Release 7

Statistics and Graphics Guide

"The real voyage of discovery consists not in seeking new landscapes, but in having new eyes."

Marcel Proust

JMP, A Business Unit of SAS
SAS Campus Drive
Cary, NC 27513

JMP Statistics and Graphics Guide, **Release 7**

ISBN 978-1-59994-419-7

SAS Institute Inc., SAS Campus Drive, Cary, North Carolina 27513.

1st printing, May 2007

Contents

JMP Statistics and Graphics Guide

3 Univariate Analysis

4 Advanced Univariate Analysis

5 Capability Analyses

A Statistical Details

Credits and Acknowledgments

Origin

JMP was developed by SAS Institute Inc., Cary, NC. JMP is not a part of the SAS System, though portions of JMP were adapted from routines in the SAS System, particularly for linear algebra and probability calculations. Version 1 of JMP went into production in October, 1989.

Credits

JMP was conceived and started by John Sall. Design and development were done by John Sall, Chung-Wei Ng, Michael Hecht, Richard Potter, Brian Corcoran, Annie Dudley Zangi, Bradley Jones, Craige Hales, Chris Gotwalt, Paul Nelson, Xan Gregg, Jianfeng Ding, Eric Hill, John Schroedl, Laura Lancaster, Scott McQuiggan, and Peng Liu. .

In the SAS Institute Technical Support division, Wendy Murphrey, Toby Trott, and Rosemary Lucas provide technical support and conduct test site administration. Statistical technical support is provided by Craig DeVault, Duane Hayes, Elizabeth Edwards, Kathleen Kiernan, Tonya Mauldin, and Doug Wielenga.

Nicole Jones, Jim Borek, Kyoko Keener, Hui Di, Joseph Morgan, Wenjun Bao, Fang Chen, Susan Shao, Hugh Crews, Yusuke Ono and Kelci Miclaus provide ongoing quality assurance. Additional testing and technical support is done by Noriki Inoue, Kyoko Takenaka, and Masakazu Okada from SAS Japan.

Bob Hickey is the release engineer.

The JMP manuals were written by Ann Lehman, Lee Creighton, John Sall, Bradley Jones, Erin Vang, Melanie Drake, and Meredith Blackwelder, with contributions from Annie Dudley Zangi and Brian Corcoran. Creative services and production was done by SAS Publications. Melanie Drake implemented the help system.

Jon Weisz and Jeff Perkinson provided project management. Also thanks to Lou Valente, Ian Cox, Mark Bailey, and Malcolm Moore for technical advice.

Genomics development was led by Russ Wolfinger. Thanks to Tzu Ming Chu, Wendy Czika, and Geoff Mann for their development work on the Genomics product.

Thanks also to Georges Guirguis, Warren Sarle, Gordon Johnston, Duane Hayes, Russell Wolfinger, Randall Tobias, Robert N. Rodriguez, Ying So, Warren Kuhfeld, George MacKensie, Bob Lucas, Warren Kuhfeld, Mike Leonard, and Padraic Neville for statistical R&D support. Thanks are also due to Doug Melzer, Bryan Wolfe, Vincent DelGobbo, Biff Beers, Russell Gonsalves, Mitchel Soltys, Dave Mackie, and Stephanie Smith, who helped us get started with SAS Foundation Services from JMP.

Acknowledgments

We owe special gratitude to the people that encouraged us to start JMP, to the alpha and beta testers of JMP, and to the reviewers of the documentation. In particular we thank Michael Benson, Howard Yetter (d), Andy Mauromoustakos, Al Best, Stan Young, Robert Muenchen, Lenore Herzenberg, Ramon Leon, Tom Lange, Homer Hegedus, Skip Weed, Michael Emptage, Pat Spagan, Paul Wenz, Mike Bowen, Lori Gates, Georgia Morgan, David Tanaka, Zoe Jewell, Sky Alibhai, David Coleman, Linda Blazek, Michael Friendly, Joe Hockman, Frank Shen, J.H. Goodman, David Iklé, Barry Hembree, Dan Obermiller, Jeff Sweeney, Lynn Vanatta, and Kris Ghosh.

Also, we thank Dick DeVeaux, Gray McQuarrie, Robert Stine, George Fraction, Avigdor Cahaner, José Ramirez, Gudmunder Axelsson, Al Fulmer, Cary Tuckfield, Ron Thisted, Nancy McDermott, Veronica Czitrom, Tom Johnson, Cy Wegman, Paul Dwyer, DaRon Huffaker, Kevin Norwood, Mike Thompson, Jack Reese, Francois Mainville, John Wass, Thomas Burger, and the Georgia Tech Aerospace Systems Design Lab.

We also thank the following individuals for expert advice in their statistical specialties: R. Hocking and P. Spector for advice on effective hypotheses; Robert Mee for screening design generators; Greg Piepel, Peter Goos, J. Stuart Hunter, Dennis Lin, Doug Montgomery, and Chris Nachtsheim for advice on design of experiments; Jason Hsu for advice on multiple comparisons methods (not all of which we were able to incorporate in JMP); Ralph O'Brien for advice on homogeneity of variance tests; Ralph O'Brien and S. Paul Wright for advice on statistical power; Keith Muller for advice in multivariate methods, Harry Martz, Wayne Nelson, Ramon Leon, Dave Trindade, Paul Tobias, and William Q. Meeker for advice on reliability plots; Lijian Yang and J.S. Marron for bivariate smoothing design; George Milliken and Yurii Bulavski for development of mixed models; Will Potts and Cathy Maahs-Fladung for data mining; Clay Thompson for advice on contour plotting algorithms; and Tom Little, Damon Stoddard, Blanton Godfrey, Tim Clapp, and Joe Ficalora for advice in the area of Six Sigma; and Josef Schmee and Alan Bowman for advice on simulation and tolerance design.

For sample data, thanks to Patrice Strahle for Pareto examples, the Texas air control board for the pollution data, and David Coleman for the pollen (eureka) data.

Translations

Erin Vang coordinated localization. Noriki Inoue, Kyoko Takenaka, and Masakazu Okada of SAS Japan were indispensable throughout the project. Special thanks to Professor Toshiro Haga (retired, Science University of Tokyo) and Professor Hirohiko Asano (Tokyo Metropolitan University for reviewing our Japanese translation. Special thanks to Dr. Fengshan Bai, Dr. Xuan Lu, and Dr. Jianguo Li, professors at Tsinghua University in Beijing, and their assistants Rui Guo, Shan Jiang, Zhicheng Wan, and Qiang Zhao, for reviewing the Simplified Chinese translation. Finally, thanks to all the members of our outstanding translation teams.

Past Support

Many people were important in the evolution of JMP. Special thanks to David DeLong, Mary Cole, Kristin Nauta, Aaron Walker, Ike Walker, Eric Gjertsen, Dave Tilley, Ruth Lee, Annette Sanders, Tim Christensen, Jeff Polzin, Eric Wasserman, Charles Soper, Wenjie Bao, and Junji Kishimoto. Thanks to SAS Institute quality assurance by Jeanne Martin, Fouad Younan, and Frank Lassiter. Additional testing for Versions 3 and 4 was done by Li Yang, Brenda Sun, Katrina Hauser, and Andrea Ritter.

Also thanks to Jenny Kendall, John Hansen, Eddie Routten, David Schlotzhauer, and James Mulherin. Thanks to Steve Shack, Greg Weier, and Maura Stokes for testing JMP Version 1.

Thanks for support from Charles Shipp, Harold Gugel (d), Jim Winters, Matthew Lay, Tim Rey, Rubin Gabriel, Brian Ruff, William Lisowski, David Morganstein, Tom Esposito, Susan West, Chris Fehily, Dan Chilko, Jim Shook, Ken Bodner, Rick Blahunka, Dana C. Aultman, and William Fehlner.

Technology License Notices

The ImageMan DLL is used with permission of Data Techniques, Inc.

Chapter 1

Preliminaries

JMP Statistical Discovery

Welcome to JMP interactive statistics and graphics. The mission of JMP is to help researchers to analyze and make discoveries in their data.

JMP is designed to be a point-and-click, walk-up-and-use product that harnesses the power of interactive statistical graphics to serve the analysis needs of the researcher. JMP provides a facility for statistical analysis and exploration that enable the user to

Discover more. With graphics you see patterns and relationships in your data, and you find the points that don't fit the pattern.

Interact more. When a product is easy to use, you focus on the problem rather than on the tool to view the problem. With interactivity, you follow up on clues and try more approaches. The more angles you try, the more you likely you are to discover things.

Understand more. With graphics, you see how the data and model work together to produce the statistics. With a well-organized well-thought-out tool, you learn the structure of the discipline. With a progressively disclosed system, you learn statistics methods in the right context.

JMP provides a rich variety of statistical and graphical methods organized into a small number of interactive platforms.

Contents

What Are the Limitations of JMP?

JMP is a great tool for statistical exploration. So what kind of tasks would JMP not be suited for?

Very large jobs JMP is not for problems with millions of data values. JMP data tables must fit in main memory. JMP makes graphs of everything, but graphs get expensive, more cluttered, and less useful when they have many thousands of points.

Data processing If you need a tool for production data processing, heavy duty programming, applications development, and data management, then JMP might not be the best choice. Instead, consider using SAS software, which was designed for those types of tasks.

Specialized statistics JMP has a wide statistical charter. With JMP, you are served with the most popular and generally useful statistical methods, but there are some specialized areas that are not currently supported in JMP.

Data tables are not spreadsheets JMP has a spreadsheet grid to browse a data table, but the rows and columns are not arbitrary like a business spreadsheet.

Comparing the Capabilities of JMP and SAS

Which product would you use when you need more data handling, more programmability, and more statistical depth than JMP? The SAS System, of course. The SAS System is 40 times larger than JMP. It is not unusual for a SAS job to process 250 million records. SAS supports a number of different programming languages, and the SAS System is very deep in statistics. For example: JMP has three multiple comparison methods and SAS has over twenty; JMP has one factor analysis approach and SAS has many; JMP can handle only single random effect mixed models; SAS has more general mixed model procedure; JMP fits proportional hazard models with a simple approach for ties; SAS fits proportional hazard models with time-varying covariates and supports exact methods for handling ties. JMP provides logistic regression. JMP provides some very useful regression diagnostics such a leverage plots. SAS provides many more, including the complete BKW set; JMP provides a nonlinear regression facility. SAS provides three separate nonlinear fitting facilities in three separate products; JMP does simple flat-triangle surface fitting and contouring. SAS provides both quintic spline and thin-plate surface fitting and contouring; if your data are locked into a specialized binary format, JMP cannot retrieve the data. SAS can read almost any kind of data and even has direct links to most database systems.

SAS is a very large system that won't run out of steam when the jobs grow big. However, that size comes at a price. Compared to JMP, the SAS System is more expensive to acquire, takes a larger learning investment, and needs more computing resources. That higher price is easily justified if you have large jobs to do every day. JMP addresses smaller jobs with a product that is smaller, but still has the methods needed for most analyses. JMP is a Macintosh/Windows-bred product that is very easy to learn and use, so that you can jump right in and not be frustrated by a steep learning curve. JMP is a very willing product that is eager to quickly show you lots of angles on your data.

When you are looking for an information delivery vehicle, sometimes you need a truck that can haul anything, every day, like SAS. But if you are out for a drive to go exploring, you might find it best to drive a small car, even a sports car like JMP. Perhaps both belong in your garage, so you can use the one that best suits your immediate task.

SAS Institute will continue to support and improve both JMP and the SAS System in a way that is consistent with their respective charters and complementary to each other.

Prerequisites

About Using the Computer

This documentation assumes familiarity with standard operations and terminology such as *click*, *double-click*, *Control-click* and *Alt-click* under Windows, *⌘-click* and *Option-click* on the Macintosh, *Shift-click*, *drag*, *select*, *copy*, and *paste*. Also know how to use menu bars and scroll bars, how to move and resize windows, and how to manipulate files in the desktop. Consult the reference guides that came with the system for more information on these topics.

Note that a right-click can be simulated on a one-button mouse by substituting Control-click.

About Statistics

JMP only assumes a minimal background in formal statistics. All analysis platforms include graphical displays with options that help review and interpret results. Each platform also includes access to help windows that offer general help and some statistical details. There are tutorials in this book that provide examples of displaying and analyzing data.

About Sample Data and Sample Scripts

The sample data is installed in an appropriate place for your operating system. Below, we list the default locations, but the location may vary depending on your (or your Network Administrator's) choices during the installation process.

A journal of the sample data is available at **Help > Sample Data Directory**. In addition, the JSL path variable `$SAMPLE_DATA` holds the directory location of the sample data files. To see its value, open an empty script window and type

```
name = Get Path Variable("$SAMPLE_DATA")
```

Select **Edit > Run Script** to execute the script. Select **View > Log** (PC and Linux) or **Window > Log** (Mac) to show the log that holds the result.

Although it is set at install time, you may desire to point to a different sample data location. In this case, submit

```
Set Path Variable("$SAMPLE_DATA",name);
```

- On Windows machines, the Sample Data folder is located in the same folder as the executable. By default, this is C:\Program Files\SAS\JMP6\Support Files English\Sample Data.

- On Linux machines, the Sample Data folder is located in the same folder as the executable. This may be /usr/local/bin/JMP or /home/<*username*>/JMP, or any other location. that you have access to.
- On the Macintosh, the Sample Data folder is located at /Library/Application Support/JMP. If you installed JMP to your home Applications directory, you may find them at ~/Library/Application Support/JMP.

The Sample Scripts folder and the Sample Import Data folders are located in the same places as the Sample Data folder.

Learning About JMP

Using JMP Help

JMP uses the familiar point-and-click interface. After installing JMP, open any of the JMP files in the sample data folder and experiment with analysis tools. Help is available for most menus, options, and reports. There are several ways to see JMP Help:

- Windows users have a standard **Help** main menu with **Contents**, **Search**, and **Index** commands, which access the JMP documentation.
- On the Macintosh, select **Help Center** from the **Help** main menu. This displays all the help books in the help facility. Click the JMP Help Book to see a list of JMP documents. The Macintosh Help Center has a search facility. Or, access each document through its table of contents within the Help system.
- Click **Help > Indexes>Statistics** access a scrolling alphabetical reference that tells how to generate specific analyses using JMP and accesses further help for that topic. Other help items are found in the same menu.
- **Help** is available from buttons in dialogs and from popup menus in JMP report windows.

After generating a report, context-sensitive help is available. Select the help tool (**?**) from the **Tools** menu and click on any part of the report. Context-sensitive help tells about the items in the report window.

Using Tutorials

The *Introductory Guide* is a collection of tutorials designed to help you learn JMP strategies. Each tutorial uses a sample data file from the Sample Data folder. By following along with these step-by-step examples, you can quickly become familiar with JMP menus, options, and report windows.

Reading About JMP

The book you are reading now is the *JMP Statistics and Graphics Guide*. It contains documentation with examples of all **Analyze** and **Graph** menu commands. See the following manuals for additional information about JMP:

- The *User Guide* has complete documentation of all JMP menus, an explanation of data table manipulation, and a description of the formula editor. There are chapters that show you how to do common tasks such as manipulating files, transforming data table columns, and cutting and pasting JMP data, statistical text reports, and graphical displays.

- *Design of Experiments* documents the Design of Experiments (DOE) facility in JMP.
- The *Scripting Guide* is a reference guide to JMP Scripting Language (JSL), the programming language that lets you automate any sequence of actions. JSL is an advanced topic, which you learn after becoming familiar with the JMP application.

Other documents give technical information to site administrators for installing and maintaining network and leased versions of JMP.

...if you are a previous JMP user

If you are familiar with the JMP environment, you might just want to know what's new. "What's New in JMP 6" gives a summary of changes and additions.

Conventions and Organization

The following conventions help you relate written material to information you see on your screen.

- Most open data table names used in examples show in Helvetica font (Animals or Animals.jmp) in this document. References to variable names in data tables and items in reports also show in Helvetica according to the way they appear on the screen or in the book illustration.
- **Note:** Special information, warnings, and limitations are noted as such with a **boldface** note.
- Reference to menu names (**File** menu) or menu items (**Save** command) appear in **Helvetica bold**.
- Words or phrases that are important or have definitions specific to JMP are in *italics* the first time you see them. For example, the word *platform* is in italics the first time you see it. Most words in italics are defined when they are first used unless clear in the context of use.

Chapter 2

Understanding JMP Analyses

JMP's Analysis Platforms

This chapter introduces you to the JMP analysis platforms. It presents the basic navigational instructions on how to use them, outlines the statistical models that underlie them, and gives some conceptual background. This chapter offers you the opportunity to make a small investment in time for a large return later.

If you are new to statistics, then this chapter helps to fill in what you need to know to start analyzing data. If you are knowledgeable about statistics and want to dive right in, you can refer to chapters that document specific platforms and analyses or to the appendix "Statistical Details," p. 975, for a more detailed explanation of statistical methods.

Contents

The JMP Statistical Platforms

There are a great many statistical methods implemented in JMP, but they are organized and consolidated into the **Analyze**, **DOE**, and **Graph** commands, which are also accessible through the JMP Starter window. For more information, see the *JMP User Guide*.

Statistical methods are folded into hierarchies, rather than accessed from a large flat list. For example, you do contingency analysis from the Contingency platform. At first you may need to reference this book or the on-line help to find your way around, but as you become more familiar with the JMP approach to statistics, the natural hierarchy in JMP becomes simple and convenient to use. Methods progressively unfold before you that suit the context of your data. Many results are done automatically and others are offered through popup menus in the analysis report window.

Each **Analyze** and **Graph** command *launches a platform*. A platform is an interactive window that you can use to analyze data, work with points on plots, and save results. The platform persists until you close it. You can have any number of data tables open at the same time and any number of platforms displayed for each data table. If you have several data tables, then the platform you launch analyzes the *current data table*. This table is either the active data table window or the table that corresponds to an analysis report window. Its name always shows in a tool bar at the top of the JMP window.

The navigational instructions for the analysis component of JMP are simple, but before you begin, note some points of design philosophy.

First, JMP implements general methods consistent with its key concepts (see "Key Concepts," p. 19). All platforms produce a variety of results that handle many situations. They are consistent throughout in the way they treat data and statistical concepts.

Second, JMP methods are generic. They adapt themselves to the context at hand. The principal concept is that of *modeling type*, which gears all analyses. For example, JMP automatically analyzes a variable that has nominal values differently than it analyzes a variable with continuous values.

Third, JMP doesn't wait for options and commands. A platform gives you most everything you need at once. If this is more information than you want, you can conceal the tables or graphs you don't need. The advantage is that there is less need to *mouse around* to find commands for extra statistics.

JMP is a discovery tool rather than a presentation facility—and the more perspectives you have on your data the better off you are. The computer is fast enough to spoil you with many views of your data. You choose among the views for the best way to present results after you discover what your data have to say.

To begin an analysis of data, open a data table and follow three steps:

- Assign modeling types to variables as needed (Figure 2.1)
- Launch an analysis platform
- Select columns to play variable roles.

1. Assign Modeling Types

The modeling type of a variable can be continuous, nominal, or ordinal. You assign a modeling type either with the popup menu in the columns panel of the data table window as shown in Figure 2.1, or with the modeling type popup menu in the Column Info dialog. This modeling type assignment becomes an attribute of the column until you change it, and is saved with the data table when you close it.

Figure 2.1 Assign a Modeling Type to Columns

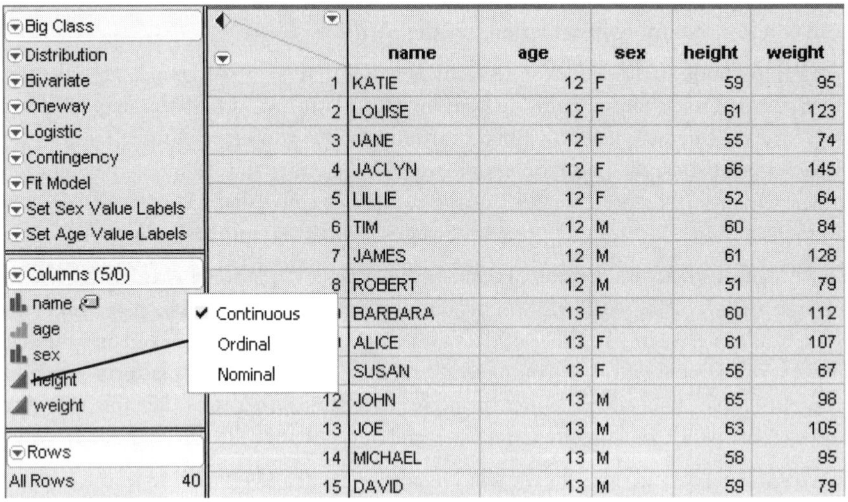

Big Class
Distribution
Bivariate
Oneway
Logistic
Contingency
Fit Model
Set Sex Value Labels
Set Age Value Labels

Columns (5/0)
name
age
sex
height
weight

Rows
All Rows 40

✔ Continuous
Ordinal
Nominal

	name	age	sex	height	weight
1	KATIE	12	F	59	95
2	LOUISE	12	F	61	123
3	JANE	12	F	55	74
4	JACLYN	12	F	66	145
5	LILLIE	12	F	52	64
6	TIM	12	M	60	84
7	JAMES	12	M	61	128
8	ROBERT	12	M	51	79
	BARBARA	13	F	60	112
	ALICE	13	F	61	107
	SUSAN	13	F	56	67
12	JOHN	13	M	65	98
13	JOE	13	M	63	105
14	MICHAEL	13	M	58	95
15	DAVID	13	M	59	79

The modeling types are not just descriptive tags. They tell each analysis platform how to analyze and graph the data. They are prescriptions for treating the values in a model. Also, there is a difference in modeling type treatment according to whether the variable is a response (Y) variable or a factor (X) variable. *Do not hesitate to change the modeling type in order to study the same variable in different ways.*

JMP uses the following modeling types:

Continuous For continuous columns, the numeric value is used directly in the model. For continuous factors, the response is a continuous function of the factor. Often the variable is measured on a continuous scale. Continuous variables are amenable to concepts like *expected value* or *mean*. In particular, continuous responses are described as having a random component that is typically thought of as a normally distributed error term.

Ordinal An ordinal value is not used directly, but only as a name. An ordinal column can be either numeric or character and have either numbers or characters as values. However, JMP interprets the column as having discrete values, which are treated as ordered categories. Numbers are ordered by their increasing numeric value, and the order of characters is the alphabetical or collating sequence.

For ordinal factors, the response steps to different values as the factor moves to different levels. For ordinal responses, JMP builds a discrete probability model for the ordered categories.

Nominal For the nominal modeling type, the values are treated as unordered categories, or names. A character column is always nominal column. The values can be either characters or numbers, but numbers are treated as unordered discrete values.

2. Choose an Analysis

You launch an analysis platform from the JMP starter window or from the **Analyze** or **Graph** menu according to the way you want *X* and *Y* variables treated together.

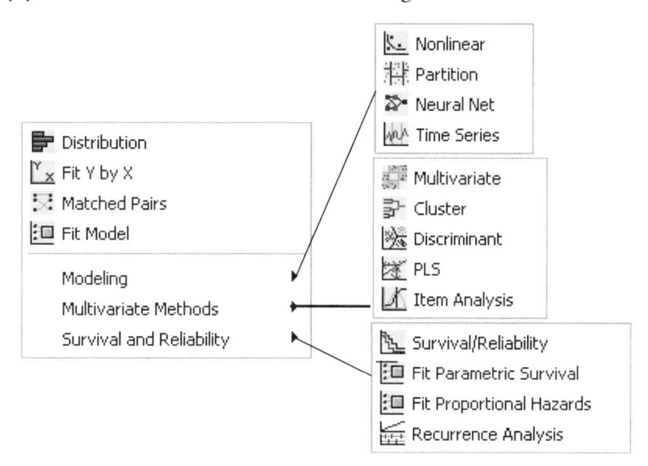

Here is a brief description of the **Analyze** platforms:

Distribution describes the distribution of each column you choose for analysis with histograms, box plots and normal plots for continuous columns, and divided (mosaic) bar charts for nominal and ordinal columns. You can also compare the computed mean and standard deviation of the distribution to a constant and examine the fit of a variety of different distributions.

Fit Y by X describes each pair of *X* and *Y* columns you specify. The displays and reports vary depending upon the modeling types (continuous, nominal, ordinal) you assign to the *X* and *Y* columns. The four combinations of response and factor modeling types lead to the four analyses: bivariate analysis, one-way analysis of variance, contingency table analysis, and logistic regression.

Matched Pairs analyzes the special situation where multiple measurements are taken on a single subject, and a paired *t*-test or repeated measures analysis is needed.

Fit Model gives the *general fitting platform* for fitting one or more *Y* columns by all the effects you create. The general fitting platform fits multiple regression models, models with complex effects, response surface models, and multivariate models including discriminant and canonical analysis. Leverage plots, least-squares means plots, and contour plots help you visualize the whole model and each effect. Special dialogs let you request contrasts, custom tests, and power details. The Fit Model platform also has special effect screening tools such as the cube plots, a prediction profiler, and a contour profiler, for models with multiple responses.

Modeling is for advanced models that are nonlinear in their parameters, or models that have correlated terms. In some cases, the form of the model is not important.

Nonlinear fits the response (*Y*) as a function of a nonlinear formula of the specified *x* variable and parameters. **Nonlinear** fits models that are nonlinear in their parameters. Nonlinear fitting begins with a formula you build in a column using the formula editor. This formula has parameters to be estimated by the nonlinear platform. When you launch **Nonlinear**, you interact with a control panel to do the fitting and plotting of results.

Partition fits classification and regression trees.

Neural implements a simple one-layer neural net.

Time Series performs simple forecasting and fits ARIMA models.

Multivariate Methods provide analysis methods for multivariate data. Cluster analysis, correlations, principal components, and discriminant analysis are the purveyance of this set of tools.

Multivariate describes relationships among response variables with correlations, a scatterplot matrix, a partial correlation matrix, the inverse correlation matrix, a multivariate outlier plot or a jackknifed outlier plot, and principal components analysis. You can also display the covariance matrix and see a three-dimensional spinning plot of the principal components.

Cluster clusters rows of a JMP table using the hierarchical, *k*-means, or EM (*expectation maximization*) method. The hierarchical cluster method displays results as a tree diagram of the clusters followed by a plot of the distances between clusters. The *k*-means cluster option, suitable for larger tables, iteratively assigns points to the number of clusters you specify. The EM mixture clustering method is for mixtures of distributions and is similar to the k-means method.

Discriminant predicts classification variables (nominal or ordinal) based on a known continuous response. Discriminant analysis can be regarded as inverse prediction from a multivariate analysis of variance.

PLS analyzes data using partial least squares.

Item Analysis fits response curves from Item Response Theory (IRT) for analyzing survey or test data.

Survival and Reliability allows analysis of univariate or multivariate survival and reliability data.

Survival/Reliability analyzes survival data using product-limit (Kaplan-Meier) life table survival computations.

Fit Parametric Survival launches a personality of the Fit Model platform to accomplish parametric fitting of censored data.

Fit Proportional Hazards launches a personality of the Fit Model platform for proportional hazard regression analysis that fits a Cox model.

Recurrence Analysis performs recurrence analysis.

3. Select Columns to Play Roles

You choose X or Y roles for the variables in your analysis depending on how you want JMP to treat them. A column with a Y role is a response or dependent variable. An *x* column is a factor variable that fits, predicts, or groups a response. Each analysis platform prompts you with a Launch dialog, similar to the one in Figure 2.2.

Figure 2.2 Assign Variable Roles

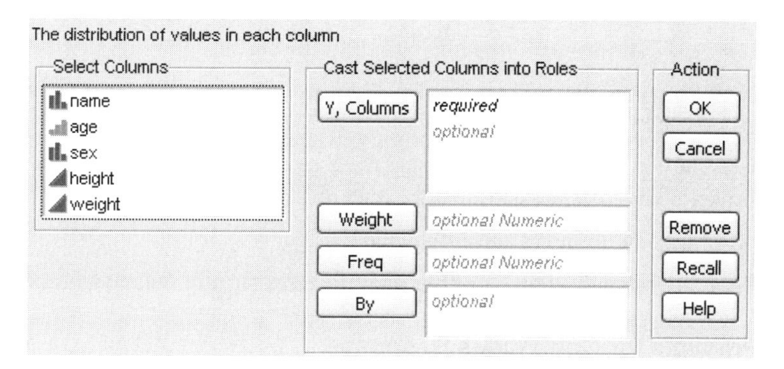

Variable roles include **X** and **Y**, as described above, and **Weight** or **Freq** when you want to assign a weight or frequency count to each row.

See the *JMP User Guide* for a more detailed discussion of types of data, modeling types, and variable roles.

Figure 2.3 is a simple illustration of the process for analyzing data with JMP.

Figure 2.3 The Three-Step General Approach to Analysis in JMP

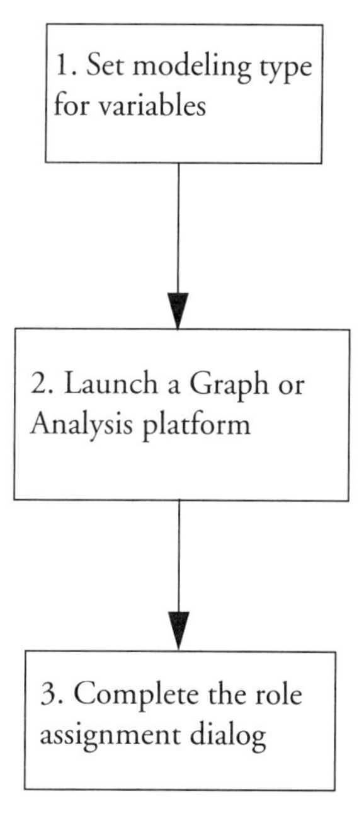

The Response Models

The models and methods available in JMP are practical, are widely used, and suit the need for a general approach in a statistical software tool. As with all statistical software, you are responsible for learning the assumptions of the models you choose to use, and the consequences if the assumptions are not met. For more information see "The Usual Assumptions," p. 18 in this chapter.

Continuous Response Model

When the response column (column assigned the Y role) is continuous, JMP fits the value of the response directly. The basic model is that for each observation,

Y = (some function of the X's and parameters) + error

Statistical tests are based on the assumption that the error term in the model is normally distributed.

Fitting Principle for Continuous Response

The Fitting principle is called *least squares*. The least squares method estimates the parameters in the model to minimize the sum of squared errors. The errors in the fitted model, called *residuals*, are the difference between the actual value of each observation and the value predicted by the fitted model.

The least squares method is equivalent to the maximum likelihood method of estimation if the errors have a normal distribution. This means that the analysis estimates the model that gives the most likely residuals. The log-likelihood is a scale multiple of the sum of squared errors for the normal distribution.

Base Model

The simplest model for continuous measurement fits just one value to predict all the response values. This value is the estimate of the *mean*. The mean is just the arithmetic average of the response values. All other models are compared to this base model.

Nominal Response Model

When the response column (column assigned the Y role) has a nominal modeling type, JMP fits the probabilities that the response is one of r different response levels given by the data values. The basic model is that for each observation,

$$\text{Probability}(Y = j\text{th response level}) = \text{some function of the } X\text{s and parameters}$$

The probability estimates must all be positive. For a given configuration of X's, the probability estimates must sum to 1 over the response levels. The function that JMP uses to predict probabilities is a composition of a linear model and a multi-response logistic function. This is sometimes called a *log-linear* model because the logs of ratios of probabilities are linear models. For example, if Y can be either 1 or 2, the log-linear model is expressed as

$$\log\left(\frac{P(Y=1)}{P(Y=2)}\right) = \text{function of } X\text{s and parameters}$$

Fitting Principle For Nominal Response

The fitting principle is called *maximum likelihood*. It estimates the parameters such that the joint probability for all the responses given by the data is the greatest obtainable by the model. Rather than reporting the joint probability (likelihood) directly, it is more manageable to report the total of the negative logs of the likelihood.

The uncertainty (–log-likelihood) is the sum of the negative logs of the probabilities attributed by the model to the responses that actually occurred in the sample data. For a sample of size n, it is often denoted as H and written

$$H = \sum_{i=1}^{n} -\log(P(y = y_i))$$

If you attribute a probability of 1 to each event that did occur, then the sum of the negative logs is zero for a perfect fit.

The nominal model can take a lot of time and memory to fit, especially if there are many response levels. JMP tracks the progress of its calculations with an *iteration history*, which shows the –log-likelihood values becoming smaller as they converge to the estimates.

Base Model

The simplest model for a nominal response is a set of constant response probabilities fitted as the occurrence rates for each response level across the whole data table. In other words, the probability that y is response level j is estimated by dividing the total sample count n into the total of each response level n_j, and is written

$$p_j = \frac{n_j}{n}$$

All other models are compared to this base model. The base model serves the same role for a nominal response as the sample mean does for continuous models.

The R^2 statistic measures the portion of the uncertainty accounted for by the model, which is

$$1 - \frac{H(\text{full model})}{H(\text{base model})}$$

However, it is rare in practice to get an R^2 near 1 for categorical models.

Ordinal Response Model

With an ordinal response (Y), as with nominal responses, JMP fits probabilities that the response is one of r different response levels given by the data.

Ordinal data have an order like continuous data. The order is used in the analysis but the spacing or distance between the ordered levels is not used. If you have a numeric response but want your model to ignore the spacing of the values, you can assign the ordinal level to that response column. If you have a classification variable and the levels are in some natural order such as low, medium, and high, you can use the ordinal modeling type. For details about the function that is fitted, see the chapters "Simple Logistic Regression," p. 167, and "Logistic Regression for Nominal and Ordinal Response," p. 403.

One way of thinking about an ordinal model is to imagine the response mechanism as a two-phase process. The x's (the linear model) first determine some variable z with continuous values. However, z is not observed. Instead, z determines the ordinal response level for y as it crosses various threshold values.

Schematically this is

z = (some function of y's and parameters) + error

y = (some level determined by the value of z)

where the error has a logistic distribution.

Specifically, the value of y ascends the ordinal levels as z crosses certain threshold values. If the levels of y are 1 through r, then

$$y = \begin{cases} r & \alpha_{r-1} \le z \\ j & \alpha_{j-1} \le z < a_j \\ 1 & z \le a_j \end{cases}$$

A different but mathematically equivalent way to envision an ordinal model is to think of a nominal model where, instead of modeling the odds, you model the cumulative probability. Instead of fitting functions for all but the last level, you fit only one function and slide it to fit each cumulative response probability.

Fitting Principle For Ordinal Response

The maximum likelihood fitting principle for an ordinal response model is the same as for a nominal response model. It estimates the parameters such that the joint probability for all the responses that occur is the greatest obtainable by the model. It uses an iterative method that is faster and uses less memory than nominal fitting.

Base Model

The simplest model for an ordinal response, like a nominal response, is a set of response probabilities fitted as the occurrence rates of the response in the whole data table.

The Factor Models

The way the x-variables (factors) are modeled to predict an expected value or probability is the subject of the factor side of the model.

The factors enter the prediction equation as a linear combination of x values and the parameters to be estimated. For a continuous response model, where i indexes the observations and j indexes the parameters, the assumed model for a typical observation, y_i, is written

$y_i = \beta_0 + \beta_1 x_{1i} + \ldots + \beta_k x_{ki} + \varepsilon_i$ where

y_i is the response

x_{ij} are functions of the data

ε_i is an unobservable realization of the random error

β_j are unknown parameters to be estimated.

The way the *x*'s in the linear model are formed from the factor terms is different for each modeling type. The linear model *x*'s can also be complex effects such as interactions or nested effects. Complex effects are discussed in detail later.

Continuous Factors

When an *x* variable is continuous, its values are used directly in the prediction formula. The *x* variables are called *regressors*, or *covariates*.

Nominal Factors

When an *x* variable has the nominal modeling type, the response can have separate prediction coefficients fit to each of its levels. The prediction coefficients are modeled to represent differences from each level to the average response across all the nominal values.

In the context of a linear model, there is an indicator variable for each nominal level except the last. When the last nominal level occurs, the indicator variables are set to –1. This causes all the indicators to be –1 so that the parameters across all levels sum to zero.

For example, suppose the nominal factor A has three levels 1, 2, and 3, and you want to fit it as a linear model. The model has the intercept and two parameters that show a value to add to the intercept if the level of A is 1 or 2. For the last level of A, you subtract both A parameters from the intercept to get the predicted value,

$$y_i = \beta_0 + \beta_1 x_1 + \beta_2 x_2 + \text{error} \qquad \text{where}$$

β_0 is the intercept parameter

β_1 and β_2 are parameters for the A nominal effect

x_1 is 1 if $A = 1$, 0 if $A = 2$, -1 if $A = 3$

x_2 is 0 if $A = 1$, 1 if $A = 2$, -1 if $A = 3$

Thus, for the Parameter Estimates report shown below, the equation to the right of the report is the prediction formula for the fitted *Y*:

Parameter Estimates

Term	Estimate
Intercept	7.9
A[1]	-2.6
A[2]	-1.8

$$7.9+\text{Match}(A)\begin{pmatrix} \text{"1"} \Rightarrow -2.6 \\ \text{"2"} \Rightarrow -1.8 \\ \text{"3"} \Rightarrow 4.4 \\ \text{else} \Rightarrow . \end{pmatrix}$$

where the value for $A = 3$ of 4.4 is the negative sum of the other two values, –2.6 and –1.8. This is just an alternative way to implement the model that fits three means to the three groups.

Ordinal Factors

When an *x* variable is assigned the ordinal modeling type, its lowest value is considered a baseline value, and each level above that baseline contributes an additional quantity used to determine the response.

In the context of a linear model, there is an indicator variable for each factor level except the first. All the indicators are set to 1 at and below the indicator for the current factor level.

Note: When you test to see if there is no effect, there is not much difference between nominal and ordinal factors for simple models. However, there are major differences when interactions are specified. See the appendix "Statistical Details," p. 975, for a discussion of complex ordinal factors. We recommend that you use nominal rather than ordinal factors for most models.

The Usual Assumptions

Before you put your faith in statistics, reassure yourself that you know both the value and the limitations of the techniques you use. Statistical methods are just tools—they cannot guard you from incorrect science (invalid statistical assumptions) or bad data.

Assumed Model

Most statistics are based on the assumption that the model is correct. You assume that the equation really does determine the response up to a normally distributed random error. To the extent that your model may not be correct, you must attenuate your credibility in the statistical reports that result from the model.

Relative Significance

Many statistical tests do not evaluate the model in an absolute sense. Significant test statistics might only be saying that the model fits better than some reduced model, such as the mean. The model can appear to fit the data but might not describe the underlying physical model well at all.

Multiple Inferences

Often the value of the statistical results is not that you believe in them directly, but rather that they provide a key to some discovery. To confirm the discovery, you may need to conduct further studies. Otherwise, you might just be sifting through the data.

For instance, if you conduct enough analyses you can find 5% significant effects in five percent of your studies by chance alone, even if the factors have no predictive value. Similarly, to the extent that you use your data to shape your model (instead of testing the correct model for the data), you are corrupting the significance levels in your report. The random error then influences your model selection and leads you to believe that your model is better than it really is.

Validity Assessment

Some of the various techniques and patterns to look for in assessing the validity of the model are as follows:

- Model validity can be checked against a saturated version of the factors with Lack of Fit tests. The Fit Model platform presents these tests automatically if you have replicated x data in a nonsaturated model.
- You can check the distribution assumptions for a continuous response by looking at plots of residuals and studentized residuals from the Fit Model platform. Or, use the **Save** commands in the platform popup menu to save the residuals in data table columns. Then use the **Analyze > Distribution** on these columns to look at a histogram with its normal curve and the normal quantile plot. The residuals are not quite independent, but you can informally identify severely nonnormal distributions.
- The best all-around diagnostic tool for continuous responses is the leverage plot because it shows the influence of each point on each hypothesis test. If you suspect that there is a mistaken value in your data, this plot helps determine if a statistical test is heavily influenced by a single point. Leverage plots are described in the chapter "Standard Least Squares: Introduction," p. 213.
- It is a good idea to scan your data for outlying values and examine them to see if they are valid observations. You can spot univariate outliers in the Distribution platform reports and plots. Bivariate outliers appear in Fit Y by X scatterplots and in the Multivariate scatterplot matrix. You can see trivariate outliers in a three-dimensional spinning plot produced by the **Graph > Spinning Plot**. Higher dimensional outliers can be found with Principal Components produced in the Spinning Plot or Multivariate platform, and with Mahalanobis and jackknifed distances computed and plotted in the Multivariate platform.

Alternative Methods

The statistical literature describes special nonparametric and robust methods, but JMP implements only a few of them at this time. These methods require fewer distributional assumptions (nonparametric), and then are more resistant to contamination (robust). However, they are less conducive to a general methodological approach, and the small sample probabilities on the test statistics can be time consuming to compute.

If you are interested in linear rank tests and need only normal large sample significance approximations, you can analyze the ranks of your data to perform the equivalent of a Wilcoxon rank-sum or Kruskal-Wallis one-way test.

If you are uncertain that a continuous response adequately meets normal assumptions, you can change the modeling type from continuous to ordinal and then analyze safely, even though this approach sacrifices some richness in the presentations and some statistical power as well.

Key Concepts

There are two key concepts that unify classical statistics and encapsulate statistical properties and fitting principles into forms you can visualize:

- a unifying concept of uncertainty
- two basic fitting machines.

These two ideas help unlock the understanding of statistics with intuitive concepts that are based on the foundation laid by mathematical statistics.

Statistics is to science what accounting is to business. It is the craft of weighing and balancing observational evidence. Statistical tests are like credibility audits. But statistical tools can do more than that. They are instruments of discovery that can show unexpected things about data and lead to interesting new ideas. Before using these powerful tools, you need to understand a bit about how they work.

Uncertainty, a Unifying Concept

When you do accounting, you total money amounts to get summaries. When you look at scientific observations in the presence of uncertainty or noise, you need some statistical measurement to summarize the data. Just as money is additive, uncertainty is additive if you choose the right measure for it.

The best measure is not the direct probability because to get a joint probability you have to assume that the observations are independent and then multiply probabilities rather than add them. It is easier to take the log of each probability because then you can sum them and the total is the log of the joint probability.

However, the log of a probability is negative because it is the log of a number between 0 and 1. In order to keep the numbers positive, JMP uses the negative log of the probability. As the probability becomes smaller, its negative log becomes larger. This measure is called uncertainty, and it is measured in reverse fashion from probability.

In business, you want to maximize revenues and minimize costs. In science you want to minimize uncertainty. Uncertainty in science plays the same role as cost plays in business. All statistical methods fit models such that uncertainty is minimized.

It is not difficult to visualize uncertainty. Just think of flipping a series of coins where each toss is independent. The probability of tossing a head is 0.5, and $-\log(0.5)$ is 2 for base 2 logarithms. The probability of tossing h heads in a row is simply

$$p = \left(\frac{1}{2}\right)^h$$

Solving for h produces

$$h = -\log_2 p$$

You can think of the uncertainty of some event as the number of consecutive “head” tosses you have to flip to get an equally rare event.

Almost everything we do statistically has uncertainty, $-\log p$, at the core. Statistical literature refers to uncertainty as *negative log-likelihood*.

The Two Basic Fitting Machines

An amazing fact about statistical fitting is that most of the classical methods reduce to using two simple machines, the spring and the pressure cylinder.

Springs

First, springs are the machine of fit for a continuous response model (Farebrother, 1987). Suppose that you have *n* points and that you want to know the expected value (mean) of the points. Envision what happens when you lay the points out on a scale and connect them to a common junction with springs (see Figure 2.4). When you let go, the springs wiggle the junction point up and down and then bring it to rest at the mean. This is what must happen according to physics.

If the data are normally distributed with a mean at the junction point where springs are attached, then the physical energy in each point's spring is proportional to the uncertainty of the data point. All you have to do to calculate the energy in the springs (the uncertainty) is to compute the sum of squared distances of each point to the mean.

To choose an estimate that attributes the least uncertainty to the observed data, the spring settling point is chosen as the estimate of the mean. That is the point that requires the least energy to stretch the springs and is equivalent to the least squares fit.

Figure 2.4 Connect Springs to Data Points

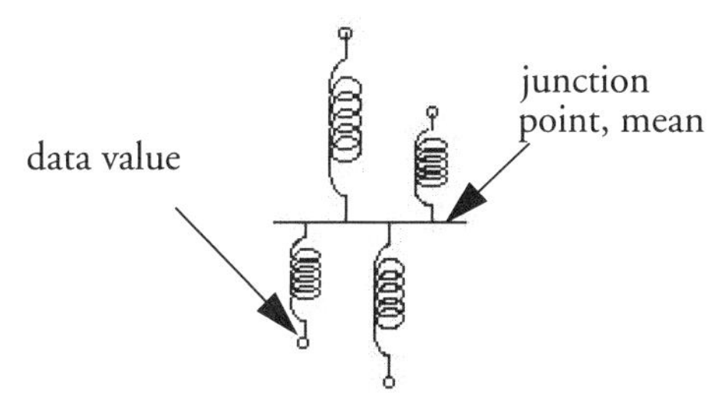

That is how you fit one mean or fit several means. That is how you fit a line, or a plane, or a hyperplane. That is how you fit almost any model to continuous data. You measure the energy or uncertainty by the sum of squares of the distances you must stretch the springs.

Statisticians put faith in the normal distribution because it is the one that requires the least faith. It is, in a sense, the most random. It has the most non-informative shape for a distribution. It is the one distribution that has the most expected uncertainty for a given variance. It is the distribution whose uncertainty is measured in squared distance. In many cases it is the limiting distribution when you have a mixture of distributions or a sum of independent quantities. It is the distribution that leads to test statistics that can be measured fairly easily.

When the fit is constrained by hypotheses, you test the hypotheses by measuring this same spring energy. Suppose you have responses from four different treatments in an experiment, and you want to test if the means are significantly different. First, envision your data plotted in groups as shown here, but with springs connected to a separate mean for each treatment. Then exert pressure against the spring force to move the individual means to the common mean. Presto! The amount of energy that constrains the means to be the same is the test statistic you need. That energy is the main ingredient in the *F*-test for the hypothesis that tests whether the means are the same.

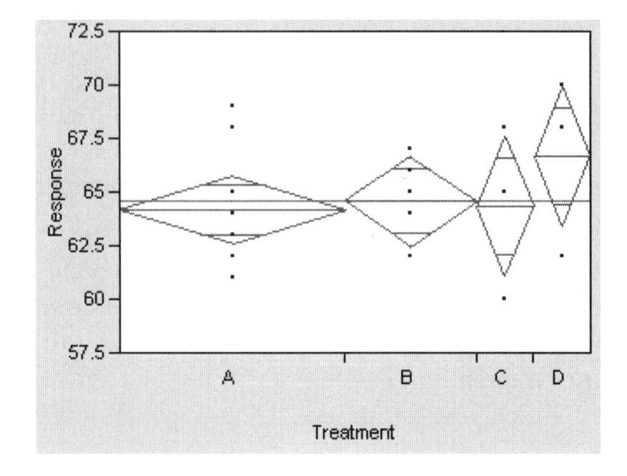

Pressure Cylinders

What if your response is categorical instead of continuous? For example, suppose that the response is the country of origin for a sample of cars. For your sample, there are probabilities for the three response levels, American, European, and Japanese. You can set these probabilities for country of origin to some estimate and then evaluate the uncertainty in your data. This uncertainty is found by summing the negative logs of the probabilities of the responses given by the data. It is written

$$H = \sum h_{y(i)} = \sum \log p_{y(i)}$$

The idea of springs illustrates how a mean is fit to continuous data. When the response is categorical, statistical methods estimate the response probabilities directly and choose the estimates that minimize the total uncertainty of the data. The probability estimates must be nonnegative and sum to 1. You can picture the response probabilities as the composition along a scale whose total length is 1. For each response observation, load into its response area a gas pressure cylinder, for example, a tire pump. Let the partitions between the response levels vary until an equilibrium of lowest potential energy is reached. The sizes of the partitions that result then estimate the response probabilities.

Figure 2.5 shows what the situation looks like for a single category such as the medium size cars (see the mosaic column from Carpoll.jmp labeled medium in Figure 2.6). Suppose there are thirteen responses (cars). The first level (American) has six responses, the next has two, and the last has five responses. The response probabilities become 6/13, 2/13, and 5/13, respectively, as the pressure against the response partitions balances out to minimize the total energy.

Figure 2.5 Effect of Pressure Cylinders in Partitions

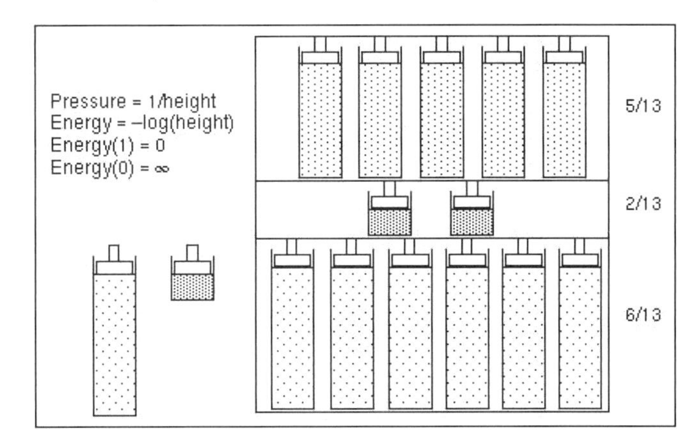

As with springs for continuous data, you can divide your sample by some factor and fit separate sets of partitions. Then test that the response rates are the same across the groups by measuring how much additional energy you need to push the partitions to be equal. Imagine the pressure cylinders for car origin probabilities grouped by the size of the car. The energy required to force the partitions in each group to align horizontally tests whether the variables have the same probabilities. Figure 2.6 shows these partitions.

Figure 2.6 A Mosaic Plot for Categorical Data

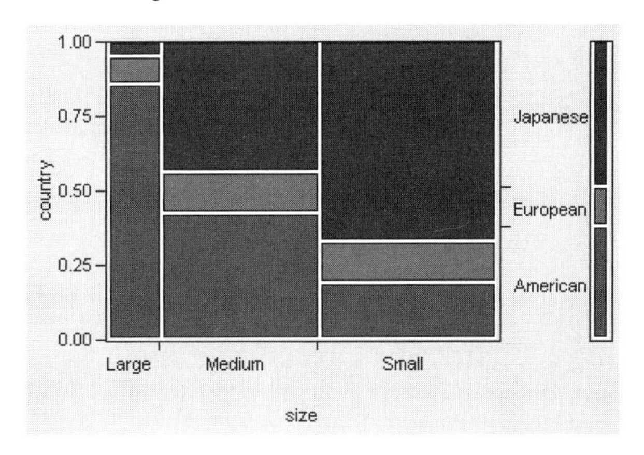

Chapter 3

Univariate Analysis

The Distribution Platform

The Distribution platform describes the distribution of values in data table columns using histograms and other graphical and text reports.

For categorical (nominal/ordinal modeling type) variables:

- The graphical display can consist of a histogram and a divided (mosaic) bar chart. The histogram shows a bar for each level of the ordinal or nominal variable.
- The text reports show counts, proportions, cumulative proportions, and Chi-Square tests.

For numeric continuous variables:

- Graphical displays begin with a histogram and outlier box plot. The histogram shows a bar for grouped values of the continuous variable. You can also request a quantile box plot, a normal quantile plot, a stem-and-leaf plot, and an overlaid nonparametric smoothing curve.
- Text reports show selected quantiles and moments. Options let you save ranks, probability scores, and normal quantile values as new columns in the data table. Optional commands test the mean and standard deviation of the column against a constant you specify, test the fit of various distributions you can specify, and perform a capability analysis for a quality control application.

Contents

Launching the Platform

To begin, choose **Distribution** from the **Analyze** menu (or tool bar), or click **Distribution** on the **Basic Stats** tab of the JMP Starter window.

When the Launch dialog appears, assign one or more numeric or character columns for a distribution analysis, as shown in Figure 3.1. To choose variables for analysis, select them in the column selector list and click the appropriate role button, or drag the variables to the role boxes on the right of the dialog.

Figure 3.1 Launch Dialog

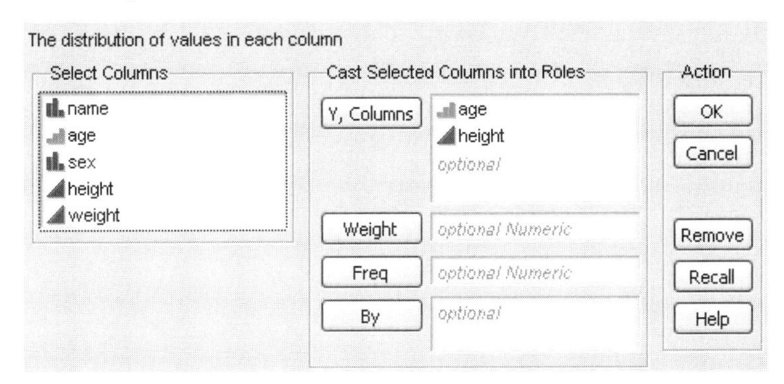

When you click **OK**, the **Distribution** command launches a platform that shows the graphical description and supporting tables for each variable entered in the **Y, Columns** list. A typical distribution layout is shown in Figure 3.2.

There is a histogram for each analysis column accompanied by plots and tables that show the distribution in different ways. Popup icons on report titles give commands appropriate for the variable's modeling type. On histogram title bars,

- the popup icon next to each continuous column name lets you test the sample mean and standard deviation against any value, fit a variety of distributions, do a capability analysis, select graph and display options, and save histogram information.
- The popup icon next to each categorical column name lets you select display options, test probabilities, and save information about the levels of the histogram.

When you add a **Freq** variable on the dialog, it affects the output by being summed in the overall count appearing in the moments table (N). Therefore, all other moment statistics (mean, standard deviation, and so on) are also affected by the **Freq** variable.

If you instead add the variable as a **Weight**, the overall count is not affected. The sum of the weights is shown in the **More Moments** section of the report. Any moment that is based on the Sum Wgts is therefore affected. Specifically, the mean, standard error of the mean, and upper and lower CIs are affected by weights.

If the variable you use for either the **Freq** or **Weight** is not an integer, the result for Sum Wgts differs, since a column designated as a Freq has all values truncated to integers.

Figure 3.2 The Distribution Platform

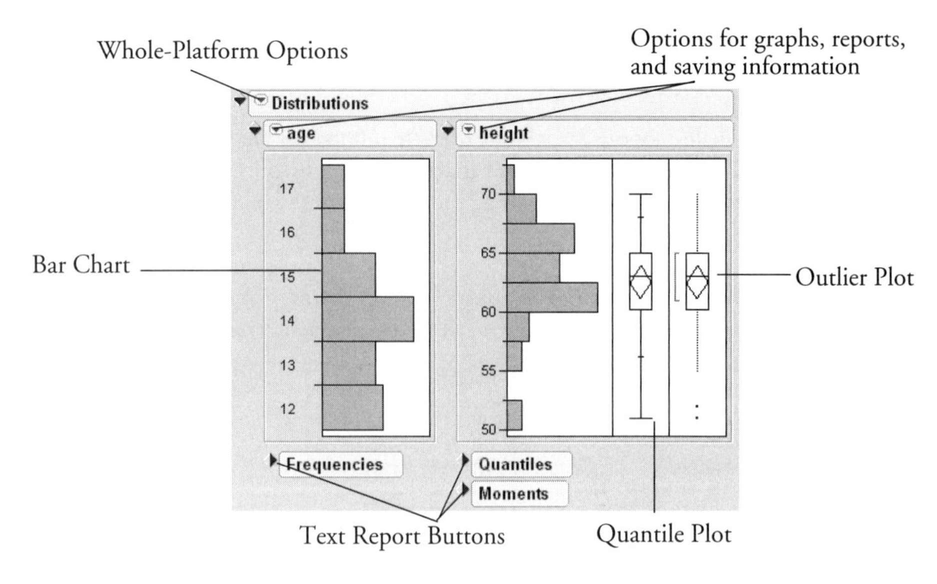

Resizing Graphs

You can resize the plot frame on any **Analyze** or **Graph** platform. To resize a graph or plot, move the cursor to the right or bottom edge or right-bottom corner of the plot frame. When it becomes a double arrow you can drag the plot frame to resize it. The heights of both a histogram and its accompanying chart or plot always adjust as one unit. However, the widths of all plots can be adjusted independently.

- Hold the Control key (Command key on the Mac) to resize all graphs together.
- The Alt (or Option) key resizes in 8-pixel increments.
- Resizing with the Shift key creates frames that maintain their aspect ratio.

Adjusting Continuous Column Histogram Bars

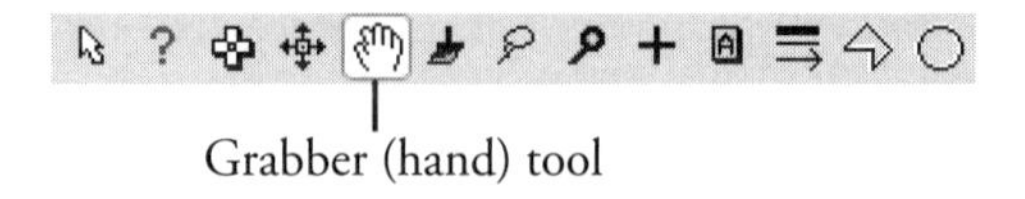

To adjust a histogram's bars, select the grabber (hand) tool from the graphics cursor tool bar and position it anywhere in the histogram plot frame. Hold down the mouse button and move the mouse. If you think of each bar as a bin that holds a number of observations, then moving the hand to the left increases the bin width and combines intervals. The number of bars decreases as the bar size increases. Moving the hand to the right decreases the bin width, producing more bars (see Figure 3.3). Moving the hand up or down shifts the bin locations on the axis, changing the contents and size of each bin.

Note: To rescale the axis of any continuous histogram, double click on the axis and complete the Axis Specification dialog.

Figure 3.3 Change Histogram Bars

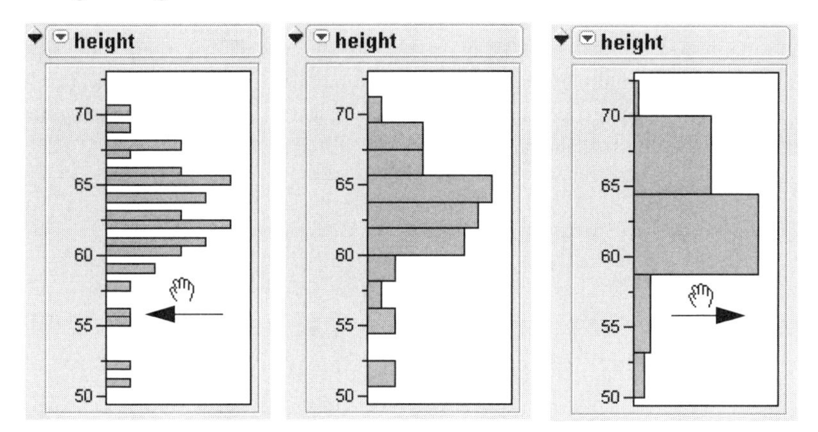

Highlighting Bars and Selecting Rows

Clicking any histogram bar with the arrow cursor highlights the bar and selects the corresponding rows in the data table. The appropriate portions of all other graphical displays also highlight the selection. Figure 3.4 shows the results of highlighting a bar in the height histogram. The corresponding rows are selected in the data table. Hold Control-click (⌘-click on the Macintosh) to deselect histogram bars.

Figure 3.4 Highlight Observations

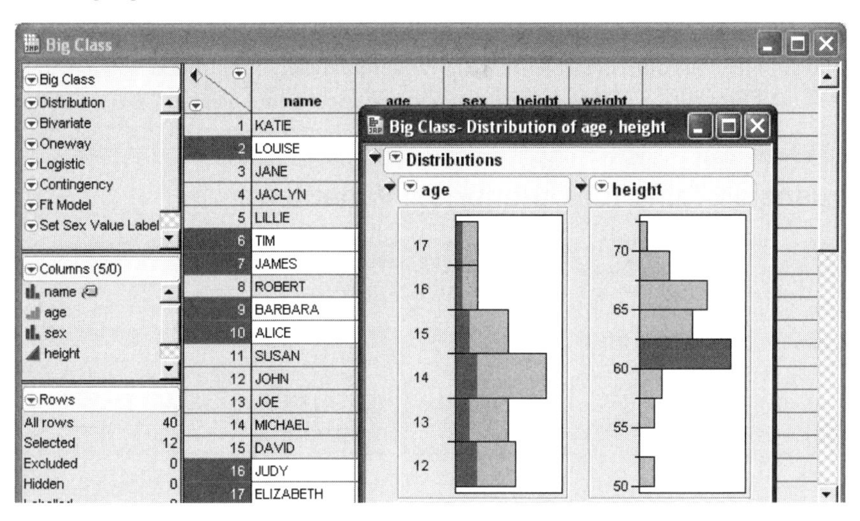

Selecting Subsets in Histograms

The Alt (Option) key is used to make selections using more than one histogram. In other words, you can select bars in one histogram, then, using the Alt key, click in another histogram to further specify the selection using another variable's value.

For example, suppose you want to select small- and medium-sized pharmaceutical companies. Use the sample data table Companies.jmp to produce histograms of the type (Type) and size (Size Co) of the listed companies. In the Size Co histogram, click in the "Small" bar, then Shift-click in the "Medium" bar. At this point, all small- and medium-sized companies are selected. To select only the Pharmaceutical companies, Alt (Option)-click in the "Pharmaceutical" bar of the Type histogram. The selection shrinks to only the pharmaceutical companies.

Figure 3.5 Selection using Alt key

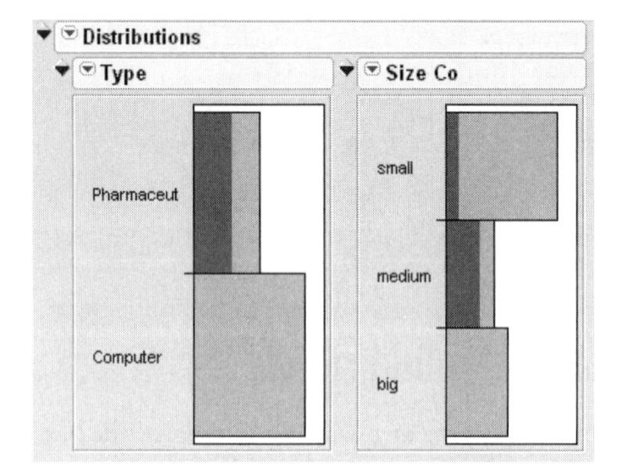

Initial Text Reports

The following sections describe tables that appear automatically (unless other preferences are in effect) as part of the Distribution analysis for the variables you select for analysis.

The Frequencies Table for Categorical Variables

Frequencies

Level	Count	Prob
A	8	0.29630
B	9	0.33333
C	5	0.18519
D	5	0.18519
Total	27	1.00000
N Missing	0	

The Frequencies table lists the following specific information for each level of a nominal or ordinal response variable. Missing values are omitted from the analysis:

Level lists each value found for a response variable.

Count lists the number of rows found for each level of a response variable.

Probability lists the probability (or proportion) of occurrence for each level of a response variable. The probability is computed as the count divided by the total frequency of the variable, shown at the bottom of the table.

StdErrProb lists the standard error of the probabilities. This column is hidden in the example above. Right-click (Control-click on the Mac) in the table and select **StdErrProb** from the **Columns** submenu to unhide it.

Cum Prob contains the cumulative sum of the column of probabilities. This column is also hidden in the example above.

The Quantiles Table for Continuous Variables

The Quantiles table lists the values of selected quantiles, sometimes called percentiles. It is a text table you can show or hide independently of the Quantile Box Plot. See “Options for Continuous Variables,” p. 39, for a discussion of the Quantile Box Plot.

Quantiles

100.0%	maximum	70.000
99.5%		70.000
97.5%		69.975
90.0%		68.000
75.0%	quartile	65.000
50.0%	median	63.000
25.0%	quartile	60.250
10.0%		56.200
2.5%		51.025
0.5%		51.000
0.0%	minimum	51.000

Quantiles are computed as follows:

To compute the *p*th quantile of N nonmissing values in a column, arrange the N values in ascending order and call these column values $y_1, y_2, ..., y_N$. Compute the rank number for the *p*th quantile as p / $100(N + 1)$.

If the result is an integer, the *p*th quantile is that rank's corresponding value.

If the result is not an integer, the *p*th quantile is found by interpolation. The *p*th quantile, denoted q_p, is

$$q_p = (1 - f)y_i + (f)y_{i+1}$$

where

n is the number of nonmissing values for a variable

$y_1, y_{2, ...,} y_n$ represents the ordered values of the variable

y_{n+1} is taken to be y_n

$np = i + f$

i is the integer part and f is the fractional part of np.

$(n + 1)p = i + f$

For example, suppose a data table has 15 rows and you want to find the 75th and 90th quantile values of a continuous column. After the column is arranged in ascending order, the ranks that contain these quantiles are computed as

$$\frac{75}{100}(15+1) = 12 \text{ and } \frac{90}{100}(15+1) = 14.4$$

The value y_{12} is the 75th quantile.

The 90th quantile is interpolated by computing a weighted average of the 14th and 15th ranked values as $y_{90} = 0.6y_{14} + 0.4y_{15}$.

The Moments Table for Continuous Variables

The Moments table displays the mean, standard deviation, and other summary statistics. If the sample is normally distributed, the data values are arranged symmetrically around the sample mean (arithmetic average).

Moments	
Mean	62.55
Std Dev	4.2423385
Std Err Mean	0.6707726
upper 95% Mean	63.906766
lower 95% Mean	61.193234
N	40

The Moments table shows the following quantities:

Mean estimates the expected value of the underlying distribution for the response variable. It is the arithmetic average of the column's values. It is the sum of the nonmissing values divided by the number of nonmissing values. If the data table has a column assigned the role of **weight**, the mean is computed by multiplying each column value by its corresponding weight, adding these values, and dividing by the sum of the weights.

Std Dev measures the spread of a distribution around the mean. It is often denoted as s and is the square root of the sample variance, denoted s^2.

$$s = \sqrt{s^2} \text{ where}$$

$$s^2 = \sum_{i=1}^{N} \frac{w_i(\bar{y}_w - y_i)^2}{N-1}$$

$$\bar{y}_w = \text{weighed mean}$$

A normal distribution is mainly defined by the mean and standard deviation. These parameters give an easy way to summarize data as the sample becomes large:

68% of the values are within one standard deviation of the mean.

95% of the values are within two standard deviations of the mean.

More than 99% of the values are within three standard deviations of the mean.

Std Err Mean the standard error of the mean, estimates the standard deviation of the distribution of mean estimators. **Std Err Mean** is computed by dividing the sample standard deviation, s, by the square root of N. If a column is assigned the role of **weight**, then the denominator is the square root of $\sum w_i$. This statistic is used in t-tests of means and in constructing confidence intervals for means. If no weight column is specified, then $w_i = 1$ for all i and $\sum w_i = N$.

upper 95% Mean and lower 95% Mean are 95% confidence limits about the mean. They define an interval that is very likely to contain the true population mean. If many random samples are drawn from the same population and each 95% confidence interval is determined, you expect 95% of the confidence intervals computed to contain the true population mean.

N is the total number of nonmissing values and is used to compute statistics in the Moments table when no column is assigned the role of **weight.**

Note that the following statistics appear when **More Moments** is selected from the **Display Options** of the platform.

Sum Wgts is the sum of a column assigned the role of **weight** and is used in the denominator for computations of the mean instead of *N*.

Sum is the sum of the response values.

Variance (the sample variance) is the square of the sample standard deviation.

Skewness measures sidedness or symmetry. It is based on the third moment about the mean and is computed

$$\sum w_i^{\frac{3}{2}} z_i^3 \frac{N}{(N-1)(N-2)}$$, where $z_i = \frac{x_i - \bar{x}}{\text{std}(x)}$ and w_i is a weight term (= 1 for equally weighted items)

Kurtosis measures peakedness or heaviness of tails. It is based on the fourth moment about the mean and is computed

$$\frac{n(n+1)}{(n-1)(n-2)(n-3)} \sum_{i=1}^{n} w_i\left(\frac{x_i - x}{s}\right)^4 - \frac{3(n-1)^2}{(n-2)(n-3)}$$, where w_i is a weight term (= 1 for equally weighted items)

CV is the percent coefficient of variation. It is computed as the standard deviation divided by the mean and multiplied by 100.

N Missing is the number of missing observations.

Whole-Platform Options

Each statistical platform has a popup menu in the outermost outline level next to the platform name. Options and commands in this menu affect all text reports and graphs on the platform.

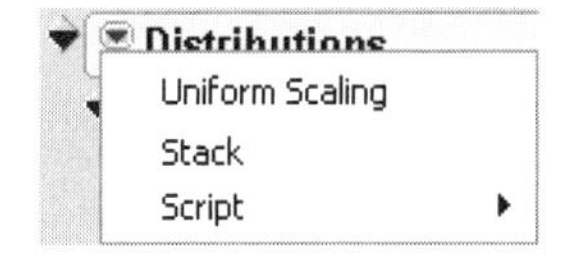

The whole-platform options for the Distribution platform include the following:

Uniform Scaling scales all axes with the same minimum, maximum, and intervals so that the distributions are easily compared. This option applies to reports for all response variables when selected.

Stack lets you orient all the output in the report window as either portrait or landscape.

Script lets you rerun or save the JSL script that produced the platform results. If the script is saved to a file, you can edit it; if it is saved with the current data table, it is available to run the next time you open the table. The JSL generated by **Save Script for All Objects** is the same as **Save Script to Script Window** if there are no By-Groups. When there are By-Groups the script includes JSL `Where` clauses that identify the By-Group levels.

Data Table Window gives a view of the underlying data table, which is especially useful when there are By-Groups.

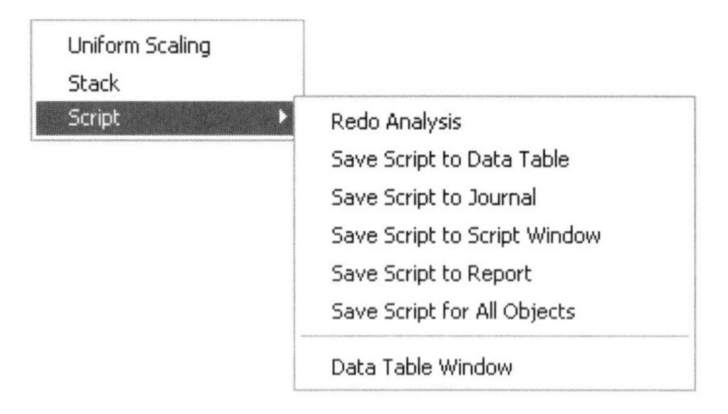

Options for Categorical Variables

All parts of the Distribution report are optional. The popup menu next to the variable name at the top of its report lists options available for that variable's modeling type. The popup menu shown here lists the commands and options for categorical (nominal/ordinal) variables. The next sections describe these options.

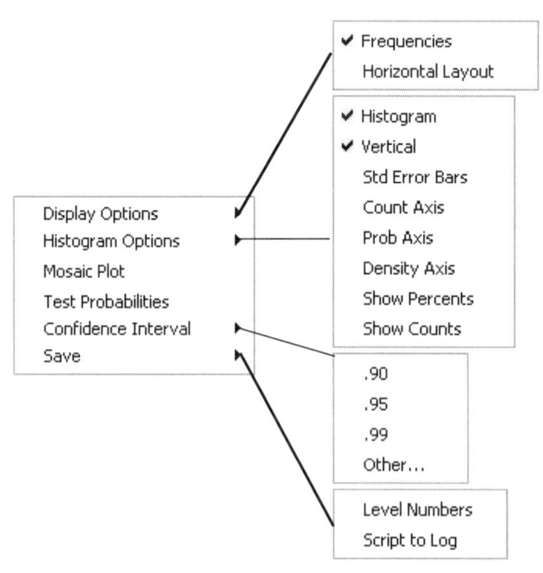

Display

The **Display Options** commands let you adjust the display of the Frequencies table (described in "The Frequencies Table for Categorical Variables," p. 30) and change the histogram from a vertical to a horizontal bar chart.

Frequencies toggles the visibility of the Frequencies table

Horizontal Layout changes both the orientation of the histogram and the relative location of the text reports.

Histogram Options

Histograms of continuous variables group the values into bars that illustrate the shape of the column's distribution. Histograms of nominal and ordinal variables are bar charts with a bar for each value (level) of the variable. **Histogram Options** has a submenu with these items:

Histogram shows or hides the histogram.

Vertical toggles the histogram (only, not the surrounding report) from a vertical to a horizontal orientation.

Std Err Bars draws the standard error bar on each level of the histogram using the pooled or proportional standard error $\sqrt{np_i(1-p_i)}$ where $p_i = n_i/n$. The standard error bar adjusts automatically when you adjust the number of bars with the hand tool.

Count Axis adds an axis that shows the frequency of column values represented by the histogram bars.

Prob Axis adds an axis that shows the proportion of column values represented by histogram bars.

Density Axis The density is the length of the bars in the histogram. Both the count and probability are based on

```
prob = (bar width)*density
count = (bar width)*density*(total count)
```

Show Percents labels the proportion of column values represented by each histogram bar.

Show Counts labels the frequency of column values represented by each histogram bar.

The following picture shows all these options in effect.

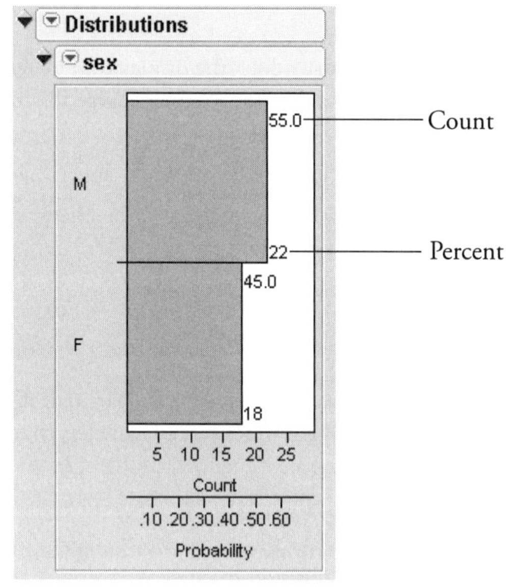

The density axis is especially useful for continuous distributions when looking at density curves that you can add with the **Fit Distribution** option. It helps show the curves' point estimates.

You can add any combination of these axes to categorical or continuous histograms.

Note: As you change the length of the bars with the hand tool, the Count and Prob axes change but the density axis remains constant.

Mosaic Plot

The **Mosaic Plot** command displays a mosaic bar chart for each nominal or ordinal response variable. A mosaic plot is a stacked bar chart where each segment is proportional to its group's frequency count. Figure 3.6 shows a mosaic plot for the nominal variable age.

Figure 3.6 Mosaic Plot

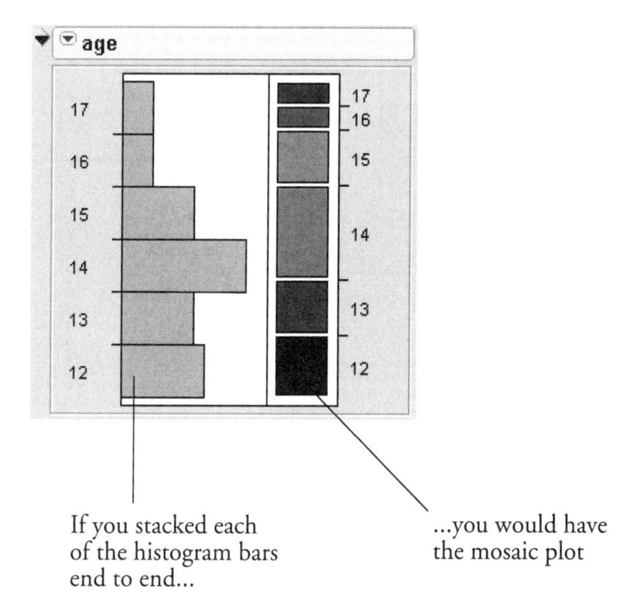

Testing Probabilities: The Test Probabilities Table

The **Test Probabilities** option displays the dialog shown here for you to enter hypothesized probabilities. The radio buttons on the dialog let you choose between rescaling hypothesized values to sum to one or using the values that you enter (rescale omitted). For cases with exactly two levels, you can choose among two different directions for the null hypothesis of an exact binomial test, or a chi-square test. When you click **Done**, Likelihood Ratio and Pearson Chi-square tests are calculated for those probabilities for the chi-square, and exact probabilities are calculated for the binomial test.

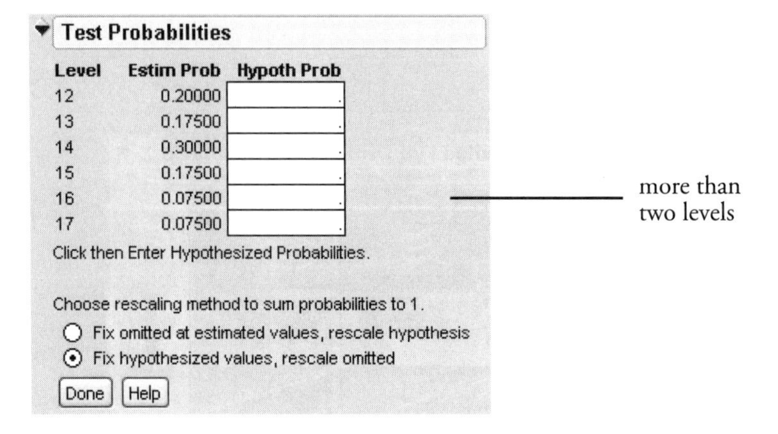

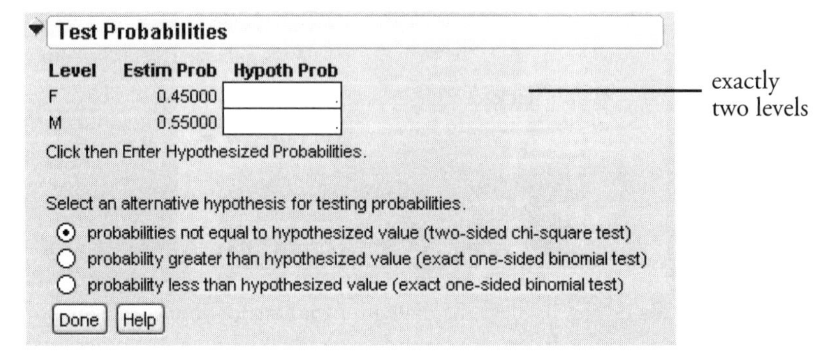

Test Probabilities scales the hypothesized values that you enter so that the probabilities sum to one. Thus, the easiest way to test that all the probabilities are equal is to enter a one in each field. If you want to test a subset of the probabilities, then leave blank the levels that are not involved. JMP then substitutes estimated probabilities.

Test Probabilities

Level	Estim Prob	Hypoth Prob
12	0.20000	0.25000
13	0.17500	0.25000
14	0.30000	0.25000
15	0.17500	0.25000
16	0.07500	0.00000
17	0.07500	0.00000

Test	ChiSquare	DF	Prob>Chisq
Likelihood Ratio	.	4	0.0000*
Pearson	.	4	0.0000*

Method: Rescaled hypothesized nonmissing values, set omitted to zero. Because a zero value was entered for one or more hypothesized probabilities and the estimated probability is greater than zero, the p-value for both Chi Square tests is zero.

Confidence Intervals: The Score Confidence Interval Table

The **Confidence Intervals** option generates the Score Confidence Intervals table, like the one shown here. The score confidence interval is recommended because it tends to have better coverage probabilities (1 – α), especially with smaller sample sizes (Agresti and Coull 1998).

Confidence Intervals

Level	Count	Prob	Lower CI	Upper CI	1-Alpha
F	18	0.45000	0.307053	0.601709	0.950
M	22	0.55000	0.398291	0.692947	
Total	40				

Note: Computed using score confidence intervals.

Save

Save options let you save information about the report as columns in the data table:

Level Numbers creates a new column in the data table called Level *colname*. The level number of each observation corresponds to the histogram bar that contains the observation. The histogram bars are numbered from low to high beginning with 1.

Script to log displays a log window that contains the script commands to generate the same report you generated. For example, **Script to log** for the example shown for the variable sex with the **Histogram** and **Mosaic Plot** option puts the following JSL script into the log:

```
Distribution(Nominal Distribution(Column( :sex), Mosaic Plot(1)))
```

Options for Continuous Variables

Features for continuous variables fall into several general categories (Figure 3.7):

- display and histogram options that affect the display of reports and charts, and commands to request additional plots
- commands to perform statistical tests and perform a capability analysis
- distribution fitting commands.

Figure 3.7 Menus of Commands and Subcommands for Continuous Variables

The next sections describe the commands in Figure 3.7.

Display

The **Display Options** command gives the following tables, which contain terms covered at the beginning of the chapter in the section "Initial Text Reports," p. 30.

Quantiles shows or hides the Quantiles table.

Moments shows or hides the Moments table, which displays the mean, standard deviation, standard error of the mean, upper and lower 95% confidence limits for the mean, and sample size.

More Moments extends the Moments table by adding Sum, Variance, Skewness, Kurtosis, and CV.

Horizontal Layout toggles between

- arranging the text reports to the right of their corresponding graphs and shows the histogram as a horizontal bar chart, and
- arranging the text reports vertically below their corresponding graphs and shows the histogram as a vertical bar chart.

Histogram

(See the previous discussion of "Options for Categorical Variables," p. 34 in this chapter.)

Normal Quantile Plot

The **Normal Quantile Plot** option adds a graph to the report that is useful for visualizing the extent to which the variable is normally distributed. If a variable is normal, the normal quantile plot approximates a diagonal straight line. This kind of plot is also called a quantile-quantile plot, or Q-Q plot.

The Normal Quantile plot also shows Lilliefors confidence bounds, reference lines, and a probability scale as illustrated in the plot shown here (Conover 1980).

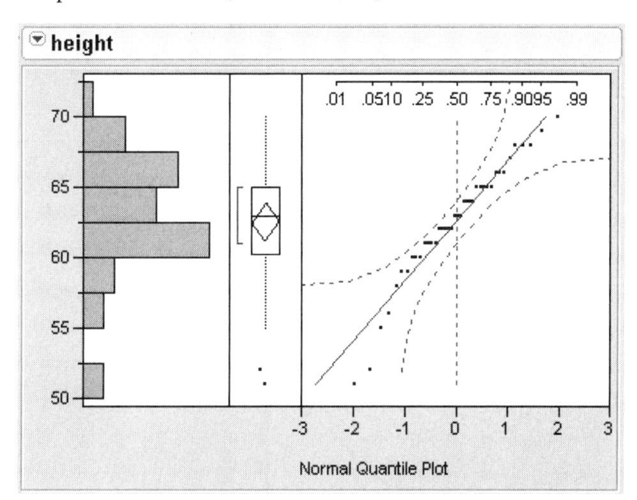

The *y*-axis of the Normal Quantile plot shows the column values. The *x*-axis shows the expected normal scores for each value. The examples in Figure 3.8 show normal quantile plots for simulations of 400 points from four different distributions:

- The plot called Normal is the normal quantile plot for a normal distribution and appears as a diagonal linear pattern.
- The second example is for a uniform distribution, a flat distribution that produces an S-shaped quantile plot. A very peaked distribution produces an inverted S-shaped quantile plot (not shown).
- Squaring a normal distribution yields a new distribution that is skewed to the right. This produces the concave normal quantile plot that is labeled Normal Squared.
- A distribution that is skewed to the left produces the convex pattern similar to the one shown in the example labeled –Normal Squared.

One other pattern not shown here is a *staircase* pattern, which is indicative of data that have been rounded or have discrete values.

Figure 3.8 Normal Quantile Plot Comparison

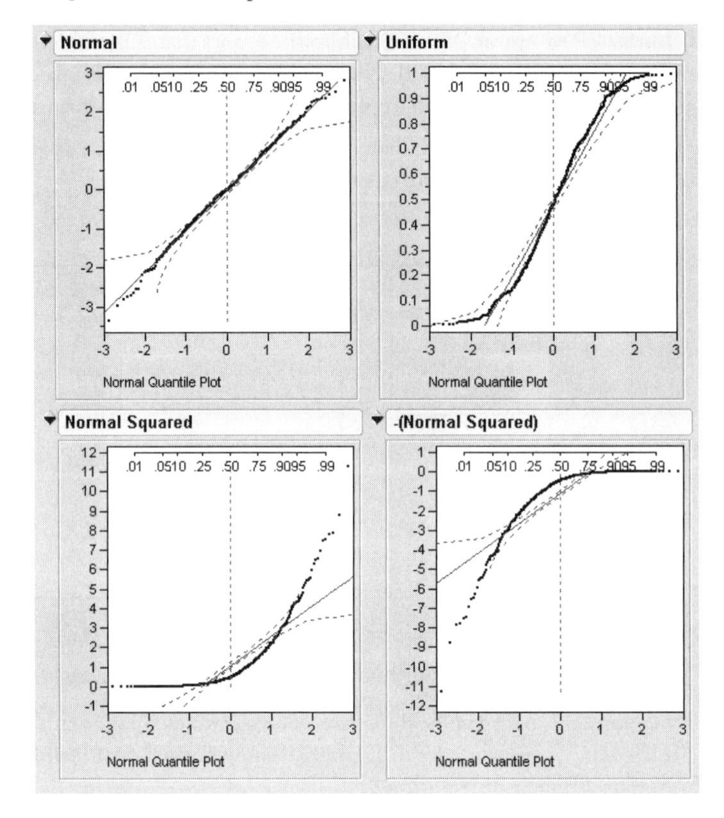

The normal quantile values are computed by the formula

$$\Phi^{-1}\left(\frac{r_i}{N+1}\right)$$

where Φ is the cumulative probability distribution function for the normal distribution, r_i is the rank of the *i*th observation, and N is the number of non-missing observations. These normal scores are Van Der Waerden approximations to the expected order statistics for the normal distribution.

Outlier Box Plot

The outlier box plot is a schematic that lets you see the sample distribution and identify points with extreme values, sometimes called *outliers*. Box Plots show selected quantiles of continuous distributions.

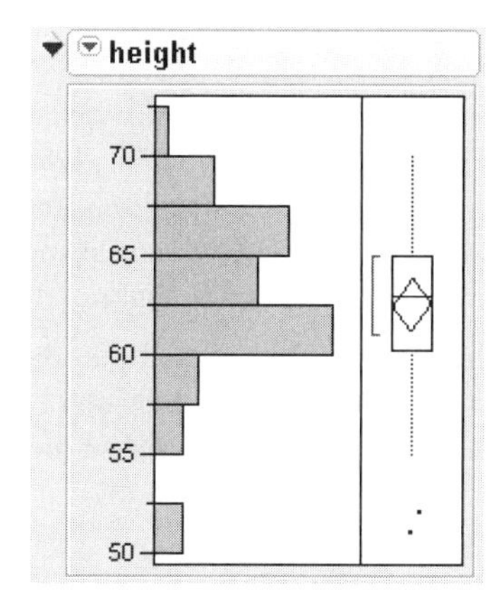

The ends of the box are the 25th and 75th quantiles, also called the *quartiles*. The difference between the quartiles is the *interquartile range*. The line across the middle of the box identifies the median sample value and the means diamond indicates the sample mean and 95% confidence interval.

The **Outlier Box Plot** command constructs a box plot with lines, sometimes called *whiskers*, extending from both ends (see Figure 3.9) that extend from each end. The whiskers extend from the ends of the box to the outer-most data point that falls within the distances computed

```
upper quartile + 1.5*(interquartile range)
lower quartile - 1.5*(interquartile range).
```

The bracket along the edge of the box identifies the *shortest half*, which is the most dense 50% of the observations (Rousseeuw and Leroy 1987).

Figure 3.9 Outlier Box Plot

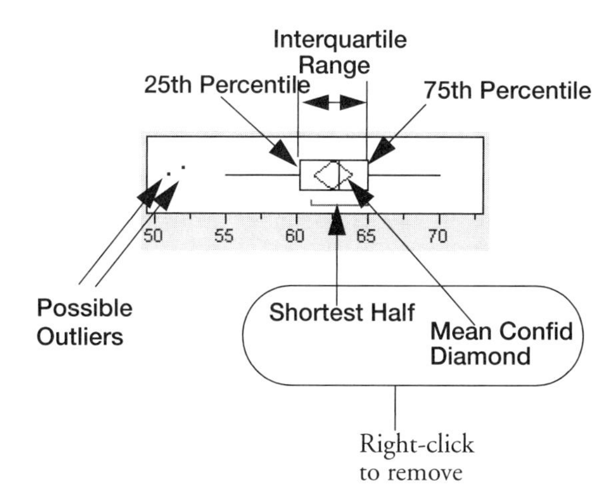

Right-clicking on the box plot reveals a menu that allows you to toggle on and off the **Mean Confid Diamond** and **Shortest Half Bracket**.

Quantile Box Plot

The **Quantile Box Plot** can sometimes show additional quantiles (90th, 95th, or 99th) on the response axis. If a distribution is normal, the quantiles shown in a box plot are approximately equidistant from each other. This lets you see at a glance if the distribution is normal. For example, if the quantile marks are grouped closely at one end but have greater spacing at the other end, the distribution is skewed toward the end with more spacing as illustrated in Figure 3.10.

Note that the quantile box plot is not the same as a Tukey outlier box plot. Quantiles are values that divide a distribution into two groups where the *p*th quantile is larger than *p*% of the values. For example, half the data are below and half are above or equal to the 50th percentile (median).

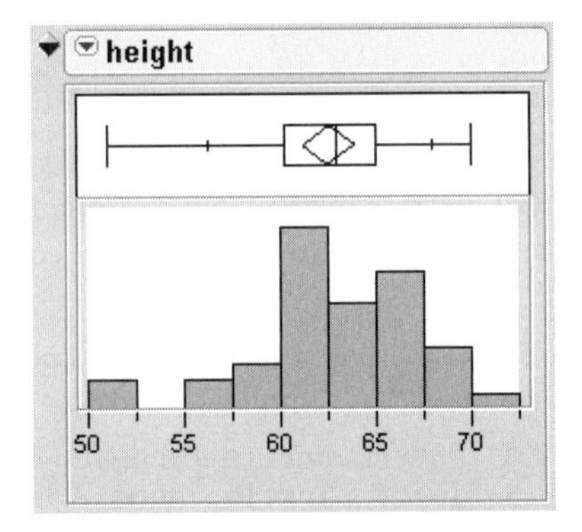

Figure 3.10 Quantile Box Plot and Quantiles Table

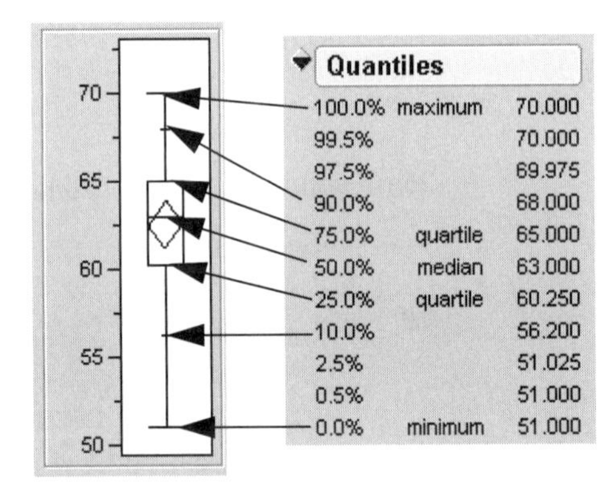

Quantiles		
100.0%	maximum	70.000
99.5%		70.000
97.5%		69.975
90.0%		68.000
75.0%	quartile	65.000
50.0%	median	63.000
25.0%	quartile	60.250
10.0%		56.200
2.5%		51.025
0.5%		51.000
0.0%	minimum	51.000

The spacing of quantiles for a normal (0,1) distribution is listed in Table 3.1 "Spacing of Quantiles in the Normal Distribution," p. 44. The two-tailed outside area, which approximately corresponds to equal normal quantile spacing, is computed `2*(1-probability)`.

Warning: Because the quantile estimates for the extreme ends of the distribution tails are highly variable, there may not be equal spacing for normal data unless the sample is large.

Table 3.1 Spacing of Quantiles in the Normal Distribution

Probability	Normal Quantile	[a]Normal Quantile / 0.674	Outside	Area (Both Tails)
0.5	0	0	1.0	

Table 3.1 Spacing of Quantiles in the Normal Distribution *(continued)*

Probability	Normal Quantile	[a]Normal Quantile / 0.674	Outside	Area (Both Tails)
0.75	[a]0.674	1	0.5	half the data are in the box, half out
0.90	1.282	1.9	0.2	80% are between the first marks
0.975	1.960	2.9	0.05	95% are between the second marks
0.995	2.576	3.8	0.01	99% are between the third marks

a. For normal distributions, the spacing between selected quantiles is approximately equal.

Stem and Leaf

The **Stem and Leaf** command constructs a plot that is a variation on the histogram. It was developed for tallying data in the days when computers were rare.

Stem and Leaf

Stem	Leaf	Count
17	2	1
16		
15		
14	25	2
13	4	1
12	3888	4
11	122223569	9
10	45567	5
9	122355899	9
8	1445	4
7	499	3
6	47	2

6|4 represents 64

Each line of the plot has a **Stem** value that is the leading digit of a range of column values. The **Leaf** values are made from the next-in-line digits of the values. You can reconstruct the data values by joining the stem and leaf by examining the legend at the bottom of the plot. The stem-and-leaf plot actively responds to clicking and the brush tool.

CDF Plot

The **CDF Plot** command forms a cumulative distribution function step plot using the observed data (with weights or frequencies if specified).

The CDF plot is related to the density curve. For a point x_i on the curve, $F(x_i)$ on the CDF plot is defined as the integral to x_i of the density function. The CDF plot estimates the areas under the density curve up to the point x_i.

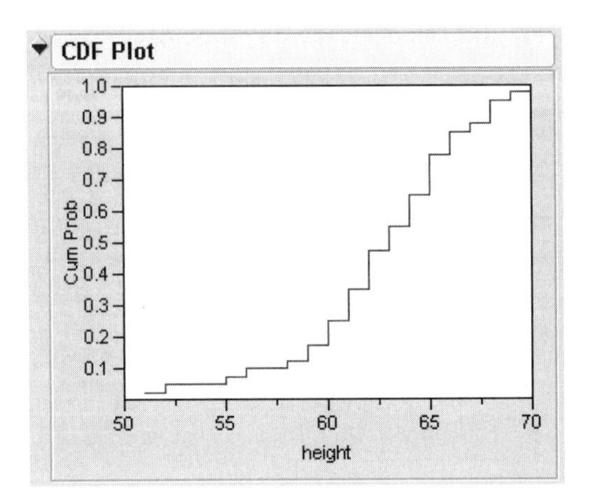

Test Mean

The **Test Mean** command prompts you to enter a test value for statistical comparison to the sample mean. If you enter a value for the standard deviation, a *z*-test is computed. Otherwise, the sample standard deviation is used to compute a *t*-statistic. Optionally, you can request the nonparametric Wilcoxon Signed-Rank test. After you click **OK**, the Test Mean table is appended to the bottom of the reports for that variable.

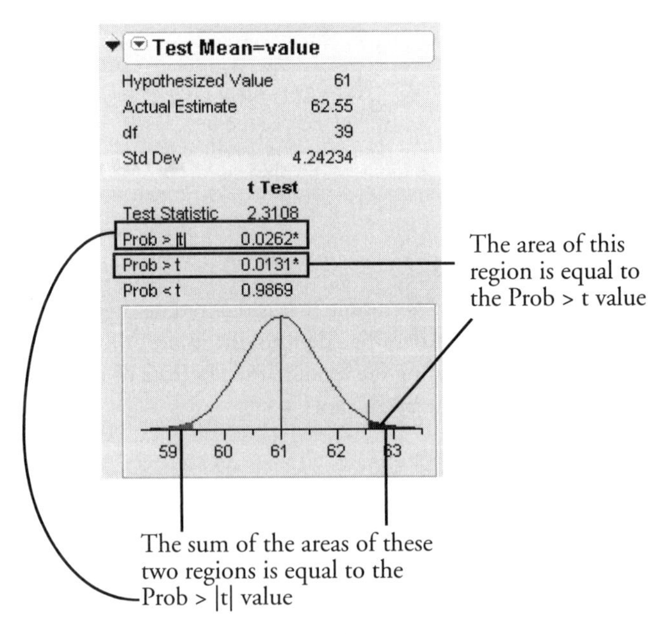

You can use the **Test Mean** command repeatedly to test different values. Each time you test the mean, a new Test Mean table is appended to the text report.

The **Test Mean** command calculates the following statistics:

t Test (or **z Test**) lists the value of the test statistic and the *p*-values for the two-sided and one-sided alternatives. The test assumes the distribution is normal.

Signed-Rank lists the value of the Wilcoxon signed-rank statistic followed by the *p*-values for the two-sided and one-sided alternatives. The test assumes only that the distribution is symmetric. The Wilcoxon signed-rank test uses average ranks for ties. The *p*-values are exact for $n \leq 20$ where n is the number of values not equal to the hypothesized value. For $n > 20$ a Student's t approximation given by Iman and Conover (1978) is used.

The probability values given in the Test Mean table are defined:

Prob>|t| is the probability of obtaining a greater absolute *t*-value by chance alone when the sample mean is not different from the hypothesized value. This is the *p*-value for observed significance of the two-tailed *t*-test.

Prob>t is the probability of obtaining a *t*-value greater than the computed sample *t*-ratio by chance alone when the sample mean is not different from the hypothesized value. This is the *p*-value for observed significance of a one-tailed *t*-test. The value of this probability is half of Prob > |t|.

Prob<t is the probability of obtaining a *t*-value less than the computed sample *t* ratio by chance alone when the sample mean is not different from the hypothesized value. This is the *p*-value for observed significance of a one-tailed *t*-test. The value of this probability is 1–Prob>t.

Test Std Dev

The **Test Std Dev** command prompts you to enter a test value for statistical comparison to the sample standard deviation (details in Neter, Wasserman, and Kutner 1990). After you click **OK**, the Test Standard Deviation table is appended to the bottom of the reports for that variable.

Test Standard Deviation=value

Hypothesized Value	4
Actual Estimate	4.24234
df	39
	ChiSquare
Test Statistic	43.8687
Min PValue	0.5454
Prob < ChiSq	0.7273
Prob > ChiSq	0.2727

You can use the **Test Std Dev** command repeatedly to test different values. Each time you test the standard deviation, a new table is appended to the text report.

$$\frac{(n-1)s^2}{\sigma^2}$$

is distributed as a Chi square variable with $n-1$ degrees of freedom. We use the critical values associated with the level of confidence desired to compute the corresponding confidence limits

$$\sqrt{\frac{(n-1)s^2}{X^2_{\alpha,\, n-1}}} \leq \sigma \leq \sqrt{\frac{(n-1)s^2}{X^2_{1-\alpha,\, n-1}}}$$

The Test Standard deviation table shows the computed chi-square statistic that tests whether the hypothesized standard deviation is the same as the computed sample standard deviation, and the probabilities associated with that Chi-square value:

Min PValue is the probability of obtaining a greater absolute Chi-square value by chance alone when the sample standard deviation is not different from the hypothesized value. This is the *p*-value for observed significance of the two-tailed test.

Prob>ChiSq is the probability of obtaining a Chi-square value *greater than* the computed sample Chi-square by chance alone when the sample standard deviation is not different from the hypothesized value. This is the *p*-value for observed significance of a one-tailed *t*-test. The value of this probability is half of Prob>|ChiSq|.

Prob<ChiSq is the probability of obtaining a Chi-square value less than the computed sample Chi-square by chance alone when the sample standard deviation is not different from the hypothesized value. This is the *p*-value for observed significance of a one-tailed *t*-test. The value of this probability is 1 – Prob>ChiSq.

Confidence Intervals

The **Confidence Intervals** option displays a popup for you to choose an α level.There is also an option to choose one-sided or two-sided confidence intervals.

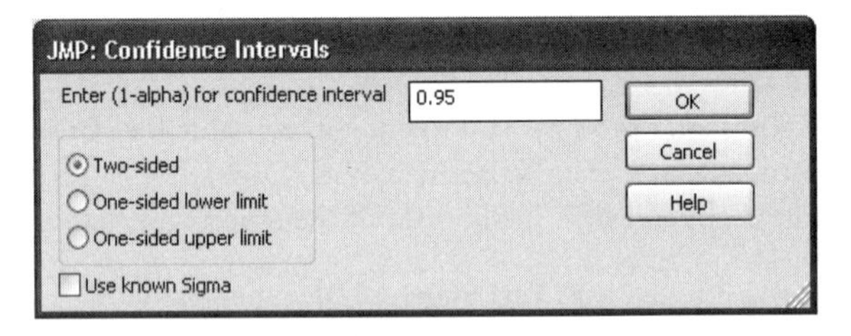

JMP then appends the Confidence Interval table to the end of the report and shows the mean and standard deviation parameter estimates with upper and lower confidence limits for $1 - \alpha$.

Confidence Intervals

Parameter	Estimate	Lower CI	Upper CI	1-Alpha
Mean	62.55	61.19323	63.90677	0.950
Std Dev	4.242338	3.475158	5.447313	

Supplying a known sigma in the confidence interval dialog yields a confidence interval about the mean based on *z*-values rather than *t*-values.

$$\mu_{1-\alpha} = \left(\bar{x} - k_{\frac{\alpha}{2}}\frac{\sigma}{\sqrt{n}}, \bar{x} + k_{\frac{\alpha}{2}}\frac{\sigma}{\sqrt{n}}\right)$$

where $k_{\frac{\alpha}{2}}$ is a standardized normal variate.

Save

You can save information computed from continuous response variables by using **Save** menu commands. Each command generates a new column in the current data table named by appending the response column name (denoted *colname* in the following definitions) to the saved statistics name

Note: You can select the **Save** commands repeatedly. This enables you to save the same statistic multiple times under different circumstances, such as before and after combining histogram bars. If you use a **Save** command multiple times, the column name for the statistic is numbered name1, name2, and so forth, to create unique column names.

The **Save** commands behave as follows:

Level Numbers creates a new column, called Level *colname*. The level number of each observation corresponds to the histogram bar that contains the observation. The histogram bars are numbered from low to high beginning with 1.

Level Midpoints creates a new column, called Midpoint *colname*. The midpoint value for each observation is computed by adding half the level width to the lower level bound.

Ranks creates a new column called Ranked *colname* that contains a ranking for each of the corresponding column's values starting at 1. Duplicate response values are assigned consecutive ranks in order of their occurrence in the spreadsheet.

Ranks Averaged creates a new column, called RankAvgd *colname*. If a value is unique, the averaged rank is the same as the rank. If a value occurs k times, the average rank is computed as the sum of the value's ranks divided by k.

Prob Scores creates a new column, called Prob *colname*. For N nonmissing scores, the probability score of a value is computed as the averaged rank of that value divided by $N + 1$. This column is similar to the empirical cumulative distribution function.

Normal Quantiles creates a new column, called N-Quantile *colname*. The normal quantile values are computed as

$$\Phi^{-1}\left(\frac{r_i}{N+1}\right)$$

where Φ is the cumulative probability distribution function for the normal distribution, r_i is the rank of the ith observation, and N is the number of nonmissing observations. These normal scores are Van Der Waerden approximations to the expected order statistics for the normal distribution. The normal quantile values can be computed using the `Normal Quantile` JMP function on a column created by the **Ranks Averaged** command. For example, the computation of the normal quantile values for a column called height is:

$$\text{Normal Quantile}\left(\frac{\text{RankAvgd height}}{(\text{Col Number}(\text{RankAvgd height})+1)}\right)$$

Standardized creates a new column, called Std *colname*. You can create standardized values for numeric columns with the formula editor using the formula shown here.

$$\frac{\text{colname} - \text{ColMean(colname)}}{\text{ColStd(colname)}}$$

Spec Limits stores the specification limits applied in a capability analysis in the current data table and automatically retrieves and displays them when you repeat the capability analysis.

Chapter 4

Advanced Univariate Analysis

Capability Analyses and Fitted Distributions

This chapter further develops the Distributions platform to include advanced options.

- Capability Analyses are used for quality control. The capability study measures the conformance of a process to certain specification limits. Several different estimates for σ are available.
- JMP can fit several statistical distributions (normal, lognormal, Weibull, extreme value, exponential, gamma, beta, and others) as well as perform goodness-of-fit tests on the fits.

Contents

Tolerance Intervals

The **Tolerance Interval** option from a Distribution report displays a dialog where the confidence (1–α) and the proportion (p) to be included in the interval are selected. JMP can produce one-sided or two-sided tolerance intervals. Select the appropriate option from the dialog box.

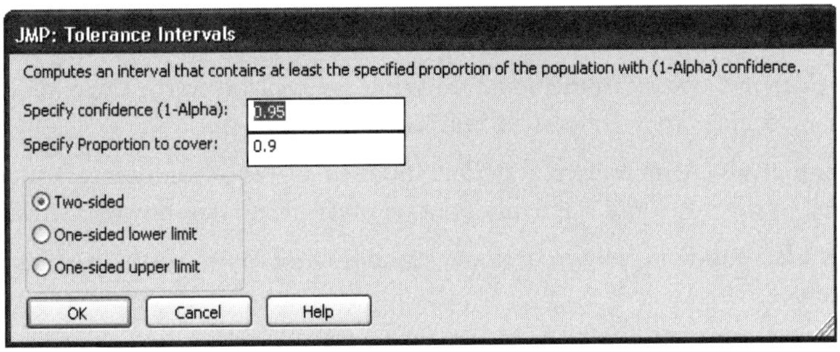

JMP: Tolerance Intervals
Computes an interval that contains at least the specified proportion of the population with (1-Alpha) confidence.
Specify confidence (1-Alpha): 0.95
Specify Proportion to cover: 0.9
Two-sided
One-sided lower limit
One-sided upper limit
OK Cancel Help

$$[\tilde{T}_{p_L}, \tilde{T}_{p_U}] = [\bar{x} - g_{(1-\alpha/2;p_L)} s, \bar{x} + g_{(1-\alpha/2;p_U)} s]$$

where

$$g_{(1-\alpha;p,n)}$$

is a constant that can be found in Table 4 of Odeh and Owen (1980).

To determine g, consider the fraction of the population captured by the tolerance interval. Tamhane and Dunlop (2000) give this fraction as

$$\Phi\left(\frac{\bar{x} + gs - \mu}{\sigma}\right) - \Phi\left(\frac{\bar{x} - gs - \mu}{\sigma}\right)$$

where Φ denotes the standard normal c.d.f. (cumulative distribution function). Therefore g solves the equation

$$P\left\{\Phi\left(\frac{\bar{X} + gS - \mu}{\sigma}\right) - \Phi\left(\frac{\bar{X} - gS - \mu}{\sigma}\right) \geq 1 - \gamma\right\} = 1 - \alpha$$

where $1-\gamma$ is the fraction of all future observations contained in the tolerance interval.

More information is given in Tables A.1a, A.1b, A.11a, and A.11b of Hahn and Meeker (1991).

A tolerance interval is an interval that one can claim contains at least a specified proportion with a specified degree of confidence—essentially, a confidence interval for a population proportion. Complete discussions of tolerance intervals are found in Hahn and Meeker (1991), and in Tamhane and Dunlop (2000).

For example, using the **height** variable from **Big Class.jmp**, the report in Figure 4.1 results from setting $\alpha = 0.05$ and $p = 0.90$.

Figure 4.1 Tolerance Interval Report

Tolerance Intervals

Parameter	Estimate	Lower TI	Upper TI	1-Alpha	Proportion
Mean	62.55	53.8309	71.2691	0.950	0.900

From this report, the researchers can say "We are 95% confident that at least 90% of the population data lie between 53.8309 and 71.2691."

Prediction Intervals

Both one-sided and two-sided Prediction Intervals are available in the Distribution platform for both the mean and standard deviation of a future sample.

Prediction intervals are intervals that, with a specified degree of confidence, contain either a single observation, or the mean and standard deviation of the next randomly selected sample.

When you select **Prediction Interval** from a variable's drop-down menu, you are presented with the following dialog.

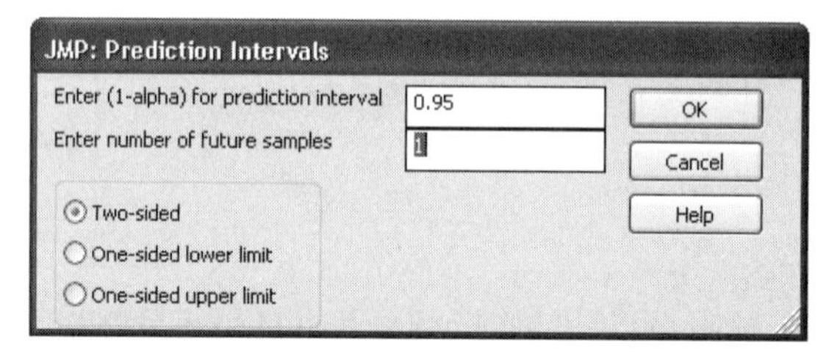

Enter the 1-α for the prediction interval in the first edit box. Since the prediction interval is dependent on the size of the future observation, you are required to enter the future sample size in the second field (we chose 2). Select lower, upper, or two-sided limits with the radio buttons and click **OK**. The following output (for a two-sided report in this case) is appended to the Distribution report.

Prediction Interval

Parameter	Estimate	Lower PI	Upper PI	1-Alpha	Future N
Mean	62.55	56.33252	68.76748	0.950	2
Std Dev	4.242338	0.133802	9.890012		

Interpretation of Prediction Intervals

If from many independent pairs of random samples, a 100(1–α)% prediction intervals is computed from the data of the first sample to contain the value(s) of the second sample, 100(1–α)% of the intervals would, in the log run, correctly bracket the future value(s). (Hahn and Meeker 1991)

Computation Formulas

The formulas JMP uses for computation of prediction intervals are as follows.

About a future individual or mean:

$$[Y_{lower}, Y_{upper}] = \bar{X} \pm t_{\left(1-\frac{\alpha}{2}, n-1\right)} \times \sqrt{\frac{1}{m}+\frac{1}{n}} \times s$$

About a future standard deviation:

$$[s_{lower}, s_{upper}] = \left[s \times \sqrt{\frac{1}{F_{\left(1-\frac{\alpha}{2};(n-1,\, m-1)\right)}}},\; s \times \sqrt{F_{\left(1-\frac{\alpha}{2};(n-1,\, m-1)\right)}} \right]$$

For references, see Hahn and Meeker (1991), pages 61-64.

Capability Analysis

The **Capability Analysis** option gives a capability analysis for quality control applications. The capability study measures the conformance of a process to given specification limits. It assumes that the data is Normally distributed. If the data is not Normally distributed, you may want to fit a Capability analysis using one of the other distributions ("The Fit Distribution Commands," p. 61)

A dialog prompts you for **Lower Spec Limit**, **Upper Spec Limit**, and **Target**. You only have to enter one of the three values. Only those fields you enter are part of the resulting Capability Analysis table. Optionally, you can enter a known value for sigma, the process standard deviation.

Capability Analyses can calculate capability indices using several different short-term estimates for σ. After requesting a **Distribution**, select **Capability Analysis** from the popup menu on the outline bar for the variable of interest. The Dialog box shown in Figure 4.2 appears, allowing specification of long-term or one or more short-term sigmas, grouped by a column or a fixed sample size and distribution type.

Figure 4.2 Capability Analysis Dialog Box

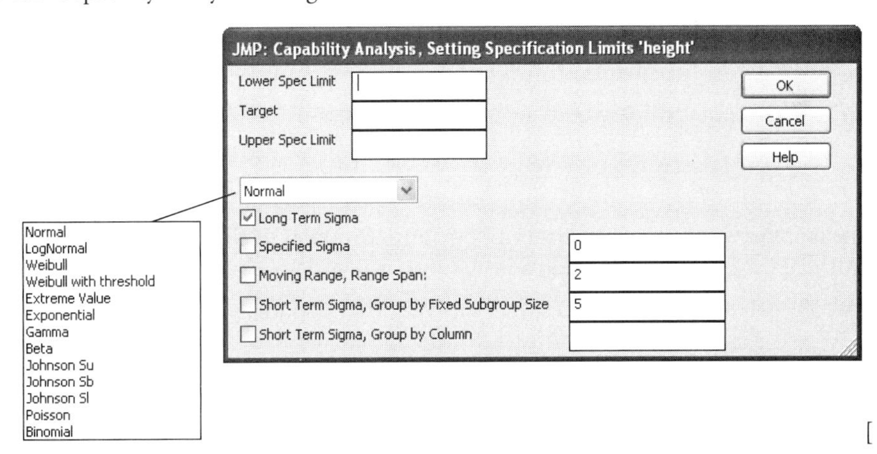

[

All capability analyses use the same formulas. The difference between the options lies in how sigma is computed. These options for sigma can be explained as:

- **Long-term** uses the overall sigma. This is the option used for P_{pk} statistics, and has sigma computed as

$$\sigma = \sqrt{\sum_{i=1}^{n} \frac{(x_i - \bar{x})^2}{n-1}}$$

- **Specified Sigma** allows the user to enter a specific, known sigma used for computing capability analyses. Sigma is, obviously, user-specified and is therefore not computed. This is the option used for control chart-generated capability analyses, where the sigma used in the chart is entered (in the dialog) as the specified sigma.
- Moving Range allows the user to enter a range span, where sigma is computed as

  $$\sigma = \frac{\bar{R}}{d_2(n)}$$

  $\bar{R}$ =the average of the moving ranges

  $d_2(n)$ = expected value of the range of n independent normally distributed variables with unit standard deviation
- **Short Term, Grouped by fixed subgroup size** computes σ using the following formula. In this case, if *r* is the number of subgroups and each *i*th subgroup is defined by the order of the data, sigma is computed as

  $$\sigma = \sqrt{\frac{\sum_{i=1}^{n} (x_{ij} - x_{i.})^2}{n - r - 1}}$$
- **Short Term, Grouped by Column** brings up a column list dialog from which you choose the grouping column. In this case, with *r* equal to the number of subgroups, sigma is computed as

  $$\sigma = \sqrt{\frac{\sum_{i=1}^{n} (x_{ij} - x_{i.})^2}{n - r - 1}}$$

(Note that this is the same formula for **Short Term, Grouped by fixed subgroup size** and is commonly referred to as the Root Mean Square Error or RMSE.)

Note: There is a preference for Distribution called **Ppk Capability Labeling** that will label the long-term capability output with P_{pk} labels. This option is found using **File > Preferences.**

When you click **OK**, the platform appends a Capability Analysis table, like the one in Figure 4.3, at the bottom of the text reports. You can remove and redo a Capability Analysis as many times as you want.

The specification limits can be stored and automatically retrieved as a column property. To do this, choose **Spec Limits** from the **Save** command menu. When you save the specification limits, they appear on the histogram when opened at a later time.

Figure 4.3 The Capability Analysis Table

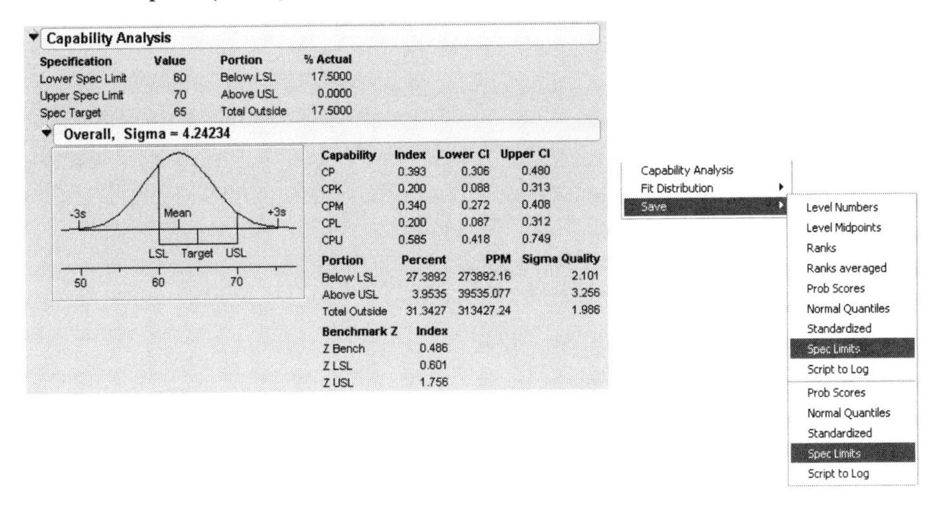

The Capability Analysis table is organized into two parts. The upper part of the table shows these quantities:

- The **Specification** column lists the names of items for which values are shown. They are **Lower Spec Limit**, **Upper Spec Limit**, and **Spec Target.**
- The **Value** column lists the values you specified for each limit and the target
- **%Actual** is the observed percent of data falling outside the specification limits.

The lower portion of the Capability Analysis table lists five basic process capability indexes, their values, and their upper and lower Confidence Intervals. It also lists the percent and PPM for areas outside the spec limits.The **PPM** column (parts per million) is the **Percent** column multiplied by 10,000.

This **Sigma Quality** measurement is frequently used in Six Sigma methods, and is also referred to as the *process sigma*.

$$\text{Sigma Quality} = \text{Normal Quantile}\left(1 - \frac{\text{Expected (\# defects)}}{n}\right) + 1.5$$

For example, if there are 3 defects in n=1,000,000 observations, the formula yields 6.03, or a 6.03 Sigma process. The above and below columns do not sum to the total column because Sigma Quality uses values from the Normal distribution, and is therefore not additive.

Table 4.1 "Capability Index Names and Computations," p. 57, describes these indices and gives computational formulas.

Table 4.1 Capability Index Names and Computations

Index	Index Name	Computation
CP	process capability ratio, C_p	(USL – LSL)/6*s* where USL is the upper spec limit LSL is the lower spec limit

Table 4.1 Capability Index Names and Computations (continued)

Index	Index Name	Computation
CIs for CP	Lower CI on CP	$CP\sqrt{\frac{\chi^2_{\frac{1-\alpha}{w},\, n-1}}{n-1}}$
	Upper CI on CP	$\sqrt{\frac{\chi^2_{\frac{1-(1-\alpha)}{w},\, n-1}}{n-1}}$
CPK (PPK for AIAG)	process capability index, C_{pk}	min(CPL, CPU)

Table 4.1 Capability Index Names and Computations (continued)

Index	Index Name	Computation
CIs for CPK	Expected Value $E(\hat{C}_{pk})$	Let $$c = \frac{\sqrt{n}\left(\mu - \frac{USL + LSL}{2}\right)}{\sigma}$$ denote the noncentrality parameter and $$d = \frac{USL - LSL}{\sigma}$$ represent the specification limit in σ units. Then the expected value is $$\frac{1}{6}\sqrt{\frac{n-1}{2n}}\frac{\Gamma\left(\frac{n-2}{2}\right)}{\Gamma\left(\frac{n-1}{2}\right)}\left(d\sqrt{n} - 2\sqrt{\frac{2}{\pi}}\exp\left(\left(\frac{-c^2}{2}\right) - 2c(1 - 2\Phi(-c))\right)\right)$$
	Variance $\mathrm{Var}(\hat{C}_{pk})$	Using c and d from above, the variance is $$\left[\frac{d^2}{36}\left(\frac{n-1}{n-3}\right)\right] - \left[\frac{d}{9\sqrt{n}}\left(\frac{n-1}{n-3}\right)\right]\left[\sqrt{\frac{2}{\pi}}\exp\left(\frac{-c^2}{2}\right) + c(1 - 2\Phi(-c))\right]$$ $$+ \frac{1}{9}\left[\frac{n-1}{n(n-3)}\right](1 + c^2) - \left(\frac{n-1}{72n}\right)\left[\frac{\Gamma\left(\frac{n-2}{2}\right)}{\Gamma\left(\frac{n-1}{2}\right)}\right]^2$$ $$\left(d\sqrt{n} - 2\sqrt{\frac{2}{\pi}}\exp\left(\left(\frac{-c^2}{2}\right) - 2c(1 - 2\Phi(-c))\right)\right)^2$$
	Lower CI	$E(\hat{C}_{pk}) - k\sqrt{\mathrm{Var}(\hat{C}_{pk})}$
	Upper CI	$E(\hat{C}_{pk}) + k\sqrt{\mathrm{Var}(\hat{C}_{pk})}$
CPM	process capability index, C_{pm}	$$\frac{\min(\text{target} - LSL, USL - \text{target})}{3\sqrt{s^2 + (\text{mean} - \text{target})^2}}$$

Table 4.1 Capability Index Names and Computations (continued)

Index	Index Name	Computation
CIs for CPM	Lower CI on CPM	$\mathrm{CPM}\sqrt{\dfrac{\chi^2_{\frac{1-\alpha}{2},\gamma}}{\gamma}}$
	Upper CI on CPM	$\mathrm{CPM}\sqrt{\dfrac{\chi^2_{\frac{1-(1-\alpha)}{2},\gamma}}{\gamma}}$
CPL	process capability ratio case of one-sided lower specification	(mean – LSL)/3*s* where *s* is the estimated standard deviation
CPU	process capability ratio of one-sided upper specification	(USL - mean)/3*s*

In Japan, a capability index of 1.33 is considered to be the minimum acceptable. For a normal distribution, this gives an expected number of nonconforming units of about 6 per 100,000.

Exact 100(1 – α)% lower and upper confidence limits for CPL are computed using a generalization of the method of Chou et al. (1990), who point out that the 100(1 – α) lower confidence limit for CPL (denoted by CPLLCL) satisfies the equation

$$\Pr\{T_{n-1}(\delta = 3\sqrt{n})\mathrm{CPLLCL} \le 3\mathrm{CPL}\sqrt{n}\} = 1-\alpha$$

where $T_{n-1}(\delta)$ has a non-central *t*-distribution with $n-1$ degrees of freedom and noncentrality parameter δ.

Exact 100(1 – α)% lower and upper confidence limits for CPU are also computed using a generalization of the method of Chou et al. (1990). who point out that the 100(1 – α) lower confidence limit for CPU (denoted CPULCL) satisfies the equation

$$\Pr\{T_{n-1}(\delta = 3\sqrt{n})(\mathrm{CPULCL} \ge 3\mathrm{CPL}\sqrt{n})\} = 1-\alpha$$

where $T_{n-1}(\delta)$ has a non-central *t*-distribution with $n-1$ degrees of freedom and noncentrality parameter δ.

At the bottom of the report, *Z* statistics are reported. *Z* represents (according to the AIAG Statistical Process Control manual) the number of standard deviation units from the process average to a value of interest such as an engineering specification. When used in capability assessment, *Z* USL is the distance to the upper specification limit and *Z* LSL is the distance to the lower specification limit.

Z USL = (USL-Xbar)/sigma = 3 * CPU

Z LSL = (Xbar-LSL)/sigma = 3 * CPL

Z Bench = Inverse Cumulative Prob(1 - P(LSL) - P(USL))

where

P(LSL) = Prob(X < LSL) = 1 - Cum Prob(*Z* LSL)

P(USL) = Prob(X > USL) = 1 - Cum Prob(*Z* USL)

Note: You can also do a non-normal capability analysis through **Fit Distribution** options, described in the next section. After you fit a distribution, you have the option to generate quantiles and a target value for the fitted distribution. If you give a target value, a capability analysis is automatically generated by using the quantile values and target you specified.

Fit Distribution

The **Fit Distribution** command gives you a list of distributions that you can fit to your analysis variable. When you select a distribution, a table that shows the fitted distribution parameter estimates is appended to the report. You can repeat this fitting process as many times as you want.

The Fit Distribution Commands

The next section discusses the distributions given by the **Fit Distribution** command.

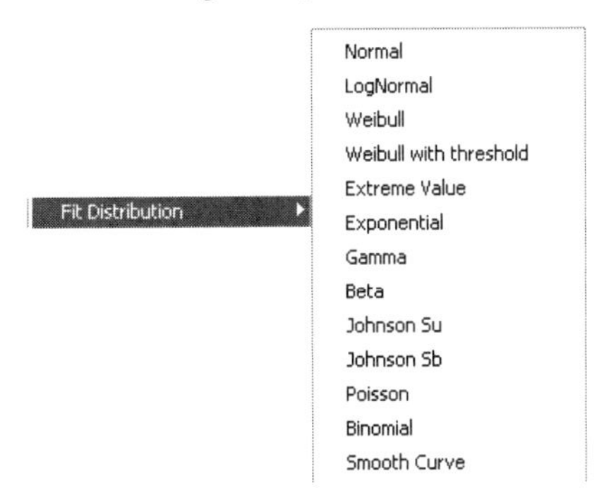

Note: Each fitted distribution has a popup menu on its title bar with additional options for that fit. These options are discussed in the next section.

Normal

The **Normal** fitting option estimates the parameters of the normal distribution based on the analysis sample. The parameters for the normal distribution are μ (the mean), which defines the location of the distribution on the *x*-axis and σ (standard deviation), which defines the dispersion or spread of the distribution. The standard normal distribution occurs when $\mu = 0$ and $\sigma = 1$. The Parameter Estimates table for the normal distribution fit shows mu (estimate of μ) and sigma (estimate of σ), with upper and lower 95% confidence limits.

The next figure shows a normal (bell-shaped) curve drawn by the **Normal** option. The data were generated by the `Random Normal()` random number function. The normal curve is often used to model measures that are symmetric with most of the values falling in the middle of the curve. You can choose **Normal** fitting for any set of data and test how well a normal distribution fits your data.

The Shapiro-Wilk test for normality is reported when the sample size is ≤2000, and the KSL test is computed for samples >2000.

Note: This example shows the **Density Curve** option from the **Fitted Normal** popup menu. This option overlays the density curve on the histogram using the parameter estimates from the sample data. See the next section for options available for each distribution fit.

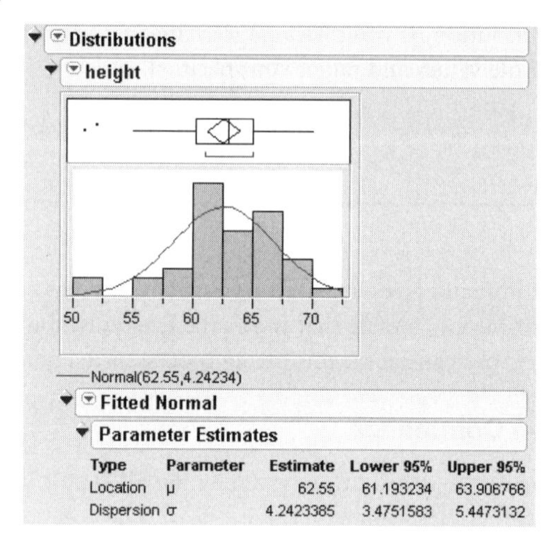

LogNormal

The **LogNormal** fitting option estimates the parameters μ and σ for the two-parameter lognormal distribution for a variable Y where Y is lognormal if and only if $X = \ln(Y)$ is normal. The data must be positive.

Weibull, Weibull with threshold, and Extreme Value

The Weibull distribution has different shapes depending on the values of α (a scale parameter) and β (a shape parameter). It often provides a good model for estimating the length of life, especially for mechanical devices and in biology. The two-parameter Weibull is the same as the three-parameter Weibull with a threshold parameter of zero.

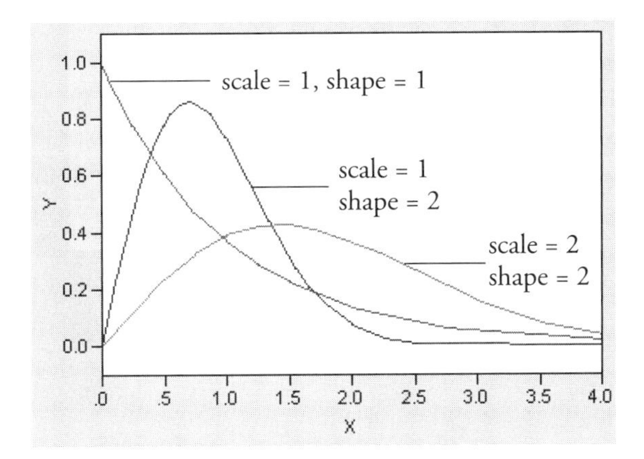

Weibull with threshold adds a threshold centrality parameter to the distribution. JMP estimates this third parameter, but if you know what it should be, you can set it through the **Fix Parameters** option.

The **Extreme Value** distribution is a two parameter Weibull (α, β) distribution with the transformed parameters $\delta = 1 / \beta$ and $\lambda = \ln(\alpha)$.

Exponential

The exponential distribution is a special case of the Weibull and gamma distributions when the shape parameter, β, is 1. The exponential distribution is especially useful for describing events that randomly occur over time, such as survival data. Another example is elapsed time between the occurrence of non-overlapping events such as the time between a user's computer query and response of server, arrival of customers at a service desk, or calls coming in at a switchboard.

Devore (1995) notes that an exponential distribution is *memoryless*, meaning that if you check a component after *t* hours and it is still working, the distribution of additional lifetime (the conditional probability of additional life given having lived) is the same as the original distribution.

Gamma

The **Gamma** fitting option estimates the gamma distribution parameters, $\alpha > 0$ and $\sigma > 0$, from the analysis sample. The parameter α, called alpha in the fitted gamma report, describes shape or curvature. The parameter σ, called sigma, is the scale parameter of the distribution.

Note: A third parameter, called Threshold, is the lower endpoint parameter and must be greater than the minimum data value in the data. It is set to zero by default. However, you can set its value with the **Fix Parameters** option in the report popup menu for the fitted distribution. It is a constant used in the density formula to adjust each sample value:

- The *standard* gamma distribution has $\sigma = 1$. Beta is called the scale parameter because values other than 1 stretch or compress the distribution along the *x*-axis.
- The χ^2-distribution is the family of gamma curves that occur when $\sigma = 2$ and $\theta = 0$.
- The exponential distribution is the family of gamma curves that occur when $\alpha = 1$ and $\theta = 0$.

Figure 4.4 shows the shape of gamma probability density functions for various values of α. The data for the histograms were generated using the `Random Gamma()` function with arguments 1, 3, and 5. The standard gamma density function is strictly decreasing when $\alpha \leq 1$. When $\alpha > 1$, the density function begins at zero when $x = 0$, increases to a maximum, and then decreases.

Figure 4.4 Histograms of Gamma Distributions for $\alpha = 1$, $\alpha = 3$, and $\alpha = 5$

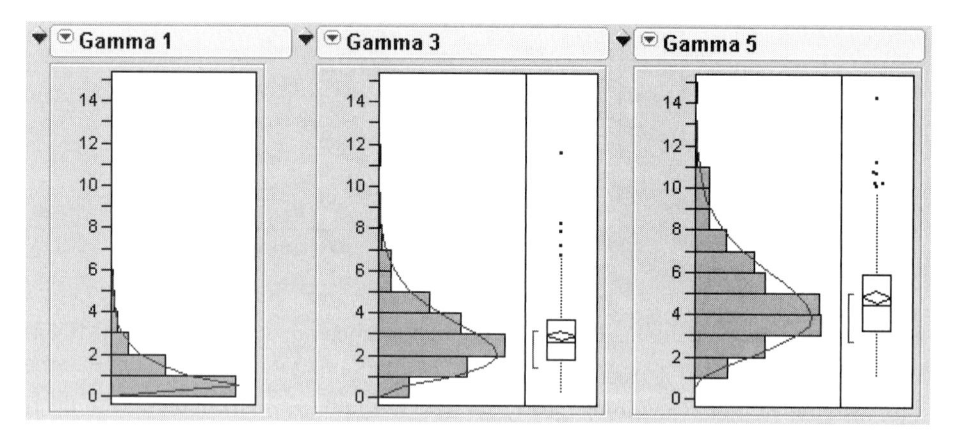

Beta

The **Beta** distribution fitting option estimates two beta distribution shape parameters, $\alpha > 0$ and $\beta > 0$ from the analysis sample. There are also a lower threshold parameter $\theta = A$, and σ, which are used to define the upper threshold as $\theta + \sigma = B$. The beta distribution has values only for the interval defined by $A \leq X \leq B$. The standard beta distribution occurs when $A = 0$ and $B = 1$.

You can set parameters to fixed values with the **Fix Parameters** option in the fitted distribution report popup menu, but the upper threshold must be greater than the maximum data value and lower thresholds must be less than the minimum data value.

The standard beta distribution is sometimes useful for modeling the probabilistic behavior of random variables such as proportions constrained to fall in the interval 0, 1.

Johnson Su, Johnson Sb

The Johnson system of distributions contains three distributions which are all based on a transformed normal distribution. These three distributions are the Johnson Su, which is unbounded for Y, the Johnson Sb, which is bounded on both tails ($0 < Y < 1$), and the Johnson Sl, leading to the lognormal family of distributions.

Johnson distributions are popular because of their flexibility. In particular, the Johnson distribution system is noted for its data-fitting capabilities because it supports every possible combination of skewness and kurtosis.

If Z is a standard normal variate, then the system is defined by

$$Z = \gamma + \delta f(y)$$

where, for the Johnson Su

$$f(y) = \ln\left(Y + \sqrt{1 + Y^2}\right) = \sinh^{-1} Y$$

$$Y = \frac{X-\theta}{\sigma} \qquad -\infty < X < \infty$$

for the Johnson Sb

$$f(Y)) = \ln\left(\frac{Y}{1-Y}\right)$$

$$Y = \frac{X-\theta}{\sigma} \qquad \theta < X < \theta + \sigma$$

and for the Johnson Sl, where $\sigma = \pm 1$.

$$f(Y) = \ln(Y)$$

$$Y = \frac{X-\theta}{\theta} \qquad \begin{array}{ll} \theta < X < \infty & \text{if } \sigma = 1 \\ -\infty < X < \theta & \text{if } \sigma = -1 \end{array}$$

Note: The S refers to system, the subscript of the range. Although we implement a different method, information on selection criteria for a particular Johnson system can be found in Slifker and Shapiro (1980).

As an example, examine the JohnsonSuExample.jmp data set. In this distribution, two-thirds of the data is between zero and five, with the minimum at –18.81. A Johnson Su gives a good fit of the data.

Figure 4.5 Johnson Su Fit

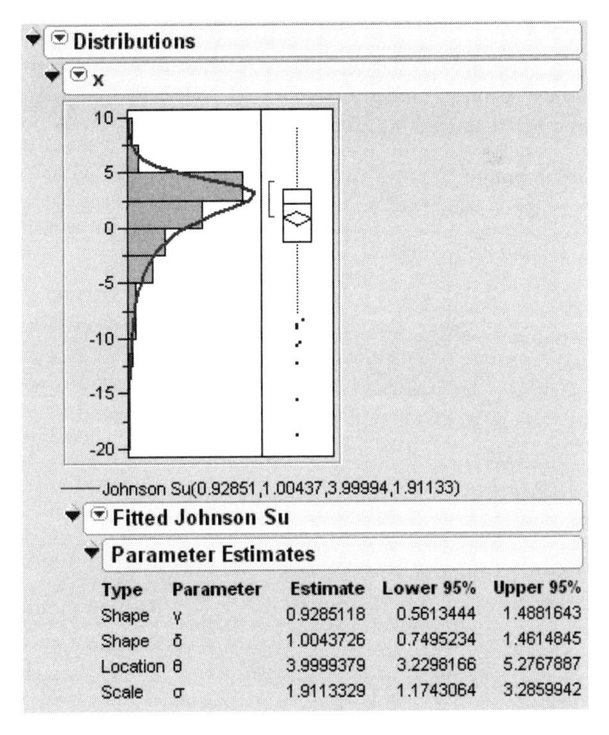

Type	Parameter	Estimate	Lower 95%	Upper 95%
Shape	γ	0.9285118	0.5613444	1.4881643
Shape	δ	1.0043726	0.7495234	1.4614845
Location	θ	3.9999379	3.2298166	5.2767887
Scale	σ	1.9113329	1.1743064	3.2859942

Poisson

The **Poisson** distribution estimates a single scale parameter $\lambda > 0$ from the analysis sample. You can fix the value of this parameter using the **Fix Parameters** option in the fitted distribution report.

The confidence intervals for the Poisson are formed using Maximum Likelihood techniques.

For sample sizes less than or equal to 100, the goodness-of-fit test is formed using two one-sided exact Kolmogorov tests combined to form a near exact test (see Conover 1972 for details). For sample sizes greater than 100, a Pearson Chi-Squared goodness-of-fit test is performed.

Note: Since the Poisson distribution is a discrete distribution, its smooth curve (or PDF) is actually a step function, with jumps occurring at every integer.

Binomial

The Binomial distribution fits a binomial distribution to data in two formats. When Binomial is selected, a dialog appears allowing you to specify one of the following:

- Successes are in a column and there are a constant number of trials, which you enter in the dialog
- The number of successes are in one column and the number of trials are in a separate column.

Smooth Curve

The **Smooth Curve** command fits a smooth curve to the continuous variable histogram using nonparametric density estimation. The smooth curve displays with a slider beneath the plot. You can use the slider to set the kernel standard deviation. The estimate is formed by summing the normal densities of the kernel standard deviation located at each data point. By changing the kernel standard deviation you can control the amount of smoothing.

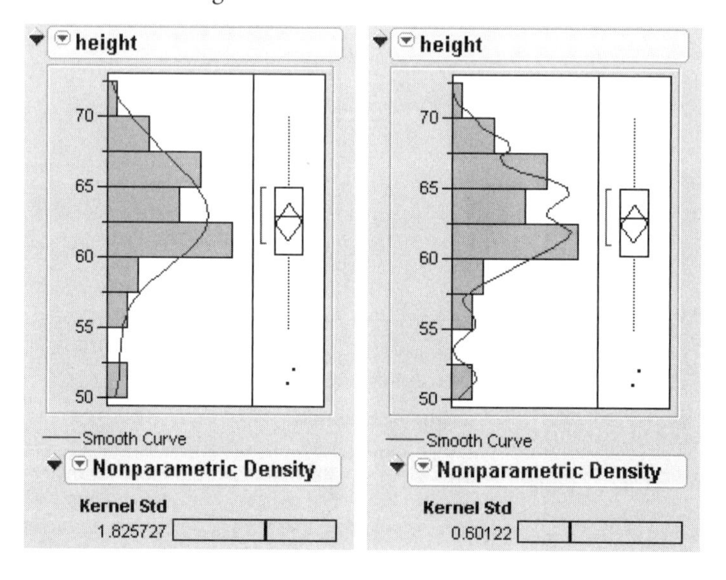

Fit Distribution Options

Each fitted distribution table also has the popup menu, as shown here, that gives options to enhance the fitted distribution report.

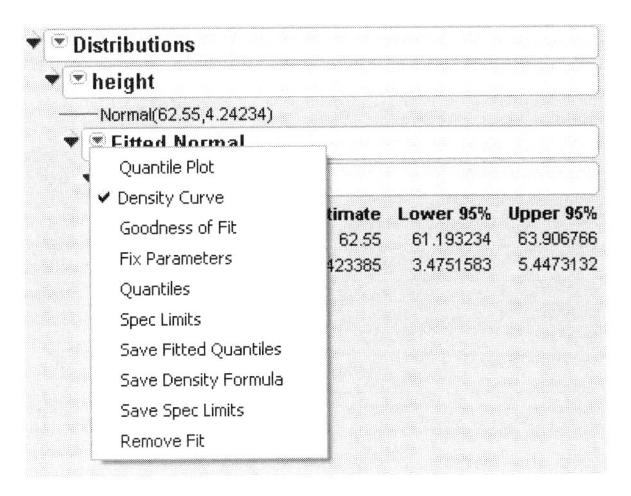

Quantile Plot generates a quantile plot with the selected distribution quantiles on the *y*-axis, and variable values on the *x*-axis.

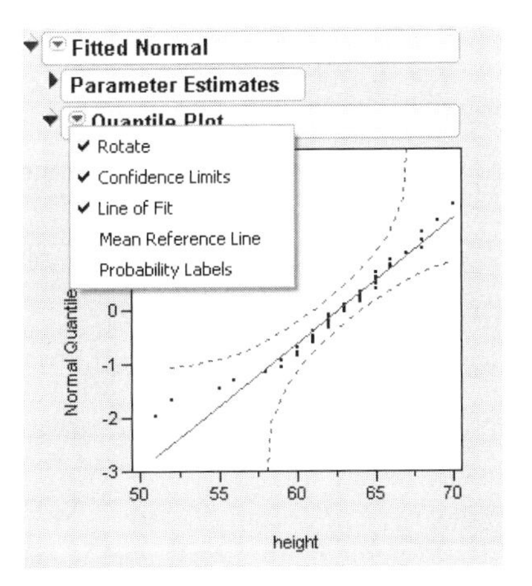

The Quantile plot has its own set of options:

- **Rotate** reverses the *x*- and *y*-axes.
- **Confidence Limits** draws Lilliefors 95% confidence limits for the Normal Quantile plot, and 95% equal precision bonds with $a = 0.001$ and $b = 0.99$ for all other quantile plots (Meeker and Escobar (1998)).
- **Line of Fit** draws the straight diagonal reference line. If a variable fits the selected distribution, the values fall approximately on the reference line.
- **Mean Reference Line** draws a horizontal line at the mean of the response.
- **Probability Labels** adds a probabilities axis on the right side of the quantile plot.

Density Curve uses the estimated parameters of the distribution to overlay a density curve on the histogram. The descriptions of the Normal and the Beta distributions, described previously, show examples of overlaid density curves.

Goodness of Fit computes the goodness of fit test for the fitted distribution. Note that the goodness-of-fit tests are not Chi-square tests, but rather are EDF (Empirical Distribution Function) tests. The EDF tests offer advantages over the Chi-square, including improved power and invariance with respect to histogram midpoints.

The following table shows which test JMP conducts for each distribution.

Table 4.2 JMP Goodness of Fit Tests

Distribution	Parameters	JMP gives this test
Beta	α and β are known	Kolmogorov's *D*
	either α or β unknown	(none)
Exponential	σ known or unknown	Kolmogorov's *D*
Gamma	α and σ known	Cramer von Mises W^2
	either α or σ unknown	(none)
Normal	μ and σ are unknown	Shapiro-Wilk (for less than 2000 data points) Kolmogorov-Smirnoff-Lillefors (For 2000 data points or more.)
	μ and σ are both known	Kolmogorov-Smirnoff-Lillefors
	either μ or σ is known	(none)
LogNormal	μ and σ are known or unknown	Kolmogorov's *D*
Binomial	*p* and *n* are known	Kolmogorov's *D*

Fix Parameters appends a dialog to the report, which you use to key specific parameter values into a User-defined Value column.

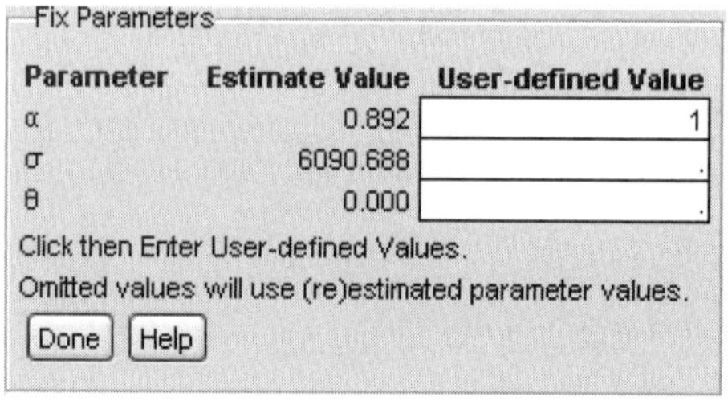

Fix Parameters

Parameter	Estimate Value	User-defined Value
α	0.892	1
σ	6090.688	.
θ	0.000	.

Click then Enter User-defined Values.
Omitted values will use (re)estimated parameter values.
Done Help

When you click **Done**, a fixed parameters table is generated that shows the parameter values you entered. All parameters that remain non-fixed are re-estimated considering the fixed constraints. Confidence limits for the other parameters are recomputed based on the fixed parameters. The null hypothesis for the adequacy test is that the new restricted parameterization adequately describes the data. This means that small *p*-values reject this hypothesis and conclude the new parameterization is *not* adequate.

Fitted Gamma

Fixed Parameters

Parameter	Value	Fixed	Lower 95%	Upper 95%
α	1.000	*	.	.
σ	6090.688		3664.9074	11501.607
θ	0.000		.	.

Adequacy LR Test

ChiSquare	p-Value
12.9754	0.0015220

Note: H0=New restricted parameterization adequately describes data. Small p-values are evidence against new parameterization.

Quantiles returns the unscaled and uncentered quantiles for the specific upper and lower that you specify in a dialog displayed by the **Quantile** option.

Spec Limits computes generalizations of the standard capability indices using the fact that for the normal distribution, 3σ is both the distance from the lower 0.135 percentile to median (or mean) and the distance from the median (or mean) to the upper 99.865 percentile. These percentiles are estimated from the fitted distribution, and the appropriate percentile-to-median distances are substituted for 3σ in the standard formulas.

Writing T for the target, LSL and USL for the lower and upper specification limits and P_α for the 100th percentile, the generalized capability indices are as follows.

$$C_{pl} = \frac{P_{0.5} - LSL}{P_{0.5} - P_{0.00135}}$$

$$C_{pu} = \frac{USL - P_{0.5}}{P_{0.99865} - P_{0.5}}$$

$$C_p = \frac{USL - LSL}{P_{0.99865} - P_{0.00135}}$$

$$C_{pk} = \min\left(\frac{P_{0.5} - LSL}{P_{0.5} - P_{0.00135}}, \frac{USL - P_{0.5}}{P_{0.99865} - P_{0.5}}\right)$$

$$K = 2 \times \frac{\left|\frac{1}{2}(USL + LSL) - P_{0.5}\right|}{USL - LSL}$$

$$C_{pm} = \frac{\min\left(\frac{T - LSL}{P_{0.5} - P_{0.00135}}, \frac{USL - T}{P_{0.99865} - P_{0.5}}\right)}{\sqrt{1 + \left(\frac{\mu - T}{\sigma}\right)^2}}$$

If the data are normally distributed, these formulas reduce to the formulas for standard capability indices (Table 4.1 "Capability Index Names and Computations," p. 57)

Save Fitted Quantiles saves the fitted quantile values as a new column in the current data table.

Save Density Formula creates a new column in the current data table that has fitted values computed by the density formula using the estimated parameters from the response variable as arguments.

Save Spec Limits saves the spec limits given in a Capability Analysis to the column.

Remove Fit removes the distribution fit information from the report window.

Chapter 5

Capability Analyses

The Capability Platform

Capability analysis, used in quality control, measures the conformance of a process to given specification limits. Using these limits, you can compare a current process to specific tolerances and maintain consistency in production. Graphical tools such as the goalpost plot and box plot give you quick visual ways of observing within-spec behaviors.

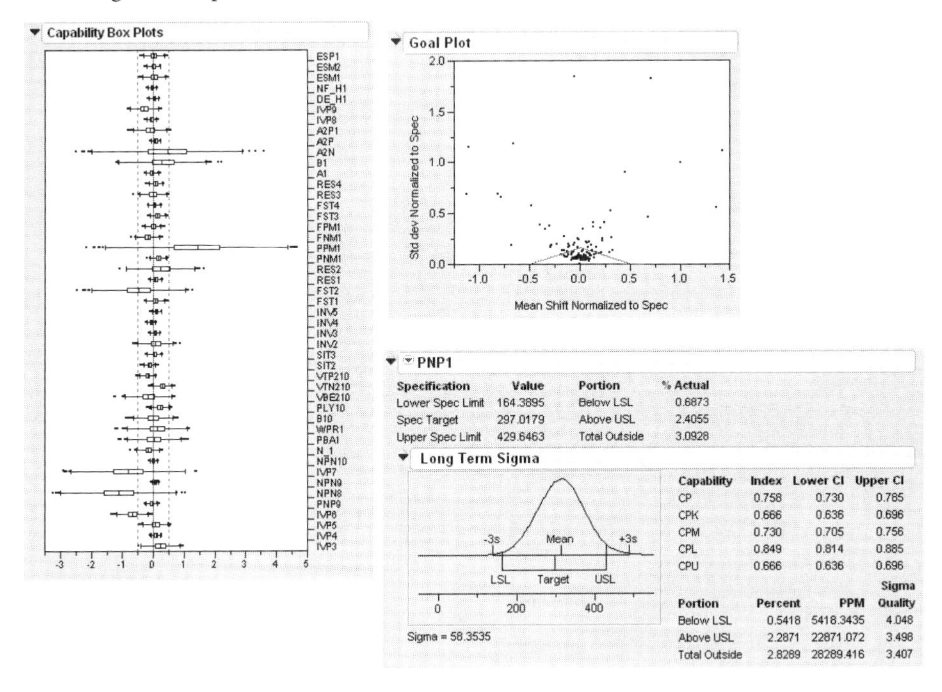

Contents

Launch the Platform

We use Cities.jmp and Semiconductor Capability.jmp in the following examples. To launch the Capability platform, select Capability from the Graph menu. This presents you with the following dialog box.

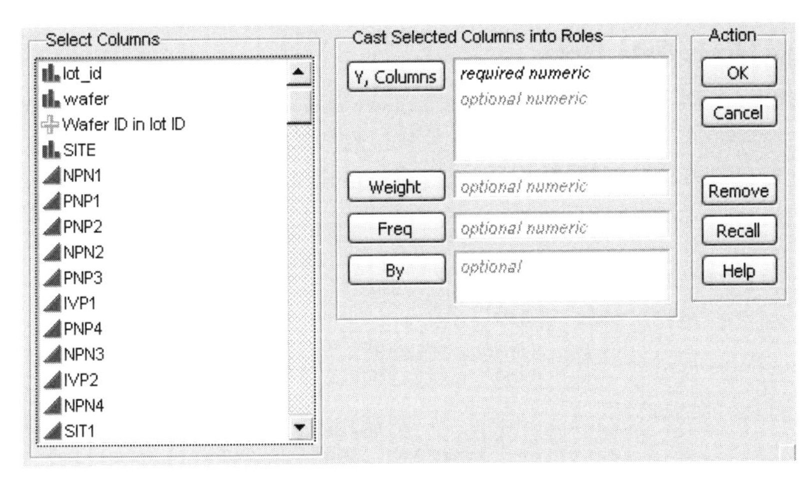

Here, you select the variables that you want to analyze. After assigning the desired variables to **Y, Columns**, click **OK** to bring up the Specification Limits dialog. Columns selected in the launch dialog are listed here, with entry fields for the lower specification limit (LSL), target, and upper specification limit (USL).

Entering Limits

At this point, specification limits should be entered for each variable. Note that manually adding the limits at this point is only one of the available methods for entering them.

1 If the limits are already stored in the data table, they can be imported using the **Import Spec Limits** command.

2 You can enter the limits as Column Properties, thereby bypassing the spec limits dialog

3 You can enter the limits on the dialog.

4 You can enter them using JSL.

Using JSL

Spec limits can be read from JSL statements or from a spec limits table.

As an example of reading in spec limits from JSL, consider the following code snippet, which places the spec limits inside a `Spec Limits()` clause.

```
// JSL for reading in spec limits
Capability(
  Y( :OZONE, :CO, :SO2, :NO ),
  Capability Box Plots( 1 ),
  Spec Limits(
```

```
      OZONE( LSL( 0 ), Target( 0.05 ), USL( 0.1 ) ),
      CO( LSL( 5 ), Target( 10 ), USL( 20 ) ),
      SO2( LSL( 0 ), Target( 0.03 ), USL( 0.08 ) ),
      NO( LSL( 0 ), Target( 0.025 ), USL( 0.6 ) )
   )
);
```

Using a Limits Data Table

A spec limits table can be in two different formats: *wide* or *tall.* Figure 5.1 shows an example both types.

Figure 5.1 Tall (top) and Wide (bottom) Spec Limit Tables

	Column 1	_LSL	_Target	_USL
1	OZONE	0.075	0.15	0.25
2	CO	5	7	12
3	SO2	0.01	0.04	0.09
4	NO	0.01	0.025	0.04

	_LimitsKey	Max deg. F Jan	OZONE	CO	SO2
1	_LSL	15	0	•	0
2	_Target	40	0.1	7	0.05
3	_USL	•	0.3	20	0.1

A tall table has one row for each column analyzed in Capability, with four columns. The first holds the column names. The other three columns need to be named, _LSL, _USL, and _TARGET.

A wide table has one column for each column analyzed in Capability, with three rows plus a _LimitsKey column. In the _LimitsKey column, the three rows need to contain the identifiers _LSL, _USL, and _TARGET.

Either of these formats can be read using the **Import Spec Limits** command.

Using a Limits Table and JSL

There is no extra syntax needed to differentiate between the two table types when they are read using JSL. The following syntax works for either table. It places the spec limits inside an `Import Spec Limits()` clause.

```
// JSL for reading in a spec limits file
Capability(
   Y( :OZONE, :CO, :SO2, :NO ),
   Capability Box Plots( 1 ),
   Spec Limits(
      Import Spec Limits(
         "<path>/filename.JMP"
);
```

Using a Dialog

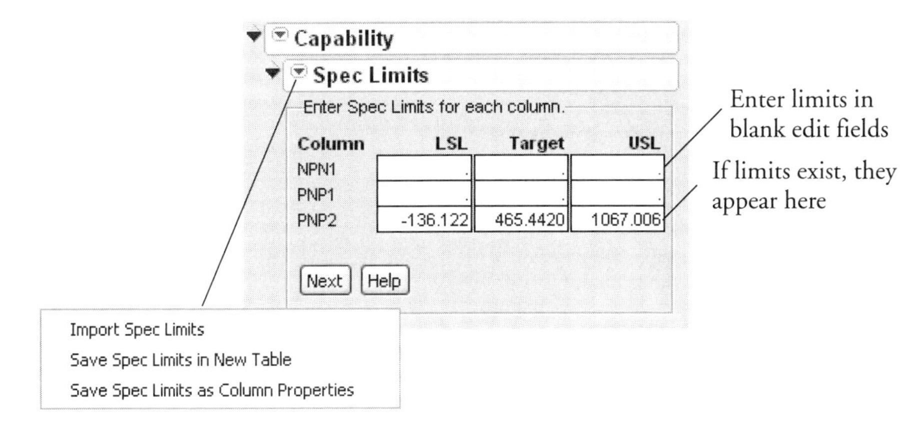

After entering or loading the specification limits, you can save them to the data table using the **Save Spec Limits as Column Properties** command.

You can also save the specification limits to a new data table with the **Save Spec Limits in New Table** command.

Capability Charts and Graphs

By default, JMP shows a goalpost plot and capability box plots. Using the platform menu, you can additionally add Normalized box plots, capability indices, and a summary table, as well as display a capability report for each individual variable in the analysis.

Goal Plot

The Goal plot shows, for each variable, the spec-normalized mean shift on the *x*-axis, and the spec-normalized standard deviation on the *y*-axis.

For each column with LSL, Target, and LSL, these quantities are defined as

Mean Shift Normalized to Spec = Mean(Col[*i*] - Target) / (USL - LSL)

Standard Deviation Normalized to Spec = Standard Deviation(Col[*i*])/ (USL - LSL)

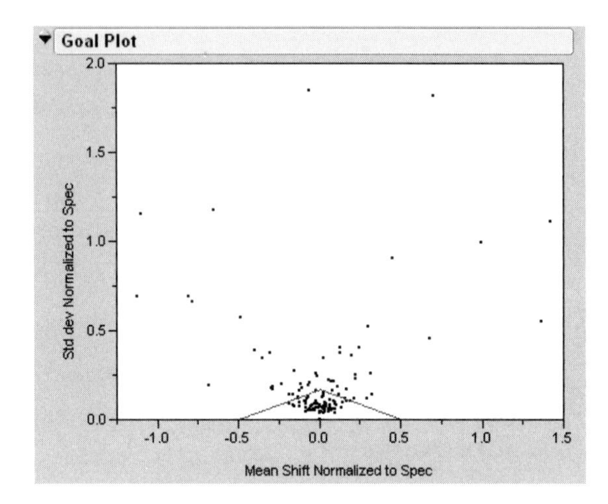

The red goal line represents a Cpk = 1, which approximates a defect rate contour line of 0.01. Points (which represent columns in this case) appearing outside the goal have Cpk < 1, which is likely problematic. Points above have a higher standard deviation. Points to the right or left are simply off-target.

Normalized Box Plots

When drawing Normalized box plots, JMP first standardizes each column by subtracting off the mean and dividing by the standard deviation. Next, quantiles are formed for each standardized column. The box plots are formed for each column from these standardized quantiles.

Figure 5.2 Normalized Box Plot

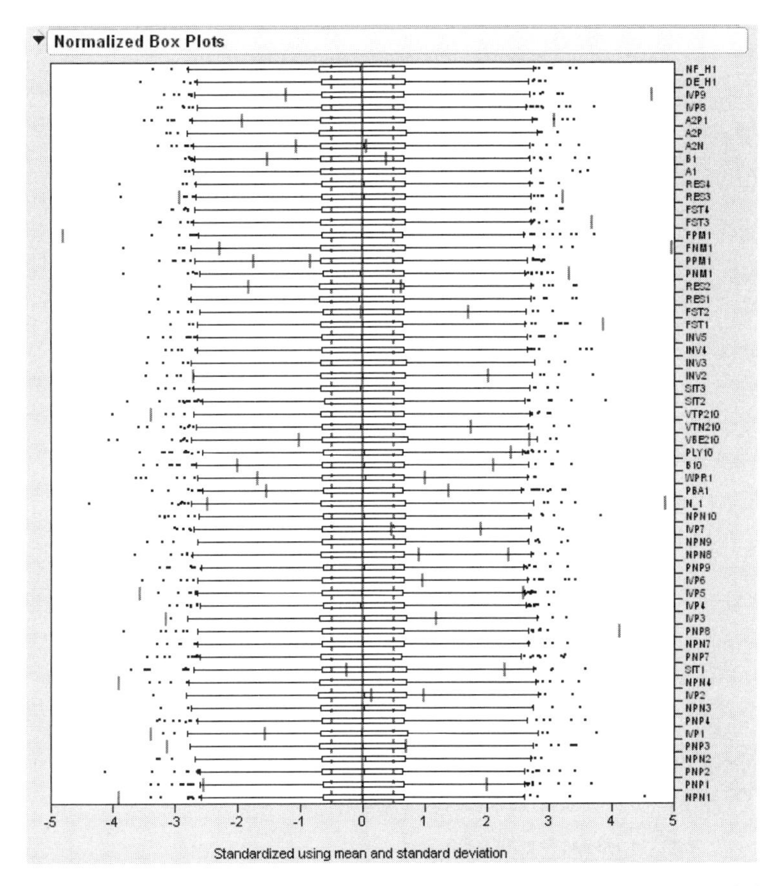

The green vertical lines represent the spec limits normalized by the mean and standard deviation. The gray vertical lines are drawn at ± 0.5 , since the data has all been standardized to a standard deviation of 1.

Capability Box Plots

Capability box plots also show a box plot for each variable in the analysis. The values for each column are centered by their target value and scaled by the difference between the specification limits. That is, for each column *Yj*,

$$Y_j = \frac{Y_{ij} - \text{Target}_j}{\text{USL}_j - \text{LSL}_j}$$

Figure 5.3 Capability Box Plot

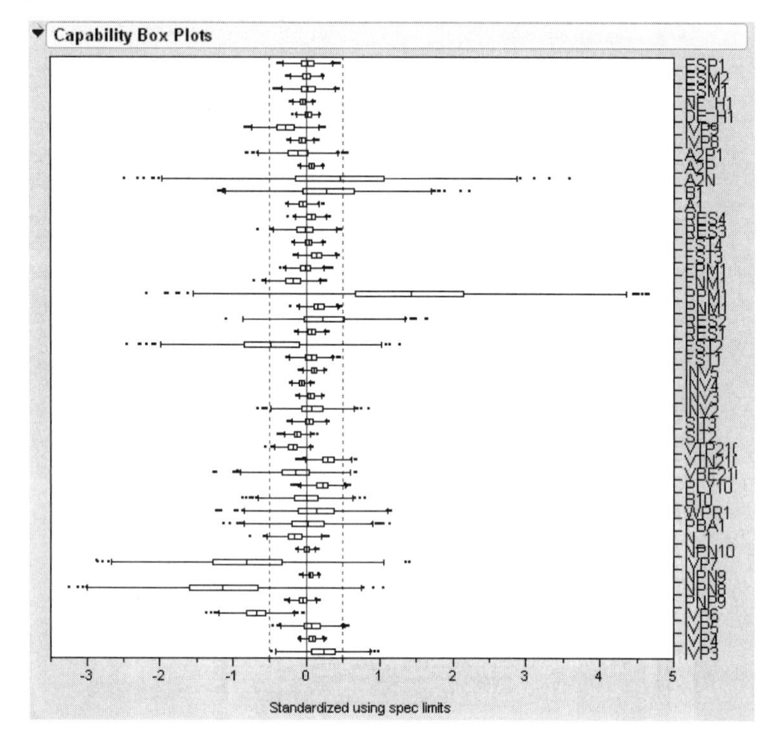

When a spec limit is missing for one or more columns, separate box plots are produced for those columns, with gray threshold lines, as shown here.

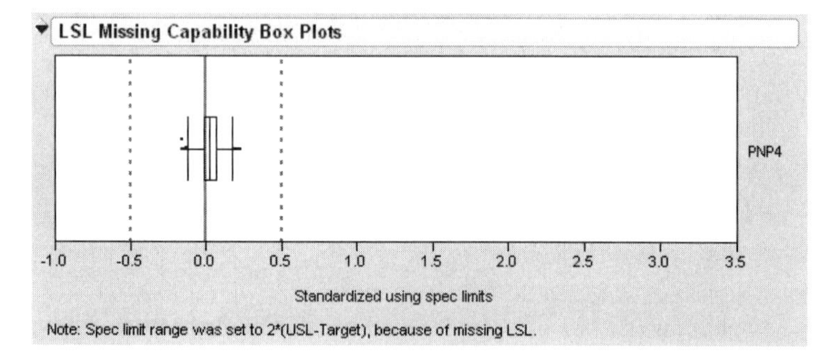

Individual Detail Reports

The Individual Detail Reports command shows a capability report for each variable in the analysis. This report is identical to the one from the Distribution platform, detailed in “Capability Analysis,” p. 55 in the “Advanced Univariate Analysis” chapter.

Figure 5.4 Individual Detail Report

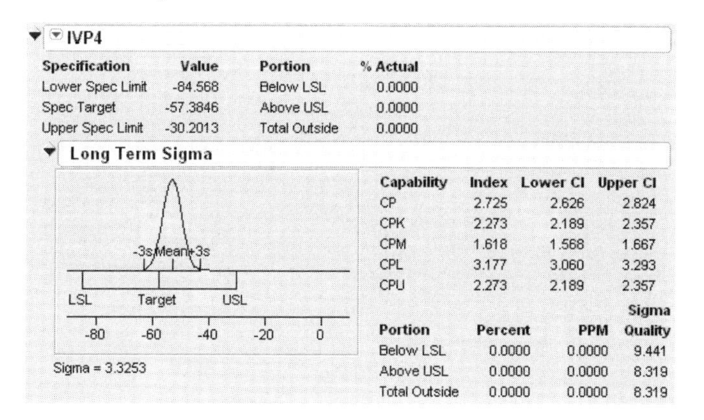

Platform Commands

In addition to the various graphs available in the platform, there are several other options on the report's main menu.

Make Summary Table makes a summary table that includes the variable's name, its spec-normalized mean shift, and its spec-normalized standard deviation.

	Variable	Mean Shift Normalized to Spec	Stdev Normalized to Spec
1	PLG 2	0.03737445	0.09481514
2	PNP5	0.05729886	-0.2081417
3	NPN6	-1.1031025	1.15507268
4	PNP6	•	•
5	PNP7	-0.0272153	0.05189182
6	NPN7	0.00932515	0.04226333
7	PNP8	0.08657827	0.10020075

Capability Indices Report shows or hides a table showing each variable's LSL, USL, target, mean, standard deviation, Cp, Cpk, and PPM. Optional columns for this report are CPM, CPL, CPU, Ppm Below LSL and Ppm Above USL. To reveal these optional columns, right-click on the report and select the column names from the **Columns** submenu.

Capability Indices

Columns	LSL	Target	USL	Mean	Standard Deviation	Cp	CpK	PPM
PLG 2	119.4695	131.7178	143.966	132.6333	2.322635	1.757807	1.6264	0.539725
PNP5	-41.9397	-46.6145	-51.7771	-47.1782	2.047573	-0.80074	-0.8528	1982391
NPN6	43.8949	44.39073	44.88655	43.29683	1.145429	0.144291	-0.1740	781798.2
PNP6	0	0	0	0	0	.	.	.
PNP7	-18.5253	-15.7487	-12.9721	-15.8998	0.288163	3.21181	3.0370	4.08e-14
NPN7	8.346467	24.42629	40.50612	24.72618	1.359174	3.943529	3.8700	9.55e-28

Goal Labels shows or hides all the tables for the points in the Goal plot.

Chapter 6

Bivariate Scatterplot and Fitting

The Fit Y by X Platform

The Bivariate platform lets you explore how the distribution of one continuous variable relates to another continuous variable. The Bivariate platform is the *continuous by continuous* personality of the **Fit Y by X** command. The analysis begins as a scatterplot of points, to which you can interactively add other types of fits:

- simple linear regression—fits a straight line through the data points
- polynomial regression of selected degree—fits a parabola or high order line
- smoothing spline—with a choice of the degree of smoothing
- bivariate normal density ellipses—graphically shows the correlation
- bivariate nonparametric density contours
- multiple fits over groups defined by a grouping variable.

Note: You can launch the Bivariate platform from the **Fit Y by X** main menu command (or tool bar) or with the **Bivariate** button on the JMP Starter. When you choose **Fit Y by X** from the **Analyze** menu, JMP performs an analysis by context, which is the analysis appropriate for the modeling types you specify for the response and factor variables. When you use the **Bivariate** button on the JMP Starter, the launch dialog expects both factor and response variables to be continuous and advises you if you specify other modeling types.

See the *JMP User Guide* for a discussion of data types and modeling types. For examples of more complex models, see in this volume the chapters "Standard Least Squares: Introduction," p. 213, "Standard Least Squares: Perspectives on the Estimates," p. 249, "Standard Least Squares: Exploring the Prediction Equation," p. 275, and "Standard Least Squares: Random Effects," p. 339, and "Multiple Response Fitting," p. 427.

Contents

Launch the Platform

When you choose **Fit Y by X** from the **Analyze** menu or tool bar or click the **Bivariate** button on the JMP Starter, you first see a Launch dialog (see Figure 6.2). After you select analysis variables and click **OK** on this dialog, **Fit Y by X** launches a platform with plots and accompanying text reports for each specified combination of response (*Y*) and factor (*X*) variables.

The type of platform varies according to the modeling types of the variables. Figure 6.1 illustrates the Fit Y by X report for continuous *X* and *Y* columns.

Figure 6.1 Fit Y by X Regression Platform

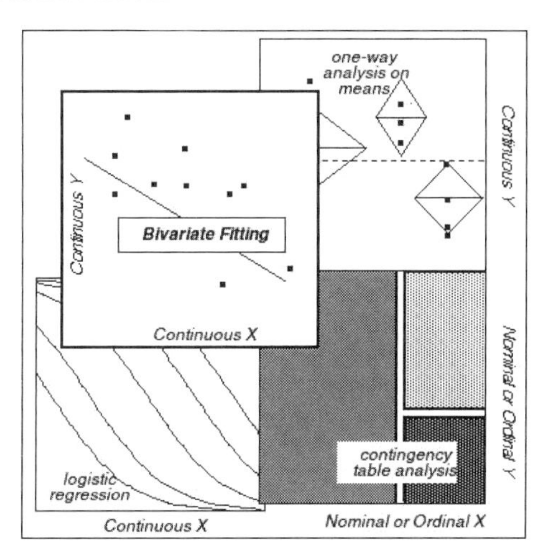

Select Columns into Roles

You assign columns for analysis with the dialog in Figure 6.2. The selector list at the left of the dialog shows all columns in the current table. To cast a column into a role, select one or more columns in the column selector list and click a role button. Or, drag variables from the column selector list to one of the following role boxes:

- **X, Factor** for the *x*-axis, one or more factor variables
- **Y, Response** for the *y*-axis, one or more response variables.
- **Block** is not used for a continuous by continuous analysis.
- **Weight** is an optional role that identifies one column whose values supply weights for the response in each *y* by *x* combination.
- **Freq** is an optional role that identifies one column whose values assign a frequency to each row for the analysis.

To remove an unwanted variable from an assigned role, select it in the role box and click **Remove**. After assigning roles, click **OK** to see the analysis for each combination of *Y* by *X* variables.

Note: Any column or combination of columns can be assigned as labels by selecting them in the Columns panel to the left of the spreadsheet. Then choose **Label** from the **Columns** popup menu at the top of the Columns panel. When you click on a point in a plot, it is identified by the value of the column(s) you assigned as a label variable.

Figure 6.2 Fit Y by X Launch Dialog

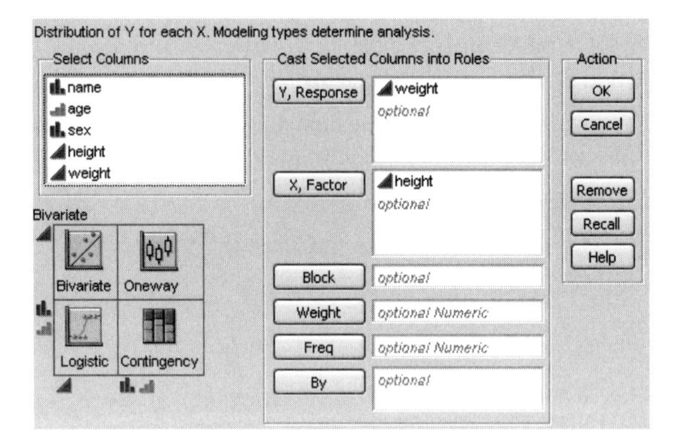

The Scatterplot

The Bivariate platform begins with a scatterplot for each pair of *X* and *Y* variables. The scatterplot, like other plots in JMP, has features to resize plots, highlight points with the cursor or brush tool, and label points.

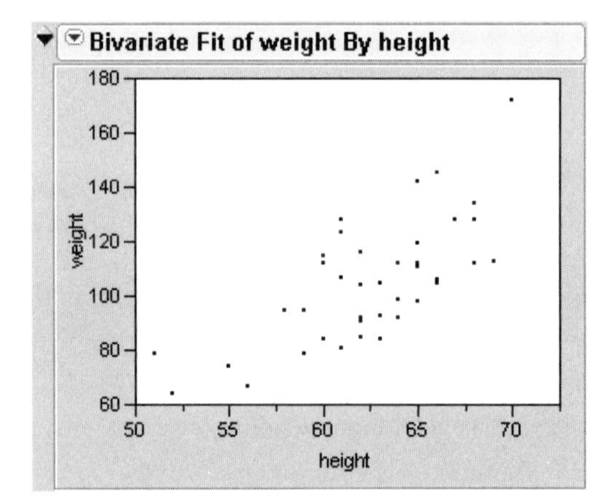

See the *JMP User Guide* for a discussion of these features.

The popup icon on the title of the scatterplot accesses a menu with fitting commands and options to overlay curves on the scatterplot and present statistical tables.

The data for the next examples are in the Big Class.jmp table found in the Sample Data folder.

Fitting Commands

The illustration here shows the popup menu next to the scatterplot name (the Fitting menu).

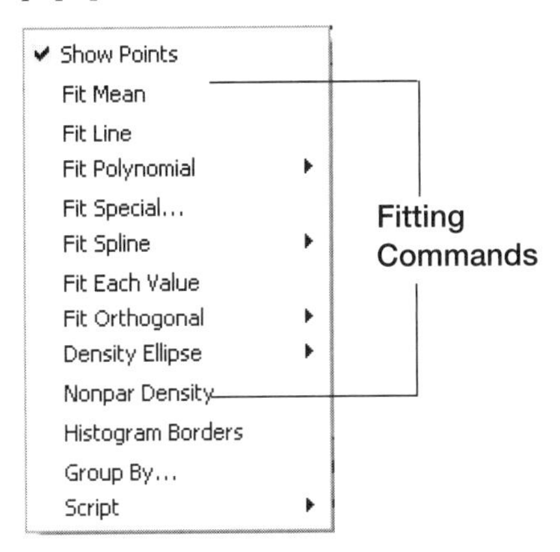

You can select fitting commands repeatedly. Each new fit appears overlaid on the scatterplot. This enables you to try fits with different points excluded and compare them on the same plot. Each time you select a fit, a text report is added to the platform. The type of report varies depending on the kind of fit you request.

Note: When you hold down the Control key (⌘ key on the Mac) and select a fitting option, the selected fit is done simultaneously for all Bivariate analyses in that window. If you Control-click (⌘-click) and drag a scatterplot resize box, all plots resize together.

The fitting commands come in two varieties:

Regression Fits Regression methods fit a curve through the points. The curve is an equation (a linear model) that is estimated by least squares. The least-squares method minimizes the sum of squared differences from each point to the line (or curve). The fit assumes that the *Y* variable is distributed as a random scatter above and below a line of fit.

Density Estimates Density estimation fits a bivariate distribution to the points. You can choose either a bivariate normal density, characterized by elliptical contours or a general nonparametric density.

The following sections describe the options and fits and their statistical results, and discuss the additional platform features.

Show Points

Show Points alternately hides or displays the points in the plot.

Fit Mean

The **Fit Mean** command adds a horizontal line to the plot at the mean of the response variable (*Y*). You can add confidence curves with the **Confid Curves Fit** and **Confid Curves Indiv** options, discussed in the section "Display Options and Save Commands," p. 104 in this chapter. It is often useful to first fit the mean as a reference line for other fits. Look ahead at Figure 6.3 for an example that compares the mean line to a straight line fit and a polynomial of degree 2. **Fit Mean** is equivalent to a polynomial fit of degree 0.

Fit Mean	
Mean	105
Std Dev [RMSE]	22.20187
Std Error	3.510424
SSE	19224

The Fit Mean Fit table shows the following quantities that summarize information about the response variable:

Mean is the sample mean (arithmetic average) of the response variable. It is the predicted response when there are no specified effects in the model.

Std Dev [RMSE] is the standard deviation of the response variable. It is the square root of the mean square of the response variable, also referred to as the root mean square error (RMSE).

Std Error is the standard deviation of the response mean. It is found by dividing the RMSE by the square root of the number of nonmissing values.

SSE is the error sum of squares for the simple mean model. It is the sum of the squared distances of each point from the mean. It appears as the sum of squares for **Error** in the analysis of variance tables for each model fit.

Fit Line and Fit Polynomial

The **Fit Line** command adds a straight-line fit to the plot using least squares regression. See the section "Display Options and Save Commands," p. 104 in this chapter for a discussion of how to add confidence curves to the plot. There are also commands to save predicted values and residuals for the linear fit as new columns in the current data table.

Fit Line finds the parameters β_0 and β_1 for the straight line that fits the points to minimize the residual sum of squares. The model for the *i*th row is written $y_i = \beta_0 + \beta_1 x_i + \varepsilon_i$.

If the confidence area of this line includes the response mean, then the slope of the line of fit is not significantly different from zero at the 0.05 significance level. The **Fit Line** command is equivalent to a polynomial fit of degree 1. Figure 6.3 shows an example of a linear fit compared to the simple mean line and to a degree 2 polynomial fit.

The **Fit Polynomial** command fits a polynomial curve of the degree you select from the **Fit Polynomial** submenu. After you select the polynomial degree, the curve is fit to the data points using least squares regression. You can add curves representing 95% confidence limits with the **Confid Curves Fit** and **Confid Curves Indiv** display options.

You can select the **Fit Polynomial** option multiple times and choose different polynomial degrees for comparison. As with the linear fit, options let you save predicted values and residuals as new columns in the current data table for each polynomial fit.

A polynomial of degree 2 is a parabola; a polynomial of degree 3 is a cubic curve. For degree *k*, the model for the *i*th observation

$$y_i = \sum_{j=0}^{k} \beta_j x_i^j + \varepsilon_i$$

The fitting commands are not toggles and can be selected repeatedly. This enables you to try fits with different points excluded and compare them on the same plot.

Figure 6.3 Example of Selected Fits

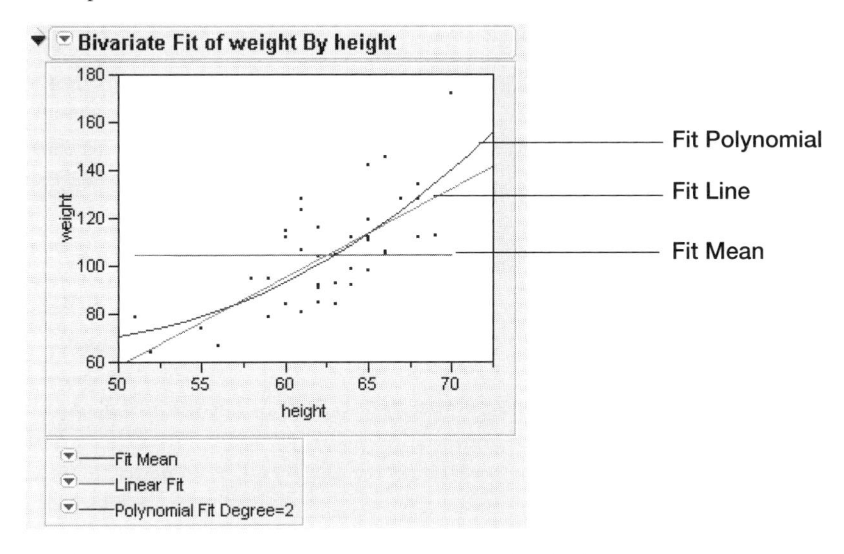

Each time you choose a linear or polynomial fit, three additional tables are appended to the report. A fourth table, Lack of Fit, also appears if there are replicates.

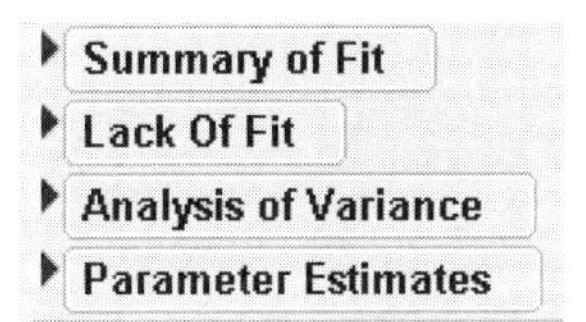

These tables are described next.

The Summary of Fit Table

The Summary of Fit tables in Figure 6.4 show the numeric summaries of the response for the linear fit and polynomial fit of degree 2 for the same data. You can compare Summary of Fit tables to see the improvement of one model over another as indicated by a larger Rsquare value and smaller Root Mean Square Error.

Figure 6.4 Summary of Fit Tables for Linear and Polynomial Fits

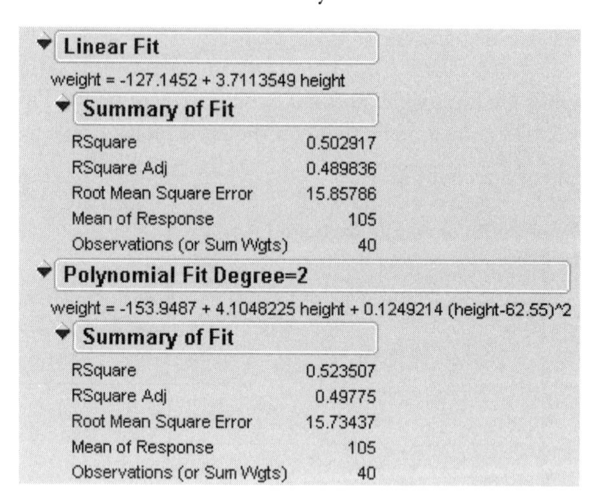

The fitting results begin by showing the equation of fit (Figure 6.4). Note that this equation is editable. The Summary of Fit table follows with these quantities:

RSquare measures the proportion of the variation around the mean explained by the linear or polynomial model. The remaining variation is attributed to random error. Rsquare is 1 if the model fits perfectly. An Rsquare of 0 indicates that the fit is no better than the simple mean model. Using quantities from the corresponding analysis of variance table, the Rsquare for any interval response fit is always calculated as

$$\frac{\text{Sum of Squares for Model}}{\text{Sum of Squares for CTotal}}$$

The Rsquare values seen in Figure 6.4 indicate that the polynomial fit of degree 2 gives only a small improvement over the linear fit.

RSquare Adj adjusts the Rsquare value to make it more comparable over models with different numbers of parameters by using the degrees of freedom in its computation. It is a ratio of mean squares instead of sums of squares and is calculated

$$1 - \frac{\text{Mean Square for Error}}{\text{Mean Square for C. Total}}$$

where mean square for Error is found in the Analysis of Variance table and the mean square for **C. Total** can be computed as the **C. Total** Sum of Squares for **C. Total** divided by its respective degrees of freedom (see Figure 6.4 and Figure 6.5).

Root Mean Square Error estimates the standard deviation of the random error. It is the square root of the mean square for Error in the Analysis of Variance tables in Figure 6.5.

Mean of Response is the sample mean (arithmetic average) of the response variable. This is the predicted response when no model effects are specified.

Observations is the number of observations used to estimate the fit. If there is a weight variable, this is the sum of the weights.

The Lack of Fit Table

The Lack of Fit table shows a special diagnostic test and appears only when the data and the model provide the opportunity. The idea is that sometimes you can estimate the error variance independently of whether you have the right form of the model. This occurs when observations are exact replicates of each other in terms of the *x* variable. The error that you measure for these exact replicates is called *pure error*. This is the portion of the sample error that cannot be explained or predicted no matter what form the model uses for the *x* variable. However, a lack of fit test may not be of much use if it has only a very few degrees of freedom for it (few replicated *x* values).

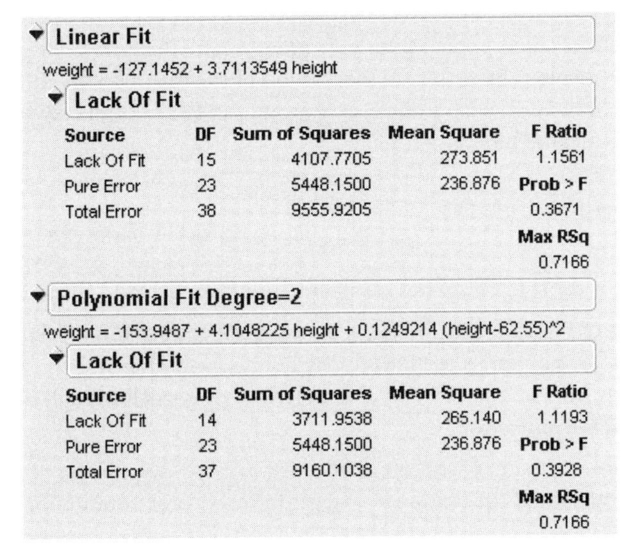
Linear Fit

weight = -127.1452 + 3.7113549 height

Lack Of Fit

Source	DF	Sum of Squares	Mean Square	F Ratio
Lack Of Fit	15	4107.7705	273.851	1.1561
Pure Error	23	5448.1500	236.876	Prob > F
Total Error	38	9555.9205		0.3671
				Max RSq
				0.7166

Polynomial Fit Degree=2

weight = -153.9487 + 4.1048225 height + 0.1249214 (height-62.55)^2

Lack Of Fit

Source	DF	Sum of Squares	Mean Square	F Ratio
Lack Of Fit	14	3711.9538	265.140	1.1193
Pure Error	23	5448.1500	236.876	Prob > F
Total Error	37	9160.1038		0.3928
				Max RSq
				0.7166

The difference between the residual error from the model and the pure error is called *lack of fit error*. Lack of fit error can be significantly greater than pure error if you have the wrong functional form of the regressor. In that case, you should try a different kind of model fit.

The Lack of Fit table shows information about the error terms:

Source lists the three sources of variation called Lack of Fit, Pure Error, and Total Error.

DF records an associated *degrees of freedom* (DF) for each source of error.

The **Total Error** DF is the degrees of freedom found on the Error line of the Analysis of Variance table (shown under "Analysis of Variance Table," p. 90). It is the difference between the **Total** DF and the **Model** DF found in that table. The **Error** DF is partitioned into degrees of freedom for lack of fit and for pure error.

The **Pure Error** DF is pooled from each group where there are multiple rows with the same values for each effect. For example, there are multiple instances in the Big Class.jmp sample data table where more than one subject has the same value of height. In general, if there are *g* groups having multiple rows with identical values for each effect, the pooled DF, denoted DF_p, is

$$DF_p = \sum_{i=1}^{g} (n_i - 1)$$

where n_i is the number of subjects in the ith group.

The **Lack of Fit** DF is the difference between the **Total Error** and **Pure Error** DF.

Sum of Squares records an associated sum of squares (SS for short) for each source of error.

The **Total Error** SS is the sum of squares found on the **Error** line of the corresponding Analysis of Variance table, shown under "Analysis of Variance Table," p. 90.

The **Pure Error** SS is pooled from each group where there are multiple rows with the same value for the x variable. This estimates the portion of the true random error that is not explained by model x effect. In general, if there are g groups having multiple rows with the same x value, the pooled SS, denoted SS_p, is written

$$SS_p = \sum_{i=1}^{g} SS_i$$

where SS_i is the sum of squares for the ith group corrected for its mean.

The **Lack of Fit SS** is the difference between the **Total Error** and **Pure Error** sum of squares. If the lack of fit SS is large, the model may not be appropriate for the data. The F-ratio described below tests whether the variation due to lack of fit is small enough to be accepted as a negligible portion of the pure error.

Mean Square is a sum of squares divided by its associated degrees of freedom. This computation converts the sum of squares to an average (mean square). F-ratios for statistical tests are the ratios of mean squares.

F Ratio is the ratio of mean square for lack of fit to mean square for Pure Error. It tests the hypothesis that the lack of fit error is zero.

Prob > F is the probability of obtaining a greater F-value by chance alone if the variation due to lack of fit variance and the pure error variance are the same. A high p value means that an insignificant proportion of error is explained by lack of fit.

Max RSq is the maximum R^2 that can be achieved by a model using only the variables in the model. Because Pure Error is invariant to the form of the model and is the minimum possible variance, Max RSq is calculated

$$1 - \frac{SS(\text{Pure error})}{SS(\text{Total for whole model})}$$

Analysis of Variance Table

Analysis of variance (ANOVA) for a linear regression partitions the total variation of a sample into components. These components are used to compute an F-ratio that evaluates the effectiveness of the model. If the probability associated with the F-ratio is small, then the model is considered a better statistical fit for the data than the response mean alone. The Analysis of Variance tables in Figure 6.5 compare a linear fit (**Fit Line**) and a second degree (**Fit Polynomial**). Both fits are statistically different from a horizontal line at the mean.

Figure 6.5 Analysis of Variance Tables

Linear Fit

weight = -127.1452 + 3.7113549 height

Analysis of Variance

Source	DF	Sum of Squares	Mean Square	F Ratio
Model	1	9668.079	9668.08	38.4460
Error	38	9555.921	251.47	**Prob > F**
C. Total	39	19224.000		<.0001*

Polynomial Fit Degree=2

weight = -153.9487 + 4.1048225 height + 0.1249214 (height-62.55)^2

Analysis of Variance

Source	DF	Sum of Squares	Mean Square	F Ratio
Model	2	10063.896	5031.95	20.3253
Error	37	9160.104	247.57	**Prob > F**
C. Total	39	19224.000		<.0001*

The Analysis of Variance table displays the following quantities:

Source lists the three sources of variation, called **Model**, **Error**, and **C. Total**.

DF records the associated degrees of freedom (DF) for each source of variation:

A degree of freedom is subtracted from the total number of nonmissing values (*N*) for each parameter estimate used in the computation. The computation of the total sample variation uses an estimate of the mean; therefore, one degree of freedom is subtracted from the total, leaving 39. The total corrected degrees of freedom are partitioned into the **Model** and **Error** terms.

One degree of freedom from the total (shown on the **Model** line) is used to estimate a single regression parameter (the slope) for the linear fit. Two degrees of freedom are used to estimate the parameters (β_1 and β_2) for a polynomial fit of degree 2.

The **Error** degrees of freedom is the difference between total DF and model DF.

Sum of Squares records an associated sum of squares (SS for short) for each source of variation:

In this example, the total (**C. Total**) sum of squared distances of each response from the sample mean is 19224, as shown in Figure 6.5. That is the sum of squares for the base model (or simple mean model) used for comparison with all other models.

For the linear regression, the sum of squared distances from each point to the line of fit reduces from 19924 to 9555.9. This is the unexplained **Error** (*residual*) SS after fitting the model. The residual SS for a second degree polynomial fit is 9160.1, accounting for slightly more variation than the linear fit. That is, the model accounts for more variation because the model SS are higher for the second degree polynomial than the linear fit.The total SS less the error SS gives the sum of squares attributed to the model.

Mean Square is a sum of squares divided by its associated degrees of freedom. The *F*-ratio for a statistical test is the ratio of the following mean squares:

The **Model** mean square for the linear fit is 9668.09 and estimates the variance, but only under the hypothesis that the model parameters are zero.

The **Error** mean square is 251.47 and estimates the variance of the error term.

F Ratio is the model mean square divided by the error mean square. The underlying hypothesis of the fit is that all the regression parameters (except the intercept) are zero. If this hypothesis is true, then both the mean square for error and the mean square for model estimate the error variance, and their ratio has an *F*-distribution. If a parameter is a significant model effect, the *F*-ratio is usually higher than expected by chance alone.

Prob > F is the observed significance probability (*p*-value) of obtaining a greater *F*-value by chance alone if the specified model fits no better than the overall response mean. Observed significance probabilities of 0.05 or less are often considered evidence of a regression effect.

Parameter Estimates Table

The terms in the Parameter Estimates table for a linear fit (Figure 6.6) are the intercept and the single *x* variable. Below the linear fit, Figure 6.6 shows the parameter estimates for a polynomial model of degree 2. For a polynomial fit of order *k*, there is an estimate for the model intercept and a parameter estimate for each of the *k* powers of the *x* variable.

Figure 6.6 Parameter Estimates Tables for Linear and Polynomial Fits

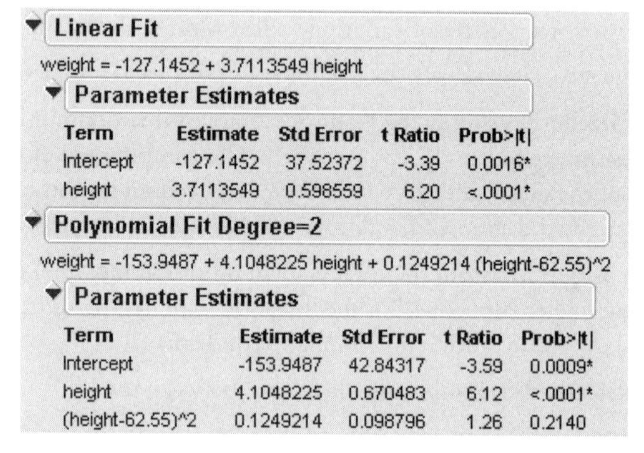

Linear Fit

weight = -127.1452 + 3.7113549 height

Parameter Estimates

Term	Estimate	Std Error	t Ratio	Prob>|t|
Intercept	-127.1452	37.52372	-3.39	0.0016*
height	3.7113549	0.598559	6.20	<.0001*

Polynomial Fit Degree=2

weight = -153.9487 + 4.1048225 height + 0.1249214 (height-62.55)^2

Parameter Estimates

Term	Estimate	Std Error	t Ratio	Prob>|t|
Intercept	-153.9487	42.84317	-3.59	0.0009*
height	4.1048225	0.670483	6.12	<.0001*
(height-62.55)^2	0.1249214	0.098796	1.26	0.2140

The Parameter Estimates table displays the following:

Term lists the name of each parameter in the requested model. The intercept is a constant term in all models.

Estimate lists the parameter estimates of the linear model. The prediction formula is the linear combination of these estimates with the values of their corresponding variables.

Std Error lists the estimates of the standard errors of the parameter estimates. They are used in constructing tests and confidence intervals.

t Ratio lists the test statistics for the hypothesis that each parameter is zero. It is the ratio of the parameter estimate to its standard error. If the hypothesis is true, then this statistic has a Student's *t*-distribution. Looking for a *t*-ratio greater than 2 in absolute value is a common rule of thumb for judging significance because it approximates the 0.05 significance level.

Prob>|t| lists the observed significance probability calculated from each *t*-ratio. It is the probability of getting, by chance alone, a *t*-ratio greater (in absolute value) than the computed value, given a true hypothesis. Often, a value below 0.05 (or sometimes 0.01) is interpreted as evidence that the parameter is significantly different from zero.

Fit Special

The **Fit Special** command displays a dialog with choices for both the *Y* and *X* transformation. Transformations include log, square root, square, reciprocal, and exponential.

Figure 6.7 The **Fit Special** Menu and Dialog

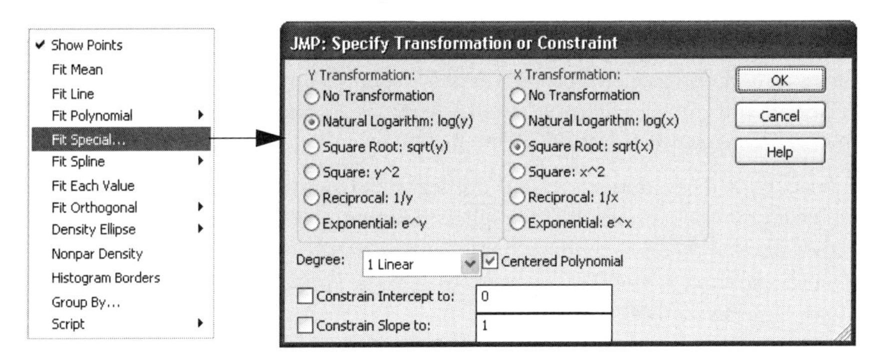

The fitted line is plotted on the original scale as shown in Figure 6.8. The regression report is shown with the transformed variables, but an extra report shows measures of fit transformed in the original *Y* scale if there is a *Y* transformation.

Figure 6.8 Fitting Transformed Variables

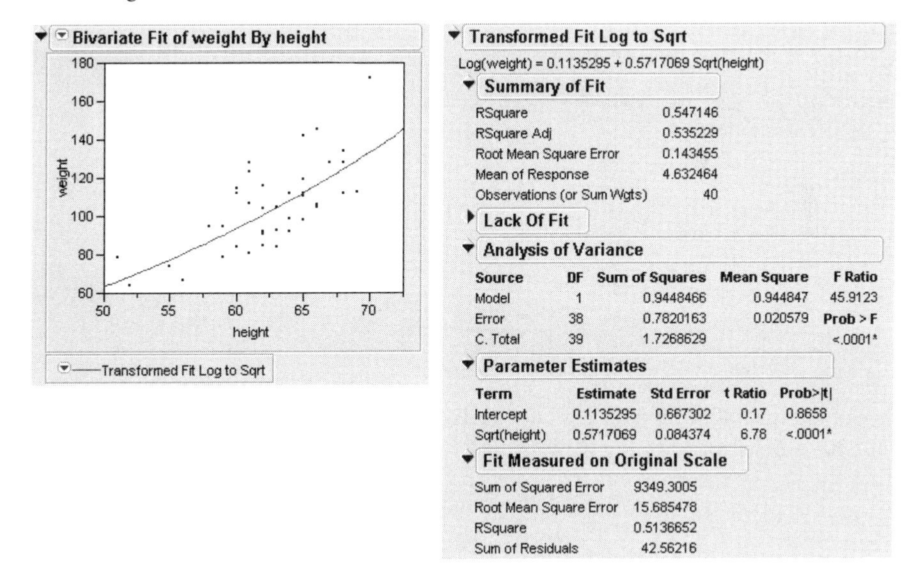

Summary of Fit	
RSquare	0.547146
RSquare Adj	0.535229
Root Mean Square Error	0.143455
Mean of Response	4.632464
Observations (or Sum Wgts)	40

Source	DF	Sum of Squares	Mean Square	F Ratio
Model	1	0.9448466	0.944847	45.9123
Error	38	0.7820163	0.020579	Prob > F
C. Total	39	1.7268629		<.0001*

Term	Estimate	Std Error	t Ratio	Prob>|t|
Intercept	0.1135295	0.667302	0.17	0.8658
Sqrt(height)	0.5717069	0.084374	6.78	<.0001*

Fit Measured on Original Scale	
Sum of Squared Error	9349.3005
Root Mean Square Error	15.685478
RSquare	0.5136652
Sum of Residuals	42.56216

Polynomial Centering

Fit Special provides a way to turn off polynomial centering, as shown in Figure 6.7. Uncheck the **Centered Polynomial** check box to turn off centering.

Fit Spline

The **Fit Spline** command fits a smoothing spline that varies in smoothness (or flexibility) according to lambda (λ), a tuning parameter in the spline formula. You select Lambda from the **Fit Spline** submenu. If you want to use a λ value not listed on the submenu, choose the **Other** selection and enter the value when prompted. In addition, you can specify whether the *x*-values should be standardized by selecting a checkbox in this dialog.

As the value of λ decreases, the error term of the spline model has more weight and the fit becomes more flexible and curved. Higher values of λ make the fit stiff (less curved), approaching a straight line. Figure 6.9 compares a stiff and a flexible spline fit.

The cubic spline method uses a set of third-degree polynomials spliced together such that the resulting curve is continuous and smooth at the splices (knot points). The estimation is done by minimizing an objective function that is a combination of the sum of squares error and a penalty for curvature integrated over the curve extent. See the paper by Reinsch (1967) or the text by Eubank (1988) for a description of this method.

The smoothing spline is an excellent way to get an idea of the shape of the expected value of the distribution of *y* across *x*. The points closest to each piece of the fitted curve have the most influence on it. The influence increases as you set the tuning parameter to produce a highly flexible curve (low value of lambda). You may find yourself trying several λ values, or using the **Lambda** slider beneath the Smoothing Spline table, before you settle on the spline curve that seems best for your data. However, λ is not invariant to the scaling of the data.

Figure 6.9 Spline Fits for Selected Values of Lambda

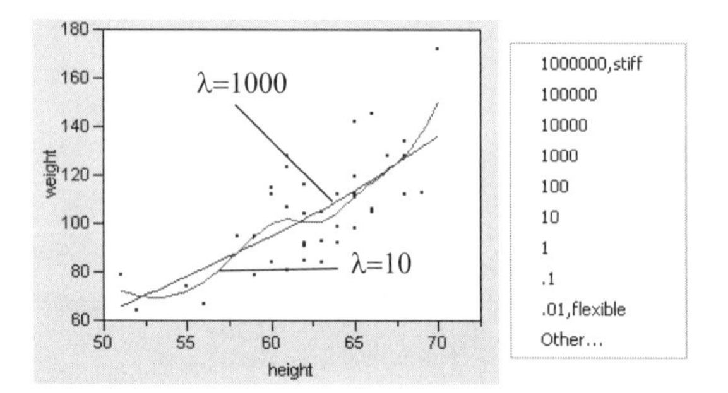

The Smoothing Spline Fit table has only the R-Square for the spline fit and the Sum of Squares Error. It also has a slider for adjusting the value of λ.

Smoothing Spline Fit, lambda=1000	
R-Square	0.520554
Sum of Squares Error	9216.87
Change Lambda:	1000

These values are useful for comparing the spline fit to other fits or for comparing different spline fits to each other:

R-Square measures the proportion of variation accounted for by the spline model. In the example above, fitting a spline with λ = 1000 explains 52.06% of the total variation.

Sum of Squares Error is the sum of squared distances from each point to the fitted spline. It is the unexplained error (*residual*) after fitting the spline model.

Note: The popup menu for the Spline Fit has the **Save Coefficients** command, which saves the spline coefficients as a new data table, with columns called X, A, B, C, and D. X gives the knot points. A, B, C, and D are the intercept, linear, quadratic, and cubic coefficients for the cubic spline slices.

Fit Each Value

The **Fit Each Value** command fits a value to each unique *x* value. Fitting each value is like doing a one-way analysis of variance, but in the continuous-by-continuous bivariate platform you can compare it to other fitted lines and see the concept of lack of fit.

Fit Orthogonal

The **Fit Orthogonal** command fits lines that adjust for variability in *X* as well as *Y*. Standard least square fitting assumes that the *x* variable is fixed and the *Y* variable is a function of *X*, plus error. If there is random variation in the measurement of *X*, then instead of fitting a standard regression line, you should fit a line that is perpendicular to the line of fit as illustrated by the figure below. But the perpendicular distance depends on how *X* and *Y* are scaled, and the scaling for the perpendicular is reserved as a statistical issue, not a graphical one.

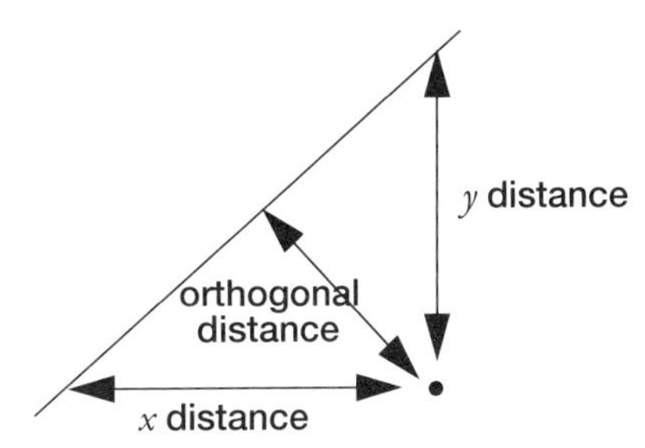

The fit requires that you specify the ratio of the variance of the error in *X* to the error in *Y*. This is the variance of the error, not the variance of the sample points, so you must choose carefully. The ratio $(\sigma_y^2)/(\sigma_x^2)$ is infinite in standard least squares because σ_x^2 is zero. If you do an orthogonal fit with a large error ratio, the fitted line approaches the standard least squares line of fit. If you specify a ratio of zero, the fit is equivalent to the regression of *X* on *Y*, instead of *Y* on *X*.

The most common use of this technique is in comparing two measurement systems that both have errors in measuring the same value. Thus, the *Y* response error and the *x* measurement error are both the same kind of measurement error. Where do you get the measurement error variances? You can't get them from bivariate data because you can't tell which measurement system produces what proportion of the error. So, you either must blindly assume some ratio like 1, or you must rely on separate repeated measurements of the same unit by the two measurement systems.

An advantage to this approach is that the computations give you predicted values for both *Y* and *X*; the predicted values are the point on the line that is closest to the data point, where closeness is relative to the variance ratio.

Confidence limits are calculated as described in Tan and Iglewicz (1999).

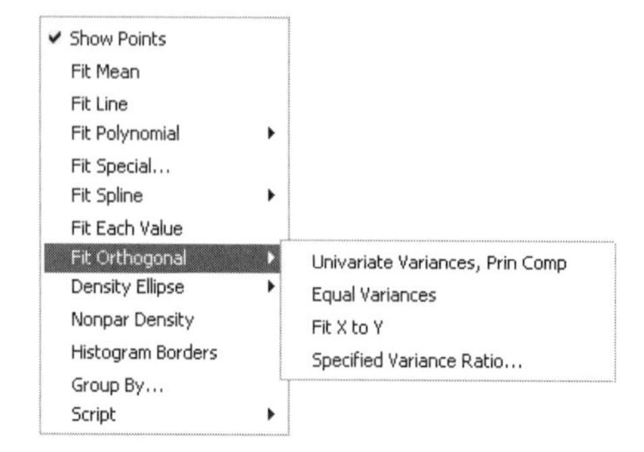

The **Fit Orthogonal** command has the following options to specify a variance ratio:

Univariate Variances, Prin Comp uses the univariate variance estimates computed from the samples of *X* and *Y*. This turns out to be the standardized first principal component.

Note: Be aware that this is not a good choice in a measurement systems application because the error variances are not likely to be proportional to the population variances.

Equal Variances uses 1 as the variance ratio, which assumes that the error variances are the same. Using equal variances is equivalent to the non-standardized first principal component line. If the scatterplot is scaled the same in both *x* and *y* directions and you show a normal density ellipse, you see that this line is the longest axis of the ellipse (Figure 6.10).

Fit X to Y uses variance ratio of zero, which indicates that *Y* effectively has no variance (see Figure 6.10).

Specified Variance Ratio lets you enter any ratio you want, giving you the ability to make use of known information about the measurement error in *X* and response error in *Y*.

The scatterplot in Figure 6.10 shows standardized height and weight values with various line fits that illustrate the behavior of the orthogonal variance ratio selections. The standard linear regression occurs when the variance of the *X* variable is considered to be very small. **Fit X to Y** is the opposite extreme, when the variation of the *Y* variable is ignored. All other lines fall between these two extremes and shift as the variance ratio changes. As the variance ratio increases, the variation in the *Y* response dominates and the slope of the fitted line shifts closer to the *Y* by *X* fit. Likewise, when you decrease the ratio, the slope of the line shifts closer to the *X* by *Y* fit.

In extreme cases, lines go through the horizontal and vertical tangents to the ellipse. The intermediate cases represent where the line goes through the longest axis of the ellipse, where longest depends on relative scaling. That is, you can resize the plot to change the aspect ratio to make the line go through the longest axis.

Figure 6.10 Illustration of Orthogonal Fitting Options

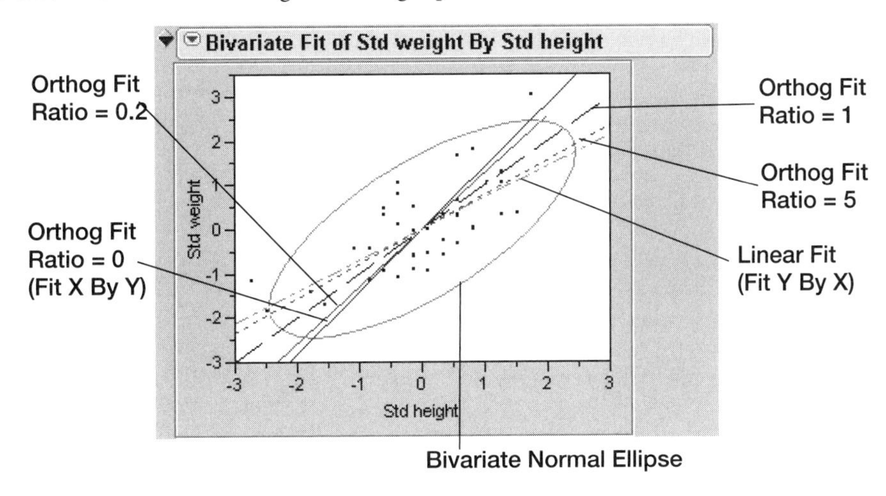

Example situation

Suppose you have two weighing scales, a good one called A, and a bad one, called B. Scale A is known to be unbiased and has a standard deviation in repeated measurements of 21 grams. Scale B is known to be biased (but linearly so) and has a standard deviation of 23 grams in repeated measurements. You have weight measurements of your experimental responses from both scales. It would be tempting to just use the measurements from Scale A, the good scale. But you can do better. The readings from Scale B still have value. Do an orthogonal regression of the B on the A weight measurements and use a variance ratio of $(23/21)^2$. Then you get predicted values for both A and B. Since A is unbiased, use the predicted values of A as your new-improved measurement, with a measurement error variance of nearly half of what you would have gotten with just Scale A.

Practical Equivalence

If you are trying to establish that the slope of the line is some value within certain bounds with a 1-alpha confidence, then this can be done, following Tan and Iglewicz (1999) using the Two One-Sided Tests (TOST) approach described further in Berger and Hsu (1996). In this case all you have to do is enter `2*alpha` into the **Confidence Interval Alpha** column, and check that the confidence limits are contained by the bounds you are testing for. For example, it is often desired to confirm that the slope is practically close to 1, where 0.8 to 1.25 are the practical limits around 1. These should be symmetric limits in the sense that if you interchange the *X* and *Y* variables, the test will be the same. Enter an alpha of 0.10 and see if the confidence limits are both within the range 0.8 to 1.25, in which case they would be declared as practical average equivalent.

Fit Density Ellipse

The **Density Ellipse** command draws an ellipse that contains the specified mass of points as determined by the probability you choose from the **Density Ellipse** submenu. The density ellipsoid is computed from the bivariate normal distribution fit to the *X* and *Y* variables. The bivariate normal density is a function of the means and standard deviations of the *X* and *Y* variables and the correlation between them. The **Other** selection lets you specify any probability greater than zero and less than or equal to one.

These ellipsoids are both density contours and confidence curves. As confidence curves they show where a given percentage of the data is expected to lie, assuming the bivariate normal distribution. Figure 6.11 compares ellipses with different probabilities.

Figure 6.11 Selected Density Ellipses

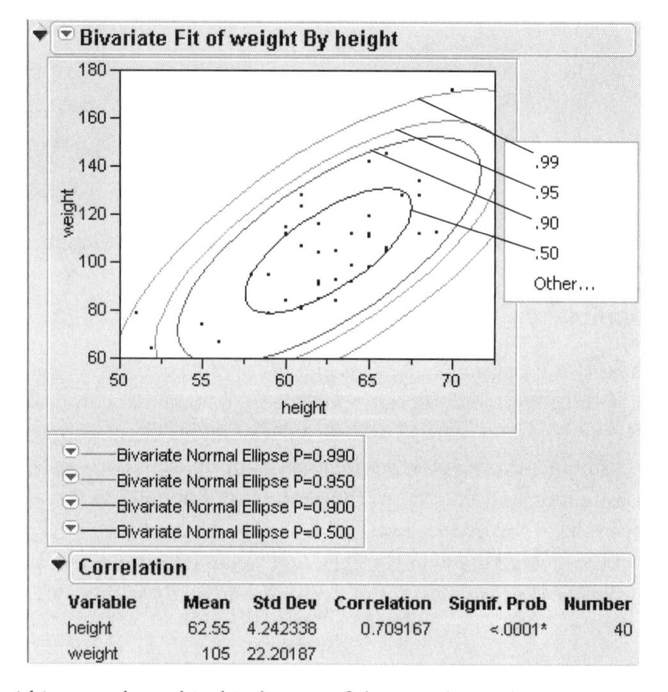

Variable	Mean	Std Dev	Correlation	Signif. Prob	Number
height	62.55	4.242338	0.709167	<.0001*	40
weight	105	22.20187			

The density ellipsoid is a good graphical indicator of the correlation between two variables. The ellipsoid collapses diagonally as the correlation between the two variables approaches either 1 or –1. The ellipsoid is more circular (less diagonally oriented) if the two variables are uncorrelated.

The Density Ellipse table (Figure 6.11) that accompanies each **Density Ellipse** fit shows the correlation coefficient for the *X* and *Y* variables and the probability that the correlation between the variables is significant. To see a matrix of ellipsoids and correlations for many pairs of variables, use the **Multivariate** command in the **Analyze > Multivariate Methods** menu.

The Bivariate table gives these quantities:

Mean gives the arithmetic average of both the *X* and *Y* variable. The mean is computed as the sum of the nonmissing values divided by the number of nonmissing values.

Std Dev gives the standard deviation of both the *X* and *Y* variable. It measures the spread of a distribution around the mean. It is often denoted as s and is the square root of the sample variance, denoted s^2. For a normal distribution, about 68% of the distribution is within one standard deviation of the mean.

A discussion of the mean and standard deviation are in the section “The Moments Table for Continuous Variables,” p. 32 in the “Univariate Analysis” chapter.

Correlation is the Pearson correlation coefficient, denoted *r*, and is computed as

$$r_{xy} = \frac{s_{xy}^2}{\sqrt{s_x^2 s_y^2}} \text{ where } s_{xy}^2 = \frac{\sum w_i(x_i - \bar{x}_i)(y_i - \bar{y}_i)}{df}$$

where w_i is either the weight of the *i*th observation if a weight column is specified, or 1 if no weight column is assigned. If there is an exact linear relationship between two variables, the correlation is 1 or –1 depending on whether the variables are positively or negatively related. If there is no relationship, the correlation tends toward zero.

Note: The correlation coefficient is also part of the statistical text report given by the Multivariate platform discussed in the chapter “Correlations and Multivariate Techniques,” p. 497. The Multivariate platform only uses rows that have values for all specified *Y* variables. If multiple *Y*’s are specified and there are missing values, an observation is eliminated from the computation even though a specific pair of variables both have values. The Multivariate platform also gives pairwise correlations.

Signif. Prob is the probability of obtaining, by chance alone, a correlation with greater absolute value than the computed value if no linear relationship exists between the *X* and *Y* variables.

The following commands are available for density ellipses

Shaded Contour shades the area inside the density ellipse.

Select Points Inside selects the points inside the ellipse.

Select Points Outside selects the points outside the ellipse.

Nonpar Density: Nonparametric Density Estimation

The **Nonpar Density** feature is designed for the situation when you have thousands of points and the scatterplot is so darkened by points that it is hard to see patterns in the point density.

Bivariate density estimation models a smooth surface that describes how dense the data points are at each point in that surface. The plot adds a set of contour lines showing the density (Figure 6.12). Optionally, the contour lines are quantile contours in 5% intervals with thicker lines at the 10% quantile intervals. This means that about 5% of the points are below the lowest contour, 10% are below the next contour, and so forth. The highest contour has about 95% of the points below it.

Figure 6.12 Contour Lines Showing Bivariate Density Information

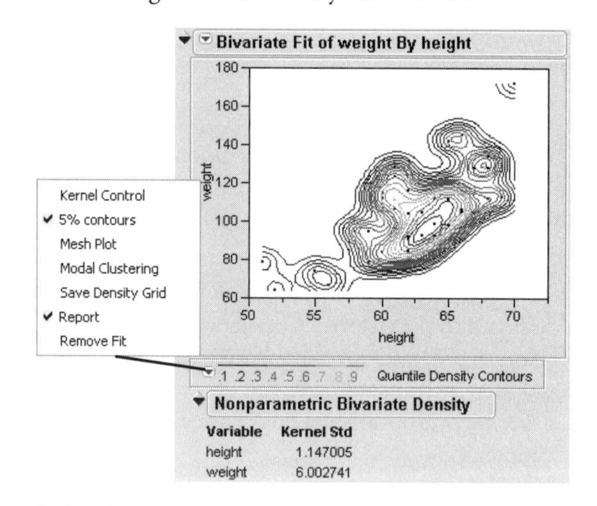

The popup menu beneath the plot gives you these optional commands:

Kernel Control displays a slider for each variable, which lets you change the kernel standard deviation used to define the range of *X* and *Y* values for determining the density of contour lines.

5% Contours shows or hides the 5% contour lines and shows only the 10% contours.

Mesh Plot is a three-dimensional plot of the density estimated over a grid of the two analysis variables (Figure 6.13).

Save Density Grid saves the density estimates and the quantiles associated with them in a new data table. The grid data can be used to visualize the density in other ways, such as with the Spinning Plot or the Contour Plot platforms.

Modal Clustering creates a new column in the current data table and fills it with cluster values.

Note: If you save the modal clustering values first and then save the density grid, the grid table will also contain the cluster values. The cluster values are useful for coloring and marking points in plots.

Figure 6.13 Mesh Plot

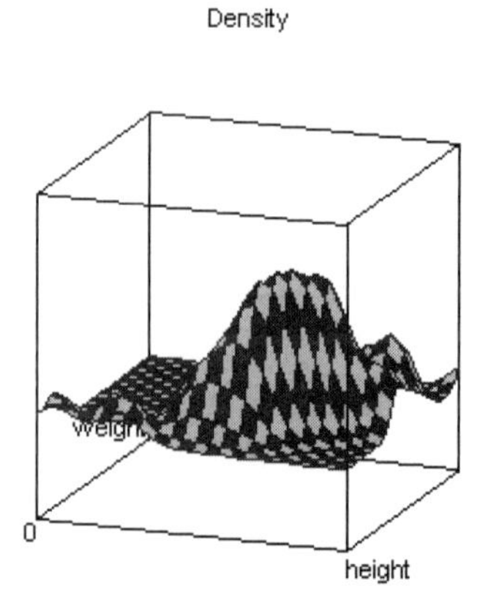

This nonparametric density method is efficient for thousands of points but relatively expensive for a small number of points. The steps in the method are as follows:

1 Divide each axis into 50 binning intervals, for a total of 2,500 bins over the whole surface.
2 Count the points in each bin.
3 Decide the smoothing kernel standard deviation using the recommendations of Bowman and Foster (1992).
4 Run a bivariate normal kernel smoother using an FFT and inverse FFT to do the convolution.
5 Create a contour map on the 2,500 bins using a bilinear surface patch model.

Histogram Borders

The **Histogram Borders** command appends histograms to the *x*- and *y*-axes of the scatterplot.

Figure 6.14 Histogram Borders

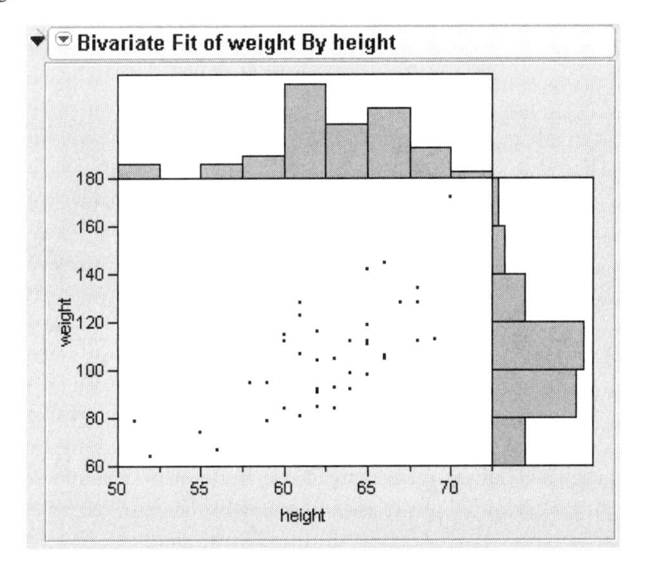

Paired *t*-test

In version 3 of JMP, the paired *t*-test was part of the Bivariate platform. The two paired variables were plotted as the *x*- and *y*- axes of a bivariate scatterplot, and a 45-degree line was superimposed on the plot. JMP 4 enhanced this analysis and its plot, promoting it into its own platform (see "Paired Data," p. 181 OF *JMP Statistics and Graphics Guide* for details on the Matched Pairs platform). However, if you prefer the version 3 form of this test, it can be accessed by *holding down the Shift key* while selecting the popup menu on the title bar of the Bivariate Report. **Paired t test** appears in the menu. Figure 6.15 shows a paired *t*-test analysis of two variables from the Dogs.jmp data table.

The plot contains a gray 45 degree line, representing where the two columns are equal. It also contains a red line, parallel to the gray line, displaced by a value equal to the difference in means between the two responses. The dashed lines around the red line of fit show the 95% confidence interval for the difference in means.

Figure 6.15 Version 3 Paired *t*-test

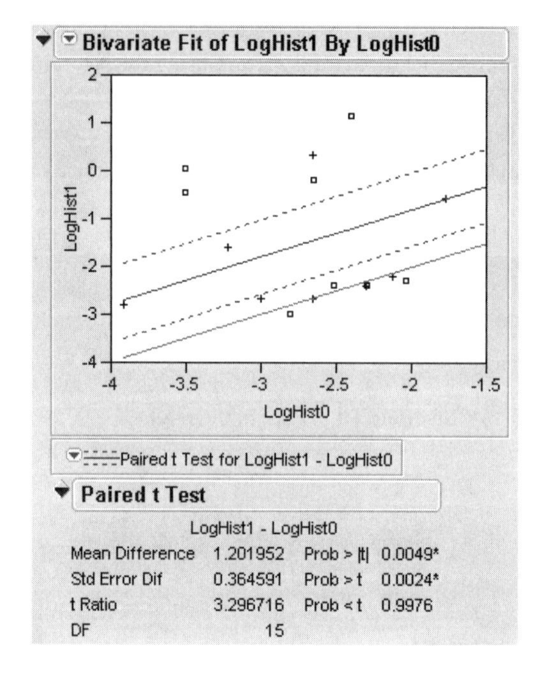

Group By

The **Group By** command in the Fitting menu displays a dialog that prompts you to select a classification (grouping) variable. When a grouping variable is in effect, the Fit Y by X platform computes a separate analysis for each level of the grouping variable and overlays the regression curves or ellipses on the scatterplot. The fit for each level of the grouping variable is identified beneath the scatterplot, with individual popup menus to save or remove fitting information.

The **Group By** command is checked in the Fitting menu when a grouping variable is in effect. You can change the grouping variable by first selecting the **Group By** command to remove (uncheck) the existing variable. Then, select the **Group By** command again and respond to its dialog as before.

An overlay of linear regression lines lets you compare slopes visually. An overlay of density ellipses can show clusters of points by levels of a grouping variable. For example, the ellipses in Figure 6.16 show clearly how the different types of hot dogs cluster with respect to the cost variables.

Note: The plot in Figure 6.16 uses the Hot dogs.jmp sample data table, and plots the variables $/oz by $/lb Protein. The **Group By** variable is Type, which gives the three density ellipses.

Figure 6.16 Density Ellipses with Grouping Variable

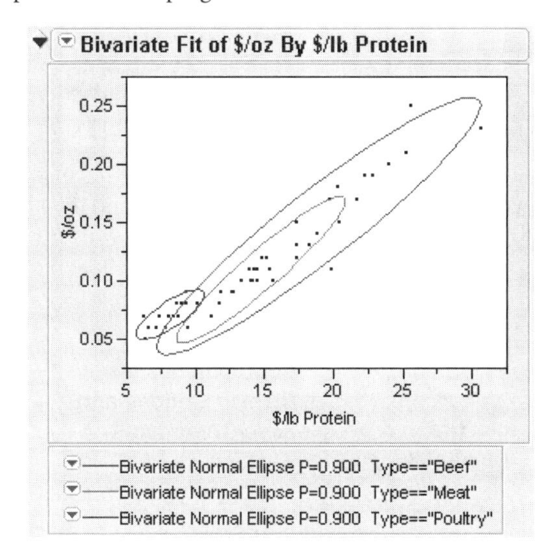

Another use for grouped regression is overlaying lines to compare slopes of different groups. The scatterplot to the left in Figure 6.17 has a single, straight-line regression that relates company profits to number of employees for all companies in a data table. This single regression shows a negative linear relationship between the two variables. The regression lines on the scatterplot to the right illustrate the same relationship separately for each type of company (computer and pharmaceutical). Statistics computed separately for each type of company indicate that profitability decreases as the number of employees increases in pharmaceutical companies but does not change in computer companies.

Figure 6.17 Regression Analysis for Whole Sample and Grouped Sample

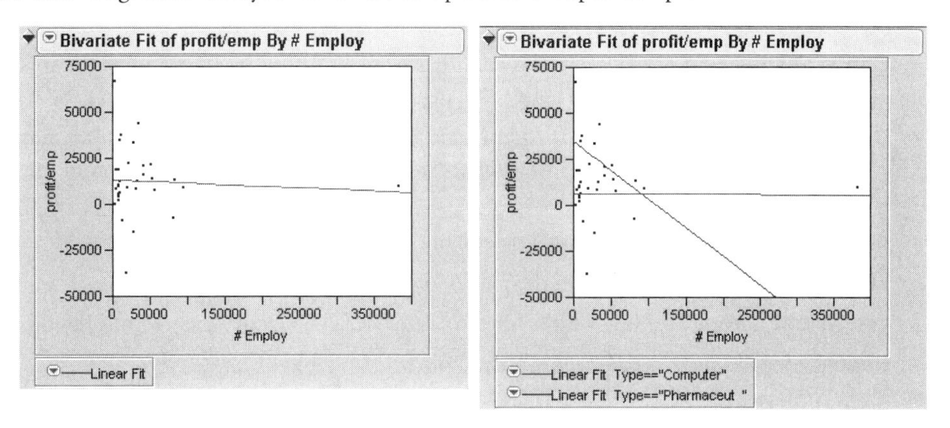

Note: The plots in Figure 6.17 are from the Companies.jmp sample data table.

Display Options and Save Commands

Each fit is drawn on the scatterplot with a solid line. You can add confidence curves to the mean, linear fit, and polynomial fit using the **Confid Curves Fit** and **Confid Curves Indiv** display popup menu options for each fit showing beneath the scatterplot (Figure 6.18).

Each fit you select also adds a text report to the analysis window. For example, the report titles shown in Figure 6.18 result when you select **Fit Mean**, then **Fit Line**, and then **Fit Polynomial** with degree 3.

The popup menu to the right in Figure 6.18 shows the display and save commands available for each requested fit.

Figure 6.18 Display Options

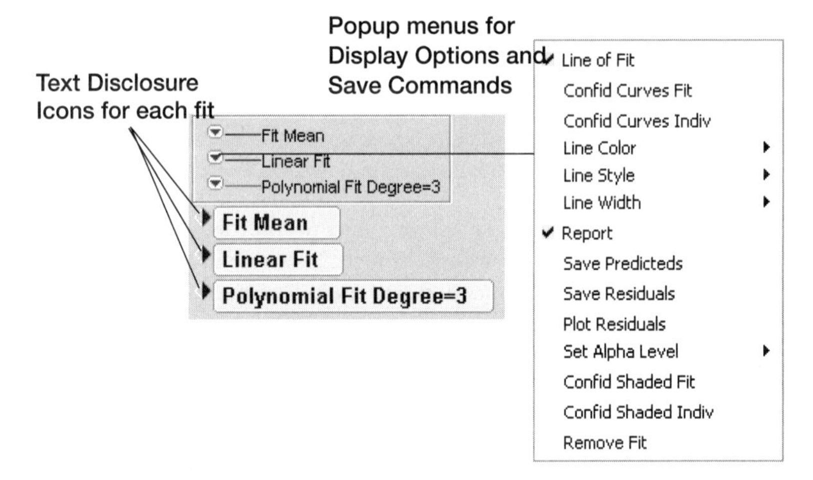

Line of Fit displays or hides the line of fit.

Confid Curves Fit displays or hides confidence curves for mean, line, and polynomial fits. This option is not available for the spline and density ellipse fits and is dimmed on those menus.

Confid Curves Indiv displays the upper and lower 95% confidence limits for an individual predicted value. The confidence limits reflect variation in the error and variation in the parameter estimates.

Line Color lets you choose from a palette of colors for assigning a color to each fit.

Line Style lets you choose from the palette of line styles for each fit.

Line Width gives three line widths for the line of fit. The default line width is the thinnest line.

Report toggles the fit's text report on and off.

Save Predicteds creates a new column in the current data table called Predicted *colname* where *colname* is the name of the *Y* variable. This column includes the prediction formula and the computed sample predicted values. The prediction formula computes values automatically for rows you add to the table.

Save Residuals creates a new column in the current data table called Residuals *colname* where *colname* is the name of the *Y* variable. Each value is the difference between the actual (observed) value and its predicted. This option is not available for the mean fit or density ellipses and is dimmed on those menus.

Plot Residuals plots the predicted and residual values.

Set Alpha Level prompts you to enter the alpha level to compute and display confidence levels for line fits, polynomial fits, and special fits.

Confid Shaded Fit draws the same curves as the **Confid Curves Fit** command, and shades the area between the curves

Confid Shaded Indiv draws the same curves as the Confid Curves Indiv command, and shades the area between the curves.

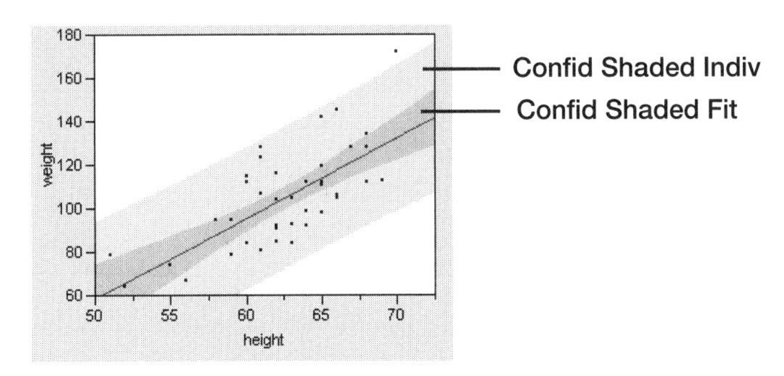

Remove Fit removes the fit from the graph and removes its text report.

Note: You can use the **Save Predicteds** and **Save Residuals** commands for each fit. If you use these commands multiple times or with a grouping variable, it is best to rename the resulting columns in the data table to reflect each fit.

Chapter 7

One-Way ANOVA

The Fit Y by X Platform

The Oneway platform analyzes how the distribution of a continuous *Y* variable differs across groups defined by a categorical *x* variable. Group means can be calculated and tested, as well as other statistics and tests. The Oneway platform is the *continuous by nominal/ordinal* personality of the **Fit Y by X** command. This platform begins with a *Y* by *X* scatterplot to show the distribution of values within each group and can

- perform a one-way analysis of variance to fit means and to test that they are equal
- perform nonparametric tests
- test for homogeneity of variance
- do multiple comparison tests on means, with means comparison circles
- present outlier box plots overlaid on each group
- show power details for the one-way layout.

Note: You can launch the Oneway platform from either the **Fit Y by X** main menu command (or tool bar), or with the **Oneway** button on the JMP Starter. When you choose **Fit Y by X** from the **Analyze** menu, JMP performs an analysis by context, which is the analysis appropriate for the modeling types of the response and factor variables you specify. When you use the **Oneway** button on the JMP Starter, the Launch dialog (role dialog) expects a continuous response and a nominal or ordinal factor variable; it advises you if you specify other modeling types.

For examples of more complex models see the chapters "Standard Least Squares: Introduction," p. 213, "Standard Least Squares: Perspectives on the Estimates," p. 249, "Standard Least Squares: Exploring the Prediction Equation," p. 275, and "Standard Least Squares: Random Effects," p. 339, which cover the Fit Model platform for standard least squares model fitting.

Also, the Variability Charts platform, discussed in the chapter "Variability Charts," p. 861, generalizes the one-way analysis in a different direction.

Contents

Launch the Platform

When you choose **Fit Y by X** from the **Analyze** menu or tool bar or click the **Oneway** button on the **Basic Stats** tab of the JMP Starter, you first see the Launch dialog shown here and described in the previous chapter, "Bivariate Scatterplot and Fitting," p. 81. The Oneway role dialog supports the **Block** selection.

The following picture uses the Analgesics.jmp data set.

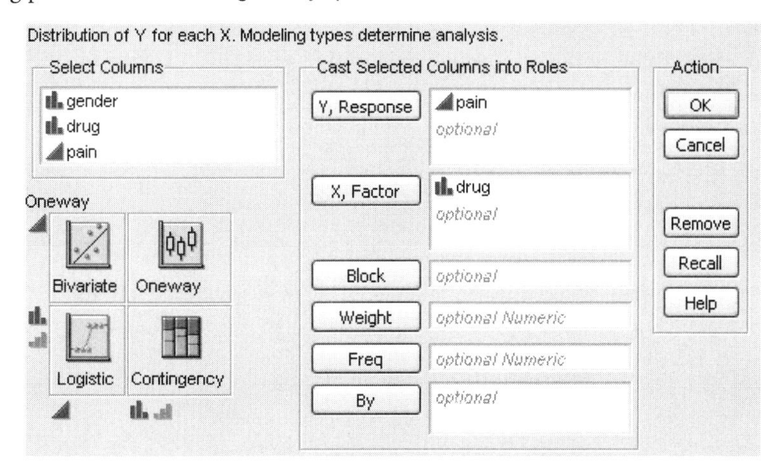

A block variable can be used to specify a second factor, which forms a two-way analysis without interaction. If there is a block variable, the analysis expects the data to be balanced and have equal counts in each block by group cell.

Figure 7.1 Fit Y By X One-Way ANOVA Platform

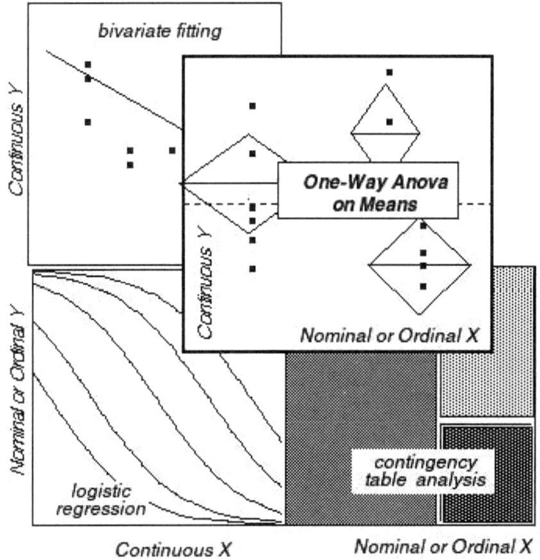

After you assign columns as response and factor variables and click **OK**, each combination of selected *x* and *Y* variables produces a separate analysis.

The data for the next example is in the Analgesics.jmp sample data table.

The Plot for a One-Way Analysis of Variance

The plot for a continuous by nominal/ordinal analysis shows the vertical distribution of response points for each factor (*X*) value (Figure 7.2). The distinct values of *x* are sometimes called levels. The basic fitting approach is to compute the mean response within each group. If the group response variance is assumed to be the same across all groups, then a one-way analysis of variance tests whether the means of the groups are equal.

Figure 7.2 The Continuous by Nominal/Ordinal Plot

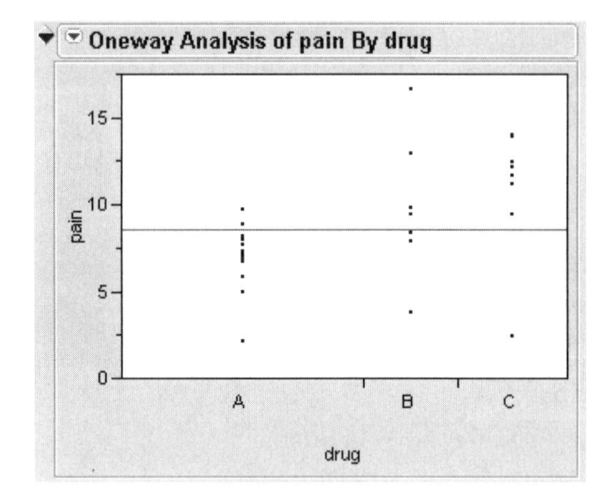

A Quick Example

Suppose the example above represents pain level scores recorded for three drugs called A, B, and C. The pain level scores are the numeric *Y* response and the three drugs are levels of the categorical *X* factor. The Oneway platform can look at the difference in group means and perform a one-way analysis of variance to determine if they are significantly different.

The initial scatterplot in Figure 7.2 indicates that one drug has consistently lower scores than the others. Also, the *x*-axis ticks are unequally spaced, proportional to the number of scores (observations) in each group.

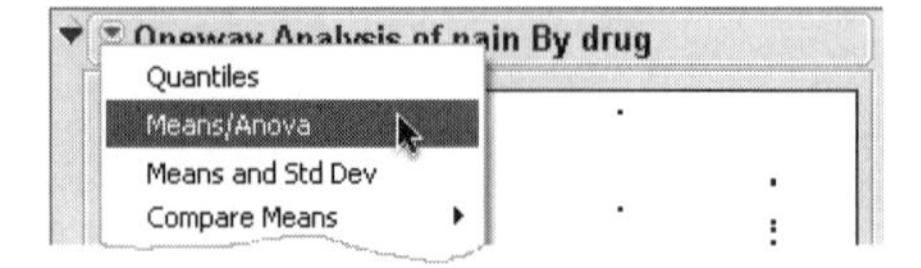

The **Means/Anova**command in the platform menu on the Oneway title bar performs an analysis of variance on the analgesics data, which automatically displays confidence interval means diamonds as shown in Figure 7.3. (Note that if there were only two levels, **Means/Anova** would show as **Means/Anova/Pooled t,** adding a pooled *t*-test to the output.)

The line across each diamond represents the group mean. The vertical span of each diamond represents the 95% confidence interval for each group. Means diamonds are covered in detail later in the chapter under "Means Diamonds and x-Axis Proportional," p. 122.

Figure 7.3 The Continuous by Nominal/Ordinal Scatterplot with Means Diamonds

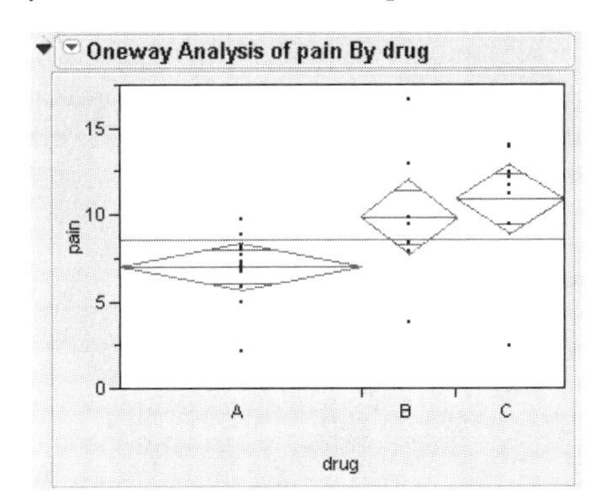

The results of the analysis of variance is in a report beneath the scatterplot that consists of the three tables in Figure 7.4.

Figure 7.4 Oneway Analysis of Variance report

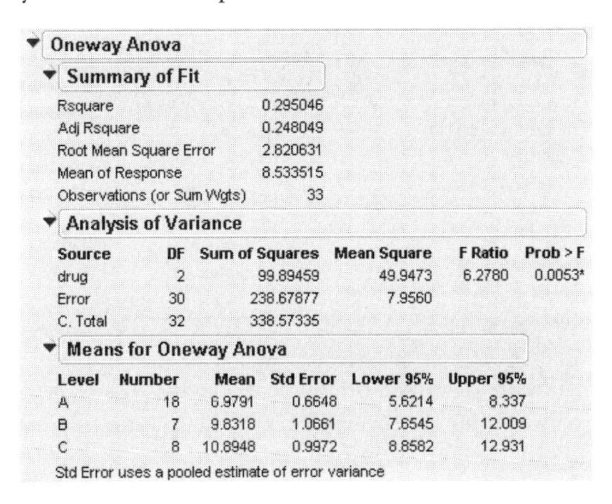

Oneway Anova

Summary of Fit

Rsquare	0.295046
Adj Rsquare	0.248049
Root Mean Square Error	2.820631
Mean of Response	8.533515
Observations (or Sum Wgts)	33

Analysis of Variance

Source	DF	Sum of Squares	Mean Square	F Ratio	Prob > F
drug	2	99.89459	49.9473	6.2780	0.0053*
Error	30	238.67877	7.9560		
C. Total	32	338.57335			

Means for Oneway Anova

Level	Number	Mean	Std Error	Lower 95%	Upper 95%
A	18	6.9791	0.6648	5.6214	8.337
B	7	9.8318	1.0661	7.6545	12.009
C	8	10.8948	0.9972	8.8582	12.931

Std Error uses a pooled estimate of error variance

- The Summary of Fit table gives overall summary information about the analysis.
- The Analysis of Variance table gives the standard ANOVA table with the probability that indicates whether or not there is a difference between the mean group scores. Note here that the Prob>F (the *p*-value) is 0.0053, which supports the visual conclusion that there is a significant difference among the groups.
  Note: If there are only two groups the analysis of variance table also shows a *t*-test.
- The Means for Oneway Anova table gives the mean, sample size, and standard error for each level of the categorical factor.

You can continue to investigate data interactively with commands that do multiple comparisons, non-parametric analysis, and other tests of interest.

The One-Way Platform Layout

Figure 7.5 is an illustration of the commands and options accessed by the popup menu icon on the title bar at the top of the scatterplot. As platform analysis commands are chosen, display options are triggered to add more elements to the graphical display as well. For example, the **Quantiles** command adds box plots; **Means/Anova/t Test** adds means diamonds; **Means/Std Dev/Std Err** adds error bars and standard deviation marks; **Compare Means** adds comparison circles.

Figure 7.5 Platform Commands and Display Options

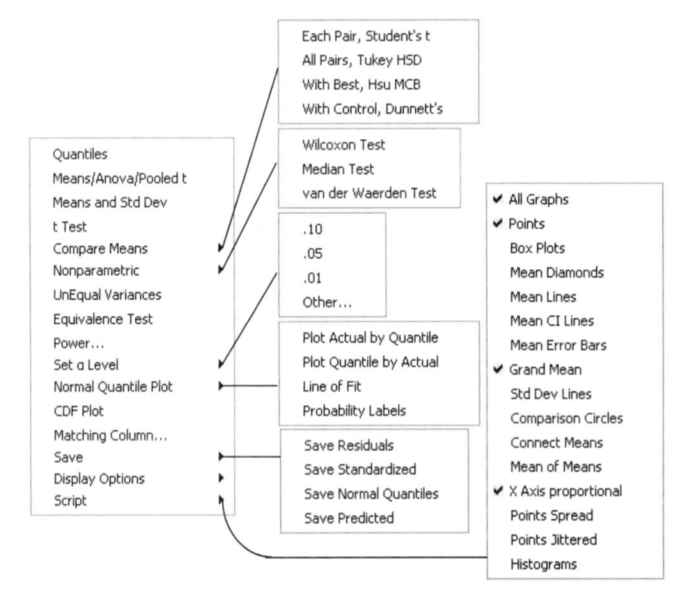

The platform commands are described briefly here. Details and examples of the platform commands and their related display options are covered later in the chapter.

Platform Commands

The popup menu icon on the title bar at the top of the scatterplot displays the platform commands and options shown in Figure 7.5. When you select a platform command, its corresponding display options automatically shows. However, you can uncheck any display option at any time.

Quantiles displays the Quantiles table, which lists the following quantiles for each group: 0% (minimum), 10%, 25%, 50% (median), 75%, 90%, and 100% (maximum), and activates **Box Plots** from the **Display Options** menu. See "Quantiles and Box Plots," p. 117 for more information.

Means/Anova/Pooled t fits means for each group and performs a one-way analysis of variance to test if there are differences among the means. Three tables are produced: a summary table, a one-way analysis of variance, and a table that lists group frequencies, means, and standard errors computed with the pooled estimate of the error variance. If there are only two groups, a *t*-test also shows. This option activates the **Means Diamonds** display option.

Means and Std Dev fits means for each group, but uses standard deviations computed within each group rather than the pooled estimate of the standard deviation used to calculate the standard errors of the means. This command displays **Mean Lines**, **Error Bars**, and **Std Dev Lines** display options.

t test produces a *t*-test report assuming the variances are not equal. The report is otherwise identical to the one shown in Figure 7.8.

Compare Means has a submenu that provides the following four multiple comparison methods for comparing sets of group means. See the sections "Multiple Comparisons," p. 124, and "Details of Comparison Circles," p. 144, for a discussion and examples of these **Compare Means** options.

Each Pair, Student's t displays a table with Student's *t*-statistics for all combinations of group means and activates the **Comparison Circles** display option.

All Pairs, Tukey HSD displays a table that shows the Tukey-Kramer HSD (honestly significant difference) comparisons of group means and corresponding comparison circles.

With Best, Hsu's MCB displays a table that shows Hsu's MCB (Multiple Comparison with the Best) comparisons of group means to the best (max or min) group mean and the appropriate comparison circles.

With Control, Dunnett's displays a table showing Dunnett's comparisons of group means with a control group that you specify, with the appropriate comparison circles.

Nonparametric has a submenu that provides the following three nonparametric comparisons of group means. See the section "Nonparametric Tests," p. 134 for statistical details.

Wilcoxon Test displays a Wilcoxon rank sums test if there are two groups and a Kruskal-Wallis nonparametric one-way analysis of variance if there are more than two groups. Note that the Wilcoxon test is equivalent to the Mann-Whitney *U*-statistic.

Median displays a two-sample median test or a *k*-sample (Brown-Mood median test) to compare group means.

Van der Waerden displays a Van der Waerden test to compare group means.

Unequal Variances command displays a table with four tests for equality of group variances: O'Brien's test, the Brown-Forsythe test, Levene's test, and Bartlett's test. When the variances across groups are not equal, the usual analysis of variance assumptions are not satisfied, the analysis of variance *F*-test is not valid, and a Welch ANOVA is preferred. See the section "Homogeneity of Variance Tests," p. 137 for statistical details about these tests.

Equivalence Test tests that a difference is less than a threshold value.

Power gives calculations of statistical power and other details about a given hypothesis test. See the section "Power Details," p. 140 for information about power computations. Also, the Power Details dialog and reports are the same as those in the general fitting platform launched by the **Fit**

Model command. Further discussion and examples for the power calculations can be found in the chapter "Standard Least Squares: Perspectives on the Estimates," p. 249.

Set Alpha Level has a submenu that allows you to choose from the most common alpha levels or specify any level with the **Other** selection. Changing the alpha level causes any confidence limits to be recalculated, adjusts the means diamonds on the plot if they are showing, and modifies the upper and lower confidence level values in reports.

Normal Quantile Plot shows overlaid normal quantile plots for each level of the *x* variable. This plot shows both the differences in the means (vertical position) and the variances (slopes) for each level of the categorical *x* factor.

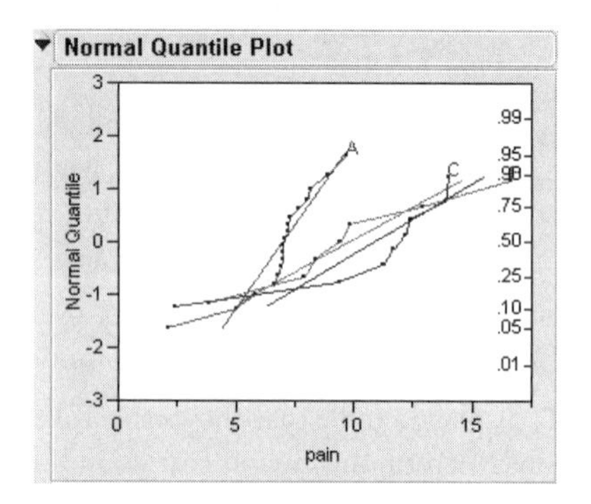

Normal Quantile plot has these additional options:

Plot Actual by Quantile generates a quantile plot with the response variable on the *y*-axis and quantiles on the *x*-axis. The plot shows quantiles computed within each level of the categorical *X* factor.

Plot Quantile by Actual reverses the *x*- and *y*-axes, as shown here.

Line of Fit draws the straight diagonal reference lines for each level of the *x* variable.

Probability Labels shows probabilities on the right axis of the Quantile by Actual plot and on the top axis of the Actual by Quantile plot.

CDF Plot plots the cumulative distribution function for all the groups in the Oneway report. Figure 7.6 shows the original analysis and the CDF plot for the **Big Class** data set's variables **height** (cast as the *Y* variable) and **sex** (cast as the *X* variable). Levels of the *x* variables in the Oneway analysis are shown as different curves. The *y* value in the Oneway analysis is the horizontal axis of the CDF plot.

Figure 7.6 Oneway CDF Plot

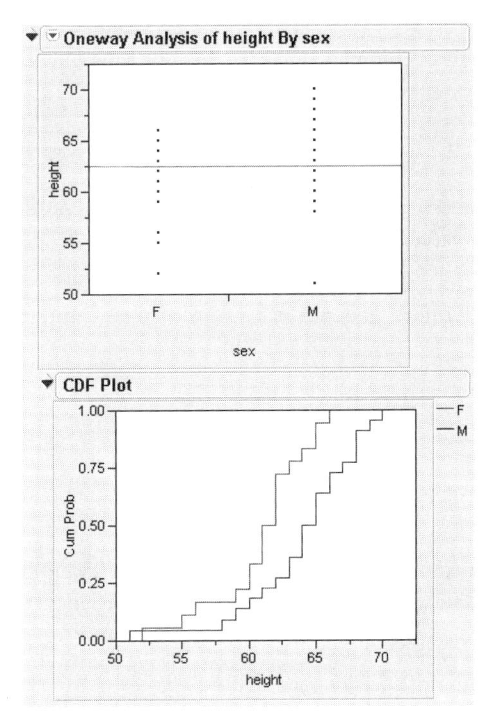

Compare Densities shows the kernel density estimate for each group.

Composition of Densities shows the summed densities, weighted by each group's counts. At each *x*-value, the Composition of Densities plot shows how each level contributes to the total.

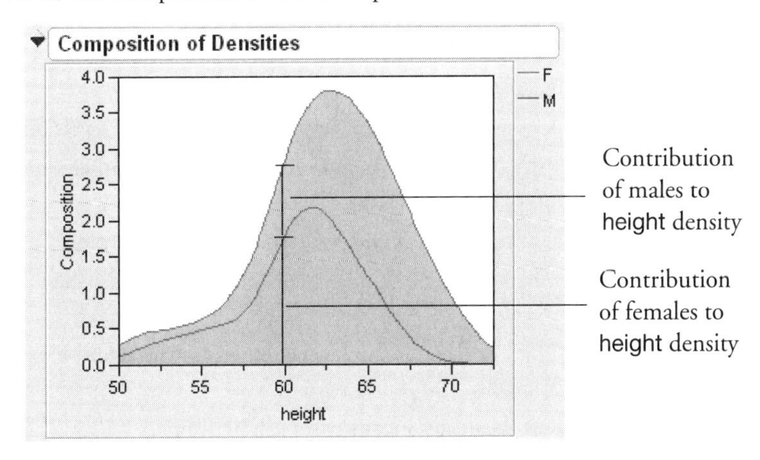

Matching Column lets you specify a matching variable to perform a matching model analysis. The **Matching Column** command addresses the case when the data in a one-way analysis come from matched (paired) data, as when observations in different groups come from the same subject. **Matching Column** activates **Show Lines** display options to connect the matching points.

You can toggle the lines off if you don't want them. See "Matching Model Analysis," p. 143 for details and an example.

Save has a submenu of commands to save the following quantities as new columns in the current data table:

Save Residuals saves values computed as the response variable minus the mean of the response variable within each level of the factor variable.

Save Standardized saves standardized values of the response variable computed within each level of the factor variable. This is the centered response divided by the standard deviation within each level.

Save Normal Quantiles saves normal quantile values computed within each level of the categorical factor variable.

Save Predicted saves the predicted mean of the response variable for each level of the factor variable.

Display Options

Display Options let you add or remove elements from the scatterplot. Note that menu items only display when they are relevant, so you may not see all the items in every situation.

All Graphs shows or hides all graphs.

Points shows data points on the scatterplot.

Box Plots shows outlier box plots for each group.

Means Diamonds draws a horizontal line through the mean of each group proportional to its x-axis.

The top and bottom points of the means diamond show the upper and lower 95% confidence points for each group.

Mean Lines draws a line at the mean of each group.

Mean CI Lines draws lines at the upper and lower 95% confidence levels for each group.

Mean Error Bars identifies the mean of each group with a large marker and shows error bars one standard error above and below the mean.

Grand Mean draws the overall mean of the Y variable on the scatterplot.

Std Dev Lines shows dotted lines one standard deviation above and below the mean of each group.

Comparison Circles show comparison circles computed for the multiple comparison method selected in the platform menu.

Connect Means connects the group means with a straight line.

X-Axis Proportional makes spacing on the x-axis proportional to the sample size of each level.

3Mean of Means draws a line at the mean of the group means.

Points Spread spreads points over the width of the interval.

Points Jittered adds random horizontal jitter so that you can see points that overlay on the same *y* value. The horizontal adjustment of points varies from 0.375 to 0.625 with a 4*(Uniform–0.5)^5 distribution.

Matching Lines connects matched points that are selected using the **Matching Columns** command.

Matching Dotted Lines draws dotted lines to connect cell means from missing cells in the table. The values used as the endpoints of the lines are obtained using a two-way ANOVA model.

Histograms draws side-by-side histograms to the right of the main graph.

Quantiles and Box Plots

The Quantiles table and outlier box plots appear automatically when you select the **Quantiles** option from the platform popup menu. However, you can show or hide them at any time either with or without platform commands or other display options in effect.

The Quantiles table lists selected percentiles for each level of the response variable, as shown here. The median is the 50th percentile, and the 25th and 75th percentiles are called the *quartiles.*

Quantiles

Level	Minimum	10%	25%	Median	75%	90%	Maximum
A	2.130667	4.713067	6.771667	7.036	7.765	8.969133	9.693333
B	3.838	3.838	7.84	9.418	12.93	16.636	16.636
C	2.404667	2.404667	9.861833	11.92	13.50967	14.01467	14.01467

The **Box Plots** option superimposes side-by-side outlier box plots on each factor group as shown in Figure 7.7. The box plots summarize the distribution of points at each factor level. The ends of the box are the 25th and 75th quantiles. The difference between the quartiles is the *interquartile range*. The line across the middle of the box identifies the median sample value. Each box has lines, sometimes called *whiskers*, that extend from each end. The whiskers extend from the ends of the box to the outermost data point that falls within the distances computed

```
upper quartile + 1.5*(interquartile range)
lower quartile - 1.5*(interquartile range).
```

Figure 7.7 Outlier Box Plots

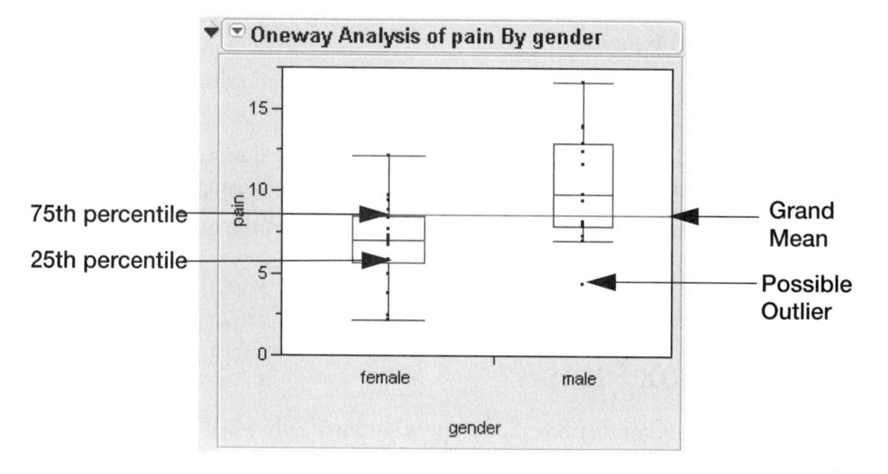

One-Way Analysis of Variance and *t*-Test: Tables and Plots

A one-way analysis of variance is the attribution and test that part of the total variability in a response is due to the difference in mean responses among the factor groups. The analysis of variance technique computes the mean within each group and the sum of squared distances from each point to the group mean. These sums of squares are totaled (pooled) to form an overall estimate of the variation unaccounted for by grouping the response. The total response variation less this residual determines the amount of variation explained by the analysis of variance model.

The **Means/Anova/t Test** command does an analysis of variance (Figure 7.4) and presents the results in the set of tables shown here. Note that when there are only two levels (as with gender) the layout includes a *t*-test table that shows tests assuming pooled (equal) variances. The t-test for unequal variances is obtained by using the **t test** command.

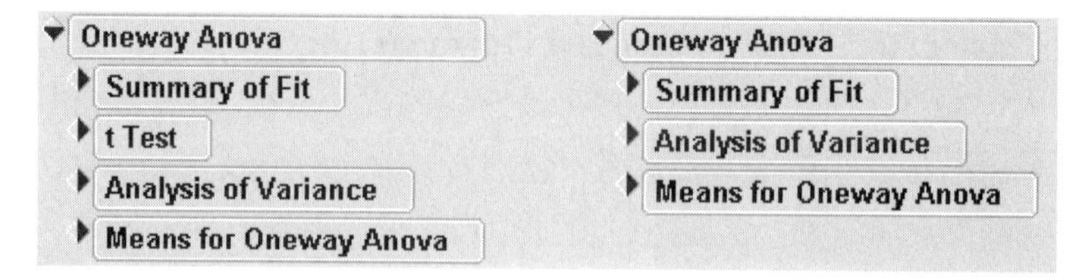

Individual tables have a disclosure button to open or close them. You can change the default layout with Preferences to display the tables you want.

The Summary of Fit Table

The Summary of Fit table shows a summary for a one-way analysis of variance, listing the following quantities.

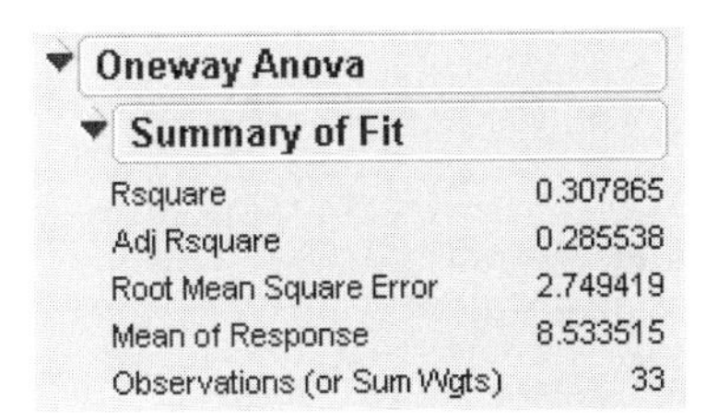

Oneway Anova	
Summary of Fit	
Rsquare	0.307865
Adj Rsquare	0.285538
Root Mean Square Error	2.749419
Mean of Response	8.533515
Observations (or Sum Wgts)	33

Rsquare measures the proportion of the variation accounted for by fitting means to each factor level. The remaining variation is attributed to random error. The R^2 value is 1 if fitting the group means accounts for all the variation with no error. An R^2 of 0 indicates that the fit serves no better as a prediction model than the overall response mean. Using quantities from the Analysis of Variance table for the model, the R^2 for any continuous response fit is always calculated:

$$\frac{\text{Sum of Squares (Model)}}{\text{Sum of Squares (C Total)}}$$

Note: R^2 is also called the *coefficient of determination.*

Rsquare Adj adjusts R^2 to make it more comparable over models with different numbers of parameters by using the degrees of freedom in its computation. It is a ratio of mean squares instead of sums of squares and is calculated:

$$1 - \frac{\text{Mean Square (Error)}}{\text{Mean Square (C Total)}}$$

where mean square for Error is found in the Analysis of Variance table and where the mean square for **C. Total** can be computed as the **C. Total** Sum of Squares divided by its respective degrees of freedom (see the Analysis of Variance Table shown next).

Root Mean Square Error estimates the standard deviation of the random error. It is the square root of the mean square for Error found in the Analysis of Variance table.

Mean of Response is the overall mean (arithmetic average) of the response variable. This is the predicted response for the base model.

Observations is the number of observations used in estimating the fit. If weights are used, this is the sum of the weights.

The *t*-test Table

If your analysis has only two *x* levels, two *t*-tests are performed in addition to the *F*-test shown in the Analysis of Variance table.

Figure 7.8 *t*-test Table

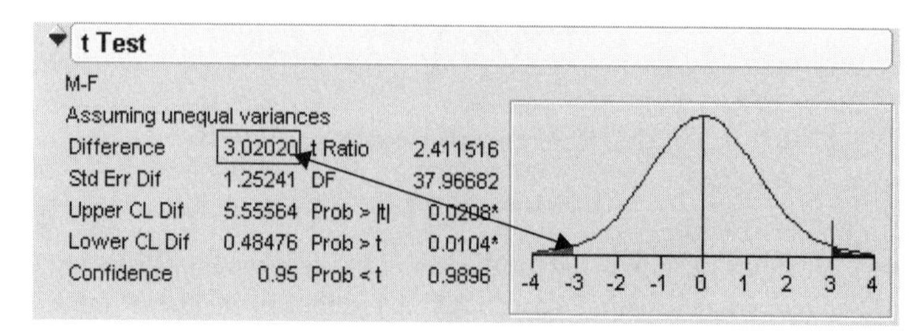

t Test

M-F

Assuming unequal variances

Difference	3.02020	t Ratio	2.411516
Std Err Dif	1.25241	DF	37.96682
Upper CL Dif	5.55564	Prob > |t|	0.0208*
Lower CL Dif	0.48476	Prob > t	0.0104*
Confidence	0.95	Prob < t	0.9896

The *t*-test Table shows these quantities for unequal-variance assumptions:

Difference shows the estimated difference between the two *x*-levels, the standard error of this difference, and a 95% confidence interval on the difference.

t-test shows the value of the *t*-test.

DF shows the degrees of freedom used in the *t*-test.

Prob > |t| shows the *p*-value associated with the *t*-test.

The plots on the right of the report show the Difference value (red line) that compares the two levels.

Note: To get a *t*-test that does assume equal variances, select **Means/Anova/Pooled t** from the platform menu.

The Analysis of Variance Table

The Analysis of Variance table *partitions* the total variation of a sample into components. The ratio of the mean square components forms an *F*-ratio that evaluates the effectiveness of the model fit. If the probability associated with the *F*-ratio is small, then the model is a statistically better fit than the overall response mean.

Analysis of Variance

Source	DF	Sum of Squares	Mean Square	F Ratio	Prob > F
gender	1	104.23492	104.235	13.7890	0.0008*
Error	31	234.33844	7.559		
C. Total	32	338.57335			

The Analysis of Variance table shows these quantities:

Source lists the three sources of variation, which are the model source (gender in the example shown here), **Error**, and **C. Total** (corrected total).

DF records an associated degrees of freedom (DF for short) for each source of variation:

A degree of freedom is subtracted from the total number of nonmissing values *N* for each parameter estimate used in a model computation. For example, an estimate of the mean parameter is used to compute the total sample variation. Thus, the DF shown on the **C. Total** line above is 32 after subtracting 1 from the total number of observations. The total degrees of freedom are *partitioned* (divided) into the **Model** (gender) and **Error** terms.

If the factor has k levels, then k - 1 degrees of freedom are used to compute the variation attributed to the analysis of variance model. In the example shown above, there are two nominal levels, which results in one degrees of freedom for the **Model**.

The **Error** degrees of freedom is the difference between the **C. Total** degrees of freedom and the **Model** degrees of freedom.

Sum of Squares records a sum of squared distances (SS for short) for each source of variation:

The total (**C. Total**) sum of squares of each response from the overall response mean is 338.57 in this example. The **C. Total** sum of squares is the base model used for comparison with all other models.

The sum of squared distances from each point to its respective group mean is 234.3338 in this example. This is the remaining unexplained **Error** (*residual*) SS after fitting the analysis of variance model.

The total SS less the error SS (104.234) gives the sum of squares attributed to the model. This tells you how much of the total variation is explained by the model.

Mean Square is a sum of squares divided by its associated degrees of freedom:

The **Model** mean square for this example is 49.947. It estimates the variance of the error but only under the hypothesis that the group means are equal.

The **Error** mean square for this example is 7.96. It estimates the variance of the error term independently of the model mean square and is unconditioned by any model hypothesis.

F Ratio is the model mean square divided by the error mean square. If the hypothesis that the group means are equal (there is no real difference between them) is true, both the mean square for error and the mean square for model estimate the error variance. Their ratio has an *F*-distribution. If the analysis of variance model results in a significant reduction of variation from the total, the *F*-ratio is higher than expected.

Prob>F is the probability of obtaining (by chance alone) an *F*-value greater than the one calculated if, in reality, fitting group means accounts for no more variation than using the overall response mean. Observed significance probabilities of 0.05 or less are often considered evidence that an analysis of variance model fits the data.

If you have specified a Block column, then the report is modified to show the contribution of the Blocks, and an additional report of Block means is shown.

Note: The **Power** popup command on the Analysis of Variance Table accesses the JMP power details facility so you can look at the power of the analysis.

The Means for Oneway Anova Table

The Means for Oneway Anova table summarizes response information for each level of the nominal or ordinal factor.

Means for Oneway Anova

Level	Number	Mean	Std Error	Lower 95%	Upper 95%
female	18	6.9111	0.64804	5.5894	8.233
male	15	10.4804	0.70990	9.0326	11.928

Std Error uses a pooled estimate of error variance

The Means for Oneway Anova table lists these quantities:

Level lists the values of the ordinal or nominal variable that name or describe the categories for fitting each group mean.

Number lists the number of observations used to calculate each group mean estimate.

Mean lists the estimates of the mean response (sample mean) of each group.

Std Error lists the estimates of the standard deviations for the group means. This standard error is estimated assuming that the variance of the response is the same in each level. It is the root mean square error found in the Summary of Fit table divided by the square root of the number of values used to compute the group mean.

Lower 95% and Upper 95% list the lower and upper 95% confidence interval for the group means.

Means Diamonds and *x*-Axis Proportional

A means diamond illustrates a sample mean and 95% confidence interval, as shown by the schematic in Figure 7.9. The line across each diamond represents the group mean. The vertical span of each diamond represents the 95% confidence interval for each group. Overlap marks are drawn $((\sqrt{2}\mathrm{CI})/2)$ above and below the group mean. For groups with equal sample sizes, overlapping marks indicate that the two group means are not significantly different at the 95% confidence level.

If the **X-Axis Proportional** display option is in effect, the horizontal extent of each group along the *x*-axis (the horizontal size of the diamond) is proportional to the sample size of each level of the *x* variable. It follows that the narrower diamonds are usually the taller ones because fewer data points yield a less precise estimate of the group mean.

The confidence interval computation assumes that variances are equal across observations. Therefore, the height of the confidence interval (diamond) is proportional to the reciprocal of the square root of the number of observations in the group.

The means diamonds automatically show when you select the **Means/Anova/t Test** option from the platform menu. However, you can show or hide them at any time with or without the **Means, Anova/t Test** command or other display options in effect.

Figure 7.9 Means Diamonds and X-Proportional Options

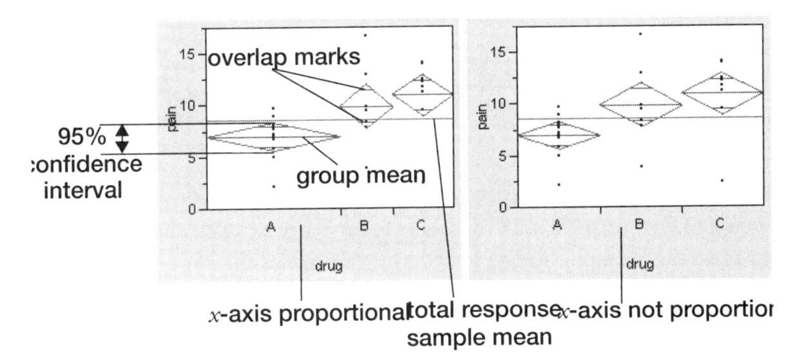

The following sections describe these platform options and their related display options.

Mean Lines, Error Bars, and Standard Deviations Lines

When you select **Mean Error Bars** from the display options submenu, the scatterplot shows a line at the mean of each group with error bars one standard error above and below each group mean. If you also select the **Std Dev Lines** option, you see dashed lines that identify one standard deviation above and below the group means, as shown here.

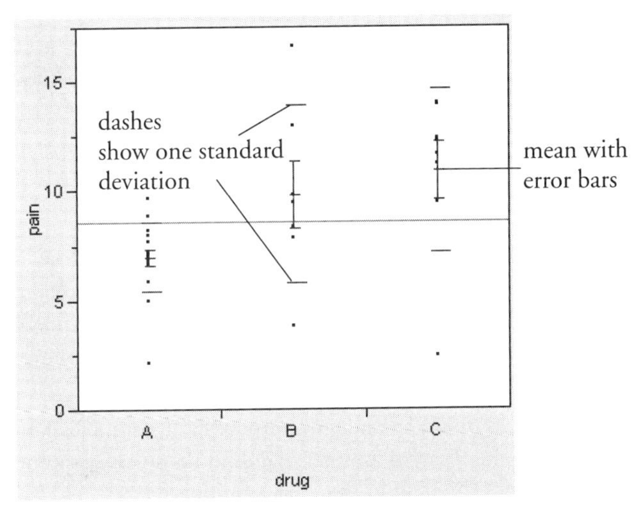

The **Mean Lines**, **Mean Error Bars**, and **Std Dev Lines** display options appear automatically when you select the **Means/Std Dev/Std Err** platform option. However, you can show or hide them at any time without platform commands or other display options in effect.

Multiple Comparisons

There are a variety of methods to test differences in group means (multiple comparisons) that vary in detail about how to size the test to accommodate different kinds of multiple comparisons. **Fit Y by X** automatically produces the standard analysis of variance and optionally offers the following four multiple comparison tests:

Each Pair, Student's t computes individual pairwise comparisons using Student's *t*-tests. This test is sized for individual comparisons. If you make many pairwise tests, there is no protection across the inferences, and thus the alpha-size (Type I) error rate across the hypothesis tests is higher than that for individual tests.

All Pairs, Tukey HSD gives a test that is sized for all differences among the means. This is the *Tukey* or *Tukey-Kramer* HSD (honestly significant difference) test. (Tukey 1953, Kramer 1956). This test is an exact alpha-level test if the sample sizes are the same and conservative if the sample sizes are different (Hayter 1984).

With Best, Hsu's MCB tests whether means are less than the unknown maximum (or greater than the unknown minimum). This is the Hsu MCB test (Hsu 1981).

With Control, Dunnett's tests whether means are different from the mean of a control group. This is Dunnett's test (Dunnett 1955).

The four multiple comparisons tests are the ones recommended by Hsu (1989) as level-5 tests for the three situations: MCA (Multiple Comparisons for *All* pairs), MCB (Multiple Comparisons with the *Best*), and MCC (Multiple Comparisons with *Control*).

If you have specified a Block column, then the multiple comparison methods are performed on data that has been adjusted for the Block means.

Visual Comparison of Group Means

Each multiple comparison test begins with a *comparison circles* plot, which is a visual representation of group mean comparisons. The plot follows with a reveal table of means comparisons. The illustration in Figure 7.10 shows the alignment of comparison circles with the confidence intervals of their respective group means for the Student's t comparison. Other comparison tests widen or shorten the radii of the circles.

Overlap marks show for each diamond and are computed as (group mean ± CI). Overlap marks in one diamond that are closer to the mean of another diamond than that diamond's overlap marks indicate that those two groups are not different at the 95% confidence level.

Figure 7.10 Visual Comparison of Group Means

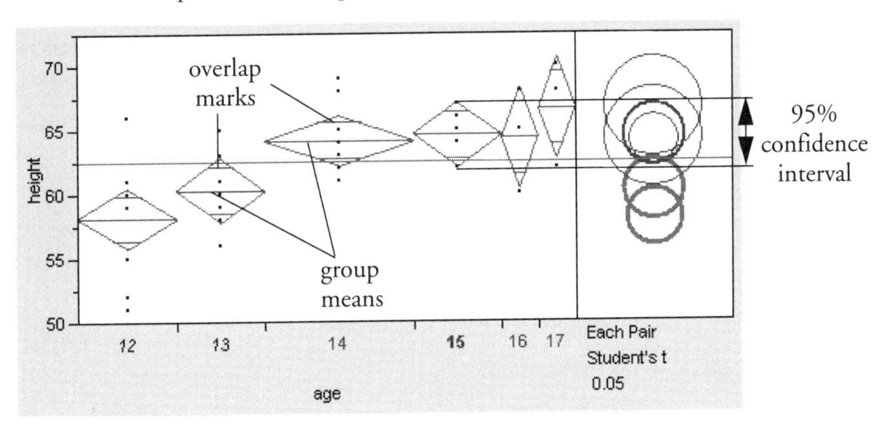

You can compare each pair of group means visually by examining how the comparison circles intersect. The outside angle of intersection tells you whether group means are significantly different (see Figure 7.11). Circles for means that are significantly different either do not intersect or intersect slightly so that the outside angle of intersection is less than 90 degrees. If the circles intersect by an angle of more than 90 degrees or if they are nested, the means are not significantly different.

Figure 7.11 Angle of Intersection and Significance

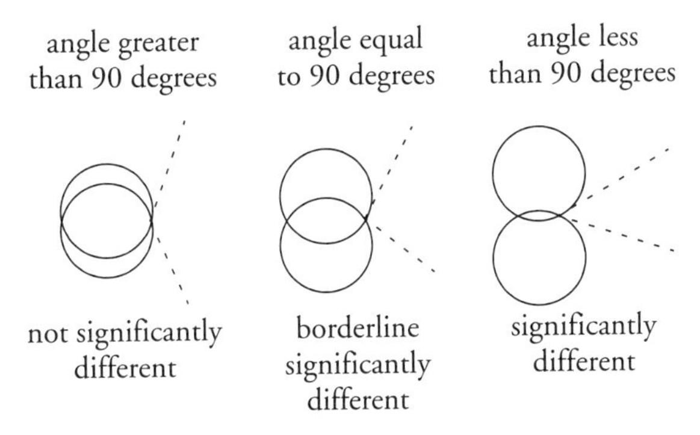

If the intersection angle is close to 90 degrees, you can verify whether the means are significantly different by clicking on the comparison circle to highlight it. The highlighted circle appears with a thick solid line. Red circles representing means that are not significantly different from the highlighted circle show with thin lines (see Figure 7.12). Circles representing means that are significantly different show with a thick gray pattern. To deselect circles, click in the graph outside the circles.

Figure 7.12 Highlighting Comparison Circles

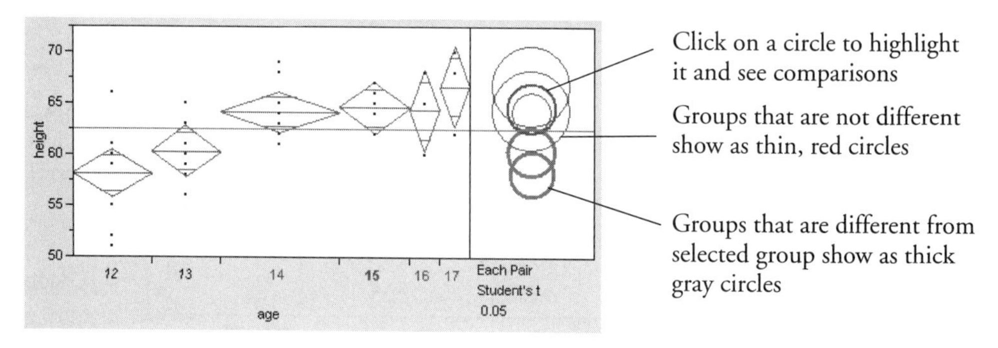

For more information about comparison circles, see "Details of Comparison Circles," p. 144.

In addition, means comparison reports contain several options that allow it to be customized. All of these methods use pooled variance estimates for the means.

Confid Quantile shows the *t*-statistic or other corresponding quantiles used for confidence intervals.

t	Alpha
2.03224	0.05

Difference Matrix shows a table of all differences of means.

Dif=Mean[i]-Mean[j]

	17	15	16	14	13	12
17	0.0000	2.0952	2.3333	2.5000	6.3810	8.5417
15	-2.0952	0.0000	0.2381	0.4048	4.2857	6.4464
16	-2.3333	-0.2381	0.0000	0.1667	4.0476	6.2083
14	-2.5000	-0.4048	-0.1667	0.0000	3.8810	6.0417
13	-6.3810	-4.2857	-4.0476	-3.8810	0.0000	2.1607
12	-8.5417	-6.4464	-6.2083	-6.0417	-2.1607	0.0000

LSD Threshold Matrix shows a matrix showing if a difference exceeds the least significant difference for all comparisons.

Abs(Dif)-LSD

	17	15	16	14	13	12
17	-5.6127	-2.6484	-3.2794	-1.9373	1.6373	3.8878
15	-2.6484	-3.6744	-4.5055	-2.8646	0.6113	2.8887
16	-3.2794	-4.5055	-5.6127	-4.2706	-0.6960	1.5545
14	-1.9373	-2.8646	-4.2706	-2.8064	0.6116	2.9040
13	1.6373	0.6113	-0.6960	0.6116	-3.6744	-1.3970
12	3.8878	2.8887	1.5545	2.9040	-1.3970	-3.4371

Connecting Letters Report shows the traditional letter-coded report where means not sharing a letter are significantly different.

Level		Mean
17	A	66.666667
15	A	64.571429
16	A B	64.333333
14	A	64.166667
13	B C	60.285714
12	C	58.125000

Levels not connected by same letter are significantly different.

Ordered Differences Report shows all the positive-side differences with their confidence interval in sorted order. For the Student's t and Tukey-Kramer comparisons, an Ordered Difference report appears below the text reports. The following report comes from the Snapdragon.jmp data table.

Figure 7.13 Ordered Differences Report

Level	- Level	Difference	Lower CL	Upper CL	p-Value
wabash	compost	6.300000	4.01814	8.581858	<.0001*
wabash	clinton	5.666667	3.38481	7.948525	0.0002*
knox	compost	5.233333	2.95148	7.515192	0.0003*
wabash	webster	4.866667	2.58481	7.148525	0.0006*
knox	clinton	4.600000	2.31814	6.881858	0.0009*
o'neill	compost	4.133333	1.85148	6.415192	0.0019*
wabash	clarion	3.800000	1.51814	6.081858	0.0035*
knox	webster	3.800000	1.51814	6.081858	0.0035*
o'neill	clinton	3.500000	1.21814	5.781858	0.0059*
knox	clarion	2.733333	0.45148	5.015192	0.0228*
o'neill	webster	2.700000	0.41814	4.981858	0.0242*
clarion	compost	2.500000	0.21814	4.781858	0.0343*
wabash	o'neill	2.166667	-0.11519	4.448525	0.0608
clarion	clinton	1.866667	-0.41519	4.148525	0.1000
o'neill	clarion	1.633333	-0.64852	3.915192	0.1448
webster	compost	1.433333	-0.84852	3.715192	0.1962
knox	o'neill	1.100000	-1.18186	3.381858	0.3143
wabash	knox	1.066667	-1.21519	3.348525	0.3285
clarion	webster	1.066667	-1.21519	3.348525	0.3285
webster	clinton	0.800000	-1.48186	3.081858	0.4597
clinton	compost	0.633333	-1.64852	2.915192	0.5566

This report shows the ranked differences, from highest to lowest, with a confidence interval band overlaid on the plot. Confidence intervals that do not fully contain their corresponding bar are significantly different from each other.

Detailed Comparisons Report shows a detailed report for each comparison. Each section shows the difference between the levels, standard error and confidence intervals, *t*-ratios, *p*-values, and degrees of freedom. A plot illustrating the comparison is shown on the right of each report.

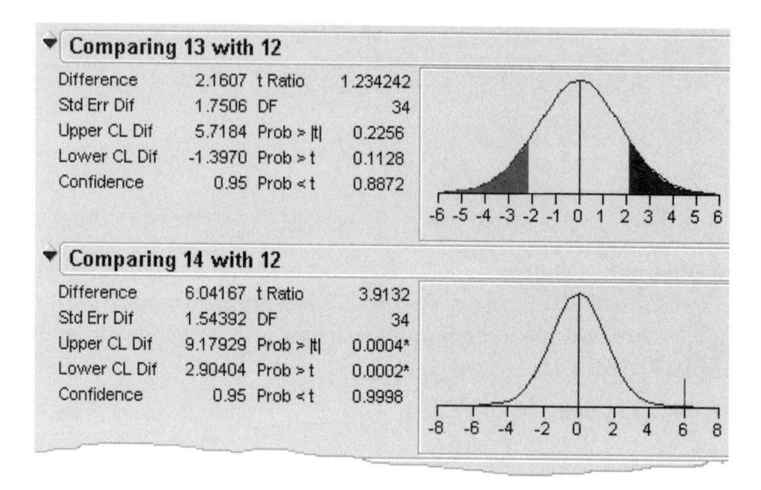

The availability of these commands differs among the comparison tests. For the Student's *t*-test, all options are available. For Tukey's HSD, all options except **Detailed Comparisons Report** are available. For both Hsu's MCB and Dunnett's test, only **Difference Matrix**, **Confid Quantile**, and **LSD Threshold Matrix** are available.

Student's *t*-Test for Each Pair

The **Each Pair, Student's t** command gives the Student's *t*-test for each pair of group levels and tests individual comparisons only. The example shown here is for a one-way layout of **weight** by **age** from the **Big Class** sample data table. It is the group comparison with comparison circles that illustrate all possible *t*-tests.

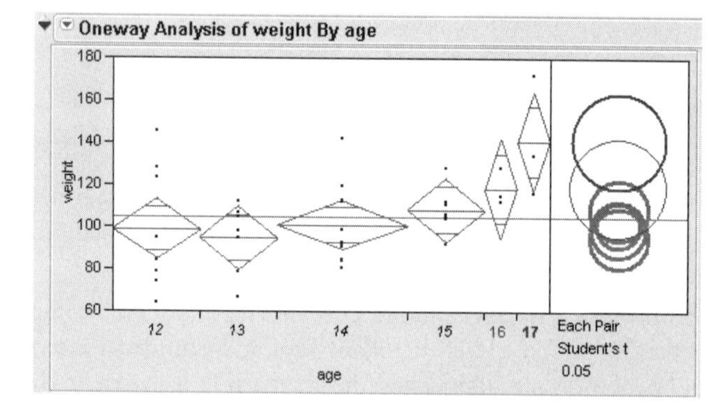

The means comparison method can be thought of as seeing if the actual difference in the means is greater than the difference that would be significant. This difference is called the LSD (least significant difference). The LSD term is used both for Student's *t* intervals and in context with intervals for other tests. In the comparison circles graph, the distance between the circles' centers represent the actual difference. The LSD is what the distance would be if the circles intersected at right angles.

Figure 7.14 shows the means comparison report with default options for Student's *t* comparisons (**Confid Quantile**, **LSD Threshold Matrix**, **Connecting Lines Report**, and **Ordered Differences Report**).

The LSD threshold table shown in Figure 7.14 is the difference between the absolute difference and the LSD that would be significantly different. If the values are positive, the difference is judged to be significant. The greatest differences are typically in the upper-right and lower-left corners, and the nonsignificant differences are generally down the diagonal of the table.

Figure 7.14 Means Comparisons for Student's *t*

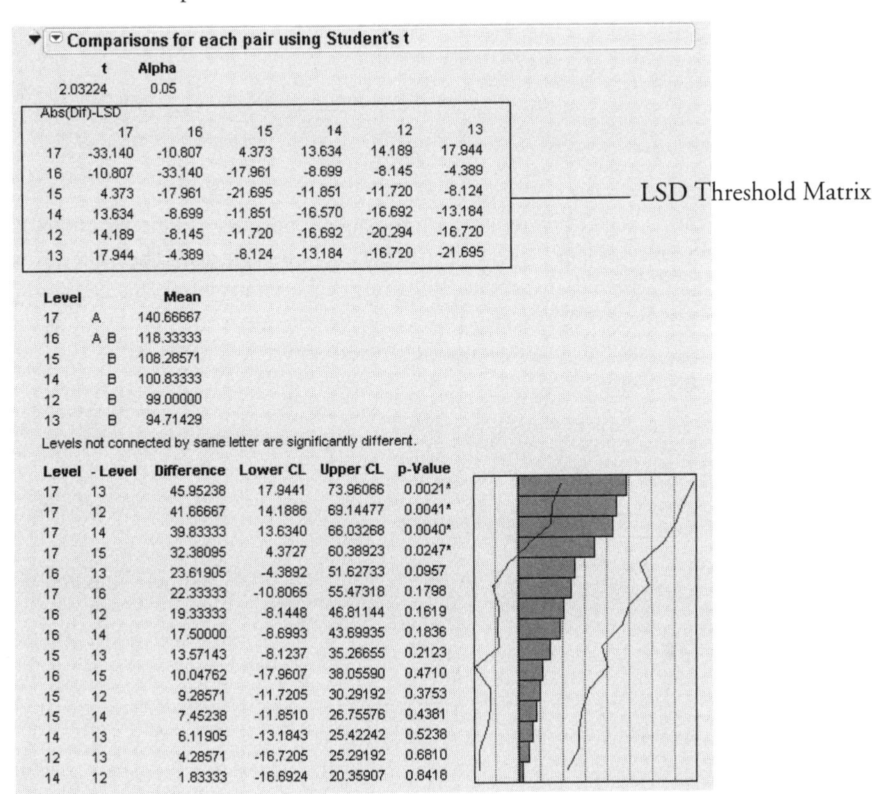

Comparisons for each pair using Student's t

t	Alpha
2.03224	0.05

Abs(Dif)-LSD

	17	16	15	14	12	13
17	-33.140	-10.807	4.373	13.634	14.189	17.944
16	-10.807	-33.140	-17.961	-8.699	-8.145	-4.389
15	4.373	-17.961	-21.695	-11.851	-11.720	-8.124
14	13.634	-8.699	-11.851	-16.570	-16.692	-13.184
12	14.189	-8.145	-11.720	-16.692	-20.294	-16.720
13	17.944	-4.389	-8.124	-13.184	-16.720	-21.695

Level			Mean
17	A		140.66667
16	A	B	118.33333
15		B	108.28571
14		B	100.83333
12		B	99.00000
13		B	94.71429

Levels not connected by same letter are significantly different.

Level	- Level	Difference	Lower CL	Upper CL	p-Value
17	13	45.95238	17.9441	73.96066	0.0021*
17	12	41.66667	14.1886	69.14477	0.0041*
17	14	39.83333	13.6340	66.03268	0.0040*
17	15	32.38095	4.3727	60.38923	0.0247*
16	13	23.61905	-4.3892	51.62733	0.0957
17	16	22.33333	-10.8065	55.47318	0.1798
16	12	19.33333	-8.1448	46.81144	0.1619
16	14	17.50000	-8.6993	43.69935	0.1836
15	13	13.57143	-8.1237	35.26655	0.2123
16	15	10.04762	-17.9607	38.05590	0.4710
15	12	9.28571	-11.7205	30.29192	0.3753
15	14	7.45238	-11.8510	26.75576	0.4381
14	13	6.11905	-13.1843	25.42242	0.5238
12	13	4.28571	-16.7205	25.29192	0.6810
14	12	1.83333	-16.6924	20.35907	0.8418

Tukey-Kramer HSD for All Comparisons

With Tukey-Kramer you protect the significance tests of all combinations of pairs, and the LSD intervals become much greater than the Student's *t* pairwise LSDs. Graphically, the comparison circles become much larger as shown here; thus, differences are less significant.

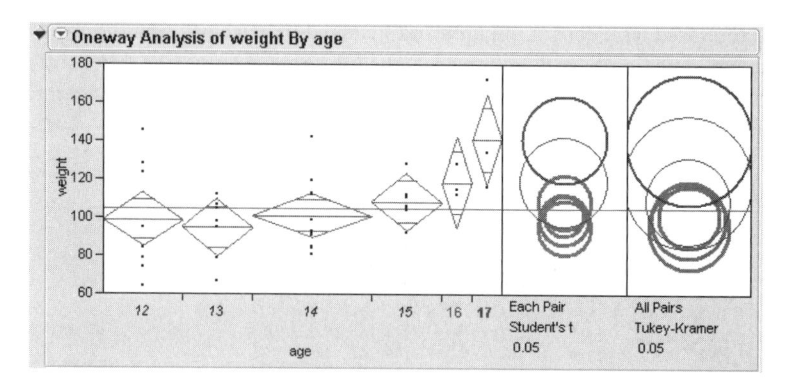

Figure 7.15 shows the Tukey's HSD report with the default options (**Confid Quantile, LSD Threshold Matrix**, **Connecting Lines Report**, and **Ordered Differences Report**). The Tukey-Kramer LSD Threshold matrix shows the actual absolute difference in the means minus the LSD, which is the difference that would be significant. Pairs with a positive value are significantly different. Because the borders of the table are sorted by the mean, the more significant differences tend toward the upper right, or symmetrically, the lower left.

Figure 7.15 Means Comparison for Tukey's HSD

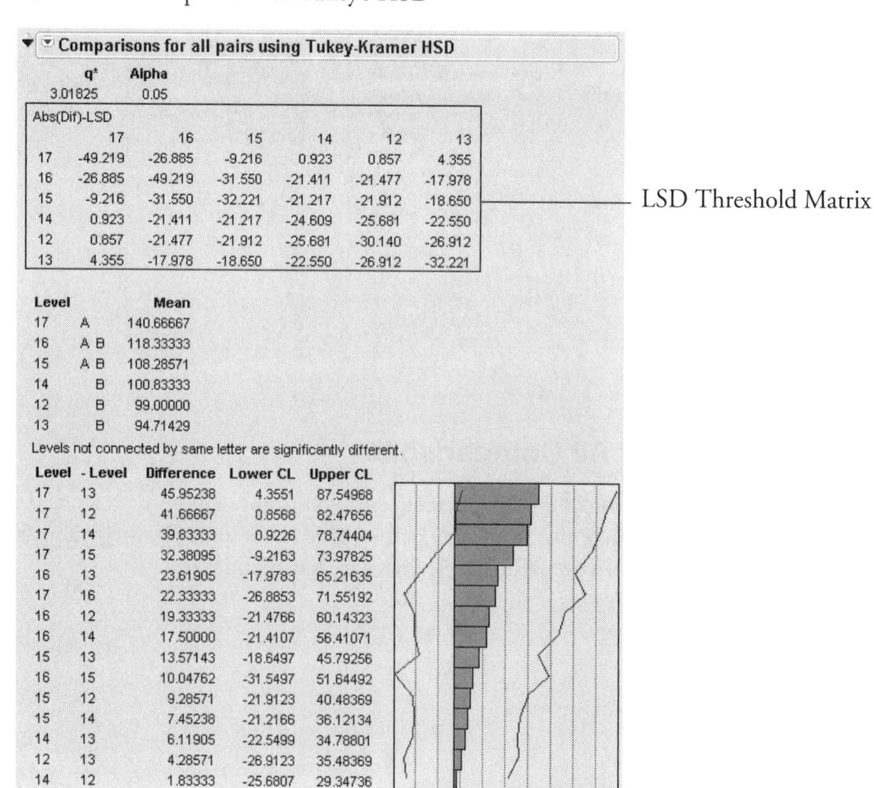

Comparisons for all pairs using Tukey-Kramer HSD

q*	Alpha
3.01825	0.05

Abs(Dif)-LSD

	17	16	15	14	12	13
17	-49.219	-26.885	-9.216	0.923	0.857	4.355
16	-26.885	-49.219	-31.550	-21.411	-21.477	-17.978
15	-9.216	-31.550	-32.221	-21.217	-21.912	-18.650
14	0.923	-21.411	-21.217	-24.609	-25.681	-22.550
12	0.857	-21.477	-21.912	-25.681	-30.140	-26.912
13	4.355	-17.978	-18.650	-22.550	-26.912	-32.221

LSD Threshold Matrix

Level			Mean
17	A		140.66667
16	A	B	118.33333
15	A	B	108.28571
14		B	100.83333
12		B	99.00000
13		B	94.71429

Levels not connected by same letter are significantly different.

Level	- Level	Difference	Lower CL	Upper CL
17	13	45.95238	4.3551	87.54968
17	12	41.66667	0.8568	82.47656
17	14	39.83333	0.9226	78.74404
17	15	32.38095	-9.2163	73.97825
16	13	23.61905	-17.9783	65.21635
17	16	22.33333	-26.8853	71.55192
16	12	19.33333	-21.4766	60.14323
16	14	17.50000	-21.4107	56.41071
15	13	13.57143	-18.6497	45.79256
16	15	10.04762	-31.5497	51.64492
15	12	9.28571	-21.9123	40.48369
15	14	7.45238	-21.2166	36.12134
14	13	6.11905	-22.5499	34.78801
12	13	4.28571	-26.9123	35.48369
14	12	1.83333	-25.6807	29.34736

The band near the diagonal tends to be negative because the means are closer. The q* (shown above the LSD Threshold Matrix table) is the quantile that is used to scale the LSDs. It has a computational role comparable to a Student's *t*.

Comparisons with the Best—Hsu's MCB

Hsu's MCB test is used to determine whether a mean can be rejected as the maximum or minimum among the means. Usually, Hsu's MCB is presented in terms of confidence intervals of the difference between each mean and the unknown maximum mean. Here, for consistency with the other JMP methods and the graphical presentation, this test is shown in terms of an LSD (least significant difference).

A mean can be considered significantly less than the maximum if there is a mean above it that is separated by more than the LSD. The quantile that scales the LSD is not the same for each mean. Graphically, the comparison circles technique does not work as well because the circles must be rescaled according to the mean being tested, unless the sample sizes are equal. The graph uses the largest quantiles value shown to make the circles. If a circle has another circle separated above it, then the two means are significantly different. If not, check the report to see if the mean is significantly less than the maximum, or click on its circle to see the differences highlighted.

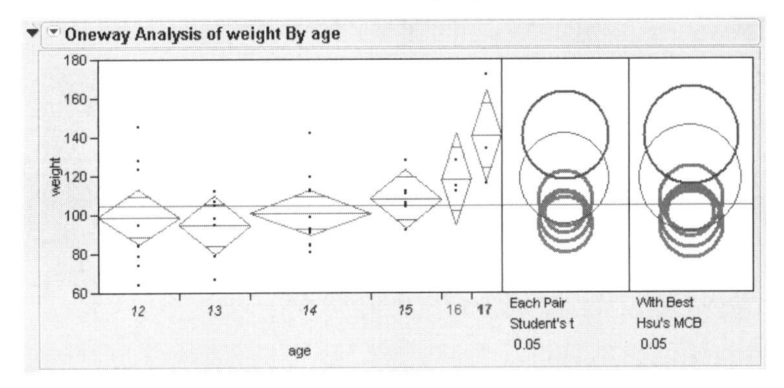

In summary, you conclude as follows. If a mean

- has means significantly separated above it, it is not regarded as the maximum.
- has means significantly separated below it, it is not regarded as the minimum.
- is significantly separated above all other means, it is regarded as the maximum.
- is significantly separated below all other means, it is regarded as the minimum.

Note: Means that are not regarded as the maximum or the minimum by MCB are also the means not contained in the selected subset of Gupta (1965) of potential maximums or minimum means.

Figure 7.16 shows the means comparison report with default options for Hsu's MCB comparisons (**Confid Quantile** and **LSD Threshold Matrix**).

Look down the columns in the LSD threshold matrix shown for each mean. For the *maximum* report, a column shows the row mean minus the column mean minus the LSD. If a column has any positive values, the means for those rows are significantly separated above the mean for the column, and the mean for the column is significantly not the maximum.

Similarly for the *minimum* report, if a column has any negative values, the means for those rows are significantly separated below the mean for the column, and the mean for the column is significantly not the minimum.

Figure 7.16 Hsu's MCB Default Report

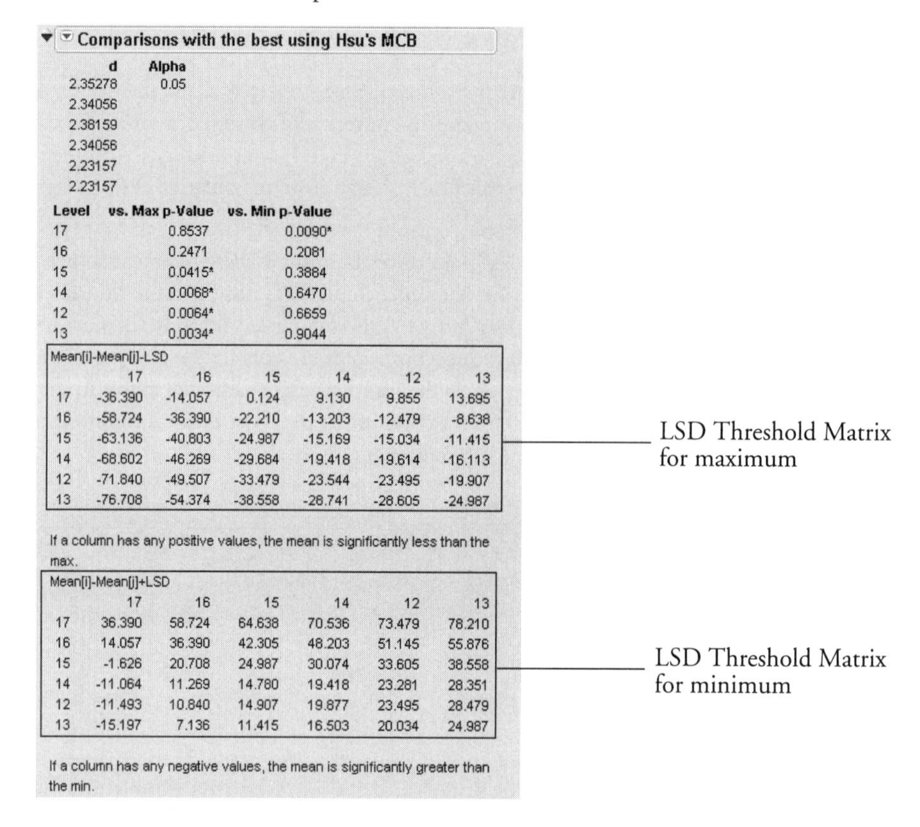

Comparisons with the best using Hsu's MCB

d	Alpha
2.35278	0.05
2.34056	
2.38159	
2.34056	
2.23157	
2.23157	

Level	vs. Max p-Value	vs. Min p-Value
17	0.8537	0.0090*
16	0.2471	0.2081
15	0.0415*	0.3884
14	0.0068*	0.6470
12	0.0064*	0.6659
13	0.0034*	0.9044

Mean[i]-Mean[j]-LSD

	17	16	15	14	12	13
17	-36.390	-14.057	0.124	9.130	9.855	13.695
16	-58.724	-36.390	-22.210	-13.203	-12.479	-8.638
15	-63.136	-40.803	-24.987	-15.169	-15.034	-11.415
14	-68.602	-46.269	-29.684	-19.418	-19.614	-16.113
12	-71.840	-49.507	-33.479	-23.544	-23.495	-19.907
13	-76.708	-54.374	-38.558	-28.741	-28.605	-24.987

If a column has any positive values, the mean is significantly less than the max.

Mean[i]-Mean[j]+LSD

	17	16	15	14	12	13
17	36.390	58.724	64.638	70.536	73.479	78.210
16	14.057	36.390	42.305	48.203	51.145	55.876
15	-1.626	20.708	24.987	30.074	33.605	38.558
14	-11.064	11.269	14.780	19.418	23.281	28.351
12	-11.493	10.840	14.907	19.877	23.495	28.479
13	-15.197	7.136	11.415	16.503	20.034	24.987

If a column has any negative values, the mean is significantly greater than the min.

Comparisons with a Control Group—Dunnett's Test

Dunnett's test compares a set of means against the mean of a *control* group. The LSDs it produces are between Student's *t* and Tukey-Kramer because they are sized to guard against an intermediate number of comparisons.

For Dunnett's test, you must first highlight a row or point on the means plot to identify that group as your *control* group. The group identified by the first highlighted row that JMP finds becomes the control group. If there are no highlighted rows, JMP presents a dialog to choose a group as the control.

The comparisons circles plot of Dunnett's test, with the control group (age 12) highlighted, is shown in Figure 7.17. The significances with the other means are coded by color and by line style. Graphically, a mean is significantly different from the control group mean if its circle is separate from the control group circle, or at least is intersecting small enough to produce an angle of intersection less than 90 degrees.

Figure 7.17 Comparison Circles for Student's *t* and Dunnett's Tests

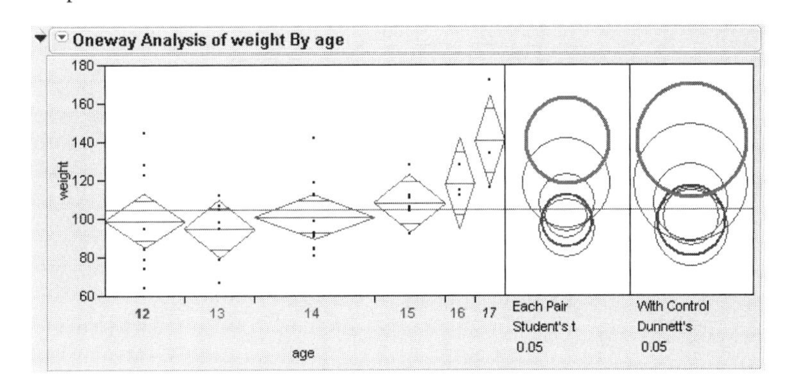

In the Dunnett's report, the |*d*| quantile is shown and can be used in a manner similar to a Student's *t*-statistic. Figure 7.18 shows the Dunnett's report with the default options selected (shows the means comparison report with default options for Student's *t* comparisons (**Confid Quantile** and **LSD Threshold Matrix**). The LSD threshold matrix shows the absolute value of the difference minus the LSD. If a value is positive, its mean is more than the LSD apart from the control group mean and is thus significantly different. The value for the control group in the report shown here (-26.596 for the 12-year-olds) is the negative of what the LSD would be for all comparisons if each group had the same number of observations as the control group, which happens to be the product of the quantile value, the square root of 2, and the standard error of the mean.

Figure 7.18 Dunnett's Test Default Report

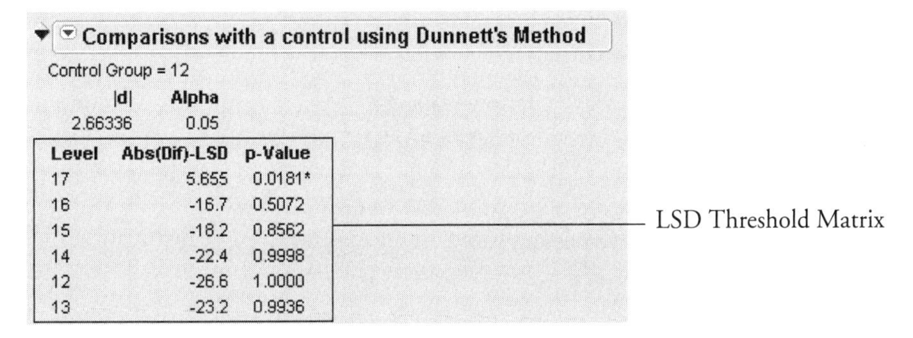

Comparisons with a control using Dunnett's Method

Control Group = 12

|d|	Alpha
2.66336	0.05

Level	Abs(Dif)-LSD	p-Value
17	5.655	0.0181*
16	-16.7	0.5072
15	-18.2	0.8562
14	-22.4	0.9998
12	-26.6	1.0000
13	-23.2	0.9936

LSD Threshold Matrix

Although the four tests all test differences between group means, different results can occur. Figure 7.19 compares the means circles for the data above, comparing the smallest group mean to the others. The student's *t*-test is least conservative and shows no difference in the smallest three means. The three more conservative tests show no difference in four group means. Even though they give the same result, the radius of the comparison circles and their angles of intersection indicate different significance levels between groups.

Figure 7.19 Comparison Circles for Four Multiple Comparison Test

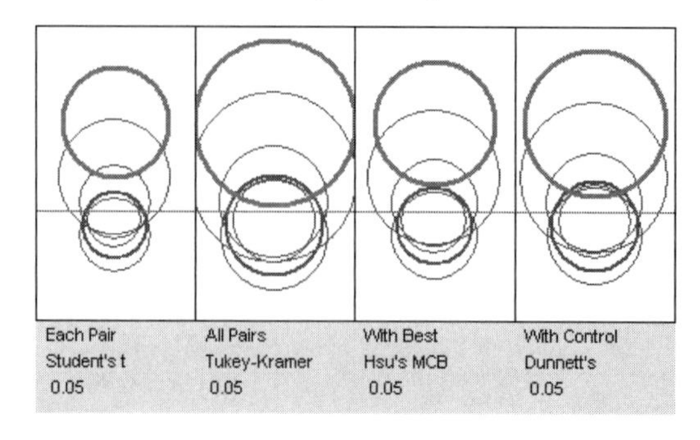

Equivalence Test

Equivalence tests try to confirm that there is no difference in means, Since this is impossible, you need to pick a threshold difference for which smaller differences is practical equivalence. The most straight-forward test to construct uses two one-sided *t*-tests from both sides of the difference interval. If both tests reject, the groups are equivalent. The **Equivalence Test** command uses this Two One-Sided Tests (TOST) approach.

For example, suppose you are interested if the difference in height between males and females is less than 6 (using the Big Class.jmp data table). After running **Fit Y by X** with height as **Y** and sex as **X**, select the **Equivalence Test** command. In the dialog that appears, enter 6 as the difference considered practically zero and click **OK** to reveal the output shown in Figure 7.20.

Figure 7.20 Practical Equivalence Output

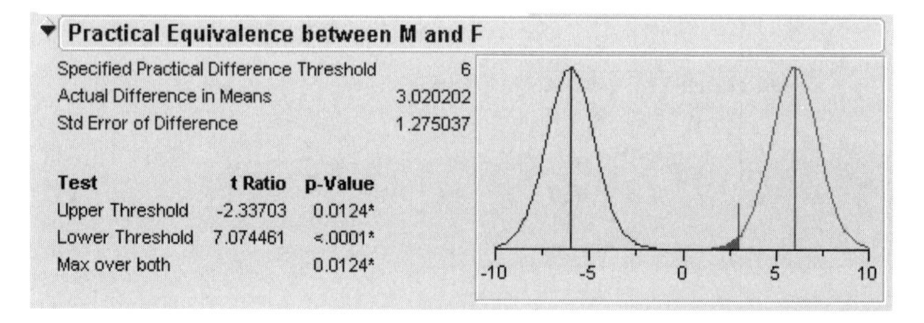

The actual difference is 3.02, and the test shows that this has a *p*-value of 0.01. Therefore, you can declare this difference to be zero for your purposes.

Nonparametric Tests

Nonparametric tests are useful to test whether group means or medians are located the same across groups. However, the usual analysis of variance assumption of normality is not made. Nonparametric tests use functions of the response variable ranks, called rank scores (Hajek 1969).

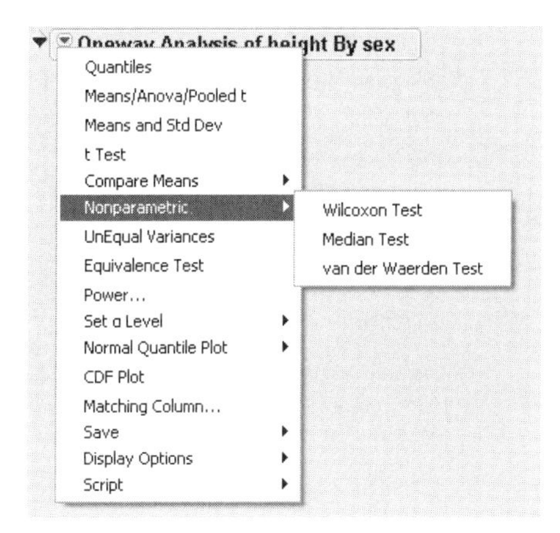

JMP offers three nonparametric tests for testing whether distributions across factor levels are centered at the same location:

- *Wilcoxon rank scores* are the simple ranks of the data. The Wilcoxon test is the most powerful rank test for errors with logistic distributions.
- *Median rank scores* are either 1 or 0 depending on whether a rank is above or below the median rank. The Median test is the most powerful rank test for errors with doubly exponential distributions.
- *Van der Waerden rank scores* are the ranks of the data divided by one plus the number of observations transformed to a normal score by applying the inverse of the normal distribution function. The Van der Waerden test is the most powerful rank test for errors with normal distributions.

If a factor has only two levels, a normal approximation to the appropriate two sample test is given as well as the Chi-square approximation to the one-way test. Table 7.1 "Guide for Using Nonparametric Tests," p. 135 lists the appropriate tests given the number of factor levels.

Table 7.1 Guide for Using Nonparametric Tests

Command	Two levels	Two or more levels	Most powerful rank-score for...
Nonpar-Wilcoxon	Wilcoxon rank-sum	Kruskal-Wallis	Logistic
Nonpar-Median	Two sample Median	*k*-sample median (Brown-Mood)	Double Exponential
Nonpar-VW	Van der Waerden	*k*-sample Van der Waerden	Normal

Oneway Analysis of height By sex

Wilcoxon / Kruskal-Wallis Tests (Rank Sums)

Level	Count	Score Sum	Score Mean	(Mean-Mean0)/Std0
F	18	275.500	15.3056	-2.538
M	22	544.500	24.7500	2.538

2-Sample Test, Normal Approximation

S	Z	Prob>|Z|
275.5	-2.53751	0.0112*

1-way Test, ChiSquare Approximation

ChiSquare	DF	Prob>ChiSq
6.5084	1	0.0107*

Median Test (Number of Points Above Median)

Level	Count	Score Sum	Score Mean	(Mean-Mean0)/Std0
F	18	4.667	0.259259	-2.815
M	22	15.333	0.696970	2.815

2-Sample Test, Normal Approximation

S	Z	Prob>|Z|
4.6666667	-2.81526	0.0049*

1-way Test, ChiSquare Approximation

ChiSquare	DF	Prob>ChiSq
7.9257	1	0.0049*

Van der Waerden Test (Normal Quantiles)

Level	Count	Score Sum	Score Mean	(Mean-Mean0)/Std0
F	18	-7.123	-0.39575	-2.436
M	22	7.123	0.32379	2.436

2-Sample Test, Normal Approximation

S	Z	Prob>|Z|
-7.123443	-2.43599	0.0149*

1-way Test, ChiSquare Approximation

ChiSquare	DF	Prob>ChiSq
5.9340	1	0.0149*

The three tables shown here have the same format and show the following:

Level lists the factor levels occurring in the data.

Count records the frequencies of each level.

Score Sum records the sum of the rank score for each level.

Score Mean records the mean rank score for each level.

(Mean-Mean0)/Std0 records the standardized score. **Mean0** is the mean score expected under the null hypothesis. **Std0** is the standard deviation of the score sum expected under the null hypothesis. The null hypothesis is that the group means or medians are in the same location across groups.

S records the test statistic for the two sample test. *S* is reported only for factors with two levels.

Z records the normal approximation to the S test statistic. *Z* is reported only for factors with two levels. This is the same as **(Mean-Mean0)/Std0** if there are only two groups.

Prob>|Z| is the two-tailed probability of obtaining by chance alone an absolute *Z* value larger than the one calculated if, in reality, the distributions of the two levels are centered at the same location. Observed significance probabilities of 0.05 or less are often considered evidence that the distributions of the two levels are not centered at the same location. **Prob>|Z|** is reported only for factors with two levels. For an upper or lower one-tail test, the observed significance level is half **Prob>|Z|**.

Chi-Square records the test statistic for the one-way test.

DF records the degrees of freedom associated with the one-way test. If a factor has k levels, the test has $k-1$ degrees of freedom.

Prob>ChiSq is the probability of obtaining by chance alone a Chi-square value larger than the one calculated if, in reality, the distributions across factor levels are centered at the same location. Observed significance probabilities of 0.05 or less are often considered evidence that the distributions across factor levels are not centered at the same location.

If you have specified a Block column, then the non-parametric tests are performed on data that has been adjusted for the Block means.

Homogeneity of Variance Tests

When the variances across groups are not equal, the usual analysis of variance assumptions are not satisfied and the ANOVA F-test is not valid. JMP gives four tests for equality of group variances and an ANOVA that is valid when the group sample variances are unequal. The concept behind the first three tests of equal variances is to perform an analysis of variance on a new response variable constructed to measure the spread in each group. The fourth test is Bartlett's test, which is similar to the likelihood-ratio test under normal distributions.

- *O'Brien's* test constructs a dependent variable so that the group means of the new variable equal the group sample variances of the original response. An ANOVA on the O'Brien variable is actually an ANOVA on the group sample variances (O'Brien 1979, Olejnik and Algina 1987).
- The *Brown-Forsythe* test shows the F-test from an ANOVA in which the response is the absolute value of the difference of each observation and the group median (Brown and Forsythe 1974a).
- *Levene's* test shows the F-test from an ANOVA in which the response is the absolute value of the difference of each observation and the group mean (Levene 1960).That is, the spread is measured as $z_{ij} = |y_{ij} - \bar{y}_i|$ (as opposed to the SAS default $z^2_{ij} = (y_{ij} - y_i)^2$).
- *Bartlett's* test compares the weighted arithmetic average of the sample variances to the weighted geometric average of the sample variances. The geometric average is always less than or equal to the arithmetic average with equality holding only when all sample variances are equal. The more variation there is among the group variances, the more these two averages differ. A function of these two averages is created, which approximates a χ^2-distribution (or, in fact, an F-distribution under a certain formulation). Large values correspond to large values of the arithmetic/geometric ratio, and hence to widely varying group variances. Dividing the Bartlett Chi-square test statistic by the degrees of freedom gives the F-value shown in the table. Bartlett's test is not very robust to violations of the normality assumption (Bartlett and Kendall 1946).

Note: If there are only two groups tested, then a standard F-test for unequal variances is performed. The F-test is the ratio of the larger to the smaller variance estimate. The p-value from the F-distribution is doubled to make it a two-sided test.

The Tests that the Variances are Equal table shows the differences between group means to the grand mean and to the median, and gives a summary of testing procedures.

If the tests of equal variances reveal that the group variances are significantly different, the Welch ANOVA for the means should be used in place of the usual ANOVA in the Fit Means table. The Welch statistic is based on the usual ANOVA F-test: however, the means have been weighted by the reciprocal of the sample variances of the group means (Welch 1951; Brown and Forsythe 1974b; Asiribo, Osebekwin, and Gurland 1990).

If there are only two levels, the Welch ANOVA is equivalent to an unequal variance *t*-test (the *t*-value is the square root of the *F*).

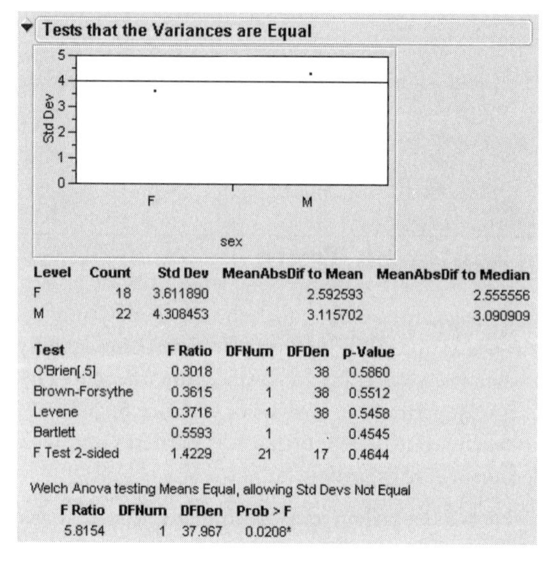

Level	Count	Std Dev	MeanAbsDif to Mean	MeanAbsDif to Median
F	18	3.611890	2.592593	2.555556
M	22	4.308453	3.115702	3.090909

Test	F Ratio	DFNum	DFDen	p-Value
O'Brien[.5]	0.3018	1	38	0.5860
Brown-Forsythe	0.3615	1	38	0.5512
Levene	0.3716	1	38	0.5458
Bartlett	0.5593	1	.	0.4545
F Test 2-sided	1.4229	21	17	0.4644

Welch Anova testing Means Equal, allowing Std Devs Not Equal

F Ratio	DFNum	DFDen	Prob > F
5.8154	1	37.967	0.0208*

The Tests that the Variances are Equal table show these quantities:

Level lists the factor levels occurring in the data.

Count records the frequencies of each level.

Std Dev records the standard deviations of the response for each factor level. The standard deviations are equal to the means of the O'Brien variable. If a level occurs only once in the data, no standard deviation is calculated.

MeanAbsDif to Mean records the mean absolute difference of the response and group mean. The mean absolute differences are equal to the group means of the Levene variable.

MeanAbsDif to Median records the absolute difference of the response and group median. The mean absolute differences are equal to the group means of the Brown-Forsythe variable.

Test lists the names of the various tests performed.

F Ratio records the following calculated *F*-statistic for each test:

O'Brien's test constructs a dependent variable so that the group means of the new variable equal the group sample variances of the original response. The O'Brien variable is computed as

$$r_{ijk} = \frac{(n_{ij}-1.5)n_{ij}(y_{ijk}-\overline{y_{ij}})^2 - 0.5s_{ij}^2(n_{ij}-1)}{(n_{ij}-1)(n_{ij}-2)}$$

where n represents the number of y_{ijk} observations.

Brown-Forsythe is the model *F*-statistic from an ANOVA on $z_{ij} = |y_{ij} - \tilde{y}_i|$ where y_i is the median response for the *i*th level. If any z_{ij} is zero, then it is replaced with the next smallest z_{ij} in the same level. This corrects for the artificial zeros occurring in levels with odd numbers of observations.

The Levene *F* is the model *F*-statistic from an ANOVA on $z_{ij} = |y_{ij} - \overline{y_{i.}}|$ where $y_{i.}$ is the mean response for the *i*th level.

Bartlett's test is calculated as

$$T = \frac{v\log\left(\sum_i \frac{v_i}{v}s_i^2\right) - \sum_i v_i \log(s_i^2)}{1 + \left(\frac{\sum_i \frac{1}{v_i} - \frac{1}{v}}{3(k-1)}\right)} \quad \text{where } v_i = n_i - 1 \text{ and } v = \sum_i v_i$$

and n_i is the count on the *i*th level and s_i^2 is the response sample variance on the *i*th level. The Bartlett statistic has a χ^2-distribution. Dividing the Chi-square test statistic by the degrees of freedom results in the reported *F*-value.

DFNum records the degrees of freedom in the numerator for each test. If a factor has *k* levels, the numerator has $k - 1$ degrees of freedom. Levels occurring only once in the data are not used in calculating test statistics for O'Brien, Brown-Forsythe, or Levene. The numerator degrees of freedom in this situation is the number of levels used in calculations minus one.

DFDen records the degrees of freedom used in the denominator for each test. For O'Brien, Brown-Forsythe, and Levene, a degree of freedom is subtracted for each factor level used in calculating the test statistic. One more degree of freedom is subtracted for the overall mean. If a factor has *k* levels, the denominator degrees of freedom is $n - k$.

Prob>F is the probability of obtaining by chance alone an *F*-value larger than the one calculated if, in reality, the variances are equal across all levels. Observed significance probabilities of 0.05 or less are often considered evidence of unequal variances across the levels.

A separate table gives the Welch ANOVA tests for equality of group means allowing for the fact that the group variances may not be equal. If there are only two levels, this is equivalent to the unequal-variance *t*-test.

Welch Anova testing Means Equal, allowing Std Devs Not Equal

F Ratio	DFNum	DFDen	Prob > F
5.8154	1	37.967	0.0208*

t Test
2.4115

The following computations are shown in the Welch ANOVA:

F Ratio is computed as

$$F = \frac{\left[\frac{\sum_i w_i(\bar{y}_i - \tilde{y}_{..})^2}{k-1}\right]}{\left\{1 + \frac{2(k-2)}{k^2-1}\left[\sum_i \frac{\left(1-\frac{w_i}{u}\right)^2}{n_i - 1}\right]\right\}} \quad \text{where } w_i = \frac{n_i}{s_i^2},\ u = \sum_i w_i,\ \tilde{y}_{..} = \sum_i \frac{w_i y_{i.}}{u},$$

and n_i is the count on the *i*th level, $\bar{y}_{i.}$ is the mean response for the *i*th level, and s_i^2 is the response sample variance for the *i*th level.

DFNum records the degrees of freedom in the numerator of the test. If a factor has *k* levels, the numerator has *k* - 1 degrees of freedom. Levels occurring only once in the data are not used in calculating the Welch ANOVA. The numerator degrees of freedom in this situation is the number of levels used in calculations minus one.

DFDen records the degrees of freedom in the denominator of the test. The Welch approximation for the denominator degrees of freedom is

$$f = \frac{1}{\left(\frac{3}{k^2-1}\right)\left[\sum_i \frac{\left(1-\frac{w_i}{u}\right)^2}{n_i - 1}\right]}$$

where w_i, n_i, and u are defined as in the *F*-ratio formula.

Prob>F is the probability of obtaining by chance alone an *F*-value larger than the one calculated if, in reality, the means are equal across all levels. Observed significance probabilities of 0.05 or less are considered evidence of unequal means across the levels.

If you have specified a Block column, then the variance tests are performed on data after it has been adjusted for the Block means.

Power Details

The **Power** command displays the Power Details Dialog (Figure 7.21). You fill in the values as described next and then click **Done**. JMP gives calculations of statistical power and other details about a given hypothesis test. (Note: Examples use Typing Data.jmp)

- *LSV* (the Least Significant Value) is the value of some parameter or function of parameters that would produce a certain *p*-value alpha. Said another way, you want to know how small an effect would be declared significant at some *p*-value alpha. The LSV provides a measuring stick for significance on the scale of the parameter, rather than on a probability scale. It shows how sensitive the design and data are.

- *LSN* (the Least Significant Number) is the total number of observations that would produce a specified *p*-value alpha given that the data has the same form. The LSN is defined as the number of observations needed to reduce the variance of the estimates enough to achieve a significant result with the given values of alpha, sigma, and delta (the significance level, the standard deviation of the error, and the effect size). If you need more data to achieve significance, the LSN helps tell you how many more. The LSN is the total number of observations that yields approximately 50% power.
- *Power* is the probability of getting significance at or below a given *p*-value alpha for a given situation. It is a function of the sample size, the effect size, the standard deviation of the error, and the significance level. The power tells you how likely your experiment is to detect a difference (effect size), at a given alpha level.

Note: When there are only two groups in a one-way layout, the LSV computed by the power facility is the same as the least significant difference (LSD) shown in the multiple comparison tables.

Figure 7.21 Power Details Dialog

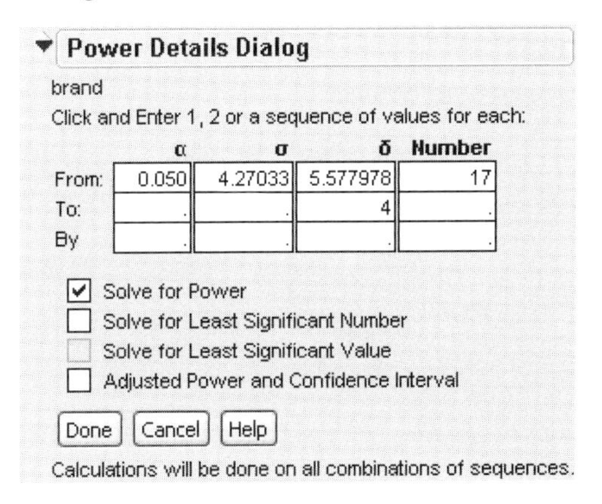

For each of four columns **Alpha**, **Sigma**, **Delta**, and **Number**, you can fill in a single value, two values, or the start, stop, and increment for a sequence of values, as shown in Figure 7.21. Power calculations are done on all possible combinations of the values you specify.

Alpha (α) is the significance level, between 0 and 1, usually 0.05, 0.01, or 0.10. Initially, a value of 0.05 shows. Click on one of the three positions to enter or edit one, two, or a sequence of values.

Sigma (σ) is the standard error of the residual error in the model. Initially, RMSE, the estimate from the square root of the mean square error is supplied here. Click on one of the three positions to enter or edit one, two, or a sequence of values.

Delta (δ) is the raw effect size. For details about effect size computations see the section "Effect Size," p. 261 in the chapter "Standard Least Squares: Perspectives on the Estimates," p. 249. The first position is initially set to the square root of the sums of squares for the hypothesis divided by n; that is, $\delta = \sqrt{SS/n}$. Click on one of the three positions to enter or edit one, two, or a sequence of values.

Number (n) is the sample size. Initially, the actual sample size is put in the first position. Click on one of the three positions to enter or edit one, two, or a sequence of values.

Click the following check boxes to request the results you want.

Solve for Power Check to solve for the power (the probability of a significant result) as a function of all four values: α, σ, δ, and n.

Solve for Least Significant Number Check to solve for the number of observations expected to be needed to achieve significance alpha given α, σ, and δ.

Solve for Least Significant Value Check this to solve for the value of the parameter or linear test that produces a *p*-value of α. This is a function of α, σ, n, and the standard error of the estimate. This feature is only available for one-degree-of-freedom tests and is usually used for individual parameters.

Adjusted Power and Confidence Interval When you look at power retrospectively, you use estimates of the standard error and the test parameters. Adjusted power is the power calculated from a more unbiased estimate of the noncentrality parameter. The confidence interval for the adjusted power is based on the confidence interval for the noncentrality estimate. Adjusted power and confidence limits are only computed for the original Delta, because that is where the random variation is.

When you finish the dialog, click **Done** to see the calculations at the bottom of the report. Note that Power and LSV are computed for the two values of δ (effect size) entered in the Power Details dialog. In this example power is high (0.881 and 0.994) for both differences. Note that significance would have been attained for the same sample results with a sample of only 8 observations (7.4 to be exact).

Power Details

Test brand

Power

α	σ	δ	Number	Power
0.0500	4.27033	4	17	0.8810
0.0500	4.27033	5.577978	17	0.9937

Popup menu

If you request a range of sample sizes (n) in the Power Details dialog, the popup menu in the Power table plots power by n.

To compute the power, you make use of the noncentral F distribution. The formula (O'Brien and Lohr 1984) is given by

$$\text{Power} = \text{Prob}(F > F_{crit}, \nu_1, \nu_2, nc)$$

where

F is distributed as the noncentral $F(nc, \nu_1, \nu_2)$ and $F_{crit} = F_{(1-\alpha, \nu 1, \nu 2)}$ is the $1 - \alpha$ quantile of the F distribution with ν_1 and ν_2 degrees of freedom.

$\nu_1 = r - 1$ is the numerator df

$\nu_2 = r(n - 1)$ is the denominator df

n is the number per group

r is the number of groups

$nc = n(CSS)/\sigma^2$ is the noncentrality parameter

$$CSS = \sum_{g=1}^{G} (\mu_g - \mu) \text{ is the corrected sum of squares}$$

μ_g is the mean of the gth group

μ is the overall mean

σ^2 is estimated by the mean squared error (MSE)

The Power Details dialog and reports are the same as those in the general fitting platform launched by the **Fit Model** command. A detailed discussion and statistical details for the power calculation reports is in the chapter "Standard Least Squares: Introduction," p. 213.

Matching Model Analysis

The **Matching Column** command lets you specify a matching (ID) variable for a matching model analysis. The **Matching Column** command addresses the case when the data in a one-way analysis come from matched (paired) data, as when observations in different groups come from the same subject.

A special case of this leads to the paired *t*-test. The **Matched Pairs** platform in JMP handles this kind of data but expects the data to be organized with the pairs in different columns, not in different observations.

The **Matching Column** command does two things:

- It fits an additive model that includes both the grouping variable (the *x* variable in the Fit Y by X analysis), and the matching variable you choose. It uses an iterative proportional fitting algorithm to do this. This algorithm makes a difference if there are hundreds of subjects because the equivalent linear model would be very slow and require huge memory resources.
- It draws lines between the points that match across the groups. If there are multiple observations with the same matching ID value, lines are drawn from the mean of the group of observations.

Matching Column activates **Show Lines** display options to connect the matching points. You can toggle the lines off if you don't want them.

As an example, consider the data in the Matching.jmp sample data table, shown in Figure 7.22. An animal (subject) is observed over time (season), which is the *x* grouping variable. Miles wandered (miles) by the same animal is the *Y* response variable, giving the kind of repeated measure situation described above. The categorical data are stacked into the season column as expected by the **Matching Column** command.

The plot on the right in Figure 7.22 shows the Fit Y by X platform plot of miles by season, with subject as the matching variable. In this example, the values of species and subject are used to label the first measurement for each subject. The analysis shows the season and subject effects with *F*-tests. These are equivalent to the tests you get with the Fit Model platform if you run two models, one with the interaction term and one without. If there are only two levels, then the *F*-test is equivalent to the paired *t*-test.

Another use for the feature is to do parallel coordinate plots when the variables are scaled similarly.

Figure 7.22 Matching Variables Example

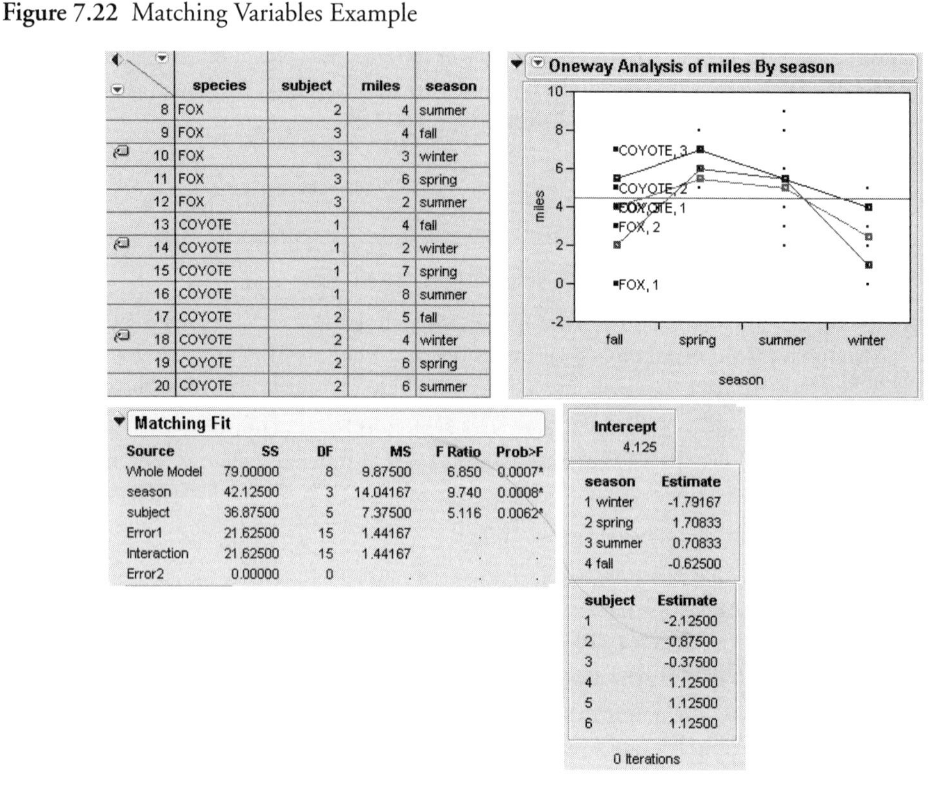

	species	subject	miles	season
8	FOX	2	4	summer
9	FOX	3	4	fall
10	FOX	3	3	winter
11	FOX	3	6	spring
12	FOX	3	2	summer
13	COYOTE	1	4	fall
14	COYOTE	1	2	winter
15	COYOTE	1	7	spring
16	COYOTE	1	8	summer
17	COYOTE	2	5	fall
18	COYOTE	2	4	winter
19	COYOTE	2	6	spring
20	COYOTE	2	6	summer

Matching Fit

Source	SS	DF	MS	F Ratio	Prob>F
Whole Model	79.00000	8	9.87500	6.850	0.0007*
season	42.12500	3	14.04167	9.740	0.0008*
subject	36.87500	5	7.37500	5.116	0.0062*
Error1	21.62500	15	1.44167	.	.
Interaction	21.62500	15	1.44167	.	.
Error2	0.00000	0	.	.	.

Intercept
4.125

season	Estimate
1 winter	-1.79167
2 spring	1.70833
3 summer	0.70833
4 fall	-0.62500

subject	Estimate
1	-2.12500
2	-0.87500
3	-0.37500
4	1.12500
5	1.12500
6	1.12500

0 Iterations

Details of Comparison Circles

One approach to comparing two means is to see if their actual difference is greater than their *least significant difference* (LSD). This least significant difference is a Student's *t*-statistic multiplied by the standard error of the difference of the two means and is written

$$\mathrm{LSD} = t_{\alpha/2}\mathrm{std}(\hat{\mu}_1 - \hat{\mu}_2)$$

The standard error of the difference of two independent means is calculated from the relationship

$$[\mathrm{std}(\hat{\mu}_1 - \hat{\mu}_2)]^2 = [\mathrm{std}(\hat{\mu}_1)]^2 + [\mathrm{std}(\hat{\mu}_2)]^2$$

When the means are uncorrelated, these quantities have the relationship

$$\mathrm{LSD}^2 = \left[t_{\alpha/2}\mathrm{std}((\hat{\mu}_1 - \hat{\mu}_2))\right]^2 = \left[t_{\alpha/2}\mathrm{std}(\hat{\mu}_1)\right]^2 + \left[t_{\alpha/2}\mathrm{std}\hat{\mu}_2\right]^2$$

These squared values form a Pythagorean relationship illustrated graphically by the right triangle shown in Figure 7.23.

Figure 7.23 Relationship of the Difference between Two Means

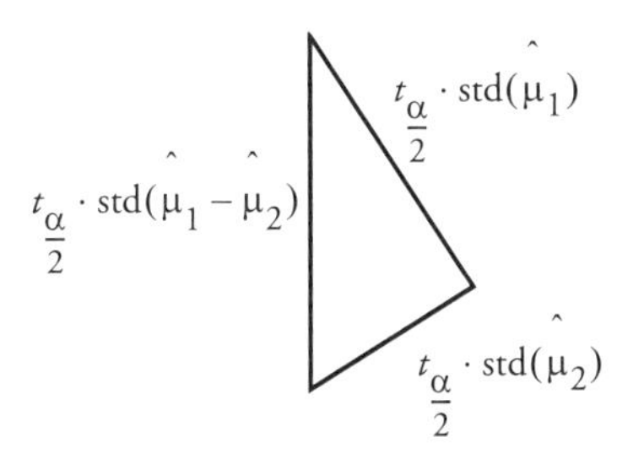

The hypotenuse of this triangle is a measuring stick for comparing means. The means are significantly different if and only if the actual difference is greater than the hypotenuse (LSD).

Now suppose that you have two means that are exactly on the borderline, where the actual difference is the same as the least significant difference. Draw the triangle with vertices at the means measured on a vertical scale. Also, draw circles around each mean such that the diameter of each is equal to the confidence interval for that mean.

Figure 7.24 Geometric Relationship of *t*-test Statistics

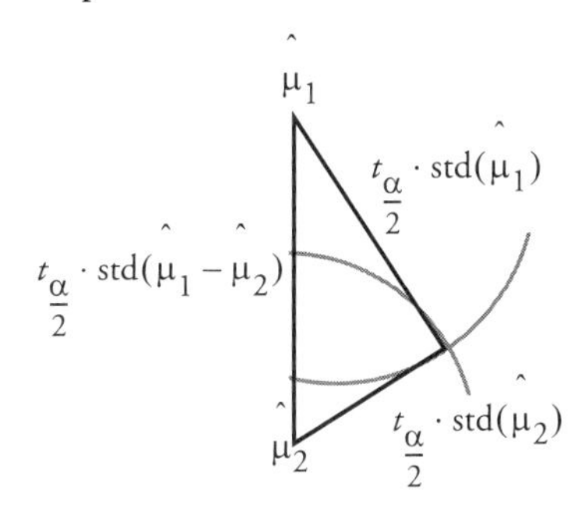

The radius of each circle is the length of the corresponding leg of the triangle, which is $t_{\alpha/2}\text{std}(\hat{\mu}_i)$.

Note that the circles must intersect at the same right angle as the triangle legs, giving the following relationship:

- If the means differ exactly by their least significant difference, then the confidence interval circles around each mean intersect at a right angle. That is, the angle of the tangents is a right angle.

Now consider the way these circles must intersect if the means are different by greater than or less than the least significant difference.

- If the circles intersect such that the outside angle is greater than a right angle, then the means *are not* significantly different. If the circles intersect such that the outside angle is less than a right angle then the means *are* significantly different. An outside angle of less than 90 degrees indicates that the means are farther apart than the least significant difference.
- If the circles do not intersect, then they are significantly different. If they nest, they are not significantly different. Figure 7.25 illustrates how to interpret comparison circles.

Figure 7.25 Angle of Intersection and Significance

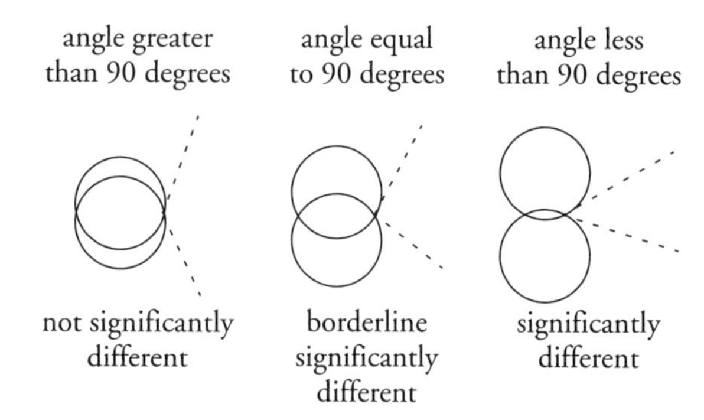

The same graphical technique works for many multiple comparison tests, substituting a different probability quantile value for the Student's *t*.

Chapter 8

Contingency Tables Analysis

The Fit Y by X Platform

The Contingency platform examines the distribution of a categorical response variable *Y* as conditioned by the values of a categorical factor *X*. The Contingency platform is the personality of the **Fit Y by X** command when both the *Y* and *X* variables are nominal or ordinal. Ordinal values are treated no differently than nominal values in this platform. For each *X* and *Y* combination, this platform presents

- a mosaic plot
- a two-way frequency (Contingency) table with frequencies, percents, and cell Chi-square (χ^2) information
- a Tests table with the Pearson (χ^2) and likelihood-ratio (G^2) Chi-square statistics. If both variables have only two levels, Fisher's exact test, relative risk, and odds ratios are also available.
- optionally, a Kappa statistic as a measure of agreement when the *X* and *Y* have the same set of values
- optionally, a correspondence analysis plot
- optionally, a Cochran-Mantel-Haenszel test for testing if there is a relationship between two categorical variables after blocking across a third classification.

Note: You can launch this platform from either the **Fit Y by X** main menu command (or toolbar), or with the **Contingency** button on the JMP Starter. When you choose **Fit Y by X** from the **Analyze** menu, JMP performs an analysis by context, which is the analysis appropriate for the modeling types of the response and factor variables you specify. When you use the **Contingency** button on the JMP Starter, the Launch dialog (role dialog) expects nominal or ordinal factor and response variables, and advises you if you specify other modeling types.

See the *JMP User Guide* for a discussion of modeling types and data types.

Contents

Launch the Platform

The **Fit Y by X** command launches a report platform for each pair of columns given factor (*X*) and response (*Y*) roles in its launch dialog. The chapter "Bivariate Scatterplot and Fitting," p. 81 discusses the launch dialog.

Each combination of *X* and *Y* variables produces a plot and accompanying text reports. The type of platform varies according to the modeling types of the *X* and *Y* variables. Figure 8.1 illustrates that the platform, when both *X* and *Y* are either nominal or ordinal columns, gives a contingency table analysis.

Figure 8.1 Fit Y By X Mosaic Plot of Contingency Table

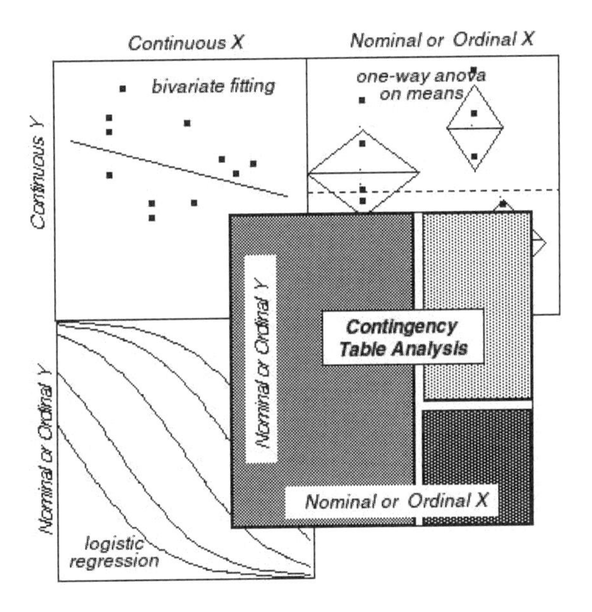

Contingency Table Analysis

When both the response and the factor are nominal or ordinal, the data are summarized by frequency counts and proportions. This information estimates the response rates of the *Y* levels for each level of the *X* factor.

Platform Commands

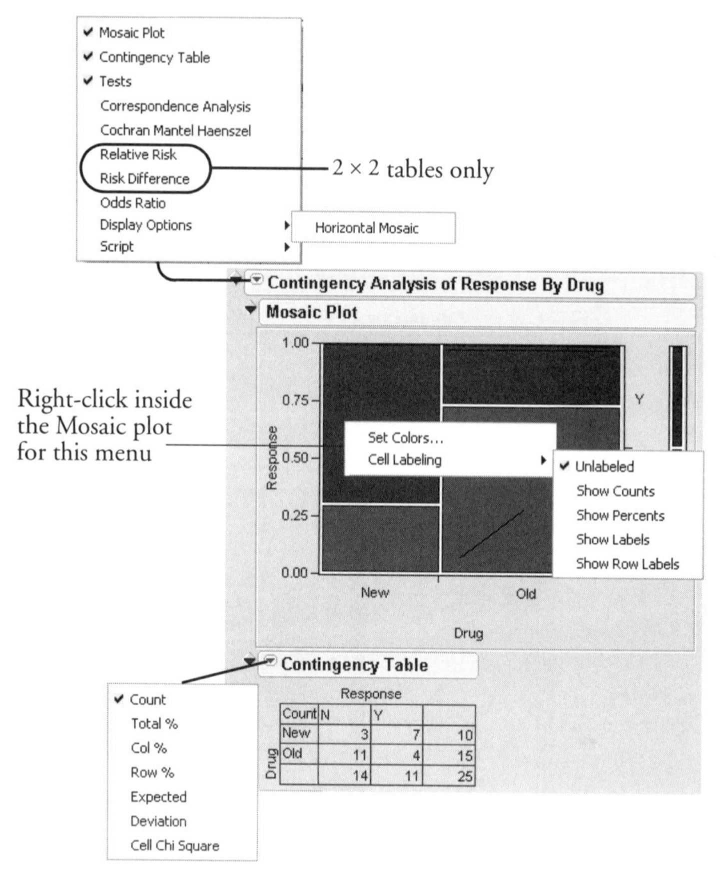

Count	N	Y	
New	3	7	10
Old	11	4	15
	14	11	25

- The popup menu on the Contingency analysis title bar gives commands for the Contingency platform, which include the frequency (Contingency) table, a statistical tests table with Chi-square tests for equal response rates, a correspondence analysis, and Cochran-Mantel-Haenszel tests. In addition, the **Display Options > Horizontal Mosaic** command on the platform menu rotates the mosaic plot 90 degrees.

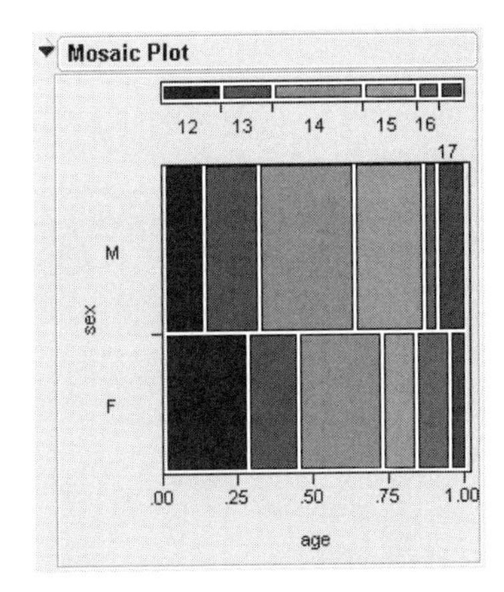

- The Mosaic plot has commands to set colors and labels, accessible by right-clicking inside the Mosaic plot itself.
- The Contingency table also has a menu to toggle the display of certain quantities, available from the red triangle drop-down menu on the Contingency table outline node. Each command is explained below.

Mosaic Plot

A two-way frequency table can be graphically portrayed in a *mosaic plot*. A mosaic plot is any plot divided into small rectangles such that the area of each rectangle is proportional to a frequency count of interest. To compute each (*X*, *Y*) proportion, JMP divides the *Y* counts within each *x* level by the total of the level. This proportion estimates the response probability for each (*X*, *Y*) level. The mosaic plot was introduced by Hartigan and Kleiner in 1981 and refined by Friendly (1994).

Figure 8.2 is a mosaic plot of age by sex from the Big Class.jmp data table:

- The proportions shown on the *x*-axis represent the relative sizes of the female and male totals.
- Likewise, the proportions shown on the right *y*-axis represent the relative sizes of each age group for the combined levels.
- The scale of the left *y*-axis shows response probability, with the whole axis being probability 1 (the total sample).

The mosaic plot is equivalent to side-by-side, divided bar charts. For each sex, the proportion of each age divides the bars vertically. To see how the response probabilities vary for the different levels of *X*, compare the heights of *Y* levels across the *X* levels. If the response divisions line up horizontally across the *X* levels, the response rates are the same.

Figure 8.2 Contingency Analysis Mosaic Plot

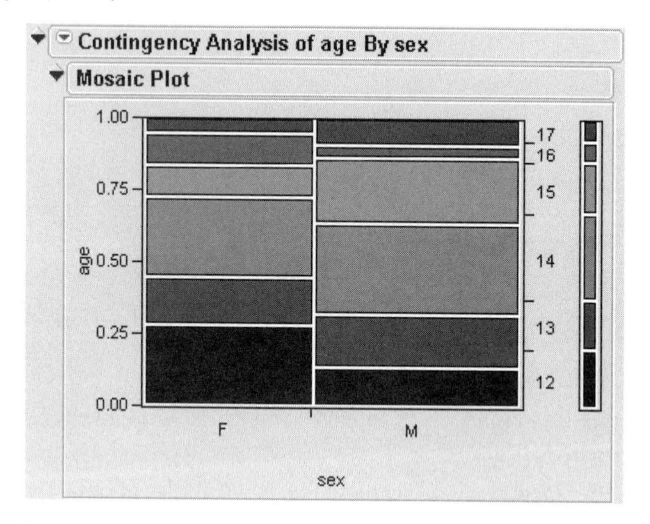

Note: When you click on a section in the mosaic plot, the section is highlighted and corresponding data table rows are selected.

There are some commands accessible by right-clicking on the Mosaic Plot itself.

Set Colors brings up a dialog that shows the current assignment of colors to levels. To change the color for any level, click on the appropriate oval in the second column of colors, which presents a dialog to pick a new color. Mosaic colors depend on whether the response column is ordinal or nominal, and whether there is an existing Value Colors column property

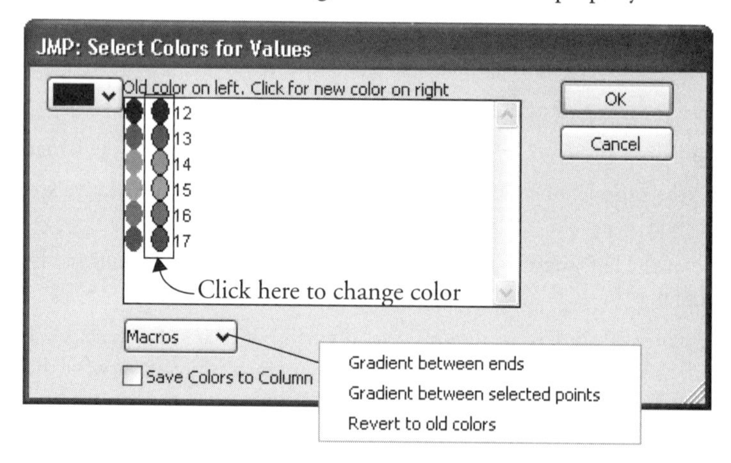

You can also have JMP compute a color gradient between any two levels using one of the **Macro** commands. If you select a range of levels (by dragging the mouse over the levels you want to select, or shift-clicking the first and last level), the **Gradient Between Selected Points** command applies a color gradient to the levels you have selected. **Gradient Between Ends** applies a gradient to all levels of the variable.

A useful way of using these macros is to drastically change the colors on two endpoints, then apply a gradient between those endpoints to give the appearance of a gradiated change.

Revert to Old Colors undoes any changes.

If you select the **Save Colors to Column** check box, a new column property **Value Colors** is added to the column. By accessing this property through the **Cols > Column Info** dialog, you can further edit the colors.

Cell Labeling allows you to specify a label to be drawn in the Mosaic plot. Options include **Show Counts** (the number of observations in each cell), **Show Percents** (the percent of observations in each cell), **Show Labels** (the levels of the x-variable corresponding to each cell), and **Show Row Labels** (the row labels for all the rows represented by the cell).

The Contingency Table

The Contingency table can appear as a simple two-way frequency table like the one shown here. There is a row for each factor level and a column for each response level. The borders of the table display the row totals, column totals, with the grand total in the lower right cell of the table.

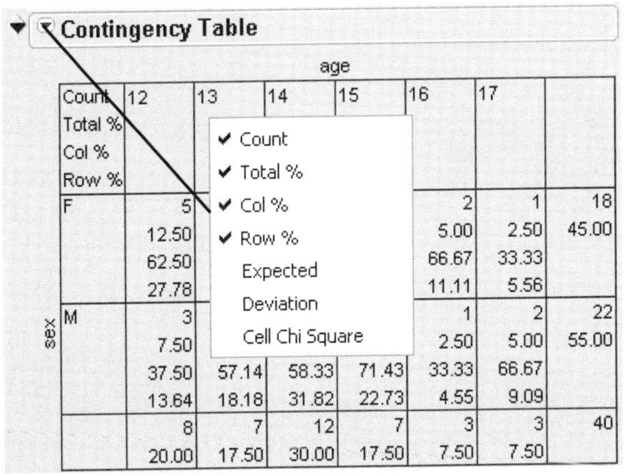

Contingency Table

sex	Count Total % Col % Row %	age 12	13	14	15	16	17	
	F	5 12.50 62.50 27.78				2 5.00 66.67 11.11	1 2.50 33.33 5.56	18 45.00
	M	3 7.50 37.50 13.64	57.14 18.18	58.33 31.82	71.43 22.73	1 2.50 33.33 4.55	2 5.00 66.67 9.09	22 55.00
		8 20.00	7 17.50	12 30.00	7 17.50	3 7.50	3 7.50	40

✔ Count
✔ Total %
✔ Col %
✔ Row %
Expected
Deviation
Cell Chi Square

You can request the following cell quantities from the popup menu next to the Contingency table name (or by right-clicking inside the table itself):

Count is the cell frequency, margin total frequencies, and grand total (total sample size).

Total% is the percent of cell counts and margin totals to the grand total.

Row% is the percent of each cell count to its row total.

Col% is the percent of each cell count to its column total.

Expected is the expected frequency (*E*) of each cell under the assumption of independence. It is computed as the product of the corresponding row total and column total, divided by the grand total.

Deviation is the observed (actual) cell frequency (*O*) minus the expected cell frequency (*E*)

Cell Chi Square is the Chi-square values computed for each cell as $(O - E)^2 / E$.

You can select or remove any of the cell options.

The Tests Report

The Tests report is analogous to the Analysis of Variance table for continuous data, and displays the two tests that the response level rates are the same across *X* levels. The negative log-likelihood for categorical data plays the same role as sums of squares in continuous data.

Tests

N	DF	-LogLike	RSquare (U)
40	5	1.2953986	0.0193

Test	ChiSquare	Prob>ChiSq
Likelihood Ratio	2.591	0.7628
Pearson	2.554	0.7683

Warning: 20% of cells have expected count less than 5, ChiSquare suspect.
Warning: Average cell count less than 5, LR ChiSquare suspect.

When both categorical variables are responses (*Y* variables), the Chi-square statistics test that they are independent. If one variable is considered as *Y* and the *X* is regarded as fixed, the Chi-square statistics test that the distribution of the *Y* variable is the same across each *X* level. This is a test of *marginal homogeneity*. The Tests table has the following information:

DF records the degrees of freedom associated with each source of uncertainty.

The total observations, **N**, in the **Big Class.jmp** table is 40. There are 35 degrees of freedom associated with the **C. Total**, which is computed as $N - (r - 1)$, where r is the number of response levels. The total degrees of freedom are partitioned into **Model** and **Error** terms.

The degrees of freedom for **Model** are used to compute the response rates for each factor level and are equal to $(s - 1)(r - 1)$, where r is the number of response levels and s is the number of factor levels.

The **Error** degrees of freedom are the difference between the **C. Total** and the **Model** degrees of freedom. There is little meaning to this value. It can become negative if the contingency table is sparse, with many cells that have no data.

–LogLike is the negative log-likelihood, which measures fit and uncertainty in the same manner that sums of squares does in continuous response situations:

–LogLike for C. Total is the negative log-likelihood (or uncertainty) when the probabilities are estimated by fixed rates for each response level. This corresponds to the total variation in an Analysis of Variance table. It is computed as

$$-\sum_{ij} n_{ij}\ln(p_j) \text{ where } p_j = \frac{N_j}{N}$$

Here n_{ij} is the count for the jth response level of the ith factor level (sample), N_j is the total of the jth response level, and $N = \sum n_{ij}$. Note that j indexes x factor levels.

Rsquare (U) is the portion of the total uncertainty attributed to the model fit. It is computed as

$$\frac{-\text{log likelihood for Model}}{-\text{log likelihood for C. Total}}$$

using the log-likelihood quantities from the Tests table. The total negative log-likelihood is found by fitting fixed response rates across the total sample. An R^2 of 1 means that the factors completely predict the categorical response. An R^2 of 0 means that there is no gain from using the model instead of fixed background response rates. In categorical data analysis, high R^2 values are somewhat rare.

Test lists two Chi-square statistical tests of the hypothesis that the response rates are the same in each sample category.

The **Likelihood Ratio** Chi-square test is computed as twice the negative log-likelihood for Model in the Tests table. Some books use the notation G^2 for this statistic. In terms of the difference of two negative log-likelihoods, one with *whole-population* response probabilities and one with *each-population* response rates, this is written:

$$G^2 = 2\left[\sum_{ij}(-n_{ij})\ln(p_j) - \sum_{ij}-n_{ij}\ln(p_{ij})\right] \text{ where } p_{ij} = \frac{n_{ij}}{N} \text{ and } p_j = \frac{N_j}{N}$$

Often textbooks write this formula more compactly as

$$G^2 = 2\sum_{ij} n_{ij}\ln\left(\frac{p_{ij}}{p_j}\right)$$

The **Pearson** Chi-square is another Chi-square test of the hypothesis that the response rates are the same in each sample category. It is calculated by summing the squares of the differences between observed and expected cell counts. The Pearson Chi-square exploits the property that frequency counts tend to a normal distribution in very large samples. The familiar form of this Chi-square statistic is

$$\chi^2 = \sum_{\text{cells}} \frac{(\text{observed} - \text{expected})^2}{\text{expected}}$$

There is no continuity correction done here, as is sometimes done in 2-by-2 tables.

Prob>ChiSq lists the probability of obtaining, by chance alone, a Chi-square value greater than the one computed if no relationship exists between the response and factor. If both variables have only two levels, Fisher's exact probabilities for the one-tailed tests and the two-tailed test are also shown.

Two-by-Two Tables

The following analyses are available if the contingency table is two-by-two.

Fisher's Exact Test is performed automatically, showing both one- and two-sided tests.

Relative Risk is a command that shows the ratios of rates from one sample to another and their confidence intervals.

Odds Ratio is a command that shows the odds ratio

$$\frac{p_{11}p_{22}}{p_{21}p_{12}}$$

and its confidence interval.

Relative Risk and Risk Difference

You can calculate Risk Ratios and Risk differences for 2×2 contingency tables using the **Relative Risk** and **Risk Difference** commands. Details of these two methods are found in Agresti (1990) section 3.4.2.

Relative Risk

For this example, use Carpoll.jmp from the Sample Data folder. Select **Analyze > Fit Y by X** and assign marital status as Y and sex as X. When the Contingency report appears, select **Relative Risk** from the report's drop-down menu.

This produces a dialog presenting you with several options.

Figure 8.3 Relative Risk Dialog Box

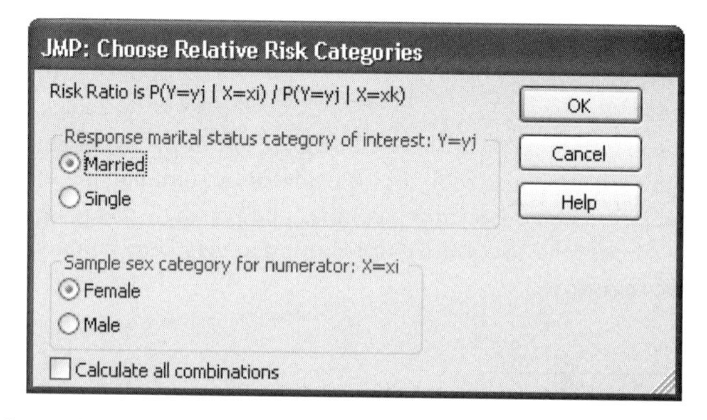

If you are interested in only a single response/factor combination, you can select them here. For example, if you clicked OK in the dialog in Figure 8.3, JMP would calculate the risk ratio of

$$\frac{P(Y = \text{Married}|X = \text{Female})}{P(Y = \text{Married}|X = \text{Male})}$$

If you would like to calculate the risk ratios for all (2×2 =4) combinations of response and factor levels, select the checkbox for **Calculate All Combinations**. Figure 8.4 shows the report when all combinations are asked for.

Figure 8.4 Risk Ratio Report

Contingency Analysis of marital status By sex

Contingency Table

marital status

Count / Total % / Col % / Row %	Married	Single	
sex: Female	95 31.35 48.47 68.84	43 14.19 40.19 31.16	138 45.54
sex: Male	101 33.33 51.53 61.21	64 21.12 59.81 38.79	165 54.46
	196 64.69	107 35.31	303

Relative Risk

Description	Relative Risk	Lower 95%	Upper 95%
P(Married|Female)/P(Married|Male)	1.124623	0.953193	1.326885
P(Married|Male)/P(Married|Female)	0.889187	0.753645	1.049105
P(Single|Female)/P(Single|Male)	0.803329	0.587179	1.099046
P(Single|Male)/P(Single|Female)	1.24482	0.90988	1.703057

To see how the relative risk is calculated, examine the first entry in the Relative Risk report.

P(Married|Female)/P(Married|Male)

These probabilities can be read off the contingency table above the Relative Risk report. We are computing probabilities concerning two levels of sex, which differs across the rows of the table. Therefore, we use the Row% to read off the probabilities:

P(Married|Female)=0.6884

P(Married|Male) = 0.6121

Therefore,

$$P(Married|Female)/P(Married|Male) = \frac{0.6884}{0.6121} = 1.1248$$

Risk Difference

Risk differences (called difference of proportions in Agresti 1990) are calculated between the levels of the factor for each level of the response. Figure 8.5 shows the calculations for the Car Poll example.

Figure 8.5 Risk Difference Report

Risk Difference

Description	Risk Difference	Lower 95%	Upper 95%
P(Married|Female)-P(Married|Male)	0.076285	-0.03095	0.183517
P(Single|Female)-P(Single|Male)	-0.07628	-0.18352	0.030948

As with risk ratios, the calculations can be read from the associated contingency table.

P(Married|Female) – P(Married|Male) = 0.6884 – 0.6121 = 0.0763

The Agreement Statistic

When there are matching levels across the two categorical variables, the **Agreement Statistic** command is available.This command displays the Kappa statistic (Agresti 1990), its standard error and Bowker's test of symmetry, also know as McNemar's test. The following example results from using raters A and B in the AttributeGage.jmp sample data table. The agreement statistic is high (0.86), reinforcing the high agreement seen by looking at the main diagonal of the contingency table.

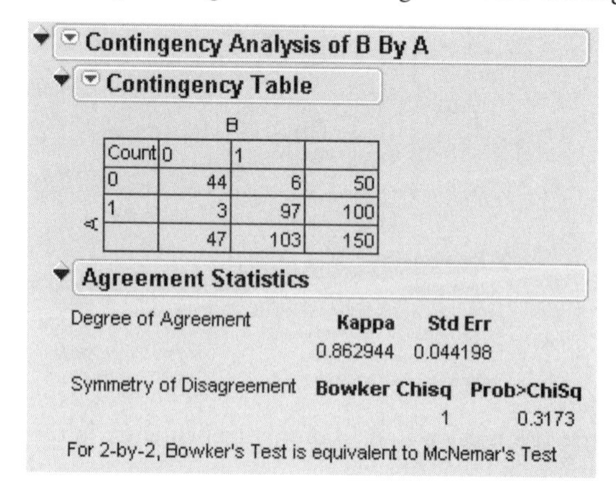

Viewing the two response variables as two independent ratings of the *n* subjects, the kappa coefficient equals +1 when there is complete agreement of the raters. When the observed agreement exceeds chance agreement, the kappa coefficient is positive, with its magnitude reflecting the strength of agreement. Although unusual in practice, kappa is negative when the observed agreement is less than chance agreement. The minimum value of kappa is between -1 and 0, depending on the marginal proportions.

Details Quantities associated with the Kappa statistic are computed as follows.

$$\hat{\kappa} = \frac{P_0 - P_e}{1 - P_e} \text{ where } P_0 = \sum_i p_{ii} \text{ and } P_e = \sum_i p_{i.}p_{.i}$$

The asymptotic variance of the simple kappa coefficient is estimated by the following:

$$\text{var} = \frac{A + B - C}{(1 - P_e)^2 n} \text{ where } A = \sum_i p_{ii}\left[1 - (p_{i.} + p_{.i})(1 - \hat{\kappa})\right],\ B = (1 - \hat{\kappa})^2 \sum_{i \neq j}\sum p_{ij}(p_{.i} + p_{j.})^2 \text{ and}$$

$$C = \left[\hat{\kappa} - P_e(1 - \hat{\kappa})\right]^2$$

For Bowker's test of symmetry, the null hypothesis is that the probabilities in the two-by-two table satisfy symmetry ($p_{12}=p_{21}$ and $p_{11}=p_{12}$).

Odds Ratio

For 2×2 contingency tables, the **Odds Ratio** command produces a report of the odds ratio, calculated as

$$\frac{p_{11} \times p_{22}}{p_{12} \times p_{21}}$$

as well as 95% confidence intervals on this ratio.s

Correspondence Analysis

Correspondence analysis is a graphical technique to show which rows or columns of a frequency table have similar patterns of counts. In the correspondence analysis plot there is a point for each row and for each column.

Define the *row profile* to be the counts in a row divided by the total count for that row. If two rows have very similar row profiles, their points in the correspondence analysis plot will be close together. Squared distances between row points are approximately proportional to Chi-square distances that test the homogeneity between the pair of rows.

Column and row profiles are alike because the problem is defined symmetrically. The distance between a row point and a column point has no meaning but the directions of columns and rows from the origin are meaningful, and the relationships help interpret the plot.

Correspondence analysis is especially popular in France (Benzecri 1973, Lebart et al. 1977) and Japan (Hayashi 1950) but has only recently become popular in the United States (Greenacre 1984).

The technique is particularly useful for tables with large numbers of levels where deriving useful information from the table can be difficult. As an example of a correspondence analysis see the Newell cheese tasting experiment as reported in McCullagh and Nelder (1989). The data are in the Cheese.jmp sample data table. The experiment records counts over nine different response levels across four different cheese additives. The variable Response levels has the ordinal modeling type, with values 1 to 9 where 9 is best tasting. The *x* factor is Cheese, and Count is a Freq (frequency) variable.

You can immediately see in the mosaic plot (Figure 8.6) that the distributions don't appear homogeneous. However, it is hard to make sense of the chart across nine levels. A correspondence analysis can help define relationships in this kind of situation. To obtain the correspondence analysis plot, select the **Correspondence Analysis** command from the popup menu on the Contingency Table title bar.

Figure 8.6 Correspondence Analysis Option

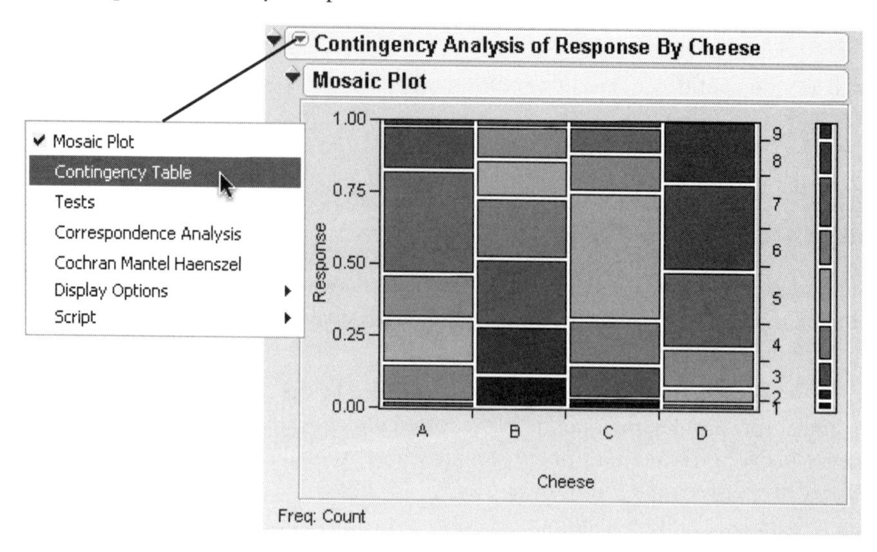

The Correspondence Analysis Plot

The **Correspondence Analysis** command first displays the plot in Figure 8.7, which shows correspondence analysis graphically with plot axes labeled c1 and c2. You read the plot like this:

- Higher response levels tend to be negative in c1 and positive in c2.
- Moderate response levels are negative in c2 and neutral in c1.
- The lowest response values are positive in c1 and c2.

The c1 axis tends to score the response levels linearly. The c2 scores indicate non-neutrality. From the directions established, it is clear that D is the winner as best-liked cheese, and in descending order the others are A, C, and B.

Figure 8.7 Correspondence Plot for the Cheese Experiment

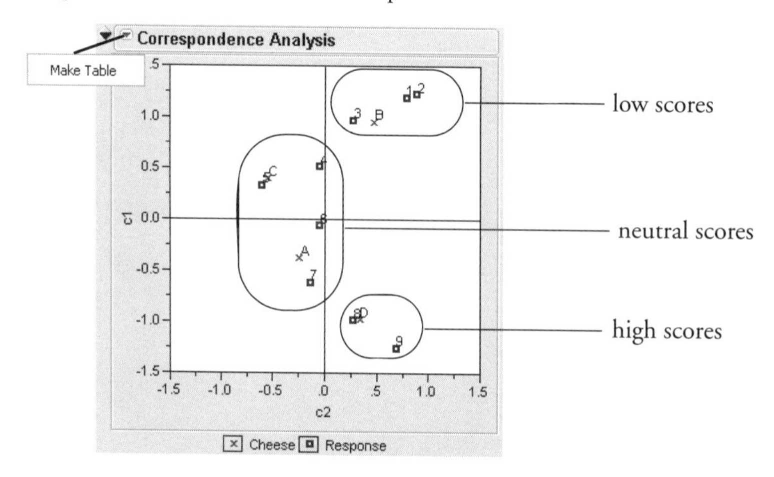

Make Table and Three Dimensional Plots

In these correspondence plots, often the points overlap and it is difficult to see all the labels. The **Make Table** command, located in the platform drop-down menu, outputs the table of both X and Y labels and scores so that you can better examine the results. The output table includes a script that produces the appropriate spinning plot so you can see a 3-D correspondence analysis.

Figure 8.8 Spinning Plot for the Cheese Example

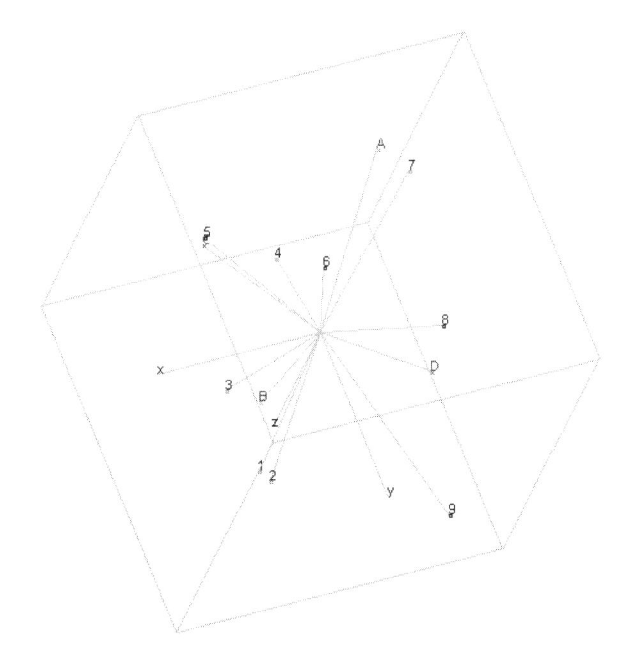

We can conclude that c1, the most prominent component, corresponds mostly with the response level, and c2 corresponds with some quality that is different between A and C cheeses, and B and D cheeses.

Statistical Details for Correspondence Analysis

You can open the **Details** table that accompanies the Correspondence Analysis plot to see statistical information about the plot.

Details

Singular Value	Inertia	Portion	Cumulative
0.73609	0.54183	0.6936	0.6936
0.42010	0.17649	0.2259	0.9195
0.25070	0.06285	0.0805	1.0000

Cheese	c1	c2	c3	Response	c1	c2	c3
A	-0.3763	-0.2528	-0.3865	1	1.190	0.7764	-0.0490
B	0.9553	0.4728	-0.0554	2	1.222	0.8811	-0.1006
C	0.3981	-0.5540	0.2467	3	0.964	0.2628	0.0900
D	-0.9771	0.3340	0.1952	4	0.507	-0.0588	-0.1693
				5	0.328	-0.6068	0.2705
				6	-0.065	-0.0617	-0.0472
				7	-0.623	-0.1480	-0.3510
				8	-0.991	0.2634	0.0443
				9	-1.259	0.6786	0.5852

The Details table has these quantities:

Singular Value lists the singular values of

$$D_r^{-0.5}(\mathrm{P} - rc')\mathrm{D}_c^{-0.5}$$

where **P** is the matrix of counts divided by the total frequency, r and c are row and column sums of P, and the D's are diagonal matrices of the values of r and c.

Inertia lists the squared singular values of the scaled table, reflecting the variation accounted for in the canonical dimensions. If the first two singular values capture the bulk of the values, then the 2-D correspondence plot is a good approximation to higher dimensional relationships in the table.

Portion is the portion of inertia with respect to the total inertia.

X variable (Cheese) c1, c2, c3 are the coordinates of the row values used to position points in the plot.

Y variable (Response) c1, c2, c2 are the coordinates of the column values used to position points in the plot.

Mail Messages Example

The following example uses an electronic mail log and looks at the pattern of senders and receivers. Figure 8.9 is a partial listing of the Mail Messages.jmp sample data with variables When (time a message was sent), From (message sender), and To (message receiver).

Figure 8.9 Mail Messages Data

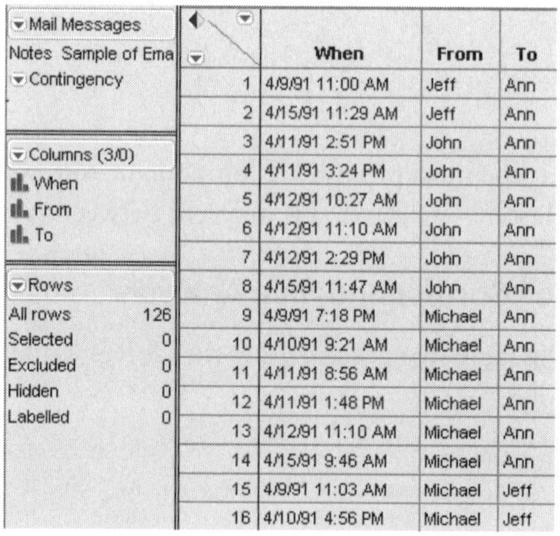

Mail Messages
Notes Sample of Ema
Contingency

Columns (3/0)
When
From
To

Rows

All rows	126
Selected	0
Excluded	0
Hidden	0
Labelled	0

	When	From	To
1	4/9/91 11:00 AM	Jeff	Ann
2	4/15/91 11:29 AM	Jeff	Ann
3	4/11/91 2:51 PM	John	Ann
4	4/11/91 3:24 PM	John	Ann
5	4/12/91 10:27 AM	John	Ann
6	4/12/91 11:10 AM	John	Ann
7	4/12/91 2:29 PM	John	Ann
8	4/15/91 11:47 AM	John	Ann
9	4/9/91 7:18 PM	Michael	Ann
10	4/10/91 9:21 AM	Michael	Ann
11	4/11/91 8:56 AM	Michael	Ann
12	4/11/91 1:48 PM	Michael	Ann
13	4/12/91 11:10 AM	Michael	Ann
14	4/15/91 9:46 AM	Michael	Ann
15	4/9/91 11:03 AM	Michael	Jeff
16	4/10/91 4:56 PM	Michael	Jeff

Fit Y by X for the nominal variables To and From gives the frequency table and Chi-square tests shown beside the data table.

Figure 8.10 Mail Log Data Table

Contingency Analysis of To By From

Contingency Table

Count	Ann	Jeff	John	Katheri	Michael	
Ann	0	0	3	0	4	7
Jeff	2	0	2	2	7	13
John	6	0	0	16	17	39
Katherine	0	0	12	0	9	21
Michael	6	12	14	14	0	46
	14	12	31	32	37	126

Tests

Source	DF	-LogLike	RSquare (U)
Model	16	55.70245	0.2907
Error	106	135.94228	
C. Total	122	191.64472	
N	126		

Test	ChiSquare	Prob>ChiSq
Likelihood Ratio	111.405	<.0001*
Pearson	79.480	<.0001*

Warning: 20% of cells have expected count less than 5, ChiSquare suspect.

The Chi-square tests in Figure 8.10 are highly significant, which means the To and From variables are not independent. Looking at the frequency table shows obvious differences. For example,

- Jeff sends messages to everyone else but only receives messages from Michael.
- Michael and John send many more messages then the others.
- Michael sends messages to everyone.
- John sends messages to everyone except Jeff.
- Katherine and Ann only send messages to Michael and John.

You can visualize the results of the contingency table with a correspondence analysis. The Details table gives the results of a correspondence analysis.

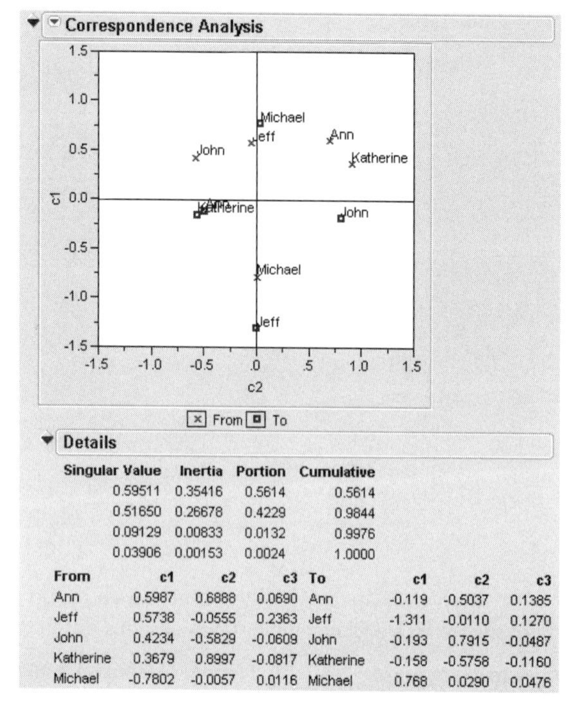

Singular Value	Inertia	Portion	Cumulative
0.59511	0.35416	0.5614	0.5614
0.51650	0.26678	0.4229	0.9844
0.09129	0.00833	0.0132	0.9976
0.03906	0.00153	0.0024	1.0000

From	c1	c2	c3	To	c1	c2	c3
Ann	0.5987	0.6888	0.0690	Ann	-0.119	-0.5037	0.1385
Jeff	0.5738	-0.0555	0.2363	Jeff	-1.311	-0.0110	0.1270
John	0.4234	-0.5829	-0.0609	John	-0.193	0.7915	-0.0487
Katherine	0.3679	0.8997	-0.0817	Katherine	-0.158	-0.5758	-0.1160
Michael	-0.7802	-0.0057	0.0116	Michael	0.768	0.0290	0.0476

The **Portion** column in the table shows that the bulk of the variation (56% + 42%) of the mail sending pattern is summarized by c1 and c2 for the To and From groups.

The Correspondence Analysis plot of c1 and c2 shows the pattern of mail distribution among the mail group:

- Katherine and Ann have similar sending and receiving patterns; they both send to Michael and John and receive from Michael, John, and Jeff.
- Jeff and Michael lie in a single dimension but have opposite sending and receiving patterns; Jeff sends to everyone and receives only from Michael. Michael sends to and receives from everyone.
- John's patterns are different than the others. He sends to Ann, Katherine, and Michael, and receives from everyone.

You can consider the effect of the third variate in the canonical dimension, c3, by plotting c1, c2 and c3 as illustrated in the previous example.

Follow these steps to create the Mail 3D table for a spin plot:

1. Right-click (Control-click on Macintosh) on the To portion of the Details table, and select the **Make Into Data Table** command to create a JMP table.
2. Create a new column and append the word "To" to the sender's name. Give the column a new name, To/From.
3. Right-click (Control-click on Macintosh) on the From portion of the Details table and select the **Make Into Data Table** command to create a second new JMP table.
4. Create a new column and append the word "From" to the receiver's name. Give this column the same name you used in the other data table (To/From).
5. Use the **Concatenate** command in the **Tables** menu to append the two tables and create the table

shown here.

	To/From	c1	c2	c3
1	From Ann	0.59870069	0.68883707	0.06901165
2	From Jeff	0.57379738	-0.0555449	0.23632648
3	From John	0.42343594	-0.5829035	-0.0608683
4	From Katherine	0.3678502	0.89971905	-0.0816566
5	From Michael	-0.7801984	-0.0056661	0.01159409
6	To Ann	-0.1191826	-0.5037318	0.13849547
7	To Jeff	-1.3110086	-0.0109701	0.12700136
8	To John	-0.193234	0.79147148	-0.0487185
9	To Katherine	-0.157544	-0.5757993	-0.1160173
10	To Michael	0.76844111	0.02902291	0.04756416

Now use the Spinning Plot platform to look at c1, c2 and c3. When the points are projected onto the (x, y) plane, the pattern is the same as the Correspondence plot shown previously. When you rotate the spinning plot as in Figure 8.11, you can clearly see an additional dimension in the relationship between John, Katherine, and Michael:

- Katherine sends messages only to Michael and John.
- John sends many more messages to Michael and Katherine than the others.
- Michael sends more messages to John and Katherine than the others.

Figure 8.11 Spinning Three Mail Pattern Correspondence Coordinates

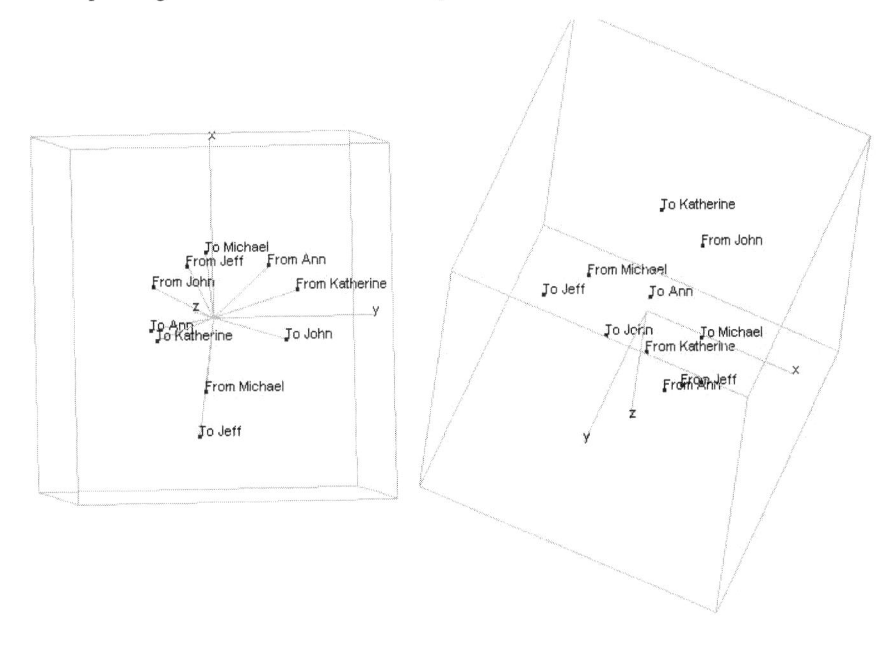

Cochran-Mantel-Haenszel Test

The contingency analysis platform offers the Cochran-Mantel-Haenszel test for testing if there is a relationship between two categorical variables after blocking across a third classification. When you select the **Cochran Mantel Haenszel** command, a dialog asks you to identify a grouping variable.

The example in Figure 8.12 uses the Hot dogs.jmp sample data table. The contingency table analysis for Type by Taste (not shown) reveals a marginally significant Chi-square probability of about 0.0799.

However, if you stratify on protein to fat ratio (values 1 to 5), Taste by Type shows no relationship at all. The Cochran-Mantel-Haenszel Chi-square test for the adjusted correlation between them has a probability of 0.857, and the Chi-square probability associated with the general association of categories is 0.282.

Figure 8.12 Cochran-Mantel-Haenszel Chi-square Test

Cochran-Mantel-Haenszel Tests

Stratified by Protein/Fat

CMH Test	ChiSquare	DF	Prob>Chisq
Correlation of Scores	0.0324	1	0.8572
Row Score by Col Categories	1.4812	2	0.4768
Col Score by Row Categories	0.4204	2	0.8104
General Assoc. of Categories	5.0559	4	0.2816

Frequency Counts

The Cochran-Mantel-Haenszel Chi-square statistic tests the alternative hypothesis that there is a linear association between the row and column variable given a third strata factor that you define:

Correlation of Scores is used when either the *Y* or *X* is ordinal. The alternative hypothesis is that there is a linear association between *Y* and *X* in at least one stratum.

Row Score by Col Categories is used when *Y* is ordinal or interval. The alternative hypothesis is that, for at least one stratum, the mean scores of the *r* rows are unequal.

Col Score by Row Categories is used when *X* is ordinal or interval. The alternative hypothesis is that, for at least one stratum, the mean scores of the *c* columns are unequal.

General Assoc. of Categories tests that for at least one stratum there is some kind of association between *X* and *Y*.

Chapter 9

Simple Logistic Regression

The Fit Y by X Platform

The Logistic platform fits the probabilities for response categories to a continuous *x* predictor. The fitted model estimates probabilities attributed to each *x* value. The logistic platform is the *nominal/ordinal* by *continuous* personality of the **Fit Y by X** command. There is a distinction between nominal and ordinal responses on this platform:

- Nominal logistic regression estimates a set of curves to partition the attributed probability among the responses.
- Ordinal logistic regression models the probability of being less than or equal to a given response. This has the effect of estimating a single logistic curve, which is shifted horizontally to produce probabilities for the ordered categories. This model is less general but more parsimonious, and is recommended for ordered responses.

Logistic regression has a long tradition with widely varying applications such as modeling dose-response data and purchase-choice data. Unfortunately, many introductory statistics courses do not cover this fairly simple method. Many texts in categorical statistics cover it (Agresti 1998), in addition to texts on logistic regression (Hosmer and Lemeshow 1989). Some analysts use the method with a different distribution function, the normal, in which case it is called *probit analysis*. Some analysts use discriminant analysis instead of logistic regression because they prefer to think of the continuous variables as *Y*'s and the categories as *X*'s and work backwards; however, discriminant analysis assumes that the continuous data are normally distributed random responses, rather than fixed regressors.

Simple logistic regression is a more graphical and simplified version of the general facility for categorical responses in the Fit Model platform. For examples of more complex logistic regression models, see the chapter "Logistic Regression for Nominal and Ordinal Response," p. 403.

Note: You can launch Logistic from either the **Fit Y by X** main menu command (or toolbar) or with the **Logistic** button on the JMP Starter. When you choose **Fit Y by X**, JMP performs an analysis by context, which is the analysis appropriate for the modeling types of the response and factor variables you specify. When you use the **Logistic** button on the JMP Starter, the Launch dialog (role dialog) expects a categorical response and a continuous factor variable, and advises you if you specify other modeling types.

Contents

Launch the Platform

Launch the Logistic platform, with either the **Fit Y by X** command on the **Analyze** menu (or toolbar) or with the **Logistic** button on the **Basic Stats** tab page of the JMP Starter. Then specify continuous factors for *x* and nominal or ordinal responses for *Y*. Each combination of *X* and *Y* variables produces a plot and accompanying text reports. See the chapter "Bivariate Scatterplot and Fitting," p. 81, for a discussion of selecting columns into roles with the Launch dialog.

Figure 9.1 shows how simple logistic regression fits within the scheme of the Fit Y by X platform for the different modeling type combinations.

Figure 9.1 Fit Y By X Logistic Regression Platform

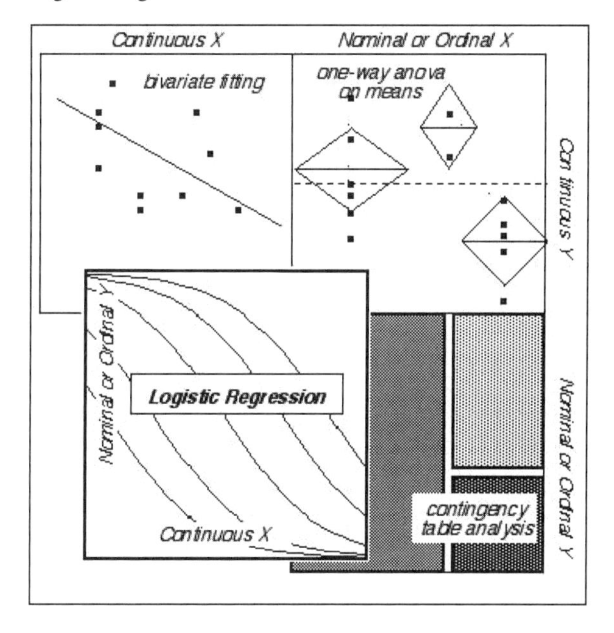

Nominal Logistic Regression

Nominal logistic regression estimates the probability of choosing one of the response levels as a smooth function of the factor. The fitted probabilities must be between 0 and 1, and must sum to 1 across the response levels for a given factor value.

In a logistic probability plot, the *y*-axis represents probability. For *k* response levels, *k* - 1 smooth curves partition the total probability (=1) among the response levels. The fitting principle for a logistic regression minimizes the sum of the negative logarithms of the probabilities fitted to the response events that occur–that is, maximum likelihood.

The following examples use the Big Class.jmp sample data table with age as a nominal variable to illustrate a logistic regression. Suppose you want to guess the value of age for a known value of weight. The age column is assigned as the categorical response, and weight is the continuous factor. The logistic regression analysis begins with the plot shown in Figure 9.2. The left *y*-axis represents probability, and the proportions shown on the right *y*-axis represent the relative sizes of each age group in the total sample.

The Cumulative Logistic Probability Plot

The cumulative logistic probability plot gives a complete picture of what the logistic model is fitting. At each *x* value, the probability scale in the *y* direction is divided up (partitioned) into probabilities for each response category. The probabilities are measured as the vertical distance between the curves, with the total across all *Y* category probabilities sum to 1.

Figure 9.2 The Nominal by Continuous Fit Y by X Platform

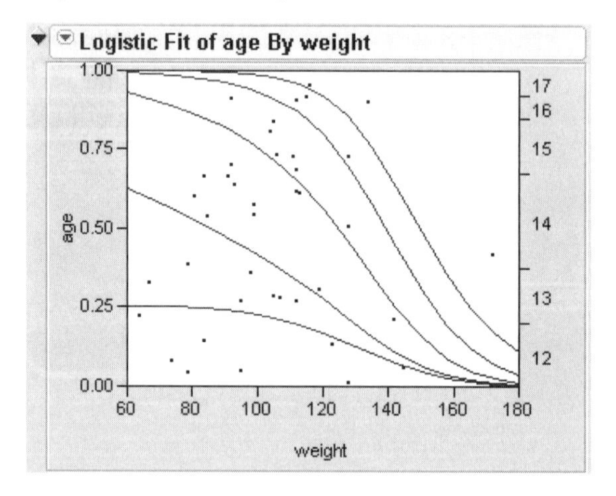

(Note: To reproduce figures in this section, make age a Nominal variable and use **Fit Y by X** with age as Y and weight as X. Don't forget to change age back to ordinal after this section!)

In Figure 9.3, the first (bottom) curve shows the probability attributed to age=12 as weight varies. The next higher curve shows the probability attributed to age≤13. Thus, the distance between the first two curves is the probability for age=13. The distance from the top curve to the top of the graph is the probability attributed to age=17. Note that as weight increases, the model gives more probability to the higher ages. At weight=180, most of the probability is attributed to age 17.

Markers for the data are drawn at their *x*-coordinate, with the *y* position jittered randomly within the range corresponding to the response category for that row. You can see that the points tend to push the lines apart and make vertical space where they occur in numbers, and allow the curves to get close together where there is no data. The data pushes the curves because the criterion that is maximized is the product of the probabilities fitted by the model. The fit tries to avoid points attributed to have a small probability, which are points crowded by the curves of fit. See "Nominal Responses," p. 977 in the appendix "Statistical Details," p. 975, for more about computational details.

Figure 9.3 Interpretation of a Logistic Probability Plot

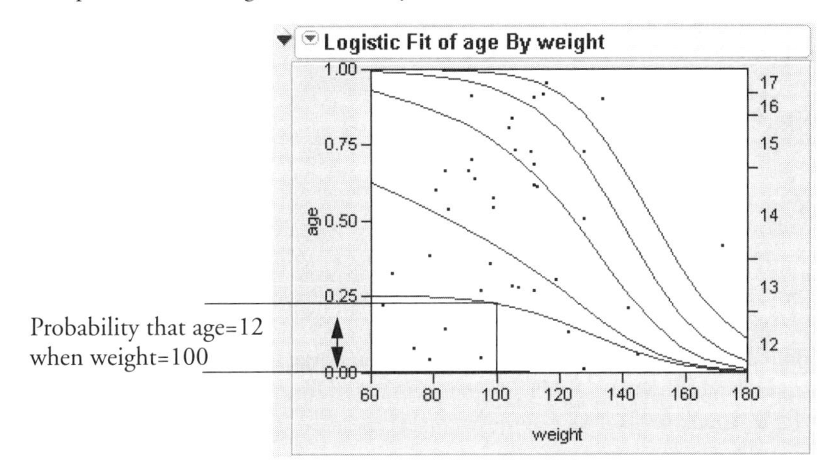

If the *x* -variable has no effect on the response, then the fitted lines are horizontal and the probabilities are constant for each response across the continuous factor range. Figure 9.4 shows a logistic probability plot for selected ages that each have the same weight distribution. The weight factor has no predictive value. Note that the right *y*-axis is partitioned according to this reduced model.

Figure 9.4 Logistic Probability Plot Showing No *Y* by *X* Relationship

age weight
Logistic

Columns (2/0)
age
weight

Rows
All rows 8
Selected 0

	age	weight
1	12	59
2	12	63
3	12	59
4	12	63
5	13	59
6	13	63
7	14	59
8	14	63

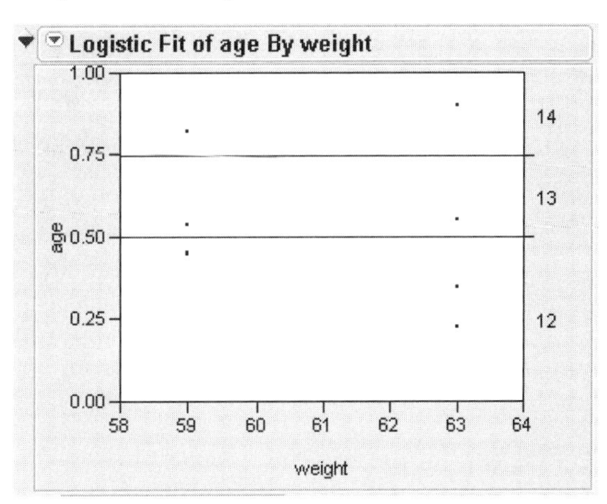

If the response is completely predicted by the value of the factor, then the logistic curves are effectively vertical. In this case, the parameter estimates become very large and are marked *unstable* in the regression report. The prediction of a response is certain (the probability is 1) at each of the factor levels, as illustrated in Figure 9.5. It is ironic that the parametric model breaks down for the perfect-fitting case, but although the parameters are unstable, the fitted probabilities are well behaved.

Figure 9.5 Logistic Plot Showing an Almost Perfect *Y* by *X* Relationship

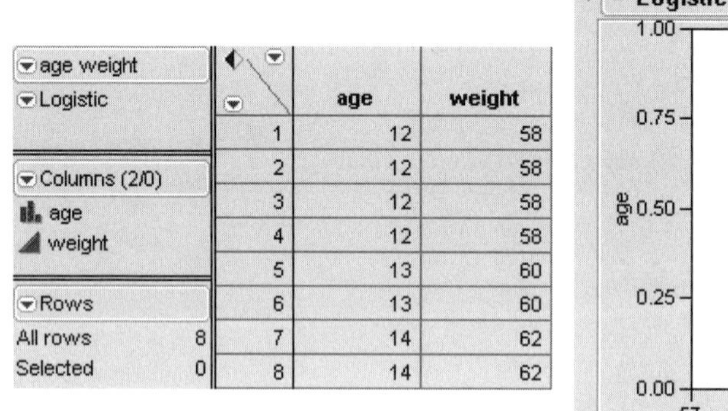

	age	weight
1	12	58
2	12	58
3	12	58
4	12	58
5	13	60
6	13	60
7	14	62
8	14	62

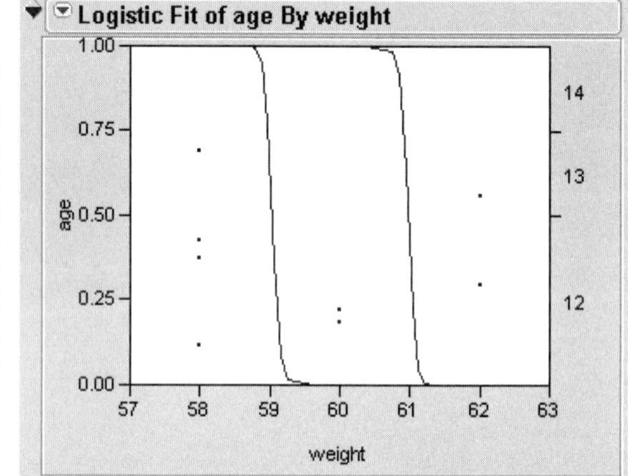

The nominal by continuous platform has two text reports, the Whole Model Test table and the Parameter Estimates table, that are similar to those described for categorical models in the chapter "Contingency Tables Analysis," p. 147.

Plot Options

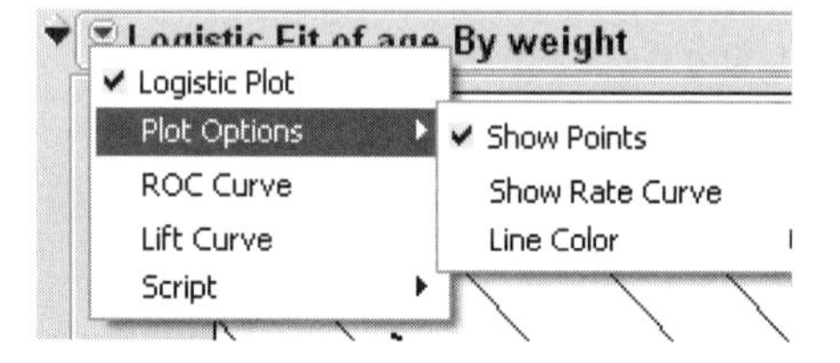

Plot options are accessible from the platform's drop-down menu.

Show Points toggles the points on or off.

Show Rate Curve is only useful if you have several points for each *x*-value. In these cases, you can get reasonable estimates of the rate at each value, and compare this rate with the fitted logistic curve. To prevent too many degenerate points, usually at zero or one, JMP only shows the rate value if there are at least three points at the *x*-value.

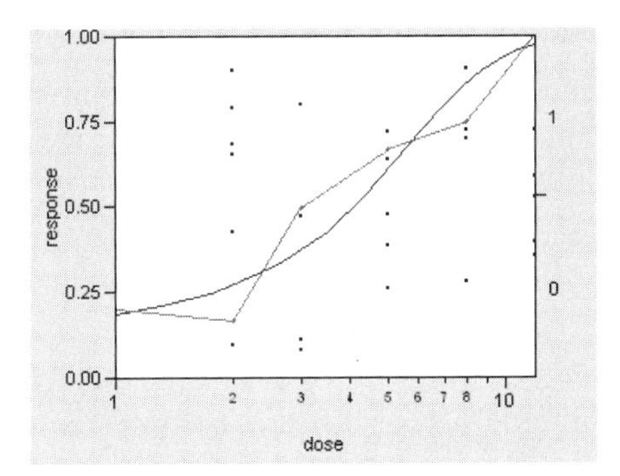

This example uses Dose Response.jmp, with a logarithmic dose axis.

Line Color allows you to pick the color of the plot curves

The Whole Model Test Table

The Whole Model Test table shows if the model fits better than constant response probabilities. This table is analogous to the Analysis of Variance table for a continuous response model. It is a specific likelihood-ratio Chi-square test that evaluates how well the categorical model fits the data. The negative sum of logs of the observed probabilities is called the negative log-likelihood (–LogLikelihood). The negative log-likelihood for categorical data plays the same role as sums of squares in continuous data. The difference in the log-likelihood from the model fitted by the data and the model with equal probabilities is a Chi-square statistic. This test statistic examines the hypothesis that the *x* variable has no effect on the response.

Logistic Fit of age By weight

Whole Model Test

Model	-LogLikelihood	DF	ChiSquare	Prob>ChiSq
Difference	6.168277	5	12.33655	0.0305*
Full	61.098073			
Reduced	67.266350			
RSquare (U)		0.0917		
Observations (or Sum Wgts)		40		

Converged by Gradient

Values of the **Rsquare (U)**, sometimes denoted as R^2, range from 0 to 1, where 1 is a perfect fit and 0 means there is no gain by using the model over using fixed background response rates. In Figure 9.5, all probabilities of response events are 1, and R^2 is 1. In Figure 9.4, all fitted lines are horizontal and Rsquare is 0. High R^2 values are rare in categorical models. An R^2 of 1 means that the factor completely predicts the categorical response.

The Whole Model Test table has these items:

Model (sometimes called Source) and **DF** in the Whole Model Test table are discussed previously in the chapter "Contingency Tables Analysis," p. 147.

–LogLikelihood measures variation, sometimes called *uncertainty*, in the sample:

Full (the full model) is the negative log-likelihood (or uncertainty) calculated after fitting the model. The fitting process involves predicting response rates with a linear model and a logistic response function. This value is minimized by the fitting process.

Reduced is the negative log-likelihood (or uncertainty) for the case when the probabilities are estimated by fixed background rates. This is the background uncertainty when the model has no effects.

The difference of these two negative log-likelihoods is the reduction due to fitting the model. Two times this value is the likelihood-ratio Chi-square test statistic.

Chi-Square is the likelihood-ratio Chi-square test of the hypothesis that the model fits no better than fixed response rates across the whole sample. It is twice the **–LogLikelihood** for the Difference Model. It is the difference of two negative log-likelihoods, one with whole-population response probabilities and one with each-population response rates. This statistic is sometimes denoted G^2 and is written

$$G^2 = 2(\sum -\ln p(\text{background}) - \sum -\ln p(\text{model}))$$

where the summations are over all observations instead of all cells.

Prob>ChiSq is the observed significance probability, often called the p value, for the Chi-square test. It is the probability of getting, by chance alone, a Chi-square value greater than the one computed. Models are often judged significant if this probability is below 0.05.

Rsquare (U) is the proportion of the total uncertainty that is attributed to the model fit. The difference between the log-likelihood from the fitted model and the log-likelihood that uses horizontal lines is a test statistic to examine the hypothesis that the factor variable has no effect on the response. The ratio of this test statistic to the background log-likelihood is subtracted from 1 to calculate R^2. More simply, it is computed as

$$\frac{-\text{loglikelihood for Difference}}{-\text{loglikelihood for Reduced}}$$

using quantities from the Whole Model Test table.

Observations (or **Sum Wgts**) is the total sample size used in computations.

The Parameter Estimates Table

The nominal logistic model fits a parameter for the intercept and slope for each of $k-1$ logistic comparisons, where k is the number of response levels. The Parameter Estimates table, shown next, lists these estimates. Each parameter estimate can be examined and tested individually, although this is seldom of much interest.

Parameter Estimates

Term	Estimate	Std Error	ChiSquare	Prob>ChiSq
Intercept[12]	13.4041298	5.8347741	5.28	0.0216*
weight[12]	-0.1054371	0.0468175	5.07	0.0243*
Intercept[13]	14.6094734	5.9440578	6.04	0.0140*
weight[13]	-0.1192624	0.0485206	6.04	0.0140*
Intercept[14]	13.2340182	5.6838347	5.42	0.0199*
weight[14]	-0.0996764	0.0448932	4.93	0.0264*
Intercept[15]	10.3190474	5.6576149	3.33	0.0682
weight[15]	-0.0769464	0.0443084	3.02	0.0825
Intercept[16]	6.26425316	5.7788496	1.18	0.2784
weight[16]	-0.0486128	0.044568	1.19	0.2754

For log odds of 12/17, 13/17, 14/17, 15/17, 16/17

Term lists each parameter in the logistic model. There is an intercept and a slope term for the factor at each level of the response variable.

lists the parameter estimates given by the logistic model.

Std Error lists the standard error of each parameter estimate. They are used to compute the statistical tests that compare each term to zero.

Chi-Square lists the Wald tests for the hypotheses that each of the parameters is zero. The Wald Chi-square is computed as $(\text{Estimate}/\text{Std Error})^2$.

Prob>ChiSq lists the observed significance probabilities for the Chi-square tests.

Inverse Prediction

Inverse prediction is the opposite of prediction. It is the prediction of *x* values from given *y* values. But in logistic regression, instead of a *y* value, you have the probability attributed to one of the *Y* levels. This feature only works when there are two response categories (a binary response). For example, in a laboratory study, you might want to know what dose of a toxin results in a 0.5 probability of killing a rat. In this case, the inverse prediction for 0.5 is called the *LD50*, the lethal dose corresponding to a 50% death rate.

The inverse prediction is the *x* value that corresponds to a given probability of the lower response category. You can use the crosshair tool to visually approximate an inverse prediction, as shown here. This example shows the logistic plot that predicts sex for given height values. But if you want to know which value of height is equally likely to be that of either a male or female, you can place the horizontal crosshair line at 0.5 on the vertical (sex) probability axis, move the cross-hair intersection to the prediction line, and read the height value that shows on the horizontal axis. In this example, a student with a height of approximately 61.529 is as likely to be male as female.

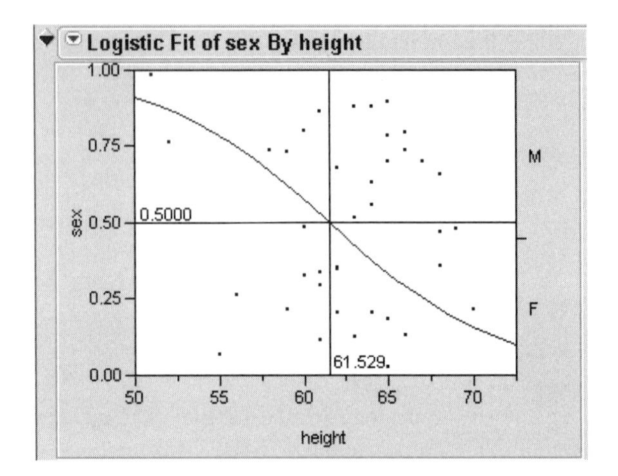

If your response has exactly two levels, the **Inverse Prediction** command enables you to request an exact inverse prediction as shown in Figure 9.6. If there is a strong relationship between the variables, the Inverse Prediction table displays upper and lower confidence limits for the Inverse Prediction.

The Fit Model platform also has an option that gives an inverse prediction with confidence limits. The chapters "Standard Least Squares: Perspectives on the Estimates," p. 249, and "Logistic Regression for Nominal and Ordinal Response," p. 403, give more information about inverse prediction.

Figure 9.6 Inverse Prediction

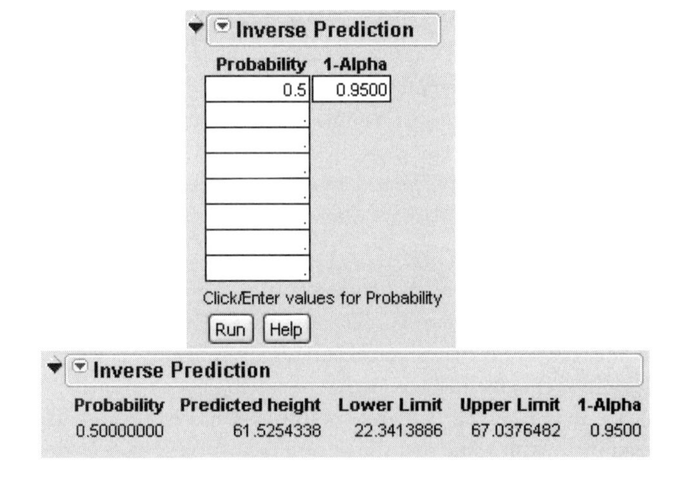

ROC (Receiver Operating Characteristic) Curves

Suppose that an *x* value is a diagnostic measurement and you want to determine some threshold value of *x* which indicates a condition exists or does not exist if the *x* value is greater than or less than the threshold. For example, you could measure a blood component level as a diagnostic test to predict a type of cancer. Now consider the diagnostic test as you vary the threshold and, thus, cause more or fewer false positives and false negatives. You then plot those rates. The ideal is to have a very narrow range of *x* criterion values that best divides true negatives and true positives. The ROC curve shows how rapidly this transition happens, with the goal being to have diagnostics that maximize the area under the curve.

Two standard definitions used in medicine are

- *Sensitivity*, the probability that a given *x* value (a test or measure) correctly predicts an existing condition. For a given *x*, the probability of incorrectly predicting the existence of a condition is 1 – sensitivity.
- *Specificity*, the probability that a test correctly predicts that a condition does not exist.

A Receiver Operating Characteristic curve (ROC) is a plot of sensitivity by (1 – specificity) for each value of *x*. The area under the ROC curve is a common index used to summarize the information contained in the curve.

When you do a simple logistic regression with a binary outcome, there is a platform option to request an ROC curve for that analysis. After selecting the ROC Curve option, a dialog box asks for a specification of which level is to be used as "positive".

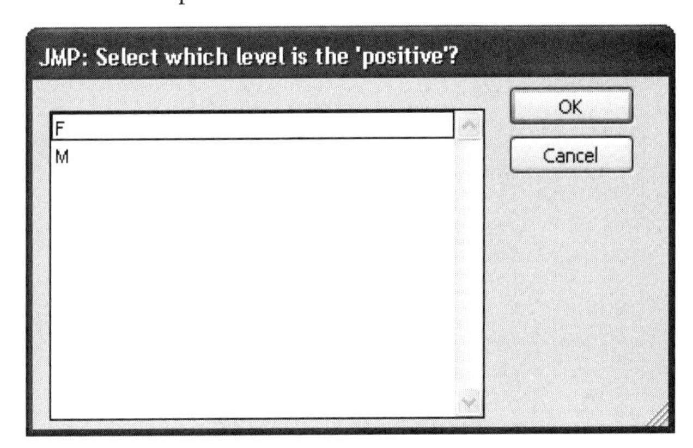

The results for the sex by height example are shown here. The ROC curve plots the probabilities described above, for predicting sex.

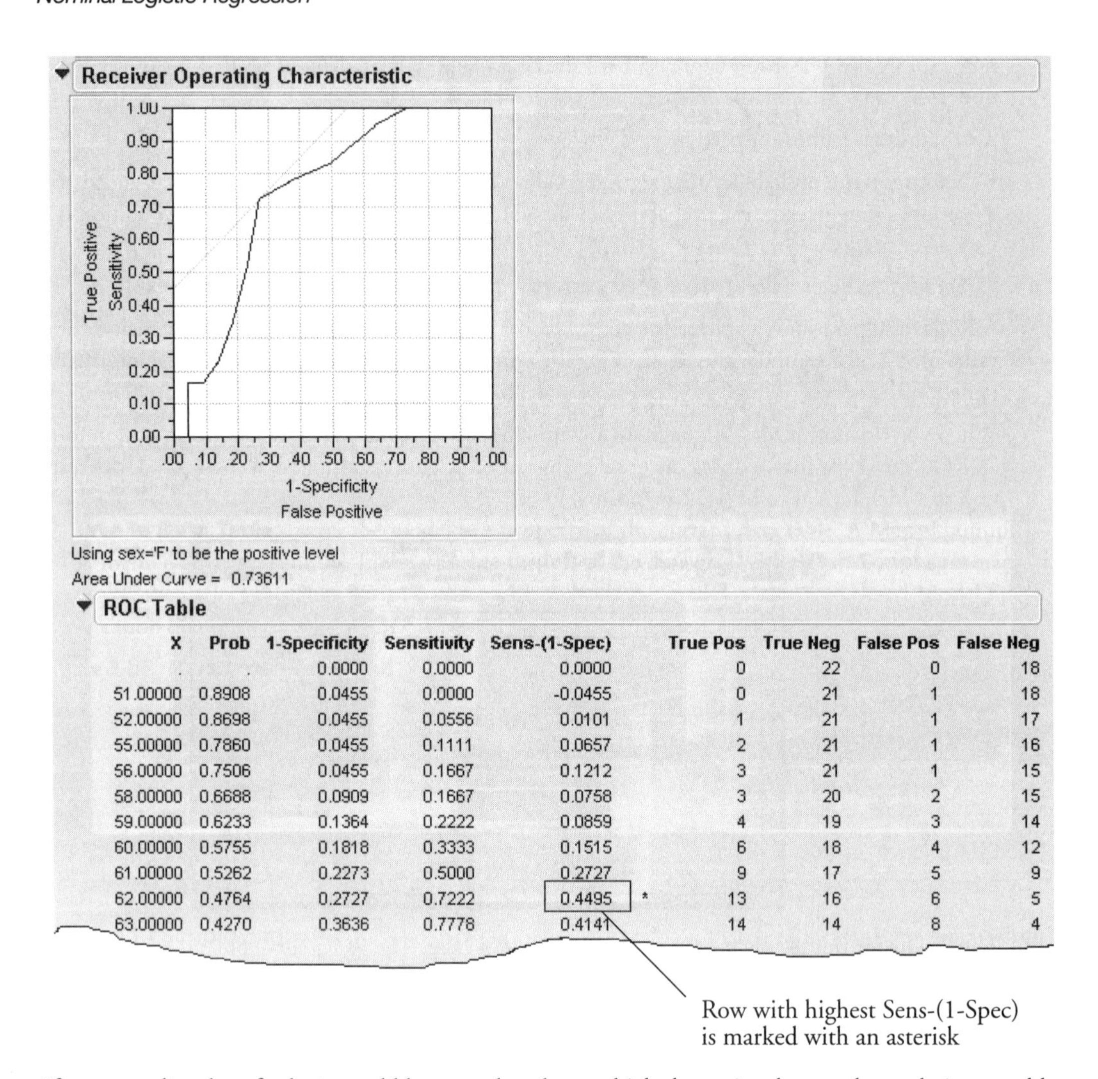

X	Prob	1-Specificity	Sensitivity	Sens-(1-Spec)		True Pos	True Neg	False Pos	False Neg
.	.	0.0000	0.0000	0.0000		0	22	0	18
51.00000	0.8908	0.0455	0.0000	-0.0455		0	21	1	18
52.00000	0.8698	0.0455	0.0556	0.0101		1	21	1	17
55.00000	0.7860	0.0455	0.1111	0.0657		2	21	1	16
56.00000	0.7506	0.0455	0.1667	0.1212		3	21	1	15
58.00000	0.6688	0.0909	0.1667	0.0758		3	20	2	15
59.00000	0.6233	0.1364	0.2222	0.0859		4	19	3	14
60.00000	0.5755	0.1818	0.3333	0.1515		6	18	4	12
61.00000	0.5262	0.2273	0.5000	0.2727		9	17	5	9
62.00000	0.4764	0.2727	0.7222	0.4495	*	13	16	6	5
63.00000	0.4270	0.3636	0.7778	0.4141		14	14	8	4

If a test predicted perfectly, it would have a value above which the entire abnormal population would fall and below which all normal values would fall. It would be perfectly sensitive and then pass through the point (0,1) on the grid. The closer the ROC curve comes to this ideal point, the better its discriminating ability. A test with no predictive ability produces a curve that follows the diagonal of the grid (DeLong, et al. 1988).

The ROC curve is a graphical representation of the relationship between false-positive and true-positive rates. A standard way to evaluate the relationship is with the area under the curve, shown below the plot. In the plot, a yellow line is drawn at a 45 degree angle tangent to the ROC Curve. This marks a good cutoff point under the assumption that false negatives and false positives have similar costs.

Lift Curve

Produces a lift curve for the model. A lift curve shows the same information as an ROC curve, but in a way to dramatize the richness of the ordering at the beginning. The *Y*-axis shows the ratio of how rich that portion of the population is in the chosen response level compared to the rate of that response level as a whole. See "Lift Curves," p. 674 IN THE "RECURSIVE PARTITIONING" CHAPTER for more details on lift curves.

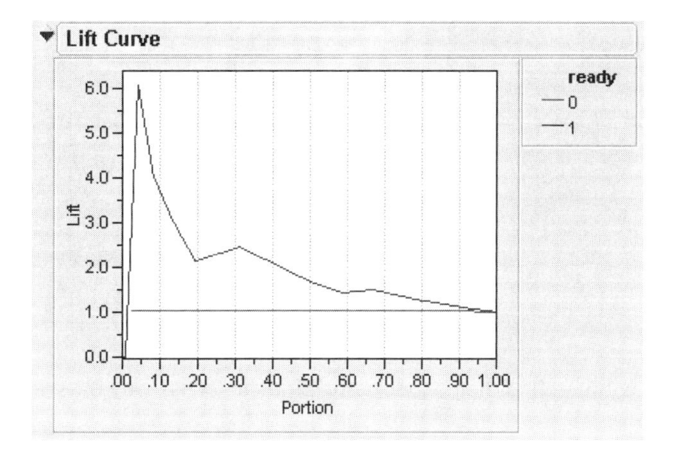

Odds Ratios

Odds ratios are implemented in the same way as they are in the Fit Model Platform. See "Odds Ratios (Nominal Responses Only)," p. 411 IN THE "LOGISTIC REGRESSION FOR NOMINAL AND ORDINAL RESPONSE" CHAPTER for details.

Ordinal Logistic Regression

When *Y* is ordinal, a modified version of logistic regression is used for fitting. The cumulative probability of being at or below each response level is modeled by a curve. The curves are the same for each level except that they are shifted to the right or left.

The ordinal logistic model fits a different intercept, but the same slope, for each of $r - 1$ cumulative logistic comparisons, where r is the number of response levels. Each parameter estimate can be examined and tested individually, although this is seldom of much interest.

The ordinal model is preferred to the nominal model when it is appropriate because it has fewer parameters to estimate. In fact, it is practical to fit ordinal responses with hundreds of response levels. See "Ordinal Responses," p. 977 in the appendix "Statistical Details," p. 975, for the logistic regression model formulation.

To illustrate an ordinal logistic regression, suppose you want to guess the value of age for a given weight value. In the plot shown here, age is the ordinal and weight is continuous. You interpret this platform the same way as the nominal by continuous platform discussed previously under "Nominal Logistic Regression," p. 170.

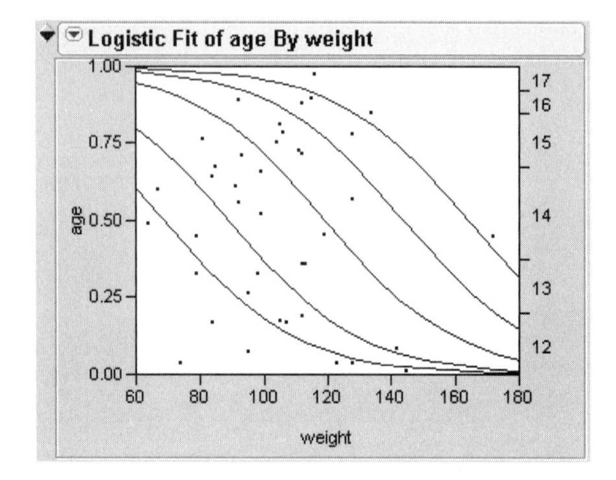

The ordinal by continuous Whole Model test tables, shown next, are the same as those for a nominal by continuous analysis.

Whole Model Test

Model	-LogLikelihood	DF	ChiSquare	Prob>ChiSq
Difference	5.037911	1	10.07582	0.0015*
Full	62.228439			
Reduced	67.266350			

RSquare (U)	0.0749
Observations (or Sum Wgts)	40

Converged by Objective

Parameter Estimates

Term	Estimate	Std Error	ChiSquare	Prob>ChiSq
Intercept[12]	3.32387173	1.5005519	4.91	0.0268*
Intercept[13]	4.29921448	1.535908	7.84	0.0051*
Intercept[14]	5.76150124	1.6417195	12.32	0.0004*
Intercept[15]	6.96581366	1.7511209	15.82	<.0001*
Intercept[16]	7.93933219	1.8612781	18.19	<.0001*
weight	-0.0483808	0.014842	10.63	0.0011*

See the section "Nominal Logistic Regression," p. 170, for a discussion of the Whole Model Test table and the Parameter Estimates table. In the Parameter Estimates table, an intercept parameter is estimated for every response level except the last, but there is only one slope parameter. The intercept parameters show the spacing of the response levels. They always increase monotonically.

Chapter 10

Paired Data

The Matched Pairs Platform

The Matched Pairs platform compares means between two response columns. For example, a matched pair might be a blood pressure measurement taken on the same subject before a treatment and again after the treatment. The responses are correlated, and the statistical method called the *paired t-test* takes that into account.

The platform produces a graph of the difference by the mean, and the Student's paired *t*-test results for all three alternative hypotheses. Additional features provide for more than two matched responses and for a grouping column to test across samples, in a simple version of repeated measures analysis.

Contents

Introduction

The Matched Pairs platform compares means between two response columns using a paired *t*-test. Often the two columns represent measurements on the same subject before and after some treatment.

If you have paired data arranged in two data table columns, then you are ready to use the Matched Pairs platform. If all the measurements are in a single column, there is an alternate approach to doing a paired analysis, mentioned below. Or, you can use the **Split** command in the **Tables** menu to split the column of measurements into two columns.

Approaches to Analyzing Matched Pairs

The paired *t*-test is equivalent to several other approaches, including the following:

- For two response columns, create a third column that calculates the difference between the two responses. Then test that the mean of the difference column is zero with the Distribution platform.
- For the two responses stored in separate rows of a single column identified by an ID column, you can do a two-way analysis of variance. One factor (column) identifies the two responses and the other factor identifies the subject. Use the Fit Y by X Oneway platform with a matching column, or use the Fit Model platform with a Block column. The test on the ID factor is equivalent to the paired *t*-test.

Note: If the data are paired, don't do a two-group unpaired *t*-test. Don't stack the data into one column and use the Fit Y by X One-way Anova on the ID without specifying a block or matching column. To do this has the effect of ignoring the correlation between the responses, which causes the test to overestimate the effect if responses are negatively correlated, or to underestimate the effect if responses are positively correlated.

Graphics for Matched Pairs

The primary graph in the platform is a plot of the difference of the two responses on the *y*-axis, and the mean of the two responses on the *x*-axis. This graph is the same as a scatterplot of the two original variables, but turned 45 degrees. A 45 degree rotation turns the original coordinates into a difference and a sum. By rescaling, this plot can show a difference and a mean, as illustrated in Figure 10.1.

Figure 10.1 Transforming to Difference by Sum is a Rotation by 45-degrees

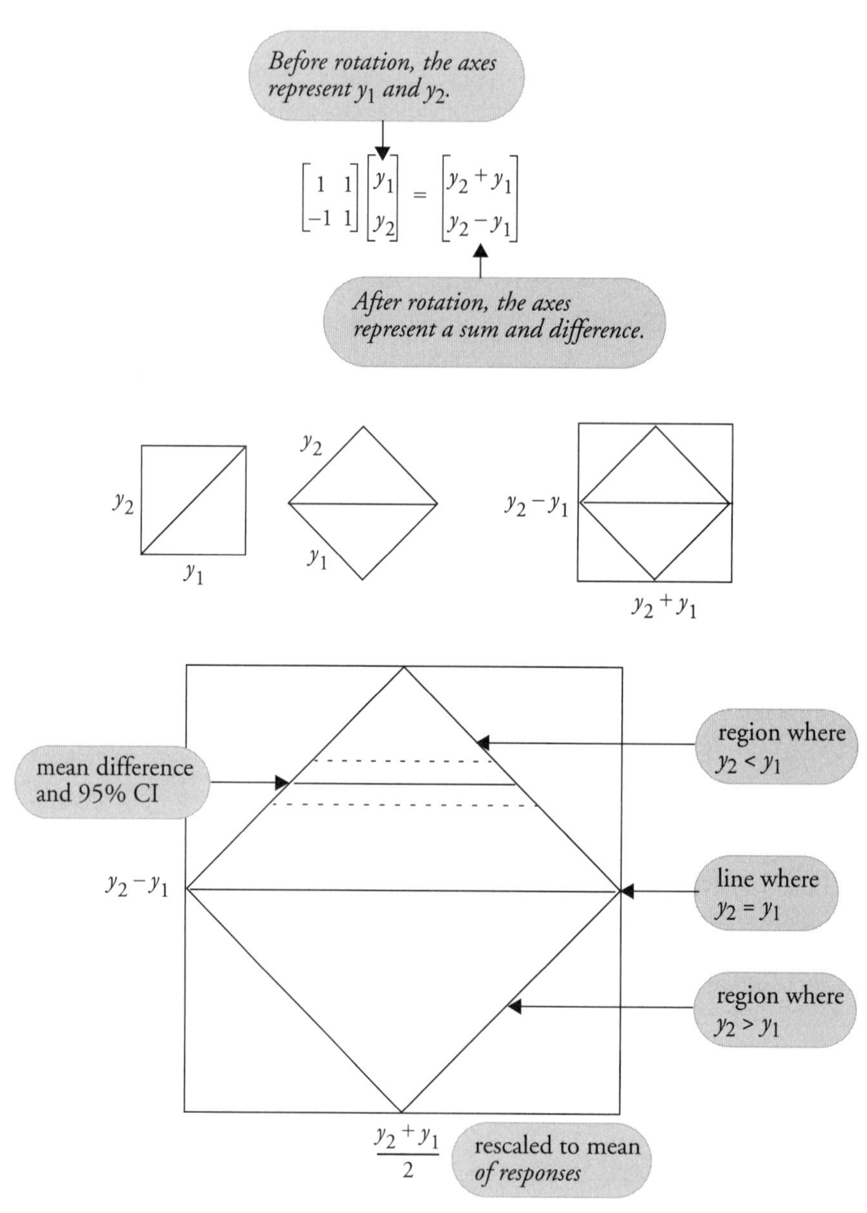

The following example uses the Dogs.jmp sample data set. Suppose you want to study a pair of measurements, called LogHist0 and LogHist1, taken on the same subject at different times. To illustrate the study method, the Dogs.jmp data has two new columns created with the formula editor, which are the difference and the mean of the two response columns.

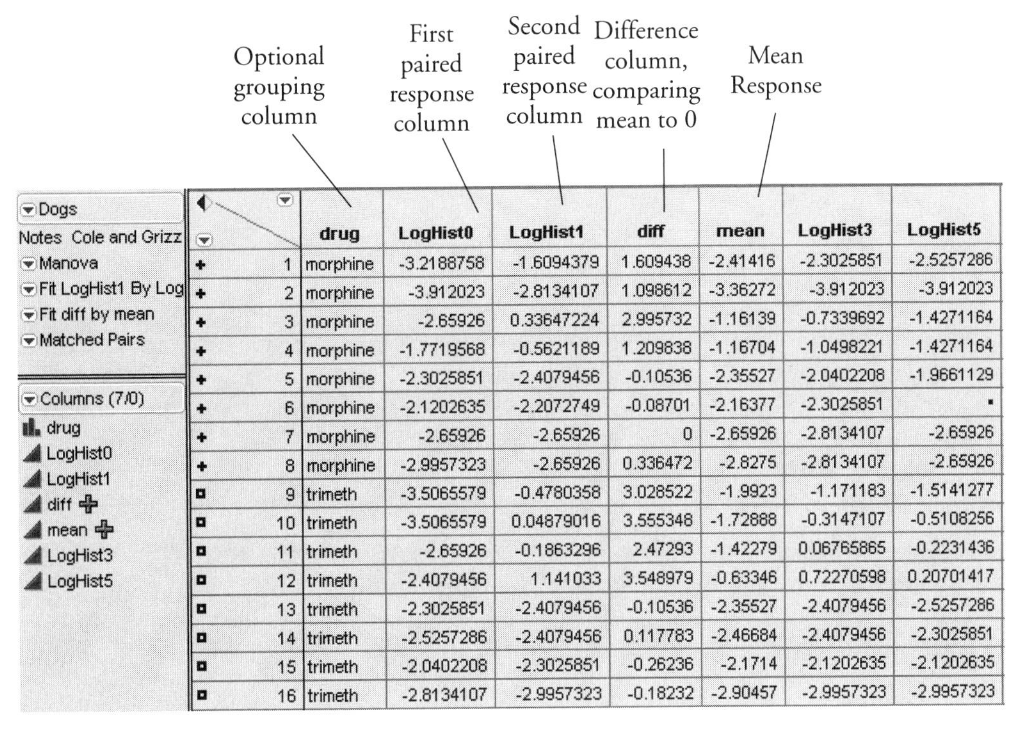

		drug	LogHist0	LogHist1	diff	mean	LogHist3	LogHist5
+	1	morphine	-3.2188758	-1.6094379	1.609438	-2.41416	-2.3025851	-2.5257286
+	2	morphine	-3.912023	-2.8134107	1.098612	-3.36272	-3.912023	-3.912023
+	3	morphine	-2.65926	0.33647224	2.995732	-1.16139	-0.7339692	-1.4271164
+	4	morphine	-1.7719568	-0.5621189	1.209838	-1.16704	-1.0498221	-1.4271164
+	5	morphine	-2.3025851	-2.4079456	-0.10536	-2.35527	-2.0402208	-1.9661129
+	6	morphine	-2.1202635	-2.2072749	-0.08701	-2.16377	-2.3025851	•
+	7	morphine	-2.65926	-2.65926	0	-2.65926	-2.8134107	-2.65926
+	8	morphine	-2.9957323	-2.65926	0.336472	-2.8275	-2.8134107	-2.65926
□	9	trimeth	-3.5065579	-0.4780358	3.028522	-1.9923	-1.171183	-1.5141277
□	10	trimeth	-3.5065579	0.04879016	3.555348	-1.72888	-0.3147107	-0.5108256
□	11	trimeth	-2.65926	-0.1863296	2.47293	-1.42279	0.06765865	-0.2231436
□	12	trimeth	-2.4079456	1.141033	3.548979	-0.63346	0.72270598	0.20701417
□	13	trimeth	-2.3025851	-2.4079456	-0.10536	-2.35527	-2.4079456	-2.5257286
□	14	trimeth	-2.5257286	-2.4079456	0.117783	-2.46684	-2.4079456	-2.3025851
□	15	trimeth	-2.0402208	-2.3025851	-0.26236	-2.1714	-2.1202635	-2.1202635
□	16	trimeth	-2.8134107	-2.9957323	-0.18232	-2.90457	-2.9957323	-2.9957323

Now, using the Bivariate platform (**Fit Y by X**), look at scatterplots of LogHist1 by LogHist0, and their difference by their mean in Figure 10.2. Note that they are the same configuration of points rotated 45 degrees.

Figure 10.2 Bivariate Plots Show Rotation

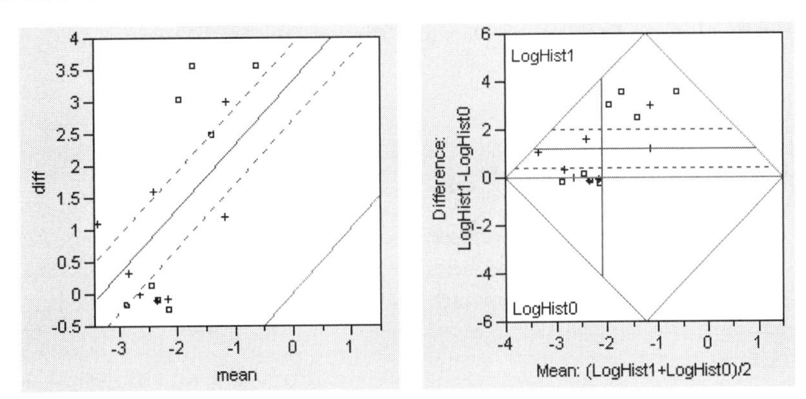

Matched Pairs Platform Output

The Matched Pairs platform combines the ideas presented in the previous sections. The graph is the scatterplot of difference by mean, which is the scatterplot on the columns turned 45 degrees. The paired *t*-test is the same as a test on the difference column.

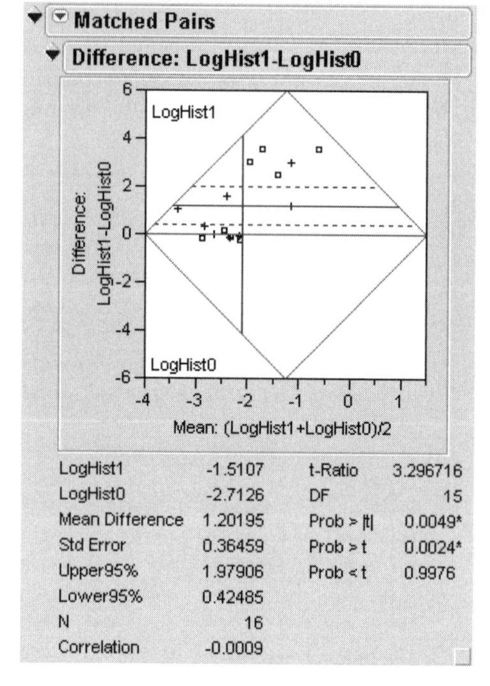

In the graph, notice the following:

- The 45-degree tilted square shows the frame of the tilted scatterplot of the original columns.
- The mean difference is shown as the horizontal line, with the 95% confidence interval above and below. If the confidence region includes the horizontal line at zero, then the means are not significantly different at the 0.05 level. In the example here, the difference is significant.
- The mean of the mean of pairs is shown by the vertical line.

Launching the Matched Pairs Platform

To see the results shown in the previous section, choose **Matched Pairs** from the **Analyze** menu. Then complete the Launch dialog as shown here. Two or more response columns must be chosen. One grouping variable can be selected, in which case the paired difference is analyzed across the grouping.

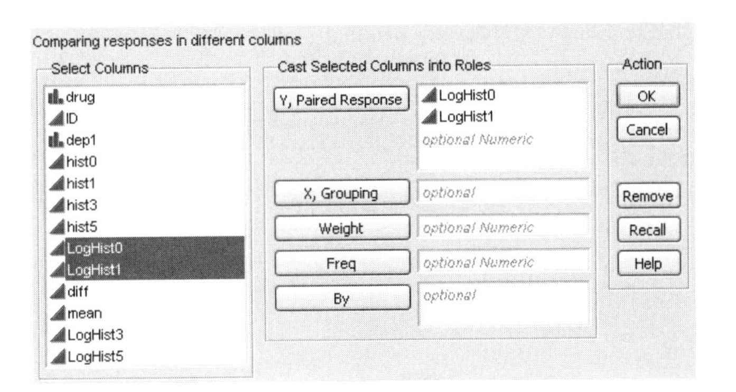

Grouping Variable

If the matched pair produces different means in different groups, you can specify that variable as **X, Grouping** when you launch the Matched Pairs Launch dialog. When there is a grouping variable specified, the platform estimates means and tests both between and among the pairs.

In the following example, the Drug variable, with values trimeth and morphine, is specified as a grouping column.

Interaction Figure 10.3 shows the results of the grouped analysis of paired data. The Mean Difference column in the report table shows the mean across rows in each group of the difference between the two paired columns. In this example, Mean Difference is the mean of (LogHist1-LogHist0) for each group, morphine and trimeth. In other terms, this is the within-subject by across-subject interaction, or split-plot by whole-plot interaction.

Figure 10.3 Matched Pairs Analysis of Grouped Data

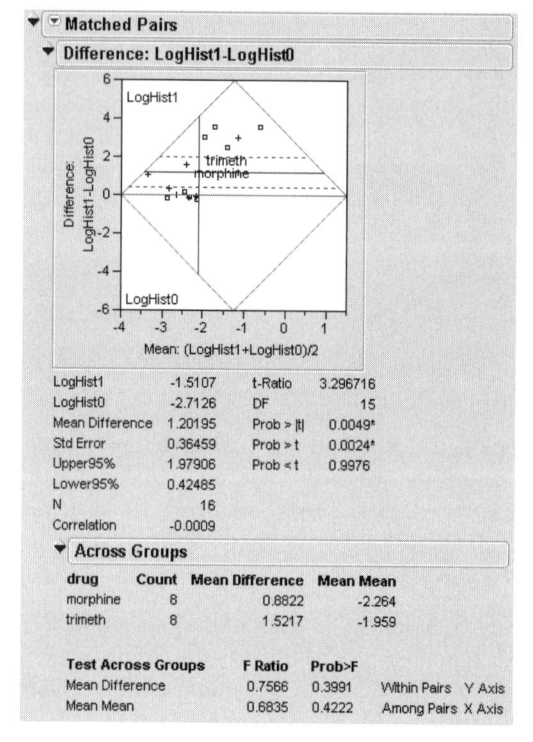

Across Subjects The **Mean Mean** report column in the Across Groups table shows the Mean across rows in each group of the Mean across the two paired columns. In this example, **Mean Mean** is mean of (LogHist1+LogHist0)/2 for each group, morphine and trimeth. In other terms, this is the across-subject or whole-plot effect.

Across Groups There are two *F*-tests that test the across group values are different. These tests correspond to the **Mean Difference** and **Mean Mean** values:

Mean Difference is testing that the change across the pair of responses is different in different groups.

Mean Mean is testing that the average response for a subject is different in different groups.

The analysis shown previously in Figure 10.3 corresponds exactly to a simple repeated measures analysis. If you use the **Manova** personality of the Fit Model platform to do a repeated measures analysis with the paired columns as two responses, one drug factor, and specify the repeated measures response function, you get the same test results.

The MANOVA results are shown here. The drug effect test on the Difference is the Within Subjects time*drug interaction *F*-test. The drug effect test on the mean is the Between Subjects *F*-test on drug.

Manova Fit

Between Subjects

drug

Test	Value	Exact F	NumDF	DenDF	Prob>F
F Test	0.048824	0.6835	1	14	0.4222

Within Subjects

Time*drug

Test	Value	Exact F	NumDF	DenDF	Prob>F
F Test	0.0540433	0.7566	1	14	0.3991
Univar unadj Epsilon=	1	0.7566	1	14	0.3991
Univar G-G Epsilon=	1	0.7566	1	14	0.3991
Univar H-F Epsilon=	1	0.7566	1	14	0.3991

These results can also be obtained using a random-effects model in the Fit Model platform using the EMS approach. In that case, the responses have to be stacked into one column that is identified by an ID column.

Correlation of Responses

In general, if the responses are positively correlated, the paired *t*-test gives a more significant *p* value than the *t*-test for independent means. If responses are negatively correlated, then the paired *t*-test will be less significant than the independent *t*-test. In most cases where the pair of measurements are taken from the same individual at different times, they are positively correlated. However, if they represent competing responses, the correlation can be negative.

Figure 10.4 shows how the positive correlation of the two responses becomes the small variance on the difference (the *y*-axis). If the correlation is negative, the ellipse is oriented in the other direction and the variance of the rotated graph is large on the *y*-axis.

Figure 10.4 Positive Correlation Before and After Rotation Shows That the Variance of the Difference Is Small

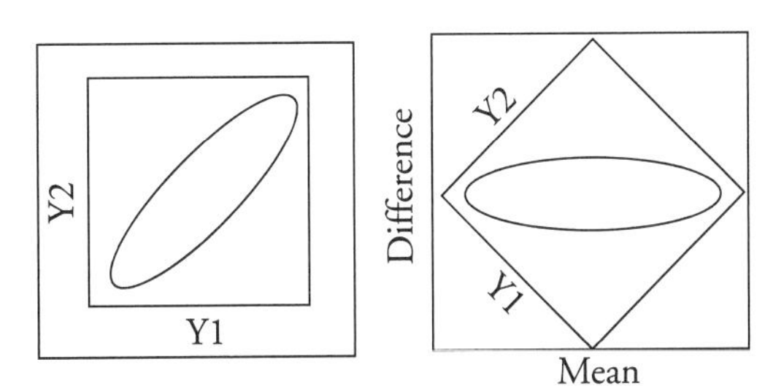

Comparison of Paired *t*-test and *t*-Test of Independent Means

Figure 10.5 shows the results of three *t*-tests on the Dogs data. Matched pairs analysis and comparing the difference between the paired columns to zero give the same correct results. Doing a *t*-test for independent groups on stacked data gives incorrect results because it ignores the correlation between the responses.

Figure 10.5 Compare Matched Pairs Analysis and Independent *t*-Test

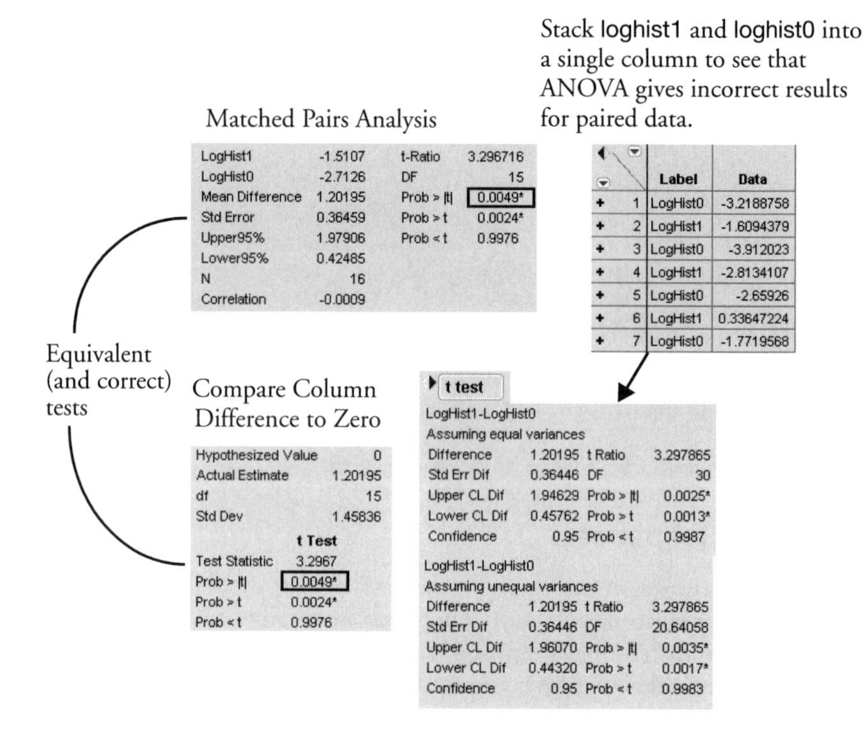

Multiple Y Columns

You can have more than two matched pairs. The matched pairs platform presents a separate analysis with plots on all possible pairings of the response columns. For an example, see the following matrix of analyses arrangement:

Figure 10.6

Y2 by Y1	Y3 by Y1	Y4 by Y1
	Y3 by Y2	Y4 by Y2
		Y4 by Y3

If there are an even number of responses, the Matched Pairs platform prompts you and asks if you would rather do a series of pairs instead of all possible pairs, which gives the following arrangement:

Figure 10.7

Y2 by Y1	Y4 by Y3

Situations for the Paired *t*-Test Plot

There are many possibilities for making statements regarding the patterns to be discovered in the new coordinates. The examples in Figure 10.8, through Figure 10.12, show five different situations and their interpretations.

Figure 10.8 No Change

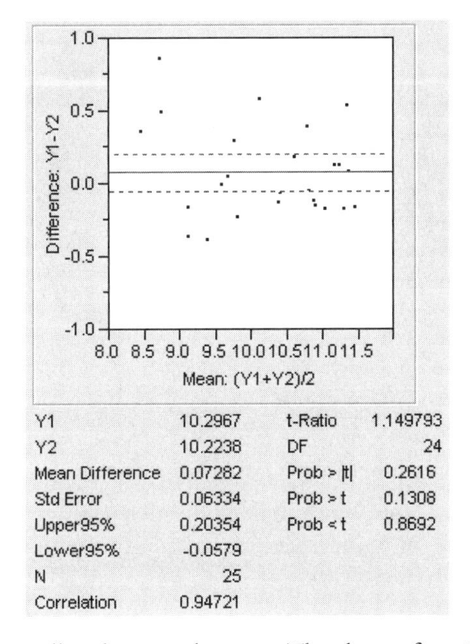

Y1	10.2967	t-Ratio	1.149793
Y2	10.2238	DF	24
Mean Difference	0.07282	Prob > |t|	0.2616
Std Error	0.06334	Prob > t	0.1308
Upper95%	0.20354	Prob < t	0.8692
Lower95%	-0.0579		
N	25		
Correlation	0.94721		

The distribution vertically is small and centered at zero. The change from Y1 to Y2 is not significant. This is the high-positive-correlation pattern of the original scatterplot and is typical.

Figure 10.9 Highly Significant Shift Down

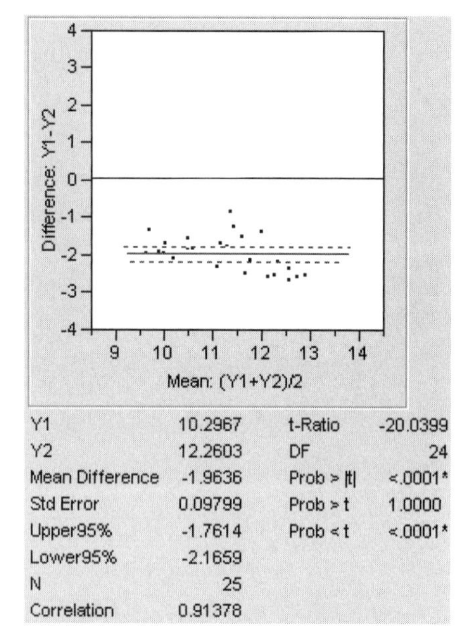

Y1	10.2967	t-Ratio	-20.0399
Y2	12.2603	DF	24
Mean Difference	-1.9636	Prob > |t|	<.0001*
Std Error	0.09799	Prob > t	1.0000
Upper95%	-1.7614	Prob < t	<.0001*
Lower95%	-2.1659		
N	25		
Correlation	0.91378		

The Y2 score is consistently lower than Y1 across all subjects.

Figure 10.10 No Average Shift, But Amplified Relationship

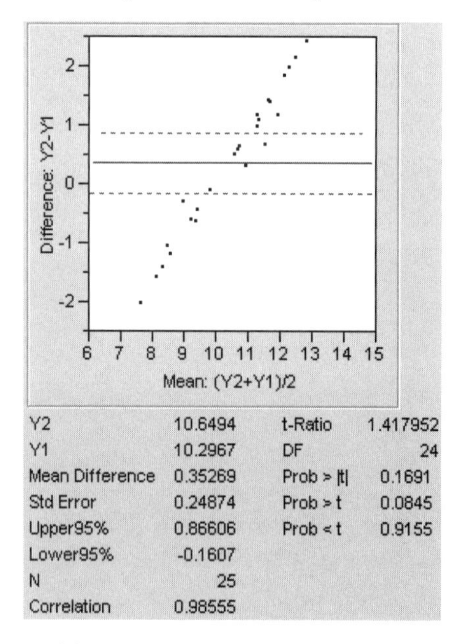

Y2	10.6494	t-Ratio	1.417952
Y1	10.2967	DF	24
Mean Difference	0.35269	Prob > |t|	0.1691
Std Error	0.24874	Prob > t	0.0845
Upper95%	0.86606	Prob < t	0.9155
Lower95%	-0.1607		
N	25		
Correlation	0.98555		

Low variance of the difference, and high variance of the mean of the two values within a subject. Overall the mean is the same from Y1 to Y2, but individually, the high scores got higher and the low scores got lower.

Figure 10.11 No Average Shift, But Reverse Relationship

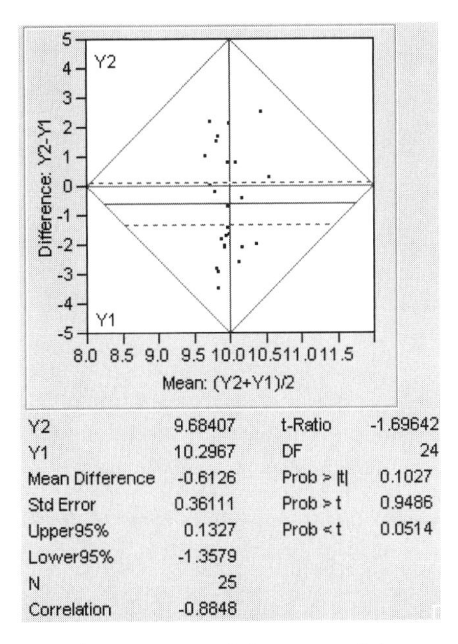

Y2	9.68407	t-Ratio	-1.69642
Y1	10.2967	DF	24
Mean Difference	-0.6126	Prob > |t|	0.1027
Std Error	0.36111	Prob > t	0.9486
Upper95%	0.1327	Prob < t	0.0514
Lower95%	-1.3579		
N	25		
Correlation	-0.8848		

High variance of the difference, and low variance of the mean of the two values within a subject. Overall the mean was the same from Y1 to Y2, but the high Y1s got low Y2s and vice-versa. This is due to the high-negative-correlation pattern of the original scatterplot and is unusual.

Figure 10.12 No Average Shift, Positive Correlation, but Damped Instead of Accelerated

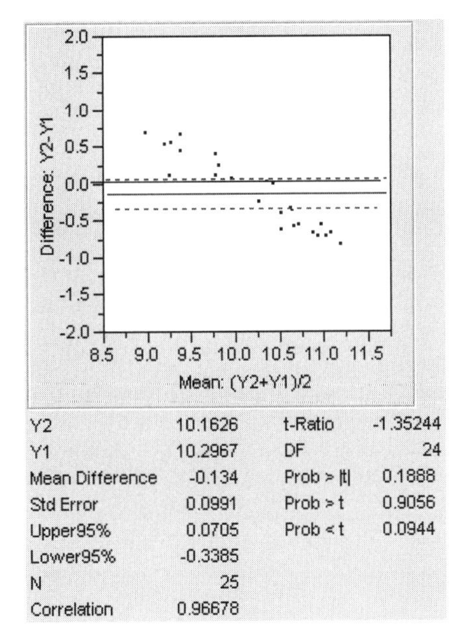

Y2	10.1626	t-Ratio	-1.35244
Y1	10.2967	DF	24
Mean Difference	-0.134	Prob > |t|	0.1888
Std Error	0.0991	Prob > t	0.9056
Upper95%	0.0705	Prob < t	0.0944
Lower95%	-0.3385		
N	25		
Correlation	0.96678		

Overall the mean is the same from Y1 to Y2, but the high scores drop a little, and low scores increase a little.

Chapter 11

Introduction to Model Fitting

The Model Specification Dialog

Analyze > Fit Model launches the *general fitting platform*, which lets you construct linear models that have more complex effects than those assumed by other JMP statistical platforms. **Fit Model** displays the Model Specification dialog (sometimes called the Fit Model dialog for short) that lets you define complex models. You choose specific model effects and error terms and add or remove terms from the model specification as needed.

The Model Specification dialog is a unified launching pad for a variety of fitting *personalities* such as:

- standard least squares fitting of one or more continuous responses (multiple regression)
- screening analysis for experimental data where there are many effects but few observations
- stepwise regression
- logistic regression for nominal or ordinal response
- multivariate analysis of variance (MANOVA) for multiple continuous responses
- log-linear variance fitting to fit both means and variances
- proportional hazard fits for censored survival data.

This chapter describes the Model Specification dialog in detail and defines the report surface display options and save commands available with various statistical analyses. The chapters "Standard Least Squares: Introduction," p. 213, "Standard Least Squares: Perspectives on the Estimates," p. 249, "Standard Least Squares: Exploring the Prediction Equation," p. 275, "Standard Least Squares: Random Effects," p. 339, "Stepwise Regression," p. 385, and "Logistic Regression for Nominal and Ordinal Response," p. 403, and "Multiple Response Fitting," p. 427, discuss standard least squares and give details and examples for each Fit Model personality.

Contents

The Model Specification Dialog: A Quick Example

The **Fit Model** command first displays the Model Specification dialog, shown in Figure 11.1. You use this dialog to tailor a model for your data. The Model Specification dialog is *nonmodal*, which means it persists until you explicitly close it. This is useful to change the model specification and fit several different models before closing the window.

The example in Figure 11.1, uses the Fitness.jmp (SAS Institute 1987) data table in the Sample Data folder. The data are results of an aerobic fitness study. The Oxy variable is a continuous response (dependent) variable. The variables RunTime, RunPulse, and RstPulse that show in the Construct Model Effects list are continuous effects (also called regressors, factors, or independent variables). The popup menu at the upper-right of the dialog shows **Standard Least Squares**, which defines the fitting method or fitting *personality*. The various fitting personalities are described later in this chapter.

Figure 11.1 The Model Specification Dialog Completed for a Multiple Regression

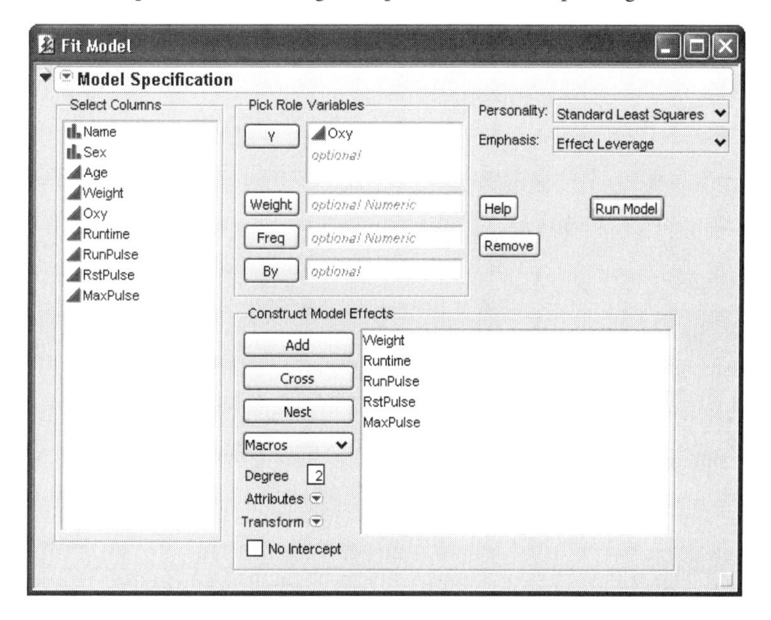

This standard least squares example, with a single continuous *Y* variable and several continuous *X* variables, specifies a multiple regression.

After a model is defined in the Model Specification dialog, click **Run Model** to perform the analysis and generate a report window with the appropriate tables and supporting graphics.

A standard least squares analysis such as this multiple regression begins by showing you the Summary of Fit table, the Parameter Estimates table, the Effect Tests table, Analysis of Variance, Residual by Predicted plots, and leverage plots. If you want, you can open additional tables and plots, such as those shown in Figure 11.2, to see details of the analysis. Or, if a screening or response surface analysis seems more appropriate, you can choose commands from the Effect Screening and Factor Profiling menus at the top of the report.

All tables and graphs available on the Fit Model platform are discussed in detail in the chapters "Standard Least Squares: Introduction," p. 213, and "Standard Least Squares: Perspectives on the Estimates," p. 249, and "Standard Least Squares: Exploring the Prediction Equation," p. 275.

See Table 11.1 "Types of Models," p. 198, and Table 11.2 "Clicking Sequences for Selected Models," p. 199, in the next section for a description of models and the clicking sequences needed to enter them into the Model Specification dialog.

Figure 11.2 Partial Report for Multiple Regression

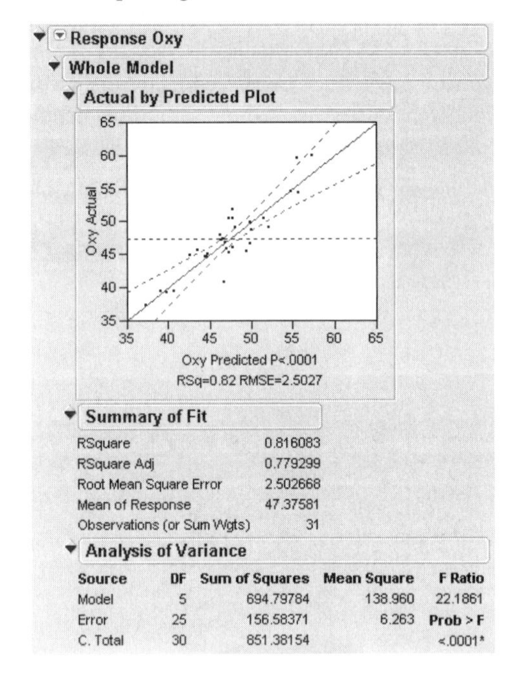

Types of Models

The list in Table 11.1 "Types of Models," p. 198 is a catalog of model examples that can be defined using the Model Specification dialog, where the effects *X* and *Z* have continuous values, and *A*, *B*, and *C* have nominal or ordinal values. This list is not exhaustive.

When you correctly specify the type of model, the model effects show in the **Construct Model Effects** list on the Model Specification dialog. Refer to Table 11.2 "Clicking Sequences for Selected Models," p. 199 to see the clicking sequences that produce some of these sets of model effects.

Table 11.1 Types of Models

Type of Model	Model Effects
simple regression	*X*
polynomial (*x*) to Degree 2	X, X*X
polynomial (*x*, *z*) to Degree 2	X, X*X, Z, Z*Z

Table 11.1 Types of Models

Type of Model	Model Effects
cubic polynomial (*x*)	*X, X*X, X*X*X*
multiple regression	*X, Z*, …, other continuous columns
one-way analysis of variance	*A*
two-way main effects analysis of variance	*A, B*
two-way analysis of variance with interaction	*A, B, A*B*
three-way full factorial	*A, B, A*B, C, A*C, B*C, A*B*C*
analysis of covariance, same slopes	*A, X*
analysis of covariance, separate slopes	*A, X, A*X*
simple nested model	*A, B[A]*
compound nested model	*A, B[A], C[A B]*
simple split-plot or repeated measure	*A, B[A]&Random, C, A*C*
response surface (*x*) model	*X&RS, X*X*
response surface (*x, z*) model	*X&RS, Z&RS, X*X, Z*Z, X*Z*
MANOVA	multiple *Y* variables

The following convention is used to specify clicking:

- If a column name is in plain typeface, click the name in the column selection list.
- If a column name is **bold**, then select that column in the dialog model effects list.
- The name of the button to click is shown in all CAPITAL LETTERS.

Table 11.2 Clicking Sequences for Selected Models

Type of Model	Clicking Sequence
simple regression	X, ADD
polynomial (*x*) to Degree 2	X, **Degree** 2, select **Polynomial to Degree** in the **Macros** popup menu
polynomial (*x, z*) to Degree 3	X, Z, **Degree** 3, select **Polynomial to Degree** in the **Macros** popup menu
multiple regression	X, ADD, Z, ADD,… or X, Z, …, ADD
one-way analysis of variance	A, ADD
two-way main effects analysis of variance	A, ADD B, ADD, or A, B, ADD
two-way analysis of variance with interaction	A, B, ADD, A, B, CROSS or A, B, select **Full Factorial** in **Macros** popup menu
three-way full factorial	A, B, C, select **Full Factorial** in **Macros** popup menu
analysis of covariance, same slopes	A, ADD, X, ADD or A, X, ADD
analysis of covariance, separate slopes	A, ADD, X, ADD, A, X, CROSS
simple nested model	A, ADD, B, ADD, A, **B**, NEST, or A, B, ADD, A, **B**, NEST

Table 11.2 Clicking Sequences for Selected Models (continued)

Type of Model	Clicking Sequence
compound nested model	A, B, ADD, A, **B**, NEST, C, ADD, A, B, **C**, NEST
simple split-plot or repeated measure	A, ADD, B, ADD, A, **B**, NEST, select **Random** from the **Effect Attributes** popup menu, C, ADD, C, A, CROSS
two-factor response surface	X, Z, select **Response Surface** in the **Macros** popup menu

The Response Buttons (Y, Weight, and Freq)

The column selection list in the upper left of the dialog lists the columns in the current data table. When you click a column name, it highlights and responds to the action that you choose with other buttons on the dialog. Either drag across columns or Shift-click to extend a selection of column names, and Control-click (⌘-click on the Macintosh) to select non-adjacent names.

To assign variables to the **Y**, **Weight**, or **Freq** roles, select them and click the corresponding role button.

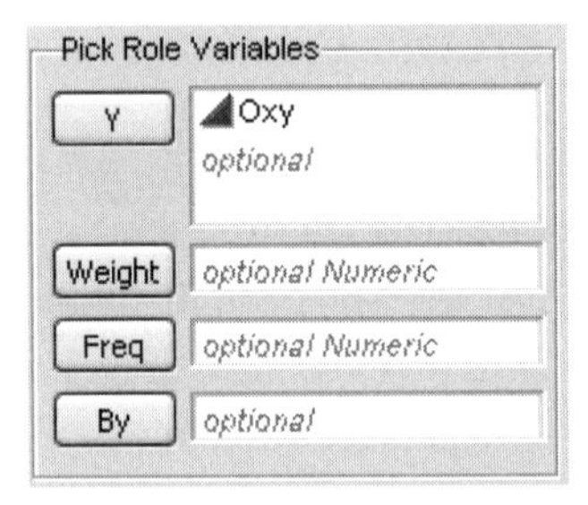

- **Y** identifies one or more response variables (the dependent variables) for the model.
- **Weight** is an optional role that identifies one column whose values supply weights for the response. Weighting affects the importance of each row in the model.
- **Freq** is an optional role that identifies one column whose values assign a frequency to each row for the analysis. The values of this variable determine how degrees of freedom are counted.

If you want to remove variables from roles, highlight them and click **Remove** or hit the backspace key.

Model Effects Buttons

Suppose that a data table contains the variables *X* and *Z* with continuous values, and *A*, *B*, and *C* with nominal values. The following paragraphs describe the buttons in the Model Specification dialog that specify model effects.

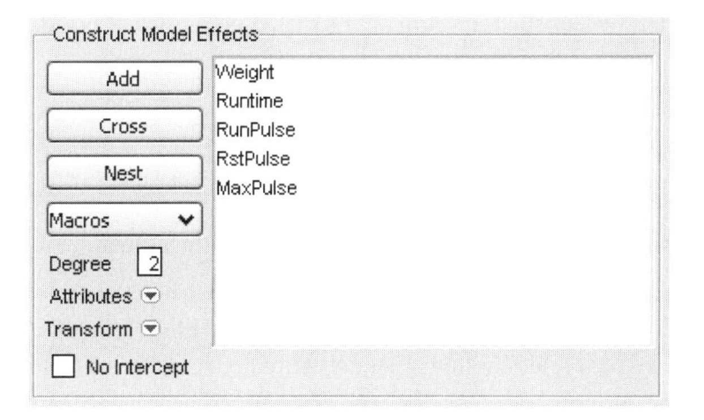

Add

To add a simple regressor (continuous column) or a main effect (nominal or ordinal column) as an x effect to any model, select the column from the column selection list and click **Add**. That column name appears in the model effects list. As you add effects, be aware of the modeling type declared for that variable and the consequences that modeling type has when fitting the model:

- Variables with continuous modeling type become simple regressors.
- Variables with nominal modeling types are represented with dummy variables to fit separate coefficients for each level in the variable.
- Nominal and ordinal modeling types are handled alike, but with a slightly different coding. See the appendix "Statistical Details," p. 975, for details on coding of effects.

If you mistakenly add an effect, select it in the model effects list and click **Remove**.

Cross

To create a compound effect by crossing two or more variables, Shift-click in the column selection list to select them. Control-click (⌘-click on the Macintosh) if the variables are not contiguous in the column selection list. Then click **Cross**:

- Crossed nominal variables become interaction effects.
- Crossed continuous variables become multiplied regressors.
- Crossing a combination of nominal and continuous variables produces special effects suitable for testing homogeneity of slopes.

If you select both a column name in the column selection list and an effect in the model effects list, the **Cross** button crosses the selected column with the selected effect and adds this compound effect to the effects list.

See the appendix "Statistical Details," p. 975, for a discussion of how crossed effects are parameterized and coded.

Nest

When levels of an effect B only occur within a single level of an effect A, then B is said to be *nested* within A and A is called the *outside effect*. The notation B[A] is read "B within A" and means that you want to fit a B effect within each level of A. The B[A] effect is like pooling B+A*B. To specify a nested effect

- select the outside effects in the column selection list and click **Add** or **Cross**
- select the nested effect in the column selection list and click **Add** or **Cross**
- select the outside effect in the column selection list
- select the nested variable in the model effects list and click **Nest**

For example, suppose that you want to specify A*B[C]. Highlight both A and B in the column selection list. Then click **Cross** to see A*B in the model effects list. Highlight the A*B term in the model effects list and C in the column selection list. Click **Nest** to see A*B[C] as a model effect.

Note: The nesting terms must be specified in order, from outer to inner. If B is nested within A, and C is nested within B, then the model is specified as: A, B[A], C[B,A].

JMP allows up to ten terms to be combined as crossed and nested.

Macros Popup Menu

Commands from the **Effect Macros** popup menu, shown here, automatically generate the effects for commonly used models. You can add or remove terms from an automatic model specification as needed.

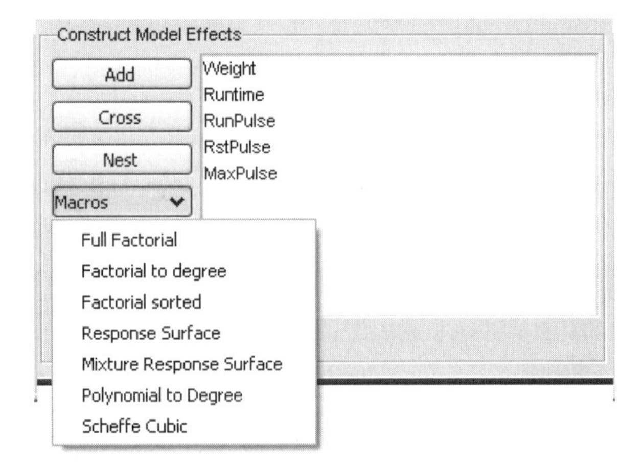

Full Factorial To look at many crossed factors, such as in a factorial design, use **Full Factorial.** It creates the set of effects corresponding to all crossings of all variables you select in the columns list. For example, if you select variables A, B, and C, the **Full Factorial** selection generates A, B, A*B, C, A*C, B*C, A*B*C in the effect lists. To remove unwanted model interactions, highlight them and click **Remove**.

Factorial to a Degree To create a limited factorial, enter the degree of interactions that you want in the **Degree** box, then select **Factorial to a Degree**.

Factorial Sorted The **Factorial Sorted** selection creates the same set of effects as **Full Factorial** but lists them in order of degree. All main effects are listed first, followed by all two-way interactions, then all three-way interactions, and so forth.

Response Surface In a response surface model, the object is to find the values of the terms that produce a maximum or minimum expected response. This is done by fitting a collection of terms in a quadratic model. The critical values for the surface are calculated from the parameter estimates and presented with a report on the shape of the surface.

To specify a response surface effect, select two or more variables in the column selection list. When you choose **Response Surface** from the **Effect Macros** popup menu, response surface expressions appear in the model effects list. For example, if you have variables A and B

Surface(A) fits $\beta_1 A + \beta_{11} A^2$

Surface(A,B) fits $\beta_1 A + \beta_{11} A^2 + \beta_2 B + \beta_{22} B^2 + \beta_{12} AB$

The response surface effect attribute, prefixed with an ampersand (&), is automatically appended to the effect name in the model effects list. The next section discusses the **Attributes** popup menu.

Mixture Response Surface The **Mixture Response Surface** design omits the squared terms and the intercept and attaches the **&RS** effect attribute to the main effects. For more information see the *Design of Experiments* Guide.

Polynomial to Degree A polynomial effect is a series of terms that are powers of one variable. To specify a polynomial effect:

- click one or more variables in the column selection list;
- enter the degree of the polynomial in the **Degree** box;
- use the **Polynomial to Degree** command in the **Macros** popup menu.

For example, suppose you select variable x. A 2nd polynomial in the effects list fits parameters for x and x^2. A 3rd polynomial in the model dialog fits parameters for x, x^2, and x^3.

Scheffe Cubic When you fit a 3rd degree polynomial model to a mixture, it turns out that you need to take special care not to introduce even-powered terms, because they are not estimable. When you get up to a cubic model, this means that you can't specify an effect like X1*X1*X2. However, it turns out that a complete polynomial specification of the surface should introduce terms of the form:

X1*X2*(X1 – X2)

which we call *Scheffe Cubic* terms.

In the Fit Model dialog, this macro creates a complete cubic model, including the Scheffe Cubic terms if you either (a) enter a 3 in the Degree box, then do a “Mixture Response Surface” command on a set of mixture columns, or(b) use the **Scheffe Cubic** command in the Macro button.

Effect Attributes and Transformations

The **Effect Attributes** popup menu has five special attributes that you can assign to an effect in a model.

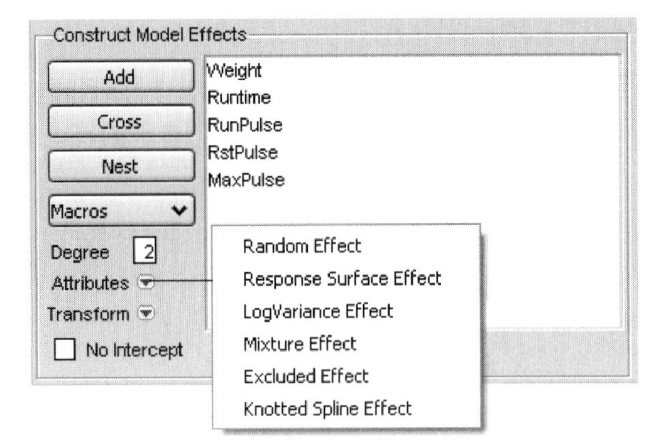

Random Effect If you have multiple error terms or random effects, as with a split-plot or repeated measures design, you can highlight them in the model effects list and select **Random Effect** from the **Attributes** popup menu. See the chapter "Standard Least Squares: Random Effects," p. 339, for a detailed discussion and example of random effects.

Response Surface Effect If you have a response surface model, you can use this attribute to identify which factors participate in the response surface. This attribute is automatically assigned if you use the **Response Surface** effects macro. You need only identify the main effects, not the crossed terms, to obtain the additional analysis done for response surface factors.

Log Variance Effect Sometimes the goal of an experiment is not just to maximize or minimize the response itself but to aim at a target response and achieve minimum variability. To analyze results from this kind of experiment:

1. Assign the response (*Y*) variable and choose **LogLinear Variance** as the fitting personality

2. Specify loglinear variance effects by highlighting them in the model effects list and select **LogVariance Effect** from the **Attributes** popup menu. The effect appears in the model effects list with **&LogVariance** next to its name.

If you want an effect to be used for both the mean and variance of the response, you must specify it as an effect twice, once with the **LogVariance Effect** attribute.

Mixture Effect You can use the **Mixture Effect** attribute to specify a mixture model effect without using the **Mixture Response Surface** effects macro. If you don't use the effects macro you have to add this attribute to the effects yourself so that the model understands which effects are part of the mixture.

Excluded Effect Use the **Excluded Effect** attribute when you want to estimate least squares means of an interaction, or include it in lack of fit calculations, but don't want it in the model.

Knotted Spline Effect Use the **Knotted Spline Effect** attribute to have JMP fit a segmentation of smooth polynomials to the specified effect. When you select this attribute, a dialog box appears that allows you to specify the number of knot points.

Note: Knotted splines are only implemented for main-effect continuous terms.

JMP follows the advice in the literature in positioning the points. If there are 100 or fewer points, the first and last knot are the fifth point inside the minimum and maximum, respectively. Otherwise, the first and last knot are placed at the 0.05 and 0.95 quantile if there are 5 or fewer knots, or the 0.025 and 0.975 quantile for more than 5 knots. The default number of knots is 5 unless there are less than or equal to 30 points, in which case the default is 3 knots.

Knotted splines are similar to smoothing splines. Figure 11.3, using Growth.jmp, contrasts the use of two fitted smoothing splines and a knotted spline.

Figure 11.3 Smoothing Spline vs. Knotted Spline

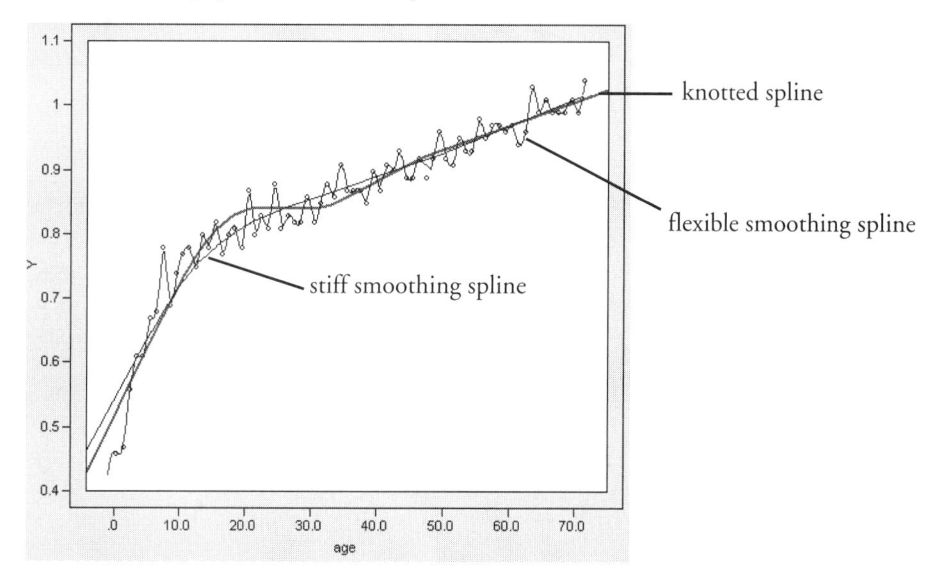

Knotted splines have the following properties in contrast to smoothing splines:

1 Knotted splines work inside of general models with many terms, where smoothing splines are for bivariate regressions.
2 The regression basis is not a function of the response.
3 Knotted splines are parsimonious, adding only $k - 2$ terms for curvature for k knot points.
4 Knotted splines are conservative compared to pure polynomials in the sense that the extrapolation outside the range of the data is a straight line, rather than a (curvy) polynomial.
5 There is an easy test for curvature.

To test for curvature, select **Estimates > Custom Test** and add a column for each knotted effect. To duplicate the curve and test for curvature for the curve in Figure 11.3,

1 Select **Analyze > Fit Model**.
2 Assign ratio to **Y** and add age as an effect to the model.
3 Select the age variable in the Construct Model Effects box and select **Attributes > Knotted Spline**

Effect.

4 Accept the default 5 knots.

5 Click **Run Model.**

When the report appears,

6 Select **Estimates > Custom Test** and notice that there are three knotted columns. Therefore, click the **Add Column** button twice to produce a total of three columns.

7 Fill in the columns so they match Figure 11.4.

Figure 11.4 Construction of Custom Test for Curvature

Custom Test

Parameter			
Intercept	0	0	0
age&Knotted	0	0	0
age&Knotted@4.5	1	0	0
age&Knotted@17.1	0	1	0
age&Knotted@29.7	0	0	1
=	0	0	0

Click and Type Above to form hypothesis test.

Done | Add Column | Help

8 Click **Done.**

This produces the report shown in Figure 11.5. The low Prob > F value indicates that there is indeed curvature to the data.

Figure 11.5 Curvature Report

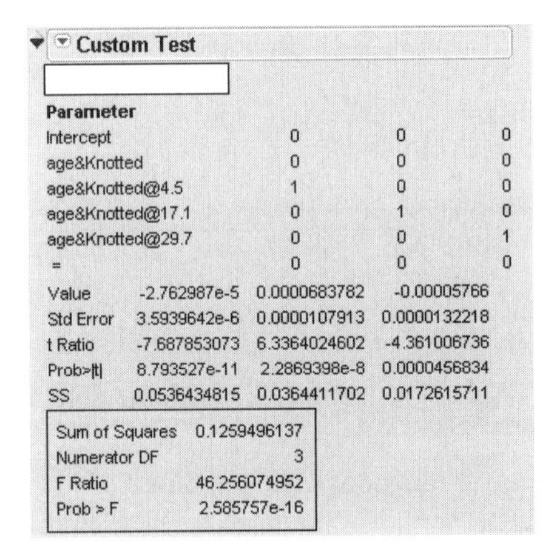

Custom Test

Parameter			
Intercept	0	0	0
age&Knotted	0	0	0
age&Knotted@4.5	1	0	0
age&Knotted@17.1	0	1	0
age&Knotted@29.7	0	0	1
=	0	0	0
Value	-2.762987e-5	0.0000683782	-0.00005766
Std Error	3.5939642e-6	0.0000107913	0.0000132218
t Ratio	-7.687853073	6.3364024602	-4.361006736
Prob>|t|	8.793527e-11	2.2869398e-8	0.0000456834
SS	0.0536434815	0.0364411702	0.0172615711

Sum of Squares	0.1259496137
Numerator DF	3
F Ratio	46.256074952
Prob > F	2.585757e-16

Transformations (Standard Least Squares Only)

The **Transformations** popup menu has eight functions available to transform selected continuous effects or *Y* columns for standard least squares analyses only. The available transformations are **None, Log, Sqrt, Square, Recipocal, Exp, Arrhenius, ArrheniusInv.** These transformations are only supported for single-column continuous effects.

The Arrhenius transformation is $T^* = \frac{11605}{Temp + 273.15}$

Fitting Personalities

The Model Specification dialog in JMP serves many different fitting methods. Rather than have a separate dialog for each method, there is one dialog with a choice of *fitting personality* for each method. Usually the personality is chosen automatically from the context of the response and factors you enter, but you can change selections from the **Fitting Personality** popup menu to alternative methods.

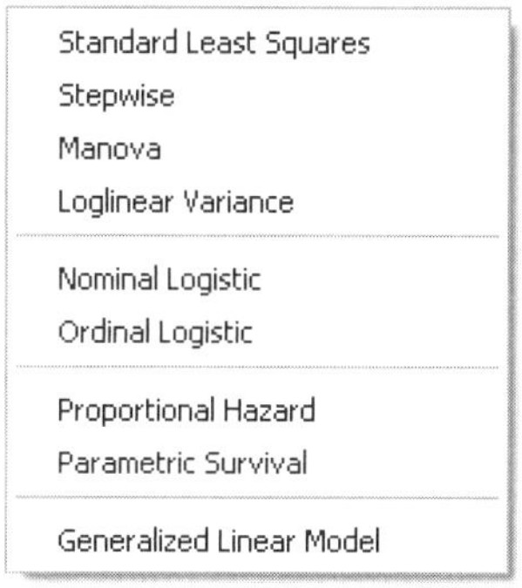

Standard Least Squares
Stepwise
Manova
Loglinear Variance

Nominal Logistic
Ordinal Logistic

Proportional Hazard
Parametric Survival

Generalized Linear Model

The following list briefly describes each type of model. Details about text reports, plots, options, special commands, and example analyses are found in the individual chapters for each type of model fit:

Standard Least Squares JMP models one or more continuous responses in the usual way through fitting a linear model by least squares. The standard least squares report platform offers two flavors of tables and graphs:

Traditional statistics are for situations where the number of error degrees of freedom allows hypothesis testing. These reports include leverage plots, least squares means, contrasts, and output formulas.

Screening and Response Surface Methodology analyze experimental data where there are many effects but few observations. Traditional statistical approaches focus on the residual error. However, because in near-saturated designs there is little or no room for estimating residuals, the focus is on the prediction equation and the effect sizes. Of the many effects that a screening design can fit, you expect a few important terms to stand out in comparison to the others. Another example is when the goal of the experiment is to optimize settings rather than show statistical significance, the factor combinations that optimize the predicted response are of overriding interest.

Stepwise Stepwise regression is an approach to selecting a subset of effects for a regression model. The **Stepwise** feature computes estimates that are the same as those of other least squares platforms, but it facilitates searching and selecting among many models. The **Stepwise** personality allows only one continuous *Y*.

Manova When there is more than one *Y* variable specified for a model with the **Manova** fitting personality selected, the Fit Model platform fits the *Y*'s to the set of specified effects and provides multivariate tests.

LogLinear Variance LogLinear Variance is used when one or more effects model the variance rather than the mean. The **LogLinear Variance** personality must be used with caution. See "Fitting Dispersion Effects with the LogLinear Variance Model," p. 375 for more information on this feature.

Nominal Logistic If the response is nominal, the Fit Model platform fits a linear model to a multilevel logistic response function using maximum likelihood.

Ordinal Logistic If the response variable has an ordinal modeling type, the Fit Model platform fits the cumulative response probabilities to the logistic distribution function of a linear model by using maximum likelihood.

Proportional Hazard The proportional hazard (Cox) model is a special fitting personality that lets you define models having failure time as the response variable with right-censoring and time-independent covariates. Often, you specify the covariate of interest as a grouping variable that defines subpopulations or strata.

Parametric Survival Parametric Survival performs the same analysis as the **Fit Parametric Survival** command on the **Analyze > Survival and Reliability** menu. See the chapter "Survival and Reliability Analysis I," p. 605, for details.

Generalized Linear Model fits generalized linear models with various distribution and link functions. See the chapter "Generalized Linear Models," p. 357 for a complete discussion of generalized linear models

Table 11.3 Characteristics of Fitting Personalities

Personality	Response (Y) Type	Notes
Standard Least Squares	≥ 1 continuous	all effect types
Stepwise	1 continuous	all effect types
MANOVA	> 1 continuous	all effect types
LogLinear Variance	1 continuous	variance effects
Nominal Logistic	1 nominal	all effect types
Ordinal Logistic	1 ordinal	all effect types
Proportional Hazard	1 continuous	survival models only
Generalized Linear Model	continuous, nominal, ordinal	all effect types

Other Model Dialog Features

Emphasis Choices

The **Emphasis** popup menu controls the type of plots you see as part of the initial analysis report.

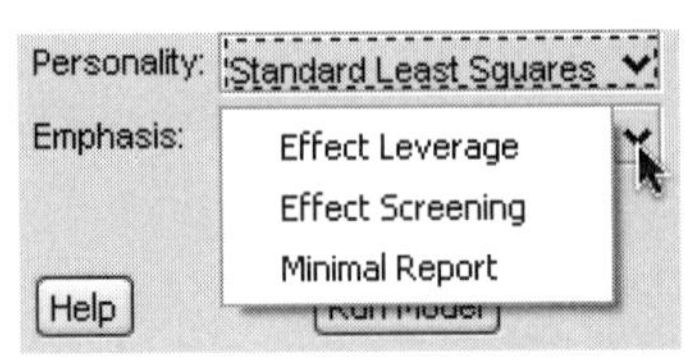

Effect Leverage begins with leverage and residual plots for the whole model. You can then request effect details and other statistical reports.

Effect Screening shows whole model information followed by a scaled parameter report with graph and the Prediction Profiler.

Minimal Report suppresses all plots except the regression plot. You request what you want from the platform popup menu.

Run Model

The **Run Model** button submits the model to the fitting platform. Because the Model Specification dialog is nonmodal, the **Run Model** button does not close the dialog window. You can continue to use the dialog and make changes to the model for additional fits. You can also make changes to the data and then refit the same model.

Validity Checks

Fit Model checks your model for errors such as duplicate effects or missing effects in a hierarchy. If you get an alert message, you can either **Continue** the fitting despite the situation, or click **Cancel** in the alert message to stop the fitting process.

In addition, your data may have missing values. The default behavior is to exclude rows if any *Y* or *X* value is missing. This can be wasteful of rows for cases when some *Y*'s have non-missing values and other *Y*'s are missing. Therefore, you may consider fitting each *Y* separately.

When this situation occurs, you are alerted with the dialog shown in Figure 11.6. This gives you the option to fit each *Y* separately. A consequence of fitting different *Y*'s separately is that the profilers then only handle one *Y* at a time, and saved scripts only run one *Y* at a time. In addition, calculations for larger problems may not be quite as fast.

Figure 11.6 Missing Values Alert

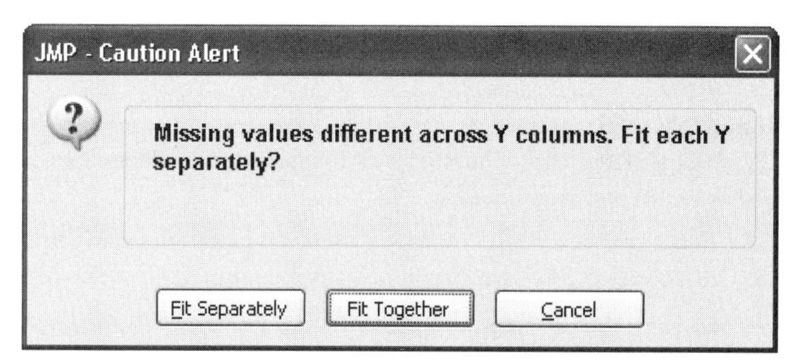

Note: This dialog only appears when the model is run interactively. Scripts continue to use the default behavior, unless `Fit Separately` is placed inside the `Run Model` command.

Other Model Specification Options

The popup menu on the title bar of the Model Specification window gives additional options.

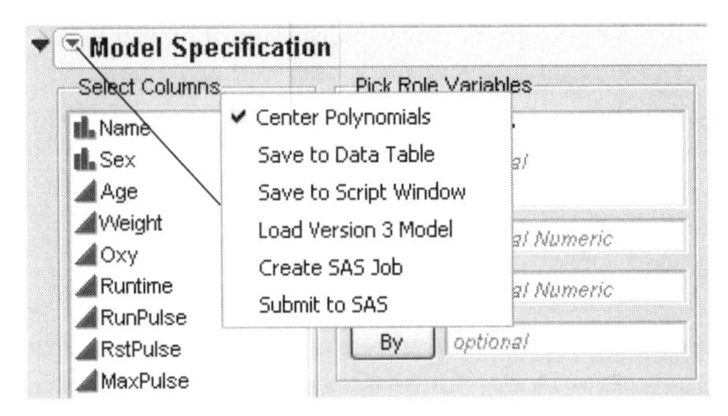

Center Polynomials causes a continuous term participating in a crossed term to be centered by its mean. Exceptions to centering are effects with coded or mixture properties. This option is important to make main effects tests be meaningful hypotheses in the presence of continuous crossed effects.

Save to Data Table saves the model as a property of the current data table. A **Model** popup menu icon appears in the Tables panel at the left of the data grid with a **Run Script** command. If you then select this **Run Script** command to submit the model, a new completed Model Specification dialog appears. The *JMP Scripting Guide* is the reference for JSL statements.

Figure 11.7 **Run Script** Command

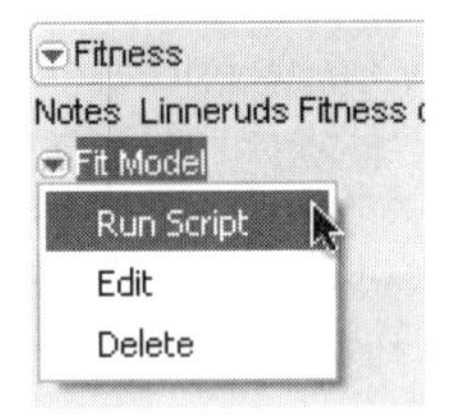

Save to Script window saves the JSL commands for the completed model in a new open script window. You can save the script window and recreate the model at any time by running the script.

Load Version 3 Model presents an Open File dialog for you to select a text file that contains JMP Version 3 model statements. The model then appears in the Model Specification dialog, and can be saved as a current-version model.

Create SAS Job creates a SAS program in a script window that can recreate the current analysis in SAS. Once created, you have several options for submitting the code to SAS.

1. Copy and Paste the resulting code into the SAS Program Editor. This method is useful if you are running an older version of SAS (pre-version 8.2).

2. Select **Edit > Run Script**. This brings up a dialog (shown here) that allows you to enter the name of an accessible SAS server (reached through IOM). If you leave the dialog blank and click **OK**, a local copy of SAS is used if it exists.

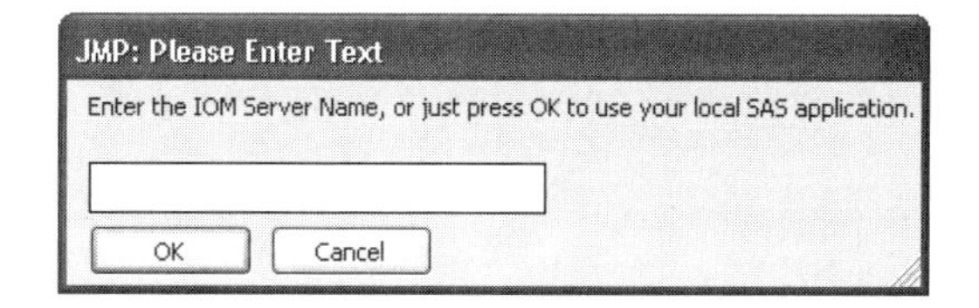

After clicking **OK**, the script is sent to SAS via IOM and the results are returned to JMP. Results are displayed in a separate log window.

3. Save the file and double-click it to open it in a local copy of SAS. This method is useful if you would like to take advantage of SAS ODS options, *e.g.* generating HTML or PDF output from the SAS code.

Submit to SAS sends the script to SAS via IOM and the results are returned to JMP. Results are displayed in a separate log window.

Chapter 12

Standard Least Squares: Introduction

The Fit Model Platform

This fitting facility can fit many different kinds of models (regression, analysis of variance, analysis of covariance, and mixed models), and you can explore the fit in many ways. So, even though this is a sophisticated platform, you can do simple things very easily.

This Standard Least Squares fitting personality is for continuous-responses fit to a linear model of factors using least squares. The results are presented in detail, including leverage plots and least squares means. There are many additional features to consider, such as making contrasts and saving output formulas.

But don't get too bogged down in this chapter if you need to skip to others:

- If you haven't learned how to specify your model in the Model Specification dialog, you might need to jump back to the chapter "Introduction to Model Fitting," p. 195.
- If you have response surface effects or want retrospective power calculations, see the chapter "Standard Least Squares: Perspectives on the Estimates," p. 249.
- For screening applications, skip ahead to the chapter "Standard Least Squares: Exploring the Prediction Equation," p. 275.
- If you have random effects, see the chapter "Standard Least Squares: Random Effects," p. 339.
- If you need the details on how JMP parameterizes its models, see the appendix "Statistical Details," p. 975.

Standard Least Squares is just one fitting personality of Fit Model platform. The other seven personalities are covered in the later chapters "Standard Least Squares: Perspectives on the Estimates," p. 249, "Standard Least Squares: Exploring the Prediction Equation," p. 275, "Standard Least Squares: Random Effects," p. 339, "Stepwise Regression," p. 385, and "Logistic Regression for Nominal and Ordinal Response," p. 403, and "Multiple Response Fitting," p. 427.

Contents

Launch the Platform: A Simple Example

To introduce the Fit Model platform, consider a simple one-way analysis of variance to test if there is a difference in the mean response among three drugs. The example data (Snedecor and Cochran 1967) is a study that measured the response of 30 subjects after treatment by each of three drugs labeled a, d, and f. The results are in the Drug.jmp sample data table.

To begin, specify the model in the Model Specification dialog for Fit Model as shown in Figure 12.1. See the chapter "Introduction to Model Fitting," p. 195, for details about how to use this dialog. To add **Drug** as an effect, select it in the column selection list and click the **Add**. When you select the column y to be the *Y* response, the Fitting Personality becomes **Standard Least Squares** and the Emphasis is **Effect Leverage**. You can change these choices in other situations. Click **Run Model** to see the initial results.

Figure 12.1 The Model Specification Dialog For a One-Way Analysis of Variance

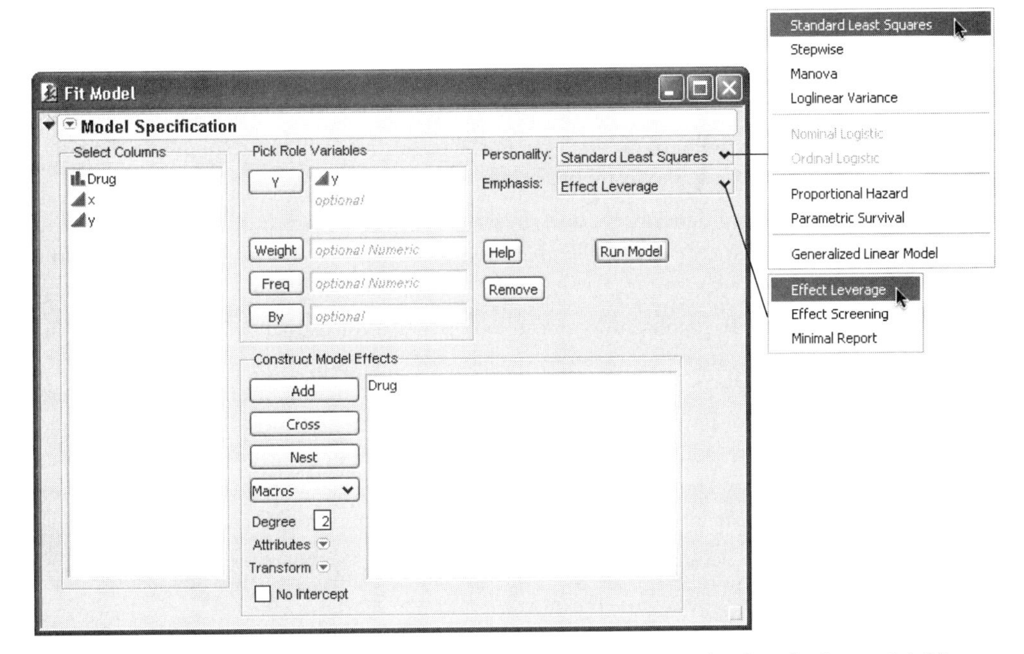

At the top of the output are the graphs in Figure 12.2 that show how the data fit the model. These graphs are called *leverage plots*, because they convey the idea of the data points pulling on the lines representing the fitted model. Leverage plots have these useful properties:

- The distance from each point to the line of fit is the error or residual for that point.
- The distance from each point to the horizontal line is what the error would be if you took out effects in the model.

Thus, strength of the effect is shown by how strongly the line of fit is suspended away from the horizontal by the points. Confidence curves are on the graph so you can see at a glance whether an effect is significant. In each plot, if the 95% confidence curves cross the horizontal reference line, then the effect is significant; if the curves do not cross, then it is not significant (at the 5% level).

Figure 12.2 Whole Model Leverage Plot and Drug Effect Leverage Plot.

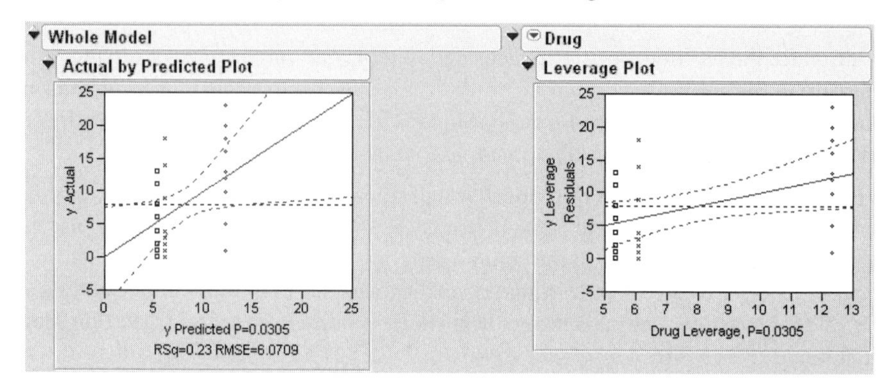

In this simple case where predicted values are simple means, the leverage plot for Drug shows a regression of the actual values on the means for the drug level. (Levin, Serlin, and Webne-Behrman (1989) showcase this idea.) The drugs a, d and f are labeled by a square, cross, and triangle respectively in Figure 12.2. Because there is only one effect in the model, the leverage plot for Whole Model and for the effect Drug are equivalent. They differ only in the scaling of the *x*-axis. The leverage plot for the Whole Model is always a plot of the actual response versus the predicted response, so the points that fall on the line are those that are perfectly predicted.

In this example, the confidence curve does cross the horizontal line; thus, the drug effect is marginally significant, even though there is considerable variation around the line of fit (in this case around the group means).

After you examine the fit graphically, you can look below it for textual details. In this case there are three levels of Drug, giving two parameters to characterize the differences among them. Text reports show the estimates and various test statistics and summary statistics concerning the parameter estimates. The Summary of Fit table and the Analysis of Variance table beneath the whole-model leverage plot show the fit as a whole. The test statistic on the two parameters of the Drug effect appears in the Effect Tests summary and in the detail reports for Drug. Because Drug is the only effect, the same test appears in the model line in the Analysis of Variance table and in the Effects Tests table.

Summary of Fit

RSquare	0.227826
RSquare Adj	0.170628
Root Mean Square Error	6.070878
Mean of Response	7.9
Observations (or Sum Wgts)	30

Analysis of Variance

Source	DF	Sum of Squares	Mean Square	F Ratio
Model	2	293.6000	146.800	3.9831
Error	27	995.1000	36.856	**Prob > F**
C. Total	29	1288.7000		0.0305*

The Analysis of Variance table shows that the Drug effect has an observed significance probability (Prob>F) of 0.0305, which is just barely significant, compared to 0.05. The RSquare reports that 22.8% of the variation in the response can be absorbed by fitting the model.

Whenever there are nominal effects (such as **Drug**), it is interesting to compare how well the levels predict the response. Rather than use the parameter estimates directly, it is usually more meaningful to compare the predicted values at the levels of the nominal values. These predicted values are called the *least squares means* (LSMeans), and in this simple case, they are the same as the ordinary means. Least squares means can differ from simple means when there are other effects in the model, as seen in later examples. The Least Squares Means table for a nominal effect shows beneath the effect's leverage plot (Figure 12.3).

The popup menu for an effect also lets you request the **LSMeans Plot**, as shown in Figure 12.3. The plot graphically shows the means and their associated 95% confidence interval.

Figure 12.3 Table of Least Squares Means and LS Means Plot

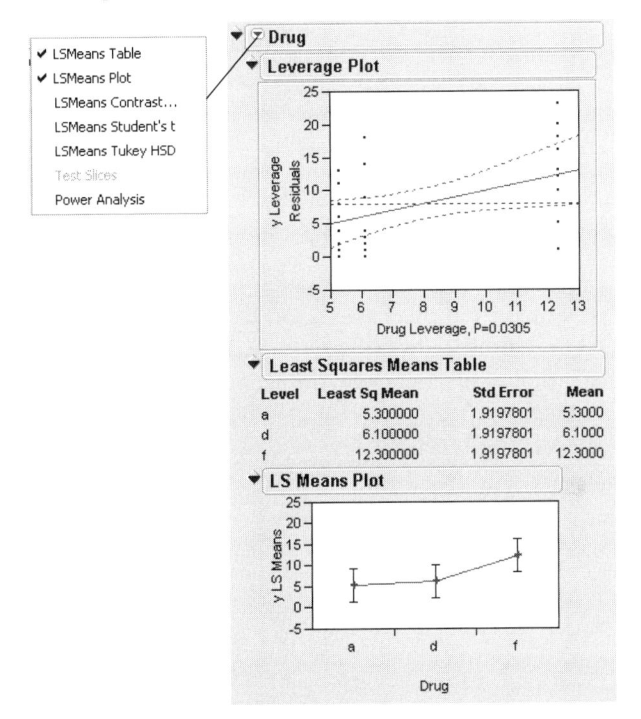

The later section "Examples with Statistical Details," p. 238, continues the drug analysis with the Parameter Estimates table and looks at the group means with the **LSMeans Contrast**.

Regression Plot

If there is exactly one continuous term in a model, and no more than one categorical term, then JMP plots the regression line (or lines), which is associated with the Plot Regression command. Selecting this command turns the plot off and on.

Figure 12.4 Regression Plot

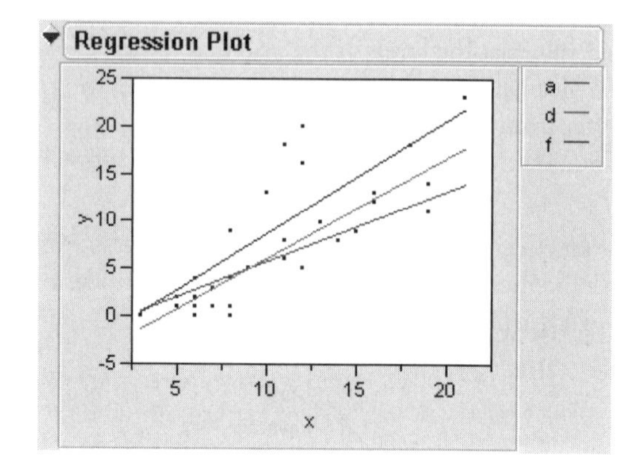

Option Packages for Emphasis

The model fitting process can produce a wide variety of tables, plots and graphs. For convenience, three standard option packages let you choose the report layout that best corresponds to your needs. Just as automobiles are made available with luxury, sport, and economy option packages, linear model results are made available in choices to adapt to your situation. Table 12.1 "Standard Least Squares Report Layout Defined by Emphasis," p. 218, describes the types of report layout for the standard least squares analysis.

The chapter "Introduction to Model Fitting," p. 195, introduces the **Emphasis** menu. The default value of **Emphasis** is the platform's best guess on what you will likely want, given the number of effects that you specify and the number of rows (observations) in the data table.

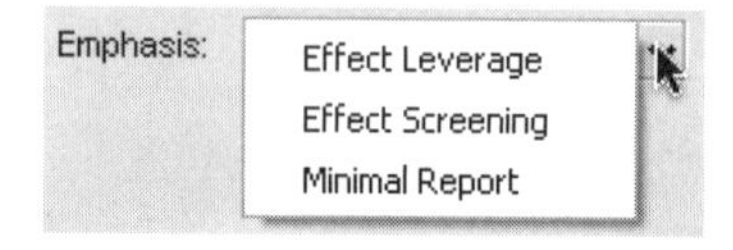

Table 12.1 Standard Least Squares Report Layout Defined by Emphasis

Emphasis	Description of Reports
Effect Leverage	Choose **Effect Leverage** when you want details on the significance of each effect. The initial reports for this emphasis features leverage plots, and the reports for each effects are arranged horizontally.
Effect Screening	The **Effect Screening** layout is better when you have many effects, and don't want details until you see which effects are stronger and which are weaker. This is the recommended for screening designs, where there are many effects and few observations, and the quest is to find the strong effects, rather than to test significance. This arrangement initially displays whole model information followed by effect details, scaled estimates, and the prediction profiler.

Table 12.1 Standard Least Squares Report Layout Defined by Emphasis

Minimal Report	**Minimal Report** starts simple; you customize the results with the tables and plots you want to see. Choosing **Minimal Report** suppresses all plots (except the regression plot), and arranges whole model and effect detail tables vertically.

Whole-Model Statistical Tables

This section starts the details on the standard statistics inside the reports. You might want to skim some of these details sections and come back to them later.

Regression reports can be turned on and off with the Regression Reports menu, accessible from the report's red triangle menu.

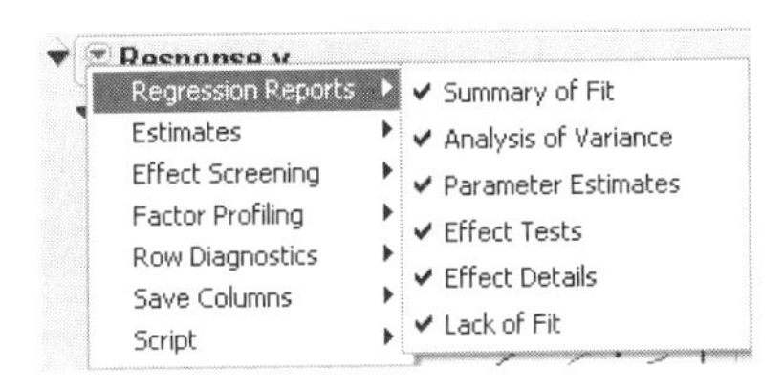

The Whole Model section shows how the model fits as a whole. The initial reports appear as illustrated in Figure 12.5. The next sections describe these tables:

- Summary of Fit
- Analysis of Variance
- Lack of Fit
- Parameter Estimates
- Effect Tests

Discussion of tables for effects and leverage plots are described under the sections "The Effect Test Table," p. 226, and "Leverage Plots," p. 227.

Figure 12.5 Arrangement of Whole Model Reports By Emphasis

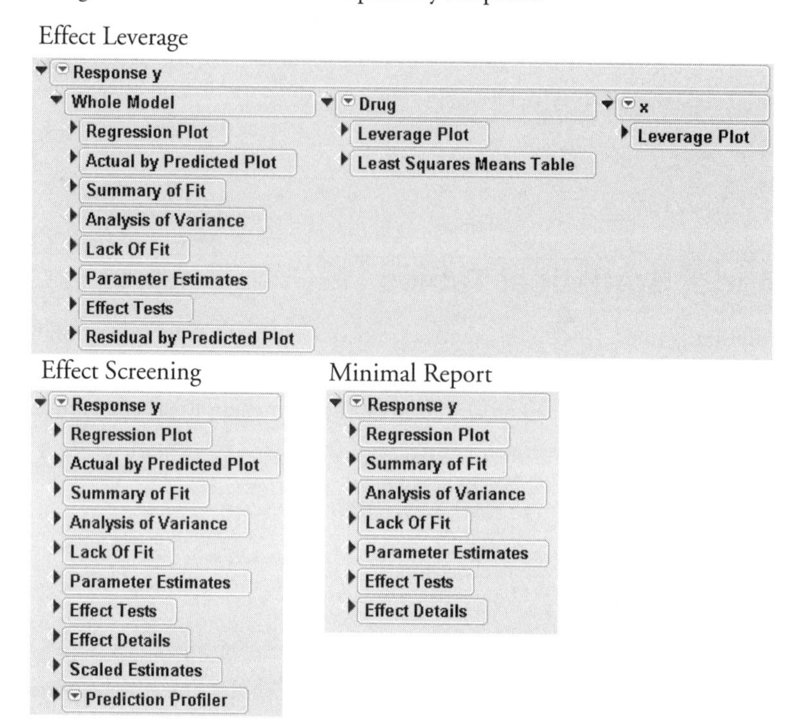

The following examples use the Drug.jmp data table from the Sample Data folder. The model specifies y as the response variable (*Y*) and drug and x as effects, where age has the ordinal modeling type.

The Summary of Fit Table

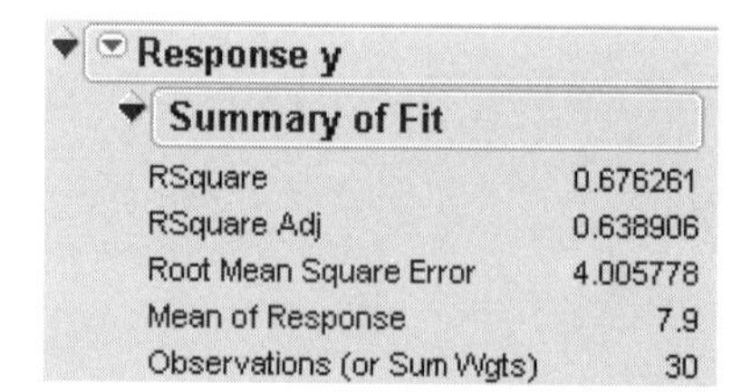

Response y

Summary of Fit

RSquare	0.676261
RSquare Adj	0.638906
Root Mean Square Error	4.005778
Mean of Response	7.9
Observations (or Sum Wgts)	30

The Summary of Fit table appears first and shows the following numeric summaries of the response for the multiple regression model:

RSquare estimates the proportion of the variation in the response around the mean that can be attributed to terms in the model rather than to random error. Using quantities for the corresponding Analysis of Variance table, R^2 is calculated:

$$\frac{\text{Sum of Squares(Model)}}{\text{Sum of Squares(CTotal)}}$$

It is also the square of the correlation between the actual and predicted response. An R^2 of 1 occurs when there is a perfect fit (the errors are all zero). An R^2 of 0 means that the fit predicts the response no better than the overall response mean.

Rsquare Adj adjusts R^2 to make it more comparable over models with different numbers of parameters by using the degrees of freedom in its computation. It is a ratio of mean squares instead of sums of squares and is calculated

$$1 - \frac{\text{Mean Square(Error)}}{\text{Mean Square(C Total)}}$$

where mean square for Error is found in the Analysis of Variance table (shown next under "The Analysis of Variance Table," p. 221) and the mean square for **C. Total** can be computed as the **C. Total** sum of squares divided by its respective degrees of freedom.

Root Mean Square Error estimates the standard deviation of the random error. It is the square root of the mean square for error in the corresponding Analysis of Variance table, and it is commonly denoted as σ.

Mean of Response is the overall mean of the response values. It is important as a base model for prediction because all other models are compared to it. The variance measured around this mean is the *corrected total* (**C. Total**) mean square in the Analysis of Variance table.

Observations (or Sum of Weights) records the number of observations used in the fit. If there are no missing values and no excluded rows, this is the same as the number of rows in the data table. If there is a column assigned the role of weight, this is the sum of the weight column values.

The Analysis of Variance Table

Response y

Analysis of Variance

Source	DF	Sum of Squares	Mean Square	F Ratio
Model	3	871.4974	290.499	18.1039
Error	26	417.2026	16.046	**Prob > F**
C. Total	29	1288.7000		<.0001*

The Analysis of Variance table shows the basic calculations for a linear model. The table compares the model fit to a simple fit to a single mean:

Source lists the three sources of variation, called **Model**, **Error**, and **C. Total**.

DF records an associated *degrees of freedom* (DF) for each source of variation.

The **C. Total** degrees of freedom is for the simple mean model. There is only one degree of freedom used (the estimate of the mean parameter) in the calculation of variation so the **C. Total** DF is always one less than the number of observations.

The total degrees of freedom are partitioned into the **Model** and **Error** terms:

The **Model** degrees of freedom is the number of parameters (except for the intercept) used to fit the model.

The **Error** DF is the difference between the **C. Total** DF and the **Model** DF.

Sum of Squares records an associated sum of squares (SS for short) for each source of variation. The SS column accounts for the variability measured in the response. It is the sum of squares of the differences between the fitted response and the actual response.

The total (**C. Total**) SS is the sum of squared distances of each response from the sample mean, which is 1288.7 in the example shown at the beginning of this section. That is the base model (or simple mean model) used for comparison with all other models.

The **Error** SS is the sum of squared differences between the fitted values and the actual values, and is 417.2 in the previous example. This sum of squares corresponds to the unexplained **Error** (*residual*) after fitting the regression model.

The total SS less the error SS gives the sum of squares attributed to the **Model.**

One common set of notations for these is SSR, SSE, and SST for sum of squares due to regression (model), error, and total, respectively.

Mean Square is a sum of squares divided by its associated degrees of freedom. This computation converts the sum of squares to an average (mean square).

The **Model** mean square is 290.5 in the table at the beginning of this section.

The **Error** mean square is 16.046 and estimates the variance of the error term. It is often denoted as MSE or s^2.

F Ratio is the model mean square divided by the error mean square. It tests the hypothesis that all the regression parameters (except the intercept) are zero. Under this whole-model hypothesis, the two mean squares have the same expectation. If the random errors are normal, then under this hypothesis the values reported in the SS column are two independent Chi-squares. The ratio of these two Chi-squares divided by their respective degrees of freedom (reported in the DF column) has an *F*-distribution. If there is a significant effect in the model, the **F Ratio** is higher than expected by chance alone.

Prob>F is the probability of obtaining a greater *F*-value by chance alone if the specified model fits no better than the overall response mean. Significance probabilities of 0.05 or less are often considered evidence that there is at least one significant regression factor in the model.

Note that large values of **Model** SS as compared with small values of **Error** SS lead to large *F*-ratios and low *p* values, which is what you want if the goal is to declare that terms in the model are significantly different from zero. Most practitioners check this *F*-test first and make sure that it is significant before delving further into the details of the fit. This significance is also shown graphically by the whole-model leverage plot described in the previous section.

The Lack of Fit Table

Response y

Lack Of Fit

Source	DF	Sum of Squares	Mean Square	F Ratio
Lack Of Fit	18	254.86926	14.1594	0.6978
Pure Error	8	162.33333	20.2917	Prob > F
Total Error	26	417.20260		0.7507
				Max RSq
				0.8740

The Lack of Fit table shows a special diagnostic test and appears only when the data and the model provide the opportunity. The idea is that sometimes you can estimate the error variance independently of whether you have the right form of the model. This occurs when observations are exact replicates of each other in terms of the *X* variables. The error that you can measure for these exact replicates is called *pure error*. This is the portion of the sample error that cannot be explained or predicted no matter which form the model uses for the *X* variables. However, a lack of fit test may not be of much use if it has only a very few degrees of freedom, *i.e.*, few replicated *x* values.

The difference between the residual error from the model and the pure error is called *lack of fit error*. A lack of fit error can be significantly greater than pure error if you have the wrong functional form of some regressor, or too few interaction effects in an analysis of variance model. In that case, you should consider adding interaction terms, if appropriate, or try to better capture the functional form of a regressor.

There are two common situations where there is no lack of fit test:

1 There are no exactly replicated points with respect to the *X* data, and therefore there are no degrees of freedom for pure error.

2 The model is *saturated*, meaning that the model itself has a degree of freedom for each different *x* value; therefore, there are no degrees of freedom for lack of fit.

The Lack of Fit table shows information about the error terms:

Source lists the three sources of variation called Lack of Fit, Pure Error, and Total Error.

DF records an associated *degrees of freedom* (DF) for each source of error.

The Total Error DF is the degrees of freedom found on the Error line of the Analysis of Variance table. It is the difference between the Total DF and the Model DF found in that table. The Error DF is partitioned into degrees of freedom for lack of fit and for pure error.

The Pure Error DF is pooled from each group where there are multiple rows with the same values for each effect. There is one instance in the Big Class.jmp data table where two subjects have the same values of age and weight (Chris and Alfred are both 14 and have a weight of 99), giving 1(2 - 1) = 1 DF for Pure Error. In general, if there are *g* groups having multiple rows with identical values for each effect, the pooled DF, denoted DF_p, is

$$DF_p = \sum_{i=1}^{g} (n_i - 1)$$

where n_i is the number of subjects in the *i*th group.

a. The Lack of Fit DF is the difference between the Total Error and Pure Error degrees of freedom.

Sum of Squares records an associated sum of squares (SS for short) for each source of error.

a. The Total Error SS is the sum of squares found on the Error line of the corresponding Analysis of Variance table.

a. The Pure Error SS is pooled from each group where there are multiple rows with the same values for each effect. This estimates the portion of the true random error that is not explained by model effects. In general, if there are *g* groups having multiple rows with like values for each effect, the pooled SS, denoted SS_p is written

$$SS_p = \sum_{i=1}^{g} SS_i$$

where SS_i is the sum of squares for the *i*th group corrected for its mean.

The Lack of Fit SS is the difference between the Total Error and Pure Error sum of squares. If the lack of fit SS is large, the model may not be appropriate for the data. The *F*-ratio described below tests whether the variation due to lack of fit is small enough to be accepted as a negligible portion of the pure error.

Mean Square is a sum of squares divided by its associated degrees of freedom. This computation converts the sum of squares to an average (mean square). *F*-ratios for statistical tests are the ratios of mean squares.

F Ratio is the ratio of mean square for Lack of Fit to mean square for Pure Error. It tests the hypothesis that the lack of fit error is zero.

Prob > F is the probability of obtaining a greater *F*-value by chance alone if the variation due to lack of fit variance and the pure error variance are the same. This means that an insignificant proportion of error is explained by lack of fit.

Max RSq is the maximum R^2 that can be achieved by a model using only the variables in the model. Because Pure Error is invariant to the form of the model and is the minimum possible variance, Max RSq is calculated

$$1 - \frac{SS(\text{Pure Error})}{SS(\text{Total for whole model})}$$

The Parameter Estimates Table

Response y

Parameter Estimates

Term	Estimate	Std Error	t Ratio	Prob>|t|
Intercept	-2.695773	1.911085	-1.41	0.1702
Drug[a]	-1.185037	1.060822	-1.12	0.2742
Drug[d]	-1.076065	1.041298	-1.03	0.3109
x	0.9871838	0.164498	6.00	<.0001*

The Parameter Estimates table shows the estimates of the parameters in the linear model and a *t*-test for the hypothesis that each parameter is zero. Simple continuous regressors have only one parameter. Models with complex classification effects have a parameter for each anticipated degree of freedom.

The Parameter Estimates table shows these quantities:

Term names the estimated parameter. The first parameter is always the intercept. Simple regressors show as the name of the data table column. Regressors that are *dummy* indicator variables constructed from nominal or ordinal effects are labeled with the names of the levels in brackets. For nominal variables, the dummy variables are coded as 1 except for the last level, which is coded as –1 across all the other dummy variables for that effect. The parameters for ordinally coded indicators, as in this example, measure the difference from each level to the level before it.

Estimate lists the parameter estimates for each term. They are the coefficients of the linear model found by least squares.

Std Error is the standard error, an estimate of the standard deviation of the distribution of the parameter estimate. It is used to construct *t*-tests and confidence intervals for the parameter.

***t* Ratio** is a statistic that tests whether the true parameter is zero. It is the ratio of the estimate to its standard error and has a Student's *t*-distribution under the hypothesis, given the usual assumptions about the model.

Prob>|t| is the probability of getting an even greater *t*-statistic (in absolute value), given the hypothesis that the parameter is zero. This is the two-tailed test against the alternatives in each direction. Probabilities less than 0.05 are often considered as significant evidence that the parameter is not zero.

Although initially hidden, the following columns are also available. Right-click (control-click on the Macintosh) and select the desired column from the **Columns** menu.

Std Beta are the parameter estimates that would have resulted from the regression had all the variables been standardized to a mean of 0 and a variance of 1.

VIF shows the variance inflation factors. High VIFs indicate a collinearity problem.

Note that the VIF is defined as

$$VIF = \frac{1}{1 - R^2}$$

Note that the definition of R^2 changes for no-intercept models. For no-intercept and hidden-intercept models, JMP uses the R^2 from the uncorrected Sum of Squares, *i.e.* from the zero model, rather than the corrected sum of squares, from the mean model.

Lower 95% is the lower 95% confidence interval for the parameter estimate.

Upper 95% is the upper 95% confidence interval for the parameter estimate.

Design Std Error is the standard error without being scaled by sigma (RMSE), and is equal to

$$\sqrt{\mathrm{diag}(X'X)^{-1}}$$

The Effect Test Table

Response y

Effect Tests

Source	Nparm	DF	Sum of Squares	F Ratio	Prob > F
Drug	2	2	68.55371	2.1361	0.1384
x	1	1	577.89740	36.0145	<.0001*

The effect tests are joint tests that all the parameters for an individual effect are zero. If an effect has only one parameter, as with simple regressors, then the tests are no different from the *t*-tests in the Parameter Estimates table. Details concerning the parameterization and handling of singularities, which are different from the SAS GLM procedure, are discussed in the appendix "Statistical Details," p. 975.

The Effect Test table shows the following information for each effect:

Source lists the names of the effects in the model.

Nparm is the number of parameters associated with the effect. Continuous effects have one parameter. Nominal effects have one less parameter than the number of levels. Crossed effects multiply the number of parameters for each term. Nested effects depend on how levels occur.

DF is the degrees of freedom for the effect test. Ordinarily **Nparm** and **DF** are the same. They are different if there are linear combinations found among the regressors such that an effect cannot be tested to its fullest extent. Sometimes the DF will even be zero, indicating that no part of the effect is testable. Whenever DF is less than Nparm, the note **Lost DFs** appears to the right of the line in the report.

Sum of Squares is the sum of squares for the hypothesis that the listed effect is zero.

F Ratio is the *F*-statistic for testing that the effect is zero. It is the ratio of the mean square for the effect divided by the mean square for error. The mean square for the effect is the sum of squares for the effect divided by its degrees of freedom.

Prob>F is the significance probability for the *F*-ratio. It is the probability that if the null hypothesis is true, a larger *F*-statistic would only occur due to random error. Values less than 0.0005 appear as 0.0000, which is conceptually zero.

Although initially hidden, a column that displays the Mean Square is also available. Right-click (control-click on the Macintosh) and select **Mean Square** from the **Columns** menu.

Saturated Models

Screening experiments often involve fully saturated models, where there are not enough degrees of freedom to estimate error. Because of this, neither standard errors for the estimates, nor *t*-ratios, nor *p*-values can be calculated in the traditional way.

For these cases, JMP uses the relative standard error, corresponding to a residual standard error of 1. In cases where all the variables are identically coded (say, [–1,1] for low and high levels), these relative standard errors are identical.

JMP also displays a Pseudo-*t*-ratio, calculated as

$$\text{Pseudo } t = \frac{\text{estimate}}{\text{relative std error} \times PSE}$$

using Lenth's PSE (pseudo standard-error) and degrees of freedom for error (DFE) equal to one-third the number of parameters. The value for Lenth's PSE is shown at the bottom of the report.

As an example, use Reactor.jmp, and fit a full factorial model (that is, all interactions up to and including the fifth order). The parameter estimates are presented in sorted order, with smallest *p*-values listed first.

Figure 12.6 Saturated Report

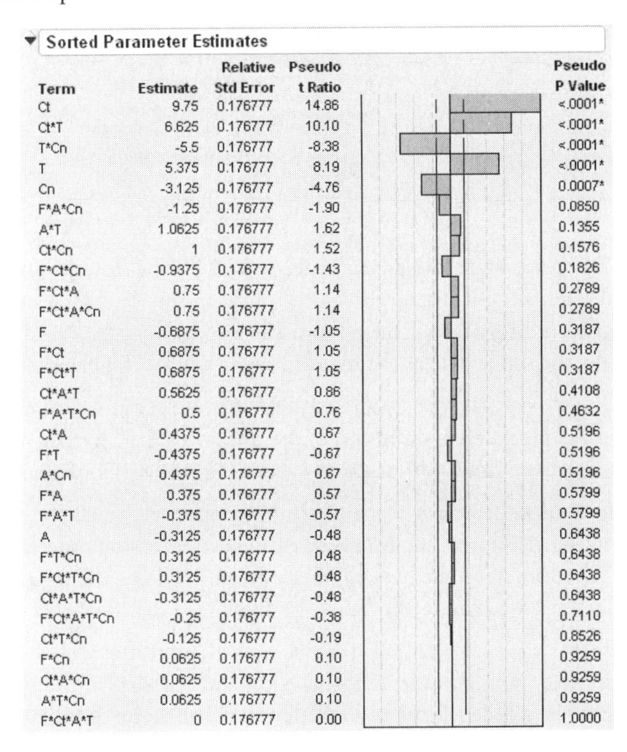

Sorted Parameter Estimates

Term	Estimate	Relative Std Error	Pseudo t Ratio	Pseudo P Value
Ct	9.75	0.176777	14.86	<.0001*
Ct*T	6.625	0.176777	10.10	<.0001*
T*Cn	-5.5	0.176777	-8.38	<.0001*
T	5.375	0.176777	8.19	<.0001*
Cn	-3.125	0.176777	-4.76	0.0007*
F*A*Cn	-1.25	0.176777	-1.90	0.0850
A*T	1.0625	0.176777	1.62	0.1355
Ct*Cn	1	0.176777	1.52	0.1576
F*Ct*Cn	-0.9375	0.176777	-1.43	0.1826
F*Ct*A	0.75	0.176777	1.14	0.2789
F*Ct*A*Cn	0.75	0.176777	1.14	0.2789
F	-0.6875	0.176777	-1.05	0.3187
F*Ct	0.6875	0.176777	1.05	0.3187
F*Ct*T	0.6875	0.176777	1.05	0.3187
Ct*A*T	0.5625	0.176777	0.86	0.4108
F*A*T*Cn	0.5	0.176777	0.76	0.4632
Ct*A	0.4375	0.176777	0.67	0.5196
F*T	-0.4375	0.176777	-0.67	0.5196
A*Cn	0.4375	0.176777	0.67	0.5196
F*A	0.375	0.176777	0.57	0.5799
F*A*T	-0.375	0.176777	-0.57	0.5799
A	-0.3125	0.176777	-0.48	0.6438
F*T*Cn	0.3125	0.176777	0.48	0.6438
F*Ct*T*Cn	0.3125	0.176777	0.48	0.6438
Ct*A*T*Cn	-0.3125	0.176777	-0.48	0.6438
F*Ct*A*T*Cn	-0.25	0.176777	-0.38	0.7110
Ct*T*Cn	-0.125	0.176777	-0.19	0.8526
F*Cn	0.0625	0.176777	0.10	0.9259
Ct*A*Cn	0.0625	0.176777	0.10	0.9259
A*T*Cn	0.0625	0.176777	0.10	0.9259
F*Ct*A*T	0	0.176777	0.00	1.0000

In cases where the relative standard errors are different (perhaps due to unequal scaling), a similar report appears. However, there is a different value for Lenth's PSE for each estimate. The PSE values are scaled by a TScale parameter, which is (I have forgotten how it is calculated).

Leverage Plots

To graphically view the significance of the model or focus attention on whether an effect is significant, you want to display the data from the point of view of the hypothesis for that effect. You might say that you want more of an X-ray picture showing the inside of the data rather than a surface view from the outside. The leverage plot is this kind of view of your data, a view calculated to give maximum insight into how the fit carries the data.

The effect in a model is tested for significance by comparing the sum of squared residuals to the sum of squared residuals of the model with that effect removed. Residual errors that are much smaller when the effect is included in the model confirm that the effect is a significant contribution to the fit.

The graphical display of an effect's significance test is called a *leverage plot* (Sall 1990). This kind of plot shows for each point what the residual would be both with and without that effect in the model. Leverage plots are found in the **Row Diagnostics** submenu of the Fit Model report.

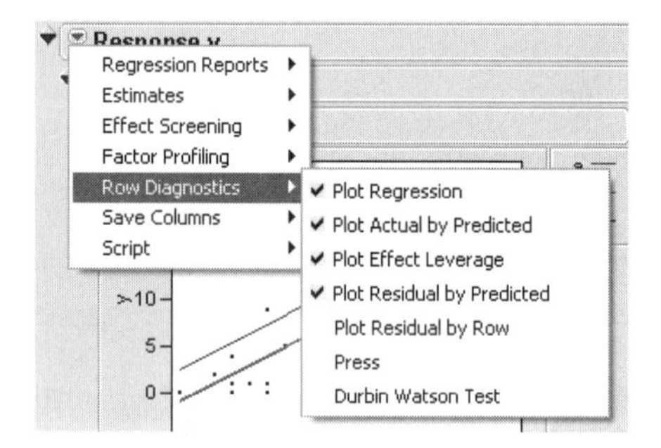

A leverage plot is constructed as illustrated in Figure 12.7. The distance from a point to the line of fit shows the actual residual. The distance from the point to the horizontal line of the mean shows what the residual error would be without the effect in the model. In other words, the mean line in this leverage plot represents the model where the hypothesized value of the parameter (effect) is constrained to zero.

Historically, leverage plots are an extension of the display called a *partial-regression residual leverage plot* by Belsley, Kuh, and Welsch (1980) or an *added variable plot* by Cook and Weisberg (1982).

The term *leverage* is used because a point exerts more influence on the fit if it is farther away from the middle of the plot in the horizontal direction. At the extremes, the differences of the residuals before and after being constrained by the hypothesis are greater and contribute a larger part of the sums of squares for that effect's hypothesis test.

The fitting platform produces a leverage plot for each effect in the model. In addition, there is one special leverage plot titled Whole Model that shows the actual values of the response plotted against the predicted values. This Whole Model leverage plot dramatizes the test that all the parameters (except intercepts) in the model are zero. The same test is reported in the Analysis of Variance report.

Figure 12.7 Illustration of a General Leverage Plot

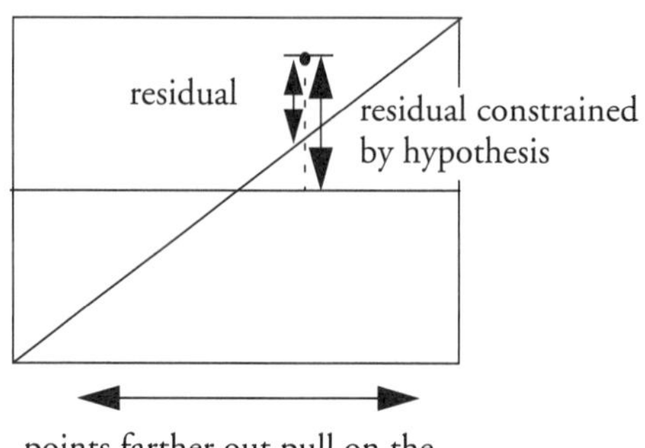

The leverage plot for the linear effect in a simple regression is the same as the traditional plot of actual response values and the regressor. The example leverage plots in Figure 12.8 result from a simple linear regression to predict the variable height with age (from Big Class.jmp). The plot on the left is the Whole Model test for all effects, and the plot on the right is the leverage plot for the effect age.

Figure 12.8 Whole Model and Effect Leverage Plots

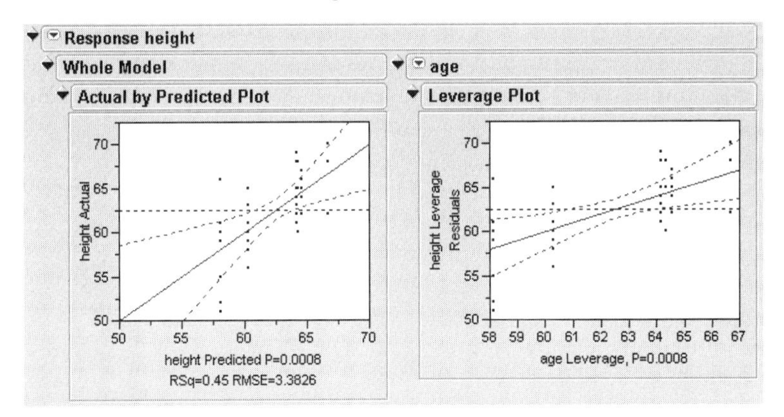

The points on a leverage plot for simple regression are actual data coordinates, and the horizontal line for the constrained model is the sample mean of the response. But when the leverage plot is for one of multiple effects, the points are no longer actual data values. The horizontal line then represents a partially constrained model instead of a model fully constrained to one mean value. However, the intuitive interpretation of the plot is the same whether for simple or multiple regression. The idea is to judge if the line of fit on the effect's leverage plot carries the points significantly better than does the horizontal line.

Figure 12.7 is a general diagram of the plots in Figure 12.8. Recall that the distance from a point to the line of fit is the actual residual and the distance from the point to the mean is the residual error if the regressor is removed from the model.

Confidence Curves

The leverage plots show with confidence curves. These indicate whether the test is significant at the 5% level by showing a confidence region for the line of fit. If the confidence region between the curves contains the horizontal line, then the effect is not significant. If the curves cross the line, the effect is significant. Compare the examples shown in Figure 12.9.

Figure 12.9 Comparison of Significance Shown in Leverage Plots

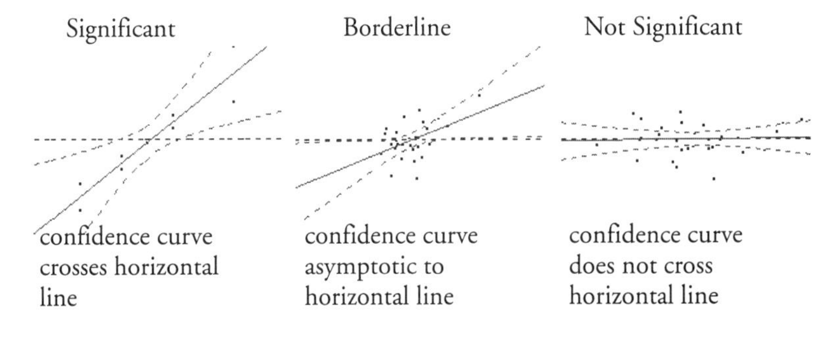

Interpretation of X Scales

If the modeling type of the regressor is continuous, then the *x*-axis is scaled like the regressor and the slope of the line of fit in the leverage plot is the parameter estimate for the regressor as illustrated to the right in Figure 12.10.

If the effect is nominal or ordinal, or a complex effect like an interaction instead of a simple regressor, then the *x*-axis cannot represent the values of the effect directly. In this case the *x*-axis is scaled like the *y*-axis, and the line of fit is a diagonal with a slope of 1. The whole model leverage plot is a version of this. The *x*-axis turns out to be the predicted response of the whole model, as illustrated by the right-hand plot in Figure 12.10.

Figure 12.10 Leverage Plots for Simple Regression and Complex Effects

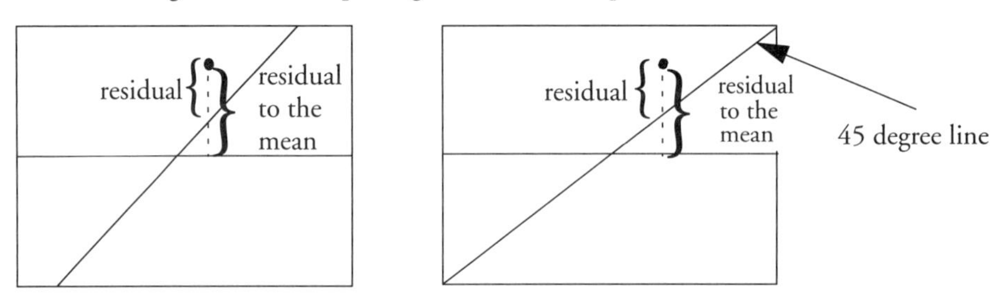

The influential points in all leverage plots are the ones far out on the *x*-axis. If two effects in a model are closely related, then these effects as a whole don't have much leverage. This problem is called *collinearity*. By scaling regressor axes by their original values, collinearity shows as shrinkage of points in the *x* direction.

See the appendix "Statistical Details," p. 975, for the details of leverage plot construction.

Effect Details

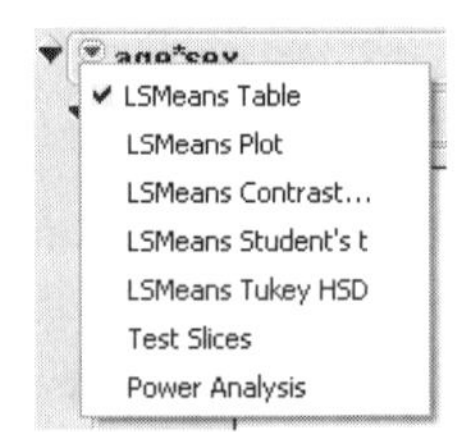

Each effect has the popup menu shown here next to its name. The effect popup menu items let you request tables, plots, and tests for that effect. The commands for an effect append results to the effect report. Optionally, you can close results or dismiss the results by deselecting the item in the menu.

The next sections describe the Effect popup menu commands.

LSMeans Table

Least squares means are predicted values from the specified model across the levels of a categorical effect where the other model factors are controlled by being set to *neutral* values. The neutral values are the sample means (possibly weighted) for regressors with interval values, and the average coefficient over the levels for unrelated nominal effects.

Least squares means are the values that let you see which levels produce higher or lower responses, holding the other variables in the model constant. Least squares means are also called *adjusted means* or *population marginal means.*

Least squares means are the statistics that are compared when effects are tested. They might not reflect typical real-world values of the response if the values of the factors do not reflect prevalent combinations of values in the real world. Least squares means are useful as comparisons in experimental situations.

Least Squares Means Table

Level	Least Sq Mean	Std Error	Mean
12	58.900157	0.7932334	58.1250
13	61.614555	0.8617459	60.2857
14	64.704970	0.6458069	64.1667
15	64.146938	0.8414738	64.5714
16	62.610762	1.3067671	64.3333
17	62.058788	1.4509920	66.6667

A Least Squares Means table with standard errors is produced for all categorical effects in the model. For main effects, the Least Squares Means table also includes the sample mean. It is common for the least squares means to be closer together than the sample means. For further details on least squares means, see the appendix "Statistical Details," p. 975.

The Least Squares Means table shows these quantities:

Level lists the names of each categorical level.

Least Sq Mean lists the least squares mean for each level of the categorical variable.

Std Error lists the standard error for each level of the categorical variable.

Mean lists the response sample mean for each level of the categorical variable. This will be different from the least squares mean if the values of other effects in the model do not balance out across this effect.

Although initially hidden, columns that display the Upper and Lower 95% confidence intervals of the mean are also available. Right-click (control-click on the Macintosh) and select **Mean Square** from the **Columns** menu.

LSMeans Plot

The **LSMeans Plot** option plots least squares means (LSMeans) plots for nominal and ordinal main effects and two-way interactions. The chapter "Standard Least Squares: Exploring the Prediction Equation," p. 275, discusses the **Interaction Plots** command in the **Factor Profiling** menu, which offers interaction plots in a different format.

Figure 12.11 shows the effect plots for main effects and two-way interactions. An interpretation of the data (Popcorn.jmp)is in the chapter "A Factorial Analysis" in the JMP Introductory Guide. In this experiment, popcorn yield measured by volume of popped corn from a given measure of kernels is compared for three conditions:

- type of popcorn (gourmet and plain)
- batch size popped (large and small)
- amount of oil used (little and lots).

Figure 12.11 LSMeans Plots for Main Effects and Interactions

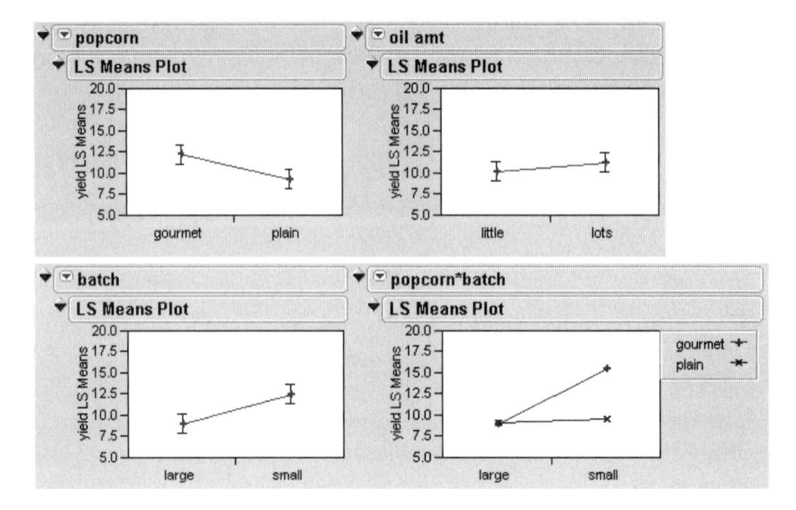

Note: To transpose the factors of the LSMeans plot for a two-factor interaction, use Shift and select the **LSMeans** option.

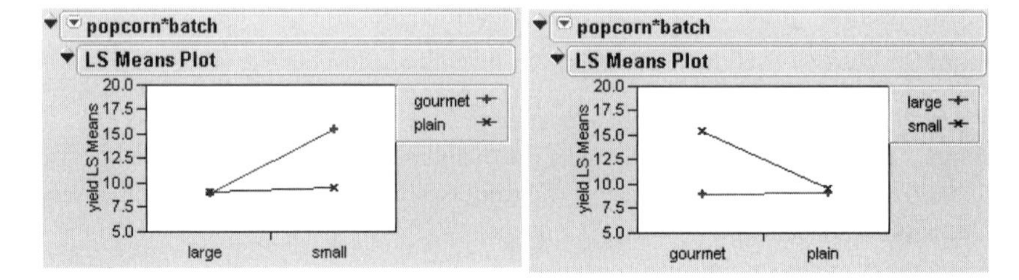

LSMeans Contrast

A *contrast* is a set of linear combinations of parameters that you want to jointly test to be zero. JMP builds contrasts in terms of the least squares means of the effect. By convention, each column of the contrast is normalized to have sum zero and an absolute sum equal to two.

To set up a contrast within an effect, use the **LSMeans Contrast** command, which displays a dialog for specifying contrasts with respect to that effect. This command is enabled only for pure classification effects.

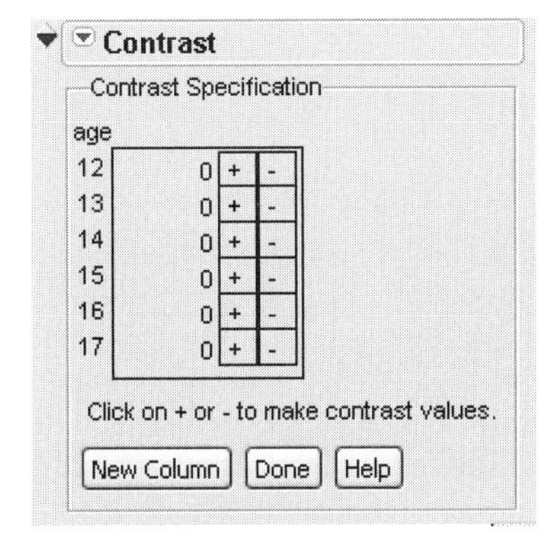

This Contrast dialog shown here is for the effect called **age**. It shows the name of the effect and the names of the levels in the effect. Beside the levels is an area enclosed in a rectangle that has a column of numbers next to boxes of + and - signs.

To construct a contrast, click the + and - boxes beside the levels that you want to compare. If possible, the dialog normalizes each time to make the sum for a column zero and the absolute sum equal to two each time you click; it adds to the plus or minus score proportionately.

For example, to form a contrast that compares the first two age levels with the second two levels, click + for the ages 12 and 13, and click - for ages 14 and 15. If you want to do more comparisons, click the **New Column** button for a new column to define the new contrast. The contrast for **age** is shown here.

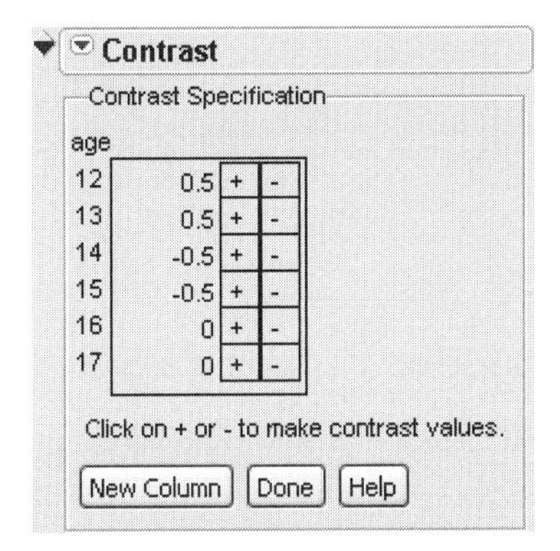

After you are through defining contrasts, click **Done**. The contrast is estimated, and the Contrast table shown here is appended to the other tables for that effect.

Contrast

Test Detail

12	0.5
13	0.5
14	-0.5
15	-0.5
16	0
17	0
Estimate	15.966
Std Error	5.7814
t Ratio	2.7616
Prob>|t|	0.0094
SS	1318.8

Sum of Squares	1318.7825474
Numerator DF	1
Denominator DF	32
F Ratio	7.6263279592
Prob > F	0.0094479824

Parameter Function

Parameter	
Intercept	0
age[13-12]	-0.5
age[14-13]	-1
age[15-14]	-0.5
age[16-15]	0
age[17-16]	0

The Contrast table shows:

- the contrast specified by the dialog as a function of the least squares means
- the estimates and standard errors of the contrast for the least squares means, and *t*-tests for each column of the contrast
- the *F*-test for all columns of the contrast tested jointly
- the Parameter Function table that shows the contrast expressed in terms of the parameters. In this example the parameters are for the ordinal variable, age.

LSMeans Student's *t*, LSMeans Tukey's HSD

The **LSMeans Student's t** and **LSMeans Tukey's HSD** commands give multiple comparison tests for model effects.

LSMeans Student's t computes individual pairwise comparisons of least squares means in the model using Student's *t*-tests. This test is sized for individual comparisons. If you make many pairwise tests, there is no protection across the inferences. Thus, the alpha-size (Type I) error rate across the hypothesis tests is higher than that for individual tests.

LSMeans Tukey's HSD gives a test that is sized for all differences among the least squares means. This is the *Tukey* or *Tukey-Kramer HSD* (Honestly Significant Difference) test. (Tukey 1953, Kramer 1956). This test is an exact alpha-level test if the sample sizes are the same and conservative if the sample sizes are different (Hayter 1984).

These tests are discussed in detail in the chapter "Multiple Comparisons," p. 124 IN THE "ONE-WAY ANOVA" CHAPTER, which has examples and a description of how to read and interpret the multiple comparison tables.

The reports from both commands have menus that allow for the display of three reports.

Crosstab Report shows or hides the crosstab report. This report is a two-way table that highlights significant differences in red.

Connecting Lines Report shows or hides a report that illustrates significant differences with letters (similar to traditional SAS GLM output). Levels not connected by the same letter are significantly different.

Ordered Differences Report shows or hides a report that ranks the differences from lowest to highest. It also plots the differences on a histogram that has overlaid confidence interval lines. See "Ordered Differences Report," p. 127 IN THE "ONE-WAY ANOVA" CHAPTER for an example of an Ordered Differences report.

Detailed Comparisons Report shows reports and graphs that compare each level of the effect with all other levels in a pairwise fashion. See "Detailed Comparisons Report," p. 127 IN THE "ONE-WAY ANOVA" CHAPTER for an example of a Detailed Comparison Report.

Equivalence Test uses the TOST method to test for practical equivalence. See "Equivalence Test," p. 134 IN THE "ONE-WAY ANOVA" CHAPTER for details.

Test Slices

The **Test Slices** command, which is enabled for interaction effects, is a quick way to do many contrasts at the same time. For each level of each classification column in the interaction, it makes comparisons among all the levels of the other classification columns in the interaction. For example, if an interaction is A*B*C then there is a slice called A=1, which tests all the B*C levels within where A=1, and so on for the levels of A, B, and C. This is a way to spot where the action is inside an interaction.

Power Analysis

Power analysis is discussed in the chapter "Standard Least Squares: Perspectives on the Estimates," p. 249.

Summary of Row Diagnostics and Save Commands

All parts of the Model Fit results are optional. When you click the triangular popup icon next to the response name at the topmost outline level, the menu shown in Figure 12.12 lists commands for the continuous response model.

Figure 12.12 Commands for Least Squares Analysis: Row Diagnostics and Save Commands

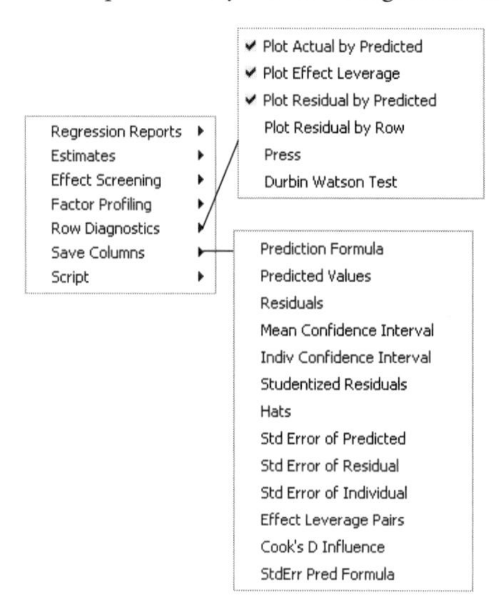

The specifics of the Fit Model platform depend on the type of analysis you do and the options and commands you ask for. Popup icons list commands and options at each level of the analysis. You always have access to all tables and plots through these menu items. Also, the default arrangement of results can be changed by using preferences or script commands.

The **Estimates** and **Effect Screening** menus are discussed in detail in the chapter "Standard Least Squares: Perspectives on the Estimates," p. 249. See the chapter "Standard Least Squares: Exploring the Prediction Equation," p. 275, for a description of the commands in the **Effect Screening** and **Factor Profiling** menus.

The next sections summarize commands in the **Row Diagnostics** and **Save** menus.

Row Diagnostics

Leverage Plots (the **Plot Actual by Predicted** and **Plot Effect Leverage** commands) are covered previously in this chapter under "Leverage Plots," p. 227.

Plot Actual by Predicted displays the observed values by the predicted values of *Y*. This is the leverage plot for the whole model.

Plot Effect Leverage produces a leverage plot for each effect in the model showing the point-by-point composition of the test for that effect.

Plot Residual By Predicted displays the residual values by the predicted values of *Y*. You typically want to see the residual values scattered randomly about zero.

Plot Residual By Row displays the residual value by the row number of its observation.

Press displays a Press statistic, which computes the residual sum of squares where the residual for each row is computed after dropping that row from the computations, and the Press RMSE, defined as $\sqrt{Press/n}$.

Durbin-Watson Test displays the Durbin-Watson statistic to test whether or not the errors have first-order autocorrelation. The autocorrelation of the residuals is also shown. The Durbin-Watson table has a popup command that computes and displays the exact probability associated with the statistic. This Durbin-Watson table is only appropriate for time series data when you suspect that the errors are correlated across time.

Note: The computation of the Durbin-Watson exact probability can be time-intensive if there are many observations. The space and time needed for the computation increase with the square and the cube of the number of observations, respectively.

Save Commands

The **Save** submenu offers the following choices. Each selection generates one or more new columns in the current data table titled as shown, where *colname* is the name of the response variable:

Prediction Formula creates a new column, called Pred Formula colname, containing the predicted values computed by the specified model. It differs from the **Save Predicted Values** column in that the prediction formula is saved with the new column. This is useful for predicting values in new rows or for obtaining a picture of the fitted model.

Use the **Column Info** command and click the **Edit Formula** button to see the prediction formula. The prediction formula can require considerable space if the model is large. If you do not need the formula with the column of predicted values, use the **Save Predicted Values** option. For information about formulas, see the *JMP User Guide*.

Predicted Values creates a new column called Predicted colname that contain the predicted values computed by the specified model.

Residuals creates a new column called Residual colname containing the residuals, which are the observed response values minus predicted values.

Mean Confidence Interval creates two new columns called Lower 95% Mean colname and Upper 95% Mean colname. The new columns contain the lower and upper 95% confidence limits for the line of fit.

Note: If you hold down the Shift key and select **Save Mean Confidence Interval**, you are prompted to enter an α-level for the computations.

Individual Confidence Interval creates two new columns called Lower95% Indiv colname and Upper95% Indiv colname. The new columns contain lower and upper 95% confidence limits for individual response values.

Note: If you hold down the Shift key and select **Save Individual Confidence Interval**, you are prompted to enter an α-level for the computations.

Studentized Residuals creates a new column called Studentized Resid colname. The new column values are the residuals divided by their standard error.

Hats creates a new column called h colname. The new column values are the diagonal values of the matrix $X(X'X)^{-1}X'$, sometimes called hat values.

Std Error of Predicted creates a new column, called StdErr Pred colname, containing the standard errors of the predicted values.

Note that if you save both a prediction formula and a standard error formula for a column, the Profiler offers (via a dialog) to use the standard error formula to make confidence intervals instead of profiling it as a separate column.

Std Error of Residual creates a new column called, StdErrResid colname, containing the standard errors of the residual values.

Std Error of Individual creates a new column, called StdErr Indiv colname, containing the standard errors of the individual predicted values.

Effect Leverage Pairs creates a set of new columns that contain the values for each leverage plot. The new columns consist of an *X* and *Y* column for each effect in the model. The columns are named as follows. If the response column name is R and the effects are X1 and X2, then the new column names are:

X Leverage of X1 for R Y Leverage of X1 for R

X Leverage of X2 for R Y Leverage of X2 for R.

Cook's D Influence saves the Cook's *D* influence statistic. Influential observations are those that, according to various criteria, appear to have a large influence on the parameter estimates.

StdErr Pred Formula creates a new column, called Pred SE colname, containing the standard error of the predicted values. It is the same as the **Std Error of Predicted** option but saves the formula with the column. Also, it can produce very large formulas.

Examples with Statistical Details

This section continues with the example at the beginning of the chapter that uses the Drug.jmp data table Snedecor and Cochran (1967, p. 422). The introduction shows a one-way analysis of variance on three drugs labeled a, d, and f, given to three groups randomly selected from 30 subjects.

Run the example again with response Y and effect Drug. The next sections dig deeper into the analysis and also add another effect, X, that has a role to play in the model.

One-Way Analysis of Variance with Contrasts

In a one-way analysis of variance, a different mean is fit to each of the different sample (response) groups, as identified by a nominal variable. To specify the model for JMP, select a continuous *Y* and a nominal *X* variable such as Drug. In this example Drug has values a, d, and f. The standard least squares fitting method translates this specification into a linear model as follows: The nominal variables define a sequence of dummy variables, which have only values 1, 0, and –1. The linear model is written

$$y_i = \beta_0 + \beta_1 x_{1i} + \beta_2 x_{2i} + \varepsilon_i$$

As shown here, the first dummy variable denotes that Drug=a contributes a value 1 and Drug=f contributes a value –1 to the dummy variable.

$$x_{1i} = \begin{bmatrix} 1 & \text{if drug=a} \\ 0 & \text{if drug=d} \\ -1 & \text{if drug=f} \end{bmatrix}$$

The second dummy variable is given values

$$x_{1i} = \begin{bmatrix} 0 & \text{if drug=a} \\ 1 & \text{if drug=d} \\ -1 & \text{if drug=f} \end{bmatrix}$$

The last level does not need a dummy variable because in this model its level is found by subtracting all the other parameters. Therefore, the coefficients sum to zero across all the levels.

The estimates of the means for the three levels in terms of this parameterization are:

$$\mu_1 = \beta_0 + \beta_1$$

$$\mu_2 = \beta_0 + \beta_2$$

$$\mu_3 = \beta_0 - \beta_1 - \beta_2$$

Solving for β yields

$$\beta_0 = \frac{(\mu_1 + \mu_2 + \mu_3)}{3} = \mu \text{ (the average over levels)}$$

$$\beta_1 = \mu_1 - \mu$$

$$\beta_2 = \mu_2 - \mu$$

$$\beta_3 = \beta_1 - \beta_2 = \mu_3 - \mu$$

Thus, if regressor variables are coded as indicators for each level minus the indicator for the last level, then the parameter for a level is interpreted as the difference between that level's response and the average response across all levels.

Response y

Whole Model

Parameter Estimates

Term	Estimate	Std Error	t Ratio	Prob>|t|
Intercept	7.9	1.108386	7.13	<.0001*
Drug[a]	-2.6	1.567494	-1.66	0.1088
Drug[d]	-1.8	1.567494	-1.15	0.2609

Effect Tests

Source	Nparm	DF	Sum of Squares	F Ratio	Prob > F
Drug	2	2	293.60000	3.9831	0.0305*

The figure here shows the Parameter Estimates and the Effect Test reports from the one-way analysis of the drug data. Figure 12.3 at the beginning of the chapter shows the Least Squares Means report and LS Means Plot for the Drug effect.

The Drug effect can be studied in more detail by using a contrast of the least squares means. To do this, click the popup icon next to the Drug effect title and select **LSMeans Contrast** to obtain the Contrast specification dialog.

Click the + boxes for drugs a and d, and the - box for drug f to define the contrast that compares drugs a and d to f (shown in Figure 12.13). Then click **Done** to obtain the Contrast report. The report shows that the Drug effect looks more significant using this one-degree-of-freedom comparison test; The LSMean for drug f is clearly significantly different from the average of the LSMeans of the other two drugs.

Figure 12.13 Contrast Example for the Drug Experiment

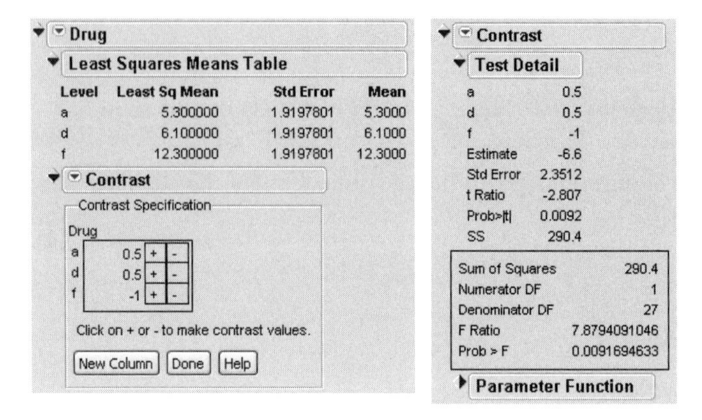

Analysis of Covariance

An analysis of variance model with an added regressor term is called an *analysis of covariance*. Suppose that the data are the same as above, but with one additional term, X_3, in the formula as a new regressor. X_1 and X_2 continue to be dummy variables that index over the three levels of the nominal effect. The model is written

$$y_i = \beta_0 + \beta_1 x_{1i} + \beta_2 x_{2i} + \beta_3 x_{3i} + \varepsilon_i$$

Now there is an intercept plus two effects, one a nominal main effect using two parameters, and the other an interval covariate regressor using one parameter.

Rerun the Snedecor and Cochran Drug.jmp example, but add the x to the model effects as a covariate. Compared with the main effects model (Drug effect only), the R^2 increases from 22.8% to 67.6%, and the standard error of the residual reduces from 6.07 to 4.0. The *F*-test significance for the whole model decreases from 0.03 to less than 0.0001.

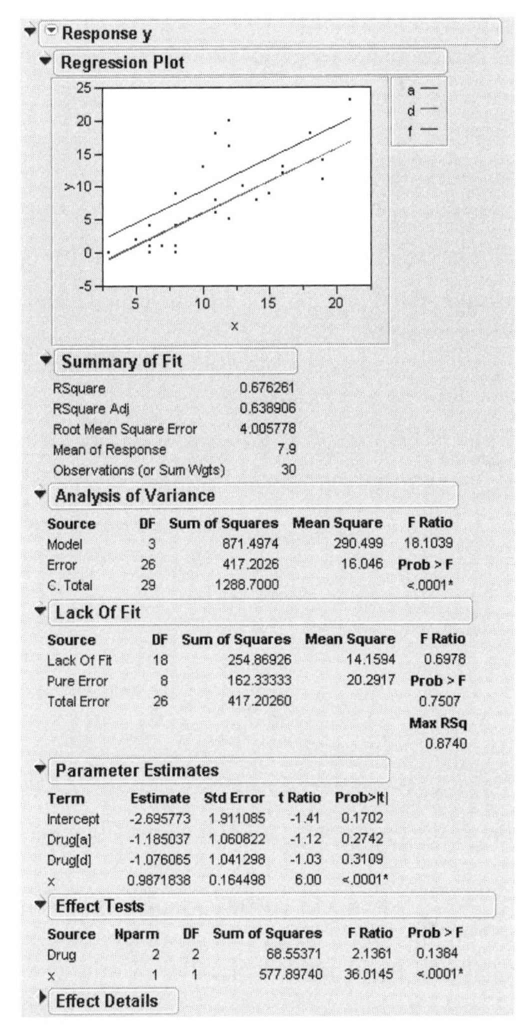

Summary of Fit	
RSquare	0.676261
RSquare Adj	0.638906
Root Mean Square Error	4.005778
Mean of Response	7.9
Observations (or Sum Wgts)	30

Source	DF	Sum of Squares	Mean Square	F Ratio
Model	3	871.4974	290.499	18.1039
Error	26	417.2026	16.046	Prob > F
C. Total	29	1288.7000		<.0001*

Source	DF	Sum of Squares	Mean Square	F Ratio
Lack Of Fit	18	254.86926	14.1594	0.6978
Pure Error	8	162.33333	20.2917	Prob > F
Total Error	26	417.20260		0.7507
				Max RSq
				0.8740

Term	Estimate	Std Error	t Ratio	Prob>|t|
Intercept	-2.695773	1.911085	-1.41	0.1702
Drug[a]	-1.185037	1.060822	-1.12	0.2742
Drug[d]	-1.076065	1.041298	-1.03	0.3109
x	0.9871838	0.164498	6.00	<.0001*

Source	Nparm	DF	Sum of Squares	F Ratio	Prob > F
Drug	2	2	68.55371	2.1361	0.1384
x	1	1	577.89740	36.0145	<.0001*

Sometimes you can investigate the functional contribution of a covariate. For example, some transformation of the covariate might fit better. If you happen to have data where there are exact duplicate observations for the regressor effects, it is possible to partition the total error into two components. One component estimates error from the data where all the *x* values are the same. The other estimates error that may contain effects for unspecified functional forms of covariates, or interactions of nominal effects. This is the basis for a lack of fit test. If the lack of fit error is significant, then the fit model platform warns that there is some effect in your data not explained by your model. Note that there is no significant lack of fit error in this example, as seen by the large probability value of 0.7507.

The covariate, X, has a substitution effect with respect to Drug. It accounts for much of the variation in the response previously accounted for by the Drug variable. Thus, even though the model is fit with much less error, the Drug effect is no longer significant. The effect previously observed in the main effects model now appears explainable to some extent in terms of the values of the covariate.

The least squares means are now different from the ordinary mean because they are adjusted for the effect of *X*, the covariate, on the response, *Y*. Now the least squares means are the predicted values that you expect for each of the three values of Drug given that the covariate, *X*, is held at some constant value. The constant value is chosen for convenience to be the mean of the covariate, which is 10.7333.

So, the prediction equation gives the least squares means as follows:

fit equation: `-2.696 - 1.185*`Drug[a - f] `- 1.0761*`Drug[d - f] `+ 0.98718*x`

for a: `-2.696 - 1.185*(1) -1.0761*(0) + 0.98718*(10.7333) = 6.71`

for d: `-2.696 - 1.185*(0) -1.0761*(1) + 0.98718*(10.7333) = 6.82`

for f: `-2.696 - 1.185*(-1) -1.0761*(-1) + 0.98718*(10.7333) = 10.16`

With two effects there is now a leverage plot for the whole model and a leverage plot for each effect. If the covariate makes no difference, the leverage plot for Drug would not be much different than shown in Figure 12.14. However, because the covariate makes a big difference, the leverage values for Drug are dispersed a bit from their least-squares means.

Figure 12.14 Comparison of Leverage Plots for Drug Test Data

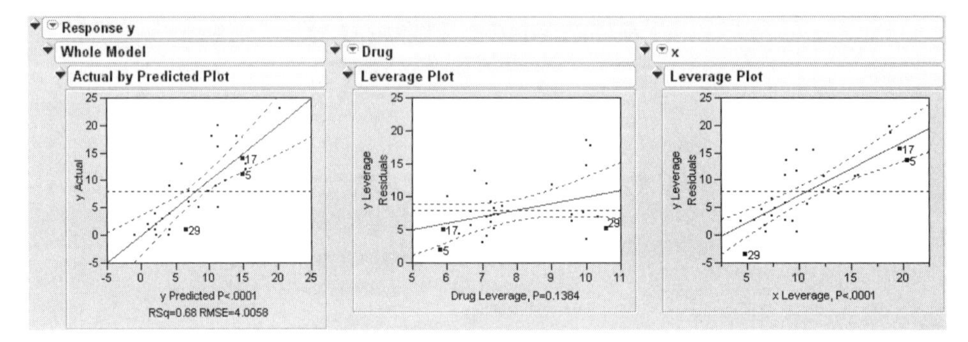

In the leverage plots above, the three points 5, 17, and 29 are highlighted to illustrate the following situations:

- Point 29 makes Drug appear less significant and x more significant. It pulls the right end of the leverage line fit for Drug towards the mean line, which tends to decrease its slope. It pulls the left end of the leverage line fit x away from the mean line, which tends to increase its slope.
- In the same manner, point 5 makes Drug appear more significant, but x less significant.
- Point 17 is important to the significance test for both Drug and x. It has a small residual but is far out on the *x*-axis. This means that if either effect or both effects were removed from the model, its residual would be large.

The **Save Prediction Formula** command from the Save menu creates a new column in the data table called Pred Formula Y using the prediction formula shown here.

```
-2.695772906127
                 ( "a" ⇒-1.1850365373806 )
+Match( Drug )   ( "d" ⇒-1.0760652051714 )
                 ( "f" ⇒2.261101742552   )
                 ( else⇒.                )
+0.98718381112985*x
```

When using this command, JMP first attempts to find a Response Limits property containing desirability functions, first from the profiler (if that option has been used), otherwise from the response columns. If you reset the desirabilities later, it affects subsequent saves, but not saved columns that were already made.

Analysis of Covariance with Separate Slopes

This example is a continuation of the Drug.jmp example presented in the previous section. The example uses data from Snedecor and Cochran (1967, p. 422). A one-way analysis of variance for a variable called Drug with values a, d, and f shows a difference in the mean response between the Drug main effect with a significance probability of 0.03.

The lack of fit test for the model with main effect Drug and covariate x is not significant. However, for the sake of illustration, this example includes the main effects and the Drug*x effect. This model tests whether the regression on the covariate has separate slopes for different Drug levels.

This specification adds two columns to the linear model (call them X_4 and X_5) that allow the slopes for the covariate to be different for each Drug level. The new variables are formed by multiplying the dummy variables for Drug by the covariate values giving

$$y_i = \beta_0 + \beta_1 x_{1i} + \beta_2 x_{2i} + \beta_3 x_{3i} + \beta_4 x_{4i} + \beta_5 x_{5i} + \varepsilon_i$$

Table 12.2 "Coding of Analysis of Covariance with Separate Slopes," p. 243, shows the coding of this Analysis of Covariance with Separate Slopes. **Note:** The mean of X is 10.7333.

Table 12.2 Coding of Analysis of Covariance with Separate Slopes

Regressor	Effect	Values
X_1	`Drug[a]`	+1 if a, 0 if d, –1 if f
X_2	`Drug[d]`	0 if a, +1 if d, –1 if f
X_3	X	the values of X
X_4	`Drug[a]*(X - 10.733)`	$+X - 10.7333$ if a, 0 if d, $-(X - 10.7333)$ if f
X_5	`Drug[d]*(X - 10.733)`	0 if a, $+X - 10.7333$ if d, $-(X - 10.7333)$ if f

The R^2 increases from 67.63% for the analysis of covariance with same slopes (see the previous example) to 69.2% for the model with separate slopes (Summary of Fit table not shown here). The separate slopes model shifts two degrees of freedom from Lack Of Fit error (Figure 12.15) to the model, increasing the DF for Model from 3 to 5 as shown in Figure 12.16.

Figure 12.15 Comparison of Lack of Fit Tables

Models with Same Slopes

Lack Of Fit

Source	DF	Sum of Squares	Mean Square	F Ratio
Lack Of Fit	18	254.86926	14.1594	0.6978
Pure Error	8	162.33333	20.2917	Prob > F
Total Error	26	417.20260		0.7507
				Max RSq
				0.8740

Models with Different Slopes

Lack Of Fit

Source	DF	Sum of Squares	Mean Square	F Ratio
Lack Of Fit	16	235.22462	14.7015	0.7245
Pure Error	8	162.33333	20.2917	Prob > F
Total Error	24	397.55795		0.7231
				Max RSq
				0.8740

The Pure Error seen in both Lack of Fit tables in Figure 12.15 is the same because there are no new variables in the separate slopes covariance model. The new effect in the separate slopes model is constructed from effects in the original analysis of covariance model.

Figure 12.16 Comparison of Analysis of Variance Tables

Analysis of Variance

Source	DF	Sum of Squares	Mean Square	F Ratio
Model	3	871.4974	290.499	18.1039
Error	26	417.2026	16.046	Prob > F
C. Total	29	1288.7000		<.0001*

Analysis of Variance

Source	DF	Sum of Squares	Mean Square	F Ratio
Model	5	891.1420	178.228	10.7594
Error	24	397.5580	16.565	Prob > F
C. Total	29	1288.7000		<.0001*

The remaining reports are the Least Squares Means table and other details. In the Effect Test reports, neither the Drug nor Drug*x effect is significant. Fitting different slopes results in adjusted least squares means for each drug and increases their standard errors slightly.

Leverage plots in Figure 12.17 illustrate the significance of the model and the x effect. In this figure, drug 'a' is labeled as a square, drug 'd' is labeled as a triangle, and drug 'f' is labeled as a diamond. The Drug main effect is less significant than it was, with the separate slopes covariate absorbing some of its effect. You can highlight points 5 and 25, which are the only ones that have much influence on the separate slopes effect.

Figure 12.17 Leverage Plots for the Model with Separate Slopes

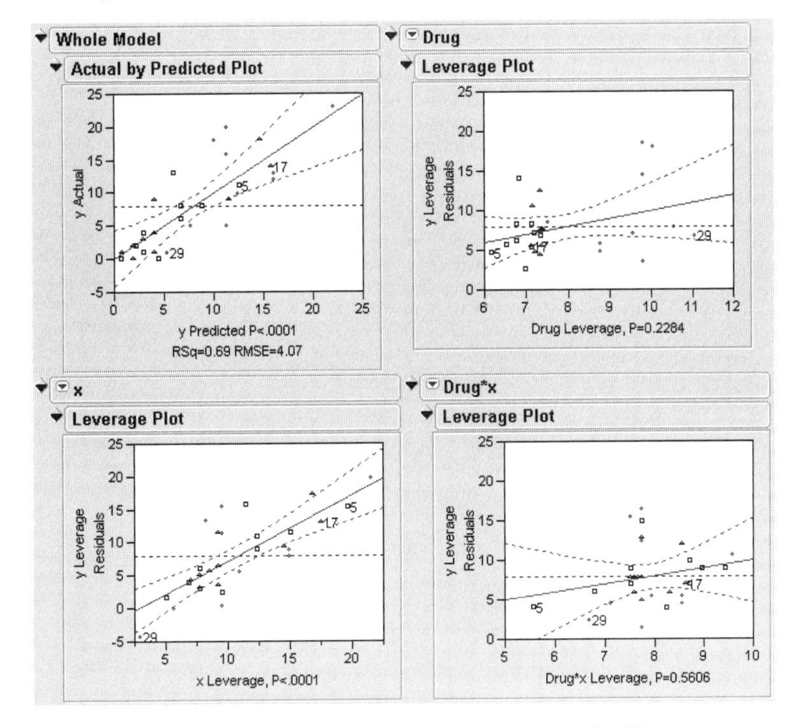

To summarize the fits made on the drug data, the following three graphs illustrate

- the simple main effects model (Figure 12.18)
- the model with a covariate, but with a common slope (Figure 12.19)
- the model with a covariate with separate slopes (Figure 12.20).

Construction of the least squares means is also shown. As an exercise, you can construct these plots by using the grouping feature Fit Y by X platform and the **Stack** table operation found in the **Tables** menu.

Figure 12.18 shows a scatterplot that illustrates the model y=Drug with lines drawn at the mean of each drug group.

Figure 12.18 Illustration of the Simple Mean Model

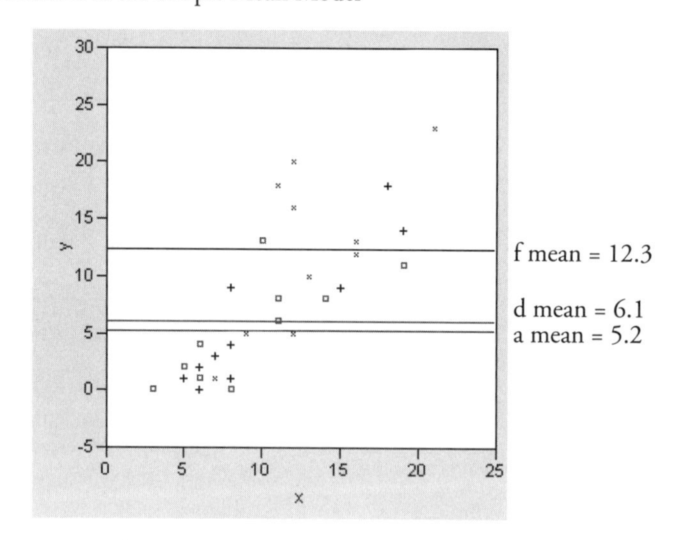

Figure 12.19 shows a scatterplot that illustrates the model y=Drug, x with lines drawn having the same slope for each drug group.

Figure 12.19 Illustration of the Analysis of Covariance Model

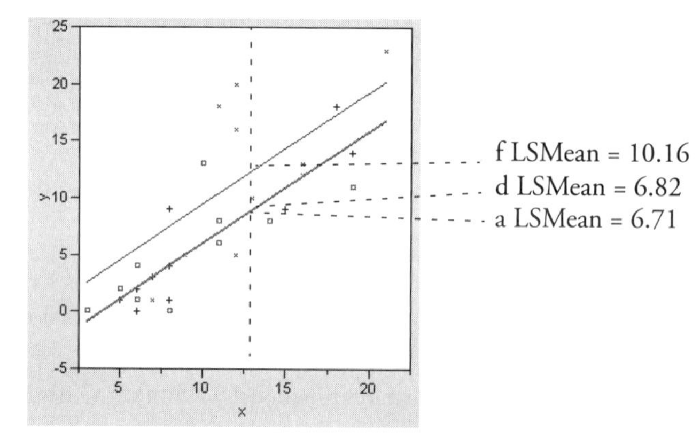

Figure 12.20 shows a scatterplot with lines drawn having different slopes for each drug group and illustrates the model y=Drug, x, and Drug*x.

Figure 12.20 Illustration of the Covariance Model with Different Slopes

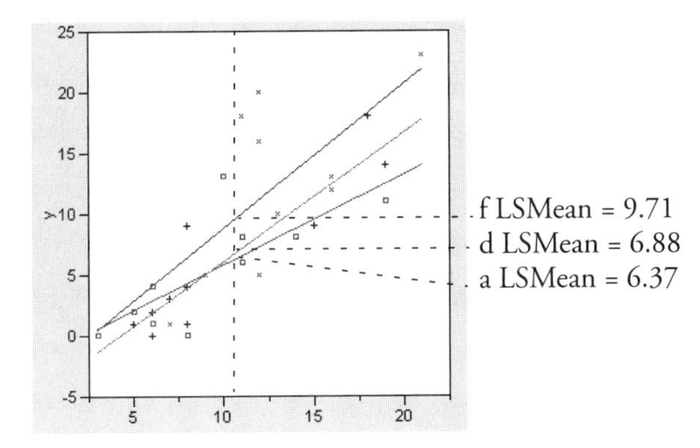

Chapter 13

Standard Least Squares: Perspectives on the Estimates

Fit Model Platform

Though the fitting platform always produces a report on the parameter estimates and tests on the effects, there are many options available to make these more interpretable. The following sections address these questions:

• How do I interpret estimates for nominal factors? How can I get the missing level coefficients?	**Expanded Estimates**
• How can I measure the size of an effect in a scale-invariant fashion?	**Scaled Estimates**
• If I have a screening design with many effects but few observations, how can I decide which effects are sizable and active?	**Effect Screening Options**
• How can I get a series of tests of effects that are independent tests whose sums of squares add up to the total, even though the design is not balanced?	**Sequential Tests**
• How can I test some specific combination?	**Custom Test**
• How can I predict (backwards) which x value led to a given y value?	**Inverse Prediction**
• How likely is an effect to be significant if I collect more data, have a different effect size, or have a different error variance?	**Power Analysis**
• How strongly are estimates correlated?	**Correlation of Estimates**

Contents

Estimates and Effect Screening Menus

Most parts of the Model Fit results are optional. When you click the popup icon next to the response name at the topmost outline level, the menu shown in Figure 13.1 lists commands for the continuous response model.

Figure 13.1 Commands for a Least Squares Analysis: the **Estimates** and **Effect Screening** Menus

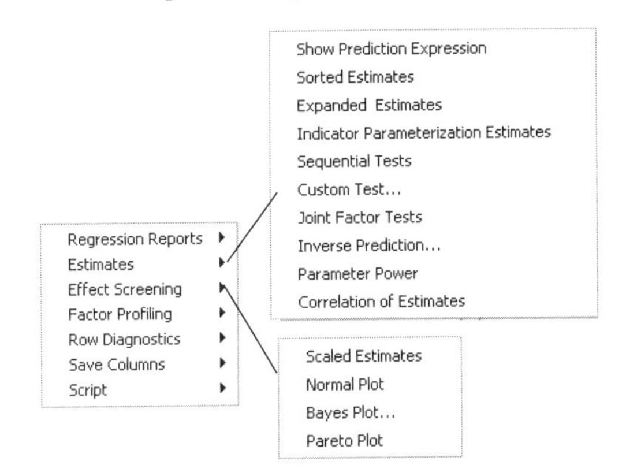

The specifics of the Fit Model platform depend on the type of analysis you do and the options and commands you ask for. The popup icons list commands and options at each level of the analysis. You always have access to all tables and plots through menu items. Also, the default arrangement of results can be changed by using preferences or script commands.

The focus of this chapter is on items in the **Estimates** and **Effect Screening** menus, which includes inverse prediction and a discussion of parameter power.

Show Prediction Expression

The **Show Prediction Expression** command places the prediction expression in the report.

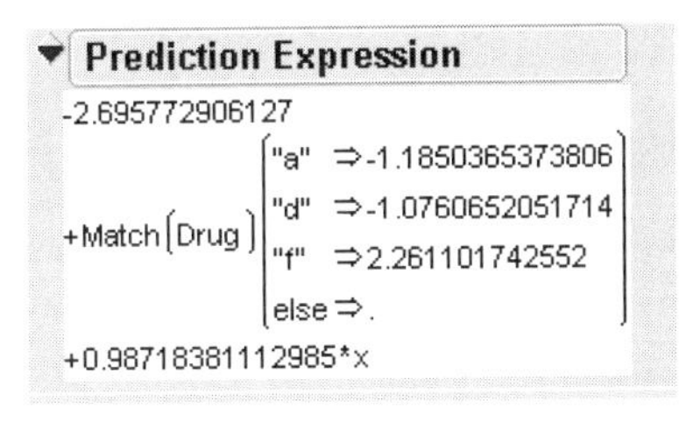

Sorted Estimates

The sorted Estimates command produces a different version of the Parameter Estimates report that is more useful in screening situations. This version of the report is especially useful if the design is saturated, when typical reports are less informative.

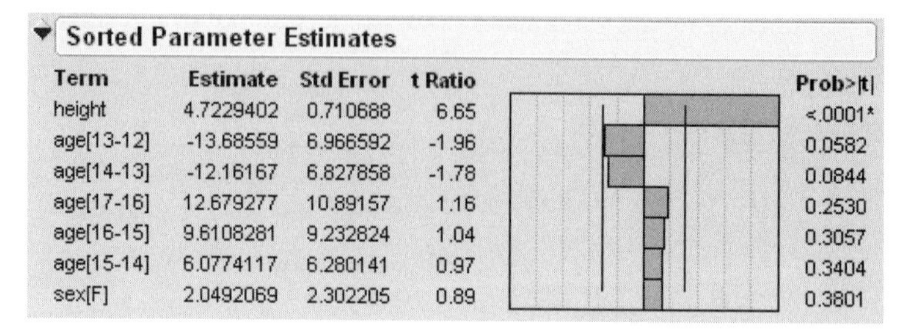

Sorted Parameter Estimates

Term	Estimate	Std Error	t Ratio	Prob>|t|
height	4.7229402	0.710688	6.65	<.0001*
age[13-12]	-13.68559	6.966592	-1.96	0.0582
age[14-13]	-12.16167	6.827858	-1.78	0.0844
age[17-16]	12.679277	10.89157	1.16	0.2530
age[16-15]	9.6108281	9.232824	1.04	0.3057
age[15-14]	6.0774117	6.280141	0.97	0.3404
sex[F]	2.0492069	2.302205	0.89	0.3801

This report is shown automatically if all the factors are two-level. It is also shown if the emphasis is screening and all the effects have only one parameter, in which case it is done instead of the Scaled Estimates report.

Note the following differences between this report and the Parameter Estimates report.

- This report does not show the intercept.
- The effects are sorted by the absolute value of the *t*-ratio, showing the most significant effects at the top.
- A bar graph shows the t ratio, with a line showing the 0.05 significance level.
- If JMP cannot obtain standard errors for the estimates, relative standard errors are used and notated.
- If there are no degrees of freedom for residual error, JMP constructs *t*-ratios and *p*-values using Lenth's Pseudo-Standard Error. These quantities are labeled with Pseudo in their name. A note explains the change and shows the PSE. To calculate *p*-values, JMP uses a DFE of $m/3$, where m is the number of parameter estimates excluding the intercept.
- If the parameter estimates have different standard errors, then the PSE is defined using the *t*-ratio rather than a common standard error.

Expanded Estimates and the Coding of Nominal Terms

Expanded Estimates is useful when there are categorical (nominal) terms in the model and you want a full set of effect coefficients.

When you have nominal terms in your model, the platform needs to construct a set of dummy columns to represent the levels in the classification. (Full details are shown in the appendix "Statistical Details," p. 975.) For n levels, there are n - 1 dummy columns. Each dummy variable is a zero-or-one indicator for a particular level, except for the last level, which is coded –1 for all dummy variables. For example, if column *A* has levels *A*1, *A*2, *A*3, then the dummy columns for *A*1 and *A*2 are as shown here.

A	*A*1 dummy	*A*2 dummy
*A*1	1	0
*A*2	0	1
*A*3	-1	-1

These columns are not displayed. They are just for conceptualizing how the fitting is done. The parameter estimates are the coefficients fit to these columns. In this case, there are two of them, labeled *A*[*A*1] and *A*[*A*2].

This coding causes the parameter estimates to be interpreted as how much the response for each level differs from the average across all levels. Suppose, however, that you want the coefficient for the last level, *A*[*A*3]. The coefficient for the last level is the negative of the sum across the other levels, because the sum across all levels is constrained to be zero. Although many other codings are possible, this coding has proven to be practical and interpretable.

However, you probably don't want to do hand calculations to get the estimate for the last level. The **Expanded Estimates** command in the **Estimates** menu calculates these missing estimates and shows them in a text report. You can verify that the mean (or sum) of the estimates across a classification is zero.

Keep in mind that the **Expanded Estimates** option with high-degree interactions of two-level factors produces a lengthy report. For example, a five-way interaction of two-level factors produces only one parameter but has $2^5 = 32$ expanded coefficients, which are all the same except for sign changes.

If you choose **Expanded Estimates** for the Drug.jmp example from the last chapter, you can see how the last level is expanded (see Figure 13.2), and that the estimates are expressions of how the least squares means differ from their average.

Figure 13.2 Comparison of Parameter Estimates and Expanded Estimates

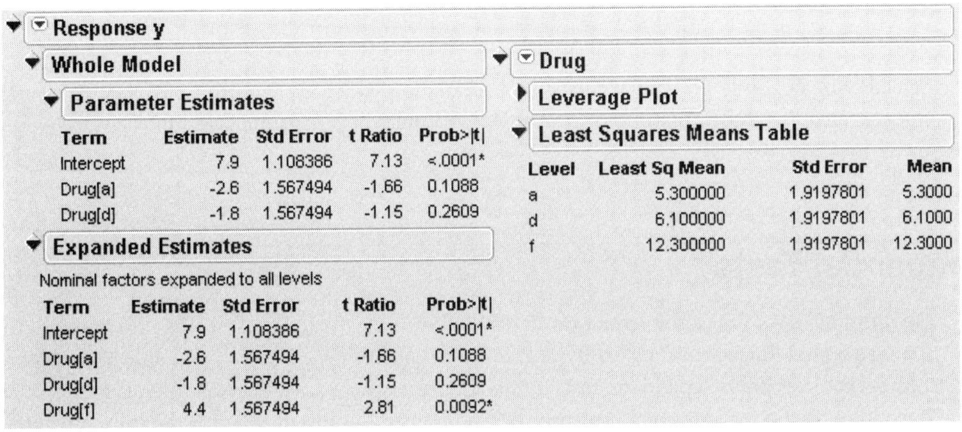

Response y

Whole Model

Parameter Estimates

Term	Estimate	Std Error	t Ratio	Prob>|t|
Intercept	7.9	1.108386	7.13	<.0001*
Drug[a]	-2.6	1.567494	-1.66	0.1088
Drug[d]	-1.8	1.567494	-1.15	0.2609

Expanded Estimates

Nominal factors expanded to all levels

Term	Estimate	Std Error	t Ratio	Prob>|t|
Intercept	7.9	1.108386	7.13	<.0001*
Drug[a]	-2.6	1.567494	-1.66	0.1088
Drug[d]	-1.8	1.567494	-1.15	0.2609
Drug[f]	4.4	1.567494	2.81	0.0092*

Drug

Leverage Plot

Least Squares Means Table

Level	Least Sq Mean	Std Error	Mean
a	5.300000	1.9197801	5.3000
d	6.100000	1.9197801	6.1000
f	12.300000	1.9197801	12.3000

To compute expanded estimates subtract the least squares category means from the grand mean giving:

```
 5.3 - 7.9 = -2.6
 6.1 - 7.9 = -1.8
12.3 - 7.9 =  4.4
```

Scaled Estimates and the Coding Of Continuous Terms

The parameter estimates are highly dependent on the scale of the factor. If you convert a factor from grams to kilograms, the parameter estimates change by a multiple of a thousand. If the same change is applied to a squared (quadratic) term, the scale changes by a multiple of a million. If you are interested in the effect size, then you should examine the estimates in a more scale-invariant fashion. This means converting from an arbitrary scale to a meaningful one so that the sizes of the estimates relate to the size of the effect on the response. There are many approaches to doing this. In JMP, the **Scaled Estimates** command on the **Effect Screening** menu gives coefficients corresponding to factors that are scaled to have a mean of zero and a range of two. If the factor is symmetrically distributed in the data then the scaled factor will have a range from –1 to 1. This corresponds to the scaling used in the design of experiments (DOE) tradition. Thus, for a simple regressor, the scaled estimate is half the predicted response change as the regression factor travels its whole range.

Scaled estimates are important in assessing effect sizes for experimental data in which the uncoded values are used. If you use coded values (–1 to 1), then the scaled estimates are no different than the regular estimates.

Also, you do not need scaled estimates if your factors have the **Coding** column property. In that case, they are converted to uncoded form when the model is estimated and the results are already in an interpretable form for effect sizes.

When you select **Scaled Estimates**, a table of estimates like the one shown here appears. As noted in the report, the estimates are parameter centered by the mean and scaled by range/2.

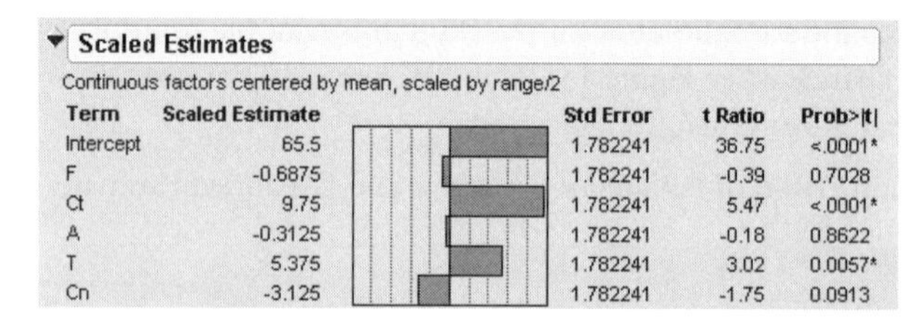

Scaled Estimates

Continuous factors centered by mean, scaled by range/2

Term	Scaled Estimate	Std Error	t Ratio	Prob>|t|
Intercept	65.5	1.782241	36.75	<.0001*
F	-0.6875	1.782241	-0.39	0.7028
Ct	9.75	1.782241	5.47	<.0001*
A	-0.3125	1.782241	-0.18	0.8622
T	5.375	1.782241	3.02	0.0057*
Cn	-3.125	1.782241	-1.75	0.0913

Scaled estimates also take care of the issues for polynomial (crossed continuous) models even if they are not centered by the parameterized **Center Polynomials** default launch option.

Sequential Tests

Sequential Tests shows the reduction in residual sum of squares as each effect is entered into the fit. The sequential tests are also called Type I sums of squares (Type I SS). A desirable property of the Type I SS is that they are independent and sum to the regression SS. An undesirable property is that they depend on the order of terms in the model. Each effect is adjusted only for the preceding effects in the model.

The following models are considered appropriate for the Type I hypotheses:

- balanced analysis of variance models specified in proper sequence (that is, interactions do not precede main effects in the effects list, and so forth)
- purely nested models specified in the proper sequence

- polynomial regression models specified in the proper sequence.

Custom Test

If you want to test a custom hypothesis, select **Custom Test** from the **Estimates** popup menu. This displays the dialog shown to the left in Figure 13.3 for constructing the test in terms of all the parameters. After filling in the test, click **Done**. The dialog then changes to a report of the results, as shown on the right in Figure 13.3.

The space beneath the **Custom Test** title bar is an editable area for entering a test name. Use the Custom Test dialog as follows:

Parameter lists the names of the model parameters. To the right of the list of parameters are columns of zeros corresponding to these parameters. Click a cell here, and enter a new value corresponding to the test you want.

Add Column adds a column of zeros so that you can test jointly several linear functions of the parameters. Use the **Add Column** button to add as many columns to the test as you want.

The last line in the Parameter list is labeled **=**. Enter a constant into this box to test the linear constraint against. For example, to test the hypothesis $\beta_0=1$, enter a 1 in the = box. In Figure 13.3, the constant is equal to zero.

When you finish specifying the test, click **Done** to see the test performed. The results are appended to the bottom of the dialog.

When the custom test is done, the report lists the test name, the function value of the parameters tested, the standard error, and other statistics for each test column in the dialog. A joint *F*-test for all columns shows at the bottom.

Warning: The test is always done with respect to residual error. If you have random effects in your model, this test may not be appropriate if you use EMS instead of REML.

Note: If you have a test within a classification effect, consider using the contrast dialog instead of a custom test, which tests hypotheses about the least squares means.

Figure 13.3 The Custom Test Dialog and Test Results for Age Variable

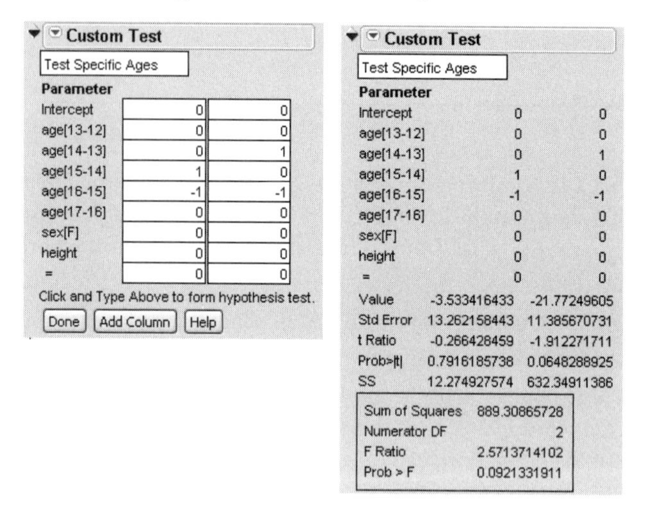

Joint Factor Tests

This command appears when interaction effects are present. For each term in the model, JMP produces a joint test on all the parameters involving that factor.

In the following example, we look at a model from Big Class.jmp, with weight as the response and height, age, and sex (with their second-order interactions) as effects. The joint test for these terms is shown in Figure 13.4.

Figure 13.4 Joint Factor Tests for Big Class model

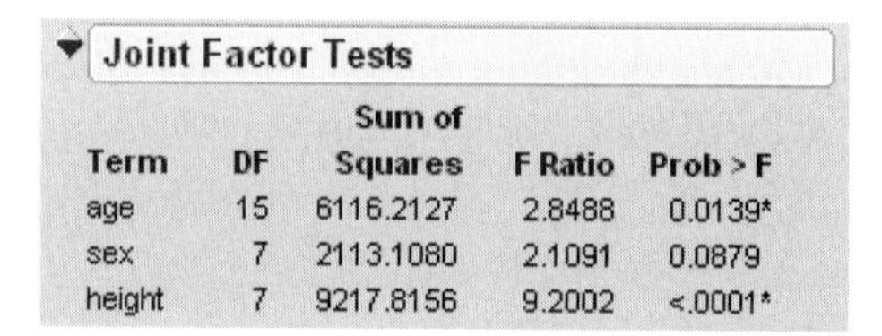

Joint Factor Tests

Term	DF	Sum of Squares	F Ratio	Prob > F
age	15	6116.2127	2.8488	0.0139*
sex	7	2113.1080	2.1091	0.0879
height	7	9217.8156	9.2002	<.0001*

Note age has 15 DFs because it is testing the five parameters for age, the five parameters for age*sex, and the five parameters for height*age, all tested to be zero.

Inverse Prediction

To find the value of x for a given y requires *inverse prediction*, sometimes called *calibration*. The **Inverse Prediction** command on the **Estimates** menu displays a dialog (Figure 13.5) that lets you ask for a specific value of one independent (X) variable, given a specific value of a dependent variable and other x

values. The inverse prediction computation includes confidence limits (fiducial limits) on the prediction.

As an example, use the Fitness.jmp file in the Sample Data folder and specify Oxy as *Y* and Runtime as *X*. When there is only a single *X*, as in this example, the Fit Y by X platform can give you a visual estimate of inverse prediction values. The figure shown here is the Fit Y by X scatterplot with the **Fit Line** option in effect. The crosshair tool approximates inverse prediction. To find which value of Runtime gives an Oxy value of 50, position the crosshair tool with its horizontal line crossing the Oxy axis at 50, and its intersection with the vertical line positioned on the prediction line. The vertical crosshair then intersects the Runtime axis at about 9.75, which is the inverse prediction.

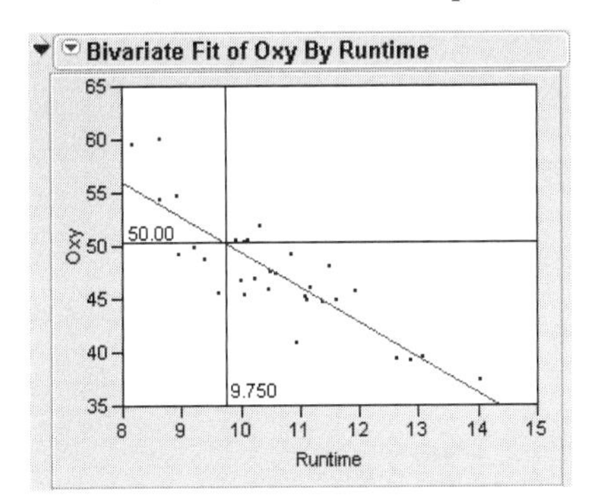

However, to see the exact prediction of Runtime, use the Fit Model dialog, assign Oxy as the *Y* variable, and assign Runtime as the single model effect. Click **Run Model** and choose **Inverse Prediction** from the **Estimates** menu. When the Inverse Prediction dialog appears, enter one or more values for Oxy, as shown in Figure 13.5. When you click **Run**, the dialog disappears and the Inverse Prediction table in Figure 13.6 gives the exact predictions for each Oxy value specified, with upper and lower 95% confidence limits. The exact prediction for Runtime when Oxy is 50 is 9.7935, which is close to the approximate prediction of 9.75 found using the Fit Y by X platform.

Figure 13.5 Inverse Prediction Given by the Fit Model Platform

Inverse Prediction

Oxy	1-Alpha
40	0.9500
45	
50	
55	
.	
.	
.	

Click/Enter values for Oxy

☐ Confid interval with respect to individual rather than expected response

Run Help

Figure 13.6 Inverse Prediction Given by the Fit Model Platform

Inverse Prediction

Oxy	Predicted Runtime	Lower Limit	Upper Limit	1-Alpha
40.000000	12.8140955	12.3214517	13.5402388	0.9500
45.000000	11.3037749	10.9863914	11.6963705	
50.000000	9.7934543	9.3882316	10.1156016	
55.000000	8.2831337	7.5378659	8.7870387	

Confidence Interval with respect to an expected response

Note: The fiducial confidence limits are formed by Fieller's method. Sometimes this method results in a degenerate (outside) interval, and JMP reports missing values. See the appendix "Statistical Details," p. 975, for information about computing confidence limits.

The inverse prediction command also predicts a single *x* value when there are multiple effects in the model. To predict a single *x*, you supply one or more *y* values of interest and set the *x* value you want to predict to be missing. By default, the other *x* values are set to the regressor's means but can be changed to any desirable value. The example in Figure 13.7–Figure 13.8 shows the dialog and results to predict Runtime where Oxy is *Y* and Runtime, RunPulse, and RstPulse are regressors.

Figure 13.7 Inverse Prediction Dialog for a Multiple Regression Model

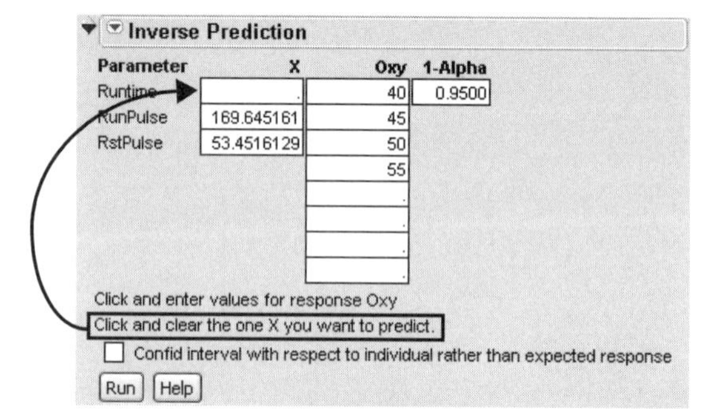

Figure 13.8 Inverse Prediction for a Multiple Regression Model

Inverse Prediction

Oxy	Predicted Runtime	Lower Limit	Upper Limit	1-Alpha
40.000000	12.9080094	12.3435031	13.8245494	0.9500
45.000000	11.3340253	10.9978860	11.7835576	
50.000000	9.7600413	9.2920075	10.1028271	
55.000000	8.1860572	7.2435302	8.7646954	

X Values

Confidence Interval with respect to an expected response

X Values

1	.	169.6452	53.45161

Parameter Power

Suppose that you want to know how likely your experiment is to detect some difference at a given α-level. The probability of getting a significant test result is termed the power. The power is a function of the unknown parameter values tested, the sample size, and the unknown residual error variance.

Or, suppose that you already did an experiment and the effect was not significant. If you think that it might have been significant if you had more data, then you would like to get a good guess of how much more data you need.

JMP offers the following calculations of statistical power and other details relating to a given hypothesis test.

- LSV, the *least significant value*, is the value of some parameter or function of parameters that would produce a certain *p*-value alpha.
- LSN, the *least significant number*, is the number of observations that would produce a specified *p*-value alpha if the data has the same structure and estimates as the current sample.
- *Power* is the probability of getting at or below a given *p*-value alpha for a given test.

The LSV, LSN, and power values are important measuring sticks that should be available for all test statistics. They are especially important when the test statistics do not show significance. If a result is not significant, the experimenter should at least know how far from significant the result is in the space of the estimate (rather than in the probability) and know how much additional data is needed to confirm significance for a given value of the parameters.

Sometimes a novice confuses the role of the null hypotheses, thinking that failure to reject the null hypothesis is equivalent to proving it. For this reason, it is recommended that the test be presented in these other aspects (power and LSN) that show how sensitive the test is. If an analysis shows no significant difference, it is useful to know the smallest difference the test is likely to detect (LSV).

The power details provided by JMP can be used for both prospective and retrospective power analyses. A prospective analysis is useful in the planning stages of a study to determine how large your sample size must be in order to obtain a desired power in tests of hypothesis. See the section "Prospective Power Analysis," p. 264, for more information and a complete example. A retrospective analysis is useful during the data analysis stage to determine the power of hypothesis tests already conducted.

Technical details for power, LSN, and LSV are covered in the section "Power Calculations," p. 996 in the appendix "Statistical Details," p. 975.

> Calculating retrospective power at the actual sample size and estimated effect size is somewhat non-informative, even controversial [Hoenig and Heisey, 2001]. Certainly, it doesn't give additional information to the significance test, but rather shows the test in just a different perspective. However we believe that many studies fail due to insufficient sample size to detect a meaningful effect size, and there should be some facility to help guide for the next study, for specified effect sizes and sample sizes.
> For more information, see John M. Hoenig and Dinnis M. Heisey, (2001) "The Abuse of Power: The Pervasive Fallacy of Power Calculations for Data Analysis.", *American Statistician* (v55 No 1, 19-24).

Power commands are only available for continuous-response models. Power and other test details are available in the following contexts:

- If you want the 0.05 level details for all parameter estimates, use the **Parameter Power** command in the **Estimates** menu. This produces the LSV, LSN, and adjusted power for an alpha of 0.05 for each parameter in the linear model.
- If you want the details for an *F*-test for a certain effect, find the **Power Analysis** command in the popup menu beneath the effect details for that effect.
- If you want the details for a Contrast, create the contrast from the popup menu next to the effect's title and select **Power Analysis** in the popup menu next to the contrast.

- If you want the details for a custom test, first create the test you want with the **Custom Test** command from the platform popup menu and then select **Power Analysis** command in the popup menu next to the Custom Test.

In all cases except the first, a Power Analysis dialog lets you enter information for the calculations you want.

The Power Analysis Dialog

The Power Analysis dialog (Figure 13.9) displays the contexts and options for test detailing. You fill in the values as described next, and then click **Done**. The results are appended at the end of the report.

Examples that follow use the Big Class.jmp data table from the Sample Data folder. The **Fit Model** script generated the report from which the Power examples were taken.

Figure 13.9 Power Analysis Dialog

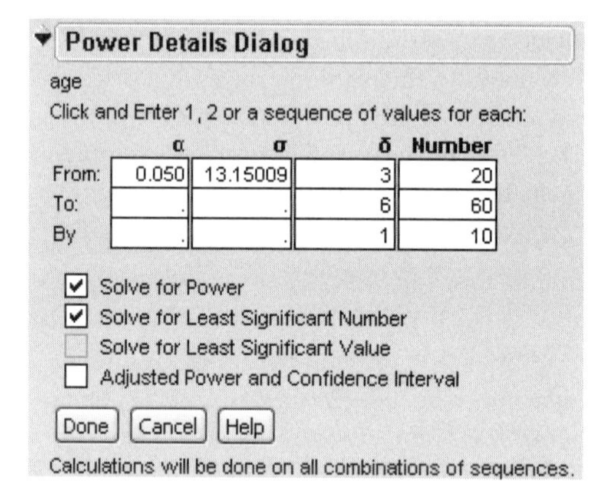

For each of four columns **Alpha**, **Sigma**, **Delta**, and **Number**, you can fill in a single value, two values, or the start, stop, and increment for a sequence of values, as shown in the dialog in Figure 13.9. Power calculations are done on all possible combinations of the values you specify.

Alpha (α) is the significance level, between 0 and 1, usually 0.05, 0.01, or 0.10. Initially, **Alpha** automatically has a value of 0.05. Click on one of the three positions to enter or edit one, two, or a sequence of values.

Sigma (σ) is the standard error of the residual error in the model. Initially, RMSE, the square root of the mean square error, is supplied here. Click on one of the three positions to enter or edit one, two, or a sequence of values.

Delta (δ) is the raw effect size. See "Effect Size," p. 261, for details. The first position is initially set to the square root of the sum of squares for the hypothesis divided by n. Click on one of the three positions to enter or edit one, two, or a sequence of values.

Number (n) is the sample size. Initially, the actual sample size is in the first position. Click on one of the three positions to enter or edit one, two, or a sequence of values.

Click the following check boxes to request the results you want:

Solve for Power Check to solve for the power (the probability of a significant result) as a function of α, σ, δ, and n.

Solve for Least Significant Number Check to solve for the number of observations expected to be needed to achieve significance alpha given α, σ, and δ.

Solve for Least Significant Value Check to solve for the value of the parameter or linear test that produces a *p*-value of alpha. This is a function of α, σ, and n. This feature is only available for one-degree-of-freedom tests and is used for individual parameters.

Adjusted Power and Confidence Interval To look at power retrospectively, you use estimates of the standard error and the test parameters. Adjusted power is the power calculated from a more unbiased estimate of the noncentrality parameter. The confidence interval for the adjusted power is based on the confidence interval for the noncentrality estimate. Adjusted power and confidence limits are only computed for the original δ, because that is where the random variation is.

Effect Size

The power is the probability that an F achieves its α-critical value given a noncentrality parameter related to the hypothesis. The noncentrality parameter is zero when the null hypothesis is true—that is, when the effect size is zero. The noncentrality parameter λ can be factored into the three components that you specify in the JMP Power dialog as

$$\lambda = (n\delta^2)/s^2 .$$

Power increases with λ, which means that it increases with sample size n and raw effect size δ and decreases with error variance σ^2. Some books (Cohen 1977) use standardized rather than raw Effect Size, $\Delta = \delta/\sigma$, which factors the noncentrality into two components $\lambda = n\Delta^2$.

Delta (δ) is initially set to the value implied by the estimate given by the square root of SSH/n, where SSH is the sum of squares for the hypothesis. If you use this estimate for delta, you might want to correct for bias by asking for the Adjusted Power.

In the special case for a balanced one-way layout with k levels

$$\delta^2 = \frac{\sum(\alpha_i - \bar{\alpha})^2}{k}$$

Because JMP's parameters are of the form

$$\beta_i = (\alpha_i - \bar{\alpha}) \text{ with } \beta_k = -\sum_{m=1}^{k-1} \alpha_m$$

the delta for a two-level balanced layout is

$$\delta^2 = \frac{\beta_1^2 + (-\beta)_1^2}{2} = \left|\beta_1^2\right|$$

Text Reports for Power Analysis

The power analysis facility calculates power as a function of every combination of the α, σ, δ, and n values you specify in the Power Analysis Dialog.

- For every combination of α, σ, and δ in the Power Analysis dialog, it calculates the least significant number.
- For every combination of α, σ, and *n* it calculates the least significant value.

For example, if you run the request shown in Figure 13.9, you get the tables shown in Figure 13.10.

Figure 13.10 The Power Analysis Tables

Power Details

Test age

Power

α	σ	δ	Number	Power
0.0500	13.15009	3	20	0.0828
0.0500	13.15009	3	30	0.1117
0.0500	13.15009	3	40	0.1426
0.0500	13.15009	3	50	0.1755
0.0500	13.15009	3	60	0.2099
0.0500	13.15009	4	20	0.1117
0.0500	13.15009	4	30	0.1694
0.0500	13.15009	4	40	0.2317
0.0500	13.15009	4	50	0.2969
0.0500	13.15009	4	60	0.3630
0.0500	13.15009	5	20	0.1524
0.0500	13.15009	5	30	0.2515
0.0500	13.15009	5	40	0.3554
.0500	13.15009	5		

Least Significant Number

α	σ	δ	Number(LSN)
0.0500	13.15009	3	216.8578
0.0500	13.15009	4	123.8892
0.0500	13.15009	5	80.93391
0.0500	13.15009	6	57.6814

If you check **Adjusted Power and Confidence Interval** in the Power Analysis dialog, the Power report includes the **AdjPower**, **LowerCL**, and **UpperCL** columns. The tables in Figure 13.11 are the result of requesting least significant number, and power and confidence limits for **Number** values 10 to 20 by 2.

Figure 13.11 The Power Analysis Tables with Confidence Limits

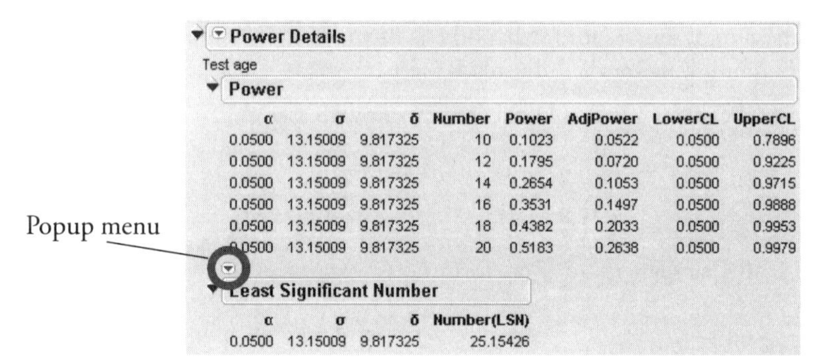

Power Details

Test age

Power

α	σ	δ	Number	Power	AdjPower	LowerCL	UpperCL
0.0500	13.15009	9.817325	10	0.1023	0.0522	0.0500	0.7896
0.0500	13.15009	9.817325	12	0.1795	0.0720	0.0500	0.9225
0.0500	13.15009	9.817325	14	0.2654	0.1053	0.0500	0.9715
0.0500	13.15009	9.817325	16	0.3531	0.1497	0.0500	0.9888
0.0500	13.15009	9.817325	18	0.4382	0.2033	0.0500	0.9953
0.0500	13.15009	9.817325	20	0.5183	0.2638	0.0500	0.9979

Least Significant Number

α	σ	δ	Number(LSN)
0.0500	13.15009	9.817325	25.15426

Plot of Power by Sample Size

The popup command located at the bottom of the Power report gives you the command **Power Plot**, which plots the Power by N columns from the Power table. The plot on the right in Figure 13.12 shows the result when you plot the example table in Figure 13.10. This plot is enhanced with horizontal and vertical grid lines on the major tick marks. Double-click on the axes and complete the dialog to change tick marks and add grid lines on an axis.

Figure 13.12 Plot of Power by Sample Size

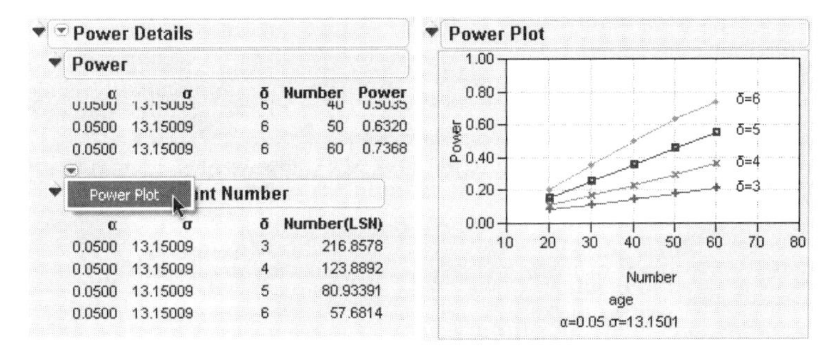

The Least Significant Value (LSV)

After a single-degree of freedom hypothesis test is performed, you often want to know how sensitive the test was. Said another way, you want to know how small an effect would be declared significant at some *p*-value alpha. The LSV provides a significance measuring stick on the scale of the parameter, rather than on a probability scale. It shows how sensitive the design and data are. It encourages proper statistical intuition concerning the null hypothesis by highlighting how small a value would be detected as significant by the data.

- The LSV is the value that the parameter must be greater than or equal to in absolute value to give the *p*-value of the significance test a value less than or equal to alpha.
- The LSV is the radius of the confidence interval for the parameter. A 1–alpha confidence interval is derived by taking the parameter estimate plus or minus the LSV.
- The absolute value of the parameter or function of the parameters tested is equal to the LSV, if and only if the *p*-value for its significance test is exactly alpha.

Compare the absolute value of the parameter estimate to the LSV. If the absolute parameter estimate is bigger, it is significantly different from zero. If the LSV is bigger, the parameter is not significantly different from zero.

The Least Significant Number (LSN)

The LSN or *least significant number* is defined to be the number of observations needed to drive down the variance of the estimates enough to achieve a significant result with the given values of alpha, sigma, and delta (the significance level, the standard deviation of the error, and the effect size, respectively). If you need more data points (a larger sample size) to achieve significance, the LSN helps tell you how many more.

Note: LSN is not a recommendation of how large a sample to take because the probability of significance (power) is only about 0.5 at the LSN.

The LSN has these characteristics:

- If the LSN is less than the actual sample size *n*, then the effect is significant. This means that you have more data than you need to detect the significance at the given alpha level.

- If the LSN is greater than n, the effect is not significant. In this case, if you believe that more data will show the same standard errors and structural results as the current sample, the LSN suggests how much data you would need to achieve significance.
- If the LSN is equal to n, then the p-value is equal to the significance level alpha. The test is on the border of significance.
- Power (described next) calculated when n = LSN is always greater than or equal to 0.5.

The Power

The power is the probability of getting a significant result. It is a function of the sample size n, the effect size δ, the standard deviation of the error σ, and the significance level α. The power tells you how likely your experiment is to detect a difference at a given α-level. Power has the following characteristics:

- If the true value of the parameter is the hypothesized value, the power should be alpha, the size of the test. You do not want to reject the hypothesis when it is true.
- If the true value of the parameters is not the hypothesized value, you want the power to be as great as possible.
- The power increases with the sample size. The power increases as the error variance decreases. The power increases as the true parameter gets farther from the hypothesized value.

The Adjusted Power and Confidence Intervals

Because power is a function of population quantities that are not known, it is usual practice to substitute sample estimates in power calculations (Wright and O'Brien 1988). If you regard these sample estimates as random, you can adjust them to have a more proper expectation. You can also construct a confidence interval for this adjusted power; however, the confidence interval is often very wide. The adjusted power and confidence intervals can only be computed for the original δ because that is where the random variation is. For details about adjusted power see "Computations for Adjusted Power," p. 998 in the appendix "Statistical Details," p. 975.

Prospective Power Analysis

Prospective analysis helps you answer the question, "Will I detect the group differences I am looking for given my proposed sample size, estimate of within-group variance, and alpha level?" In a prospective power analysis, you must provide estimates of the group means and sample sizes in a data table and an estimate of the within-group standard deviation σ in the Power Analysis dialog.

The **Sample Size and Power** command found on the **DOE** menu offers a facility that computes power, sample size, or the effect size you want to detect, for a given alpha and error standard deviation. You supply two of these values and the Sample Size and Power feature computes the third. If you supply only one of these values, the result is a plot of the other two. This feature is available for the single sample, two sample, and k sample situations, and assumes equal sample sizes. For unequal sample sizes, you can do as shown in the following example

This example is from Wright and O'Brien (1988). Dr. Noah Decay, a dentist, is planning an experiment in which he wishes to compare two treatments that are believed to reduce dental plaque. Dr. Decay believes that the means for the control group and two treatments are 32, 26, and 24, respectively, and that the within-group standard deviation is between 6 and 9. The control group has twice as many patients as each treatment group.

To run a prospective power analysis on the test for the group effect, Dr. Decay must first create the data table shown here, with three columns Group, Mean, and N, that contain the group names, group means, and proposed group sizes. You can open the Noah Decay.jmp sample data to see this table.

Untitled

Columns (3/0): grp, mean, n

	grp	mean	n
1	Contro	32	20
2	T1	26	10
3	T2	24	10

Dr. Decay uses the Fit Model platform to set up a one-way analysis of variance (as shown here). Note that the sample size variable, N, is declared as **Freq**. Also, the **Minimal Reports** emphasis option is in effect.

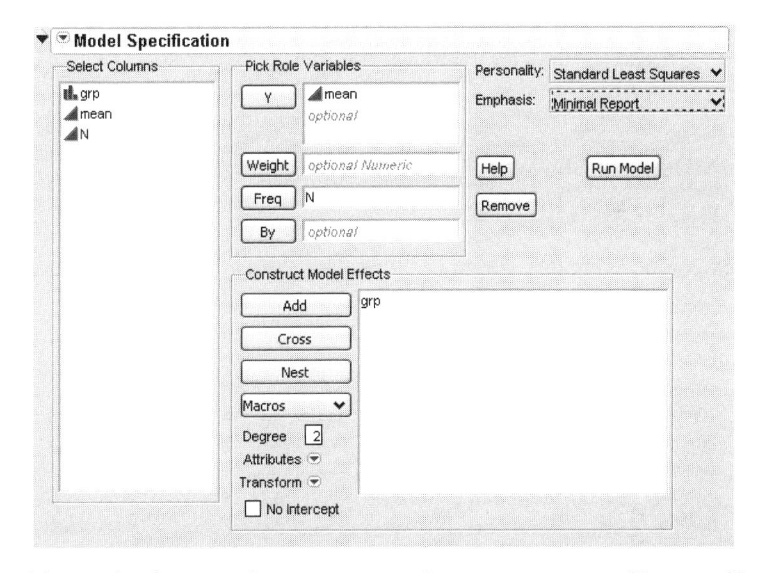

The analysis yields zero for the sum of squares error and mean square error. However, Dr. Decay suspects that the standard deviation is between 6 and 9. He now chooses **Power Analysis** from the Grp report popup menu beneath Effect Details and checks **Solve for Power** and **Solve for Least Significant Number**. Then, the Power Analysis dialog as shown to the left in Figure 13.13 gives calculations for all combinations of alpha = 0.05, sample sizes of 40 to 60, and standard deviations (sigma) of 6 to 9 by 1.

The **Delta** shown was calculated from the specified mean values. A prospective power analysis uses population values, which give an unbiased noncentrality parameter value, so adjusted power and confidence limits are not relevant.

Figure 13.13 Power Details and Power by N Plot for Prospective Analysis

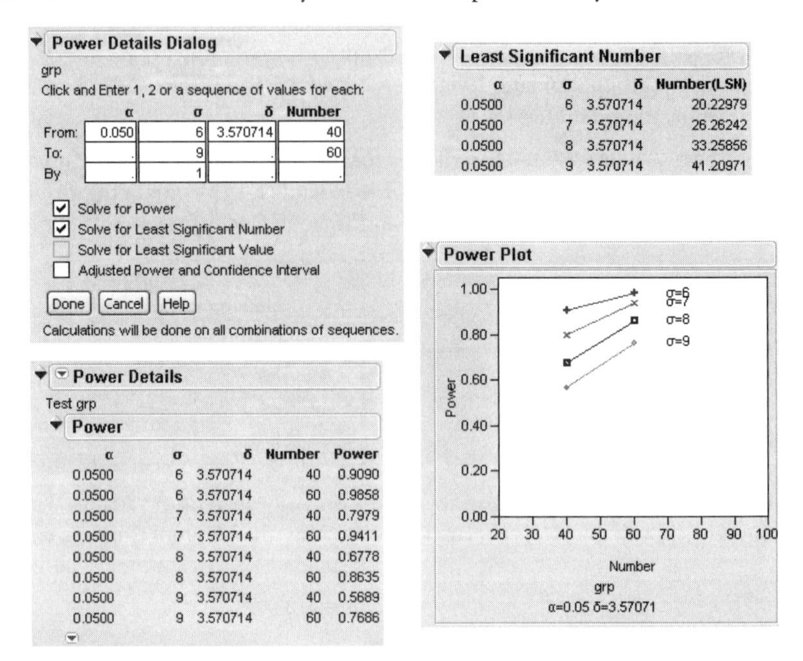

Correlation of Estimates

The **Correlation of Estimates** option in the **Estimates** platform menu produces a correlation matrix for the all effects in a model. The example shown here uses the Tiretread data with ABRASON as the response, two factors SILICA and SILANE, and their interaction.

Figure 13.14 Correlation of Estimates Report

Response ABRASION

Whole Model

Correlation of Estimates

Corr

	Intercept	SILICA	SILANE	(SILICA-1.2)*(SILANE-50)
Intercept	1.0000	-0.4281	-0.8919	0.0000
SILICA	-0.4281	1.0000	0.0000	0.0000
SILANE	-0.8919	0.0000	1.0000	0.0000
(SILICA-1.2)*(SILANE-50)	0.0000	0.0000	0.0000	1.0000

Effect Screening

The **Effect Screening** commands help examine the sizes of the effects. The **Scaled Estimates** command was covered earlier under "Scaled Estimates and the Coding Of Continuous Terms," p. 254, which dealt with scaling issues. The other **Effect Screening** commands are discussed here. These commands correct for scaling and for correlations among the estimates. The features of these **Effect Screening** commands are derived by noticing three things:

1. The process of fitting can be thought of as converting one set of realizations of random values (the response values) into another set of realizations of random values (the parameter estimates). If the design is balanced with an equal number of levels per factor, these estimates are independent and identically distributed, just as the responses are. If there are as many parameters as data points, they even have the same variance as the responses.
2. If you are fitting a screening design with many effects and only a few runs, you expect that only a few effects are active. That is, a few effects have sizable impact and the rest of them are inactive (they are estimating zeroes). This is called the assumption of *effect sparsity.*
3. Given points 1 and 2 above, you can think of screening as a way to determine which effects are inactive with random values around zero and which ones are outliers, not part of the distribution of inactive effects.

Thus, you treat the estimates themselves as a set of data to help you judge which effects are active and which are inactive. If there are few runs, with little or no degrees of freedom for error, then there are no classical significance tests, and this approach is especially needed.

The last three **Effect Screening** commands look at parameters in this way. There are two default reports, Lenth's PSE and the Parameter Estimates Population table with significant factors highlighted. The **Effect Screening** command also has a submenu for looking at the model parameters from different angles using scaled estimates and three plots.

Figure 13.15 Effect Screening

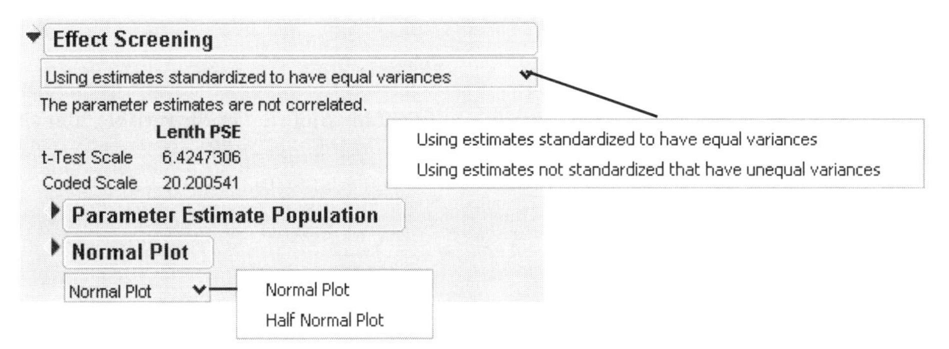

The following descriptions are for an analysis on the Reactor.jmp data using four factors, Ct, A, T, and Cn, with all two-factor interactions included in the model.

Lenth's Method

An estimate of standard error is calculated using the method of Lenth (1989) and shows in the Effect Screening table (shown above). This estimate, called the *pseudo standard error*, is formed by taking 1.5

times the median absolute value of the estimates after removing all the estimates greater than 3.75 times the median absolute estimate in the complete set of estimates.

Parameter Estimates Population

Most inferences about effect size first assume that the estimates are uncorrelated and have equal variances. This is true for fractional factorials and many classical experimental designs. However, for some designs it is not true. The last three **Effect Screening** commands display the Parameter Estimates Population report, which first finds the correlation of the estimates and tells you whether or not the estimates are uncorrelated and have equal variances.

If the estimates are correlated, a normalizing transformation can be applied to make them uncorrelated and have equal variances. If there is an estimate of error variance, the transformed estimates have variances of near 1 and are the same as the *t*-ratios. If an estimate of error variance is unavailable, the transformed estimates have the same variance as the error.

If the estimates are uncorrelated and have equal variances, then the following notes appear and the analysis is straightforward.

- The parameter estimates have equal variances
- The parameter estimates are not correlated.

If the estimates are correlated and/or have unequal variances, then each of these two notes may change into a popup menu showing that it has transformed the estimates, and giving you the option to undo the transformation.

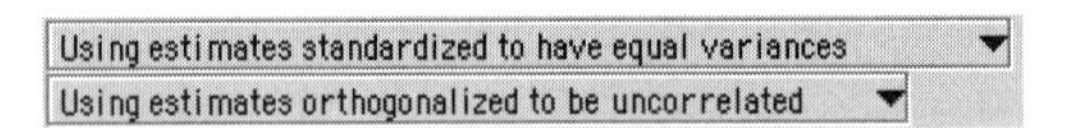

If you click these popup menus and undo both transformations, the lines look like this:

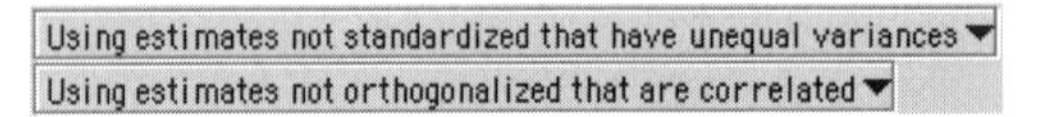

The transformation to make the estimates uncorrelated is the same as that used to calculate sequential sums of squares. The estimates measure the additional contribution of the variable after all previous variables have been entered into the model.

Parameter Estimate Population

Term	Original	t Ratio	Orthog Coded	Orthog t Test	Prob>|t|
Intercept	65.5000	118.5903	65.5000	118.5903	<.0001*
Ct	9.7500	17.6527	9.7500	17.6527	<.0001*
A	-0.3125	-0.5658	-0.3125	-0.5658	0.5775
T	5.3750	9.7316	5.3750	9.7316	<.0001*
Cn	-3.1250	-5.6579	-3.1250	-5.6579	<.0001*
Ct*A	0.4375	0.7921	0.4375	0.7921	0.4372
Ct*T	6.6250	11.9948	6.6250	11.9948	<.0001*
Ct*Cn	1.0000	1.8105	1.0000	1.8105	0.0845
A*T	1.0625	1.9237	1.0625	1.9237	0.0681
A*Cn	0.4375	0.7921	0.4375	0.7921	0.4372
T*Cn	-5.5000	-9.9580	-5.5000	-9.9580	<.0001*

The table here shows a Parameter Estimate Population table, which has these columns:

Original lists the parameter estimates for the fitted linear model. These estimates can be compared in size only if the X's are scaled the same.

t ratio is the t test associated with this parameter.

Orthog Coded contains the orthogonalized estimates. They are used in the Pareto plot because this plot partitions the sum of the effects, which requires orthogonality. The orthogonalized values are computed by premultiplying the column vector of the Original estimates by the Cholesky root of $X'X$. These estimates depend on the model order unless the original design is orthogonal.

If the design was orthogonal and balanced, then these estimates will be identical to the original estimates. If they are not, then each effect's contribution is measured after it is made orthogonal to the effects before it.

Orthog t-Test lists the parameter estimates after a transformation that makes them independent and identically distributed. These values are used in the Normal Plot (discussed next), which requires uncorrelated estimates of equal variance. The *p*-values associated with Orthog *t*-test estimates (given in the Prob>|t| column) are equivalent to Type I sequential tests. This means that if the parameters of the model are correlated, the estimates and their *p*-values depend on the order of terms in the model. The Orthog t-test estimates are computed by dividing the orthogonalized estimates (Orthog Coded estimates) by their standard errors. The Orthog t-test values let JMP treat the estimates as if they were from a random sample for use in Normal plots or Bayes plots.

Prob>|t| is the significance level or *p*-value associated with the values in the Orthog t-Test column.

Normal Plot

The **Normal Plot** command displays the Parameter Estimates Population table (discussed in the previous section) and shows a normal plot of these parameter estimates (Daniel 1959). The estimates are on the vertical axis and the normal quantiles on the horizontal axis. The normal plot for *Y* in the Reactor.jmp screening model is shown in Figure 13.16. If all effects are due to random noise, they tend to follow a straight line with slope σ, the standard error. The line with slope equal to the Lenth's PSE estimate is shown in blue.

The Normal plot helps you pick out effects that deviate from the normal lines. Estimates that deviate substantially are labeled. In this example, The A factor does not appear important. But not only are the other three factors (T, Ct, and Cn) active, T (temperature) appears to interact with both Ct and Cn.

Figure 13.16 Normal Plot for the Reactor Data with Two-Factor Interactions

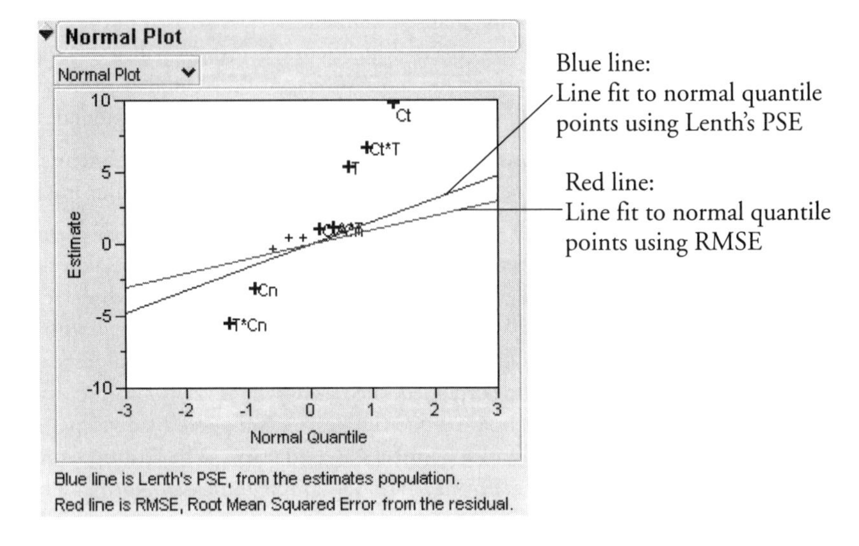

Half-Normal Plot

At the top of the Normal plot is a popup menu as shown in Figure 13.15 "Effect Screening," p. 267.

Click this and select **Half Normal Plot** to obtain a plot of the absolute values of the estimates against the normal quantiles for the absolute value normal distribution, as shown in Figure 13.17. Some analysts prefer this to the Normal Plot.

Figure 13.17 Half Normal Plot

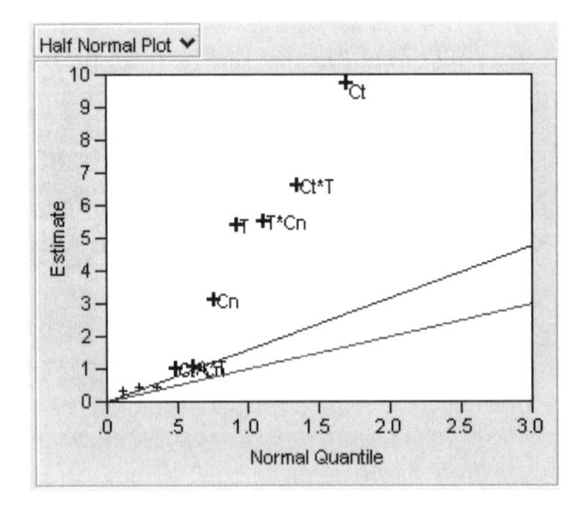

Bayes Plot

Another approach to resolving which effects are important (sometimes referred to as which contrasts are active) is to compute posterior probabilities using a Bayesian approach. This method, due to Box and Meyer (1986), assumes that the estimates are a mixture from two distributions. Some portion of the effects is assumed to come from pure random noise with a small variance. The remaining terms are assumed to come from a *contaminating* distribution that has a variance *K* times larger than the error variance.

The prior probability for an effect is the chance you give that effect of being nonzero, (or being in the contaminating distribution). These priors are usually set to equal values for each effect, and 0.2 is a commonly recommended prior probability value. The *K* contamination coefficient is often set at 10, which says the contaminating distribution has a variance that is 10 times the error variance.

The Bayes plot is done with respect to normalized estimates (JMP lists as Orthog *t*-Test), which have been transformed to be uncorrelated and have equal variance. To see a Box-Meyer Bayesian analysis for the reactor example, select **Bayes Plot** from the **Effect Screening** menu. The dialog panel, shown to the left in Figure 13.18, asks you to fill in the prior probabilities and the *K* coefficient. You can also edit the other values. It is not necessary to have a nonzero DFE. Click **Go** to start the calculation of the posterior probabilities.

Figure 13.18 Bayes Plot Dialog and Plot

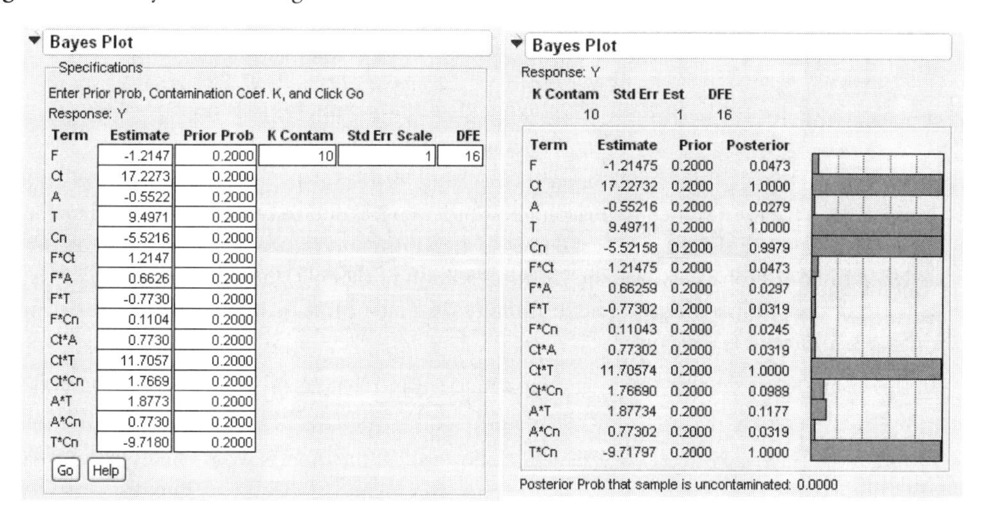

The **Std Err Scale** field is set to 0 for a saturated model with no estimate of error. If there is an estimate of standard error (the root mean square error), this field is set to 1 because the estimates have already been transformed and scaled to unit variance. If you edit the field to specify a different value, it should be done as a scale factor of the RMSE estimate.

The resulting posterior probabilities are listed and plotted with bars as shown in Figure 13.18. The literature refers to this as the Bayes Plot. An overall posterior probability is also listed for the outcome that the sample is uncontaminated. This is formed from the posteriors for each sample β_i by

$$\text{P(uncontam)} = \prod_{i=1}^{n} (1 - \beta_i(\sigma))$$

Bayes Plot for Factor Activity

JMP includes a script that allows you to determine which factors are active in the design. Found in the Sample Scripts folder, BayesPlotforFactors.jsl can be used to produce the Factor Activity plot. Open and run the script (**Edit > Run Script**) on the Reactor data, using Y as Y and F, Ct, A, T, and Cn as factors. This produces the following plot.

Figure 13.19 Bayes Plot for Factor Activity

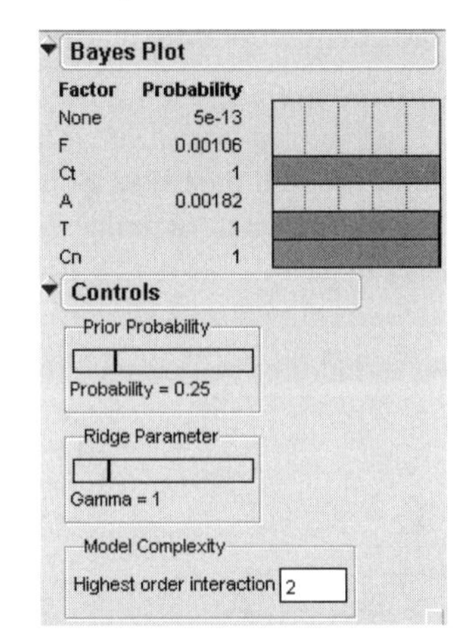

In this case, we have specified that the highest order interaction to consider is two. Therefore, all possible models that include (up to) second order interactions are constructed, and, based on the value assigned to Prior Probability (see the Controls section of the plot), a posterior probability is computed for each of the possible models. The probability for a factor is the sum of the probabilities for each of the models where it was involved.

In this example, we see that Ct, T, and Cn are active and that A and F are not. This agrees with Figure 13.18.

Note: If the ridge parameter were zero (not allowed), all the models would be fit by least-squares. As the ridge parameter gets large the parameter estimates for any model shrink towards zero. Details on the ridge parameter (and why it cannot be zero) are explained in Box and Meyer (1993).

Pareto Plot

The **Pareto Plot** selection gives plots of the absolute values of the orthogonalized estimates showing their composition relative to the sum of the absolute values. The Pareto plot is done with respect to the unorthogonalized estimates.

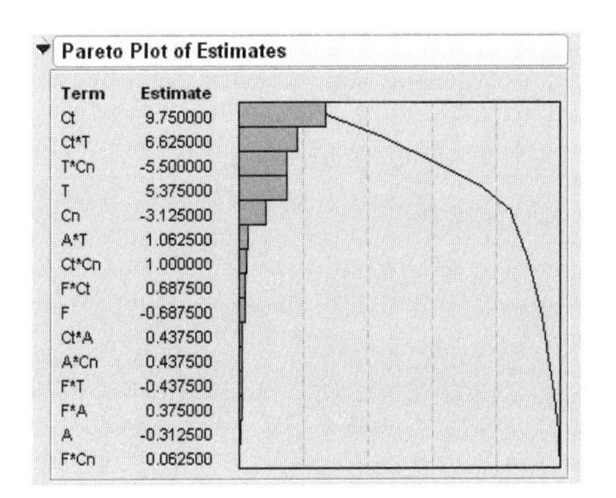
Pareto Plot of Estimates
Term Estimate
Ct 9.750000
Ct*T 6.625000
T*Cn -5.500000
T 5.375000
Cn -3.125000
A*T 1.062500
Ct*Cn 1.000000
F*Ct 0.687500
F -0.687500
Ct*A 0.437500
A*Cn 0.437500
F*T -0.437500
F*A 0.375000
A -0.312500
F*Cn 0.062500

Chapter 14

Standard Least Squares: Exploring the Prediction Equation

The Fit Model Platform

Assuming that the prediction equation is estimated well, there is still work in exploring the equation itself to answer a number of questions:

- What kind of curvature does the response surface have?
- What are the predicted values at the corners of the factor space?
- Would a transformation on the response produce a better fit?

The tools described in this chapter explore the prediction equation to answer these questions in the context of assuming that the equation is correct enough to work with.

Contents

Exploring the Prediction Equation

Figure 14.1 illustrates the relevant commands available for the least squares fitting personality. This chapter covers the commands that are not involved in profiling (See "The Profiler," p. 291). The examples use the Reactor.jmp sample data file.

Figure 14.1 Commands and Options for a Least Squares Analysis

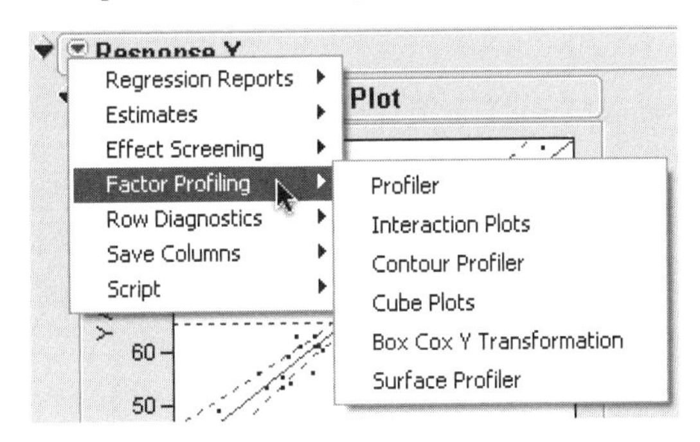

Here is a quick description of each feature to help you decide which section you need.

Profiler, Contour Profiler, and Surface Profiler display cuts through response surfaces for analyzing and optimizing a model. For complete details on using the Profilers, see the chapter "Profiling," p. 287.

Interaction plots are multiple profile plots across one factor under different settings of another factor. The traces are not parallel when there is a sizable interaction.

Cube Plots show predicted values in the corners of the factor space.

Box Cox Y Transformation finds a power transformation of the response that fits best.

The Profiler

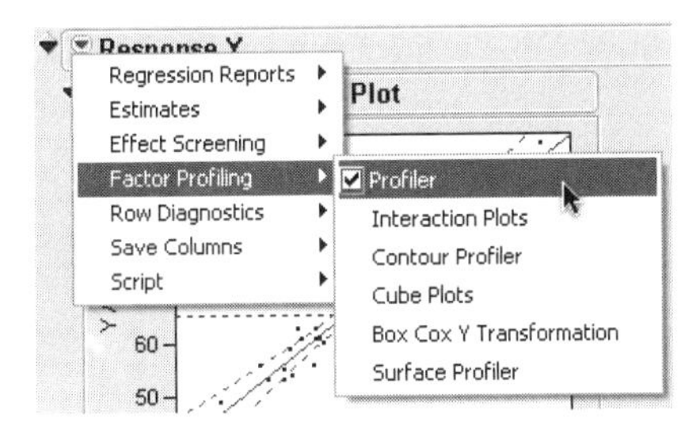

The **Profiler** displays prediction traces (see Figure 14.2) for each *X* variable. The vertical dotted line for each *X* variable shows its *current value* or *current setting*.

If the variable is nominal, the *x*-axis identifies categories.

For each *X* variable, the value above the factor name is its current value. You change the current value by clicking in the graph or by dragging the dotted line where you want the new current value to be.

- The horizontal dotted line shows the *current predicted value* of each *Y* variable for the current values of the *X* variables.
- The blue lines within the plots show how the predicted value changes when you change the current value of an individual *X* variable. The 95% confidence interval for the predicted values is shown by a dotted curve surrounding the prediction trace (for continuous variables) or an error bar (for categorical variables).

The Prediction Profiler is a way of changing one variable at a time and looking at the effect on the predicted response.

Figure 14.2 Illustration of Prediction Traces

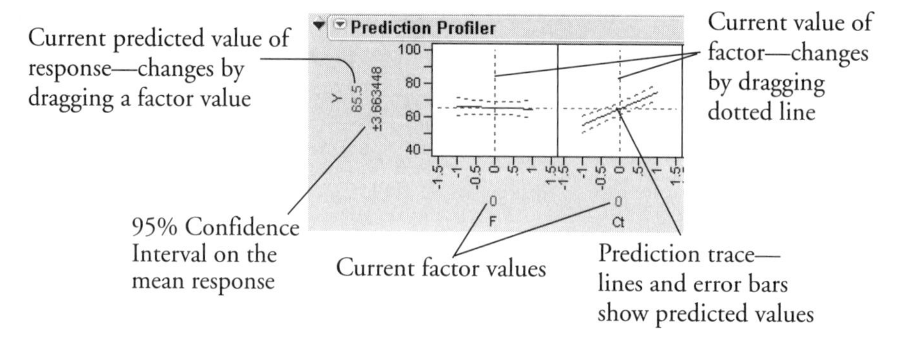

For details on using the profiler, see the chapter "The Profiler," p. 291.

Contour Profiler

The **Contour Profiler** option in the **Factor Profiling** submenu brings up an interactive contour profiling facility. This is useful for optimizing response surfaces graphically.

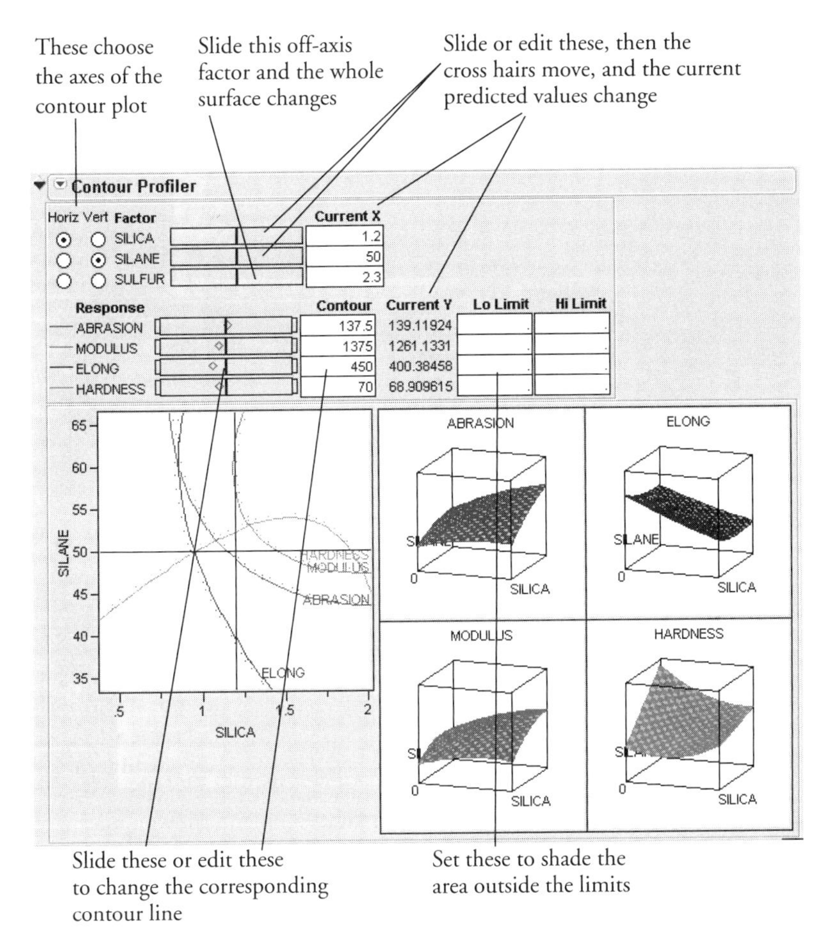

Details on the Contour profiler are found in "Contour Profiler," p. 306 in the "Profiling" chapter.

Surface Profiler

The Surface Profiler shows a three-dimensional surface plot of the response surface. Details of Surface Plots are found in the "Plotting Surfaces" chapter.

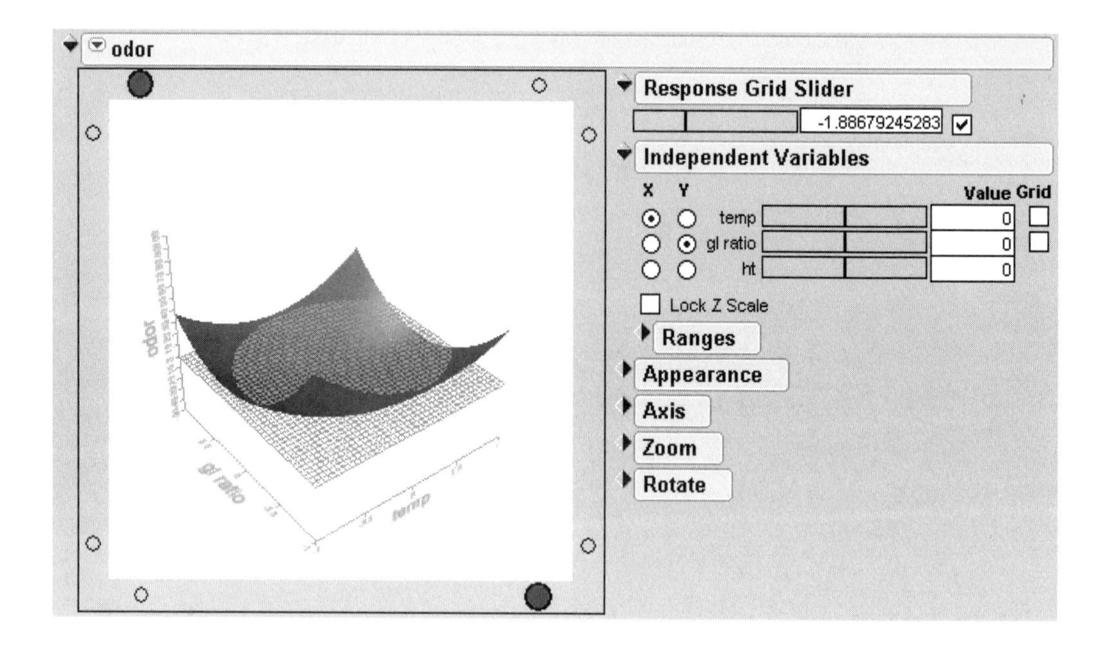

Interaction Plots

The **Interaction Plots** selection in the **Factor Profiling** submenu shows a matrix of interaction plots when there are interaction effects in the model. As an example, use the **Reactor.jmp** sample data with factors **Ct**, **A**, **T**, and **Cn**, and all two-factor interactions. You can then see the interaction effects with the **Interaction Plots** command.

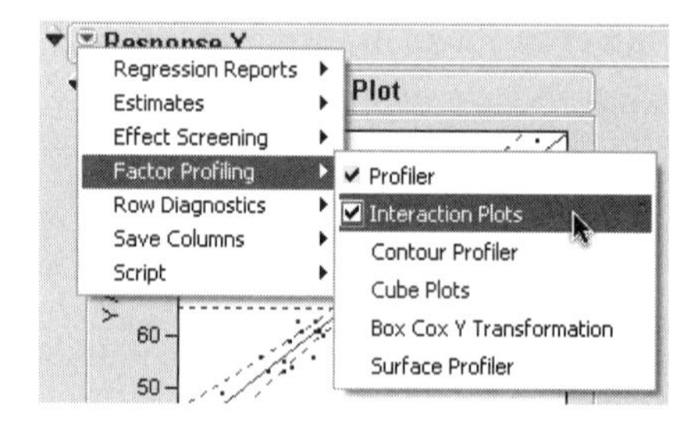

In an interaction plot, evidence of interaction shows as nonparallel lines. For example, in the **T*Cn** plot in the bottom row of plots (Figure 14.3) the effect of Cn is very small at the low values of temperature, but it diverges widely for the high values of temperature. The interaction of **Cn** with **T** tends to mask the effect of **Cn** as a main effect.

Figure 14.3 Interaction Plots for the Reactor Example

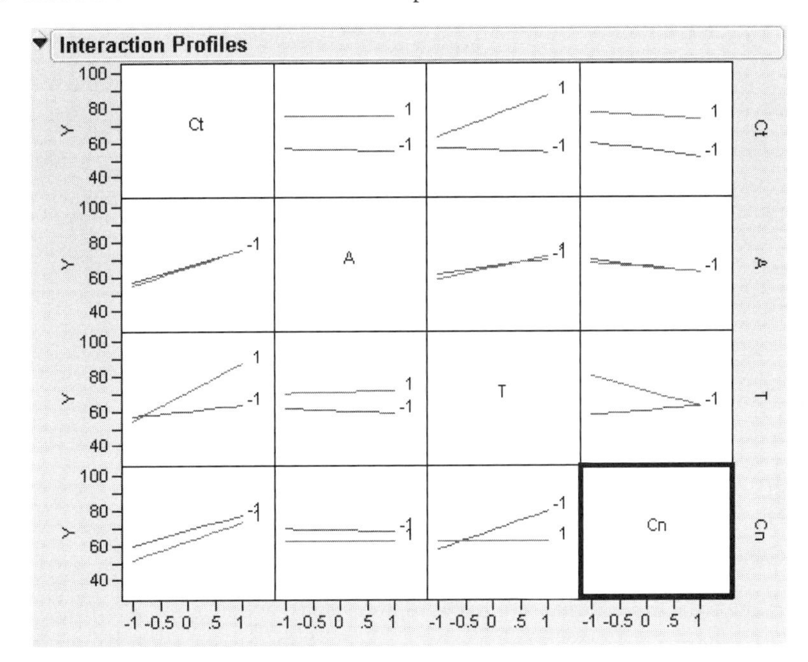

Note that the lines of a cell in an interaction plot appear dotted rather than solid when there is no corresponding interaction term in the model.

Cube Plots

The **Cube Plots** option in the **Factor Profiling** submenu displays a set of predicted values for the extremes of the factor ranges, laid out on the vertices of cubes, as illustrated in Figure 14.4. If a factor is nominal, the vertices are the first and last level. To resize the cube plots, click at the lower right corner of the plot frame and drag when the cursor becomes a double arrow.

If you want to change the layout so that terms are mapped to different cube coordinates, click one of the term names in the first cube and drag it to the coordinate you want. In the example below, if you click T and drag it to Cn=–1, then T identifies the cubes and Cn identifies the front-to-back coordinate. If there is more than one response, the multiple responses are shown stacked at each vertex.

Figure 14.4 Cube Plot Examples

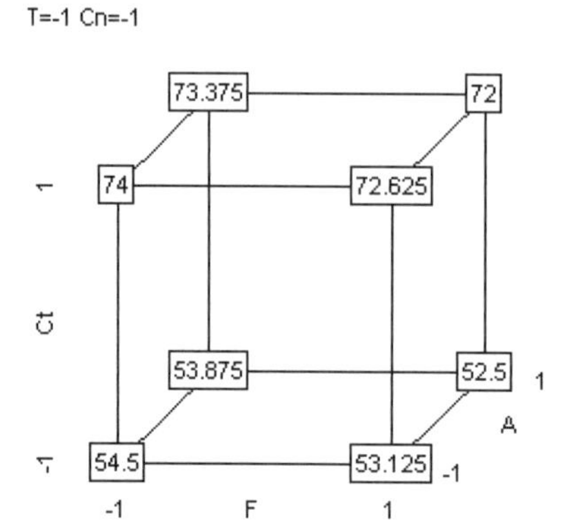

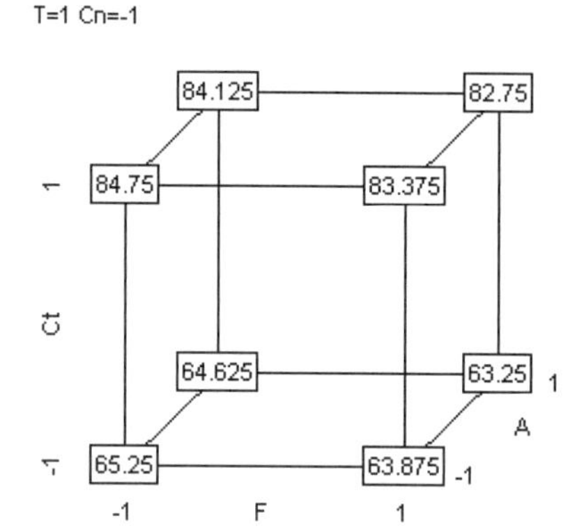

Response Surface Curvature

Often in industrial experiments, the goal is to find values for one or more factors that maximize or minimize the response. JMP provides surface modeling with special reports that show the critical values, the surface curvature, and a response contour plot.

The **Response Surface** selection in the **Effect Macros** popup menu automatically constructs all the linear, quadratic, and cross product terms needed for a response surface model.

The same model can be specified using the **Add** and **Cross** buttons to create model effects in the Model Specification dialog. You then select a model term and assign it the **Response Surface** effect attribute found in the **Attributes** popup menu. The response surface effects show with **&RS** after their name in the **Construct Model Effects** list, as shown in Figure 14.5.

After the parameters are estimated, critical values for the factors in the estimated surface can be found. Critical points occur at either maximums, minimums, or saddle points in the surface. The eigenvalues and eigenvectors of the calculated quadratic parameters determine the type of curvature. The eigenvectors show the principal directions of the surface, including the directions of greatest and smallest curvature.

The following example uses the Odor Control Original.jmp file in the Sample Data folder. The example uses data discussed in John (1971). The objective is to find the range of temperature (temp), gas-liquid ratio (gl ratio), and height (Ht) values that minimize the odor of a chemical production process. To run this response surface model, select the response variable, odor, from the variable selector list and click **Y**. Drag or Shift-click to highlight the factors **temp**, **gl ratio**, and **ht**, and choose **Response Surface** from the **Macros** popup menu. Note that the main effects automatically appear in the **Construct Model Effects** list with **&RS** after their name (see Figure 14.5).

Figure 14.5 Model Specification dialog For Response Surface Analysis

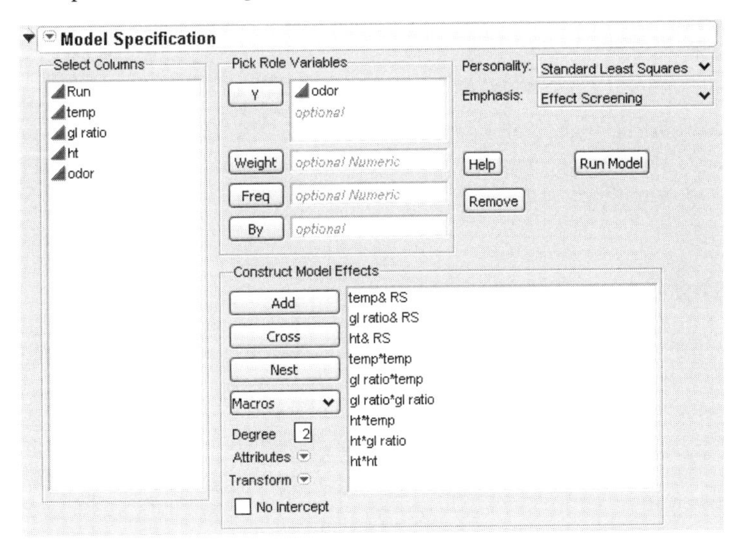

Parameter Estimates

The Parameter Estimates table shows estimates of the linear terms, the quadratic terms for each factor, and the crossed effect of the two factors.

Parameter Estimates

Term	Estimate	Std Error	t Ratio	Prob>|t|
Intercept	-30.66667	12.97797	-2.36	0.0645
temp	-12.125	7.947353	-1.53	0.1876
gl ratio	-17	7.947353	-2.14	0.0854
ht	-21.375	7.947353	-2.69	0.0433*
temp*temp	32.083333	11.69819	2.74	0.0407*
gl ratio*temp	8.25	11.23925	0.73	0.4959
gl ratio*gl ratio	47.833333	11.69819	4.09	0.0095*
ht*temp	1.5	11.23925	0.13	0.8990
ht*gl ratio	-1.75	11.23925	-0.16	0.8824
ht*ht	6.0833333	11.69819	0.52	0.6252

If you use the **Prediction Formula** command in the **Save Columns** submenu menu, a new data table column called Pred Formula odor saves the prediction formula using the coefficients from the Parameter Estimates table. The following formula is from the JMP calculator for the column of predicted values:

```
-30.666672+-12.125*temp+-17*gl
  ratio-21.375*ht++(temp)*((temp)*32.083337)+(gl ratio)*((temp)*8.25)++(gl
  ratio)*((gl ratio)*47.8333337)+(ht)*((temp)*1.5)++(ht)*((gl
  ratio)*-1.75)+(ht)*((ht)*6.083368)
```

The probability value of 0.0657 in the Analysis of Variance table indicates that the three-variable response surface model is only marginally better than the sample mean.

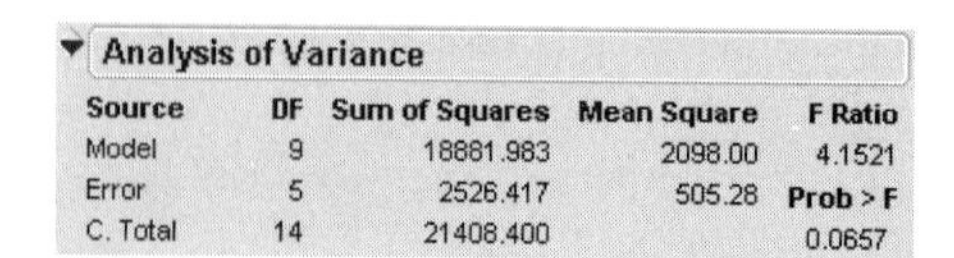

Analysis of Variance

Source	DF	Sum of Squares	Mean Square	F Ratio
Model	9	18881.983	2098.00	4.1521
Error	5	2526.417	505.28	Prob > F
C. Total	14	21408.400		0.0657

The response surface analysis also displays the Response Surface table and the Solution table (Figure 14.6). The Solution table shows the critical values for the surface variables and indicates that the surface solution point is a minimum for this example.

Canonical Curvature Table

The Canonical Curvature report, found under the Response Surface title, shows the eigenstructure (Figure 14.6), which is useful for identifying the shape and orientation of the curvature and results from the eigenvalue decomposition of the matrix of second-order parameter estimates. The eigenvalues (given in the first row of the Canonical Curvature table) are negative if the response surface shape curves back from a maximum. The eigenvalues are positive if the surface shape curves up from a minimum. If the eigenvalues are mixed, the surface is saddle shaped, curving up in one direction and down in another direction.

The eigenvectors listed beneath the eigenvalues (48.8588, 31.1035, and 6.0377) show the orientations of the principal axes. In this example the eigenvalues are positive, which indicates that the curvature bends up from a minimum. The direction where the curvature is the greatest corresponds to the largest eigenvalue (48.8588) and the variable with the largest component of the associated eigenvector (gl ratio). The direction with the eigenvalue of 31.1035 is loaded more on temp, which has nearly as much curvature as that for gl ratio.

Sometimes a zero eigenvalue occurs. This means that along the direction described by the corresponding eigenvector, the fitted surface is flat.

Figure 14.6 Basic Reports for Response Surface Model

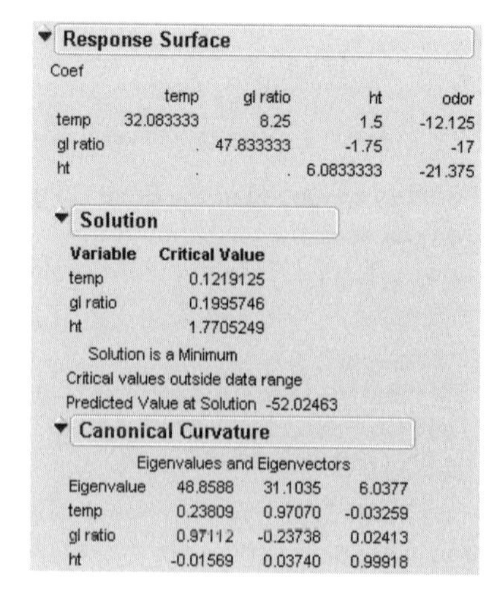

Response Surface

Coef

	temp	gl ratio	ht	odor
temp	32.083333	8.25	1.5	-12.125
gl ratio	.	47.833333	-1.75	-17
ht	.	.	6.0833333	-21.375

Solution

Variable	Critical Value
temp	0.1219125
gl ratio	0.1995746
ht	1.7705249

Solution is a Minimum
Critical values outside data range
Predicted Value at Solution -52.02463

Canonical Curvature

Eigenvalues and Eigenvectors

Eigenvalue	48.8588	31.1035	6.0377
temp	0.23809	0.97070	-0.03259
gl ratio	0.97112	-0.23738	0.02413
ht	-0.01569	0.03740	0.99918

Box Cox *Y* Transformations

Sometimes a transformation on the response fits the model better than the original response. A commonly used transformation raises the response to some power. Box and Cox (1964) formalized and described this family of power transformations. The **Factor Profiling** menu has the **Box Cox Y Transformation** command. The formula for the transformation is constructed so that it provides a continuous definition and the error sums of squares are comparable.

$$Y^{(\lambda)} = \begin{cases} \dfrac{y^{\lambda} - 1}{\lambda \dot{y}^{\lambda - 1}} & \text{if } \lambda \neq 0 \\ \dot{y}\ln(y) & \text{if } \lambda = 0 \end{cases} \quad \text{where } \dot{y} \text{ is the geometric mean}$$

The plot shown here illustrates the effect of this family of power transformations on *Y*.

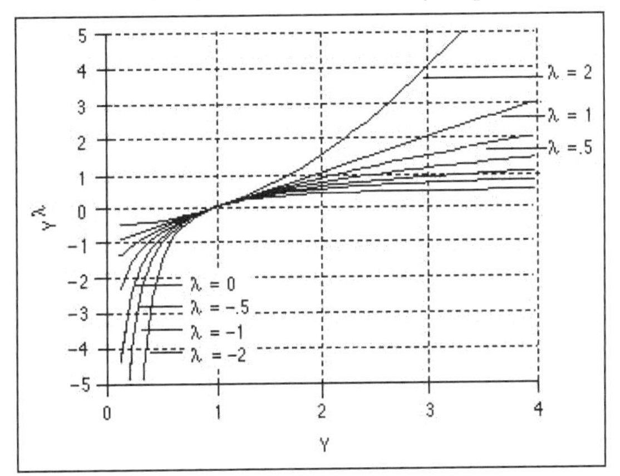

The **Box-Cox Y Transformation** command fits transformations from λ = –2 to 2 in increments of 0.2, and it plots the sum of squares error (SSE) across the λ power. The plot below shows the best fit when λ is between 1.0 and 1.5 for the **Reactor.jmp** data using the model with effects F, Ct, A, T, and Cn and all two-factor interactions. The best transformation is found on the plot by finding the lowest point on the curve.

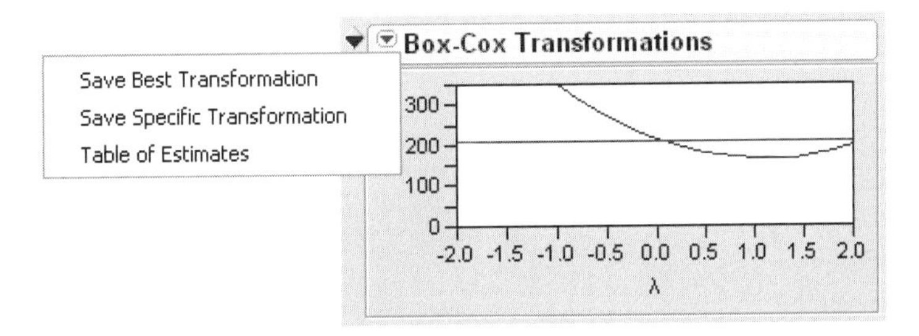

The **Save Best Transformation** command accessed by the popup icon next to the table reveal button creates a new column in the data table and saves the formula for the best transformation.

The **Save Specific Transformation** command behaves in the same way, but first prompts for a lambda value.

The **Table of Estimates** command creates a new data table of parameter estimates for all values of λ from -2 to 2.

Chapter 15

Profiling

Response Surface Visualization, Optimization, and Simulation

Profiling is an approach to visualizing response surfaces by seeing what would happen if you change just one or two factors at a time. Essentially, a profile is a cross-section view.

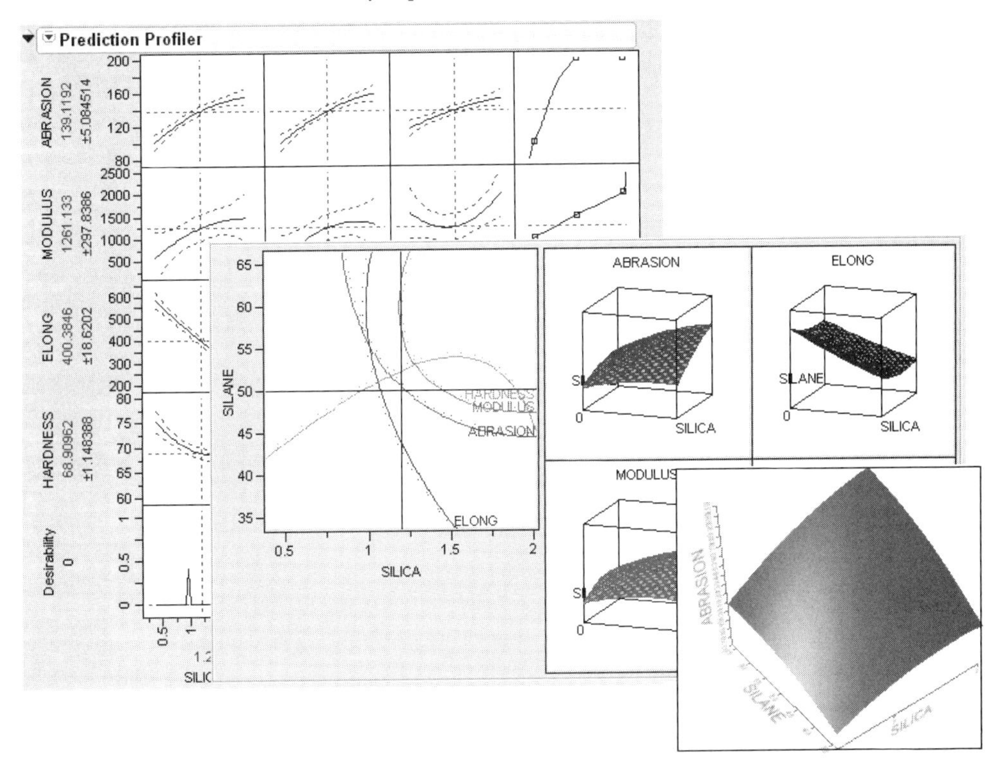

Contents

Introduction to Profiling

It is easy to visualize a response surface with one input factor *X* and one output factor *Y*. It becomes harder as more factors and responses are added. The profilers provide a number of highly interactive cross-sectional views of any response surface.

Desirability profiling and optimization features are available to help find good factor settings and produce desirable responses.

Simulation and defect profiling features are available for when you need to make responses that are robust and high-quality when the factors have variation.

Profiling Features in JMP

There are four profiler facilities in JMP, accessible from a number of fitting platforms and the main menu.They are used to profile data column formulas.

Table 15.1 Profiler Features Summary

Facility	Description	Features
Profiler	Shows vertical slices across each factor, holding other factors at current values	Desirability, Optimization, Simulator, Propagation of Error
Contour Profiler	Horizontal slices show contour lines for two factors at a time	Simulator
Surface Profiler	3-D plots of responses for 2 factors at a time, or a contour surface plot for 3 factors at a time	
Custom Profiler	A non-graphical profiler and numerical optimizer.	General Optimization, Simulator

In addition to these, in certain situations a Categorical Profiler or Interaction Profiler is available.Profiler availability is shown in Table 15.2.

Table 15.2 Profiler Facilities in JMP

Location	Profiler	Contour Profiler	Surface Profiler	Others
Graph Menu (as a Platform)	✓	✓	✓	Custom Profiler
Fit Model: Least Squares	✓	✓	✓	Interaction Profiler
Fit Model: Logistic				Categorical Profiler
Fit Model: LogVariance	✓	✓	✓	
Fit Model: Generalized Linear	✓	✓	✓	
Nonlinear: Factors and Response	✓	✓	✓	
Nonlinear: Parameters and SSE	✓	✓	✓	
Neural Net	✓	✓	✓	Categorical Profiler
Gaussian Process	✓	✓	✓	
Custom Design Prediction Variance	✓		✓	

In this chapter, we use the following terms interchangeably:

factor, input variable, *X* column, independent variable, setting

response, output variable, *Y* column, dependent variable, outcome

The Profiler (with a capital P) is one of several profilers (lowercase p). Sometimes, to distinguish the Profiler from other Profilers, we call it the Prediction Profiler.

When the profiler is invoked as (main menu) platforms, rather than through a fitting platform, you provide columns with formulas as the **Y, Prediction Formula** columns. These formulas could have been saved from the fitting platforms.

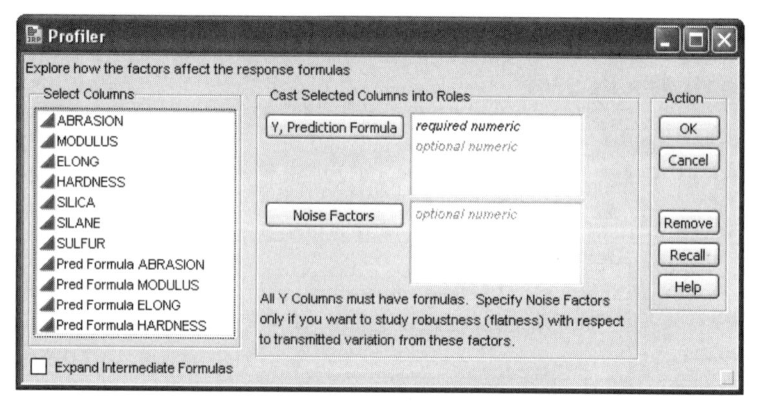

The columns referenced in the formulas become the *X* columns (unless the column is also a *Y*).

Y, Prediction Formula are the response columns containing formulas

Noise Factors are only used in special cases for modeling derivatives. Details are in "Noise Factors (Robust Engineering)," p. 332.

Expand Intermediate Formulas tells JMP that if an ingredient column to a formula is a column that itself has a formula, to substitute the inner formula, as long as it refers to other columns.

The Surface Plot platform is discussed in a separate chapter. The Surface Profiler is very similar to the Surface Plot platform, except Surface Plot has more modes of operation. Neither the Surface Plot platform nor the Surface Profiler have some of the capabilities common to other profilers

The Profiler

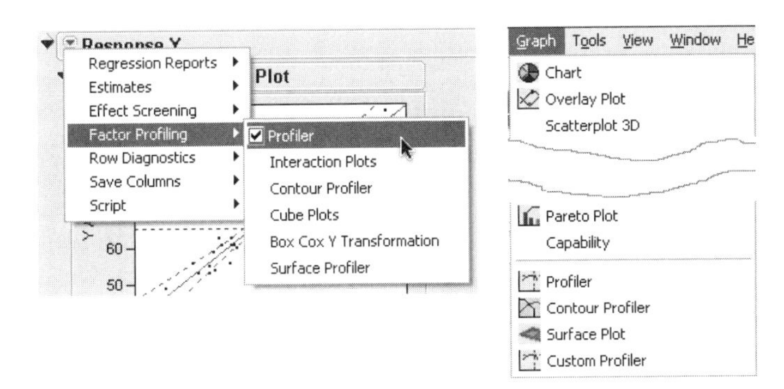

The **Profiler** displays profile traces (see Figure 15.1) for each *X* variable. A *profile trace* is the predicted response as one variable is changed while the others are held constant at the current values. The Profiler recomputes the profiles and predicted responses (in real time) as you vary the value of an *X* variable.

- The vertical dotted line for each *X* variable shows its *current value* or *current setting*. If the variable is nominal, the *x*-axis identifies categories. See "Interpreting the Profiles," p. 292, for more details. For each *X* variable, the value above the factor name is its current value. You change the current value by clicking in the graph or by dragging the dotted line where you want the new current value to be.
- The horizontal dotted line shows the *current predicted value* of each *Y* variable for the current values of the *X* variables.
- The black lines within the plots show how the predicted value changes when you change the current value of an individual *X* variable. In fitting platforms, the 95% confidence interval for the predicted values is shown by a dotted blue curve surrounding the prediction trace (for continuous variables) or the context of an error bar (for categorical variables).

The Profiler is a way of changing one variable at a time and looking at the effect on the predicted response.

Figure 15.1 Illustration of Traces

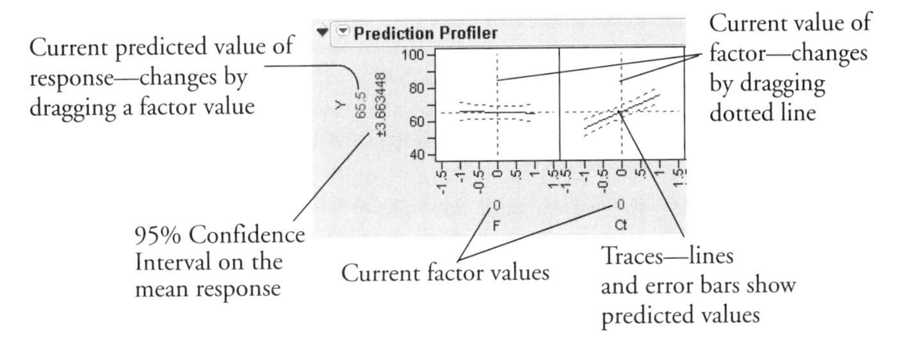

The Profiler in some situations computes confidence intervals for each profiled column. If you have saved both a standard error formula and a prediction formula for the same column, the Profiler offers to use the standard errors to produce the confidence intervals rather than profiling them as a separate column.

Interpreting the Profiles

The illustration in Figure 15.2 describes how to use the components of the profiler. There are several important points to note when interpreting a prediction profile:

- The importance of a factor can be assessed to some extent by the steepness of the prediction trace. If the model has curvature terms (such as squared terms), then the traces may be curved.
- If you change a factor's value, then its prediction trace is not affected, but the prediction traces of all the other factors can change. The *Y* response line must cross the intersection points of the prediction traces with their current value lines.

Note: If there are interaction effects or cross-product effects in the model, the prediction traces can shift their slope and curvature as you change current values of other terms. That is what interaction is all about. If there are no interaction effects, the traces only change in height, not slope or shape.

Figure 15.2 Changing one Factor From 0 to 0.75

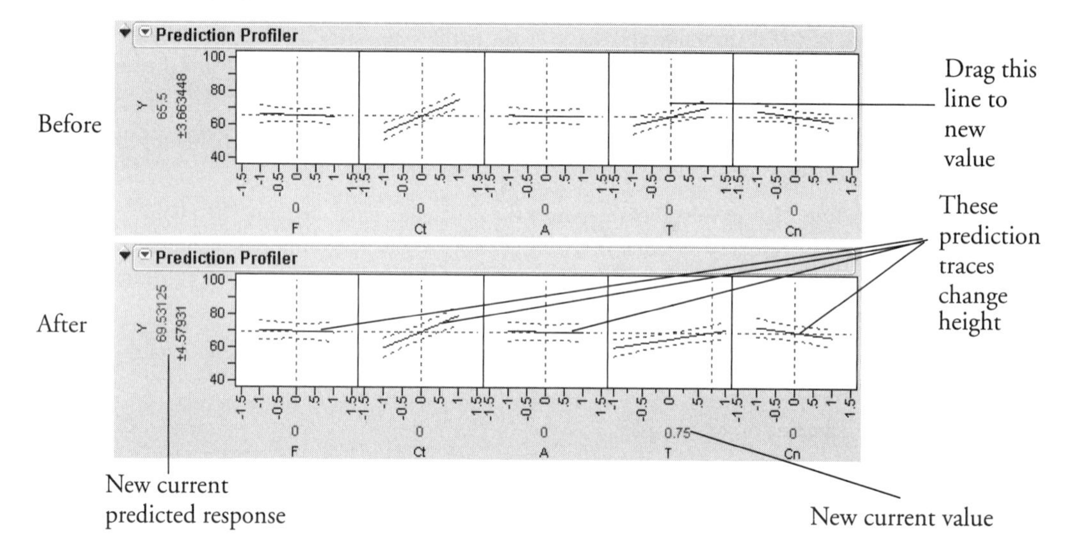

Prediction profiles are especially useful in multiple-response models to help judge which factor values can optimize a complex set of criteria.

Click on a graph or drag the current value line right or left to change the factor's current value. The response values change as shown by a horizontal reference line in the body of the graph. When you click in the vertical axis, the *X* value at that point displays. Double-click in an axis to bring up a dialog that changes its settings.

Thinking about Profiling as Cross-sectioning

In the following example using Tiretread.jmp, look at the response surface of the expression for Modulus as a function of Sulfur and Silane (holding Silica constant). Now look at how a grid that cuts across Silane at the Sulfur value of 2.25. Note how the slice intersects the surface at the blue line. If you transfer that blue line down below, it becomes the profile for Silane. Similarly, note the grid across Sulfur at the Silane value of 50. The intersection is the other blue line, which transferred down to the Sulfur graph becomes the profile for Sulfur.

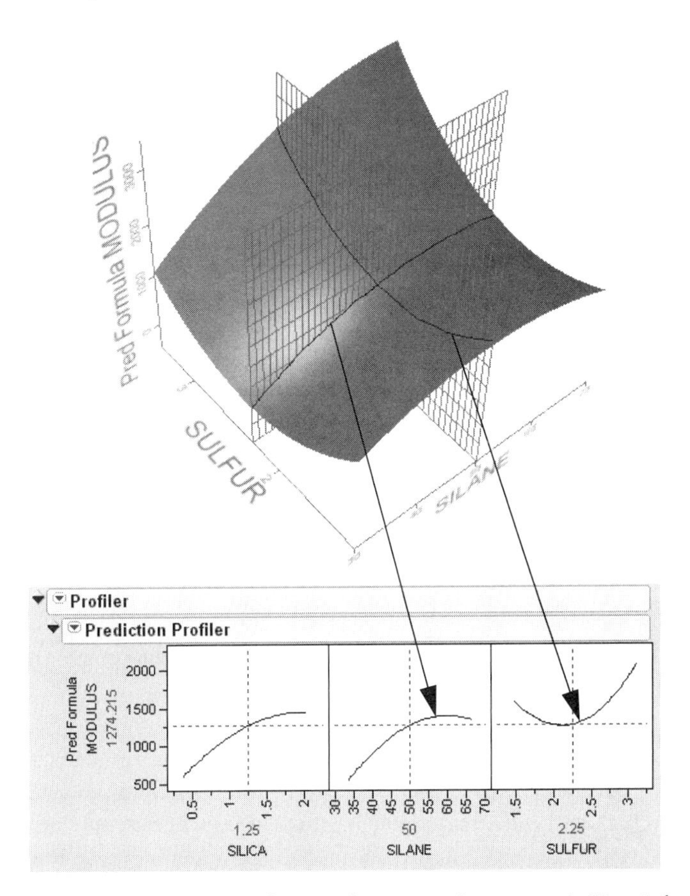

Now consider changing the current value of Sulfur from 2.25 down to 1.5. Here is how the grid cuts the response surface line, with the new blue line showing the new profile.

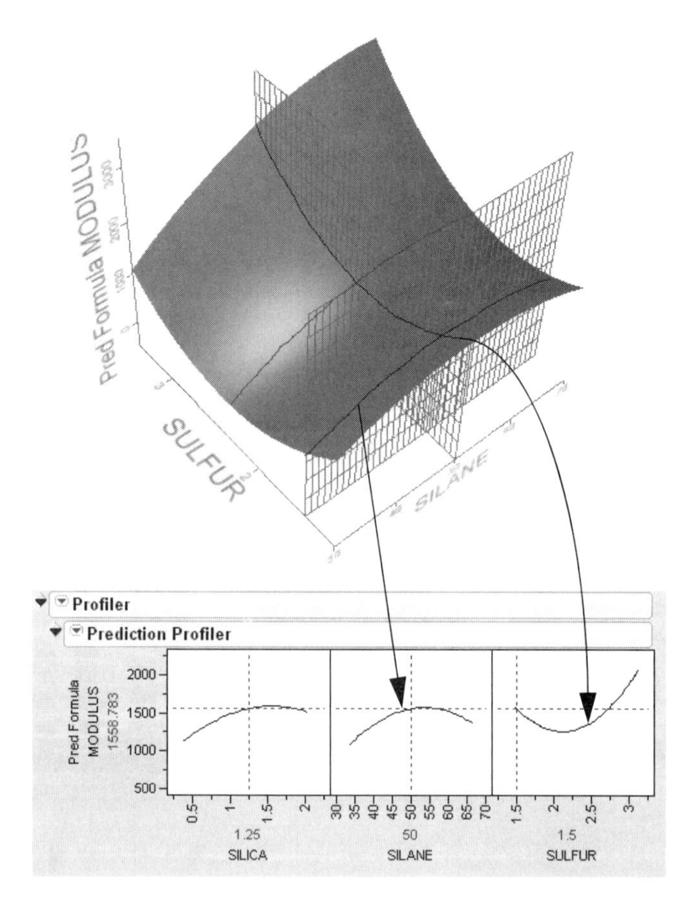

In the profiler, note the new value just moves along the same curve for Sulfur, the Sulfur curve itself doesn't change. But the profile for Silane is now taken at a different cut for Sulfur, and is a little higher and reaches its peak in the different place, closer to the current Silane value of 50.

Setting or Locking a Factor's Values

If you Alt-click (Option-click on the Macintosh) in a graph, a dialog prompts you to enter specific settings for the factor.

Figure 15.3 Continuous Factor Settings Dialog

JMP: Current Settings for x

Current Value: 10.733
Minimum Setting: 3
Maximum Setting: 21
Number of Plotted Points: 41
Show
Lock Factor Setting:

OK
Cancel

For continuous variables, you can specify

Current Value is the value used to calculate displayed values in the profiler, equivalent to the red vertical line in the graph.

Minimum Setting is the minimum value of the factor's axis

Maximum Value is the maximum value of the factor's axis

Number of Plotted Points specifies the number of points used in plotting the factor's prediction traces

Show is a checkbox that allows you to show or hide the factor in the profiler.

Lock Factor Setting locks the value of the factor at its current setting.

Profiler Options

The popup menu on the Prediction Profiler title bar has the following options to help you set the profiler axes and desirability functions:

Confidence Intervals shows or hides the confidence bars on the prediction traces of continuous factors. These are available only when the profiler is used inside certain fitting platforms.

Sensitivity Indicator displays a purple triangle whose height and direction correspond to the value of the derivative of the profile function at its current value. This is useful in large profiles to be able to quickly spot the sensitive cells.

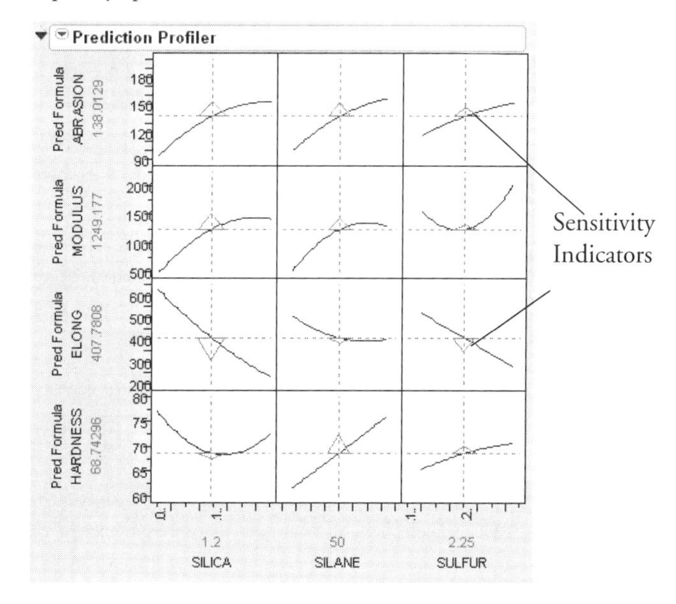

Propagation of Error bars appear under certain circumstances (see "Propagation of Error Bars," p. 303).

Desirability Functions shows or hides the desirability functions, as illustrated by Figure 15.6 and Figure 15.7. Desirability is discussed in "Desirability Profiling and Optimization," p. 298.

Maximize Desirability sets the current factor values to maximize the desirability functions.

Maximization Options allows you to refine the optimization settings through a dialog.

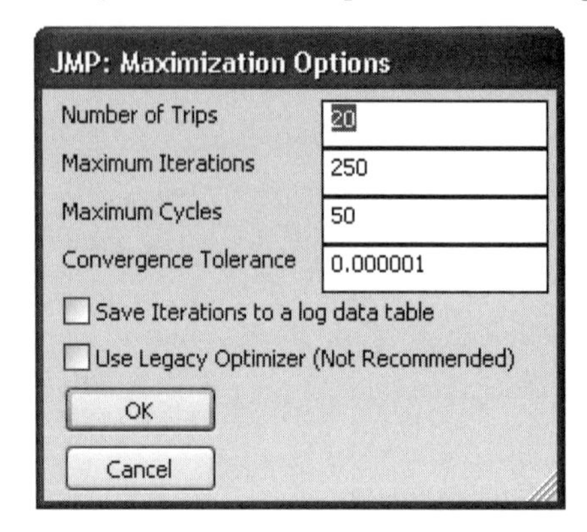

Save Desirabilities saves the three desirability function settings for each factor, and the associated desirability values, as a Response Limits column property in the data table. These correspond to the coordinates of the handles in the desirability plots.

Set Desirabilities brings up a dialog where specific desirability values can be set.

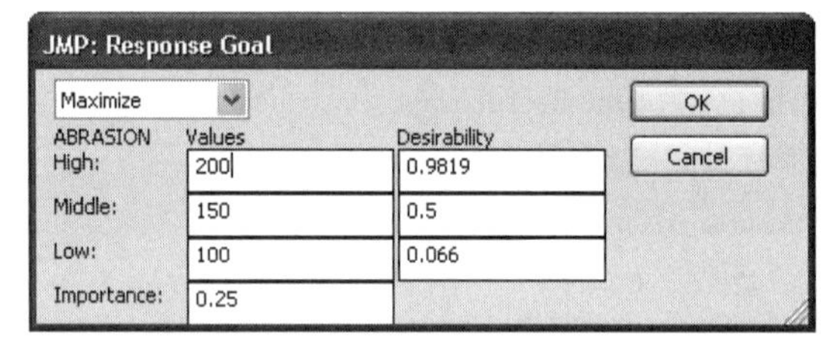

Save Desirability Formula creates a column in the data table with a formula for Desirability. The formula uses the fitting formula when it can, or the response variables when it can't access the fitting formula.

Reset Factor Grid displays a dialog for each value allowing you to enter specific values for a factor's current settings. See the section "Setting or Locking a Factor's Values," p. 294 for details on these dialog boxes.

Factor Settings is a submenu that consists of the following options:

Remember Settings adds an outline node to the report that accumulates the values of the current settings each time the **Remember Settings** command is invoked. Each remembered setting is preceded by a radio button that is used to reset to those settings.

Set To Data in Row assigns the values of a data table row to the Profiler.

Copy Settings Script and **Paste Settings Script** allow you to move the current Profiler's settings to a Profiler in another report.

Append Settings to Table appends the current profiler's settings to the end of the data table. This is useful if you have a combination of settings in the Profiler that you want to add to an experiment in order to do another run.

Link Profilers links all the profilers together, so that a change in a factor in one profiler causes that factor to change to that value in all other profilers, including Surface Plot, where that factor appears as a factor. This is a global option, set or unset for all profilers.

Set Script sets a script that is called each time a factor changes. The set script receives a list of arguments of the form

```
{factor1 = n1, factor2 = n2, ...}
```

For example, to write this list to the log, first define a function

```
ProfileCallbackLog = Function({arg},show(arg));
```

Then enter `ProfileCallbackLog` in the **Set Script** dialog.

Similar functions convert the factor values to global values:

```
ProfileCallbackAssign = Function({arg},evalList(arg));
```

Or access the values one at a time:

```
ProfileCallbackAccess = Function({arg},f1=arg["f1"];f2=arg["f2"]);
```

Output Grid Table produces a new data table with columns for the factors that contain grid values, columns for each of the responses with computed values at each grid point, and the desirability computation at each grid point.

If you have a lot of factors, it is impractical to use the **Output Grid Table** command, since it produces a large table. In such cases, you should lock some of the factors, which are held at locked, constant values. To get the dialog to specify locked columns, Alt- or Option-click inside the profiler graph to get a dialog that has a **Lock Factor Setting** checkbox.

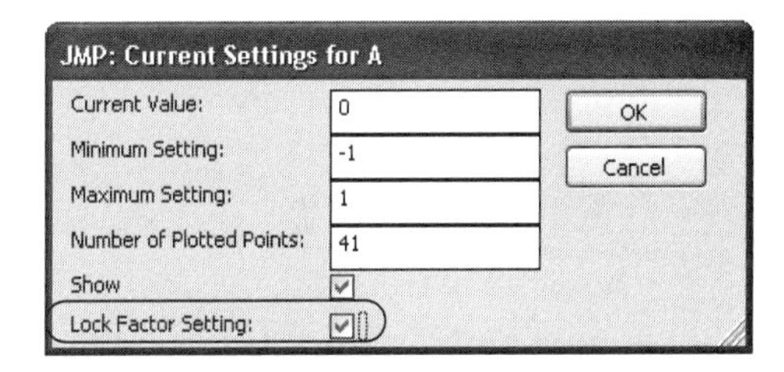

Output Random Table prompts for a number of runs and creates an output table with that many rows, with random factor settings and predicted values over those settings. This is equivalent to (but much simpler than) opening the Simulator, resetting all the factors to a random uniform

distribution, then simulating output. This command is similar to **Output Grid Table**, except it results in a random table rather than a sequenced one.

The prime reason to make uniform random factor tables is to explore the factor space in a multivariate way using graphical queries. This technique is called *Filtered Monte Carlo.*

Suppose you want to see the locus of all factor settings that produce a given range to desirable response settings. By selecting and hiding the points that don't qualify (using graphical brushing or the Data Filter), you see the possibilities of what is left: the opportunity space yielding the result you want.

Interaction Profiler brings up interaction plots that are interactive with respect to the profiler values. This option can help visualize third degree interactions by seeing how the plot changes as current values for the terms are changed. The cells that change for a given term are the cells that do not involve that term directly.

Default N Levels allows you to set the default number of levels for each continuous factor. This option is useful when the Profiler is especially large. When calculating the traces for the first time, JMP measures how long it takes. If this time is greater than three seconds, you are alerted that decreasing the Default N Levels speeds up the calculations.

Conditional Predictions appears when random effects are included in the model. The random effects are shown and used in formulating the predicted value.

Simulator launches the Simulator. The Simulator enables you to create Monte Carlo simulations using random noise added to factors and predictions for the model. A typical use is to set fixed factors at their optimal settings, and uncontrolled factors and model noise to random values and find out the rate that the responses are outside the specification limits. See the section on the Simulator for details on its use.

Desirability Profiling and Optimization

Often there are multiple responses measured for each set of experimental conditions, and the desirability of the outcome involves several or all of these responses. For example, you might want to maximize one response, minimize another, and keep a third response close to some target value. In desirability profiling, you specify a desirability function for each response. The overall desirability can be defined as the geometric mean of the desirability for each response.

To use desirability profiling, click the popup icon on the Prediction Profile title bar and select **Desirability Functions**.

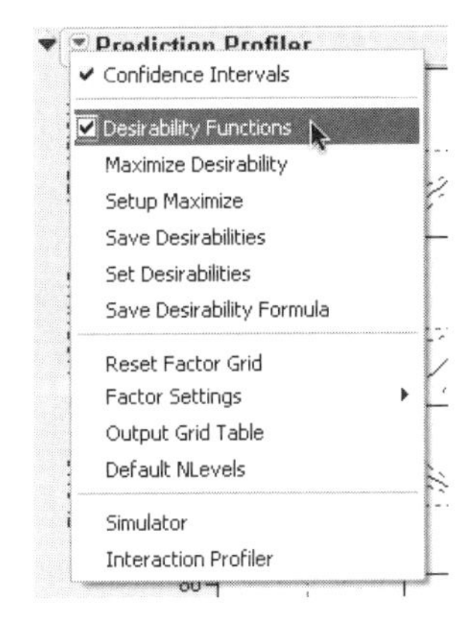

Note: If the response column has a Response Limits property, desirability functions are turned on by default.

This command appends a new row to the bottom of the plot matrix, dedicated to graphing desirability. The row has a plot for each factor showing its *desirability trace*, as illustrated in Figure 15.4. It also adds a column that has adjustable desirability function for each *Y* variable. The overall desirability measure shows on a scale of zero to one at the left of the row of desirability traces.

Figure 15.4 The Desirability Profiler

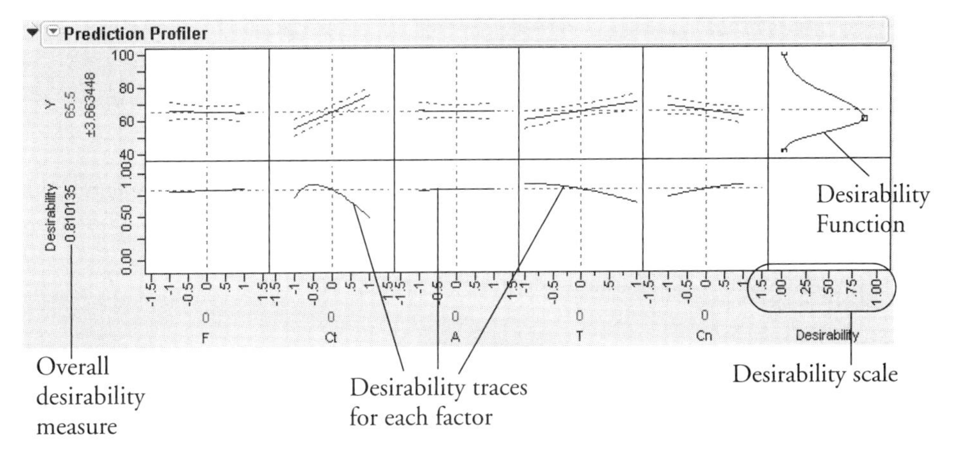

The desirability profiler components and examples of desirability function settings are discussed next.

The Desirability Function

To use a variable's desirability function, drag the function handles to represent a response value.

As you drag these handles, the changing response value shows in the area labeled **Desirability** to the left of the plots. The dotted line is the response for the current factor settings. The overall desirability shows to the left of the row of desirability traces. Alternatively, you can select **Set Desirabilities** to enter specific values for the points.

On Desirability Functions

The desirability functions are smooth piecewise functions that are crafted to fit the control points.

- The minimize and maximize functions are three-part piecewise smooth functions that have exponential tails and a cubic middle.
- The target function is a piecewise function that is a scale multiple of a normal density on either side of the target (with different curves on each side), which is also piecewise smooth and fit to the control points.

These choices give the functions good behavior as the desirability values switch between the maximize, target, and minimize values. For completeness, we implemented the upside-down target also.

JMP doesn't use the Derringer and Suich functional forms. Since they are not smooth, they do not always work well with JMP's optimization algorithm.

The control points are not allowed to reach all the way to zero or one at the tail control points.

The next illustration shows steps to create desirability settings.

Maximize The default desirability function setting is maximize ("higher is better"). The top function handle is positioned at the maximum Y value and aligned at the high desirability, close to 1. The bottom function handle is positioned at the minimum Y value and aligned at a low desirability, close to 0.

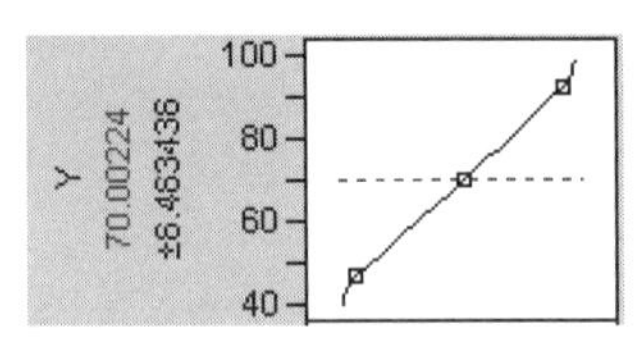

Target You can designate a target value as "best." In this example, the middle function handle is positioned at Y = 55 and aligned with the maximum desirability of 1. Y becomes less desirable as its value approaches either 70 or 42. The top and bottom function handles at Y = 70 and Y = 42 are positioned at the minimum desirability close to 0.

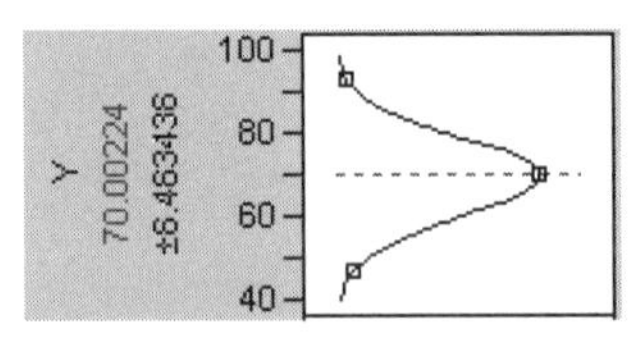

Minimize The minimize ("smaller is better") desirability function associates high response values with low desirability and low response values with high desirability. The curve is the maximization curve flipped around a horizontal line at the center of plot.

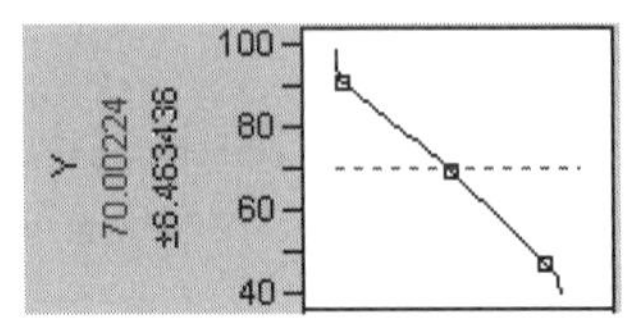

Note: Dragging the top or bottom point of a maximize or minimize desirability function across the *y*-value of the middle point results in the opposite point reflecting, so that a Minimize becomes a Maximize, and vice versa.

The Desirability Profile

The last row of plots shows the desirability trace for each response. The numerical value beside the word Desirability on the vertical axis is the geometric mean of the desirability measures. This row of plots shows both the current desirability and the trace of desirabilities that result from changing one factor at a time.

For example, suppose the desirability function in Figure 15.5 is for a target value of 60. The desirability plots indicate that you could reach the optimum desirability for the target by either lowering Ct toward its low value or by lowering A The profile in Figure 15.5 shows when Ct is -0.74. The desirability is at a maximum and Y is exactly at the target of 60.

Figure 15.5 Prediction Profile Plot with Adjusted Desirability and Factor Values

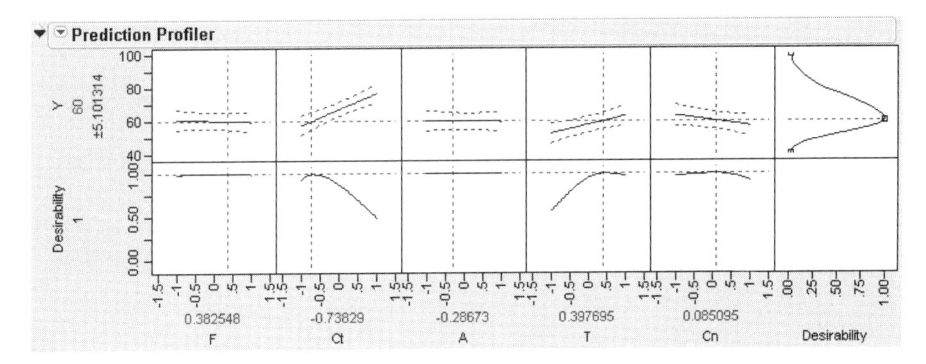

Desirability Profiling for Multiple Responses

A desirability index becomes especially useful when there are multiple responses. The idea was pioneered by Derringer and Suich (1980), who give the following example. Suppose there are four responses, ABRASION, MODULUS, ELONG, and HARDNESS. Three factors, SILICA, SILANE, and SULPHUR, were used in a central composite design.

The data are in the Tiretread.jmp table in the Sample Data folder. Use the **RSM for 4 responses** script in the data table, which defines a model for the four responses with a full quadratic response surface. The summary tables and effect information appear for all the responses, followed by the prediction profiler shown in Figure 15.6. The default profiles have all *X* variables set at their midpoint factor values.

Figure 15.6 Profiler for Multiple Responses Before Optimization

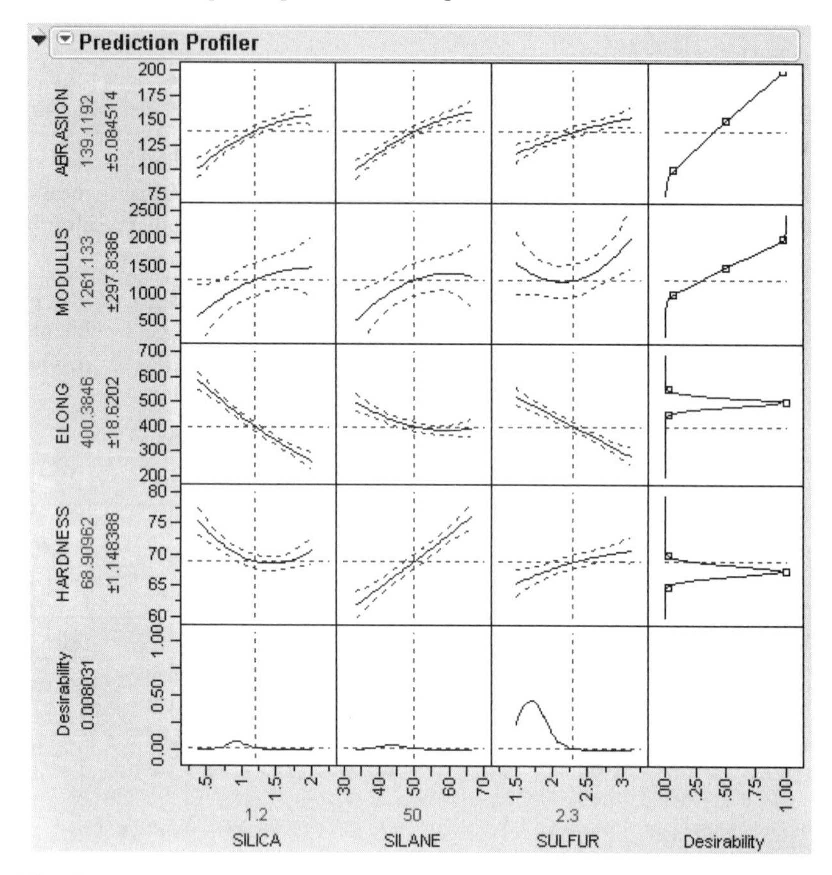

The desirability functions in Figure 15.7 are set for the targets, and the *x* values are adjusted to achieve at least a local optimum. The desirability traces at the bottom decrease everywhere except the current values of the effects, which indicates that any further adjustment could decrease the overall desirability.

1 Maximum ABRASION and maximum MODULUS are most desirable.

2 ELONG target of 500 is most desirable.

3 HARDNESS target of 67.5 is most desirable.

Figure 15.7 Profiler After Optimization

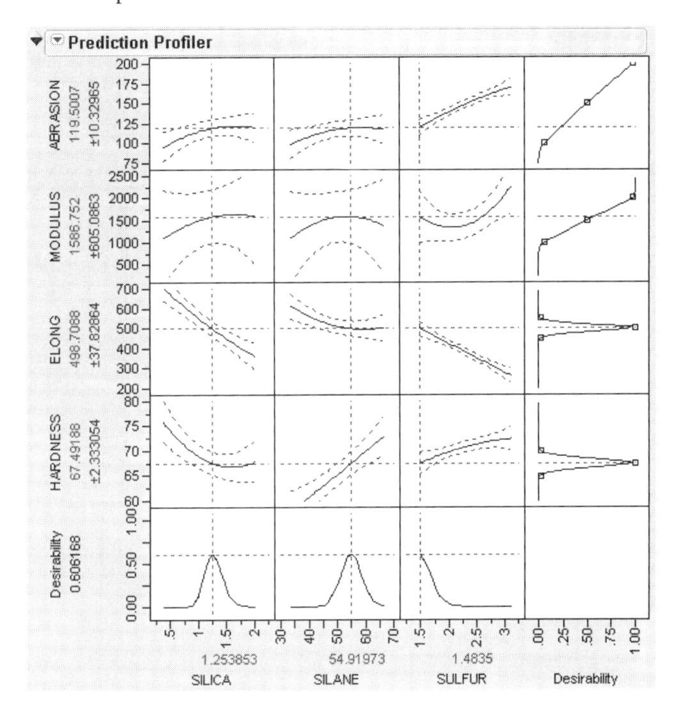

Questions and Answers for Profiling

What if I want to put desirability functions on the factors as well as on the responses? In Fit Model, you can include *X* factors also as *Y* responses in the Fit Model dialog. The fit will be perfect, and you can add desirability functions. If you are working with column factors, you need a new column whose formula is just the factor.

What if I want to fit a different model to each response? Do the fits individually and use the **Save Prediction Formula** to make new columns in the data table. Then use the **Profiler** plot in the **Graph** menu to plot the prediction columns.

What if my responses have different missing value patterns? JMP omits a row if any response is missing. This means that I lose data if I fit several responses simultaneously. Can I still use the Prediction Profiler? See the answer to the previous question.

Special Profiler Topics

Propagation of Error Bars

Propagation of error (POE) is important when attributing the variation of the response in terms of variation in the factor values when the factor values are not very controllable.

In JMP's implementation, the Profiler first looks at the factor and response variables to see if there is a **Sigma** column property (a specification for the process standard deviation of the column, accessed through the **Cols > Column Info** dialog box). If the property exists, then the **Prop of Error Bars** command becomes accessible in the Profiler drop-down menu. This displays the 3σ interval that is implied on the response due to the variation in the factor.

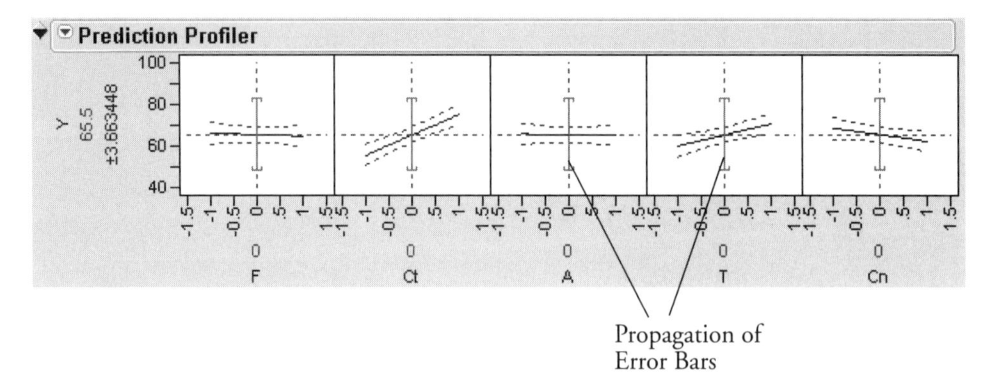

The interval is calculated (assuming that the variation among the factors is uncorrelated) by

$$\sum_{i=1}^{\#\ \text{factors}} \left(\sigma_{x_i}^2 \times \left(\frac{\partial f}{\partial x_i}\right)^2\right) + \sigma_y^2$$

where f is the prediction function and x_i is the ith factor.

Currently, these partial derivatives are calculated by numerical derivatives (centered, with δ=xrange/10000)

POE limits increase dramatically in response surface models when you are over a more sloped part of the response surface. One of the goals of robust processes is to operate in flat areas of the response surface so that variations in the factors do not amplify in their effect on the response.

Customized Desirability Functions

It is possible to use a customized desirability function based on predicted values of other Y variables. For example, suppose you want to maximize using the function

$$\frac{\textit{Pred Formula ABRASION}}{96} + \frac{\textit{Pred Formula MODULUS}}{700} + \frac{33}{(|\textit{Pred Formula ELONG} - 450| + 1)} + \frac{2}{(|\textit{Pred Formula HARDNESS} - 67| + 1)}$$

First, create a column called MyDesire that contains the above formula. Then, launch the Profiler using **Graph > Profiler** and including all the Pred Formula columns and the MyDesire column. All the desirability functions for the individual effects must be turned off by double-clicking on the desirability trace and selecting **None** in the dialog that appears (Figure 15.8).

Figure 15.8 Selecting No Desirability Goal

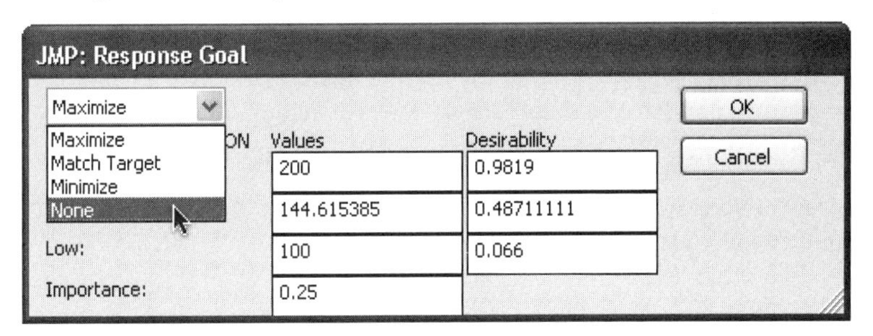

At this point, selecting **Maximize Desirability** uses only the custom myDesire function.

Figure 15.9 Maximized Custom Desirability

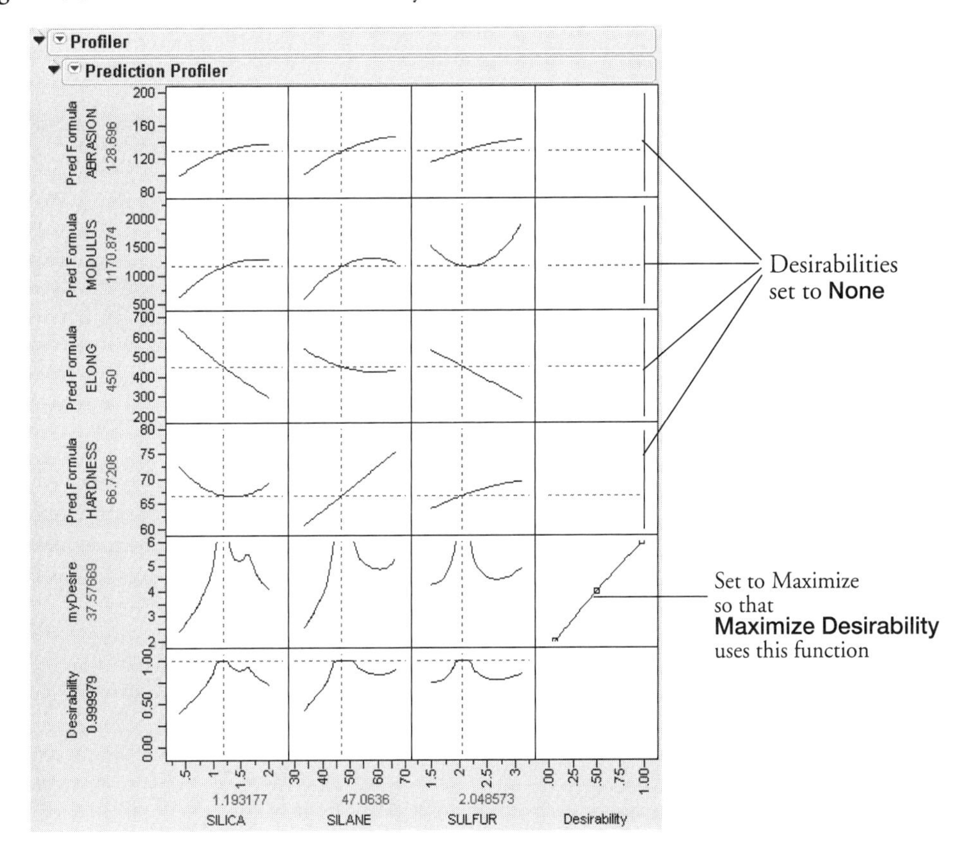

Mixture Designs

When analyzing a mixture design, JMP constrains the ranges of the factors so that settings outside the mixture constraints are not possible. This is why, in some mixture designs, the profile traces appear to turn abruptly.

In addition, two new options appear in the Profiler menu when there are mixture components that have constraints, other than the usual zero-to-one constraint.

Turn At Boundaries lets the settings continue along the boundary of the restraint condition.

Stop At Boundaries truncates the prediction traces to the region where strict proportionality is maintained.

Expanding Intermediate Formulas

The Profiler launch dialog has an **Expand Intermediate Formulas** checkbox. When this is checked, then when the formula is examined for profiling, if it references another column that has a formula containing references to other columns, then it substitutes that formula and profiles with respect to the end references—not the intermediate column references.

In addition, using the **Expand Intermediate Formulas** checkbox enables the **Save Expanded Formulas** command in the platform pop-up menu, enabling you to save the expanded formulas to the original data table.

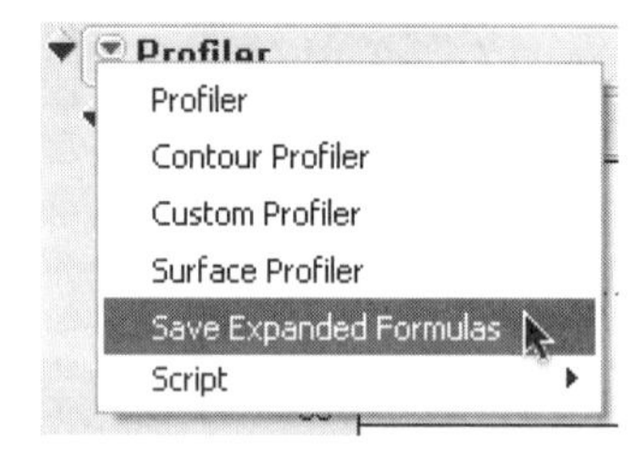

For example, when Fit Model fits a logistic regression for two levels and formulas are saved, there is a linear model column called Lin[xxx] which has the formula in terms of the regression terms, and there are end formulas Prob[xxx] which are functions of the Lin[xxx] column. Only the formulas that themselves reference other columns are considered for "digging". Columns that are expressions only of constants or random functions, or row calculations are not considered.

Contour Profiler

The **Contour Profiler** shows response contours for two factors at a time. The interactive contour profiling facility is useful for optimizing response surfaces graphically.

Here is a contour surface for the ABRASION response by SILANE and SILICA (holding SULFUR constant). Note the grid slicing the surface at the ABRASION value of 145. The cut defines a contour line, which can be graphed in 3-D.

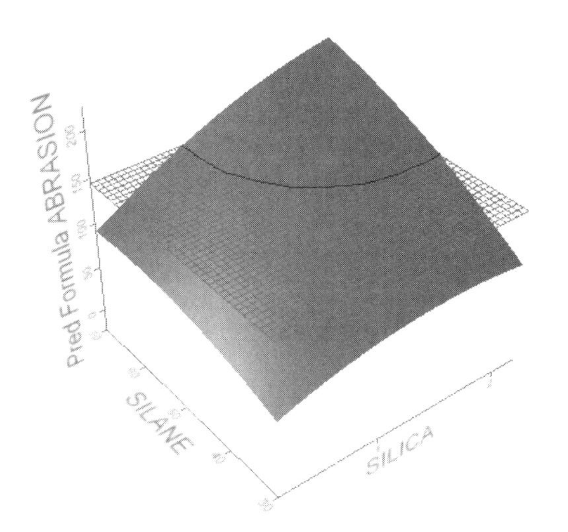

Here is the 2-D Contour profiler for that cut.

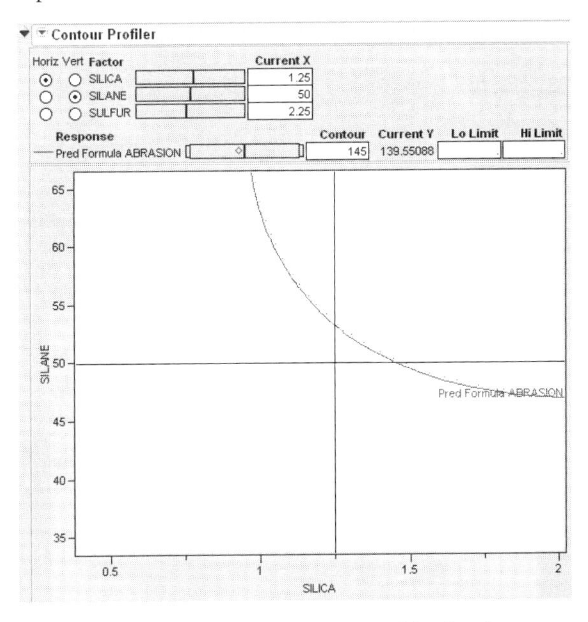

Figure 15.10 shows an annotated example of the Contour Profiler for the Tiretread sample data. To see this example, run a response surface model with ABRASION, MODULUS, ELONG, and HARDNESS as response variables, and SILICA, SILANE, and SULFUR as factors. The script attached to the data table accomplishes this. The following features are shown in Figure 15.10:

- There are slider controls and edit fields for both the *X* and *Y* variables.
- The Current X values generate the Current Y values. The Current X location is shown by the crosshair lines on the graph. The Current Y values are shown by the small red diamonds in the slider control.

- The other lines on the graph are the contours for the responses set by the *Y* slider controls or by entering values in the Contour column. There is a separately colored contour for each response (4 in this example for the response variables).
- You can enter low and high limits to the responses, which results in a shaded region. To set the limits, you can click and drag from the side zones of the *Y* sliders or enter values in the Lo Limit or Hi Limit columns.
- If you have more than two factors, use the check box in the upper left of the plot to switch the graph to other factors.
- Option (Alt) click on the slider control changes the scale of the slider (and the plot for an active *X* variable).
- For each contour, there is a dotted line in the direction of higher response values, so that you get a sense of direction.

Figure 15.10 The Contour Profiler and Popup Menu Commands

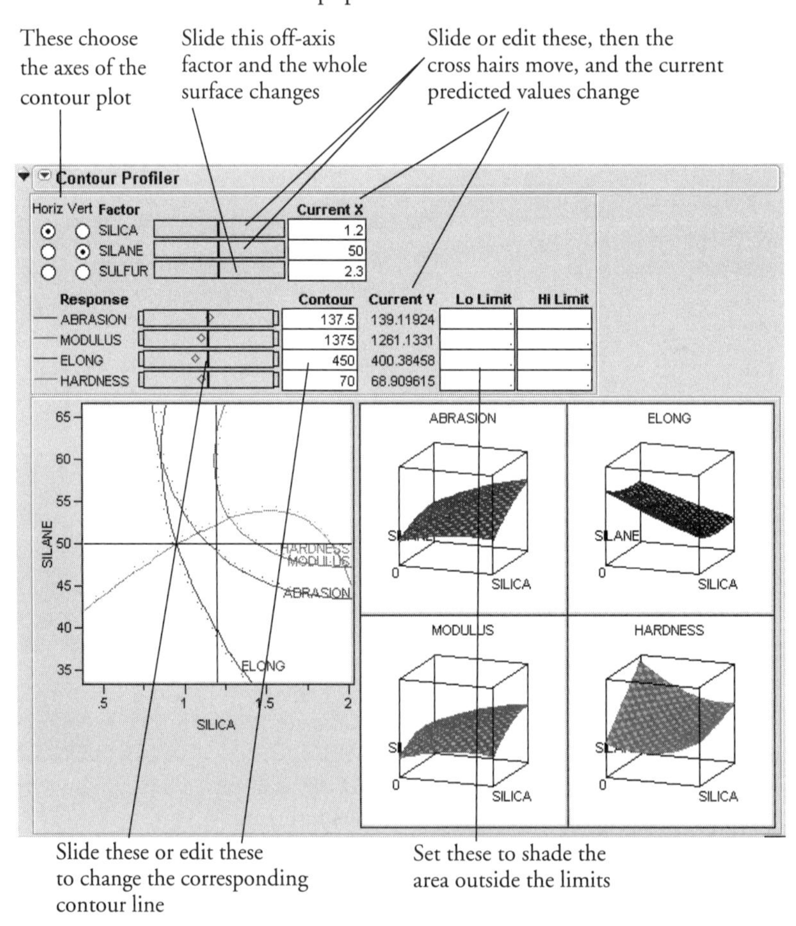

Grid Density lets you set the density of the mesh plots (Surface Plots).

Graph Updating gives you the options to update the Contour Profiler **Per Mouse Move**, which updates continuously, or **Per Mouse Up**, which waits for the mouse to be released to update. (The difference might not be noticeable on a fast machine.)

Surface Plot hides or shows mesh plots with the Contour Profiler, as shown here. The mesh plots are displayed by default.

Contour Label hides or shows a label for the contour lines. The label colors match the contour colors.

Contour Grid draws contours on the Contour Profiler plot at intervals you specify.

Factor Settings is a submenu of commands that allows you to save and transfer the Contour Profiler's settings to other parts of JMP. Details on this submenu are found in the discussion of the profiler on p. 309.

Mixtures

For mixture designs, a Lock column appears in the Contour Profiler (Figure 15.11). This column allows you to lock settings for mixture values so that they are not changed when the mixture needs to be adjusted due to other mixture effects being changed. When locked columns exist, the shaded area for a mixture recognizes the newly restricted area.

Figure 15.11 Lock Columns

Horiz	Vert	Factor		Current X	Lock
⦿	○	X1		0.25	☐
○	⦿	X2		0.25	☐
○	○	X3		0.25	☐
○	○	X4		0.25	☐

Constraint Shading

Specifying limits to the Y's shades the areas outside the limits as shown in Figure 15.12. The unshaded white area becomes the feasible region. To maximize ABRASION subject to the limits,

- Move the ABRASION slider up until it reached the upper right corner of the white area, then
- Select another combination of factors using the radio buttons
- Repeat using other combinations of the two factors.

Figure 15.12 Settings for Contour Shading

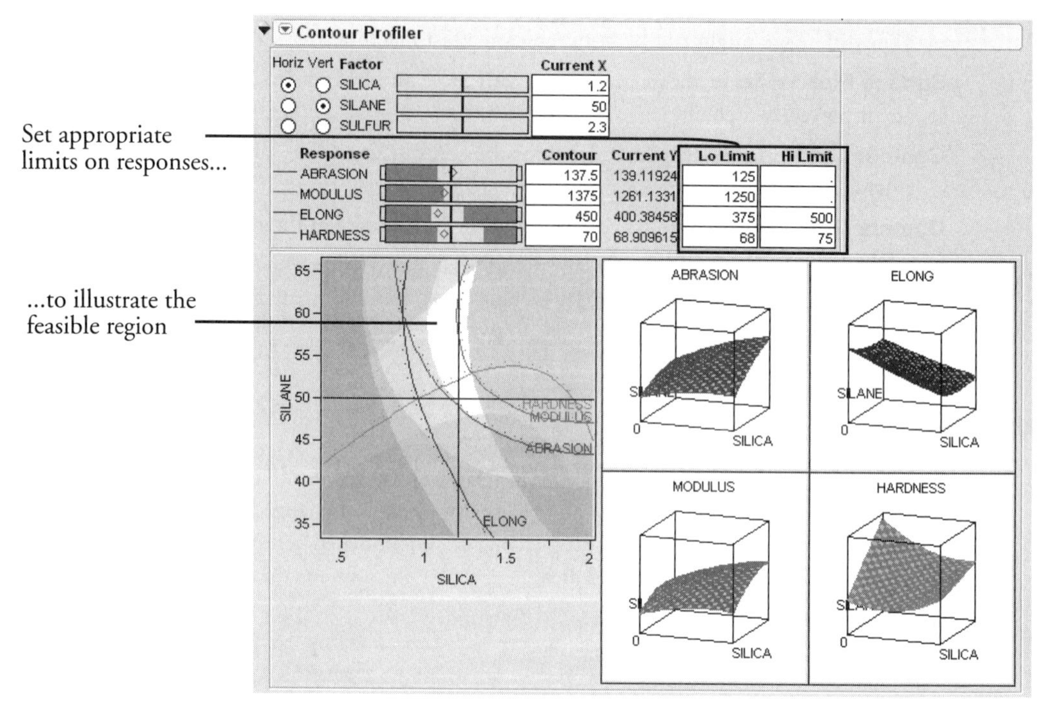

The unshaded white space becomes the feasible region for responses.

Surface Profiler

The Surface Profiler shows a three-dimensional surface plot of the response surface. Details of Surface Plots are found in the “Plotting Surfaces” chapter.

The **Response Grid Slider** adjusts an optional grid that moves along the response axis.

Note that the Surface Profiler can show either continuous or categorical factors.

Figure 15.13 Surface Profiler

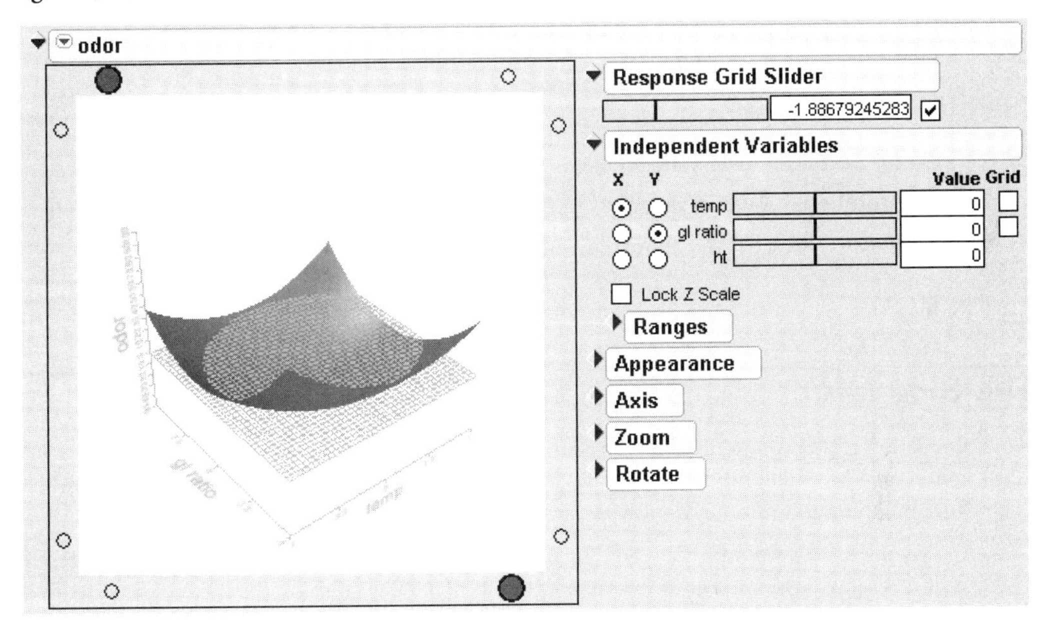

The Custom Profiler

The Custom Profiler allows you to optimize factor settings computationally, without graphical output. This is used for large problems that would have too many graphs to visualize well.

It has many fields in common with other profilers. Its upper half closely resembles the Contour Profiler (p. 306). The Benchmark field represents the value of the prediction formula based on current settings. Click Reset Benchmark to update the results.

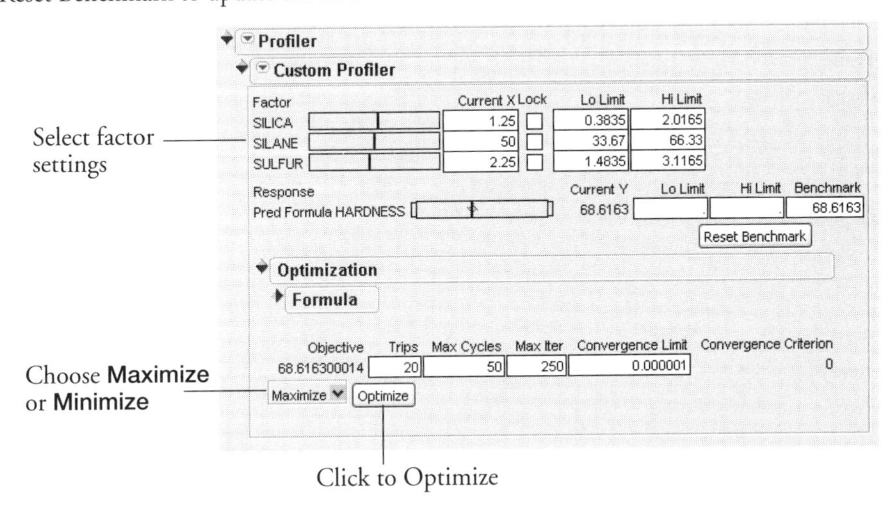

The Optimization outline node allows you to specify the formula to be optimized and specifications about the optimization iterations. Click the **Optimize** button to optimize based on current settings.

Custom Profiler Options

Factor Settings is a submenu identical to the one covered on p. 297.

Log Iterations outputs iterations to the log window.

Simulator launches the Simulator.

The Simulator

Simulation allows you to discover the distribution of model outputs as a function of the random variation in the factors and model noise. The simulation facility in the profilers provides a way to set up the random inputs and run the simulations, producing an output table of simulated values.

An example of this facility's use would be to find out the defect rate of a process that has been fit, and see if it is robust with respect to variation in the factors. If specification limits have been set in the responses, they are carried over into the simulation output, allowing a prospective capability analysis of the simulated model using new factors settings.

In the Profiler, the Simulator is integrated into the graphical layout. Factor specifications are aligned below each factor's profile. A simulation histogram is shown on the right for each response.

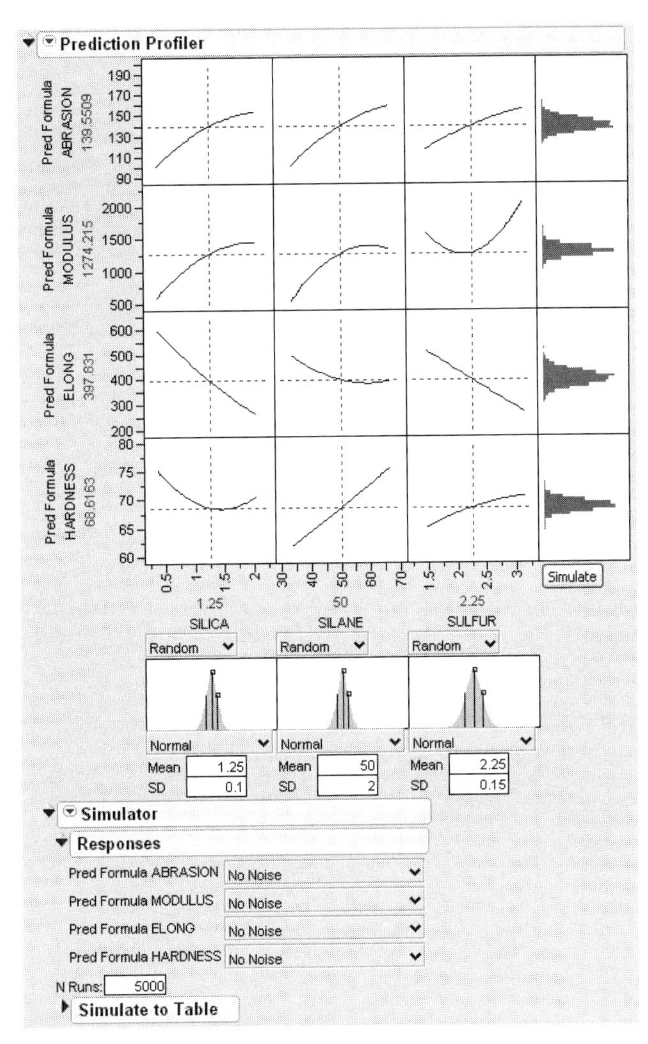

In the other profilers, the Simulator is less graphical, and kept separate. There are no integrated histograms, and the interface is textual. However, the internals and output tables are the same.

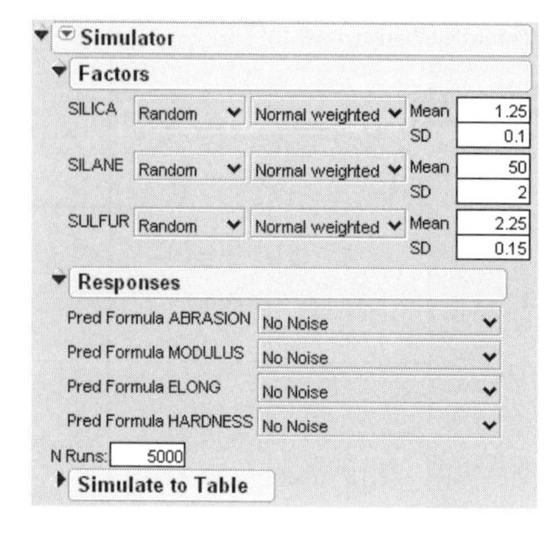

Specification of Factors

Factors (inputs) and responses (outputs) are already given roles by being in the Profiler. Additional specifications for the simulator are on how to give random values to the factors, and add random noise to the responses.

For each factor, the choices of how to give values are as follows:

Fixed fixes the factor at the specified value. The initial value is the current value in the profiler, which may be a value obtained through optimization.

Random gives the factor a random value with the specified distribution and distributional parameters.

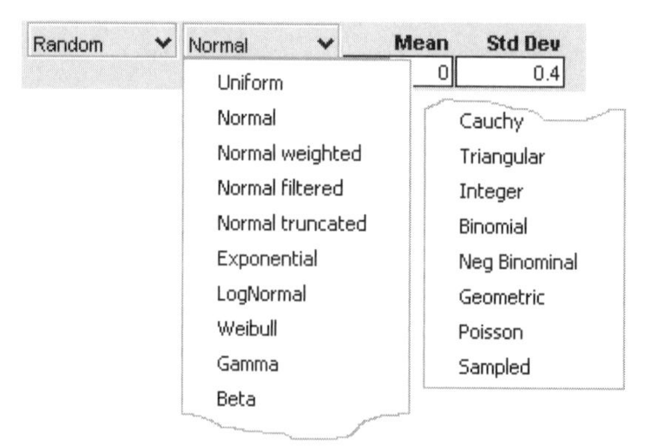

See the *JMP User's Guide* for descriptions of most of these random functions.If the factor is categorical, then the distribution is characterized by probabilities specified for each category, with the values normalized to sum to 1.

Sampled means that JMP randomly picks values from that column in the data table.

Normal weighted is Normally distributed with the given mean and standard deviation, but a special stratified and weighted sampling system is used to simulate very rare events far out into the tails of the distribution. This is a good choice when you want to measure very low defect rates accurately.

Normal truncated is a normal distribution limited by lower and upper limits. Any random realization that exceeds these limits is discarded and the next variate within the limits is chosen. This is used to simulate an inspection system where inputs that do not satisfy specification limits are discarded or sent back.

Normal censored is a normal distribution limited by lower and upper limits. Any random realization that exceeds a limit is just set to that limit, putting a density mass on the limits. This is used to simulate a re-work system where inputs that do not satisfy specification limits are reworked until they are at that limit.

In the Profiler, a graphical specification shows the form of the density for the several of the distributions (uniform, all normals, triangular, and multinomial), and provides control points that can be dragged to change the distribution. The drag points for the Normal are the mean and the mean plus one standard deviation. The filtered and truncated add points for the lower and upper limits. The uniform and triangular have limit control points, and the triangular adds the mode.

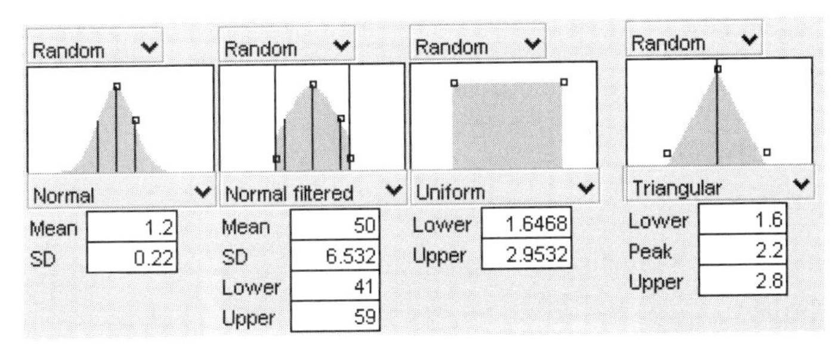

Expression allows you to write your own expression in JMP Scripting Language (JSL) form into a field. This gives you flexibility to make up a new random distribution. For example, you could create a truncated normal distribution that guaranteed non-negative values with an expression like `Max(0,RandomNormal(5,2))`. In addition, character results are supported, so `If(Random Uniform() < 2, "M", "F")` works fine.

Multivariate allows you to generate a multivariate normal for when you have correlated factors. Specify the mean and standard deviation with the factor, and a correlation matrix separately.

Multivariate	
Mean	1.2
SD	0.6532

Specifying the Response

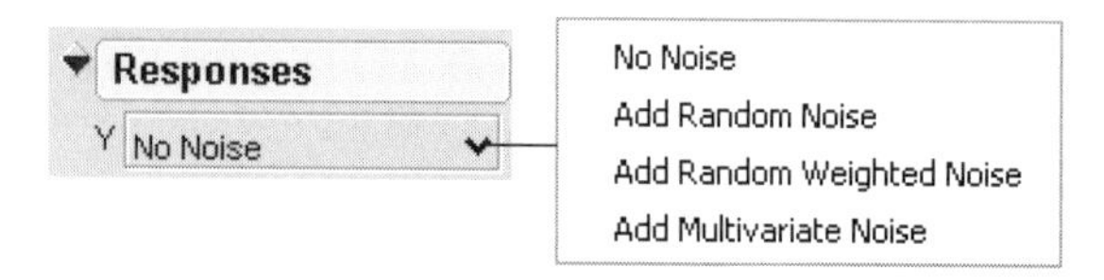

If the model is only partly a function of the factors, and the rest of the variation of the response is attributed to random noise, then you will want to specify this with the responses. The choices are:

No Noise just evaluates the response from the model, with no additional random noise added.

Add Random Noise obtains the response by adding a normal random number with the specified standard deviation to the evaluated model

Add Random Weighted Noise is distributed like Add Random Noise, but with weighted sampling to enable good extreme tail estimates.

Add Multivariate Noise yields a response as follows: A multivariate random normal vector is obtained using a specified correlation structure, and it is scaled by the specified standard deviation and added to the value obtained by the model.

The Simulator Menu

Spec Limits shows or edits specification limits

Defect Profiler shows the defect rate as an isolated function of each factor. This command is enabled when spec limits are available, as described below.

Defect Parametric Profile shows the defect rate as an isolated function of the parameters of each factor's distribution. It is enabled when the Defect Profiler is launched.

Simulation Experiment is used to run a designed simulation experiment on the locations of the distribution. Two dialogs appear, allowing you to specify the number of runs, and the portion of factor space (where 1 is the full space) to be used in the experiment. There is one row for each set of simulation runs at one set of locations. The responses include the defect rate for each response with spec limits. The experimental design is a Latin Hypercube. After the experiment, it would be appropriate to fit a Gaussian Process model on the overall defect rate, or a root of it, or a logarithm of it.

Using Specification Limits

The profilers support specification limits on the responses, providing a number of features

- In the Profiler, if you don't have the **Response Limits** property set up in the input data table to provide desirability coordinates, JMP looks for a **Spec Limits** property and constructs desirability functions appropriate to those **Spec Limits.**
- If you use the Simulator to output simulation tables, JMP copies **Spec Limits** to the output data tables, making accounting for defect rates and capability indices easy.
- Adding **Spec Limits** enables a feature called the **Defect Profiler.**

In the following example, we assume that the following Spec Limits have been specified.

Table 15.3 Spec Limits for Tiretread.jmp data table

Response	LSL	Target	USL
Abrasion	110		
Modulus			2000
Elong	350	450	
Hardness	66	70	

To set these limits in the data table, highlight a column and select **Cols > Column Info**. Then, click the **Column Properties** button and select the **Spec Limits** property.

If you are already in the Simulator in a profiler, another way to enter them is to use the **Spec Limits** command in the Simulator.

Spec Limits

Response	LSL	USL
ABRASION	110	.
MODULUS	.	2000
ELONG	350	550
HARDNESS	66	.

With these specification limits, and the random specifications shown, click the Simulate button. Notice the spec limit lines in the output histograms.

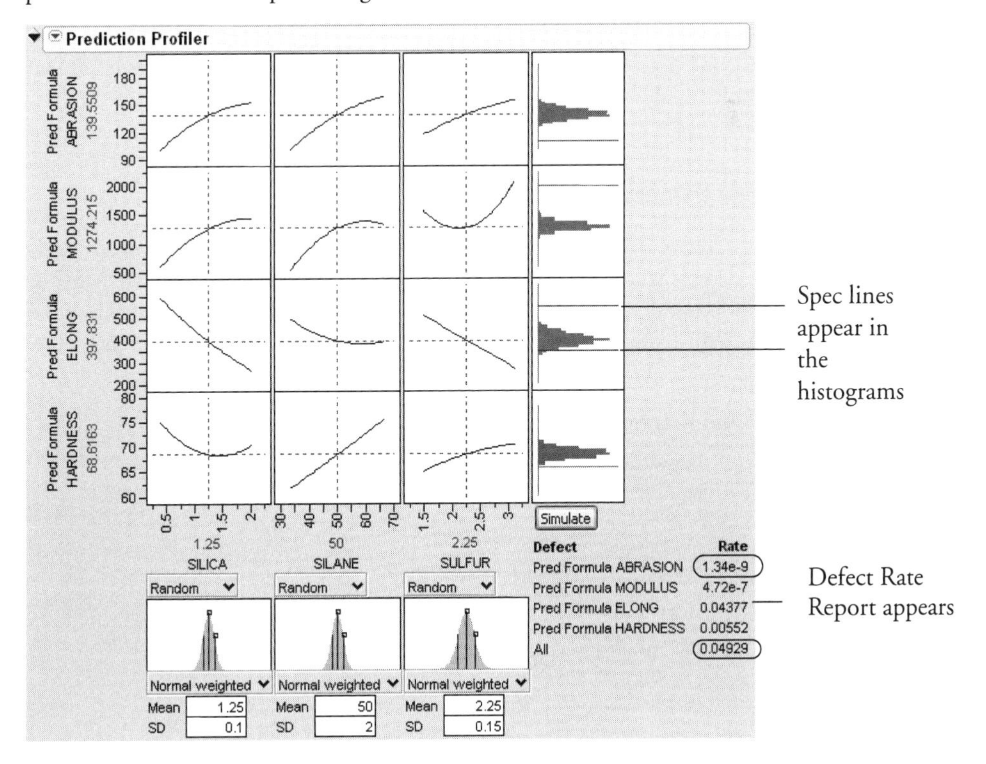

Look at the histogram for Abrasion. The upper spec limit is far above the distribution, yet the Simulator is able to estimate a defect rate for it: 1.34e-9. This despite only having 5000 runs in the simulation. It can do this rare-event estimation when you use a **Normal weighted** distribution.

Note that the overall defect rate is 0.05, and that most of these defects are in the Elong variable.

To see this weighted simulation in action, click the **Make Table** button and examine the Weight column.

	Weight	Freq	Y In Spec
1	0.2	2e+11	1
2	0.2	2e+11	1
3	0.2	2e+11	1
4	0.2	2e+11	1
5	0.175	1.75e+11	1
6	0.021875	2.1875e+10	0
7	0.00273438	2734375000	0
8	0.0003418	341796875	0
9	4.27246e-5	42724609.4	1
10	5.34058e-6	5340576.17	0
11	6.67572e-7	667572.022	1
12	8.34465e-8	83446.5027	1
13	1.04308e-8	10430.8128	1
14	1.30385e-9	1303.85158	0
15	1.6298e-10	162.981517	0
16	2.3283e-11	23.2830422	1

JMP generates extreme values for the later observations, using very small weights to compensate. Since the Distribution platform handles frequencies better than weights, there is also a column of frequencies, which is simply the weights multiplied by 10^{12}.

The output data set contains a Distribution script appropriate to analyze the simulation data completely with a capability analysis.

Simulating General Formulas

Though the profiler and simulator are designed to work from formulas stored from a model fit, it works from any formulas. A typical application of simulation from spreadsheet-based what-if simulators is to exercise financial models under certain probability scenarios to obtain the distribution of the objectives. This can be done in JMP—the key is to store the formulas into columns, set up ranges, and then conduct the simulation.

Table 15.4 Factors and Responses for a Financial Simulation

Inputs (Factors)	Unit Sales	random uniform between 1000 and 2000
	Unit Price	fixed
	Unit Cost	random normal with mean 2.25 and std dev 0.1
Outputs (Responses)	Revenue	formula: Unit Sales*Unit Price
	Total Cost	formula: Unit Sales*Unit Cost + 1200
	Profit	formula: Revenue – Total Cost

The following JSL script creates the data table with some initial scaling data and stores formulas into the output variables. It also launches the Custom Profiler.

```
dt = newTable("Sales Model");
dt<<newColumn("Unit Sales",Values({1000,2000}));
dt<<newColumn("Unit Price",Values({2,4}));
dt<<newColumn("Unit Cost",Values({2,2.5}));
dt<<newColumn("Revenue",Formula(:Unit Sales*:Unit Price));
dt<<newColumn("Total Cost",Formula(:Unit Sales*:Unit Cost + 1200));
dt<<newColumn("Profit",Formula(:Revenue-:Total Cost), Set Property("Spec
  Limits",{LSL(0)}));
Profiler(Y(:Revenue,:Total Cost,:Profit), Objective Formula(Profit));
```

Sales Model

Columns (6/0): Unit Sales, Unit Price, Unit Cost, Revenue, Total Cost, Profit

Rows: All rows 2

	Unit Sales	Unit Price	Unit Cost	Revenue	Total Cost	Profit
1	1000	2	2	2000	3200	-1200
2	2000	4	2.5	8000	6200	1800

Unit Sales*Unit Price

Unit Sales*Unit Cost+1200

Revenue- Total Cost

Once they are created, Select the Simulator from the Profiler. Use the specifications from Table 15.4 "Factors and Responses for a Financial Simulation," p. 318 to fill in the Simulator.

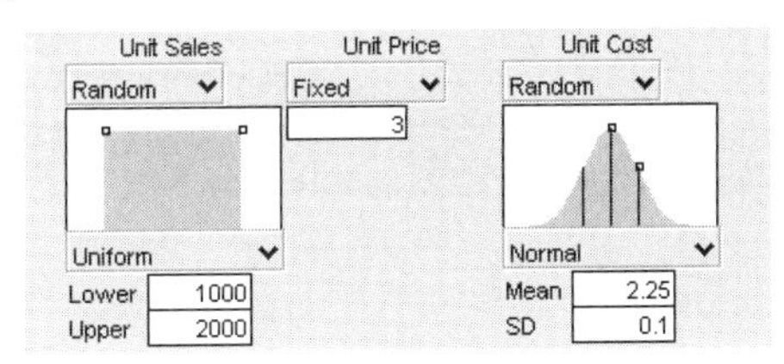

Now, run the simulation and the script in the resulting table, which produce the following histograms in the Profiler.

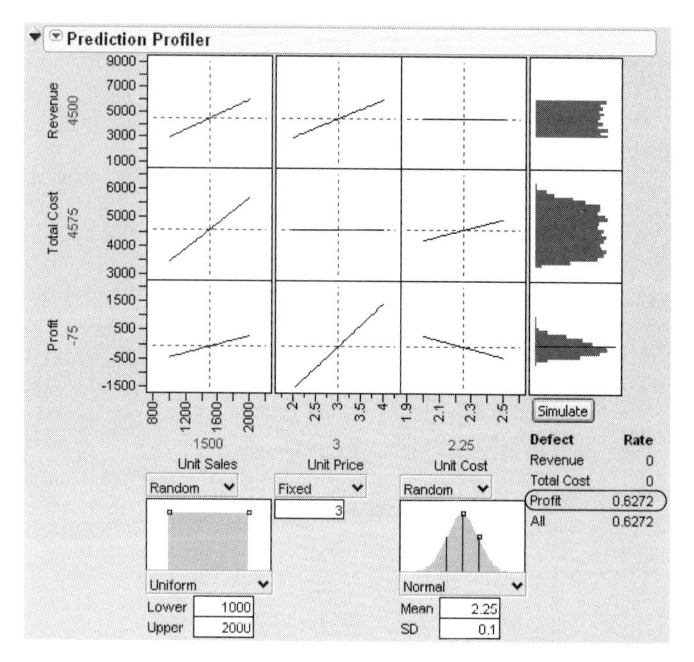

It looks like we are not very likely to be profitable. By putting specification limits on Profit of a lower limit of zero, the capability report would say that the probability of being unprofitable is 63%.

So we raise the price to $3.25 (that is, change Unit Price from 3 to 3.25) and rerun the analysis. Again, we ask for a capability analysis with a lower spec limit of zero, and we see that the probability of being unprofitable is down to about 21%.

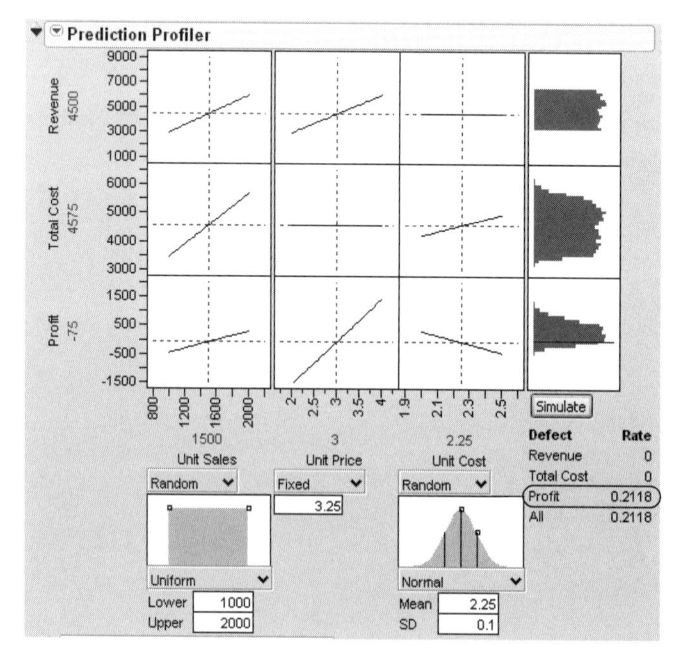

The Defect Profiler

The defect profiler shows the probability of an out-of-spec output defect as a function of each factor, while the other factors vary randomly. This is used to help visualize which factor's distributional changes the process is most sensitive to, in the quest to improve quality and decrease cost.

Specification limits define what is a defect, and random factors provide the variation to produce defects in the simulation. Both need to be present for a Defect Profile to be meaningful.

At least one of the Factors must be declared **Random** for a defect simulation to be meaningful, otherwise the simulation outputs would be constant. These are specified though the simulator Factor specifications.

Important: If you need to estimate very small defect rates, use **Normal weighted** instead of just **Normal**. This allows defect rates of just a few parts per million to be estimated well with only a few thousand simulation runs.

> *Tolerance Design* is the investigation of how defect rates on the outputs can be controlled by controlling variability in the input factors.
>
> The input factors have variation. Specification limits are used to tell the supplier of the input what range of values are acceptable. These input factors then go into a process producing outputs, and the customer of the outputs then judges if these outputs are within an acceptable range.
>
> Sometimes, a Tolerance Design study shows that spec limits on input are unnecessarily tight, and loosening these limits results in cheaper product without a meaningful sacrifice in quality. In these cases, Tolerance Design can save money.
>
> In other cases, a Tolerance Design study may find that either tighter limits or different targets result in higher quality. In all cases, it is valuable to learn which inputs the defect rate in the outputs are most sensitive to.

This graph shows the defect rate as a function of each factor as if it were a constant, but all the other factors varied according to their random specification. If there are multiple outputs with Spec Limits, then there is a defect rate curve color-coded for each output. A black curve shows the defect rate curve for any output being defective—this curve is above all the colored curves.

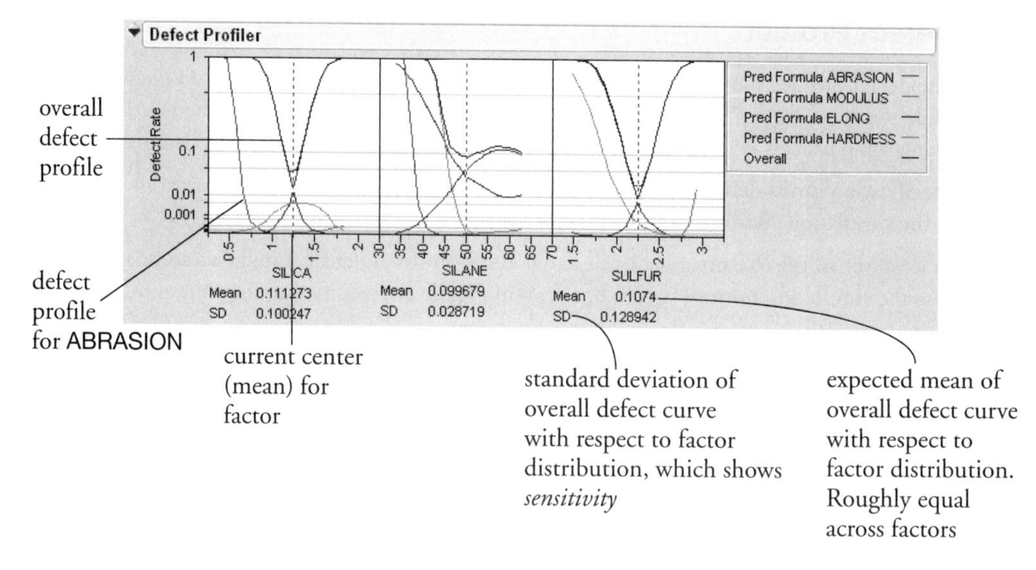

Graph Scale

Defect rates are shown on a cubic root scale, so that small defect rates are shown in some detail even though large defect rates may be visible. A log scale is not used because zero rates are not uncommon and need to be shown.

Expected Defects

Reported below each defect profile plot is the mean and standard deviation (SD). The mean is the overall defect rate, calculated by integrating the defect profile curve with the specified factor distribution.

In this case, the mean estimates the same defect rate that is reported below the other factors, and also the rate estimated for the overall simulation below the histograms (*i.e.* if you clicked the **Simulate** button). Since each estimate of the rate is obtained in a different way, they may be a little different. If they are very different, you may need to use more simulation runs. In addition, check that the range of the factor scale is wide enough to cover the distribution well.

The standard deviation a good measure of the sensitivity of the defect rates to the factor. It is quite small if either the factor profile were flat, or the factor distribution has a very small variance. Comparing SD's across factors is a good way to know which factor should get more attention to reducing variation.

The mean and SD are updated when you change the factor distribution. This is one way to explore how to reduce defects as a function of one particular factor at a time. You can click and drag a handle point on the factor distribution, and watch the mean and SD change as you drag. However, changes are not updated across factors until you click the ReRun button to do another set of simulation runs

The defect rate will generally agree with rates for other factors and for the rate shown under the histograms for an overall simulation.

Simulation Method and Details

Assume we want a defect profile for factor X1, in the presence of random variation in X2 and X3. A series of *n* simulation runs is done at each of *k* points in a grid of equally spaced values of X1. (*k* is generally set at17). At each grid point, of the *n* runs, suppose that there are *m* defects due to outputs the specification limits. At that grid point, the defect rate is *m/n*. With normal weighted, these are done in a weighted fashion. These defect rates are connected and plotted as a continuous function of X1.

Notes

Recalculation The profile curve is not recalculated automatically when distributions change, though the expected value is. It is done this way because the simulations could take a while to run.

Limited goals Profiling does not address the general optimization problem, that of optimizing the value of quality balanced against the costs given functions that represent all aspects of the problem. This more general problem would benefit from a surrogate model and space filling design to explore this space and optimize to it.

Jagged Defect Profiles The defect profiles tend to get uneven when they are low. This is due to exaggerating the differences for low values of the cubic scale. If the overall defect curve (black line) is smooth, and the defect rates are somewhat consistent, then you are probably taking enough runs. If the Black line is jagged and not very low, then increase the number of runs. 20,000 runs is often enough to stabilize the curves.

Defect Profiler Workflow

To show a common workflow with the Defect profiler, we use Tiretread.jmp with the specification limits from earlier in the chapter, and the saved predictions from the **RSM for 4 Responses** script. We also give the following random specifications to the three factors.

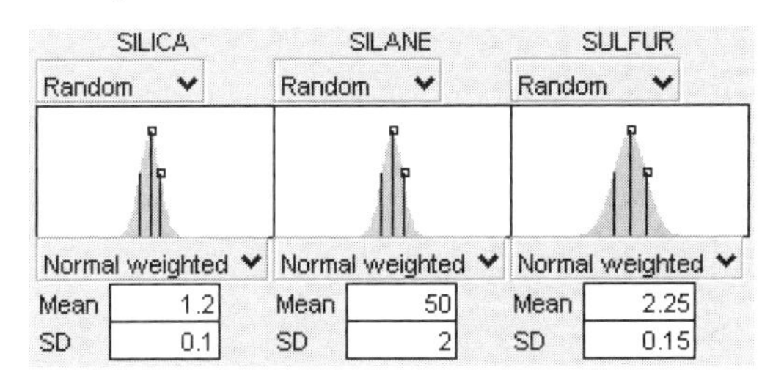

Select **Defect Profiler** to see the defect profiles.

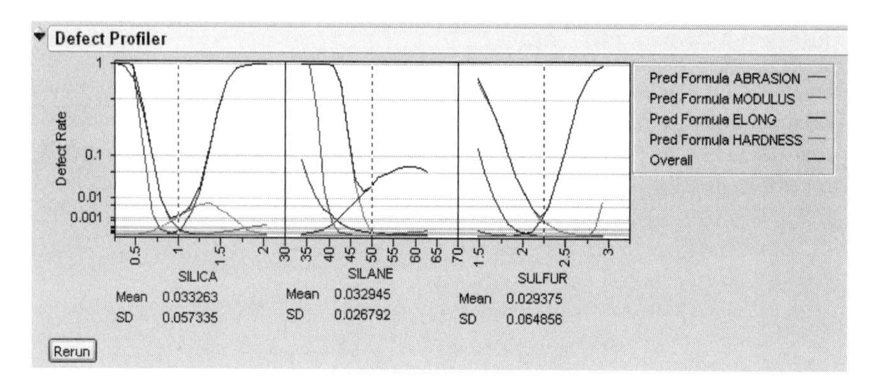

The black curve on each factor shows the defect rate if you could fix that factor at the *x*-axis value, but leave the other features random.

Look at the curve for SILICA. As its values vary, its defect rate goes from the lowest 0.005 at SILICA=0.9, quickly up to a defect rate of 1 at SILICA=0.4. or 1.8. However, SILICA is itself random. If you imagine integrating the density curve of SILICA with its defect profile curve, you could estimate the average defect rate: 0.033. This is actually estimating the same value as above in the defect rate near the histograms (the 0.03567) by numerically integrating two curves, rather than by the overall simulation. The numbers are not exactly the same. However, we now also get an estimate of the standard deviation of the defect rate with respect to the variation in SILICA. This value (labeled SD) is 0.573. The standard deviation is intimately related to the sensitivity of the defect rate with respect to the distribution of that factor.

Looking at the SDs across the three factors, we see that the SD for SULFUR is much higher than the SD for SILICA, which is in turn much higher than the SD for SILANE. This means that to improve the defect rate, improving the distribution in SULFUR should have the greatest effect. A distribution can be improved in three ways: changing its mean, changing its standard deviation, or by chopping off the distribution by rejecting parts that don't meet certain specification limits.

In order to visualize all these changes, there is another command in the Simulator, **Defect Parametric Profile**, which shows how single changes in the factor distributions affects the defect rate.

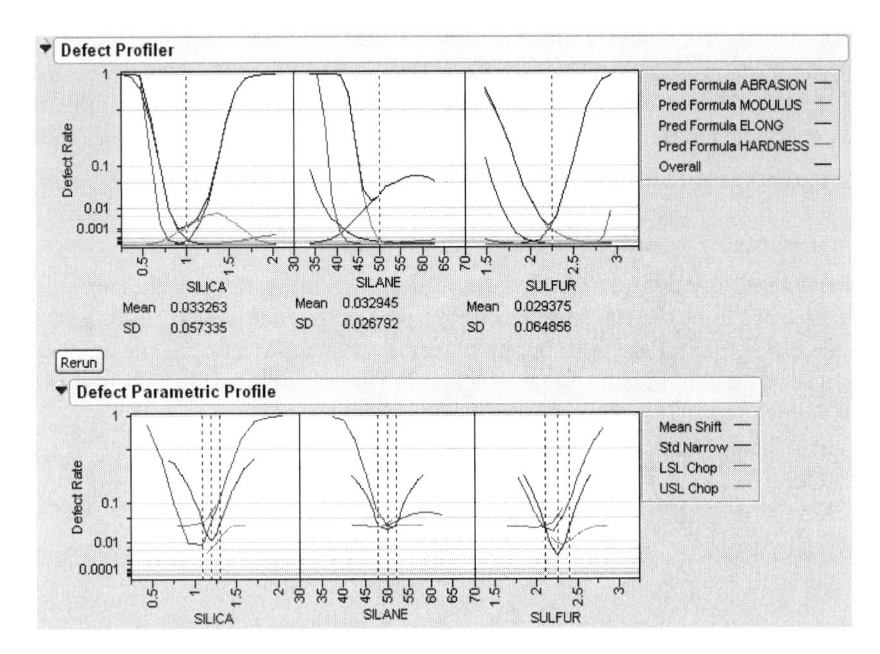

Let's look closely at the **SULFUR** situation.

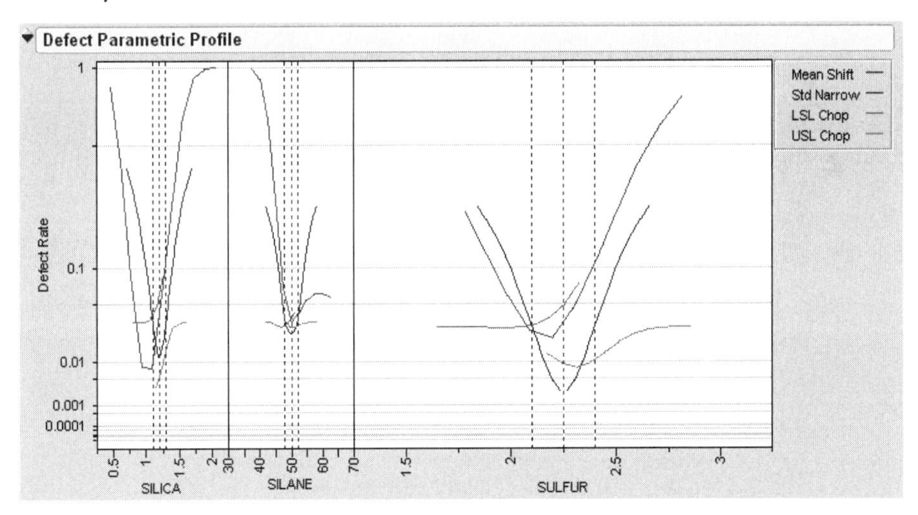

First, note that the current defect rate (0.03) is represented in four ways corresponding to each of the four curves.

For the red curve, mean shift, the current rate is where the red solid line intersects the vertical red dotted line. One opportunity to reduce the defect rate is to shift the mean slightly to the left. If you use the crosshair tool on this plot, you see that a mean shift reduces the defect rate to 0.037.

For the blue curve, Std Narrow, the current rate represents where the solid blue line intersects the two dotted blue lines. The Std Narrow curves represent the change in defect rate by changing the standard deviation of the factor. The dotted blue lines represent one standard deviation below and above the current mean. The solid blue lines are drawn symmetrically around the center. At the center, the blue line typically reaches a minimum, representing the defect rate for a standard deviation of zero. That is, if we totally eliminate variation in SULFUR, the defect rate is still around 0.024. This is much better than 0.03. If you look at the other Defect parametric profile curves, you can see that this is better than reducing variation in the other factors, something we suspected by the SD value for SULFUR.

For the green curve, LSL Chop, there are no interesting opportunities in this example, because the green curve is above current defect rates for the whole curve. This means that reducing the variation by rejecting parts with too-small values for SULFUR will not help.

For the orange curve, USL Chop, there are good opportunities. Reading the curve from the right, the curve starts out at the current defect rate (0.03), then as you start rejecting more parts by decreasing the USL for SULFUR, the defect rate improves—almost as much as reducing the variation to zero. However, moving a spec limit to the center is equivalent to throwing away half the parts, which may not be a practical solution.

Looking at all the opportunities over all the factors, it now looks like the best first move is to change the mean of SILICA to 1.

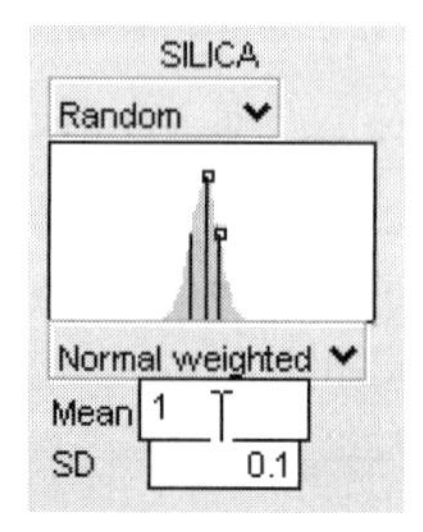

After changing the mean of SILICA, all the defect curves become invalid and need to be rerun. After clicking **Rerun**, we get a new perspective on defect rates.

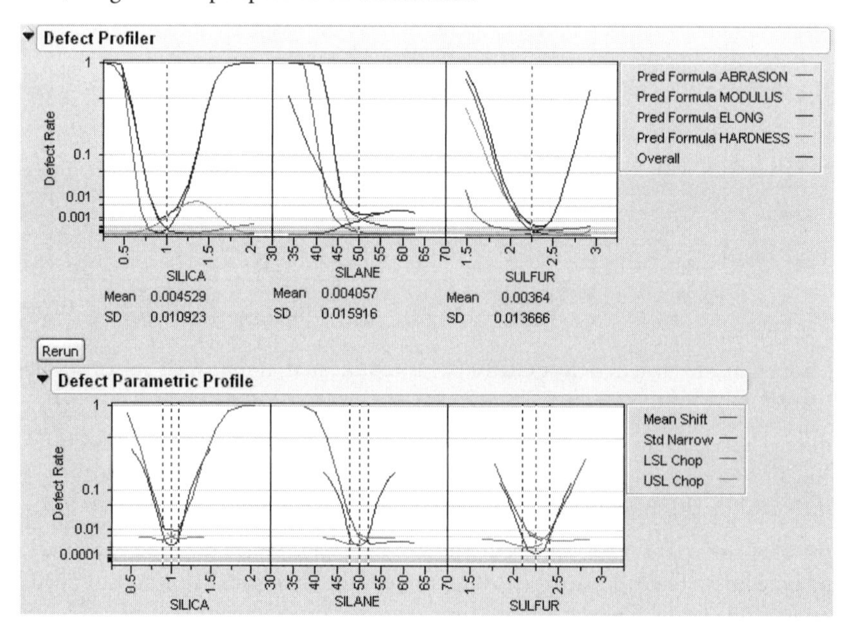

Now, the defect rate is down to about 0.004, much improved. The SD values suggest that SILANE might be the next variable to consider, but the SD values aren't that different from each other.

Shifting the mean of SILANE from 50 to 52 looks promising, so we do that and **Rerun**.

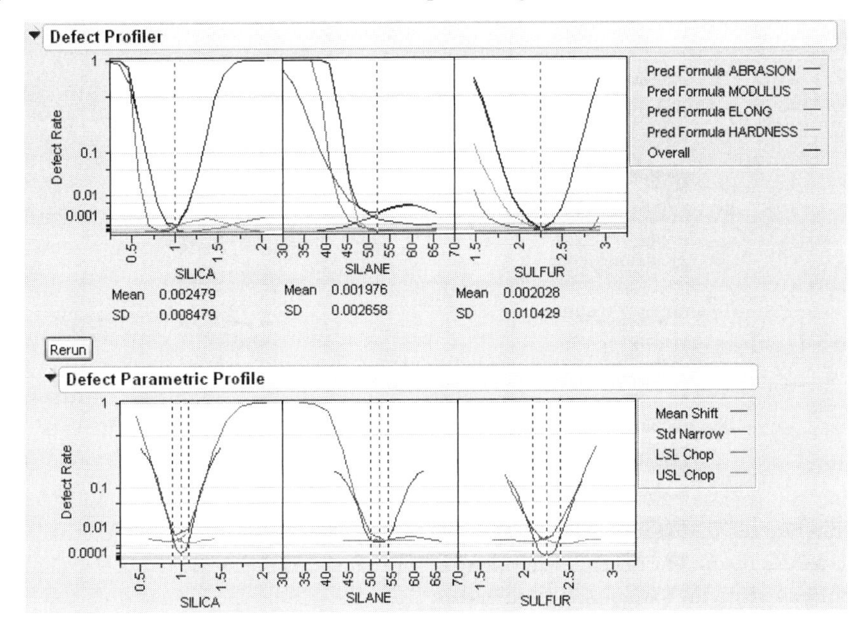

Now it is clear that reducing variation in SULFUR is the best prospect for further defect reduction. Suppose we can reduce the standard deviation from 0.15 to 0.08 without much sacrifice in cost.

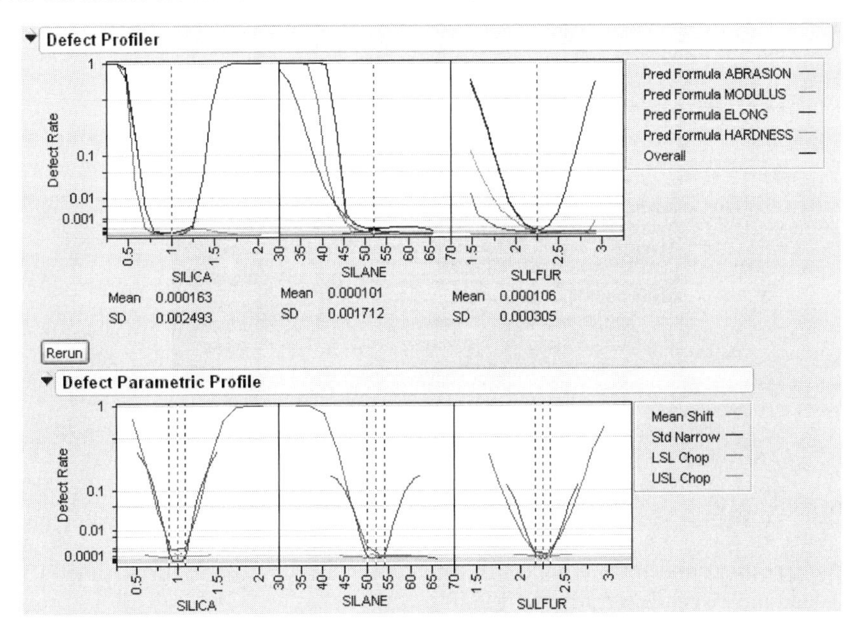

After rerunning, the defect rate has dropped to about 0.0001: one defect in 10,000. To see the opportunities better, double click to change the vertical scale to get more detail.

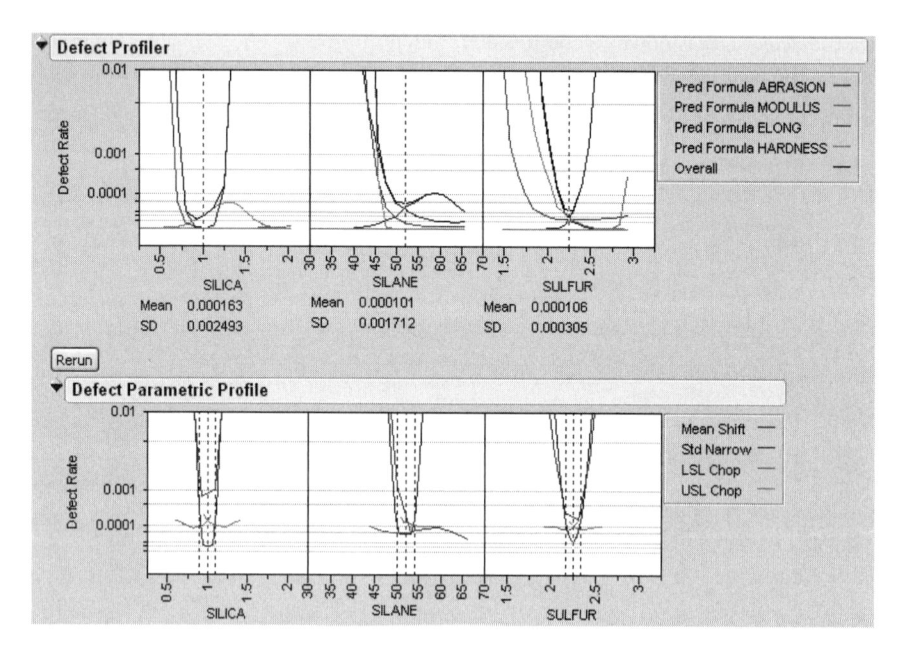

We can proceed in this fashion as far as is practical. The opportunities for further improvement would be to reduce the mean of SILICA slightly to 0.95, to increase the mean of SILANE slightly to 52.5, and to further reduce the variation in SULFUR to 0.05.

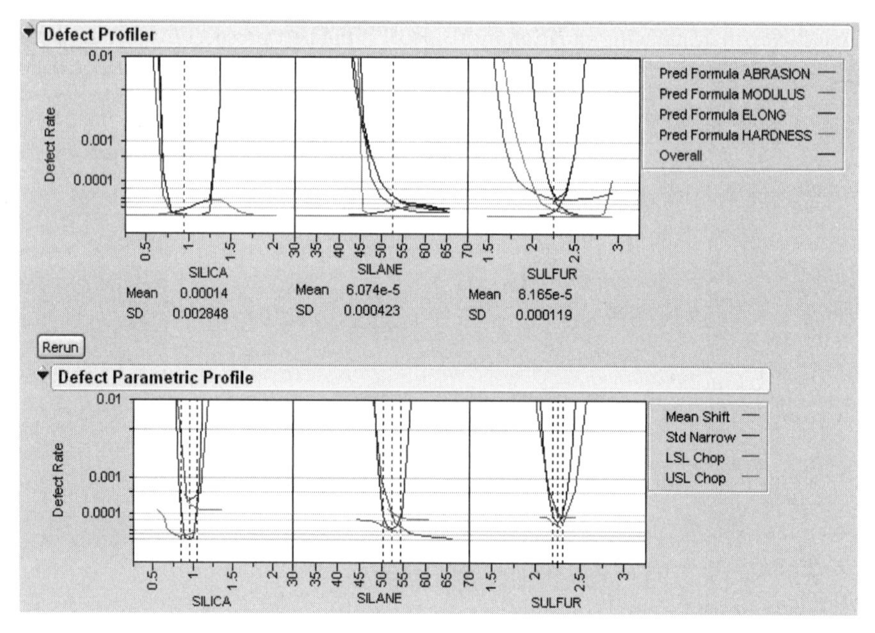

Now we see opportunities by reducing the variation in SILICA to 0.08, increasing the value of SILANE to 54, and increasing the mean of SULFUR to 2.3.

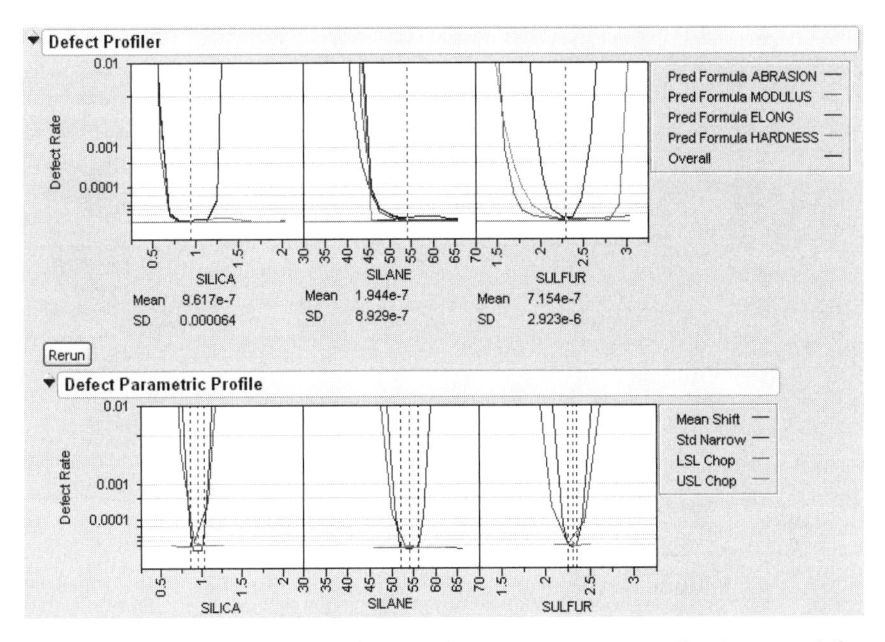

After rerunning with these settings, the defect rate decreases to 3e-8. Actually, the mean defect rates across the factors become relatively less reliable at this small rate. The accuracy could be improved by reducing the ranges of the factors in the profiler a little so that it integrates the distributions better.

This level of fine-tuning is probably not practical, because the experiment that estimated the response surface is probably not at this high level of accuracy. Once the ranges have been refined, you may need to conduct another experiment focusing on the area that you know is closer to the optimum.

Bank Profile Example

This example, developed by Schmee and Bowman (2004), examines bank processing time for loan approvals. This is the sum of four other processing times: Receive Application, Review, Credit Check, and Loan Finalization. The sample data table Bank Loan.jmp has two versions of the response, one with just an upper spec limit (Time) and one with both upper and Lower limits (Time2sided). Figure 15.14 shows the distribution of the four constituent processing times and the response with limits 42 and 76.

Figure 15.14 Total Processing Time as Sum of Four Distributions.

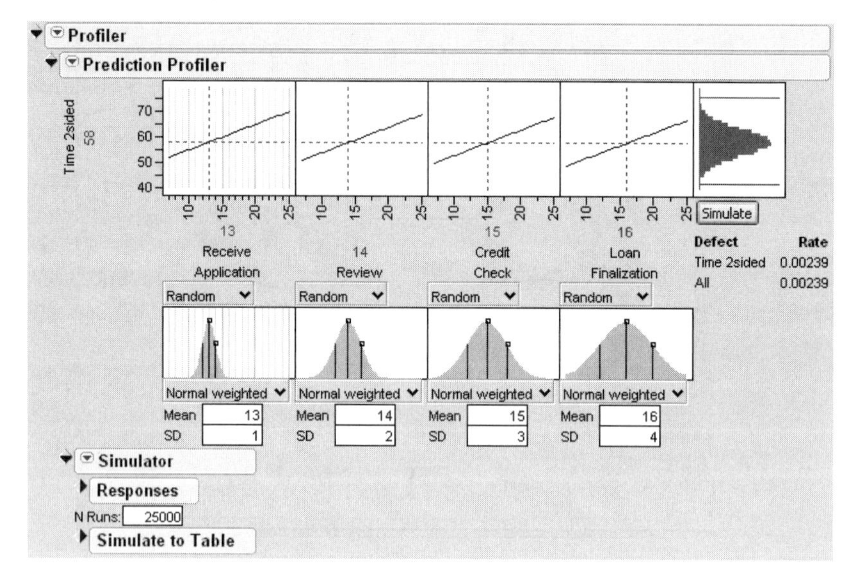

Using 25000 (rather than the default 5000) runs with these distributions, and click the **Simulate** button to see the defect rate for this process. It appears to be around 0.2 percent.

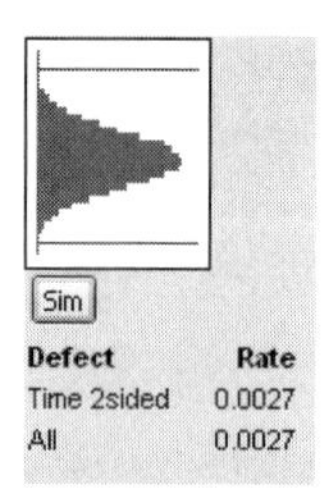

Invoking the Defect Profiler (**Simulator > Defect Profiler**) produces the following report. You may need to change the *Y*-axis so that its range is 0 to 0.1 to match this picture.

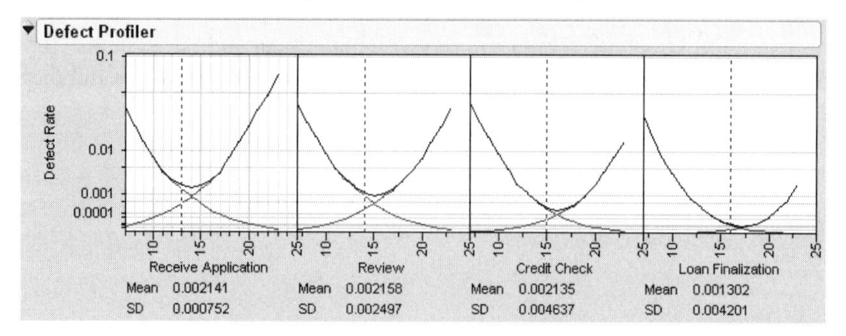

The defect rates vary around 0.002, but the rate for **Loan Finalization** is much smaller. The rates should be about the same. The difference is due to the range of that factor in the profiler not being wide enough to capture enough of the tails of the distribution for the numerical integration to be accurate.

Double-click on the axis for Loan Finalization in the upper profiler and set the range to be from 0 to 30 and click the **Rerun** button. The mean defect rates are now closer to agreement.

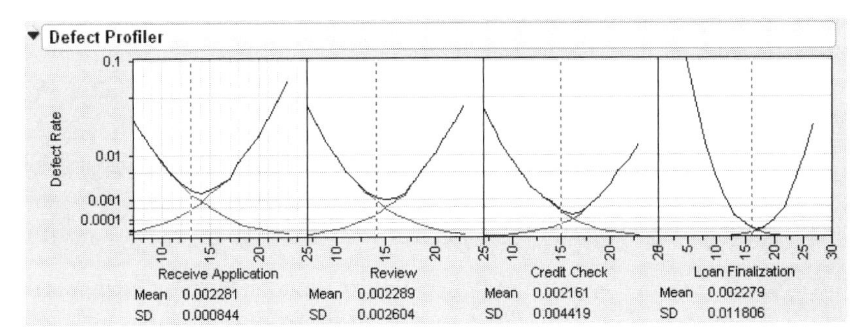

Looking at the defect profiles, we see that the current centers are reasonably close to the minimums. Looking a the standard deviations shows that the last factor has the most sensitivity overall. This is what we expect, given the larger variance of that factor.

Let's explore this further by reducing the standard deviation of Loan Finalization from 4 to 3.

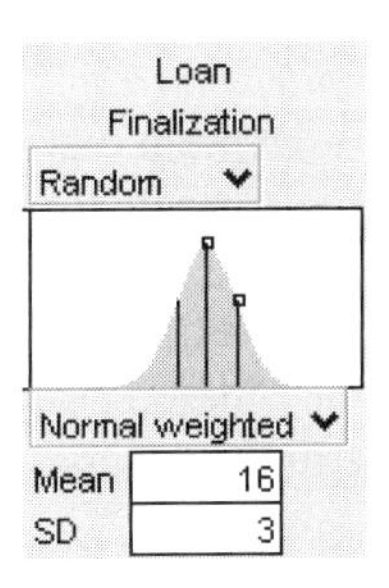

Click **Simulate** to see the defect rate decrease to around 0.0005.

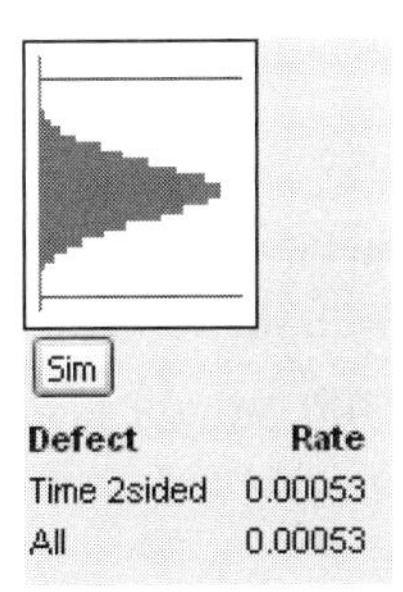

With this new defect rate, click **Rerun** to update the simulations and get new defect curves for all the factors.

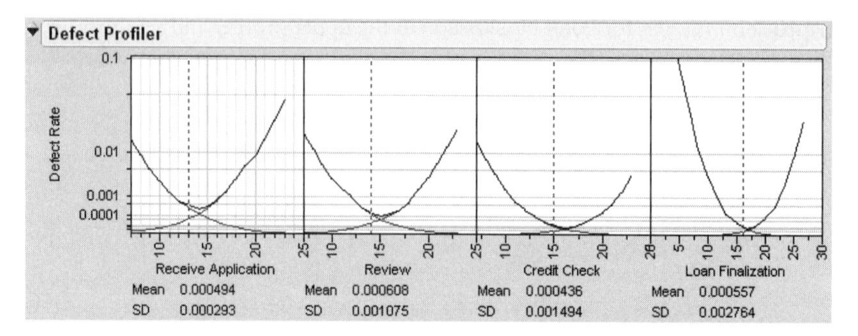

Another approach to reducing variation is to eliminate variation by rejecting inputs that are outside spec limits. So, we could set **Credit Check** to **Normal Truncated** with limits of 8 and 21. By switching to a **Normal Truncated** distribution, this reduces defect rates further.

Noise Factors (Robust Engineering)

Robust process engineering enables you to produce acceptable products reliably, despite variation in the process variables. Even when your experiment has controllable factors, there is a certain amount of uncontrollable variation in the factors that affects the response. This is called *transmitted variation*. Factors with this variation are called *noise factors*. Some factors you can't control at all, like environmental noise factors. Some factors can have their mean controlled, but not their standard deviation. This is often the case for intermediate factors that are output from a different process or manufacturing step.

A good approach to making the process robust is to match the target at the flattest place of the noise response surface so that the noise has little influence on the process. Mathematically, this is the value where the first derivatives of each response with respect to each noise factor are zero. JMP, of course, computes the derivatives for you.

To analyze a data set with noise factors,

- Run the appropriate model (using, for example the **Fit Model** platform).
- Save the model to the data table with the **Save > Prediction Formula** command.
- Launch the **Profiler** (from the **Graph** menu).
- Assign the prediction formula to the **Y, Prediction Formula** role and the noise factors to the **Noise Factor** role.
- Click **OK**.

The resulting profiler shows response functions and their appropriate derivatives with respect to the noise factors, with the derivatives set to have maximum desirability at zero.

- Select **Maximize Desirability** from the Profiler menu.

This finds the best settings of the factors, balanced with respect to minimizing transmitted variation from the noise factors.

Example

As an example, use the Tiretread.jmp sample data set. This data set shows the results of a tire manufacturer's experiment whose objective is to match a target value of hardness = 67.5 based on three factors: silica, silane, and sulfur content. Suppose the silane and sulfur content are easily (and precisely) controllable, but silica expresses variability that is worth considering.

- Run the RSM For Four Responses script attached to the data table by clicking on its red triangle and selecting **Run Script**.
- Click **Run Model** in the menu that appears.

When the analysis report appears,

- Scroll down to the **Response Hardness** outline node and select **Save > Prediction Formula** from the node's drop-down menu.

This saves the prediction formula in the original data table with the name Pred Formula HARDNESS.

For comparison, first maximize the factors for hardness without considering variation from the noise factor.

- Select **Graph > Profiler** to launch the profiler.
- Select **Maximize Desirability** to find the optimum factor settings for our target value of HARDNESS.

We get the following profiler display. Notice that the SILICA factor's optimum value is on a sloped part of a profile curve. This means that variations in SILICA are transmitted to become variations in the response, HARDNESS.

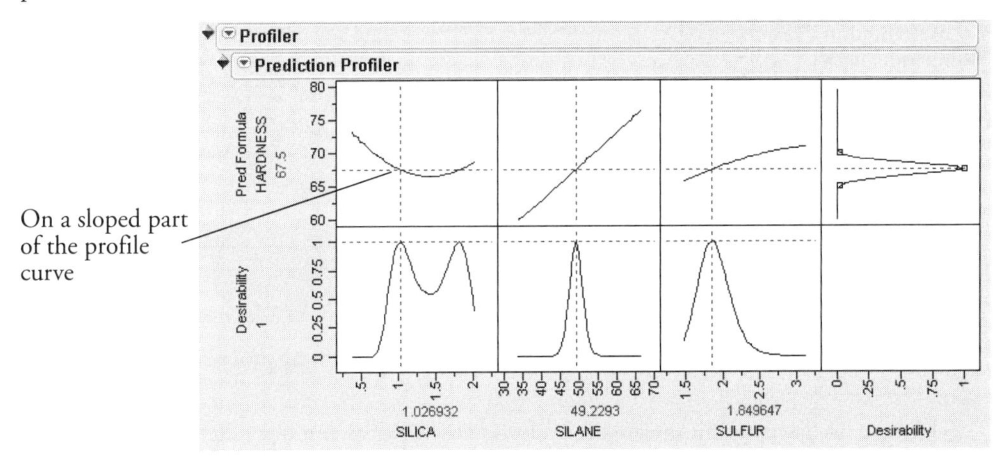

Note: You may get different results from these because different combinations of factor values can all hit the target.

Now, we would like to not just optimize for a specific target value of HARDNESS, but also be on a flat part of the curve. So, repeat the process and add SILICA as a noise factor.

- Select **Graph > Profiler.**
- Assign Pred Formula HARDNESS to **Y, Prediction Formula** variable and SILICA to **Noise Factors**, then click **OK.**

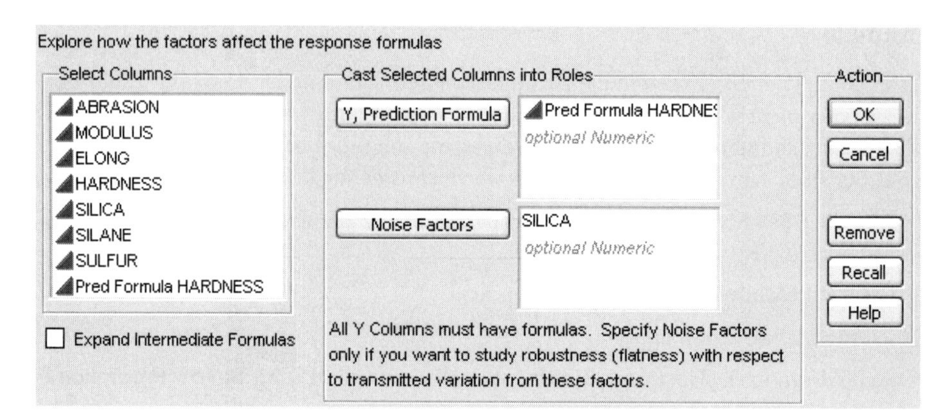

The resulting profiler has the appropriate derivative of the fitted model with respect to the noise factor, set to be maximized at zero, its flattest point.

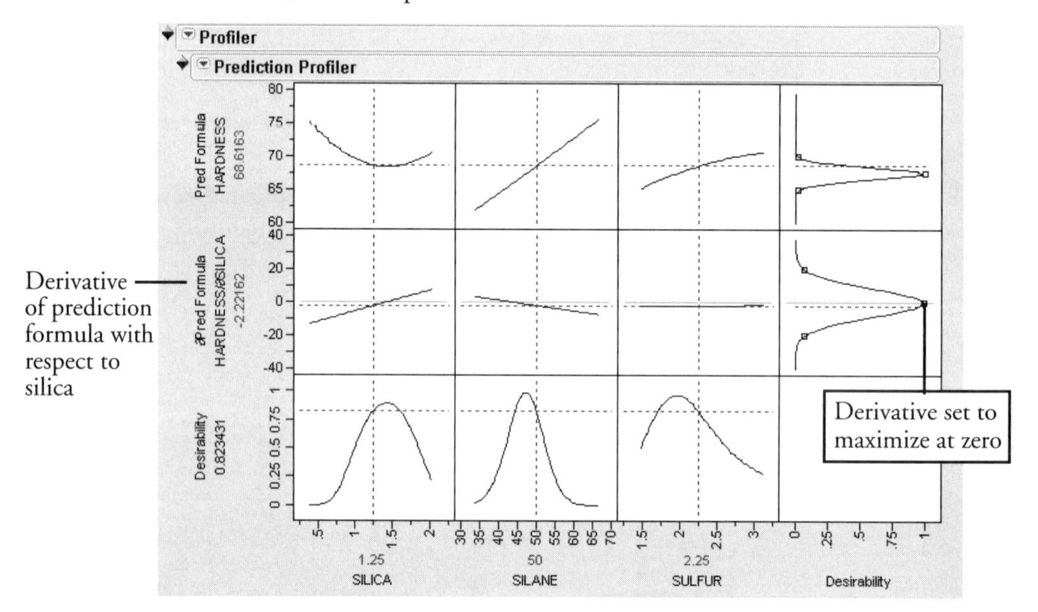

- Select **Maximize Desirability** to find the optimum values for the process factor, balancing for the noise factor.

This time, we have also hit the targeted value of HARDNESS, but our value of SILICA is on its flatter region.

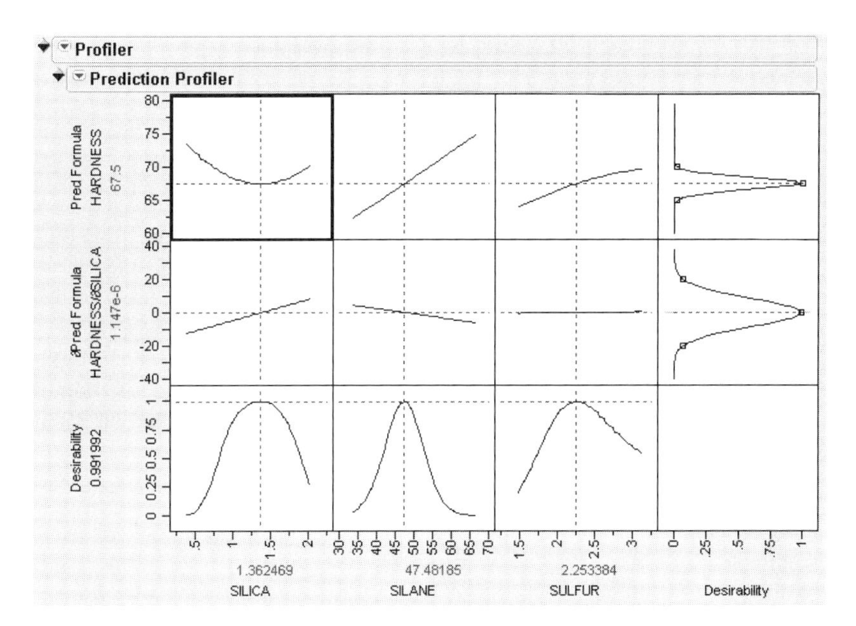

You can easily see the effect this has on the variance of the predictions. From each Profiler (one without the noise factor, one with),

- Select **Simulator** from the platform menu.
- Assign **SILICA** to have a random Normal distribution with a standard deviation of 0.05.

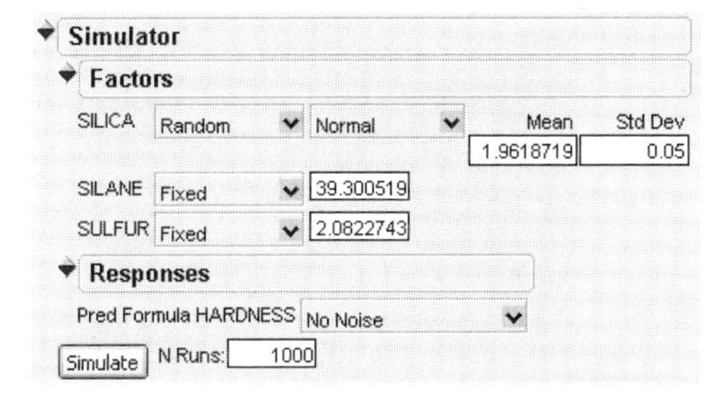

- Click **Simulate**.

Doing these steps for both the original and noise-factor-optimal simulations results in two similar data tables, each holding a simulation. In order to make two comparable histograms of the predictions, we need the two prediction columns in a single data table.

- Copy one **Pred Formula HARDNESS** column into the other data table. Name them **Without Noise Factor** and **With Noise Factor**, respectively.
- Select **Analyze > Distribution** and assign both prediction columns as **Y**.

When the histograms appear,

- Select **Uniform Scaling** from the **Distribution** main title bar.

The resulting histograms show the dramatic difference.

It is also interesting to note the shape of the robust response's transmitted variation (seen in the profiler with the noise factor). It is skewed because it straddles the minima of HARDNESS by SILICA.

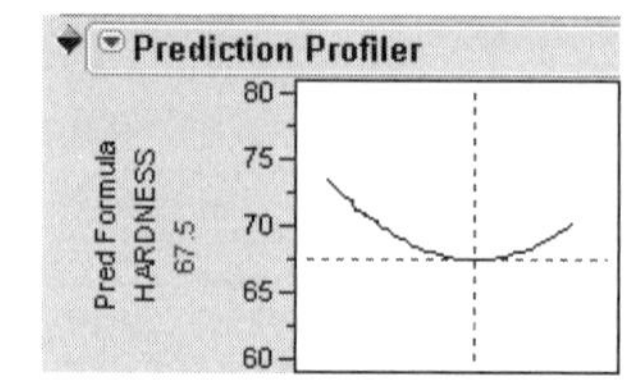

Therefore, transmitted variation from the center can only make hardness increase. When the non-robust solution is used, the variation could be transmitted either way. In the comparison histograms above, note that the With Noise Factor distribution only has error trailing off in one direction.

Other Platforms

Noise factor optimization is also available in the Contour Profiler (p. 306) and the Custom Profiler.

Statistical Details

Normal Weighted Distribution

JMP uses the multivariate radial strata method for each factor that uses the **Normal Weighted** distribution. This seems to work better than a number of Importance Sampling methods, as a multivariate Normal Integrator accurate in the extreme tails.

First, define strata and calculate corresponding probabilities and weights. For d random factors, the strata are radial intervals as follows.

Figure 15.15 Strata Intervals

Strata Number	Inside Distance	Outside Distance
0	0	$\sqrt{d}$
1	$\sqrt{d}$	$\sqrt{d+\sqrt{2d}}$
2	$\sqrt{d+\sqrt{2d}}$	$\sqrt{d+2\sqrt{2d}}$

Figure 15.15 Strata Intervals

Strata Number	Inside Distance	Outside Distance
i	$\sqrt{d+(i-1)\sqrt{2d}}$	$\sqrt{d+i\sqrt{2d}}$
$N_{Strata}-1$	previous	∞

The default number of strata is 12. To change the number of strata, a hidden command **N Strata** is available if you hold the Shift key down when accessing the drop down menu for the Simulator.

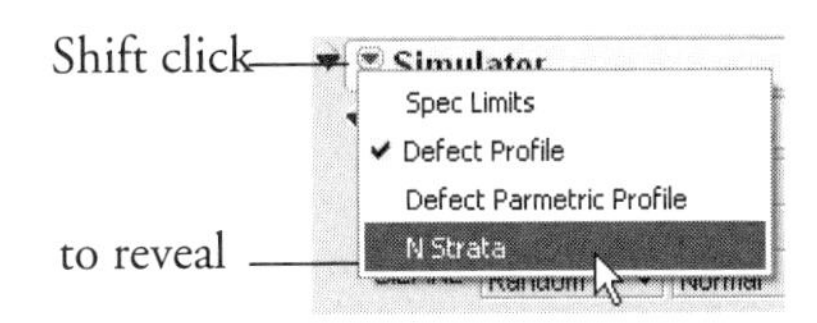

Increase the sample size as needed to maintain an even number of strata.

For each simulation run,

1 Select a strata as mod(i – 1, N_{Strata}) for run i.

2 Determine a random n-dimensional direction by scaling multivariate Normal (0,1) deviates to unit norm.

3 Determine a random distance using a chi-square quantile appropriate for the strata of a random uniform argument.

4 Scale the variates so the norm is the random distance.

5 Scale and re-center the variates individually to be as specified for each factor.

The resulting factor distributions are multivariate normal with the appropriate means and standard deviations when estimated with the right weights. Note that you cannot use the Distribution standard deviation with weights, because it does not estimate the desired value. However, multiplying the weight by a large value, like 10^{12}, and using that as a Freq value results in the correct standard deviation.

Chapter 16

Standard Least Squares: Random Effects

The Fit Model Platform

Random effects are those where the effect levels are chosen randomly from a larger population of levels, which they represent as a sample. In contrast, the levels of fixed effects are of direct interest rather than representative. If you have both random and fixed (nonrandom) effects in a model, it is called a *mixed model.*

It is very important to declare random effects. Otherwise, the test statistics produced from the fitted model are calculated with the wrong assumptions.

Typical random effects are

- subjects in a *repeated measures* experiment, where the subject is measured at several times.
- plots in a *split plot experiment*, where an experimental unit is subdivided and multiple measurements are taken, usually with another treatment applied to the subunits.
- *measurement* studies, where multiple measurements are taken in order to study measurement variation.
- random coefficients models, where the random effect is built with a continuous term crossed with categories.

The Fit Model platform in JMP fits mixed models using modern methods now generally regarded as best practice:

- REML estimation method (REstricted Maximum Likelihood)
- Kenward-Roger tests

For historical interest only, the platform also offers the Method of Moments (EMS), but this is no longer a recommended method except in special cases where it is equivalent to REML.

If you have a model where all effects are random, you can also fit it in the Variability Chart platform.

Contents

Topics in Random Effects

Introduction to Random Effects

Levels in random effects are randomly selected from a larger population of levels. For the purpose of testing hypotheses, the distribution of the effect on the response over the levels is assumed to be normal, with mean zero and some variance (called a *variance component*).

In one sense, every model has at least one random effect, which is the effect that makes up the residual error. The units making up individual observations are assumed to be randomly selected from a much larger population, and the effect sizes are assumed to have a mean of zero and some variance, σ^2.

The most common model that has random effects other than residual error is the repeated measures or split plot model. Table 16.1 "Types of Effects in a Split plot Model," p. 341, lists the types of effects in a split plot model. In these models the experiment has two layers. Some effects are applied on the whole plots or subjects of the experiment. Then these plots are divided or the subjects are measured at different times and other effects are applied within those subunits. The effects describing the whole plots or subjects are one random effect, and the subplots or repeated measures are another random effect. Usually the subunit effect is omitted from the model and absorbed as residual error.

Table 16.1 Types of Effects in a Split plot Model

Split Plot Model	Type of Effect	Repeated Measures Model
whole plot treatment	fixed effect	across subjects treatment
whole plot ID	random effect	subject ID
subplot treatment	fixed effect	within subject treatment
subplot ID	random effect	repeated measures ID

Each of these cases can be treated as a layered model, and there are several traditional ways to fit them in a fair way. The situation is treated as two different experiments:

1 The whole plot experiment has whole plot or subjects as the experimental unit to form its error term.

2 Subplot treatment has individual measurements for the experimental units to form its error term (left as residual error).

The older, traditional way to test whole plots is to do any one of the following:

- Take means across the measurements and fit these means to the whole plot effects.
- Form an F-ratio by dividing the whole plot mean squares by the whole plot ID mean squares.
- Organize the data so that the split or repeated measures form different columns and do a MANOVA model, and use the univariate statistics.

While these approaches work if the structure is simple and the data is complete and balanced, there is a more general model that works for any structure of random effects. This more generalized model is called the *mixed model*, because it has both fixed and random effects.

Another common situation that involves multiple random effects is in measurement systems where there are multiple measurements with different parts, different operators, different gauges, and different repetitions. In this situation, all the effects are regarded as random.

Generalizability

Random effects are randomly selected from a larger population, where the distribution of their effect on the response is assumed to be a realization of a normal distribution with a mean of zero and a variance that can be estimated.

Often, the exact effect sizes are not of direct interest. It is the fact that they represent the larger population that is of interest. What you learn about the mean and variance of the effect tells you something about the general population from which the effect levels were drawn. That is different from fixed effects, where you only know about the levels you actually encounter in the data.

The REML Method

The REML (REstricted or REsidual Maximum Likelihood) method for fitting mixed models is now the mainstream, state-of-the-art method, supplanting older methods.

In the days before availability of powerful computers, researchers needed to restrict their interest to situations in which there were computational short cuts to obtain estimates of variance components and tests on fixed effects in a mixed model. Most books today introduce mixed models using these short cuts that work on balanced data. The Method of Moments provided a way to calculate what the expected value of Mean Squares (EMS) were in terms of the variance components, and then back-solve to obtain the variance components. It was also possible using these techniques to obtain expressions for test statistics that had the right expected value under the null hypotheses that were synthesized from mean squares.

If your model satisfies certain conditions (*i.e.* it has random effects that contain the terms of the fixed effects they provide random structure for) then you can use the EMS choice to produce these traditional analyses. However, since the newer REML method produces identical results to these models, but is considerably more general, the EMS method is never recommended.

The REML approach was pioneered by Patterson and Thompson in 1974. See also Wolfinger, Tobias, and Sall (1994) and Searle, Casella, and McCulloch(1992). The reason to prefer REML is that it works without depending on balanced data, or shortcut approximations, and it gets all the tests right, even contrasts that work across interactions. Most packages that use the traditional EMS method are either not able to test some of these contrasts, or compute incorrect variances for them.

Unrestricted Parameterization for Variance Components in JMP

Note: Read this section only if you are concerned about matching the results of certain textbooks.

There are two different statistical traditions for parameterizing the variance components: the unrestricted and the restricted approaches. JMP and SAS use the unrestricted approach. In this approach, while the estimated effects always sum to zero, the true effects are not assumed to sum to zero over a particular random selection made of the random levels. This is the same assumption as for residual error. The estimates make the residual errors have mean zero, and the true mean is zero. But a random draw of data using the true parameters will be some random event that might not have a mean of exactly zero.

You need to know about this assumption because many statistics textbooks use the restricted approach. Both approaches have been widely taught for 50 years. A good source that explains both sides is Cobb (1998), section 13.3.

Negative Variances

Note: Read this section only when you are concerned about negative variance components.

Though variances are always positive, it is possible to have a situation where the unbiased estimate of the variance is negative. This happens in experiments when an effect is very weak, and by chance the resulting data causes the estimate to be negative. This usually happens when there are few levels of a random effect that correspond to a variance component.

JMP can produce negative estimates for both REML and EMS. For REML, there is a checkbox in the model dialog to constrain the estimate to be non-negative. We recommend that you do not check this. Constraining the variance estimates leads to bias in the tests for the fixed effects.

If you remain uncomfortable about negative estimates of variances, please consider that the random effects model is statistically equivalent to the model where the variance components are really covariances across errors within a whole plot. It is not hard to think of situations in which the covariance estimate can be negative, either by random happenstance, or by a real process in which deviations in some observations in one direction would lead to deviations in the other direction in other observations. When random effects are modeled this way, the covariance structure is called *compound symmetry*.

So, consider negative variance estimates as useful information. If the negative value is small, it can be considered happenstance in the case of a small true variance. If the negative value is larger (the variance ratio can get as big as 0.5), it is a troubleshooting sign that the rows are not as independent as you had assumed, and some process worth investigating is happening within blocks.

Random Effects *BLUP* Parameters

Random effects have a dual character. In one perspective, they appear like residual error, often the error associated with a whole-plot experimental unit. In another respect, they are like fixed effects with a parameter to associate with each effect category level. As parameters, you have extra information about them-they are derived from a normal distribution with mean zero and the variance estimated by the variance component. The effect of this extra information is that the estimates of the parameters are shrunk towards zero. The parameter estimates associated with random effects are called *BLUPs* (Best Linear Unbiased Predictors). Some researchers consider these BLUPs as parameters of interest, and others consider them by-products of the method that are not interesting in themselves. In JMP, these estimates are available, but in an initially-closed report.

BLUP parameter estimates are used to estimate random-effect least squares means, which are therefore also shrunken towards the grand mean, at least compared to what they would be if the effect were treated as a fixed effect. The degree of shrinkage depends on the variance of the effect and the number of observations per level in the effect. With large variance estimates, there is little shrinkage. If the variance component is small, then more shrinkage takes place. If the variance component is zero, the effect levels are shrunk to exactly zero. It is even possible to obtain highly negative variance components where the shrinkage is reversed. You can consider fixed effects as a special case of random effects where the variance component is very large.

If the number of observations per level is large, the estimate will shrink less. If there are very few observations per level, the estimates will shrink more. If there are infinite observations, there is no shrinkage and the answers are the same as fixed effects.

The REML method balances the information on each individual level with the information on the variances across levels.

As an example, suppose that you have batting averages for different baseball players. The variance component for the batting performance across player describes how much variation is usual between players in their batting averages. If the player only plays a few times and if the batting average is unusually small or large, then you tend not to trust that estimate, because it is based on only a few at-bats; the estimate has a high standard error. But if you mixed it with the grand mean, that is, shrunk the estimate towards the grand mean, you would trust the estimate more. For players that have a long batting record, you would shrink much less than those with a short record.

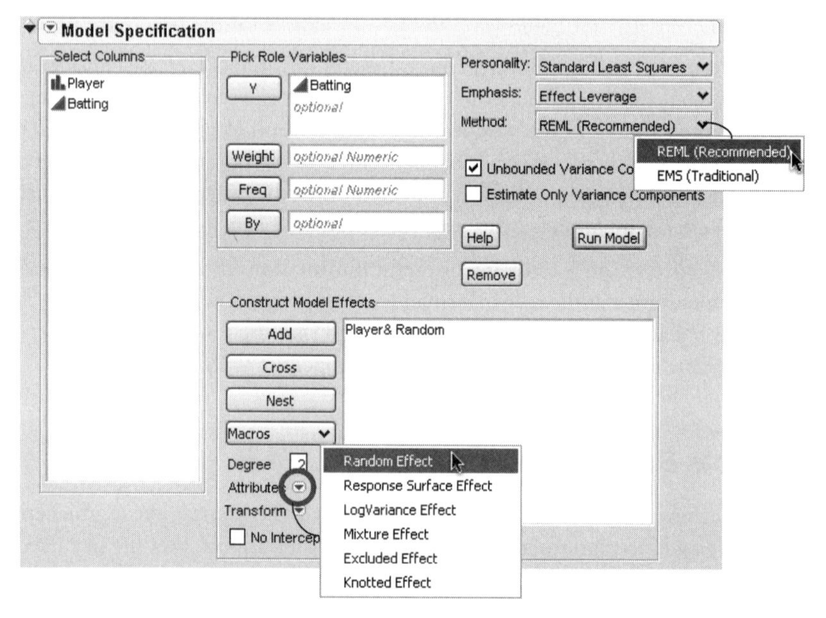

You can run this example and see the results for yourself. The example batting average data are in the **Baseball.jmp** sample data file, with variables **Player** and **Batting**. To compare the Method of Moments (EMS) and REML, run the model twice. **Batting** is *Y* and **Player** is the random effect, as shown here.

Run the model and select **REML (Recommended)** from the **Method** popup menu.

Run the model again with **EMS (Traditional)** as **Method**.

Table 16.2 "Comparison of Estimates Between Method of Moments and REML," p. 345, summarizes the estimates between Method of Moments and REML across a set of baseball players in this simulated example. Note that Suarez, with only 3 at-bats, is shrunk more than the others with more at-bats.

Table 16.2 Comparison of Estimates Between Method of Moments and REML

	Method of Moments	REML	N
Variance Component	0.01765	0.019648	
Anderson	0.29500000	0.29640407	6
Jones	0.20227273	0.20389793	11
Mitchell	0.32333333	0.32426295	6
Rodriguez	0.55000000	0.54713393	6
Smith	0.35681818	0.35702094	11
Suarez	0.55000000	0.54436227	3
Least Squares Means	same as ordinary means	shrunken from means	

REML and Traditional Methods Agree on the Standard Cases

It turns out that in balanced designs, the REML *F*-test values for fixed effects will be the same as with the Method of Moments (Expected Means Squares) approach. The degrees of freedom could differ in some cases. There are a number of methods of obtaining the degrees of freedom for REML *F*-tests; the one that JMP uses is the smallest degrees of freedom associated with a containing effect.

F-Tests in Mixed Models

Note: This section details the tests produced with REML

The REML method obtains the variance components and parameter estimates, but there are a few additional steps to obtain tests on fixed effects in the model. The objective is to construct the F statistic and associated degrees of freedom to obtain a p-value for the significance test.

Historically, in simple models using the Method of Moments (EMS), standard tests were derived by construction of quadratic forms that had the right expectation under the null hypothesis. Where a mean square had to be synthesized from a linear combination of mean squares to have the right expectation, Satterthwaite's method could be used to obtain the degrees of freedom to get the *p*-value. Sometimes these were fractional degrees of freedom, just as you might find in a modern (Aspin-Welch) Student's *t*-test.

With modern computing power and recent methods, we have much improved techniques to obtain the tests. First, Kackar and Harville (1984) found a way to estimate a bias-correction term for small samples. This was refined by Kenward and Roger (1997) to correct further and obtain the degrees of freedom that gave the closest match of an *F*-distribution to the distribution of the test statistic. These are not easy calculations, consequently they can take some time to perform for larger models.

If you have a simple balanced model, the results from REML-Kenward-Roger will agree with the results from the traditional approach, provided that the estimates aren't bounded at zero.

- These results do not depend on analyzing the syntactic structure of the model. There are no rules about finding containing effects. The method doesn't care if your whole plot fixed effects are purely nested in whole plot random effects-it will get the right answer regardless.

- These results do not depend on having categorical factors. It handles continuous (random coefficient) models just as easily.
- These methods will produce different (and better) results than older versions of JMP (that is, earlier than JMP 6) that implemented older, less precise, technology to do these tests.
- These methods do not depend on having positive variance components. Negative variance components are not only supported, but need to be allowed in order for the tests to be unbiased.

Our goal in implementing these methods was not just to handle general cases, but to handle cases without the user needing to know very much about the details. Just declare which effects are random, and everything else is automatic. It is particularly important that engineers learn to declare random effects, because they have a history of performing inadvertent split-plot experiments where the structure is not identified.

Specifying Random Effects

Models with Random Effects use the same Model Specification dialog as other models. To identify a random effect, highlight it in the model effects list and select **Random Effect** from the Attributes popup menu. This appends &Random to the effect name in the model effect list.

Split Plot Example

The most common type of layered design is a balanced split plot, often in the form of repeated measures across time. One experimental unit for some of the effects is subdivided, (sometimes by time period) and other effects are applied to these subunits.

As an example, consider the Animals.jmp data found in the Sample Data folder (the data are fictional). Figure 16.1, shows a listing of the table. The study collected information about the difference in seasonal hunting habits of foxes and coyotes. Three foxes and three coyotes were marked and observed periodically (each season) for a year. The average number of miles (rounded to the nearest mile) they wandered from their dens during different seasons of the year was recorded. The model is defined by

- the continuous response variable called miles
- the species effect with values fox or coyote
- the season effect with values fall, winter, spring, and summer
- an animal identification code called subject, with nominal values 1, 2, and 3 for both foxes and coyotes.

There are two layers to the model.

1. The top layer is the between-subject layer, in which the effect of being a fox or coyote (species effect) is tested with respect to the variation from subject to subject. The bottom layer is the within-subject layer, in which the repeated-measures factor for the four seasons (season effect) is tested with respect to the variation from season to season within a subject. The within-subject variability is reflected in the residual error.
2. The season effect can use the residual error for the denominator of its *F*-statistics. However, the between-subject variability is not measured by residual error and must be captured with the subject

within species (subject[species]) effect in the model. The *F*-statistic for the between-subject effect species uses this nested effect instead of residual error for its *F*-ratio denominator.

Figure 16.1 Data Table for Repeated Measure Experiment

Animals
Notes Example Data- Repeate
Repeated Measures Model

Columns (4/0)
species
subject
miles
season

Rows
All Rows 24
Selected 0
Excluded 0
Hidden 0
Labelled 0

	species	subject	miles	season
1	FOX	1	0	fall
2	FOX	1	0	winter
3	FOX	1	5	spring
4	FOX	1	3	summer
5	FOX	2	3	fall
6	FOX	2	1	winter
7	FOX	2	5	spring
8	FOX	2	4	summer
9	FOX	3	4	fall
10	FOX	3	3	winter
11	FOX	3	6	spring
12	FOX	3	2	summer
13	COYOTE	1	4	fall
14	COYOTE	1	2	winter
15	COYOTE	1	7	spring
16	COYOTE	1	8	summer
17	COYOTE	2	5	fall
18	COYOTE	2	4	winter
19	COYOTE	2	6	spring
20	COYOTE	2	6	summer
21	COYOTE	3	7	fall
22	COYOTE	3	5	winter
23	COYOTE	3	8	spring
24	COYOTE	3	9	summer

The Model Dialog

The **Fit Model** command lets you construct model terms and identify error terms. To run the nested model Animals data with the correct *F*-statistics, specify the response column and add terms to the Construct Model Effects list as follows to see the completed dialog in Figure 16.2.

- Select miles from the column selector and click **Y**.
- Select species from the column selector list and click **Add**.
- Select subject from the column selector list and click **Add**.
- Select species from the column selector list again, and also select subject in the Construct Model Effects list. Click **Nest** to add the subject within species (subject[species]) effect to the model.
- Select the nested effect, subject[species], and choose **Random Effect** from the **Attributes** popup menu. This nested effect is now identified as an error term for the species effect and shows as subject[species]&Random.
- Select season from the column selector list and click **Add**.

When you assign any effect as random from the **Attributes** popup menu, the Method options (REML and EMS) appear at the top-right of the dialog, with REML selected as the default.

Figure 16.2 Model Specification Dialog for Repeated Measures Experiment

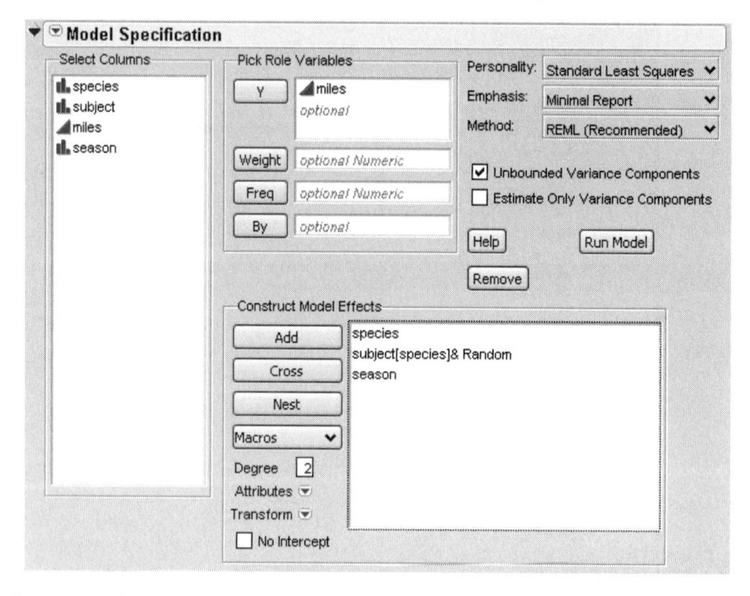

Note: JMP only creates design columns for those combinations that actually occur in the data.

Note: You do not have to have a completely contained effects. For example, if A*B*C is in the model, then A*B, A*C, and B*C do not necessarily have to be in the model.

REML Results

A nice feature of REML is that the report doesn't need qualification (Figure 16.3). The estimates are all properly shrunk and the standard errors are properly scaled (SAS Institute Inc. 1996). The variance components are shown as a ratio to the error variance, and as a portion of the total variance.

There is no special table of synthetic test creation, because all the adjustments are automatically taken care of by the model itself. There is no table of expected means squares, because the method does not need this.

If you have random effects in the model, the analysis of variance report is not shown. This is because the variance does not partition in the usual way, nor do the degrees of freedom attribute in the usual way, for REML-based estimates. You can obtain the residual variance estimate from the REML report rather than from the analysis of variance report.

Figure 16.3 Report of REML Analysis

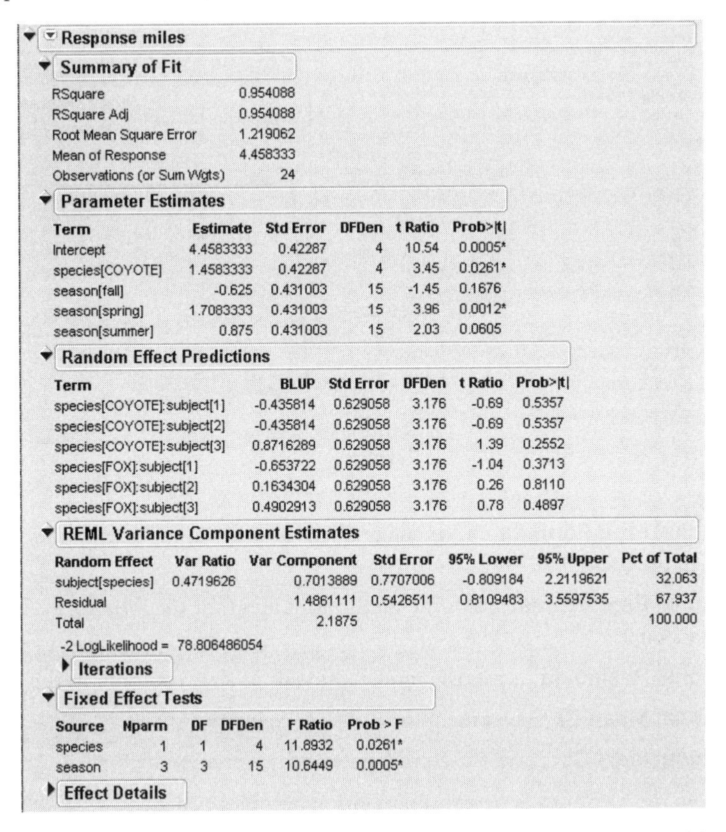

Response miles

Summary of Fit

RSquare	0.954088
RSquare Adj	0.954088
Root Mean Square Error	1.219062
Mean of Response	4.458333
Observations (or Sum Wgts)	24

Parameter Estimates

Term	Estimate	Std Error	DFDen	t Ratio	Prob>|t|
Intercept	4.4583333	0.42287	4	10.54	0.0005*
species[COYOTE]	1.4583333	0.42287	4	3.45	0.0261*
season[fall]	-0.625	0.431003	15	-1.45	0.1676
season[spring]	1.7083333	0.431003	15	3.96	0.0012*
season[summer]	0.875	0.431003	15	2.03	0.0605

Random Effect Predictions

Term	BLUP	Std Error	DFDen	t Ratio	Prob>|t|
species[COYOTE]:subject[1]	-0.435814	0.629058	3.176	-0.69	0.5357
species[COYOTE]:subject[2]	-0.435814	0.629058	3.176	-0.69	0.5357
species[COYOTE]:subject[3]	0.8716289	0.629058	3.176	1.39	0.2552
species[FOX]:subject[1]	-0.653722	0.629058	3.176	-1.04	0.3713
species[FOX]:subject[2]	0.1634304	0.629058	3.176	0.26	0.8110
species[FOX]:subject[3]	0.4902913	0.629058	3.176	0.78	0.4897

REML Variance Component Estimates

Random Effect	Var Ratio	Var Component	Std Error	95% Lower	95% Upper	Pct of Total
subject[species]	0.4719626	0.7013889	0.7707006	-0.809184	2.2119621	32.063
Residual		1.4861111	0.5426511	0.8109483	3.5597535	67.937
Total		2.1875				100.000

-2 LogLikelihood = 78.806486054

Iterations

Fixed Effect Tests

Source	Nparm	DF	DFDen	F Ratio	Prob > F
species	1	1	4	11.8932	0.0261*
season	3	3	15	10.6449	0.0005*

Effect Details

You can Right-click on the Variance Components Estimates table to toggle in the additional columns to show a 95% confidence interval for the variance components using the Satterthwaite (1946) approximation).

Note: If there is a random effect in the model and you invoke the Profiler, the random term is suppressed in the Profiler output. The predictions are integrated over Random effects, so they would appear flat (*i.e.* uninteresting). You can still make them show with the **Reset Factor Grid** command.

REML Save Menu

When REML is used, the following extra commands appear in the Save submenu. All these commands allow the Random effects to participate in the prediction, rather than using their expected value (zero).

In the following figure, the prediction formula for the above example is presented without participation of random effects (left) and with their participation (right).

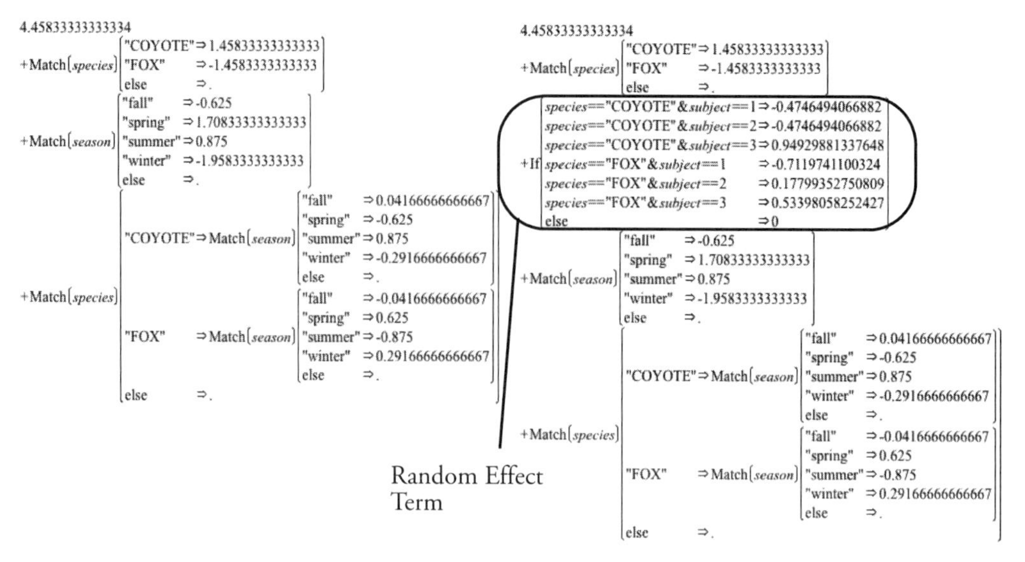

Conditional Pred Formula saves the prediction formula, including random effects, to a new column in the data table.

Conditional Pred Values saves the predicted values (not the formula itself) to a new column in the data table.

Conditional Residuals saves the model residuals to a new column in the data table.

Conditional Mean CI saves the mean confidence interval

Conditional Indiv CI saves the confidence interval for an individual.

In addition to these options, a new profiler option becomes available

Conditional Predictions shows the random effects and uses them in formulating the predicted value. In the following example, the random effects are not shown when the profiler is first shown. After selecting **Conditional Predictions**, the levels of the random effect variable appear. Note that the confidence intervals for the other factors change when the random effect is included.

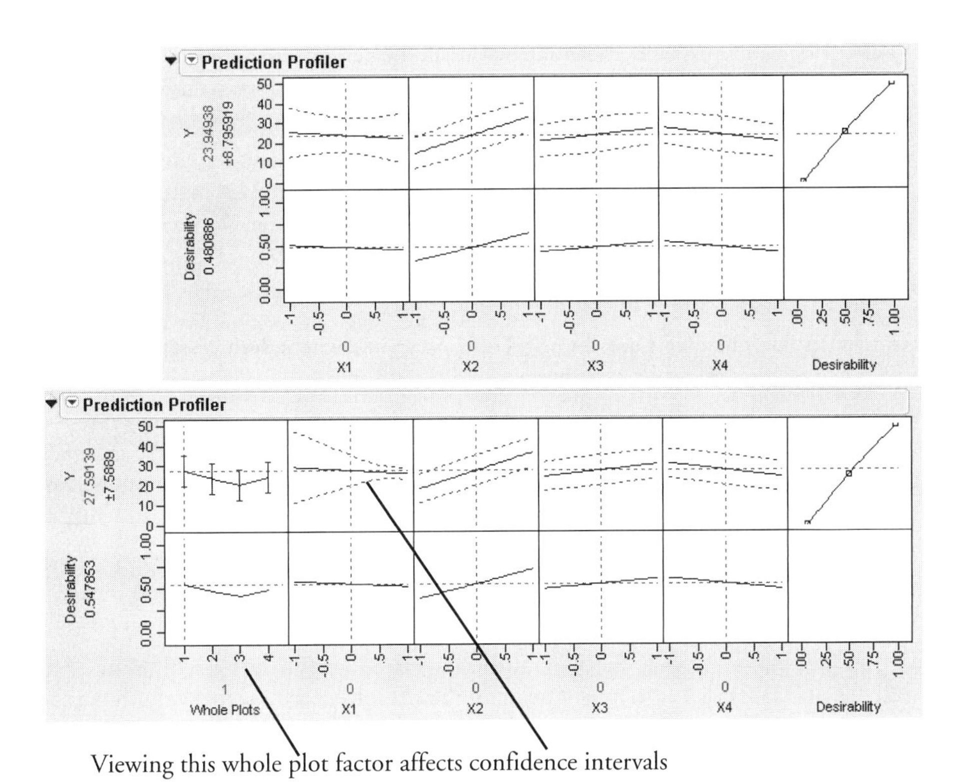

Viewing this whole plot factor affects confidence intervals

Method of Moments Results

Note: This section is only of use in matching historical results

---

We no longer recommend the Method of Moments, but we understand the need to support it for teaching use, in order to match the results of many textbooks still in use.

You have the option of choosing the traditional Method of Moments (EMS, Expected Mean Square) approach from the **Method** popup menu on the Model Specification dialog, as shown here.

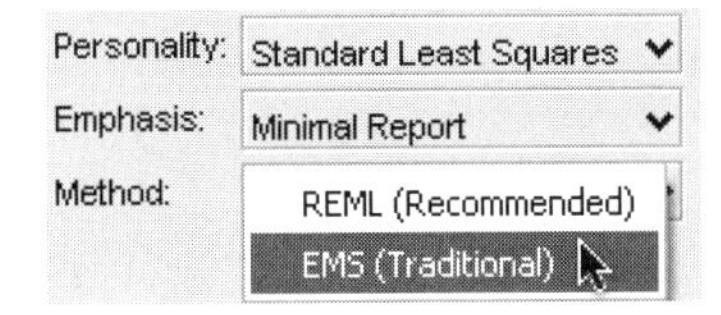

Results from the steps for the Method of Moments are as follows:

- For each effect, the coefficients of the expected mean squares for that effect are calculated. This is a linear combination of the variance components and fixed effect values that describes the expected value of the mean square for that effect. All effects also have a unit coefficient on the residual variance.

- The coefficients of expected mean squares for all the random effects, including the residual error, are gathered into a matrix, and this is used to solve for variance components for each random effect.
- For each effect to be tested, a denominator for that effect is synthesized using the terms of the linear combination of mean squares in the numerator that don't contain the effect to be tested or other fixed effects. Thus, the expectation is equal for those terms common to the numerator and denominator. The remaining terms in the numerator then constitute the effect test.
- Degrees of freedom for the synthesized denominator are constructed using Satterthwaite's method.
- The effect tests use the synthetic denominator.

JMP handles random effects like the SAS GLM procedure with a `Random` statement and the `Test` option. Figure 16.4, shows example results.

Warning: Standard errors for least squares means and denominators for contrast *F*-tests also use the synthesized denominators. Contrasts using synthetic denominators might not be appropriate, especially in crossed effects compared at common levels. The leverage plots and custom tests are done with respect to the residual, so they might not be appropriate.

Warning: Crossed and nested relationships must be declared explicitly. For example, if knowing a subject ID also identifies the group that contains the subject, (that is, if each subject is in only one group), then subject must be declared as nested within group. In that situation, the nesting must be explicitly declared to define the design structure.

Limitation: JMP cannot fit a layered design if the effect for a layer's error term cannot be specified under current effect syntax. An example of this is a design with a Latin Square on whole plots for which the error term would be `Row*Column–Treatment`. Fitting such special cases with JMP requires constructing your own *F*-tests using sequential sums of squares from several model runs.

For the Animals example above, the standard method of moments reports are as follows.

Figure 16.4 Report of Method of Moments Analysis for Animals Data

Response miles

Summary of Fit

RSquare	0.838417
RSquare Adj	0.75224
Root Mean Square Error	1.219062
Mean of Response	4.458333
Observations (or Sum Wgts)	24

Analysis of Variance

Source	DF	Sum of Squares	Mean Square	F Ratio
Model	8	115.66667	14.4583	9.7290
Error	15	22.29167	1.4861	Prob > F
C. Total	23	137.95833		0.0001*

Parameter Estimates

Term	Estimate	Std Error	t Ratio	Prob>|t|
Intercept	4.4583333	0.24884	17.92	<.0001*
species[COYOTE]	1.4583333	0.24884	5.86	<.0001*
season[fall]	-0.625	0.431003	-1.45	0.1676
season[spring]	1.7083333	0.431003	3.96	0.0012*
season[summer]	0.875	0.431003	2.03	0.0605
species[COYOTE]:subject[1]	-0.666667	0.49768	-1.34	0.2003
species[COYOTE]:subject[2]	-0.666667	0.49768	-1.34	0.2003
species[FOX]:subject[1]	-1	0.49768	-2.01	0.0628
species[FOX]:subject[2]	0.25	0.49768	0.50	0.6227

Expected Mean Squares

The Mean Square per row by the Variance Component per column

EMS

	Intercept	species	subject[species]&Random	season
Intercept	0	0	0	0
species	0	12	4	0
subject[species]&Random	0	0	4	0
season	0	0	0	6

plus 1.0 times Residual Error Variance

Variance Component Estimates

Component	Var Comp Est	Percent of Total
subject[species]&Random	0.701389	32.063
Residual	1.486111	67.937
Total	2.1875	100.000

These estimates based on equating Mean Squares to Expected Value.

Test Denominator Synthesis

Source	MS Den	DF Den	Denom MS Synthesis
species	4.29167	4	subject[species]&Random
subject[species]&Random	1.48611	15	Residual
season	1.48611	15	Residual

Tests wrt Random Effects

Source	SS	MS Num	DF Num	F Ratio	Prob > F
species	51.0417	51.0417	1	11.8932	0.0261*
subject[species]&Random	17.1667	4.29167	4	2.8879	0.0588
season	47.4583	15.8194	3	10.6449	0.0005*

Effect Details

The random submatrix from the EMS table is inverted and multiplied into the mean squares to obtain variance component estimates. These estimates are usually (but not necessarily) positive. The variance component estimate for the residual is the Mean Square Error.

Note that the CV of the variance components is initially hidden in the Variance Components Estimates report. To reveal it, right-click (Control-click on the Macintosh) and select **Columns > CV** from the menu that appears.

Unbalanced Example

The Machine.jmp data table in the Sample Data folder records 44 responses (rating) given by six subjects for three different machines. The 44 responses are distributed as shown here.

Contingency Table

Count (machine \ person)	1	2	3	4	5	6	
1	1	2	1	2	3	3	12
2	1	3	2	3	2	3	14
3	3	3	3	3	3	3	18
	5	8	6	8	8	9	44

You define the model in the Model Specification dialog as shown in Figure 16.5. This design declares two random effects, person and machine*person.

Figure 16.6 shows the variance components estimates and effect tests for this analysis.

Figure 16.5 Model Specification Dialog for Model with Two Random Effects

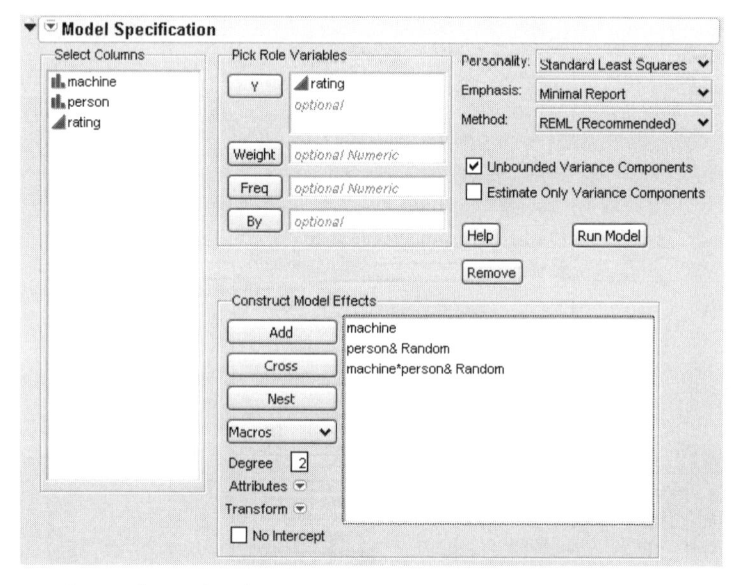

Figure 16.6 Report for Analysis of Unbalanced Design with Two Random Effects

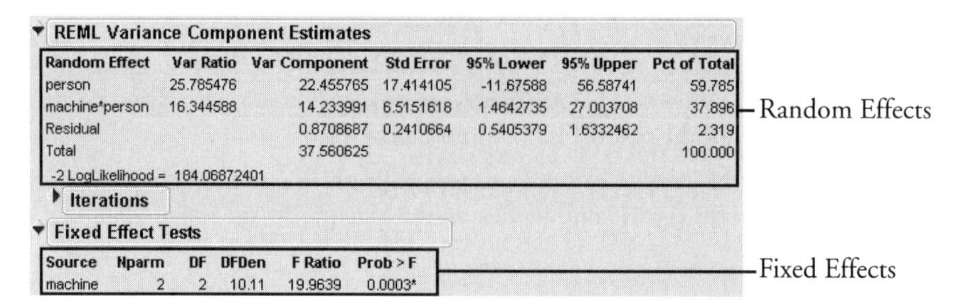

REML Variance Component Estimates

Random Effect	Var Ratio	Var Component	Std Error	95% Lower	95% Upper	Pct of Total
person	25.785476	22.455765	17.414105	-11.67588	56.58741	59.785
machine*person	16.344588	14.233991	6.5151618	1.4642735	27.003708	37.896
Residual		0.8708687	0.2410664	0.5405379	1.6332462	2.319
Total		37.560625				100.000

-2 LogLikelihood = 184.06872401

Iterations

Fixed Effect Tests

Source	Nparm	DF	DFDen	F Ratio	Prob > F
machine	2	2	10.11	19.9639	0.0003*

The variance ratio (Var Ratio) for a random effect is its variance component (Var Component) value divided by the variance component for the residual. Thus a variance ratio of one (1) for the random effect indicates that the random effect is no different than the residual, and the residual could be used as the denominator for the *F*-ratio. A larger variance ratio indicates that the random effect is a more appropriate denominator in the *F*-ratio.

Chapter 17

Generalized Linear Models

The Fit Model Platform

Generalized Linear Models provide a unified way to fit responses that don't fit the usual requirements of least-squares fits. In particular, frequency counts, which are characterized as having a Poisson distribution indexed by a model, are easily fit by a Generalized Linear Model.

The technique, pioneered by Nelder and Wedderburn (1972) involves a set of iteratively reweighted least-squares fits of a transformed response.

Additional features of JMP's Generalized Linear Model personality are

- likelihood ratio statistics for user-defined contrasts, that is, linear functions of the parameters, and *p*-values based on their asymptotic chi-square distributions
- estimated values, standard errors, and confidence limits for user-defined contrasts and least-squares means
- graphical profilers for examining the model
- confidence intervals for model parameters based on the profile likelihood function

Contents

Generalized Linear Models

While traditional linear models are used extensively in statistical data analysis, there are types of problems for which they are not appropriate.

- It may not be reasonable to assume that data are normally distributed. For example, the normal distribution (which is continuous) may not be adequate for modeling counts or measured proportions that are considered to be discrete.
- If the mean of the data is naturally restricted to a range of values, the traditional linear model may not be appropriate, since the linear predictor can take on any value. For example, the mean of a measured proportion is between 0 and 1, but the linear predictor of the mean in a traditional linear model is not restricted to this range.
- It may not be realistic to assume that the variance of the data is constant for all observations. For example, it is not unusual to observe data where the variance increases with the mean of the data.

A generalized linear model extends the traditional linear model and is, therefore, applicable to a wider range of data analysis problems. See the section "Examples of Generalized Linear Models," p. 361 for the form of a probability distribution from the exponential family of distributions.

As in the case of traditional linear models, fitted generalized linear models can be summarized through statistics such as parameter estimates, their standard errors, and goodness-of-fit statistics. You can also make statistical inference about the parameters using confidence intervals and hypothesis tests. However, specific inference procedures are usually based on asymptotic considerations, since exact distribution theory is not available or is not practical for all generalized linear models.

The Generalized Linear Model Personality

Generalized linear models are fit as a personality of the Fit Model Dialog. After selecting **Analyze > Fit Model**, select **Generalized Linear Model** from the drop-down menu before or after assigning the effects to the model.

Figure 17.1 Generalized Linear Model Launch Dialog

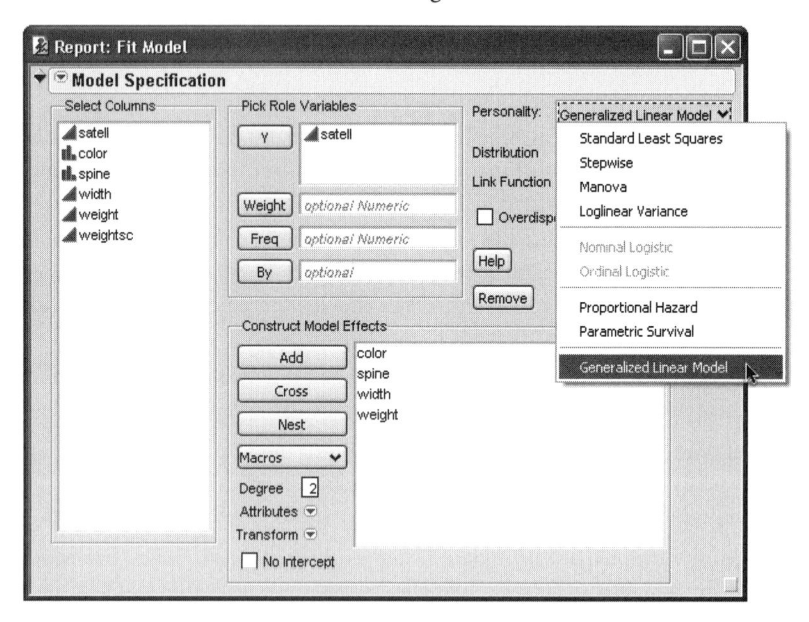

When you specify that you are fitting a generalized linear model, the Fit Model dialog changes to allow you to select a Distribution and a Link Function. In addition, an Offset button and an option for over-dispersion tests and intervals appears.

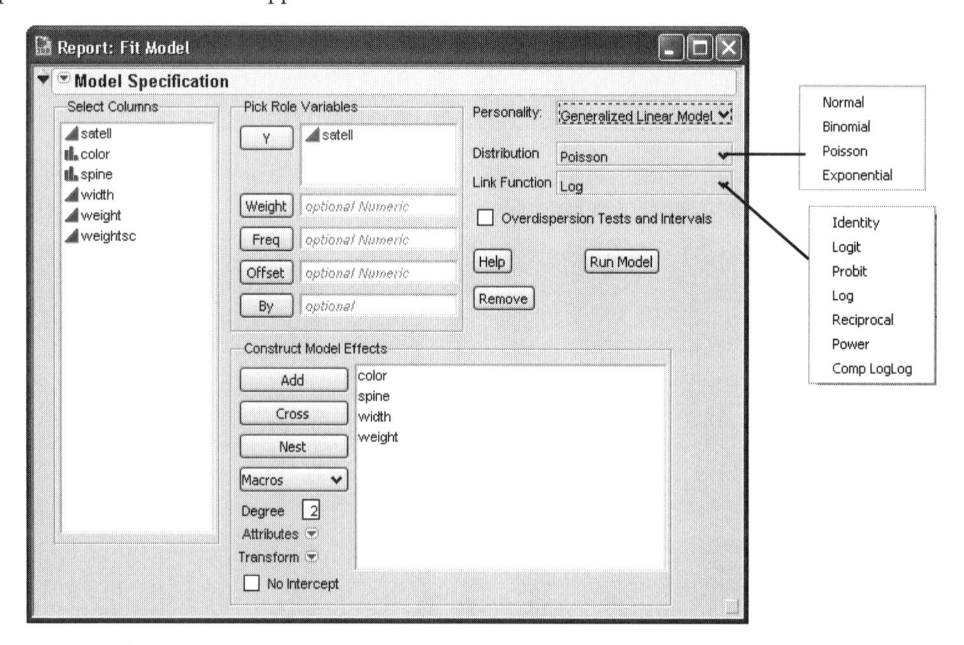

Examples of Generalized Linear Models

You construct a generalized linear model by deciding on response and explanatory variables for your data and choosing an appropriate link function and response probability distribution. Explanatory variables can be any combination of continuous variables, classification variables, and interactions.

Table 17.1 Examples of Generalized Linear Models

Model	Response Variable	Distribution	Canonical Link Function
Traditional Linear Model	continuous	Normal	identity, $g(\mu) = \mu$
Logistic Regression	a count or a binary random variable	Binomial	logit, $g(\mu) = \log\left(\frac{\mu}{1-\mu}\right)$
Poisson Regression in Log Linear Model	a count	Poisson	log, $g(\mu) = \log(\mu)$
Exponential Regression	positive continuous	Exponential	$\frac{1}{\mu}$

JMP fits a generalized linear model to the data by maximum likelihood estimation of the parameter vector. There is, in general, no closed form solution for the maximum likelihood estimates of the parameters. JMP estimates the parameters of the model numerically through an iterative fitting process. The dispersion parameter ϕ is also estimated by dividing the Pearson goodness-of-fit statistic by its degrees of freedom. Covariances, standard errors, and confidence limits are computed for the estimated parameters based on the asymptotic normality of maximum likelihood estimators.

A number of link functions and probability distributions are available in JMP. The built-in link functions are

identity: $g(\mu) = \mu$

logit: $g(\mu) = \log\left(\frac{\mu}{1-\mu}\right)$

probit: $g(\mu) = \Phi^{-1}(\mu)$, where Φ is the standard normal cumulative distribution function

log: $g(\mu) = \log(\mu)$

reciprocal: $g(\mu) = \frac{1}{\mu}$

power: $g(\mu) = \begin{cases} \mu^{\lambda} & \text{if } (\lambda \neq 0) \\ \log(\mu) & \text{if } \lambda = 0 \end{cases}$

complementary log-log: $g(m) = \log(-\log(1 - \mu))$

The available distributions and associated variance functions are

normal: $V(\mu) = 1$

binomial (proportion): $V(\mu) = \mu(1 - \mu)$

Poisson: $V(\mu) = \mu$

Note: When you select the **Binomial** distribution, a second **Y** variable specifying the number of trials is required.

Model Selection and Deviance

An important aspect of generalized linear modeling is the selection of explanatory variables in the model. Changes in goodness-of-fit statistics are often used to evaluate the contribution of subsets of explanatory variables to a particular model. The *deviance*, defined to be twice the difference between the maximum attainable log likelihood and the log likelihood at the maximum likelihood estimates of the regression parameters, is often used as a measure of goodness of fit. The maximum attainable log likelihood is achieved with a model that has a parameter for every observation. The following table displays the deviance for each of the probability distributions available in JMP.

Figure 17.2 Deviance Functions

Distribution	Deviance
normal	$\sum_i w_i(y_i - \mu_i)^2$
Poisson	$2\sum_i w_i\left[y_i \log\left(\frac{y_i}{\mu_i}\right) - (y_i - \mu_i)\right]$
binomial[a]	$2\sum_i w_i m_i\left[y_i \log\left(\frac{y_i}{\mu_i}\right) + (1 - y_i)\log\left(\frac{1 - y_i}{1 - \mu_i}\right)\right]$

a. In the binomial case, $y_i = r_i / m_i$, where r_i is a binomial count and m_i is the binomial number of trials parameter

The Pearson chi-square statistic is defined as

$$X^2 = \sum_i \frac{w_i(y_i - \mu_i)^2}{V(\mu_i)}$$

where V is the variance function.

One strategy for variable selection is to fit a sequence of models, beginning with a simple model with only an intercept term, and then include one additional explanatory variable in each successive model. You can measure the importance of the additional explanatory variable by the difference in deviances or fitted log likelihoods between successive models. Asymptotic tests computed by JMP enable you to assess the statistical significance of the additional term.

Examples

The following examples illustrate how to use JMP's generalized linear models platform.

Poisson Regression

This example uses data from a study of nesting horseshoe crabs. Each female crab had a male crab resident in her nest. This study investigated whether there were other males, called satellites, residing nearby. The data set CrabSatellites.jmp contains a response variable listing the number of satellites, as well as variables describing the female crab's color, spine condition, weight, and carapace width. The data are shown in Figure 17.3.

Figure 17.3 Crab Satellite Data

CrabSatellites
Notes Horseshoe cr
Model
Columns (6/0)
satell
color *
spine *
width
weight
weightsc +
Rows
All Rows 173

	satell	color	spine	width	weight	weightsc
1	8	Medium	Both Worn/Broken	28.3	3050	1.06234565
2	0	Dark Med	Both Worn/Broken	22.5	1550	-1.5371951
3	9	Light Med	Both Good	26	2300	-0.2374247
4	0	Dark Med	Both Worn/Broken	24.8	2100	-0.5840302
5	4	Dark Med	Both Worn/Broken	26	2600	0.28248343
6	0	Medium	Both Worn/Broken	23.8	2100	-0.5840302
7	0	Light Med	Both Good	26.5	2350	-0.1507734
8	0	Dark Med	One Worn/Broken	24.7	1900	-0.9306356
9	0	Medium	Both Good	23.7	1950	-0.8439842
10	0	Dark Med	Both Worn/Broken	25.6	2150	-0.4973788
11	0	Dark Med	Both Worn/Broken	24.3	2150	-0.4973788
12	0	Medium	Both Worn/Broken	25.8	2650	0.36913479

To fit the Poisson loglinear model,

- Select **Analyze > Fit Model**
- Assign satell as **Y**
- Assign color, spine, width, and weight as **Effects**
- Choose the **Generalized Linear Model** Personality
- Choose the **Poisson** Distribution

The **Log** Link function should be selected for you automatically.

- Click **Run Model**.

The results are shown in Figure 17.4.

Figure 17.4 Crab Satellite Results

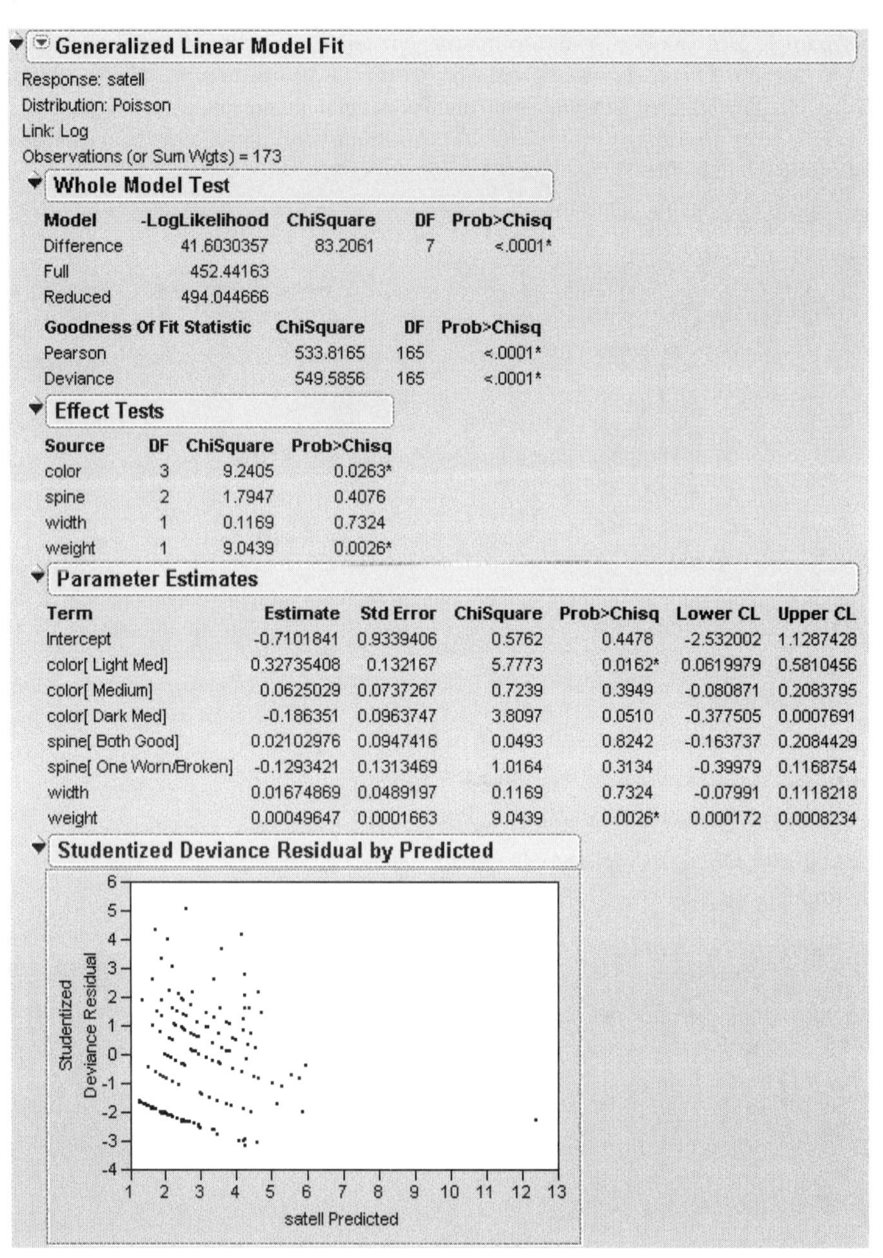

Generalized Linear Model Fit

Response: satell
Distribution: Poisson
Link: Log
Observations (or Sum Wgts) = 173

Whole Model Test

Model	-LogLikelihood	ChiSquare	DF	Prob>Chisq
Difference	41.6030357	83.2061	7	<.0001*
Full	452.44163			
Reduced	494.044666			

Goodness Of Fit Statistic	ChiSquare	DF	Prob>Chisq
Pearson	533.8165	165	<.0001*
Deviance	549.5856	165	<.0001*

Effect Tests

Source	DF	ChiSquare	Prob>Chisq
color	3	9.2405	0.0263*
spine	2	1.7947	0.4076
width	1	0.1169	0.7324
weight	1	9.0439	0.0026*

Parameter Estimates

Term	Estimate	Std Error	ChiSquare	Prob>Chisq	Lower CL	Upper CL
Intercept	-0.7101841	0.9339406	0.5762	0.4478	-2.532002	1.1287428
color[Light Med]	0.32735408	0.132167	5.7773	0.0162*	0.0619979	0.5810456
color[Medium]	0.0625029	0.0737267	0.7239	0.3949	-0.080871	0.2083795
color[Dark Med]	-0.186351	0.0963747	3.8097	0.0510	-0.377505	0.0007691
spine[Both Good]	0.02102976	0.0947416	0.0493	0.8242	-0.163737	0.2084429
spine[One Worn/Broken]	-0.1293421	0.1313469	1.0164	0.3134	-0.39979	0.1168754
width	0.01674869	0.0489197	0.1169	0.7324	-0.07991	0.1118218
weight	0.00049647	0.0001663	9.0439	0.0026*	0.000172	0.0008234

Studentized Deviance Residual by Predicted

The Whole Model Test gives information on the model as a whole.

- First, comparisons among full and reduced models are presented. The Reduced model is a model containing only an intercept. The Full model contains all of the effects. The Difference model is the difference of the log likelihood of the full and reduced models. The Prob>Chisq is analogous to a whole-model *F*-test.

- Second, goodness-of-fit statistics are presented. Analogous to lack-of-fit tests, they test for adequacy of the model. Low values of the goodness-of-fit statistics indicate that you may need to add higher-order terms to the model, add more covariates, change the distribution, or (in Poisson and binomial cases especially) consider adding an overdispersion parameter.

The Effect Tests table shows joint tests that all the parameters for an individual effect are zero. If an effect has only one parameter, as with simple regressors, then the tests are no different from the tests in the Parameter Estimates table.

The Parameter Estimates table shows the estimates of the parameters in the model and a *test* for the hypothesis that each parameter is zero. Simple continuous regressors have only one parameter. Models with complex classification effects have a parameter for each anticipated degree of freedom. Confidence limits are also displayed.

Poisson Regression with Offset

The sample data table Ship Damage.JMP is adapted from one found in McCullugh and Nelder (1983). It contains information on a certain type of damage caused by waves to the forward section of the hull. Hull construction engineers are interested in the risk of damage associated with three variables: ship Type, the year the ship was constructed (Yr Made) and the block of years the ship saw service (Yr Used, with 60=1960-74).

In this analysis we use the variable Service, the log of the aggregate months of service, as an *offset variable*. An offset variable is one that is treated like a regression covariate whose parameter is fixed to be 1.0.

These are most used often to scale the modeling of the mean in Poisson regression situations with log link. In this example, we use log(months of service) since one would expect that the number of repairs to be proportional to the number of months in service. To see how this works, assume the linear component of the GLM is called eta. Then with a log link function, the model of the mean with the offset included is:

exp(Log(months of service) + eta) = (months of service) * exp(eta).

To run this example, assign

- **Generalized Linear Model** as the **Personality**
- Poisson as the **Distribution**, which automatically selects the **Log** link function
- N to **Y**
- Service to **Offset**
- Type, Yr Made, Yr Used as effects in the model
- **Overdispersion Tests and Intervals** with a check mark

The Fit Model dialog should appear like the one shown in Figure 17.5.

Figure 17.5 Ship Damage Fit Model Dialog

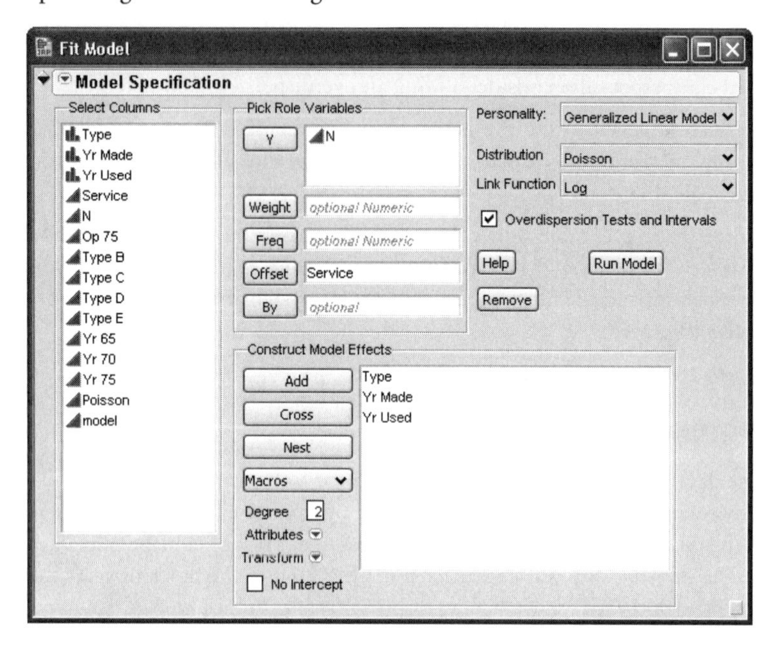

When you click Run Model, you see the report shown in Figure 17.6. Notice that all three effects (Type, Yr Made, Yr Used) are significant.

Figure 17.6 Ship Damage Report

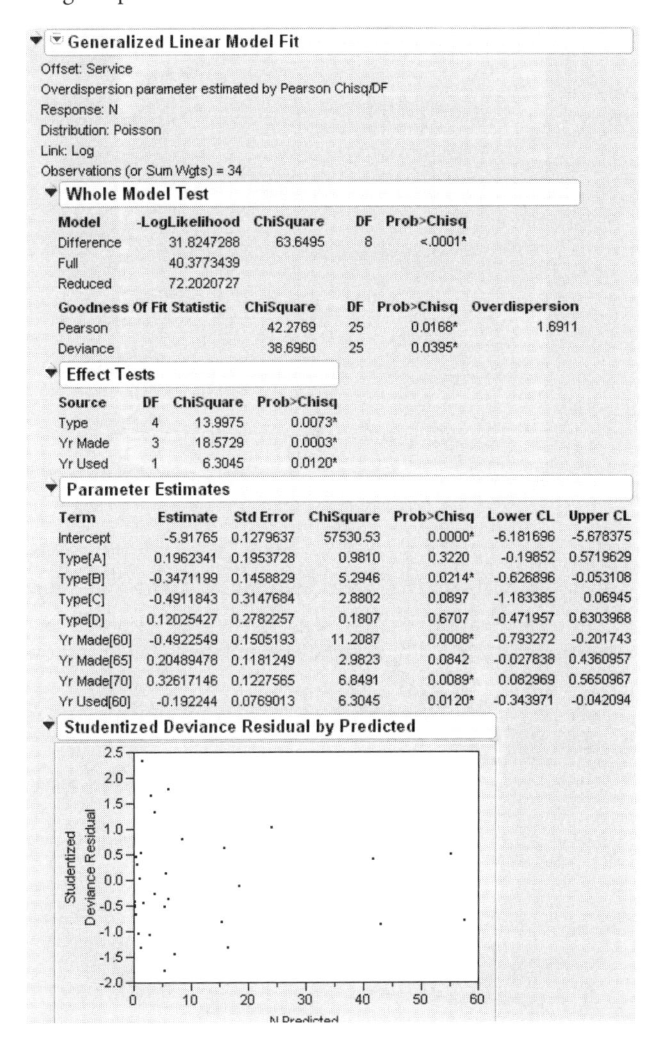

Generalized Linear Model Fit

Offset: Service
Overdispersion parameter estimated by Pearson Chisq/DF
Response: N
Distribution: Poisson
Link: Log
Observations (or Sum Wgts) = 34

Whole Model Test

Model	-LogLikelihood	ChiSquare	DF	Prob>Chisq
Difference	31.8247288	63.6495	8	<.0001*
Full	40.3773439			
Reduced	72.2020727			

Goodness Of Fit Statistic	ChiSquare	DF	Prob>Chisq	Overdispersion
Pearson	42.2769	25	0.0168*	1.6911
Deviance	38.6960	25	0.0395*	

Effect Tests

Source	DF	ChiSquare	Prob>Chisq
Type	4	13.9975	0.0073*
Yr Made	3	18.5729	0.0003*
Yr Used	1	6.3045	0.0120*

Parameter Estimates

Term	Estimate	Std Error	ChiSquare	Prob>Chisq	Lower CL	Upper CL
Intercept	-5.91765	0.1279637	57530.53	0.0000*	-6.181696	-5.678375
Type[A]	0.1962341	0.1953728	0.9810	0.3220	-0.19852	0.5719629
Type[B]	-0.3471199	0.1458829	5.2946	0.0214*	-0.626896	-0.053108
Type[C]	-0.4911843	0.3147684	2.8802	0.0897	-1.183385	0.06945
Type[D]	0.12025427	0.2782257	0.1807	0.6707	-0.471957	0.6303968
Yr Made[60]	-0.4922549	0.1505193	11.2087	0.0008*	-0.793272	-0.201743
Yr Made[65]	0.20489478	0.1181249	2.9823	0.0842	-0.027838	0.4360957
Yr Made[70]	0.32617146	0.1227565	6.8491	0.0089*	0.082969	0.5650967
Yr Used[60]	-0.192244	0.0769013	6.3045	0.0120*	-0.343971	-0.042094

Studentized Deviance Residual by Predicted

Normal Regression, Log Link

Consider the following data set, where x is an explanatory variable and y is the response variable.

Figure 17.7 Nor.jmp data set

nor
Model
Overlay Plot
Columns (3/0)
X
Y
Rows

Rows	
All rows	16
Selected	0
Excluded	0
Hidden	0
Labelled	0

	X	Y
1	0	5
2	0	7
3	0	9
4	1	7
5	1	10
6	1	8
7	2	11
8	2	9
9	3	16
10	3	13
11	3	14
12	4	25
13	4	24
14	5	34
15	5	32
16	5	30

Using Fit Y By X, you can easily see that y varies nonlinearly with x and that the variance is approximately constant. A normal distribution with a log link function is chosen to model these data; that is, $\log(\mu_i) = \mathbf{x}_i'\beta$ so that $\mu_t = \exp(\mathbf{x}_i'\beta)$. The completed Fit Model dialog is shown in Figure 17.8.

Figure 17.8 Nor Fit Model Dialog

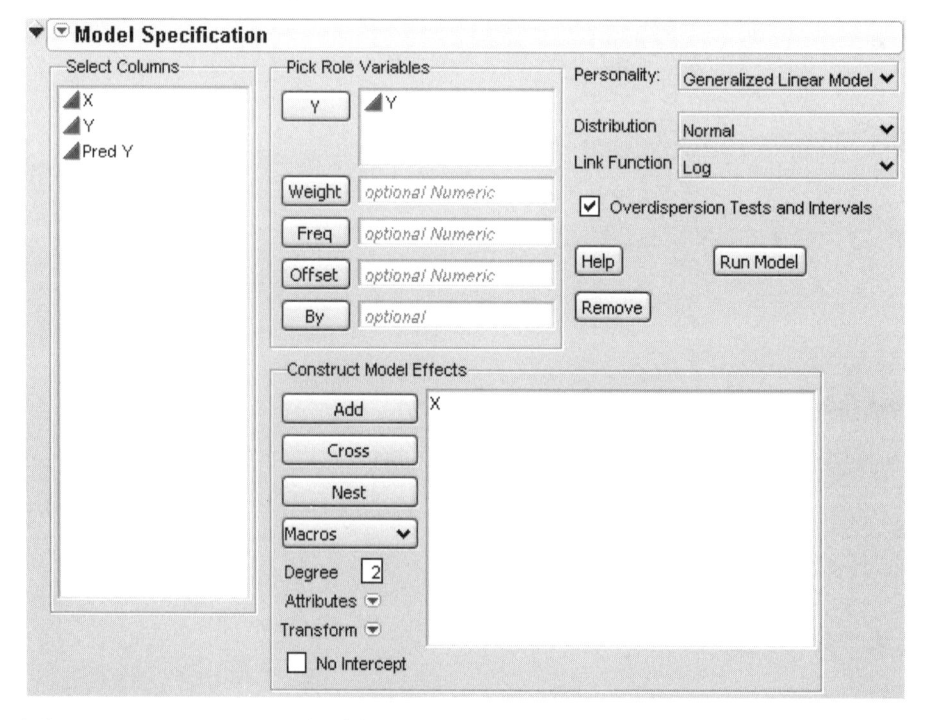

After clicking **Run Model**, you get the following report.

Figure 17.9 Nor results

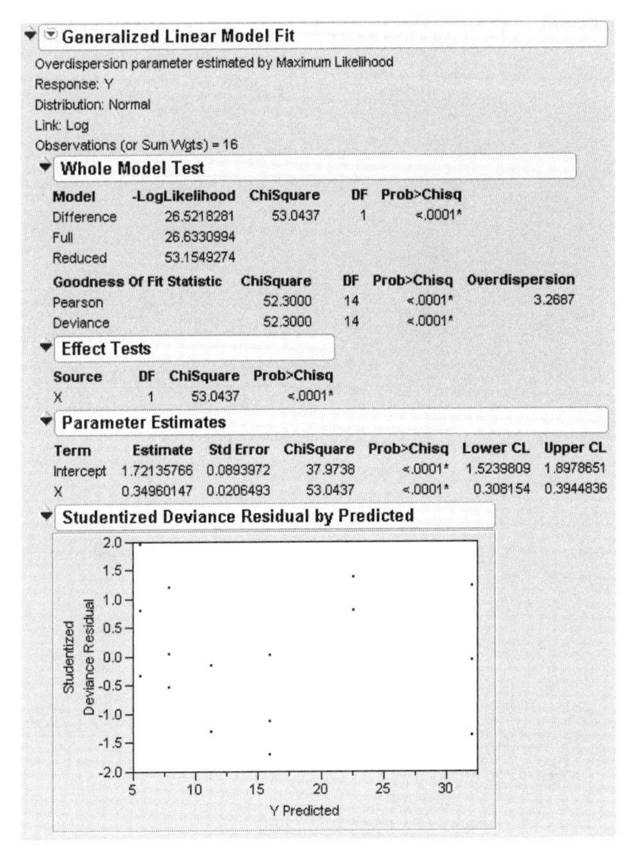

Generalized Linear Model Fit

Overdispersion parameter estimated by Maximum Likelihood
Response: Y
Distribution: Normal
Link: Log
Observations (or Sum Wgts) = 16

Whole Model Test

Model	-LogLikelihood	ChiSquare	DF	Prob>Chisq
Difference	26.5218281	53.0437	1	<.0001*
Full	26.6330994			
Reduced	53.1549274			

Goodness Of Fit Statistic	ChiSquare	DF	Prob>Chisq	Overdispersion
Pearson	52.3000	14	<.0001*	3.2687
Deviance	52.3000	14	<.0001*	

Effect Tests

Source	DF	ChiSquare	Prob>Chisq
X	1	53.0437	<.0001*

Parameter Estimates

Term	Estimate	Std Error	ChiSquare	Prob>Chisq	Lower CL	Upper CL
Intercept	1.72135766	0.0893972	37.9738	<.0001*	1.5239809	1.8978651
X	0.34960147	0.0206493	53.0437	<.0001*	0.308154	0.3944836

Studentized Deviance Residual by Predicted

You can get a plot of the prediction curve on the original values by

- selecting **Save Columns > Predicted Values**
- Selecting **Graph > Overlay Plot** with Y and Pred Y as **Y** and X as **X**.
- In the resulting plot, right-click on Pred Y to de-select **Show Points**, then repeat to select **Connect Points**.

Figure 17.10 Prediction Curve and Original Points

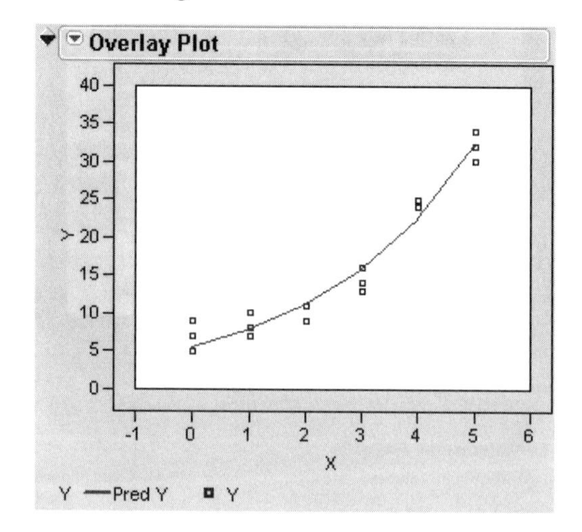

Because the distribution is normal, the Studentized Deviance residuals and the Deviance residuals are the same. To see this, select **Diagnostic Plots > Deviance Residuals by Predicted** from the platform drop-down menu.

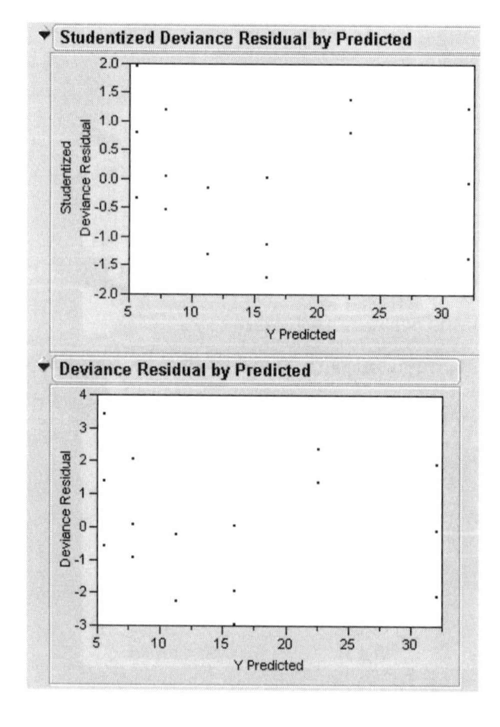

Platform Commands

The following commands are available in the Generalized Linear Model report.

Custom Test allows you to test a custom hypothesis. Refer to "Custom Test," p. 255 in the "Standard Least Squares: Perspectives on the Estimates" chapter for details on custom tests.

Contrast allows you to test for differences in levels within a variable. In the Crab Sattelite example, suppose you want to test whether the dark-colored crabs attracted a different number of sattelites than the medium-colored crabs. Selecting **Contrast** brings up the following dialog.

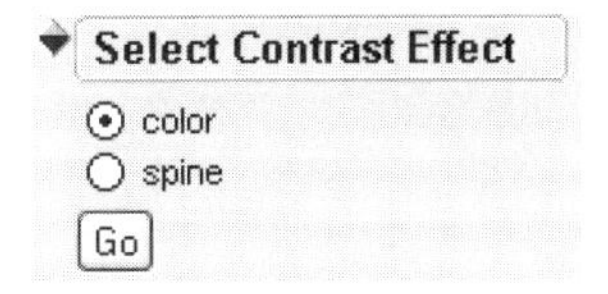

Here you choose color, the variable of interest. When you click **Go**, you are presented with a Contrast Specification dialog.

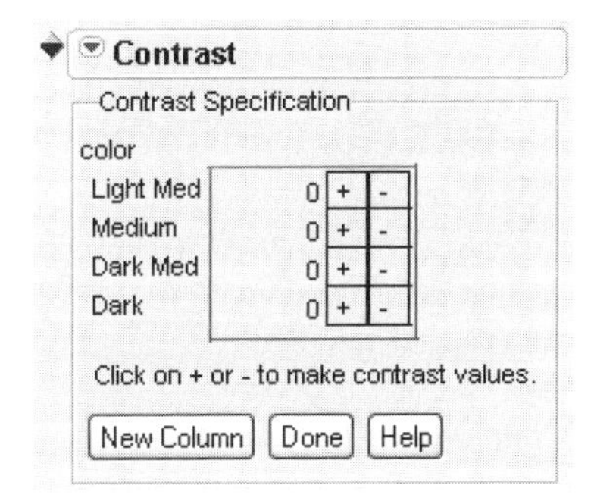

To compare the dark-colored to the medium-colored, click the + button beside Dark and Dark Med, and the - button beside Medium and Light Medium.

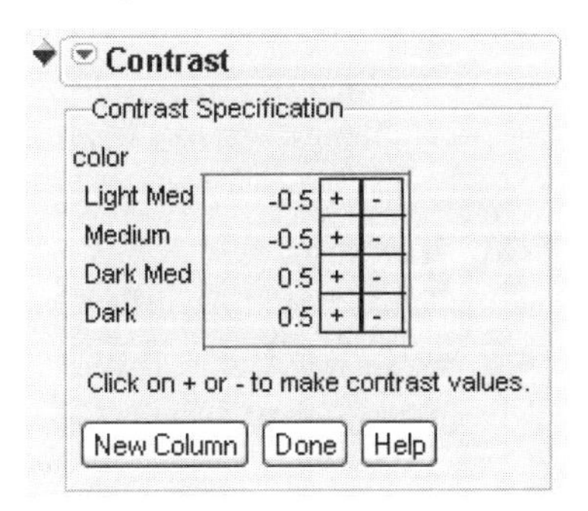

Click **Done** to get the Contrast report shown here.

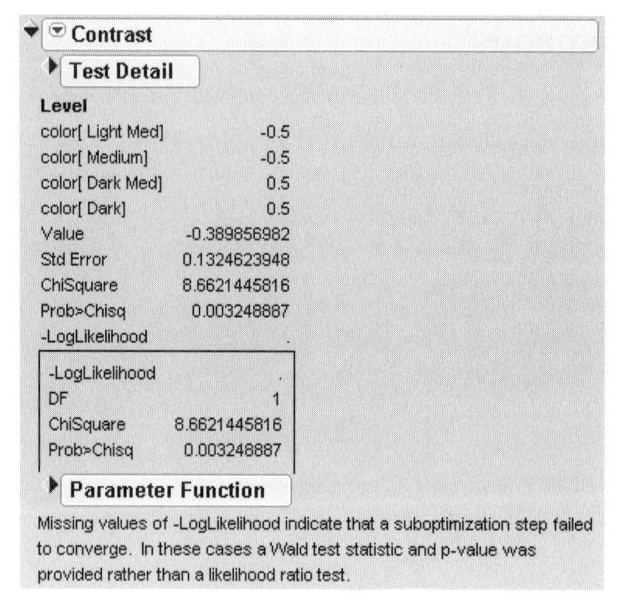

Since the Prob>Chisq is less than 0.05, we have evidence that there is a difference in satellite attraction based on color.

Covariance of Estimates produces a covariance matrix for all the effects in a model. The estimated covariance matrix of the parameter estimator is given by

$$\Sigma = -\mathbf{H}^{-1}$$

where **H** is the Hessian (or second derivative) matrix evaluated using the parameter estimates on the last iteration. Note that the dispersion parameter, whether estimated or specified, is incorporated into **H**. Rows and columns corresponding to aliased parameters are not included in Σ.

Correlation of Estimates produces a correlation matrix for all the effects in a model. The correlation matrix is the normalized covariance matrix. That is, if σ_{ij} is an element of Σ, then the corressponding element of the correlation matrix is $\sigma_{ij}/\sigma_i\sigma_j$, where $\sigma_i = \sqrt{\sigma_{ii}}$

Profiler brings up the Profiler for examining prediction traces for each X variable. Details on the profiler are found in "The Profiler," p. 277 IN THE "STANDARD LEAST SQUARES: EXPLORING THE PREDICTION EQUATION" CHAPTER.

Contour Profiler brings up an interactive contour profiler. Details are found in "Contour Profiler," p. 306 IN THE "PROFILING" CHAPTER.

Surface Profiler brings up a 3-D surface profiler. Details of Surface Plots are found in the "Plotting Surfaces" chapter.

Diagnostic Plots is a submenu containing commands that allow you to plot combinations of residuals, predicted values, and actual values to search for outliers and determine the adequacy of your model. Deviance is discussed above in "Model Selection and Deviance," p. 362. The following plots are available:

Studentized Deviance Residuals by Predicted

Studentized Pearson Residuals by Predicted

Pearson Residuals By Predicted

Deviance Residuals by Predicted

Actual by Predicted

Save Columns is a submenu that lets you save certain quantities in the data table.

Prediction Formula saves the formula that predicts the current model

Predicted Values saves the values predicted by the current model

Mean Confidence Interval saves the 95% confidence limits for the prediction equation. The confidence limits reflect variation in the parameter estimates.

Save Indiv Confidence Limits saves the confidence limits for a given individual value. The confidence limits reflect variation in the error and variation in the parameter estimates.

Deviance Residuals saves the deviance residuals

Pearson Residuals saves the Pearson residuals

Studentized Deviance Residuals saves the studentized deviance residuals

Studentized Pearson Residuals saves the studentized Pearson residuals

Chapter 18

Fitting Dispersion Effects with the LogLinear Variance Model

The Fit Model Platform

This fitting platform allows you to model both the expected value and the variance of a response using regression models. The log of the variance is fit to one linear model simultaneously with the expected response fit to a different linear model.

> The estimates are demanding in their need for a lot of well-designed, well-fitting data. You need more data to fit variances than you do means.

For many engineers, the goal of an experiment is not to maximize or minimize the response itself, but to aim at a target response and achieve minimum variability. The loglinear variance model provides a very general and effective way to model variances, and can be used for unreplicated data, as well as data with replications.

Modeling dispersion effects is not very widely covered in textbooks, with the exception of the Taguchi framework. In a Taguchi-style experiment, this is handled by taking multiple measurements across settings of an outer array, constructing a new response which measures the variability off-target across this outer array, and then fitting the model to find out the factors that produce minimum variability. This kind of modeling requires a specialized design that is a complete cartesian product of two designs. The method of this chapter models variances in a more flexible, model-based approach. The particular performance statistic that Taguchi recommends for variability modeling is $STD = -\log(s)$. In JMP's methodology, the $\log(s^2)$ is modeled and combined with a model with a mean. The two are basically equivalent, since $\log(s^2)=2\log(s)$.

Contents

The Loglinear Variance Model

The loglinear-variance model (Harvey 1976, Cook and Weisberg 1983, Aitken 1987, Carroll and Ruppert 1988) provides a neat way to model the variance simply through a linear model. In addition to regressor terms to model the mean response, there are regressor terms in a linear model to model the log of the variance:

mean model: $E(y) = X\beta$

variance model: $\log(Variance(y)) = Z\lambda$,

or equivalently

$Variance(y) = \exp(Z\lambda)$

where the columns of X are the regressors for the mean of the response, and the columns of Z are the regressors for the variance of the response. β has the regular linear model parameters, and λ has the parameters of the variance model.

Estimation Method

Log-linear variance models are estimated using REML.

Loglinear Variance Models in JMP

This section introduces a new kind of effect, a *dispersion* effect, labeled as a *log-variance* effect, that can model changes in the variance of the response. This is implemented in the Fit Model platform by a fitting personality called the *log-variance* personality.

Model Specification

Log-linear variance effects are specified in the Fit Model dialog by highlighting them and selecting **LogVariance Effect** from the **Attributes** drop-down menu. **&LogVariance** appears at the end of the effect. When you use this attribute, it also changes the fitting **Personality** at the top to **LogLinear Variance.** If you want an effect to be used for both the mean and variance of the response, then you must specify it twice, once with the **LogVariance** option.

The effects you specify with the log-variance attribute become the effects that generate the *Z*'s in the model, and the other effects become the *X*'s in the model.

Example

The data table InjectionMolding.jmp contains the experimental results from a 7-factor 2^{7-3} fractional factorial design with four added centerpoints [from Myers and Montgomery, 1995, page 519, originally Montgomery, 1991]. Preliminary investigation determined that the mean response only seemed to vary with the first two factors, Mold Temperature, and Screw Speed, and the variance seemed to be affected by Holding Time.

Figure 18.1 Injection Molding Data

InjectionMolding
Model
Columns (9/0)
Pattern
MoldTemp
Screw Speed
Hold Time
Gate Size
Cycle Time
Moisture
Pressure
Shrinkage
Rows
All rows 20
Selected 0

	Pattern	MoldTemp	Screw Speed	Hold Time	Gate Size	Cycle Time	Moisture	Pressure	Shrinkage
1	-------	-1	-1	-1	-1	-1	-1	-1	6
2	+----++	1	-1	-1	-1	1	-1	1	10
3	-+--+-+	-1	1	-1	-1	1	1	-1	32
4	++--++-	1	1	-1	-1	-1	1	1	60
5	--+-++-	-1	-1	1	-1	1	1	1	4
6	+-+-+-+	1	-1	1	-1	-1	1	-1	15
7	-++--++	-1	1	1	-1	-1	-1	1	26
8	+++----	1	1	1	-1	1	-1	-1	60
9	---++++	-1	-1	-1	1	-1	1	1	8
10	+--++--	1	-1	-1	1	1	1	-1	12
11	-+-+-+-	-1	1	-1	1	1	-1	1	34
12	++-+--+	1	1	-1	1	-1	-1	-1	60
13	--++--+	-1	-1	1	1	1	-1	-1	16
14	+-++-+-	1	-1	1	1	-1	-1	1	5
15	-++++--	-1	1	1	1	-1	1	-1	37
16	+++++++	1	1	1	1	1	1	1	52
17	0000000	0	0	0	0	0	0	0	25
18	0000000	0	0	0	0	0	0	0	29
19	0000000	0	0	0	0	0	0	0	24
20	0000000	0	0	0	0	0	0	0	27

To proceed with the analysis, select **Analyze > Fit Model** and complete the Fit Model dialog as shown in Figure 18.2. After Hold Time was added to the model, it was selected and changed to a LogVariance effect through the **Attributes** popup menu. This also forced the fitting personality to change to **Loglinear Variance**. Click the **Run Model** button to start the fitting.

Figure 18.2 Fit Model Dialog

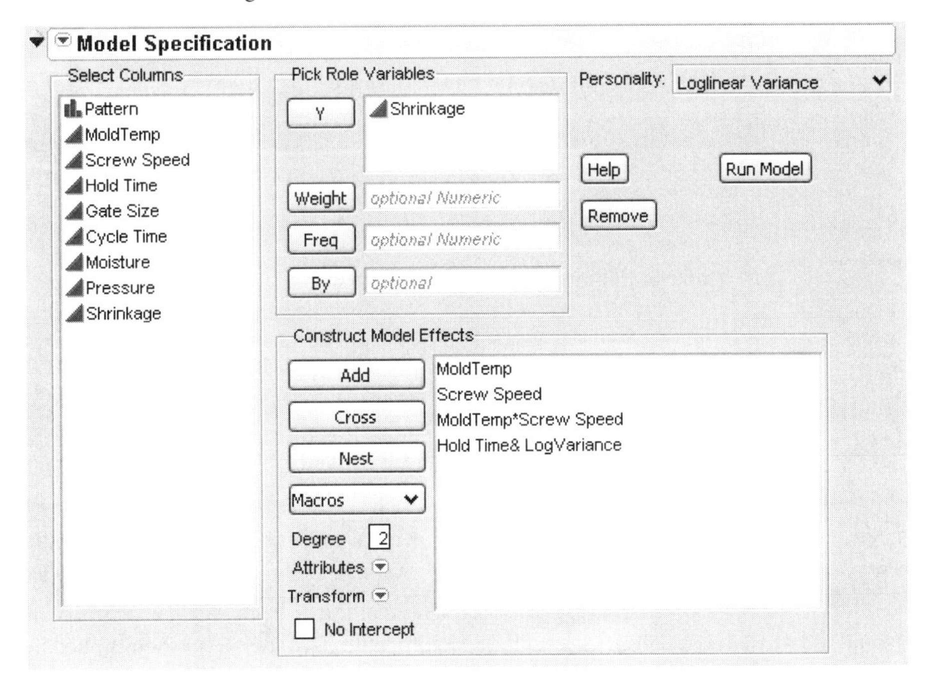

Displayed Output

The top portion of the resulting report shows the fitting of the Expected response, with reports similar to standard least squares, though actually derived from restricted maximum likelihood (REML).

Figure 18.3 Mean Model Output

Mean Model for Shrinkage

Parameter Estimates

Term	Estimate	Std Error	t Ratio	Prob>|t|
Intercept	27.582516	0.380697	72.45	<.0001*
MoldTemp	7.683713	0.392907	19.56	<.0001*
Screw Speed	18.673515	0.392907	47.53	<.0001*
MoldTemp*Screw Speed	5.765297	0.392907	14.67	<.0001*

Fixed Effect Tests

Source	Nparm	DF	DFDen	F Ratio	Prob > F
MoldTemp	1	1	5.758	382.4387	<.0001*
Screw Speed	1	1	5.758	2258.768	<.0001*
MoldTemp*Screw Speed	1	1	5.758	215.3094	<.0001*

Figure 18.4 Variance Model Output

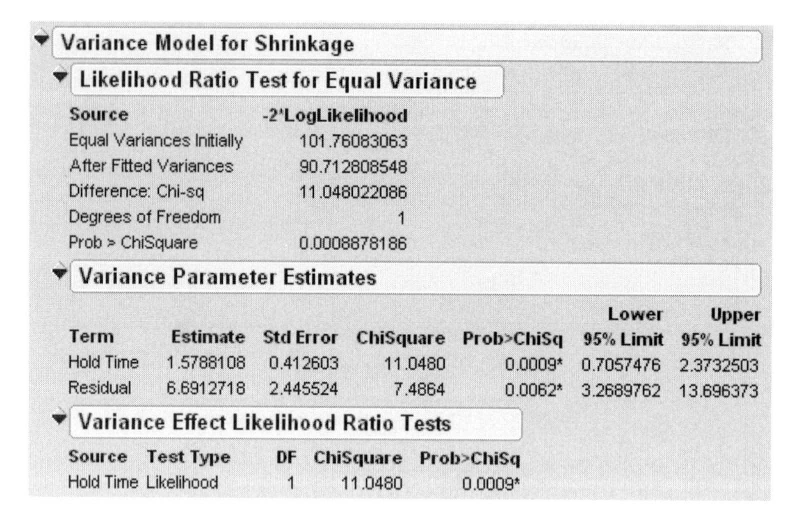

Variance Model for Shrinkage

Likelihood Ratio Test for Equal Variance

Source	-2*LogLikelihood
Equal Variances Initially	101.76083063
After Fitted Variances	90.712808548
Difference: Chi-sq	11.048022086
Degrees of Freedom	1
Prob > ChiSquare	0.0008878186

Variance Parameter Estimates

Term	Estimate	Std Error	ChiSquare	Prob>ChiSq	Lower 95% Limit	Upper 95% Limit
Hold Time	1.5788108	0.412603	11.0480	0.0009*	0.7057476	2.3732503
Residual	6.6912718	2.445524	7.4864	0.0062*	3.2689762	13.696373

Variance Effect Likelihood Ratio Tests

Source	Test Type	DF	ChiSquare	Prob>ChiSq
Hold Time	Likelihood	1	11.0480	0.0009*

The second portion of the report shows the fit of the variance model. The **Variance Parameter Estimates** report shows the estimates and relevant statistics.

The hidden column **exp(Estimate)** is the exponential of the estimate. So, if the factors are coded to have +1 and -1 values, then the +1 level for a factor would have the variance multiplied by the **exp(Estimate)** value and the -1 level would have the variance multiplied by the reciprocal of this column. To see a hidden column, right-click on the report and select the name of the column from the **Columns** menu that appears.

The hidden column labeled **exp(2|Estimate|)**is the ratio of the higher to the lower variance if the regressor has the range -1 to +1. The report also shows the standard error, chi-square, *p*-value, and profile likelihood confidence limits of each estimate. The residual parameter is the overall estimate of the variance, given all other regressors are zero.

Does the variance model fit significantly better than the original model? The likelihood ratio test for this question compares the fitted model with the model where all parameters are zero except the intercept, the model of equal-variance. In this case the *p*-value is highly significant. Changes in **Hold Time** change the variance.

The **Variance Effect Likelihood Ratio Tests** refit the model without each term in turn to create the likelihood ratio tests. These are generally more trusted than Wald tests.

Platform Options

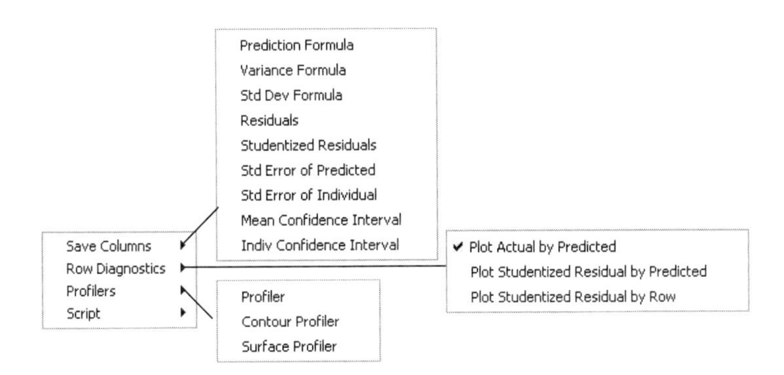

Save Columns

Each of these commands creates one or more columns in the data table.

Prediction Formula creates a new column, called colname mean, containing the predicted values for the mean computed by the specified model.

Variance Formula creates a new column, called colname variance, containing the predicted values for the variance computed by the specified model.

Std Dev Formula creates a new column, called colname Std Dev, containing the predicted values for the standard deviation computed by the specified model.

Residuals creates a new column called colname Residual containing the residuals, which are the observed response values minus predicted values.

Studentized Residuals creates a new column called colname Studentized Resid. The new column values are the residuals divided by their standard error.

Std Error of Predicted creates a new column, called StdErr Pred colname, containing the standard errors of the predicted values.

Std Error of Individual creates a new column, called StdErr Indiv colname, containing the standard errors of the individual predicted values.

Mean Confidence Interval creates two new columns, Lower 95% Mean colname and Upper 95% Mean colname that are the bounds for a confidence interval for the prediction mean.

Individ Confidence Interval creates two new columns, Lower 95% Indiv colname and Upper 95% Indiv colname that are the bounds for a confidence interval for the prediction mean.

Row Diagnostics

Plot Actual by Predicted displays the observed values by the predicted values of *Y*. This is the leverage plot for the whole model.

Plot Studentized Residual by Predicted displays the Studentized residuals by the predicted values of *Y*.

Plot Studentized Residual by Actual displays the Studentized residuals by the actual values of *Y*.

Profilers

Profiler, **Contour Profiler**, and **Surface Profiler** are the standard JMP profilers, detailed in the chapter "Standard Least Squares: Exploring the Prediction Equation," p. 275.

Examining the Residuals

To see the dispersion effect, we invoked the Oneway platform on **ShrinkageResidual** (produced using the **Save Columns > Residual** command) by **Hold Time**: With this plot it is easy to see the variance go up as Hold Time is increased. This is treating **Hold Time** as a nominal factor, though it is originally continuous in the fit above.

Figure 18.5 Residual by Dispersion Effect

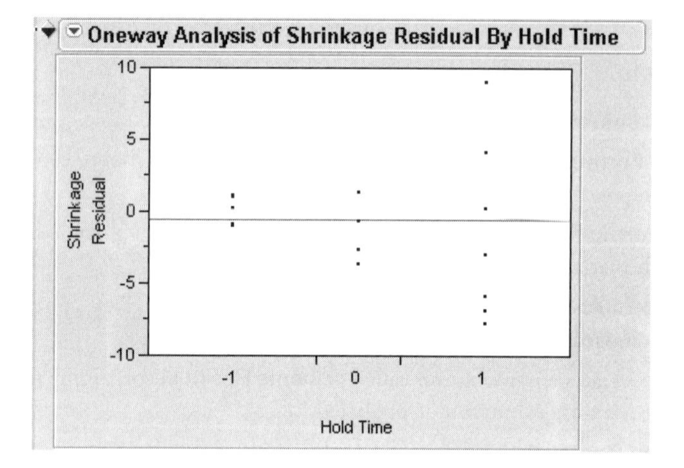

Profiling the Fitted Model

The **Profiler**, **Contour Profiler**, or **Surface Profiler** can be used to gain more insight on the fitted model. Each can be selected from the platform drop-down menu under the **Profilers** submenu.For example, suppose that the goal was to find the factor settings that achieved a target of 36.35 for the response, but at the smallest variance. Fit the models and choose Profiler from the report menu. For example, Figure 18.6 shows the Profiler set up to match a target value for a mean and to minimize variance.

One of the best ways to see the relationship between the mean and the variance (both modeled with the LogVariance personality) is through looking at the individual prediction confidence intervals about the mean. To see prediction intervals in the Profiler, select **Prediction Intervals** from its drop-down menu. Regular confidence intervals (those shown by default in the Profiler) do not show information about the variance model as well as individual prediction confidence intervals do. Prediction intervals show both the mean and variance model in one graph.

If Y is the modeled response, and you want a prediction interval for a new observation at x_{new}, then

$$s^2|x_{new} = s_Y^2|x_{new} + s_{\hat{Y}}^2|x_{new}$$

where

$s^2|x_{new}$ is the variance for the individual prediction at x_{new}

$s^2_Y|x_{new}$ is the variance of the distribution of Y at x_{new}

$s^2_{\hat{Y}}\big|x_{new}$ is the variance of the sampling distribution of $\hat{Y}$, and is also the variance for the mean.

Because the variance of the individual prediction contains the variance of the distribution of *Y*, the effects of the changing variance for *Y* can be seen. Not only are the individual prediction intervals wider, but they can change shape with a change in the variance effects. Figure 18.7 shows prediction intervals for the situation in Figure 18.6.

Figure 18.6 Profiler to Match Target and Minimize Variance

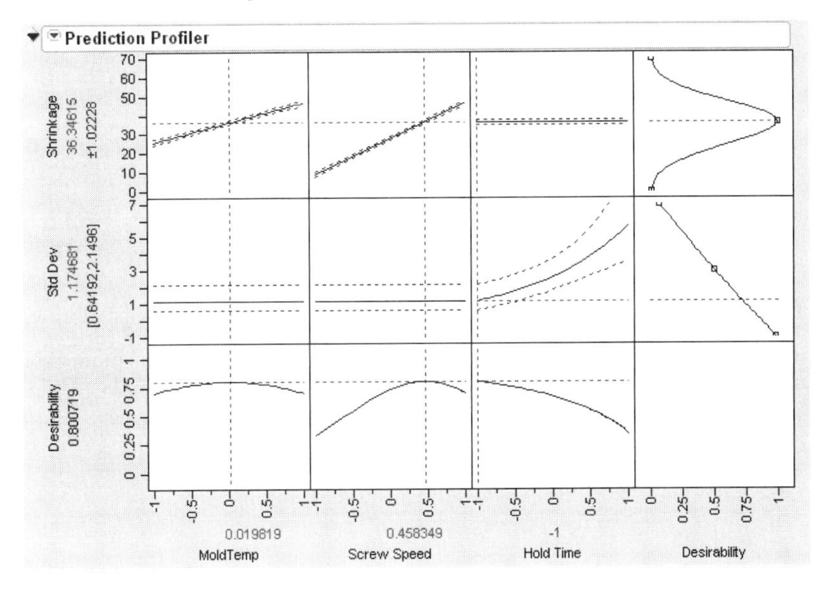

Figure 18.7 Prediction Intervals

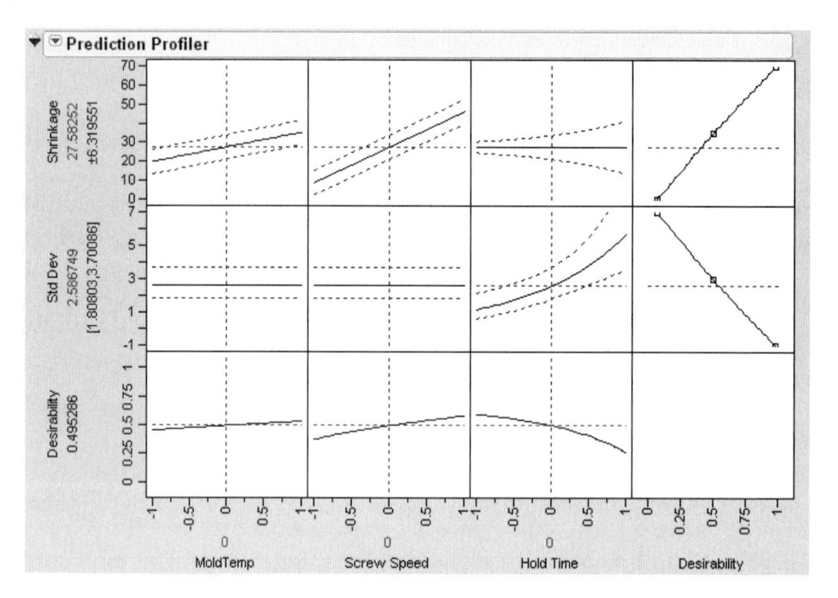

Comments

Every time another parameter is estimated for the mean model, at least one more observation is needed, and preferably more. But with variance parameters, several more observations for each variance parameter is needed. It takes more data to estimate variances than it does means.

The log-linear variance model is a very flexible way to fit dispersion effects, and the method deserves much more attention than it has received so far in the literature.

Chapter 19

Stepwise Regression

The Fit Model Platform

In JMP, stepwise regression is a *personality* of the Model Fitting platform; it is one of the selections in the Fitting Personality popup menu on the Model Specification dialog.

Stepwise regression is an approach to selecting a subset of effects for a regression model. It is used when there is little theory to guide the selection of terms for a model and the modeler, in desperation, wants to use whatever seems to provide a good fit.

The approach is somewhat controversial. The significance levels on the statistics for selected models violate the standard statistical assumptions because the model has been selected rather than tested within a fixed model. On the positive side, the approach has been of practical use for 30 years in helping to trim out models to predict many kinds of responses. The book *Subset Selection in Regression*, by A. J. Miller (1990), brings statistical sense to model selection statistics.

This chapter uses the term "significance probability" in a mechanical way to represent that the calculation would be valid in a fixed model, recognizing that the true significance probability could be nowhere near the reported one.

Contents

Introduction to Stepwise Regression

In JMP, stepwise regression is a *personality* of the Model Fitting platform—it is one of the selections in the Fitting Personality popup menu on the Model Specification dialog (see Figure 19.2). The **Stepwise** feature computes estimates that are the same as those of other least squares platforms, but it facilitates searching and selecting among many models.

A Multiple Regression Example

As an example, consider the Fitness.jmp (SAS Institute Inc. 1987) data table in the Sample Data folder, which is the result of an aerobic fitness study. Figure 19.1 shows a partial listing of the Fitness.jmp data table.

Aerobic fitness can be evaluated using a special test that measures the oxygen uptake of a person running on a treadmill for a prescribed distance. However, it would be more economical to find a formula that uses simpler measurements that evaluate fitness and predict oxygen uptake. To identify such an equation, measurements of age, weight, runtime, and pulse were taken for 31 participants who ran 1.5 miles.

To find a good oxygen uptake prediction equation, you need to compare many different regression models. The Stepwise platform lets you search through models with combinations of effects and choose the model you want.

Figure 19.1 The Fitness Data Table

Fitness
Notes Linneruds Fitn
Fit Model
Stepwise Fit
Columns (9/0)
Name
Sex
Age
Weight
Oxy
Runtime
RunPulse
RstPulse
MaxPulse
Rows
All rows 31

	Name	Sex	Age	Weight	Oxy	Runtime	RunPulse	RstPulse	MaxPulse
1	Donna	F	42	68.15	59.57	8.17	166	40	172
2	Gracie	F	38	81.87	60.06	8.63	170	48	186
3	Luanne	F	43	85.84	54.30	8.65	156	45	168
4	Mimi	F	50	70.87	54.63	8.92	146	48	155
5	Chris	M	49	81.42	49.16	8.95	180	44	185
6	Allen	M	38	89.02	49.87	9.22	178	55	180
7	Nancy	F	49	76.32	48.67	9.40	186	56	188
8	Patty	F	52	76.32	45.44	9.63	164	48	166
9	Suzanne	F	57	59.08	50.55	9.93	148	49	155
10	Teresa	F	51	77.91	46.67	10.00	162	48	168
11	Bob	M	40	75.07	45.31	10.07	185	62	185
12	Harriett	F	49	73.37	50.39	10.08	168	67	168
13	Jane	F	44	73.03	50.54	10.13	168	45	168
14	Harold	M	48	91.63	46.77	10.25	162	48	164
15	Sammy	M	54	83.12	51.86	10.33	166	50	170

Note: For purposes of illustration, certain values of MaxPulse and RunPulse have been changed from data reported by Rawlings (1988, p.105).

Figure 19.2 Model Specification Dialog for a Stepwise Model

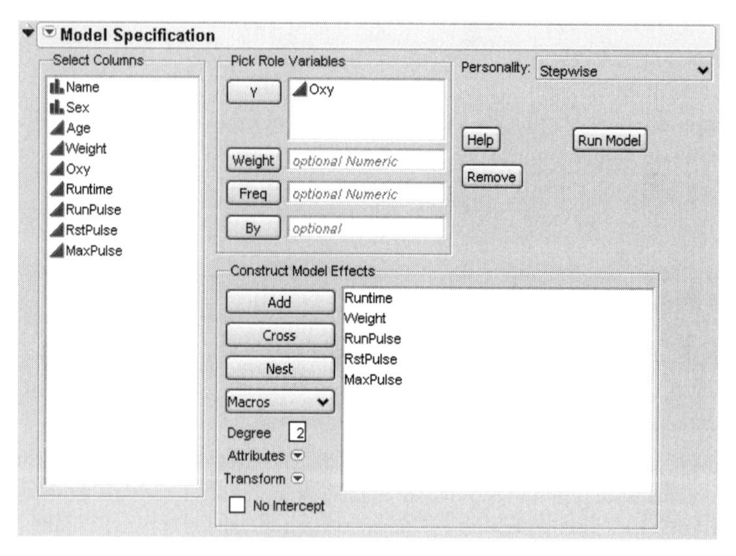

To do stepwise regression, first select **Fit Model** in the **Analyze** menu. In the Model Specification dialog, choose a *Y* response (Oxy) and a number of candidate *X* regressor columns.

Then, select **Stepwise** from the Fitting Personality popup menu in the top-right corner of the dialog, as shown in Figure 19.2, and click **Run Model.**

When the model runs, it displays a window that shows three areas:

- The Stepwise Regression Control panel, which is an interactive control panel for operating the platform
- The Current Estimates table, which shows the current status of the specified model, and provides additional control features
- The Step History table, which lists the steps in the stepwise selection.

The following sections describe the components of these areas and tell how to use them. If you have categorical responses, see "Logistic Stepwise Regression," p. 400.

Stepwise Regression Control Panel

The Stepwise Regression Control Panel (Control Panel for short), shown next, has editable areas, buttons and a popup menu. You use these dialog features to limit regressor effect probabilities, determine the method of selecting effects, begin or stop the selection process, and create a model.

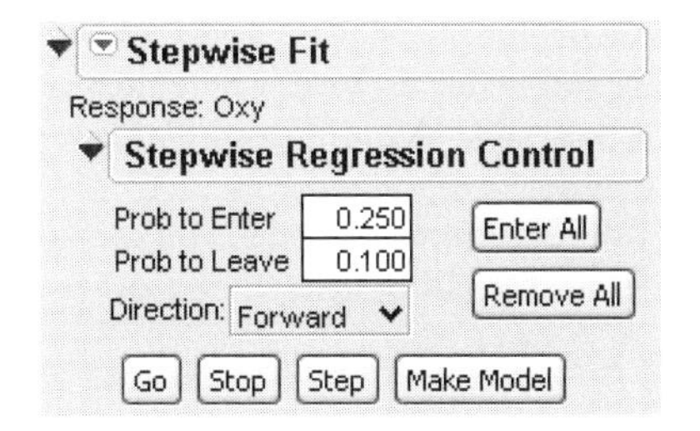

You use the Control Panel as follows:

Prob to Enter is the significance probability that must be attributed to a regressor term for it to be considered as a forward step and entered into the model. Click the field to enter a value.

Prob to Leave is the significance probability that must be attributed to a regressor term in order for it to be considered as a backward step and removed from the model. Click the field to enter a value.

Direction accesses the popup menu shown here, which lets you choose how you want variables to enter the regression equation.

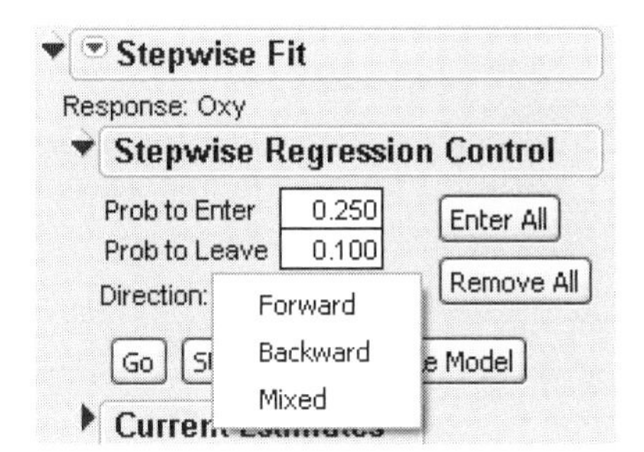

Forward brings in the regressor that most improves the fit, given that term is significant at the level specified by **Prob to Enter.**

Backward removes the regressor that affects the fit the least, given that term is not significant at the level specified in **Prob to Leave**.

Mixed alternates the forward and backward steps. It includes the most significant term that satisfies Prob to Enter and removes the least significant term satisfying **Prob to Leave**. It continues removing terms until the remaining terms are significant and then it changes to the forward direction.

Buttons on the controls panel let you control the stepwise processing:

Go starts the selection process. The process continues to run in the background until the model is finished.

Stop stops the background selection process.

Step stops after each step of the stepwise process

Enter All enters all unlocked terms into the model.

Remove All removes all terms from the model.

Make Model forms a model for the Model Specification Dialog from the model currently showing in the Current Estimates table. In cases where there are nominal or ordinal terms, **Make Model** can create new data table columns to contain terms that are needed for the model.

Tip: If you have a hierarchy of terms in your model, you can specify their entry rules using the **Rules** drop-down menu. See "Rules for Including Related Terms," p. 397 for details.

Current Estimates Table

The Current Estimates table lets you enter, remove, and lock in model effects. The platform begins with no terms in the model except for the intercept, as is shown here. The intercept is permanently locked into the model.

Current Estimates

SSE	DFE	MSE	RSquare	RSquare Adj	Cp	AIC
851.38154	30	28.379385	0.0000	0.0000	106.93073	104.6991

Lock	Entered	Parameter	Estimate	nDF	SS	"F Ratio"	"Prob>F"
☑	☑	Intercept	47.3758065	1	0	0.000	1.0000
☐	☐	Runtime	0	1	632.9001	84.008	0.0000
☐	☐	Weight	0	1	22.55181	0.789	0.3817
☐	☐	RunPulse	0	1	134.8447	5.457	0.0266
☐	☐	RstPulse	0	1	135.7828	5.503	0.0260
☐	☐	MaxPulse	0	1	47.71646	1.722	0.1997

You use check boxes to define the stepwise regression process:

Lock locks a term in or out of the model. **Lock** does not permit a term that is checked to be entered or removed from the model. Click an effect's check box to change its lock status.

Entered shows whether a term is currently in the model. You can click a term's check box to manually bring an effect into or out of the model.

Parameter lists the names of the regressor terms (effects).

Estimate is the current parameter estimate. It is missing (.) if the effect is not currently in the model.

nDF is the number of degrees of freedom for a term. A term has more than one degree of freedom if its entry into a model also forces other terms into the model.

SS is the reduction in the error (residual) SS if the term is entered into the model or the increase in the error SS if the term is removed from the model. If a term is restricted in some fashion, it could have a reported SS of zero.

"F Ratio" is the traditional test statistic to test that the term effect is zero. It is the square of a *t*-ratio. It is in quotation marks because it does not have an *F*-distribution for testing the term because the model was selected as it was fit.

"Prob>F" is the significance level associated with the *F*-statistic. Like the "F Ratio," it is in quotation marks because it is not to be trusted as a real significance probability.

Statistics for the current model appear above the list of effects:

SSE, DFE, MSE are the sum of squares, degrees of freedom, and mean square error (residual) of the current model.

RSquare is the proportion of the variation in the response that can be attributed to terms in the model rather than to random error.

RSquareAdj adjusts R^2 to make it more comparable over models with different numbers of parameters by using the degrees of freedom in its computation. The adjusted R^2 is useful in stepwise procedure because you are looking at many different models and want to adjust for the number of terms in the model.

Cp is Mallow's C_p criterion for selecting a model. It is an alternative measure of total squared error defined as

$$C_p = \left(\frac{\mathrm{SSE}_p}{s^2}\right) - (N - 2p)$$

where s^2 is the MSE for the full model and SSE_p is the sum-of-squares error for a model with p variables, including the intercept. Note that p is the number of x-variables+1. If C_p is graphed with p, Mallows (1973) recommends choosing the model where C_p first approaches p.

AIC is Akaike's Information Criterion defined as

$$\mathrm{AIC} = n\ln\left(\frac{\mathrm{SSE}}{n}\right) + 2p$$

where n is the number of observations and p is the number of model parameters including the intercept. This is a general criterion for choosing the best number of parameters to include in a model. The model that has the smallest value of AIC is considered the best (Akaike 1974).

Forward Selection Example

The default method of selection is the **Forward** selection. You can proceed with the Fitness.jmp data example using the **Step** button on the Control Panel. You see that after one step, the most significant term Runtime is entered into the model (top Current Estimates table in Figure 19.3). Click **Go** to see the stepwise process run to completion.

The bottom table in Figure 19.3 shows that all the terms have been added except RstPulse and Weight which are not significant at the **Prob to Enter** value of 0.25 specified in the Stepwise Regression Control Panel.

Figure 19.3 Current Estimates Table

After one step

Current Estimates

SSE	DFE	MSE	RSquare	RSquare Adj	Cp	AIC
218.48144	29	7.5338429	0.7434	0.7345	7.8825312	64.53413

Lock	Entered	Parameter	Estimate	nDF	SS	"F Ratio"	"Prob>F"
☑	☑	Intercept	82.4217727	1	0	0.000	1.0000
☐	☑	Runtime	-3.3105554	1	632.9001	84.008	0.0000
☐	☐	Weight	0	1	1.323628	0.171	0.6827
☐	☐	RunPulse	0	1	15.36208	2.118	0.1567
☐	☐	RstPulse	0	1	0.130138	0.017	0.8981
☐	☐	MaxPulse	0	1	1.567361	0.202	0.6563

After all steps

Current Estimates

SSE	DFE	MSE	RSquare	RSquare Adj	Cp	AIC
161.77233	27	5.9915679	0.8100	0.7889	2.8284105	59.21829

Lock	Entered	Parameter	Estimate	nDF	SS	"F Ratio"	"Prob>F"
☑	☑	Intercept	80.9007896	1	0	0.000	1.0000
☐	☑	Runtime	-2.9701867	1	443.2028	73.971	0.0000
☐	☐	Weight	0	1	4.989591	0.827	0.3714
☐	☑	RunPulse	-0.3751142	1	55.14175	9.203	0.0053
☐	☐	RstPulse	0	1	0.350744	0.056	0.8140
☐	☑	MaxPulse	0.35421891	1	41.34703	6.901	0.0140

Step History Table

As each step is taken, the Step History table records the effect of adding a term to the model. The Step History table for the Fitness data example shows the order in which the terms entered the model and shows the effect as reflected by R^2 and C_p.

Step History

Step	Parameter	Action	"Sig Prob"	Seq SS	RSquare	Cp	p
1	Runtime	Entered	0.0000	632.9001	0.7434	7.8825	2
2	RunPulse	Entered	0.1567	15.36208	0.7614	7.4298	3
3	MaxPulse	Entered	0.0140	41.34703	0.8100	2.8284	4

If you use Mallow's C_p as a model selection criterion, select the model where C_p approaches p, the number of parameters in the model. In this example, three or four variables appear to be a good choice for a regression model.

Backwards Selection Example

In backwards selection, terms are entered into the model and then least significant terms are removed until all the remaining terms are significant.

To do a backwards selection stepwise regression in JMP, use the Control Panel and click **Enter All**, which displays the Current Estimates table as shown in Figure 19.4. Then select **Backward** from the Direction popup menu and click either **Step** or **Go**.

Figure 19.4 Backwards Selection Example in Stepwise Regression

Stepwise Fit

Response: Oxy

Stepwise Regression Control

Prob to Enter 0.250
Prob to Leave 0.100
Direction: Backward

Enter All | Remove All

Go | Stop | Step | Make Model

Current Estimates

SSE	DFE	MSE	RSquare	RSquare Adj	Cp	AIC
156.58371	25	6.2633484	0.8161	0.7793	6	62.20771

Lock	Entered	Parameter	Estimate	nDF	SS	"F Ratio"	"Prob>F"
☑	☑	Intercept	82.3936054	1	0	0.000	1.0000
☐	☑	Runtime	-2.9518165	1	366.3375	58.489	0.0000
☐	☑	Weight	-0.0509071	1	4.83788	0.772	0.3878
☐	☑	RunPulse	-0.3970425	1	59.51519	9.502	0.0049
☐	☑	RstPulse	0.01239004	1	0.199033	0.032	0.8600
☐	☑	MaxPulse	0.38479281	1	45.83023	7.317	0.0121

Step History

The first backward step removes **RstPulse** and the second step removes **Weight**. No further terms meet the **Prob to Leave** probability specified in the Control Panel. The Current Estimates and Step History tables shown in Figure 19.5 summarize the backwards stepwise selection process.

Figure 19.5 Current Estimates with Terms Removed and Step History Table

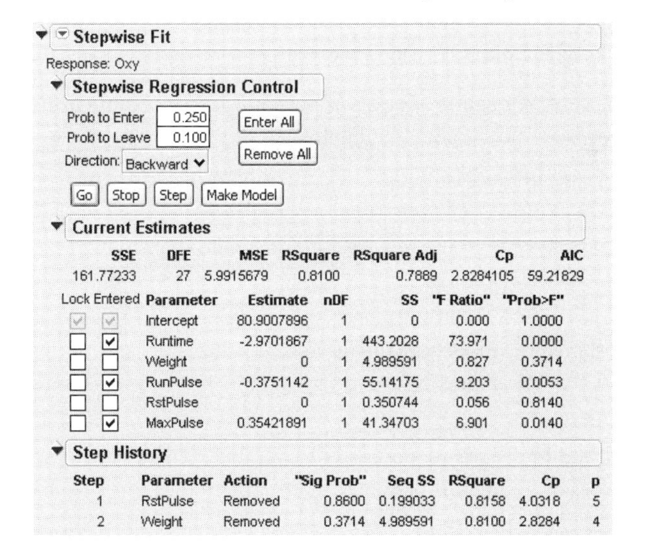

Stepwise Fit

Response: Oxy

Stepwise Regression Control

Prob to Enter 0.250
Prob to Leave 0.100
Direction: Backward

Enter All | Remove All

Go | Stop | Step | Make Model

Current Estimates

SSE	DFE	MSE	RSquare	RSquare Adj	Cp	AIC
161.77233	27	5.9915679	0.8100	0.7889	2.8284105	59.21829

Lock	Entered	Parameter	Estimate	nDF	SS	"F Ratio"	"Prob>F"
☑	☑	Intercept	80.9007896	1	0	0.000	1.0000
☐	☑	Runtime	-2.9701867	1	443.2028	73.971	0.0000
☐	☐	Weight	0	1	4.989591	0.827	0.3714
☐	☑	RunPulse	-0.3751142	1	55.14175	9.203	0.0053
☐	☐	RstPulse	0	1	0.350744	0.056	0.8140
☐	☑	MaxPulse	0.35421891	1	41.34703	6.901	0.0140

Step History

Step	Parameter	Action	"Sig Prob"	Seq SS	RSquare	Cp	p
1	RstPulse	Removed	0.8600	0.199033	0.8158	4.0318	5
2	Weight	Removed	0.3714	4.989591	0.8100	2.8284	4

Make Model

When you click **Make Model**, the model seen in the Current Estimates table appears in the Model Specification dialog. For example, if you click **Make Model** after the backward selection in Figure 19.5, the Model Specification dialog appears as shown in Figure 19.6 without a fitting personality selection.

Figure 19.6 New Model Specification dialog from Forward Stepwise Procedure

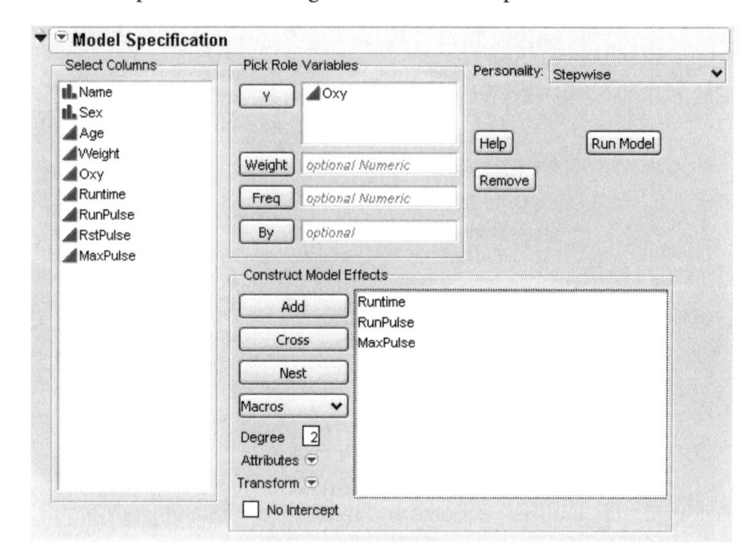

All Possible Regression Models

Stepwise includes an **All Possible Models** command. It is accessible from the red triangle drop-down menu on the stepwise control panel (see Figure 19.7).

Figure 19.7 All Possible Models

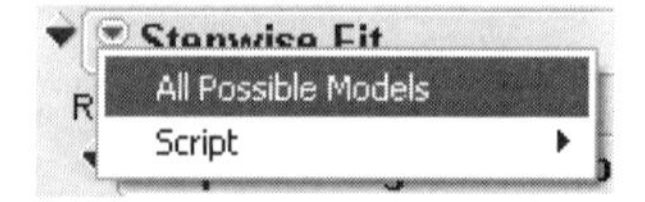

When selected, all possible models of the regression parameters are run, resulting in the report seen in Figure 19.8. Note that this report is for a three-variable model consisting of Runtime, RunPulse, and MaxPulse.

The models are listed in decreasing order of the number of parameters they contain. The model with the highest R^2 for each number of parameters is highlighted.

We suggest that no more than about 15 variables be used with this platform. More may be possible, but can strain computer memory (and human patience).

Figure 19.8 All Models Report

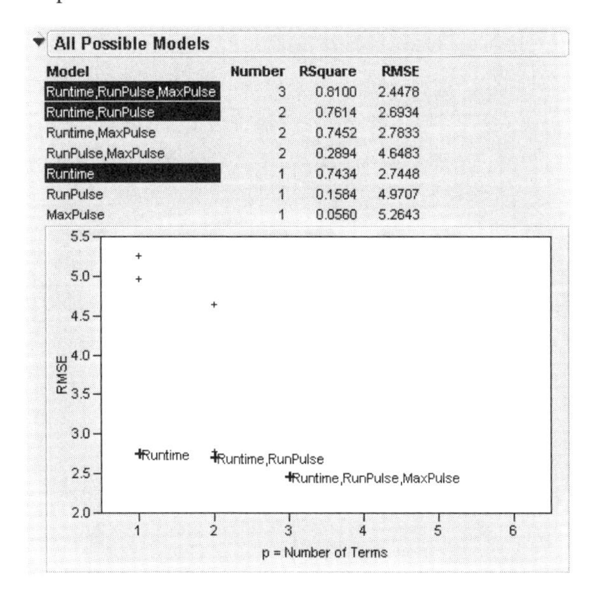

All Possible Models

Model	Number	RSquare	RMSE
Runtime,RunPulse,MaxPulse	3	0.8100	2.4478
Runtime,RunPulse	2	0.7614	2.6934
Runtime,MaxPulse	2	0.7452	2.7833
RunPulse,MaxPulse	2	0.2894	4.6483
Runtime	1	0.7434	2.7448
RunPulse	1	0.1584	4.9707
MaxPulse	1	0.0560	5.2643

Note: Mallow's C_p statistic is computed, but initially hidden in the table. To make it visible, Right-click (Control-click on the Macintosh) and select **Columns > Cp** from the menu that appears.

Models with Crossed, Interaction, or Polynomial Terms

Often with models from experimental designs, you have cross-product or interaction terms. For continuous factors, these are simple multiplications. For nominal and ordinal factors, the interactions are outer products of many columns. When there are crossed terms, you usually want to impose rules on the model selection process so that a crossed term cannot be entered unless all its subterms (terms that contain it) are in the model.

The next example uses the Reactor.jmp sample data (Box, Hunter, and Hunter 1978). The response is Y and the variables are F, Ct, A, T, and Cn.

The model shown here is composed of all factorial terms up to two-factor interactions for the five continuous factors. Note that some terms have more than one degree of freedom (nDF) due to the restrictions placed on some of the terms. Under the model selection rules described above, a crossed term cannot be entered into the model until all its subterms are also in the model. For example, if the stepwise process enters F*Ct, then it must also enter F and Ct, which gives F*Ct an nDF of 3.

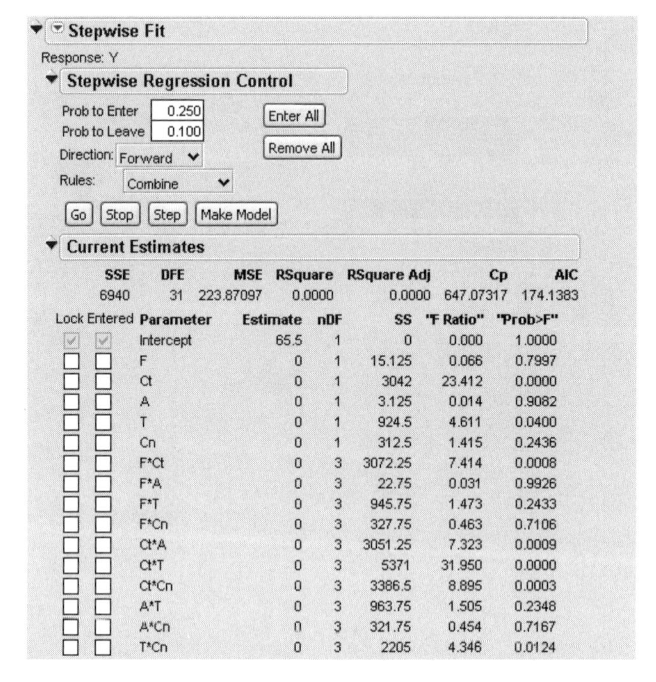

Stepwise Fit

Response: Y

Stepwise Regression Control

Prob to Enter 0.250
Prob to Leave 0.100
Direction: Forward
Rules: Combine

Enter All | Remove All

Go | Stop | Step | Make Model

Current Estimates

SSE	DFE	MSE	RSquare	RSquare Adj	Cp	AIC
6940	31	223.87097	0.0000	0.0000	647.07317	174.1383

Lock	Entered	Parameter	Estimate	nDF	SS	"F Ratio"	"Prob>F"
☑	☑	Intercept	65.5	1	0	0.000	1.0000
☐	☐	F	0	1	15.125	0.066	0.7997
☐	☐	Ct	0	1	3042	23.412	0.0000
☐	☐	A	0	1	3.125	0.014	0.9082
☐	☐	T	0	1	924.5	4.611	0.0400
☐	☐	Cn	0	1	312.5	1.415	0.2436
☐	☐	F*Ct	0	3	3072.25	7.414	0.0008
☐	☐	F*A	0	3	22.75	0.031	0.9926
☐	☐	F*T	0	3	945.75	1.473	0.2433
☐	☐	F*Cn	0	3	327.75	0.463	0.7106
☐	☐	Ct*A	0	3	3051.25	7.323	0.0009
☐	☐	Ct*T	0	3	5371	31.950	0.0000
☐	☐	Ct*Cn	0	3	3386.5	8.895	0.0003
☐	☐	A*T	0	3	963.75	1.505	0.2348
☐	☐	A*Cn	0	3	321.75	0.454	0.7167
☐	☐	T*Cn	0	3	2205	4.346	0.0124

The progress of multiterm inclusion is a balance between numerator degrees of freedom and opportunities to improve the fit. When there are significant interaction terms, often several terms enter at the same step. After one step in this example, Ct*T is entered along with its two contained effects Ct and T. However, a step back is not symmetric because a crossed term can be removed without removing its two component terms. Note that Ct now has 2 degrees of freedom because if Stepwise removes Ct, it also removes Ct*T.

Current Estimates

SSE	DFE	MSE	RSquare	RSquare Adj	Cp	AIC
1569	28	56.035714	0.7739	0.7497	129.07317	132.5587

Lock	Entered	Parameter	Estimate	nDF	SS	"F Ratio"	"Prob>F"
☑	☑	Intercept	65.5	1	0	0.000	1.0000
☐	☐	F	0	1	15.125	0.263	0.6124
☐	☑	Ct	9.75	2	4446.5	39.676	0.0000
☐	☐	A	0	1	3.125	0.054	0.8182
☐	☑	T	5.375	2	2329	20.781	0.0000
☐	☐	Cn	0	1	312.5	6.715	0.0152
☐	☐	F*Ct	0	2	30.25	0.256	0.7764
☐	☐	F*A	0	3	22.75	0.123	0.9459
☐	☐	F*T	0	2	21.25	0.178	0.8376
☐	☐	F*Cn	0	3	327.75	2.200	0.1130
☐	☐	Ct*A	0	2	9.25	0.077	0.9260
☐	☑	Ct*T	6.625	1	1404.5	25.064	0.0000
☐	☐	Ct*Cn	0	2	344.5	3.657	0.0398
☐	☐	A*T	0	2	39.25	0.334	0.7194
☐	☐	A*Cn	0	3	321.75	2.150	0.1192
☐	☐	T*Cn	0	2	1280.5	57.700	0.0000

Rules for Including Related Terms

You can change the rules that are applied when there is a hierarchy of terms in the model. Notice that when terms are related, an extra popup menu called **Rules** appears.

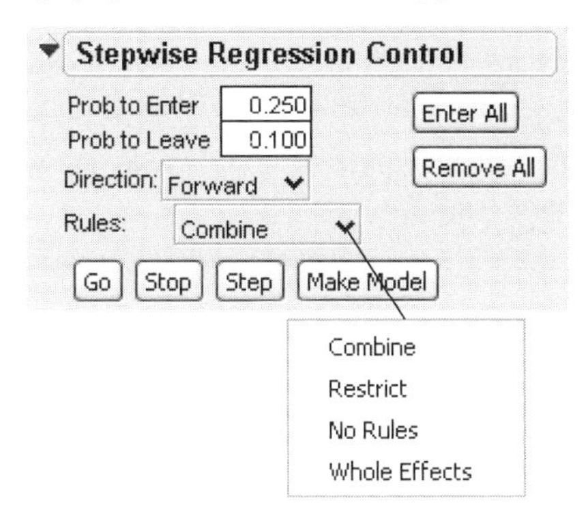

The **Rules** choices are used for related terms:

Combine groups a term with its precedent terms and calculates the group's significance probability for entry as a joint *F*-test. **Combine** is the default rule.

Restrict restricts the terms that have precedents so that they cannot be entered until their precedents are entered.

No Rules gives the selection routine complete freedom to choose terms, regardless of whether the routine breaks a hierarchy or not.

Whole Effects enters only whole effects, when all terms in the effect are significant.

Models with Nominal and Ordinal Terms

Traditionally, stepwise regression has not addressed the situation when there are categorical terms in the model. When nominal or ordinal terms are in regression models, they are carried as sets of dummy or indicator columns. When there are only two levels, there is no problem because they generate only a single column. However, for more than two levels, multiple columns must be handled. The convention in JMP for nominal variables in standard platforms is to model these terms so that the parameter estimates average out to zero across all the levels.

In the stepwise platform, categorical variables (nominal and ordinal) are coded in a hierarchical fashion, which is different from the other least squares fitting platforms. In hierarchical coding, the levels of the categorical variable are considered in some order and a split is made to make the two groups of levels that most separate the means of the response. Then, each group is further subdivided into its most separated subgroups, and so on, until all the levels are distinguished into k - 1 terms for k levels.

For example, consider the Football.jmp data used to predict Speed from Weight and the variable Position2. Position2 is a nominal variable with values representing football positions with one- and two-letter abbreviations.

- Considering only **Position2** and the response variable, **Speed**, the method first splits **Position** into two groups with the term **Position2{wr&db&o&lb&te–ki&l}**. The slower group consists of the wide receivers (**wr**), defensive backs (**db**), and offensive backs (**o**), linebackers (**lb**), and tight ends (**te**). The faster group is the kickers (**ki**), and linemen (**l**), which splits as **Position2{ki–&l}**.
- The next split subdivides the slower group into wide receivers (**wr**), defensive backs (**db**), and offensive backs(**o**) versus linebackers (**lb**) and tight ends (**te**), shown as **Position2{wr&db&o–lb&te}**. The linebackers (**lb**) and tight ends split (**te**) to form **Position2{lb–te}**.
- The slower group divides again giving **Position2{wr&db-o}**, wide receivers (**wr**) and defensive backs (**db**) versus offensive backs (**o**).
- The last subdivision is the wide receivers (**wr**) and defensive backs (**db**), **Position2{wr–db}**.

These terms can be illustrated by the tree hierarchy shown at the top in Figure 19.9. The Current Estimates table at the bottom lists the terms, with their estimates.

Figure 19.9 Tree Structure of Terms and Corresponding Current Estimates Table

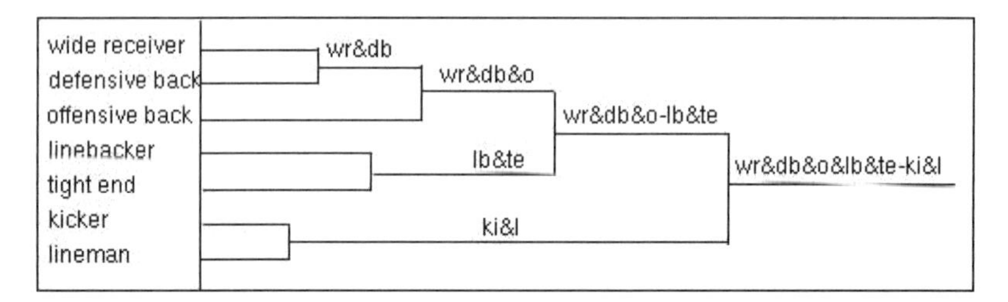

Current Estimates

SSE	DFE	MSE	RSquare	RSquare Adj	Cp	AIC
1489.5446	110	13.541314	0.0000	0.0000	137.9381	290.2332

Lock	Entered	Parameter	Estimate	nDF	SS	"F Ratio"	"Prob>F"
☑	☑	Intercept	61.0852252	1	0	0.000	1.0000
☐	☐	Weight	0	1	619.6596	77.646	0.0000
☐	☐	Position2{wr&db&o&lb&te-ki&l}	0	1	723.3538	102.906	0.0000
☐	☐	Position2{wr&db&o-lb&te}	0	2	818.9823	65.952	0.0000
☐	☐	Position2{wr&db-o}	0	3	824.4063	44.207	0.0000
☐	☐	Position2{wr-db}	0	4	826.0589	32.993	0.0000
☐	☐	Position2{lb-te}	0	3	819.667	43.642	0.0000
☐	☐	Position2{ki-l}	0	2	749.098	54.631	0.0000

Using the default **Combine** rule for terms to enter a model or the **Restrict** rule, a term cannot enter the model unless all the terms above it in the hierarchy have been entered. Thus, it is simple to bring in the term **Position2{wr&db&o-lb&te&ki&l}** because there is nothing above it in the hierarchy. But to enter **{te-ki}** requires that the three other terms above it in the hierarchy are entered: **Position2{lb-te&ki}**, **Position2{lb&te&ki-l}**, and **Position2{wr&db&o-lb&te&ki&l}**.

There are several reasons to choose a hierarchical coding:

1. It is an orthonormal coding, so that the transformation on the *X* columns is the same as the transformation on the Beta columns. In matrix form,

   $$XTT^{-1}\beta = XTT'\beta \text{ because } T^{-1} = T'.$$

2. The hierarchy leads to natural stepping rules.
3. The groupings that make the greatest initial separation enter the model early.

For nominal terms, the order of levels is determined by the means of the Ys. For ordinal terms, the order is fixed.

Make Model Command for Hierarchical Terms

When you click **Make Model** for a model with nominal or ordinal terms, **Fit Model** creates a new set of columns in the data table that it needs. The model appears in a new Fit Model window for the response variable. The next example uses the Hotdogs2 sample data to illustrate how **Stepwise** constructs a model with hierarchical effects.

A simple model (Figure 19.10) looks at the cost per ounce as a function of Type (Meat, Beef, Poultry) and Size (Jumbo, Regular, Hors d'oeuvre). **Stepwise** constructs the hierarchical tree of the model effects listed in Figure 19.10. The tree is formed using the combinations of factor levels that have the greatest influence on the model.

In this example, a forward **Stepwise** process enters effects using the **Restrict** from the **Rules** popup menu, and includes three of the four terms in the model.

Figure 19.10 Stepwise Platform for Model with Hierarchical Effects

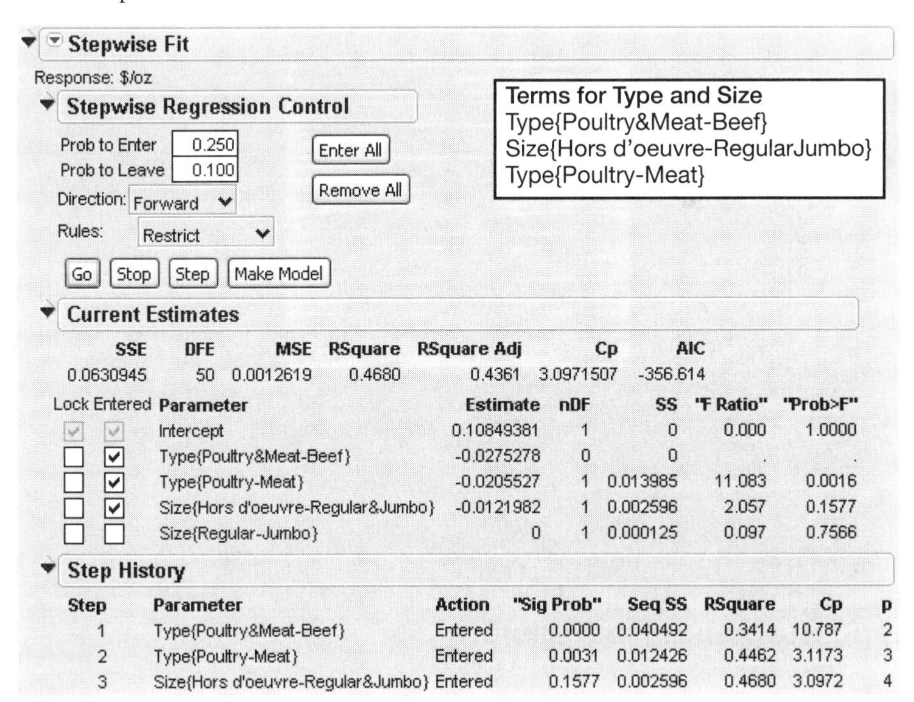

When you choose **Make Model** in the Stepwise Regression Control Panel, two actions occur:

- The indicator variables listed in the Current Estimates table appear in the Hotdogs2 data table, as shown in Figure 19.11.

- A new Model Specification dialog opens. The effects are the new variables that were selected by the stepwise analysis process for further data analysis.

Figure 19.11 Indicator Variables Constructed by the Stepwise Analysis

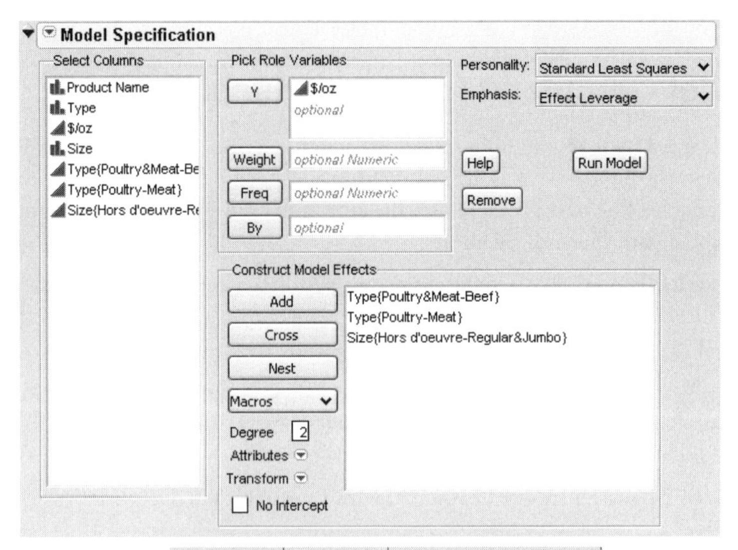

Type{Poultry& Meat-Beef}	Type{Poultry-Meat}	Size{Hors d'oeuvre-Regular&Jumbo}
-1	0	-1
-1	0	-1
-1	0	-1
-1	0	-1
-1	0	-1
-1	0	-1
-1	0	-1
-1	0	-1
-1	0	-1
-1	0	-1
-1	0	-1
-1	0	-1
-1	0	-1
-1	0	-1
-1	0	-1
-1	0	-1
-1	0	-1

Logistic Stepwise Regression

JMP performs logistic stepwise regression in a similar way to standard least-squares logistic regression. To run a logistic stepwise regression, simply add terms to the model as usual and choose Stepwise from the personality drop-down menu.

The difference in the report when the response is categorical is in the **Current Estimates** section of the report. Wald/Score chi-square statistics appear, and the overall fit of the model is shown as -LogLikelihood. An example is shown in Figure 19.12.

Figure 19.12 Logistic Stepwise Report

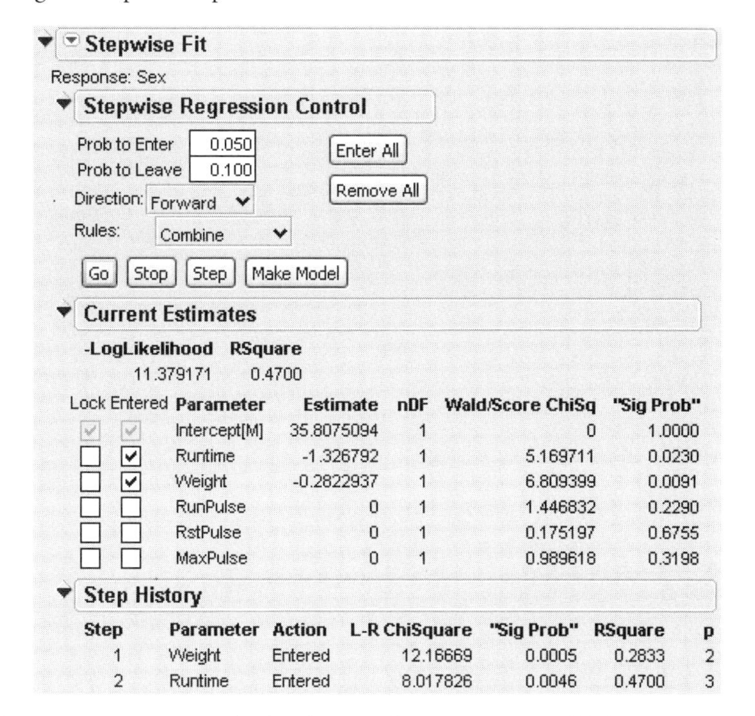

The enter and remove statistics are calculated using cheap Score or Wald chi-square tests respectively, but the regression estimates and log-likelihood values are based on the full iterative maximum likelihood fit. If you want to compare the Wald/Score values, look at the Step History report.

Chapter 20

Logistic Regression for Nominal and Ordinal Response

The Fit Model Platform

If the model response is nominal, the Fit Model platform fits a linear model to a multi-level logistic response function using maximum likelihood. Likelihood-ratio test statistics are computed for the whole model. Lack of Fit tests and Wald test statistics are computed for each effect in the model. Options give likelihood-ratio tests for effects and confidence limits and odds ratios for the maximum likelihood parameter estimates.

If the response variable is ordinal, the platform fits the cumulative response probabilities to the logistic distribution function of a linear model using maximum likelihood. Likelihood-ratio test statistics are provided for the whole model and lack of fit.

For simple main effects, you can use the Fit Y by X platform described in the chapter "Simple Logistic Regression," p. 167, to see a cumulative logistic probability plot for each effect. Details for these model are in discussed in the chapter "Understanding JMP Analyses," p. 7, and the appendix "Statistical Details," p. 975.

Contents

Introduction to Logistic Models

Logistic regression fits nominal *Y* responses to a linear model of *X* terms. To be more precise, it fits probabilities for the response levels using a logistic function. For two response levels the function is

$$\text{Prob}\ (Y = \text{1st response}) = (1 + e^{-Xb})^{-1}$$

or equivalently

$$\log\left(\frac{\text{Prob}(Y = \text{1st response})}{\text{Prob}(Y = \text{2nd response})}\right) = Xb$$

For r nominal responses, where $r > 2$, it fits $r - 1$ sets of linear model parameters of the form

$$\log\left(\frac{\text{Prob}(Y = j)}{\text{Prob}(Y = r)}\right) = X_j b$$

The fitting principal of maximum likelihood means that the βs are chosen to maximize the joint probability attributed by the model to the responses that did occur. This fitting principal is equivalent to minimizing the negative log-likelihood (–LogLikelihood)

$$\text{Loss} = -\text{loglikelihood} = -\log\left(\text{Prob}\left(\sum_{i=1}^{n} i\text{th row has the } y_j\text{th response}\right)\right)$$

as attributed by the model.

As an example, consider an experiment that was performed on metal ingots prepared with different heating and soaking times. The ingots were then tested for readiness to roll (Cox 1970). The Ingots.jmp data table in the Sample Data folder has the experimental results.

Ingots
Notes Metal ingots prepar
Logistic Regression

Columns (4/0)
heat
soak
ready
count

Rows
All rows 38

	heat	soak	ready	count
1	7	1.0	1	0
2	7	1.0	0	10
3	7	1.7	1	0
4	7	1.7	0	17
5	7	2.2	1	0
6	7	2.2	0	7
7	7	2.8	1	0
8	7	2.8	0	12
9	7	4.0	1	0
10	7	4.0	0	9
11	14	1.0	1	0
12	14	1.0	0	31

The categorical variable called ready has values 1 and 0 for readiness and not readiness to roll, respectively.

The Fit Model platform fits the probability of the *not readiness* (0) response to a logistic cumulative distribution function applied to the linear model with regressors heat and soak:

$$\text{Probability(not ready to roll)} = \frac{1}{1 + e^{-(\beta_0 + \beta_1 \text{heat} + \beta_2 \text{soak})}}$$

The parameters are estimated by minimizing the sum of the negative logs of the probabilities attributed to the observations by the model (maximum likelihood).

To analyze this model, select **Analyze > Fit Model**. The ready variable is Y, the response, and heat and soak are the model effects. The count column is the **Freq** variable. When you click **Run Model**, iterative calculations take place. When the fitting process converges, the nominal/ordinal regression report appears. The following sections discuss the report layout and statistical tables, and show examples.

The Statistical Report

Initially the logistic platform produces the reports shown here. Lack of fit tests show only if they are applicable and Likelihood-ratio tests are done only on request.

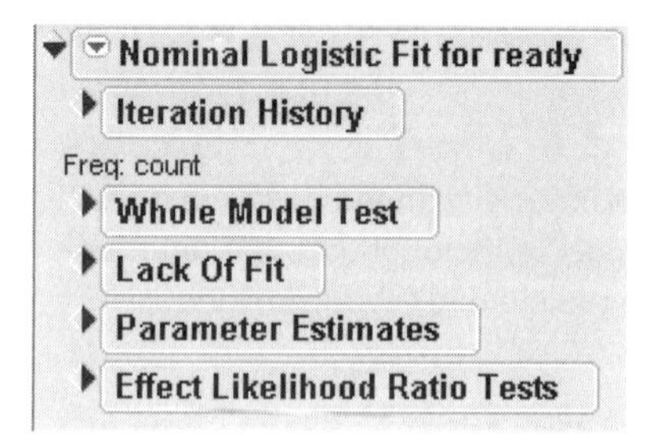

All tests compare the fit of the specified model with subset or superset models, as illustrated in Figure 20.1. If a test shows significance, then the higher order model is justified.

- Whole model tests: if the specified model is significantly better than a reduced model without any effects except the intercepts
- Lack of Fit tests: if a saturated model is significantly better than the specified model
- Effect tests: if the specified model is significantly better than a model without a given effect.

Figure 20.1 Relationship of Statistical Tables

Tests are a comparison of model fits

reduced model
(with only intercepts)

Whole Model Test

(default)
Wald Effect Test

specified model

model without
i^{th} effect

Likelihood-ratio
Effect Test
(optional)

Lack-of-Fit Test
(Goodness of Fit G^2)

saturated model
(a parameter for each unique
combination of x values)

Logistic Plot

If your model contains a single continuous effect, then a logistic report similar to the one in Fit Y By X appears. See Figure 9.3 "Interpretation of a Logistic Probability Plot," p. 171 for an interpretation of these plots.

Iteration History

After launching Fit Model, an iterative estimation process begins and is reported iteration by iteration. After the fitting process completes, you can open the Iteration History report and see the iteration steps. If the fitting takes too long, you can cancel by pressing the Escape key (⌘-Period on the Macintosh) at any time. Otherwise the iteration process stops when either the log-likelihood doesn't change by more than a very small relative amount (Obj-Criterion), the parameter estimates don't change by a small relative amount (Delta-Criterion), or 15 iterations have been performed.

Nominal Logistic Fit for ready

Iteration History

Iter	LogLikelihood	Step	Delta-Criterion	Obj-Criterion
1	-268.2479589	Initial	2159406105	.
2	-76.29480707	Newton	2.20769391	2.51561021
3	-53.3803292	Newton	1.49888471	0.4291878
4	-48.34608553	Newton	0.66510118	0.10410776
5	-47.69191265	Newton	0.13286898	0.01371377
6	-47.67282965	Newton	0.0058697	0.00040021
7	-47.67280663	Newton	9.5523e-6	4.82788e-7

Whole Model Test

The Whole Model table shows tests that compare the whole-model fit to the model that omits all the regressor effects except the intercept parameters. The test is analogous to the Analysis of Variance table for continuous responses. The negative log-likelihood corresponds to the sums of squares, and the Chi-square test corresponds to the *F*-test.

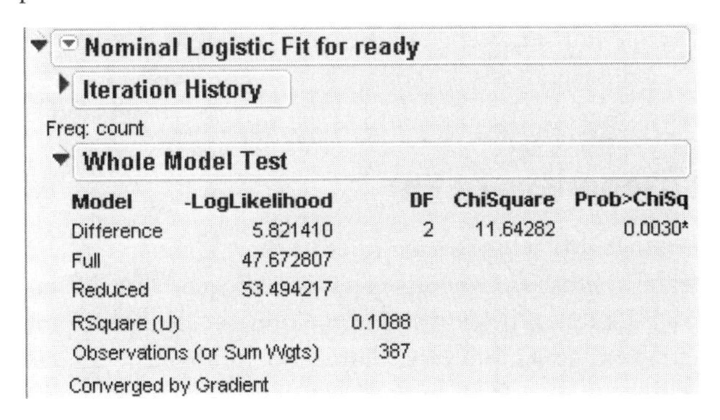

Nominal Logistic Fit for ready

Iteration History

Freq: count

Whole Model Test

Model	-LogLikelihood	DF	ChiSquare	Prob>ChiSq
Difference	5.821410	2	11.64282	0.0030*
Full	47.672807			
Reduced	53.494217			

RSquare (U)	0.1088
Observations (or Sum Wgts)	387

Converged by Gradient

The Whole Model table shows these quantities:

Model lists the model labels called Difference (difference between the Full model and the Reduced model), Full (model that includes the intercepts and all effects), and Reduced (the model that includes only the intercepts).

–LogLikelihood records an associated negative log-likelihood for each of the models.

Difference is the difference between the Reduced and Full models. It measures the significance of the regressors as a whole to the fit.

Full describes the negative log-likelihood for the complete model.

Reduced describes the negative log-likelihood that results from a model with only intercept parameters. For the ingot experiment, the –LogLikelihood for the reduced model that includes only the intercepts is 53.49.

DF records an associated degrees of freedom (DF) for the Difference between the Full and Reduced model. For the ingots experiment, there are two parameters that represent different heating and soaking times, so there are 2 degrees of freedom.

Chi-Square is the Likelihood-ratio Chi-square test for the hypothesis that all regression parameters are zero. It is computed by taking twice the difference in negative log-likelihoods between the fitted model and the reduced model that has only intercepts.

Prob>ChiSq is the probability of obtaining a greater Chi-square value by chance alone if the specified model fits no better than the model that includes only intercepts.

RSquare (U) shows the R^2, which is the ratio of the **Difference** to the **Reduced** negative log-likelihood values. It is sometimes referred to as U, the uncertainty coefficient. **RSquare** ranges from zero for no improvement to 1 for a perfect fit. A **Nominal** model rarely has a high **Rsquare**, and it has an **Rsquare** of 1 only when all the probabilities of the events that occur are 1.

Observations (or **Sum Wgts**) is the total number of observations in the sample.

After fitting the full model with two regressors in the ingots example, the –LogLikelihood on the Difference line shows a reduction to 5.82 from the Reduced –LogLikelihood of 53.49. The ratio of Difference to Reduced (the proportion of the uncertainty attributed to the fit) is 10.8% and is reported as the Rsquare (U).

To test that the regressors as a whole are significant (the Whole Model test), a Chi-square statistic is computed by taking twice the difference in negative log-likelihoods between the fitted model and the reduced model that has only intercepts. In the ingots example, this Chi-square value is $2 \times 5.82 = 11.64$, and is significant at 0.003.

Lack of Fit Test (Goodness of Fit)

The next questions that JMP addresses are whether there is enough information using the variables in the current model or whether more complex terms need to be added. The Lack of Fit test, sometimes called a Goodness of Fit test, provides this information. It calculates a pure-error negative log-likelihood by constructing categories for every combination of the regressor values in the data (Saturated line in the Lack Of Fit table), and it tests whether this log-likelihood is significantly better than the Fitted model.

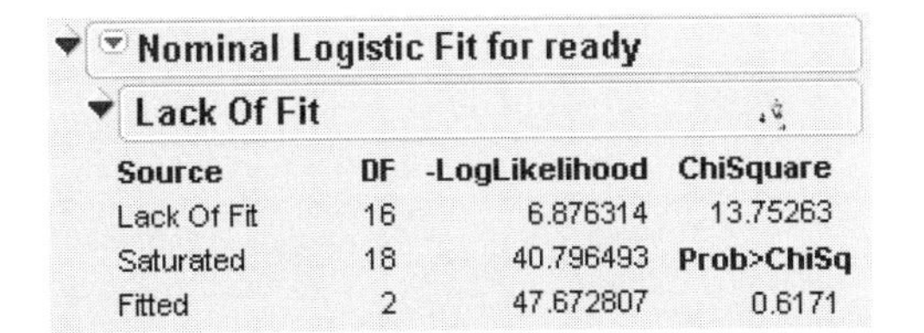
Nominal Logistic Fit for ready

Lack Of Fit

Source	DF	-LogLikelihood	ChiSquare
Lack Of Fit	16	6.876314	13.75263
Saturated	18	40.796493	Prob>ChiSq
Fitted	2	47.672807	0.6171

The Saturated degrees of freedom is m–1, where m is the number of unique populations. The Fitted degrees of freedom is the number of parameters not including the intercept. For the Ingots example, these are 18 and 2 DF, respectively. The Lack of Fit DF is the difference between the Saturated and Fitted models, in this case 18–2=16.

The Lack of Fit table lists the negative log-likelihood for error due to Lack of Fit, error in a Saturated model (pure error), and the total error in the Fitted model. Chi-square statistics test for lack of fit.

In this example, the lack of fit Chi-square is not significant (Prob>ChiSq = 0.617) and supports the conclusion that there is little to be gained by introducing additional variables, such as using polynomials or crossed terms.

Likelihood-ratio Tests

The **Likelihood Ratio Tests** command produces a table like the one shown here. The Likelihood-ratio Chi-square tests are calculated as twice the difference of the log-likelihoods between the full model and the model constrained by the hypothesis to be tested (the model without the effect). These tests can take time to do because each test requires a separate set of iterations.

This is the default test if the fit took less than ten seconds to complete.

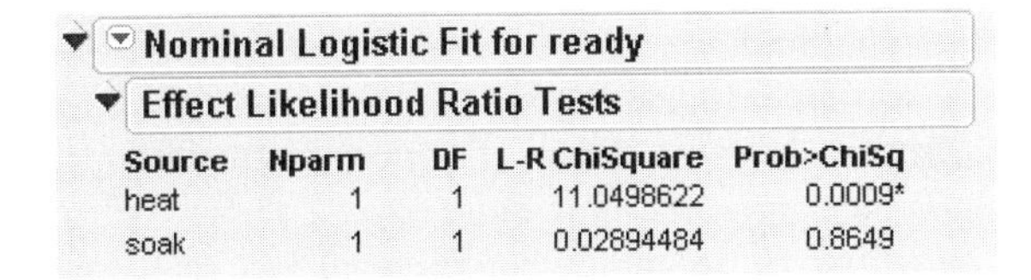
Nominal Logistic Fit for ready

Effect Likelihood Ratio Tests

Source	Nparm	DF	L-R ChiSquare	Prob>ChiSq
heat	1	1	11.0498622	0.0009*
soak	1	1	0.02894484	0.8649

Platform Options

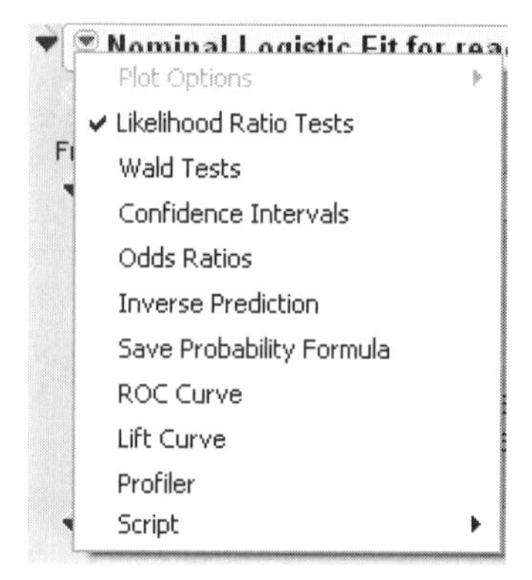

The popup menu next to the analysis name gives you the six additional commands that are described next.

Wald Tests for Effects

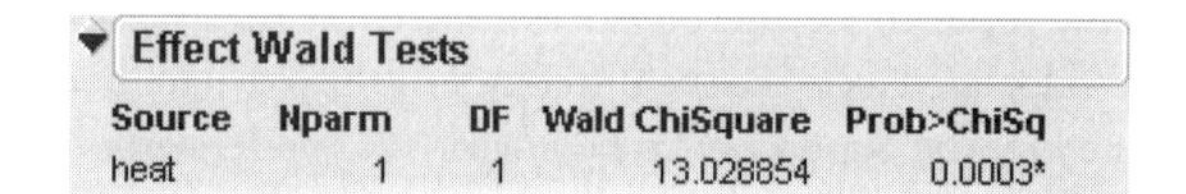

Effect Wald Tests

Source	Nparm	DF	Wald ChiSquare	Prob>ChiSq
heat	1	1	13.028854	0.0003*

One downside to likelihood ratio tests is that they involve refitting the whole model, which uses another series of iterations. Therefore, they could take a long time for big problems. The logistic fitting platform gives an optional simpler test that serves the same function. The Wald Chi-square is a one-step linear approximation to the likelihood-ratio test, and it is a by-product of the calculations. Though Wald tests are considered less trustworthy, they do provide an adequate significance indicator for screening effects. Each parameter estimate and effect is shown with a Wald test. This is the default test if the fit takes more than ten seconds to complete.

Likelihood-ratio tests are a platform option discussed under “Likelihood-ratio Tests,” p. 409, and we highly recommend using this option.

Confidence Intervals

You can also request profile likelihood confidence intervals for the model parameters.When you select the **Confidence Intervals** command, a dialog prompts you to enter α to compute the $1 - \alpha$ confidence intervals, or you can use the default of $\alpha = 0.05$. Each confidence limit requires a set of iterations in the model fit and can be expensive. Further, the effort does not always succeed in finding limits.

Parameter Estimates

Term	Estimate	Std Error	ChiSquare	Prob>ChiSq	Lower 95%	Upper 95%
Intercept	5.55915959	1.1196934	24.65	<.0001*	3.45325843	7.90106457
heat	-0.0820307	0.0237345	11.95	0.0005*	-0.1292562	-0.0348569
soak	-0.0567708	0.3312129	0.03	0.8639	-0.6674285	0.6628852

For log odds of 0/1

Odds Ratios (Nominal Responses Only)

When you select **Odds Ratios**, a report appears showing **Unit Odds Ratios** and **Range Odds Ratios**, as shown in Figure 20.2.

Figure 20.2 Odds Ratios

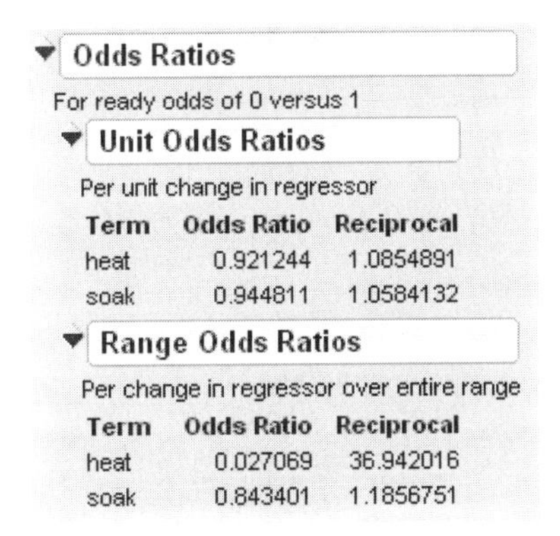

Odds Ratios

For ready odds of 0 versus 1

Unit Odds Ratios

Per unit change in regressor

Term	Odds Ratio	Reciprocal
heat	0.921244	1.0854891
soak	0.944811	1.0584132

Range Odds Ratios

Per change in regressor over entire range

Term	Odds Ratio	Reciprocal
heat	0.027069	36.942016
soak	0.843401	1.1856751

From the introduction (for two response levels), we had

$$\log\left(\frac{\text{Prob}(Y = \text{1st response})}{\text{Prob}(Y = \text{2nd response})}\right) = \text{X}b$$

so the odds ratio

$$\frac{\text{Prob}(Y = \text{1st response})}{\text{Prob}(Y = \text{2nd response})} = \exp(X\beta) = \exp(\beta_0) \cdot \exp(\beta_1 X_1) \cdots \exp(\beta_i X_i)$$

Note that $\exp(\beta_i(X_i + 1)) = \exp(\beta_i X_i)\exp(\beta_i)$. This shows that if X_i changes by a unit amount, the odds is multiplied by $\exp(\beta_i)$, which we label the unit odds ratio. As X_i changes over its whole range, the odds is multiplied by $\exp((X_{high} - X_{low})\beta_i)$ which we label the range odds ratio. For binary responses, the log odds ratio for flipped response levels involves only changing the sign of the parameter, so you may want the reciprocal of the reported value to focus on the last response level instead of the first.

Two-level nominal effects are coded 1 and -1 for the first and second levels, so range odds ratios or their reciprocals would be of interest.

Dose Response Example

In the Dose Response.jmp sample data table, the dose varies between 1 and 12. Using Fit Model, we got the following report after requesting **Odds Ratio** from the platform drop-down menu.

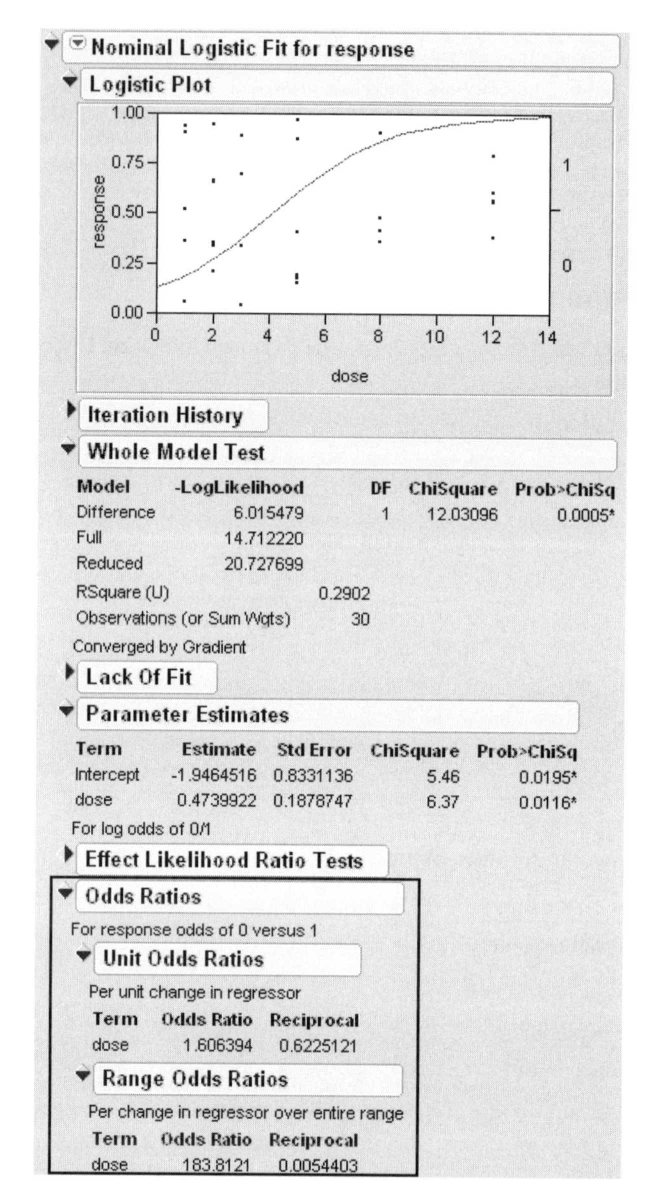

The unit odds ratio for dose is 1.606 (which is exp(0.473)) and indicates that the odds of getting a $Y = 0$ rather than $Y = 1$ improves by a factor of 1.606 for each increase of one unit of dose. The range odds ratio for dose is 183.8 (exp((12-1)*0.473)) and indicates that the odds improve by a factor of 183.8 as dose is varied between 1 and 12.

Inverse Prediction

For a two-level response, the **Inverse Prediction** command finds the *x* value that results in a specified probability. For example, using the **Ingots.jmp** data, ignore the **Soak** variable, and fit the probability of ready by heat using the **Fit Y by X** command for a simple logistic regression (Count is **Freq**). The cumulative logistic probability plot shows the result.

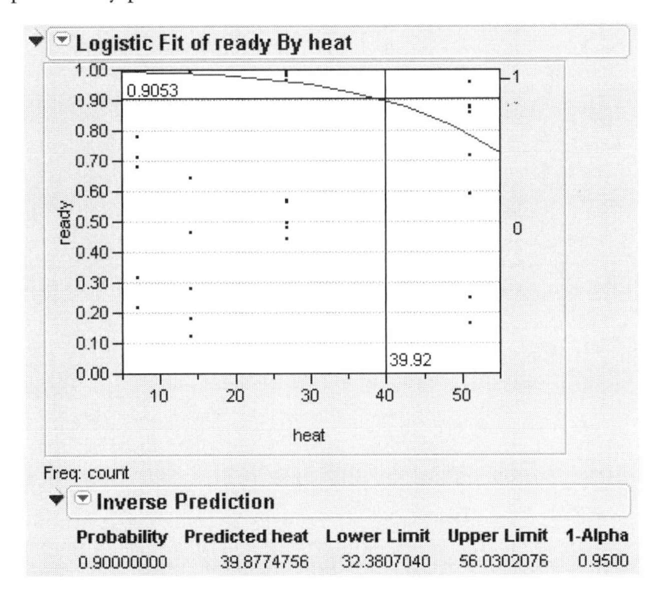

Probability	Predicted heat	Lower Limit	Upper Limit	1-Alpha
0.90000000	39.8774756	32.3807040	56.0302076	0.9500

Note that the fitted curve crosses the 0.9 probability level at a heat level of about 39.5, which is the inverse prediction. To be more precise and to get a fiducial confidence interval, you can use the **Inverse Prediction** command for **Fit Y by X** (see the chapter "Simple Logistic Regression," p. 167, for more discussion of the Fit Y by X simple logistic analysis). When you specify exactly 0.9 in the Inverse Prediction dialog, the predicted value (inverse prediction) for heat is 39.8775, as shown in the table above.

However, if you have another regressor variable (Soak), you must use the Fit Model platform. Then the **Inverse Prediction** command displays the Inverse Prediction dialog shown in Figure 20.3, for requesting the probability of obtaining a given value for one independent variable. To complete the dialog, click and type values in the editable X and Probability columns. Enter a value for a single *X* (heat or soak) and the probabilities you want for the prediction. Set the remaining independent variable to missing by clicking in its X field and deleting. The missing regressor is the value that it will predict.

Figure 20.3 The Inverse Prediction Dialog and Table

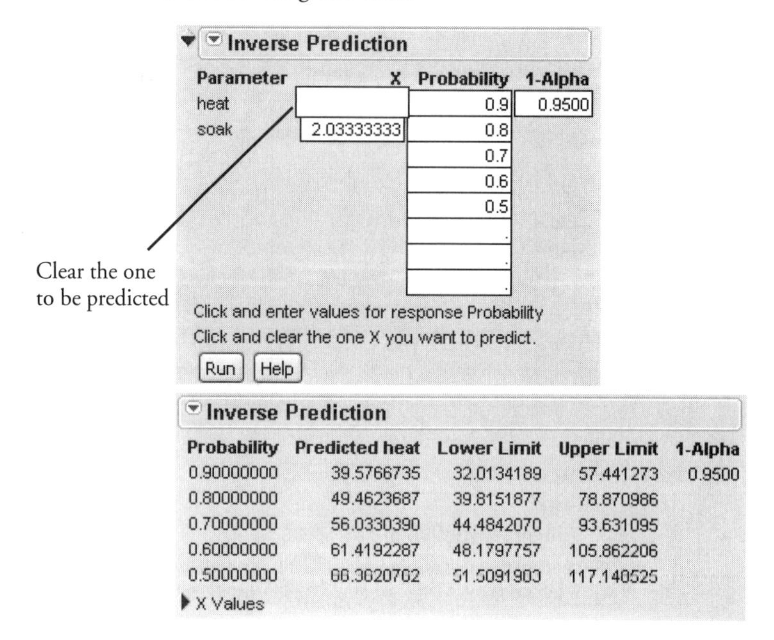

See the appendix "Statistical Details," p. 975, for more details about inverse prediction.

Save Commands

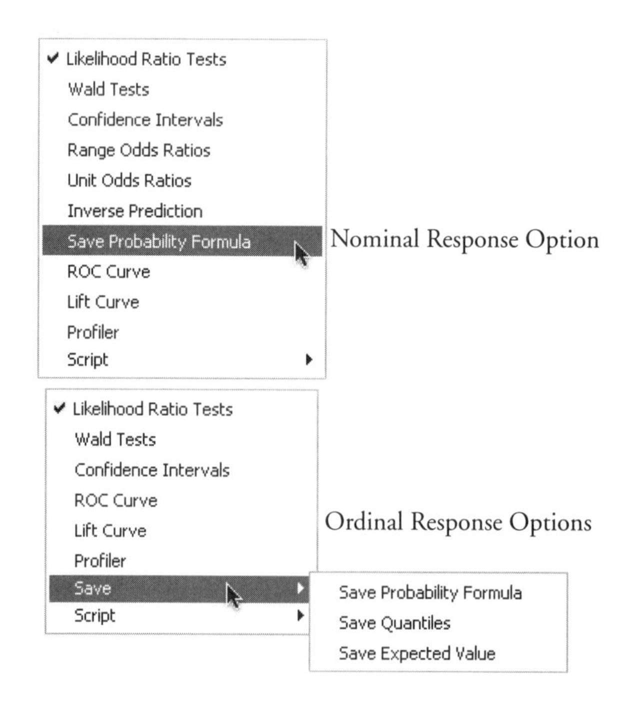

If you have ordinal or nominal response models, the **Save Probability Formula** command creates new data table columns.

If the response is numeric and has the ordinal modeling type, the **Save Quantiles** and **Save Expected Values** commands are also available.

The **Save** commands create the following new columns:

Save Probability Formula creates columns in the current data table that save formulas for linear combinations of the response levels, prediction formulas for the response levels, and a prediction formula giving the most likely response.

For a nominal response model with *r* levels, JMP creates

- columns called Lin[*j*] that contain a linear combination of the regressors for response levels $j = 1, 2, \ldots r - 1$
- a column called Prob[*r*], with a formula for the fit to the last level, *r*
- columns called Prob[*j*] for $j < r$ with a formula for the fit to level *j*
- a column called MostLikely responsename that picks the most likely level of each row based on the computed probabilities.
  For an ordinal response model with *r* levels, JMP creates
- a column called Linear that contains the formula for a linear combination of the regressors without an intercept term
- columns called Cum[*j*], each with a formula for the cumulative probability that the response is less than or equal to level *j*, for levels $j = 1, 2, \ldots r - 1$. There is no Cum[$j = 1, 2, \ldots r - 1$] that is 1 for all rows
- columns called Prob[$j = 1, 2, \ldots r - 1$], for $1 < j < r$, each with the formula for the probability that the response is level *j*. Prob[*j*] is the difference between Cum[*j*] and Cum[*j* –1]. Prob[1] is Cum[1], and Prob[*r*] is 1–Cum[*r* –1].
- a column called MostLikely responsename that picks the most likely level of each row based on the computed probabilities.

Save Quantiles creates columns in the current data table named OrdQ.05, OrdQ.50, and OrdQ.95 that fit the quantiles for these three probabilities.

Save Expected Values creates a column in the current data table called Ord Expected that is the linear combination of the response values with the fitted response probabilities for each row and gives the expected value.

ROC

Receiver Operating Characteristic (ROC) curves measure the sorting efficiency of the model's fitted probabilities to sort the response levels. ROC curves can also aid in setting criterion points in diagnostic tests. The higher the curve from the diagonal, the better the fit. An introduction to ROC curves is found in the section “ROC (Receiver Operating Characteristic) Curves,” p. 176 in the “Simple Logistic Regression” chapter. If the logistic fit has more than two response levels, it produces a generalized ROC curve (identical to the one in the Partition platform). In such a plot, there is a curve for each response level, which is the ROC curve of that level versus all other levels. Details on these ROC curves are found in “Graphs for Goodness of Fit,” p. 671 in the “Recursive Partitioning” chapter.

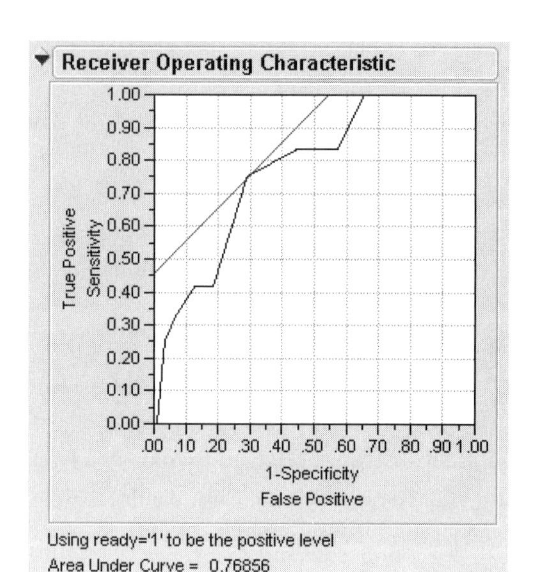

Lift Curve

Produces a lift curve for the model. A lift curve shows the same information as an ROC curve, but in a way to dramatize the richness of the ordering at the beginning. The *Y*-axis shows the ratio of how rich that portion of the population is in the chosen response level compared to the rate of that response level as a whole. See “Lift Curves,” p. 674 IN THE “RECURSIVE PARTITIONING” CHAPTER for more details on lift curves.

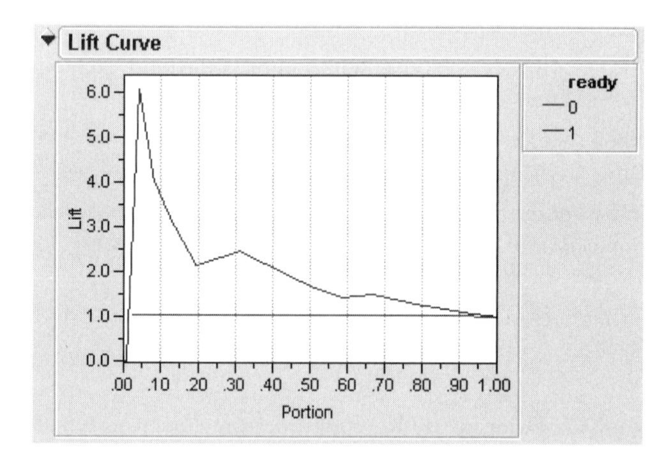

Profiler

Brings up the prediction profiler, showing the fitted values for a specified response probability as the values of the factors in the model are changed. This is currently available for nominal responses, but not ordinal responses.

Nominal Logistic Model Example: The Detergent Data

A market research study was undertaken to evaluate preference for a brand of detergent (Ries and Smith 1963). The results are in the Detergent.jmp sample data table.

Detergent
Notes Ries and Smith (1963) Ch
Nominal Logistic Regression
Columns (5/0)
brand
softness
previous use
temperature
count
Rows
All rows 24

	brand	softness	previous use	temperature	count
1	x	soft	yes	high	19
2	x	soft	yes	low	57
3	x	soft	no	high	29
4	x	soft	no	low	63
5	m	soft	yes	high	29
6	m	soft	yes	low	49
7	m	soft	no	high	27
8	m	soft	no	low	53
9	x	med	yes	high	23
10	x	med	yes	low	47

The model is defined by

- the response variable, brand with values m and x
- an effect called softness (water softness) with values soft, medium, and hard
- an effect called previous use with values yes and no
- an effect called temperature with values high and low
- a count variable, count, which gives the frequency counts for each combination of effect categories.

The study begins by specifying the full three-factor factorial model as shown by the Model Specification dialog in Figure 20.4. To specify a factorial model, highlight the three main effects in the column selector list. Then select **Full Factorial** from the **Macros** popup menu.

Figure 20.4 A Three-Factor Factorial Model with Nominal Response

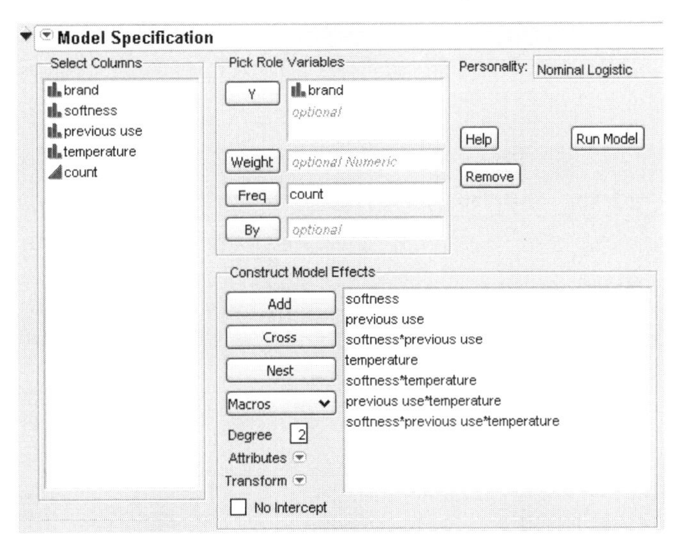

The tables in Figure 20.5 show the three-factor model as a whole to be significant (Prob>ChiSq = 0.0006) in the Whole Model table. The Effect Likelihood Ratio Tests table shows that the effects that include softness do not contribute significantly to the model fit.

Figure 20.5 Tables for Nominal Response Three-Factor Factorial

Nominal Logistic Fit for brand

Iteration History

Freq: count

Whole Model Test

Model	-LogLikelihood	DF	ChiSquare	Prob>ChiSq
Difference	16.41281	11	32.82562	0.0006*
Full	682.24780			
Reduced	698.66061			

RSquare (U)	0.0235
Observations (or Sum Wgts)	1008

Converged by Gradient

Parameter Estimates

Effect Likelihood Ratio Tests

Source	Nparm	DF	L-R ChiSquare	Prob>ChiSq
softness	2	2	0.09804239	0.9522
previous use	1	1	22.1316677	<.0001*
softness*previous use	2	2	3.78609668	0.1506
temperature	1	1	3.63914019	0.0564
softness*temperature	2	2	0.19617686	0.9066
previous use*temperature	1	1	2.26089207	0.1327
softness*previous use*temperature	2	2	0.73731691	0.6917

Next, use the Model Specification dialog again to remove the softness factor and its interactions because they don't appear to be significant. You can do this easily by double-clicking the softness factor in the Fit Model dialog. A dialog then appears asking if you want to remove the other factors that involve softness (you do).This leaves the two-factor factorial model in Figure 20.6.

Figure 20.6 A Two-factor Factorial Model with Nominal Response

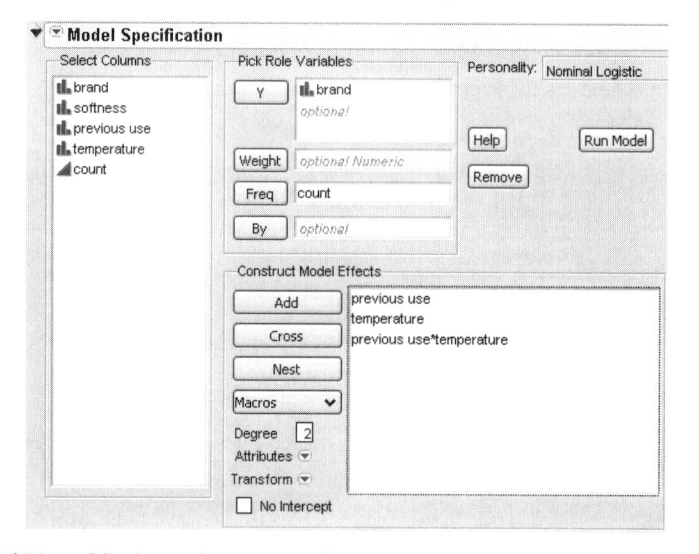

The Whole Model Test table shows that the two-factor model fits as well as the three-factor model. In fact, the three-factor Whole Model table in Figure 20.5 shows a larger Chi-square value (32.83) than the Chi-square value for the two-factor model (27.17). This results from the change in degrees of freedom used to compute the Chi-square values and their probabilities.

Whole Model Test

Model	-LogLikelihood	DF	ChiSquare	Prob>ChiSq
Difference	13.58479	3	27.16958	<.0001*
Full	685.07582			
Reduced	698.66061			

RSquare (U)	0.0194
Observations (or Sum Wgts)	1008

Converged by Objective

The report shown in Figure 20.7, supports the conclusion that previous use of a detergent brand has a strong effect on detergent preference, water temperature has a borderline significant effect, but water softness does not seem to matter.

Figure 20.7 Report for Nominal Response Two-Factor Factorial

Parameter Estimates

Term	Estimate	Std Error	ChiSquare	Prob>ChiSq
Intercept	0.03042324	0.0666826	0.21	0.6482
previous use[no]	-0.3148401	0.0666826	22.29	<.0001*
temperature[low]	-0.13311	0.0666826	3.98	0.0459*
previous use[no]*temperature[low]	0.11105724	0.0666826	2.77	0.0958

For log odds of m/x

Effect Likelihood Ratio Tests

Source	Nparm	DF	L-R ChiSquare	Prob>ChiSq
previous use	1	1	22.6511053	<.0001*
temperature	1	1	3.99625472	0.0456*
previous use*temperature	1	1	2.78794928	0.0950

Ordinal Logistic Example: The Cheese Data

If the response variable has an ordinal modeling type, the platform fits the cumulative response probabilities to the logistic function of a linear model using maximum likelihood. Likelihood-ratio test statistics are provided for the whole model and lack of fit. Wald test statistics are provided for each effect.

If there is an ordinal response and a single continuous numeric effect, the ordinal logistic platform in Fit Y by X displays a cumulative logistic probability plot.

Details of modeling types are discussed in the chapter "Understanding JMP Analyses," p. 7, and the details of fitting appear in the appendix "Statistical Details," p. 975. The method is discussed in Walker and Duncan (1967), Nelson (1976), Harrell (1986), and McCullagh and Nelder (1983).

Note: If there are many response levels, the ordinal model is much faster to fit and uses less memory than the nominal model.

As an example of ordinal logistic model fitting, McCullagh and Nelder (1983) report an experiment by Newell to test whether various cheese additives (A to D) had an effect on taste. Taste was measured by a tasting panel and recorded on an ordinal scale from 1 (strong dislike) to 9 (excellent taste). The data are in the Cheese.jmp sample data table.

Cheese
Notes Data from McCulloug
Fit Model
Contingency

Columns (4/0)
Cheese
Response
Count
Score

Rows
All rows 36

	Cheese	Response	Count	Score
1	A	1	0	-0.8613
2	A	2	0	-0.8613
3	A	3	1	-0.8613
4	A	4	7	-0.8613
5	A	5	8	-0.8613
6	A	6	8	-0.8613
7	A	7	19	-0.8613
8	A	8	8	-0.8613
9	A	9	1	-0.8613
10	B	1	6	2.4885
11	B	2	9	2.4885

To run the model, assign **Response** as **Y**, **Cheese** as the effect, and **Count** as **Freq.**

The method in this example required only six iterations to reduce the background **–LogLikelihood** of 429.9 to 355.67. This reduction yields a likelihood-ratio Chi-square for the whole model of 148.45 with 3 degrees of freedom, showing the difference in perceived cheese taste to be highly significant.

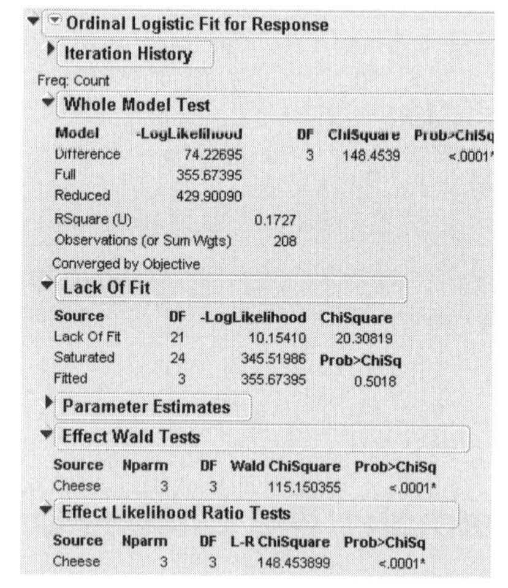
Ordinal Logistic Fit for Response
Iteration History
Freq: Count
Whole Model Test

Model	-LogLikelihood	DF	ChiSquare	Prob>ChiSq
Difference	74.22695	3	148.4539	<.0001*
Full	355.67395			
Reduced	429.90090			

RSquare (U) 0.1727
Observations (or Sum Wgts) 208
Converged by Objective

Lack Of Fit

Source	DF	-LogLikelihood	ChiSquare
Lack Of Fit	21	10.15410	20.30819
Saturated	24	345.51986	Prob>ChiSq
Fitted	3	355.67395	0.5018

Parameter Estimates
Effect Wald Tests

Source	Nparm	DF	Wald ChiSquare	Prob>ChiSq
Cheese	3	3	115.150355	<.0001*

Effect Likelihood Ratio Tests

Source	Nparm	DF	L-R ChiSquare	Prob>ChiSq
Cheese	3	3	148.453899	<.0001*

The Lack of Fit test happens to be testing the ordinal response model compared to the nominal model. This is because the model is saturated if the response is treated as nominal rather than ordinal, giving 21 additional parameters, which is the **Lack of Fit** degrees of freedom. The nonsignificance of **Lack of Fit** leads one to believe that the ordinal model is reasonable.

There are eight intercept parameters because there are nine response categories. There are only three structural parameters. As a nominal problem, there are $8 \times 3 = 24$ structural parameters.

When there is only one effect, its test is equivalent to the Likelihood-ratio test for the whole model. The Likelihood-ratio Chi-square is 148.45, different than the Wald Chi-square of 115.12, which illustrates the point that Wald tests are to be regarded with some skepticism.

Figure 20.8

A	–0.8622	2nd place
B	2.4895	least liked
C	0.8477	3rd place
D	–2.4755	most liked

To see if a cheese additive is preferred, look for the most negative values of the parameters (Cheese D's effect is the negative sum of the others).

You can also use the Fit Y by X platform for this model, which treats nominal and ordinal responses the same and shows a contingency table analysis. The error log-likelihood, shown here, agrees with the lack-of-fit saturated log-likelihood shown previously.

Contingency Analysis of Response By Cheese

Freq: Count

Tests

Source	DF	-LogLike	RSquare (U)
Model	24	84.38105	0.1963
Error	176	345.51986	
C. Total	200	429.90090	
N	208		

Test	ChiSquare	Prob>ChiSq
Likelihood Ratio	168.762	<.0001*
Pearson	162.482	<.0001*

Warning: 20% of cells have expected count less than 5, ChiSquare suspect.

If you want to see a graph of the response probabilities as a function of the parameter estimates for the four cheeses, you can create a new continuous variable (call it **Score**) with the formula shown in Figure 20.9. Use the new variable as a response surface effect in the Model Specification dialog as shown. To create the model in Figure 20.9, select **Score** in the column selector list, and then select **Response Surface** from the Macros popup menu on the Model Specification dialog.

Figure 20.9 Model Dialog For Ordinal Logistic Regression

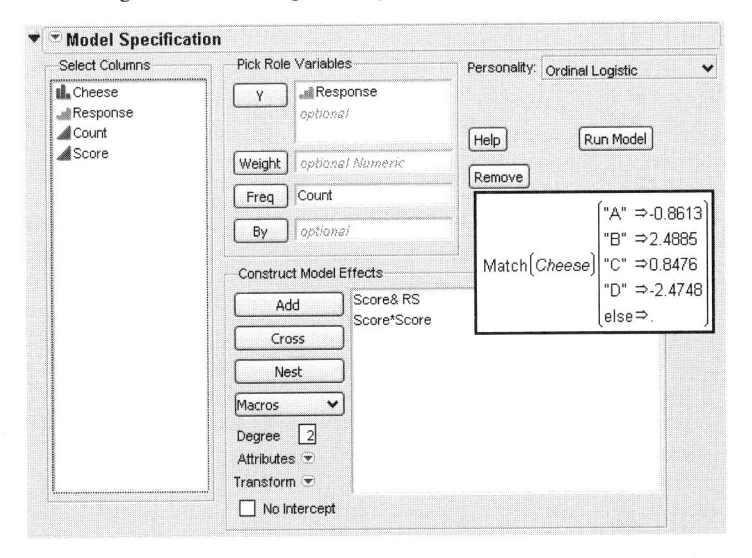

Click **Run Model** to see the analysis report and the cumulative logistic probability plot in Figure 20.10. The distance between each curve is the fitted response probability for the levels in the order for the levels on the right axis of the plot.

Figure 20.10 Cumulative Probability Plot for Ordinal Logistic Regression

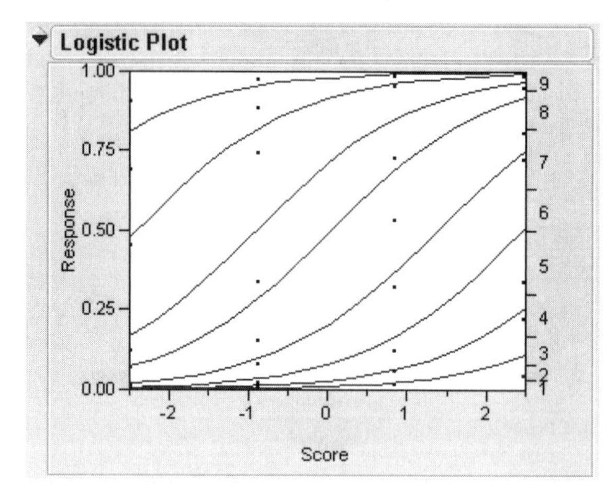

Quadratic Ordinal Logistic Example: Salt in Popcorn Data

The Ordinal Response Model can fit a quadratic surface to optimize the probabilities of the higher or lower responses. The arithmetic in terms of the structural parameters is the same as that for continuous responses. Up to five factors can be used, but this example has only one factor, for which there is a probability plot.

Consider the case of a microwave popcorn manufacturer who wants to find out how much salt consumers like in their popcorn. To do this, the manufacturer looks for the maximum probability of a favorable response as a function of how much salt is added to the popcorn package. An experiment controls salt amount at 0, 1, 2, and 3 teaspoons, and the respondents rate the taste on a scale of 1=low to 5=high. The optimum amount of salt is the amount that maximizes the probability of more favorable responses. The ten observations for each of the salt levels are shown in Table 20.1.

Table 20.1 Salt in Popcorn

Salt Amount	Salt Rating Response									
no salt	1	3	2	4	2	2	1	4	3	4
1 tsp.	4	5	3	4	5	4	5	5	4	5
2 tsp.	4	3	5	1	4	2	5	4	3	2
3 tsp.	3	1	2	3	1	2	1	2	1	2

Use **Fit Model** with the Salt in Popcorn.jmp sample data to fit the ordinal taste test to the surface effect of salt. Complete the Model Specification dialog as shown in the previous example.

The report shows how the quadratic model fits the response probabilities. The curves, instead of being shifted logistic curves, become a folded pile of curves where each achieves its optimum at the same point. The critical value is at –0.5 b1/b2 where b1 is the linear coefficient and b2 is the quadratic. From the Parameter Estimates table you can compute the optimum as

$$-0.5 \times (-3.4855)/1.3496 = 1.29$$

teaspoons of salt.

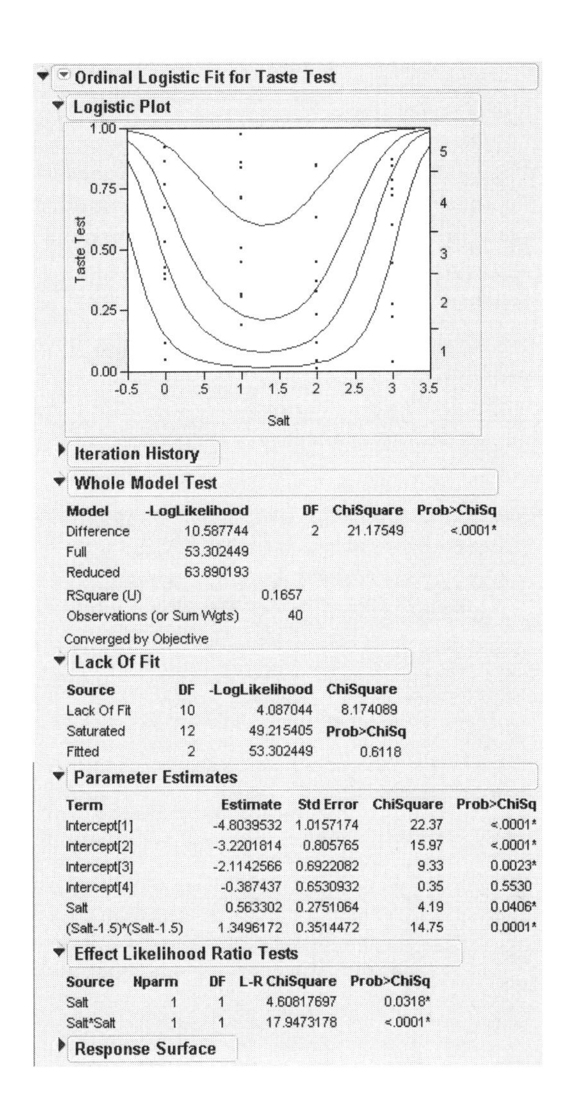

Model	-LogLikelihood	DF	ChiSquare	Prob>ChiSq
Difference	10.587744	2	21.17549	<.0001*
Full	53.302449			
Reduced	63.890193			

RSquare (U)	0.1657
Observations (or Sum Wgts)	40

Converged by Objective

Lack Of Fit

Source	DF	-LogLikelihood	ChiSquare
Lack Of Fit	10	4.087044	8.174089
Saturated	12	49.215405	Prob>ChiSq
Fitted	2	53.302449	0.6118

Parameter Estimates

Term	Estimate	Std Error	ChiSquare	Prob>ChiSq
Intercept[1]	-4.8039532	1.0157174	22.37	<.0001*
Intercept[2]	-3.2201814	0.805765	15.97	<.0001*
Intercept[3]	-2.1142566	0.6922082	9.33	0.0023*
Intercept[4]	-0.387437	0.6530932	0.35	0.5530
Salt	0.563302	0.2751064	4.19	0.0406*
(Salt-1.5)*(Salt-1.5)	1.3496172	0.3514472	14.75	0.0001*

Effect Likelihood Ratio Tests

Source	Nparm	DF	L-R ChiSquare	Prob>ChiSq
Salt	1	1	4.60817697	0.0318*
Salt*Salt	1	1	17.9473178	<.0001*

Response Surface

The distance between each curve measures the probability of each of the five response levels. The probability for the highest response level is the distance from the top curve to the top of the plot rectangle. This distance reaches a maximum when the amount of salt is about 1.3 teaspoons. All curves share the same critical point.

The parameter estimates for Salt and Salt*Salt become the coefficients used to find the critical value. Although it appears as a minimum, it is only a minimum with respect to the probability curves. It is really a maximum in the sense of maximizing the probability of the highest response.

What to Do If Your Data Are Counts in Multiple Columns

Data that are frequencies (counts) listed in several columns of your data table are not the form you need for logistic regression. For example, the Ingots2.jmp data table in the data folder (left-hand table in Figure 20.11) has columns Nready and Nnotready that give the number of ready and number of not ready ingots for each combination of Heat and Soak values. To do a logistic regression, you need the data organized like the right-hand table in Figure 20.11.

To make a the new table, select the **Stack** command from the **Tables** menu. Complete the Stack dialog by choosing Nready and NNotReady as the columns to stack, and then click **OK** in the Stack dialog. This creates the new table to the right in Figure 20.11. If you use the default column names, Label is the response (*Y*) column and Data is the frequency column.

The example in the section "Introduction to Logistic Models," p. 405, shows a logistic regression using a sample data table Ingots.jmp. It has a frequency column called count (equivalent to the Data column in the table below) and a response variable called Ready, with values 1 to represent ingots that are ready and 0 for not ready.

Figure 20.11 Changing the Form of a Data Table

Heat	Soak	Nnotready	Nready	Ntotal
7	1	10	0	10
7	1.7	17	0	17
7	2.2	7	0	7
7	2.8	12	0	12
7	4	9	0	9
14	1	31	0	31
14	1.7	43	0	43
14	2.2	31	2	33
14	2.8	31	0	31
14	4	19	0	19
27	1	55	1	56
27	1.7	40	4	44
27	2.2	21	0	21

Heat	Soak	Ntotal	Label	Data
7	1	10	Nnotready	10
7	1	10	Nready	0
7	1.7	17	Nnotready	17
7	1.7	17	Nready	0
7	2.2	7	Nnotready	7
7	2.2	7	Nready	0
7	2.8	12	Nnotready	12
7	2.8	12	Nready	0
7	4	9	Nnotready	9
7	4	9	Nready	0
14	1	31	Nnotready	31
14	1	31	Nready	0
14	1.7	43	Nnotready	43

Note: Some columns in the data tables are hidden.

Chapter 21

Multiple Response Fitting

The Fit Model Platform

When more than one *Y* is specified for a model in the Fit Model platform, you can choose the **Manova** personality for multivariate fitting. Multivariate models fit several *Y*'s to a set of effects. Functions across the Y's can be tested with appropriate response designs. In addition to standard MANOVA models, you can do the following techniques with this facility:

- Repeated measures analysis when repeated measurements are taken on each subject and you want to analyze effects both between subjects and within subjects across the measurements. This multivariate approach is especially important when the correlation structure across the measurements is arbitrary.
- Canonical correlation to find the linear combination of the *X* and *Y* variables that has the highest correlation.
- Discriminant analysis to find distance formulas between points and the multivariate means of various groups so that points can be classified into the groups that they are most likely to be in. A more complete implementation of discriminant analysis is in the **Discriminant** platform.

Multivariate fitting has a different flavor from univariate fitting. The multivariate fit begins with a rudimentary preliminary analysis that shows parameter estimates and least squares means. You can then specify a response design across the Y's, and multivariate tests are performed.

Contents

Multiple Response Model Specification

To form a multivariate model, choose **Fit Model** from the **Analyze** menu and assign multiple Y's in the dialog. Then select **Manova** from the fitting personality popup menu, as shown below, and click **Run Model**.

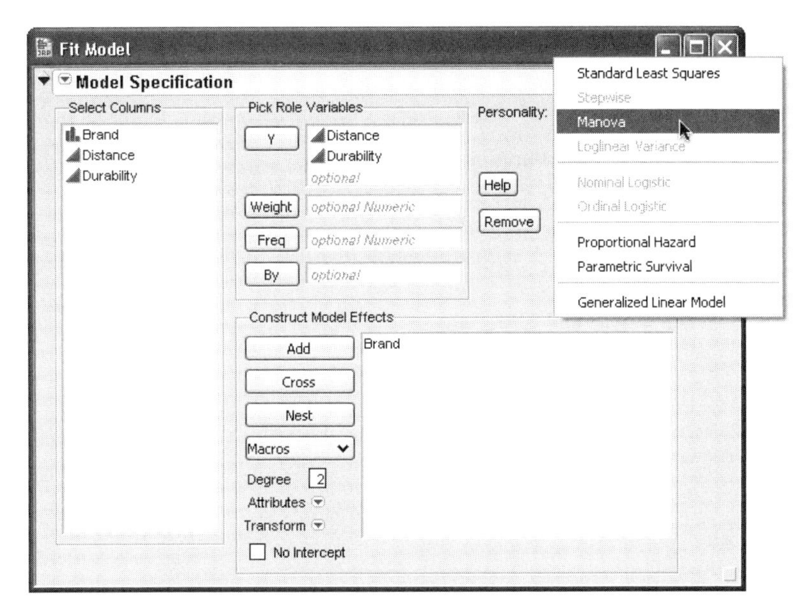

The next example uses the Golf Balls.jmp sample data table (McClave and Dietrich, 1988). The data are a comparison of distances traveled and a measure of durability for three brands of golf balls. A robotic golfer hit a random sample of ten balls for each brand in a random sequence. The hypothesis to test is that distance and durability are the same for the three golf ball brands.

Initial Fit

The multivariate fitting platform fits the responses to the effects using least squares. The figure below shows the MANOVA report for results from the initial fit. All tables are shown closed here, but some are usually open by default. These results might not be very interesting in themselves, because no response design has been specified yet. After you specify a response model, the multivariate platform displays tables of multivariate estimates and tests. The following sections discuss the initial tables, response design specification, and tables of multivariate details.

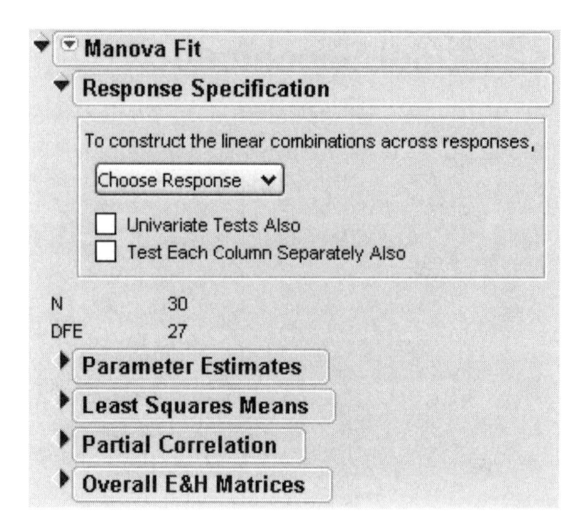

Parameter Estimates Table

The Parameter Estimates table includes only the parameter estimates for each response variable, without details like standard errors or *t*-tests. There is a column for each response variable.

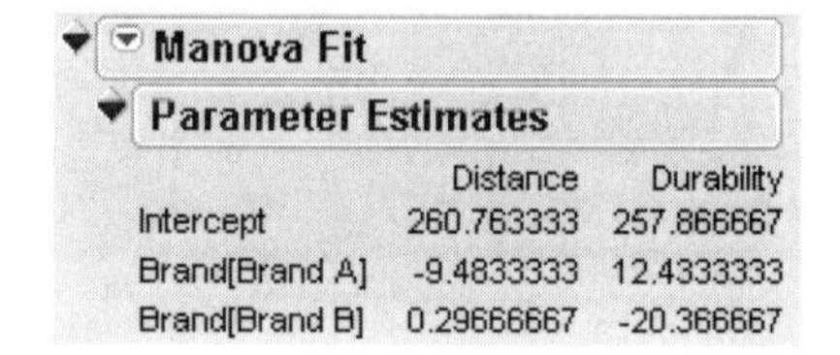

Manova Fit

Parameter Estimates

	Distance	Durability
Intercept	260.763333	257.866667
Brand[Brand A]	-9.4833333	12.4333333
Brand[Brand B]	0.29666667	-20.366667

Least Squares Means Report

For each pure nominal effect, the Least Squares Means table reports the overall least squares means of all the response columns, least squares means of each nominal level, and LS means plots of the means. Figure 21.1 shows the LS Mean plot of the golf ball brands and the table of least squares means. The same plot and table are available for the overall means, as seen by the closed reveal/conceal button.

Figure 21.1 Least Squares Means Report

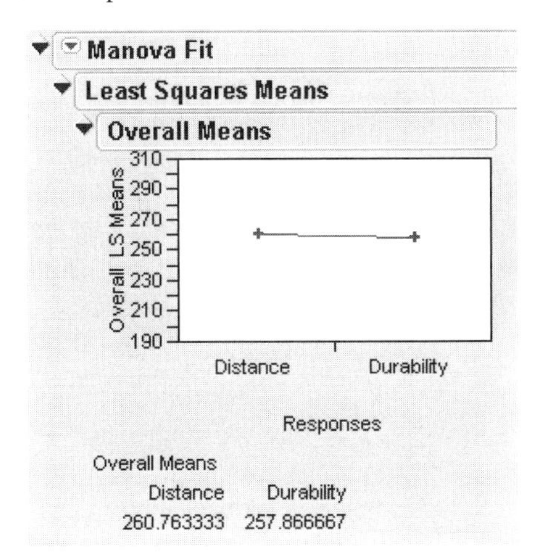

Partial Covariance and Correlation Tables

The Partial Correlation table shows the covariance matrix and the partial correlation matrix of residuals from the initial fit, adjusted for the *X* effects.

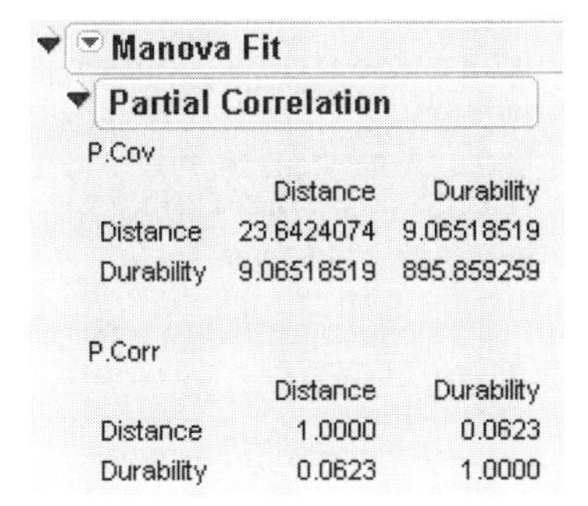
Manova Fit

Partial Correlation

P.Cov

	Distance	Durability
Distance	23.6424074	9.06518519
Durability	9.06518519	895.859259

P.Corr

	Distance	Durability
Distance	1.0000	0.0623
Durability	0.0623	1.0000

Overall E&H Matrices

The main ingredients of multivariate tests are the **E** and the **H** matrices:

- The elements of the **E** matrix are the cross products of the residuals.
- The **H** matrices correspond to hypothesis sums of squares and cross products.

There is an **H** matrix for the whole model and for each effect in the model. Diagonal elements of the **E** and **H** matrices correspond to the hypothesis (numerator) and error (denominator) sum of squares for the univariate *F* tests. New **E** and **H** matrices for any given response design are formed from these initial matrices, and the multivariate test statistics are computed from them.

Manova Fit

Overall E&H Matrices

E

	Distance	Durability
Durability	244.76	24188.2

Whole Model H

	Distance	Durability
Distance	1744.16467	-510.70667
Durability	-510.70667	6323.26667

Intercept

	Distance	Durability
Distance	2039925.48	2017265.15
Durability	2017265.15	1994856.53

Brand

	Distance	Durability
Distance	1744.16467	-510.70667
Durability	-510.70667	6323.26667

Specification of the Response Design

You use the Response Specification dialog at the top of the report to specify the response designs for various tests.

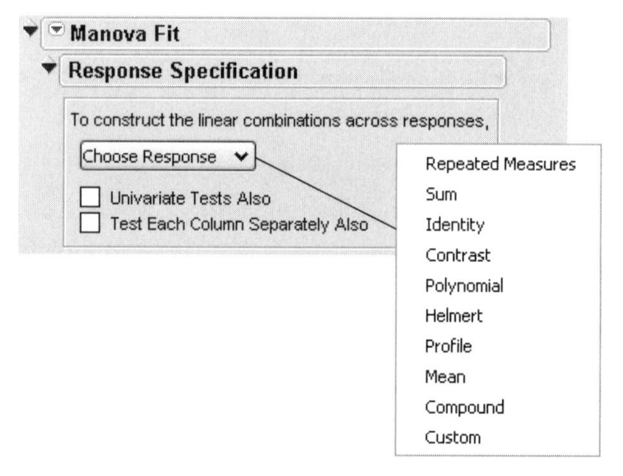

The dialog has these optional check boxes:

- The **Univariate Tests Also** check box is used in repeated measures models to obtain adjusted and unadjusted univariate repeated measures tests as well as multivariate tests.
- The **Test Each Column Separately Also** check box is used in obtaining univariate ANOVA tests on each response as well as multivariate tests.

The response design forms the **M** matrix. The columns of an **M** matrix define a set of transformation variables for the multivariate analysis. The **Choose Response** popup menu lists the choices for **M** shown below. The popup menu has the following response matrices; it also allows you the ability to build your own response with the **Custom** command.

Repeated Measures constructs and runs both Sum and Contrast responses.

Sum is the sum of the responses, giving a single value.

Identity uses each separate response, the identity matrix.

Contrast compares each response and the first response.

Polynomial constructs a matrix of orthogonal polynomials.

Helmert compares each response with the combined responses listed below it.

Profile compares each response with the following response.

Mean compares each response versus the mean of the others.

Compound creates and runs several response functions that are appropriate if the responses are compounded from two effects.

Custom uses any custom M matrix you enter.

Note: Initially in this chapter the matrix names E and H refer to the error and hypothesis cross products. After specification of a response design, E and H refer to those cross products transformed by the response design, which are actually $M'EM$ and $M'HM$.

The most typical response designs are **Repeated Measures** and **Identity** for multivariate regression. There is little difference in the tests given by **Contrast**, **Helmert**, **Profile**, and **Mean** because they span the same space. However, the tests and details in the Least Squares means and Parameter Estimates tables for them show correspondingly different highlights.

The **Repeated Measures** and the **Compound** response selections display dialogs to specify response effect names. They then continue on to fit several response functions without waiting for further user input.

Otherwise, selections expand the control panel and give you more opportunities to refine the specification. The response function is shown in the form of the M matrix, and control buttons appear at the bottom of the dialog (see the figure below). You use this part of the dialog in the following ways:

- Check the **Univariate Tests Also** check box to see the adjusted and unadjusted univariate tests as well as the multivariate tests in a repeated measures situation. Univariate tests are done with respect to an orthonormalized version of M.
- Click any value in the M matrix text edit boxes to change a value.
- Use the **Orthogonalize** button to orthonormalize the matrix. Orthonormalization is done after the column contrasts (sum to zero) for all response types except **Sum** and **Identity**.
- Use **Delete Last Column** to reduce the dimensionality of the transformation.

The figure here shows the **Identity** transformation selected from the **Response Design** popup menu and the default M matrix for the *Y* variables distance and durability. The Response Specification dialog stays open to specify further response designs for estimation and testing.

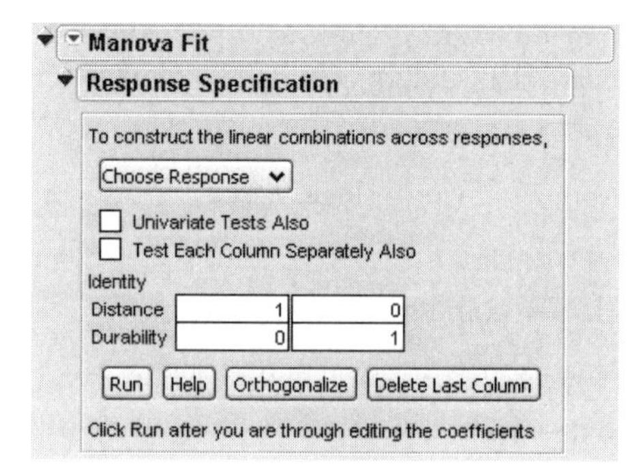

Multivariate Tests

After you complete the Response Specification dialog, click Run at the bottom of the dialog to see multivariate estimations and tests. The analysis reports will appear appended to the initial tables.

Figure 21.2 Multivariate Test Menus

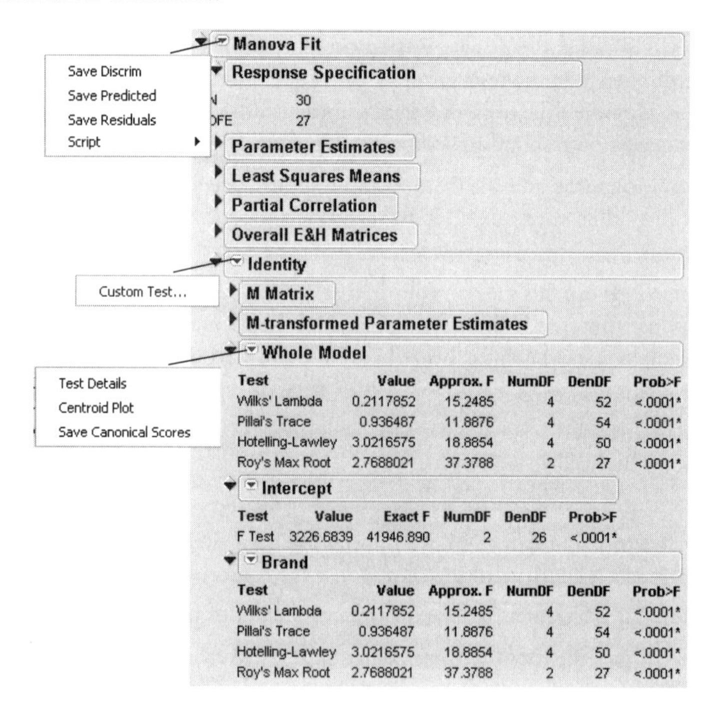

Each time you run a response design, JMP extends the report with additional tables for that design. The example to the right shows the outline for the Identity design and these extended report tables:

- the **M** Matrix response design
- the **M**-transformed parameter estimates

- whole model tests for all effects jointly except the intercept tests for the Intercept and each effect in the model.

In addition, the popup command next to the response matrix name lets you construct custom tests. Menus next to the model effects also give additional options, described later.

The Extended Multivariate Report

In multivariate fits, the sums of squares due to hypothesis and error are matrices of squares and cross products instead of single numbers. And there are lots of ways to measure how large a value the matrix for the hypothesis sums of squares and cross products (called **H** or **SSCP**) is compared to that matrix for the residual (called **E**). JMP reports the four multivariate tests that are commonly described in the literature. If you are looking for a test at an exact significance level, you will have to go hunting for tables in reference books. Fortunately, all four tests can be transformed into an approximate *F*-test. If the response design yields a single value, or if the hypothesis is a single degree of freedom, the multivariate tests are equivalent and yield the same exact *F*-test. JMP labels the text Exact F; otherwise, JMP labels it Approx F.

In the golf balls example described at the beginning of this chapter, there is only one effect so the Whole Model test and the test for Brand are the same, which show the four multivariate tests with approximate F tests. There is only a single intercept with two DF (one for each response), so the *F*-test for it is exact and is labeled Exact F.

The popup menu icons next to the Whole Model, Intercept, and Brand table names access commands to generate additional information, which includes eigenvalues, canonical correlations, a list of centroid values, a centroid plot, and a **Save** command that lets you save canonical variates. These options are discussed in the section "Commands for Response Type and Effects," p. 443.

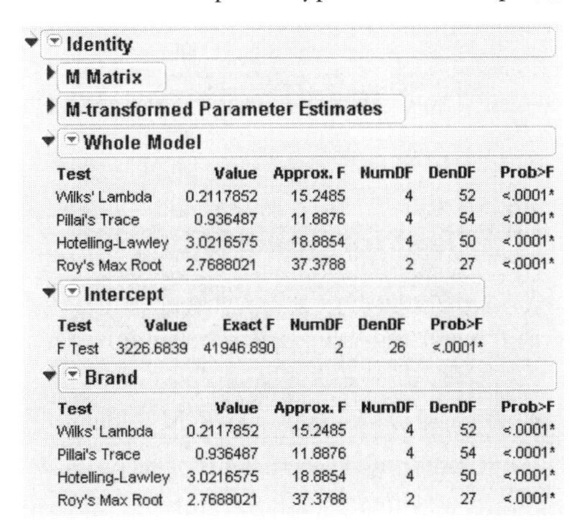

Identity

M Matrix

M-transformed Parameter Estimates

Whole Model

Test	Value	Approx. F	NumDF	DenDF	Prob>F
Wilks' Lambda	0.2117852	15.2485	4	52	<.0001*
Pillai's Trace	0.936487	11.8876	4	54	<.0001*
Hotelling-Lawley	3.0216575	18.8854	4	50	<.0001*
Roy's Max Root	2.7688021	37.3788	2	27	<.0001*

Intercept

Test	Value	Exact F	NumDF	DenDF	Prob>F
F Test	3226.6839	41946.890	2	26	<.0001*

Brand

Test	Value	Approx. F	NumDF	DenDF	Prob>F
Wilks' Lambda	0.2117852	15.2485	4	52	<.0001*
Pillai's Trace	0.936487	11.8876	4	54	<.0001*
Hotelling-Lawley	3.0216575	18.8854	4	50	<.0001*
Roy's Max Root	2.7688021	37.3788	2	27	<.0001*

The effect (Brand in this example) popup menu also includes the option to specify contrasts.

The custom test and contrast features are the same as those for regression with a single response. See the chapters "Standard Least Squares: Introduction," p. 213, and "Standard Least Squares: Perspectives on the Estimates," p. 249, for a description and examples of contrasts and custom tests.

"Multivariate Details," p. 994 in the appendix "Statistical Details," p. 975, shows formulas for the MANOVA table tests.

Each MANOVA test table, except the Sphericity Test table, has these elements:

Test labels each statistical test in the table. If the number of response function values (columns specified in the **M** matrix) is 1 or if an effect has only one degree of freedom per response function, the exact *F*-test is presented. Otherwise, the standard four multivariate test statistics are given with approximate F tests: Wilk's Lambda (Λ), Pillai's Trace, the Hotelling-Lawley Trace, and Roy's Maximum Root.

Value the value of each multivariate statistical test in the report.

Approx. F (or Exact F) the *F*-values corresponding to the multivariate tests. If the response design yields a single value or if the test is one degree of freedom, this will be an exact *F*-test.

NumDF the numerator degrees of freedom.

DenDF the denominator degrees of freedom.

Prob>F the significance probability corresponding to the *F*-value.

Comparison of Multivariate Tests

Although the four standard multivariate tests often give similar results, there are situations where they differ, and one may have advantages over another. Unfortunately, there is no clear winner. In general, the order of preference in terms of power is

1 Pillai's Trace
2 Wilk's Lambda
3 Hotelling-Lawley Trace
4 Roy's Maximum Root.

When there is a large deviation from the null hypothesis and the eigenvalues differ widely, the order of preference is the reverse (Seber 1984).

Univariate Tests and the Test for Sphericity

There are cases, such as a repeated measures model, that allow transformation of a multivariate problem into a univariate problem (Huynh and Feldt 1970). Using univariate tests in a multivariate context is valid

- if the response design matrix **M** is orthonormal (**M′M** = Identity).
- if **M** yields more than one response the coefficients of each transformation sum to zero.
- if the *sphericity* condition is met. The sphericity condition means that the M-transformed responses are uncorrelated and have the same variance. **M′ΣM** is proportional to an identity matrix, where Σ is the covariance of the Y's.

If these conditions hold, the diagonal elements of the **E** and **H** test matrices sum to make a univariate sums of squares for the denominator and numerator of an *F*-test.

When you check the **Univariate Tests Also** check box in the Response Specification dialog, you see a Sphericity Test table, like the one shown here, and adjusted univariate *F* tests in the multivariate report tables.

Sphericity Test	
Mauchly Criterion	0.1752641
ChiSquare	16.930873
DF	5
Prob >Chisq	0.0046328

The sphericity test checks the appropriateness of an unadjusted univariate *F*-test for the within-subject effects using the Mauchly criterion to test the sphericity assumption (Anderson 1958). The sphericity test and the univariate tests are always done using an orthonormalized **M** matrix. You interpret the sphericity test as follows:

- If the sphericity Chi-square test is not significant, you can use the unadjusted univariate F tests.
- If the sphericity test is significant, use the multivariate or the adjusted univariate tests.

The univariate *F*-statistic has an approximate *F*-distribution even without sphericity, but the degrees of freedom for numerator and denominator are reduced by some fraction epsilon (ε) Box (1954), Geisser and Greenhouse (1958), and Huynh-Feldt (1976) offer techniques for estimating the epsilon degrees-of-freedom adjustment. Muller and Barton (1989) recommend the Geisser-Greenhouse version, based on a study of power.

The epsilon adjusted tests in the multivariate report are labeled G-G (Greenhouse-Geisser) or H-F (Huynh-Feldt), with the epsilon adjustment shown in the value column.

Multivariate Model with Repeated Measures

One common use of multivariate fitting is to analyze data with repeated measures, also called *longitudinal data*. A subject is measured repeatedly across time, and the data are arranged so that each of the time measurements form a variable. Because of correlation between the measurements, data should not be stacked into a single column and analyzed as a univariate model unless the correlations form a pattern termed *sphericity*. See the previous section, "Univariate Tests and the Test for Sphericity," p. 436, for more details about this topic.

With repeated measures, the analysis is divided into two layers:

- Between-subject (or across-subject) effects are modeled by fitting the sum of the repeated measures columns to the model effects. This corresponds to using the **Sum** response function, which is an M-matrix that is a single vector of 1's.
- Within-subjects effects (repeated effects, or time effects) are modeled with a response function that fits differences in the repeated measures columns. This analysis can be done using the **Contrast** response function or any of the other similar differencing functions: **Polynomial**, **Helmert**, **Profile**, or **Mean**. When you model differences across the repeated measures, think of the differences as being a new within-subjects effect, usually time. When you fit effects in the model, interpret them as the interaction with the within-subjects effect. For example, the effect for Intercept becomes the Time (within-subject) effect, showing overall differences across the repeated measures. If you have

an effect A, the within-subjects tests are interpreted to be the tests for the A*Time interaction, which model how the differences across repeated measures vary across the A effect.

Table 21.1 "Corresponding Multivariate and Univariate Tests," p. 439, shows the relationship between the response function and the model effects compared with what a univariate model specification would be. Using both the **Sum** (between-subjects) and **Contrast** (within-subjects) models, you should be able to reconstruct the tests that would have resulted from stacking the responses into a single column and obtaining a standard univariate fit.

There is a direct and an indirect way to perform the repeated measures analyses:

- The direct way is to use the popup menu item Repeated Measures. This prompts you to name the effect that represents the within-subject effect across the repeated measures. Then it fits both the Contrast and the Sum response functions. An advantage of this way is that the effects are labeled appropriately with the within-subjects effect name.
- The indirect way is to specify the two response functions individually. First, do the Sum response function and second, do either Contrast or one of the other functions that model differences. You will have to remember to associate the within-subjects effect with the model effects in the contrast fit.

Repeated Measures Example

As an example, consider a study by Cole and Grizzle (1966). The results are in the Dogs.jmp table in the sample data folder. Sixteen dogs are assigned to four groups defined by variables drug and dep1, each having two levels. The dependent variable is the blood concentration of histamine at 0, 1, 3, and 5 minutes after injection of the drug. The log of the concentration is used to minimize the correlation between the mean and variance of the data.

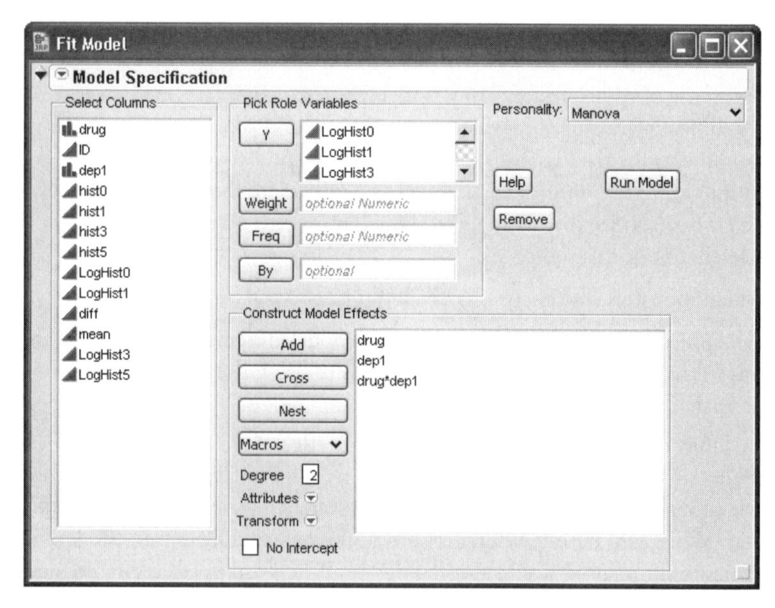

For repeated measures, a repeated measures design on the control panel popup menu (shown previously) is specified, which then displays a dialog like the one here.

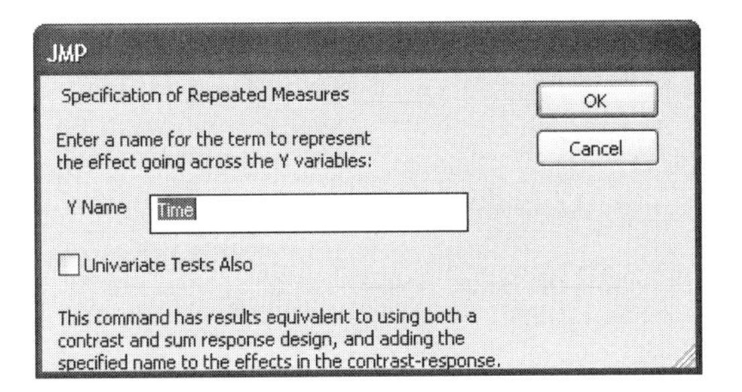

If you check the **Univariate Tests Also** check box, the report includes tests corresponding to the univariate tests shown in Figure 21.2. Univariate tests are calculated as if the responses were stacked into a single column.

Table 21.1 "Corresponding Multivariate and Univariate Tests," p. 439, shows how the multivariate tests for a **Sum** and **Contrast** response designs correspond to how univariate tests would be labeled if the data for columns LogHist0, LogHist1, LogHist3, and LogHist5 were stacked into a single *Y* column, with the new rows identified with a nominal grouping variable, Time.

Table 21.1 Corresponding Multivariate and Univariate Tests

Sum M-Matrix Between Subjects		Contrast M-Matrix Within Subjects	
Multivariate Test	**Univariate Test**	**Multivariate Test**	**Univariate Test**
intercept	intercept	intercept	time
drug	drug	drug	time*drug
depl	depl	depl	time*depl

The between-subjects analysis is produced first. This analysis is the same (except titling) as it would have been if **Sum** had been selected on the popup menu.

The within-subjects analysis is produced next. This analysis is the same (except titling) as it would have been if **Contrast** had been selected on the popup menu, though the within-subject effect name (Time) has been added to the effect names in the report. Note that the position formerly occupied by Intercept is Time, because the intercept term is estimating overall differences across the repeated measurements.

Figure 21.3 Repeated Measures Analysis of Dogs Data

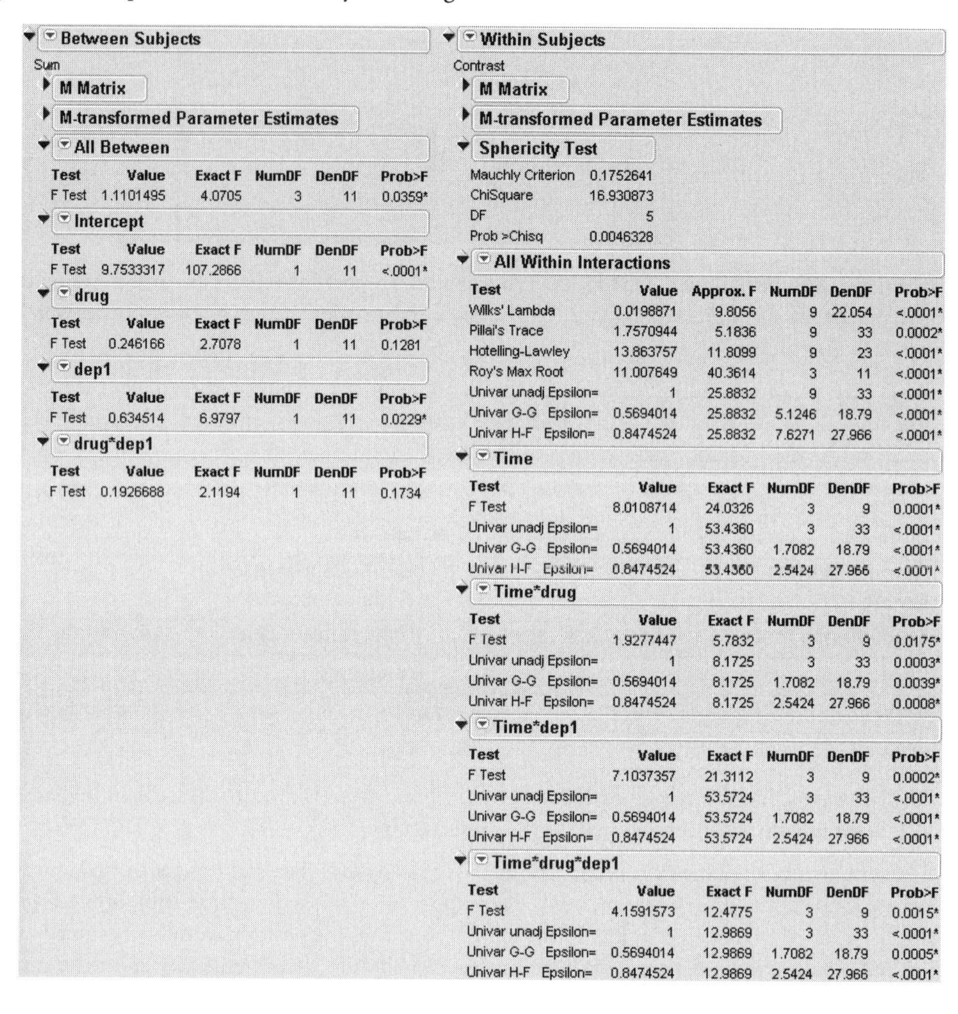

Between Subjects

Sum

M Matrix

M-transformed Parameter Estimates

All Between

Test	Value	Exact F	NumDF	DenDF	Prob>F
F Test	1.1101495	4.0705	3	11	0.0359*

Intercept

Test	Value	Exact F	NumDF	DenDF	Prob>F
F Test	9.7533317	107.2866	1	11	<.0001*

drug

Test	Value	Exact F	NumDF	DenDF	Prob>F
F Test	0.246166	2.7078	1	11	0.1281

dep1

Test	Value	Exact F	NumDF	DenDF	Prob>F
F Test	0.634514	6.9797	1	11	0.0229*

drug*dep1

Test	Value	Exact F	NumDF	DenDF	Prob>F
F Test	0.1926688	2.1194	1	11	0.1734

Within Subjects

Contrast

M Matrix

M-transformed Parameter Estimates

Sphericity Test

Mauchly Criterion	0.1752641
ChiSquare	16.930873
DF	5
Prob >Chisq	0.0046328

All Within Interactions

Test	Value	Approx. F	NumDF	DenDF	Prob>F
Wilks' Lambda	0.0198871	9.8056	9	22.054	<.0001*
Pillai's Trace	1.7570944	5.1836	9	33	0.0002*
Hotelling-Lawley	13.863757	11.8099	9	23	<.0001*
Roy's Max Root	11.007649	40.3614	3	11	<.0001*
Univar unadj Epsilon=	1	25.8832	9	33	<.0001*
Univar G-G Epsilon=	0.5694014	25.8832	5.1246	18.79	<.0001*
Univar H-F Epsilon=	0.8474524	25.8832	7.6271	27.966	<.0001*

Time

Test	Value	Exact F	NumDF	DenDF	Prob>F
F Test	8.0108714	24.0326	3	9	0.0001*
Univar unadj Epsilon=	1	53.4360	3	33	<.0001*
Univar G-G Epsilon=	0.5694014	53.4360	1.7082	18.79	<.0001*
Univar H-F Epsilon=	0.8474524	53.4360	2.5424	27.966	<.0001*

Time*drug

Test	Value	Exact F	NumDF	DenDF	Prob>F
F Test	1.9277447	5.7832	3	9	0.0175*
Univar unadj Epsilon=	1	8.1725	3	33	0.0003*
Univar G-G Epsilon=	0.5694014	8.1725	1.7082	18.79	0.0039*
Univar H-F Epsilon=	0.8474524	8.1725	2.5424	27.966	0.0008*

Time*dep1

Test	Value	Exact F	NumDF	DenDF	Prob>F
F Test	7.1037357	21.3112	3	9	0.0002*
Univar unadj Epsilon=	1	53.5724	3	33	<.0001*
Univar G-G Epsilon=	0.5694014	53.5724	1.7082	18.79	<.0001*
Univar H-F Epsilon=	0.8474524	53.5724	2.5424	27.966	<.0001*

Time*drug*dep1

Test	Value	Exact F	NumDF	DenDF	Prob>F
F Test	4.1591573	12.4775	3	9	0.0015*
Univar unadj Epsilon=	1	12.9869	3	33	<.0001*
Univar G-G Epsilon=	0.5694014	12.9869	1.7082	18.79	0.0005*
Univar H-F Epsilon=	0.8474524	12.9869	2.5424	27.966	<.0001*

A Compound Multivariate Model

JMP can handle data with layers of repeated measures. For example, see the table called Cholesterol.jmp in the Sample Data folder (Figure 21.4). Groups of five subjects belong to one of four treatment groups called A, B, Control, and Placebo. Cholesterol was measured in the morning and again in the afternoon once a month for three months (the data are fictional). In this example the response columns are arranged chronologically with time of day within month. The columns could have been ordered with AM in the first three columns followed by three columns of PM measures.

Figure 21.4 Data for a Compound Multivariate Design

Cholesterol
Notes Fictional data-Exar
Fit Model
Columns (7/0)
treatment
April AM
April PM
May AM
May PM
June AM
June PM
Rows
All rows 20

	treatment	April AM	April PM	May AM	May PM	June AM	June PM
1	A	278	280	204	208	171	175
2	A	278	281	195	199	185	189
3	A	276	280	213	219	179	181
4	A	276	281	201	211	183	188
5	A	279	285	188	192	170	174
6	B	266	270	220	224	180	184
7	B	280	284	228	232	200	204
8	B	284	288	233	237	175	179
9	B	273	277	215	219	202	207
10	B	281	285	237	241	199	205
11	Control	278	282	273	277	281	285
12	Control	273	277	274	278	280	282
13	Control	282	285	276	281	279	282

To analyze this experimental design, complete the Model Specification dialog by declaring treatment as the only effect and all six response columns as *Y* variables. When you click **Run Model**, the four tables for the initial fit appear.

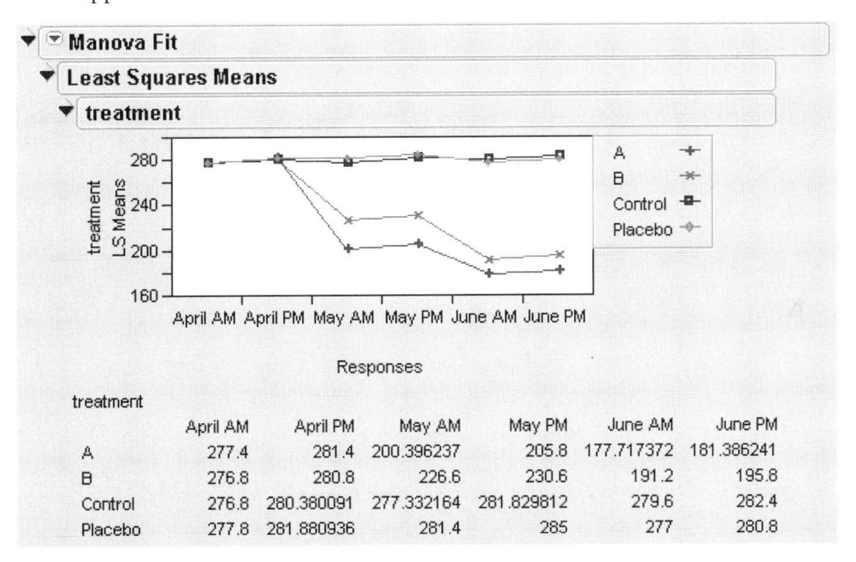

treatment	April AM	April PM	May AM	May PM	June AM	June PM
A	277.4	281.4	200.396237	205.8	177.717372	181.386241
B	276.8	280.8	226.6	230.6	191.2	195.8
Control	276.8	280.380091	277.332164	281.829812	279.6	282.4
Placebo	277.8	281.880936	281.4	285	277	280.8

You can see in the Least Squares Means table that the four treatment groups began the study with very similar mean cholesterol values. The A and B treatment groups appear to have lower cholesterol values at the end of the trial period. The control and placebo groups remain unchanged.

Next, choose **Compound** from the response design popup menu. You complete this dialog to tell JMP how the responses are arranged in the data table and the number of levels of each response. In the cholesterol example, the time of day columns are arranged within month. Therefore, you name time of day in the first box and the month effect in the second box (see Figure 21.5). Note that testing the interaction effect is optional.

Figure 21.5 Compound Dialog

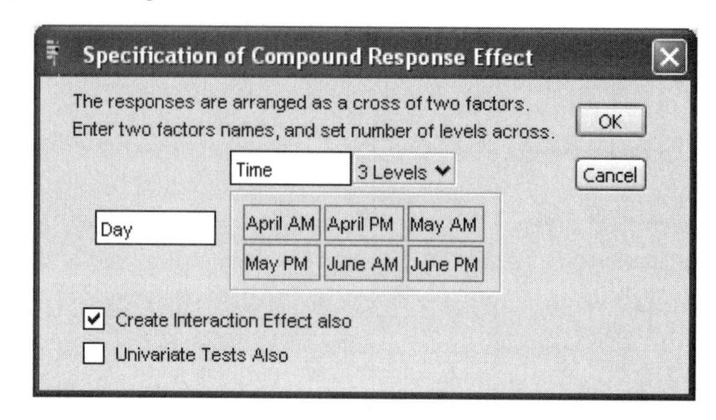

When you click **OK**, the tests for each effect display. Parts of the statistical report are shown in Figure 21.6.

- The time of day that the measurement is taken has no significant effect. Note that because there are only two levels of time measurement, the exact *F*-test is shown.
- The second table in Figure 21.6, shows a significant difference in monthly cholesterol levels among the treatment groups.
- The interaction between time of day and treatment, shown by the table at the bottom of the figure, is not significant.

Figure 21.6 Cholesterol Study Results

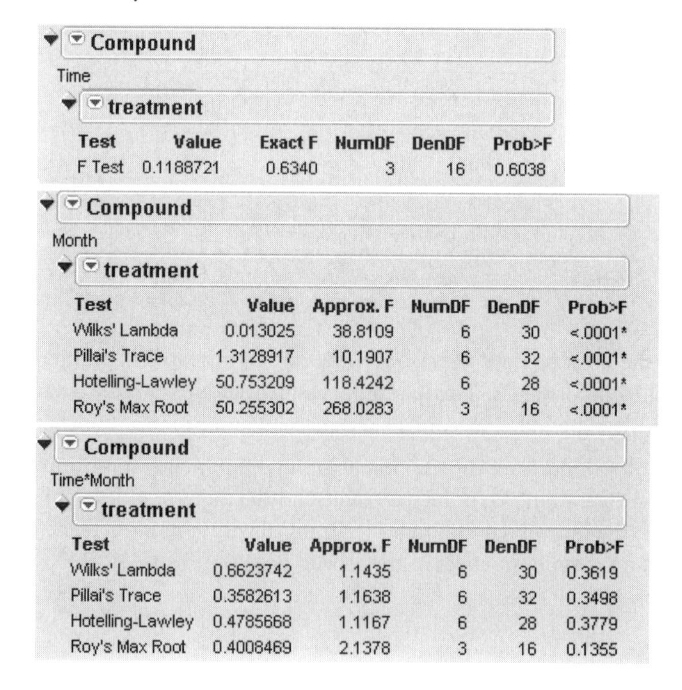

Compound

Time

treatment

Test	Value	Exact F	NumDF	DenDF	Prob>F
F Test	0.1188721	0.6340	3	16	0.6038

Compound

Month

treatment

Test	Value	Approx. F	NumDF	DenDF	Prob>F
Wilks' Lambda	0.013025	38.8109	6	30	<.0001*
Pillai's Trace	1.3128917	10.1907	6	32	<.0001*
Hotelling-Lawley	50.753209	118.4242	6	28	<.0001*
Roy's Max Root	50.255302	268.0283	3	16	<.0001*

Compound

Time*Month

treatment

Test	Value	Approx. F	NumDF	DenDF	Prob>F
Wilks' Lambda	0.6623742	1.1435	6	30	0.3619
Pillai's Trace	0.3582613	1.1638	6	32	0.3498
Hotelling-Lawley	0.4785668	1.1167	6	28	0.3779
Roy's Max Root	0.4008469	2.1378	3	16	0.1355

Commands for Response Type and Effects

The **Custom Test** popup command (Figure 21.2 of "Multivariate Tests," p. 434) next to the name of the response type name displays the Custom Test dialog for setting up custom tests of effect levels. See "Custom Test," p. 255 IN THE "STANDARD LEAST SQUARES: PERSPECTIVES ON THE ESTIMATES" CHAPTER for a description of how to use the Custom Test dialog to create custom tests.

The popup menu icon beside each effect name gives you the commands shown here, to request additional information about the multivariate fit:

Test Details displays the eigenvalues and eigenvectors of the $E^{-1}H$ matrix used to construct multivariate test statistics.

Centroid Plot plots the centroids (multivariate least-squares means) on the first two canonical variables formed from the test space.

Save Canonical Scores saves variables called canon[1], canon[2], and so on, as columns in the current data table. These columns have both the values and their formulas.

Contrast performs the statistical contrasts of treatment levels that you specify in the contrasts dialog. **Note:** The **Contrast** command is the same as for regression with a single response. See the chapter "Standard Least Squares: Perspectives on the Estimates," p. 249, for a description and examples of the **Custom Test** and **Contrast** commands.

Test Details (Canonical Details)

As an example, open Fisher's Iris data, Iris.jmp, found in the Sample Data folder (Mardia, Kent, and Bibby 1979). The Iris data have three levels of Species named Virginica, Setosa, and Versicolor. There are four measures (Petal length, Petal width, Sepal length, and Sepal width) taken on each sample. Fit a multivariate model (**Identity**) for the Species effect, with the four petal and sepal measures assigned as responses (Y). Then select the **Test Details** command from the Species popup menu.

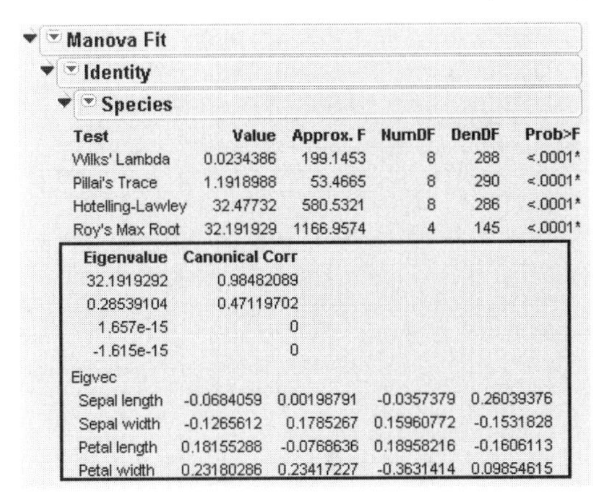

Manova Fit
Identity
Species

Test	Value	Approx. F	NumDF	DenDF	Prob>F
Wilks' Lambda	0.0234386	199.1453	8	288	<.0001*
Pillai's Trace	1.1918988	53.4665	8	290	<.0001*
Hotelling-Lawley	32.47732	580.5321	8	286	<.0001*
Roy's Max Root	32.191929	1166.9574	4	145	<.0001*

Eigenvalue	Canonical Corr
32.1919292	0.98482089
0.28539104	0.47119702
1.657e-15	0
-1.615e-15	0

Eigvec				
Sepal length	-0.0684059	0.00198791	-0.0357379	0.26039376
Sepal width	-0.1265612	0.1785267	0.15960772	-0.1531828
Petal length	0.18155288	-0.0768636	0.18958216	-0.1606113
Petal width	0.23180286	0.23417227	-0.3631414	0.09854615

The **Test Details** command appends the canonical details for an effect to its test report (as shown here outlined in black). This additional information lists the eigenvalues and eigenvectors of the $E^{-1}H$ matrix used for construction of the multivariate test statistics.

EigenValue lists the eigenvalues of the $E^{-1}H$ matrix used in computing the multivariate test statistics.

Canonical Corr lists the canonical correlations associated with each eigenvalue. This is the canonical correlation of the transformed responses with the effects, corrected for all other effects in the model.

Eigvec lists the eigenvectors of the $E^{-1}H$ matrix, or equivalently of $(E + \mathbf{H})^{-1}H$.

The Centroid Plot

The **Centroid Plot** command plots the centroids (multivariate least-squares means) on the first two canonical variables formed from the test space, as in Figure 21.7. The first canonical axis is the vertical axis so that if the test space is only one dimensional the centroids align on a vertical axis. The centroid points appear with a circle corresponding to the 95% confidence region (Mardia, Kent, and Bibby, 1980). When centroid plots are created under effect tests, circles corresponding to the effect being tested appear in red. Other circles appear blue. Biplot rays show the directions of the original response variables in the test space.

Click the **Centroid Val** disclosure icon to append additional information (shown in Figure 21.7) to the text report.

In the centroid plot in Figure 21.7, the first canonical axis with an eigenvalue accounts for much more separation than does the second axis. The means are well separated (discriminated), with the first group farther apart than the other two. The first canonical variable seems to load the petal length variables against the petal width variables. Relationships among groups of variables can be verified with Biplot Rays and the associated eigenvectors.

Figure 21.7 Centroid Plot and Centroid Values

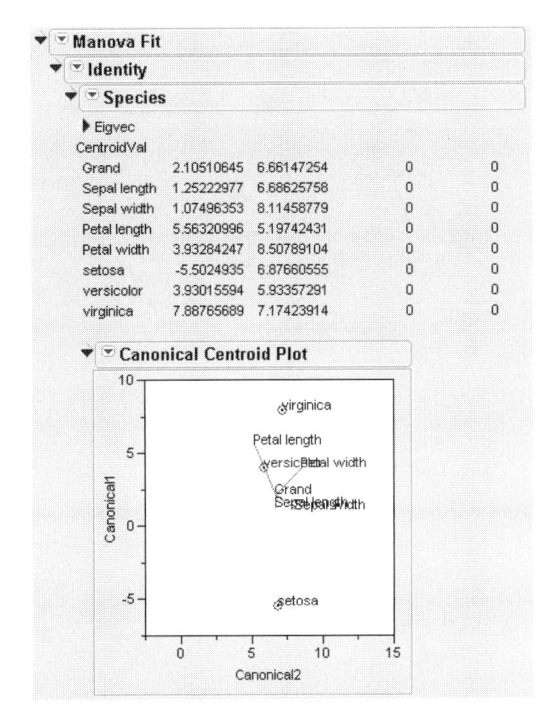

CentroidVal				
Grand	2.10510645	6.66147254	0	0
Sepal length	1.25222977	6.68625758	0	0
Sepal width	1.07496353	8.11458779	0	0
Petal length	5.56320996	5.19742431	0	0
Petal width	3.93284247	8.50789104	0	0
setosa	-5.5024935	6.87660555	0	0
versicolor	3.93015594	5.93357291	0	0
virginica	7.88765689	7.17423914	0	0

Save Canonical Scores (Canonical Correlation)

Canonical correlation analysis is not a specific command, but it can be done by a sequence of commands in the multivariate fitting platform:

- Choose the **Fit Model** command.
- Specify Y's and X's, select **Manova** from the **Personality** drop-down menu, and click **Run Model**.
- Choose the **Identity** as your multivariate function in the Choose Response Selection dialog.
- Select **Test Details** and then **Save Canonical Scores** from the Whole Model popup commands. The details list the canonical correlations (Canonical Corr) next to the eigenvalues. The saved variables are called canon[1], canon[2], and so on. These columns contain both the values and their formulas.
- To obtain the canonical variables for the *X* side, repeat the same steps but interchange the *X* and *Y* variables. If you already have the columns canon[n], the new columns are called canon[1]2 (or another number) appended makes the name unique.

For example, try the Linnerud data from Rawlings (1988) called Exercise.jmp found in the Sample Data folder. It has the physiological *X* variables weight, waist, and pulse, and the exercise *Y* variables chins, situps, and jumps. The table below shows how the Whole Model table looks after the details are requested.

Whole Model

Test	Value	Approx. F	NumDF	DenDF	Prob>F
Wilks' Lambda	0.3503905	2.0482	9	34.223	0.0635
Pillai's Trace	0.6784815	1.5587	9	48	0.1551
Hotelling-Lawley	1.7719415	2.4938	9	38	0.0238*
Roy's Max Root	1.7247387	9.1986	3	16	0.0009*

Eigenvalue	Canonical Corr
1.72473874	0.79560815
0.0419084	0.20055604
0.00529433	0.07257029

Eigvec

chins	0.02503681	-0.016636	0.05641878
situps	0.00637953	0.0004622	-0.004547
jumps	-0.0052909	0.0048507	0.0018787

The output canonical variables use the eigenvectors shown as the linear combination of the *Y* variables. For example, the formula for canon[1] is

```
0.025036802*chins + 0.0063795*situps + -0.0052909*jumps
```

This canonical analysis does not produce a standardized variable with mean 0 and standard deviation 1, but it is easy to define a new standardized variable with the calculator that has these features.

Interchanging the *X* and *Y* variables gives the corresponding report, shown here. Note that the canonical correlations are the same as those in the table above, but the eigenvectors concern the other variables.

Whole Model

Test	Value	Approx. F	NumDF	DenDF	Prob>F
Wilks' Lambda	0.3503905	2.0482	9	34.223	0.0635
Pillai's Trace	0.6784815	1.5587	9	48	0.1551
Hotelling-Lawley	1.7719415	2.4938	9	38	0.0238*
Roy's Max Root	1.7247387	9.1986	3	16	0.0009*

Eigenvalue	Canonical Corr
1.72473874	0.79560815
0.0419084	0.20055604
0.00529433	0.07257029

Eigvec

weight	-0.0118927	0.01787202	-0.0017792
waist	0.18678646	-0.0863452	0.03635125
pulse	-0.003105	0.00750573	0.03352166

You can use the **Multivariate** command in the **Analyze > Multivariate Methods** menu to look at a correlation table for the canonical variables. The Multivariate platform shows zero correlations between each canonical variable except for the correlations between the two sets, which are the canonical correlations.

Multivariate

Correlations

	Canon[1]	Canon[2]	Canon[3]	Canon[1] 2	Canon[2] 2	Canon[3] 2
Canon[1]	1.0000	0.0000	0.0000	-0.7956	0.0000	0.0000
Canon[2]	0.0000	1.0000	0.0000	0.0000	-0.2006	0.0000
Canon[3]	0.0000	0.0000	1.0000	0.0000	0.0000	-0.0726
Canon[1] 2	-0.7956	0.0000	0.0000	1.0000	0.0000	0.0000
Canon[2] 2	0.0000	-0.2006	0.0000	0.0000	1.0000	0.0000
Canon[3] 2	0.0000	0.0000	-0.0726	0.0000	0.0000	1.0000

The signs of the correlations are arbitrary because the signs of their respective eigenvectors are arbitrary.

Discriminant Analysis

Discriminant analysis is a method of predicting some level of a one-way classification based on known values of the responses. The technique is based on how close a set of measurement variables are to the multivariate means of the levels being predicted. Discriminant analysis is more fully implemented using the Discriminant Platform("Discriminant Analysis," p. 593).

In JMP you specify the measurement variables as *Y* effects and the classification variable as a single *X* effect. The multivariate fitting platform gives estimates of the means and the covariance matrix for the data, assuming the covariances are the same for each group. You obtain discriminant information with the **Save Discrim** command in the popup menu next to the MANOVA platform name. This command saves distances and probabilities as columns in the current data table using the initial **E** and **H** matrices.

For a classification variable with *k* levels, JMP adds *k* distance columns, *k* classification probability columns, the predicted classification column, and two columns of other computational information to the current data table.

Again use Fisher's Iris data (Iris.jmp in the Sample Data folder) as found in Mardia, Kent, and Bibby. There are *k* = 3 levels of species and four measures on each sample. The **Save Discrim** command adds the following nine columns to the Iris.jmp data table.

Dist[0] is the quadratic form needed in the Mahalanobis distance calculations.

Dist[SETOSA] is the Mahalanobis distance of the observation from the Setosa centroid.

Dist[VERSICOLOR] is the Mahalanobis distance of the observation from the Versicolor centroid.

Dist[VIRGINICA] is the Mahalanobis distance of the observation from the Virginica centroid.

Prob[0] is the sum of the negative exponentials of the Mahalanobis distances, used below.

Prob[SETOSA] is posterior probability of being in the Setosa category.

Prob[VERSICOLOR] is posterior probability of being in the Versicolor category.

Prob[VIRGINICA] is posterior probability of being in the Virginica category.

Pred Species is the species that is most likely from the probabilities.

You can use the new columns in the data table with other JMP platforms to summarize the discriminant analysis with reports and graphs. For example, if you use the Fit Y by X platform and specify Pred Species as *X* and Species as *Y* variable, the Contingency Table report shown here summarizes the discriminant classifications.

	Pred Species				
	Count Total %	setosa	versicolor	virginica	
Species	setosa	50 33.33	0 0.00	0 0.00	50 33.33
	versicolor	0 0.00	48 32.00	2 1.33	50 33.33
	virginica	0 0.00	1 0.67	49 32.67	50 33.33
		50 33.33	49 32.67	51 34.00	150

If you want to plot the posterior probabilities for each species, use the Fit Y by X platform.

The *Y* variables are the probability scores for each group and the *X* variable is the classification variable. The example shown here plots the probability of classification as Versicolor for each species. The Pred Species column is the label and identifies three misclassifications.

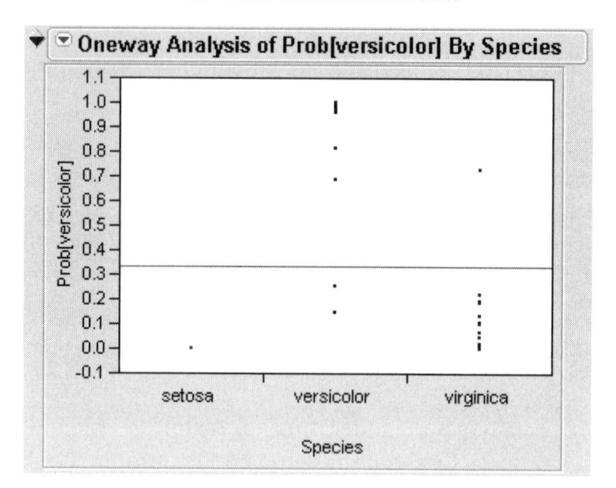

Chapter 22

Analyzing Screening Designs

The Screening Platform

The Screening platform helps select a model to fit to a two-level screening design by showing which effects are large.

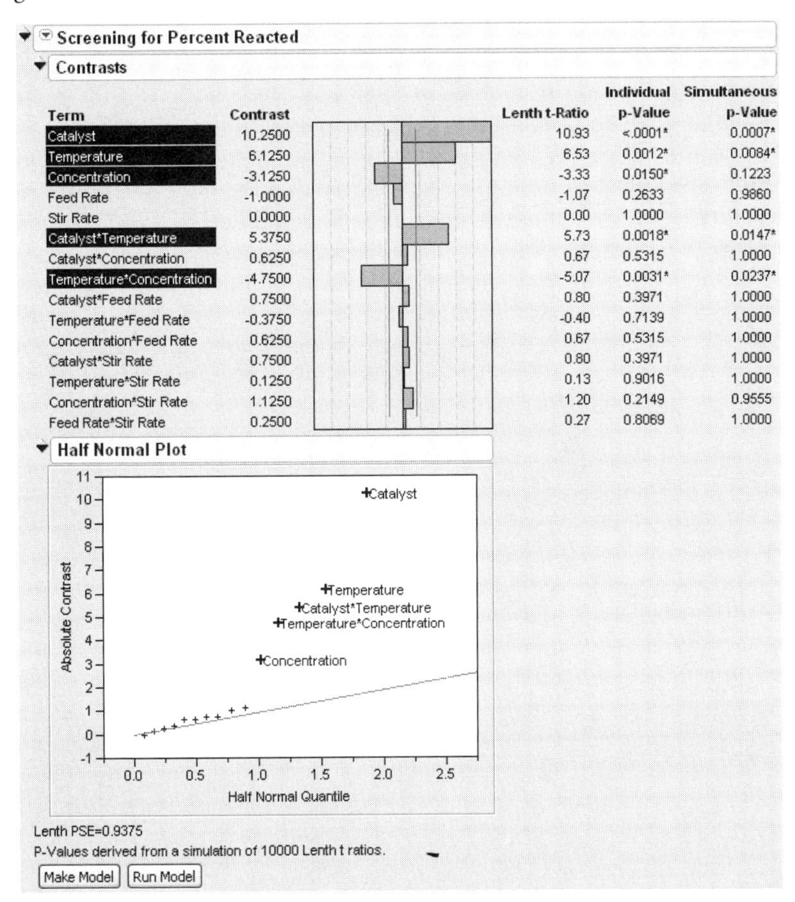

Term	Contrast	Lenth t-Ratio	Individual p-Value	Simultaneous p-Value
Catalyst	10.2500	10.93	<.0001*	0.0007*
Temperature	6.1250	6.53	0.0012*	0.0084*
Concentration	-3.1250	-3.33	0.0150*	0.1223
Feed Rate	-1.0000	-1.07	0.2633	0.9860
Stir Rate	0.0000	0.00	1.0000	1.0000
Catalyst*Temperature	5.3750	5.73	0.0018*	0.0147*
Catalyst*Concentration	0.6250	0.67	0.5315	1.0000
Temperature*Concentration	-4.7500	-5.07	0.0031*	0.0237*
Catalyst*Feed Rate	0.7500	0.80	0.3971	1.0000
Temperature*Feed Rate	-0.3750	-0.40	0.7139	1.0000
Concentration*Feed Rate	0.6250	0.67	0.5315	1.0000
Catalyst*Stir Rate	0.7500	0.80	0.3971	1.0000
Temperature*Stir Rate	0.1250	0.13	0.9016	1.0000
Concentration*Stir Rate	1.1250	1.20	0.2149	0.9555
Feed Rate*Stir Rate	0.2500	0.27	0.8069	1.0000

Contents

The Screening Platform

For two-level screening designs, the goal is to search for large effects. The screening platform looks for these big effects, helping you formulate a model for fitting the screening design.

Screening situations depend on *effect sparsity*, where most effects are assumed to be inactive. Using this assumption, the smaller estimates can be used to help estimate the error in the model and determine if the larger effects are real. Basically, if all the effects are inactive, they should vary randomly with none sticking too far out from the rest.

Using the Screening Platform

If your data is all two-level and orthogonal, then all the statistics in this platform should work well.

If categorical terms have more than two levels, then the Screening platform is not appropriate for the design. JMP produces a warning message and treats the level numbers as a continuous regressor. The variation across the factor is scattered across main and polynomial effects for that term.

For highly supersaturated main effect designs, the Screening platform is effective in selecting factors, but is not as effective at estimating the error or the significance. The Monte Carlo simulation to produce *p*-values doesn't use assumptions which are not valid for this case.

If your data is not orthogonal, then the constructed estimates are different than standard regression estimates. JMP can pick out the big effects, but does not effectively test each effect, since later effects are artificially orthogonalized, probably making earlier effects look more significant at the expense of later ones.

Note that the Screening platform is not appropriate for mixture designs.

Comparing Screening and Fit Model

Consider Reactor Half Fraction.jmp, from the Sample Data folder. The data is derived from a design in Box, Hunter, and Hunter (1978), and we are interested in a model with main effects and two-way interactions—that is, a model with fifteen parameters for a design with sixteen runs.

Running this model with the Fit Model platform results in the report shown in Figure 22.1, and illustrates why the Screening platform is needed.

Figure 22.1 Traditional Saturated Reactor Half Fraction Design Output

Response Percent Reacted

Summary of Fit

RSquare	1
RSquare Adj	.
Root Mean Square Error	.
Mean of Response	65.25
Observations (or Sum Wgts)	16

Parameter Estimates

Term	Estimate	Std Error	t Ratio	Prob>|t|
Intercept	65.25	.	.	.
Feed Rate(10,15)	-1	.	.	.
Catalyst(1,2)	10.25	.	.	.
Stir Rate(100,120)	0	.	.	.
Temperature(140,180)	6.125	.	.	.
Concentration(3,6)	-3.125	.	.	.
Feed Rate*Catalyst	0.75	.	.	.
Feed Rate*Stir Rate	0.25	.	.	.
Feed Rate*Temperature	-0.375	.	.	.
Feed Rate*Concentration	0.625	.	.	.
Catalyst*Stir Rate	0.75	.	.	.
Catalyst*Temperature	5.375	.	.	.
Catalyst*Concentration	0.625	.	.	.
Stir Rate*Temperature	0.125	.	.	.
Stir Rate*Concentration	1.125	.	.	.
Temperature*Concentration	-4.75	.	.	.

JMP can calculate parameter estimates, but with no degrees of freedom for error, standard errors, *t*-ratios, and *p*-values are all missing. Rather than use Fit Model, it is better to use the Screening platform, which specializes in getting the most information out of these situations, leading to a better model.

Here is the output from Screening for the same data.

Figure 22.2 Reactor Half Fraction Screening Design

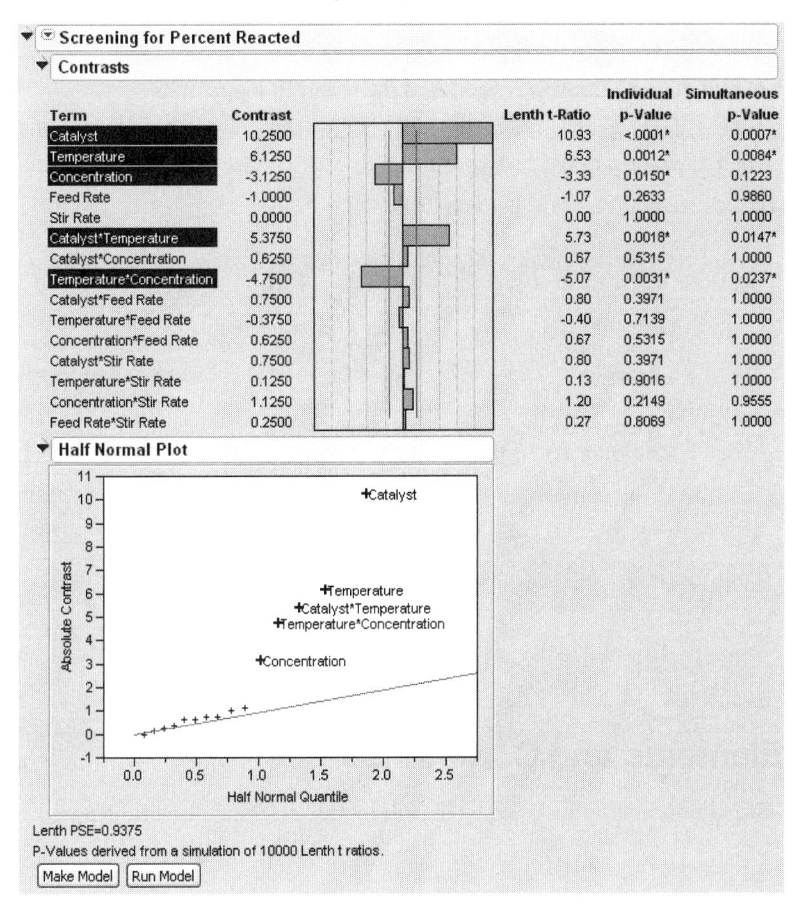

Screening for Percent Reacted

Contrasts

Term	Contrast	Lenth t-Ratio	Individual p-Value	Simultaneous p-Value
Catalyst	10.2500	10.93	<.0001*	0.0007*
Temperature	6.1250	6.53	0.0012*	0.0084*
Concentration	-3.1250	-3.33	0.0150*	0.1223
Feed Rate	-1.0000	-1.07	0.2633	0.9860
Stir Rate	0.0000	0.00	1.0000	1.0000
Catalyst*Temperature	5.3750	5.73	0.0018*	0.0147*
Catalyst*Concentration	0.6250	0.67	0.5315	1.0000
Temperature*Concentration	-4.7500	-5.07	0.0031*	0.0237*
Catalyst*Feed Rate	0.7500	0.80	0.3971	1.0000
Temperature*Feed Rate	-0.3750	-0.40	0.7139	1.0000
Concentration*Feed Rate	0.6250	0.67	0.5315	1.0000
Catalyst*Stir Rate	0.7500	0.80	0.3971	1.0000
Temperature*Stir Rate	0.1250	0.13	0.9016	1.0000
Concentration*Stir Rate	1.1250	1.20	0.2149	0.9555
Feed Rate*Stir Rate	0.2500	0.27	0.8069	1.0000

Compare the following differences between the Fit Model report and the Screening report.

- Estimates labeled **Contrast**. Effects whose individual *p*-value is less than 0.2 are highlighted.
- A *t*-ratio is calculated using Lenth's PSE (pseudo-standard error). The PSE is shown below the Half Normal Plot.
- Both individual and simultaneous *p*-values are shown. Those that are less than 0.05 are shown with an asterisk.
- A Half Normal plot allows for quick visual examination of the effects. Effects initially highlighted in the effects list are also labeled in this plot.
- Buttons at the bottom of the report also operate on the highlighted variables. The **Make Model** button opens the Fit Model dialog using the current highlighted factors. The **Run Model** button runs the model immediately.

For this example, Catalyst, Temperature, and Concentration, along with two of their two-factor interactions, are selected.

Launch the Platform

The Screening platform is launched via **Analyze > Modeling > Screening**. The completed launch dialog is shown in Figure 22.3, which produced the report in Figure 22.2.

In the launch dialog, just choose the factors. JMP constructs interactions automatically, unlike Fit Model, where the interactions are added distinctly.

Figure 22.3 Screening Platform Launch Dialog

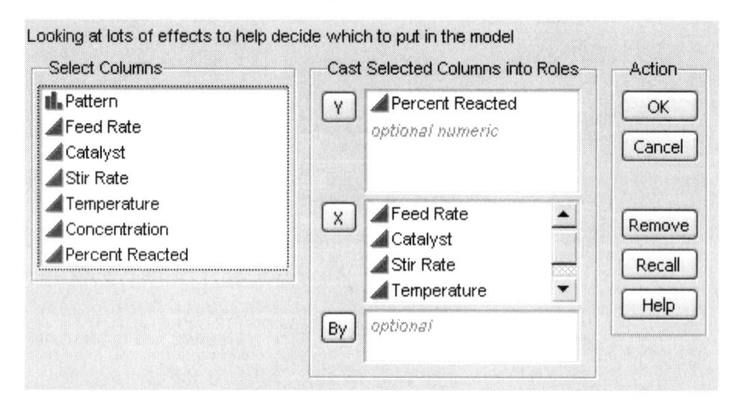

Report Elements and Commands

The following information is shown by default in the Screening platform report.

Contrasts

The **Contrasts** section shows the following columns.

Term is the factor name.

Contrast is the estimate for the factor. For orthogonal designs, this number is the same as the regression parameter estimate. This is not the case for non-orthogonal designs. An asterisk may appear next to the contrast, indicating lack of orthogonality.

Bar Graph shows the *t*-ratios with blue lines marking a critical value at 0.05 significance.

Lenth t Ratio is Lenth's *t*-ratio, calculated as $\frac{\text{Contrast}}{\text{PSE}}$.

Individual p-value is analogous to the standard *p*-values for a linear model. Small values of this value indicate a significant effect. Refer to "Statistical Details," p. 456 for details.

Simultaneous p-value is used like the individual *p*-value, but is multiple-comparison adjusted.

Aliases only appears if there are exact aliases of later effects to earlier effects.

Half Normal Plot

The Half Normal plot shows the absolute value of the contrasts against the normal quantiles for the absolute value normal distribution. Significant effects show as being separated from the line toward the upper right of the graph.

Note that this plot is interactive. If you would like to select a different set of effects to use in the model, drag a rectangle around those that you would like to select, or Control-click the effect names in the report.

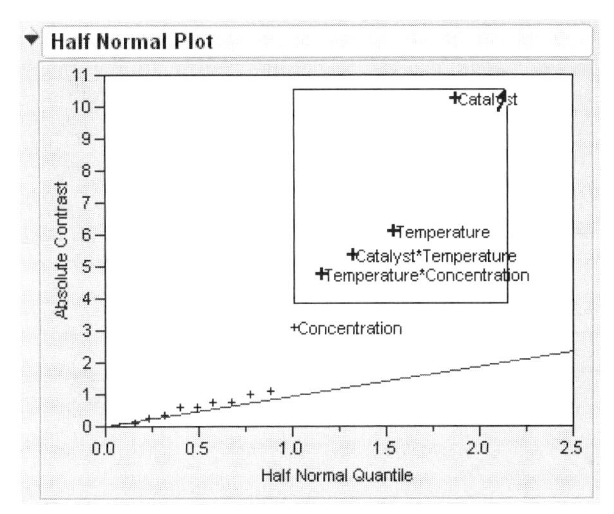

Launching a Model

The two buttons at the bottom of the plot construct a model using the currently selected effects.

Make Model launches the Fit Model dialog populated with selected effects and responses.

Run Model launches the Fit Model dialog, but also runs the model immediately.

Tips on Using the Platform

Higher-Order Effects. Control-click to select more or fewer effects. The effect selection is not constrained by hierarchy, so if, for example, A*B is selected, check to make sure the lower-order effects (the A and B main effects) are also selected. Control-click to select them if they are not already selected.

Re-running the Analysis. Do not expect the *p*-values to be exactly the same if the analysis is re-run. The Monte Carlo method should give similar, but not identical, values if the same analysis was repeated.

Statistical Details

Operation

The Screening platform has a carefully defined order of operations.

- First, the main effect terms enter according to the absolute size of their contrast. All effects are orthogonalized to the effects preceding them in the model. The method assures that their order is the same as it would be in a forward stepwise regression. Ordering by main effects also helps in selecting preferred aliased terms later in the process.
- After main effects, all second-order interactions are brought in, followed by third-order interactions, and so on. The second-order interactions cross with all earlier terms before bringing in a new term. For example, with size-ordered main effects A, B, C, and D, B*C enters before A*D. If a factor has more than two levels, square and higher-order polynomial terms are also considered.
- An effect that is an exact alias for an effect already in the model shows in the alias column. Effects that are a linear combination of several previous effects are not displayed. If there is partial aliasing, *i.e.* a lack of orthogonality, the effects involved are marked with an asterisk.
- The process continues until n effects are obtained, where n is the number of rows in the data table, thus fully saturating the model. If complete saturation is not possible with the factors, JMP generates random orthogonalized effects to absorb the rest of the variation. They are labeled Null n where *n* is a number. This situation occurs, for example, if there are exact replicate rows in the design.

Screening as an Orthogonal Rotation

Mathematically, the Screening platform takes the n values in the response vector and rotates them into n new values that are mapped by the space of the factors and their interactions.

$$\text{Contrasts} = \mathbf{T}' \times \text{Responses}$$

where T is an orthonormalized set of values starting with the intercept, main effects of factors, two-way interactions, three-way interactions, and so on, until n values have been obtained. Since the first column of T is an intercept, and all the other columns are orthogonal to it, these other columns are all contrasts, *i.e.* they sum to zero. Since T is orthogonal, it can serve as X in a linear model, but it doesn't need inversion, since $\mathbf{T}'$ is also $\mathbf{T}^{-1}$ and also $(\mathbf{T}'\mathbf{T})\mathbf{T}'$, so the contrasts are the parameters estimated in a linear model.

If no effect in the model is active after the intercept, the contrasts are just an orthogonal rotation of random independent variates into different random independent variates with the same variance. To the extent that some effects are active, the inactive effects still represent the same variation as the error in the model. The hope is that the effects and the design are strong enough to separate which are active from which are random error.

Lenth's Pseudo-Standard Error

At this point, Lenth's method (Lenth, 1989) identifies inactive effects from which it constructs an estimate of the residual standard error—the *Lenth Pseudo Standard Error* (*PSE*).

The value for Lenth's PSE is shown at the bottom of the Screening report. From the PSE, t-ratios are obtained. To generate p-values, a Monte Carlo simulation of 10,000 runs of $n - 1$ purely random values is created and Lenth t-ratios are produced from each set. The p-value is the interpolated fractional position among these values in descending order. The simultaneous *p*-value is the interpolation along the max(|*t*|) of the $n - 1$ values across the runs. This technique is similar to that in Yee and Hamada (2000).

If you want to run more or less than the 10,000 default runs, assign a value to a global JSL variable named `LenthSimN`. JMP carries out that number of runs if the variable is defined. If `LenthSimN=0`, the standard *t*-distribution is used (not recommended).

Analyzing a Plackett-Burman Design

Plackett-Burman designs are an alternative to fractional-factorial screening designs. Two-level fractional factorial designs must, by their nature, have a number of runs that are a power of two. Plackett-Burman designs exist for 12-, 24-, and 28-run designs.

Weld-Repaired Castings.jmp from the Sample Data folder uses a Plackett-Burman design, and is found in textbooks like Giesbrecht and Gumpertz (2004) and Box, Hunter, and Hunter (2005). Seven factors thought to be influential on the quality of a weld (Initial Structure, Bead Size, Pressure Treatment, Heat Treatment, Cooling Rate, Polish, and Final Treatment) are investigated using a 12-run Plackett-Burman design. The response is $100 \times \log(\text{lifetime})$. (There are also four terms that were used to model error that are not used in this analysis.)

Using the Screening platform, select the seven effects as **X** and **Log Life** as **Y**. The following report appears, showing only a single significant effect.

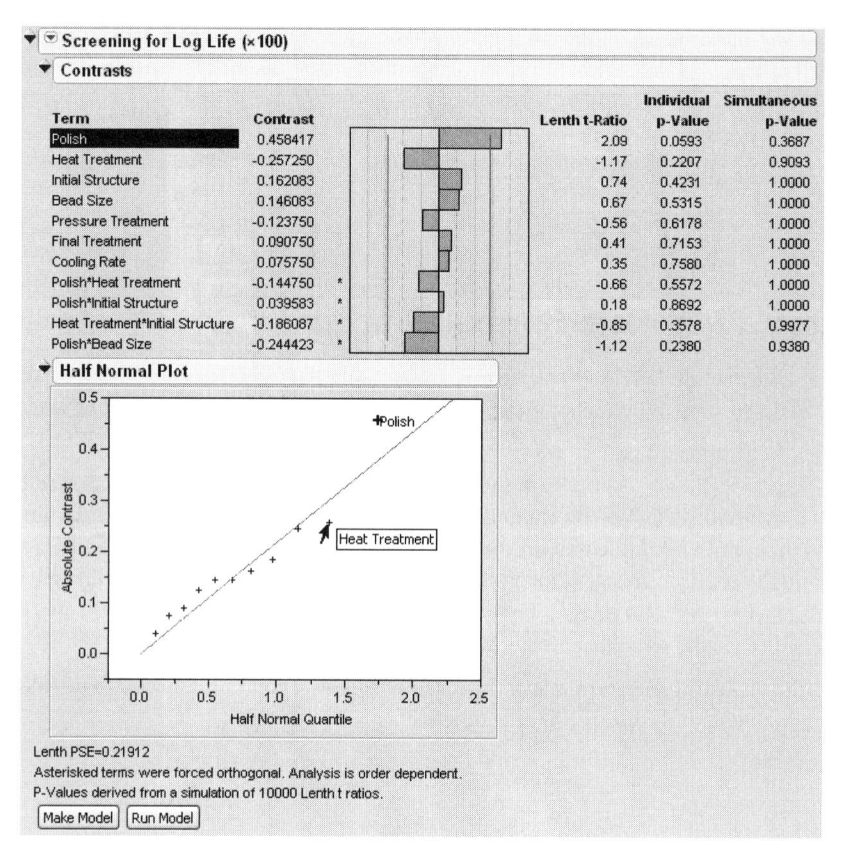
Screening for Log Life (×100)
Contrasts

Term	Contrast		Lenth t-Ratio	Individual p-Value	Simultaneous p-Value
Polish	0.458417		2.09	0.0593	0.3687
Heat Treatment	-0.257250		-1.17	0.2207	0.9093
Initial Structure	0.162083		0.74	0.4231	1.0000
Bead Size	0.146083		0.67	0.5315	1.0000
Pressure Treatment	-0.123750		-0.56	0.6178	1.0000
Final Treatment	0.090750		0.41	0.7153	1.0000
Cooling Rate	0.075750		0.35	0.7580	1.0000
Polish*Heat Treatment	-0.144750	*	-0.66	0.5572	1.0000
Polish*Initial Structure	0.039583	*	0.18	0.8692	1.0000
Heat Treatment*Initial Structure	-0.186087	*	-0.85	0.3578	0.9977
Polish*Bead Size	-0.244423	*	-1.12	0.2380	0.9380

Lenth PSE=0.21912
Asterisked terms were forced orthogonal. Analysis is order dependent.
P-Values derived from a simulation of 10000 Lenth t ratios.
Make Model Run Model

Note asterisks mark four terms, indicating that they are not orthogonal to effects preceding them, and the obtained contrast value was after orthogonalization. So, they would not match corresponding regression estimates.

Analyzing a Supersaturated Design

Supersaturated designs have more factors than runs. The objective is to determine which effects are active. They rely heavily on effect sparsity for their analysis, so the Screening platform is ideal for their analysis.

As an example, look at Supersaturated.jmp, from the Sample Data folder, a simulated data set with18 factors but only 12 runs. Y is generated by

$$Y = 2(X7) + 5(X10) - 3(X15) + \varepsilon$$

where $\varepsilon \sim N(0,1)$. So, Y has been constructed with three active factors.

To detect the active factors, run the Screening platform with X1–X18 as **X** and Y as **Y**. The following report appears.

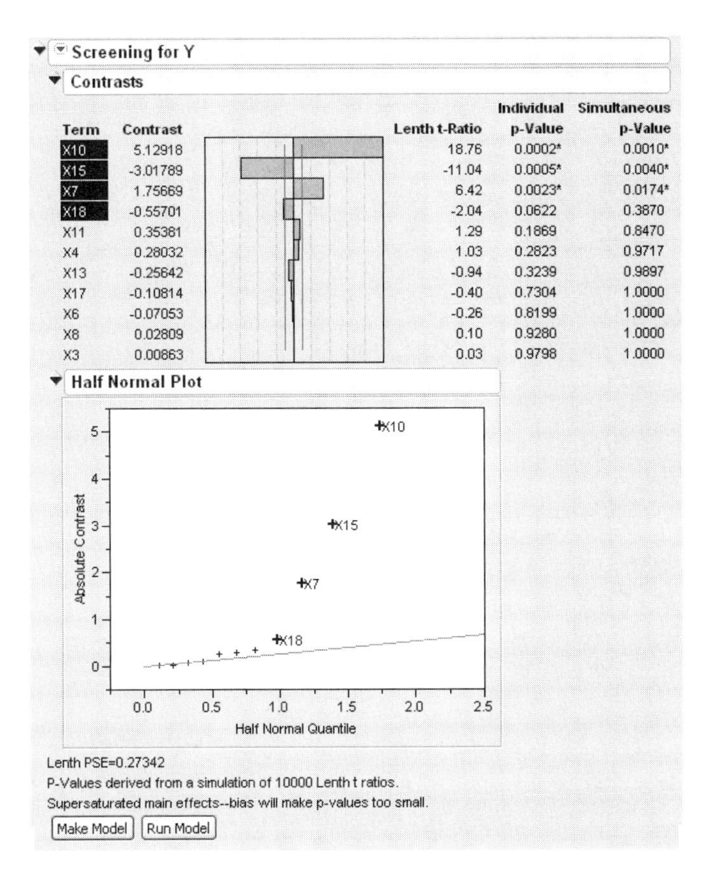

Note that the three active factors have been highlighted. One other factors, X18, has also been highlighted. It shows in the Half Normal plot quite near to the blue line, indicating that they are close to the 0.1 cutoff significance value. The 0.1 critical value is generous in its selection of factors so you don't miss those that are possibly active.

The contrasts of 5.1, –3, and 1.8 are close to their simulated values (5, –3, 1), but will be closer if you run the selected model as a regression, without the effect of orthogonalization.

The *p*-values, while useful, are not entirely valid statistically, since they are based on a simulation that assumes orthogonal designs, which is not the case for supersaturated designs.

Chapter 23

Nonlinear Regression

The Nonlinear Platform

The **Nonlinear** command fits models that are nonlinear in their parameters, using least-squares or a custom loss function. You do not need to use this platform if the model is nonlinear in the variables but linear in the parameters. Similarly, you can use the linear platforms if you can transform your model to be linear in the parameters.

Nonlinear models are more difficult to fit than linear models. They require more preparation with the specification of the model and initial guesses for parameter values. Iterative methods are used to search for the least-squares estimates, and there is no guarantee that a solution will be found. Indeed it is possible to diverge, or even to converge on a local solution that is not the least-squares solution. Nonlinear fits do not have some of the nice properties that linear models have, and the results must be interpreted with caution. For example, there is no well defined R^2 statistic; the standard errors of the estimates are approximations; and leverage plots are not provided.

Nonlinear fitting begins with a formula for the model that you build in a column using the formula editor or select using JMP's built-in model library. The formula is specified with parameters to be estimated. After entering the formula, select the **Nonlinear** command and work with its interactive iteration Control Panel to do the fitting.

The platform itself is easy to use. The iterative techniques employed in JMP are variations of the - Gauss-Newton method with a line search. This method employs derivatives of the model with respect to each parameter, but these derivatives are obtained automatically by the platform. The Newton-Raphson method is also available by requesting second derivatives. There are features for finding profile confidence limits of the parameters and for plotting the fitted function if the model is a function of just one column.

It is also possible to define special loss functions to use instead of least-squares. The loss function can be a function of the model, other variables, and parameters. When a custom loss function is defined, the platform minimizes the sum of the loss function across rows; thus, if you want maximum likelihood, specify something that sums to the negative of the log-likelihood as the loss.

Contents

The Nonlinear Fitting Process

The **Nonlinear** command on the **Analyze > Modeling** submenu launches an interactive nonlinear fitting facility. You orchestrate the fitting process as a coordination of three important parts of JMP: the data table, the formula editor, and the Nonlinear platform.

- You define the column and its prediction formula with the formula editor. The formula is specified with parameters to be estimated. Use the formula editor to define parameters and give them initial values.
- Launch the Nonlinear platform with the response variable in the Y role and the model column with fitting formula in the Prediction Column. If no *Y* column is given, then the Prediction Column formula is for residuals. If you have a loss column specified, then you might not need either a Prediction column or *Y* column specification.
- Interact with the platform through the iteration Control Panel.

A Simple Exponential Example

The US Population.jmp table in the Nonlinear Examples of the Sample Data folder illustrates a simple nonlinear exponential model.

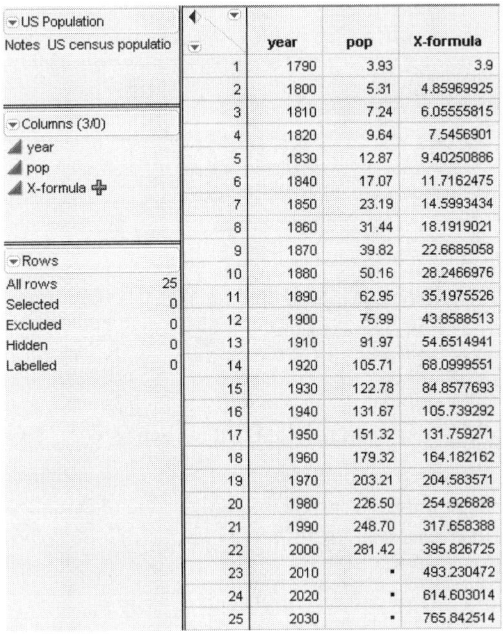

US Population
Notes US census populatio

Columns (3/0)
year
pop
X-formula

Rows
All rows 25
Selected 0
Excluded 0
Hidden 0
Labelled 0

	year	pop	X-formula
1	1790	3.93	3.9
2	1800	5.31	4.85969925
3	1810	7.24	6.05555815
4	1820	9.64	7.5456901
5	1830	12.87	9.40250886
6	1840	17.07	11.7162475
7	1850	23.19	14.5993434
8	1860	31.44	18.1919021
9	1870	39.82	22.6685058
10	1880	50.16	28.2466976
11	1890	62.95	35.1975526
12	1900	75.99	43.8588513
13	1910	91.97	54.6514941
14	1920	105.71	68.0999551
15	1930	122.78	84.8577693
16	1940	131.67	105.739292
17	1950	151.32	131.759271
18	1960	179.32	164.182162
19	1970	203.21	204.583571
20	1980	226.50	254.926828
21	1990	248.70	317.658388
22	2000	281.42	395.826725
23	2010	•	493.230472
24	2020	•	614.603014
25	2030	•	765.842514

The pop (population) values are modeled as a nonlinear function of year. To see the nonlinear formula, highlight the X-formula column (which contains the model). Then select the **Formula** command in the **Columns** menu. The column formula editor appears with a formula that describes an exponential model for the growth of population in the U. S. between 1790 and 1990.

Creating a Formula with Parameters

When you fit a nonlinear formula to data, the first step is to create a column in the data table using the formula editor to build a prediction formula, which includes parameters to be estimated. The formula contains the parameters' initial values.

Begin in the formula editor by defining the parameters for the nonlinear formula. Select **Parameters** from the popup menu above the column selector list as shown to the left in Figure 23.1. The list changes from the column selector list to a parameter selector list. When **New Parameter** appears in the selector list, click on it and respond to the New Parameter definition dialog, as shown in Figure 23.1. Use this dialog to name the parameter and assign it an initial value. When you click **OK**, the new parameter name appears in the selector list. Continue this process to create additional parameters. The population example uses two parameters. You can now build the formula you need by using data table columns and parameters.

Figure 23.1 Creating Parameters for a Formula

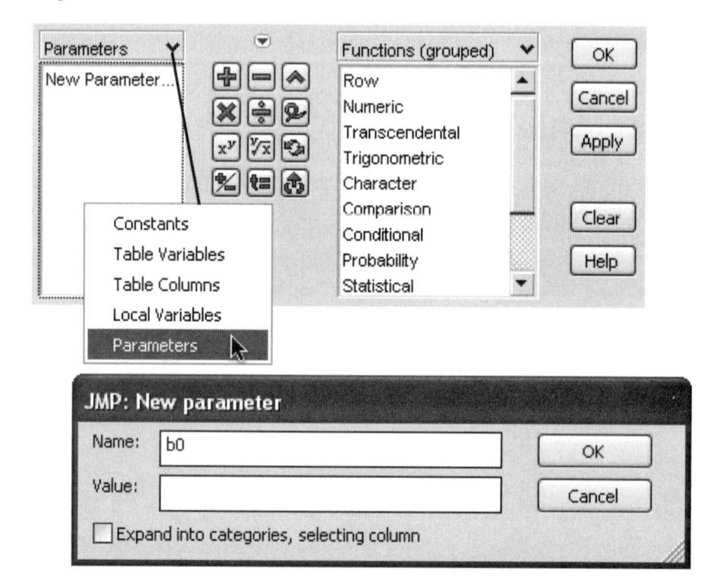

The parameter names are arbitrary and in this example were given initial values chosen as follows:

- **B0** is the prediction of population at year 1790, which should be near 3.93, the actual value of the first recorded population value in the data table. Therefore, 3.9 seems to be a reasonable initial value for B0.
- The **B1** growth rate parameter is given the initial value of 0.022, which is the estimate of the slope you get when you fit the natural log of pop to year with a straight line (or **Fit Special** and specify ln transformation for pop). This initial value seems reasonable because the nonlinear exponential model can have similar final parameter estimates.

You can now build a formula using data table columns and parameters.

```
B0*Exp(B1*(year - 1790))
```

Enter the formula by clicking on column names, parameter names, operator keys, and function list items in the formula editor panel. Alternatively, double-click on the formula in the formula editor to change it into text-edit mode. The formula can be typed in as text.

See the *JMP User Guide* for information about building formulas.

Launch the Nonlinear Platform

When the formula is complete, choose **Analyze > Modeling > Nonlinear** and complete the Launch dialog as shown in Figure 23.2. Select the pop as **Y, Response**, and the column with the fitting formula (X-formula) as **X, Predictor Formula.**

Figure 23.2 The Nonlinear Launch Dialog

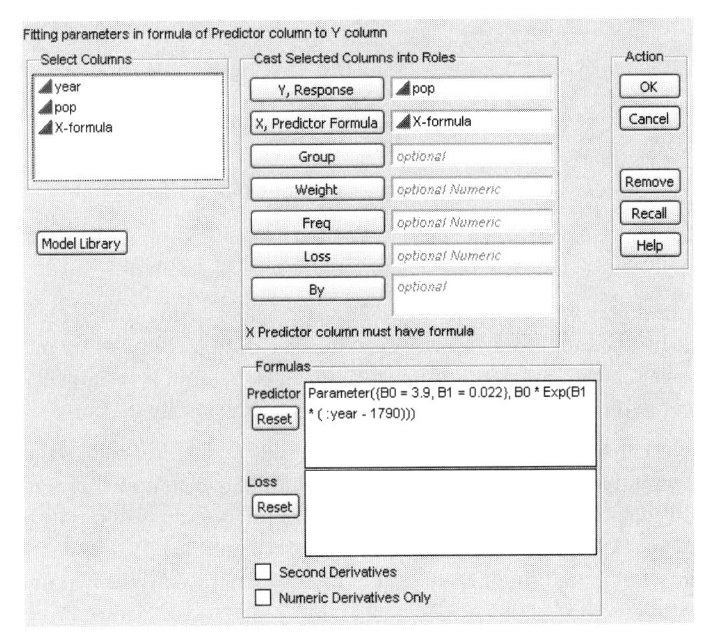

The model formula (and formula for a loss function if there is one) appear in text form in the **Formulas** area at the bottom of the Nonlinear Launch dialog (see Figure 23.2).

A *Y* variable is not needed if the fitting formula calculates residuals instead of predicted values or if there is a custom loss function. In some cases with loss functions, even the **X, Predictor Formula** column (the model) is not necessary.

Note: You can also select **Weight**, **Freq**, **By,** and **Group** columns. If the **Weight** column contains a formula, it is recalculated for each iteration. This also causes the column with the predicted value to update at each iteration. If you specify a **Group** variable, then when you do the plot, JMP overlays a curve for each value of the **Group** column. This assumes that the model is only a function of the **Group** column and one other continuous column.

Drive the Iteration Control Panel

When you complete the Launch dialog, click **OK** to see the Iteration Control Panel. The Control Panel lets you tailor the nonlinear fitting process. The parameters' initial values from the formula construction show in the Control Panel. Edit the **Stop Limit** fields if you need more or less accuracy for the converged solutions. If you edit the **Current Values** of the parameters, click **Reset** to set them for the next fit. Click **Go** to start the fitting iterations. You can watch the iterations progress in the Control Panel.

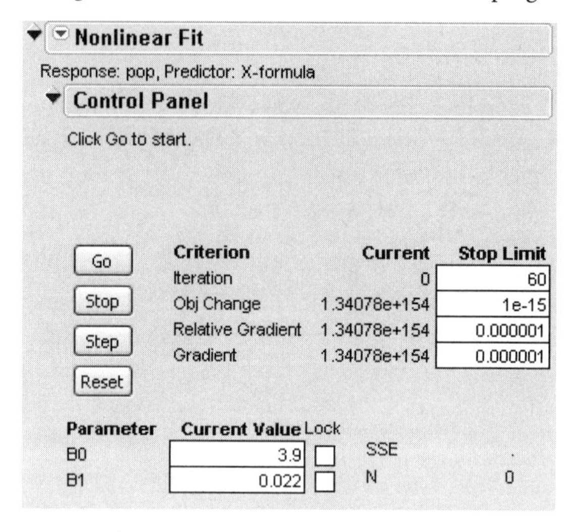

If the iterations do not converge, then try other parameter starting values; each time you click **Reset** and rerun the fit, the sum of squares error is recalculated, so it is easy to try many values. Sometimes locking some parameters and iterating on the rest can be helpful. If the model is a function of only one column, then use the sliders on the output **Plot** to visually modify the curve to fit better.

If the iterations converge, then a solution report appears and additional fields appear in the Control Panel, as shown in Figure 23.3. Click **Confidence Limits**, which now shows at the bottom of the Iteration Control Panel, to calculate profile-likelihood confidence intervals on the parameter estimates. These confidence limits show in the Solution table (see Figure 23.4). The confidence limit calculations involve a new set of iterations for each limit of each parameter, and the iterations often do not find the limits successfully.

When there is a single factor, the results include a graph of the response by the factor as shown in Figure 23.4. The solution estimates for the parameters are listed beneath the plot with sliders to alter the values and modify the fitted curve accordingly.

Figure 23.3 Nonlinear Iteration Control Panel After Fitting Process Is Complete

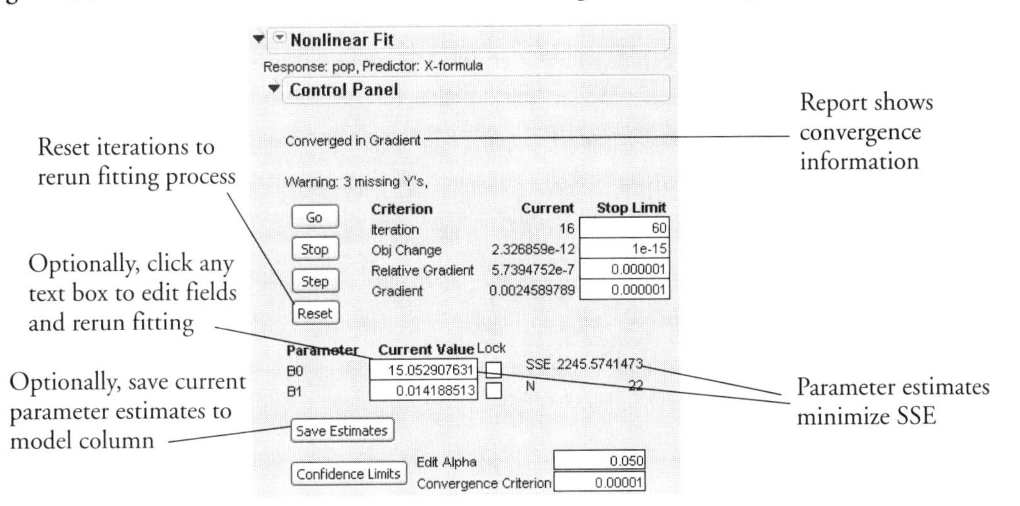

Click the **Confidence Limits** button to generate confidence intervals for the parameters.

Figure 23.4 Results for Exponential Population Growth Example

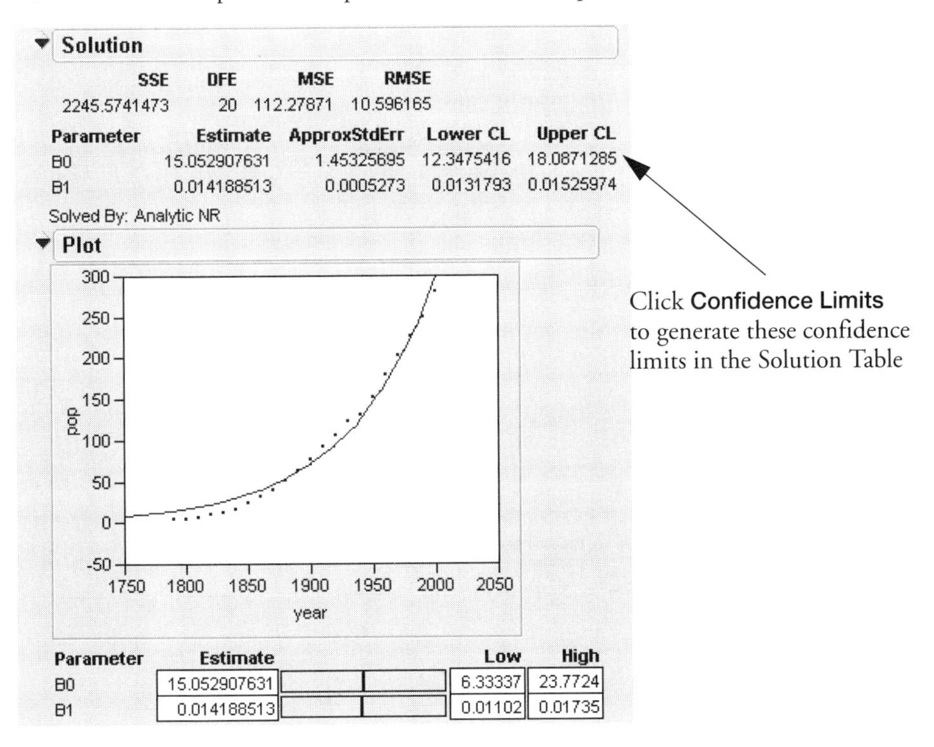

Using the Model Library

The built-in model library is invoked by clicking the **Model Library** button on the Nonlinear launch dialog. Once it is invoked, a window called Nonlinear Model Library appears.

Figure 23.5 Nonlinear Model Library Dialog

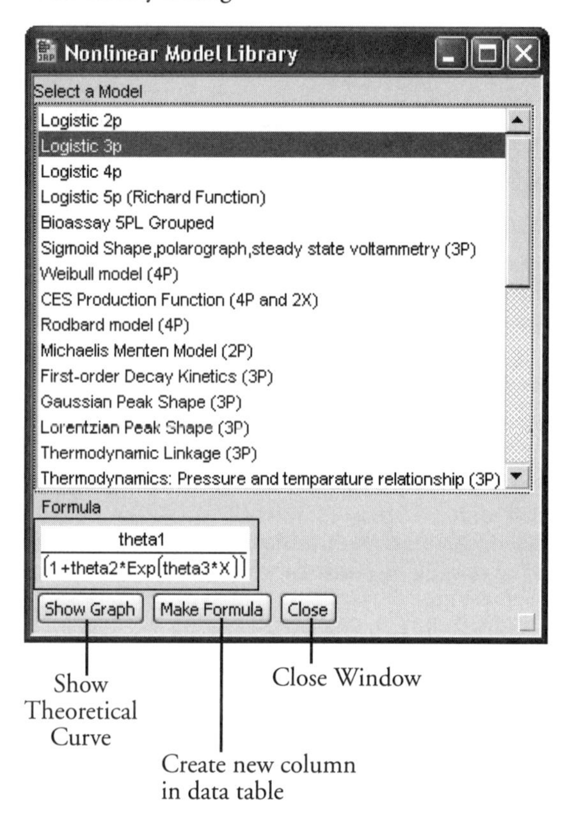

There are about 33 models included in this library. Users can select one of the models and click either **Show Graph** or **Make Formula**.

Show Graph shows a 2-D theoretical curve for one-parameter models and a 3-D surface plot for two-parameter models. No graph is available for models with more than two explanatory (*X*)variables. On the graph window, change the default initial values of parameters using the slider, or clicking and typing values in directly (see Figure 23.6).

Figure 23.6 Example Graph in Model Library

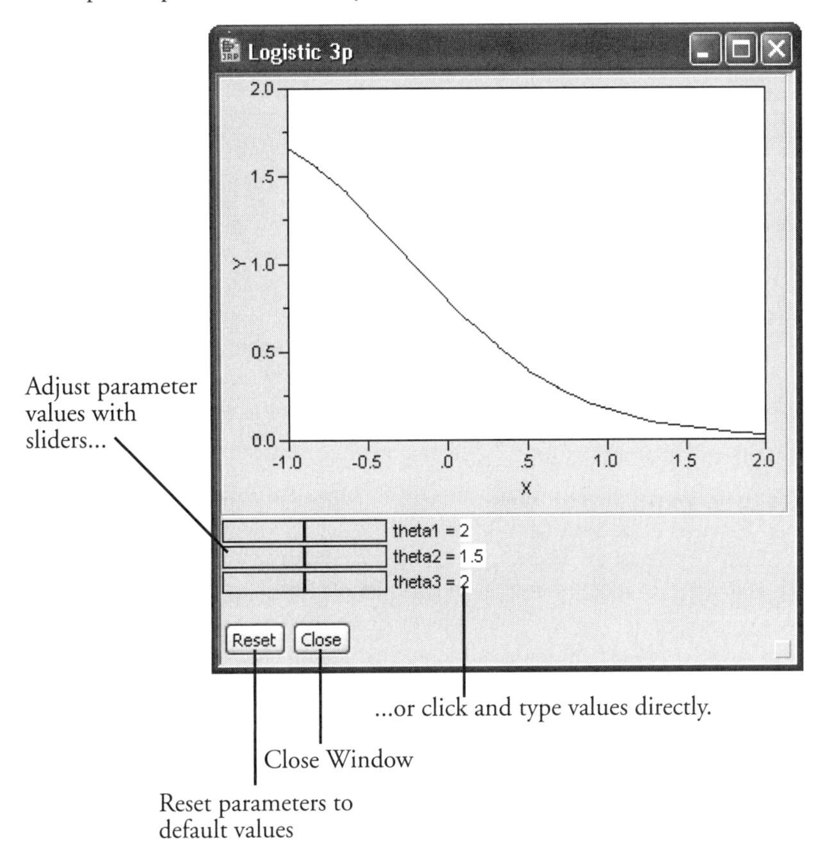

The **Reset** button on the graph window sets the initial values of parameters back to their default values.

Clicking **Make Formula** opens a window that allows you to select the *x*-variables for the model. Select a column from the data table and click X (or X1, or X2, depending on the number of variables in the model). Then, click **OK** to make a new column in the data table which has the latest initial values of parameters.

If the data table already has columns called X1 and X2, a new column with a formula is created without the Select Columns window being launched. Once the formula is created in the data table, continue the analysis in the nonlinear platform, assigning the newly-created column as the **X, Predictor Formula** in the Nonlinear launch dialog.

Many of the functions are named Model A, Model B, etc., whose definitions are shown in Table 23.2 "Guide to Nonlinear Modeling Templates," p. 494.

Customizing the Nonlinear Model Library

The Model Library is created by a built-in script named `nonlinLib.jsl`, located in the **Builtin Scripts** subdirectory of the folder that contains JMP (Windows and Linux) or the Application Package (Macintosh). You can customize the nonlinear library script by modifying this script.

To add a model, you must add three lines to the list named `Listofmodellist#`. These three lines are actually a list themselves which consists of the following three parts.

- Model name, a quoted string
- Model formula, an expression
- Model scale.

For example, suppose you want to add a model called "Simple Exponential Growth" that has the form

$$y = b_1 e^{kx}$$

Add the following lines to the `nonlibLib.jsl` script

```
{//Simple Exponential Growth
    "Simple Exponential Growth",
    Expr(Parameter({b1=2, k=0.5}, b1*exp(k * :X)),
    lowx = -1; highx = 2; lowy = 0;  highy = 2},
```

Some things to note:

- The first line is simply an open bracket (starting the list) and an optional comment. The second line is the string that is displayed in the model library window.
- The values of `lowx`, `highx`, `lowy`, and `highy` specify the initial window for the theoretical graph.
- There is a comma as the last character in the example above. If this is the final entry in the `Listofmodellist#` list, the comma can be omitted.
- If the model uses more than two parameters, replace the last line (containing the graph limits) with the quoted string "String Not Available".
  To delete a model, delete the corresponding three-lined list from the `Listofmodellist#` list.

Details for Formula Editor

In the formula editor, when you add a parameter, note the checkbox for **Expand Into Categories, selecting column**. This command is used to add several parameters (one for each level of a variable) at once. When you select this option, a dialog appears that allows you to select a column. After selection, a new parameter appears in the Parameters list with the name *D_column*. This is, in a sense, a macro that inserts a Match expression containing a separate parameter for each level of the selected column.

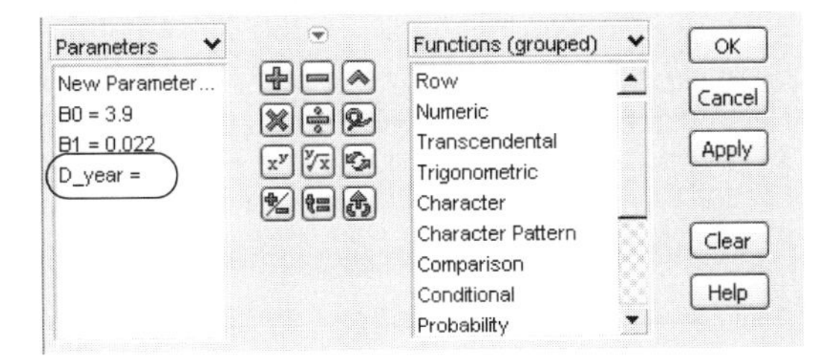

Once you decide where the list of parameters should be inserted, click the new parameter name to insert the group of parameters into the prediction formula.

Details of the Iteration Control Panel

The Nonlinear platform appears with the Iteration Control Panel shown in Figure 23.7. The initial Control Panel has these features:

- a processing messages area, which shows the progress of the fitting processing
- editable parameter fields for starting values for the parameters. Click **Reset** after you edit these.
- buttons to start (**Go**), stop (**Stop**), and step (**Step**) through the fitting process
- a **Reset** button to reset the editable values into the formula, reset the iteration values, and calculate the SSE at these new values.
- initial values for convergence criteria and step counts
- a popup menu with fitting options to specify computational methods and save confidence limits. These popup menu items are described in the next section.

Figure 23.7 Iteration Control Panel Before Fitting

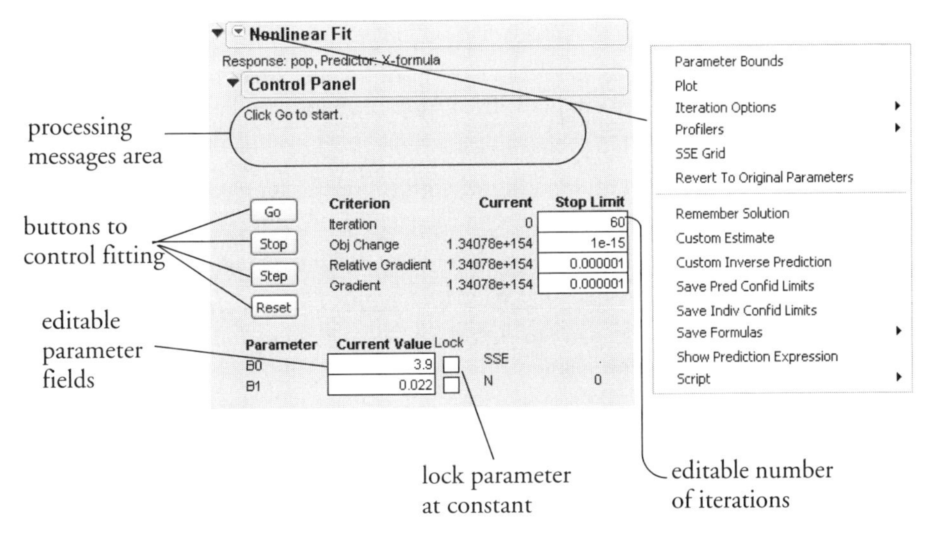

After the fitting process converges, controls appear at the bottom of the Control Panel that let you specify the alpha level and convergence criterion for computation of confidence limits and predictions (Figure 23.8). When you click **Confidence Limits**, the results appear in the Solution table as shown previously in Figure 23.4.

Figure 23.8 Control Panel After Fitting

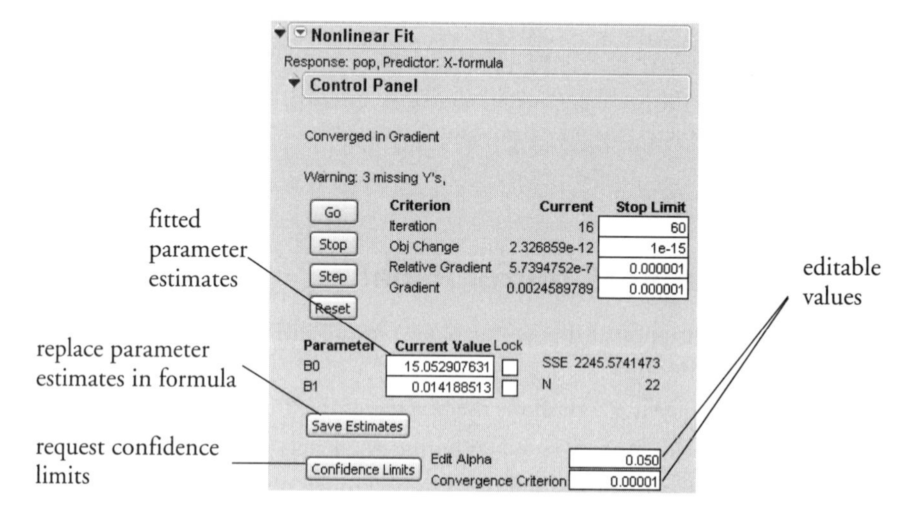

The next sections describe the features of the iteration Control Panel in more detail.

Panel Buttons

Go starts an iterative fitting process. The process runs in the background so that you can do other things while it runs. If you want to re-run the model, you must first click the **Reset** button. A message in the status bar reminds you to do so if you click **Go** multiple times.

Stop stops the iterative process. Because the platform only listens to events between steps, it might take a few seconds to get a response from **Stop**. Iterations can be restarted by clicking **Go** again.

Step takes a single iteration in the process.

Reset resets the iteration counts and convergence criteria, copies the editable values into the internal formulas, and then calculates the SSE for these new values. Use **Reset** if you want to

- take extra steps after the iterations have stopped. For example, extra steps are needed if you set the Stop Limit in order to solve for greater accuracy.
- edit the parameter estimates to clear the previous fitting information and begin iterations using these new starting values.
- calculate the SSE with the current parameter values.

The Current Parameter Estimates

The Control Panel also shows the **Current Value** of the parameter estimates and the **SSE** values as they compute. At the start, these are the initial values given in the formula. Later, they are values that result from the iterative process. The SSE is missing before iterations begin and is the first SSE displayed after the nonlinear fit.

If you want to change the parameter values and start or continue the fitting process, click and edit the parameters in the **Current Value** column of the parameter estimates. Then click **Reset**, followed by **Go**.

Note: Nonlinear makes its own copy of the column formula, so if you open the formula editor for the model column, the changes you make there will not affect an active Nonlinear platform. The parameter values are not stored back into the column formula until you click **Save Estimates**.

Save Estimates

After the nonlinear fitting is complete, the **Save Estimates** button appears on the Control Panel. When you click **Save Estimates**, the parameter estimates are stored into the column formula, causing the column to be recalculated by the data table. If you are using By-groups, this does not work because the formula is in the source table, instead of the By-groups.

Confidence Limits

The **Confidence Limits** button is available above the Solution Table after fitting. It computes confidence intervals for each parameter estimate. These intervals are profile-likelihood confidence limits, and each limit for each parameter involves a new set of iterations, so it can take a lot of time and result in missing values if the iterations fail to find solutions. The calculations do not run in the background but you can cancel them by typing the Escape key (⌘-period key on the Mac). See "The Solution Table," p. 478, for details about confidence limits.

The Nonlinear Fit Popup Menu

The popup menu icon on the Nonlinear Fit title bar lets you tailor the fitting process, request additional information about the results, and save confidence limits as new columns in the current data table.

The following list describes each of the popup menu items:

Parameter Bounds

To set bounds on the parameters, select the **Parameter Bounds** option to reveal new editable fields next to the parameter values, into which you can enter lower and upper bounds to the parameters. Unbounded parameters are signified by leaving the field as a missing value (.).

If you save a script, the bounds are saved in this manner:

```
Parameter Bounds(paramName(lower, upper)...)
```

Plot

If the prediction formula is a function of exactly one other continuous variable and you specify the response separately as the *Y* variable, then **Plot** shows a plot of the function and the observed data (see Figure 23.4). If you specify a Group variable at launch, then a curve shows for each group. The initial estimates of the parameters show beneath the plot, with sliders to adjust the values. You can use these sliders to adjust the parameters for and obtain an empirical fit of the prediction function and use the resulting parameter estimates and starting values for the nonlinear fitting process.

Iteration Options

Iteration options let you specify options for the optimization algorithm.

Iteration Log opens an additional window (named **Iteration Log**) and records each step of the fitting process. The iteration log lets you see parameter estimates at each step and the objective criterion as they change at each step during the fitting process.

Numeric Derivatives Only is useful when you have a model that is too messy to take analytic derivatives for. It can also be valuable in obtaining convergence in tough cases.

Second Deriv Method uses second derivatives as well as first derivatives in the iterative method to find a solution. With second derivatives, the method is called Newton-Raphson rather than Gauss-Newton. This method is only useful if the residuals are unusually large or if you specify a custom loss function and your model is not linear in its parameters.

Note: You can also request the second derivative method with a check box on the launch dialog.

Newton chooses whether Gauss-Newton (for regular least squares) or Newton-Raphson (for models with loss functions) as the optimization method.

QuasiNewton SR1 chooses QuasiNewton SR1 as the optimization method.

QuasiNewton BFGS chooses QuasiNewton BFGS as the optimization method.

Accept Current Estimates tells JMP to produce the solution report with the current estimates, even if the estimates did not converge.

Show Derivatives shows the derivatives of the nonlinear formula in the JMP log. Use the **Log** command in the **View** or **Window** menu to see the JMP log if it is not currently open. See “Notes Concerning Derivatives,” p. 491, for technical information about derivatives.

Unthreaded runs the iterations in the main computational thread. In most cases, JMP does the computations in a separate computational thread. This improves the responsiveness of JMP while doing other things during the nonlinear calculations. However, there are some isolated cases (models that have side effects that call display routines, for example) that should be run in the main thread, so this option should be turned on.

Profilers

The Profilers submenu brings up three facilities for viewing the prediction surface. Details on profiling are discussed in “Profiling Features in JMP,” p. 289.

Profiler brings up the JMP Profiler. The Profiler lets you view vertical slices of the surface across each *x*-variable in turn, as well as find optimal values of the factors. Details of the Profiler are discussed in “The Profiler,” p. 277.

Contour Profiler brings up the JMP Contour Profiler. The Contour profiler lets you see two-dimensional contours as well as three dimensional mesh plots. Details of the Contour Profiler are discussed in "Contour Profiler," p. 306.

Surface Profiler creates a three-dimensional surface plot. There must be two continuous factors for this facility to work. Details of surface plots are discussed in the "Plotting Surfaces" chapter.

Parameter Profiler brings up the JMP Profiler and profiles the SSE or loss as a function of the parameters. Details of the Profiler are discussed in "The Profiler," p. 277.

Parameter Contour Profiler brings up the JMP Contour Profiler and profiles the SSE or loss as a function of the parameters. Details of the Contour Profiler are discussed in "Contour Profiler," p. 306.

Parameter Surface Profiler creates a three-dimensional surface plot and profiles the SSE or loss as a function of the parameters. Details of surface plots are discussed in the "Plotting Surfaces" chapter.

SSE Grid

To explore the sensitivity of the fit with respect to the parameter estimates, you can create a grid around the solution estimates and compute the error sum of squares for each value. The solution estimates should have the minimum SSE.

To create a grid table, use the **SSE Grid** command, which opens the Specify Grid for Output dialog, shown here. Edit the values in the dialog to customize the grid.

Specify Grid for Output

Parameter	Min	Max	N Points
B0	11.419765247	18.686050014	11
B1	0.0128702655	0.0155067604	11

Go

The Specify Grid for Output table shows these items:

Parameter lists the parameters in the fitting model (B0 and B1 in the figure above).

Min displays the minimum parameter value used in the grid calculations. By default, Min is the solution estimate minus 2.5 times the ApproxStdErr.

Max displays the maximum parameter value used in the grid calculations. By default, Max is the solution estimate plus 2.5 times the ApproxStdErr.

N Points gives the number of points to create for each parameter. To calculate the total number of points in the new grid table, multiply all the **N Points** values. Initially **N Points** is 11 for the first two parameters and 3 for the rest. If you specify new values, use odd values to ensure that the grid table includes the solution estimates. Setting **N Points** to 0 for any parameter records only the solution estimate in the grid table.

When you click **Go**, JMP creates a new untitled table with grid coordinates, like the table shown here. A highlighted X marks the solution estimate row if the solution is in the table. You can use the **Spinning Plot** command in the **Graph** menu to explore the SSE or likelihood surface. See Figure 23.13 for a spinning plot example.

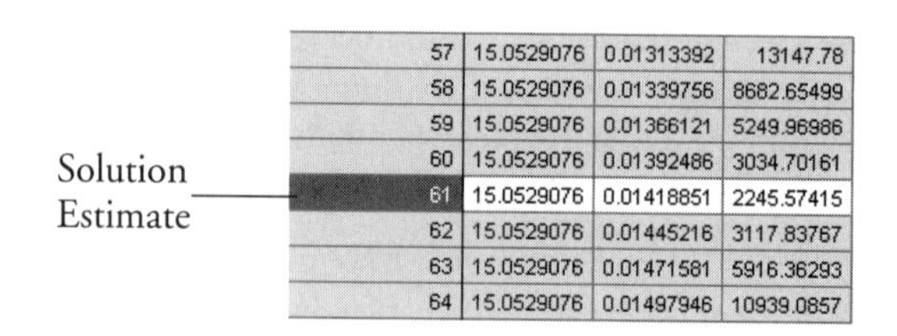

57	15.0529076	0.01313392	13147.78
58	15.0529076	0.01339756	8682.65499
59	15.0529076	0.01366121	5249.96986
60	15.0529076	0.01392486	3034.70161
61	15.0529076	0.01418851	2245.57415
62	15.0529076	0.01445216	3117.83767
63	15.0529076	0.01471581	5916.36293
64	15.0529076	0.01497946	10939.0857

Revert to Original Parameters

The Revert to Original Parameters command resets the Nonlinear platform to use the originally-specified parameters (*i.e.* the parameters that existed before any fitting).

Remember Solution

The **Remember Solution** command is used to stage a series of restricted estimates for use in constructing tests on the restrictions. This command is only used after you finish estimating for each model.

For example, suppose you want to test an unrestricted exponential model with a model where one parameter is constrained. First, set the parameter to its restricted value and check the **Lock** box for it. Then, click **Go** to obtain the restricted results. Click the **Remember Solution** command to remember this solution. Now, uncheck the lock and repeat the procedure, clicking **Reset**, then clicking **Go**, then **Remember Solution**. JMP then shows the fitting statistics for each solution, and a test comparing the solutions.Figure 23.9 shows an example using **US Population.jmp**, where the **b1** parameter is constrained to be 0.1 in the restricted model.

Figure 23.9 Remembered Models

Remembered Models

Model	SSE	DFE	MSE	Restrictions
Unrestricted Model	2245.5741	20	112.2787	
Restricted Model	277552.85	21	13216.8	B1=1

Hypothesized	Alternative	Denominator	SS	NDF	DDF	F Ratio	Prob > F
Restricted Model	Unrestricted Model	Unrestricted Model	275307.28	1	20	2451.999	<.0001*

Parameter	Unrestricted Model
B0	15.052907631
B1	0.014188513

Custom Estimate

The **Custom Estimate** command lets you provide an expression involving only parameters. JMP calculates the expression using the current estimates, and also calculate a standard error of the expression using a first-order Taylor series approximation.

Custom Inverse Prediction

Given a *y*-value, the **Custom Inverse Prediction** command calculates the *x*-value yielding this *y*-value for the estimated model. It also calculates a standard error. JMP must be able to invert the model to obtain an inverse expression, and it can only do this if the model is a function of one variable mentioned one time in the expression. The standard error is based on the first-order Taylor series approximation using the inverted expression. The confidence interval uses a *t*-quantile with the standard error.

Save Pred Confid Limits
Save Indiv Confid Limits

The **Save** options create new columns in the current data table called LowerM, UpperM, LowerI, and UpperI that contain the estimated lower- and upper-95% confidence limits computed using the asymptotic linear approximation. **Save Pred Confid Limits** saves confidence limits of the nonlinear fit, i.e. the expected or fitted response, which involves the variances in the estimates, but does not involve the variance of the error term itself. **Save Indiv Confid Limits** saves confidence limit values for the individual predicted values, which adds the variance of the error term to the variance of prediction involving the estimates, to form the interval.

Save Formulas

Save Prediction Formula Creates a new column containing the prediction model as its formula, with all the parameters replaced by their estimates. The values produced will be the same as in the Model column after you click the Save Estimates button. The resulting column will be called "Fitted name" where "name" is the name of the column specifying the model.

Save Std Error of Predicted Creates a new column containing the formula for the standard error of the prediction, with all the parameters replaced by their estimates. This standard error accounts for all the uncertainty in the parameter estimates, but does not add the uncertainty in predicting individual *Y*'s. The resulting column will be called StdError Fitted *name* where *name* is the name of the column specifying the model. The formula is of the form `Sqrt(VecQuadratic(matrix1,vector1))` where `matrix1` is the covariance matrix associated with the parameter estimates, and `vector1` is a composition of the partial derivatives of the model with respect to each parameter.

Save Std Error of Individual Creates a new column containing the formula for the standard error of the individual value, with all the parameters replaced by their estimates. This standard error accounts for the uncertainty in both the parameter estimates and in the error for the model. The resulting column will be called StdError Indiv *name* where *name* is the name of the column specifying the model. The formula is of the form `Sqrt(VecQuadratic(matrix1,vector1)+mse)` where `matrix1` is the covariance matrix associated with the parameter estimates, `vector1` is a composition of the partial derivatives of the model with respect to each parameter, and *mse* is the estimate of error variance.

Save Residual Formula Creates a column containing a formula to calculate the residuals from the current model. The new column is named Resuduals *name* where *name* is the name of the Y column.

Save Inverse Prediction Formula Creates a column containing a formula to predict the *x* value given the *y* value. The resulting column is called InvPred *name* where *name* is the name of the *Y* column. Two additional columns are created with formulas for the standard error of prediction and the standard error of individual, named StdError InvPred *name* and StdErrIndiv InvPred *name* respectively.

The internal implementation of this uses the same routine as the JSL operator `InvertExpr`, and has the same limitations, that the expression contain invertible expressions only for the expression path leading to the inversion target, that the target be in the formula only once, and that it use favorable assumptions to limit functions to an invertible region.

Save Specific Solving Formula is, in simple cases, equivalent to **Save Inverse Prediction Formula.** However, this command allows the formula to be a function of several variables and allows expressions to be substituted. This feature only works for solving easily invertible operators and functions that occur just once in the formula.

After selecting this command, a dialog appears that allows you to select the variable to solve for. Here, you can also edit the names of the columns in the resulting table.

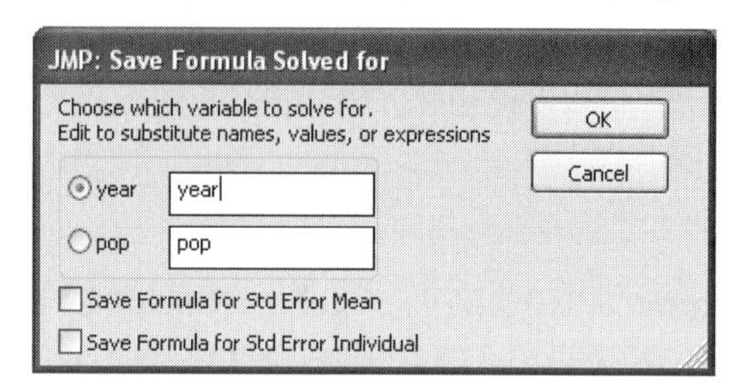

You can also substitute values for the names in the dialog. In these cases, the formula is solved for those values.

Show Prediction Expression

The **Show Prediction Expression** command toggles the display of the fitting expression at the top of the report.

Details of Solution Results

When you click **Go** on the iteration Control Panel, the fitting process begins. The iterations run in the background until the method either converges or fails to converge. During fitting, process messages on the Control Panel monitor the iterative process. When the iteration process completes, these messages show whether the convergence criteria are met.

If the convergence criteria are met, reports are appended to the iteration Control Panel, arranged as shown here. The **Solution** table and **Correlation of Estimates** table always appear. The graph of the fitted function is available only if the prediction formula is a function of one other variable (see Figure 23.4).

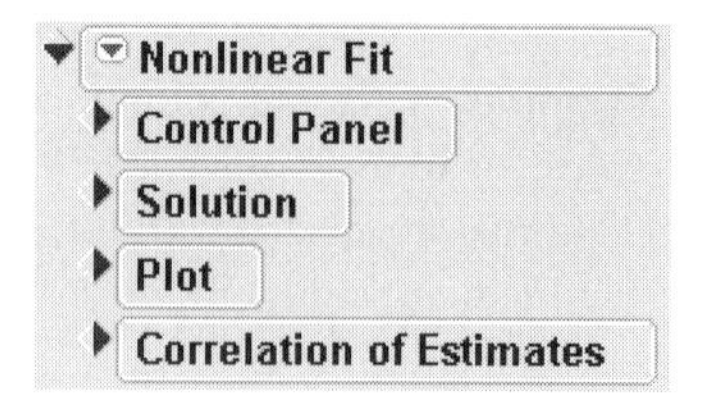

The Solution Table

When the iterations are complete, the results display in the Solution table like the one below.

Solution

SSE	DFE	MSE	RMSE
2245.5741473	20	112.27871	10.596165

Parameter	Estimate	ApproxStdErr
B0	15.052907631	1.45325695
B1	0.014188513	0.0005273

Solved By: Analytic NR

SSE shows the residual sum of squares error. SSE is the objective that is to be minimized. If a custom loss function is specified, this is the sum of the loss function.

DFE is the degrees of freedom for error, which is the number of observations used minus the number of parameters fitted.

MSE shows the mean squared error. It is the estimate of the variance of the residual error, which is the SSE divided by the DFE.

RMSE estimates the standard deviation of the residual error, which is square root of the MSE described above.

Parameter lists the names that you gave the parameters in the fitting formula.

Estimate lists the parameter estimates produced. Keep in mind that with nonlinear regression, there may be problems with this estimate even if everything seems to work.

ApproxStdErr lists the approximate standard error, which is computed analogously to linear regression. It is formed by the product of the RMSE and the square root of the diagonals of the derivative cross-products matrix inverse.

Lower CL and Upper CL are the lower and upper $100(1 - \alpha)$ percent confidence limits for the parameters. They are missing until you click the **Confidence Limits** on the Control Panel.

The upper and lower confidence limits are based on a search for the value of each parameter after minimizing with respect to the other parameters, that produces a SSE greater by a certain amount than the solution's minimum SSE. The goal of this difference is based on the *F*-distribution. The intervals are sometimes called *likelihood confidence intervals* or *profile likelihood confidence intervals* (Bates and Watts 1988; Ratkowsky 1990).

Solution

SSE	DFE	MSE	RMSE
2245.5741473	20	112.27871	10.596165

Parameter	Estimate	ApproxStdErr	Lower CL	Upper CL
B0	15.052907631	1.45325695	12.3475416	18.0871285
B1	0.014188513	0.0005273	0.0131793	0.01525974

Solved By: Analytic NR

Excluded Points

Sometimes, data is held back as a set for cross-validation of the results. The Nonlinear platform now calculates statistics of fit for excluded data. You can use this feature in conjunction with **Remember Solution** to change the exclusions, and get a new report reflecting the different exclusions.

For example, the following report uses US Population.jmp. Ten percent of the data points were excluded, then a nonlinear fit was run. This solution was remembered as "First Excluded", then another ten percent of the data points were excluded. This solution was remembered as "Second Excluded.

Figure 23.10 Excluded and Remembered Models

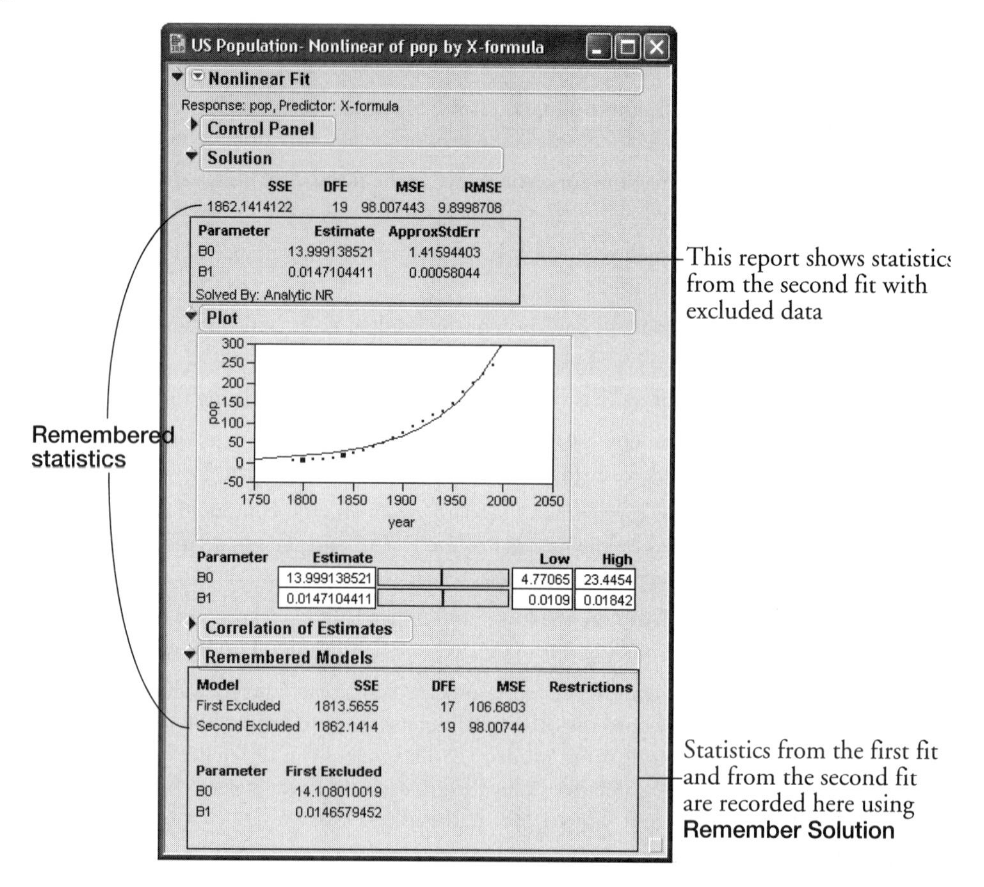

Profile Confidence Limits

Profile confidence limits all start with a *goal SSE*, which is a sum of squared errors (or sum of loss function) that an F-test considers significantly different from the solution SSE at the given alpha level. If the loss function is specified to be a negative log-likelihood, then a Chi-square quantile is used instead of an *F* quantile. For each parameter's upper confidence limit, the parameter value is moved up until the SSE reaches the goal SSE, but as the parameter value is moved up, all the other parameters are adjusted to be least squares estimates subject to the change in the profiled parameter. Conceptually, this is a compounded set of nested iterations, but internally there is a way to do this with one set of iterations developed by Johnston and DeLong (see SAS/Stat 9.1 vol. 3 pp. 1666-1667).

The diagram in Figure 23.11, shows the contour of the goal SSE or negative likelihood, with the least squares (or least loss) solution inside the shaded region:

- The asymptotic standard errors produce confidence intervals that approximate the region with an ellipsoid and take the parameter values at the extremes (at the horizontal and vertical tangents).
- Profile confidence limits find the parameter values at the extremes of the true region, rather than the approximating ellipsoid.

Figure 23.11 Diagram of Confidence Region

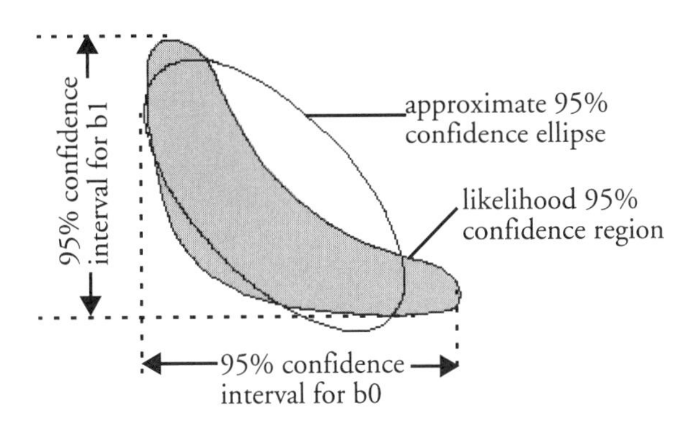

Likelihood confidence intervals are more trustworthy than confidence intervals calculated from approximate standard errors. If a particular limit cannot be found, computations begin for the next limit. When you have difficulty obtaining convergence, try the following:

- use a larger alpha, resulting in a shorter interval, more likely to be better behaved
- use the option for second derivatives
- relax the confidence limit criteria.

Fitted Function Graph

If the prediction formula is a function of exactly one other variable and you specify the response separately as the *Y* variable, then the **Plot** disclosure button displays a plot of the function and the observed data as shown here.

The sliders to the right of the estimates beneath the graph can be used to experiment with the parameter values and observe the effect on the curve. As you change parameter values, the graph changes accordingly.

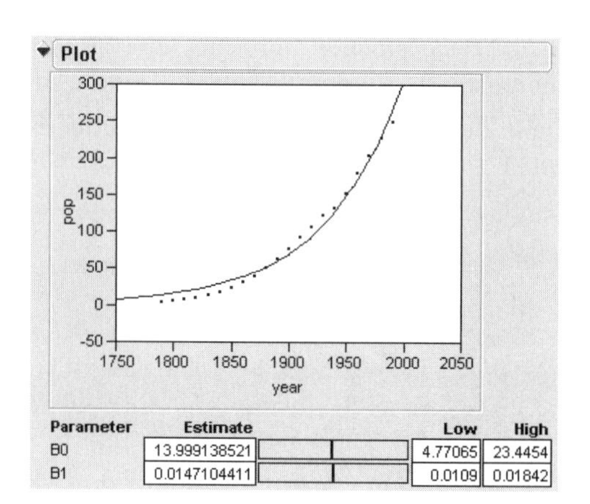

Chemical Kinetics Example

In pharmacology, the relationship between the concentration of an available dissolved organic substrate and the rate of uptake of the substrate is often modeled by the Michaelis-Menten equation, of the form

$$Y = \frac{V_{max}x}{K+x}$$

where Y is the velocity of uptake and x is the concentration. The parameters V_{max} and K estimate the maximum velocity and transport constant, respectively. The **Chemical Kinetics.jmp** data table (found in the **Nonlinear Examples** folder of the sample data) records the uptake velocity of a glucose-type substrate in an incubated sediment sample. The data table stores the Michaelis-Menten model with starting parameter values of Vmax=0.04 and K=0.

1 Select **Nonlinear** from the **Analyze > Modeling** menu and assign Model(x) as the **X, Predictor Formula** and Velocity(y) as **Y, Response**. Click **OK** and when the Control Panel appears, first use the **Plot** option, from the popup menu on the Nonlinear title bar, to see the model at the current starting values.

2 Next, adjust the **Lo** and **Hi** values of the k slider to be -2 to 2. Use the slider to change the value of K slightly, so you can get a better starting fit.

3 Click **Step** once, to see how the parameters and plot change.

4 Click **Go** to continue the iteration steps until convergence.

Figure 23.12 illustrates the fitting process described above.

Figure 23.12 Example of Plot Option to Observe Nonlinear Fitting Process

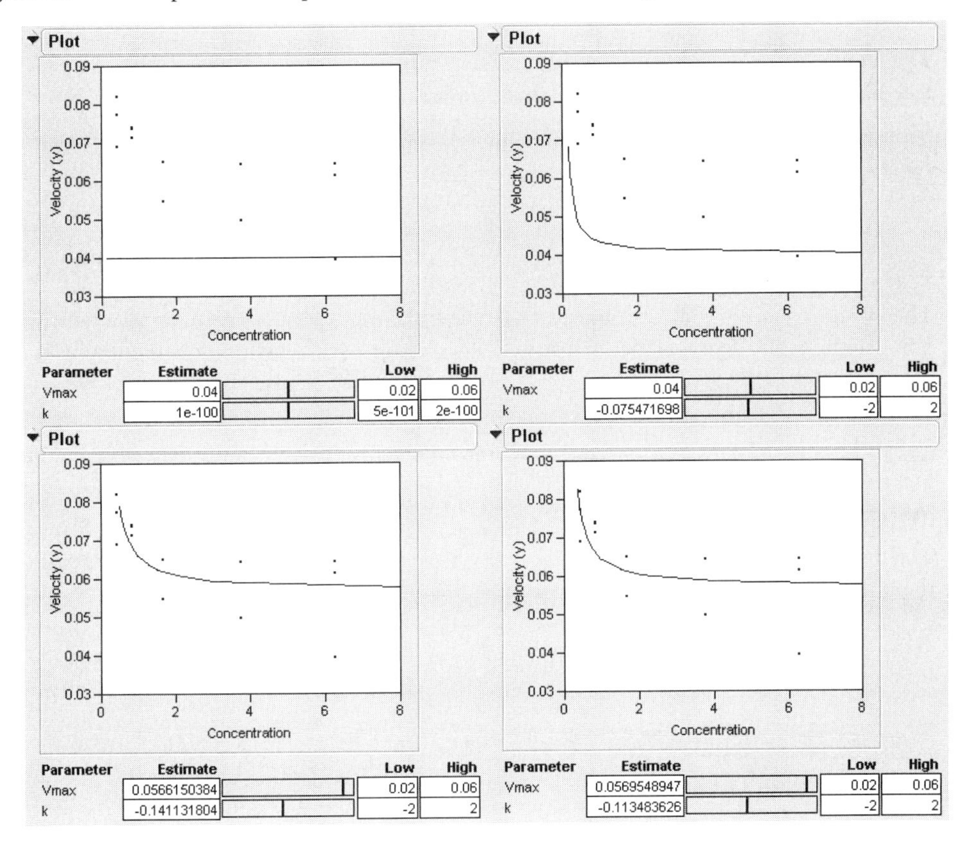

The **SSE Grid** option in the Nonlinear Fit popup menu helps you investigate the error sum of squares around the parameter estimates. An unrotated spinning plot (home position) of the new table displays a contour of the SSE as a function of the parameter estimates grid, with a highlighted x marking the solution. Alternatively, you can view the surface using the Surface Plot platform. Spin the plot to view the surface around the estimates, as shown for both cases in Figure 23.13.

Figure 23.13 Spinning SSE Grid

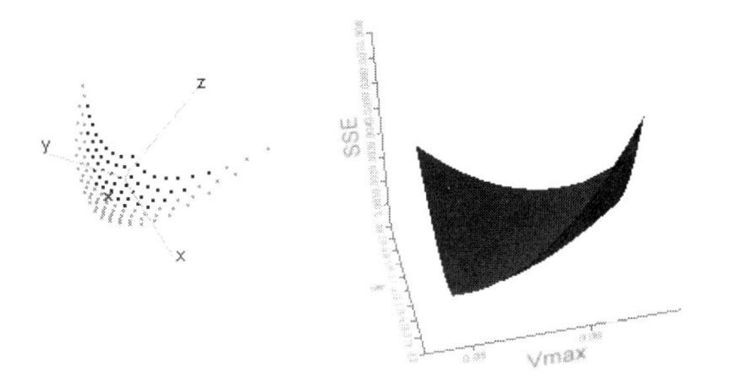

How Custom Loss Functions Work

The nonlinear facility can minimize or maximize functions other than the default sum of squares residual. This section shows the mathematics of how it is done.

Suppose that $f(\beta)$ is the model. Then the Nonlinear platform attempts to minimize the sum of the loss functions written as

$$L = \sum_{i=1}^{n} \rho(f(\beta))$$

The loss function $\rho(\bullet)$ for each row can be a function of other variables in the data table. It must have non-zero first- and second-order derivatives. The default $\rho(\bullet)$ function, squared-residuals, is

$$\rho(f(\beta)) = (y - f(\beta))^2$$

To specify a model with a custom loss function, construct a variable in the data table and build the loss function. After launching the Nonlinear platform, select the column containing the loss function as the loss variable.

The nonlinear minimization formula works by taking the first two derivatives of $\rho(\bullet)$ with respect to the model, and forming the gradient and an approximate Hessian as follows:

$$L = \sum_{i=1}^{n} \rho(f(\beta))$$

$$\frac{\partial L}{\partial \beta_j} = \sum_{i=1}^{n} \frac{\partial \rho(f(\beta))}{\partial f} \frac{\partial f}{\partial \beta_j}$$

$$\frac{\partial^2 L}{\partial \beta_j \partial \beta_k} = \sum_{i=1}^{n} \left[\frac{\partial^2 \rho(f(\beta))}{(\partial f)^2} \frac{\partial f}{\partial \beta_j} \frac{\partial f}{\partial \beta_k} + \frac{\partial \rho(f(\beta))}{\partial f} \frac{\partial^2 f}{\partial \beta_k \partial \beta_j} \right]$$

If $f(\bullet)$ is linear in the parameters, the second term in the last equation is zero. If not, you can still hope that its sum is small relative to the first term, and use

$$\frac{\partial^2 L}{\partial \beta_j \partial \beta_k} \cong \sum_{i=1}^{n} \frac{\partial^2 \rho(f(\beta))}{(\partial f)^2} \frac{\partial f}{\partial \beta_j} \frac{\partial f}{\partial \beta_k}$$

The second term will probably be small if ρ is the squared residual because the sum of residuals is small—zero if there is an intercept term. For least squares, this is the term that distinguishes Gauss-Newton from Newton-Raphson. In JMP, the second term is calculated only if the option **Second Deriv. Method** is checked.

Parameters in the Loss Function

You can use parameters in loss functions but they might not be the same as the parameters in the model. JMP uses first and second derivatives for these parameters, which means it takes a full Newton-Raphson steps for these parameters.

Maximum Likelihood Example: Logistic Regression

In this example, we show several variations of minimizing a loss function, in which the loss function is the negative of a log-likelihood function, thus producing maximum likelihood estimates.

The **Logistic w Loss.jmp** data table in the **Nonlinear Examples** sample data folder has an example for fitting a logistic regression using a loss function. Figure 23.14 shows a partial listing of the **Logistic w Loss.jmp** table.

The Y column is the proportion of responses (proportion number of 1s) for equal-sized samples of x values. The **Model Y** column has the linear model formula **b0+b1*X**. However, the **Loss** column has the nonlinear formula shown beneath the model formula. It is the negative log-likelihood for each observation, which is the negative log of the probability of getting the response that you did get.

Figure 23.14 The **Logistic w Loss.jmp** Data Table and Formulas

Model Formula $b_0 + b_1X$

Loss Formula

$$\text{If}\begin{cases} Y==1 \Rightarrow -\text{Log}\left(\dfrac{1}{(1+\text{Exp}(\textit{Model Y}))}\right) \\ \text{else} \Rightarrow -\text{Log}\left(1-\dfrac{1}{(1+\text{Exp}(\textit{Model Y}))}\right) \end{cases}$$

	Y	X	Model Y	Loss	Model2 Y	Loss2
1	0.57	20	0.0000	0.6931	0.5000	0.6931
2	0.72	30	0.0000	0.6931	0.5000	0.6931
3	0.81	40	0.0000	0.6931	0.5000	0.6931
4	0.87	50	0.0000	0.6931	0.5000	0.6931
5	0.91	60	0.0000	0.6931	0.5000	0.6931
6	0.94	70	0.0000	0.6931	0.5000	0.6931

To run the model, choose **Analyze >Modeling > Nonlinear** and complete the Launch dialog by using **Model Y** as the **X, Predictor Formula**, and **Loss** as the **Loss** function. Click **OK** to see the Iteration Control Panel. The initial values of zero (or 10^{-100}, very close to zero) for the parameters **b0** and **b1** are satisfactory.

Click **Go** to run the model. The results show in the Solution table with the negative log-likelihood as the SSE.

Solution

Loss	DFE	Avg Loss	Sqrt Avg Loss
7.8070527905	18	0.4337252	0.6585781

Parameter	Estimate	ApproxStdErr	Lower CL	Upper CL
b0	5.5114510517	2.43728165	1.86656603	11.9850962
b1	-0.034755348	0.01595111	-0.0761181	-0.0098423

Solved By: Analytic NR

The same problem can be handled differently by defining a model column formula that absorbs the logistic function and a loss function that uses the model column **Model2** to form the probability for a categorical response level. The negative log-likelihood for **Loss2** is the probability for the event that occurred. This value is either P(Model occurred), or 1 – P(Model occurred)

In this situation it is best to use second derivatives because the model is nonlinear. With least squares, the second derivatives are multiplied by residuals, which are usually near zero. For custom loss functions, second derivatives can play a stronger role.

Select **Nonlinear** again and complete the role assignment dialog with **Y** as **Y, Response**, **Model2 Y** as **X, Predictor Formula**, and **Loss2** as the loss function.

Model2 Y Formula

$$\frac{1}{(1+\text{Exp}(\mathbf{b0}+\mathbf{b1}*X))}$$

Loss2 Formula

$$-\text{If}\begin{pmatrix} Y==1 \Rightarrow \text{Log}(\textit{Model2 Y}) \\ \text{else} \Rightarrow \text{Log}(1-\textit{Model2 Y}) \end{pmatrix}$$

Check the **Second Derivatives** check box on the Launch dialog and click **OK**. Then run the model again. The Solution table shows the same log-likelihood values and same parameter estimates as before, with slightly different asymptotic standard errors.

Iteratively Reweighted Least Squares Example

Iteratively Reweighted Least Squares (IRLS) is used for robust regression (Holland and Welsch 1977) and to fit maximum likelihoods for many models (McCullagh and Nelder 1983). The Nonlinear platform can use a weight column with a formula and repeatedly reevaluate the formula if it is a function of the model. These methods can serve as automatic outlier down-weighters because, depending on the weight function, large residual values can lead to very small weights. For example, Beaton and Tukey (1974) suggest minimizing the function

$$S_{\text{biweight}} = \sum_{i=1}^{n} \rho(r_i)$$

$$\rho(r_i) = \left(\frac{B^2}{2}\right)\left(1-\left(1-\left(\frac{r_i}{B}\right)^2\right)^2\right) \text{ if } |r_i| \le B \quad \text{, where}$$

$$\rho(r_i) = \frac{B^2}{2} \text{ otherwise}$$

$$r = \frac{|\text{residual}|}{\sigma}$$

σ is measure of the error, such as $1.5 \times \text{med}(|\text{LS residuals}|)$, (Myers, 1989),

B = a tuning constant, usually between 1 and 3.

The **IRLS Example.jmp** sample data table has columns constructed to illustrate an IRLS process.

IRLS Example
Notes Iteratively Rewight
Columns (6/0): pop, year, model, B, weight, weight2
Rows
All rows 24

	pop	year	model	B	weight	weight2
1	3.93	1790	-3205887.	1.9	0.8117	0.8117
2	5.31	1800	-3241795.	1.9	0.8077	0.8077
3	7.24	1810	-3277903.	1.9	0.8036	0.8036
4	9.64	1820	-3314211.	1.9	0.7995	0.7995
5	12.87	1830	-3350718.	1.9	0.7953	0.7953
6	17.07	1840	-3387423.	1.9	0.7910	0.7910
7	23.19	1850	-3424327.	1.9	0.7867	0.7867
8	31.44	1860	-3461429.	1.9	0.7824	0.7824
9	39.82	1870	-3498731.	1.9	0.7779	0.7779
10	50.16	1880	-3536230.	1.9	0.7735	0.7735

The **IRLS Example** table columns use the following formulas:

- the model column (**model**), with a model formula that uses three parameters **b0**, **b1**, and **b2**
```
pop - (b0 + b1*year + b2*year*year)
```

- *B*, a tuning constant
- the weight column (weight), with a formula that uses model, *B*, and a local variable *t1*, a measure of error:

$$t1 = 1.5 * \text{Col Quantile}\left(|model|, 0.5\right);$$

$$\text{If}\begin{cases} \left|\dfrac{model}{t1}\right| <= B \Rightarrow \left(1 - \left(\dfrac{\left|\dfrac{model}{t1}\right|}{B}\right)^2\right)^2 \\ \text{else} \Rightarrow 0 \end{cases};$$

Myers (1989) suggests limiting the tuning constant to a range of 1 to 3. If you're having trouble getting your model to converge, try moving the tuning constant closer to 2. For this example, B (the tuning factor) is equal to 1.9. Myers also proposes using a a measure of error computed as $1.5 \times \text{med}(|\text{LS residuals}|)$. In this example, the formula includes a JSL statement that defines the local variable t1 as $1.5 \times \text{quantiles}_{0.5}(|\text{model}|)$.

Note: The variable that specifies the model must appear in the weight formula so that weights are updated with each iteration. If the actual formula is used, JMP treats the parameters as different sets of parameters and doesn't update them.

To complete the IRLS example, follow these steps:

1 (Simple Least Squares Step) Select **Analyze >Modeling > Nonlinear.** Choose model as the **X, Predictor Formula** column but don't select a weight yet. The parameter estimates from this least squares analysis provide starting values for the weighted model. Click **OK** on the Launch dialog and then **Go** when the Control Panel appears. You should get the starting values shown here.

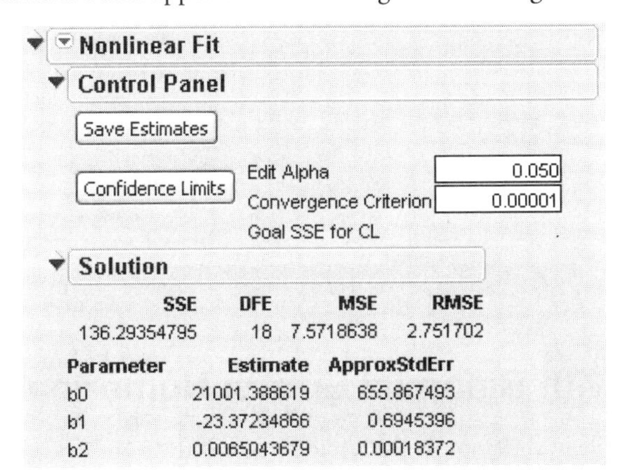

At the end of the iterations, click **Save Estimates** on the Control Panel to set the b0, b1, and b2 parameters to these estimates, which are needed for the IRLS step.

2 (IRLS Step) Again, select **Nonlinear.** Assign model as the prediction column and weight as the **Weight** variable. Click **OK** to see the Control Panel and then click **Go** to get the final parameter estimates shown in Figure 23.15. The SSE is reduced from 136.2935 for the least squares analysis to 3.0492 for the IRLS analysis.

IRLS gives small weights to extreme values, which tend to reject them from the analysis. Years 1940 and 1950 are easy to identify as outliers because their weights are very small. IRLS reduces the impact of overly influential data points and can thereby better model the rest of the data. Figure 23.16 shows a partial listing of the final data table.

Figure 23.15 Final IRLS Solution Table

Solution

SSE	DFE	MSE	RMSE
3.0488660287	18	0.1693814	0.4115598

Parameter	Estimate	ApproxStdErr
b0	21417.272994	140.711009
b1	-23.82668091	0.14899575
b2	0.0066285009	3.94146e-5

Figure 23.16 Final Data Table

IRLS Example
Notes Iteratively Rewight

Columns (6/0)
pop
year
model
B
weight
weight2

Rows
All rows 24
Selected 0
Excluded 0
Hidden 0
Labelled 0

	pop	year	model	B	weight	weight2
1	3.93	1790	-1.600658	1.9	0.3241	0.3241
2	5.31	1800	-0.004979	1.9	1.0000	1.0000
3	7.24	1810	0.841826	1.9	0.7759	0.7759
4	9.64	1820	0.855758	1.9	0.7689	0.7689
5	12.87	1830	0.397817	1.9	0.9475	0.9475
6	17.07	1840	-0.385997	1.9	0.9505	0.9505
7	23.19	1850	-0.551685	1.9	0.9003	0.9003
8	31.44	1860	0.111752	1.9	0.9958	0.9958
9	39.82	1870	-0.402683	1.9	0.9462	0.9462
10	50.16	1880	-0.255992	1.9	0.9781	0.9781
11	62.95	1890	1.044825	1.9	0.6666	0.6666
12	75.99	1900	1.299768	1.9	0.5126	0.5126
13	91.97	1910	3.184838	1.9	0.0000	0.0000
14	105.71	1920	1.529035	1.9	0.3684	0.3684
15	122.78	1930	1.899358	1.9	0.1549	0.1549
16	131.67	1940	-7.202192	1.9	0.0000	0.0000
17	151.33	1950	-6.842616	1.9	0.0000	0.0000
18	179.32	1960	0.558086	1.9	0.8980	0.8980
19	203.21	1970	2.547914	1.9	0.0000	0.0000
20	226.50	1980	2.637869	1.9	0.0000	0.0000
21	248.70	1990	0.337951	1.9	0.9620	0.9620
22	•	2000	•	1.9	0.0000	0.0000
23	•	2010	•	1.9	0.0000	0.0000
24	•	2020	•	1.9	0.0000	0.0000

Probit Model with Binomial Errors: Numerical Derivatives

The Ingots2.jmp file in the Sample Data folder records the numbers of ingots tested for readiness after different treatments of heating and soaking times.

Ingots2
Notes From D.R.Cox

Columns (7/0)
Heat
Soak
Nnotready
Nready
Ntotal
P
Loss

Rows
All rows 19
Selected 0
Excluded 0
Hidden 0
Labelled 0

	Heat	Soak	Nnotready	Nready	Ntotal	P	Loss
1	7	1	10	0	10	0.500	6.9315
2	7	1.7	17	0	17	0.500	11.7835
3	7	2.2	7	0	7	0.500	4.8520
4	7	2.8	12	0	12	0.500	8.3178
5	7	4	9	0	9	0.500	6.2383
6	14	1	31	0	31	0.500	21.4876
7	14	1.7	43	0	43	0.500	29.8053
8	14	2.2	31	2	33	0.500	22.8739
9	14	2.8	31	0	31	0.500	21.4876
10	14	4	19	0	19	0.500	13.1698
11	27	1	55	1	56	0.500	38.8162
12	27	1.7	40	4	44	0.500	30.4985
13	27	2.2	21	0	21	0.500	14.5561
14	27	2.8	21	1	22	0.500	15.2492
15	27	4	15	1	16	0.500	11.0904
16	51	1	10	3	13	0.500	9.0109
17	51	1.7	1	0	1	0.500	0.6931
18	51	2.2	1	0	1	0.500	0.6931
19	51	4	1	0	1	0.500	0.6931

The response variable, NReady, is binomial, depending on the number of ingots tested (NTotal) and the heating and soaking times. Maximum likelihood estimates for parameters from a probit model with binomial errors are obtained using

- numerical derivatives
- the negative log-likelihood as a loss function
- the Newton-Raphson method.

The average number of ingots ready is the product of the number tested and the probability that an ingot is ready for use given the amount of time it was heated and soaked. Using a probit model, the fitting equation for the model variable, P, is

```
Normal Distribution[b0+b1*Heat+b2*Soak]
```

The argument to the `Normal Distribution` function is a linear model of the treatments. Because the `Normal Distribution` function is used, the Nonlinear platform specifies numerical derivatives. For more information on numeric derivatives see "Notes Concerning Derivatives," p. 491.

To specify binomial errors, the loss function, Loss, has the formula

```
-[Nready*Log[p] + [Ntotal - Nready]*Log[1 - p]]
```

To complete the Nonlinear Launch dialog, specify P as the **X, Predictor Formula** and Loss as **Loss**. Click **Second Derivatives** on the Launch dialog. Click **OK**

Default starting values of (near) zero for the parameters should result in the solution shown in Figure 23.17. JMP used the Newton-Raphson, or Second Derivatives method, to obtain the solution.

Figure 23.17 Solution for the Ingots2 Data Using the Second Derivatives Method

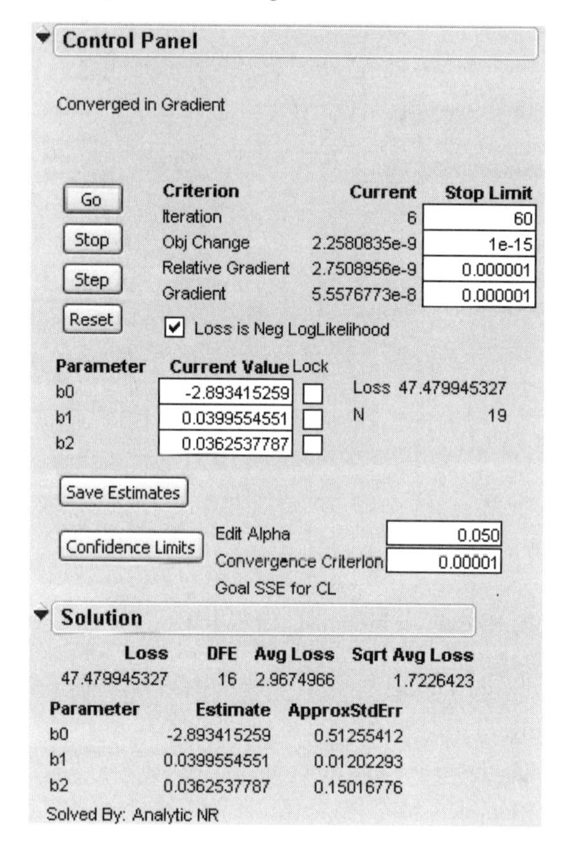

Poisson Loss Function

A Poisson distribution is often used as a model to fit frequency counts.

$$P(Y = n) = \frac{e^{-\mu}\mu^{n}}{n!}, \quad n = 0, 1, 2, \ldots$$

where μ can be a single parameter, or a linear model with many parameters. Many texts and papers show how the model can be transformed and fit with iteratively reweighted least squares (Nelder and Wedderburn 1972). However, in JMP it is more straightforward to fit the model directly. For example, McCullagh and Nelder (1989) show how to analyze the number of reported damage incidents caused by waves to cargo-carrying vessels. The data are in the Ship Damage.jmp sample data, and includes columns called model and Poisson.

The formula for the Poisson column is the loss function (or –log-likelihood) for the Poisson distribution, which is the log of the probability distribution function:

```
-(N*model-Exp(model)-Log(Gamma(N+1))
```

where Gamma is the Γ function $\Gamma(n + 1) = n!$ and e^{model} represents μ.

The formula for model is

$$\mathbf{b0} + \text{MatchMZ}(\textit{Type})\begin{pmatrix} \text{"B"} \Rightarrow \mathbf{B} \\ \text{"C"} \Rightarrow \mathbf{C} \\ \text{"D"} \Rightarrow \mathbf{D} \\ \text{"E"} \Rightarrow \mathbf{E} \\ \text{else} \Rightarrow 0 \end{pmatrix} + \text{MatchMZ}(\textit{Yr Made})\begin{pmatrix} \text{"65"} \Rightarrow \mathbf{Yr\ 65} \\ \text{"70"} \Rightarrow \mathbf{Yr\ 70} \\ \text{"75"} \Rightarrow \mathbf{Yr\ 75} \\ \text{else} \Rightarrow 0 \end{pmatrix} + \text{MatchMZ}(\textit{Yr Used})\begin{pmatrix} \text{"75"} \Rightarrow \mathbf{Used\ 75} \\ \text{else} \Rightarrow 0 \end{pmatrix} + \textit{Service}$$

To run the model, choose **Analyze > Modeling > Nonlinear**. In the Launch dialog, select model as the **X, Predictor Formula** and Poisson as **Loss**, and then click **OK**. In the Control Panel, set the initial value for B0 to 1, the other parameters to 0, and click **Go**. The table in Figure 23.18 shows the results after the model has run and you select **Confidence Limits**.

Figure 23.18 Solution Table for the Poisson Loss Example

Solution

Loss	DFE	Avg Loss	Sqrt Avg Loss
68.281234087	25	2.7312494	1.6526492

Parameter	Estimate	ApproxStdErr	Lower CL	Upper CL
b0	-6.405914771	0.21744445	-6.8430512	-5.9896833
B	-0.543353982	0.17758996	-0.881379	-0.183552
C	-0.687418388	0.32904722	-1.3764541	-0.0745299
D	-0.075979835	0.2905789	-0.6715296	0.47523936
E	0.325581683	0.23587934	-0.143451	0.78520431
Yr 65	0.6971496668	0.1496412	0.40752821	0.99512758
Yr 70	0.8184263427	0.16977314	0.48728046	1.15369087
Yr 75	0.4534435315	0.23317069	-0.0123098	0.90388895
Used 75	0.3844880718	0.11827198	0.15341509	0.61742312

Solved By: Analytic NR

Notes Concerning Derivatives

The nonlinear platform takes symbolic derivatives for formulas with most common operations. This section shows what kind of derivative expressions result.

If you open the Negative Exponential.jmp nonlinear sample data example, the actual formula looks something like this:

```
Parameter({b0=0.5, b1=0.5,}b0*(1-Exp(-b1*X)))
```

The Parameter block in the formula is hidden if you use the formula editor, but that is how it is stored in the column and how it appears in the Nonlinear Launch dialog. Two parameters named **b0** and **b1** are given initial values and used in the formula to be fit.

The Nonlinear platform makes a separate copy of the formula, and edits it to extract the parameters from the expression and maps the references to them to the place where they will be estimated. Nonlinear takes the analytic derivatives of the prediction formula with respect to the parameters. If you use the **Show Derivatives** command, you get the resulting formulas listed in the log, like this:

Prediction Model:

```
b0 * First(T#1=1-(T#2=Exp(-b1*X)), T#3=-(-1*T#2*X))
```

The Derivative of Model with respect to the parameters is:

```
{T#1, T#3*b0}
```

The derivative facility works like this:

- In order to avoid calculating subexpressions repeatedly, the prediction model is threaded with assignments to store the values of subexpressions that it will need for derivative calculations. The assignments are made to names like `T#1`, `T#2`, and so forth.
- When the prediction model needs additional subexpressions evaluated, it uses the `First` function, which returns the value of the first argument expression, and also evaluates the other arguments. In this case additional assignments that will be needed for derivatives.
- The derivative table itself is a list of expressions, one expression for each parameter to be fit. For example, the derivative of the model with respect to **b0** is `T#1`; its thread in the prediction model is `1-(Exp(-b1*X))`. The derivative with respect to **b1** is `T#3*b0`, which is `-(-1*Exp(-b1*X)*X)*b0` if you substitute in the assignments above. Although many optimizations are made, it doesn't always combine the operations optimally, as you can see by the expression for `T#3`, which doesn't remove a double negation.

If you ask for second derivatives, then you get a list of $(m(m + 1))/2$ second derivative expressions in a list, where m is the number of parameters.

If you specify a loss function, then the formula editor takes derivatives with respect to parameters, if it has any, and it takes first and second derivatives with respect to the model, if there is one.

If the derivative mechanism doesn't know how to take the analytic derivative of a function, then it takes numerical derivatives, using the `NumDeriv` function. If this occurs, then the platform shows the delta it used to evaluate the change in the function with respect to a delta change in the arguments. You may need to experiment with different delta settings to obtain good numerical derivatives.

Tips

There are always many ways to represent a given model, and some ways behave much better than other forms. Ratkowsky (1990) covers alternative forms in his text.

If you have repeated subexpressions that occur several places in a formula, then it is better to make an assignment to a temporary variable, and then refer to it later in the formula. For example, one of the model formulas above was this:

```
If(Y==0, Log(1/(1+Exp(model))), Log(1 - 1/(1 + Exp(model))));
```

This could be simplified by factoring out an expression and assigning it to a local variable:

```
temp=1/(1+Exp(model));
If(Y==0, Log(temp), Log(1-temp));
```

The derivative facility can track derivatives across assignments and conditionals.

Notes on Effective Nonlinear Modeling

We strongly encourage you to *center polynomials.*

Anywhere you have a complete polynomial term that is linear in the parameters, it is always good to center the polynomials. This improves the condition of the numerical surface for optimization. For example, if you have an expression like

$$a_1 + b_1x + c_1x^2$$

you should transform it to

$$a_2 + b_2(x - \bar{x}) + c_2(x - \bar{x})^2$$

The two models are equivalent, apart from a transformation of the parameters, but the second model is far easier to fit if the model is nonlinear.

The transformation of the parameters is easy to solve.

$$\begin{aligned} a_1 &= a_2 - b_2\bar{x} + c_2\bar{x} \\ b_1 &= b_2 - 2c_2\bar{x} \\ c_1 &= c_2 \end{aligned}$$

If the number of iterations still goes to the maximum, increase the maximum number of iterations and select **Second Deriv Method** from the red triangle menu.

There is really no one omnibus optimization method that works well on all problems. JMP has options like **Newton**, **QuasiNewton BFGS**, **QuasiNewton SR1**, **Second Deriv Method**, and **Numeric Derivatives Only** to expand the range of problems that are solvable by the Nonlinear Platform.

So if JMP's defaults are unable to converge to the solution for a particular problem, using various combinations of these settings increase the odds of obtaining convergence.

Some models are very sensitive to starting values of the parameters. Working on new starting values is often effective. Edit the starting values and click **Reset** to see the effect.

The plot often helps. Use the sliders to visually modify the curve to fit better. The parameter profilers can help, but may be too slow for anything but small data sets.

Notes Concerning Scripting

If you are using a JSL script to run the Nonlinear platform, then the following commands could be useful, along with the commands listed under the Nonlinear platform popup menu.

Table 23.1 JSL commands for Nonlinear

Model(Parameter({*name=expr*,…},*expr*)	The model to fit. If not specified, it looks in the *X* column's formula, if there is an *X* column.
Loss(*expr*) or Loss(Parameter({*name=expr*,…},*expr*)	The loss function to use. If not specified, it looks in the Loss columns formula, if there is a loss column.
Go	Like the **Go** button, it starts the iterations, but unlike the **Go** button, the iterations run in the foreground instead of the background.
Finish	Used instead of the Go command. `Finish` halts script execution until the fit is complete.
Reset	Same as the **Reset** button.
Set Parameter(*name=expression*,…)	To provide new starting values. Expression may even be a Column function like Col Maximum. name is a parameter name.
Lock Parameter(*name*,…)	To Lock or Unlock. name is a parameter name.
Save Estimates	Same as the **Save Estimates** button
Confidence Limits	Same as the **Confidence Limits** button
SSE Grid(*name*(*first*,*last*,),…)	To calculate the SSE on a Grid, on *n* values in the range specified for each parameter. Unspecified parameters are left at the solution value.

The first example below uses a By-group and specifies an expression to evaluate to get starting values. The second example specifies the model itself.

```
Nonlinear(x(Model),y(pop),By(group),
   Set Parameter(B0=ColMinimum(Pop),B1=.03),
   Go);
Nonlinear(y(pop),
   Model(
      Parameter({B0=3.9,B1=0.022},
      B0*Exp(B1*(year-1790)))),
   SSEGrid(B0(2,20,10),B1(.01,.05,6)));
```

Nonlinear Modeling Templates

This section shows examples of nonlinear models and sources of data for each model. These model formulas are in the Model Library, discussed in the section "Using the Model Library," p. 468.

Table 23.2 Guide to Nonlinear Modeling Templates

Data Reference	Formula	Model
Meyers (1988), p. 310	$\frac{\theta_1 x}{\theta_2 + x}$	A Michaelis-Menten

Table 23.2 Guide to Nonlinear Modeling Templates (continued)

Data Reference	Formula	Model
Draper and Smith (1981), p. 522, L	$\theta_1[1 - \exp(\theta_2 x)]$	B
Draper and Smith (1981), p. 476	$\theta_1 + (0.49 - \theta_1)\exp[-\theta_2(x - 8)]$	C
Draper and Smith (1981), p. 519, H	$\exp\left\{-\theta_1 x_1 \exp\left[\theta_2 \cdot \left(\frac{1}{x_2} - \frac{1}{620}\right)\right]\right\}$	D
Draper and Smith (1981), p. 519, H	$\theta_1 x^{\theta_2}$	E
Bates and Watts (1988), p. 310	$\theta_1 + \theta_2\exp(\theta_3 x)$	F Asymptotic Regression
Bates and Watts (1988), p. 310	$\frac{\theta_1}{\theta_2 \exp(\theta_3 x)}$	G Logistic
Bates and Watts (1988), p. 310	$\theta_1\exp[-\exp(\theta_2 - \theta_3 x)]$	H Gompertz Growth
Draper and Smith (1981), p. 524, N	$\theta_1[1 - (\theta_2 e)^{\theta_3 x}]$	I
Draper and Smith (1981), p. 524, N	$\theta_1 - \ln[1 + \theta_2\exp(-\theta_3 x)]$	J Log-Logistic
Draper and Smith (1981), p. 524, P	$\theta_1 + \frac{\theta_2}{x^{\theta_3}}$	K
Draper and Smith (1981), p. 524, P	$\ln[\theta_1\exp(-\theta_2 x) + (1 - \theta_1)\exp(-\theta_3 x)]$	L
Bates and Watts (1988), p. 271	$\frac{\theta_1\theta_3\left(x_2 - \frac{x_3}{1.632}\right)}{1 + \theta_2 x_1 + \theta_3 x_2 + \theta_4 x_3}$	M
Bates and Watts (1988), p.310	$\frac{\theta_2\theta_3 + \theta_1 x^{\theta_4}}{\theta_3 + x^{\theta_4}}$	O Morgan-Mercer-Florin
Bates and Watts (1988), p. 310	$\frac{\theta_1}{[1 + \theta_2 \exp(-\theta_3 x)]^{\frac{1}{\theta_4}}}$	P Richards Growth

Table 23.2 Guide to Nonlinear Modeling Templates (continued)

Data Reference	Formula	Model
Bates and Watts (1988), p. 274	$\frac{\theta_1}{\theta_2 + x_1} + \theta_3 x_2 + \theta_4 x_2^2 + \theta_5 x_2^3 + (\theta_6 + \theta_7 x_2^2) x_2 \exp\left(\frac{-x_1}{\theta_8 + \theta_9 x_2^2}\right)$	S

Chapter 24

Correlations and Multivariate Techniques

The Multivariate Platform

The **Multivariate** platform specializes in exploring how many variables relate to each other. The platform begins by showing a standard correlation matrix. The Multivariate platform popup menu gives the additional correlations options and other techniques for looking at multiple variables such as

- a scatterplot matrix with normal density ellipses
- inverse, partial, and pairwise correlations
- a covariance matrix
- nonparametric measures of association
- simple statistics (such as mean and standard deviation)
- an outlier analysis that shows how far a row is from the center, respecting the correlation structure.
- a principal components facility with several options for rotated components and a table showing rotated factor patterns and communalities.

All plots and the current data table are linked. You can highlight points on any scatterplot in the scatterplot matrix, or the outlier distance plot. The points are highlighted on all other plots and are selected in the data table.

Contents

Launch the Platform and Select Options

When you choose **Analyze > Multivariate Methods > Multivariate**, a standard correlation matrix and scatterplot matrix appears first. The platform popup menu shown here lists additional correlation options and other techniques for looking at multiple variables. The following sections describe the tables and plots offered by the Multivariate platform.

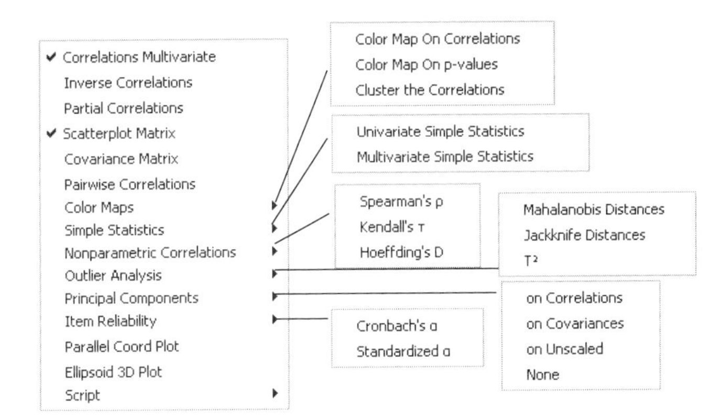

In most of the following analysis options, a missing value in an observation causes the entire observation to be deleted. The exceptions are in **Pairwise Correlations**, which exclude rows that are missing on either of the variables under consideration, and **Simple Statistics > Univariate**, which calculates its statistics column-by-column, without regard to missing values in other columns.

Many of the following examples use the **Solubility.jmp** sample data table.

Correlations Multivariate

The **Correlations Multivariate** option gives the Correlations table, which is a matrix of correlation coefficients that summarizes the strength of the linear relationships between each pair of response (*Y*) variables. This correlation matrix only uses the observations that have nonmissing values for all variables in the analysis.

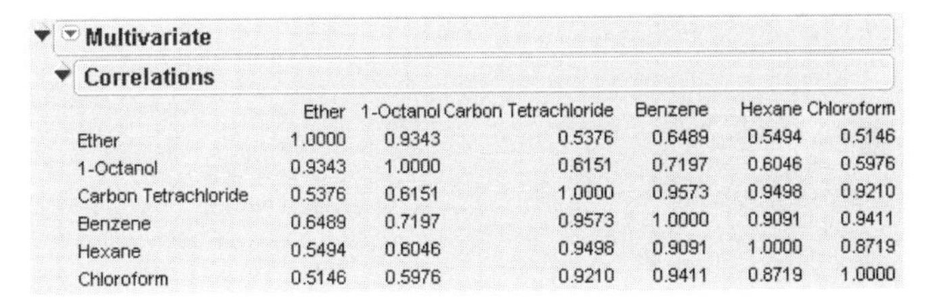

Multivariate

Correlations

	Ether	1-Octanol	Carbon Tetrachloride	Benzene	Hexane	Chloroform
Ether	1.0000	0.9343	0.5376	0.6489	0.5494	0.5146
1-Octanol	0.9343	1.0000	0.6151	0.7197	0.6046	0.5976
Carbon Tetrachloride	0.5376	0.6151	1.0000	0.9573	0.9498	0.9210
Benzene	0.6489	0.7197	0.9573	1.0000	0.9091	0.9411
Hexane	0.5494	0.6046	0.9498	0.9091	1.0000	0.8719
Chloroform	0.5146	0.5976	0.9210	0.9411	0.8719	1.0000

Inverse Correlations and Partial Correlations

The inverse correlation matrix (**Inverse Corr** table), shown at the top in the next figure, provides useful multivariate information. The diagonal elements of the matrix are a function of how closely the variable is a linear function of the other variables. In the inverse correlation, the diagonal is $1/(1 - R^2)$ for the fit of that variable by all the other variables. If the multiple correlation is zero, the diagonal inverse element is 1. If the multiple correlation is 1, then the inverse element becomes infinite and is reported missing.

Multivariate

Inverse Corr

	Ether	1-Octanol	Carbon Tetrachloride	Benzene	Hexane	Chloroform
Ether	8.5445	-7.7565	2.9732	-2.6483	-1.7262	1.4972
1-Octanol	-7.7565	9.6893	-0.2165	-2.8052	1.0470	0.1280
Carbon Tetrachloride	2.9732	-0.2165	25.2505	-15.5648	-10.8871	-0.5151
Benzene	-2.6483	-2.8052	-15.5648	27.6656	1.4066	-9.8884
Hexane	-1.7262	1.0470	-10.8871	1.4066	10.6777	-0.3445
Chloroform	1.4972	0.1280	-0.5151	-9.8884	-0.3445	10.2337

Partial Corr

	Ether	1-Octanol	Carbon Tetrachloride	Benzene	Hexane	Chloroform
Ether	.	0.8525	-0.2024	0.1722	0.1807	-0.1601
1-Octanol	0.8525	.	0.0138	0.1713	-0.1029	-0.0129
Carbon Tetrachloride	-0.2024	0.0138	.	0.5889	0.6630	0.0320
Benzene	0.1722	0.1713	0.5889	.	-0.0818	0.5877
Hexane	0.1807	-0.1029	0.6630	-0.0818	.	0.0330
Chloroform	-0.1601	-0.0129	0.0320	0.5877	0.0330	.

partialed with respect to all other variables

The partial correlation table (**Partial Corr** table) shows the partial correlations of each pair of variables after adjusting for all the other variables. This is the negative of the inverse correlation matrix scaled to unit diagonal.

Scatterplot Matrix

To help you visualize the correlations, a scatterplot for each pair of response variables displays in a matrix arrangement, as shown in Figure 24.1. The scatterplot matrix is shown by default. If the scatterplots are not showing, select **Scatterplot Matrix** from the platform popup menu. The cells of the scatterplot matrix are size-linked so that stretching a plot from any cell resizes all the scatterplot cells.

By default, a 95% bivariate normal density ellipse is imposed on each scatterplot. If the variables are bivariate normally distributed, this ellipse encloses approximately 95% of the points. The correlation of the variables is seen by the collapsing of the ellipse along the diagonal axis. If the ellipse is fairly round and is not diagonally oriented, the variables are uncorrelated.

Figure 24.1 Example of a Scatterplot Matrix

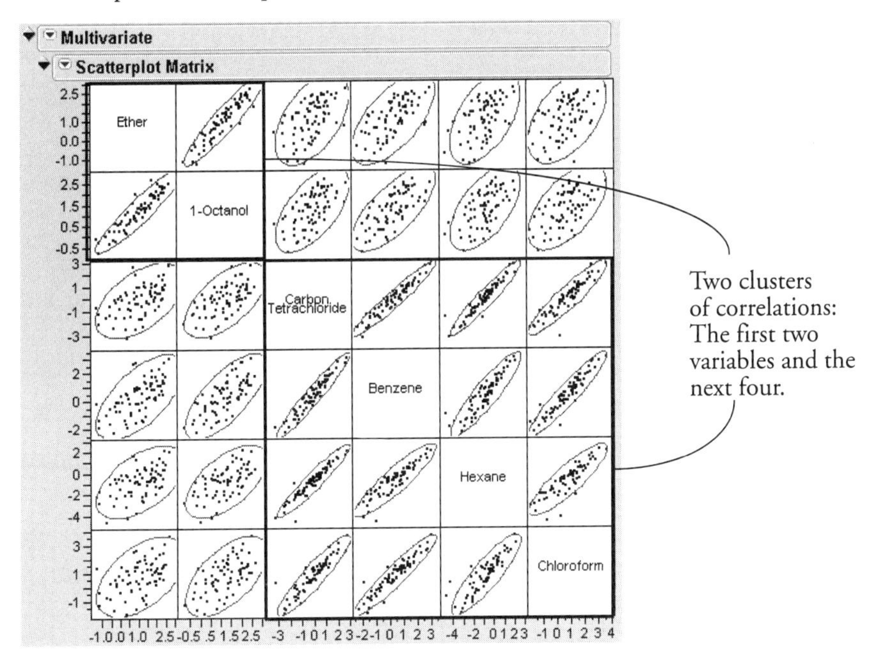

The popup menu next on the **Scatterplot Matrix** title bar button lets you tailor the matrix with color and density ellipses and by setting the α-level.

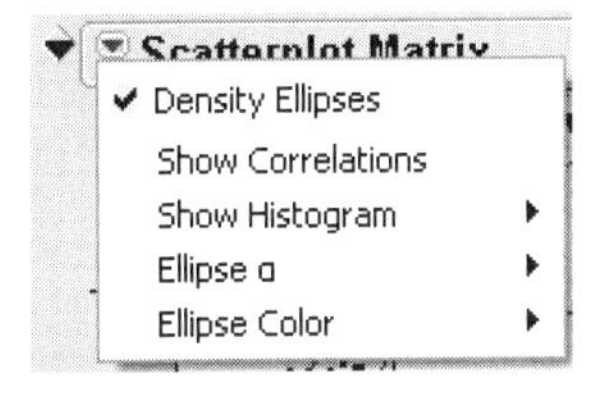

Density Ellipses toggles the display of the density ellipses on the scatterplots constructed by the α level that you choose. By default they are 95% ellipses.

Show Correlations shows the correlation of each histogram in the upper left corner of each scatterplot.

Show Histogram draws histograms in the diagonal of the scatterplot matrix. These histograms can be specified as **Horizontal** or **Vertical**. In addition, you can toggle the counts that label each bar with the **Show Counts** command.

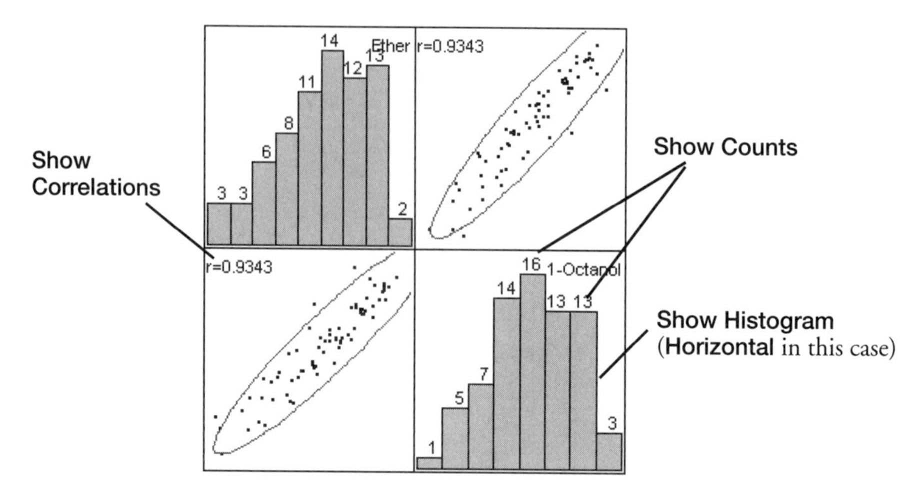

Ellipse α lets you select from a submenu of standard α-levels or select the **Other** command and specifically set the α level for the density ellipses.

Ellipse Color lets you select from a palette of colors to change the color of the ellipses.

You can reorder the scatterplot matrix columns by dragging a diagonal (label) cell to another position on the diagonal. For example, if you drag the cell of the column labeled 1-octanol diagonally down one cell, the columns reorder as shown in Figure 24.2.

When you look for patterns in the whole scatterplot matrix with reordered columns, you clearly see the variables cluster into groups based on their correlations, as illustrated previously by the two groups showing in Figure 24.1.

Figure 24.2 Reorder Scatterplot Matrix

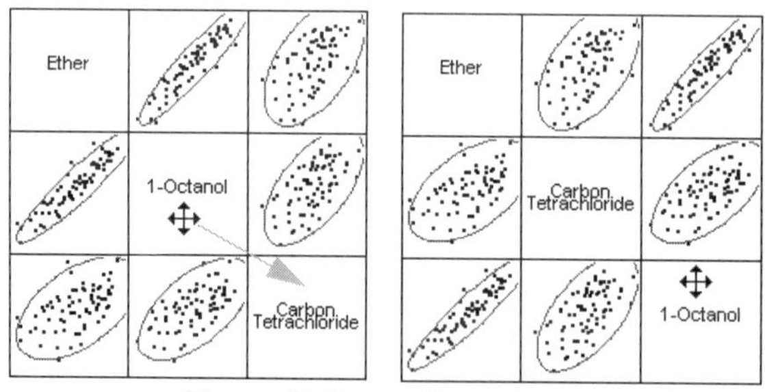

Drag cell diagonally

Covariance Matrix

The **Covariance Matrix** command displays the covariance matrix for the analysis.

Multivariate

Covariance Matrix

	Ether	1-Octanol	Carbon Tetrachloride	Benzene	Hexane	Chloroform
Ether	1.00217	0.76870	0.68356	0.79922	0.76796	0.59845
1-Octanol	0.76870	0.67553	0.64207	0.72776	0.69393	0.57055
Carbon Tetrachloride	0.68356	0.64207	1.61295	1.49577	1.68446	1.35876
Benzene	0.79922	0.72776	1.49577	1.51350	1.56172	1.34496
Hexane	0.76796	0.69393	1.68446	1.56172	1.94995	1.41442
Chloroform	0.59845	0.57055	1.35876	1.34496	1.41442	1.34948

Pairwise Correlations

The Pairwise Correlations table lists the Pearson product-moment correlations for each pair of *Y* variables, using all available values. The count values differ if any pair has a missing value for either variable. These are values produced by the **Density Ellipse** option on the Fit Y by X platform.

The Pairwise Correlations report also shows significance probabilities and compares the correlations with a bar chart, as shown in Figure 24.3.

Figure 24.3 Pairwise Correlations Report

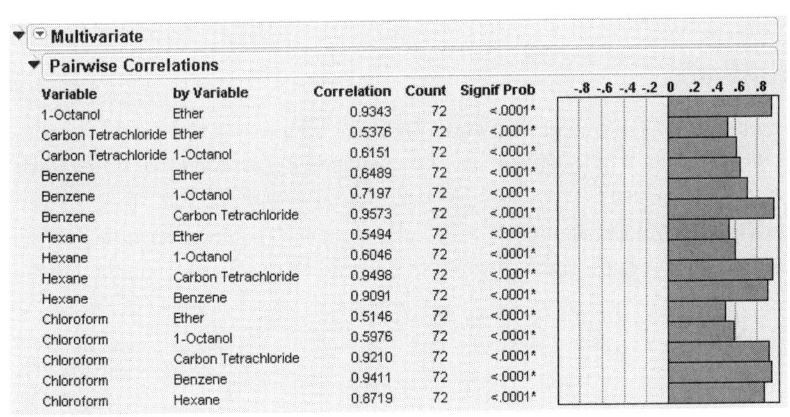

Multivariate

Pairwise Correlations

Variable	by Variable	Correlation	Count	Signif Prob
1-Octanol	Ether	0.9343	72	<.0001*
Carbon Tetrachloride	Ether	0.5376	72	<.0001*
Carbon Tetrachloride	1-Octanol	0.6151	72	<.0001*
Benzene	Ether	0.6489	72	<.0001*
Benzene	1-Octanol	0.7197	72	<.0001*
Benzene	Carbon Tetrachloride	0.9573	72	<.0001*
Hexane	Ether	0.5494	72	<.0001*
Hexane	1-Octanol	0.6046	72	<.0001*
Hexane	Carbon Tetrachloride	0.9498	72	<.0001*
Hexane	Benzene	0.9091	72	<.0001*
Chloroform	Ether	0.5146	72	<.0001*
Chloroform	1-Octanol	0.5976	72	<.0001*
Chloroform	Carbon Tetrachloride	0.9210	72	<.0001*
Chloroform	Benzene	0.9411	72	<.0001*
Chloroform	Hexane	0.8719	72	<.0001*

Color Maps

The Color Map submenu gives you three choices: **Color Map On Correlations**, **Color Map On p-values**, and **Cluster the Correlations**. The first option produces the cell plot on the left of Figure 24.4, showing the correlations among variables on a scale from red (+1) to blue (-1). **Color Map on p-values** shows the significance of the correlations on a scale from $p = 0$ (red) to $p = 1$(blue).**Cluster the Correlations** groups variables that have similar correlations together, which (in the case of Figure 24.4 looks exactly like the color map on correlations.)

Figure 24.4 Cell Plots

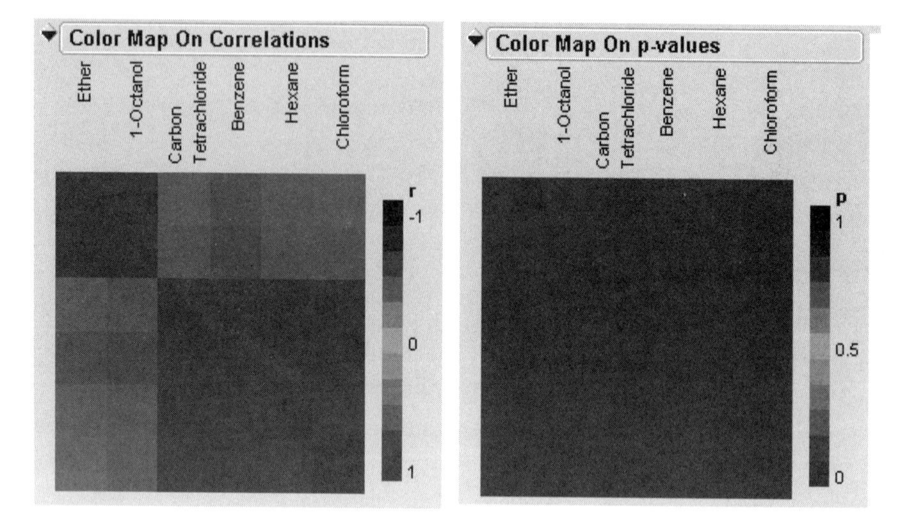

Simple Statistics

The Simple Statistics submenu allows you to display simple statistics (mean, standard deviation, and so on) for each column. These statistics can be calculated in two ways that differ when there are missing values in the data table.

Univariate Simple Statistics are calculated on each column, regardless of values in other columns. These values match the ones that would be produced using the Distribution platform.

Univariate Simple Statistics

Column	N	Mean	Std Dev	Sum	Minimum	Maximum
Ether	72	1.1010	1.0011	79.2700	-1.3000	2.7000
1-Octanol	72	1.2665	0.8219	91.1900	-0.7700	2.7000
Carbon Tetrachloride	72	-0.0746	1.2700	-5.3700	-3.3000	2.7400
Benzene	72	0.3949	1.2302	28.4300	-2.2600	3.0300
Hexane	72	-0.7817	1.3964	-56.280	-4.6000	2.1000
Chloroform	72	0.8065	1.1617	58.0700	-1.9200	3.6700

Note: Statistics were calculated for each column independently without regard for missing values in other columns.

Multivariate Simple Statistics are calculated by dropping any row that has a missing value for any column in the analysis. These are the statistics that are used by the Multivariate platform to calculate correlations.

Multivariate

Multivariate Simple Statistics

Column	N	Mean	Std Dev	Sum	Minimum	Maximum
Ether	72	1.1010	1.0011	79.2700	-1.3000	2.7000
1-Octanol	72	1.2665	0.8219	91.1900	-0.7700	2.7000
Carbon Tetrachloride	72	-0.0746	1.2700	-5.3700	-3.3000	2.7400
Benzene	72	0.3949	1.2302	28.4300	-2.2600	3.0300
Hexane	72	-0.7817	1.3964	-56.280	-4.6000	2.1000
Chloroform	72	0.8065	1.1617	58.0700	-1.9200	3.6700

Note: Rows with missing values were excluded.

Nonparametric Correlations

When you select **Nonparametric Correlations** from the platform popup menu, the Nonparametric Measures of Association table is shown. The Nonparametric submenu offers these three nonparametric measures:

Spearman's Rho is a correlation coefficient computed on the ranks of the data values instead of on the values themselves.

Kendall's Tau is based on the number of concordant and discordant pairs of observations. A pair is *concordant* if the observation with the larger value of X also has the larger value of Y. A pair is *discordant* if the observation with the larger value of X has the smaller value of Y. There is a correction for tied pairs (pairs of observations that have equal values of *X* or equal values of *Y*).

Hoeffding's D is a statistical scale that ranges from –0.5 to 1, with large positive values indicating dependence.The statistic approximates a weighted sum over observations of chi-square statistics for two-by-two classification tables, and detects more general departures from independence.

The Nonparametric Measures of Association report also shows significance probabilities for all measures and compares them with a bar chart similar to the one in Figure 24.3.

See "Computations and Statistical Details," p. 510, for computational information.

Outlier Analysis

The **Outlier Analysis** submenu toggles the display of three plots:

- Mahalanobis Distance
- Jackknife Distances
- T^2 Statistic.

These all measure distance with respect to the correlation structure, as in the normal ellipse contours shown here. Point A is an outlier because it violated the correlation structure rather than because it is an outlier in any of the coordinate directions.

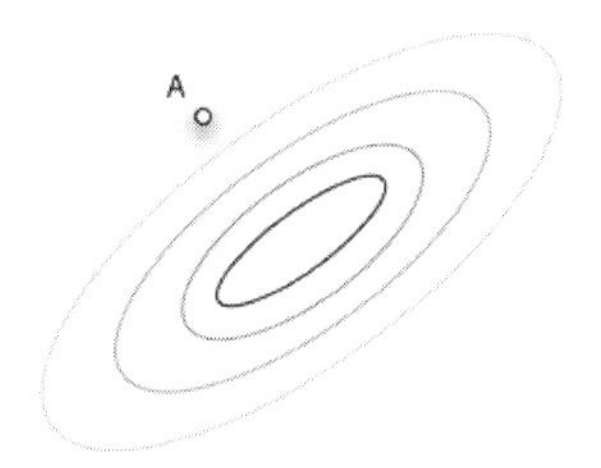

Figure 24.5 shows the *Mahalanobis distance* of each point from the multivariate mean (centroid). The standard Mahalanobis distance depends on estimates of the mean, standard deviation, and correlation for the data. The distance is plotted for each observation number. Extreme multivariate outliers can be identified by highlighting the points with the largest distance values. See "Computations and Statistical Details," p. 510, for more information.

Figure 24.5 Mahalanobis Outlier Distance Plot

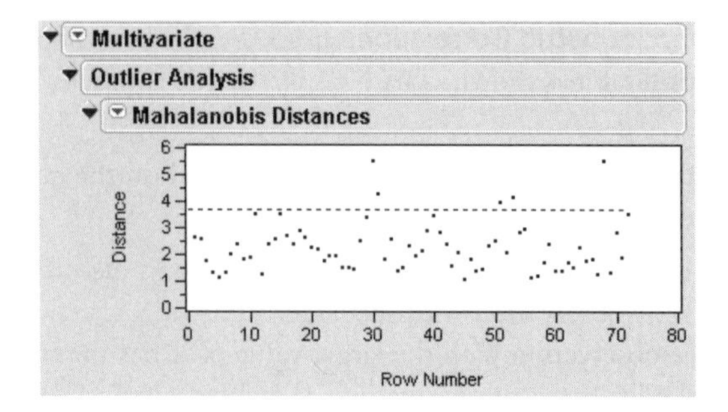

JMP can also calculate an alternate distance using a *jackknife* technique. The distance for each observation is calculated with estimates of the mean, standard deviation, and correlation matrix that do not include the observation itself. The jackknifed distances are useful when there is an outlier. In this case, the Mahalanobis distance is distorted and tends to disguise the outlier or make other points look more outlying than they are.

Figure 24.6 Jackknifed Distances Plot

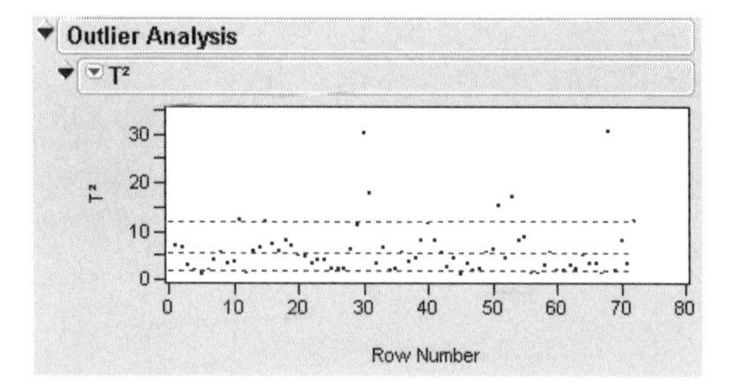

A third option is the T^2 statistic. This is the square of the Mahalanobis distance, and is preferred for multivariate control charts. The plot includes the value of the calculated T^2 statistic, as well as its upper control limit. Values that fall outside this limit may be an outlier.

Figure 24.7 T^2 Statistic Plot

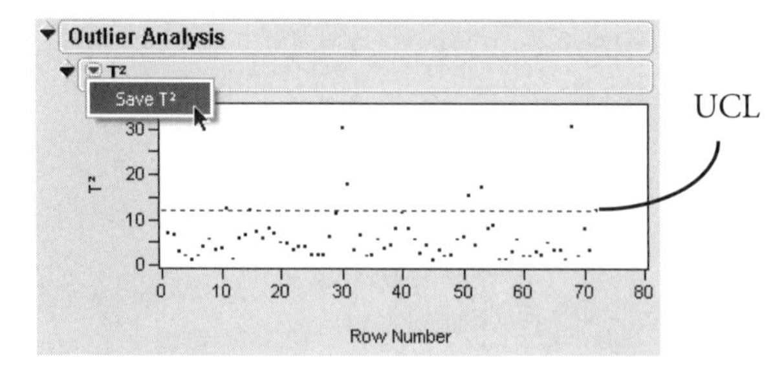

Saving Distances and Values

Each of the three outlier methods allows you to save distances back to the data table. To save the distances, use the popup menu beside the name of the distance you want to save. For example, the following figure saves values for the T^2 statistic.

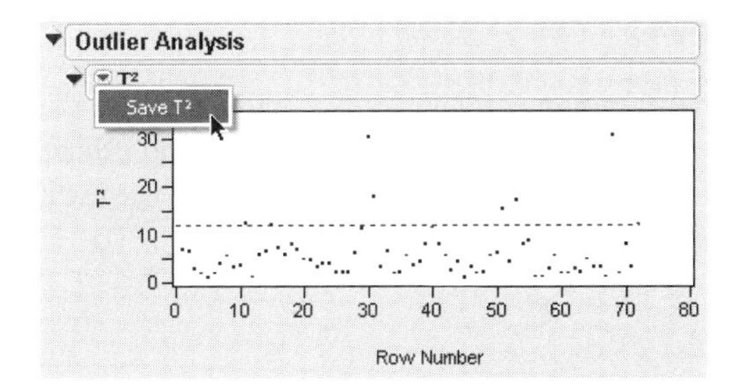

There is no formula saved with the distance column. This means the distance is not recomputed if you modify the data table. If you add or delete columns or change values in the data table, you should select **Analyze > Multivariate Methods > Multivariate** again to compute new distances.

Figure 24.8 Distances Saved in Data Table

	Labels	1-Octanol	Ether	Chloroform	Benzene	Carbon Tetrachloride	Hexane	T²
1	METHANOL	-0.770	-1.150	-1.260	-1.890	-2.100	-2.800	6.78222253
2	ETHANOL	-0.310	-0.570	-0.850	-1.620	-1.400	-2.100	6.43013185
3	PROPANOL	0.250	-0.020	-0.400	-0.700	-0.820	-1.520	2.98191944
4	BUTANOL	0.880	0.890	0.450	-0.120	-0.400	-0.700	1.71232073
5	PENTANOL	1.560	1.200	1.050	0.620	0.400	-0.400	1.27177633
6	HEXANOL	2.030	1.800	1.690	1.300	0.990	0.460	1.71265557
7	HEPTANOL	2.410	2.400	2.410	1.910	1.670	1.010	3.97104362
8	ACETIC_ACID	-0.170	-0.340	-1.600	-2.260	-2.450	-3.060	5.56524509

In addition to saving the distance values for each row, a column property is created that holds the distance value (for Mahalanobis Distance and Jackknife distance) or a list containing the UCL of the T^2 statistic.

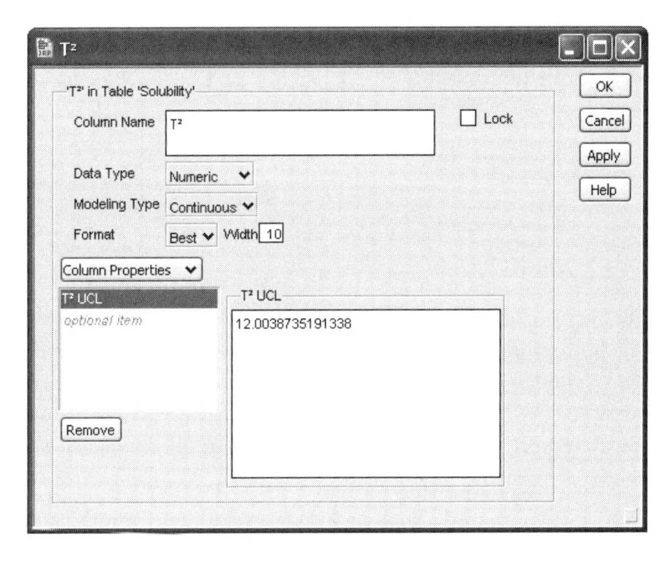

Principal Components

Principal components is a technique to take linear combinations of the original variables such that the first principal component has maximum variation, the second principal component has the next most variation subject to being orthogonal to the first, and so on. See the chapter “Principal Components,” p. 515 for details on principal components.

Item Reliability

Item reliability indicates how consistently a set of instruments measures an overall response. Cronbach’s α (Cronbach 1951) is one measure of reliability. Two primary applications for Cronbach’s α are industrial instrument reliability and questionnaire analysis.

Cronbach’s α is based on the average correlation of items in a measurement scale. It is equivalent to computing the average of all split-half correlations in the data set. The **Standardized** α can be requested if the items have variances that vary widely. The items in the scale can be continuous, as in a Likert scale, or categorical.

> Cronbach’s α is not related to a significance level α. Also, item reliability is unrelated to survival time reliability analysis.

As an example, consider the **Danger.jmp** data in the **Sample Data** folder. This table lists 30 items having some level of inherent danger. Three groups of people (students, nonstudents, and experts) ranked the items according to perceived level of danger. Note that Nuclear power is rated as very dangerous (1) by both students and nonstudents but was ranked low (20) by experts. On the other hand, motorcycles are ranked fifth or sixth by the three judging groups.

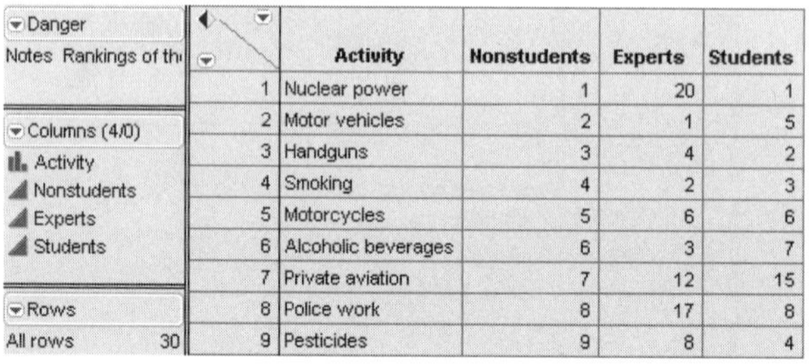

Danger
Notes Rankings of th
Columns (4/0)
Activity
Nonstudents
Experts
Students
Rows
All rows 30

	Activity	Nonstudents	Experts	Students
1	Nuclear power	1	20	1
2	Motor vehicles	2	1	5
3	Handguns	3	4	2
4	Smoking	4	2	3
5	Motorcycles	5	6	6
6	Alcoholic beverages	6	3	7
7	Private aviation	7	12	15
8	Police work	8	17	8
9	Pesticides	9	8	4

You can use Cronbach’s α to evaluate the agreement in the perceived way the groups ranked the items. The results at the bottom in Figure 24.9 show an overall α of 0.8666, which indicates a high correlation of the ranked values among the three groups. Further, when you remove the experts from the analysis, the **Nonstudents** and **Students** ranked the dangers nearly the same, with Cronbach’s α scores of 0.7785 and 0.7448, respectively. Nunnally (1979) suggests a Cronbach’s α of 0.7 as a rule-of-thumb acceptable level of agreement.

To look at the influence of an individual item, JMP excludes it from the computations and shows the effect of the Cronbach’s α value. If α increases when you exclude a variable (item), that variable is not highly correlated with the other variables. If the α decreases, you can conclude that the variable is correlated with the other items in the scale. Computations for Cronbach’s α are given in the next section, “Computations and Statistical Details,” p. 510.

Note that in this kind of example, where the values are the same set of ranks for each group, standardizing the data has no effect.

Figure 24.9 Cronbach's α Report

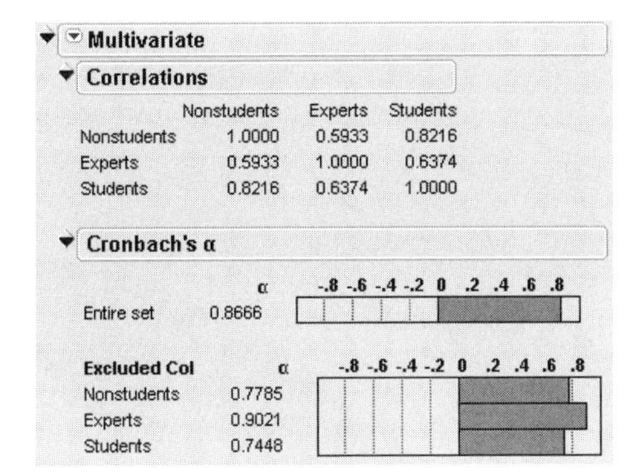

Parallel Coordinate Plot

The **Parallel Coordinate Plot** option shows or hides a parallel coordinate plot of the variables.

Figure 24.10 Parallel Coordinate Plot

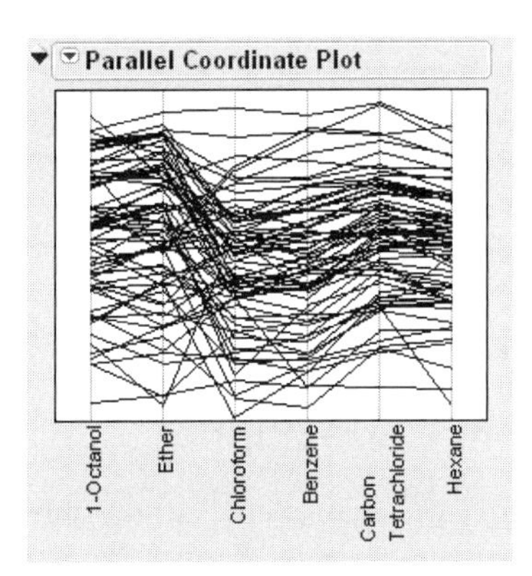

Ellipsoid 3D Plot

The Ellipsoid 3D Plot toggles a 95% confidence ellipsoid around three chosen variables. When the command is first invoked, a dialog asks which three variables to include in the plot.

Figure 24.11 Ellipsoid 3D Plot

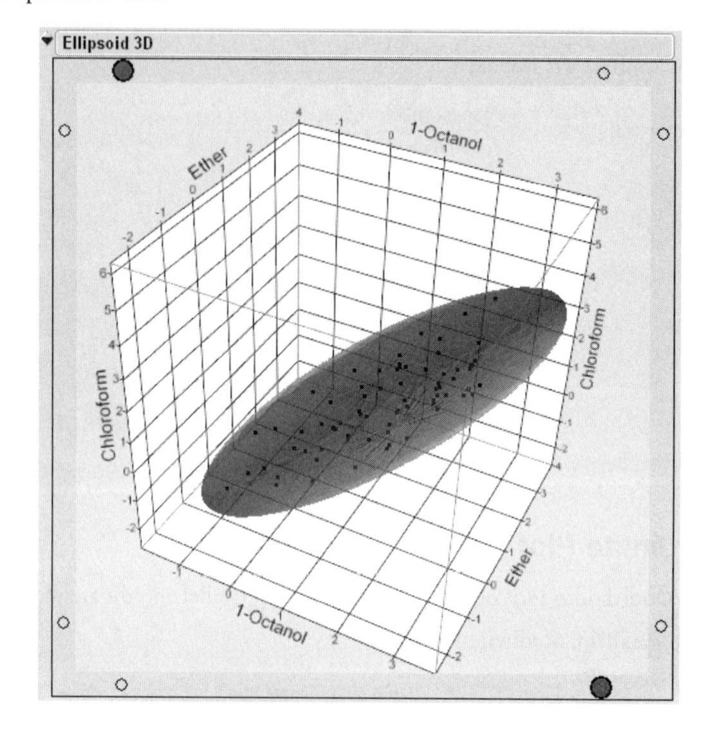

Computations and Statistical Details

Pearson Product-Moment Correlation

The Pearson product-moment correlation coefficient measures the strength of the linear relationship between two variables. For response variables *X* and *Y*, it is denoted as *r* and computed as

$$r = \frac{\sum(x-\bar{x})(y-\bar{y})}{\sqrt{\sum(x-\bar{x})^2}\sqrt{\sum(y-\bar{y})^2}}.$$

If there is an exact linear relationship between two variables, the correlation is 1 or –1, depending on whether the variables are positively or negatively related. If there is no linear relationship, the correlation tends toward zero.

Nonparametric Measures of Association

For the Spearman, Kendall, or Hoeffding correlations, the data are first ranked. Computations are then performed on the ranks of the data values. Average ranks are used in case of ties.

Spearman's ρ (rho) Coefficients

Spearman's ρ correlation coefficient is computed on the ranks of the data using the formula for the Pearson's correlation previously described.

Kendall's τ_b Coefficients

Kendall's τ_b coefficients are based on the number of concordant and discordant pairs. A pair of rows for two variables is *concordant* if they agree in which variable is greater. Otherwise they are discordant, or tied.

The formula

$$\tau_b = \frac{\sum_{i<j} \operatorname{sgn}(x_i - x_j)\operatorname{sgn}(y_i - y_j)}{\sqrt{(T_0 - T_1)(T_0 - T_2)}}$$

computes Kendall's τ_b where

$$T_0 = (n(n-1))/2,$$

$$T_1 = \sum((t_i)(t_i - 1))/2, \text{ and}$$

$$T_2 = \sum((u_i)(u_i - 1))/2,$$

Note that $\operatorname{sgn}(z)$ is equal to 1 if $z > 0$, 0 if $z = 0$, and –1 if $z < 0$.

The t_i (the u_i) are the number of tied x (respectively y) values in the ith group of tied x (respectively y) values, n is the number of observations, and Kendall's τ_b ranges from –1 to 1. If a weight variable is specified, it is ignored.

Computations proceed in the following way:

- Observations are ranked in order according to the value of the first variable.
- The observations are then re-ranked according to the values of the second variable.
- The number of interchanges of the first variable is used to compute Kendall's τ_b.

Hoeffding's *D* Statistic

The formula for Hoeffding's D (1948) is

$$D = 30\left(\frac{(n-2)(n-3)D_1 + D_2 - 2(n-2)D_3}{n(n-1)(n-2)(n-3)(n-4)}\right) \text{ where}$$

$$D_1 = S_i(Q_i - 1)(Q_i - 2)$$

$$D_2 = S_i(R_i - 1)(S_i - 1)(S_i - 2)$$

$$D_3 = (R_i - 1)(S_i - 2)(Q_i - 1)$$

The R_i and S_i are ranks of the x and y values, and the Q_i (sometimes called bivariate ranks) are one plus the number of points that have both x and y values less than the ith points. A point that is tied on its x value or y value, but not on both, contributes 1/2 to Q_i if the other value is less than the corresponding value for the ith point. A point tied on both x and y contributes 1/4 to Q_i.

When there are no ties among observations, the D statistic has values between –0.5 and 1, with 1 indicating complete dependence. If a weight variable is specified, it is ignored.

Inverse Correlation Matrix

The inverse correlation matrix provides useful multivariate information. The diagonal elements of the inverse correlation matrix, sometimes called the variance inflation factors (VIF), are a function of how closely the variable is a linear function of the other variables. Specifically, if the correlation matrix is denoted **R** and the inverse correlation matrix is denoted $\mathbf{R}^{-1}$, the diagonal element is denoted r^{ii} and is computed as

$$r^{ii} = \mathrm{VIF}_i = \frac{1}{1 - R_i^2},$$

where $\mathbf{R}_i^2$ is the coefficient of variation from the model regressing the ith explanatory variable on the other explanatory variables. Thus, a large r^{ii} indicates that the ith variable is highly correlated with any number of the other variables.

Note that the definition of R^2 changes for no-intercept models. For no-intercept and hidden-intercept models, JMP uses the R^2 from the uncorrected Sum of Squares, *i.e.* from the zero model, rather than the corrected sum of squares, from the mean model.

Distance Measures

The Outlier Distance plot shows the Mahalanobis distance of each point from the multivariate mean (centroid). The Mahalanobis distance takes into account the correlation structure of the data as well as the individual scales. For each value, the distance is denoted d_i and is computed as

$$d_i = \sqrt{(Y_i - \bar{Y})'S^{-1}(Y_i - \bar{Y})} \text{ where}$$

Y_i is the data for the ith row

$\bar{Y}$ is the row of means, and

S is the estimated covariance matrix for the data

The reference line drawn on the Mahalanobis Distance plot is computed as $\sqrt{F \times \text{nvars}}$ where *nvars* is the number of variables and the computation for F in formula editor notation is:

F Quantile(0.95, nvars, n–nvars–1, centered at 0)

If a variable is an exact linear combination of other variables, then the correlation matrix is singular and the row and the column for that variable are zeroed out. The generalized inverse that results is still valid for forming the distances.

The T^2 distance is just the square of the Mahalanobis distance, so $T_i^2 = d_i^2$. The upper control limit on the T^2 is

$$UCL = \frac{(n-1)^2}{n}\beta_{\left[a;\frac{p}{2};\frac{n-p-1}{2}\right]}$$

where

n = number of observations

p = number of variables (columns)

$\alpha = \alpha^{th}$ quantile

β= beta distribution

Multivariate distances are useful for spotting outliers in many dimensions. However, if the variables are highly correlated in a multivariate sense, then a point can be seen as an outlier in multivariate space without looking unusual along any subset of dimensions. Said another way, when data are correlated, it is possible for a point to be unremarkable when seen along one or two axes but still be an outlier by violating the correlation.

Statistical Warning: This outlier distance is not particularly robust in the sense that outlying points themselves can distort the estimate of the covariances and means in such a way that outliers are disguised. You might want to use the alternate distance command so that distances are computed with a jackknife method. The alternate distance for each observation uses estimates of the mean, standard deviation, and correlation matrix that do not include the observation itself.

Cronbach's α

Cronbach's α is defined as

$$\alpha = \frac{k\sum\frac{c}{v}}{1+(k-1)\sum\frac{c}{v}}$$

where

k = the number of items in the scale
c = the average covariance between items
v = the average variance between items.

If the items are standardized to have a constant variance, the formula becomes

$$\alpha = \frac{k(r)}{1+(k-1)r} \quad \text{where}$$

r = the average correlation between items.

The larger the overall α coefficient, the more confident you can feel that your items contribute to a reliable scale or test. The coefficient can approach 1.0 if you have many highly correlated items.

Chapter 25

Principal Components

Reducing Dimensionality

The purpose of principal component analysis is to derive a small number of independent linear combinations (principal components) of a set of variables that capture as much of the variability in the original variables as possible.

JMP also offers several types of orthogonal and oblique Factor-Analysis-Style rotations to help interpret the extracted components.

Principal components can be accessed through the **Multivariate** platform, through the **Scatterplot 3D** platform, or through the **Principal Components** command on the **Analyze > Multivariate Methods** menu. All map to the same routines, documented in this chapter.

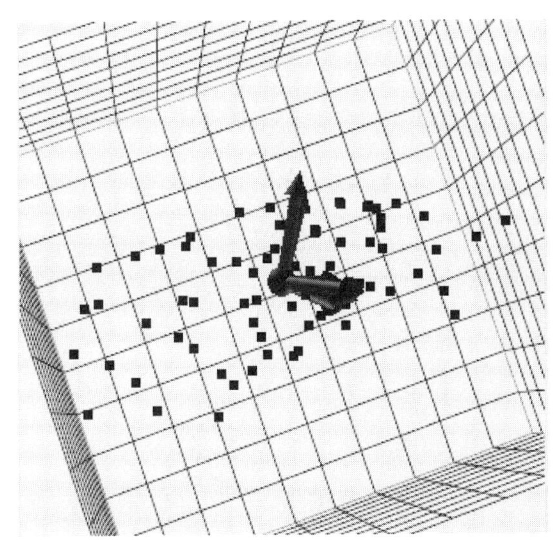

Contents

Principal Components

If you want to see the arrangement of points across many correlated variables, you can use principal component analysis to show the most prominent directions of the high-dimensional data. Using principal component analysis reduces the *dimensionality* of a set of data. Principal components is a way to picture the structure of the data as completely as possible by using as few variables as possible.

For *n* original variables, *n* principal components are formed as follows:

- The first principal component is the linear combination of the standardized original variables that has the greatest possible variance.
- Each subsequent principal component is the linear combination of the standardized original variables that has the greatest possible variance and is uncorrelated with all previously defined components.

Each principal component is calculated by taking a linear combination of an eigenvector of the correlation matrix with a standardized original variable. The eigenvalues show the variance of each component.

Principal components is important in visualizing multivariate data by reducing it to dimensionalities that are graphable.

To illustrate, suppose that you have a scatterplot for two highly correlated variables. If you compute a linear combination of the two variables that best captures the scatter of the points, then you have the combination indicated by the P1 line as shown on the left in Figure 25.1. Rotating and reflecting the plot so that P1 is the variable on the main axis, as seen on the right in Figure 25.1, gives you the best one-dimensional approximation of a two-dimensional cloud of points. The second principal component, P2, describes the remaining variation. This is an example of a principal components decomposition for two variables.

Figure 25.1 Two Principal Components

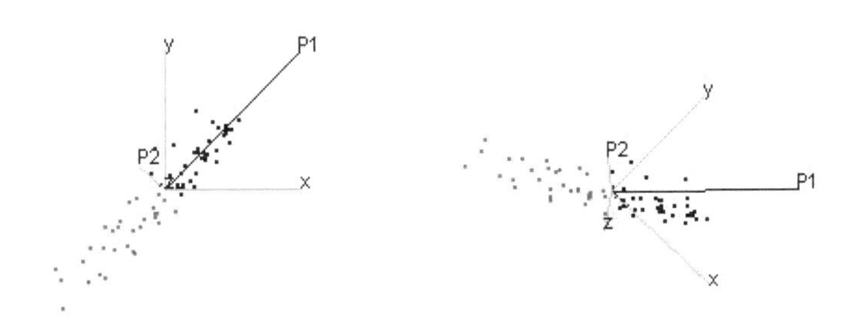

Similarly, you can find the best three-dimensional view of higher dimensional data by placing their first three principal components as the spinning axes and examining the plot.

Launch the Platform

When launched from the **Analyze > Graph** menu as a separate platform, JMP presents you with a dialog box to specify the variables involved in the analysis. In this example, we use all the continuous variables from the **Solubility.jmp** data set.

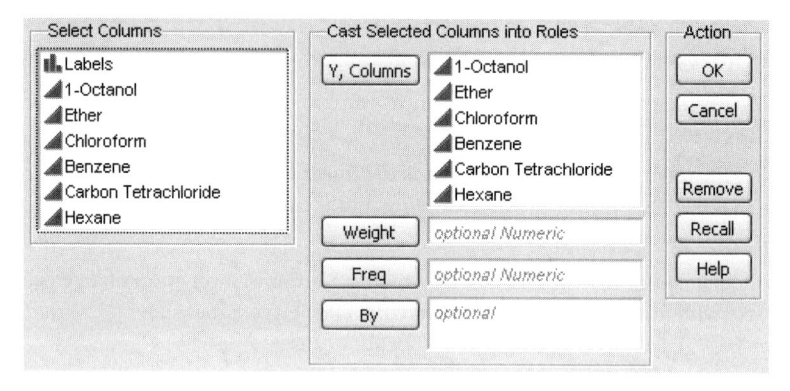

Alternatively, select the **Principal Components** menu command from the report menus of the Multivariate platform or the Scatterplot 3D platform.

Reports and Options

The initial principal components report (Figure 25.2) summarizes the variation of the specified *Y* variables with principal components. The principal components are derived from an eigenvalue decomposition of the correlation matrix, the covariance matrix, or on the unscaled and uncentered data.

The details in the report show how the principal components absorb the variation in the data. The principal component points are derived from the eigenvector linear combination of the standardized variables.

Figure 25.2 Principal Components/Factor Analysis Report

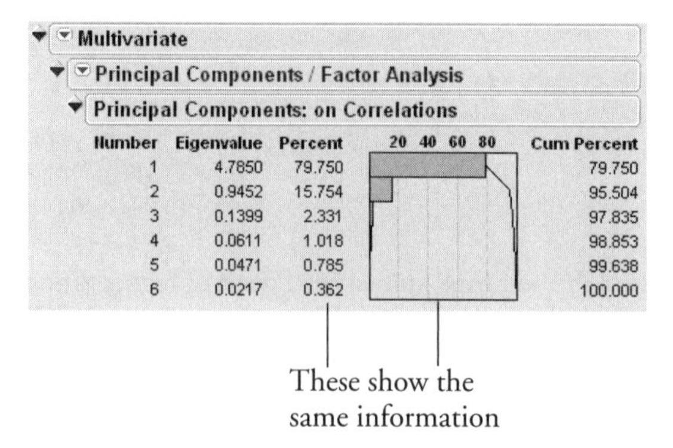

Number	Eigenvalue	Percent	Cum Percent
1	4.7850	79.750	79.750
2	0.9452	15.754	95.504
3	0.1399	2.331	97.835
4	0.0611	1.018	98.853
5	0.0471	0.785	99.638
6	0.0217	0.362	100.000

Number enumerates the principal components. There are as many principal components as there are variables in the analysis.

Eigenvalue lists the eigenvalue that corresponds to each principal component in order from largest to smallest (first principal component, second principal component, and so on). The eigenvalues represent a partition of the total variation in the multivariate sample. They sum to the number of variables when the principal components analysis is done on the correlation matrix.

The additional columns in the Principal Components report give the following information:

Percent lists each eigenvalue as a percent of the total eigenvalues. In the example above, the first principal component accounts for 79.7503% of the variation in the sample. The Pareto plot to the right of this column shows the same information in graphical form.

Cum Percent shows the cumulative percent of variation represented by the eigenvalues. In the example above, the first three principal components account for 97.8348% of the variation in the sample.

The popup menu on the **Principal Components/ Factor Analysis** title bar gives you options to see the results of rotating principal components and saving principal component scores.

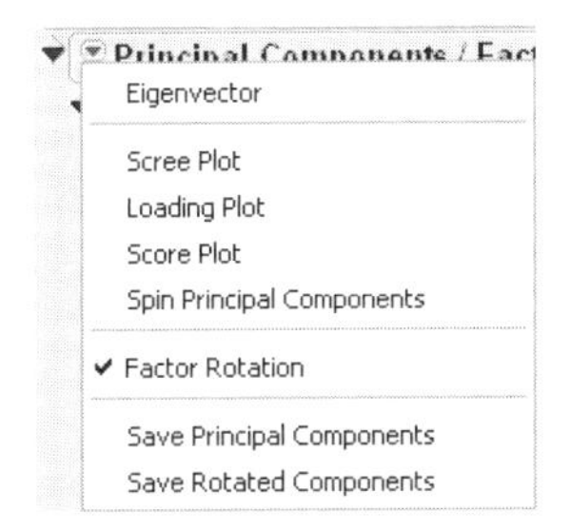

Eigenvectors shows columns of values that correspond to the eigenvectors for each of the principal components, in order, from left to right. Using these coefficients to form a linear combination of the original variables produces the principal component variables.

Eigenvector

1-Octanol	0.37441	0.55987	-0.11070	-0.65842	0.31660	0.01874
Ether	0.34834	0.64314	0.11973	0.62764	-0.20890	0.11456
Chloroform	0.41940	-0.29864	-0.64850	0.30599	0.43061	0.18793
Benzene	0.44561	-0.14756	-0.21904	-0.09455	-0.49849	-0.68865
Carbon Tetrachloride	0.43102	-0.29736	0.18487	-0.24135	-0.45965	0.64968
Hexane	0.42217	-0.27117	0.68608	0.10831	0.45926	-0.23426

Scree Plot shows or hides a scree plot of the eigenvalues vs. the number of components.

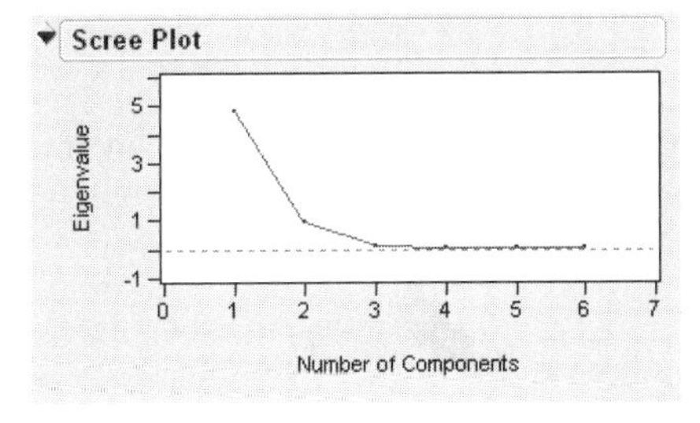

Loading Plot shows or hides a matrix of two-dimensional representations of factor loadings. After being prompted for the number of loadings to show, a matrix of plots showing all possible pairs of these loadings appears. In Figure 25.3, there are three loadings shown.

Figure 25.3 Loading Plot

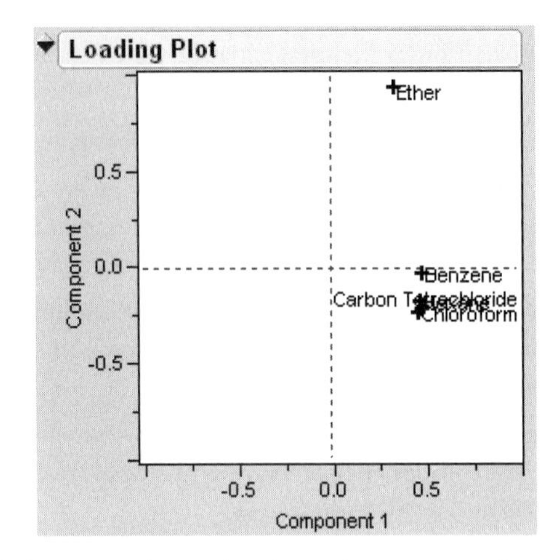

Score Plot shows or hides a two-dimensional matrix representation of the scores for each pair of principal components. After being prompted for the number of principal components to show, a matrix of plots showing all possible pairs of these component scorings appears. In Figure 25.4, there are three loadings shown.

Figure 25.4 Score Plot

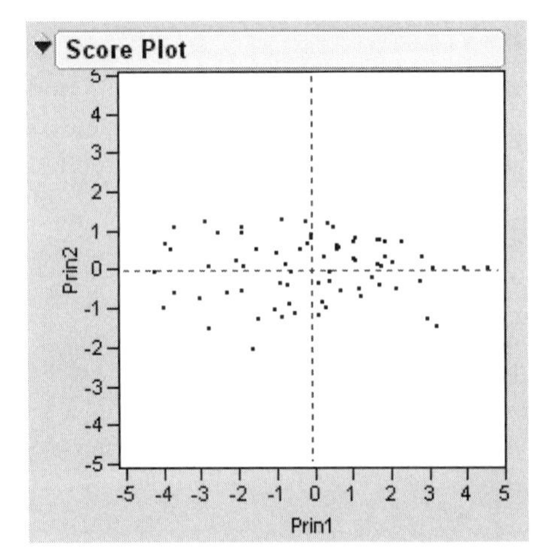

Spin Principal Components displays a 3D plot of any three principal components. When you first invoke the command, the first three principal components are presented.

Figure 25.5 Principal Components Display Controls

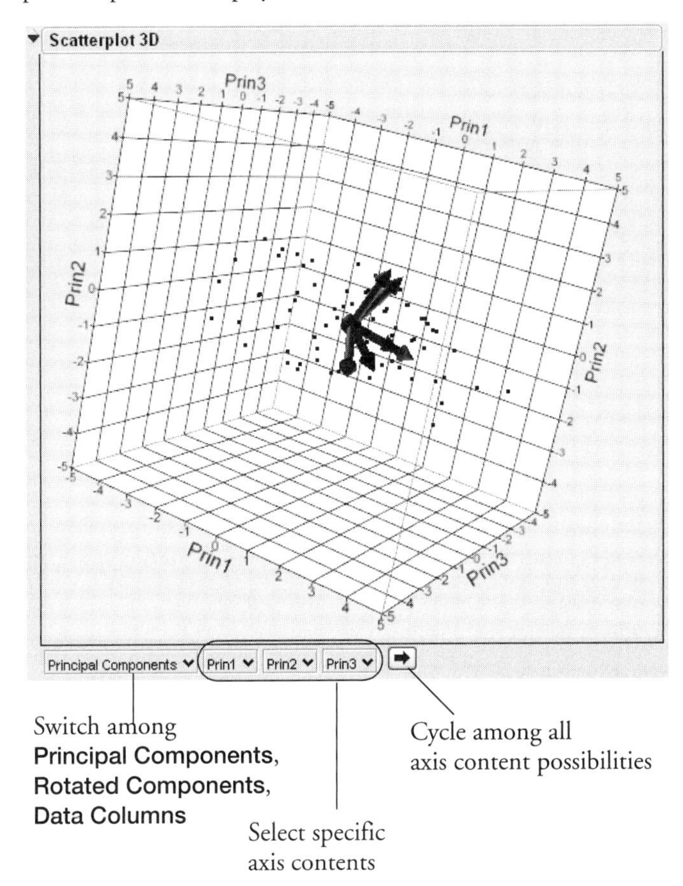

The variables show as rays in the spinning plot. These rays, called *biplot rays*, approximate the variables as a function of the principal components on the axes. If there are only two or three variables, the rays represent the variables exactly. The length of the ray corresponds to the eigenvalue or variance of the principal component. Because the first principal component captures the most variability possible, the first principal; component ray can be long compared to the subsequent components.

There are three sets of controls that allow you to choose what to display:

- Choose to show **Principal Components**, **Rotated Components**, or **Data Columns** on the axes
- Select specific things to appear on the axis (for example, choose which principal components are displayed)
- Cycle among all possible axis contents. For example, when showing principal components, the axes cycle so that all principal components appear on all axes in order.

Factor Rotation gives you factor-analysis-style rotations of the principal components. Selecting the command results in the following dialog.

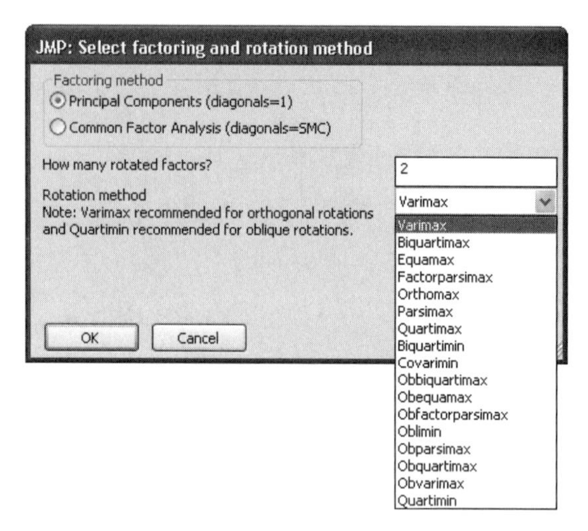

Use this dialog to specify:

1. Factoring method, **Principal Components** (where the diagonals are 1) or **Common Factor Analysis** (where the diagonals are SMC, the squared multiple correlation between X_i and the other $p - 1$ variables).

2. The number of factors to rotate

3. The rotation method.

Rotations are used to better align the directions of the factors with the original variables so that the factors may be more interpretable. You hope for clusters of variables that are highly correlated to define the rotated factors. The rotation's success at clustering the interpretability is highly dependent on the number of factors that you choose to rotate.

After the initial extraction, the factors are uncorrelated with each other. If the factors are rotated by an orthogonal transformation, the rotated factors are also uncorrelated. If the factors are rotated by an oblique transformation, the rotated factors become correlated. Oblique rotations often produce more useful patterns than do orthogonal rotations. However, a consequence of correlated factors is that there is no single unambiguous measure of the importance of a factor in explaining a variable.

The Rotated Components table shown here gives the rotated factor pattern for prior communalities equal to 1.0, two rotated factors, and the varimax rotation method. The varimax method tries to make elements of this matrix go toward 1 or 0 to show the clustering of variables. Note that the first rotated component loads on the Carbon Tetrachloride–Chloroform–Benzene–Hexane cluster and that the second rotated component loads on the ether–1-octanol cluster.

Factor Rotation: Varimax

Prior Communality Estimates:1

Unrotated Factor Pattern

Ether	0.761978	0.625283
1-Octanol	0.819019	0.544318
Carbon Tetrachloride	0.942849	-0.289099
Benzene	0.974761	-0.143460
Hexane	0.923483	-0.263641
Chloroform	0.917422	-0.290351

Rotated Factor Pattern

Ether	0.2758844	0.9462967
1-Octanol	0.3686763	0.9116749
Carbon Tetrachloride	0.9414227	0.2937115
Benzene	0.8855353	0.4319341
Hexane	0.9110652	0.3037894
Chloroform	0.9211420	0.2783219

Rotation Matrix

0.82537	0.56459
-0.56459	0.82537

Final Communality Estimates

Ether	0.97159
1-Octanol	0.96707
Carbon Tetrachloride	0.97254
Benzene	0.97074
Hexane	0.92233
Chloroform	0.92597

Std Score Coefs

Ether	-0.242051	0.635901
1-Octanol	-0.183851	0.571933
Carbon Tetrachloride	0.335313	-0.141191
Benzene	0.253827	-0.010255
Hexane	0.316767	-0.121247
Chloroform	0.331675	-0.145285

Variance	Percent	Cum Percent
3.5610	59.350	59.350
2.1692	36.154	95.504

Save Principal Components, Save Rotated Components save the number of principal components and rotated components that you specify as new columns in the current data table.

Chapter 26

Clustering

The Cluster Platform

Clustering is the technique of grouping rows together that share similar values across a number of variables. It is a wonderful exploratory technique to help you understand the clumping structure of your data. JMP provides three different clustering methods:

- Hierarchical clustering is appropriate for small tables, up to several thousand rows. It combines rows in an hierarchical sequence portrayed as a tree. In JMP, the tree, also called a dendrogram, is a dynamic, responding graph. You can choose the number of clusters you like after the tree is built.
- *K*-means clustering is appropriate for larger tables, up to hundreds of thousands of rows. It makes a fairly good guess at cluster seed points. It then starts an iteration of alternately assigning points to clusters and recalculating cluster centers. You have to specify the number of clusters before you start the process.
- Normal mixtures are appropriate when data is assumed to come from a mixture of multivariate normal distributions that overlap. Maximum likelihood is used to estimate the mixture proportions and the means, standard deviations, and correlations jointly. This approach is particularly good at estimating the total counts in each group. However each point, rather than being classified into one group, is assigned a probability of being in each group. The EM algorithm is used to obtain estimates.

After the clustering process is complete, you can save the cluster assignments to the data table or use them to set colors and markers for the rows.

Contents

Introduction to Clustering Methods

Clustering is a multivariate technique of grouping rows together that share similar values. It can use any number of variables. The variables must be numeric variables for which numerical differences makes sense. The common situation is that data are not scattered evenly through *n*-dimensional space, but rather they form clumps, locally dense areas, modes, or clusters. The identification of these clusters goes a long way towards characterizing the distribution of values.

JMP provides two approaches to clustering:

- *hierarchical clustering* for small tables, up to several thousand rows
- *k-means* and *normal mixtures* clustering for large tables, up to hundreds of thousands of rows.

Hierarchical clustering is also called *agglomerative clustering* because it is a combining process. The method starts with each point (row) as its own cluster. At each step the clustering process calculates the distance between each cluster, and combines the two clusters that are closest together. This combining continues until all the points are in one final cluster. The user then chooses the number of clusters that seems right and cuts the clustering tree at that point. The combining record is portrayed as a tree, called a *dendrogram*, with the single points as leaves, the final single cluster of all points as the trunk, and the intermediate cluster combinations as branches. Since the process starts with $n(n + 1)/2$ distances for n points, this method becomes too expensive in memory and time when n is large. For 500 points, the method needs a megabyte of memory for distances and can take many minutes to complete.

Hierarchical clustering also supports character columns. If the column is ordinal, then the data value used for clustering is just the index of the ordered category, treated as if it were continuous data. If the column is nominal, then the categories must match to contribute a distance of zero. They contribute a distance of 2 otherwise.

JMP offers five rules for defining distances between clusters: Average, Centroid, Ward, Single, and Complete. Each rule can generate a different sequence of clusters.

K-means clustering is an iterative follow-the-leader strategy. First, the user must specify the number of clusters, *k*. Then a search algorithm goes out and finds *k* points in the data, called *seeds*, that are not close to each other. Each seed is then treated as a cluster center. The routine goes through the points (rows) and assigns each point to the cluster it is closest to. For each cluster, a new cluster center is formed as the means (centroid) of the points currently in the cluster. This process continues as an alternation between assigning points to clusters and recalculating cluster centers until the clusters become stable.

Normal mixtures clustering, like *k*-means clustering, begins with a user-defined number of clusters and then selects distance seeds. JMP uses either the cluster centers chosen by *k*-means as seeds or by the values of a variable you choose to set the cluster centers. However, each point, rather than being classified into one group, is assigned a probability of being in each group.

SOMs are a variation on *k*-means where the cluster centers become a grid.

K-means, normal mixtures, and SOM clustering are doubly-iterative processes. The clustering process iterates between two steps in a particular implementation of the EM algorithm:

- The *expectation step* of mixture clustering assigns each observation a probability of belonging to each cluster.
- For each cluster, a new center is formed using every observation with its probability of membership as a weight. This is the *maximization step*.

This process continues alternating between the expectation and maximization steps until the clusters become stable.

Note: For *k*-means clustering, you can choose a variable whose values form preset fixed centers for clusters, instead of using the default random seeds for clusters.

The Cluster Launch Dialog

When you choose **Cluster** from the **Analyze > Multivariate Methods** submenu, the Cluster Launch dialog shown in Figure 26.1 appears. You can specify as many *Y* variables as you want by selecting the variables in the column selector list and clicking **Y, Columns**. Then select either the **Hierarchical** or the **KMeans** clustering method from the popup menu on the Cluster Launch dialog.

Hierarchical clustering supports character columns as follows. *K*-Means clusering only supports numeric columns.

- If the column is ordinal, then the data value used for clustering is just the index of the ordered category, treated as if it were continuous data.
- If the column is nominal, then the categories must match to contribute a distance of zero, or contribute a standardized distance of 2 otherwise.

Numeric ordinal columns are treated the same as numeric continuous columns.

By default, data are first standardized by the column mean and standard deviation. Uncheck the **Standardize Data** check box if you do not want the cluster distances computed on standardized values.

An **Ordering** column can (optionally) be specified in the Hierarchical Clustering launch dialog. In the ordering column, clusters are sorted by their mean value.

One way to utilize this feature is to complete a Principal Components analysis (using **Multivariate**) and save the first principal component to use as an **Ordering** column. The clusters are then sorted by these values.

The Cluster platform offers these two methods:

Hierarchical If you choose this, you must also choose one of five clustering distance options: average linkage, centroid, Ward's minimum variance, single linkage, and complete linkage. The clustering methods differ in how the distance between two clusters is computed. These clustering methods are discussed under "Technical Details for Hierarchical Clustering," p. 537. Hierarchical clustering runs to completion when you click **OK**.

KMeans If you choose **KMeans**, you must specify either the number of clusters to form or select a variable whose values form preset cluster centers. Other options become available when the Cluster Control Panel appears. When you click **OK** on the launch dialog, the cluster process picks seeds, but it does not run to completion until you click buttons in the Cluster Control Panel. See the later section "K-Means Clustering," p. 539, for further discussion.

Figure 26.1 The Cluster Launch Dialog

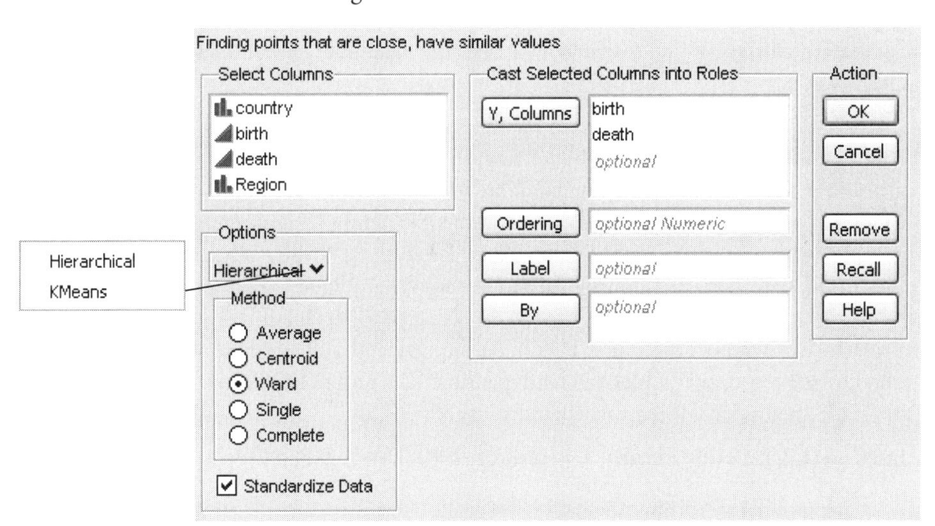

Hierarchical Clustering

The **Hierarchical** option clusters rows that group the points (rows) of a JMP table into clusters whose values are close to each other relative to those of other clusters. Hierarchical clustering is a process that starts with each point in its own cluster. At each step, the two clusters that are closest together are combined into a single cluster. This process continues until there is only one cluster containing all the points. This kind of clustering is good for smaller data sets (a few hundred observations).

To see a simple example of hierarchical clustering, select the **Birth Death Subset.jmp** table from the **Sample Data** folder. The data are the 1976 crude birth and death rates per 100,000 people.

Birth Death Subset
Notes 1976 Crude Birth an
Columns (3/0)
country
birth
death
Rows
All rows 34

	country	birth	death
1	AFGHANISTAN	52	30
2	ALGERIA	50	16
3	ARGENTINA	22	10
4	AUSTRALIA	16	8
5	AUSTRIA	12	13
6	BANGLADESH	47	19
7	BRAZIL	36	10
8	CANADA	16	7

When you select the **Cluster** command, the Cluster Launch dialog (shown previously in Figure 26.1) appears. In this example the **birth** and **death** columns are used as cluster variables in the default method, Ward's minimum variance, for hierarchical clustering.

In clustering problems, it is especially useful to assign a label variable so that you can determine which observations cluster together. In this example, **country** is the label variable.

When you click **OK**, the clustering process proceeds. When finished, the Hierarchical Cluster platform report consists of a Clustering History table, a dendrogram tree diagram, and a plot of the distances between the clusters. If you assigned a variable the label role, its values identify each observation on the dendrogram.

Open the Clustering History table to see the results shown in Figure 26.2. The number of clusters begins with 33, which is the number of rows in the data table minus one. You can see that the two closest points, the Philippines and Thailand, are joined to reduce the number of existing clusters to 32. They show as the first Leader and Joiner in the Clustering History table. The next two closest points are Bangladesh and Saudi Arabia, followed by Poland and USSR. When Brazil is joined by the Philippines in the sixth line, the Philippines had already been joined by Thailand, making it the first cluster with three points. The last single point to be joined to others is Afghanistan, which reduces the number of clusters from four to three at that join. At the very end a cluster of seven points led by Afghanistan is joined by the rest of the points, led by Argentina. The order of the clusters at each join is unimportant, essentially an accident of the way the data was sorted.

Figure 26.2 Clustering History Table for the Birth Death Subset Data

Hierarchical Clustering

Clustering History

Number of Clusters	Distance	Leader	Joiner
33	0.000000000	PHILIPPINES	THAILAND
32	0.104271097	BANGLADESH	SAUDI ARABIA
31	0.104271097	POLAND	USSR
30	0.104271097	GERMANY, FED REP OF	UNITED KINGDOM
29	0.104271097	AUSTRALIA	YUGOSLAVIA
28	0.120401892	BRAZIL	PHILIPPINES
27	0.140113736	TAIWAN	KOREA, REPUBLIC OF
26	0.140113736	FRANCE	ITALY

<portion deleted>

Number of Clusters	Distance	Leader	Joiner
5	1.501738614	ARGENTINA	AUSTRIA
4	1.903914282	BRAZIL	TAIWAN
3	2.380570164	AFGHANISTAN	ALGERIA
2	3.590473842	ARGENTINA	BRAZIL
1	6.095747490	AFGHANISTAN	ARGENTINA

The clustering sequence is easily visualized with the help of the *dendrogram*, shown in Figure 26.3. A dendrogram is a tree diagram that lists each observation, and shows which cluster it is in and when it entered its cluster.

You can drag the small diamond-shaped handle at either the top to bottom of the dendrogram to identify a given number of clusters. If you click on any cluster stem, all the members of the cluster highlight in the dendrogram and in the data table.

The scree plot beneath the dendrogram has a point for each cluster join. The ordinate is the distance that was bridged to join the clusters at each step. Often there is a natural break where the distance jumps up suddenly. These breaks suggest natural cutting points to determine the number of clusters.

Figure 26.3 Cluster Dendrogram for the Birth-Death Subset Table

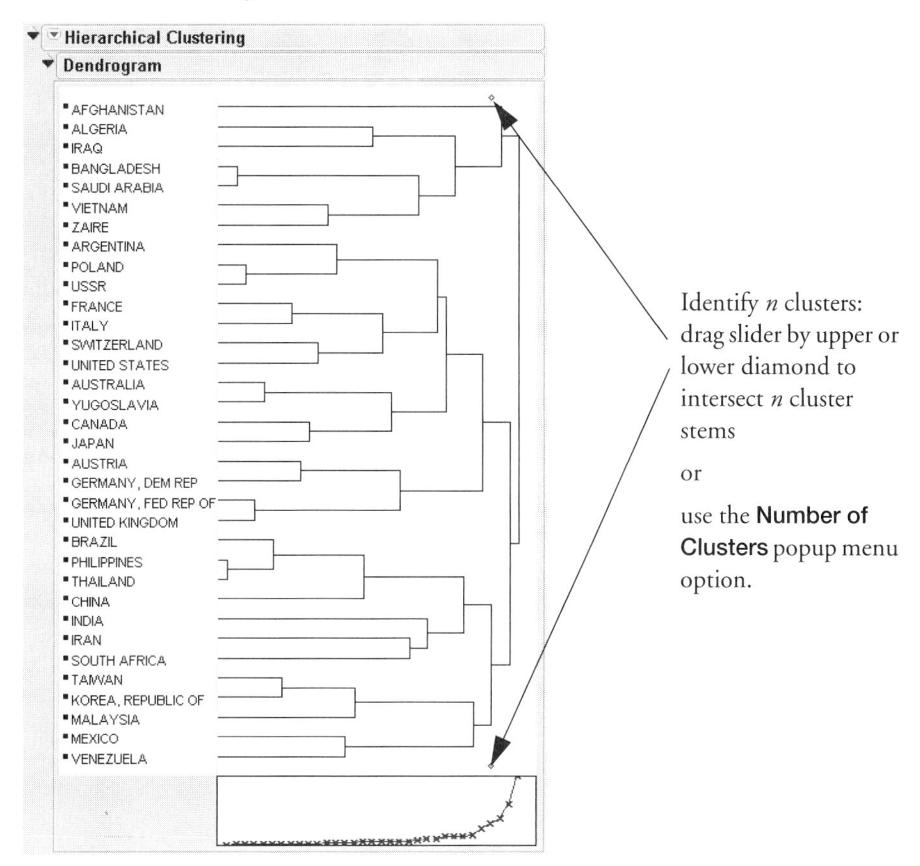

Ordering and Color Map Example

The ordering option on the cluster dialog offers the option of sorting the data by a particular column. When using the first principal component as the column by which to sort, the data is ordered by generally small values to generally large values simultaneously across all variables used.

The color map option plots the value of the data for each variable on a graduated color scale, depending on the color scheme selected. Ordering the data first across all the variables will give a color map on the dendrogram that is sorted across all variables in overall descending order.

The following example, using the **Solubility.JMP** data, consists of two runs. For the first run,

- Choose **Analyze > Multivariate Methods > Cluster**, leaving the default **Hierarchical** option.
- Choose all six continuous variables as Y's.
- Click **OK**.
- Choose **Color Map > Green to Black to Red** from the platform popup menu.

A portion of the resulting dendrogram is shown in Figure 26.4.

Figure 26.4 Run 1 Dendrogram

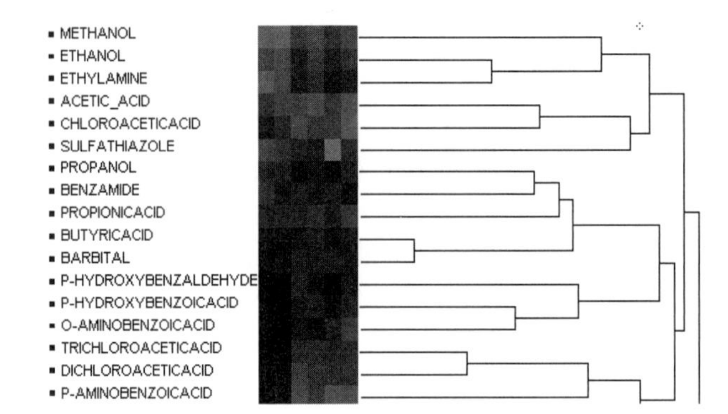

Next, run a principal components analysis on correlations and save the first principal component to the data table.

- Choose **Multivariate Methods > Multivariate.**
- Select all six continuous variables as Y's.
- Click **OK.**
- From the platform menu, choose **Principal Components > on Correlations.**
- Choose **Save Principal Components.**
- In the dialog that appears, choose to save one principal component.

Now, display the dendrogram.

- Choose **Multivariate Methods > Cluster.**
- Choose all six continuous variables as Y's.
- Choose Prin1 as the Ordering variable.
- Click **OK.**
- Choose **Color Map > Green to Black to Red.**

The result is displayed in Figure 26.5. In the second run, you should see a slightly better ordering of the colors. Fewer bright reds appear at the top end. Notice that the cluster containing Hydroquinone and Trichloroacetic Acid moves toward the top with the greens, and the cluster containing Hexanol and Dimethylphenol moves toward the bottom toward the reds.

On a larger data set with more variables and more random sorting, the difference will be more noticeable.

Figure 26.5 Run 2 Dendrogram

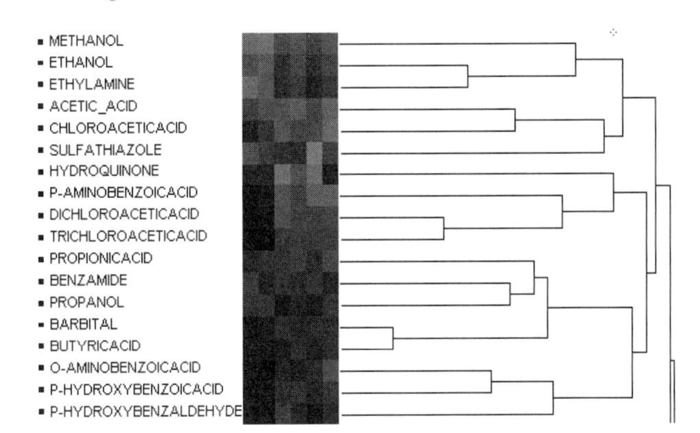

Hierarchical Cluster Options

Hierarchical clustering has the popup menu with the following commands.

Color Clusters and Mark Clusters automatically assign colors and markers to the clusters identified by the position of the sliding diamond in the dendrogram.

Save Clusters saves the cluster number of each row in a new data table column for the number of clusters identified by the slider in the dendrogram.

Save Display Order creates a new column in the data table that lists the order of each observation the cluster table.

Save Cluster Hierarchy saves information needed if you are going to do a custom dendrogram with scripting. For each clustering it outputs three rows, the joiner, the leader, and the result, with the cluster centers, size, and other information.

Number of Clusters prompts you to enter a number of clusters and positions the dendrogram slider to the that number.

Color Map is an option to add a color map showing the values of all the data colored across it's value range. There are several color theme choices in a submenu. Another term for this feature is *heat map*.

Dendrogram Scale contains options for scaling the dendrogram. **Even Spacing** shows the distance between each join point as equal. **Geometric Spacing** is useful when there are many clusters and you want the clusters near the top of the tree to be more visible than those at the bottom. (This option is the default now for more than 256 rows). **Distance Scale** shows the actual joining distance between each join point, and is the same scale used on the plot produced by the **Distance Graph** command.

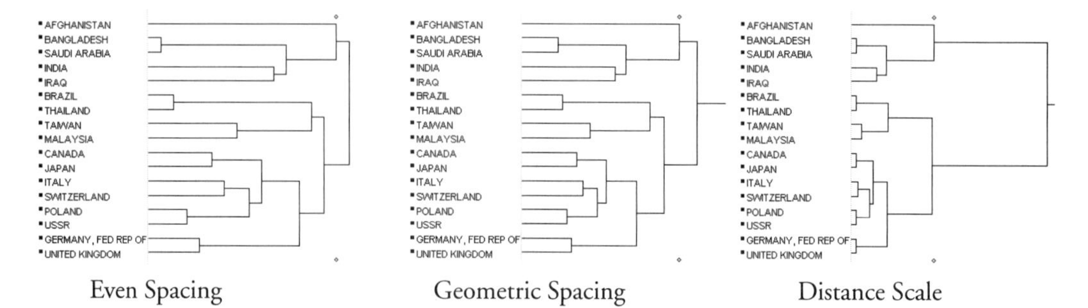

Even Spacing Geometric Spacing Distance Scale

Distance Graph shows or hides the scree plot at the bottom of the histogram.

Two way clustering adds clustering by column. A color map is automatically added with the column dendrogram at its base.

Legend only appears when a color map (produced by the **Color Map** command in the same menu) is present. The **Legend** command produces a legend showing the values represented by the color map.

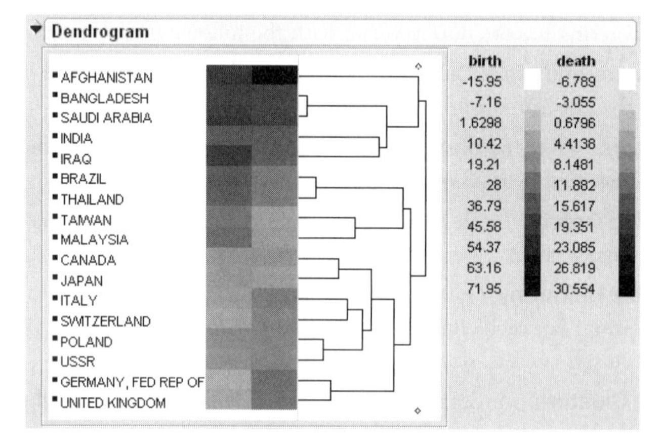

When there are a large number of clusters and the frame is squeezed down, **Cluster** adapts the cluster labels to smaller fonts, as needed.

Figure 26.6 Dendrogram with **Geometric X Scale** and **Color Map**

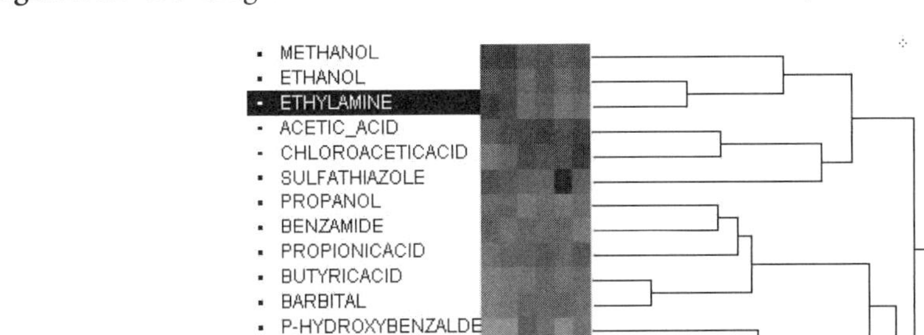

Visualizing Clusters

In the **Birth Death Subset** table you can use Fit Y by X to see a scatterplot of births and deaths, which lets you visualize possible clusters. You can use colors, markers, and the cluster numbers to identify the clusters in the scatterplot. For example, the table here shows the result after you save three clusters (identified by the dendrogram in Figure 26.3) in the **Birth Death Subset** table.

	country	birth	death	Cluster
1	AFGHANISTAN	52	30	1
2	ALGERIA	50	16	1
3	ARGENTINA	22	10	2
4	AUSTRALIA	16	8	2
5	AUSTRIA	12	13	2
6	BANGLADESH	47	19	1

Now use **Analyze > Fit Y by X** to plot birth by death and note how the points arrange within clusters. The plot on the left in Figure 26.7 shows clusters identified with markers assigned by the **Color Clusters** option. **Country** is the Label column and selected countries are labeled. In the scatterplot on the right, the new **Cluster** column is assigned as the label column and labels all the points.

Figure 26.7 Scatterplots That Show Three Clusters

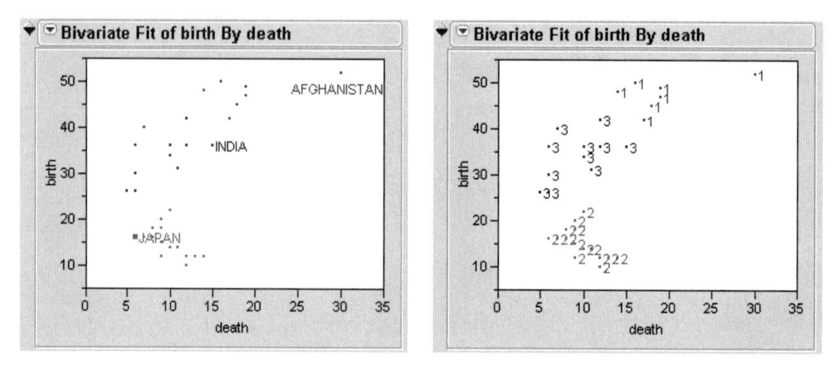

To obtain the two scatterplots in Figure 26.8:

1 Click and drag the sliding diamond on the dendrogram to the position signifying eight clusters.
2 Use **Save Clusters** to create a new data table column.
3 Give the new column, Cluster, the label role.
4 Plot birth by death with **Fit Y by X**, label all points (rows), and resize the plot.
5 Fit 95% ellipses using the Cluster column as the grouping variable.

Figure 26.8 Scatterplots That Show Eight Clusters

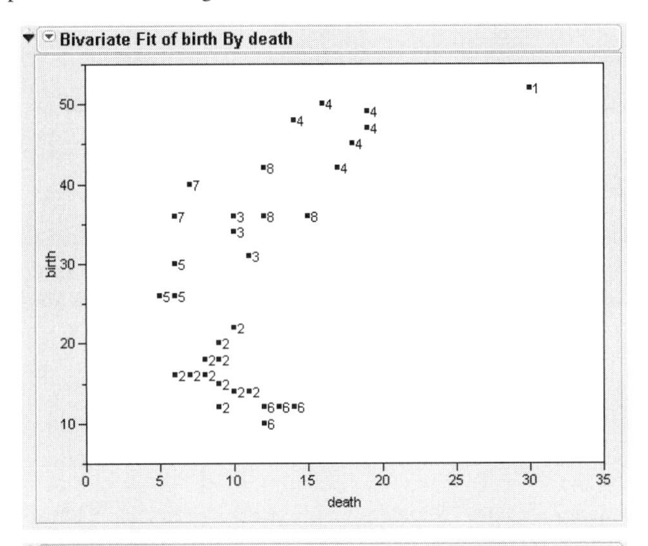

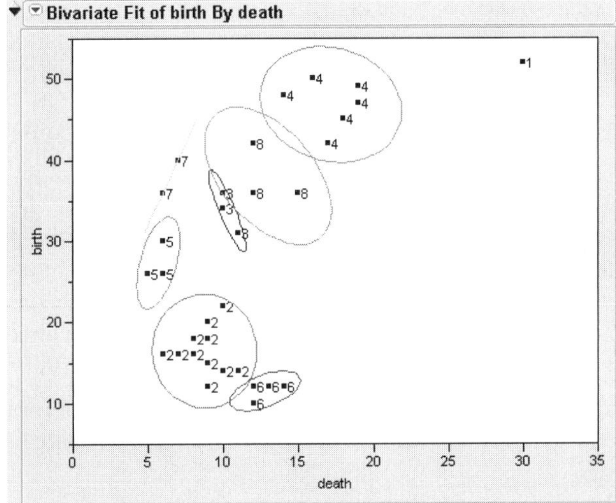

Technical Details for Hierarchical Clustering

The following description of hierarchical clustering methods gives distance formulas that use the following notation. Lowercase symbols generally pertain to observations and uppercase symbols to clusters.

n is the number of observations

v is the number of variables

x_i is the ith observation

C_K is the Kth cluster, subset of $\{1, 2,..., n\}$

N_K is the number of observations in C_K

$\bar{x}$ is the sample mean vector

$\overline{x_K}$ is the mean vector for cluster C_K

$\|x\|$ is the square root of the sum of the squares of the elements of **x** (the Euclidean length of the vector **x**)

$d(x_i, x_j)$ is $\|x\|^2$

Average Linkage In average linkage, the distance between two clusters is the average distance between pairs of observations, or one in each cluster. Average linkage tends to join clusters with small variances and is slightly biased toward producing clusters with the same variance (Sokal and Michener 1958).

Distance for the average linkage cluster method is

$$D_{KL} = \sum_{i \in C_K} \sum_{j \in C_L} \frac{d(x_i, x_j)}{N_K N_L}$$

Centroid Method In the centroid method, the distance between two clusters is defined as the squared Euclidean distance between their means. The centroid method is more robust to outliers than most other hierarchical methods but in other respects might not perform as well as Ward's method or average linkage (Milligan 1980).

Distance for the centroid method of clustering is

$$D_{KL} = \|\overline{x_K} - \overline{x_L}\|^2$$

Ward's In Ward's minimum variance method, the distance between two clusters is the ANOVA sum of squares between the two clusters added up over all the variables. At each generation, the within-cluster sum of squares is minimized over all partitions obtainable by merging two clusters from the previous generation. The sums of squares are easier to interpret when they are divided by the total sum of squares to give the proportions of variance (squared semipartial correlations).

Ward's method joins clusters to maximize the likelihood at each level of the hierarchy under the assumptions of multivariate normal mixtures, spherical covariance matrices, and equal sampling probabilities.

Ward's method tends to join clusters with a small number of observations and is strongly biased toward producing clusters with roughly the same number of observations. It is also very sensitive to outliers (Milligan 1980).

Distance for Ward's method is

$$D_{KL} = \frac{\|\overline{x_K} - \overline{x_L}\|^2}{\frac{1}{N_K} + \frac{1}{N_L}}$$

Single Linkage In single linkage the distance between two clusters is the minimum distance between an observation in one cluster and an observation in the other cluster. Single linkage has many desirable theoretical properties (Jardine and Sibson 1976); Fisher and Van Ness 1971; Hartigan 1981) but has fared poorly in Monte Carlo studies (Milligan 1980). By imposing no constraints on the shape of clusters, single linkage sacrifices performance in the recovery of compact clusters in return for the ability to detect elongated and irregular clusters. Single linkage tends to chop off the tails of distributions before separating the main clusters (Hartigan 1981). Single linkage was originated by Florek et al. (1951a, 1951b) and later reinvented by McQuitty (1957) and Sneath (1957).

Distance for the single linkage cluster method is

$$D_{KL} = \min_{i \in C_K} \min_{j \in C_L} d(x_i, x_j)$$

Complete Linkage In complete linkage, the distance between two clusters is the maximum distance between an observation in one cluster and an observation in the other cluster. Complete linkage is strongly biased toward producing clusters with roughly equal diameters and can be severely distorted by moderate outliers (Milligan 1980).

Distance for the Complete linkage cluster method is

$$D_{KL} = \max_{i \in C_K} \max_{j \in C_L} d(x_i, x_j)$$

K-Means Clustering

The *k*-means approach to clustering performs an iterative alternating fitting process to form the number of specified clusters. The *k*-means method first selects a set *n* points called *cluster seeds* as a first guess of the means of the clusters. Each observation is assigned to the nearest seed to form a set of temporary clusters. The seeds are then replaced by the cluster means, the points are reassigned, and the process continues until no further changes occur in the clusters. When the clustering process is finished, you see tables showing brief summaries of the clusters. The *k*-means approach is a special case of a general approach called the *EM algorithm*, where E stands for Expectation (the cluster means in this case) and the M stands for maximization, which means assigning points to closest clusters in this case.

The *k*-means method is intended for use with larger data tables, from approximately 200 to 100,000 observations. With smaller data tables, the results can be highly sensitive to the order of the observations in the data table.

To see an example of the *k*-means clustering, open the Cytometry.jmp sample data table. This table has eight numeric variables and 5000 observations.

Choose the **Cluster** command and select **KMeans** from the popup menu on the Launch dialog. Add the variables CD3 and CD8 as **Y, Columns** variables (see Figure 26.9). You can enter any number of clusters. However, the clustering proceeds only for the number specified (four in this example), as shown in Figure 26.9. If you want to see the results of a different number of clusters, run the cluster process again. This can be easily done from the Cluster Control Panel (discussed next).

Figure 26.9 Cluster Launch Dialog for *k*-Means Cluster with four Clusters

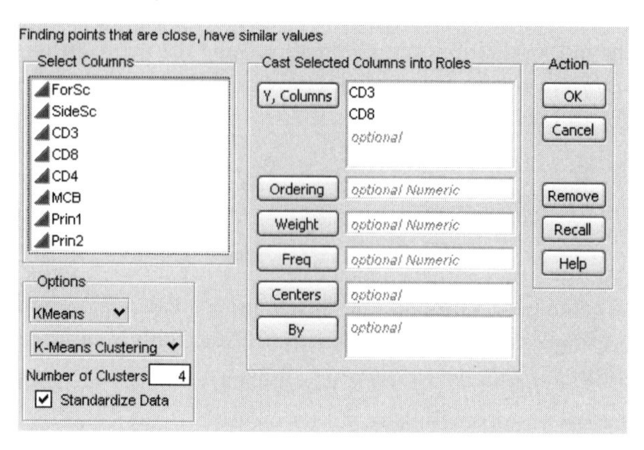

Note: If you choose a **Centers** column, JMP uses that column as the cluster identifier. The number of levels of the **Centers** variable identifies the number of clusters for the *k*-means process. The cluster means and standard deviations are then fixed at the observed mean and standard deviation of each level of the **Centers** variable.

When you click **OK**, the Control Panel appears. At this time, the initial seeding is complete and observations have been assigned to temporary clusters as shown in the Cluster Summary table beneath the Control Panel (see Figure 26.11).

Figure 26.10 Iterative Clustering Control Panel

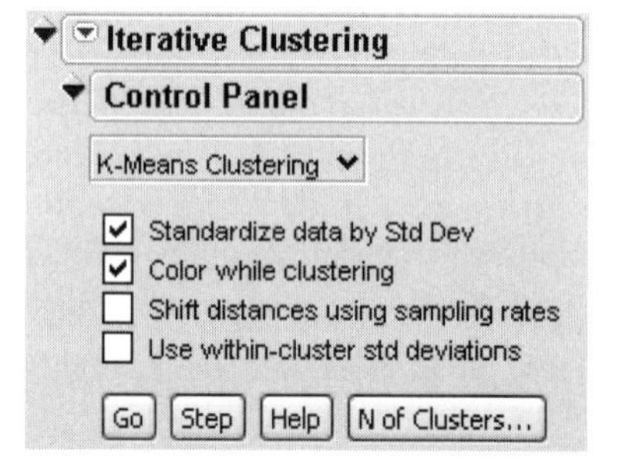

The Control Panel has these functions:

- A drop-down list to switch between *K*-Means, Normal Mixtures, and SOMs.
- The **Go** button starts the iterations of the clustering process and lets them continue.
- The **Step** button does the clustering one step at a time.
- the **N of Clusters** button allows you to change the number of clusters. If you click this, a dialog prompts you for the number, and JMP finds new cluster seeds.

Note: If you want to stop the iterations, hold down the Escape key on Windows, or hold down the Command key and type a period on the Macintosh.

The check boxes on the Control Panel have the following effects:

Standardize data by Std Dev causes distances to be calculated so that for each dimension, the difference is divided by the sample standard deviation for that variable across all rows. This does the same thing as the **Standardize Data** check box in the Cluster Launch dialog, except that in *k*-means clustering the standardize feature can be turned on or off anytime to affect the succeeding iterations.

Color while clustering assigns a different color to each cluster and colors the rows as they are assigned to clusters. This is a nice feature for large problems when you have plots open, because you can be entertained and enlightened as you watch the process at work.

Shift distances using sampling rates assumes that you have a mix of unequally sized clusters, and points should give preference to being assigned to larger clusters because there is a greater prior probability that it is from a larger cluster. This option is an advanced feature. Whether or not you use this option, the prior distances (that are subtracted from the point-to-cluster distances) are shown in the column titled Prior Dist. The calculations for this option are implied, but not shown for normal mixtures.

Use within-cluster std deviations If you don't use this option, all distances are scaled by an overall estimate of the standard deviation of each variable. If you use this option, then it will calculate distances scaled by the standard deviation estimated for each cluster. This option is not recommended for *k*-means clustering, but is turned on by default for normal mixture estimation.

When you click **Go** in the Control Panel, the clustering process begins. You can see the results of each iteration as it occurs in the summary tables below the Control Panel. When no further changes occur in cluster assignment, the process stops. Final results appear as in the right-hand table of Figure 26.11.

Figure 26.11 *K*-Means Cluster Platform Before and After Iterations

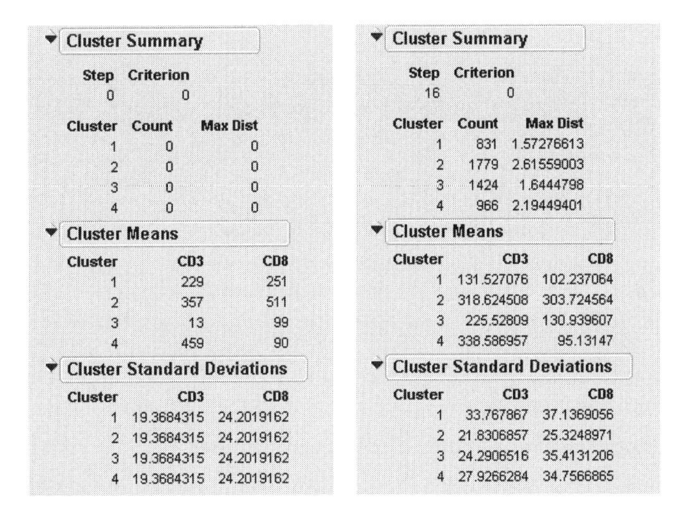

Cluster Summary

Step	Criterion
0	0

Cluster	Count	Max Dist
1	0	0
2	0	0
3	0	0
4	0	0

Cluster Means

Cluster	CD3	CD8
1	229	251
2	357	511
3	13	99
4	459	90

Cluster Standard Deviations

Cluster	CD3	CD8
1	19.3684315	24.2019162
2	19.3684315	24.2019162
3	19.3684315	24.2019162
4	19.3684315	24.2019162

Cluster Summary

Step	Criterion
16	0

Cluster	Count	Max Dist
1	831	1.57276613
2	1779	2.61559003
3	1424	1.6444798
4	966	2.19449401

Cluster Means

Cluster	CD3	CD8
1	131.527076	102.237064
2	318.624508	303.724564
3	225.52809	130.939607
4	338.586957	95.13147

Cluster Standard Deviations

Cluster	CD3	CD8
1	33.767867	37.1369056
2	21.8306857	25.3248971
3	24.2906516	35.4131206
4	27.9266284	34.7566865

Viewing a *K*-Means Clustering Process

The *k*-means platform does not produce any graphical displays. However, you can see the clustering as it occurs by watching a scatterplot of any two of the variables used to cluster. In this example there are only two variables, so a scatterplot displayed by the Fit Y by X platform gives a complete picture before the clustering takes place. The 5,000 points to the left in Figure 26.12 appear very dense, but exploring the data in this way can help you decide how many clusters to request.

During the clustering process, you can see the points highlight and assume different colors as they are assigned to different clusters. When the clustering is complete, the scatterplot in this example looks like the plot on the right in Figure 26.12, with the clusters showing in color (changed to shades of gray on black-and-white printers).

Figure 26.12 Scatterplot of Two Variables Used for *K*-Means Clustering

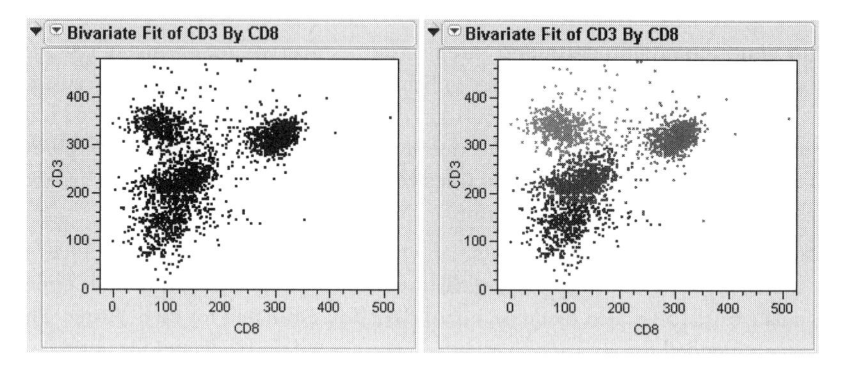

Platform Options

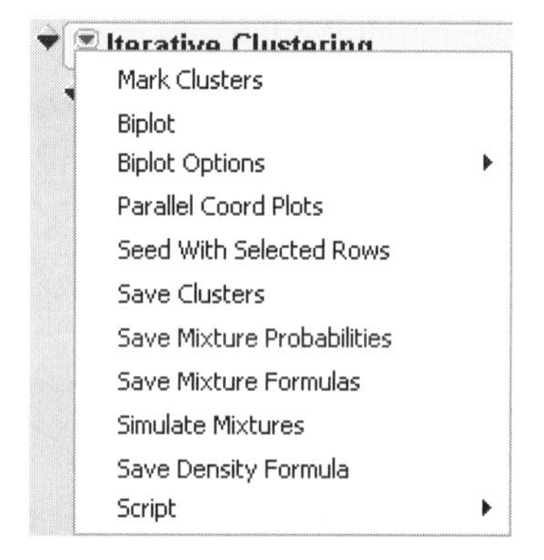

These options are accessed from the popup menu on the title. They apply to *k*-means, normal mixtures, and SOM methods.

Mark Clusters assigns a unique marker to each cluster.

Biplot shows a plot of the points and clusters in the first two principal components of the data. Circles are drawn around the cluster centers with area proportional to the number of points in the cluster. The directions of the variables in principal component space are shown with a set of axis rays emanating from a central point. The rays are initially drawn at the center of the plot, but can be dragged to a different part of the plot so as not to obscure points or clusters. To move the rays, click and drag at the point where they meet. If the biplot is selected before the iterations are performed, it shows how the clusters change at each iteration.

Biplot Options contains two options for controlling the Biplot. **Show Biplot Rays** allows you to turn off the biplot rays. **Biplot Ray Position** allows you to position the biplot ray display. This is viable since biplot rays only signify the directions of the original variables in canonical space, and there is no special significance to where they are placed in the graph.

Parallel Coord Plots plots each cluster separately, plus an additional plot showing the means for each cluster. It is updated as the platform progresses through iterations.

Seed with Selected Rows Select the rows in the data table that you want to start the cluster centers at, and then select this option. The number of clusters will change to the number of rows that are selected in the data table.

Save Clusters creates a new column with the cluster number that each row is assigned to. For normal mixtures, this is the cluster that is most likely.

Save Mixture Probabilities creates a column for each cluster and saves the probability an observation belongs to that cluster in the column. This works for *k*-means clustering only.

Save Mixture Formulas creates columns with mixture probabilities, but stores their formulas in the column and needs additional columns to hold intermediate results for the formulas. Use this feature if you want to score probabilities for excluded data, or data you add to the table.

Simulate Mixtures for normal mixtures—creates a new data table that simulates data using the estimated cluster mixing probabilities, means, and standard deviations for each cluster. The main use of this option is to analyze real data next to simulated data to see if the model adequately describes the distribution of the data.

Save Density Formula saves the density formula in the data table.

Self Organizing Maps

Self-Organizing Maps (SOMs, pronounced "somz") are implemented in the *K*-Means Clustering platform. The technique was developed by Teuvo Kohonen (1989) and further extended by a number of other neural network enthusiasts and statisticians. The original SOM was cast as a learning process, like the original neural net algorithms, but the version implemented here is done in a much more straightforward way as a simple variation on *k*-means clustering. In the SOM literature, this would be called a *batch algorithm* using a *locally weighted linear smoother.*

The goal of a SOM is to not only form clusters, but form them in a particular layout on a cluster grid, such that points in clusters that are near each other in the SOM grid are also near each other in multivariate space. In classical *k*-means clustering, the structure of the clusters is arbitrary, but in SOMs the clusters have the grid structure. This grid structure may be invaluable in interpreting the clusters in two dimensions.

Figure 26.13 Biplot with *K*-means and with SOM Option

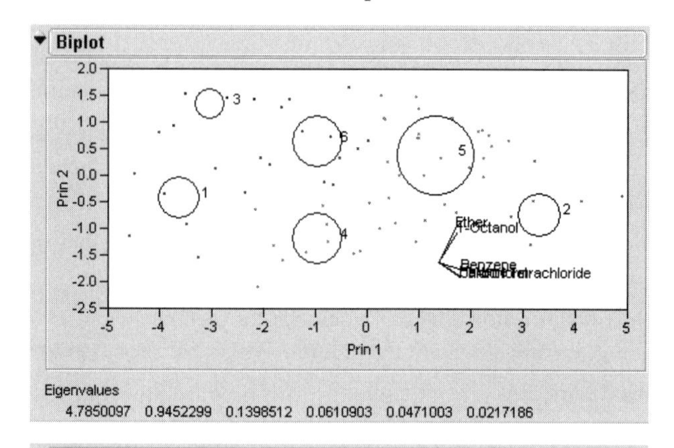

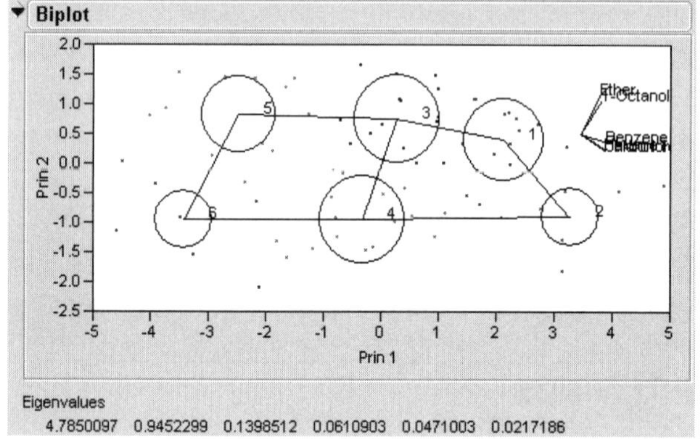

Creating SOMs in JMP

Use the Birth Death.jmp sample data table. Select **Analyze > Multivariate Methods > Cluster** and complete the launch dialog as shown in Figure 26.14. Select **K Means Cluster** and **Self Organizing Map** from the drop-down menus.

Figure 26.14 Self Organizing Map Selection

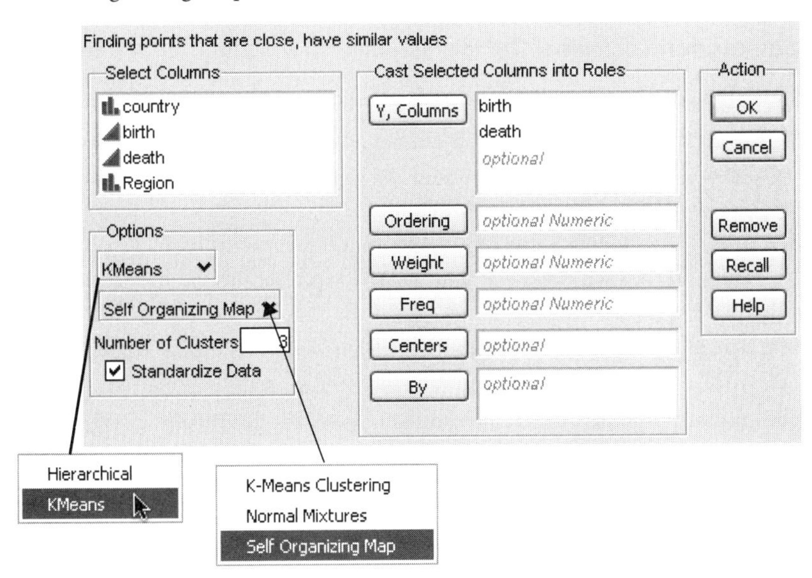

When you request an SOM, JMP needs to factor the number of clusters into grid dimensions. For example, if you specify 24 clusters, JMP factors the clusters into a 6 by 4 grid, the grid closest to a square. Therefore, the number of clusters must not be prime. If given a prime number for the number of clusters, JMP does not complete the SOM. If this happens, change **Number of Clusters** to a composite number.

After specifying variables and clicking **OK**, the iterative clustering control panel appears.

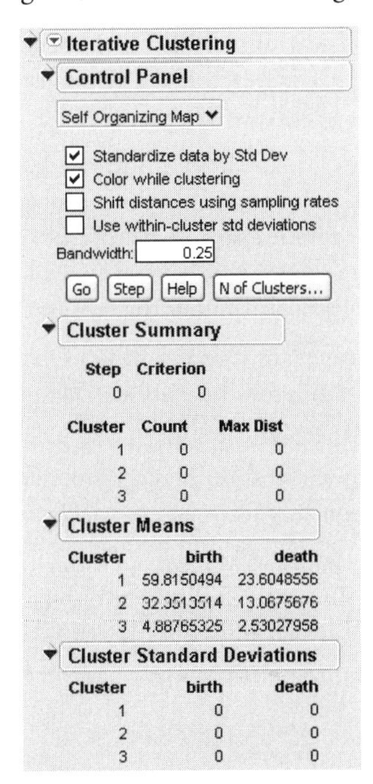

See Figure 26.10 "Iterative Clustering Control Panel," p. 540 for details on the options in this panel.

Implementation Technical Details

The SOM implementation in JMP proceeds as follows:

- The first step is to obtain good initial cluster seeds that provide a good coverage of the multidimensional space. JMP uses principal components to determine the two directions which capture the most variation in the data.
- JMP then lays out a grid in this principal component space with its edges 2.5 standard deviations from the middle in each direction. The clusters seeds are formed by translating this grid back into the original space of the variables.
- The cluster assignment proceeds as with *k*-means, with each point assigned to the cluster closest to it.
- The means are estimated for each cluster as in *k*-means. JMP then uses these means to set up a weighted regression with each variable as the response in the regression, and the SOM grid coordinates as the regressors. The weighting function uses a 'kernel' function that gives large weight to the cluster whose center is being estimated, with smaller weights given to clusters farther away from the cluster in the SOM grid. The new cluster means are the predicted values from this regression.
- These iterations proceed until the process has converged.

Normal Mixtures

Normal mixtures is an iterative technique implemented in the *k*-means clustering platform, but rather than being a clustering method to group rows, it is more of an estimation method to characterize the cluster groups. Rather than classifying each row into a cluster, it estimates the probability that a row is in each cluster. (McLachlan, G.J. and Krishnan, T. 1997)

Hierarchical and *k*-means clustering methods work well when clusters are well separated, but when clusters overlap, assigning each point to one cluster is problematic. In the overlap areas, there are points from several clusters sharing the same space. It is especially important to use normal mixtures rather than *k*-means clustering if you want an accurate estimate of the total population in each group, because it is based on membership probabilities, rather than arbitrary cluster assignments based on borders.

The algorithm for normal mixtures method is a BFGS Quasi-Newton method, and since the estimates are heavily dependent on initial guesses, the platform will go through a number of tours, each with randomly selected points as initial centers.

Doing multiple tours makes the estimation process somewhat expensive, so considerable patience is required for large problems. Controls allow you to specify the tour and iteration limits.

For an example of the use of **Normal Mixtures**, open the Iris.jmp sample data table. This data set was first introduced by Fisher (1936), and includes four different measurements: sepal length, sepal width, petal length, and petal width, performed on samples of 50 each for three species of iris.

	Sepal length	Sepal width	Petal length	Petal width	Species
1	5.1	3.5	1.4	0.2	setosa
2	4.9	3.0	1.4	0.2	setosa
3	4.7	3.2	1.3	0.2	setosa
4	4.6	3.1	1.5	0.2	setosa
5	5.0	3.6	1.4	0.2	setosa
6	5.4	3.9	1.7	0.4	setosa
7	4.6	3.4	1.4	0.3	setosa
8	5.0	3.4	1.5	0.2	setosa

The normal mixtures approach to clustering predicts the proportion of responses expected within each cluster. The assumption is that the joint probability distribution of the measurement columns can be approximated using a mixture of three multivariate normal distributions, which represent three different clusters corresponding to the three species of iris. The distributions have individual mean vectors and standard deviation vectors for each cluster.

The result will have a good estimate of the proportion from each group. To perform normal mixtures clustering with this example, choose the **Cluster** command from the **Analyze > Multivariate Methods** menu and select **KMeans** then **Normal Mixtures** from the popup menu on the Launch dialog.

Recall that the number of clusters must be selected beforehand (three in this example). To complete the launch dialog, assign Sepal length, Sepal width, Petal length, and Petal width as the *Y* columns.

When you click **OK**, the Iterative Clustering Control Panel is displayed as before, with initial cluster means and standard deviations for each *Y* column.

Click **Go**. The clusters are refined after each iteration until the process has converged. It can take hundreds of iterations for some problems and prompts whether to continue iterations. The Iris data here converged in 105 iterations, as shown in Figure 26.15.

Notice that the final results give the proportions and counts of the observations (rows) that are attributed to each cluster but that the counts are not exact integers as they would be in a *k*-means clustering. This is because normal mixtures clustering determines probable cluster membership rather than definite membership, which allows for cluster overlap. In this case the groups have estimates of 35.78% in Cluster 1, 30.88% in Cluster 2, and 33.33% in Cluster 3 giving a good approximation to the 33.33% sample of each species.

There are several ways to save cluster information:

- You can save any table in the report by context-clicking (right-click on Windows, Control-click on Macintosh) in the table and selecting the command **Make into Data Table**.
- You can save the mixture probabilities to the original data table with the **Save Mixture Probabilities** command in the popup menu on the title bar. This command creates a new column in the data table for each cluster and saves the probability that an observation (row) belongs to that cluster. Or use the **Save Mixture Formulas** to obtain these columns with formulas for scoring other data.
- Another option is to save the cluster numbers to the data table using the **Save Clusters** command in the popup menu on the Iterative Clustering title bar. This creates a new column in the data table and uses the *most probable* cluster location to assign a cluster number to each observation.

Figure 26.15 Final Results of a Normal Mixtures Clustering Example

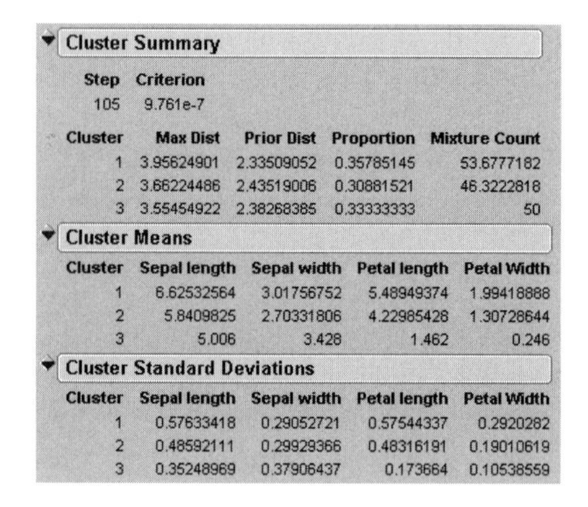

Cluster Summary

Step	Criterion
105	9.761e-7

Cluster	Max Dist	Prior Dist	Proportion	Mixture Count
1	3.95624901	2.33509052	0.35785145	53.6777182
2	3.66224486	2.43519006	0.30881521	46.3222818
3	3.55454922	2.38268385	0.33333333	50

Cluster Means

Cluster	Sepal length	Sepal width	Petal length	Petal Width
1	6.62532564	3.01756752	5.48949374	1.99418888
2	5.8409825	2.70331806	4.22985428	1.30728644
3	5.006	3.428	1.462	0.246

Cluster Standard Deviations

Cluster	Sepal length	Sepal width	Petal length	Petal Width
1	0.57633418	0.29052721	0.57544337	0.2920282
2	0.48592111	0.29929366	0.48316191	0.19010619
3	0.35248969	0.37906437	0.173664	0.10538559

Fixing Cluster Centers

There are two choices when you have known groups of observations, and you want to score clusters for other data:

1. If you are using Normal Mixtures, use the **Save Mixture Formulas** command, delete all the rows, and save a copy of the resulting table. This empty table serves as a table template. Then concatenate a data table containing your data to the table template so that the formulas from the first table compute values for the second table.
2. Make a column that identifies the clusters for the known data, and is missing or blank for the new data to be scored. Then, launch the Cluster platform and select this column in the **Centers** column list on the Cluster Launch dialog.

The following example was inspired by a case study undertaken by Do and McLachlan (1984) to determine the diet of owls in terms of the estimated proportion of several species of Malaysian rats. The researchers used two data sets. The first (classification) data set had 179 observations and contained measures of skull length, teeth row, palentine foramen, and jaw width for seven known species. The second (observational) set had 115 observations and contained the skull length, skull width, jaw width, and teeth row, but no information regarding species. The goal was to use the information from the classification data set to categorize the rats in the observational data set into species.

Both the classification and the observational data are combined (appended) into the Owl Diet.jmp data table in the Sample Data folder, leaving the Species column with missing values for rats from the observational data set (Figure 26.16).

To do this in JMP, choose **Cluster** from the **Analyze > Multivariate Methods** menu and select **K Means**, then **Normal Mixtures** on the Launch dialog. Use skull length, teeth row, palatine foramen, and jaw length as **Y** Columns and species as the **Centers** Column. Click **OK** to see the Cluster Control panel.

JMP gives as initial values the cluster centers from the classification data set. When you click **Go**, JMP classifies the observations with missing values for the Centers variable into their closest clusters. This approach is similar to discriminant analysis methods.

Figure 26.16 Owl diet Data and Distribution of Species

Owl Diet

Columns (5/0)
- skull length
- teeth row
- palatine foramen
- jaw length
- species

Rows
All rows 294

	skull length	teeth row	palatine foramen	jaw length	species
1	298.2611	49.543	48.66454	103.1874	exulans
2	299.6613	45.175	50.65786	112.806	exulans
3	324.0066	48.058	54.88021	116.5792	exulans
4	316.8173	47.411	55.43373	125.1312	exulans
5	310.8253	45.535	57.9427	119.065	exulans
6	327.4231	47.732	48.92113	121.6545	exulans
7	303.7252	46.156	49.5088	117.3014	exulans
8	303.6622	47.345	55.73164	118.7323	exulans
9	307.4479	46.127	57.93869	108.8317	exulans
10	310.8235	48.723	48.92026	115.4827	exulans

Figure 26.17 shows both the preliminary counts and the result of clustering the observational rows with the Species used as the **Centers** column.

Figure 26.17 Cluster Example with Centers Variable

Cluster Summary

Step	Criterion
0	0

Cluster	species	Proportion	Mixture Count
1	annandalfi	0.14285714	0
2	argentiventer	0.14285714	0
3	exulans	0.14285714	0
4	rajah	0.14285714	0
5	surifer	0.14285714	0
6	tiomanicus	0.14285714	0
7	whiteheadi	0.14285714	0

Cluster centers defined by species

Cluster Summary

Step	Criterion
0	0

Cluster	species	Proportion	Mixture Count
1	annandalfi	0.09457353	27.9505282
2	argentiventer	0.08229707	24.0485359
3	exulans	0.02262121	6.7279926
4	rajah	0.04323352	3.97045395
5	surifer	0.02319349	6.8527356
6	tiomanicus	0.08154074	23.3078996
7	whiteheadi	0.65254044	201.141854

Cluster centers defined by species

Details of the Estimation Process

Since the estimation is hampered by outliers, JMP uses a robust method of estimating the normal parameters. JMP computes the estimates via maximum likelihood with respect to a mixture of Huberized normal distributions (a class of modified normal distributions that was tailor-made to be more outlier resistant than the normal distribution).

The Huberized Gaussian distribution has pdf $\Phi_k(x)$.

$$\Phi_k(x) = \frac{\exp(-\rho(x))}{c_k}$$

$$\rho(x) = \begin{cases} \dfrac{x^2}{2} & \text{if } |x| \le k \\ k|x| - \dfrac{k^2}{2} & \text{if } |x| > k \end{cases}$$

$$c_k = \sqrt{2\pi}[\Phi(k) - \Phi(-k)] + 2\frac{\exp(-k^2/2)}{k}$$

So, in the limit as k becomes arbitrarily large, $\Phi_k(x)$ tends toward the normal PDF. As $k \to 0$, $\Phi_k(x)$ tends toward the exponential (Laplace) distribution.

The regularization parameter k is set so that P(Normal(x) < k) = Huber Coverage, where Normal(x) indicates a multivariate normal variate. Huber Coverage is a user field, which defaults to 0.90.

Chapter 27

Partial Least Squares

The PLS Platform

The PLS platform fits models using partial least squares (PLS). PLS balances the two objectives of explaining response variation and explaining predictor variation. It is especially useful since it is appropriate in analysis that have more *x*-variables than observations.

The number of factors to extract depends on the data. Basing the model on more extracted factors (successive linear combinations of the predictors also called *components* or *latent vectors*) optimally addresses one or both of two goals: explaining response variation and explaining predictor variation. In particular, the method of partial least squares balances the two objectives, seeking for factors that explain both response and predictor variation. Remember that extracting too many factors can cause over-fitting, (tailoring the model too much to the current data,) to the detriment of future predictions. The PLS platform enables you to choose the number of extracted factors by *cross-validation*, that is, fitting the model to part of the data and minimizing the prediction error for the unfitted part. It also allows for interactive exploration of the extracted factors through the use of the Profiler, which greatly simplifies the identification of the factors' meaning.

Contents

PLS

Ordinary least squares regression, as implemented in JMP platforms such as Fit Model and Fit Y by X, has the single goal of minimizing sample response prediction error, seeking linear functions of the predictors that explain as much variation in each response as possible. The techniques implemented in the PLS platform have the additional goal of accounting for variation in the predictors, under the assumption that directions in the predictor space that are well sampled should provide better prediction for new observations when the predictors are highly correlated. The techniques implemented in the PLS platform work by extracting successive linear combinations of the predictors, called *factors* (also called *components* or *latent vectors*), which optimally address the combined goals of explaining response variation and explaining predictor variation. In particular, the method of partial least squares balances the two objectives, seeking factors that explain both response and predictor variation.

The most important point, however, is that PLS works in cases where OLS does not. If the data have more *x*-variables than observations, OLS cannot produce estimates, where PLS can.

Launch the Platform

As an example, open the Baltic.jmp data table. The data are reported in Umetrics (1995); the original source is Lindberg, Persson, and Wold (1983). Suppose that you are researching pollution in the Baltic Sea, and you would like to use the spectra of samples of sea water to determine the amounts of three compounds present in samples from the Baltic Sea: lignin sulfonate (pulp industry pollution), humic acids (natural forest products), and detergent (optical whitener). Spectrometric calibration is a type of problem in which partial least squares can be very effective. The predictors are the spectra emission intensities at different frequencies in sample spectrum (v1–v27), and the responses are the amounts of various chemicals in the sample.

For the purposes of calibrating the model, samples with known compositions are used. The calibration data consist of 16 samples of known concentrations of lignin sulfonate, humic acids, and detergent, with spectra based on 27 frequencies (or, equivalently, wavelengths).

Launch the PLS platform by selecting **Analyze > Multivariate Methods > PLS**. The launch dialog in Figure 27.1 appears.

Figure 27.1 PLS Launch Dialog

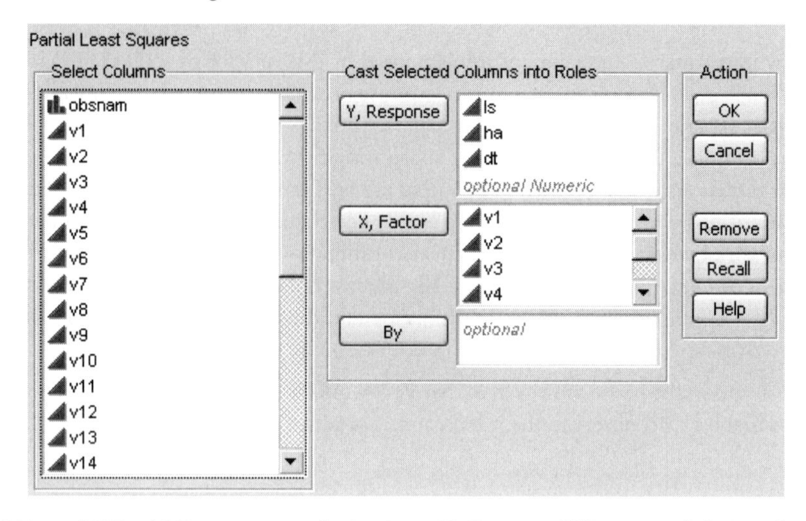

Assign LS, HA, and DT as **Y, Response** and v1–v27 as **X, Factors.** When you click **OK**, the initial PLS report appears.

At the top of this report, you see plots of each *X* score against its corresponding *Y* score, as seen in Figure 27.2. As your analysis continues, this plot updates according to the number of scores you specify.

Figure 27.2 Scores Plot

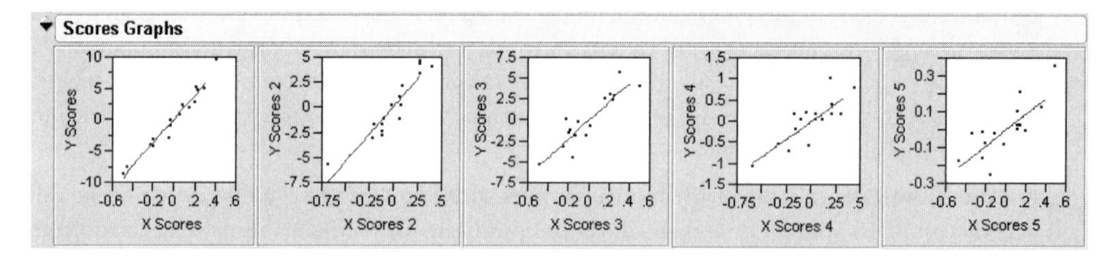

By default, JMP determines the default number of dimensions as the smallest dimension (k) which gives a slope ratio (slope of Y Score –X Score chart/slope of Y Score k–X Score k chart) greater than 30. In the Baltic example, five latent vectors are extracted by default. For the Baltic data, the percent of variation explained by each latent factor is shown in the Percent Variance Explained report (Figure 27.3). Note that most of the variation in *X* is explained by three factors; most of the variation in *Y* is explained by six or seven. Note that in this case, "factors" refers to latent variables, not columns in the data table.

Figure 27.3 Percent Variance Explained Report

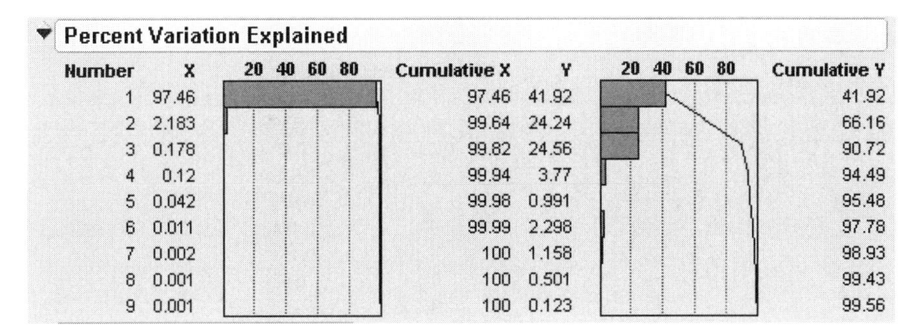

Percent Variation Explained

Number	X	20 40 60 80	Cumulative X	Y	20 40 60 80	Cumulative Y
1	97.46		97.46	41.92		41.92
2	2.183		99.64	24.24		66.16
3	0.178		99.82	24.56		90.72
4	0.12		99.94	3.77		94.49
5	0.042		99.98	0.991		95.48
6	0.011		99.99	2.298		97.78
7	0.002		100	1.158		98.93
8	0.001		100	0.501		99.43
9	0.001		100	0.123		99.56

The PLS model is incomplete until you specify the number of factors to use in the model. JMP suggests a number of factors in the N Latent Vectors control panel, shown in Figure 27.4.

Figure 27.4 N Latent Vectors Control Panel

N Latent Vectors

Number of Dimensions	5
Minimum	1
Default	5
Maximum	9
Use Cross Validation	0
User Specified	0

Go

Minimum specifies the minimum number of latent vectors that can be estimated in the model.

Default is JMP's suggestion on the number of vectors needed to explain the variation in the data.

Maximum is the maximum number of latent vectors that can be estimated in the model.

Use Cross Validation is the number of latent vectors that can be estimated in the model using cross-validation.

User Specified uses the number of latent vectors specified beside Number of Dimensions.

You can choose the number of factors by using cross-validation, in which the data set is divided into two or more groups. You fit the model to all groups except one, then you check the capability of the model to predict responses for the group omitted. Repeating this for each group, you then can measure the overall capability of a given form of the model. JMP displays the results using the prediction root mean square error for each number of factors, shown in Figure 27.5.

Figure 27.5 Cross-Validation Report

Cross Validation

Number	Prediction RMSE
1	0.865
2	0.798
3	0.57
4	0.505
5	0.638
6	0.464
7	0.413
8	0.46
9	0.458

The model with the lowest prediction RMSE has seven factors. However, notice that this is almost the same as a model with four factors. A model with four factors is preferable, so the prediction reports are based on the four-model factors.

Along with the cross-validation report, you see plots of the predicted *Y*'s and predicted *X*'s, as shown in Figure 27.6. Note that these are the scaled values of *X* and *Y*, not the original values. Below these plots are a text box to enter a value to predict and four slider boxes to adjust the values of each latent vector.

Figure 27.6 Prediction Graph and Sliders

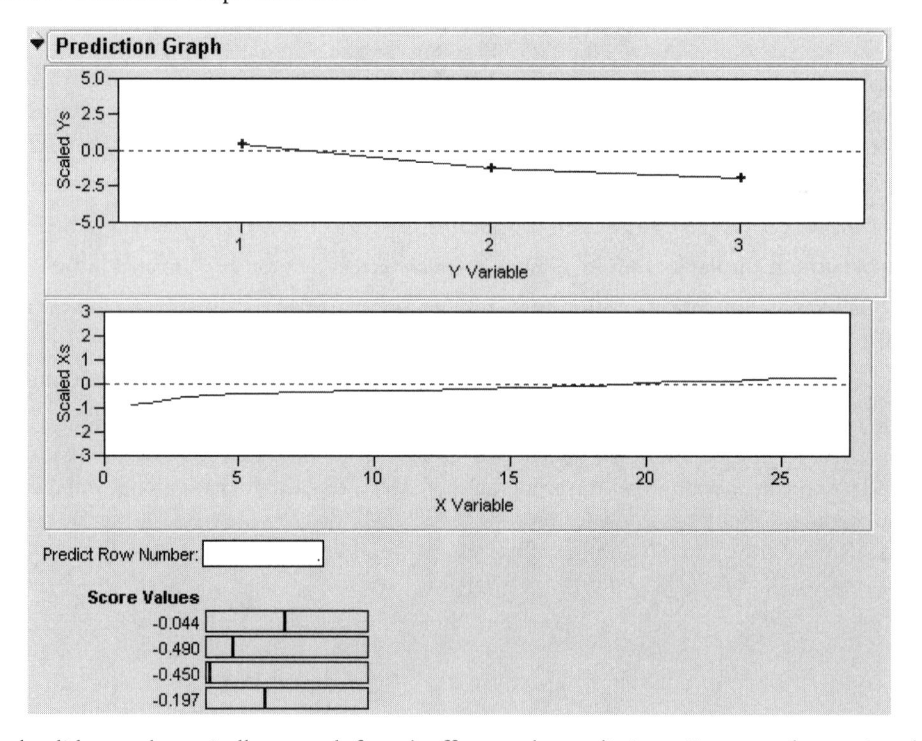

Move the sliders to dynamically see each factor's effect on the predictions. For example, moving the fourth slider in this example shows that this latent variable has a great effect on the second Y-variable (humic acid). This factor obviously represents a quantity that affects humic acid far more than lignin sulfonate or detergent. It may represent something that increases production of natural (rather than man-made) pollution.

Similarly, moving the second slider affects the amount of the 11th and 21st spectral values. Therefore, the physical quantity that this represents manifests itself through variations in those frequencies far more than the others, which remain essentially constant.

To predict the values of a specific row number, enter the row number in the text box above the score value sliders. For example, to predict the values of the first row, enter a 1 in the box. Predicted values appear on the graphs of both the *X* and *Y* factors (Figure 27.7).

Figure 27.7 Prediction of Row 1

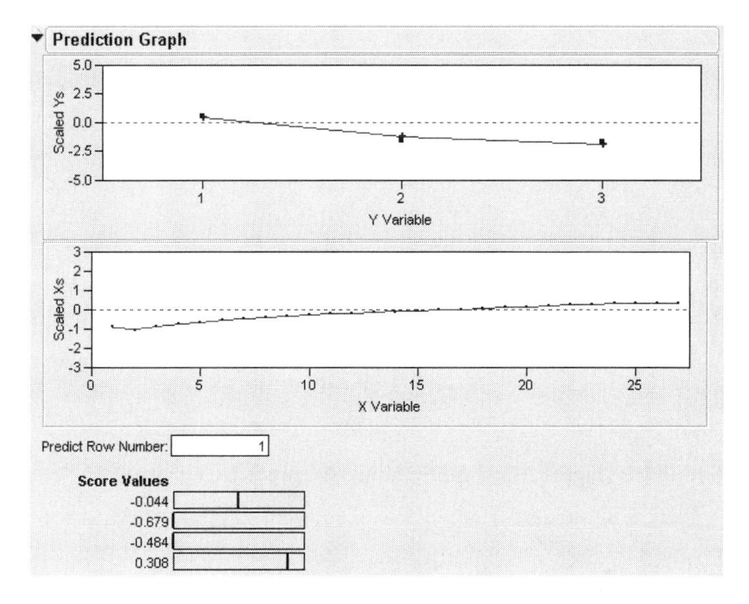

Model Coefficients and PRESS Residuals

Both the Model Coefficients report and the PRESS Residuals report are closed initially. The model coefficients appear under the prediction graph, and are shown for this example in Figure 27.8.

Figure 27.8 Model Coefficients

Model Coefficients

	ls	ha	dt
Intercept	0.2729595	0.0344186	-67.57717
v1	-0.000218	-0.000103	0.0418071
v2	-9.542e-5	-1.46e-6	0.0158662
v3	4.4942e-5	-5.931e-5	0.0060717
v4	7.2032e-5	-6.565e-5	0.0033525
v5	0.0001148	-9.051e-5	0.0016756
v6	0.0001196	-8.129e-5	0.0002634
v7	9.333e-5	-4.115e-5	-0.001371
v8	4.1874e-5	2.4943e-5	-0.00339
v9	-0.000048	0.0001269	-0.005697
v10	-0.000133	0.0002092	-0.006251
v11	-0.000172	0.0002386	-0.005494
v12	-0.000226	0.0002697	-0.003442
v13	-0.000213	0.0002483	-0.002222
v14	-0.000153	0.0001845	-0.001
v15	-0.000158	0.0001797	0.0002265
v16	-6.084e-5	9.2942e-5	0.0005355
v17	-0.000011	5.1952e-5	0.0005537
v18	9.3896e-6	4.0165e-5	0.000319
v19	0.0000707	-1.455e-5	0.0009073
v20	0.0001932	-0.000102	-0.001097
v21	0.0002602	-0.000155	-0.001278
v22	0.0003509	-0.000209	-0.003893
v23	0.0003545	-0.000194	-0.005417
v24	0.0004301	-0.000256	-0.005427
v25	0.0007435	-0.000499	-0.009134
v26	0.000799	-0.00053	-0.010708
v27	0.0008435	-0.00056	-0.01104

The PRESS residuals (Figure 27.9) appear above the prediction graph.

Figure 27.9 PRESS Residuals

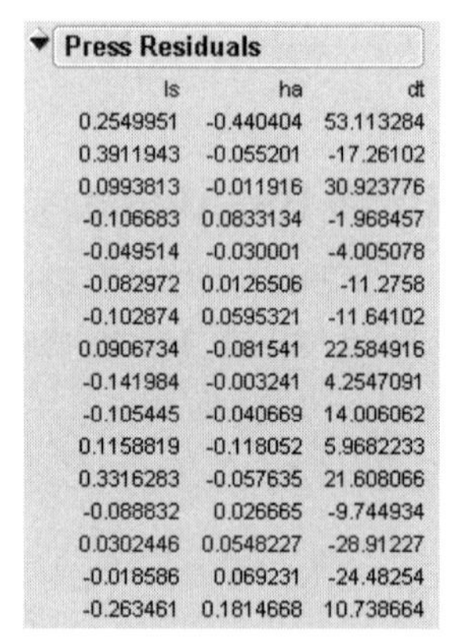

Press Residuals

ls	ha	dt
0.2549951	-0.440404	53.113284
0.3911943	-0.055201	-17.26102
0.0993813	-0.011916	30.923776
-0.106683	0.0833134	-1.968457
-0.049514	-0.030001	-4.005078
-0.082972	0.0126506	-11.2758
-0.102874	0.0595321	-11.64102
0.0906734	-0.081541	22.584916
-0.141984	-0.003241	4.2547091
-0.105445	-0.040669	14.006062
0.1158819	-0.118052	5.9682233
0.3316283	-0.057635	21.608066
-0.088832	0.026665	-9.744934
0.0302446	0.0548227	-28.91227
-0.018586	0.069231	-24.48254
-0.263461	0.1814668	10.738664

Cross-Validation

The quality of the observed data fit cannot be used to choose the number of factors to extract; the number of extracted factors must be chosen on the basis of how well the model fits observations not involved in the modeling procedure itself.

One method of choosing the number of extracted factors is to fit the model to only part of the available data (the *training set*) and to measure how well models with different numbers of extracted factors fit the other part of the data (the *test set*). This is called *test set validation.*

To do this in JMP, select the test set and exclude them from the analysis. Then, fit a PLS model and save the prediction formula to the data table. This adds a column to the data table that contains predicted values for both the training set and test set.

Platform Options

The PLS platform has several options, used to save predicted values and formulas from an analysis. All are found in the platform drop-down menu on the uppermost outline bar of the report.

Show Points shows or hides the points on the *Y* scores vs. *X* scores scatterplots.

Show Confidence Lines shows or hides the confidence lines on the *Y* scores vs. *X* scores scatterplots. Note that these should only be used for outlier detection, since the platform doesn't compute traditional confidence lines.

Save Prediction Formula saves a prediction formula for each *Y* variable in the data table.

Save Outputs saves the outputs of the platform in two ways. One is as data tables, facilitating further analyses. In addition, scripts are added to the data table containing matrices of *X* scores, *Y* scores, *X* Loadings, and *Y* loadings.

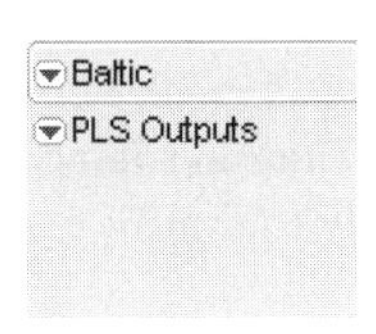

When the script is executed, the number of each of these quantities appears in the log window. For this example,

```
NRow(X Scores): 16
NRow(X Loadings):27
NRow(Y Scores): 16
NRow(Y Loadings):3
```

Statistical Details

Partial least squares (PLS) works by extracting one factor at a time. Let $X = X_0$ be the centered and scaled matrix of predictors and $Y = Y_0$ the centered and scaled matrix of response values. The PLS method starts with a linear combination $\mathbf{t} = X_0\mathbf{w}$ of the predictors, where **t** is called a *score vector* and **w** is its associated *weight vector.* The PLS method predicts both X_0 and Y_0 by regression on **t**:

$$\hat{X}_0 = \mathbf{t}\mathbf{p}', \text{ where } \mathbf{p}' = (\mathbf{t}'\mathbf{t})^{-1}\mathbf{t}'X_0$$

$$\hat{Y}_0 = \mathbf{tc}', \text{ where } \mathbf{c}' = (\mathbf{t}'\mathbf{t})^{-1}\mathbf{t}'Y_0$$

The vectors $\mathbf{p}$ and $\mathbf{c}$ are called the *X*- and *Y-loadings*, respectively.

The specific linear combination $\mathbf{t} = X_0\mathbf{w}$ is the one that has maximum covariance $\mathbf{t}_0\mathbf{u}$ with some response linear combination $\mathbf{u} = Y_0\mathbf{q}$. Another characterization is that the *X*- and *Y*-weights $\mathbf{w}$ and $\mathbf{q}$ are proportional to the first left and right singular vectors of the covariance matrix $X_0'Y_0$ or, equivalently, the first eigenvectors of $X_0'Y_0Y_0'X_0$ and $Y_0'X_0X_0'Y_0$respectively.

This accounts for how the first PLS factor is extracted. The second factor is extracted in the same way by replacing X_0 and Y_0 with the *X*- and *Y*-residuals from the first factor

$$X_1 = X_0 - \hat{X}_0$$

$$Y_1 = Y_0 - \hat{Y}_0$$

These residuals are also called the *deflated X* and *Y* blocks. The process of extracting a score vector and deflating the data matrices is repeated for as many extracted factors as are desired.

Centering and Scaling

By default, the predictors and the responses are centered and scaled to have mean 0 and standard deviation 1. Centering the predictors and the responses ensures that the criterion for choosing successive factors is based on how much variation they explain, in either the predictors or the responses or both. Without centering, both the mean variable value and the variation around that mean are involved in selecting factors. Scaling serves to place all predictors and responses on an equal footing relative to their variation in the data. For example, if Time and Temp are two of the predictors, then scaling says that a change of std(Time) in Time is roughly equivalent to a change of std(Temp) in Temp.

Missing Values

Observations with any missing independent variables are excluded from the analysis, and no predictions are computed for such observations. Observations with no missing independent variables but any missing dependent variables are also excluded from the analysis, but predictions are computed.

Chapter 28

Neural Nets

The Neural Platform

The Neural platform implements a standard type of neural network. Neural nets are used to predict one or more response variables from a flexible network of functions of input variables. Neural networks can be very good predictors when it is not necessary to know the functional form of the response surface.

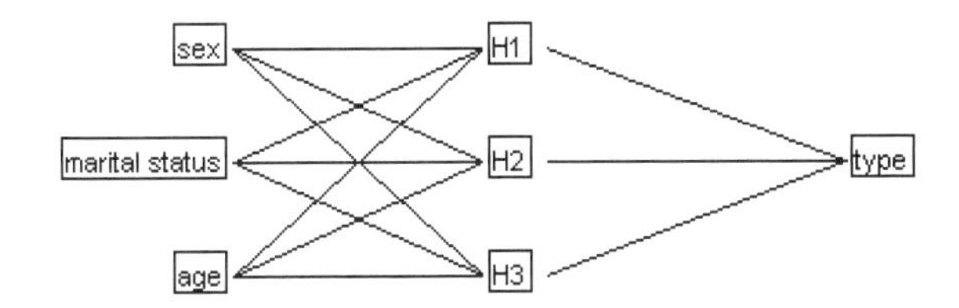

Contents

Introduction

There is a wide variety of literature on Neural Nets. The early literature suggested that the S-shaped function was an activation function that worked like nerve cells in the brain, and that by strengthening and weakening the coefficients in an iterative way, one could simulate a learning process. We prefer to think of the model as a useful nonlinear regression model, and JMP fits it using standard nonlinear least-squares regression methods.

The advantage of a neural net is that it can efficiently and flexibly model different response surfaces. Any surface can be approximated to any accuracy given enough hidden nodes.

Neural nets have disadvantages.

- The results are not as interpretable, since there is an intermediate layer rather than a direct path from the X's to the Y's as in the case of regular regression.
- It is easy to overfit a set of data so that it no longer predicts future data well.
- The fit is not parametrically stable, so that in many fits the objective function converges long before the parameters settle down.
- There are local optima, so that you can get a different answer each time you run it, and that a number of the converged estimates will not be optimal.

In consequence of the local optima, the platform will estimate the model many different times from different random starting estimates. Each starting estimate and consequent iteration is called a *tour*, and the default number of tours is 20. This should assure a good fit in simple models.

In consequence of the overfitting problem, neural nets are almost always fit with a *holdback sample* to crossvalidate the estimates on data that is not used in the estimation process. The fitting data is usually called the *training set*, and the holdback is called the *validation set*. To form a holdback set, select the rows to hold back and set the Exclude row state using (**Rows > Exclude/Unexclude**). There are features in the Rows menu to select this set randomly (**Rows > Row Selection > Select Randomly**).

A device to avoid overfitting and improve the numerical convergence is the *overfit penalty* (often called *weight decay*), which puts a penalty on the size of the parameter estimates.

Background

A neural network is just a set of nonlinear equations that predict output variables (Y's) from input variables (X's) in a flexible way using layers of linear regressions and S-shaped functions. JMP's implementation is for one hidden layer, and is specified as follows.

An S-shaped activation function used to scale continuous arguments is the familiar logistic function

$$S(x) = \frac{1}{1+e^{-x}}$$

which scales values to have mean 0 and standard deviation 1. (Nominal values are coded as in linear models).

Note: This function is implemented in JMP scripting language as `Squish()`.

Each H_j, the unobserved hidden nodes, are defined as

$$H_j = S_H\left(c_j + \sum_{i=1}^{N_X} (a_{ij}X_i)\right)$$

where

N_X is the number of X variables

$S_H(x)$ is the logistic function

and

X_i are the inputs, usually scaled to be in the interval [0, 1].

The outputs $\hat{Y}_k$ are then calculated as

$$\hat{Y}_k = S_Y\left(d_k + \sum_{j=1}^{N_H} (b_{jk}H_j)\right)$$

where

N_H is the number of hidden nodes.

$S_Y(x)$ is the identity function.

The $\hat{Y}_k$ values are also scaled to be in the interval [0, 1].

The *X* (input) and *Y* (output) values are JMP data columns. The hidden nodes are neither stored in the data table nor specified.

The coefficients *a*, *b*, *c*, and *d* are to be estimated.

Neural Net Example

In this example, start with Tiretread.jmp, a multiple-Y response surface experiment described in Derringer and Suich (1980). There are four Y output response variables, and three X input factors. When this data is fit using regression, a quadratic response surface model is used, with many additional terms for squares and crossproducts. When it is fit with neural nets, the nonlinear behavior is handled within the model, so only the three input variables are specified.

To begin, select **Analyze > Modeling > Neural Nets**. Specify variable roles as shown in Figure 28.1.

Figure 28.1 Launch the Platform

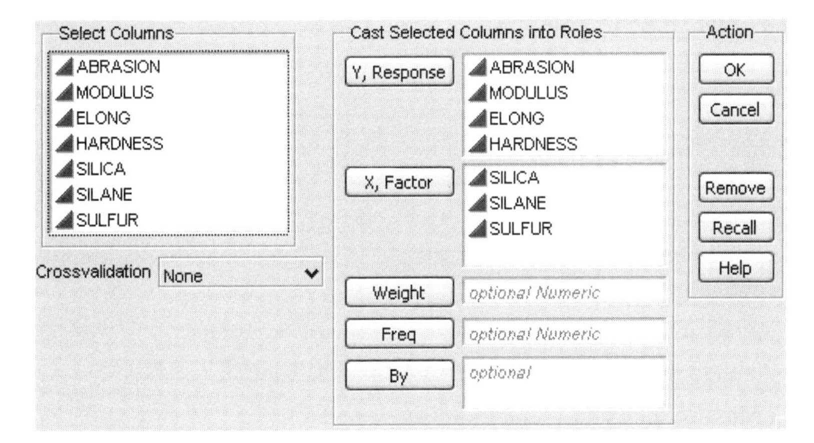

Click **OK** to see a Control Panel that allows you to change various settings before fitting the model.

Figure 28.2 Neural Net Control Panel

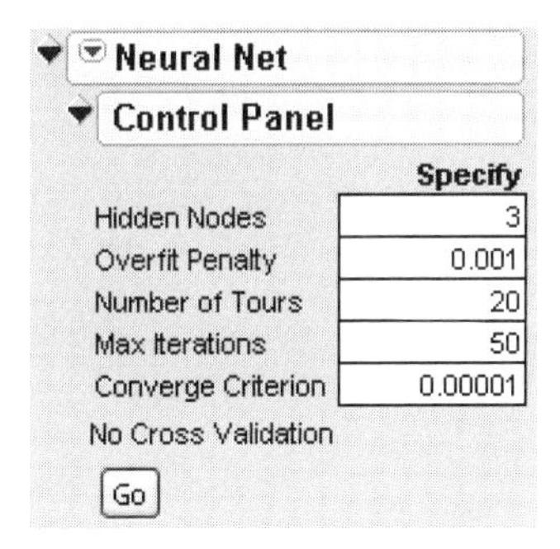

Control Panel Options

Hidden Nodes is the most important number to specify. Set this value too low and it will underfit; set it too high and it will overfit.

Overfit Penalty helps prevent the model from overfitting. The optimization criterion is formed by the sum of squared residuals plus the overfitting penalty times the sum of squares of the parameter estimates. When a neural net is overfit, the parameters are too big, and this criterion helps keep the parameters (weights) small. The penalty coefficient is often called *lambda* or *weight decay.* If you set it to zero, then not only will you tend to overfit, but the numerical iterations will tend to have a lot of trouble converging. If you have a lot of data, then the overfit penalty will have less of an effect. If you have very little data, then the overfit penalty will damp down the estimates a lot. It is recommended that you try several values between 0.0001 and 0.01.

Number of Tours sets the number of tours. Neural nets tend to have a lot of local minima, so that in order to more likely find global minima, the model is fit many times (tours) at different random starting values. Twenty tours is recommended. If this takes up too much time, then specify less. If you don't trust that you have a global optimum, then specify more.

Max Iterations is the number of iterations it will take on each tour before reporting nonconvergence.

Converge Criterion is the relative change in the objective function that an iteration must meet to decide it has converged. The value tested is

$$|\text{change}| / \min(10^{-8}, \max(10^{8}, \text{obj}))$$

where change is the change in the value of the objective function (obj) for a (non-step-shortened) iteration.

At this point, you may want to create a record of the tours, iterations, or estimates during the estimation process. The **Set Iteration Log Options** menu, available from the platform drop-down, has the following options.

Log the tours shows the best objective function value at each tour.

Log the iteration shows the objective function for each iteration.

Log the estimates shows the estimates at each iteration.

Figure 28.3 Iteration Log Options

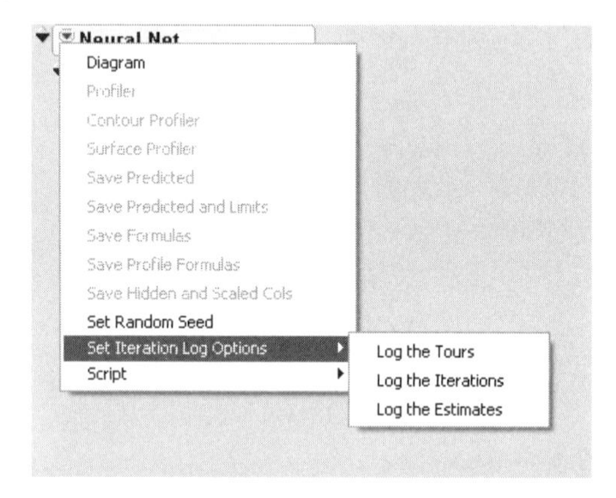

The Tiretread example uses three hidden nodes and the default value for the overfit penalty. This yields a model with 28 parameters—far fewer than the 40 parameters of a complete quadratic response surface. With only 20 rows, it isn't reasonable to have a holdback sample.

Reports and Graphs in Neural Nets

The results show the best fit found. There is no guarantee that this is actually the best fit globally, though with 20 tours, it is probably near optimal. Of the 20 tours, note that two converged to the objective value. Eleven other tours converged to a worse value. Seven tours gave up after reaching the maximum number of iterations.

Figure 28.4 Initial Results

Current Fit Results

	Objective		
SSE	7.305038041	2	Converged At Best
Penalty	0.690903788	15	Converged Worse Than Best
Total	7.995941829	0	Stuck on Flat
N	20	0	Failed to Improve
Nparm	28	3	Reached Max Iter

Y	SSE	RMSE	SSE Scaled	RSME Scaled	RSquare
ABRASION	1766.9746756	10.5088495	2.9805677335	0.43160802	0.8431
MODULUS	461100.18104	169.760895	2.0907430775	0.36148505	0.8900
ELONG	13707.372475	29.2696221	1.1414049612	0.26709139	0.9399
HARDNESS	22.032571342	1.17347165	1.0923222688	0.26128556	0.9425

The parameter estimates are often not very interesting in themselves. JMP doesn't even calculate standard errors for them.

Figure 28.5 Parameter Estimates Table

Parameter Estimates

Parameter	Estimate
H1:Intercept	2.060528299
H2:Intercept	0.2401238033
H3:Intercept	-1.018455532
H1:SILICA	1.075430135
H1:SILANE	-0.374990429
H1:SULFUR	0.4099397546
H2:SILICA	0.2451288425
H2:SILANE	0.304649963
H2:SULFUR	0.2852337204
H3:SILICA	-4.719331234
H3:SILANE	-0.181576735
H3:SULFUR	-4.284836694
ABRASION:Intercept	-9.739108521
MODULUS:Intercept	-19.50169102
ELONG:Intercept	5.5566263963
HARDNESS:Intercept	3.4166062206
ABRASION:H1	3.7005885859
ABRASION:H2	10.554509418
ABRASION:H3	2.0967755979
MODULUS:H1	10.507175045
MODULUS:H2	14.967453748
MODULUS:H3	6.5064548003
ELONG:H1	-2.883701397
ELONG:H2	-5.720434067
ELONG:H3	0.1268115398
HARDNESS:H1	-7.569585465
HARDNESS:H2	6.0834357947
HARDNESS:H3	-1.260515245

Random Seed
1969868217

Note the value of the Random Seed at the bottom of the report. This is the value seeded to the random number generator that gave the results in the current report. If you want to duplicate results for a single tour exactly, specify this seed as the starting value for iterations using the `Random Reset (seed)` JSL command and specify the number of tours as 1.

There are plots that show the actual variable vs. the predicted value (Figure 28.6), and plots to show residuals by predicted values (Figure 28.7).

Figure 28.6 Actual by Predicted Plot

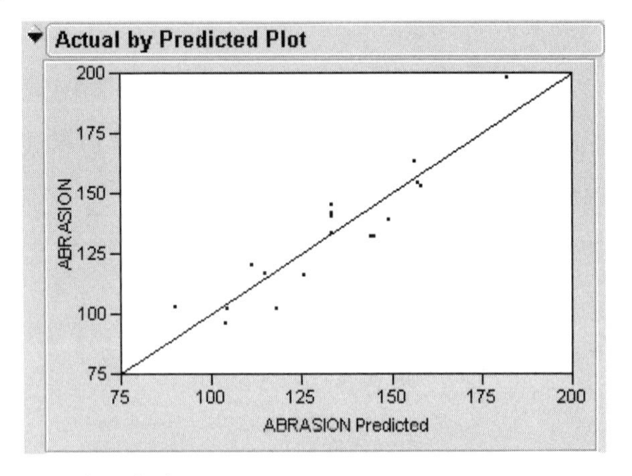

Figure 28.7 Residual By Predicted Plot

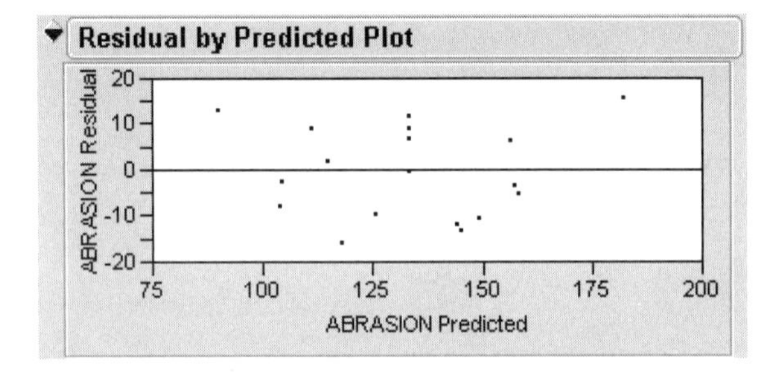

The shape of the slices of the response surface are interesting. The **Profiler** option shows these in the same interactive form used in other platforms. (See the "The Profiler," p. 277 in the "Standard Least Squares: Exploring the Prediction Equation" chapter for complete documentation of the Profiler).

Figure 28.8 shows the Prediction Profiler for two runs of the Neural Net platform, using three different numbers of hidden nodes.

- The first fit shows two hidden nodes with default values for other settings.
- The second shows six hidden nodes—too many for this situation, resulting in overfitting. The curves are obviously too flexible.
- The third shows six hidden nodes, but with an increased overfit penalty (0.1 instead of the default 0.001). By increasing the overfit penalty, the curves are stiffer.

Figure 28.8 Prediction Profiler

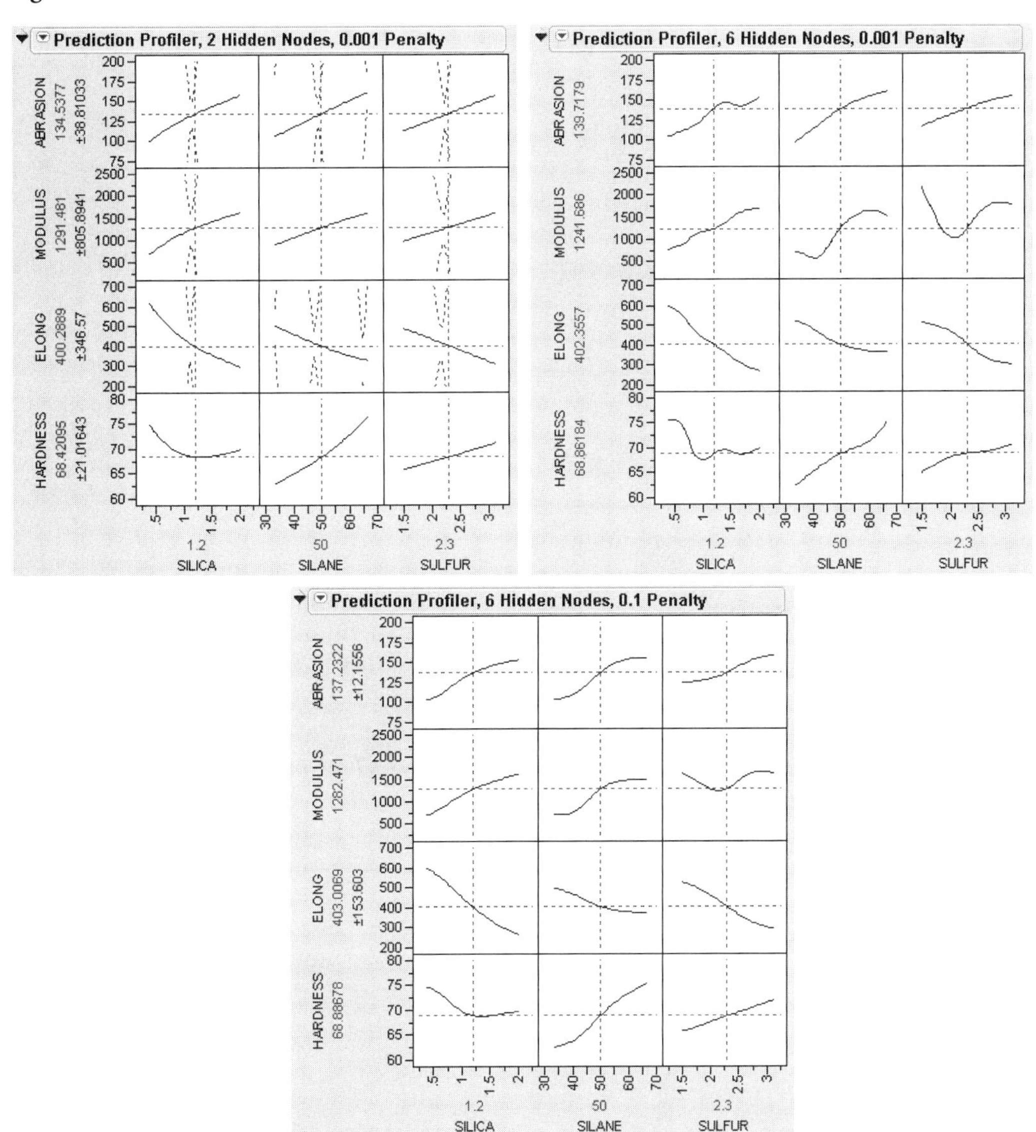

The Fit History control is a compilation of all the neural nets fit during a session. Radio buttons on the control allow you to switch among previously-fit models, changing the graphs and reports to fit the selected model. For example, Figure 28.9 shows the Fit History for the three models shown in Figure 28.8.

Figure 28.9 Fit History Control

Fit History

Nodes	Penalty	RSquare	
2	0.001	0.80305	○
6	0.001	0.98163	○
6	0.1	0.93389	⊙

Categorical Graphs and Options

When analyzing a categorical response, additional graphs and options appear.

The **Categorical Profiler** shows the response probabilities in one row of graphs. The distance between the separating curves are the probabilities.

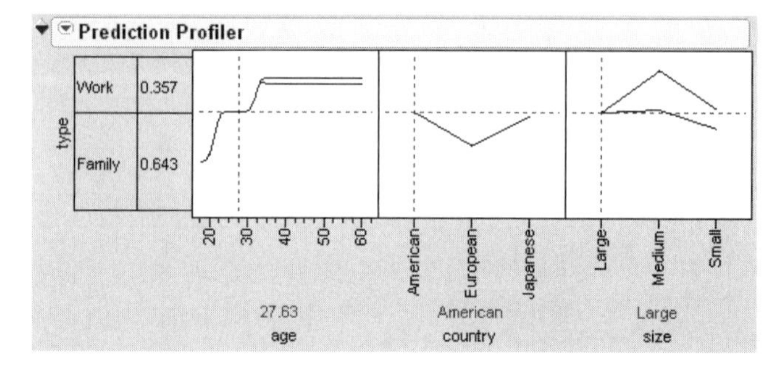

Predicted Value Histograms show the probability that each response value is predicted by the model.

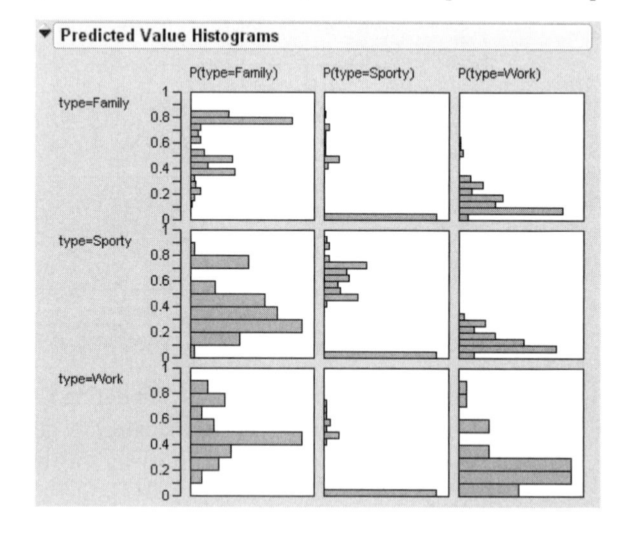

Using Cross-validation

Generally speaking, neural network models are highly overparameterized, so that models that seem to offer the best fit of the (training) data are overfit for other data.

Models chosen by goodness-of-fit type methods alone often appear to be accurate predictors on the training data, but are in fact very poor predictors on a new data set. In other words, the model does not generalize well to new data.

To find models which do generalize well to new data, use cross-validation. This reserves a portion of the existing data set that is then used to verify the model generated by the non-held-back, training data. JMP has three options for cross-validation, all available on the launch dialog: **None**, **K Fold**, or **Random Holdback**.

Figure 28.10 Cross-Validation Options

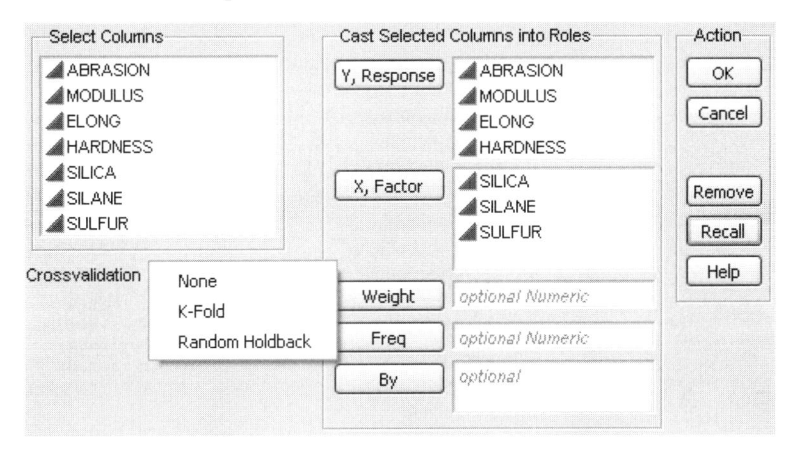

Note: These methods generate random samples, so your numbers may not agree with those shown here. The **Set Random Seed** command (Figure 28.3) can set the random seed to provide consistent results for a single tour if desired.

Random Holdback

Random Holdback is the simpler of the two methods. Before any models are fit to the data, the data are partitioned into two groups: a training group used for fitting the models, and a testing (holdback) group for comparing models that have been fit to the training data.

Holdback-type cross-validation is best suited for large data sets where there is enough data to reasonably fit complex models using 75% or less of the data (In fact, 66.67% is JMP's default holdback percentage).

The models are fit using only the training data. Once the models are fit, JMP computes R^2s for the training data (denoted RSquare in JMP's report) and the testing data (denoted CV RSquare in the report). In practice, the R^2 values for the training data increase monotonically with the number of nodes in the network, and as the penalty parameter decreases. However, the testing R^2s (CV RSquare) initially increase with the number of nodes, but then begin to decrease. Once a certain number of nodes are reached, the model begins to overfit the data.

Example

Figure 28.11 shows the Tire tread example from Figure 28.1 with the default random holdback of 66.667% (and all other default options). This data set is actually too small for a holdback, but is used here for consistency in examples.

Figure 28.11 Neural Net Example with Random Holdback

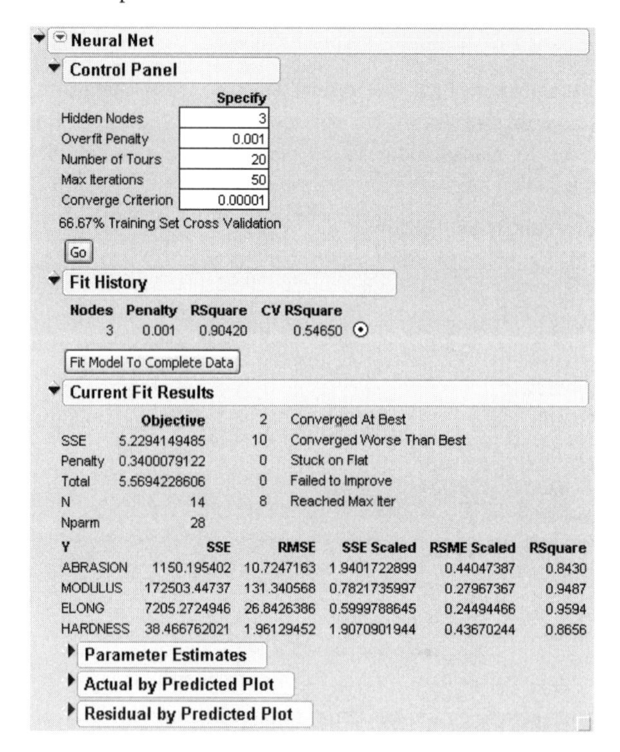

Note the new Fit History outline node that records the results of all the models that have been fit during the current session. It records the number of nodes used in each model, the penalty parameters, fit R^2s for the training data, and the cross-validation R^2s for the testing (holdback) data. There is also a radio button on the right hand side that is used for selecting models. Selecting a model using this radio button allows the user to make predictions using the selected model, recalculates all the statistics for the model that were originally in the "Results" menu, and redraws all the profilers using this model, if the profilers are enabled.

To see this in action, change the number of hidden nodes from three to 5 in the Neural Net Control Panel and hit **Go**. The model is recalculated, and a second line is added to the Fit History table.

Figure 28.12 Second Neural Net Model

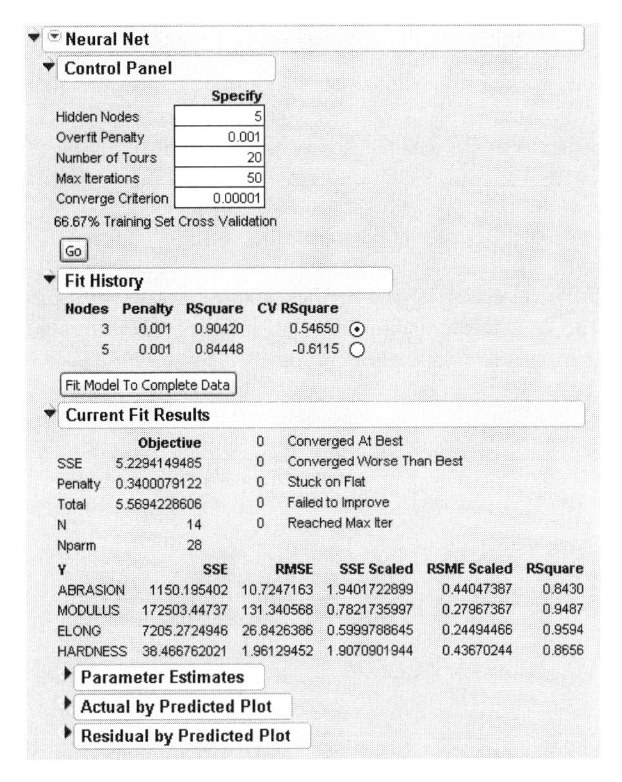

This second model is a worse fit to the data than the first model. If desired, more models can be run with different settings for other items in the Control Panel.

Of these two models, the first is the better. We would therefore like to fit this model with all the data. To do so, press the **Fit Model To Complete Data** button. Pressing this button causes JMP to fit the currently selected model to the complete data set (training+testing data), and overwrites the training parameter values with the new ones.

K-Fold Cross-validation

K-Fold cross-validation is a technique that is suited for smaller data sets. In this type of cross-validation, the data are partitioned randomly into a number of groups that you specify (*K*) and the following procedure is followed:

1 The model is fit to the complete data set.

2 Then, using the overall fit for starting values, for $i = 1, 2, ..., K$, the ith group is treated as a holdback set the model is fit to the remaining $N-N/K$ observations and an R^2 is calculated using the ith group as a holdback

3 After iterating through the K starting values, the CVRsquare is set to be the average of the R^2 values for the K iterations.

K-Fold cross-validation is about *K* times as computationally intensive as Holdback cross-validation, because of the number of times the model is refit. Thus, if the model is complicated (large number of nodes) and the data set is large (more than 1000 observations, say) this method should not be used. However, for smaller data sets the CVRsquares should be less variable and much more reliable than they would be for the corresponding (1-1/*K*)*100% Holdback cross-validation fit statistics. *K*-Fold cross-validation generally works well for K between 5 and 20.

Example

Again using the tire tread example from Figure 28.1, we pick K-Fold cross-validation from the launch dialog. This presents us with the preliminary launch dialog shown in Figure 28.13. This is where you specify *K*, the number of cross-validation groups.

Figure 28.13 K-Fold Launch Dialog

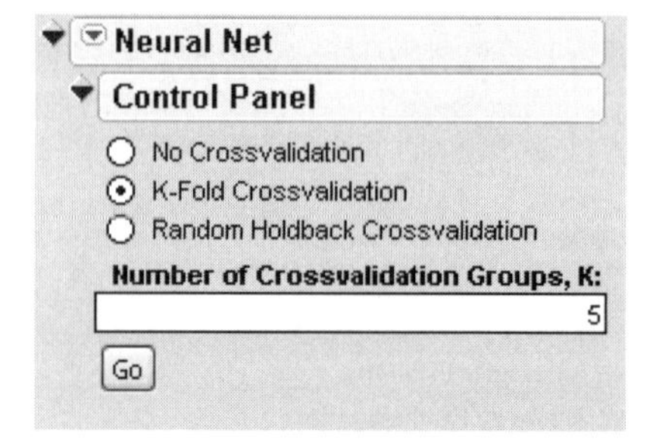

After clicking **Go**, you are presented with the standard Neural Net control panel (Figure 28.2) and analysis proceeds as in previous examples. The Fit History outline node works as described in the Random Holdback section, except that the **Fit Model To Complete Data** button is not available.

Platform Options

The following options are found in the platform popup menu from a Neural Net report.

Diagram is an option to produce a picture of the neural network, showing how each layer is connected to the layer before it. Select **Diagram** from the Control Panel popup menu.

Figure 28.14 Neural Net Diagram

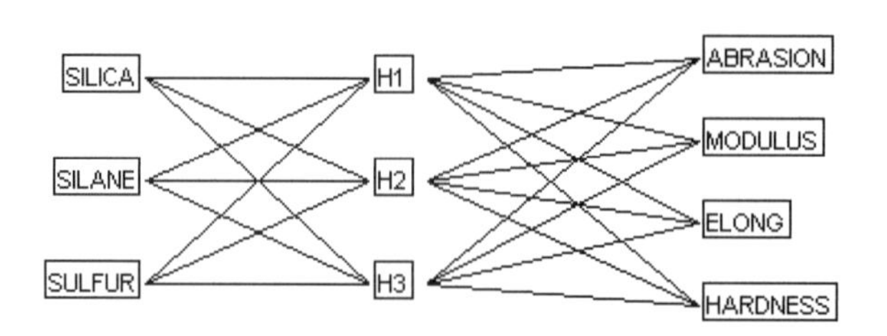

Sequence of Fits brings up a dialog that lets you fit several models sequentially.

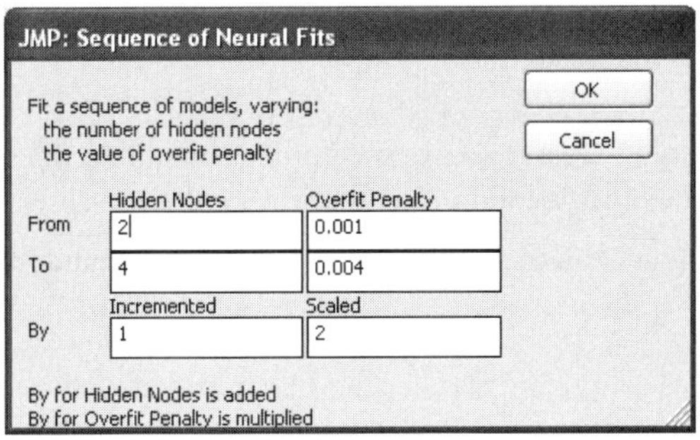

JMP: Sequence of Neural Fits

Fit a sequence of models, varying:
the number of hidden nodes
the value of overfit penalty

OK
Cancel

	Hidden Nodes	Overfit Penalty
From	2	0.001
To	4	0.004
	Incremented	Scaled
By	1	2

By for Hidden Nodes is added
By for Overfit Penalty is multiplied

Enter lower and upper bounds for both the number of hidden nodes and the overfit penalty in the corresponding From: and To: field.

Note: When cycling through the sequence of models, the number of hidden nodes is incremented (additively) by the amount in the **By** field, while the overfit penalty is incremented by multiplying by the scale found in the **By** field. For example, the following sequence of fits is the result of the settings in the dialog box above (2 to 4 hidden nodes, incremented by 1, with overfit penalty 0.001 to 0.004, scaled by 2):

Fit History

Nodes	Penalty	RSquare	
2	0.001	0.80301	○
2	0.002	0.80142	○
2	0.004	0.79919	○
3	0.001	0.90357	○
3	0.002	0.89991	○
3	0.004	0.89473	○
4	0.001	0.93573	○
4	0.002	0.93295	○
4	0.004	0.93067	⦿

Profiler shows the Prediction profiler (Figure 28.8). See the section "The Profiler," p. 291 IN THE "PROFILING" CHAPTER for details on the Profiler.

Contour Profiler activates the Contour Profiler. See the section "Contour Profiler," p. 306 IN THE "PROFILING" CHAPTER for details on the Contour Profiler.

Surface Profiler activates the Surface Profiler. See "Plotting Surfaces," p. 763 for details on the Surface Profiler.

Save Predicted creates a new column in the data table for each response variable, holding the predicted values.

Save Predicted and Limits creates new columns in the data table for upper and lower 95% confidence intervals and a column holding the predicted response for each row.

Save Formulas creates a new column in the data table for each response variable. This column holds the prediction formula for each response, so predicted values are calculated for each row. This option is useful if rows are added to the data table, since predicted values are automatically calculated. Use **Save Predicted** if formulas are not desired.

Save Profile Formulas saves formulas almost equivalent to the formulas made with the **Save Formulas** command, but the hidden layer calculations are embedded into the final predictor formulas rather than made through an intermediate hidden layer. This allows the use of the Profiler and Contour Profiler.

Save Hidden and Scaled Cols makes new columns containing the scaled data used to make the estimates, the values of the hidden columns, and of the predicted values. No formulas are saved.

Set Random Seed allows you to specify the seed used by the random number generator.

Chapter 29

Categorical Response Analysis

The Categorical Platform

The Categorical platform does tabulation and summarization of categorical response data, including multiple response data, and calculates test statistics. It is designed to handle survey and other categorical response data, including multiple response data like defect records, side effects, and so on.

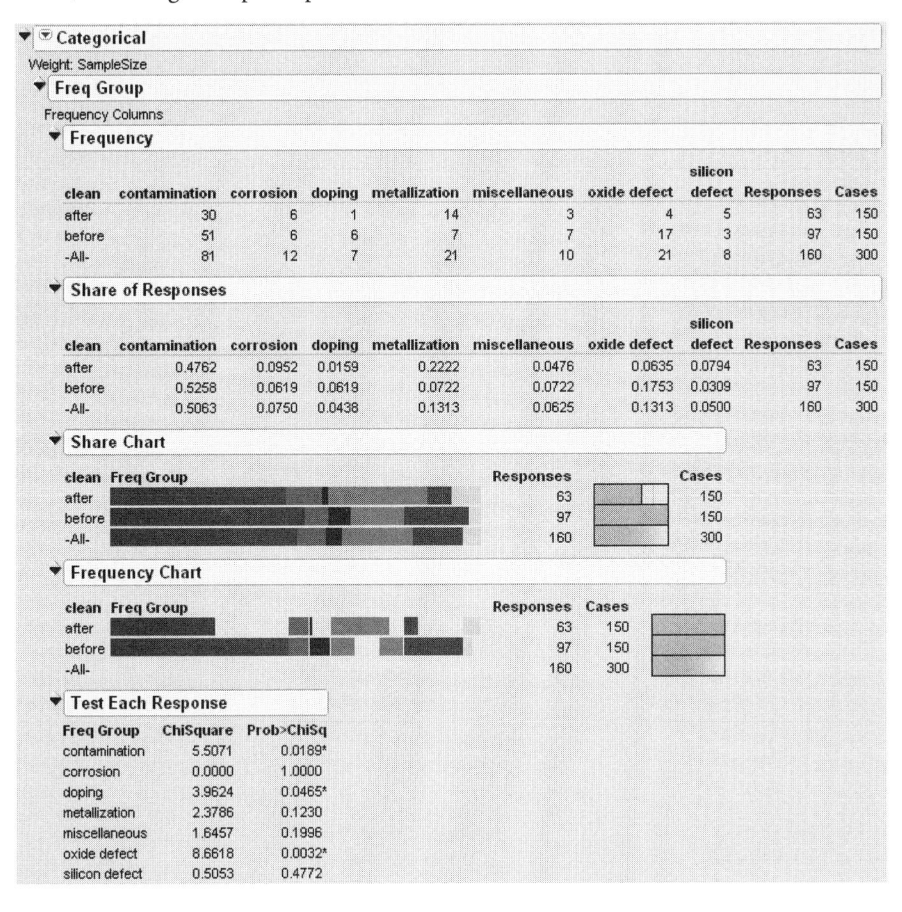

Categorical

Weight: SampleSize

Freq Group

Frequency Columns

Frequency

clean	contamination	corrosion	doping	metallization	miscellaneous	oxide defect	silicon defect	Responses	Cases
after	30	6	1	14	3	4	5	63	150
before	51	6	6	7	7	17	3	97	150
-All-	81	12	7	21	10	21	8	160	300

Share of Responses

clean	contamination	corrosion	doping	metallization	miscellaneous	oxide defect	silicon defect	Responses	Cases
after	0.4762	0.0952	0.0159	0.2222	0.0476	0.0635	0.0794	63	150
before	0.5258	0.0619	0.0619	0.0722	0.0722	0.1753	0.0309	97	150
-All-	0.5063	0.0750	0.0438	0.1313	0.0625	0.1313	0.0500	160	300

Share Chart

clean	Freq Group	Responses	Cases
after		63	150
before		97	150
-All-		160	300

Frequency Chart

clean	Freq Group	Responses	Cases
after		63	150
before		97	150
-All-		160	300

Test Each Response

Freq Group	ChiSquare	Prob>ChiSq
contamination	5.5071	0.0189*
corrosion	0.0000	1.0000
doping	3.9624	0.0465*
metallization	2.3786	0.1230
miscellaneous	1.6457	0.1996
oxide defect	8.6618	0.0032*
silicon defect	0.5053	0.4772

Contents

The Categorical Platform

The Categorical platform has capabilities similar to other platforms. The choice of platforms depends on your focus, the shape of your data, and the desired level of detail. Table 29.1 shows several of JMP's analysis platforms, and their strengths.

Table 29.1 Comparing JMP's Categorical Analyses

Platform	Specialty
Distribution	Separate, ungrouped categorical responses
Fit Y By X: Contingency	Two-way situations, including chi-square tests, correspondence analysis, agreement.
Pareto Plot	Graphical analysis of multiple-response data, especially multiple-response defect data, with more rate tests than Fit Y By X.
Variability Chart: Attribute	Attribute gauge studies, with more detailed on rater agreement.
Fit Model	Logistic categorical responses and generalized linear models.
Partition, Neural Net	Specific categorical response models

The strength of the categorical platform is that it can handle responses in a wide variety of formats without needing to reshape the data.

Launching the Platform

The platform has a unique launch dialog. The buttons select groups of columns (as opposed to single columns like other JMP dialogs) for each of the situations.

Figure 29.1 Categorical Platform Launch Dialog

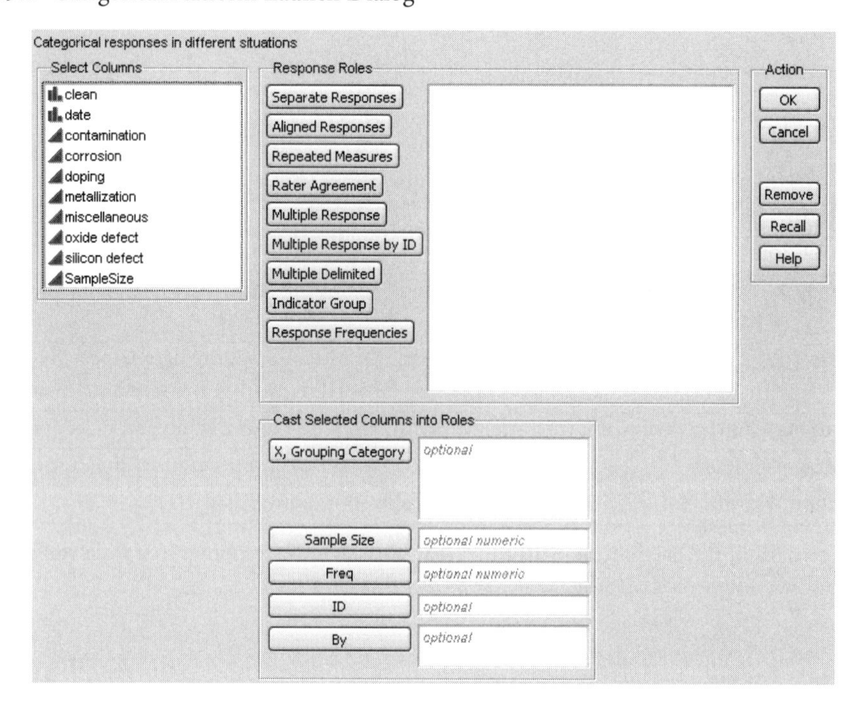

Response Roles

Use the response roles buttons to choose selected columns as responses with specified roles. The response roles are summarized in Table 29.2.

Table 29.2 Response Roles

Response Role	Description	Example Data
Separate Responses	Separate responses are in each column, resulting in a separate analysis for each response	**ID Drink Entre** John Coffee Chicken Jane Tea Veggie
Aligned Responses	Responses share common categories, resulting in better-organized reports	**ID Coffee Tea** John Like Dislike Jane Dislike Like
Repeated Measures	Aligned responses from an individual across different times or situations	**ID Morning Noon Night** John Coffee Coffee Water Jane Tea Water Tea
Rater Agreement	Aligned responses from different raters evaluating the same unit, to study agreement across raters	**Drink John Jane** Coffee Like Dislike Tea Dislike Like Water Like Like

Table 29.2 Response Roles (continued)

Response Role	Description	Example Data
Multiple Response	Aligned responses, where multiple responses are entered across several columns, but treated as one grouped response	ID Drink1 Drink2 Drink3 John Coffee Milk Water Jane Tea Water
Multiple Response by ID	Multiple responses across rows that have the same ID values	ID Drinks John Coffee John Milk John Water Jane Tea Jane Water
Multiple Delimited	Several responses in a single cell, separated by commas	ID Drinks John Coffee,Milk,Water Jane Tea,Water
Indicator Group	Binary responses across columns, like checked or not, yes or no, but all in a related group	ID Coffee Milk Tea Water John Y Y N Y Jane N N Y Y
Response Frequencies	Columns containing frequency counts for each response level, all in a related group	Group Coffee Milk Tea Water A 12 15 8 19 B 9 20 6 22

Other Launch Dialog Options

The Launch dialog has standard JMP options (like **By** variables) and additionally:

X, Grouping Categories defines sample groups to break the counts into.

Sample Size For multiple response roles with summarized data, defines the number of individual units in the group for which that frequency is applicable to. For example, a Freq column might say there were 50 defects, where the sample size variable would say that they reflect the defects for a batch of 100 units.

ID is only required and used when Multiple Response by ID is used.

Freq is for presummarized data, specifying the column containing frequency counts for each row.

Failure Rate Examples

The following examples come from testing a fabrication line on three different occasions under two different conditions. Each set of operating conditions yielded 50 data points. Inspectors recorded the following kinds of defects:

1 contamination

2 corrosion

3 doping

4 metallization
5 miscellaneous
6 oxide defect
7 silicon defect

Each unit could have several defects or even several defects of the same kind. We illustrate the data in a variety of different shapes all supported directly by the Categorical platform.

Response Frequencies

Suppose the data has columns containing frequency counts for each batch, and a column showing the total number of units of the batch (as in Failure3Freq.jmp).

	clean	date	contamination	corrosion	doping	metallization	miscellaneous	oxide defect	silicon defect	SampleSize
1	after	OCT 1	12	2	0	4	2	1	2	50
2	after	OCT 2	10	1	1	5	1	2	3	50
3	after	OCT 3	8	3	0	5	0	1	0	50
4	before	OCT 1	14	2	1	2	3	8	1	50
5	before	OCT 2	15	2	2	1	4	6	0	50
6	before	OCT 3	22	2	3	4	0	3	2	50

To generate the appropriate analysis, highlight the frequency variables and click **Response Frequencies**. The **X, Grouping Category** columns are clean and date, and SampleSize is **Sample Size**.

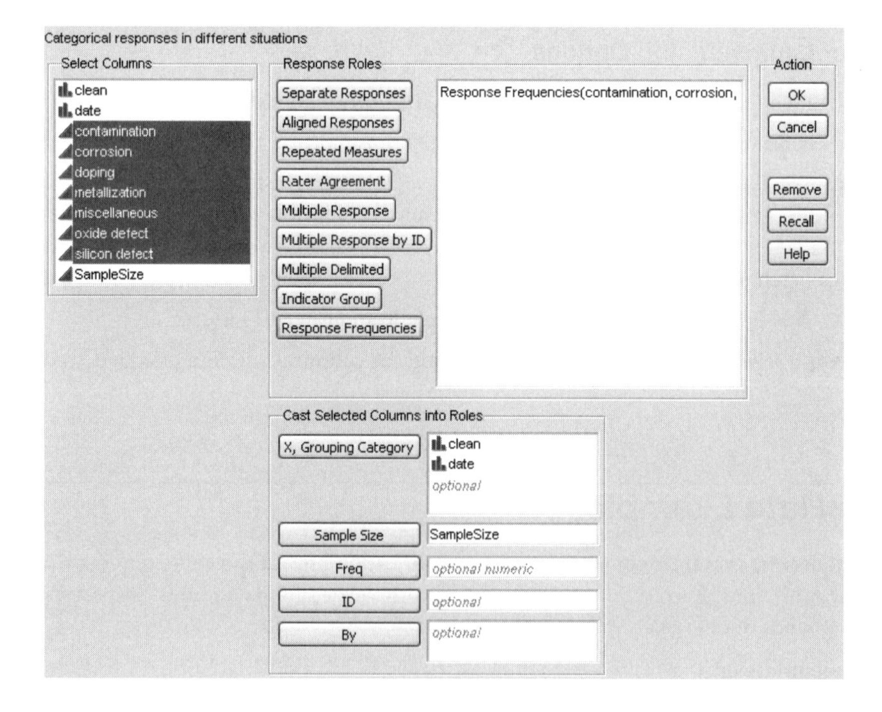

In the resulting output, a frequency count table shows the total number of defects for each defect type. There is a separate row for each of the six batches. The last two columns show the total number of defects (**Responses**) and the total number of units (**Cases**).

Figure 29.2 Defect Rate Output

Frequency

clean	date	contamination	corrosion	doping	metallization	miscellaneous	oxide defect	silicon defect	Responses	Cases
after	OCT 1	12	2	0	4	2	1	2	23	50
	OCT 2	10	1	1	5	1	2	3	23	50
	OCT 3	8	3	0	5	0	1	0	17	50
before	OCT 1	14	2	1	2	3	8	1	31	50
	OCT 2	15	2	2	1	4	6	0	30	50
	OCT 3	22	2	3	4	0	3	2	36	50
-All-	-All-	81	12	7	21	10	21	8	160	300

Indicator Group

In some cases, the data is not yet summarized, so there are individual records for each unit. We illustrate this situation in failure3Indicators.jmp.

	clean	date	ID	ID Label	contamination	corrosion	doping	metallization	miscellaneous	oxide defect	silicon defect
1	before	OCT 1	1	OCT 1 before	0	0	0	0	0	0	0
2	before	OCT 1	1	OCT 1 before	0	0	0	0	0	1	0
3	before	OCT 1	1	OCT 1 before	1	0	0	0	0	1	0
4	before	OCT 1	1	OCT 1 before	0	0	0	0	0	0	0
5	before	OCT 1	1	OCT 1 before	1	0	0	0	0	0	0
6	before	OCT 1	1	OCT 1 before	0	0	0	0	0	1	0
7	before	OCT 1	1	OCT 1 before	1	0	0	0	0	0	0
8	before	OCT 1	1	OCT 1 before	0	0	0	0	0	0	0
9	before	OCT 1	1	OCT 1 before	0	0	0	0	0	0	0
10	before	OCT 1	1	OCT 1 before	1	0	0	1	0	0	0
11	before	OCT 1	1	OCT 1 before	0	0	0	0	0	0	0
12	before	OCT 1	1	OCT 1 before	0	0	0	0	0	0	0
13	before	OCT 1	1	OCT 1 before	0	0	0	0	0	0	0
14	before	OCT 1	1	OCT 1 before	1	0	0	0	0	0	0
15	before	OCT 1	1	OCT 1 before	0	0	0	0	0	0	0
16	before	OCT 1	1	OCT 1 before	0	0	0	0	1	0	0

With data like this, specify all the defect columns with the **Indicator Group** button. clean and date are in the **X, Grouping Category** role. When you click **OK**, you get the same output as in the Response Group example (Figure 29.2).

Multiple Delimited

Suppose that for each unit, the inspector entered the observed defects for each unit. The defects are listed in a single column, delimited by a comma. (failure3Delimited.jmp).

	failures	clean	date	ID	ID Label
1		before	OCT 1	1	OCT 1 before
2	oxide defect	before	OCT 1	1	OCT 1 before
3	contamination,oxide defect	before	OCT 1	1	OCT 1 before
4		before	OCT 1	1	OCT 1 before
5	contamination	before	OCT 1	1	OCT 1 before
6	oxide defect	before	OCT 1	1	OCT 1 before
7	contamination	before	OCT 1	1	OCT 1 before
8		before	OCT 1	1	OCT 1 before
9		before	OCT 1	1	OCT 1 before
10	metallization,contamination	before	OCT 1	1	OCT 1 before
11		before	OCT 1	1	OCT 1 before
12		before	OCT 1	1	OCT 1 before
13		before	OCT 1	1	OCT 1 before
14	contamination	before	OCT 1	1	OCT 1 before

To get the appropriate analysis, specify the delimited defect columns with the **Multiple Delimited** button. The results are identical to those in Figure 29.2.

Note: If more than one delimited column is specified, separate analyses are produced for each column.

Multiple Response By ID

Suppose each failure type is a separate record, with an ID column that can be used to link together different defect types for each unit, as in failure3ID.jmp.

	failure	N	clean	date	SampleSize	ID
1	contamination	14	before	OCT 1	50	OCT 1 before
2	corrosion	2	before	OCT 1	50	OCT 1 before
3	doping	1	before	OCT 1	50	OCT 1 before
4	metallization	2	before	OCT 1	50	OCT 1 before
5	miscellaneous	3	before	OCT 1	50	OCT 1 before
6	oxide defect	8	before	OCT 1	50	OCT 1 before
7	silicon defect	1	before	OCT 1	50	OCT 1 before
8	doping	0	after	OCT 1	50	OCT 1 after
9	corrosion	2	after	OCT 1	50	OCT 1 after
10	metallization	4	after	OCT 1	50	OCT 1 after
11	miscellaneous	2	after	OCT 1	50	OCT 1 after
12	oxide defect	1	after	OCT 1	50	OCT 1 after
13	contamination	12	after	OCT 1	50	OCT 1 after

Launch the Categorical platform with columns specified as in Figure 29.3. Column roles are similar to the other defect examples, but note that N is used as **Freq** and SampleSize is used as **Sample Size**.

Figure 29.3 **Multiple Response By ID** Launch Dialog

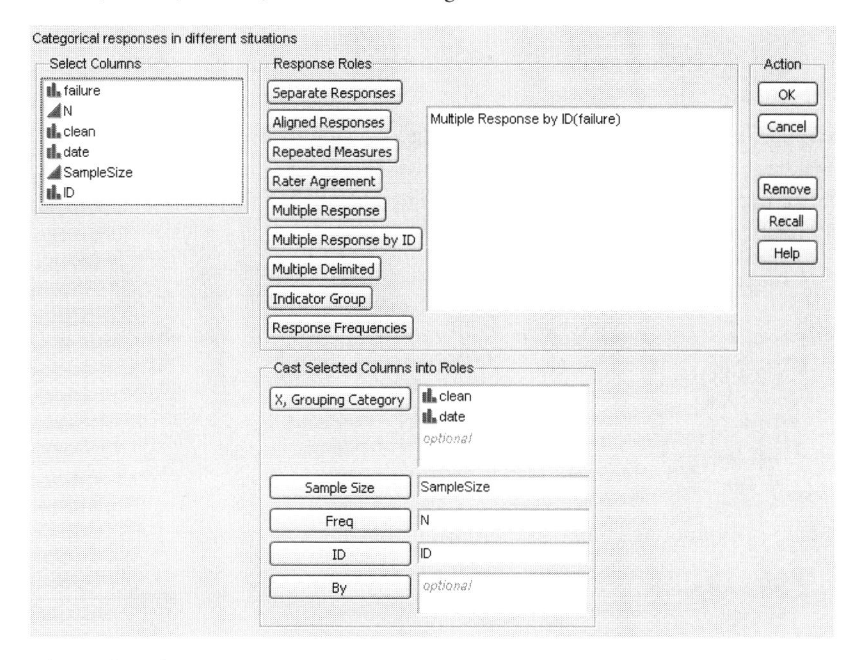

Results are identical to those in Figure 29.2.

Multiple Response

Suppose that the defects for each unit are entered via a web page, but since each unit rarely has more than three defect types, the form has three fields to enter any of the defect types for a unit, as in failure3MultipleField.jmp.

	clean	date	ID	ID Label	Failure1	Failure2	Failure3
1	before	OCT 1	1	OCT 1 before			
2	before	OCT 1	1	OCT 1 before	oxide defect		
3	before	OCT 1	1	OCT 1 before	contamination	oxide defect	
4	before	OCT 1	1	OCT 1 before			
5	before	OCT 1	1	OCT 1 before	contamination		
6	before	OCT 1	1	OCT 1 before	oxide defect		
7	before	OCT 1	1	OCT 1 before	contamination		
8	before	OCT 1	1	OCT 1 before			
9	before	OCT 1	1	OCT 1 before			
10	before	OCT 1	1	OCT 1 before	metallization	contamination	
11	before	OCT 1	1	OCT 1 before			
12	before	OCT 1	1	OCT 1 before			
13	before	OCT 1	1	OCT 1 before			
14	before	OCT 1	1	OCT 1 before	contamination		

Select the three columns containing defect types, and click the **Multiple Response** button. After specifying clean and date as **X, Grouping** variables, click **OK** to see the analysis results. As in the other examples, it is the same as in Figure 29.2.

Report Content

The Categorical platform produces analyses that are built from several reports. We illustrate them with the defect data.

The topmost report is a **Frequency** count table, showing the frequency counts for each category with the total frequency (**Responses**) and total units (**Cases**) on the right. In this example, we have total defects for each defect type by the 6 batches. The last two columns show the total number of defects (**Responses**) and the total number of units (**Cases**).

Frequency

clean	date	contamination	corrosion	doping	metallization	miscellaneous	oxide defect	silicon defect	Responses	Cases
after	OCT 1	12	2	0	4	2	1	2	23	50
	OCT 2	10	1	1	5	1	2	3	23	50
	OCT 3	8	3	0	5	0	1	0	17	50
before	OCT 1	14	2	1	2	3	8	1	31	50
	OCT 2	15	2	2	1	4	6	0	30	50
	OCT 3	22	2	3	4	0	3	2	36	50
-All-	-All-	81	12	7	21	10	21	8	160	300

The **Share of Responses** report is built by dividing each count by the total number of responses.

Share of Responses

clean	date	contamination	corrosion	doping	metallization	miscellaneous	oxide defect	silicon defect	Responses	Cases
after	OCT 1	0.5217	0.0870	0.0000	0.1739	0.0870	0.0435	0.0870	23	50
after	OCT 2	0.4348	0.0435	0.0435	0.2174	0.0435	0.0870	0.1304	23	50
after	OCT 3	0.4706	0.1765	0.0000	0.2941	0.0000	0.0588	0.0000	17	50
before	OCT 1	0.4516	0.0645	0.0323	0.0645	0.0968	0.2581	0.0323	31	50
	OCT 2	0.5000	0.0667	0.0667	0.0333	0.1333	0.2000	0.0000	30	50
	OCT 3	0.6111	0.0556	0.0833	0.1111	0.0000	0.0833	0.0556	36	50
-All-	-All-	0.5063	0.0750	0.0438	0.1313	0.0625	0.1313	0.0500	160	300

As an example, examine the second row of the table (corresponding to October 2, after cleaning). The 12 contamination defects were (12/23)=52.17% of all defects.

The **Rate Per Case** report divides each count in the frequency table by the total number of cases.

Rate Per Case

clean	date	contamination	corrosion	doping	metallization	miscellaneous	oxide defect	silicon defect	Responses	Cases
after	OCT 1	0.2400	0.0400	0.0000	0.0800	0.0400	0.0200	0.0400	23	50
	OCT 2	0.2000	0.0200	0.0200	0.1000	0.0200	0.0400	0.0600	23	50
	OCT 3	0.1600	0.0600	0.0000	0.1000	0.0000	0.0200	0.0000	17	50
before	OCT 1	0.2800	0.0400	0.0200	0.0400	0.0600	0.1600	0.0200	31	50
	OCT 2	0.3000	0.0400	0.0400	0.0200	0.0800	0.1200	0.0000	30	50
	OCT 3	0.4400	0.0400	0.0600	0.0800	0.0000	0.0600	0.0400	36	50
-All-	-All-	0.2700	0.0400	0.0233	0.0700	0.0333	0.0700	0.0267	160	300

For example, in the first row of the table (October 1, after cleaning), the 12 contamination defects are from 50 units, making the rate per unit 24%.

The **Share Chart** section presents two bar chart columns. The one on the left shows the **Share of Responses** as a divided bar chart. The bar length is proportional to the percentage of responses for each type. Since a bar in one sample might reflect a much higher frequency than a bar in another sample, the bar chart on the right shows the number of responses. The bar filled with a gradient means the the number of responses for that bar is more than the width of the bar chart allows.

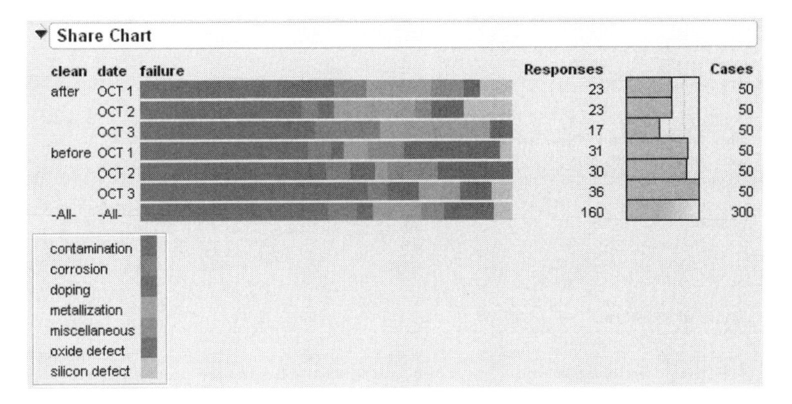

The **Frequency Chart** shows response frequencies. The bars in the chart on the left reflect the frequency count on the same scale. The bar chart on the right shows the number of units.

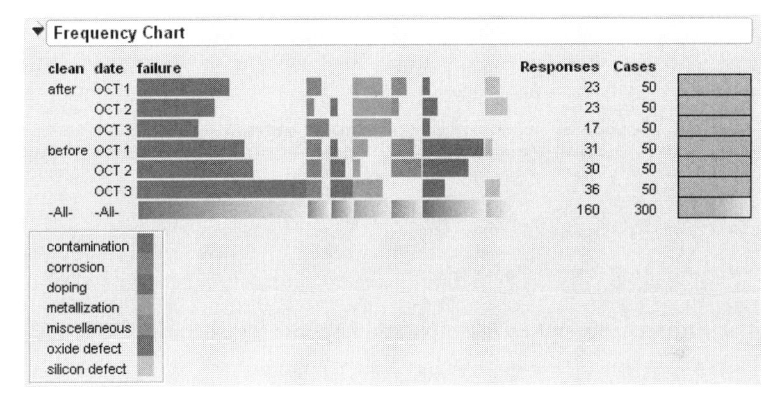

Note again that the gradient-filled bars are used when the number of responses or units is more than the width of the bar chart allows.

Report Format

The default report format is the **Table Format**.

Freq Group

Frequency Columns

Frequency

clean	date	contamination	corrosion	doping	metallization	miscellaneous	oxide defect	silicon defect	Responses	Cases
after	OCT 1	12	2	0	4	2	1	2	23	50
	OCT 2	10	1	1	5	1	2	3	23	50
	OCT 3	8	3	0	5	0	1	0	17	50
before	OCT 1	14	2	1	2	3	8	1	31	50
	OCT 2	15	2	2	1	4	6	0	30	50
	OCT 3	22	2	3	4	0	3	2	36	50
-All-	-All-	81	12	7	21	10	21	8	160	300

Share of Responses

clean	date	contamination	corrosion	doping	metallization	miscellaneous	oxide defect	silicon defect	Responses	Cases
after	OCT 1	0.5217	0.0870	0.0000	0.1739	0.0870	0.0435	0.0870	23	50
	OCT 2	0.4348	0.0435	0.0435	0.2174	0.0435	0.0870	0.1304	23	50
	OCT 3	0.4706	0.1765	0.0000	0.2941	0.0000	0.0588	0.0000	17	50
before	OCT 1	0.4516	0.0645	0.0323	0.0645	0.0968	0.2581	0.0323	31	50
	OCT 2	0.5000	0.0667	0.0667	0.0333	0.1333	0.2000	0.0000	30	50
	OCT 3	0.6111	0.0556	0.0833	0.1111	0.0000	0.0833	0.0556	36	50
-All-	-All-	0.5063	0.0750	0.0438	0.1313	0.0625	0.1313	0.0500	160	300

Rate Per Case

clean	date	contamination	corrosion	doping	metallization	miscellaneous	oxide defect	silicon defect	Responses	Cases
after	OCT 1	0.2400	0.0400	0.0000	0.0800	0.0400	0.0200	0.0400	23	50
	OCT 2	0.2000	0.0200	0.0200	0.1000	0.0200	0.0400	0.0600	23	50
	OCT 3	0.1600	0.0600	0.0000	0.1000	0.0000	0.0200	0.0000	17	50
before	OCT 1	0.2800	0.0400	0.0200	0.0400	0.0600	0.1600	0.0200	31	50
	OCT 2	0.3000	0.0400	0.0400	0.0200	0.0800	0.1200	0.0000	30	50
	OCT 3	0.4400	0.0400	0.0600	0.0800	0.0000	0.0600	0.0400	36	50
-All-	-All-	0.2700	0.0400	0.0233	0.0700	0.0333	0.0700	0.0267	160	300

To gather all three statistics for each sample and response together, use the **Crosstab** format.

Freq Group

clean,date: Freq Share Rate	contamination	corrosion	doping	metallization	miscellaneous	oxide defect	silicon defect	
after,OCT 1	12 0.522 0.240	2 0.087 0.040	0 0.000 0.000	4 0.174 0.080	2 0.087 0.040	1 0.043 0.020	2 0.087 0.040	23 50.000
after,OCT 2	10 0.435 0.200	1 0.043 0.020	1 0.043 0.020	5 0.217 0.100	1 0.043 0.020	2 0.087 0.040	3 0.130 0.060	23 50.000
after,OCT 3	8 0.471 0.160	3 0.176 0.060	0 0.000 0.000	5 0.294 0.100	0 0.000 0.000	1 0.059 0.020	0 0.000 0.000	17 50.000
before,OCT 1	14 0.452 0.280	2 0.065 0.040	1 0.032 0.020	2 0.065 0.040	3 0.097 0.060	8 0.258 0.160	1 0.032 0.020	31 50.000
before,OCT 2	15 0.500 0.300	2 0.067 0.040	2 0.067 0.040	1 0.033 0.020	4 0.133 0.080	6 0.200 0.120	0 0.000 0.000	30 50.000
before,OCT 3	22 0.611 0.440	2 0.056 0.040	3 0.083 0.060	4 0.111 0.080	0 0.000 0.000	3 0.083 0.060	2 0.056 0.040	36 50.000

Both **Table Format** and **Crosstab Format** have transposed versions (**Table Transposed** and **Crosstab Transposed**). These are useful when there are a lot of response categories but not a lot of samples.

Statistical Methods

The following commands appear in the platform menu depending on context.

Table 29.3 Categorical Platform Commands

	Command	Supported Response Contexts	Question	Details
These require multiple X columns	Test Response Homogeneity	**Separate Responses** **Aligned Responses** **Repeated Measures** **Response Frequencies** (if no **Sample Size**)	Are the probabilities across the response categories the same across sample categories?	Marginal Homogeneity (Independence) Test, both Pearson and Chi-square likelihood ratio chi-square
These require multiple X columns	Test Each Response	**Multiple Response** **Multiple Response by ID** (with **Sample Size**) **Multiple Delimited** **Response Frequence** with **Sample Size**	For each response category, are the rates the same across sample categories?	Poisson regression on sample for each defect frequency.
	Agreement Statistic	**Rater Agreement**	How closely raters agree, and is the lack of agreement symmetrical?	Kappa for agreement, Bowker/McNemar for symmetry.
	Transition Report	**Repeated Measures**	Show the change in categories across time	Transition counts and rates matrices.

Test Response Homogeneity

In this situation, there is typically one categorical response variable, and one categorical sample variable. Multiple sample variables are treated as a single variable.

The test is the chi-square test for marginal homogeneity, testing that the response probabilities are the same across samples. This is equivalent to a test for independence when the sample category is like a response. There are two versions of this test, the Pearson form and the Likelihood Ratio form, both with chi-square statistics.

As an example, open Car Poll.jmp and launch the Categorical platform. Choose country as a **Separate Response** and marital status as an **X, Grouping Variable**. When the report appears, select **Test Reponse Homogeneity** from the platform menu.

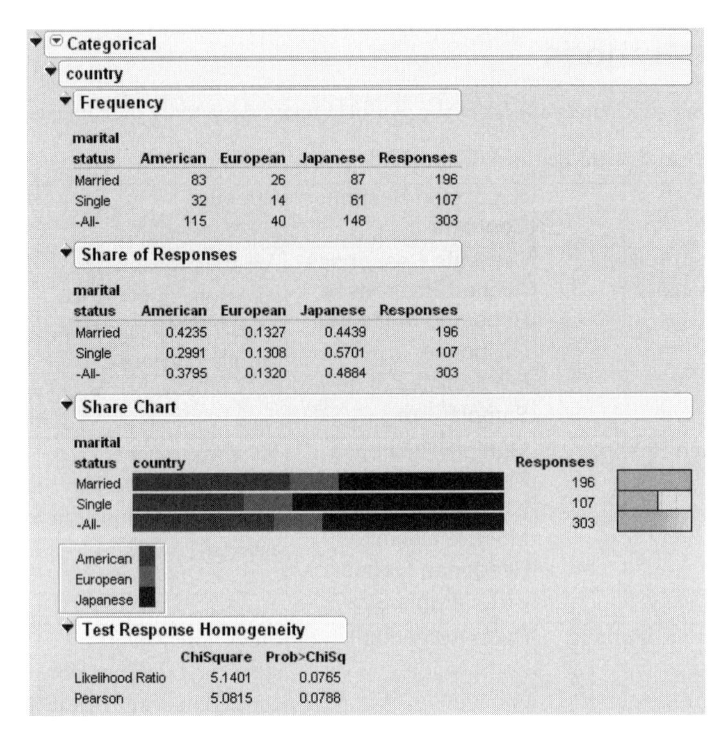

Categorical

country

Frequency

marital status	American	European	Japanese	Responses
Married	83	26	87	196
Single	32	14	61	107
-All-	115	40	148	303

Share of Responses

marital status	American	European	Japanese	Responses
Married	0.4235	0.1327	0.4439	196
Single	0.2991	0.1308	0.5701	107
-All-	0.3795	0.1320	0.4884	303

Share Chart

Test Response Homogeneity

	ChiSquare	Prob>ChiSq
Likelihood Ratio	5.1401	0.0765
Pearson	5.0815	0.0788

The Share chart seems to indicate that the married group seems has higher probability to buy American cars, and the single group has higher probability to buy Japanese cars, but the statistical test only shows a significance of 0.07.Therefore, the difference in response probabilities across marital status is not statistically significant.

Test Each Response

When there are multiple responses, each response category can be modeled separately. The question is whether the response rates are the same across samples. For each response category, we assume the frequency count has a random Poisson distribution. The rate test is obtained using a Poisson regression (through generalized linear models) of the frequency per unit modeled by the sample categorical variable. The result is a likelihood ratio chi-square test of whether the rates are different across samples.

This test can also be done by the **Pareto** platform, as well as in the **Generalized Linear Model** personality of the **Fit Model** platform.

As an example, open failure3Freq.jmp. We want to compare the samples across the clean treatment variable, so launch the Categorical platform and assign all the defect columns as **Response Frequencies**, with clean as an **X, Grouping** variable and SampleSize as **Sample Size**. Then, select **Test Each Response** from the platform menu.

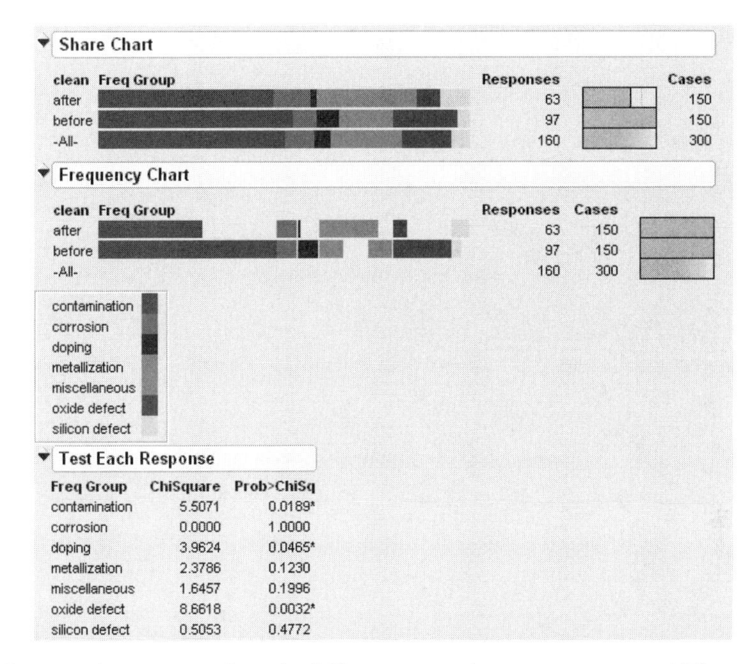

Freq Group	ChiSquare	Prob>ChiSq
contamination	5.5071	0.0189*
corrosion	0.0000	1.0000
doping	3.9624	0.0465*
metallization	2.3786	0.1230
miscellaneous	1.6457	0.1996
oxide defect	8.6618	0.0032*
silicon defect	0.5053	0.4772

For which defects are the rates significantly different across the clean treatments? The *p*-values show that oxide defect is the most significantly different, followed by contamination, then doping. The other defects are not significantly different with this amount of data.

Rater Agreement

The Rater agreement analysis answers the questions of how closely raters agree with one another, and if the lack of agreement is symmetrical.

As an example, open AttributeGage.jmp. The Attribute Chart script runs the Variability Chart platform, which has a test for agreement among raters.

Agreement Comparisons

Rater	Compared with Rater	Kappa	Standard Error
A	B	0.8629	0.0442
A	C	0.7761	0.0547
B	C	0.7880	0.0537

Launch the categorical Platform and designate the three raters (A, B, and C) as Rater Agreement in the launch dialog. In the resulting report, you have a similar test for agreement that is augmented by a symmetry test that the lack of agreement is symmetric.

Agreement Statistics

Degree of Agreement

Response1	Response2	Kappa	Std Err	Bowker Symmetry	Bowker PValue
A	B	0.862944	0.044198	1	0.3173
A	C	0.776119	0.054671	0.066667	0.7963
B	C	0.788007	0.053702	1.142857	0.2850

Chapter 30

Discriminant Analysis

The Discriminant Platform

Discriminant Analysis seeks to find a way to predict a classification (*X*) variable (nominal or ordinal) based on known continuous responses (*Y*). It can be regarded as inverse prediction from a multivariate analysis of variance (MANOVA). In fact, the **Manova** personality of Fit Model also provides some discriminant features.

Features include:

- A stepwise selection dialog to help choose variables that discriminate well
- Choice among Linear, Quadratic, or Regularized-Parameter analyses
- The discriminant scores, showing which groups each point is near.
- A misclassification summary.
- Plots of the points and means in a canonical space, which is the space that shows the best separation of means.
- Options to save prediction distances and probabilities to the data table.

Contents

Introduction

Discriminant Analysis is an alternative to logistic regression. In logistic regression, the classification variable is random and predicted by the continuous variables, whereas in discriminant analysis the classifications are fixed, and the Y's are realizations of random variables. However, in both cases, the categorical value is predicted by the continuous.

There are several varieties of discriminant analysis. JMP implements linear and quadratic Discriminant Analysis, along with a method that blends both types. In linear discriminant analysis, it is assumed that the Y's are normally distributed with the same variances and covariances, but that there are different means for each group defined by X. In quadratic Discriminant Analysis, the covariances can be different across groups.Both methods measure the distance from each point in the data set to each group's multivariate mean (often called a *centroid*) and classifies the point to the closest group. The distance measure used is the Mahalanobis distance, which takes into account the variances and covariances between the variables.

Discriminating Groups

Fisher's Iris data set is the classic example of discriminant analysis. Four measurements are taken from a sample consisting of three different species. The goal is to identify the species accurately using the values of the four measurements. Open Iris.jmp, and select **Analyze > Multivariate Methods > Discriminant** to launch the Discriminant Analysis platform. The launch dialog in Figure 30.1 appears.

Figure 30.1 Discriminant Launch Dialog

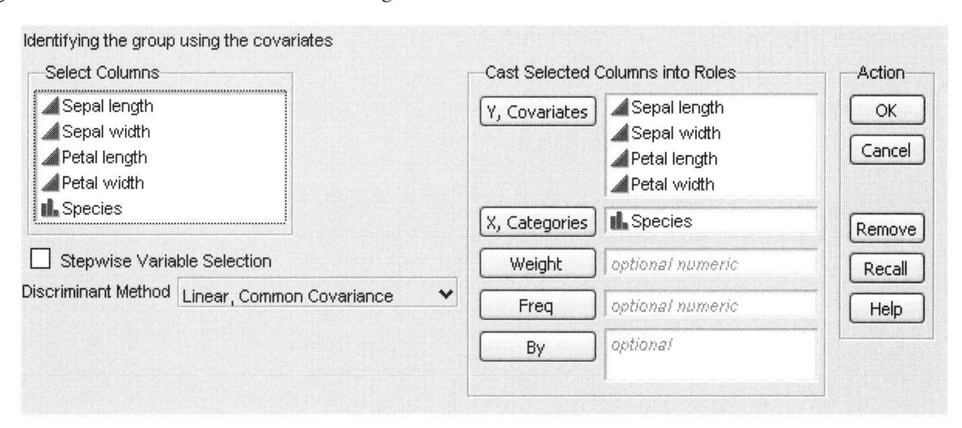

If you want to find which variables discriminate well, click the checkbox for **Stepwise Variable Selection.** Otherwise, the platform uses all the variables you specify. In this example, specify the four continuous variables as **Y, Covariates** and Species as **X, Categories.**

Discriminant Method

JMP offers three kinds of Discriminant Analysis. All three calculate distances as the Mahalanobis distance from each point to each group's multivariate mean. The difference in the methods is only in the covariance matrix used in the computations.

Linear Discriminant Analysis uses a common (*i.e.* within-) covariance matrix for all groups.

Quadratic Discriminant Analysis uses a separate covariance matrix for each group.

Quadratic discriminant suffers in small data sets because it does not having enough data to make nicely invertible and stable covariance matrices. Regularized discriminant ameliorates these problems and still allows for differences among groups.

Regularized Discriminant Analysis is a compromise between the linear and quadratic methods, governed by two arguments. When you choose **Regularized Discriminant Analysis**, a dialog appears allowing specification of these two parameters.

The first parameter (**Lambda, Shrinkage to Common Covariance**) specifies how to mix the individual and group covariance matrices. For this parameter, 1 corresponds to Linear Discriminant Analysis and 0 corresponds to Quadratic Discriminant Analysis.

The second parameter (**Gamma, Shrinkage to Diagonal**)specifies whether to deflate the non-diagonal elements, that is, the covariances across variables. If you choose 1, then the covariance matrix is forced to be diagonal.

Therefore, assigning 0,0 to these parameters is identical to requesting quadratic discriminant analysis. Similarly, a 1,0 assignment requests linear discriminant analysis. These cases, along with a Regularized Discriminant Analysis example with l=0.4 and g=0.4 are shown in Figure 30.2.

Use this table to help decide on the regularization.

Use lower Lambda (Gamma) when	**Use higher Lambda (Gamma) when**
Covariances are different (Variables are correlated)	Covariances are the same (Variables are uncorrelated)
lots of data	little data
small number of variables	many variables

Figure 30.2 Linear, Quadratic, and Regularized Discriminant Analysis

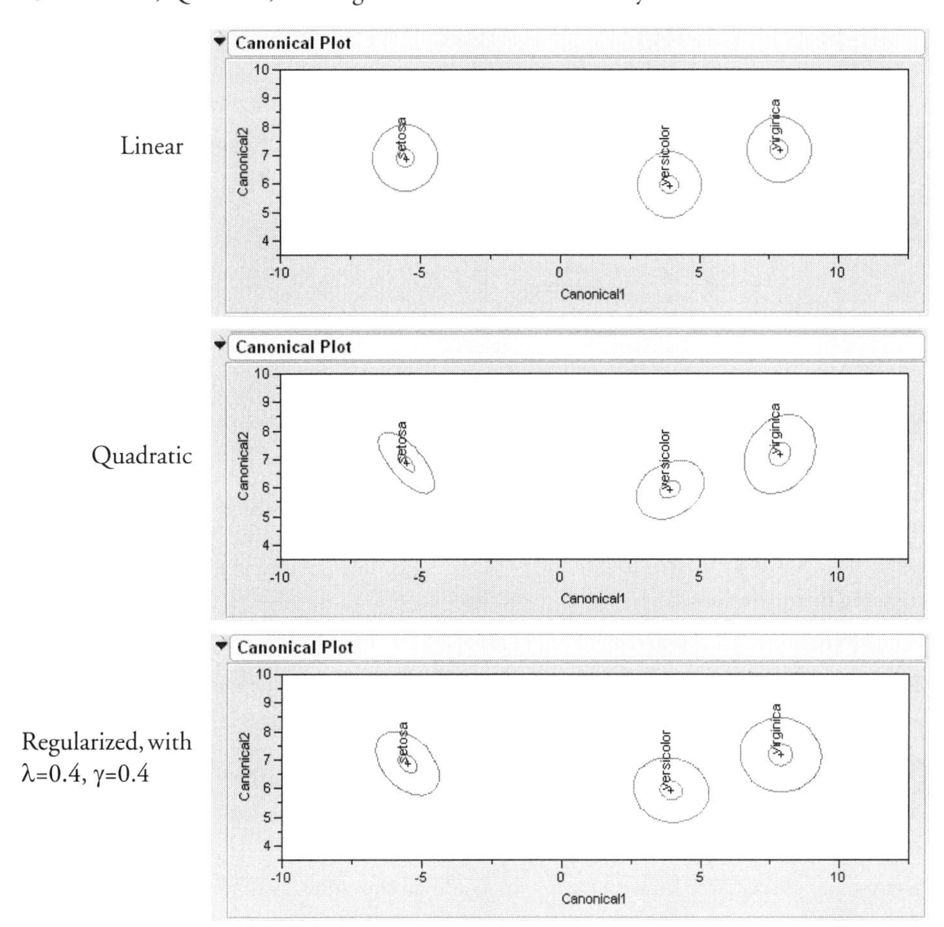

Stepwise Selection

If you choose **Stepwise Variable Selection**, a dialog appears (Figure 30.3) to select variables. You can review which columns have large *F* ratios or small *p*-values and control which columns are entered into the discriminant model. In addition, the dialog displays how many columns are currently in and out of the model, and the largest and smallest *p*-values to be entered or removed.

Note: Stepwise only supports Linear Discriminant Analysis

Figure 30.3 Stepwise Control Panel

Discriminant Analysis

Column Selection

Click to select columns into discriminant model

Columns In	0	Smallest P to Enter	0.0000000
Columns Out	4	Largest P to Remove	.

Step Forward | Enter All

Step Backward | Remove All | Apply This Model

Lock	Entered	Column	F Ratio	Prob>F
☐	☐	Sepal length	119.265	0.0000000
☐	☐	Sepal width	49.160	0.0000000
☐	☐	Petal length	1180.16	0.0000000
☐	☐	Petal width	960.007	0.0000000

Entered checkboxes show which columns are currently in the model. You can manually click columns in or out of the model.

Lock checkboxes are used when you want to force a column to stay in its current state regardless of any stepping by the buttons.

Step Forward adds the most significant column not already in the model.

Step Backward removes the least significant column already in the model.

Enter All enters all the columns into the model

Remove All removes all the columns from the model

Apply This Model is used when you are finished deciding the columns to include in the analysis, and want to proceed to estimation and scoring.

Figure 30.4 shows three forward steps, which add all the columns to the model except **Sepal length.**

Figure 30.4 Stepped Model

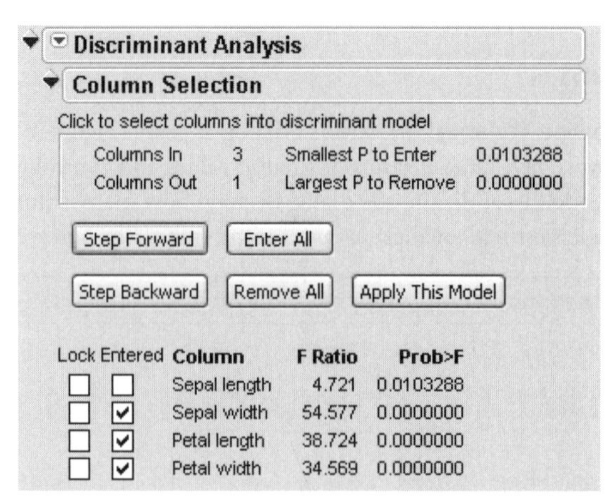

Click **Apply This Model** to estimate the model. After estimation and scoring are done, two reports are produced: a Canonical Plot (Figure 30.5), and a Scoring Report.

Canonical Plot

The Canonical Plot shows the points and multivariate means in the two dimensions that best separate the groups.

Figure 30.5 Canonical Plot

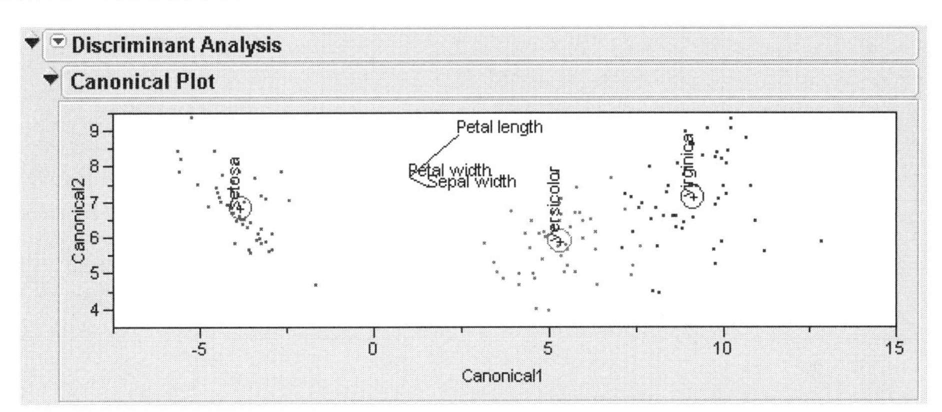

- Each row in the data set is a point, controlled by the **Canonical Options > Show Points** option
- Each multivariate mean is a labeled circle. The size of the circle corresponds to a 95% confidence limit for the mean. Groups that are significantly different tend to have non-intersecting circles. This is controlled by the **Canonical Options > Show Means CL Ellipses** option.
- The directions of the variables in the canonical space is shown by labeled rays emanating from the grand mean. This is controlled by the **Canonical Options > Show Biplot Rays** option. You can drag the center of the biplot rays to other places in the graph.
- The option **Show Normal 50% Contours** shows areas that contain roughly 50% of the points for that group if the assumptions are correct. Under linear discriminant analysis, they are all the same size and shape.

In order to have the points color-coded like the centroid circles, use the **Color Points** option or button. This is equivalent to **Rows > Color or Mark by Column**, coloring by the classification column.

The canonical plot can also be referred to as a *biplot* when both the points and the variable direction rays are shown together, as in Figure 30.5. It is identical to the Centroid plot produced in the **Manova** personality of the Fit Model platform.

Discriminant Scores

The scores report shows how well each point is classified. The first five columns of the report represent the actual (observed) data values, showing row numbers, the actual classification, the distance to the mean of that classification, and the associated probability. JMP graphs the -log(prob) to show the loss in log-likelihood when a point is predicted poorly. When the red bar is large, the point is being poorly predicted.

On the right of the graph, the report shows the category with the highest prediction probability, and, if they exist, other categories that have a predicted probability of over 0.1. An asterisk(*) indicates which points are misclassified.

In Figure 30.6, the option **Show Interesting Rows Only** option is set so that only those rows that have fitted probabilities between 0.05 and 0.95 or are misclassified are shown.

Figure 30.6 Show Interesting Rows Only

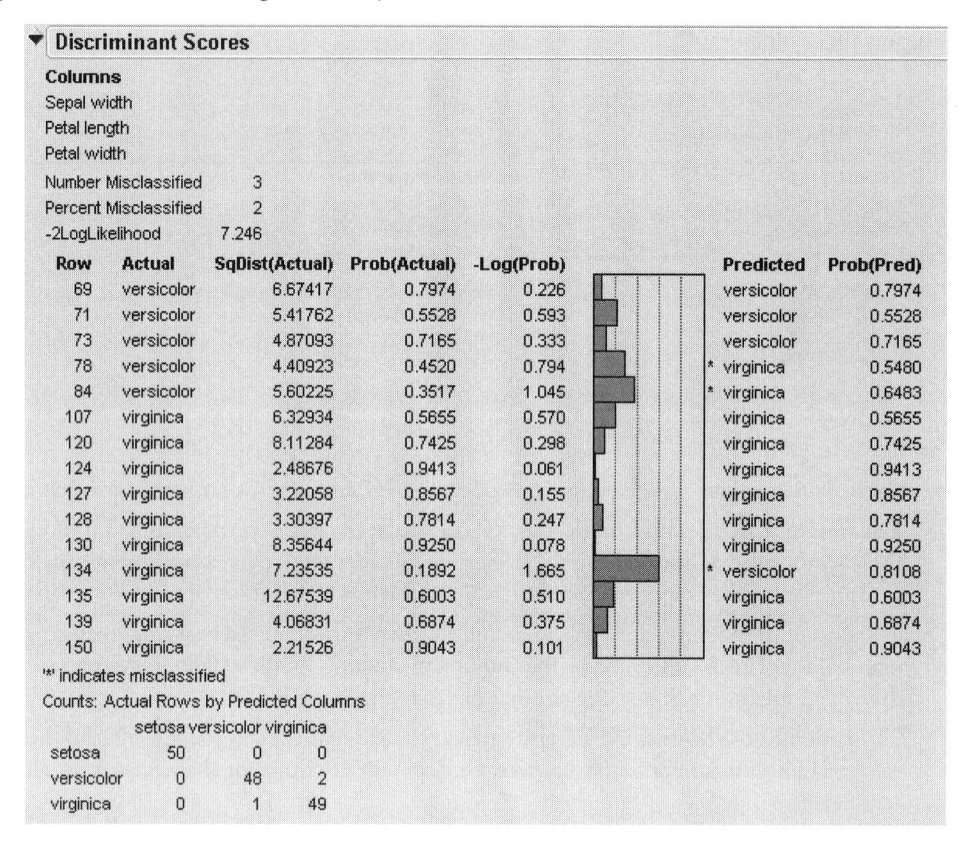

Discriminant Scores

Columns
Sepal width
Petal length
Petal width

Number Misclassified	3
Percent Misclassified	2
-2LogLikelihood	7.246

Row	Actual	SqDist(Actual)	Prob(Actual)	-Log(Prob)		Predicted	Prob(Pred)
69	versicolor	6.67417	0.7974	0.226		versicolor	0.7974
71	versicolor	5.41762	0.5528	0.593		versicolor	0.5528
73	versicolor	4.87093	0.7165	0.333		versicolor	0.7165
78	versicolor	4.40923	0.4520	0.794	*	virginica	0.5480
84	versicolor	5.60225	0.3517	1.045	*	virginica	0.6483
107	virginica	6.32934	0.5655	0.570		virginica	0.5655
120	virginica	8.11284	0.7425	0.298		virginica	0.7425
124	virginica	2.48676	0.9413	0.061		virginica	0.9413
127	virginica	3.22058	0.8567	0.155		virginica	0.8567
128	virginica	3.30397	0.7814	0.247		virginica	0.7814
130	virginica	8.35644	0.9250	0.078		virginica	0.9250
134	virginica	7.23535	0.1892	1.665	*	versicolor	0.8108
135	virginica	12.67539	0.6003	0.510		virginica	0.6003
139	virginica	4.06831	0.6874	0.375		virginica	0.6874
150	virginica	2.21526	0.9043	0.101		virginica	0.9043

'*' indicates misclassified

Counts: Actual Rows by Predicted Columns

	setosa	versicolor	virginica
setosa	50	0	0
versicolor	0	48	2
virginica	0	1	49

The Counts report appears at the bottom of Figure 30.6. The counts are zero off the diagonal if everything is classified correctly.

Commands and Options

The following commands are available from the platform popup menu.

Stepwise Variable Selection returns to the stepwise control panel.

Discriminant Method chooses the discriminant method. Details are shown in the section "Discriminant Method," p. 596.

Linear, Common Covariance assumes a common covariance.

Quadratic, Different Covariances which estimates different covariances.

Regularized, Compromise Method a compromise method.

Score Data shows or hides the listing of the scores by row in the **Discriminant Scores** portion of the report.

Score Options deal with the scoring of the observations.

Show Interesting Rows Only shows rows that are misclassified and those where p>.05 and p<0.95) for any p, the attributed probability.

Show Classification Counts shows a matrix of actual by predicted counts for each category. When the data are perfectly predicted, the off-diagonal elements are zero.

Select Misclassified Rows selects the misclassified rows in the original data table.

Select Uncertain Rows selects the rows which have uncertain classifications in the data table. When this option is selected, a dialog box appears so you can specify the difference (0.1 is the default) to be marked as uncertain.

Save Formulas saves formulas to the data table. The distance formulas are Dist[0], needed in the Mahalanobis distance calculations, and a Dist[ ] column for each X-level's Mahalanobis distance. Probability formulas are Prob[0], the sum of the exponentials of -0.5 times the Mahalanobis distances, and a Prob[ ] column for each X-level's posterior probability of being in that category. The Pred column holds the most likely level for each row.

Show Distances to each group appends a table to show the squared distance to each group.

Show Probabilities to each group appends a table to show the probabilities to each group.

ROC Curve appends an ROC curve to the report.

Canonical Plot shows or hides the Canonical Plot.

Canonical Options all affect the Canonical Plot.

Show Points shows or hides the points in the plot

Show Means CL Ellipses shows or hides the 95% confidence ellipse of each mean. The ellipses appear as a circle because of scaling. Categories with more observations have smaller circles.

Show Normal 50% Contours shows or hides the normal contours which estimate the region where 50% of the level's points lie.

Show Biplot Rays shows or hides the biplot rays. These rays indicate the directions of the variables in the canonical space.

Biplot Ray Position allows you to specify the position and radius scaling (default = 1) of the biplot rays in the canonical plot.

Color Points colors the points based on levels of the X variable. This statement is equivalent to selecting **Rows > Color or Mark by Column** and selecting the X variable.

Show Canonical Details shows or hides the Canonical details. Details for the Iris data set are shown in Figure 30.8. The matrices at the bottom of the report are opened by clicking on the blue triangle beside their name and closed by clicking on the name of the matrix.

Save Canonical Scores creates columns in the data table holding the canonical score for each observation. The new columns are named Canon[ ].

Canonical 3D Plot is only available when there are four or more groups (Figure 30.7). It shows a three-dimensional version of the Canonical Plot and respects other Canonical Options.

Figure 30.7 Canonical 3D Plot

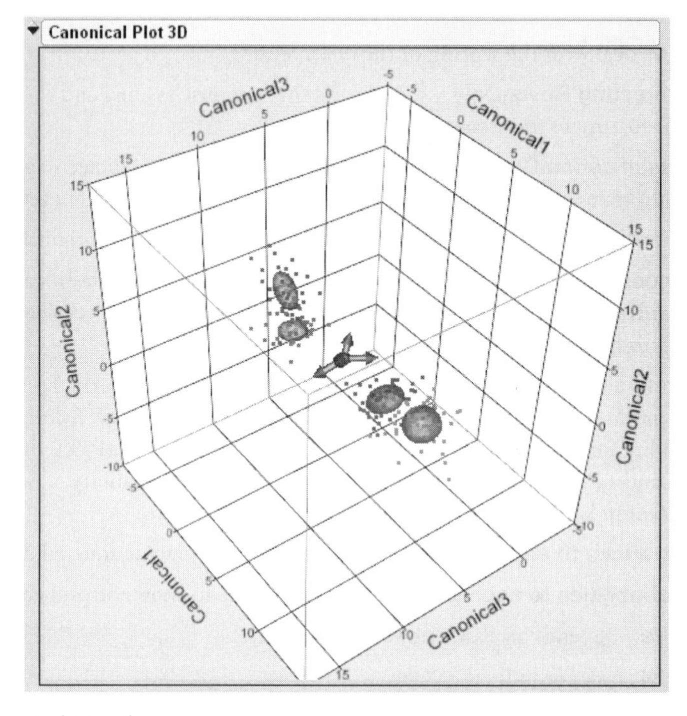

Figure 30.8 Canonical Details

Eigenvalue	Percent	Cum Percent	Canonical Corr
30.1498095	99.0624	99.0624	0.9838176
0.28537471	0.9376	100.0000	0.47118653
-8.963e-15	0.0000	100.0000	0

Test	Value	Approx. F	NumDF	DenDF	Prob>F
Wilks' Lambda	0.0249755	257.5032	6	290	<.0001*
Pillai's Trace	1.1899138	71.4852	6	292	<.0001*
Hotelling-Lawley	30.435184	730.4444	6	288	<.0001*
Roy's Max Root	30.149809	1467.2907	3	146	<.0001*

▶ Within Matrix

Open matrix — Between Matrix

	Sepal width	Petal length	Petal width
Sepal width	0.0771764	-0.389385	-0.156005
Petal length	-0.389385	2.9734884	1.2705714
Petal width	-0.156005	1.2705714	0.5470295

Closed Matrix — ▶ Eigenvectors

▶ Scoring Coefs

Specify Priors lets you specify the prior probabilities for each level of the X variable.

Equal Probabilities assigns an equal probability to each level.

Proportional to Occurrence assigns probabilities to each level according to their frequency in the data.

Other brings up a dialog to allow custom specification of the priors, shown in Figure 30.9. By default, each level is assigned equal probabilities.

Figure 30.9 Specify Prior Probabilities Dialog

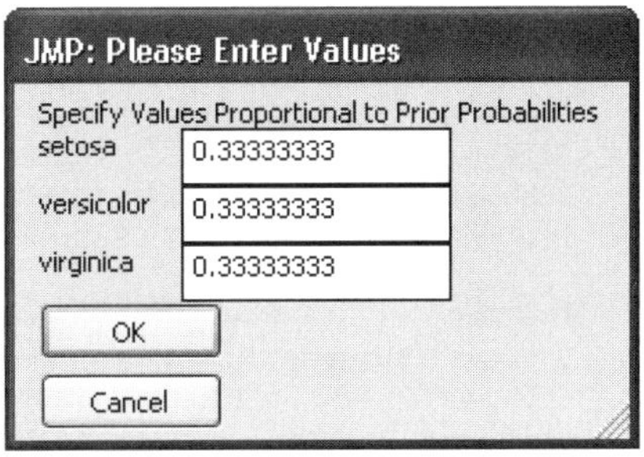

JMP: Please Enter Values
Specify Values Proportional to Prior Probabilities
setosa 0.33333333
versicolor 0.33333333
virginica 0.33333333
OK
Cancel

Consider New Levels is used when you have some points that may not fit any known group, but instead may be from an unscored, new group.

Save Discrim Matrices creates a global list (`DiscrimResults`) for use in JMP's scripting language. The list contains a list of `YNames`, a list of `XNames`, a list of `XValues`, a matrix of `YMeans`, and a matrix of `YPartialCov` (covariances). An example from the iris data `DiscrimResults` is

```
{
  YNames = {"Sepal length", "Sepal width", "Petal length", "Petal Width"},
  XName = "Species", XValues = {"setosa", "versicolor", "virginica"},
  YMeans = [5.005999999999999 3.428000000000001 1.462
  0.2459999999999999,5.936 2.77 4.26 1.326, 6.587999999999998 2.974 5.552
  2.026],
  YPartialCov = [0.2650081632653061 0.09272108843537413 0.1675142857142858
  0.03840136054421769, 0.09272108843537413 0.1153877551020407
  0.055243537414966 0.03271020408163266, 0.1675142857142858
  0.055243537414966 0.1851877551020409 0.04266530612244898,
  0.03840136054421769 0.03271020408163266 0.04266530612244898
  0.04188163265306122]
  }
```

`Get Discrim Matrices`,only available through scripting, obtains the same values as **Save Discrim Matrices**, but returns them to the caller rather than storing them in the data table.

Show Within Covariances shows or hides the Covariance Matrix report. The report for this example is shown in Figure 30.10.

Figure 30.10 Covariance Matrix Report

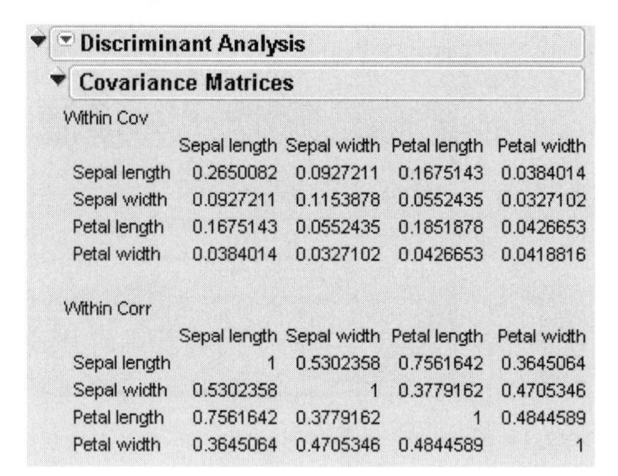

Discriminant Analysis

Covariance Matrices

Within Cov

	Sepal length	Sepal width	Petal length	Petal width
Sepal length	0.2650082	0.0927211	0.1675143	0.0384014
Sepal width	0.0927211	0.1153878	0.0552435	0.0327102
Petal length	0.1675143	0.0552435	0.1851878	0.0426653
Petal width	0.0384014	0.0327102	0.0426653	0.0418816

Within Corr

	Sepal length	Sepal width	Petal length	Petal width
Sepal length	1	0.5302358	0.7561642	0.3645064
Sepal width	0.5302358	1	0.3779162	0.4705346
Petal length	0.7561642	0.3779162	1	0.4844589
Petal width	0.3645064	0.4705346	0.4844589	1

Show Group Means shows or hides a table with the means of each variable. Means are shown for each level and for all levels of the X variable.

Chapter 31

Survival and Reliability Analysis I

Univariate Survival

Survival data contain duration times until the occurrence of a specific event and are sometimes referred to as *event-time* response data. The event is usually failure, such as the failure of an engine or death of a patient. If the event does not occur before the end of a study for an observation, the observation is said to be *censored*.

The **Survival and Reliability** submenu accesses four kinds of survival analysis.

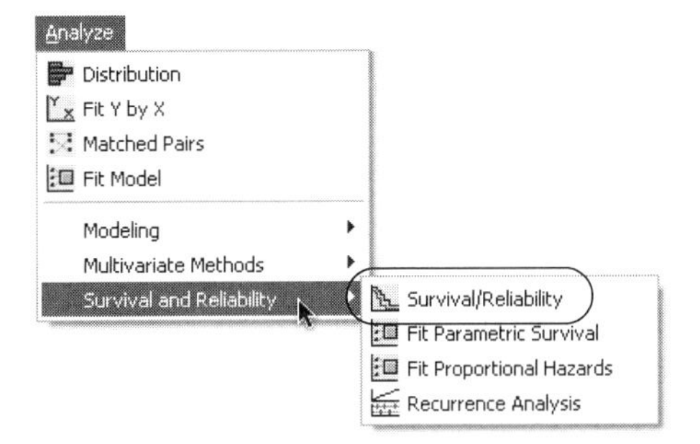

This chapter focuses on univariate survival, the first item on the Survival and Reliability Menu. **Survival/Reliability** calculates estimates of survival functions using the product-limit (Kaplan-Meier) method for one or more groups of right-censored data. It gives an overlay plot of estimated survival function for each group and for the whole sample. JMP also computes the log rank and Wilcoxon statistics to test homogeneity between the groups. Optionally, there are diagnostic plots and fitting options for exponential, Weibull, and lognormal survival distributions, and an analysis of competing causes for the Weibull model. Interval censoring is supported by Turnbull estimates.

In the "Survival and Reliability Analysis II" chapter, parametric and proportional hazards models are discussed. "Recurrence Analysis," p. 649 gives details on recurrence analysis.

Contents

Introduction to Survival Analysis

Survival data needs to be analyzed with specialized methods for two reasons:

1 The survival times have specialized non-normal distributions, like the exponential, Weibull, and lognormal.

2 Some of the data could be censored—you don't know the exact survival time but you know that it is greater than the specified value. This is called *right-censoring*. Censoring happens when the study ends without all the units failing, or when a patient has to leave the study before it is finished. The censored observations cannot be ignored without biasing the analysis.

The elements of a survival model are

- a time indicating how long until the unit (or patient) either experienced the event or was censored. Time is the model response (*Y*).
- a censoring indicator that denotes whether an observation experienced the event or was censored. JMP uses the convention that the code for a censored unit is 1 (or any nonzero value) and the code for a non-censored event is zero.
- explanatory variables if a regression model is used.
- If interval censoring is needed, then two y variables hold the lower and upper limits bounding the event time.

The terms failure time, survival time, and duration are used interchangeably.

Univariate Survival Analysis

To do a univariate survival analysis, choose **Analyze > Survival and Reliability** and select **Survival/ Reliability** from its submenu.

After you complete the launch dialog (shown here), and click **OK**, the **Survival and Reliability** command produces product-limit (also called Kaplan-Meier) survival estimates, exploratory plots with optional parameter estimates, and a comparison of survival curves when there is more than one group.

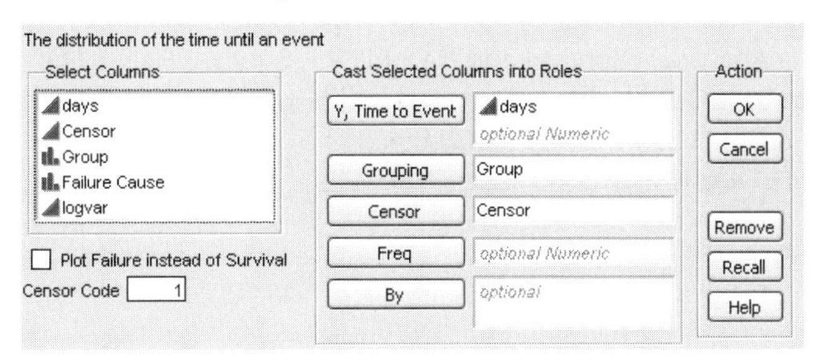

Selecting Variables for Univariate Survival Analysis

The Survival platform requires only a time (*Y*) variable, which must be duration or survival times. The censor, grouping, and frequency variables are optional. The sort-order of the data doesn't matter.

Y, Time to Event is the only required variable, which contains the time to event or time to censoring. If you have interval censoring, then you specify two *Y* variables, the lower and upper limits.

Censor is for a column in your table that has the censor code, zero for uncensored, non-zero for censored.

Grouping is for a column to classify the data into groups, which are fit separately.

Freq is for a column whose values are the frequencies of observations for each row when there are multiple units recorded.

By default, censor column values of 0 indicate the event (*e.g.* failure), and a non-zero code is a censored value. To support data sets that are coded differently, you can enter a difference code for the censoring (*e.g.* 0) in the launch dialog in the Censor Code edit box. To provide compatibility with previous versions of JMP, a code of 1 is interpreted to mean that any positive value indicates censoring.

Example: Fan Reliability

The failure of diesel generator fans was studied by Nelson (1982 p. 133, also Meeker and Escobar, 1998, appendix C1). A partial listing of the data is shown in Figure 31.1.

Figure 31.1 Fan Data

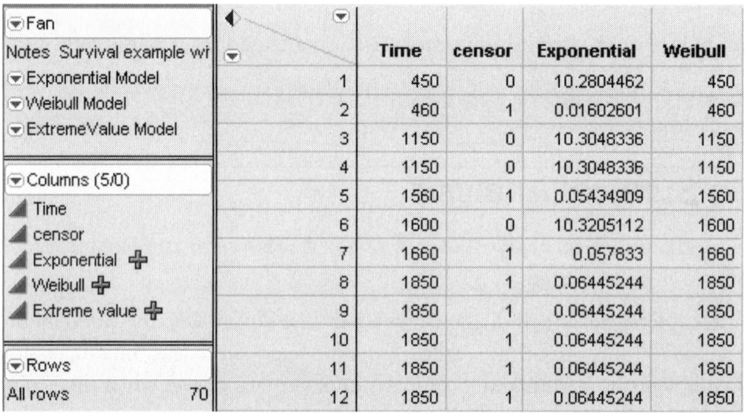

Fan
Notes Survival example wi
Exponential Model
Weibull Model
ExtremeValue Model

Columns (5/0)
Time
censor
Exponential
Weibull
Extreme value

Rows
All rows 70

	Time	censor	Exponential	Weibull
1	450	0	10.2804462	450
2	460	1	0.01602601	460
3	1150	0	10.3048336	1150
4	1150	0	10.3048336	1150
5	1560	1	0.05434909	1560
6	1600	0	10.3205112	1600
7	1660	1	0.057833	1660
8	1850	1	0.06445244	1850
9	1850	1	0.06445244	1850
10	1850	1	0.06445244	1850
11	1850	1	0.06445244	1850
12	1850	1	0.06445244	1850

After launching **Analyze > Survival and Reliability > Survival/Reliability**, specify Time as **Y, Time to Event** and censor as **Censor**. Also, check the checkbox for **Plot Failure instead of Survival**, since it is more conventional to show a failure probability plot instead of its reverse (a survival probability plot). The completed dialog is shown in Figure 31.2.

Figure 31.2 Fan Launch Dialog

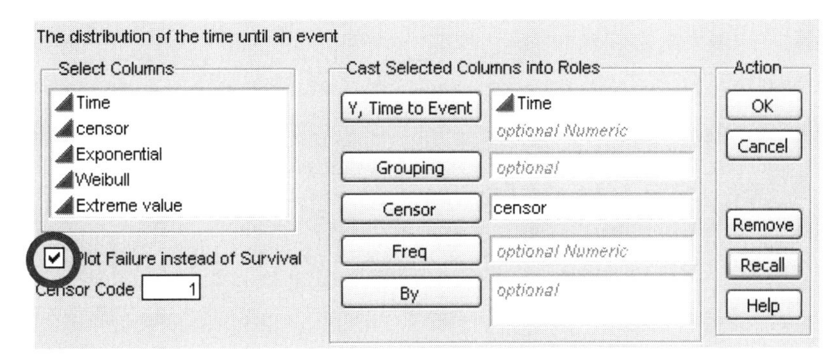

Figure 31.3 shows the Failure plot. Notice the increasing failure probability as a function of time.

Figure 31.3 Fan Initial Output

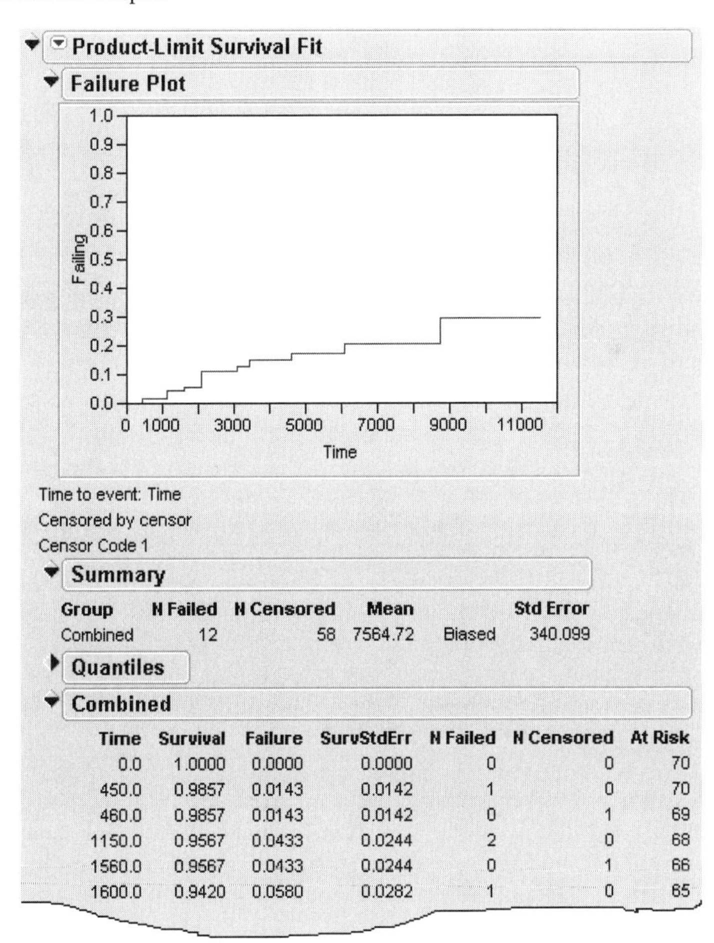

Group	N Failed	N Censored	Mean		Std Error
Combined	12	58	7564.72	Biased	340.099

Time	Survival	Failure	SurvStdErr	N Failed	N Censored	At Risk
0.0	1.0000	0.0000	0.0000	0	0	70
450.0	0.9857	0.0143	0.0142	1	0	70
460.0	0.9857	0.0143	0.0142	0	1	69
1150.0	0.9567	0.0433	0.0244	2	0	68
1560.0	0.9567	0.0433	0.0244	0	1	66
1600.0	0.9420	0.0580	0.0282	1	0	65

Usually, the next step is to explore distributional fits, such as a Weibull model, using the **Plot** and **Fit** options for that distribution.

Figure 31.4 Weibull Output for Fan Data

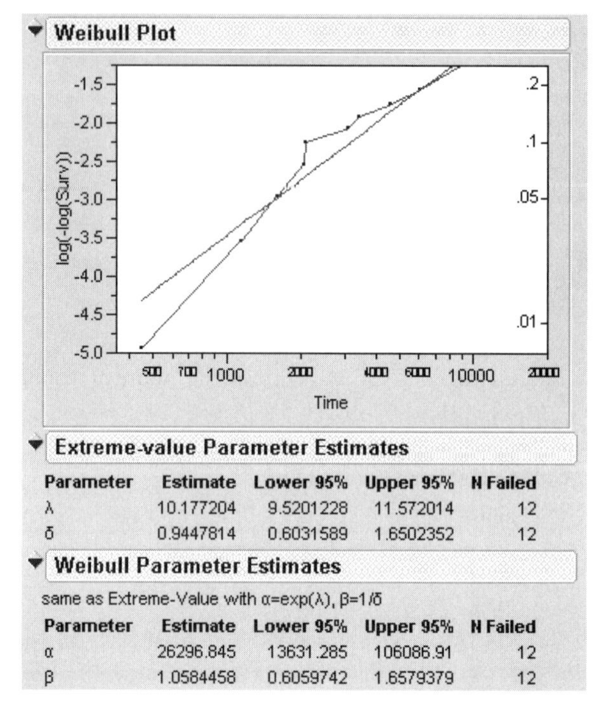

Parameter	Estimate	Lower 95%	Upper 95%	N Failed
λ	10.177204	9.5201228	11.572014	12
δ	0.9447814	0.6031589	1.6502352	12

Parameter	Estimate	Lower 95%	Upper 95%	N Failed
α	26296.845	13631.285	106086.91	12
β	1.0584458	0.6059742	1.6579379	12

Since the fit is reasonable and the Beta estimate is near 1, you can conclude that this looks like an exponential distribution, which has a constant hazard rate. The **Fitted Distribution Plots** option produces the three views of the distributional fit.

Figure 31.5 Fitted Distribution Plots

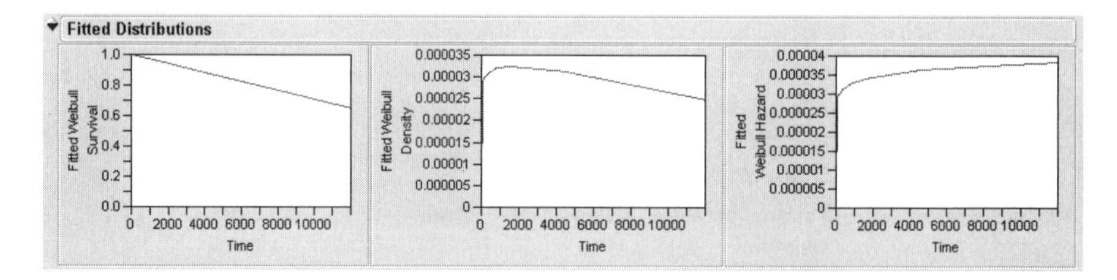

Example: Rats Data

An experiment was undertaken to characterize the survival time of rats exposed to a carcinogen in two treatment groups. The data are in the Rats.jmp table found in the sample data folder. The days variable is the survival time in days. Some observations are censored. The event in this example is death. The objective is to see if rats in one treatment group live longer (more days) than rats in the other treatment group.

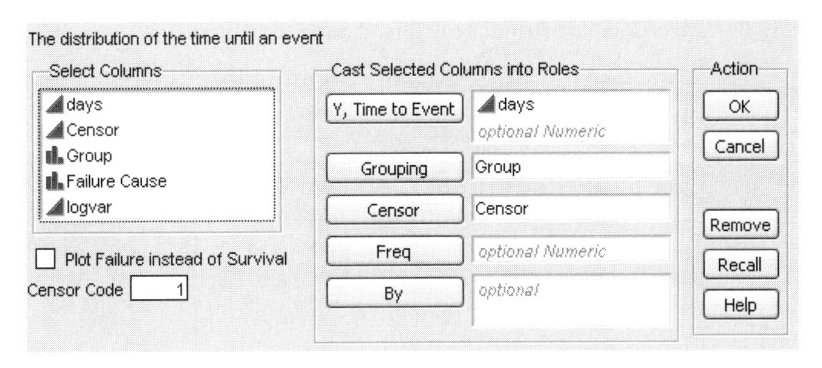

Use the Survival launch dialog to assign columns to the roles as shown in the example dialog above.

Overview of the Univariate Survival Platform

The **Survival** platform computes product-limit (Kaplan-Meier) survival estimates for one or more groups. It can be used as a complete analysis or is useful as an exploratory analysis to gain information for more complex model fitting.

The Kaplan-Meier Survival platform

- shows a plot of the estimated survival function for each group and, optionally, for the whole sample.
- calculates and lists survival function estimates for each group and for the combined sample.
- optionally, displays exponential, Weibull, and lognormal diagnostic failure plots to graphically check the appropriateness of using these distributions for further regression modeling. Parameter estimates are available on request.
- computes the Log Rank and Wilcoxon Chi-square statistics to test homogeneity of the estimated survival function across the groups.
- optionally, performs analysis of competing causes, prompting for a cause of failure variable, and estimating a Weibull failure time distribution for censoring patterns corresponding to each cause.

Initially, the Survival platform displays overlay step plots of estimated survival functions for each group as shown in Figure 31.6. A legend identifies groups by color and line type.

Tables beneath the plot give summary statistics and quantiles for survival times, list the estimated survival time for each observation computed within groups and survival times computed from the combined sample. When there is more than one group, statistical tests are given that compare the survival curves.

Figure 31.6 Survival Plot and Report Structure of Survival Platform

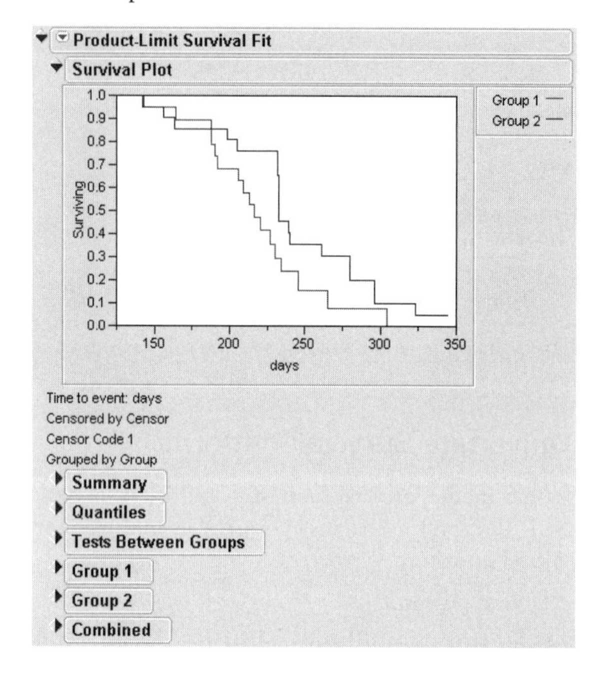

Statistical Reports for the Univariate Analysis

Initial reports are the Summary table and Quantiles table (shown in Figure 31.7). The Summary table shows the number of failed and number of censored observations for each group (when there are groups) and for the whole study, and the mean and standard deviations adjusted for censoring. For computational details on these statistics, see the *SAS/Stat User's Guide* (2001).

The quantiles table shows time to failure statistics for individual groups and combined groups. These include the median survival time, with upper and lower 95% confidence limits. The median survival time is the time (number of days) at which half the subjects have failed. The quartile survival times (25% and 75%) are also included.

Figure 31.7 Summary Statistics for the Univariate Survival Analysis

Product-Limit Survival Fit

Summary

Group	N Failed	N Censored	Mean		Std Error
Group 1	17	2	218.757		9.40318
Group 2	19	2	240.795	Biased	11.206
Combined	36	4	230.729	Biased	7.57346

Quantiles

Group	Median Time	Lower95%	Upper95%	25% Failures	75% Failures
Group 1	216	190	234	190	234
Group 2	233	232	280	232	280
Combined	232	213	239	198	261

When there are multiple groups, the Tests Between Groups table, shown below, gives statistical tests for homogeneity between the groups (Kalbfleisch and Prentice 1980).

Tests Between Groups

Test	ChiSquare	DF	Prob>ChiSq
Log-Rank	3.1227	1	0.0772
Wilcoxon	2.6510	1	0.1035

Test names two statistical tests of the hypothesis that the survival functions are the same across groups.

Chi-Square gives the Chi-square approximations for the statistical tests.

The **Log-Rank** test places more weight on larger survival times and is most useful when the ratio of hazard functions in the groups being compared is approximately constant. The hazard function is the instantaneous failure rate at a given time. It is also called the *mortality rate* or *force of mortality*.

The **Wilcoxon** test places more weight on early survival times and is the optimum rank test if the error distribution is logistic (Kalbfleisch and Prentice, 1980).

DF gives the degrees of freedom for the statistical tests.

Prob>ChiSq lists the probability of obtaining, by chance alone, a Chi-square value greater than the one computed if the survival functions are the same for all groups.

Figure 31.8 shows an example of the product-limit survival function estimates for one of the groups.

Figure 31.8 Example of Survival Estimates Table

Product-Limit Survival Fit

Group 1

days	Survival	Failure	SurvStdErr	N Failed	N Censored	At Risk
0.000	1.0000	0.0000	0.0000	0	0	19
143.000	0.9474	0.0526	0.0512	1	0	19
164.000	0.8947	0.1053	0.0704	1	0	18
188.000	0.7895	0.2105	0.0935	2	0	17
190.000	0.7368	0.2632	0.1010	1	0	15
192.000	0.6842	0.3158	0.1066	1	0	14
206.000	0.6316	0.3684	0.1107	1	0	13
209.000	0.5789	0.4211	0.1133	1	0	12
213.000	0.5263	0.4737	0.1145	1	0	11
216.000	0.4737	0.5263	0.1145	1	1	10
220.000	0.4145	0.5855	0.1145	1	0	8
227.000	0.3553	0.6447	0.1124	1	0	7
230.000	0.2961	0.7039	0.1082	1	0	6
234.000	0.2368	0.7632	0.1015	1	0	5
244.000	0.2368	0.7632	0.1015	0	1	4
246.000	0.1579	0.8421	0.0934	1	0	3
265.000	0.0789	0.9211	0.0728	1	0	2
304.000	0.0000	1.0000	0.0000	1	0	1

Note: When the final time recorded is a censored observation, the report indicates a *biased* mean estimate. The biased mean estimate is a lower bound for the true mean.

Platform Options

The popup menu on the Product-Limit Survival Fit title bar lists options for the Survival platform.

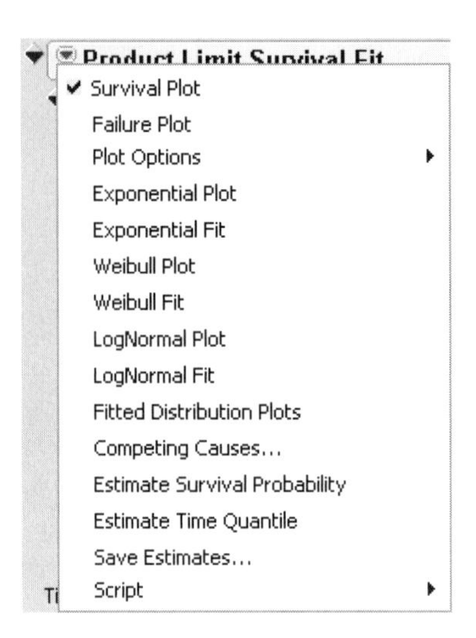

All of the platform options alternately hide or display information. The following list summarizes these options:

Survival Plot displays the overlaid survival plots for each group, as shown in Figure 31.6.

Failure Plot displays the overlaid failure plots (proportion failing over time) for each group in the tradition of the Reliability literature. A Failure Plot reverses the *y*-axis to show the number of failures rather than the number of survivors. The difference is easily seen in an example. Both plots from the Rats.jmp data table appear in Figure 31.9.

Note that **Failure Plot** replaces the **Reverse Y Axis** command found in older versions of JMP (which is still available in scripts).

Figure 31.9 Survival Plot and Failure Plot of the Rats data

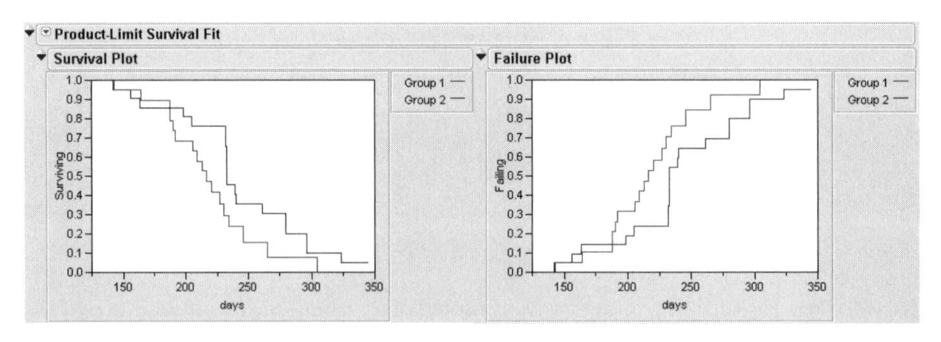

Plot Options is a submenu that contains the following options. Note that the first four options (**Show Points**, **Show Combined**, **Show Confid Interval**, and **Show Simultaneous CI**) pertain to the initial survival plot, while the last two (**Midstep Quantile Points** and **Connect Quantile Points**) only pertain to the distributional plots.

Show Points hides or shows the sample points at each step of the survival plot. Failures are shown at the bottom of the steps, and censorings are indicated by points above the steps.

Show Kaplan Meier hides or shows the Kaplan-Meier curves.

Show Combined displays the survival curve for the combined groups in the Survival Plot.

Show Confid Interval shows 95% confidence on the survival plot for groups and for the combined plot when it is displayed with the **Show Combined** option.

When the plot has the **Show Points** and **Show Combined** options in effect, the survival plot for the total or combined sample shows as a gray line and the points show at the plot steps of each group.

Show Simultaneous CI toggles the confidence intervals for all groups on the plot.

Midstep Quantile Points changes the plotting positions to use the *modified Kaplan-Meier* plotting positions, which are equivalent to taking mid-step positions of the Kaplan-Meier curve, rather than the bottom-of-step positions. A consequence of this is that there is an additional point in the plot. This option is recommended, so by default it is turned on.

Connect Quantile Points toggles the lines in the plot on and off. By default, this option is on.

Fitted Quantile toggles the straight-line fit on the fitted Weibull, lognormal, or Exponential Quantile plot.

Fitted Quantile CI Lines toggles the 95% confidence bands for the fitted Weibull, lognormal, or Exponential Quantile plot.

Fitted Quantile CI Shaded toggles the display of the 95% confidence bands for a fit as a shaded area or dashed lines.

Fitted Survival CI toggles the confidence intervals (on the survival plot) of the fitted distribution.

Fitted Failure CI toggles the confidence intervals (on the failure plot) of the fitted distribution.

Exponential Plot when checked, plots –log(Surv) by time for each group where Surv is the product-limit estimate of the survival distribution. Lines that are approximately linear empirically indicate the appropriateness of using an exponential model for further analysis. In Figure 31.11, the lines for Group 1 and Group 2 in the Exponential Plot are curved rather than straight, which indicates that the exponential distribution is not appropriate for this data.

Exponential Fit produces the Exponential Parameters table shown in Figure 31.11. The parameter Theta corresponds to the mean failure time.

Weibull Plot plots log(–log(Surv)) by log(time) for each group, where Surv is the product-limit estimate of the survival distribution. A Weibull plot that has approximately parallel and straight lines indicates a Weibull survival distribution model might be appropriate to use for further analysis.

For both the Weibull Fit and the Lognormal Fit, if you hold down the Shift key while selecting the command, the following dialog box appears.

JMP

- [x] Confidence Level for Limits, e.g. .95 — 0.95 — OK
- [ ] Enter constrained value for σ: — Cancel
- [ ] Confidence Contour Plot

Here you can set the confidence level for limits, a constrained value for sigma (see "WeiBayes Analysis," p. 621 for details on using this option), and request a Confidence Contour Plot.

Figure 31.10 Confidence Contours Plot

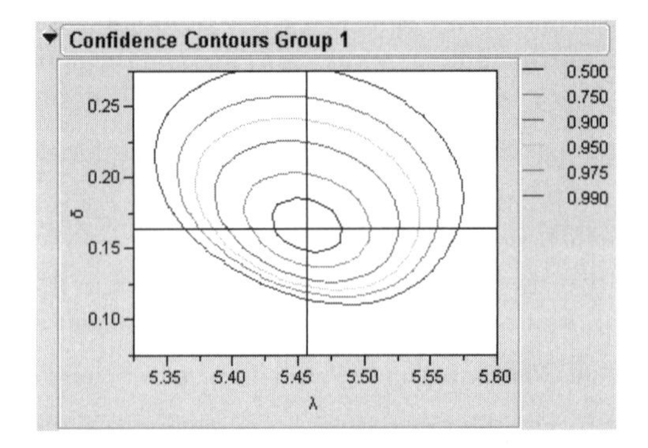

Weibull Fit produces the two popular forms of estimates shown in the Extreme Value Parameter Estimates table and the Weibull Parameter Estimates tables (Figure 31.11). The Alpha parameter is the 63.2 percentile of the failure-time distribution. The Extreme-value table shows a different parameterization of the same fit, where Lambda = ln(Alpha) and Delta = 1/Beta.

Lognormal Plot plots the inverse cumulative normal distribution of 1 minus the survival (denoted Probit(1-Surv)) by the log of time for each group, where Surv is the product-limit estimate of the survival distribution. A lognormal plot that has approximately parallel and straight lines indicates a lognormal distribution is appropriate to use for further analysis. Hold down the Shift key for more options, explained in the Weibull Plot section above.

Lognormal Fit produces the LogNormal Parameter Estimates table shown in Figure 31.11. Mu and Sigma correspond to the mean and standard deviation of a normally distributed natural logarithm of the time variable.

Fitted Distribution Plots is available in conjunction with the fitted distributions to show three plots corresponding to the fitted distributions: Survival, Density, and Hazard. No plot will appear if you haven't done a Fit command. An example is shown in the next section.

Competing Causes prompts for a column in the data table that contains labels for causes of failure. Then for each cause, the estimation of the Weibull model is performed using that cause to indicate a failure event and other causes to indicate censoring. The fitted distribution is shown by the dashed line in the Survival Plot.

Estimate Survival Probability brings up a dialog allowing you to enter up to ten times. The survival probabilities are estimated for the entered times.

Estimate Time Quantile brings up a dialog allowing you to enter up to ten probabilities. The time quantiles are estimated for each entered probability.

Save Estimates creates a JMP data table that has all the information in the Survival Estimates table.

Fitting Distributions

For each of the three distributions JMP supports, there is a plot command and a fit command. Use the plot command to see if the event markers seem to follow a straight line—which they will tend to do when the distribution is right. Then, use the fit commands to estimate the parameters.

Figure 31.11 Exponential, Weibull, and Lognormal Plots and Tables

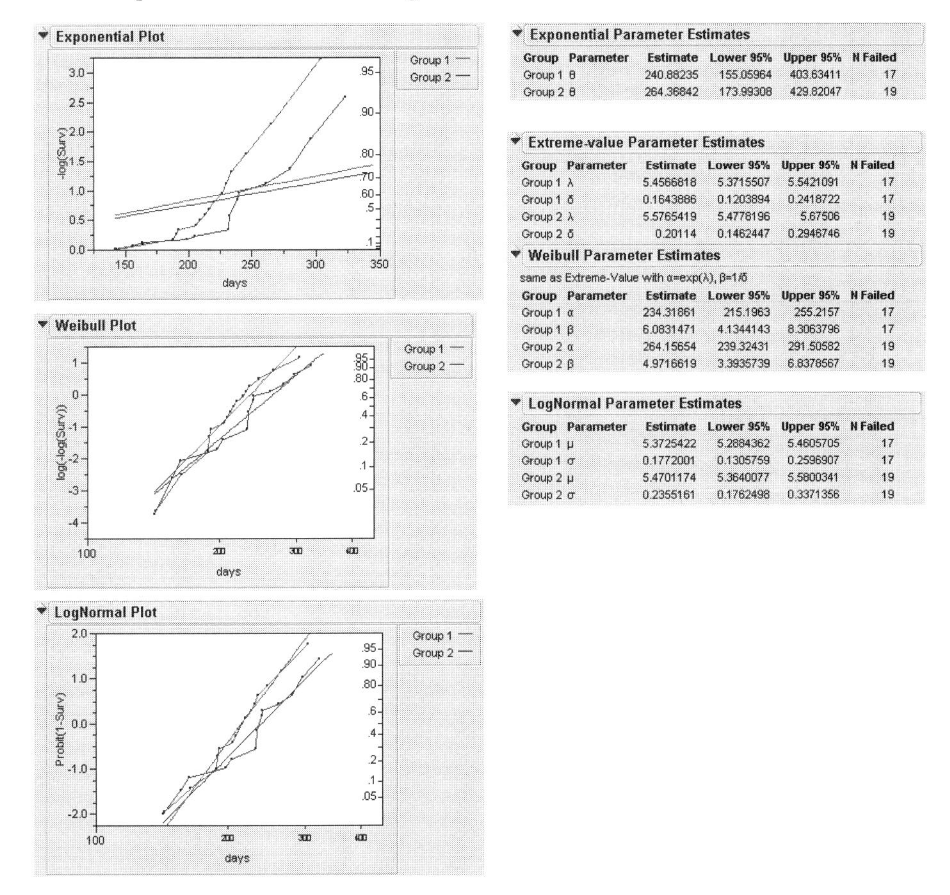

Exponential Parameter Estimates

Group	Parameter	Estimate	Lower 95%	Upper 95%	N Failed
Group 1	θ	240.88235	155.05964	403.63411	17
Group 2	θ	264.36842	173.99308	429.82047	19

Extreme-value Parameter Estimates

Group	Parameter	Estimate	Lower 95%	Upper 95%	N Failed
Group 1	λ	5.4566818	5.3715507	5.5421091	17
Group 1	δ	0.1643886	0.1203894	0.2418722	17
Group 2	λ	5.5765419	5.4778196	5.67506	19
Group 2	δ	0.20114	0.1462447	0.2946746	19

Weibull Parameter Estimates

same as Extreme-Value with α=exp(λ), β=1/δ

Group	Parameter	Estimate	Lower 95%	Upper 95%	N Failed
Group 1	α	234.31861	215.1963	255.2157	17
Group 1	β	6.0831471	4.1344143	8.3063796	17
Group 2	α	264.15654	239.32431	291.50582	19
Group 2	β	4.9716619	3.3935739	6.8378567	19

LogNormal Parameter Estimates

Group	Parameter	Estimate	Lower 95%	Upper 95%	N Failed
Group 1	μ	5.3725422	5.2884362	5.4605705	17
Group 1	σ	0.1772001	0.1305759	0.2596907	17
Group 2	μ	5.4701174	5.3640077	5.5800341	19
Group 2	σ	0.2355161	0.1762498	0.3371356	19

The following table shows what to plot that makes a straight line fit for that distribution:

Distribution Plot	X Axis	YAxis	Interpretation
Exponential	time	-log(S)	slope is 1/theta
Weibull	log(time)	log(-log(S))	slope is beta
Log-Normal	log(time)	Probit(1-S)	slope is 1/sigma

The exponential distribution is the simplest, with only one parameter, which we call theta. It is a constant-hazard distribution, with no memory of how long it has survived to affect how likely an event is. The parameter theta is the expected lifetime.

The Weibull distribution is the most popular for event-time data. There are many ways in which different authors parameterize the distribution (as shown in Table 31.1 "Various Weibull parameters in terms of JMP's alpha and beta," p. 619). JMP reports two parameterizations—labeled the lambda-delta extreme value parameterization, and the Weibull alpha-beta parameterization. The alpha-beta parameterization is used in the Reliability literature (Nelson 1990). Alpha has an interpretation of the quantile at which 63.2% of the units fail. Beta is interpreted as whether the hazard increases (beta>1) or decreases (beta<1) with time, with beta=1 being a constant hazard exponential distribution.

The Log-Normal distribution is also very popular. It is the distribution in which if you take the log of the values, the distribution would be normal. If you want to fit data to a normal distribution, you can take the exp() of it and analyze it as log-normal, as is done later in the Tobit example.

The option **Fitted Distribution Plots** will show the fitted distributions. Here you see 9 plots, the Survival, Density, and Hazard plots for the three distributions, Exponential, Weibull, and LogNormal. The plots share the same axis scaling so that the distributions can be compared easily.

Figure 31.12 Fitted Distribution Plots for Three Distributions.

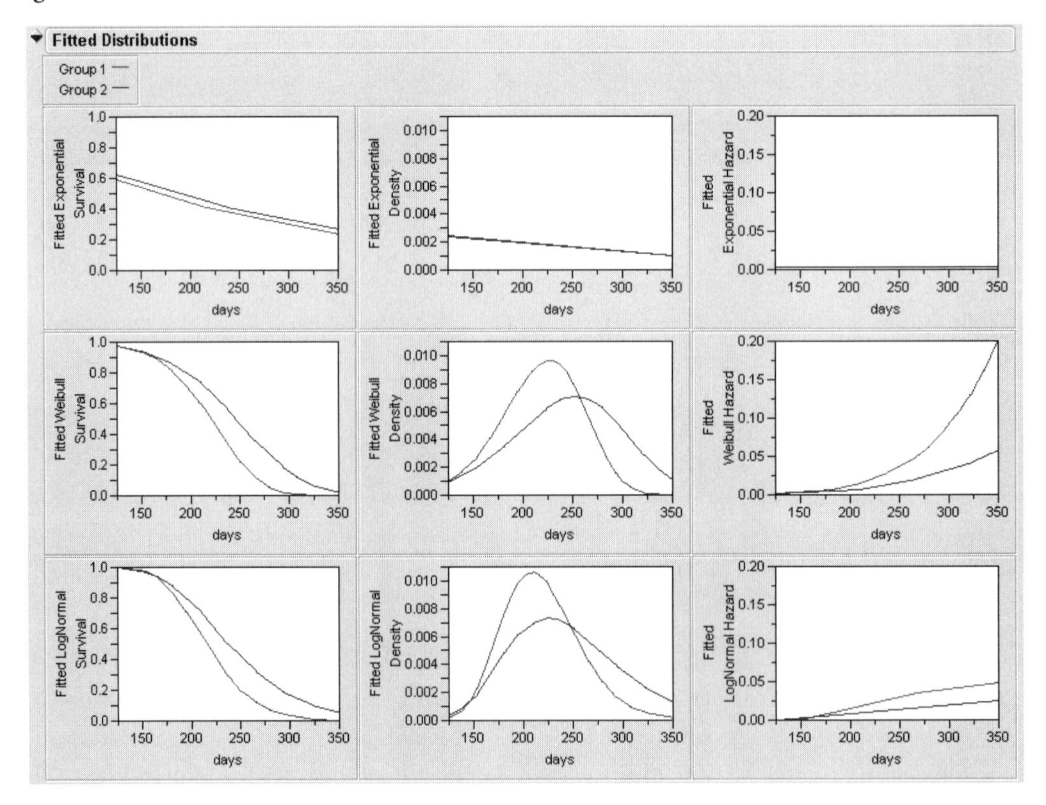

These plots can be transferred to other graphs through the use of graphics scripts. To copy the graph, context-click on the plot to be copied and select **Edit Graphics Script**. Highlight the desired script and copy it to the clipboard. On the destination plot, context-click and select **Add Graphics Script**. Paste the script from the clipboard into the window that results.

Table 31.1 Various Weibull parameters in terms of JMP's *alpha* and *beta*

JMP Weibull	*alpha*	*beta*
Wayne Nelson	alpha=*alpha*	beta=*beta*
Meeker and Escobar	eta=*alpha*	beta=*beta*
Tobias and Trindade	c = *alpha*	m = *beta*
Kececioglu	eta=*alpha*	beta=*beta*
Hosmer and Lemeshow	exp(X beta)=*alpha*	lambda=*beta*
Blishke and Murthy	beta=*alpha*	alpha=*beta*
Kalbfleisch and Prentice	lambda = 1/*alpha*	p = *beta*
JMP Extreme Value	lambda=log(*alpha*)	delta=1/*beta*
Meeker and Escobar s.e.v.	mu=log(*alpha*)	sigma=1/*beta*

Interval Censoring

With interval censored data, you only know that the events occurred in some time interval. The Turnbull method is used to obtain non-parametric estimates of the survival function.

In this example (Nelson 1990 p. 147), microprocessor units are tested and inspected at various times and the failed units counted. Missing values in one of the columns indicate that you don't know the lower or upper limit, and therefore the event is left or right censored, respectively. The data is shown in Figure 31.13.

Figure 31.13 Microprocessor Data

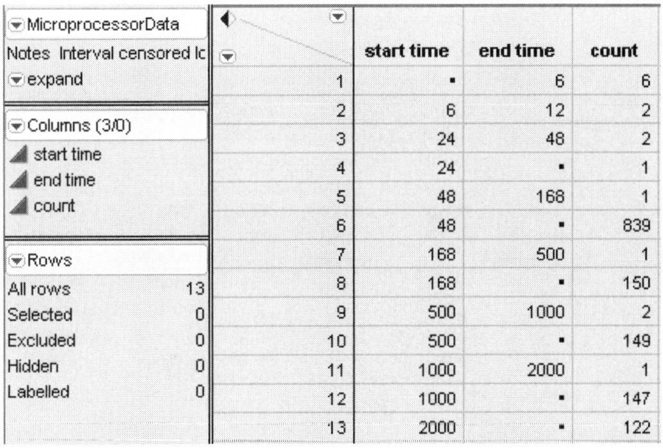

MicroprocessorData
Notes Interval censored lc
expand
Columns (3/0)
start time
end time
count
Rows
All rows 13
Selected 0
Excluded 0
Hidden 0
Labelled 0

	start time	end time	count
1	•	6	6
2	6	12	2
3	24	48	2
4	24	•	1
5	48	168	1
6	48	•	839
7	168	500	1
8	168	•	150
9	500	1000	2
10	500	•	149
11	1000	2000	1
12	1000	•	147
13	2000	•	122

When you launch the Survival platform, specify the lower and upper time limits as two Y columns, as shown in Figure 31.14.

Figure 31.14 Interval Censoring Launch Dialog

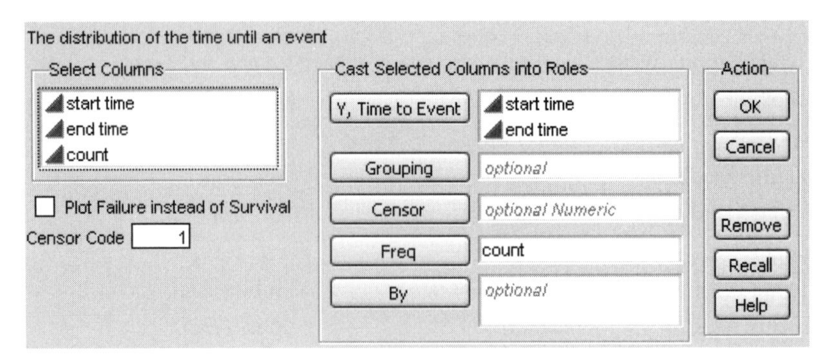

The resulting Turnbull estimates are shown. Turnbull estimates may have gaps in time where the survival probability is not estimable, as seen here between, for example, 12 and 24.

At this point, select a distribution to see its fitted estimates —in this case, a Lognormal distribution is fit. The failure plot has been resized to be smaller, since the plot does not show much with very small failure rates.

Figure 31.15 Interval Censoring Output

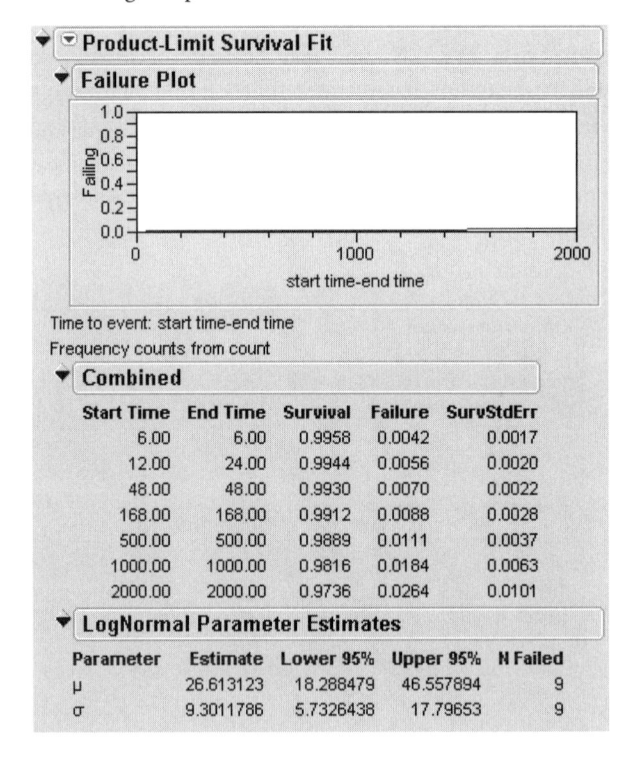

Start Time	End Time	Survival	Failure	SurvStdErr
6.00	6.00	0.9958	0.0042	0.0017
12.00	24.00	0.9944	0.0056	0.0020
48.00	48.00	0.9930	0.0070	0.0022
168.00	168.00	0.9912	0.0088	0.0028
500.00	500.00	0.9889	0.0111	0.0037
1000.00	1000.00	0.9816	0.0184	0.0063
2000.00	2000.00	0.9736	0.0264	0.0101

Parameter	Estimate	Lower 95%	Upper 95%	N Failed
μ	26.613123	18.288479	46.557894	9
σ	9.3011786	5.7326438	17.79653	9

WeiBayes Analysis

JMP can now constrain the values of the Beta (Weibull) and Sigma (LogNormal) parameters when fitting these distributions. This feature is needed in *WeiBayes* situations (Abernethy, 1996), such as

- where there are few or no failures,
- there are existing historical values for beta, and
- there is still a need to estimate alpha.

With no failures, the standard technique is to add a failure at the end and the estimates would reflect a kind of lower bound on what the alpha value would be, rather than a real estimate. This feature allows for a true estimation.

To use this feature, hold down the Shift key when selecting **Weibull** (or **LogNormal**) **Fit**, and enter a value for the parameter as prompted (for example, beta or sigma, respectively).

Estimation of Competing Causes

Sometimes there are multiple causes of failure in a system. For example, suppose that a manufacturing process has several stages and the failure of any stage causes a failure of the whole system. If the different causes are independent, the failure times can be modeled by an estimation of the survival distribution for each cause. A censored estimation is undertaken for a given cause by treating all the event times that are not from that cause as censored observations.

Appliance
Notes From Wayne Nelson's
Survival
Columns (3/0)
Group
Time Cycles
Cause Code
Rows
All rows 36

	Group	Time Cycles	Cause Code
1	2	11	1
2	2	2223	9
3	2	4329	9
4	2	3112	9
5	2	13403	0
6	2	6367	0
7	2	2451	5
8	2	381	6
9	2	1062	5

Nelson (1982) discusses the failure times of a small electrical appliance that has a number of causes of failure. One group (Group 2) of the data is in the JMP data table **Appliance.jmp** sample data, in the **Reliability** subfolder. The table above shows a partial listing of the appliance failure data.

To specify the analysis you only need to enter the time variable (**Time Cycles**) in the Survival dialog. Then use the **Competing Causes** menu command, which prompts you to choose a column in the data table to label the causes of failure. For this example choose **Cause Code** as the label variable.

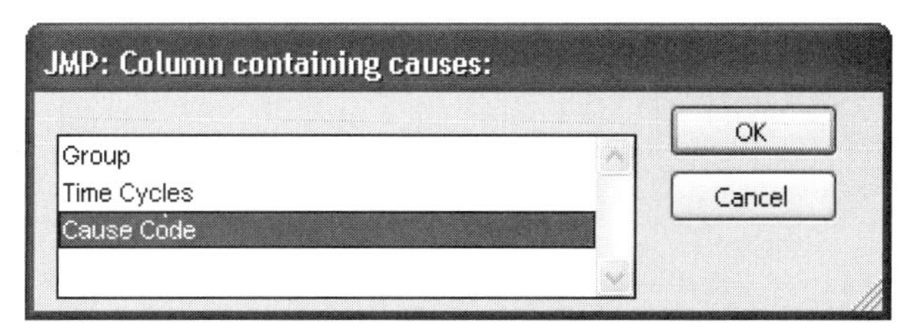

The survival distribution for the whole system is just the product of the survival probabilities. The Competing Causes table gives the estimates of Alpha and Beta for each failure cause. It is shown with the hazard plot in Figure 31.16.

In this example, most of the failures were due to cause 9. Cause 1 only occurred once and couldn't produce good Weibull estimates. Cause 15 only happened for very short times and resulted in a small beta and large alpha. Remembering that alpha is the estimate of the 63.2% quantile of failure time, causes of early failures often have very large alphas; if they don't cause an early failure, they tend to not cause later failures.

Figure 31.16 Competing Causes Plot and Table

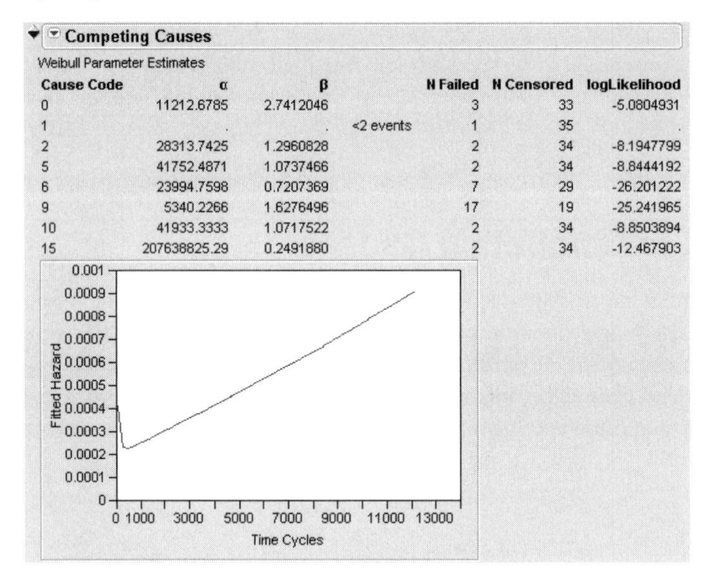

Competing Causes

Weibull Parameter Estimates

Cause Code	α	β		N Failed	N Censored	logLikelihood
0	11212.6785	2.7412046		3	33	-5.0804931
1	.	.	<2 events	1	35	.
2	28313.7425	1.2960828		2	34	-8.1947799
5	41752.4871	1.0737466		2	34	-8.8444192
6	23994.7598	0.7207369		7	29	-26.201222
9	5340.2266	1.6276496		17	19	-25.241965
10	41933.3333	1.0717522		2	34	-8.8503894
15	207638825.29	0.2491880		2	34	-12.467903

Figure 31.17 shows the (Fit Y by X) plot of Time Cycles by Cause Code with the quantiles option in effect. This plot gives an idea of how the alphas and betas relate to the failure distribution.

Figure 31.17 Fit Y by X Plot of Time Cycles by Cause Code with Box Plots

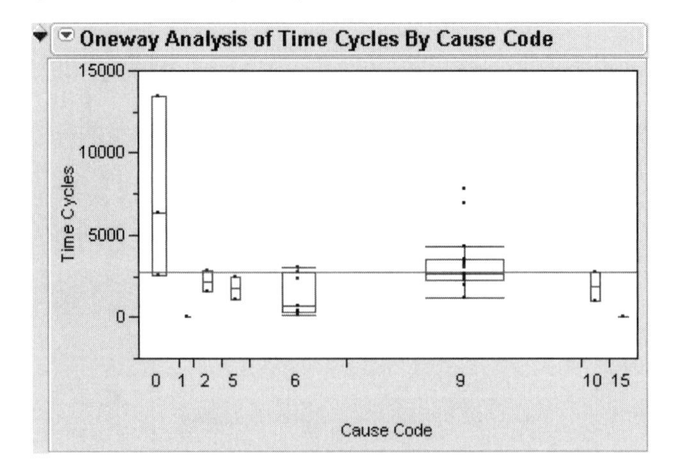

Omitting Causes

If cause 9 was corrected, how would that affect the survival due to the remaining causes?

The popup menu icon on the Competing Causes title bar accesses the menu shown here. The **Omit Causes** command prompts you for one or more cause values to omit and then calculates competing causes again.

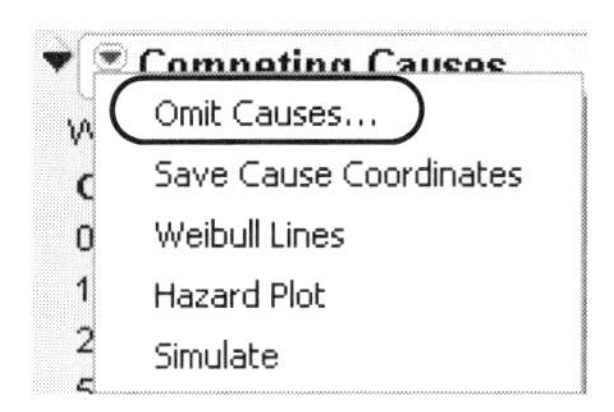

In this example, the competing risk without cause 9 gives survival estimates and the revised survival plot shown first in Figure 31.18. Note that the survival rate (as shown by the dashed line) doesn't improve much until 2,000 cycles, but then becomes much better even out as far as 10,000 cycles.

Figure 31.18 Survival Plots with Omitted Causes

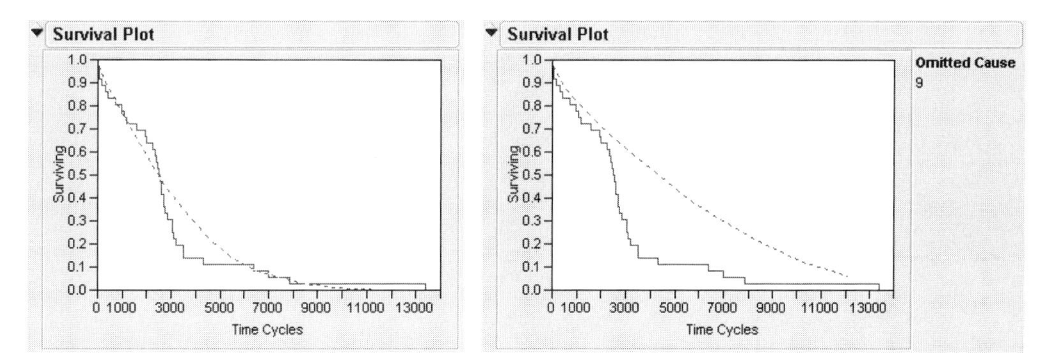

Saving Competing Causes Information

The Competing Causes table popup menu has commands to save estimates and to save fail-by-cause coordinates.

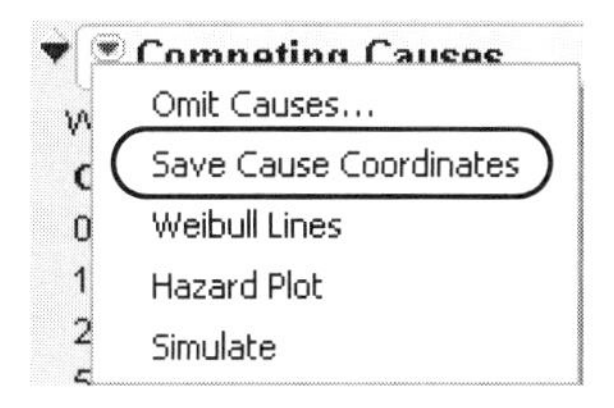

The **Save Cause Coordinates** command adds a new column to the current table called log(–log(Surv)). This information is often used to plot against the time variable, with a grouping variable, such as the code for type of failure.

Simulating Time and Cause Data

The Competing Causes table popup menu contains the **Simulate** command.

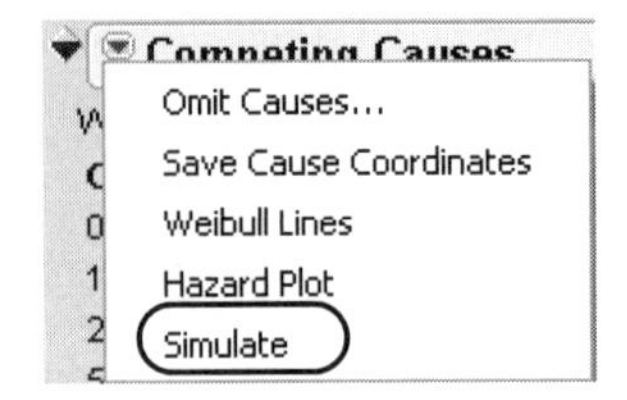

This command creates a new data table with Time and Cause data from the weibull distribution as estimated by the data.

Chapter 32

Survival and Reliability Analysis II

Regression Fitting

This chapter focuses on the second and third entries in the **Survival and Reliability** submenu.

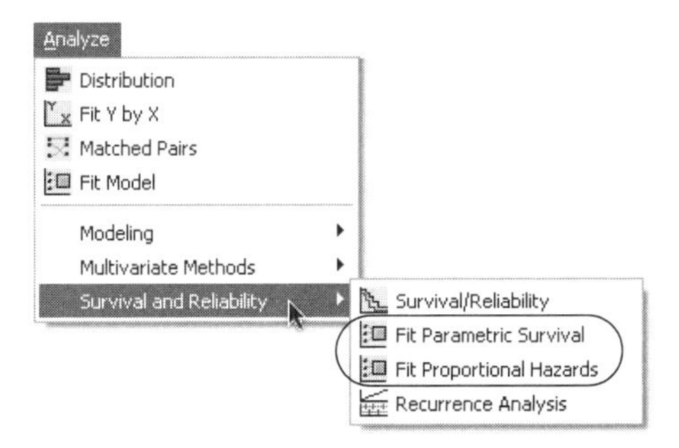

Fit Parametric Survival launches a regression platform that fits a survival distribution scaled to a linear model. The distributions to choose from are exponential, Weibull, and lognormal. The regression uses an iterative maximum likelihood method. Accelerated Failure models are one kind of regression model. This chapter also shows how to fit parametric models using the Nonlinear platform.

Fit Proportional Hazards launches a regression platform that uses Cox's proportional hazards method to fit a linear model. Proportional hazards models (Cox 1972) are popular regression models for survival data with covariates. This model is semi parametric; the linear model is estimated, but the form of the hazard function is not. Time-varying covariates are not supported.

Analyze > Fit Model also accesses the Parametric Regression and Proportional Hazard survival techniques as fitting personalities in the Fit Model dialog.

Contents

Parametric Regression Survival Fitting

If you have survival times in which the hazard rate is proportional to a function of one or more variables you need a regression platform that fits a linear regression model but takes into account the survival distribution and censoring. You can do this kind of analysis with the **Fit Parametric Survival** command in the **Survival** submenu, or use the Fit Model fitting personality called **Fit Parametric Survival**.

Example: Computer Program Execution Time

The data table comptime.jmp, from Meeker and Escobar (p. 434), is data on the analysis of computer program execution time whose log-normal distribution depends on the regressor Load. It is found in the Reliability subfolder of the sample data.

Figure 32.1 Comptime.jmp Data

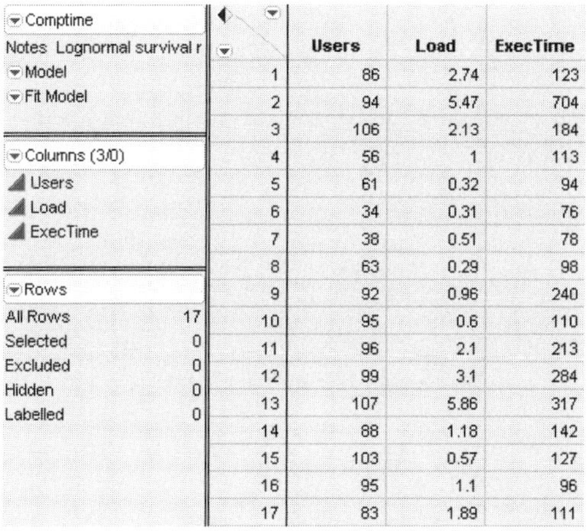

Comptime
Notes Lognormal survival r
Model
Fit Model

Columns (3/0)
Users
Load
ExecTime

Rows

All Rows	17
Selected	0
Excluded	0
Hidden	0
Labelled	0

	Users	Load	ExecTime
1	86	2.74	123
2	94	5.47	704
3	106	2.13	184
4	56	1	113
5	61	0.32	94
6	34	0.31	76
7	38	0.51	78
8	63	0.29	98
9	92	0.96	240
10	95	0.6	110
11	96	2.1	213
12	99	3.1	284
13	107	5.86	317
14	88	1.18	142
15	103	0.57	127
16	95	1.1	96
17	83	1.89	111

To begin the analysis, select **Analyze > Survival and Reliability > Fit Parametric Survival**. When the launch dialog appears, select ExecTime as the **Time to Event** and add Load as an Effect in the model. Also, change the **Distrib** from the default Weibull to LogNormal. The completed dialog should appear as in Figure 32.2.

Figure 32.2 Computing Time Dialog

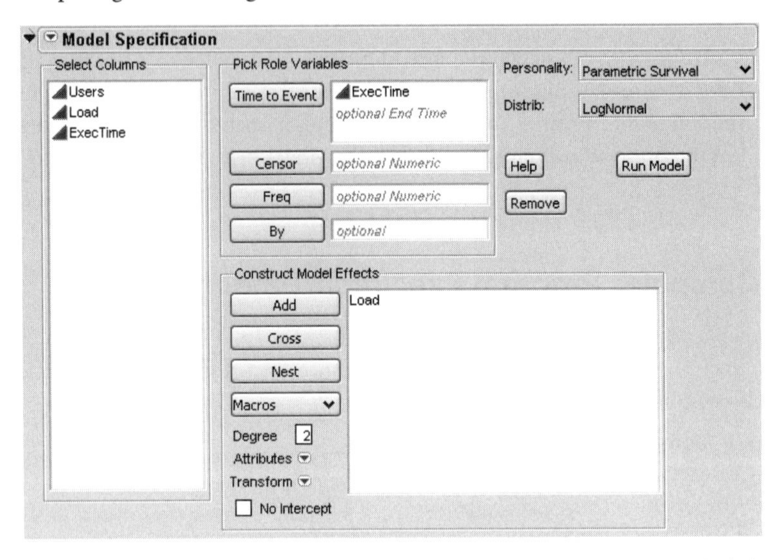

When there is only one regressor, a plot of the survival curves for three survival probabilities are shown as a function of the regressor.

Figure 32.3 Computing Time Output

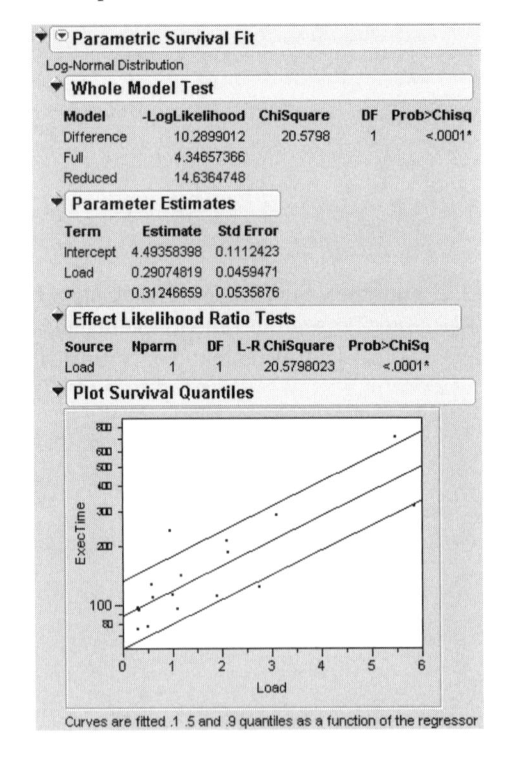
Parametric Survival Fit
Log-Normal Distribution

Whole Model Test

Model	-LogLikelihood	ChiSquare	DF	Prob>Chisq
Difference	10.2899012	20.5798	1	<.0001*
Full	4.34657366			
Reduced	14.6364748			

Parameter Estimates

Term	Estimate	Std Error
Intercept	4.49358398	0.1112423
Load	0.29074819	0.0459471
σ	0.31246659	0.0535876

Effect Likelihood Ratio Tests

Source	Nparm	DF	L-R ChiSquare	Prob>ChiSq
Load	1	1	20.5798023	<.0001*

Curves are fitted .1 .5 and .9 quantiles as a function of the regressor

Time quantiles, as described on page 438 of Meeker and Escobar, are desired for when 90% of jobs are finished under a system load of 5. Select the **Estimate Time Quantile** command, which brings up a dialog as shown in Figure 32.4. Enter 5 as the Load, and 0.1 as the **Survival Prob**.

Figure 32.4 **Estimate Time Quantile** Dialog

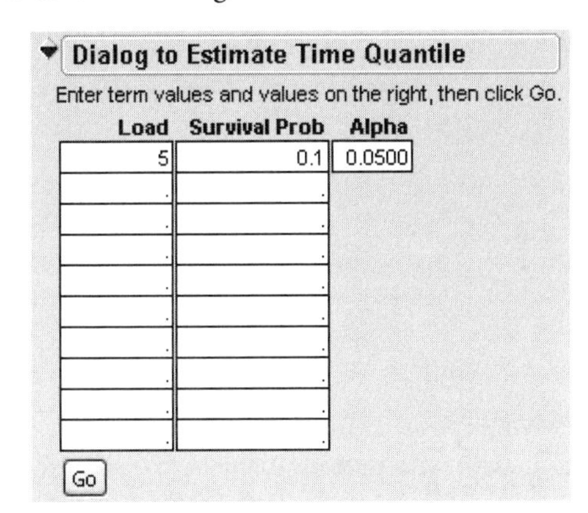

Click **Go** to produce the quantile estimates and a confidence interval.

Figure 32.5 Estimates of Time Quantile

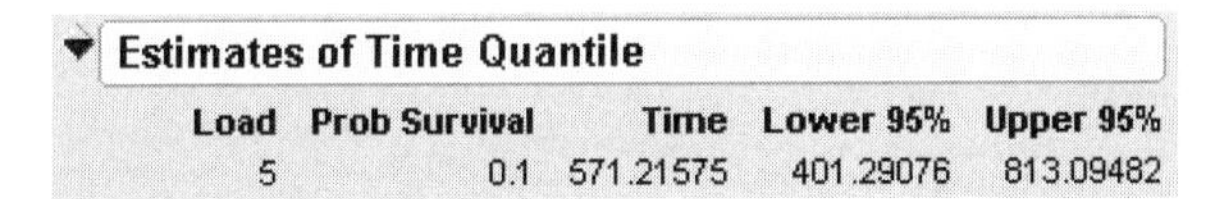

Estimates of Time Quantile

Load	Prob Survival	Time	Lower 95%	Upper 95%
5	0.1	571.21575	401.29076	813.09482

This estimates that 90% of the jobs will be done by 571 seconds of execution time under a system load of 5.

Options and Reports

As an example to illustrate the components of this platform, use the VA Lung Cancer.jmp table from the Sample Data folder (Kalbfleisch and Prentice 1980). The response, Time, is the survival time in days of a group of lung cancer patients. The regressors are age in years (Age) and time in months from diagnosis to entry into the trial (Diag Time). Censor = 1 indicates censored values.

VA Lung Cancer
Notes Nonparamteric survi

Columns (12/0)
Time
Cell Type
Treatment
Prior
Age
Diag Time
KPS
censor
Model
Weibull loss
LogNormal loss
Exponential

Rows
All Rows 137

	Time	Cell Type	Treatment	Prior	Age	Diag Time	KPS	censor
1	3	Adeno	Standard	No	43	3	30	0
2	7	Adeno	Test	No	58	4	40	0
3	8	Adeno	Standard	Yes	61	19	20	0
4	8	Adeno	Test	No	66	5	50	0
5	12	Adeno	Standard	Yes	63	4	50	0
6	18	Adeno	Test	Yes	69	5	40	0
7	19	Adeno	Test	No	42	10	50	0
8	24	Adeno	Test	No	60	2	40	0
9	31	Adeno	Test	No	39	3	80	0
10	35	Adeno	Standard	No	62	6	40	0
11	36	Adeno	Test	No	61	8	70	0
12	45	Adeno	Test	No	69	3	40	0
13	48	Adeno	Test	No	81	4	10	0
14	51	Adeno	Test	No	62	5	60	0
15	52	Adeno	Test	No	43	3	60	0
16	73	Adeno	Test	No	70	3	60	0
17	80	Adeno	Test	No	63	4	40	0

The **Fit Parametric Survival** survival command launches the Fit Model dialog shown in Figure 32.6. It is specialized for survival analysis with buttons that label the time (**Time to Event**) and censor (**Censor**) variables. Also, the fitting personality shows as **Parametric Survival** and an additional popup menu lets you choose the type of survival distribution (**Weibull**, **LogNormal**, or **Exponential**) that you think is appropriate for your data.

Figure 32.6 Fit Model Dialog for Parametric Survival Analysis

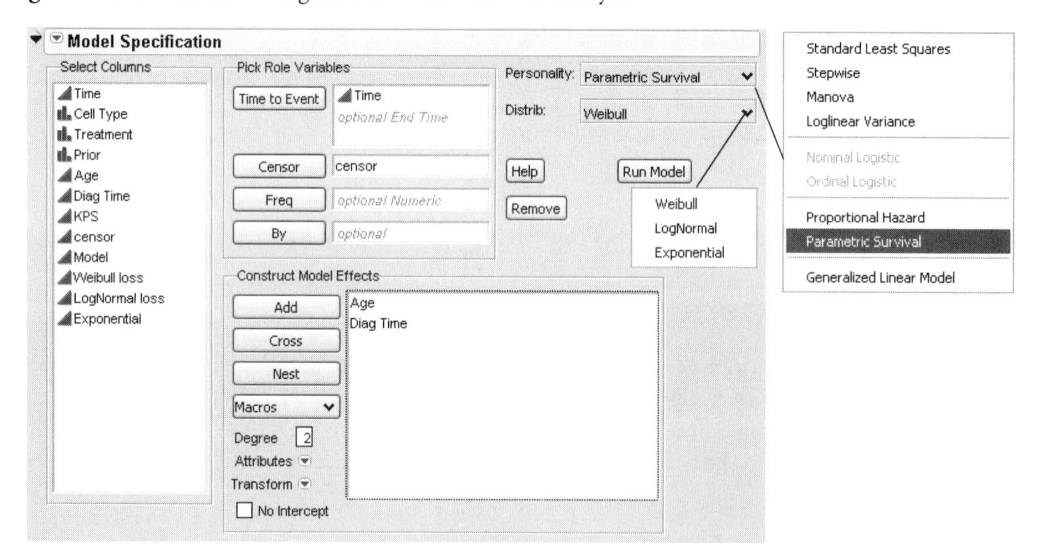

When you click **Run Model** on the Model Specification dialog, the survival analysis report shows the Whole Model Test and Parameter Estimates. The Whole Model test compares the complete fit with what would have been fit if no regressor variables were in the model and only an intercept term were fit.

The case of only having an intercept term yields a fit that is the same as that from the univariate survival platform. Delta (or beta) is the slope parameter of the Weibull distribution. The other Weibull parameter (lambda or alpha) is equivalent to the whole linear model.

The following parametric survival options (see Figure 32.7) are available:

Likelihood Ratio Tests produces tests that compare the log-likelihood from the fitted model to one that removes each term from the model individually. The likelihood-ratio test is appended to the text reports.

Confidence Intervals calculates a profile-likelihood 95% confidence interval on each parameter and lists them in the Parameter Estimates table.

Correlation of Estimates produces a correlation matrix for the model effects with each other and with the parameter of the fitting distribution (Weibull or LogNormal).

Covariance of Estimates produces a covariance matrix for the model effects with each other and with the parameter of the fitting distribution (Weibull or LogNormal).

Estimate Survival Probability brings up a dialog where you specify regressor values and one or more time values. JMP then calculates the survival and failure probabilities with 95% confidence limits for all possible combinations of the entries.

Estimate Time Quantile brings up a dialog where you specify regressor values and one or more survival values. It then calculates the time quantiles and 95% confidence limits for all possible combinations of the entries.

Residual Quantile Plot shows a plot with the residuals on the *x*-axis and the Kaplan-Meier estimated quantiles on the *y*-axis. In cases of interval censoring, the midpoint is used. The residuals are the simplest form of Cox-Snell residuals, which convert event times to a censored standard Weibull or other standard distribution.

Save Residuals creates a new column to hold the residuals.

Figure 32.7 Parametric Survival Model Reports

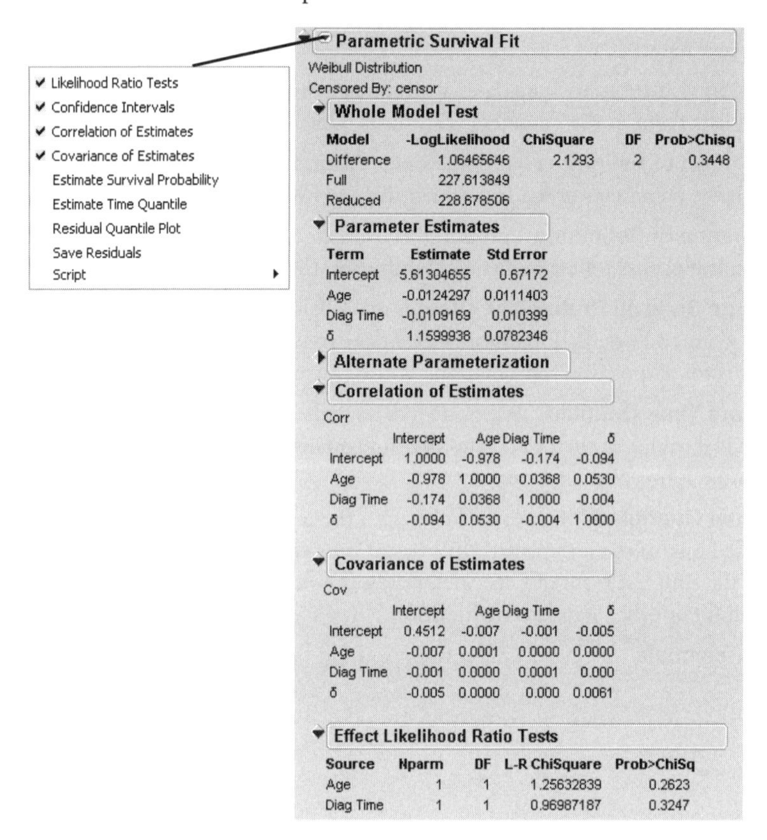

See the section "Nonlinear Parametric Survival Models," p. 641, for further details on the statistical model.

Example: Arrhenius Accelerated Failure Log-Normal Model

The Devalt.jmp data set (also known as the Device A data) is described by Meeker and Escobar (1998 p. 493) as originating from Hooper and Amster (1990). The data set is found in the Reliability sample data folder. A partial listing of the data set is in Figure 32.8.

Figure 32.8 Devalt Data Table

Devalt
Notes Meeker and Escobar
Bivariate
Survival
Model
Fit Model

Columns (6/0)
Hours
Status
Weight
Temp
Censor
x

Rows
All Rows 37

		Hours	Status	Weight	Temp	Censor	x
z	1	5000	Censored	30	10	1	40.9853435
	2	1298	Failed	1	40	0	37.0589175
	3	1390	Failed	1	40	0	37.0589175
	4	3187	Failed	1	40	0	37.0589175
	5	3241	Failed	1	40	0	37.0589175
	6	3261	Failed	1	40	0	37.0589175
	7	3313	Failed	1	40	0	37.0589175
	8	4501	Failed	1	40	0	37.0589175
	9	4568	Failed	1	40	0	37.0589175
	10	4841	Failed	1	40	0	37.0589175
	11	4982	Failed	1	40	0	37.0589175
z	12	5000	Censored	90	40	1	37.0589175
	13	581	Failed	1	60	0	34.8341588

Units are stressed by heating in order to make them fail soon enough to obtain enough failures to fit the distribution.

Use the Bivariate platform to see a plot of hours by temperature using the log scale for time. [figure 19.1 in Meeker and Escobar]. This can be accomplished manually (**Analyze > Fit Y by X** with Hours as Y and Temp as X) or by running the Bivariate script on the left side of the data table.

Figure 32.9 Bivariate Plot of Hours by log Temp

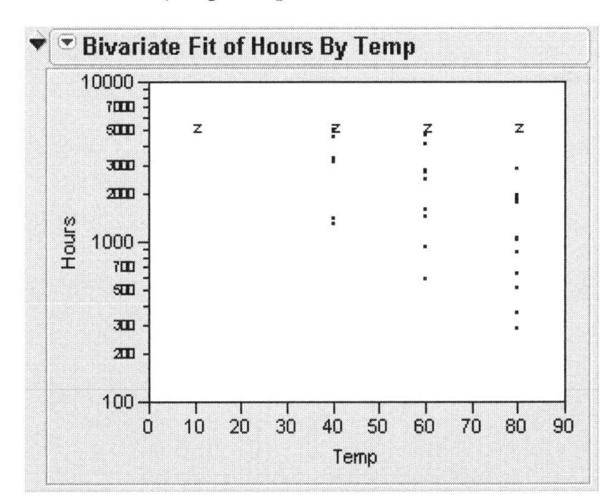

Use the univariate survival platform to produce a LogNormal plot of the data for each temperature. To do so, select **Analyze > Survival and Reliability > Survival/Reliability** with Hours as **Y**, Censor as **Censor**, Temp as **grouping**, and Weight as **Freq**. Alternatively, run the Survival script attached to the data table. Either method produces the plot shown in Figure 32.10.

Figure 32.10 Lognormal Plot

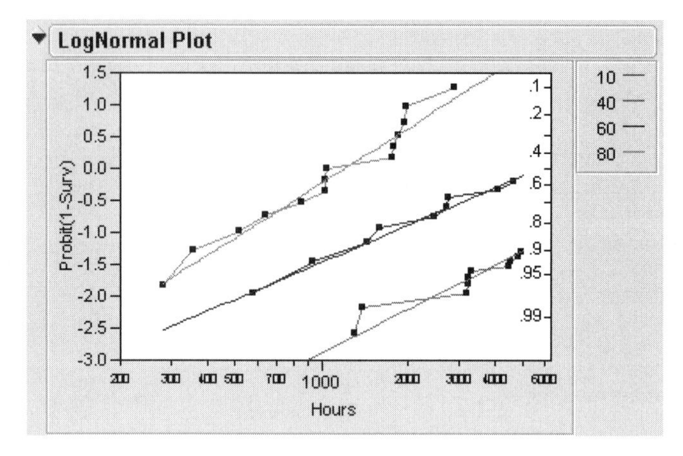

Then, fit one model using a regressor for temperature. The regressor x, also known as the *activation energy*, is the Arrhenius transformation of temperature calculated by the formula stored in the X column of the data table:

$$\frac{11605}{(Temp+273.15)}$$

Hours is the failure or censoring time. The log-normal distribution is fit to the distribution of this time. To do so, select **Analyze > Survival and Reliability > Fit Parametric Survival**, assign Hours as **Time to Event** , x as a model effect, Censor as **Censor** and Weight as **Freq**. Also, change the Distribution type to **LogNormal**.

After clicking **Run Model**, the result shows the regression fit of the data. If there is only one regressor and it is continuous, then a plot of the survival as a function of the regressor is shown, with lines at 0.1, 0.5, and 0.9 survival probabilities. If the regressor column has a formula in terms of one other column, as in this case, the plot is done with respect to the inner column. In this case the regressor was the column x, but the plot is done with respect to Temp, which x is a function of.

Figure 32.11 Devalt Parametric Output

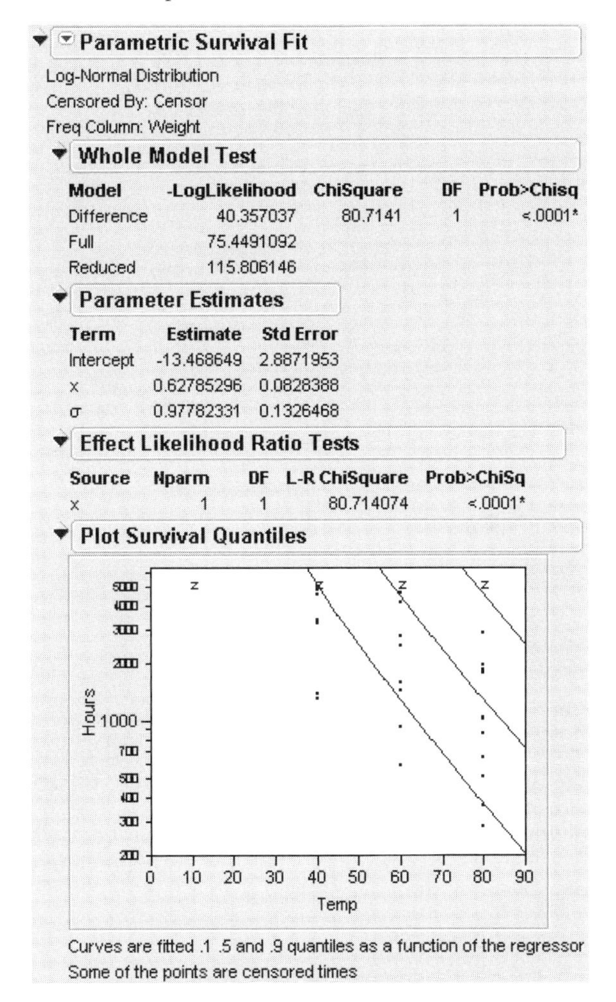
Parametric Survival Fit

Log-Normal Distribution
Censored By: Censor
Freq Column: Weight

Whole Model Test

Model	-LogLikelihood	ChiSquare	DF	Prob>Chisq
Difference	40.357037	80.7141	1	<.0001*
Full	75.4491092			
Reduced	115.806146			

Parameter Estimates

Term	Estimate	Std Error
Intercept	-13.468649	2.8871953
x	0.62785296	0.0828388
σ	0.97782331	0.1326468

Effect Likelihood Ratio Tests

Source	Nparm	DF	L-R ChiSquare	Prob>ChiSq
x	1	1	80.714074	<.0001*

Plot Survival Quantiles

Finally, we illustrate how to get estimates of survival probabilities extrapolated to a temperature of 10 degrees celsius for the times 10000 and 30000 hours. Select the **Estimate Survival Probabillity** command, and enter the following values into the dialog. The Arrhenius transformation of 10 degrees is 40.9853, the regressor value.

Figure 32.12 Estimating Survival Probabilities

Dialog to Estimate Survival

Enter term values and values on the right, then click Go.

x	Time	Alpha
40.9853	30000	0.0500
.	10000	
.	.	
.	.	
.	.	
.	.	
.	.	
.	.	
.	.	
.	.	

Go

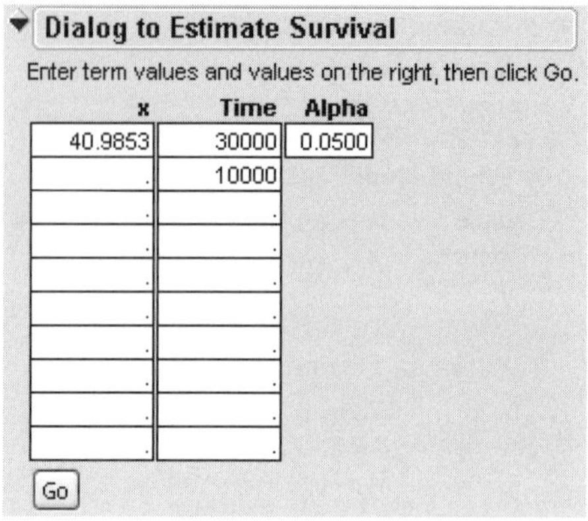

After clicking **Go**, the report shows the estimates and a confidence interval, as in Meeker and Escobar's Example 19.8.

Figure 32.13 Survival Probabilities

Estimates of Survival

x	Time	Prob Failure	Std Error	Lower 95%	Upper 95%	Prob Survival
40.9853	30000	0.02278	0.02251	0.00320	0.14473	0.97722
40.9853	10000	0.00090	0.00124	0.00006	0.01347	0.99910

Example: Interval-Censored Accelerated Failure Time Model

Continuing with another example from Meeker and Escobar (1998 p 508), icdevice02.jmp shows data in which failures were found to have happened between inspection intervals. The data, found in the Reliability sample data folder, is illustrated in Figure 32.14.

Figure 32.14 ICDevice02 Data

ICdevice02
Notes Interval censor
Survival
Model
Columns (6/0)
HoursL
HoursU
Status
Count
DegreesC
x
Rows
All rows 11
Selected 0

	HoursL	HoursU	Status	Count	DegreesC	x
1	1536	•	Right	50	150	27.4252629
2	1536	•	Right	50	175	25.8953475
3	96	•	Right	50	200	24.5271056
4	384	788	Interval	1	250	22.1829303
5	788	1536	Interval	3	250	22.1829303
6	1536	2304	Interval	5	250	22.1829303
7	2304	•	Right	41	250	22.1829303
8	192	384	Interval	4	300	20.2477536
9	384	788	Interval	27	300	20.2477536
10	788	1536	Interval	16	300	20.2477536
11	1536	•	Right	3	300	20.2477536

The model uses two *y*-variables, containing the upper and lower bounds on the failure times. Right-censored times are shown with missing upper bounds. To perform the analysis, select **Analyze >**

Survival and Reliability > Fit Parametric Survival with both HoursL and HoursU as **Time to Event**, Count as **Freq**, and DegreesC as an effect in the model. The resulting regression has a plot of time by degrees.

Figure 32.15 ICDevice Output

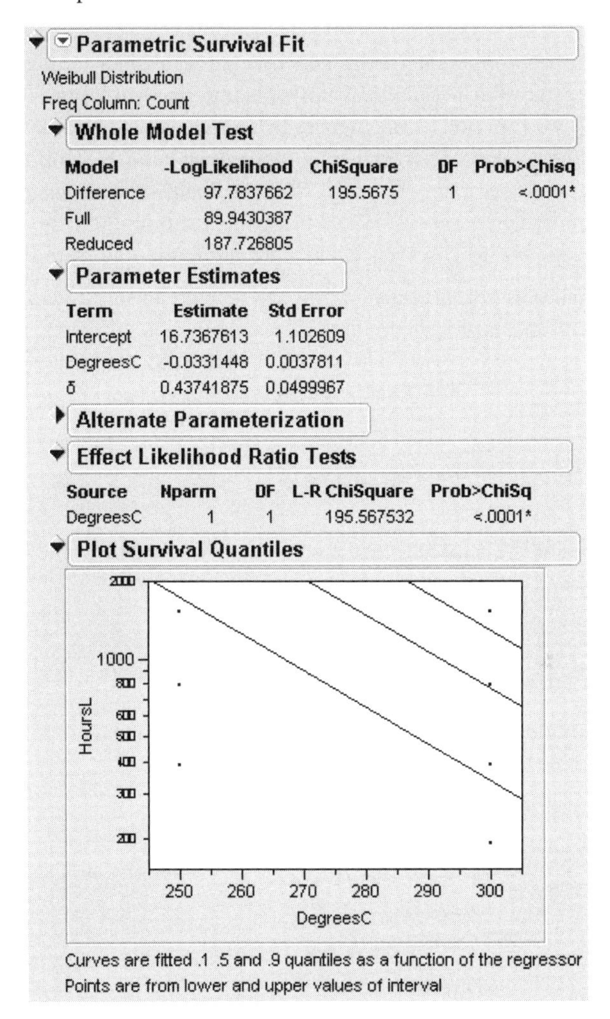

Example: Right-Censored Data; Tobit Model

The Tobit model is normal, truncated at zero. However, you can take the exponential of the response and set up the intervals for a right-censored problem. In Tobit2.jmp, found in the Reliability folder of the sample data, run the attached script to estimate the lognormal.

Proportional Hazards Model

The proportional hazards model is a special semi-parametric regression model proposed by D. R. Cox (1972) to examine the effect of explanatory variables on survival times. The survival time of each member of a population is assumed to follow its own hazard function.

Proportional hazards is nonparametric in the sense that it involves an unspecified arbitrary baseline hazard function. It is parametric because it assumes parametric form for the covariates. The baseline hazard function is scaled by a function of the model's (time-independent) covariates to give a general hazard function. Unlike the Kaplan-Meier analysis, proportional hazards computes parameter estimates and standard errors for each covariate. The regression parameters (β) associated with the explanatory variables and their standard errors are estimated using the maximum likelihood method. A conditional risk ratio (or hazard ratio) and its confidence limits are also computed from the parameter estimates.

The survival estimates in proportional hazards are generated using an empirical method (Lawless, 1982) and represent the empirical cumulative hazard function estimates, $H(t)$, of the survivor function, $S(t)$, and can be written as $S_0 = \exp(-H(t))$, with the hazard function

$$\mathrm{H}(t) = \sum_{j:t_j < t} \frac{d_j}{\sum_{l \in R_j} e^{x_l \hat{\beta}}}$$

When there are ties in the response, that is when there is more than one failure at a given time event, the Breslow likelihood is used.

Example: Rats Data

The following example uses the Rats.jmp sample data. To define a proportional hazard survival model, choose **Proportional Hazards** from the **Survival** submenu.

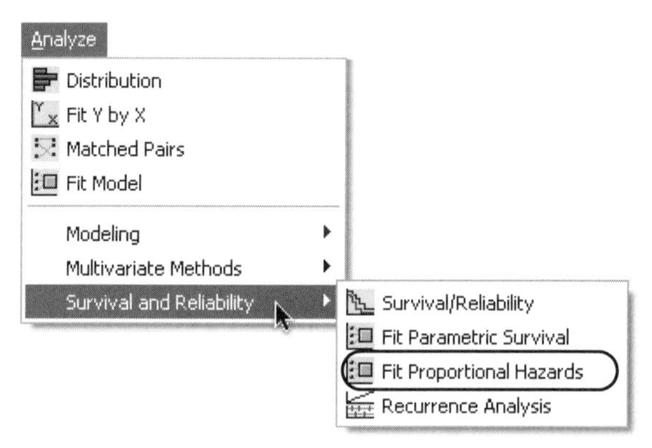

This launches the Fit Model dialog for survival analysis, with **Proportional Hazard** showing as the Fitting Personality. Alternatively, you can use the **Fit Model** command and specify the **Proportional Hazard** fitting personality (Figure 32.16).

Next, assign Days as the **Time to Event** variable, Censor as **Censor** and add Group to the model effects list. The next section describes the proportional hazard analysis results.

Figure 32.16 Fit Model Dialog for Proportional Hazard Analysis

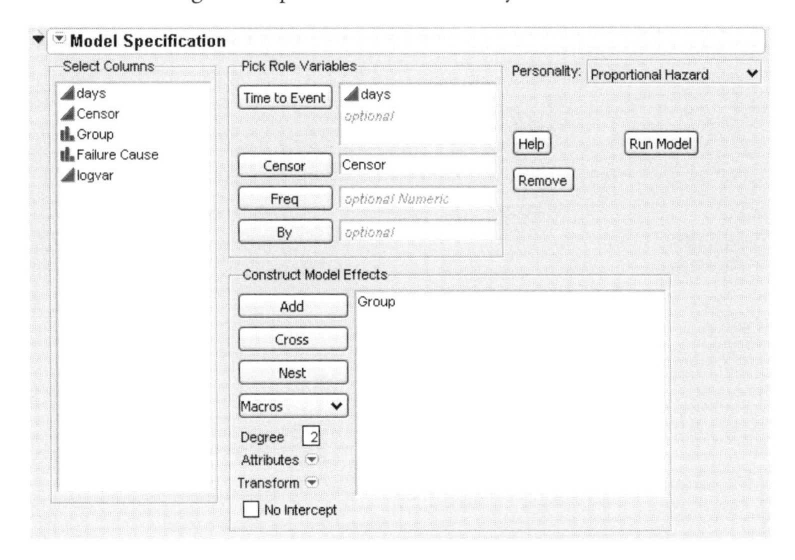

Statistical Reports for the Proportional Hazard Model

Finding parameter estimates for a proportional hazards model is an iterative procedure that displays during the model calculations. The Iteration History table lists results of the first 15 iterations. When the fitting is complete, the report in Figure 32.17 appears.

- The Whole Model Test table shows the negative of the natural log of the likelihood function (–LogLikelihood) for the model with and without the grouping covariate.
- Twice the positive difference between them gives a Chi-square test of the hypothesis that there is no difference in survival time between the groups. The degrees of freedom (DF) are equal to the change in the number of parameters between the full and reduced models.
- The Parameter Estimates table gives the parameter estimate for Group, its standard error, and 95% upper and lower confidence limits. A positive parameter estimate corresponds to a risk ratio that is greater than 1. A confidence interval that does not include zero is evidence that there is an alpha-level significant difference between groups.
- The Effect Likelihood-Ratio Tests shows the likelihood-ratio Chi-square test on the null hypothesis that the parameter estimate for the Group covariate is zero. Because Group only has two values, the test of the null hypothesis for no difference between the groups shown in the Whole Model Test table is the same as the null hypothesis that the regression coefficient for Group is zero.

Figure 32.17 Reports for Proportional Hazard Analysis

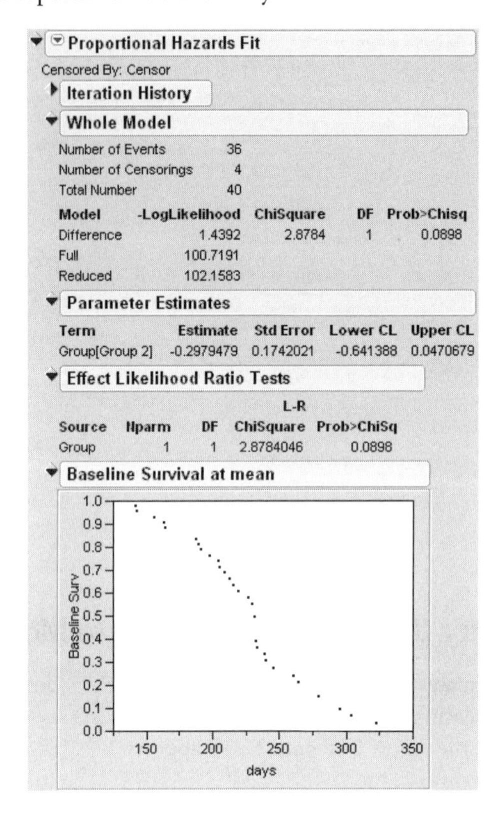
Proportional Hazards Fit

Censored By: Censor

Iteration History

Whole Model

Number of Events	36
Number of Censorings	4
Total Number	40

Model	-LogLikelihood	ChiSquare	DF	Prob>Chisq
Difference	1.4392	2.8784	1	0.0898
Full	100.7191			
Reduced	102.1583			

Parameter Estimates

Term	Estimate	Std Error	Lower CL	Upper CL
Group[Group 2]	-0.2979479	0.1742021	-0.641388	0.0470679

Effect Likelihood Ratio Tests

Source	Nparm	DF	L-R ChiSquare	Prob>ChiSq
Group	1	1	2.8784046	0.0898

Baseline Survival at mean

Risk Ratios

There are two risk ratios available as options from the Proportional Hazards report.

Unit Risk Ratio is Exp(estimate) and is related to unit changes in the regressor.

Range Risk Ratio is Exp[estimate($x_{Max} - x_{Min}$)]and shows the change over the whole range of the regressor.

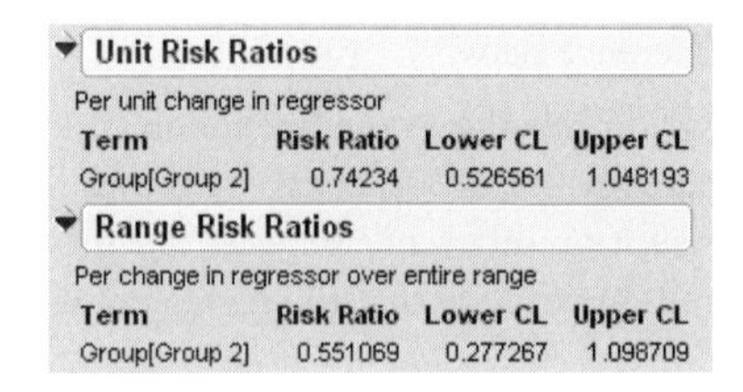
Unit Risk Ratios

Per unit change in regressor

Term	Risk Ratio	Lower CL	Upper CL
Group[Group 2]	0.74234	0.526561	1.048193

Range Risk Ratios

Per change in regressor over entire range

Term	Risk Ratio	Lower CL	Upper CL
Group[Group 2]	0.551069	0.277267	1.098709

Nonlinear Parametric Survival Models

This section shows how to use the Nonlinear platform for survival models. You only need to learn the techniques in this section if

- the model is nonlinear
- you need a distribution other than exponential, Weibull, or lognormal
- you have censoring that is not the usual right, left, or interval censoring.

With the ability to estimate parameters in specified loss functions, the Nonlinear platform becomes a powerful tool for fitting maximum likelihood models. See the chapter "Nonlinear Regression," p. 461, for complete information about the Nonlinear platform.

To fit a nonlinear model when data are censored, you first use the formula editor to create a parametric equation that represents a loss function adjusted for censored observations. Then use the **Nonlinear** command in the **Analyze > Modeling** menu, which estimates the parameters using maximum likelihood.

As an example, suppose that you have a table with the variable time as the response. First, create a new column, model, that is a linear model. Use the calculator to build a formula for model as the natural log of time minus the linear model—that is, `ln(time) - B + B*z` where `z` is the regressor.

Then, because the nonlinear platform minimizes the loss and you want to maximize the likelihood, create a loss function as the negative of the loglikelihood. The loglikelihood formula must be a conditional formula that depends on the censoring of a given observation (if some of the observations are censored).

Loss Formulas for Survival Distributions

The following formulas are for the negative log-likelihoods to fit common parametric models. Each formula uses the calculator `if` conditional function with the uncensored case of the conditional first and the right-censored case as the `Else` clause. You can copy these formulas from tables in the Loss Function Templates folder in Sample Data and paste them into your data table. Table 32.1 "Formulas for Loss Functions," p. 642, shows the loss functions as they appear in the columns created by the formula editor.

Exponential Loss Function

The exponential loss function is shown in Table 32.1 "Formulas for Loss Functions," p. 642, where sigma represents the mean of the exponential distribution and Time is the age at failure.

A characteristic of the exponential distribution is that the instantaneous failure rate remains constant over time. This means that the chance of failure for all subjects during a given length of time is the same regardless of how long a subject has been in the study.

Weibull Loss Function

The Weibull density function often provides a good model for the lifetime distributions. You can use the Univariate Survival platform for an initial investigation of data to determine if the Weibull loss function is appropriate for your data.

There are examples of a one-parameter, two-parameter, and extreme-value functions in the Loss Function Templates folder

Lognormal Loss Function

The formula shown below is the lognormal loss function where `Normal Distribution(model/sigma)` is the standard normal distribution function. The hazard function has value 0 at $t = 0$, increases to a maximum, then decreases and approaches zero as t becomes large.

Loglogistic Loss Function

The loglogistic function has a symmetric density with mean 0 and slightly heavier tails than the normal density function. Once you have selected a loss function, choose the **Nonlinear** command and complete the dialog. If the response is included in the model formula, no Y variable is needed. The model is the prediction column and a loss function column is the loss column.

Table 32.1 Formulas for Loss Functions

Exponential Loss Function
$$-\mathrm{If}\begin{pmatrix} \mathit{Censor}{=}{=}0 \Rightarrow -\mathrm{Log}(\mathbf{sigma}) - \dfrac{\mathit{Time}}{\mathbf{sigma}} \\ \mathrm{else} \Rightarrow -\left(\dfrac{\mathit{Time}}{\mathbf{sigma}}\right) \end{pmatrix}$$
Weibull Loss Function
$$-\mathrm{If}\begin{pmatrix} \mathit{censor}{=}{=}0 \Rightarrow \dfrac{\mathit{Model}}{\mathbf{sigma}} - \mathrm{Exp}\left(\dfrac{\mathit{Model}}{\mathbf{sigma}}\right) - \mathrm{Log}(\mathbf{sigma}) \\ \mathrm{else} \Rightarrow -\mathrm{Exp}\left(\dfrac{\mathit{Model}}{\mathbf{sigma}}\right) \end{pmatrix}$$
Lognormal Loss Function
$$-\mathrm{If}\begin{pmatrix} \mathit{censor}{=}{=}0 \Rightarrow -0.5*\left(\dfrac{\mathit{Model}}{\mathbf{sigma}}\right)^2 - 0.5*\mathrm{Log}(2*\mathrm{Pi}) - \mathrm{Log}(\mathbf{sigma}) \\ \mathrm{else} \Rightarrow \mathrm{Log}\left(1 - \mathrm{Normal\ Distribution}\left(\dfrac{\mathit{Model}}{\mathbf{sigma}}\right)\right) \end{pmatrix}$$
Loglogistic Loss Function
$$-\mathrm{If}\begin{pmatrix} \mathit{censor}{=}{=}0 \Rightarrow \left(\dfrac{\mathit{Model}}{\mathbf{sigma}} - 2*\mathrm{Log}\left(1+\mathrm{Exp}\left(\dfrac{\mathit{Model}}{\mathbf{sigma}}\right)\right)\right) - \mathrm{Log}(\mathbf{sigma}) \\ \mathrm{else} \Rightarrow -\mathrm{Log}\left(1+\mathrm{Exp}\left(\dfrac{\mathit{Model}}{\mathbf{sigma}}\right)\right) \end{pmatrix}$$

Weibull Loss Function Example

This example uses the VA Lung Cancer.jmp table. Models are fit to the survival time using the Weibull, lognormal, and exponential distributions. Model fits include a simple survival model containing only two regressors, a more complex model with the all regressors and some covariates, and the creation of dummy variables for the covariate Cell Type to be included in the full model.

Open the data table VA Lung Cancer.jmp. The first model and all the loss functions have already been created as formulas in the data table. The model column has the formula

```
Log(Time) - (B0 + b1*Age + b2*Diag Time)
```

Initial Parameter Estimates

Nonlinear model fitting is often sensitive to the initial values you give to the model parameters. In this example, one way to find reasonable initial values is to first use the Nonlinear platform to fit only the linear model. When the model converges, the solution values for the parameters become the initial parameter values for the nonlinear model.

To do this select **Analyze >Modeling > Nonlinear**. Assign model as **X, Predictor Formula**, and click **OK** to see the Nonlinear Control Panel, shown to the left in Figure 32.18. When you click **Go**, the platform computes the least squares parameter estimates for this model, as shown in the right-hand Control Panel.

Click **Save Estimates** to set the parameter estimates in the column formulas to those estimated by this initial nonlinear fitting process.

Figure 32.18 Initial Parameter Values in the Nonlinear Fit Control Panel

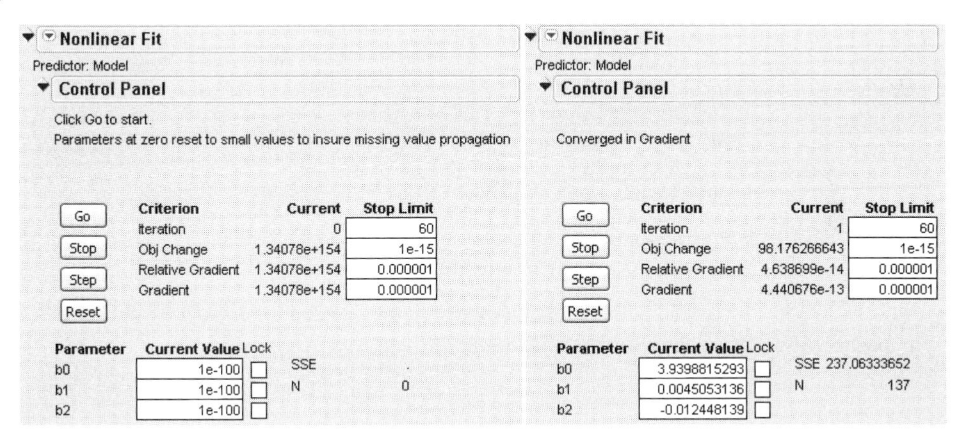

Estimates with Respect to the Loss Function

The Weibull column has the Weibull formula previously shown. To continue with the fitting process, Choose **Nonlinear** again, select model as **X, Predictor Formula** column and Weibull loss as **Loss**. When you click **OK**, the Nonlinear Fit Control Panel on the left in Figure 32.19 appears. There is now the additional parameter called sigma in the loss function. Because it is in the denominator of a fraction, a starting value of 1 is reasonable for sigma. When using any loss function other than the default, the **Loss is –LogLikelihood** box on the Control Panel is checked by default. When you click **Go**, the fitting process converges as shown in the right-hand control panel.

Figure 32.19 Nonlinear Model with Custom Loss Function

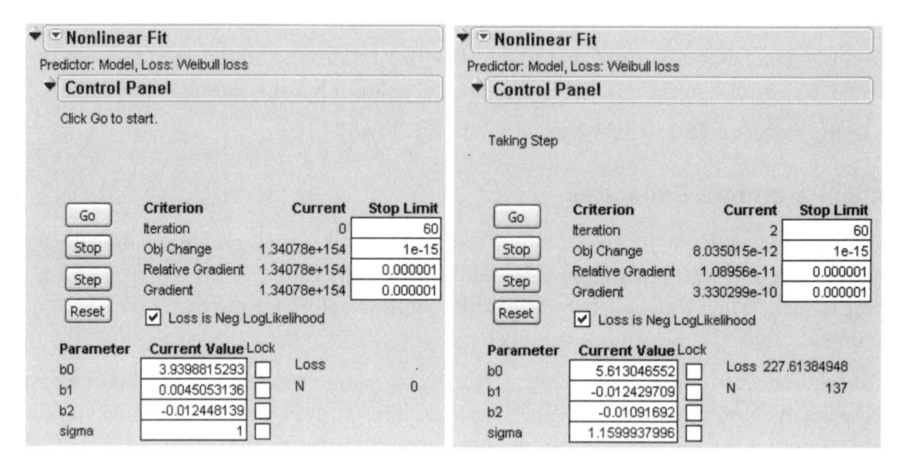

The fitting process estimates the parameters by maximizing the negative log of the Weibull likelihood function. After the Solution table appears, you can click the **Confidence Limits** button on the Control Panel to get lower and upper 95% confidence limits for the parameters, as shown in the Solution table.

Solution

Loss	DFE	Avg Loss	Sqrt Avg Loss
227.61384948	133	1.7113823	1.3081981

Parameter	Estimate	ApproxStdErr	Lower CL	Upper CL
b0	5.613046552	0.67171997	4.31169667	6.96484024
b1	-0.012429709	0.01114031	-0.0346969	0.00930037
b2	-0.01091692	0.010399	-0.0289695	0.01220355
sigma	1.1599937996	0.07823465	1.0209347	1.33040517

Note: Because the confidence limits are profile likelihood confidence intervals instead of the standard asymptotic confidence intervals, they can take time to compute.

You can also run the model with the predefined exponential and lognormal loss functions. Before you fit another model, reset the parameter estimates to the least squares estimates, as they may not converge otherwise.

Tobit Model Example

You can also analyze left-censored data in the Nonlinear platform. For the left-censored data, zero is the censored value because it also represents the smallest known time for an observation. The tobit model is popular in economics for responses that must be positive or zero, with zero representing a censor point.

The Tobit2.jmp data table in the Reliability sample data folder can be used to illustrate a Tobit model. The response variable is a measure of the durability of a product and cannot be less than zero (Durable, is left-censored at zero). Age and Liquidity are independent variables. The table also includes the model and tobit loss function. The model in residual form is `durable-(b0+b1*age+b2*liquidity)`. Tobit Loss has the formula shown here.

$$-\text{If}\begin{pmatrix} durable{=}{=}0 \Rightarrow \text{Log}\left(1\text{-Normal Distribution}\left(\frac{-Model}{\textbf{Sigma}}\right)\right) \\ \text{else} \quad \Rightarrow -\text{Log}(\textbf{Sigma})-0.5*\left(\frac{Model}{\textbf{Sigma}}\right)^2-0.5*\text{Log}(2*\text{Pi}) \end{pmatrix}$$

To run the model choose **Nonlinear**, select Model as **X, Predictor Formula** and Tobit Loss as **Loss**, and click **OK**. When the Nonlinear Fit Control Panel appears, the model parameter starting values are set to (near) zero, the loss function parameter Sigma is set to 1, and the **Loss is –LogLikelihood** is checked. Click **Go**. If you also click **Confidence Limits** in the Control Panel after the model converges, you see the solution table shown here.

Solution

Loss	DFE	Avg Loss	Sqrt Avg Loss
28.92596097	16	1.8078726	1.3445715

Parameter	Estimate	ApproxStdErr	Lower CL	Upper CL
b0	15.277120063	16.0327167	-22.059913	54.7012259
b1	-0.134007501	0.21893135	-0.7365041	0.32841861
b2	-0.045135584	0.05826851	-0.1858906	0.09330941
Sigma	5.5693492202	1.72814365	3.31613117	11.5291865

Solved By: Analytic NR

Fitting Simple Survival Distributions

The following examples show how to use maximum likelihood methods to estimate distributions from time-censored data when there are no effects other than the censor status. The Loss Function Templates folder has templates with formulas for an exponential, one-and two-parameter Weibull, extreme value, normal, and lognormal loss functions. To use these loss functions, copy your time and censor values into the Time and censor columns of the loss function template.

To run the model, select **Nonlinear** and assign the loss column as the **Loss** variable. Because both the response model and the censor status are included in the loss function and there are no other effects, you do not need prediction column (model variable).

Exponential, Weibull and Extreme-Value Loss Function Examples

The Fan.jmp data table in the Reliability sample data folder illustrates the Exponential, Weibull, and Extreme value loss functions discussed in Nelson (1982). The data are from a study of 70 diesel fans that accumulated a total of 344,440 hours in service. The fans were placed in service at different times. The response is failure time of the fans or run time, if censored. A partial listing of the Fan.jmp table is shown here.

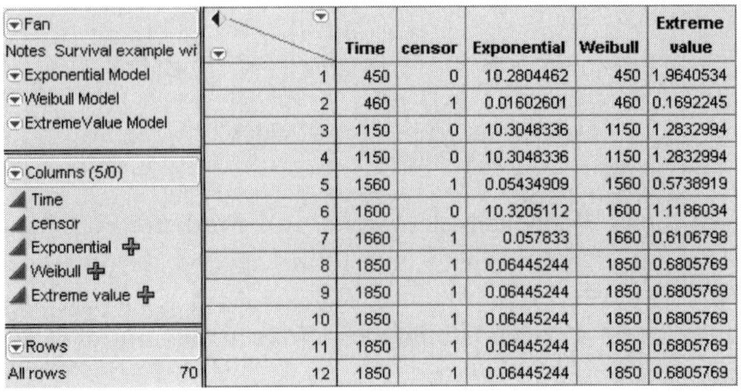

Fan
Notes Survival example wi
Exponential Model
Weibull Model
ExtremeValue Model

Columns (5/0)
Time
censor
Exponential
Weibull
Extreme value

Rows
All rows 70

	Time	censor	Exponential	Weibull	Extreme value
1	450	0	10.2804462	450	1.9640534
2	460	1	0.01602601	460	0.1692245
3	1150	0	10.3048336	1150	1.2832994
4	1150	0	10.3048336	1150	1.2832994
5	1560	1	0.05434909	1560	0.5738919
6	1600	0	10.3205112	1600	1.1186034
7	1660	1	0.057833	1660	0.6106798
8	1850	1	0.06445244	1850	0.6805769
9	1850	1	0.06445244	1850	0.6805769
10	1850	1	0.06445244	1850	0.6805769
11	1850	1	0.06445244	1850	0.6805769
12	1850	1	0.06445244	1850	0.6805769

Table 32.2 "Formulas for Loss Functions," p. 646, shows the formulas for the loss functions as they appear in the formula editor.

Table 32.2 Formulas for Loss Functions

Two-parameter exponential	$-\text{If}\begin{cases} censor==0 \Rightarrow -\text{Log}(\textbf{sigma}) - \dfrac{Time}{\textbf{sigma}} \\ \text{else} \Rightarrow -\left(\dfrac{Time}{\textbf{sigma}}\right) \end{cases}$
Weibull	$-\text{If}\begin{cases} censor!=0 \Rightarrow -\text{Exp}\left(\textbf{Beta}*\text{Log}\left(\dfrac{Time}{\textbf{Alpha}}\right)\right) \\ \text{else} \Rightarrow \text{Log}(\textbf{Beta})+(\textbf{Beta}-1)*\text{Log}(Time)-\textbf{Beta}*\text{Log}(\textbf{Alpha}) +- \text{Exp}\left(\textbf{Beta}*\text{Log}\left(\dfrac{Time}{\textbf{Alpha}}\right)\right) \end{cases}$
Extreme Value	$-\text{If}\begin{cases} censor==0 \Rightarrow -\left(\text{Log}(\textbf{delta})+\text{Exp}\left(\dfrac{\text{Log}\left(\frac{Time}{1000}\right)-\textbf{lambda}}{\textbf{delta}}\right)\right) + \dfrac{\text{Log}\left(\frac{Time}{1000}\right)-\textbf{lambda}}{\textbf{delta}} \\ \text{else} \Rightarrow -\text{Exp}\left(\dfrac{\text{Log}\left(\frac{Time}{1000}\right)-\textbf{lambda}}{\textbf{delta}}\right) \end{cases}$

To use the exponential loss function, select **Nonlinear**, choose **Exponential** as the **Loss** function, and click **OK**. In the Nonlinear Fit Control Panel, enter 1 as the starting value for **Sigma**, and click **Go**. After the model converges, click **Confidence Limits** to see the results shown here.

Exponential Solution

Loss	DFE	Avg Loss	Sqrt Avg Loss
135.17751124	69	1.9590944	1.3996765

Parameter	Estimate	ApproxStdErr	Lower CL	Upper CL
sigma	28505.121855	8172.09114	17113.5957	53602.4745

Solved By: Analytic NR

Weibull Solution

Loss	DFE	Avg Loss	Sqrt Avg Loss
135.15271994	68	1.98754	1.4098014

Parameter	Estimate	ApproxStdErr	Lower CL	Upper CL
Alpha	26296.817205	12251.3882	13629.9746	106086.918
Beta	1.0584462448	0.26825073	0.60609377	1.65792545

Solved By: Analytic NR

Extreme Value Solution

Loss	DFE	Avg Loss	Sqrt Avg Loss
42.247996552	68	0.6212941	0.7882221

Parameter	Estimate	ApproxStdErr	Lower CL	Upper CL
lambda	3.2694489763	0.46588966	2.61236736	4.66416689
delta	0.9447814453	0.23944403	0.60315891	1.6501854

Solved By: Analytic NR

To use the Weibull loss function with two parameters or the extreme value loss function, again select **Nonlinear** and choose the loss function you want. Use starting values of 1 for the alpha and beta parameters in the Weibull function and for delta and lambda in the extreme-value function. The results are shown above.

Note: Be sure to check the Loss is –LogLikelihood check box before you click **Go**.

Lognormal Loss Function Example

The Locomotive.jmp data in the Reliability sample data folder can be used to illustrate a lognormal loss. The lognormal distribution is useful when the range of the data is several powers of 10. The logNormal column in the table has the formula

$$\text{If}\left(\begin{array}{ll} \textit{censor}!=0 \Rightarrow -\text{Log}\left(1-\text{Normal Distribution}\left(\dfrac{\text{Log}_{10}(\textit{Time})-\mathbf{Mu}}{\mathbf{sigma}}\right)\right) \\ \text{else} \quad \Rightarrow \text{Log}(\textit{Time}^{*}\mathbf{sigma})+0.5^{*}\text{Log}(2^{*}\text{Pi})+0.5^{*}\left(\dfrac{\text{Log}_{10}(\textit{Time})-\mathbf{Mu}}{\mathbf{sigma}}\right)^{2}-\text{Log}(0.4343) \end{array}\right)$$

The lognormal loss function can be very sensitive to starting values for its parameters. As the lognormal is similar to the normal distribution, you can create a new variable that is the $\log_{10}$ of Time and use **Distribution** to find the mean and standard deviation of this column. Then, use those values as starting values for the Nonlinear platform. In this example the mean of $\log_{10}$ of Time is 2.05 and the standard deviation is 0.15.

Run this example as described in the previous examples. Assign lognormal as the **Loss** function. In the Nonlinear Fit Control Panel give Mu and Sigma the starting values 2.05 and 0.15 and click **Go**. After the Solution is found, you can click **Confidence Limits**, on the Control Panel and see the table shown here.

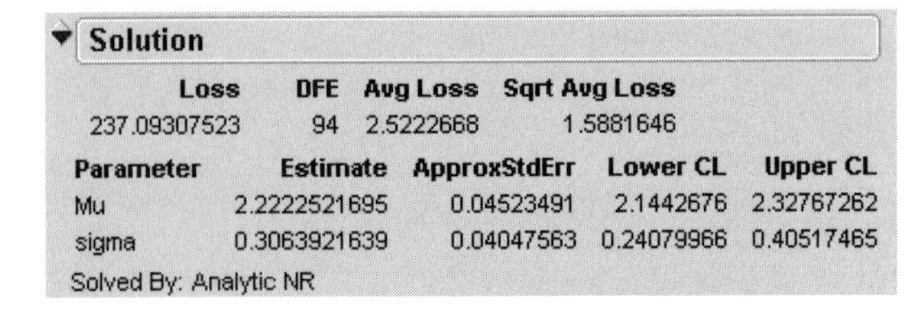
Solution

Loss	DFE	Avg Loss	Sqrt Avg Loss
237.09307523	94	2.5222668	1.5881646

Parameter	Estimate	ApproxStdErr	Lower CL	Upper CL
Mu	2.2222521695	0.04523491	2.1442676	2.32767262
sigma	0.3063921639	0.04047563	0.24079966	0.40517465

Solved By: Analytic NR

Note: Remember to **Save Estimates** before requesting confidence limits.

The maximum likelihood estimates of the lognormal parameters are 2.2223 for Mu and 0.3064 for Sigma (in base 10 logs). The corresponding estimate of the median of the lognormal distribution is the antilog of 2.2223 ($10^{2.2223}$), which is approximately 167,000 miles. This represents the typical life for a locomotive engine.

Chapter 33

Recurrence Analysis

The Recurrence Platform

This chapter focuses on the final item in the **Survival and Reliability** submenu, **Recurrence Analysis**.

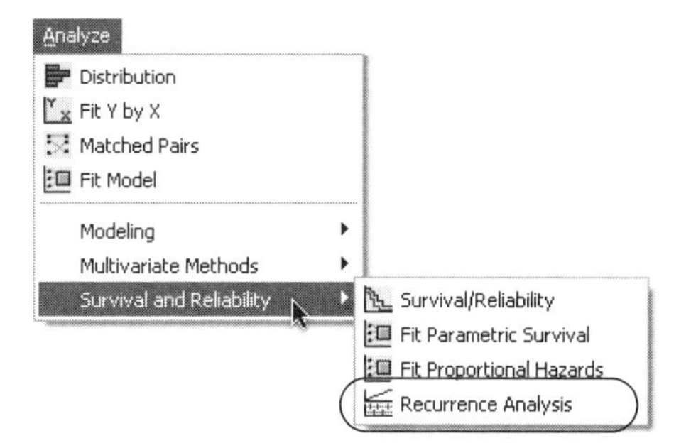

Recurrence Analysis analyzes event times like survival or reliability analysis, but they are events that can recur several times for each unit. Typically, these events are when a unit breaks down and needs to be repaired, then the unit is put back into service after the repair. The units are followed until they are ultimately taken out of service. Similarly, recurrence analysis can be used to analyze data from continuing treatements of a long-term disease, such as the recurrence of tumors in patients receiving treatment for bladder cancer. The goal of the analysis is to obtain the MCF, the mean cumulative function, which shows the total cost per unit as a function of time, where cost can be just the number of repairs, or it can be the actual cost of repair.

Contents

Recurrence Data Analysis

Recurrent event data involves the cumulative frequency or cost of repairs as units age. In JMP, the **Recurrence Analysis** platform analyzes recurrent events data.

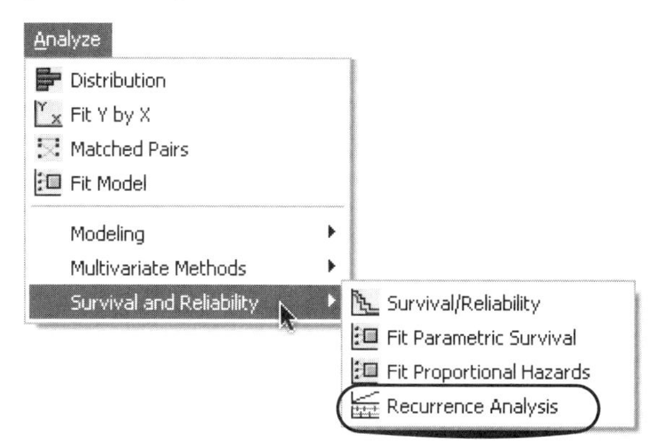

The data for recurrence analysis have one row for each observed event and a closing row with the last observed age of a unit and can include any number of units or systems.

Launching the Platform

To launch the platform, select **Analyze > Survival and Reliability > Recurrence Analysis**. The dialog contains the following roles.

Y, Age at Event contains the unit's age at the time of an event, or at the time of reference for the end of service. This is a required numeric column.

Grouping is an optional column if you want to produce separate MCF estimates for different groups, identified by this column.

Cost, End-of-Service is a column that must contain one of the following

- a 1, indicating an event has occurred (*e.g.* a unit was repaired)
- a cost for the event (*e.g.* the cost of the repair)
- a zero, indicating that the unit went out-of-service, or is no longer being studied. All units (each System ID) must have one row with a zero for this column, with the **Y, Age at Event** column containing the final observed age.

If costs are given here, the MCF is a mean cumulative cost per unit as a function of age. If indicators (1's) are used here, then the MCF is the mean cumulative count of events per unit as a function of age

Label, System ID is a required column identifying the unit for each event and censoring age.

If each unit does not have exactly one last observed age in the table (where the Cost column cell is zero), then JMP gives an error message to that effect.

Examples

Valve Seat Repairs Example

A typical unit might be a system (say. a component of an engine or appliance). For example, consider the sample data table Engine Valve Seat.jmp (Meeker and Escobar 1998, p. 395 and Nelson, 2003), which records valve seat replacements in locomotive engines. The EngineID column identifies a specific locomotive unit. Age is time in days from beginning of service to replacement of the engine valve seat. Note that an engine can have multiple rows with its age at each replacement and its cost, corresponding to multiple repairs. Here, Cost=1 indicates the last observed age of a locomotive.

Engine Valve Seat
Notes Recurrence data fro
Recurrence Analysis
Columns (3/0)
EngineID
Age
Cost
Rows
All rows 89

	EngineID	Age	Cost
1	251	761	0
2	252	759	0
3	327	667	0
4	327	98	1
5	328	667	0
6	328	326	1
7	328	653	1
8	328	653	1

Complete the launch dialog as shown here.

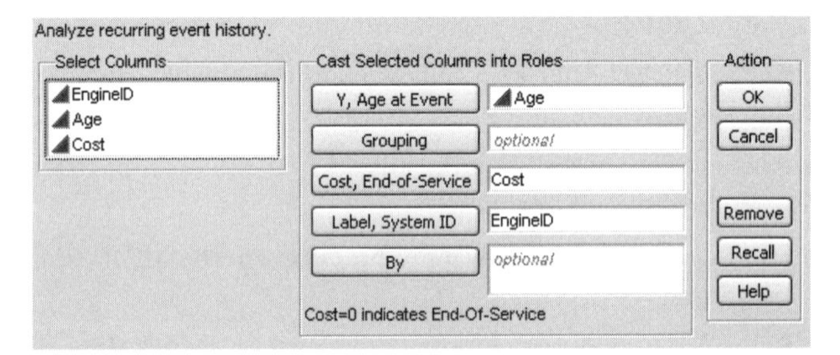

When you click **OK**, the Recurrence platform shows the reports in Figure 33.1 and Figure 33.2. The MCF plot shows the sample mean cumulative function. For each age, this is the nonparametric estimate of the mean cumulative cost or number of events per unit. This function goes up as the units get older and total costs grow.

The plot in Figure 33.1 shows that about 580 days is the age that averages (approximately) one repair event.

Figure 33.1 MCF Plot and Partial Table For Recurrence Analysis

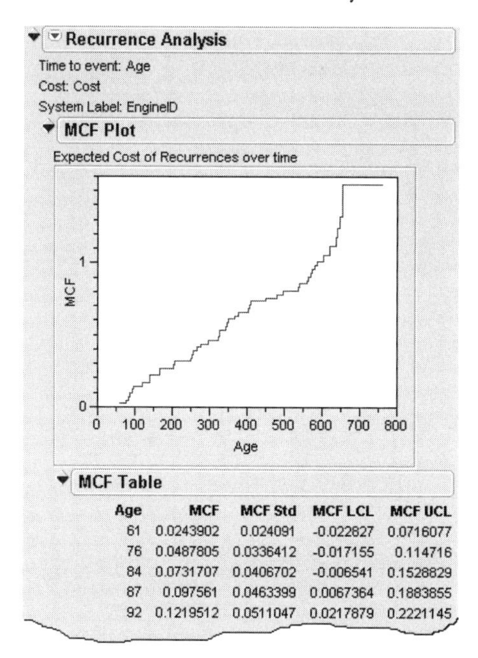

Age	MCF	MCF Std	MCF LCL	MCF UCL
61	0.0243902	0.024091	-0.022827	0.0716077
76	0.0487805	0.0336412	-0.017155	0.114716
84	0.0731707	0.0406702	-0.006541	0.1528829
87	0.097561	0.0463399	0.0067364	0.1883855
92	0.1219512	0.0511047	0.0217879	0.2221145

The event plot in Figure 33.2,shows a time line for each unit. There are markers at each time of repair, and each line extends to that unit's last observed age. For example, unit 409 was observed at 389 days and had three valve replacements.

Figure 33.2 Event Plot For Valve Seat Replacements

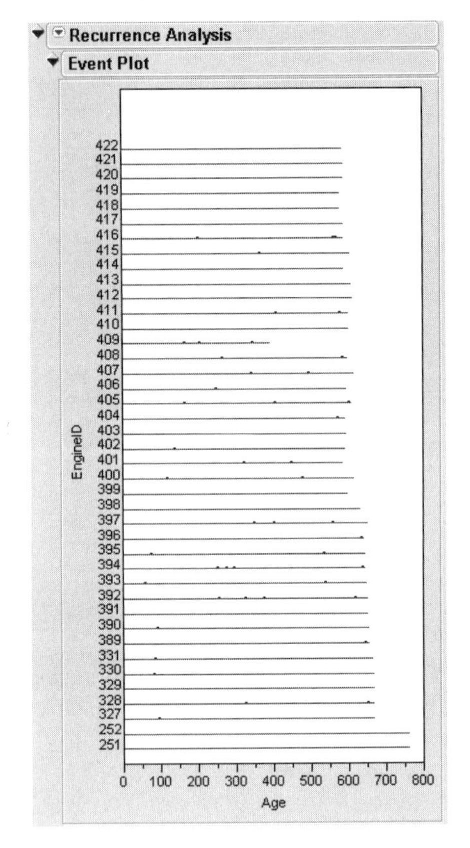

Options

MCF Plot toggles on and off the MCF plot

MCF Confidence Limits toggles on and off the lines corresponding to the approximate 95% confidence limits of the MCF.

Event Plot toggles on and off the Event plot.

Plot MCF Differences If you have a grouping variable, this option will offer to create a plot of the difference of MCF's, including a 95% confidence interval for that difference. The MCF's are significantly different where the confidence interval lines do not cross the zero line.

Bladder Cancer Recurrences Example

The sample data file Bladder Cancer.jmp (Andrews and Herzberg, 1985, table 45) contains data on bladder tumor recurrences from the Veteran's Administration Co-operative Urological Research Group. All patients presented with superficial bladder tumors which were removed upon entering the trial. Each patient was then assigned to one of three treatment groups: placebo pills, pyridoxine (vitamin B6) pills, or periodic chemotherapy with thiotepa. The following analysis of tumor recurrence explores the progression of the disease, and whether there is a difference among the three treatments.

Launch the platform with the options shown in Figure 33.3.

Figure 33.3 Bladder Cancer Launch Dialog

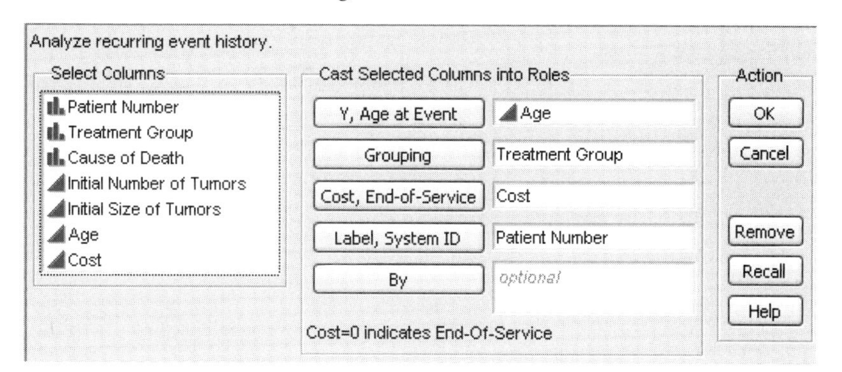

The resulting report shows MCF plots for the three treatments.

Figure 33.4 Bladder Cancer MCF Plot

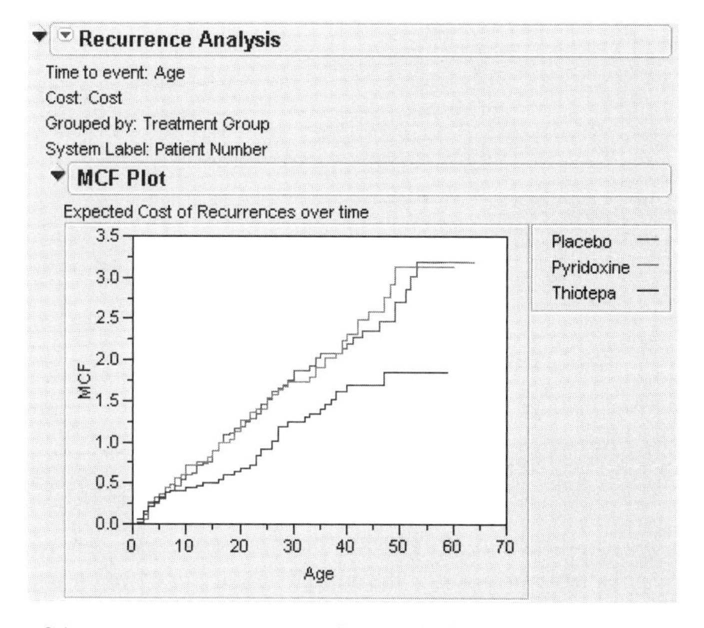

Note that all three of the MCF curves are essentially straight lines. The slopes (rates of recurrence) are therefore constant over time, implying that patients do not seem to get better or worse as the disease progresses.

To examine if there are differences among the treatments, select the **Plot MCF Differences** command from the platform drop-down menu to get the following plots.

Figure 33.5 MCF Differences

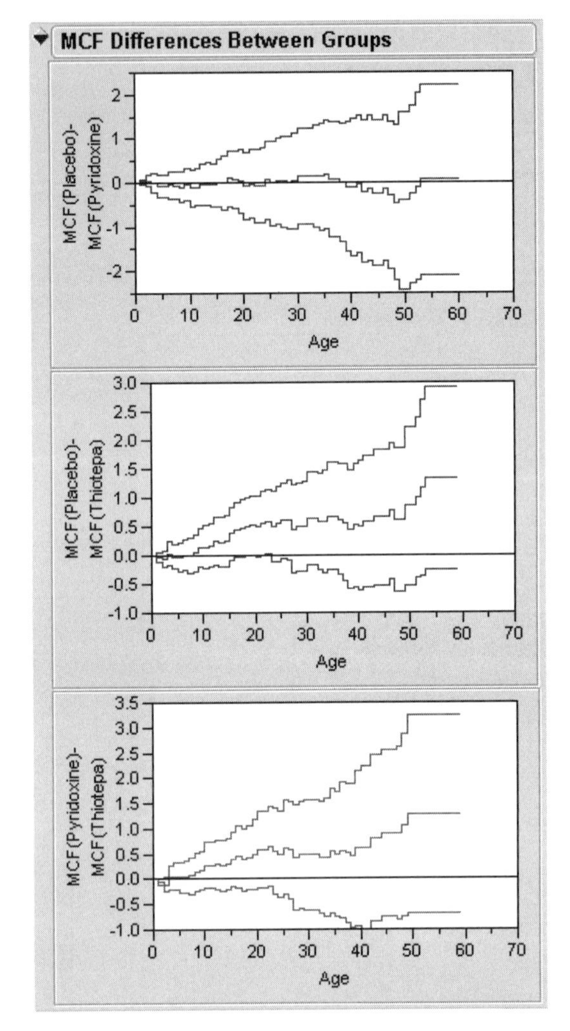

To determine whether there is a statistically significant difference between treatments, examine the confidence limits on the differences plot. If the limits do not include zero, the treatments are convincingly different. The graphs in Figure 33.5 show there is no significant difference among the treatments.

Chapter 34

Recursive Partitioning

The Partition Platform

The **Partition** platform recursively partitions data according to a relationship between the *X* and *Y* values, creating a tree of partitions. It finds a set of cuts or groupings of X values that best predict a Y value. It does this by exhaustively searching all possible cuts or groupings. These splits (or *partitions*) of the data are done recursively forming a tree of decision rules until the desired fit is reached.

Variations of this technique go by many names and brand names: decision trees, CART^{TM}, CHAID^{TM}, C4.5, C5, and others. The technique is often taught as a data mining technique, because

- it is good for exploring relationships without having a good prior model
- it handles large problems easily, and
- the results are very interpretable.

The classic application is where you want to turn a data table of symptoms and diagnoses of a certain illness into a hierarchy of questions to ask new patients in order to make a quick initial diagnosis.

The factor columns (*X*'s) can be either continuous or categorical (nominal or ordinal). If an *X* is continuous, then the splits (partitions) are created by a *cutting value*. The sample is divided into values below and above this cutting value. If the *X* is categorical, then the sample is divided into two groups of levels.

The response column (*Y*) can also be either continuous or categorical (nominal or ordinal). If *Y* is continuous, then the platform fits means, and creates splits which most significantly separate the means by examining the sums of squares due to the means differences. If *Y* is categorical, then the response rates (the estimated probability for each response level) become the fitted value, and the most significant split is determined by the largest likelihood-ratio chi-square statistic. In either case, the split is chosen to maximize the difference in the responses between the two branches of the split. The resulting tree has the appearance of artificial intelligence by machine learning.

This is a powerful platform, since it examines a very large number of possible splits and picks the optimum one.

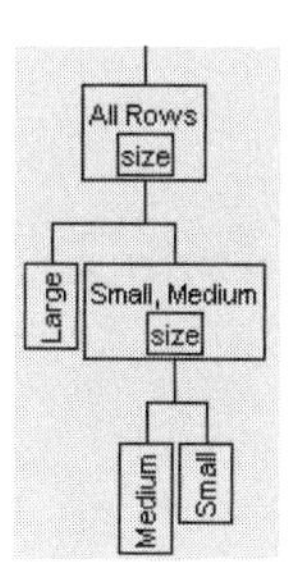

Contents

The Partition Graph

The Partition platform displays slightly different outputs depending on whether the *y* variables in the model are continuous or categorical. Output for the partition report is shown in Figure 34.1 and Figure 34.2.[1]

Figure 34.1 Partition Output for Categorical Variables

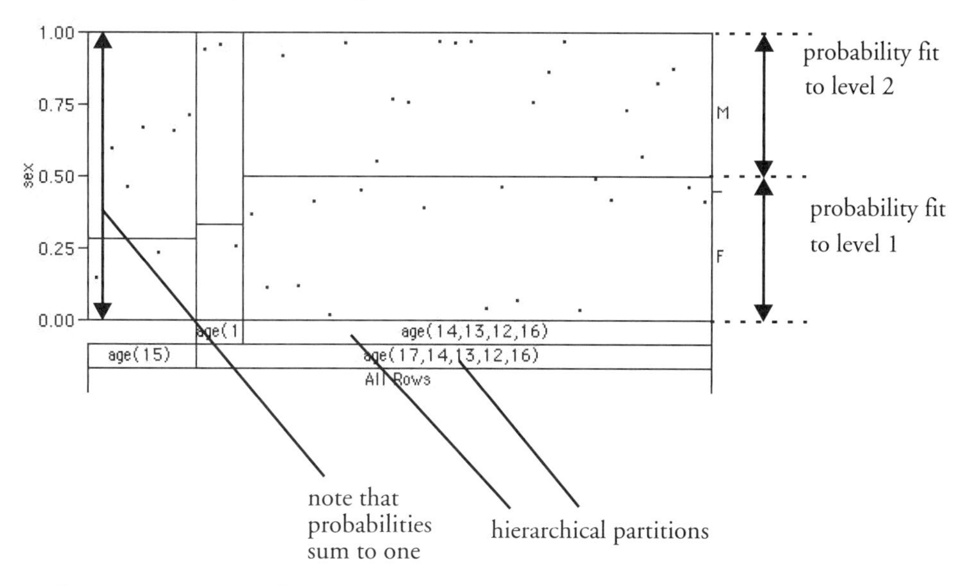

Figure 34.2 Partition Output for Continuous Variables

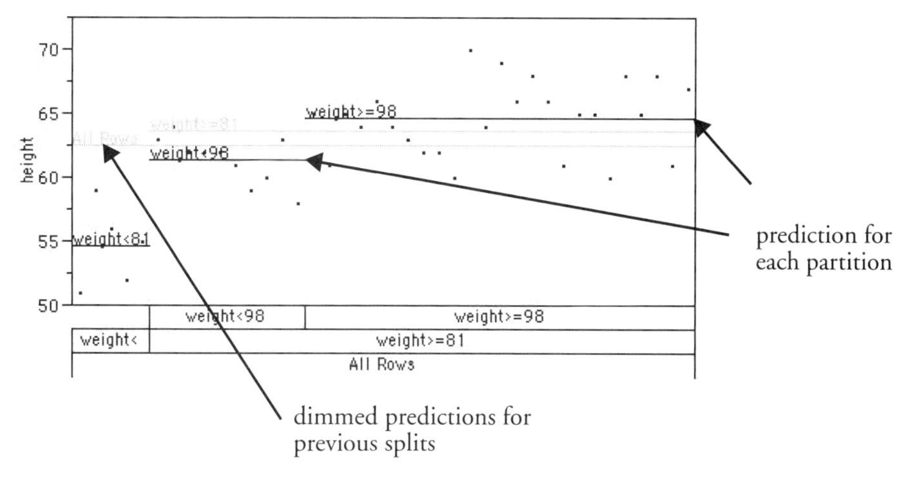

Each marker represents a response from the category and partition it is in. The *y* position is random within the *y* partition, and the *x* position is a random permutation so the points will be in the same rectangle but at different positions in successive analyses.

1. The categorical version has been referred to as a *double-decker plot* (Hoffman, 2001).

Categorical Response Example

Open Car Poll.jmp and invoke **Partition** with the country as the *Y* variable, and the other columns (sex, marital status, age, size, and type) as *X* variables (Figure 34.3). The goal of this example is to investigate which factors affect the country of purchased cars. Specify the *Y* column and *X* column in the launch dialog. **Weight** or **Freq** columns may be used if you have pre-summarized data.

Figure 34.3 Partition Platform

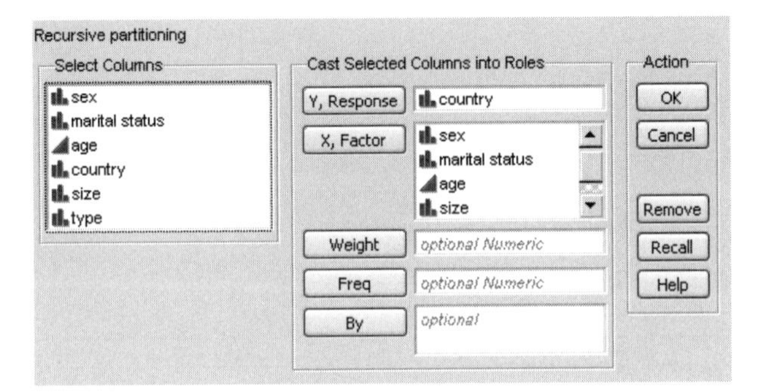

The platform starts with one group containing the nonmissing *y* values. On the first **Split** command, the platform calculates a splitting value (or groups) for the *X* column that optimally splits the group into two groups. This splitting continues with each step, choosing the best split.

Initially, everything is in one group. In order to determine the first, optimal split, each *X* column must be considered. The candidate columns are shown in the **Candidates** report, which you can open and browse to see which split to make next. As shown in Figure 34.4, the Size column has the largest G^2, and is therefore the optimum split to make.

The optimum split is noted by an asterisk. However, there are cases where the test statistic is higher for one variable, but the logworth is higher for a different variable. In this case > and < are used to point in the best direction for each variable. (The asterisk corresponds to the condition where they agree).

Figure 34.4 Initial Partition Report

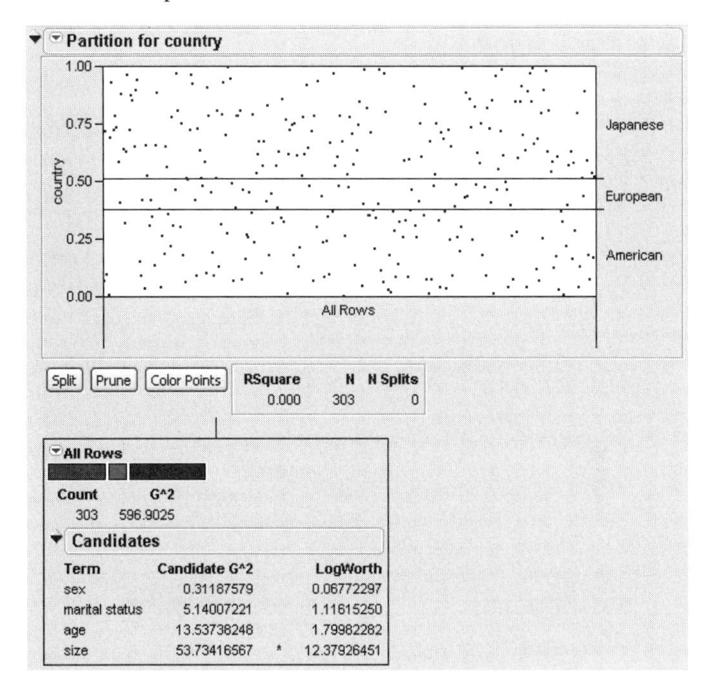

Note the three reported statistics to the right of the buttons below the plot.

RSquare is the current R^2 value.

N is the number of observations (if no Freq variable is used) or the number of rows with non-zero frequency.

NSplits is the current number of splits.

Whenever there is a missing value on the splitting statistic, JMP must randomly assign those points into one side or the other of the split. When this happens, **Imputes** appears beside the graph and tells you the number of times this has happened.

Click the **Split** button and it splits on the Size column, dividing the population into two groups, one for “Large”, and one for “Small, Medium”.

Figure 34.5 Partition after First Split

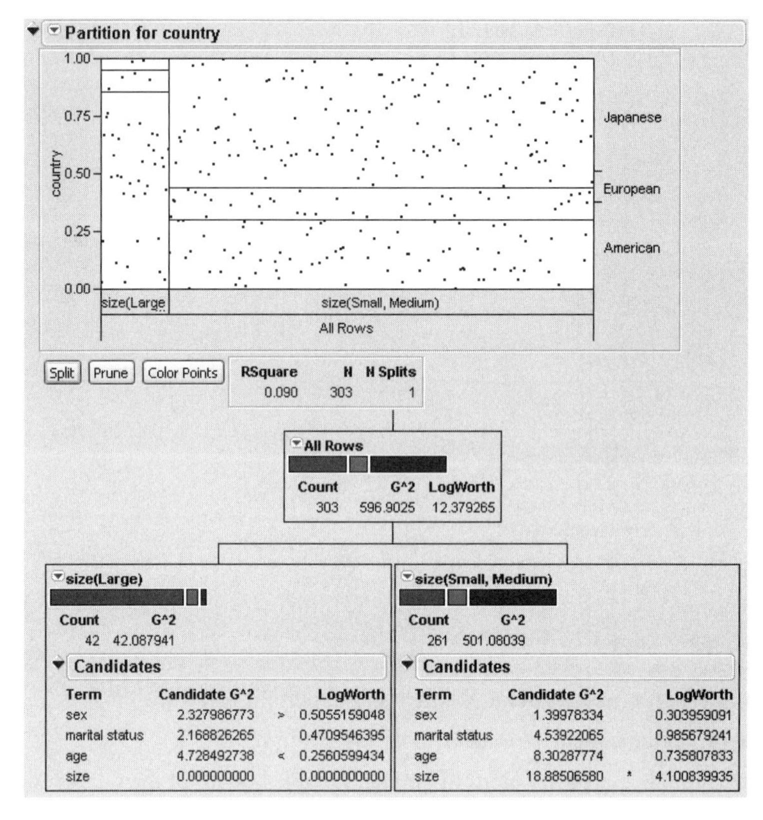

Notice that the buyers of large cars have much different country-brand rates than buyers of small and medium cars. This data was gathered in the early 1980's when large cars were more strongly associated with American brands.

As before, an examination of the candidates shows what the next split will be—in this case, another split on the Size variable.

Split again, and JMP chooses Size again, dividing into "Small" on one split, and "Medium" on the other.

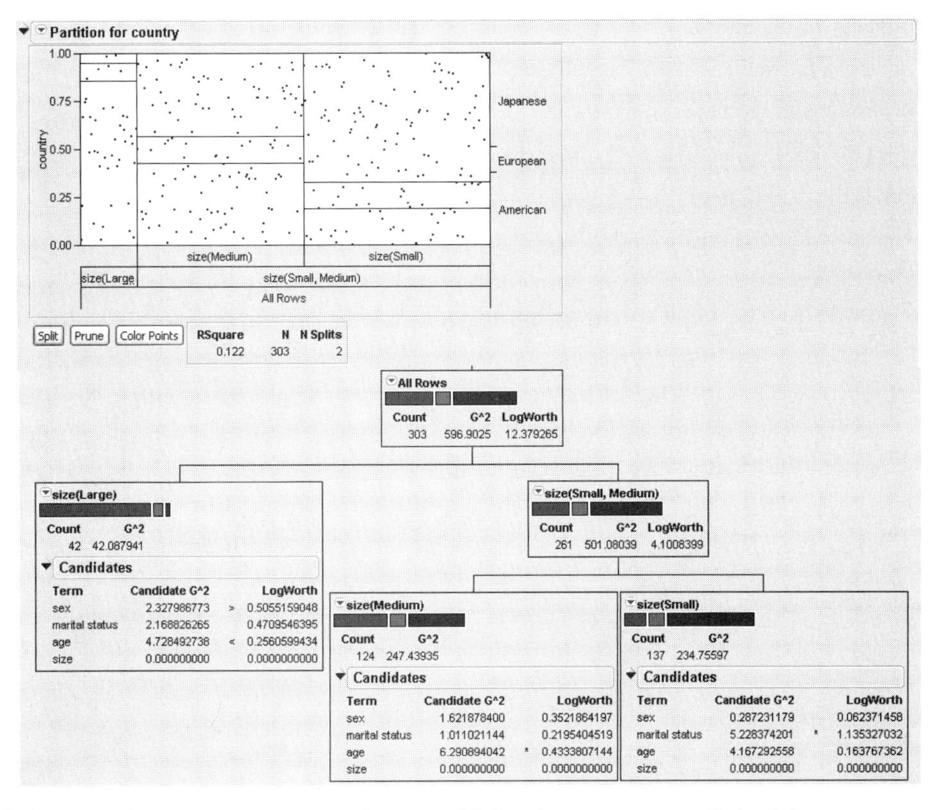

Click the **Prune** button to reverse operations, combining the most recent split back into one group.

Continuous Response Example

To illustrate the Partition platform with a continuous-response example, open the **Boston Housing.jmp** data table. This data, from Belsley, Kuh, and Welsch (1980) comes from the 1970 census, has the following variables.

mvalue is the median value of the homes

crim is the per capita crime rate

zn is the proportion of the town's residential land zoned for lots greater than 25,000 square feet

indus is the proportion of nonretail business acres per town

chas is a dummy variable with the value of 1 if the tract bounds on the Charles River

nox is the nitrogen oxide concentration in parts per hundred million squared

rooms are the average number of rooms squared

age is the proportion of owner-occupied units built prior to 1940

distance is the logarithm of the weighted distances to five employment centers in the Boston region radial is the logarithm of index of accessibility to radial highways

tax is the full-value property tax rate per $10,000

pt it the pupil-teacher ratio by town

b is $(B - 0.63)^2$ where B is the proportion of blacks in the population

lsat is the logarithm of the proportion of the population that is lower status.

Figure 34.6 shows the initial display with mvalue as the Y variable and all others as effects, showing that rooms is the best candidate.

Figure 34.6 Initial Boston Housing Display

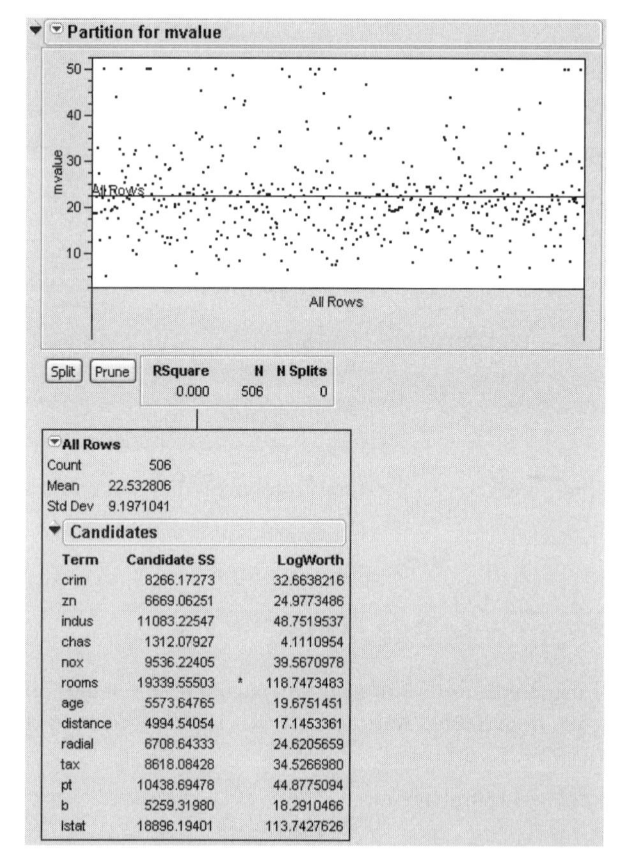

After this first split, a horizontal line appears for each group involved in the split. These lines are labeled with the variable and the level of the split. In this case, the split is on the rooms variable at 6.94. Nodes that are not leaves (such as the line for “All Rows” after the first split) are shown as light gray lines.

Basically, census tracts with large median houses have larger mean values than census tracts with small median houses.

Figure 34.7 After First Split

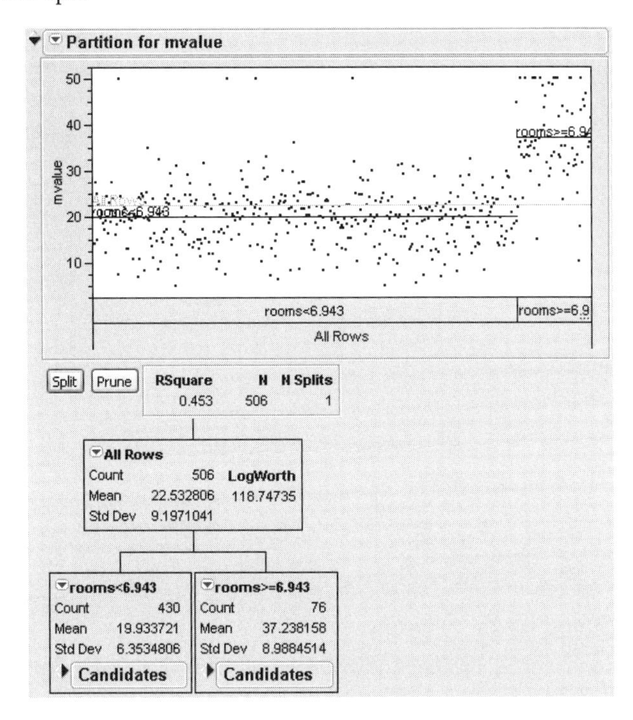

Figure 34.8 shows the plot after three splits with the **Small Tree View** option turned on. The small tree view uses the Small font, found in JMP's preferences, which has in this case been increased to 8 point type. Note that when the number of rows in a partition is small, the label may not have enough room. You can drag downward from the lower corner or lower limit of the labeling box to give it more room to wrap the labels across several lines.

Figure 34.8 After Three Splits

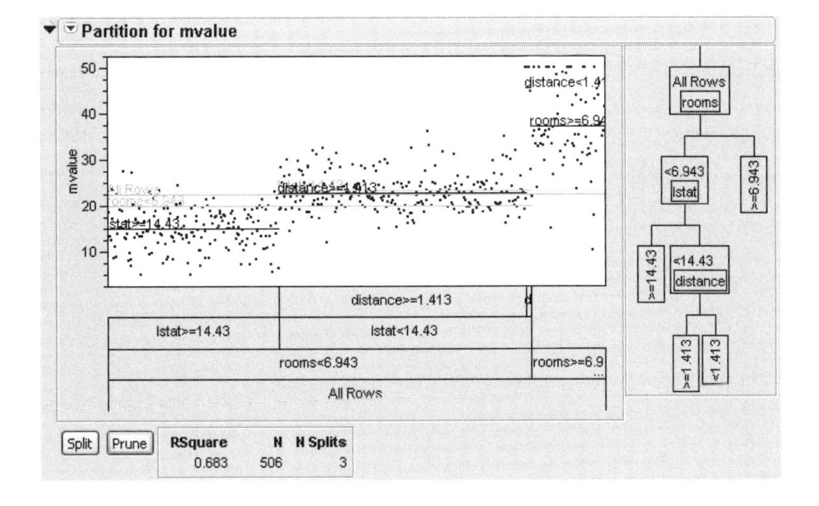

Platform Commands

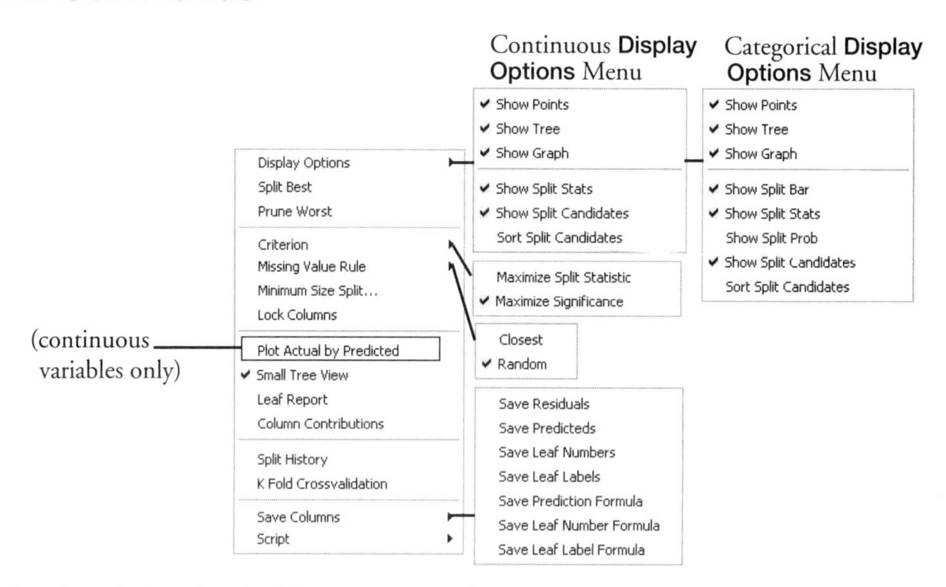

The Partition platform has the following commands.

Display Options gives a submenu consisting of items that toggle report elements on and off.

Show Points shows points rather than colored panels.

Show Tree toggles the large tree of partitions.

Show Graph toggles the partition graph.

Splits with continuous responses have the following two options.

Show Split Stats shows the split statistics. *e.g.* the G^2 or the mean and standard deviation

Show Split Candidates shows the split candidates in an outline node, giving G^2 or SS and log worth for each *X*-variable.

Sort Spit Candidates sorts the candidates report by the statistic or the log(worth), whichever is appropriate. This option can be turned on and off. When off, it doesn't change any reports, but new candidate reports are sorted in the order the *X* terms are specified, rather than by a statistic.

Splits with categorical responses have the following four options.

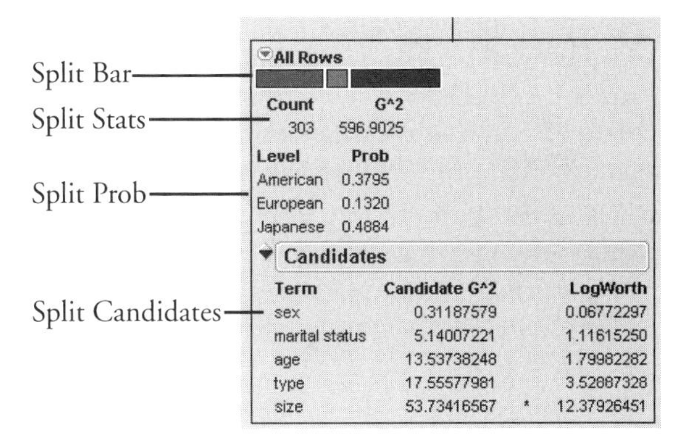

Show Split Bar toggles the colored bars showing the split proportions in each leaf.

Show Split Stats toggles the split statistics.

Show Split Prob, off by default, shows the probabilities for each level of the response variable.

Show Split Candidates toggles the candidates outline node.

Split Best splits the tree at the optimal split point. This is the same action as the **Split** button.

Prune Worst removes the terminal split that has the least discrimination ability. This is equivalent to hitting the Prune Button.

Criterion offers two choices for splitting. **Maximize Split Statistic** splits based on the raw value of sum of squares (continuous cases) or G^2(categorical cases). **Maximize Significance** calculates significance values for each split candidate, and uses these rather than the raw values to determine the best split. This is the default choice for Criterion.

When **Maximize Significance** is chosen, the candidate report includes a column titled **LogWorth**. This is the negative log of the adjusted p-value. That is,

LogWorth = $-\log_{10}(p\text{-value})$

The adjusted log-p-value is calculated in a complex manner that takes into account the number of different ways splits can occur. This calculation is very fair compared to the unadjusted p-value, which is biased to favor X's with many levels, and the Bonferroni p-value, which overadjusts to the effects that it is biased to favor—those X's with small numbers of levels. Details on the method are discussed in a white paper "Monte Carlo Calibration of Distributions of Partition Statistics" found on the jmp website www.jmp.com.

It is reported on the logarithmic scale because many significance values could easily underflow the numbers representable in machine floating-point form. It is reported as a negative log so that large values are associated with significant terms.

The essential difference between the two methods is that **Maximize Significance** tends to choose splits involving fewer levels, compared to the **Maximize Split Statistic** method, which allows a significant result to happen by chance when there are many opportunities.

For continuous responses, **Maximize Significance** tends to choose groups with small within-sample variances, whereas **Maximize Split Statistic** takes effect size and sample size into account, rather than residual variance.

Missing Value Rule is a submenu with two choices. **Closest** assigns a missing value to the group that the response fits best with. Unfortunately, **Closest** gives too optimistic a view on how good the split is. **Random**, the default, is an option used when a more realistic measure of the goodness of fit is needed. It is also appropriate for situations involving a lot of missing values. With the **Random** option, JMP assigns the missing values randomly with probability based on the non-missing sample sizes in the partition being split.

Minimum Size Split presents a dialog box where you enter a number or a fractional portion of the total sample size which becomes the minimum size split allowed. The default is 5. To specify a fraction of the sample size, enter a value less than 1. To specify an actual number, enter a value greater than or equal to 1.

Lock Columns reveals a check box table to allow you to interactively lock columns so that they are not considered for splitting. You can toggle off the check box to display without affecting the individual locks.

Plot Actual by Predicted produces a plot of actual values by predicted values.

Small Tree View displays a smaller version of the partition tree to the right of the scatterplot.

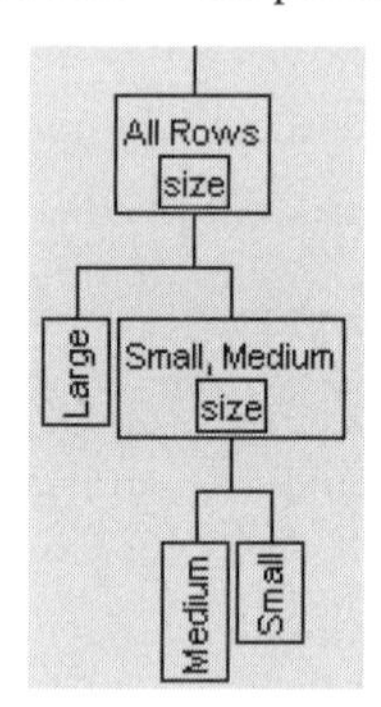

Leaf Report gives the mean and count or rates for the bottom-level leaves of the report.

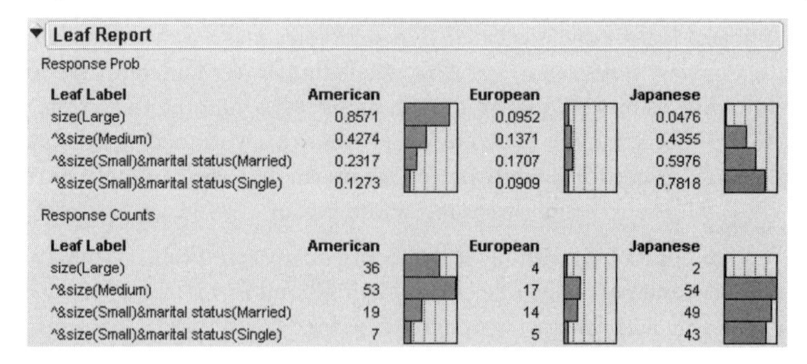
Leaf Report

Response Prob

Leaf Label	American	European	Japanese
size(Large)	0.8571	0.0952	0.0476
^&size(Medium)	0.4274	0.1371	0.4355
^&size(Small)&marital status(Married)	0.2317	0.1707	0.5976
^&size(Small)&marital status(Single)	0.1273	0.0909	0.7818

Response Counts

Leaf Label	American	European	Japanese
size(Large)	36	4	2
^&size(Medium)	53	17	54
^&size(Small)&marital status(Married)	19	14	49
^&size(Small)&marital status(Single)	7	5	43

Save Columns is a submenu, detailed below.

Column Contributions brings up a report showing how each input column contributed to the fit, including how many times it was split and the total G^2 or Sum of Squares attributed to that column.

Split History shows the Split History graph. This graph displays the R^2 vs. the number of splits. If you use *K*-fold or excluded row cross-validation, separate curves are drawn for the *K*-Fold R^2 or the Excluded R^2. The following picture shows a Split History graph with excluded points.

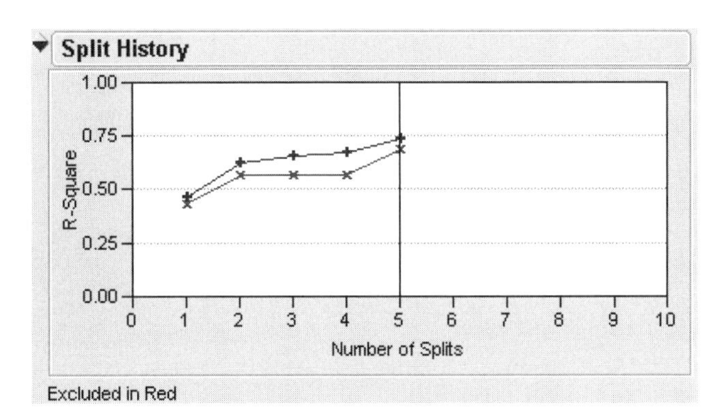

K Fold Crossvalidation randomly assigns all the (nonexcluded) rows to one of *k* groups, then calculates the error for each point using means or rates estimated from all the groups except the group to which the point belongs. The result is a cross-validated R^2.

Lift Curve is described in the section “Lift Curves,” p. 674.

ROC Curve is described in the section “ROC Curve,” p. 672.

Color Points colors the points based on their response classification. This option is only available for categorical predictors. To make this helpful command easily accessible, a button for it is also provided. The button is removed once the command is performed.

Script is the typical JMP script submenu, used to repeat the analysis or save a script to various places.

The following commands appear under the **Save Columns** submenu.

Save Residuals saves the residual values from the model to the data table.

Save Predicted saves the predicted values from the model to the data table. These are the leaf means for continuous responses, or a set of response level rates for categorical responses.

Save Leaf Numbers saves the leaf numbers of the tree to a column in the data table.

Save Leaf Labels saves leaf labels of the tree to the data table. The labels document each branch that the row would trace along the tree, with each branch separated by “&”. A label in the above analysis could be represented as “size(Small,Medium)&size(Small)”. However, JMP does not include redundant information in the form of category labels that are repeated. When a category label for a leaf references an inclusive list of categories in a higher tree node, JMP places a caret (‘^”) where the tree node with redundant labels occurs. Therefore, “size(Small,Medium)&size(Small)” is presented as ^&size(Small).

Save Prediction Formula saves the prediction formula to a column in the data table. The formula is made up of nested conditional clauses. For categorical responses, there is a new column for each response level.

Save Leaf Number Formula saves a formula that computes the leaf number to a column in the data table.

Save Leaf Label Formula saves a formula that computes the leaf label to a column in the data table.

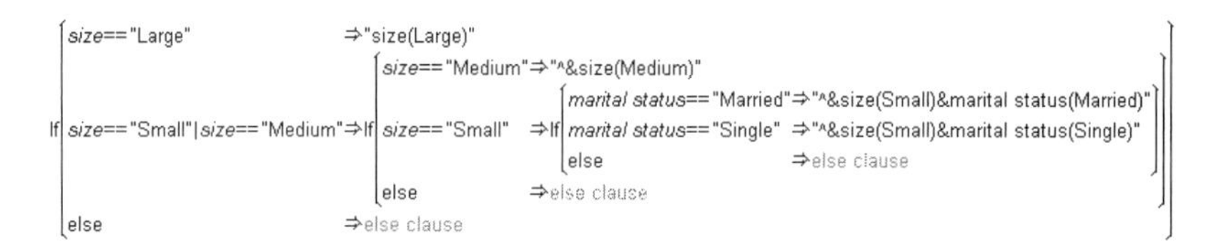

	sex	marital status	age	country	size	type	Leaf Label Formula
4	Male	Single	29	American	Large	Family	size(Large)
5	Male	Married	39	American	Medium	Family	^&size(Medium)
6	Male	Single	34	Japanese	Medium	Family	^&size(Medium)
7	Female	Married	42	American	Large	Family	size(Large)
8	Female	Married	40	European	Medium	Family	^&size(Medium)
9	Male	Married	28	American	Medium	Sporty	^&size(Medium)
10	Female	Married	26	American	Medium	Family	^&size(Medium)
11	Female	Married	26	European	Small	Sporty	^&size(Small)&marital status(Married)
12	Male	Single	26	European	Medium	Sporty	^&size(Medium)
13	Female	Married	37	American	Medium	Sporty	^&size(Medium)
14	Male	Single	25	Japanese	Small	Sporty	^&size(Small)&marital status(Single)
15	Female	Single	27	Japanese	Medium	Family	^&size(Medium)
16	Male	Married	41	American	Large	Family	size(Large)
17	Male	Single	35	American	Small	Sporty	^&size(Small)&marital status(Single)
18	Male	[illegible]	[illegible]	[illegible]	Medium	Work	^&size(Medium)
19	[illegible]	[illegible]	[illegible]	[illegible]	[illegible]	[illegible]	[illegible]

At each split, the submenu has these commands

Split Best finds and executes the best split at or below this node.

Split Here splits at the selected node on the best column to split by.

Split Specific lets you specify where a split takes place. This is useful in showing what the criterion is as a function of the cut point, as well as in determining custom cut points. The resulting table has a table script that allows for easy graphing of the data.

After selecting this command, the following dialog appears.

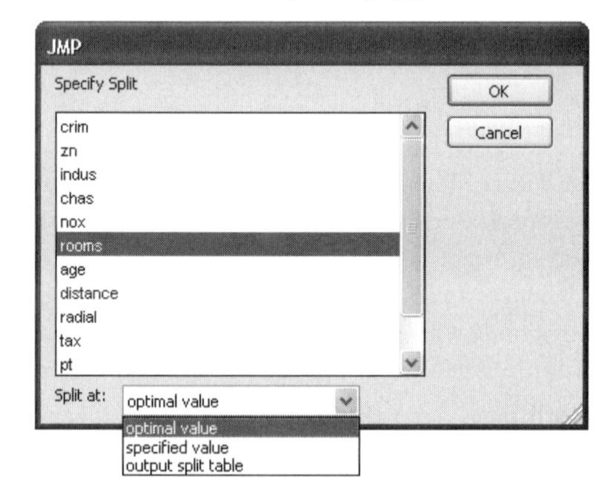

Optimal Value splits at the optimal value of the selected variable.

Specified Value allows you to specify the level where the split takes place.

Output Split Table produces a data table showing all possible splits and their associated split value.

Prune Below eliminates the splits below the selected node.

Prune Worst finds and removes the worst split below the selected node.

Select Rows selects the data table rows corresponding to this node. You can extend the selection by holding down the Shift key and choosing this command from another node. With the Shift key held down, the current node's rows are not deselected.

Show Details produces a data table that shows the split criterion for a selected variable. The data table, composed of split intervals and their associated criterion values, has an attached script that produces an appropriate graph for the criterion.

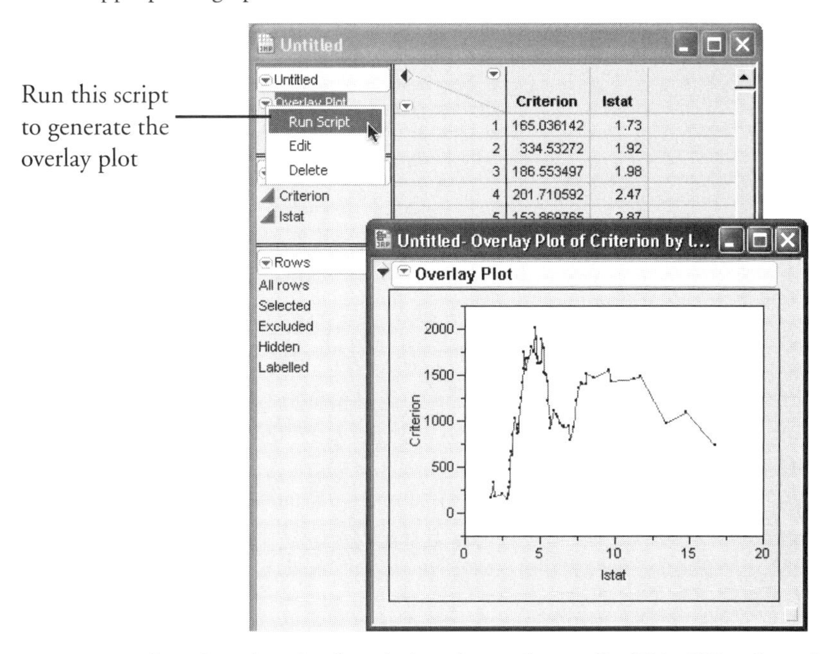

Lock prevents a node or its subnodes from being chosen for a split. This differs from the **Lock Columns** command, which prevents certain columns from being chosen for future splits. When **Lock** is checked, a lock icon is shown in the node title.

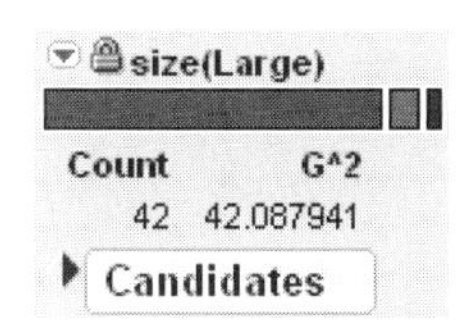

Graphs for Goodness of Fit

The graph for goodness of fit depends on which type of response you use.

Actual by Predicted Plot

For continuous responses, the Actual by Predicted plot shows how well it is fit. Each leaf is predicted with its mean, so the x-coordinates are these means. The actual values form a scatter of points around each leaf mean. A diagonal line represents the locus of where predicted and actual values are the same. For a perfect fit, all the points would be on this diagonal.

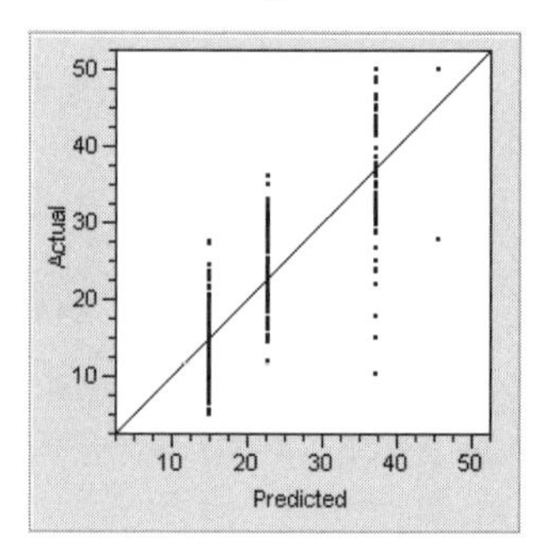

ROC Curve

The classical definition of ROC curve involves the count of True Positives by False Positives as you accumulate the frequencies across a rank ordering. The True Positive *y*-axis is labeled "Sensitivity" and the False Positive *X*-axis is labeled "1-Specificity". The idea is that if you slide across the rank ordered predictor and classify everything to the left as positive and to the right as negative, this traces the trade-off across the predictor's values.

To generalize for polytomous cases (more than 2 response levels), Partition creates an ROC curve for each response level versus the other levels. If there are only two levels, one is the diagonal reflection of the other, representing the different curves based on which is regarded as the "positive" response level.

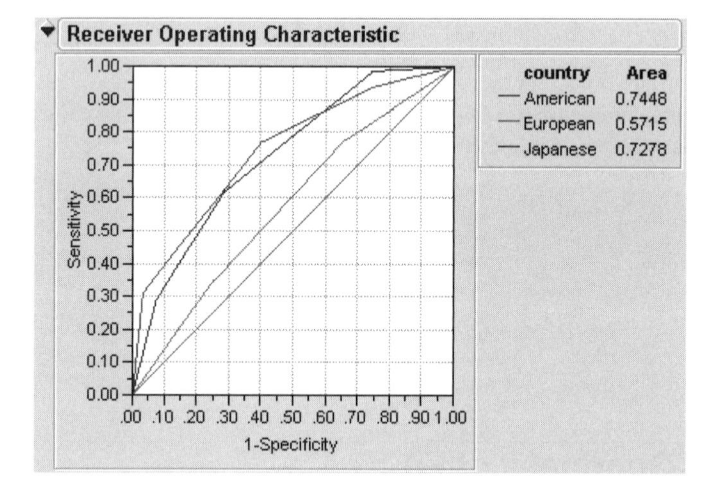

ROC curves are nothing more than a curve of the sorting efficiency of the model. The model rank-orders the fitted probabilities for a given *Y*-value, then starting at the lower left corner, draws the curve *up* when the row comes from that category, and to the right when the *Y* is another category.

In the following picture, n1= the number of *Y*'s where *Y*=1 and n0 is the number of *Y*'s where *Y*=0

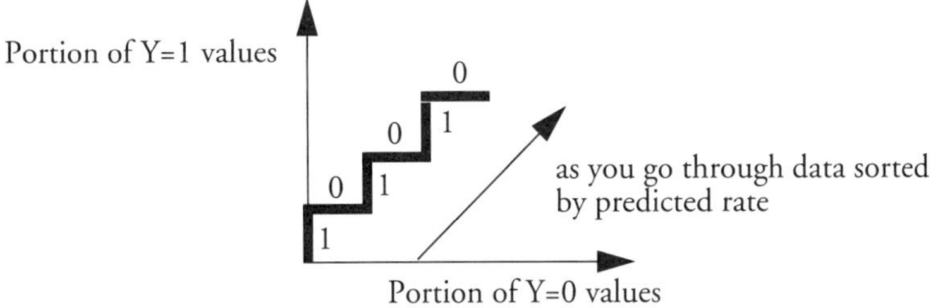

If the model perfectly rank-orders the response values, then the sorted data has all the targeted values first, followed by all the other values. The curve moves all the way to the top before it moves at all to the right.

Figure 34.9 ROC for Perfect Fit

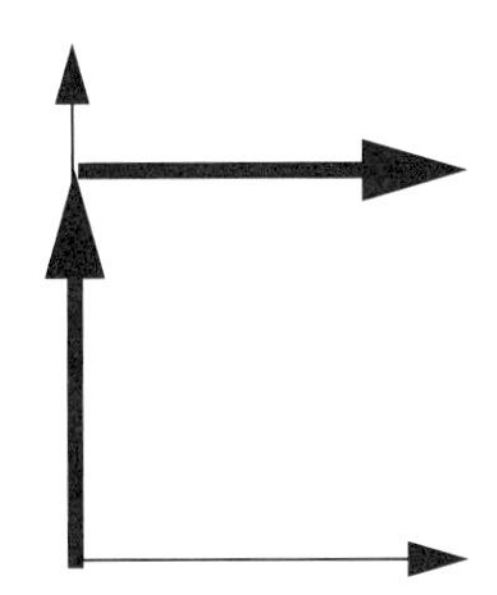

If the model does not predict well, it wanders more or less diagonally from the bottom left to top right.

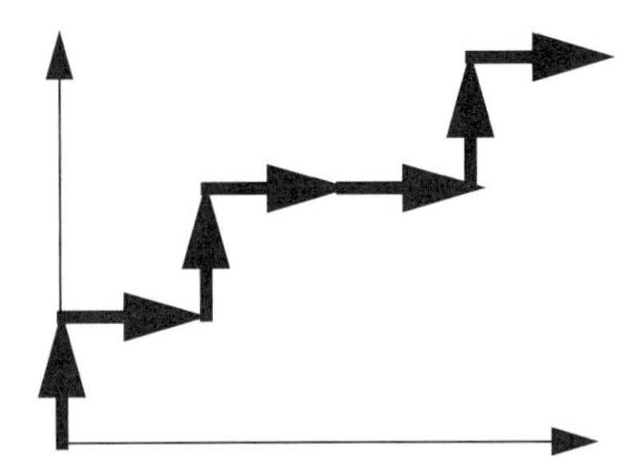

In practice, the curve lifts off the diagonal. The area under the curve is the indicator of the goodness of fit, and varies between 0.5 (no predictive value) and 1 (perfect fit).

If a partition contains a section that is all or almost all one response level, then the curve lifts almost vertically at the left for a while. This means that a sample is almost completely sensitive to detecting that level. If a partition contains none or almost none of a response level, the curve at the top will cross almost horizontally for a while. This means that there is a sample that is almost completely specific to not having that response level.

Because partitions contain clumps of rows with the same (*i.e.* tied) predicted rates, the curve actually goes slanted, rather than purely up or down.

For polytomous cases, you get to see which response categories lift off the diagonal the most. In the CarPoll example above, the European cars are being identified much less than the other two categories.

The American's start out with the most sensitive response (Size(Large)) and the Japanese with the most negative specific (Size(Large)'s small share for Japanese).

Lift Curves

A lift curve shows the same information as an ROC curve, but in a way to dramatize the richness of the ordering at the beginning. The Y-axis shows the ratio of how rich that portion of the population is in the chosen response level compared to the rate of that response level as a whole. For example, if the top-rated 10% of fitted probabilities have a 25% richness of the chosen response compared with 5% richness over the whole population, the lift curve would go through the *X*-coordinate of 0.10 at a Y-coordinate of 25% / 5%, or 5. All lift curves reach (1,1) at the right, as the population as a whole has the general response rate.

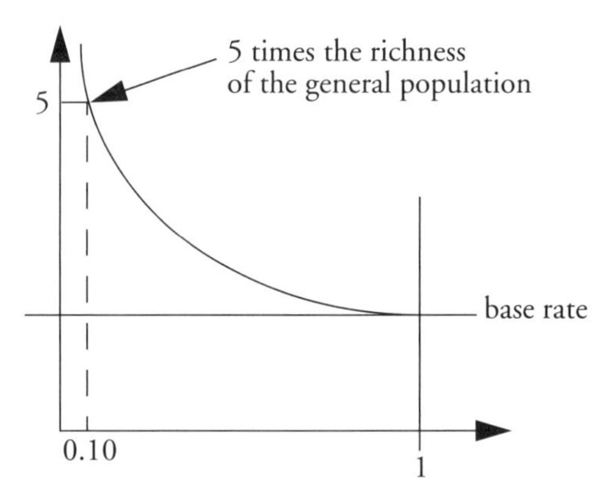

Sorted from highest predicted rate to lowest predicted rate

In problem situations where the response rate for a category is very low anyway (for example, a direct mail response rate), the lift curve explains things with more detail than the ROC curve.

Figure 34.10 Lift Curve

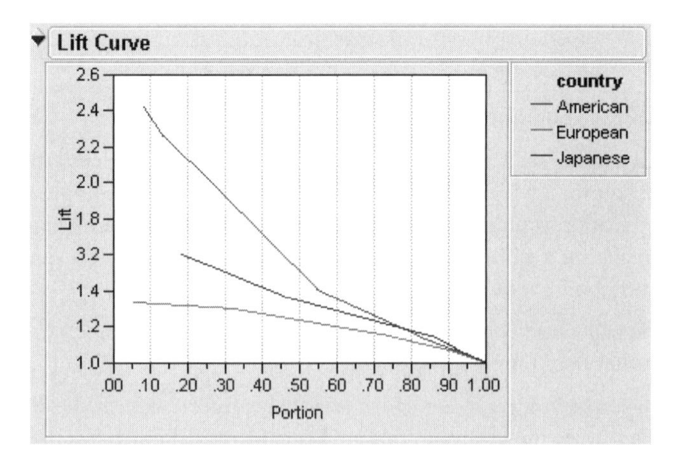

Cross-validation

cross-validation can be accomplished in two ways in the partition platform.

- using K-fold cross-validation
- using excluded rows.

K-Fold Cross-validation

K-fold cross-validation randomly assigns all the (nonexcluded) rows to one of *k* groups, then calculates the error for each point using means or rates estimated from all the groups except the group to which the point belongs. The result is a cross-validated R^2.

Figure 34.11 K-Fold cross-validation Reports

Categorical Response Report

Crossvalidation

k-fold		-2LogLike	RSquare
5	Folded	553.532648	0.0727
	Overall	543.168336	0.0900

Continuous Response Report

Crossvalidation

k-fold		SSE	RSquare
5	Folded	23531.012	0.4491
	Overall	23376.7404	0.4527

Excluded Rows

It is common to use part of a data set to compute a model (the training data set) and use the remaining data to validate the model (the validation data set). To use this technique in JMP, select rows to exclude and exclude them with the **Rows > Exclude/Unexclude** command. Frequently, the rows are selected with the **Rows > Row Selection > Select Randomly** command, which allows you to select a certain percentage or certain number of rows randomly.

If you run the Partition platform with excluded rows, the report changes slightly to show R^2 for both the included and excluded portions of the data set. Figure 34.12 shows this additional report.

Figure 34.12 Reports with Excluded Values

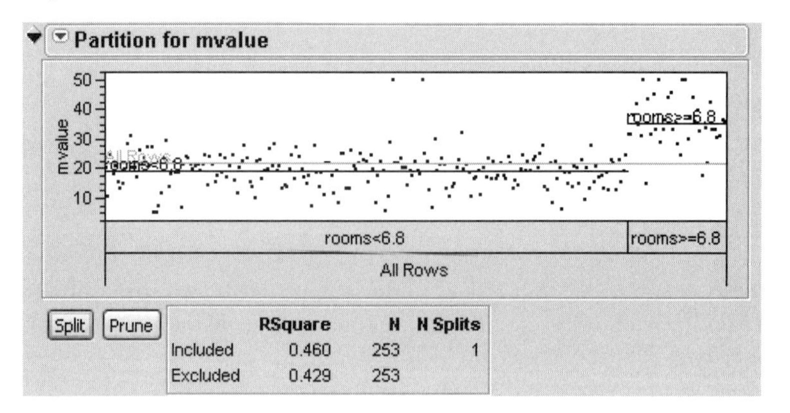

Example with Fisher's Irises

An illustrative example of partitioning is shown with the Iris.jmp sample data table. This is the classic data set containing measurements of iris flowers for three species.

After opening the data set, select **Analyze >Modeling > Partition**. Assign species the **Y** role and Petal Length and Petal Width the **X** role. The report in Figure 34.13 appears.

Figure 34.13 Initial Partition Report

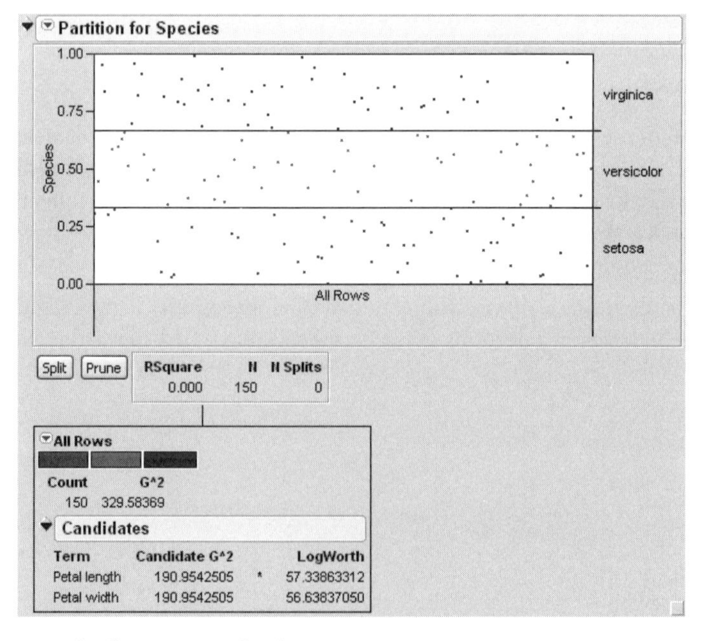

The first split (Figure 34.14) partitions the data at Petal Length = 3.

Figure 34.14 First Split of Iris Data

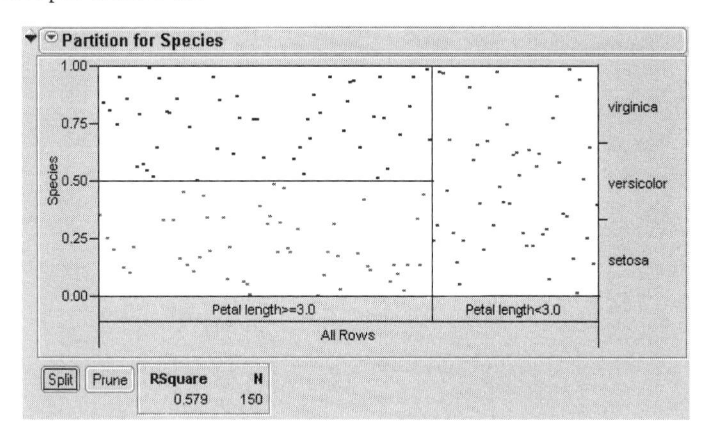

This is a good split location because all the "setosa" irises have petal length less than three, and no other species does. Therefore, this response group perfectly discriminates. A plot of Petal Width vs. Petal Length, with a reference line at Petal Length = 3, shows this clearly (Figure 34.15).

Figure 34.15 Bivariate Plot of Petal Width vs. Petal Length

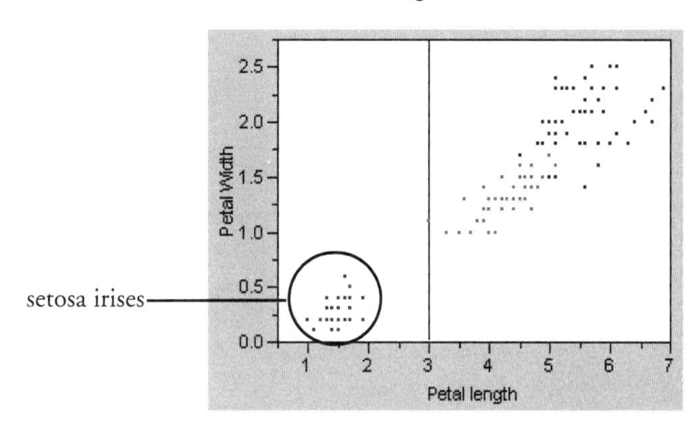

A second split results in the report shown in Figure 34.16.

Figure 34.16 Second Split Report

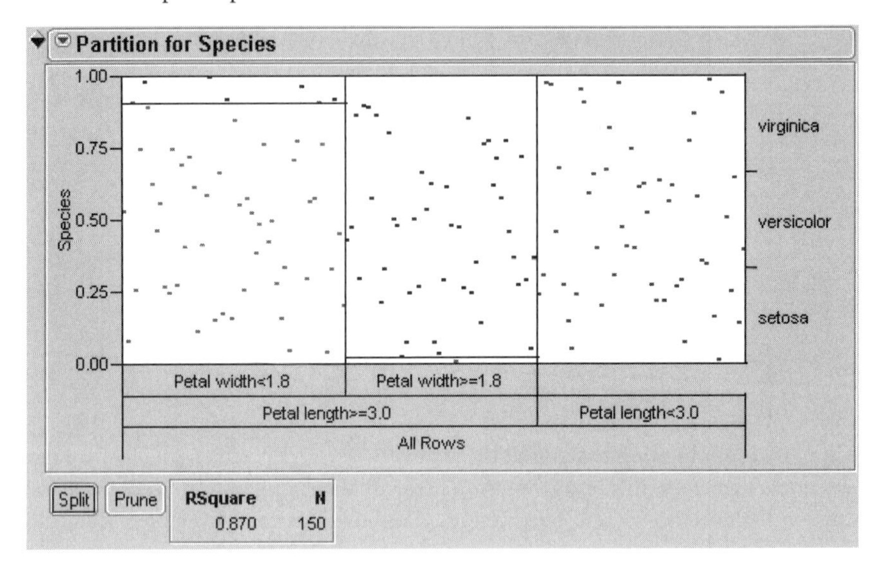

The second split partitioned at Pedal Width = 1.8. Again, a bivariate plot shows the situation.

Figure 34.17 Bivariate Plot After Second Split

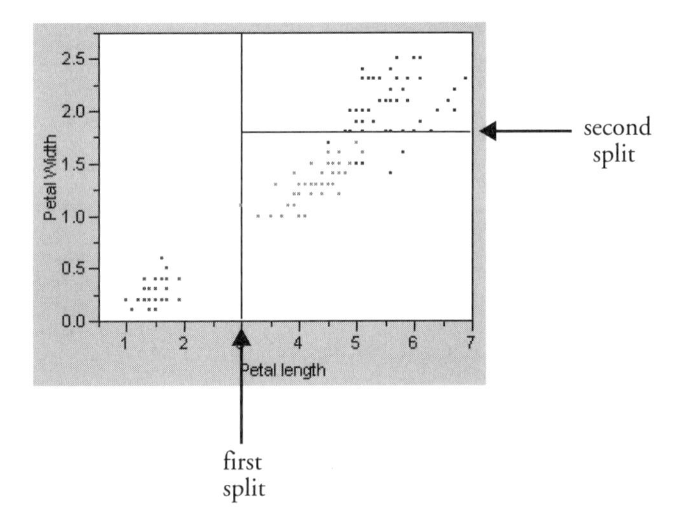

This split is the best partition of the remaining points—in this case, the species "virginica" and "versicolor".

A third split partitions the petal length again at Petal Length = 5. Examine the bivariate plot(Figure 34.19) to see why this split works well.

Figure 34.18 Third Split Report

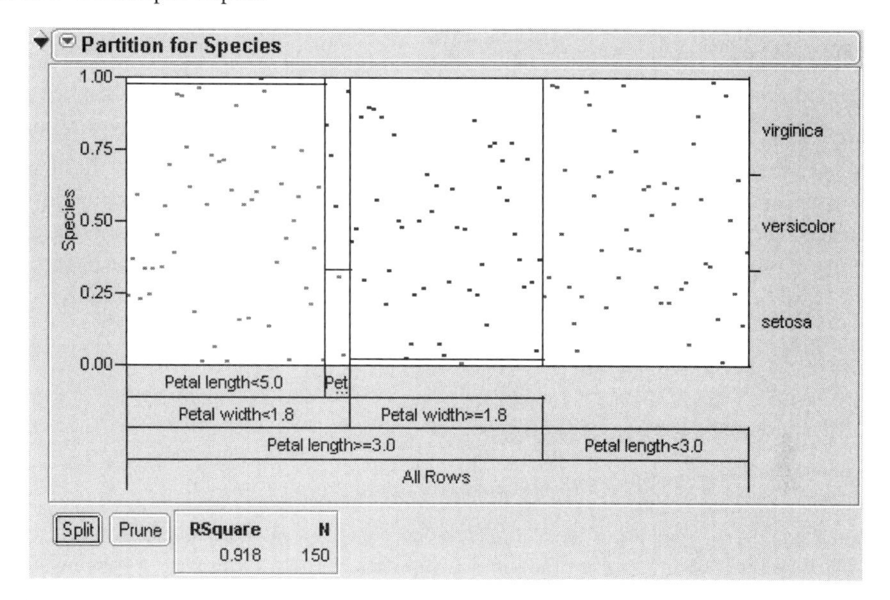

Figure 34.19 Bivariate Plot for Third Split

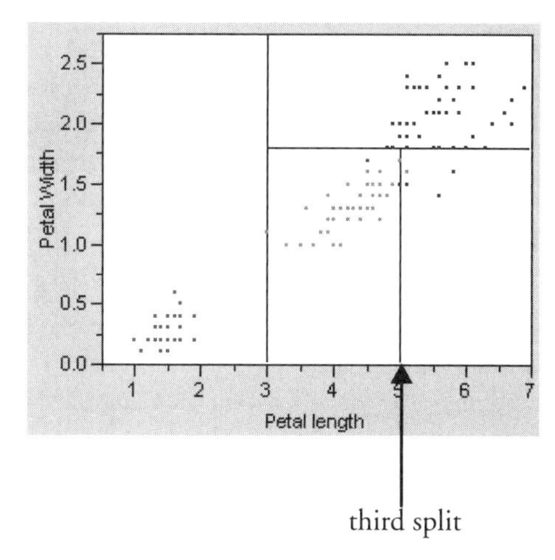

The result is that most of the data is "purified" into partitions dominated by one species, with one remaining partition in the boundary region that has a more even mixture of species

Statistical Details

The response can be either continuous, or categorical (nominal or ordinal). If *Y* is categorical, then it is fitting the probabilities estimated for the response levels, minimizing the residual log-likelihood chi-square [2*entropy]. If the response is continuous, then the platform fits means, minimizing the sum of squared errors.

The factors can be either continuous, or categorical (nominal or ordinal). If an *X* is continuous, then the partition is done according to a splitting "cut" value for *X*. If *X* is categorical, then it divides the *X* categories into two groups of levels and considers all possible groupings into two levels.

Criterion. For continuous responses, the statistic is the Sum of Squares, that is, the change in the squared errors due to the split. For categorical responses, the criterion is the likelihood-ratio chi-square, reported as **G^2** in the report. This is actually twice the [natural log] entropy or twice the change in the entropy. Entropy is Σ -log(*p*) for each observation, where *p* is the probability attributed to the response that occurred.

A candidate G^2 that has been chosen is

$$G^2_{\text{test}} = G^2_{\text{parent}} - (G^2_{\text{left}} + G^2_{\text{right}}).$$

A candidate SS that has been chosen is

$$SS_{\text{test}} = SS_{\text{parent}} - (SS_{\text{right}} + SS_{\text{left}}) \text{ where SS in a node is just } s^2(n - 1).$$

When Partition calculates a G^2 or R^2 on excluded data for a categorical response, it uses the rate value $0.25/m$ when it encounters a zero rate in a group with m rows. Otherwise, a missing statistic would be reported, since the logarithm of zero is undefined.

Performance

Although most of the platform is very fast, there is one case that is not—a categorical response with more than two levels, and a categorical *X* with more than ten levels. For *k* levels in the *X*, it searches all possible splits, which is 2^{k-1}. This quickly becomes a large number. For now, avoid this case. In a future release, heuristic ways of handling this case are planned.

Missing Values

The **Missing Value Rule > Random** option assigns a missing X to a random group rather than best-fitting group. For **Missing Value Rule > Closest**, if the *Y* is missing, the row is simply ignored. If an *X* is missing, the row is currently ignored for calculating the criterion values, but then the row becomes part of the group that it best fits into, determined by *Y*'s value.

Stopping Rules

Currently, the platform is purely interactive. Keep pushing the **Split** button until the result is satisfactory. Unlike Data Mining packages that are expected to run unattended and automatically stop, Partition is an exploratory platform intended to help you investigate relationships interactively.

35

Time Series Analysis

The Time Series Platform

The Time Series platform lets you explore, analyze, and forecast univariate time series. A time series is a set $y_1, y_2, \ldots, y_N$ of observations taken over a series of equally-spaced time periods. The analysis begins with a plot of the points in the time series. In addition, the platform displays graphs of the autocorrelations and partial autocorrelations of the series. These indicate how and to what degree each point in the series is correlated with earlier values in the series. You can interactively add

Variograms a characterization of process disturbances

AR coefficients autoregressive coefficients

Spectral Density Plots versus period and frequency, with white noise tests.

These graphs can be used to identify the type of model appropriate for describing and predicting (forecasting) the evolution of the time series. The model types include

ARIMA autoregressive integrated moving-average, often called Box-Jenkins models

Seasonal ARIMA ARIMA models with a seasonal component

Smoothing Models several forms of exponential smoothing and Winter's method.

Transfer Function Models for modeling with input series.

Note: The Time Series Launch dialog requires that one or more continuous variables be assigned as the time series. Optionally, you can specify a time ID variable, which is used to label the time axis, or one or more input series. If a time ID variable is specified, it must be continuous, sorted ascending, and without missing values.

Contents

Launch the Platform

To begin a time series analysis, choose the **Time Series** command from the **Analyze > Modeling** submenu to display the Time Series Launch dialog (Figure 35.1). This dialog allows you to specify the number of lags to use in computing the autocorrelations and partial autocorrelations. It also lets you specify the number of future periods to forecast using each model fitted to the data. After you select analysis variables and click **OK** on this dialog, a platform launches with plots and accompanying text reports for each of the time series (*Y*) variables you specified.

Select Columns into Roles

You assign columns for analysis with the dialog in Figure 35.1. The selector list at the left of the dialog shows all columns in the current table. To cast a column into a role, select one or more columns in the column selector list and click a role button. Or, drag variables from the column selector list to one of the following role boxes:

X, Time ID for the *x*-axis, one variable used for labeling the time axis

Y, Time Series for the *y*-axis, one or more time series variables.

To remove an unwanted variable from an assigned role, select it in the role box and click **Remove.** After assigning roles, click **OK** to see the analysis for each time series variable versus the time ID.

You set the number of lags for the autocorrelation and partial autocorrelation plots in the **Autocorrelation Lags** box. This is the maximum number of periods between points used in the computation of the correlations. It must be more than one but less than the number of rows. A commonly used rule of thumb for the maximum number of lags is $n/4$, where n is the number of observations. The **Forecast Periods** box allows you to set the number of periods into the future that the fitted models are forecast. By default, JMP uses 25 lags and 25 forecast periods

The data for the next examples are in the Seriesg.jmp table found in the Time Series sample data folder (Box and Jenkins 1976). The time series variable is Passengers and the time ID is Time.

Figure 35.1 The Time Series Launch Dialog

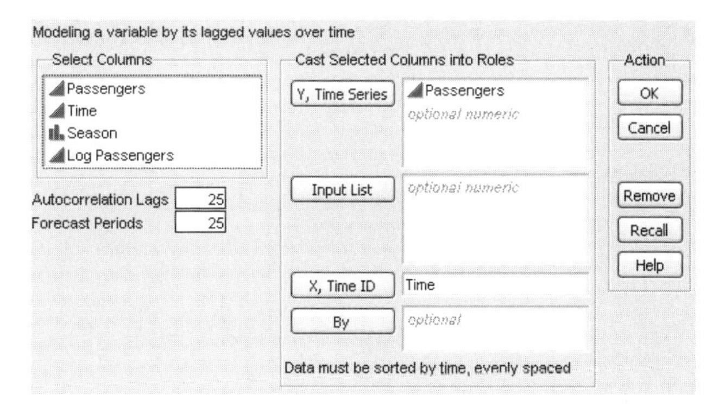

The Time Series Graph

The Time Series platform begins with a plot of each times series by the time ID, or row number if no time ID is specified (Figure 35.2). The plot, like others in JMP, has features to resize the graph, highlight points with the cursor or brush tool, and label points. See the JMP User Guide for a discussion of these features.

Figure 35.2 Time Series Plot of Seriesg (Airline Passenger) Data

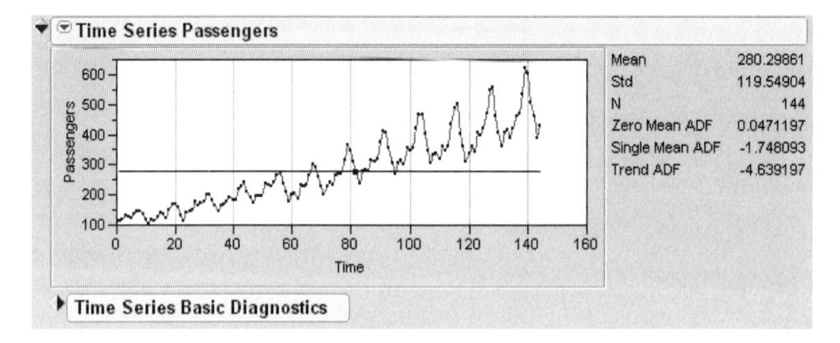

If you open **Time Series Basic Diagnostic Tables**, graphs of the autocorrelation and partial autocorrelation (Figure 35.3) of the time series are shown.

The platform popup menu, discussed next, also has fitting commands and options for displaying additional graphs and statistical tables.

Time Series Commands

The popup menu next to the time series name has the commands shown here.

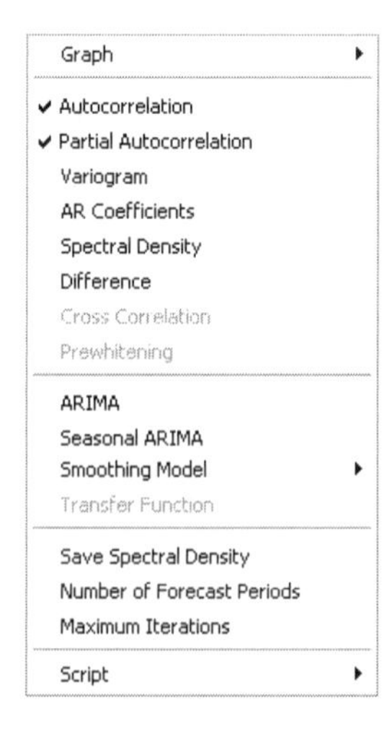

The first nine items in this menu control the descriptive and diagnostic graphs and tables. These are typically used to determine the nature of the model to be fitted to the series.

The **ARIMA**, **Seasonal ARIMA**, and **Smoothing Model** commands are for fitting various models to the data and producing forecasts. You can select the model fitting commands repeatedly. The result of each new fit is appended to the report. After the first model has been fit, a summary of all the models is inserted just above the first model report (an example is shown in "Model Comparison Table," p. 691).

The following sections describe options and model fits, discuss statistical results, and cover additional platform features.

Graph

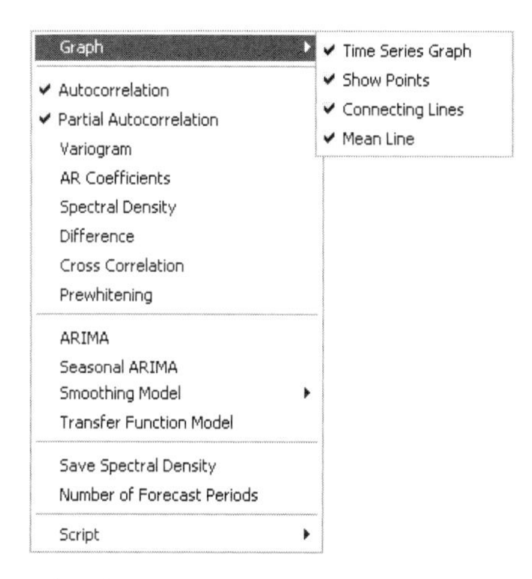

The Time Series platform begins by showing a time series plot, like the one shown previously in Figure 35.2. The **Graph** command on the platform popup menu has a submenu of controls for the time series plot with the following commands.

Time Series Graph hides or displays the time series graph.

Show Points hides or displays the points in the time series graph.

Connecting Lines hides or displays the lines connecting the points in the time series graph.

Mean Line hides or displays a horizontal line in the time series graph that depicts the mean of the time series.

Autocorrelation

The **Autocorrelation** command alternately hides or displays the autocorrelation graph of the sample, often called the *sample autocorrelation function*. This graph describes the correlation between all the pairs of points in the time series with a given separation in time or lag. The autocorrelation for the *k*th lag is

$$r_k = \frac{c_k}{c_0} \text{ where } c_k = \frac{1}{N}\sum_{t=k+1}^{N}(y_t-\bar{y})(y_{t-k}-\bar{y})$$

where $\bar{y}$ is the mean of the N nonmissing points in the time series. The bars graphically depict the autocorrelations.

By definition, the first autocorrelation (lag 0) always has length 1. The curves show twice the large-lag standard error (± 2 standard errors), computed as

$$SE_k = \sqrt{\frac{1}{N}\sum_{i=1}^{k-1} r_i^2}$$

The autocorrelation plot for the **Seriesg** data is shown on the left in Figure 35.3. You can examine the autocorrelation and partial autocorrelations plots to determine whether the time series is stationary (meaning it has a fixed mean and standard deviation over time) and what model might be appropriate to fit the time series.

In addition, the Ljung-Box Q and p-values are shown for each lag.The Q-statistic is used to test whether a group of autocorrelations is significantly different from zero or to test that the residuals from a model can be distinguished from white-noise.

Partial Autocorrelation

The **Partial Autocorrelation** command alternately hides or displays the graph of the sample partial autocorrelations. The plot on the right in Figure 35.3 shows the partial autocorrelation function for the **Seriesg** data. The solid blue lines represent ± 2 standard errors for approximate 95% confidence limits, where the standard error is computed

$$SE_k = \frac{1}{\sqrt{n}} \text{ for all } k$$

Figure 35.3 Autocorrelation and Partial Correlation Plots

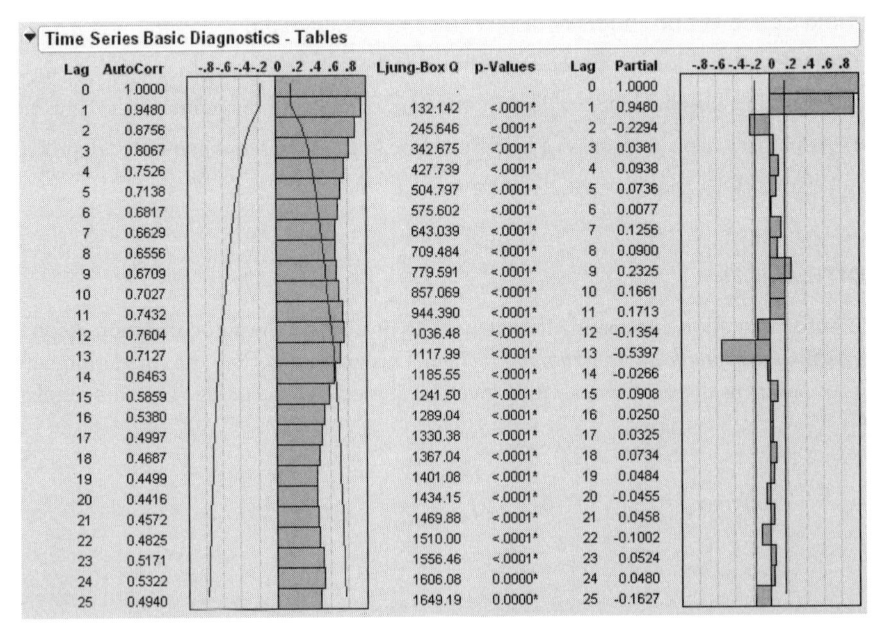

Time Series Basic Diagnostics - Tables

Lag	AutoCorr	Ljung-Box Q	p-Values	Lag	Partial
0	1.0000	.	.	0	1.0000
1	0.9480	132.142	<.0001*	1	0.9480
2	0.8756	245.646	<.0001*	2	-0.2294
3	0.8067	342.675	<.0001*	3	0.0381
4	0.7526	427.739	<.0001*	4	0.0938
5	0.7138	504.797	<.0001*	5	0.0736
6	0.6817	575.602	<.0001*	6	0.0077
7	0.6629	643.039	<.0001*	7	0.1256
8	0.6556	709.484	<.0001*	8	0.0900
9	0.6709	779.591	<.0001*	9	0.2325
10	0.7027	857.069	<.0001*	10	0.1661
11	0.7432	944.390	<.0001*	11	0.1713
12	0.7604	1036.48	<.0001*	12	-0.1354
13	0.7127	1117.99	<.0001*	13	-0.5397
14	0.6463	1185.55	<.0001*	14	-0.0266
15	0.5859	1241.50	<.0001*	15	0.0908
16	0.5380	1289.04	<.0001*	16	0.0250
17	0.4997	1330.38	<.0001*	17	0.0325
18	0.4687	1367.04	<.0001*	18	0.0734
19	0.4499	1401.08	<.0001*	19	0.0484
20	0.4416	1434.15	<.0001*	20	-0.0455
21	0.4572	1469.88	<.0001*	21	0.0458
22	0.4825	1510.00	<.0001*	22	-0.1002
23	0.5171	1556.46	<.0001*	23	0.0524
24	0.5322	1606.08	0.0000*	24	0.0480
25	0.4940	1649.19	0.0000*	25	-0.1627

Variogram

The **Variogram** command alternately displays or hides the graph of the variogram. The variogram measures the variance of the differences of points k lags apart and compares it to that for points one lag apart. The variogram is computed from the autocorrelations as

$$V_k = \frac{1 - r_{k+1}}{1 - r_1}$$

where r_k is the autocorrelation at lag k. The plot on the left in Figure 35.4 shows the Variogram graph for the Seriesg data.

AR Coefficients

The **AR Coefficients** command alternately displays or hides the graph of the least squares estimates of the autoregressive (AR) coefficients. The definition of these coefficients is given below. These coefficients approximate those that you would obtain from fitting a high-order, purely autoregressive model. The right-hand graph in Figure 35.4 shows the AR coefficients for the Seriesg data.

Figure 35.4 Variogram Graph (left) and AR Coefficient Graph (right)

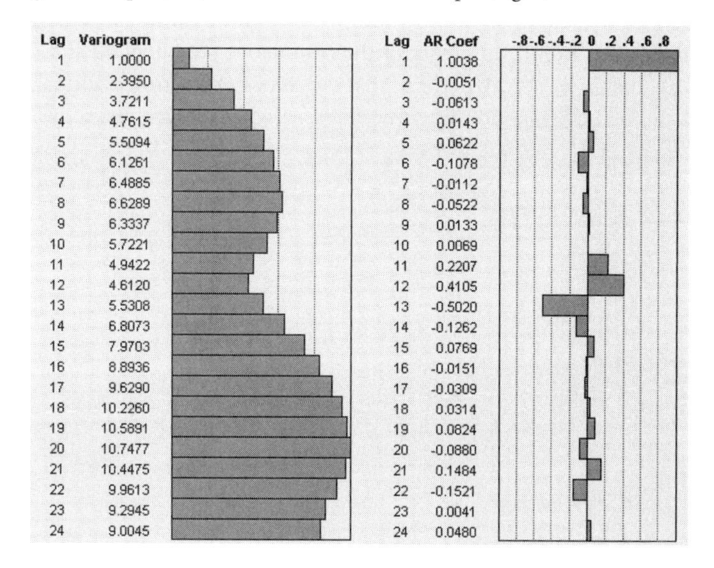

Lag	Variogram	Lag	AR Coef
1	1.0000	1	1.0038
2	2.3950	2	-0.0051
3	3.7211	3	-0.0613
4	4.7615	4	0.0143
5	5.5094	5	0.0622
6	6.1261	6	-0.1078
7	6.4885	7	-0.0112
8	6.6289	8	-0.0522
9	6.3337	9	0.0133
10	5.7221	10	0.0069
11	4.9422	11	0.2207
12	4.6120	12	0.4105
13	5.5308	13	-0.5020
14	6.8073	14	-0.1262
15	7.9703	15	0.0769
16	8.8936	16	-0.0151
17	9.6290	17	-0.0309
18	10.2260	18	0.0314
19	10.5891	19	0.0824
20	10.7477	20	-0.0880
21	10.4475	21	0.1484
22	9.9613	22	-0.1521
23	9.2945	23	0.0041
24	9.0045	24	0.0480

Spectral Density

The **Spectral Density** command alternately displays or hides the graphs of the spectral density as a function of period and frequency (Figure 35.5).

The least squares estimates of the coefficients of the Fourier series

$$a_t = \frac{2}{N} \sum_{i=1}^{N} y_t \cos(2\pi f_i t)$$

and

$$b_t = \frac{2}{N}\sum_{i=1}^{N} y_t \sin(2\pi f_i t)$$

where $f_i = i/N$ are combined to form the periodogram $I(f_i) = \frac{N}{2}(a_i^2 + b_i^2)$, which represents the intensity at frequency f_i.

The periodogram is smoothed and scaled by $1/(4\pi)$ to form the spectral density.

The *Fisher's Kappa* statistic tests the null hypothesis that the values in the series are drawn from a normal distribution with variance 1 against the alternative hypothesis that the series has some periodic component. Kappa is the ratio of the maximum value of the periodogram, $I(f_i)$, and its average value. The probability of observing a larger Kappa if the null hypothesis is true is given by

$$(k > \kappa) = 1 - \sum_{j=0}^{q} (-1)^j \binom{q}{j} \left[\max\left(1 - \frac{jk}{q}, 0\right)\right]^{q-1}$$

where $q = N / 2$ if N is even, $q = (N - 1) / 2$ if N is odd, and κ is the observed value of Kappa. The null hypothesis is rejected if this probability is less than the significance level α.

For $q - 1 > 100$, *Bartlett's Kolmogorov-Smirnov* compares the normalized cumulative periodogram to the cumulative distribution function of the uniform distribution on the interval (0, 1). The test statistic equals the maximum absolute difference of the cumulative periodogram and the uniform CDF. If it exceeds $a/(\sqrt{q})$, then reject the hypothesis that the series comes from a normal distribution. The values $a = 1.36$ and $a = 1.63$ correspond to significance levels 5% and 1% respectively.

Figure 35.5 Spectral Density Plots

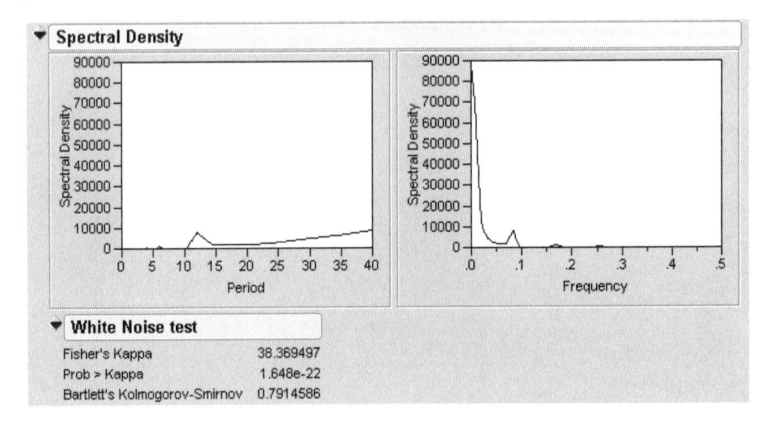

Save Spectral Density

Save Spectral Density creates a new table containing the spectral density and periodogram where the *i*th row corresponds to the frequency $f_i = 1 / N$ (that is, the *i*th harmonic of $1 / N$)

The new data table has these columns:

Period is the period of the *i*th harmonic, $1 / f_i$.

Frequency is the frequency of the harmonic, f_i.

Angular Frequency is the angular frequency of the harmonic, $2\pi f_i$

Sine is the Fourier sine coefficients, a_i

Cosine is the Fourier cosine coefficients, b_i

Periodogram is the periodogram, $I(f_i)$.

Spectral Density is the spectral density, a smoothed version of the periodogram.

Number of Forecast Periods

The **Number of Forecast Periods** command displays a dialog for you to reset the number of periods into the future that the fitted models will forecast. The initial value is set in the Time Series Launch dialog. All existing and future forecast results will show the new number of periods with this command.

Difference

Many time series do not exhibit a fixed mean. Such nonstationary series are not suitable for description by some time series models such as those with only autoregressive and moving average terms (ARMA models). However, these series can often be made stationary by differencing the values in the series. The differenced series is given by

$$w_t = (1 - B)^d (1 - B^s)^D y_t$$

where t is the time index and **B** is the backshift operator defined by $\mathbf{B}y_t = y_{t-1}$.

The **Difference** command computes the differenced series and produces graphs of the autocorrelations and partial autocorrelations of the differenced series. These graphs can be used to determine if the differenced series is stationary.

Several of the time series models described in the next sections accommodate a differencing operation (the ARIMA, Seasonal ARIMA models, and some of the smoothing models). The **Difference** command is useful for determining the order of differencing that should be specified in these models.

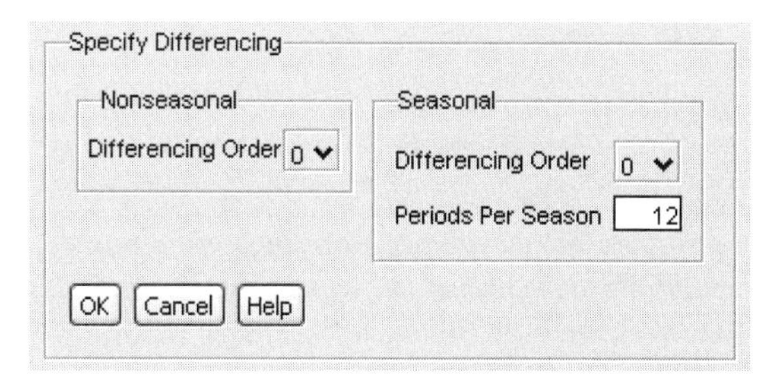

The Specify Differencing dialog appears in the report window when you select the Difference command. It allows you to specify the differencing operation you want to apply to the time series. Click **OK** to see the results of the differencing operation. The Specify Differencing dialog allows you to specify the Nonseasonal Differencing Order, d, the Seasonal Differencing Order, D, and the number of Periods Per Season, s. Selecting zero for the value of the differencing order is equivalent to no differencing of that kind.

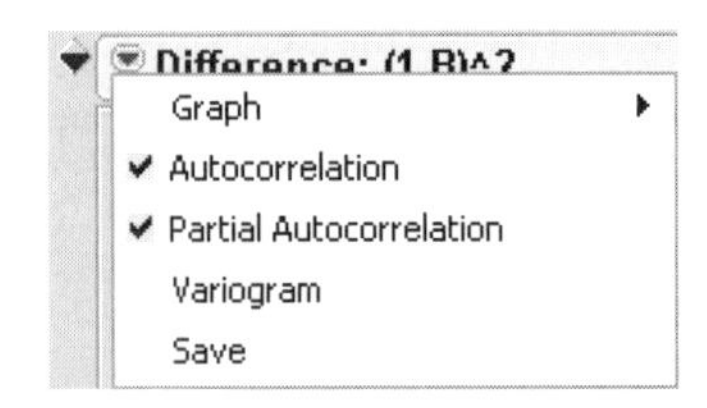

The Difference plot has the popup menu shown here, with the following options:

Graph controls the plot of the differenced series and behaves the same as those under the Time Series **Graph** menu.

Autocorrelation alternately displays or hides the autocorrelation of the differenced series.

Partial Autocorrelation alternately hides or displays the partial autocorrelations of differenced series.

Variogram alternately hides or displays the variogram of the differenced series.

Save appends the differenced series to the original data table. The leading $d + sD$ elements are lost in the differencing process. They are represented as missing values in the saved series.

Modeling Reports

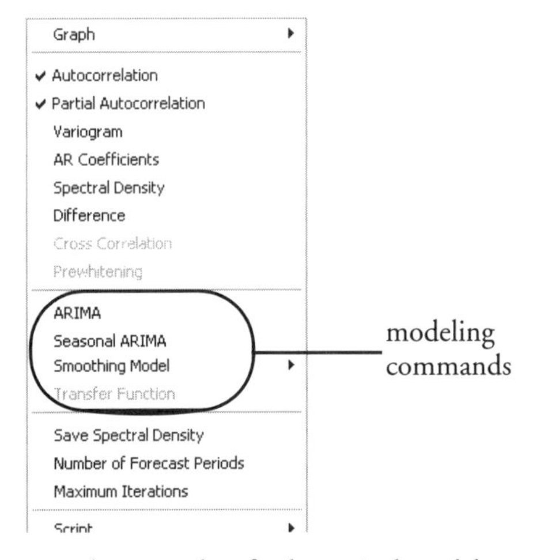

The time series modeling commands are used to fit theoretical models to the series and use the fitted model to predict (forecast) future values of the series. These commands also produce statistics and residuals that allow you to ascertain the adequacy of the model you have elected to use. You can select the modeling commands repeatedly. Each time you select a model, a report of the results of the fit and a forecast is added to the platform results.

The fit of each model begins with a dialog that lets you specify the details of the model being fit as well as how it will be fit. Each general class of models has its own dialog, as discussed previously in their respective sections. The models are fit by maximizing the likelihood function, using a Kalman filter to compute the likelihood function. The ARIMA, seasonal ARIMA, and smoothing models begin with the following report tables.

Model Comparison Table

Model Comparison

Model	DF	Variance	AIC	SBC	RSquare	-2LogLH	AIC Rank	SBC Rank	MAPE	MAE
AR(1)	142	0.0021557	-470.3323	-464.3926	0.922	-474.3323	3	3	1.728906	0.041140
MA(1)	142	0.0110938	-235.3252	-229.3855	0.682	-239.3252	4	4	3.729174	0.088265
ARMA(1, 1)	141	0.0020344	-477.6091	-468.6997	0.924	-483.6091	2	2	1.692408	0.040146
ARIMA(1, 1, 1)	140	0.0019644	-482.1410	-473.2525	0.946	-488.141	1	1	1.575766	0.037706

The Model Comparison table summarizes the fit statistics for each model. You can use it to compare several models fitted to the same time series. Each row corresponds to a different model. The numerical values in the table are drawn from the Model Summary table for each fitted model. The Model Comparison table shown above summarizes the ARIMA models (1, 0, 0), (0, 0, 1), (1, 0, 1), and (1, 1, 1) respectively.

Model Summary Table

Each model fit generates a Model Summary table, which summarizes the statistics of the fit. In the formulae below, n is the length of the series and k is the number of fitted parameters in the model.

DF is the number of degrees of freedom in the fit, $n - k$.

Sum of Squared Errors is the sum of the squared residuals.

Model: ARIMA(1, 1, 1)

Model Summary

DF	140	Stable	Yes
Sum of Squared Errors	0.27501907	Invertible	Yes
Variance Estimate	0.00196442		
Standard Deviation	0.0443218		
Akaike's 'A' Information Criterion	-482.14102		
Schwarz's Bayesian Criterion	-473.25249		
RSquare	0.94630418		
RSquare Adj	0.9455371		
MAPE	1.57576648		

Variance Estimate the unconditional sum of squares (SSE) divided by the number of degrees of freedom, SSE / $(n - k)$. This is the sample estimate of the variance of the random shocks a_t, described in the section "ARIMA Model," p. 695.

Standard Deviation is the square root of the variance estimate. This is a sample estimate of the standard deviation of a_t, the random shocks

Akaike's Information Criterion [AIC], Schwartz's Bayesian Criterion [SBC or BIC] Smaller values of these criteria indicate better fit. They are computed:

$$\text{AIC} = -2\text{loglikelihood} + 2k \text{ and } \text{SBC} = -2\text{loglikelihood} + k\ln(n)$$

RSquare RSquare is computed

$1 - \frac{\text{SSE}}{\text{SST}}$ where $\text{SST} = \sum_{i=1}^{n} (y_i - \bar{y}_i)^2$ and $\text{SSE} = \sum_{i=1}^{n} (y_i - \hat{y}_i)^2$, $\hat{y}_i$ are the one-step-ahead forecasts, and $\bar{y}_i$ is the mean y_i.

Note: If the model fits the series badly, the model error sum of squares, SSE might be larger than the total sum of squares, SST and R^2 will be negative.

RSquare Adj The adjusted R^2 is

$$1 - \left[\frac{(n-1)}{(n-k)}(1 - R^2)\right] .$$

MAPE is the Mean Absolute Percentage Error, and is computed

$$\frac{100}{n} \sum_{i=1}^{n} \left| \frac{y_i - \hat{y}_i}{y_i} \right|$$

MAE is the Mean Absolute Error, and is computed

$$\frac{1}{n} \sum_{i=1}^{n} |y_i - \hat{y}_i|$$

-2LogLikelihood is minus two times the natural log of the likelihood function evaluated at the best-fit parameter estimates. Smaller values are better fits.

Stable indicates whether the autoregressive operator is stable. That is, whether all the roots of $\phi(z) = 0$ lie outside the unit circle.

Invertible indicates whether the moving average operator is invertible. That is, whether all the roots of $\theta(z) = 0$ lie outside the unit circle.

Note: The ϕ and θ operators are defined in the section "ARIMA Model," p. 695.

Parameter Estimates Table

Model: ARIMA(1, 1, 1)

Model Summary

Parameter Estimates

Term	Lag	Estimate	Std Error	t Ratio	Prob>|t|	Constant Estimate
AR1	1	-0.4767619	0.1152473	-4.14	<.0001*	3.61907311
MA1	1	-0.864542	0.0713742	-12.11	<.0001*	
Intercept	0	2.45068156	3.2495581	0.75	0.4520	

There is a Parameter Estimates table for each selected fit, which gives the estimates for the time series model parameters. Each type of model has its own set of parameters. They are described in the sections on specific time series models. The Parameter Estimates table has these terms:

Term lists the name of the parameter. These are described below for each model type. Some models contain an *intercept* or mean term. In those models, the related *constant estimate* is also shown. The definition of the constant estimate is given under the description of ARIMA models.

Factor (Seasonal ARIMA only) lists the factor of the model that contains the parameter. This is only shown for multiplicative models. In the multiplicative seasonal models, Factor 1 is nonseasonal and Factor 2 is seasonal.

Lag (ARIMA and Seasonal ARIMA only) lists the degree of the lag or backshift operator that is applied to the term to which the parameter is multiplied.

Estimate lists the parameter estimates of the time series model.

Std Error lists the estimates of the standard errors of the parameter estimates. They are used in constructing tests and confidence intervals.

t Ratio lists the test statistics for the hypotheses that each parameter is zero. It is the ratio of the parameter estimate to its standard error. If the hypothesis is true, then this statistic has an approximate Student's *t*-distribution. Looking for a *t*-ratio greater than 2 in absolute value is a common rule of thumb for judging significance because it approximates the 0.05 significance level.

Prob>|t| lists the observed significance probability calculated from each *t*-ratio. It is the probability of getting, by chance alone, a *t*-ratio greater (in absolute value) than the computed value, given a true hypothesis. Often, a value below 0.05 (or sometimes 0.01) is interpreted as evidence that the parameter is significantly different from zero.

The Parameter Estimates table also gives the **Constant Estimate**, for models that contain an intercept or mean term. The definition of the constant estimate is given under "ARIMA Model," p. 695.

Forecast Plot

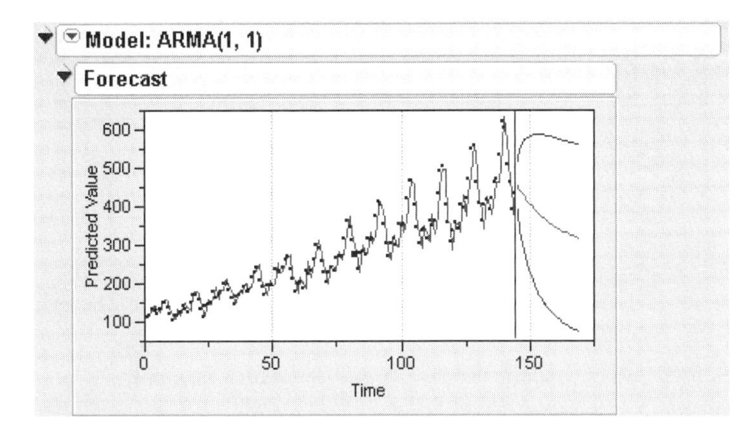

Each model has its own Forecast plot. The Forecast plot shows the values that the model predicts for the time series. It is divided by a vertical line into two regions. To the left of the separating line the one-step-ahead forecasts are shown overlaid with the input data points. To the right of the line are the future values forecast by the model and the confidence intervals for the forecasts.

You can control the number of forecast values by changing the setting of the **Forecast Periods** box in the platform launch dialog or by selecting **Number of Forecast Periods** from the Time Series drop-down menu. The data and confidence intervals can be toggled on and off using the **Show Points** and **Show Confidence Interval** commands on the model's popup menu.

Residuals

The graphs under the residuals section of the output show the values of the residuals based on the fitted model. These are the actual values minus the one-step-ahead predicted values. In addition, the autocorrelation and partial autocorrelation of these residuals are shown. These can be used to determine whether the fitted model is adequate to describe the data. If it is, the points in the residual plot should be normally distributed about the zero line and the autocorrelation and partial autocorrelation of the residuals should not have any significant components for lags greater than zero.

Iteration History

The model parameter estimation is an iterative procedure by which the log-likelihood is maximized by adjusting the estimates of the parameters. The iteration history for each model you request shows the value of the objective function for each iteration. This can be useful for diagnosing problems with the fitting procedure. Attempting to fit a model which is poorly suited to the data can result in a large number of iterations that fail to converge on an optimum value for the likelihood. The Iteration History table shows the following quantities:

Iter lists the iteration number.

Iteration History lists the objective function value for each step.

Step lists the type of iteration step.

Obj-Criterion lists the norm of the gradient of the objective function.

Model Report Options

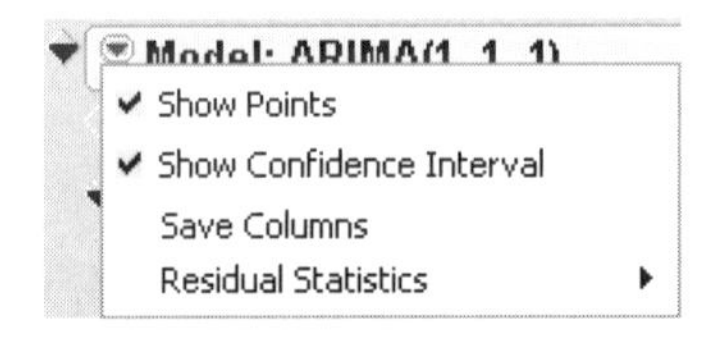

The title bar for each model you request has the popup menu shown to the right, with the following options for that model:

Show Points hides or shows the data points in the forecast graph.

Show Confidence Interval hides or shows the confidence intervals in the forecast graph.

Save Columns creates a new data table with columns representing the results of the model.

Residual Statistics controls which displays of residual statistics are shown for the model. These displays are described in the section "Time Series Commands," p. 684; however, they are applied to the residual series (the one-step-ahead model predictions minus the input series).

ARIMA Model

An **A**uto**R**egressive **I**ntegrated **M**oving **A**verage (ARIMA) model predicts future values of a time series by a linear combination of its past values and a series of errors (also known as *random shocks* or *innovations*). The **ARIMA** command performs a maximum likelihood fit of the specified ARIMA model to the time series.

For a response series $\{y_i\}$, the general form for the ARIMA model is:

$$\phi(B)(w_t - \mu) = \theta(B)a_t$$

where

t is the time index

B is the backshift operator defined as $By_t = y_{t-1}$

$w_t = (1-B)^d y_t$ is the response series after differencing

μ is the intercept or mean term.

$\phi(B)$ and $\theta(B)$, respectively, the autoregressive operator and the moving average operator and are written

$$\phi(B) = 1 - \phi_1 B - \phi_2 B^2 - \ldots - \phi_p B^p \text{ and } \theta(B) = 1 - \theta_1 B - \theta_2 B^2 - \ldots - \theta_q B^q$$

a_t are the sequence of random shocks.

The a_t are assumed to be independent and normally distributed with mean zero and constant variance. The model can be rewritten as

$\phi(B)w_t = \delta + \theta(B)a_t$ where the constant estimate δ is given by the relation

$$\delta = \phi(B)\mu = \mu - \phi_1\mu - \phi_2\mu - \ldots - \phi_p\mu .$$

The ARIMA command displays the Specify ARIMA Model dialog, which allows you to specify the ARIMA model you want to fit. The results appear when you click **Estimate**.

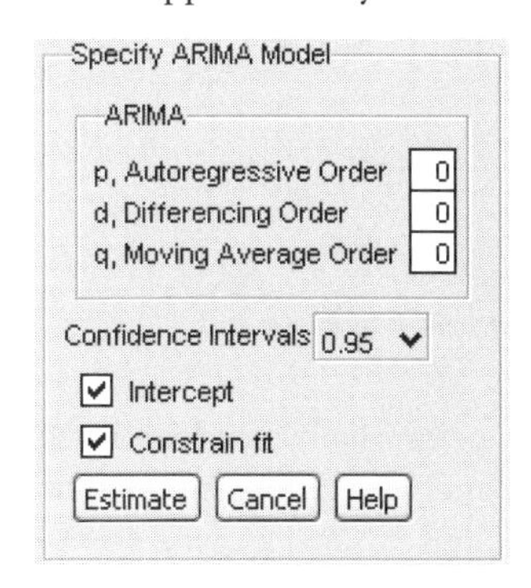

Use the Specify ARIMA Model dialog for the following three orders that can be specified for an ARIMA model:

1 The **Autoregressive Order** is the order (*p*) of the polynomial $\varphi(B)$ operator.
2 The **Differencing Order** is the order (*d*) of the differencing operator.

3 The **Moving Average Order** is the order (q) of the differencing operator $\theta(B)$.
4 An ARIMA model is commonly denoted ARIMA(p,d,q). If any of $p,d,$ or q are zero, the corresponding letters are often dropped. For example, if p and d are zero, then model would be denoted MA(q).

The **Confidence Intervals** box allows you to set the confidence level between 0 and 1 for the forecast confidence bands. The **Intercept** check box determines whether the intercept term μ will be part of the model. If the **Constrain fit** check box is checked, the fitting procedure will constrain the autoregressive parameters to always remain within the stable region and the moving average parameters within the invertible region. You might want to uncheck this box if the fitter is having difficulty finding the true optimum or if you want to speed up the fit. You can check the Model Summary table to see if the resulting fitted model is stable and invertible.

Seasonal ARIMA

In the case of **Seasonal ARIMA** modeling, the differencing, autoregressive, and moving average operators are the product of seasonal and nonseasonal polynomials:

$$w_t = (1-B)^d(1-B^s)^D y_t$$

$$\varphi(B) = (1-\varphi_{1,1}B-\varphi_{1,2}B^2-\ldots-\varphi_{1,p}B^p)(1-\varphi_{2,s}B^s-\varphi_{2,2s}B^{2s}-\ldots-\varphi_{2,Ps}B^{Ps})$$

$$\theta(B) = (1-\theta_{1,1}B-\theta_{1,2}B^2-\ldots-\theta_{1,q}B^q)(1-\theta_{2,s}B^s-\theta_{2,2s}B^{2s}-\ldots-\theta_{2,Qs}B^{Qs})$$

where s is the number of periods in a season. The first index on the coefficients is the factor number (1 indicates nonseasonal, 2 indicates seasonal) and the second is the lag of the term.

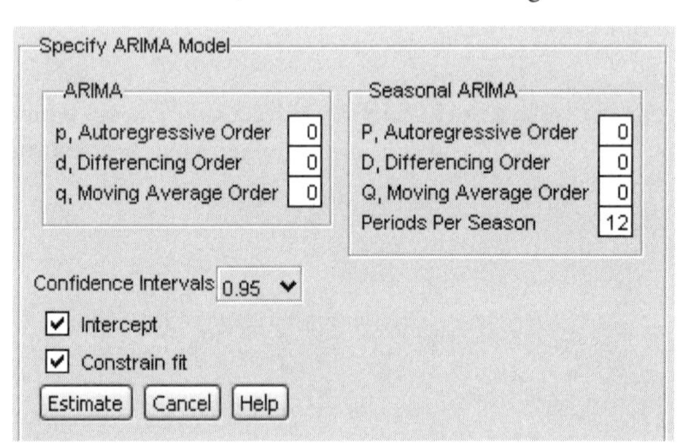

The Seasonal ARIMA dialog appears when you select the **Seasonal ARIMA** command. It has the same elements as the ARIMA dialog and adds elements for specifying the seasonal autoregressive order (P), seasonal differencing order (D), and seasonal moving average order (Q). Also, the **Periods Per Season** box lets you specify the number of periods per season (s). The seasonal ARIMA models are denoted as Seasonal ARIMA(p,d,q)(P,D,Q)s.

Transfer Functions

This example analyzes the gas furnace data (SeriesJ.jmp) from Box and Jenkins. To begin the analysis, select **Input Gas Ratio** as the **Input List** and **Output CO2** as the **Y, Time Series**. The launch dialog should appear as in Figure 35.6.

Figure 35.6 Series J Launch Dialog

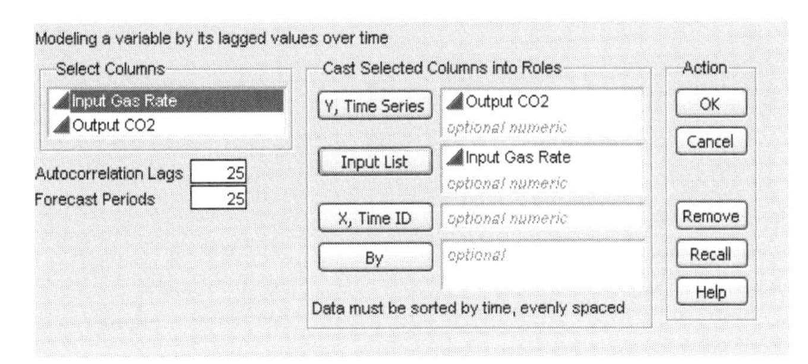

When you click **OK**, the report in Figure 35.7 appears.

Report and Menu Structure

This preliminary report shows diagnostic information and groups the analysis in two main parts. The first part, under Time Series Output CO2, contains analyses of the output series, while the Input Time Series Panel, contains analyses on the input series. The latter may include more than one series.

Figure 35.7 Series J Report

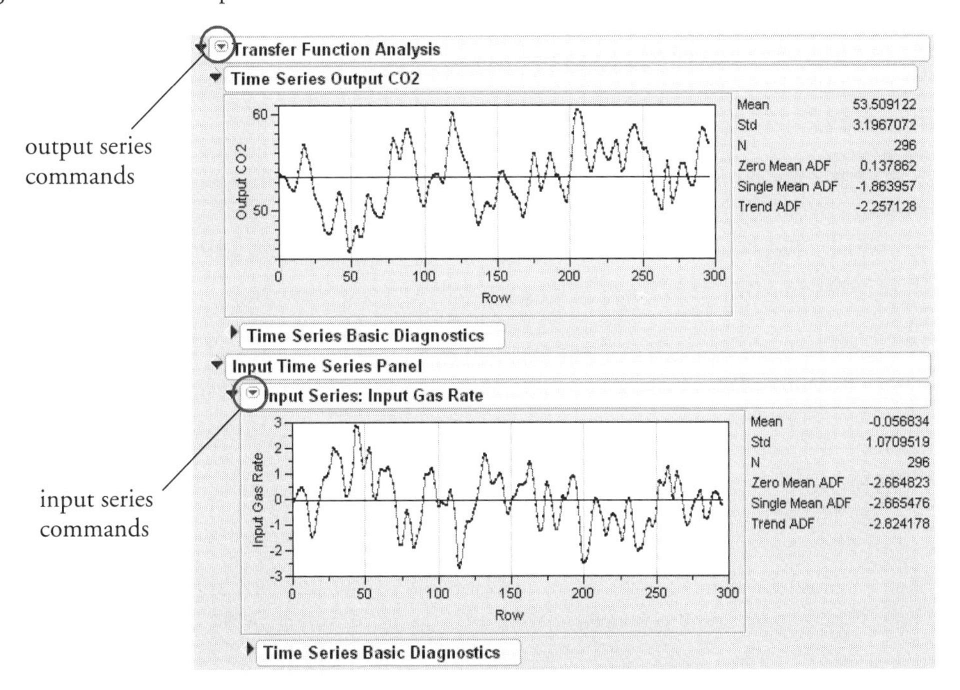

Each report section has its own set of commands. For the output (top) series, the commands are accessible from the red triangle on the outermost outline bar (**Transfer Function Analysis**). For the input (bottom) series, the red triangle is located on the inner outline bar (**Input Series: Input Gas Rate**).

Figure 35.8 shows these two command sets. Note their organization. Both start with a **Graph** command. The next set of commands are for exploration. The third set is for model building. The fourth set includes functions that control the platform.

Figure 35.8 Output and Input Series Menus

Diagnostics

Both parts give basic diagnostics, including the sample mean (**Mean**), sample standard deviation (**Std**), and series length (**N**).

In addition, the platform tests for stationarity using *Augmented Dickey-Fuller* (ADF) tests.

Zero Mean ADF tests against a random walk with zero mean, *i.e.*

$$x_t = \phi x_{t-1} + e_t$$

Single Mean ADF tests against a random walk with a non-zero mean, *i.e.*

$$x_t - \mu = \phi(x_{t-1} - \mu) + e_t$$

Trend ADF tests against a random walk with anon-zero mean and a linear trend, *i.e.*

$$x_t - \mu - \beta t = \phi[x_{t-1} - \mu - \beta(t-1)] + e_t$$

Basic diagnostics also include the autocorrelation and partial autocorrelation functions, as well as the Ljung-Box *Q*-statistic and *p*-values., found under the **Time Series Basic Diagnostics** outline node.

The **Cross Correlation** command adds a cross-correlation plot to the report. The length of the plot is twice that of an autocorrelation plot, or $2 \times \text{ACF length} + 1$.

Figure 35.9 Cross Correlation Plot

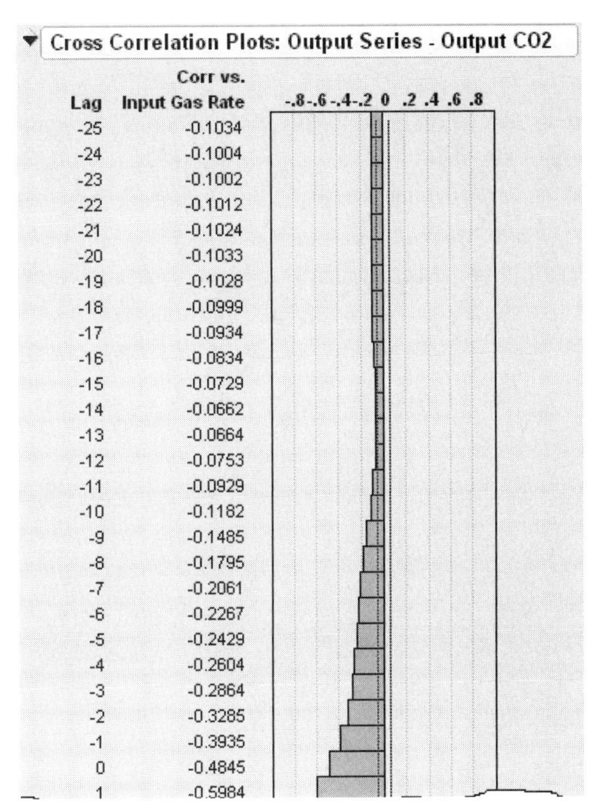

Cross Correlation Plots: Output Series - Output CO2

Lag	Corr vs. Input Gas Rate
-25	-0.1034
-24	-0.1004
-23	-0.1002
-22	-0.1012
-21	-0.1024
-20	-0.1033
-19	-0.1028
-18	-0.0999
-17	-0.0934
-16	-0.0834
-15	-0.0729
-14	-0.0662
-13	-0.0664
-12	-0.0753
-11	-0.0929
-10	-0.1182
-9	-0.1485
-8	-0.1795
-7	-0.2061
-6	-0.2267
-5	-0.2429
-4	-0.2604
-3	-0.2864
-2	-0.3285
-1	-0.3935
0	-0.4845
1	-0.5984

The plot includes plots of the output series versus all input series, in both numerical and graphical forms. The blue lines indicate standard errors for the statistics.

Model Building

Building a transfer function model is quite similar to building an ARIMA model, in that it is an iterative process of exploring, fitting, and comparing.

Before building a model and during the data exploration process, it is sometimes useful to prewhiten the data. This means find an adequate model for the input series, apply the model to the output, and get residuals from both series. Compute cross-correlations from residual series and identify the proper orders for the transfer function polynomials.

To prewhiten the input series, select the **Prewhitening** command. This brings up a dialog similar to the ARIMA dialog where you specify a stochastic model for the input series. For our SeriesJ example, we use an ARMA(2,2) prewhitening model, as shown in Figure 35.10.

Figure 35.10 Prewhitening Dialog

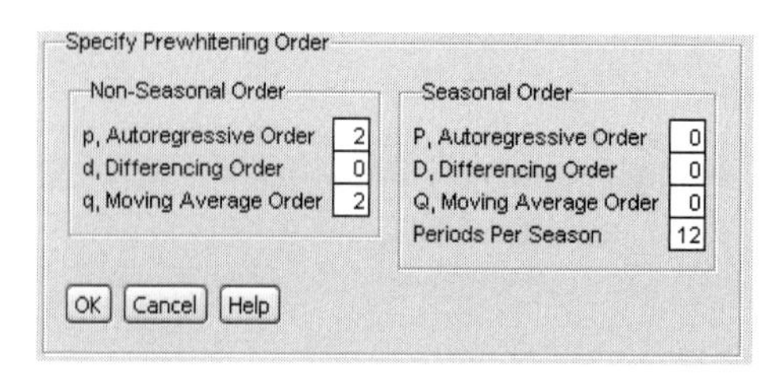

Click **OK** to reveal the Prewhitening plot.

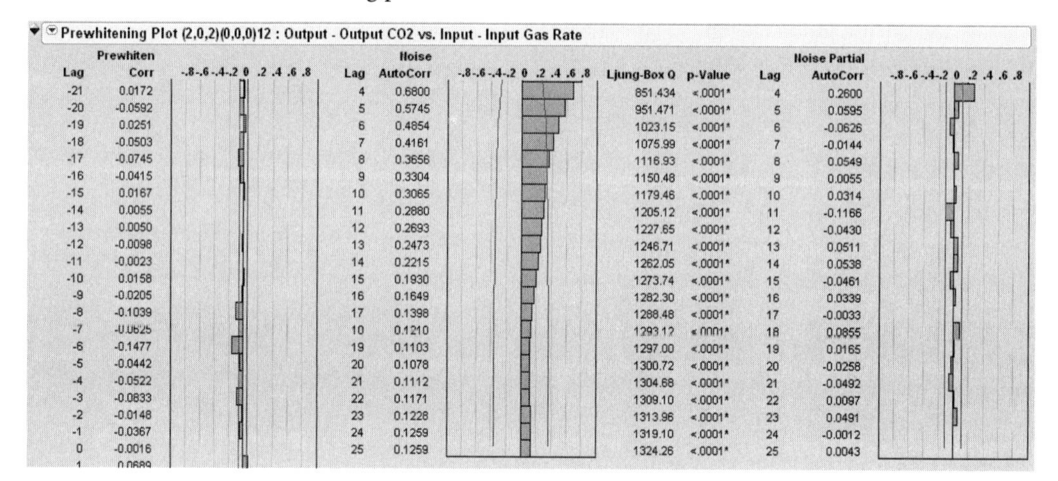
Prewhitening Plot (2,0,2)(0,0,0)12 : Output - Output CO2 vs. Input - Input Gas Rate

Lag	Prewhiten Corr
-21	0.0172
-20	-0.0592
-19	0.0251
-18	-0.0503
-17	-0.0745
-16	-0.0415
-15	0.0167
-14	0.0055
-13	0.0050
-12	-0.0098
-11	-0.0023
-10	0.0158
-9	-0.0205
-8	-0.1039
-7	-0.0625
-6	-0.1477
-5	-0.0442
-4	-0.0522
-3	-0.0833
-2	-0.0148
-1	-0.0367
0	-0.0016
1	0.0689

Lag	Noise AutoCorr	Ljung-Box Q	p-Value	Lag	Noise Partial AutoCorr
4	0.6800	851.434	<.0001*	4	0.2600
5	0.5745	951.471	<.0001*	5	0.0595
6	0.4854	1023.15	<.0001*	6	-0.0626
7	0.4161	1075.99	<.0001*	7	-0.0144
8	0.3656	1116.93	<.0001*	8	0.0549
9	0.3304	1150.48	<.0001*	9	0.0055
10	0.3065	1179.46	<.0001*	10	0.0314
11	0.2880	1205.12	<.0001*	11	-0.1166
12	0.2693	1227.65	<.0001*	12	-0.0430
13	0.2473	1246.71	<.0001*	13	0.0511
14	0.2215	1262.05	<.0001*	14	0.0538
15	0.1930	1273.74	<.0001*	15	-0.0461
16	0.1649	1282.30	<.0001*	16	0.0339
17	0.1398	1288.48	<.0001*	17	-0.0033
18	0.1210	1293.12	<.0001*	18	0.0855
19	0.1103	1297.00	<.0001*	19	0.0165
20	0.1078	1300.72	<.0001*	20	-0.0258
21	0.1112	1304.68	<.0001*	21	-0.0492
22	0.1171	1309.10	<.0001*	22	0.0097
23	0.1228	1313.96	<.0001*	23	0.0491
24	0.1259	1319.10	<.0001*	24	-0.0012
25	0.1259	1324.26	<.0001*	25	0.0043

Patterns in these plots suggest terms in the transfer function model.

Transfer Function Model

A typical transfer function model with m inputs can be represented as

$$Y_t - \mu = \frac{\omega_1(B)}{\delta_1(B)} X_{1,\, t-d1} + \ldots + \frac{\omega_m(B)}{\delta_m(B)} X_{m,\, m-dm} + \frac{\theta(B)}{\varphi(B)} e_t$$

where

Y_t denotes the output series

X_1 to X_m denote m input series

e_t represents the noise series.

$X_{1,\,t\text{–}d1}$ indicates the series X_1 is indexed by t with a $d1$-step lag.

μ represents the mean level of the model

φ(B) and θ(B) represent autoregressive and moving average polynomials from an ARIMA model

$\omega_k(B)$ and $\delta_k(B)$ represent numerator and denominator factors (or polynomials) for individual transfer functions, with k representing an index for the 1 to m individual inputs.

Each polynomial in the above model can contain two parts, either non-seasonal, seasonal, or a product of the two as in seasonal ARIMA. When specifying a model, leave the default 0 for any part that you do not want in the model.

Select **Transfer Function** to bring up the model specification dialog.

Figure 35.11 Transfer Function Specification Dialog

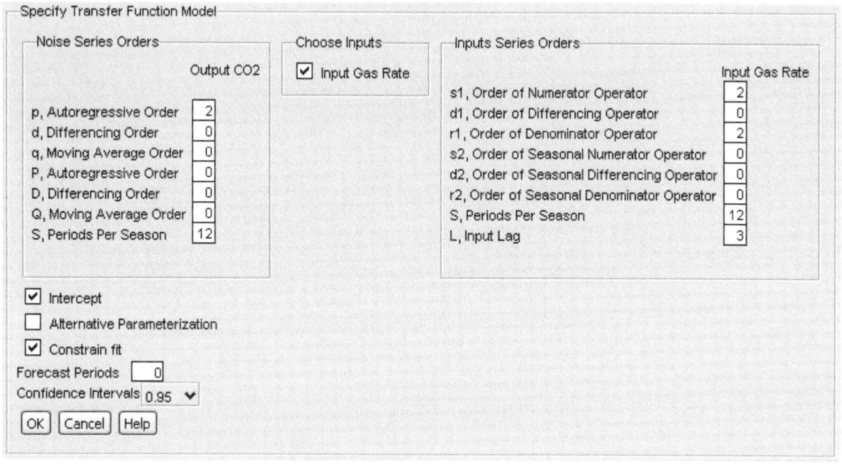

Specify Transfer Function Model
Noise Series Orders
Output CO2
p, Autoregressive Order 2
d, Differencing Order 0
q, Moving Average Order 0
P, Autoregressive Order 0
D, Differencing Order 0
Q, Moving Average Order 0
S, Periods Per Season 12
Choose Inputs
Input Gas Rate
Inputs Series Orders
Input Gas Rate
s1, Order of Numerator Operator 2
d1, Order of Differencing Operator 0
r1, Order of Denominator Operator 2
s2, Order of Seasonal Numerator Operator 0
d2, Order of Seasonal Differencing Operator 0
r2, Order of Seasonal Denominator Operator 0
S, Periods Per Season 12
L, Input Lag 3
Intercept
Alternative Parameterization
Constrain fit
Forecast Periods 0
Confidence Intervals 0.95
OK Cancel Help

The dialog consists of several parts.

Noise Series Orders contains specifications for the noise series. Lowercase letters are coefficients for non-seasonal polynomials, and uppercase letters for seasonal ones.

Choose Inputs lets you select the input series for the model.

Input Series Orders specifies polynomials related to the input series.The first three orders deal with non-seasonal polynomials. The next four are for seasonal polynomials. The final is for an input lag.

In addition, there are three options that control model fitting.

Intercept specifies whether μ is zero or not.

Alternative Parameterization specifies whether the general regression coefficient is factored out of the numerator polynomials.

Constrain Fit toggles constraining of the AR and MA coefficients.

Forecast Periods specifies the number of forecasting periods for forecasting.

Using the information from prewhitening, we specify the model as shown in Figure 35.11.

Model Reports

The analysis report is titled Transfer Function Model and is indexed sequentially. Results for the Series J example are shown in Figure 35.12.

Figure 35.12 Series J Transfer Function Reports

Transfer Function Model (1)

Model Summary

DF	283
Sum of Squared Errors	16.6190766
Variance Estimate	0.05872465
Standard Deviation	0.24233169
Akaike's 'A' Information Criterion	11.8748507
Schwarz's Bayesian Criterion	41.2614368
RSquare	0.99450574
RSquare Adj	0.9943722
MAPE	0.30783027
MAE	0.16538985
-2LogLikelihood	-4.1251438

Parameter Estimates

Variable	Term	Factor	Lag	Estimate	Std Error	t Ratio	Prob>|t|
Input Gas Rate	Num0,0	0	0	-0.53155	0.0737711	-7.21	<.0001*
Input Gas Rate	Num1,1	1	1	0.37446	0.1052303	3.56	0.0004*
Input Gas Rate	Num1,2	1	2	0.51270	0.1120626	4.58	<.0001*
Input Gas Rate	Den1,1	1	1	0.55931	0.0638098	8.77	<.0001*
Input Gas Rate	Den1,2	1	2	-0.00738	0.0362491	-0.20	0.8389
Output CO2	AR1,1	1	1	1.52719	0.0467323	32.68	<.0001*
Output CO2	AR1,2	1	2	-0.62881	0.0494885	-12.71	<.0001*
	Intercept	0	0	53.36300	0.1375902	387.84	0.0000*

$$\text{Output CO2}_t = 53.363 + \frac{(-0.5315 - 0.3745 * B - 0.5127 * B^2)}{(1 - 0.5593 * B + 0.0074 * B^2)} * \text{Input Gas Rate}_{t-3} + \frac{1}{(1 - 1.5272 * B + 0.6288 * B^2)} * e_t$$

Model Summary gathers information that is useful for comparing models.

Parameter Estimates shows the parameter estimates and is similar to the ARIMA version. In addition, the Variable column shows the correspondence between series names and parameters. The table is followed by the formula of the model. Note the notation **B** is for the backshift operator.

Residuals, Iteration History are the same as their ARIMA counterparts.

Interactive Forecasting provides a forecasting graph based on a specified confidence interval. The functionality changes based on the number entered in the Forecast Periods box.

If the number of forecast periods is less than or equal to the Input Lag, the forecasting box shows the forecast for the number of periods. A confidence interval around the prediction is shown in blue, and this confidence interval can be changed by entering a number in the Confidence Interval box above the graph.

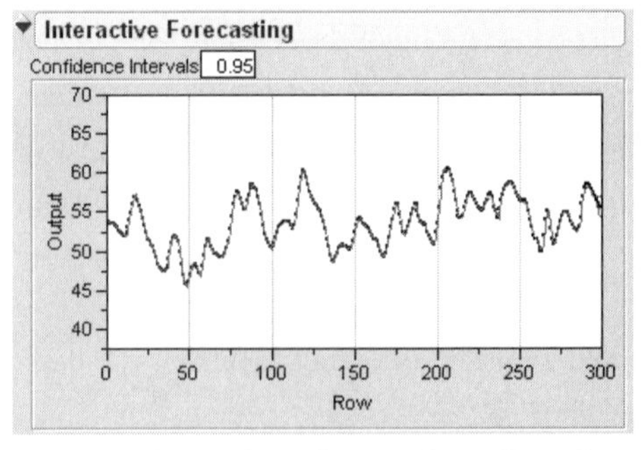

If the number of forecast periods is larger than the number of lags (say, eight in our example), the presentation is a little different.

8 forecast periods with an input lag of 3

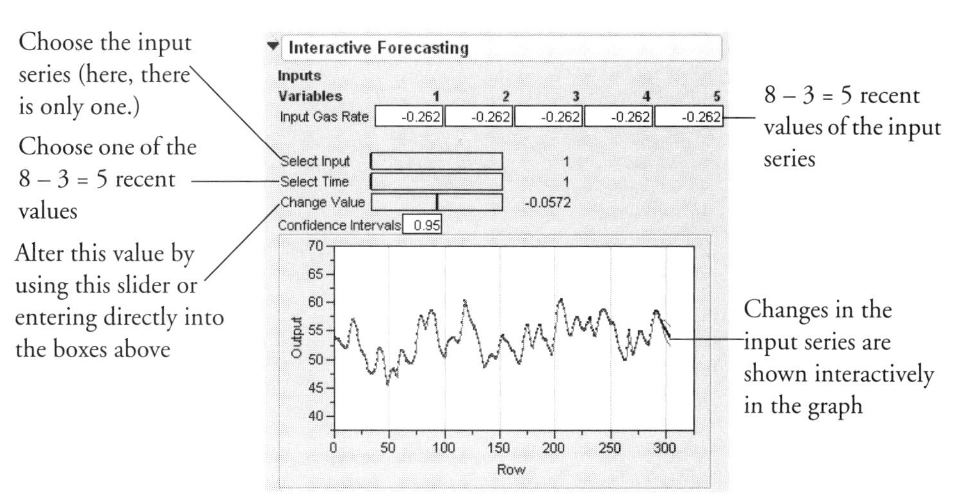

Here, you manipulate lagged values of the series by entering values into the edit boxes next to the series, or by manipulating the sliders. As before, the confidence interval can also be changed. The results of your changes are reflected in real time in the Interactive Forecasting graph.

The following commands are available from the report drop-down menu.

Save Columns creates a new data table containing the input and output series, a time column, predicted output with standard errors, residuals, and 95% confidence limits.

Create SAS Job creates PROC ARIMA code that can reproduce this model.

Submit to SAS submits PROC ARIMA code to SAS that reproduces the model.

Model Comparison Table

The model comparison table works like its ARIMA counterpart by accumulating statistics on the models you specify.

Model Comparison

Model	DF	Variance	AIC	SBC	RSquare	-2LogLH	AIC Rank	SBC Rank	MAPE	MAE
Transfer Function Model (1)	283	0.0587246	11.874851	41.261437	0.995	-4.125149	1	1	0.307830	0.165390
Transfer Function Model (2)	282	0.0588643	13.538767	46.598676	0.995	-4.461233	2	2	0.308171	0.165563
Transfer Function Model (3)	283	3.5484375	1226.6773	1256.0639	0.668	1210.6773	3	3	2.653034	1.431912

Fitting Notes

A regression model with serially correlated errors can be specified by including regressors in the model and not specifying any polynomial orders.

Intervention analysis can also be conducted, but prewhitening is no longer meaningful.

Currently, the transfer function model platform has limited capability of supporting missing values.

Smoothing Models

JMP offers a variety of smoothing techniques.

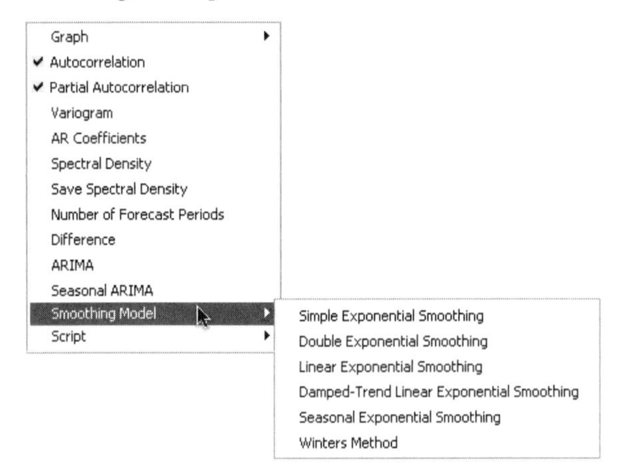

Smoothing models represent the evolution of a time series by the model:

$y_t = \mu_t + \beta_t + s(t) + a_t$ where

μ_t is the time-varying mean term,

β_t is the time-varying slope term,

$s(t)$ is one of the s time-varying seasonal terms,

a_t are the random shocks.

Models without a trend have $\beta_t = 0$ and nonseasonal models have $s(t) = 0$. The estimators for these time-varying terms are

L_t smoothed level that estimates μ_t

T_t is a smoothed trend that estimates β_t

S_{t-j} for $j = 0, 1, \ldots, s-1$ are the estimates of the $s(t)$.

Each smoothing model defines a set of recursive smoothing equations that describes the evolution of these estimators. The smoothing equations are written in terms of model parameters called *smoothing weights*. They are

α, the level smoothing weight
γ, the trend smoothing weight
φ, the trend damping weight
δ, the seasonal smoothing weight.

While these parameters enter each model in a different way (or not at all), they have the common property that larger weights give more influence to recent data while smaller weights give less influence to recent data.

Each smoothing model has an ARIMA model equivalent. These ARIMA equivalents are used to estimate the smoothing weights and provide forecasts. You may not be able to specify the equivalent ARIMA model using the **ARIMA** command because some smoothing models intrinsically constrain the ARIMA model parameters in ways the ARIMA command will not allow.

Smoothing Model Dialog

The Smoothing Model dialog appears in the report window when you select one of the smoothing model commands.

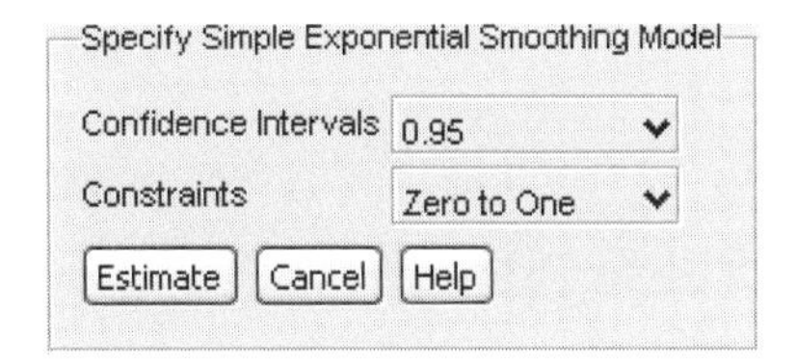

The **Confidence Intervals** box allows you to set the confidence level for the forecast confidence bands. The dialogs for seasonal smoothing models include a **Periods Per Season** box for setting the number of periods in a season. The dialog also lets you to specify what type of constraint you want to enforce on the smoothing weights during the fit. The constraints are:

Zero To One keeps the values of the smoothing weights in the range zero to one.

Unconstrained allows the parameters to range freely.

Stable Invertible constrains the parameters such that the equivalent ARIMA model is stable and invertible.

Custom expands the dialog to allow you to set constraints on individual smoothing weights.

Each smoothing weight can be **Bounded**, **Fixed**, or **Unconstrained** as determined by the setting of the popup menu next to the weight's name.

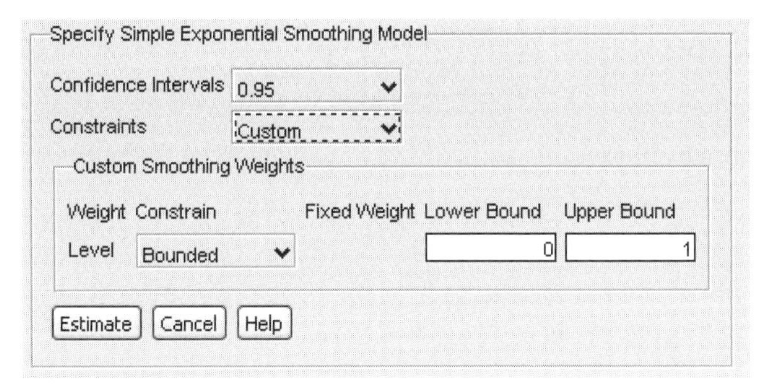

The example shown here has the Level weight (α) fixed at a value of 0.3 and the Trend weight (γ) bounded by 0 and 1. In this case, the value of the Trend weight is allowed to move within the range 0 to 1 while the Level weight is held at 0.3. Note that you can specify all the smoothing weights in advance by using these custom constraints. In that case, none of the weights would be estimated from the data although forecasts and residuals would still be computed. When you click **Estimate**, the results of the fit appear in place of the dialog.

Simple Exponential Smoothing

The model for simple exponential smoothing is $y_t = \mu_t + \alpha_t$.

The smoothing equation, $L_t = \alpha y_t + (1 - \alpha)L_{t-1}$, is defined in terms of a single smoothing weight α. This model is equivalent to an ARIMA(0, 1, 1) model where

$$(1 - B)y_t = (1 - \theta B)\alpha_t \text{ with } \theta = 1 - \alpha .$$

The moving average form of the model is

$$y_t = a_t + \sum_{j-1}^{\infty} \alpha a_{t-j}$$

Double (Brown) Exponential Smoothing

The model for double exponential smoothing is $y_t = \mu_t + \beta_1 t + a_t$.

The smoothing equations, defined in terms of a single smoothing weight α are

$$L_t = \alpha y_t + (1 - \alpha)L_{t-1} \text{ and } T_t = \alpha(L_t - L_{t-1}) + (1 - \alpha)T_{t-1} .$$

This model is equivalent to an ARIMA(0, 1, 1)(0, 1, 1)1 model

$$(1 - B)^2 y_t = (1 - \theta B)^2 a_t \text{ where } \theta_{1,1} = \theta_{2,1} \text{ with } \theta = 1 - \alpha .$$

The moving average form of the model is

$$y_t = a_t + \sum_{j=1}^{\infty} (2\alpha + (j - 1)\alpha^2)a_{t-j}$$

Linear (Holt) Exponential Smoothing

The model for linear exponential smoothing is $y_t = \mu_t + \beta_t t + a_t$.

The smoothing equations defined in terms of smoothing weights α and γ are

$$L_t = \alpha y_t + (1 - \alpha)(L_{t-1} + T_{t-1}) \text{ and } T_t = \gamma(L_t - L_{t-1}) + (1 - \gamma)T_{t-1}$$

This model is equivalent to an ARIMA(0, 2, 2) model where

$$(1 - B)^2 y_t = (1 - \theta B - \theta_2 B^2)a_t \text{ with } \theta = 2 - \alpha - \alpha\gamma \text{ and } \theta_2 = \alpha - 1 .$$

The moving average form of the model is

$$y_t = a_t + \sum_{j=1}^{\infty} (\alpha + j\alpha\gamma)a_{t-j}$$

Damped-Trend Linear Exponential Smoothing

The model for damped-trend linear exponential smoothing is $y_t = \mu_t + \beta_t t + a_t$.

The smoothing equations in terms of smoothing weights α, γ, and φ are

$$L_t = \alpha y_t + (1-\alpha)(L_{t-1} + \varphi T_{t-1}) \text{ and } T_t = \gamma(L_t - L_{t-1}) + (1-\gamma)\varphi T_{t-1}$$

This model is equivalent to an ARIMA(1, 1, 2) model where

$$(1-\varphi B)(1-B)y_t = (1-\theta_1 B - \theta_2 B^2)a_t \text{ with } \theta_1 = 1+\varphi-\alpha-\alpha\gamma\varphi \text{ and } \theta_2 = (\alpha-1)\varphi .$$

The moving average form of the model is

$$y_t = \alpha_t + \sum_{j=1}^{\infty}\left(\frac{\alpha + \alpha\gamma\varphi(\varphi^j - 1)}{\varphi - 1}\right)\alpha_{t-j}$$

Seasonal Exponential Smoothing

The model for seasonal exponential smoothing is $y_t = \mu_t + s(t) + a_t$.

The smoothing equations in terms of smoothing weights α and δ are

$$L_t = \alpha(y_t - S_{t-s}) + (1-\alpha)L_{t-1} \text{ and } S_t = \delta(y_t - L_{t-s}) + (1-\delta)\varphi S_{t-s}$$

This model is equivalent to a seasonal ARIMA(0, 1, 1)(0, 1, 0)S model where we define

$$\theta_1 = \theta_{1,1},\ \theta_2 = \theta_{2,s} = \theta_{2,s},\text{ and } \theta_3 = -\theta_{1,1}\theta_{2,s}$$

so

$$(1-B)(1-B^s)y_t = (1-\theta_1 B - \theta_2 B^2 - \theta_3 B^{s+1})a_t$$

with

$$\theta_1 = 1-\alpha,\ \theta_2 = \delta(1-\alpha),\text{ and } \theta_3 = (1-\alpha)(\delta-1) .$$

The moving average form of the model is

$$y_t = a_t + \sum_{j=1}^{\infty}\psi_j a_{t-j} \quad \text{where } \psi = \begin{cases}\alpha \text{ for } j \bmod s \neq 0 \\ \alpha + \delta(1-\alpha) \text{ for } j \bmod s = 0\end{cases}$$

Winters Method (Additive)

The model for the additive version of Winter's method is $y_t = \mu_t + \beta_t t + s(t) + a_t$.

The smoothing equations in terms of weights α, γ, and δ are

$$L_t = \alpha(y_t - S_{t-s}) + (1-\alpha)(L_{t-1} + T_{t-1}) \quad , \quad T_t = \gamma(L_t - L_{t-1}) + (1-\gamma)T_{t-1} \quad \text{, and}$$
$$S_t = \delta(y_t - L_t) + (1-\delta)S_{t-s} .$$

This model is equivalent to a seasonal ARIMA(0, 1, s+1)(0, 1, 0)s model

$$(1-B)(1-B^2)y_t = \left(1 - \sum_{i=1}^{s+1} \theta_i B^i\right) a_t$$

The moving average form of the model is

$$y_t = a_t + \sum_{j=1}^{\infty} \Psi_j a_{t-j}$$

where

$$\psi = \begin{cases} \alpha + j\alpha\gamma \text{ for } j\text{mod}s \neq 0 \\ \alpha + j\alpha\gamma + \delta(1-\alpha) \text{ for} j\text{mod}s = 0 \end{cases}$$

36

Gaussian Processes

Models for Analyzing Computer Experiments

The Gaussian Process platform is used to model the relationship between a continuous response and one or more continuous predictors. These models are common in areas like computer simulation experiments, such as the output of finite element codes, and they often perfectly interpolate the data. Gaussian processes can deal with these no-error-term models.

The Gaussian Process platform fits a spatial correlation model to the data, where the correlation of the response between two observations decreases as the values of the independent variables become more distant.

The main purpose for using this platform is to obtain a prediction formula that can be used for further analysis and optimization.

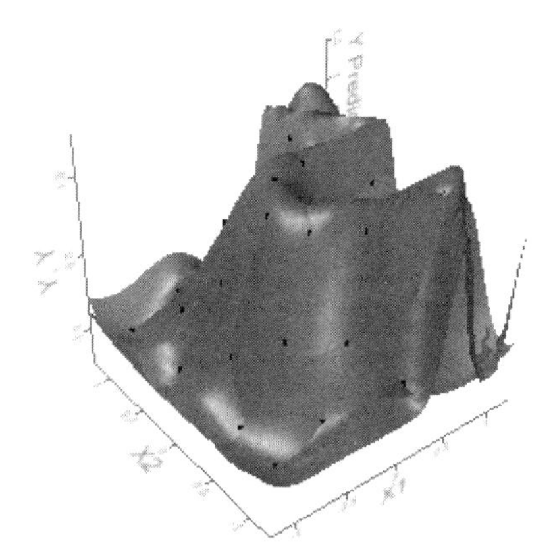

Contents

Launching the Platform

To launch the Gaussian Process platform, choose **Analyze > Modeling > Gaussian Process** from the main menu bar. Here, we illustrate the platform with 2D Gaussian Process Example.jmp data set, found in the Sample Data folder.

Figure 36.1 Gaussian Process Launch Dialog

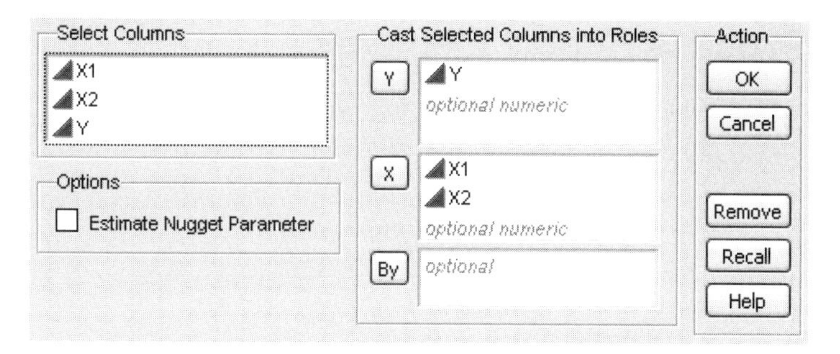

In this example, we are interested in finding the explanatory power of the two *x*-variables (X1 and X2) on Y. An overlay plot of X1 and X2 shows their even dispersal in the factor space.

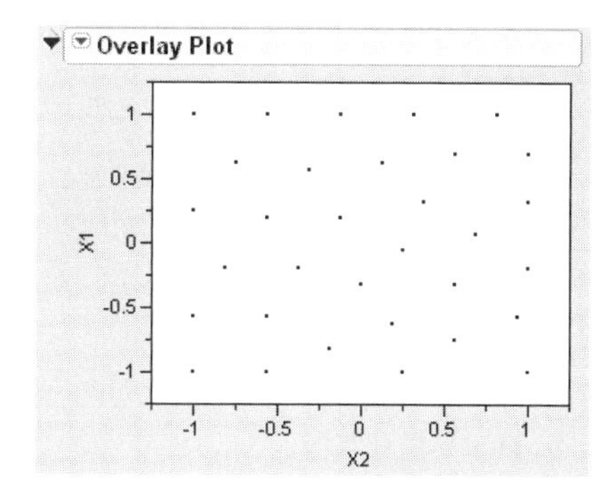

Since this is generated data, we can peek at the function that generates the Y values. It is this function that we want to estimate.

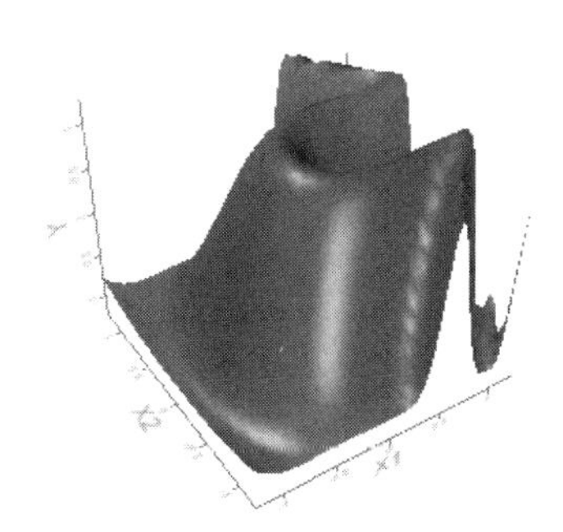

Note: If you know that there is noise or randomness in your simulation data and would like to introduce a ridge parameter into the estimation procedure (estimated by maximum likelihood), check the **Estimate Nugget Parameter** option in the launch dialog.

Gaussian Process Report Elements

After clicking **OK** from the launch dialog, the following report appears.

Figure 36.2 Gaussian Process Default Report

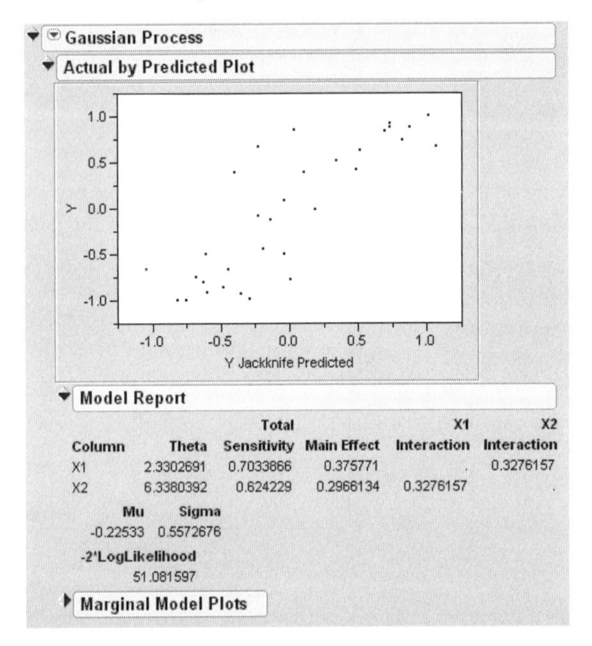

We examine each section in detail.

Actual by Predicted Plot

The Actual by Predicted plot shows the actual Y values on the *y*-axis and the jackknife predicted values on the *x*-axis. One measure of goodness-of-fit is how well the points lie along the 45 degree diagonal line.

Model Report

The Model Report shows a functional ANOVA table for the product exponential correlation function that the platform estimates. Specifically, it is an analysis of variance table, but the variation is computed using a function-driven method.

The Total Variation is the integrated variability over the entire experimental space.

For each covariate, we can create a marginal prediction formula by averaging the overall prediction formula over the values of all the other factors. The functional main effect of X1 is the integrated total variation due to X1 alone. In this case, we see that 37.6% of the variation in Y is due to X1.

The ratio of

$$\frac{\text{Functional X1 effect}}{\text{Total Variation}}$$

is the value listed as the Main Effect in the Model report. A similar ratio exists for each factor in the model.

Functional interaction effects, computed in a similar way, are also listed in the Model Report table.

Summing the value for main effect and all interaction terms gives the Total Sensitivity, the amount of influence a factor and all its two-way interactions have on the response variable.

Mu, Theta, and Sigma

JMP uses the product exponential correlation function with a power of 2 as the estimated model. This comes with the assumptions that *Y* is Normally distributed with mean μ and standard deviation $\sigma^2\mathbf{R}$. The **R** matrix is composed of elements

$$r_{ij} = \exp\left(-\sum_k \theta_k (x_{ik} - x_{jk})^2\right)$$

In the Model report, Mu is the Normal distribution mean μ, Sigma is the Normal Distribution parameter σ, and the Theta column corresponds to the values of θ_k in the definition of **R**.

These parameters are all fitted via maximum likelihood.

Note: If you see **Nugget parameters set to avoid singular variance matrix**, JMP has added a ridge parameter to the variance matrix so that it is invertible.

Marginal Model Plots

The Marginal Model plots are shown in Figure 36.3.

Figure 36.3 Marginal Model Plots

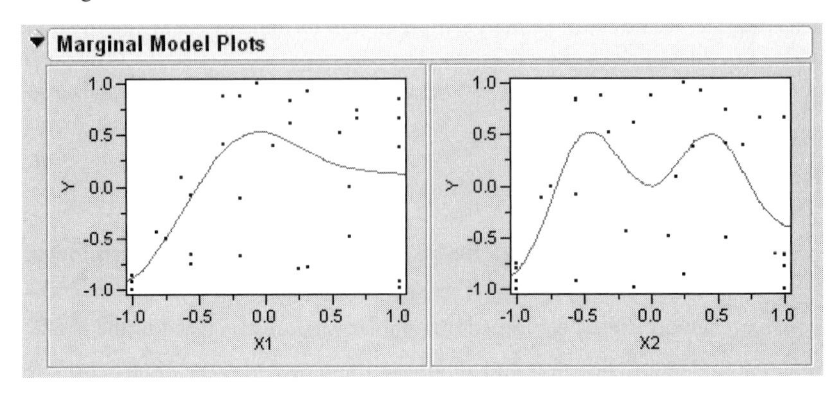

These plots show the average value each factor across all other factors. In this two-dimensional example, we examine we examine slices of X1 from –1 to 1, and plot the average value at each point.

Platform Options

The Gaussian Process platform has the following options:

Profiler brings up the standard Profiler.

Contour Profiler brings up the Contour Profiler.

Surface Profiler brings up the Surface Profiler.

Details on Profiling are found in the "Profiling" chapter on p. 287.

Save Prediction Formula creates a new column in the data table containing the prediction formula.

In Figure 36.4, we use the saved prediction formula to compare the prediction to the actual data points.

Figure 36.4 Two Views of the Prediction Surface and Actual Ys.

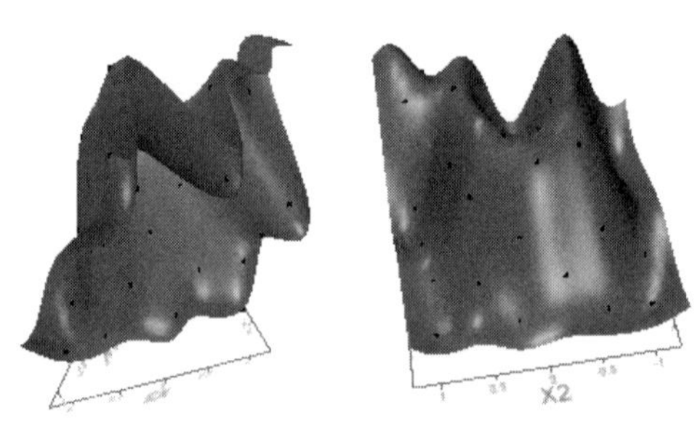

Borehole Hypercube Example

A more complicated model is seen using Borehole Latin Hypercube.jmp, found in the Design Experiments folder.

To launch the analysis, fill out the Gaussian Process dialog as shown in Figure 36.5.

Figure 36.5 Borehole Latin Hypercube Launch Dialog

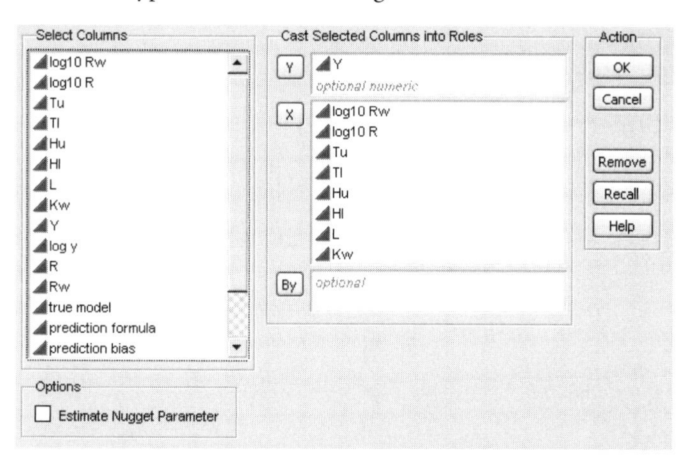

When you click OK, the following Actual by Predicted plot appears.

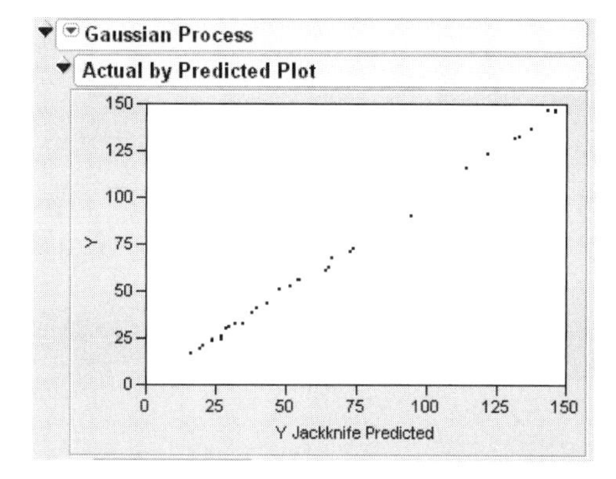

Since the points are close to the 45 degree diagonal line, we can be confident that the Gaussian process prediction model is a good approximation to the true function that generated the data.

The Model Report shows us that this is mainly due to one factor, logRw. The main effect explains 87.5% of the variation, with 90.5% explained when all second-order interactions are included.

Model Report

Column	Theta	Total Sensitivity	Main Effect	log10 Rw Interaction	log10 R Interaction	Tu Interaction	Tl Interaction	Hu Interaction	Hl Interaction	L Interaction	Kw Interaction
log10 Rw	4.644789	0.9047126	0.8751062	.	0	0	0	0.0092619	0.0102189	0.0084618	0.0016638
log10 R	0	0	0	0	.	0	0	0	0	0	0
Tu	0	0	0	0	0	.	0	0	0	0	0
Tl	0	0	0	0	0	0	.	0	0	0	0
Hu	2.1669e-6	0.0400141	0.0304916	0.0092619	0	0	0	.	3.2176e-7	0.0002417	1.8471e-5
Hl	1.9013e-6	0.0424113	0.0320666	0.0102189	0	0	0	3.2176e-7	.	0.000121	4.426e-6
L	1.4549e-7	0.0349244	0.0260815	0.0084618	0	0	0	0.0002417	0.000121	.	1.8352e-5
Kw	1.4577e-9	0.0079433	0.0062382	0.0016638	0	0	0	1.8471e-5	4.426e-6	1.8352e-5	.

Mu	Sigma	Nugget
151.25536	12454.796	0.0001

-2*LogLikelihood
205.56463

Nugget parameter set to avoid singular variance matrix.

Factors with a theta value of 0 do not impact the prediction formula at all. It is as if they have been dropped from the model.

The Marginal Model plots confirm that logRw is a highly involved participant in Y's variation.

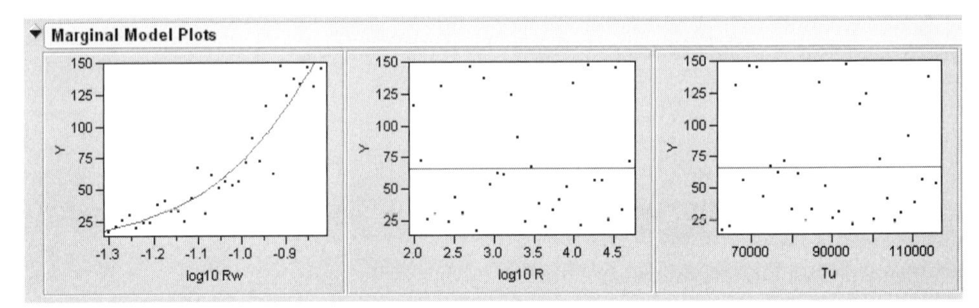

Chapter 37

Item Response Theory

The Item Analysis Platform

Item Response Theory (IRT) is a method of scoring tests. Although classical test theory methods have been widely used for a century, IRT provides a better and more scientifically-based scoring procedure. Its advantages include

- scoring tests at the item level, giving insight into the contributions of each item on the total test score
- producing scores of both the test takers and the test items on the same scale
- fitting nonlinear logistic curves, more representative of actual test performance than classical linear statistics.

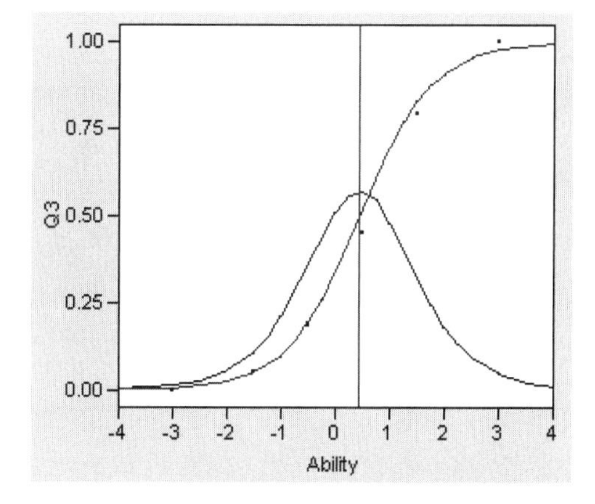

Contents

Introduction to Item Response Theory

Psychological measurement is the process of assigning quantitative values as representations of characteristics of individuals or objects, so-called *psychological constructs. Measurement theories* consist of the rules by which those quantitative values are assigned. Item response theory (IRT) is a measurement theory.

IRT utilizes a mathematical function to relate an individual's probability of correctly responding to an item to a trait of that individual. Frequently, this trait is not directly measurable, and is therefore called a *latent trait.*

To see how IRT relates traits to probabilities, first examine a test question that follows the Guttman "perfect scale" as shown in Figure 37.1. The horizontal axis represents the amount of the theoretical trait that the examinee has. The vertical axis represents the probability that the examinee will get the item correct. The curve in Figure 37.1 is called an *item characteristic curve* (ICC).

Figure 37.1 Item characteristic curve of a perfect scale item

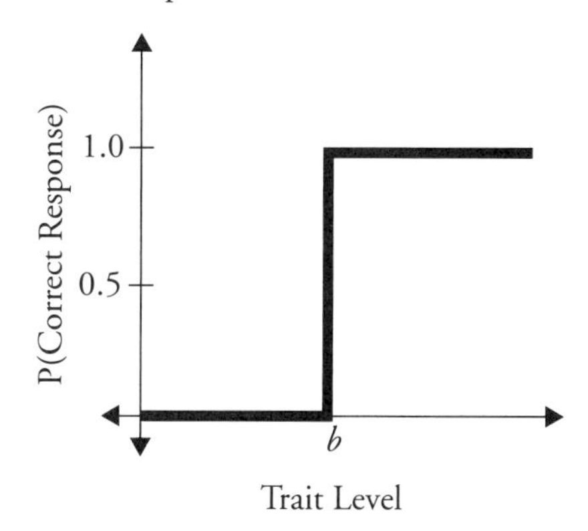

This figure shows that a person who has ability less than the value *b* has a zero percent chance of getting the item correct. A person with trait level higher than *b* has a 100 percent chance of getting the item correct.

Of course, this is an unrealistic item, but it is illustrative in showing how a trait and a question probability relate to each other. More typical is a curve that allows probabilities that very from zero to one. A typical curve found empirically is the S-shaped logistic function with a lower asymptote at zero and upper asymptote at one. It is markedly nonlinear. An example curve is shown in Figure 37.2.

Figure 37.2 Example item response curve

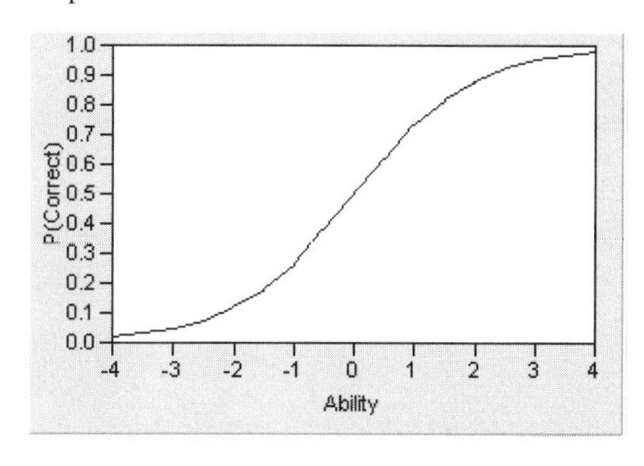

The logistic model is the best choice to model this curve, since it has desirable asymptotic properties, yet is easier to deal with computationally than other proposed models (such as the cumulative normal density function). The model itself is

$$P(\theta) = c + \frac{1-c}{1+e^{-(a)(\theta-b)}}$$

In this model, referred to as a *3PL* (three-parameter logistic) model, the variable *a* represents the steepness of the curve at its inflection point. Curves with varying values of *a* are shown in Figure 37.3. This parameter can be interpreted as a measure of the discrimination of an item—that is, how much more difficult the item is for people with high levels of the trait than for those with low levels of the trait. Very large values of *a* make the model practically the step function shown in Figure 37.1. It is generally assumed that an examinee will have a higher probability of getting an item correct as their level of the trait increases. Therefore, *a* is assumed to be positive and the ICC is monotonically increasing. Some use this positive-increasing property of the curve as a test of the appropriateness of the item. Items whose curves do not have this shape should be considered as candidates to be dropped from the test.

Figure 37.3 Logistic model for several values of *a*

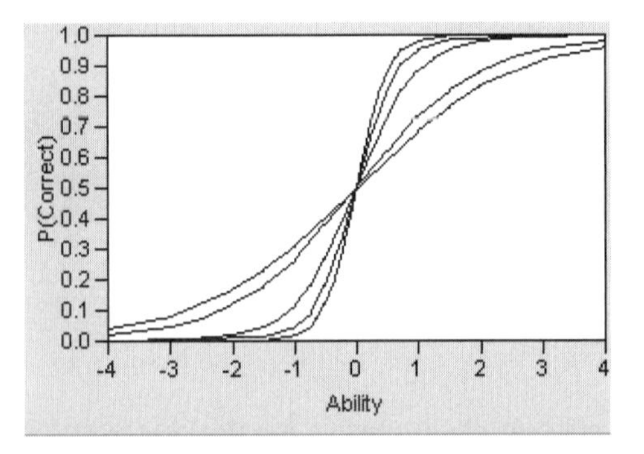

Changing the value of *b* merely shifts the curve from left to right, as shown in Figure 37.4. It corresponds to the value of θ at the point where *P*(θ)=0.5. The parameter *b* can therefore be interpreted as item difficulty where (graphically), the more difficult items have their inflection points farther to the right along their *x*-coordinate.

Figure 37.4 Logistic curve for several values of *b*

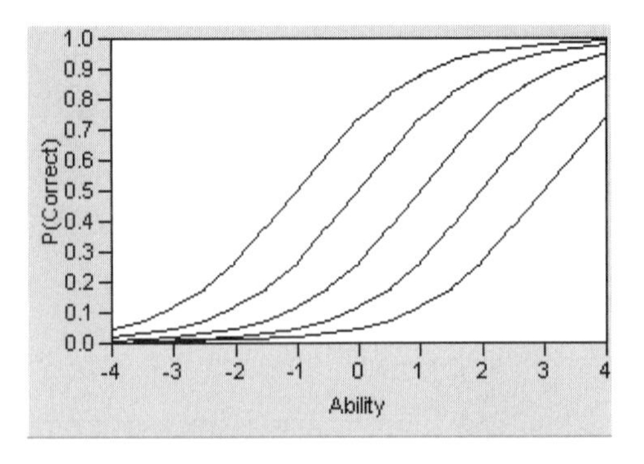

Notice that

$$\lim_{\theta \to -\infty} P(\theta) = c$$

and therefore c represents the lower asymptote, which can be non-zero. ICCs for several values of c are shown graphically in Figure 37.5. The c parameter is theoretically pleasing, since a person with no ability of the trait may have a non-zero chance of getting an item right. Therefore, c is sometimes called the *pseudo-guessing parameter.*

Figure 37.5 Logistic model for several values of c

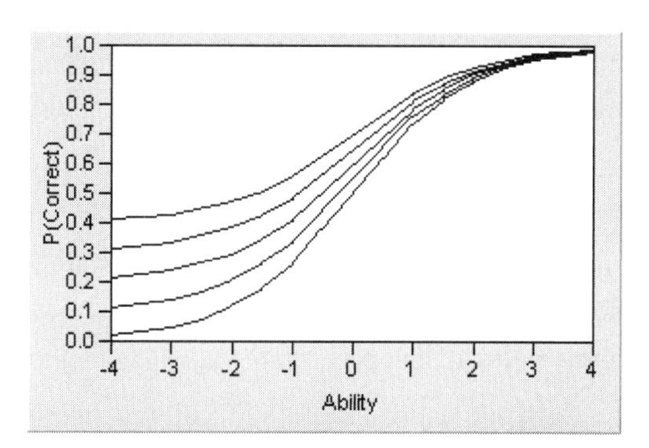

By varying these three parameters, a wide variety of probability curves are available for modeling. A sample of three different ICCs is shown in Figure 37.6. Note that the lower asymptote varies, but the upper asymptote does not. This is because of the assumption that there may be a lower guessing parameter, but as the trait level increases, there is always a theoretical chance of 100% probability of correctly answering the item.

Figure 37.6 Three item characteristic curves

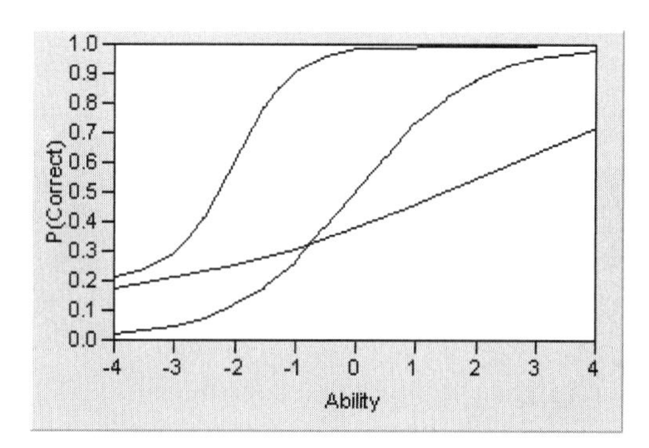

Note, however, that the 3PL model may by unnecessarily complex for many situations. If, for example, the c parameter is restricted to be zero (in practice, a reasonable restriction), there are fewer parameters to predict. This model, where only a and b parameters are estimated, is called the 2PL model.

Another advantage of the 2PL model (aside from its greater stability than the 3PL) is that b can be interpreted as the point where an examinee has a 50 percent chance of getting an item correct. This interpretation is not true for 3PL models.

A further restriction can be imposed on the general model when a researcher can assume that test items have equal discriminating power. In these cases, the parameter b is set equal to zero (essentially dropped from the model), leaving a single parameter to be estimated. This *1PL* model is frequently called the

Rasch model, named after Danish mathematician Georg Rasch, the developer of the model.The Rasch model is quite elegant, and is the least expensive to use computationally.

Caution: You must have a lot of data to produce stable parameter estimates using a 3PL model. 2PL models are frequently sufficient for tests that intuitively deserve a guessing parameter. Therefore, the 2PL model is the default and recommended model.

Launching the Platform

As an example, open the sample data file MathScienceTest.JMP. These data are a subset of the data from the Third International Mathematics and Science Study (TIMMS) conducted in 1996.

To launch the Item Analysis platform, select **Analyze > Multivariate Methods > Item Analysis.** This shows the dialog in Figure 37.7.

Figure 37.7 Item Analysis Launch Dialog

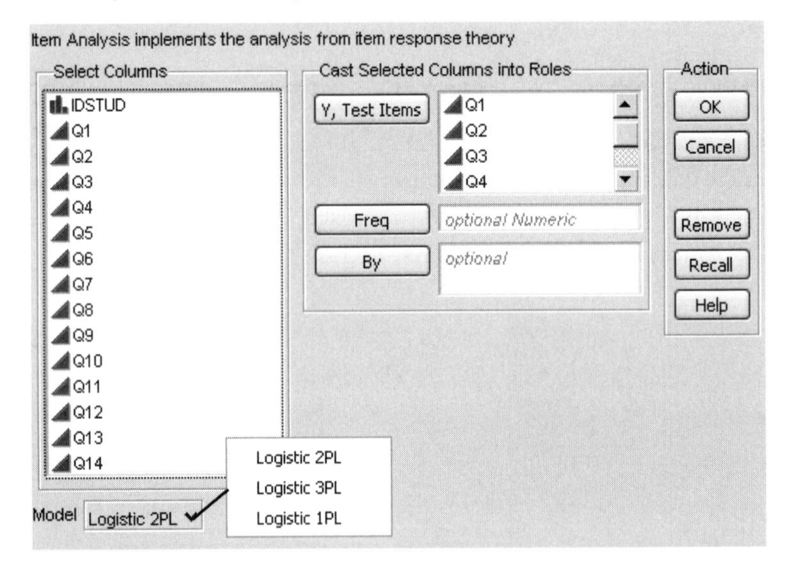

Y, Test Items are the questions from the test instrument.

Freq optionally specifies a variable used to specify the number of times each response pattern appears.

By performs a separate analysis for each level of the specified variable.

Specify the desired model (1PL, 2PL, or 3PL) by selecting it from the **Model** drop-down menu.

For this example, specify all fourteen continuous questions (Q1, Q2,..., Q14) as **Y, Test Items** and click **OK.** This accepts the default 2PL model.

Special note on 3PL Models

If you select the 3PL model, a dialog pops up asking for a penalty for the *c* parameters (thresholds). This is not asking for the threshold itself. The penalty it requests is similar to the type of penalty parameter that you would see in ridge regression, or in neural networks.

The penalty is on the sample variance of the estimated thresholds, so that large values of the penalty force the estimated thresholds' values to be closer together. This has the effect of speeding up the computations, and reducing the variability of the threshold (at the expense of some bias).

In cases where the items are questions on a multiple choice test where there are the same number of possible responses for each question, there is often reason to believe (*a priori*) that the threshold parameters would be similar across items. For example, if you are analyzing the results of a 20-question multiple choice test where each question had four possible responses, it is reasonable to believe that the guessing, or threshold, parameters would all be near 0.25. So, in some cases, applying a penalty like this has some "physical intuition" to support it, in addition to its computational advantages.

Item Analysis Output

The following plots appear in Item Analysis reports.

Characteristic Curves

Item characteristic curves for each question appear in the top section of the output. Initially, all curves are shown stacked in a single column. They can be rearranged using the **Number of Plots Across** command, found in the drop down menu of the report title bar. For Figure 37.8, four plots across are displayed.

Figure 37.8 Component Curves

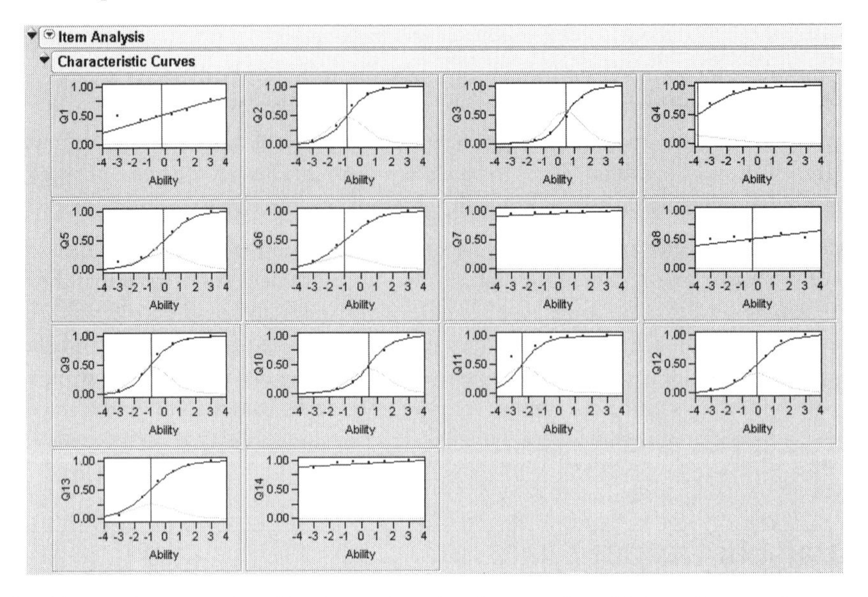

A vertical red line is drawn at the inflection point of each curve. In addition, dots are drawn at the actual proportion correct for each ability level, providing a graphical method of judging goodness-of-fit.

Gray information curves show the amount of information each question contributes to the overall information of the test. The information curve is the slope of the ICC curve, which is maximized at the inflection point.

Figure 37.9 Elements of the ICC Display

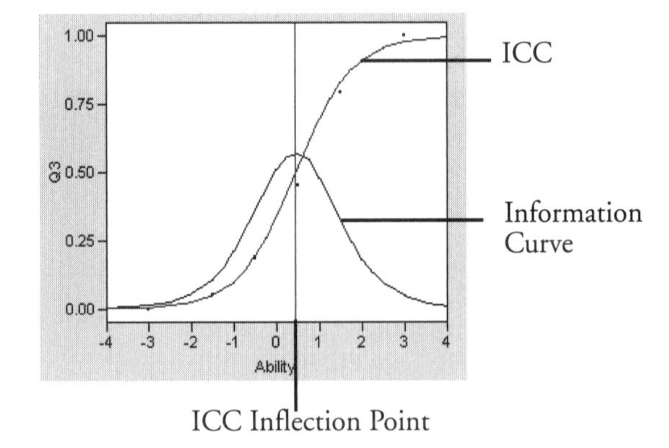

Information Curves

Questions provide varying levels of information for different ability levels. The gray information curves for each item show the amount of information that each question contributes to the total information of the test. The total information of the test for the entire range of abilities is shown in the Information Plot section of the report (Figure 37.10).

Figure 37.10 Information Plot

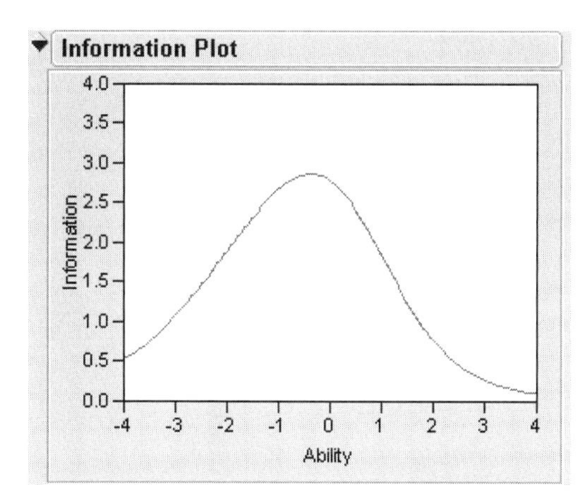

Dual Plots

The information gained from item difficulty parameters in IRT models can be used to construct an increasing scale of questions, from easiest to hardest, on the same scale as the examinees. This structure gives information on which items are associated with low levels of the trait, and which are associated with high levels of the trait.

JMP shows this correspondence with a *dual plot*. The dual plot for this example is shown in Figure 37.11.

Figure 37.11 Dual Plot

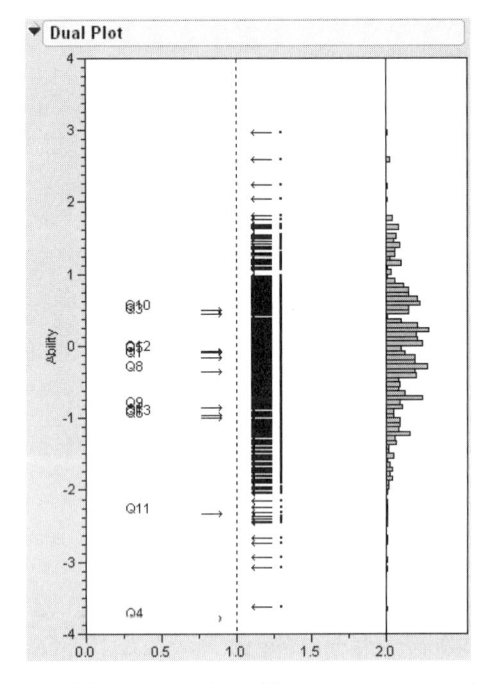

Questions are plotted to the left of the vertical dotted line, examinees on the right. In addition, a histogram of ability levels is appended to the right side of the plot.

This example shows a wide range of abilities. Q10 is rated as difficult, with an examinee needing to be around half a standard deviations above the mean in order to have a 50% chance of correctly answering the question. Other questions are distributed at lower ability levels, with Q11 and Q4 appearing as easier. There are some questions that are off the displayed scale (Q7 and Q14).

The estimated parameter estimates appear below the Dual Plot, as shown in Figure 37.12.

Figure 37.12 Parameter Estimates

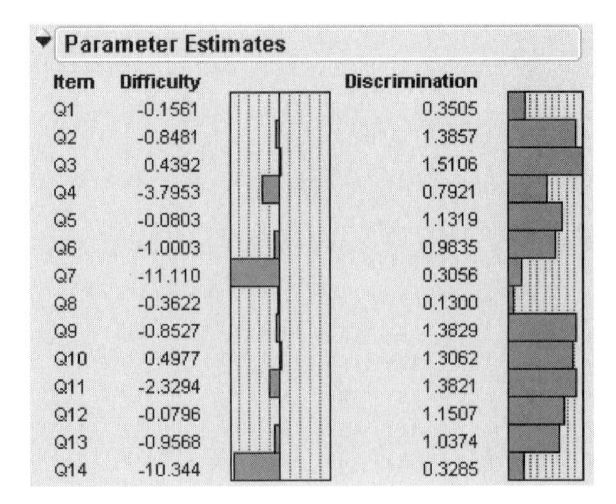

Parameter Estimates

Item	Difficulty	Discrimination
Q1	-0.1561	0.3505
Q2	-0.8481	1.3857
Q3	0.4392	1.5106
Q4	-3.7953	0.7921
Q5	-0.0803	1.1319
Q6	-1.0003	0.9835
Q7	-11.110	0.3056
Q8	-0.3622	0.1300
Q9	-0.8527	1.3829
Q10	0.4977	1.3062
Q11	-2.3294	1.3821
Q12	-0.0796	1.1507
Q13	-0.9568	1.0374
Q14	-10.344	0.3285

Item identifies the test item.

Difficulty is the *b* parameter from the model. A histogram of the difficulty parameters is shown beside the difficulty estimates.

Discrimination is the *a* parameter from the model, shown only for 2PL and 3PL models. A histogram of the discrimination parameters is shown beside the discrimination estimates.

Threshold is the *c* parameter from the model, shown only for 3PL models.

Platform Commands

The following three commands are available from the drop-down menu on the title bar of the report.

Number of Plots Across brings up a dialog to specify how many plots should be grouped together on a single line. Initially, plots are stacked one-across. Figure 37.8 "Component Curves," p. 724 shows four plots across.

Save Ability Formula creates a new column in the data table containing a formula for calculating ability levels. Since the ability levels are stored as a formula, you can add rows to the data table and have them scored using the stored ability estimates. In addition, you can run several models and store several estimates of ability in the same data table.

The ability is computed using the `IRT Ability` function. The function has the following form

```
IRT Ability (Q1, Q2,...,Qn, [a1, a2,..., an, b1, b2,..., bn, c1, c2, ...,
cn]);
```

where `Q1, Q2,...,Qn` are columns from the data table containing items, `a1, a2,..., an` are the corresponding discrimination parameters, `b1, b2,..., bn` are the corresponding difficulty parameters for the items, and `c1, c2, ..., cn` are the corresponding threshold parameters. Note that the parameters are entered as a matrix, surrounded by square brackets.

Script accesses the standard script menu for saving and re-computing an analysis.

Technical Details

Note that $P(\theta)$ does not necessarily represent the probability of a positive response from a *particular* individual. It is certainly feasible that an examinee might definitely select an incorrect answer, or that an examinee may know an answer for sure, based on the prior experiences and knowledge of the examinee, apart from the trait level. It is more correct to think of $P(\theta)$ as the probability of response for a set of individuals with ability level θ. Said another way, if a large group of individuals with equal trait levels answered the item, $P(\theta)$ predicts the proportion that would answer the item correctly. This implies that IRT models are item-invariant; theoretically, they would have the same parameters regardless of the group tested.

An assumption of these IRT models is that the underlying trait is unidimensional. That is to say, there is a single underlying trait that the questions measure which can be theoretically measured on a continuum. This continuum is the horizontal axis in the plots of the curves. If there are several traits being measured, each of which have complex interactions with each other, then these unidimensional models are not appropriate.

Chapter 38

Bar, Line, and Pie Charts

The Chart Platform

The **Chart** command in the **Graph** menu (or toolbar) charts statistics for every numeric variable you specify. **Graph > Chart** creates a bar for each level of a categorical *X* variable, and charts a count or statistic (if requested). You can have up to two *X* variables as categories. The *X* values are always treated as discrete values.

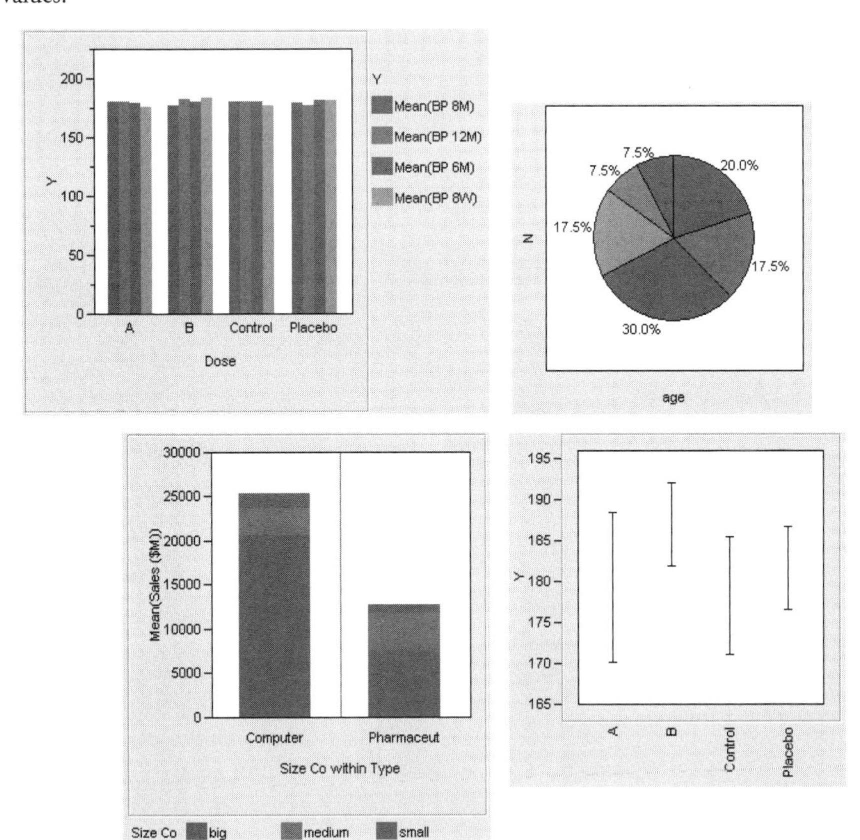

Contents

Bar, Line, Needle, Point, and Pie Charts

The **Chart** command in the **Graph** menu charts statistics or counts for every numeric variable you specify. If you specify a statistic, **Graph > Chart** computes the statistic you request and creates a bar for each level of a categorical *X* variable. If you don't request a statistic, JMP creates a chart using counts for *Y* values. You can have up to two *X* variables as categories. If you don't specify an *X* variable, JMP plots a bar for each row of the data table. All *X* values are always treated as discrete values.

The Chart platform is roughly equivalent to the **Tables > Summarize** command. It is useful for making graphical representations of summary statistics. If you want to make a plot of individual data points (rather than summaries of data points), we recommend using the Overlay Plot platform. See "Overlay Plots," p. 751 for details on overlay plots.

The Launch Dialog

When you select the **Chart** command, you first see the Chart Launch dialog (Figure 38.1). Here you can assign

- up to two *X* variables, which appear on the *x*-axis in the order you put them in the dialog
- as many numeric *Y* variables (statistics) as you want. If the data are already summarized, choose **Data** as the statistics option.

Figure 38.1 The Chart Launch Dialog

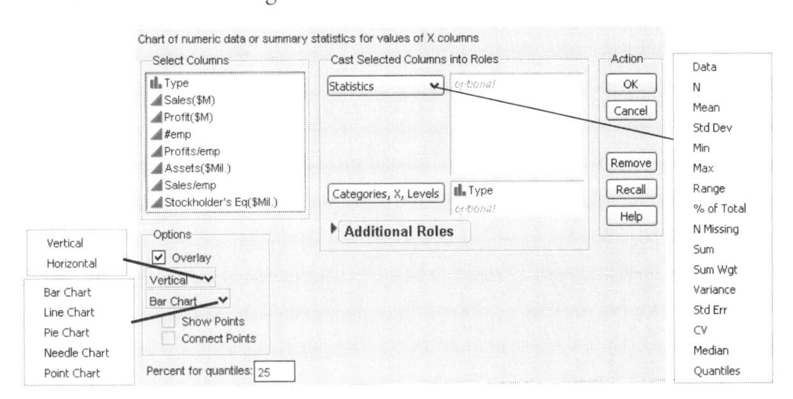

Charting options on the left of the dialog let you choose the orientation of the chart as **Vertical** or **Horizontal**. You can also select the chart types **Bar Chart**, **Line Chart**, **Pie Chart**, **Needle Chart**, and **Point Chart**. You can always change these options after the chart appears (See "Single-Chart Options," p. 742).

Figure 38.2 Charting Options

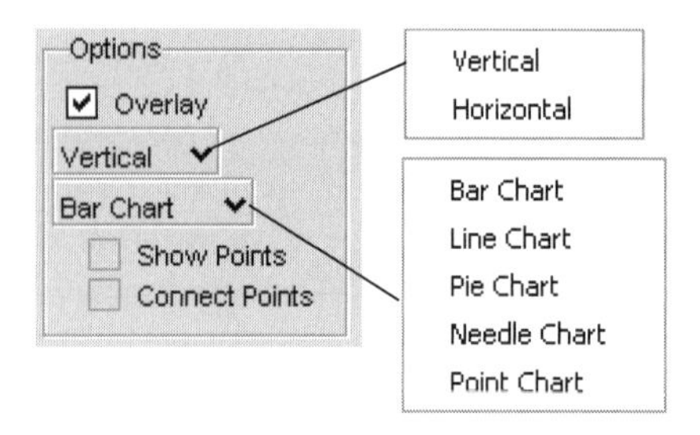

Categories, X, Levels

Optionally, you can use up to two *X* variables whose levels are categories on the *x*-axis. The Chart platform produces a bar for each level or combination of levels of the *X* variables. If you don't specify an *X* variable, the chart has a bar for each row in the data table.

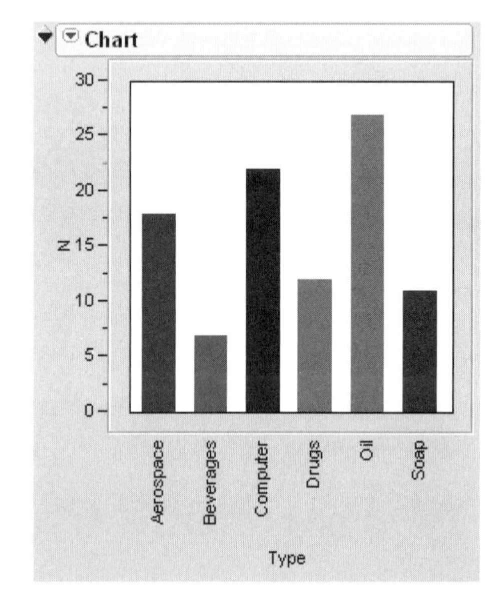

Ordering

By default, the Chart platform orders the bars using one of the common orders supported by JMP (months, days of the week, etc.). However, if the grouping column has a **Row Order Levels** column property, the levels are ordered in that order. If the grouping column has a **Value Ordering** column property, it uses that order. If both **Row Order Levels** and **Value Ordering** properties are defined, the **Row Order Levels** property has precedence.With neither property in effect, bars are drawn in alphanumeric order.

Charting Counts

You can specify an *X* variable without specifying any *Y* variables. In these cases, JMP produces bar charts that show counts for each level of the *X* variable(s) you specify. In the following example using

Big Class, age was specified as an **Categories, X, Levels** variable. No statistics variables were specified. A chart showing the number of people in each age group is the result.

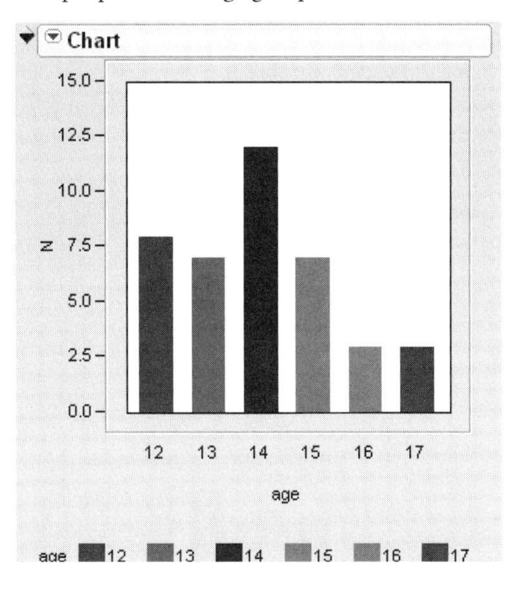

Two X-variables

If you specify two *X* -variables (with no statistics variables), JMP divides the data into groups based on the levels in the two *X*-variables, then plots the number of members in each group. In the following example, the two *X*-variables were age and sex (in that order). Since sex is listed second, it is nested within age.

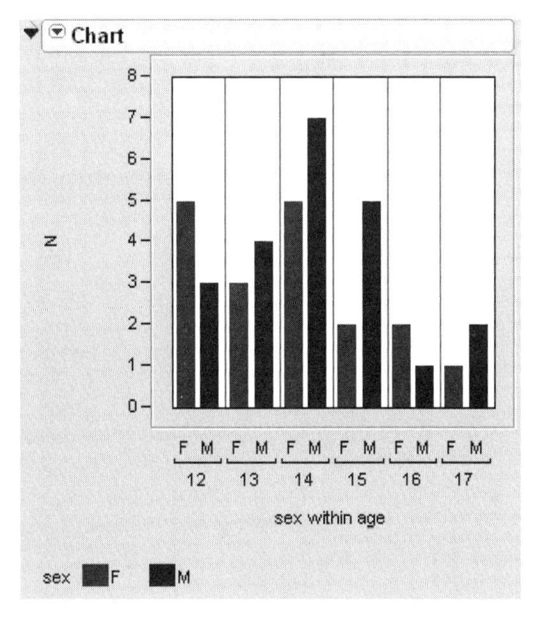

Y Variables Statistics

The two examples above did not use any statistics. However, you can plot as many statistics as you want on the *y*-axis. The **Statistics** popup menu on the launch dialog lists the available statistics. To specify the *y*-axis, highlight one or more numeric columns in the variable selection list and select from the list of statistics. The statistics to be plotted then appear in the statistics list on the right of the dialog (See Figure 38.1). If all the statistics requested are counting statistics (*e.g.*, N) for the same column, that column is the default X.

As a simple example, here we plot the mean height of students based on their age group. To do so,

1 Select age and click **Categories, X, Levels**

2 Select height and click **Statistics**. Select **Mean** from the drop-down list of statistics available.

3 Click **OK**.

This produces the following chart.

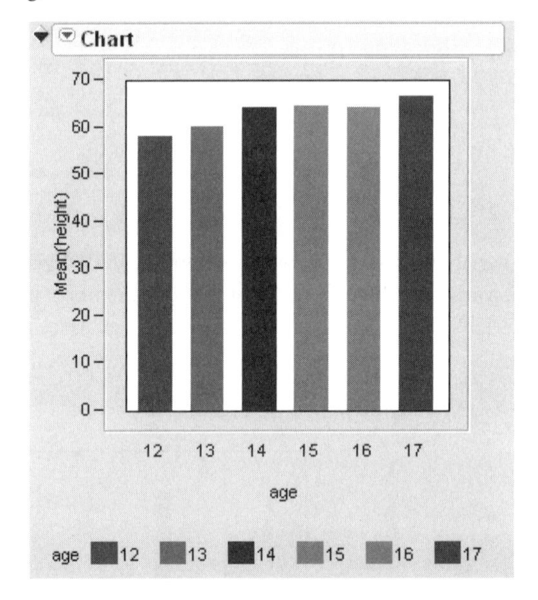

Note: Remember that the Chart platform plots a bar for each row in the data table by default. So, if you forgot to include age as an **Categories, X, Levels**, you would get a chart like the following.

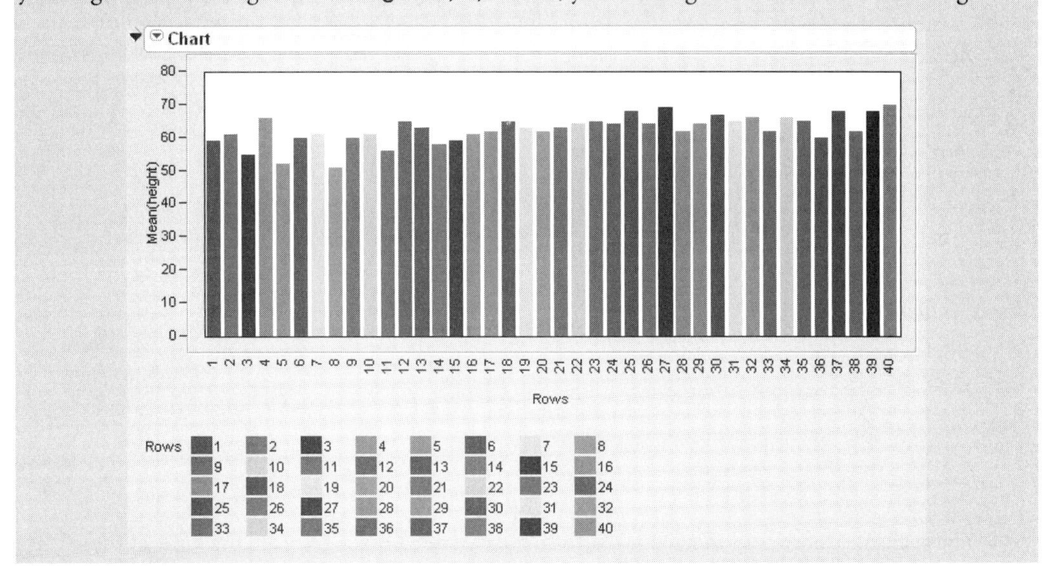

Multiple statistics (*Y* variables) show as side-by-side bars for categorical variable levels when the **Overlay** option (found on the platform drop-down menu) is in effect. The example here shows mean sales and mean profits for a categorical variable, Type. If the **Overlay** option were not in effect, two separate charts would be displayed.

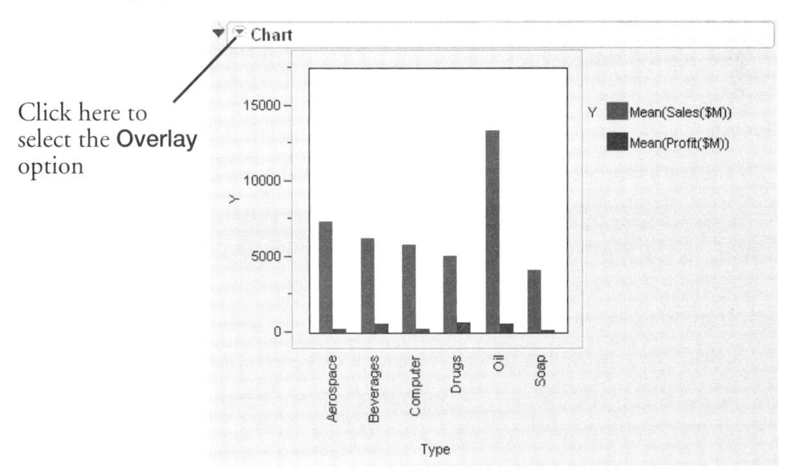

You can combine multiple *Y* statistics variables with multiple **Categories, X, Levels** variables.

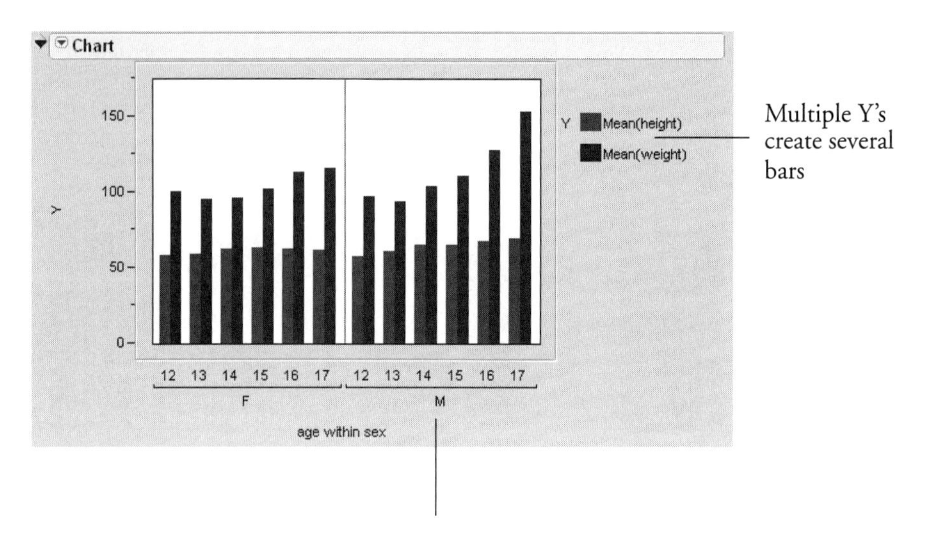

The available statistics in the Chart platform are described here. They are the same as those computed by statistical platforms in the **Analyze** menu and the **Summary** command in the **Tables** menu.

Data charts the value of each row in the data table when there is no categorical variable. If there is a categorical variable, **Data** produces a point plot within its levels.

N is the number of nonmissing values and is used to compute statistics when there is no column assigned as a weight variable. The Chart platform shows **N** for each level of a categorical variable.

% of Total is the percentage of the total number of rows that each level of a categorical value encompasses.

N Missing is the number of missing values in each level of a categorical variable.

Min is the least value, excluding missing values, in the level of a categorical variable.

Max is the greatest value in the level of a categorical variable.

Sum Wgt is the sum of all values in a column assigned the Weight role in the Launch dialog, and is used instead of **N** to compute other statistics. **Chart** shows the sum of the weight variable for each level of a categorical variable.

Sum is the sum of all values in each level of a categorical variable.

Mean is the arithmetic average of a column's values. It is the sum of nonmissing values divided by the number of nonmissing values.

Variance is the sample variance computed for each level of a categorical variable.

Std Dev is the sample standard deviation computed for each level of a categorical variable. It is the square root of the variance of the level values.

Std Err is the standard error of the mean of each level of a categorical variable. It is the standard deviation, **Std Dev**, divided by the square root of **N** for each level. If a column is assigned a weight variable, then the denominator is the square root of the sum of the weights.

CV is the coefficient of variation of a column's values. It is computed by dividing the column standard deviation by the column mean.

Range is the difference between the maximum and minimum values in each level of a categorical variable.

Median is the middle value in each level of a categorical variable. Half of the values in the level are greater than or equal to the median and half are less than the median.

Quantiles divide a data set so that n% of the data is below the nth quantile. To compute a specific quantile, enter the quantile value in the box located in the lower-left of the Chart launch dialog (see Figure 38.1) before requesting **Quantile** from the menu.

Additional Roles

The following options are found on the launch dialog by opening the Additional Roles outline node.

Grouping Variables

When there are one or more **Grouping** variables, the Chart platform produces independent results for each level or combination of levels of the grouping variables. These results are in the same window, but in separate plots. For example, the following chart was produced with the following specifications:

- (Y) **Statistic**: Mean(height)
- **Categories, X, Levels**: age
- **Group**: sex

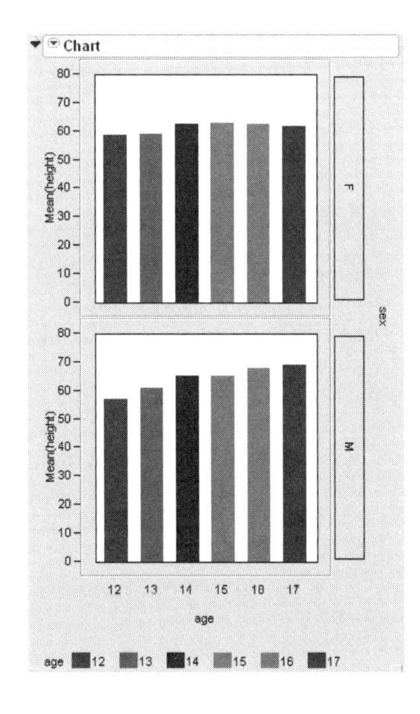

You can have multiple grouping columns as well. The following example uses the data table **Car-Poll.jmp**, with the following specifications:

- (Y) **Statistic**: Mean(age)
- **Categories, X, Levels**: sex and type
- **Group**: marital status and country

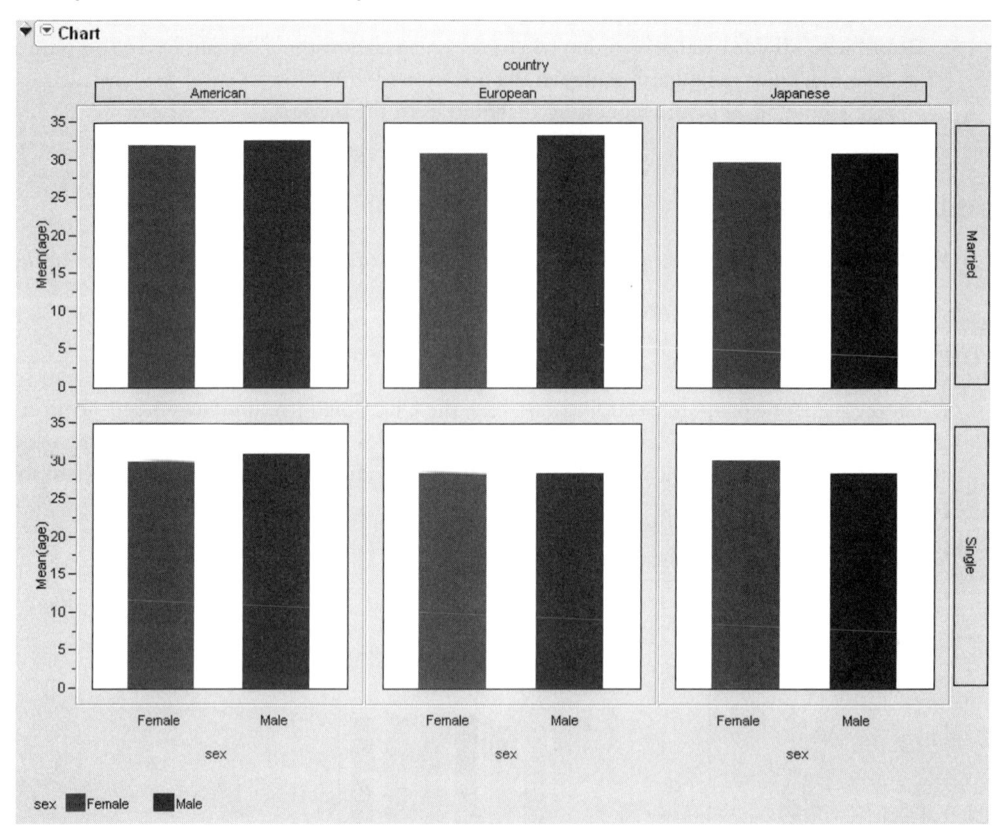

Grouping with multiple X's works as expected. The multiple group labels appear around the borders of the charts, and the multiple X's cause divisions within charts.

- (Y) **Statistic**: Mean(age)
- **Categories, X, Levels**: sex
- **Group**: marital status and country

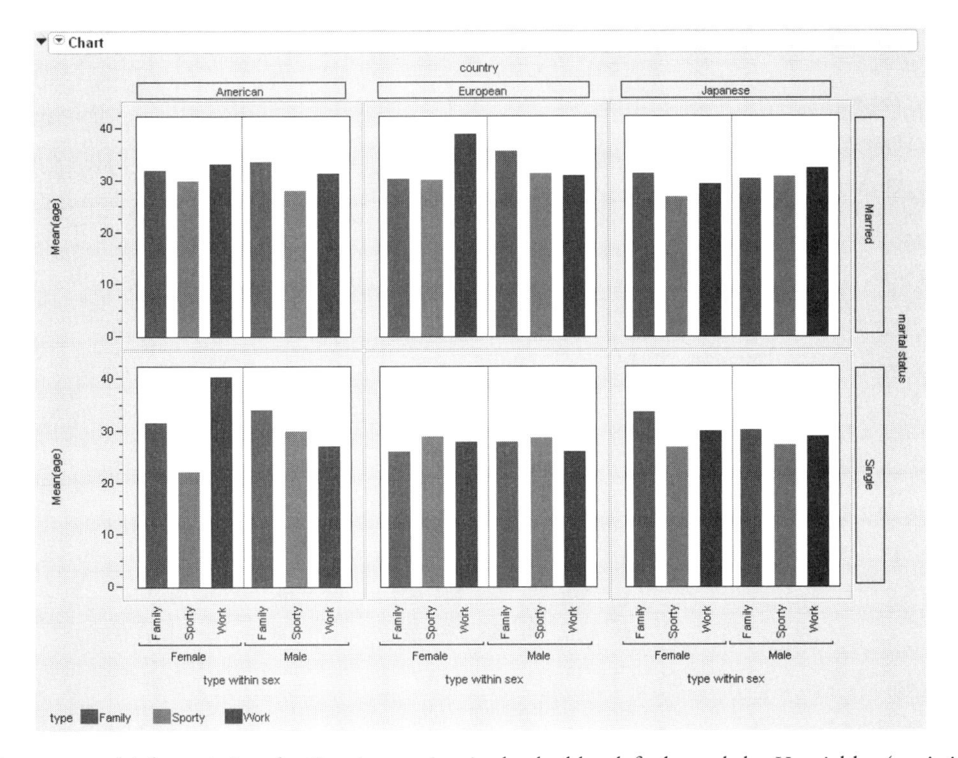

If there are multiple statistics, the **Overlay** option is checked by default, and the *Y* variables (statistics) are overlaid (*i.e.*, plotted on the same chart) for each level of the grouping variable, as in the following picture.

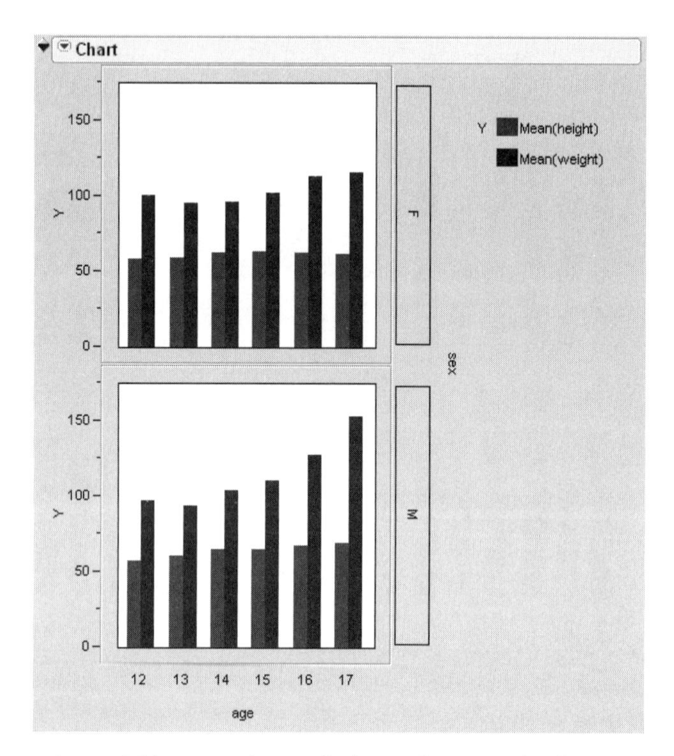

The levels of the grouping variable cannot be overlaid into the same plot frame. For this example, the groups are M and F, and the **Overlay** option cannot be used to superimpose the two M and F graphs. To see that kind of result, use **Categories, X, Levels** instead of **Grouping** variables.

By Variables

By variables cause plots to be created in separate outline nodes. Compare the following two Charts, both created with an **Categories, X, Levels** of age. One uses the **By** role, the other uses the **Group** role.

The **By** option creates separate outline nodes, one for M, one for F

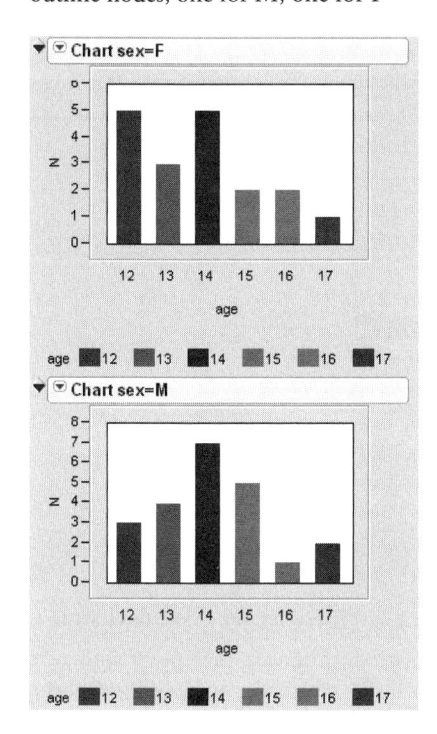

The **Group** option creates separate charts for M and F, but only one outline node

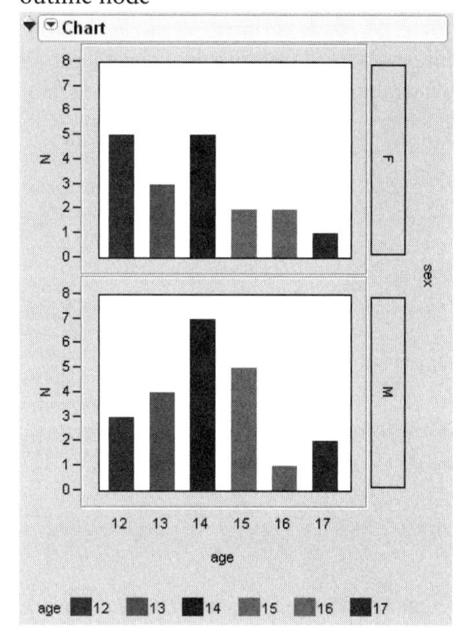

Charting Options

Options on the Launch dialog and in the platform popup menu let you choose various chart options. Platform options affect all charts in the report window. However, many options can be applied to individual charts (single-chart options), as described next.

Coloring Bars in a Chart

Bars can be colored interactively or automatically after the chart has been created.

- To manually set the color of all the bars, make sure none of them are selected, choose **Level Options > Color** and select from the color palette.
- To set the color of a single bar, select the bar in the chart, choose **Level Options > Color** and select from the color palette.
- To automatically assign a color to a level, assign the color to the **Value Colors** column property, accessible through the **Cols > Column Info** dialog box

Single-Chart Options

You can apply the options in the popup menu shown here to individual charts. When there is a single chart, right-click in the plot frame to see the popup menu shown here. If the overlay option is in effect, context-click on an individual chart legend to see the menu. When you access this menu through the platform popup menu, without any legends highlighted, the commands apply to all charts.

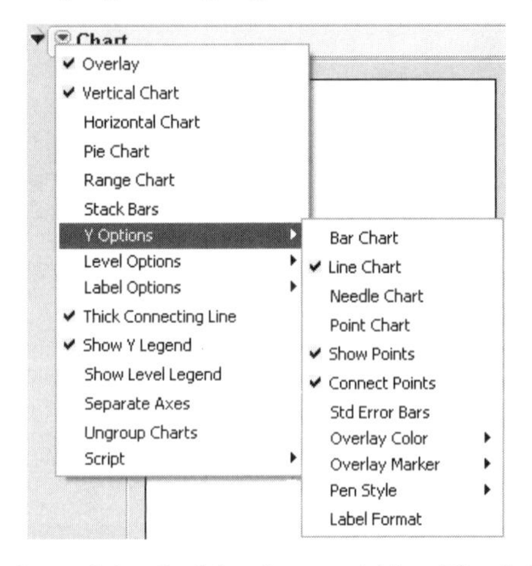

Bar Chart displays a bar for each level of the chart variables. The default chart is a bar chart.

Line Chart replaces a bar chart with a line chart and connects each point with a straight line. You can also choose the **Show Points** option to show or hide the points. **Line Chart** is also available as a platform option, which then applies to all charts at once.

Needle Chart replaces each bar with a line drawn from the axis to the plotted value. **Needle Chart** is also available as a platform option, which then applies to all charts at once.

Point Chart shows only the plotted points, without connecting them.

Show Points toggles the point markers on a line or needle chart on or off.

Connect Points toggles the line connecting points on or off.

Std Error Bars toggles the standard error bar on plots of means on or off. Note that this option is only available for plots that involve means of variables.

Overlay Color associates color to statistics (*y*-axis). This is useful to identify its points when overlaid with other charts.

Overlay Marker associates a marker to statistics, to identify them in overlaid charts, as shown here.

Pen Style lets you choose a line style from the palette, as shown here.

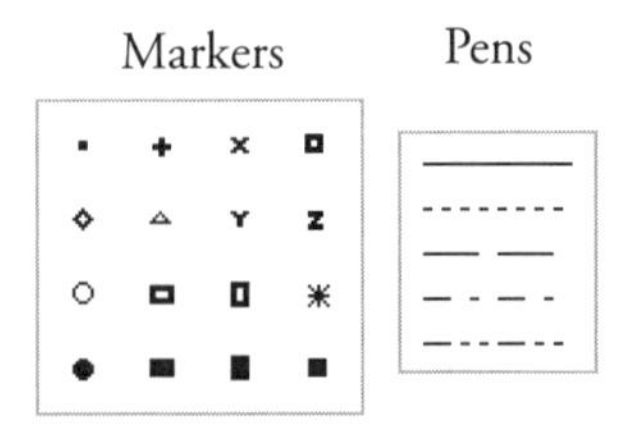

Label Format brings up a dialog that lets you specify the field width and number of decimals for labels.

As an example of these options, open Stock Prices.jmp. The variables represent the dates and values of a stock over time. Note that the variable YearWeek is a computed column representing the year and week in a single variable. An ideal plot for this data is a range chart showing the high, low, and average close values for each stock. For those weeks where data exists for multiple days, the average of the values is plotted.

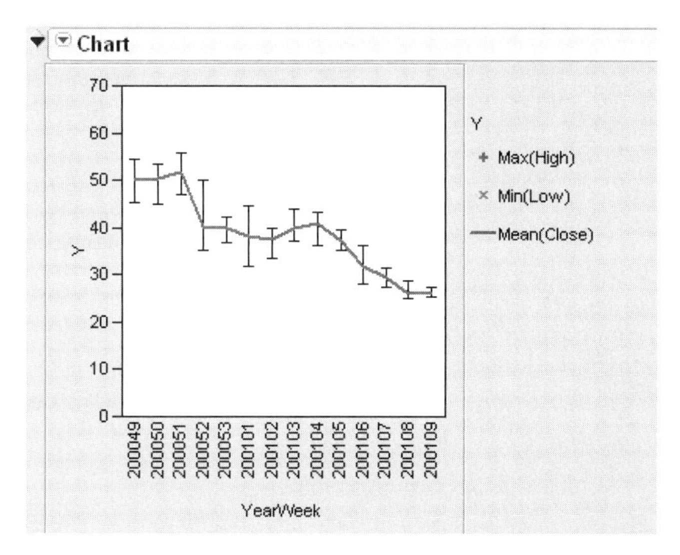

The plot can be reproduced with the JSL script Chart, attached to the data table.

```
Chart(X( :YearWeek), Y(Max( :High), Min( :Low), Mean( :Close)), Range
  Chart(1), Y[3] << Connect Points(1))
```

The script also shows the statistics and options used in the plot—the mean of High, Low, and Close as Y variables, YearWeek as the X variable, with the **Range Chart** and **Connect Points** options turned on.

Individual-Level Options

You can click on the legend for a level within a plot to highlight it. If you right-click on a highlighted legend, the commands to access the **Colors** and **Markers** palettes appear; the commands then affect only the highlighted level and its associated rows in the data table.

Platform Options

After you specify chart variables and click **OK** on the Chart launch dialog, the charts appear with a platform popup menu with several options.

✔ Overlay
✔ Vertical Chart
Horizontal Chart
Pie Chart
Range Chart
Stack Bars
Y Options ▸
Level Options ▸
Label Options ▸
✔ Thick Connecting Line
✔ Show Y Legend
Show Level Legend
Separate Axes
Ungroup Charts
Script ▸

Overlay displays a single overlaid chart when you have more than one *Y* (statistics) variable. You can assign any chart type to the individual charts and overlay them. For example, the chart shown below has three overlaid variables; two are bar charts, and one is a line chart. When **Overlay** is not checked, the platform shows duplicate axis notation for each chart. By default, the axis notation only shows for the last chart displayed if the charts are not overlaid.

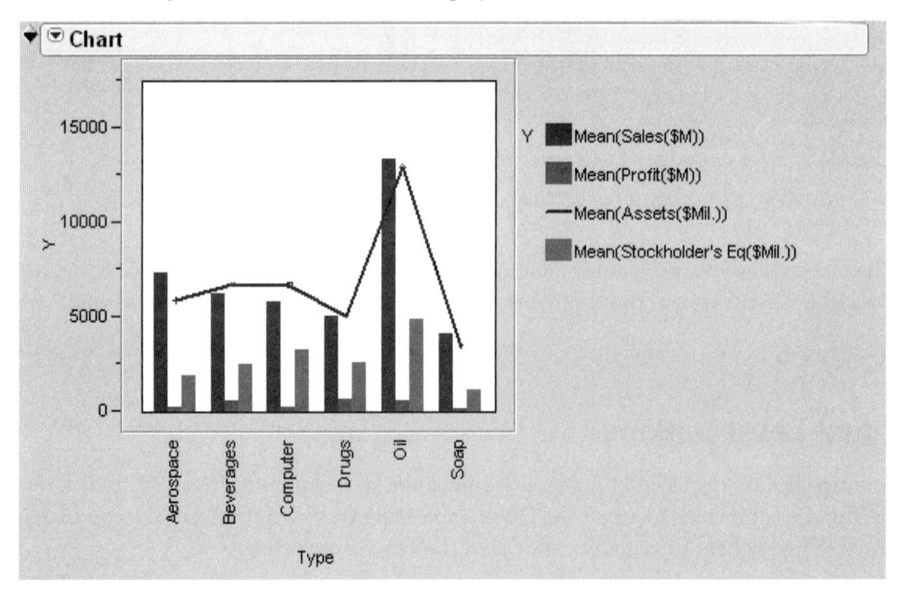

Vertical, Horizontal changes a horizontal bar chart or a pie chart to a vertical bar chart (**Vertical**), or a vertical bar chart or a pie chart to a horizontal bar chart (**Horizontal**).

Pie changes a horizontal or vertical chart type to a pie chart.

Range Chart displays a range chart. See "Create a Range Chart," p. 750, for an example of a range chart.

Figure 38.3 Range Chart

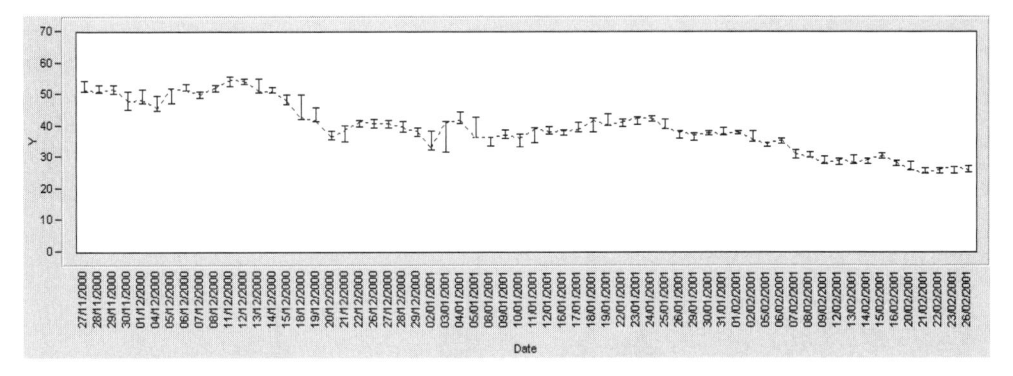

Stack Bars stacks the bars from levels of subgroup end-to-end. To use this option, you need two **Categories, X, Levels** variables and a statistic. See "Plot a Stacked Bar Chart," p. 747, for an example of stacking bars.

Y Options lets you access the options described previously under "Single-Chart Options," p. 742. These options affect the chart whose legend you highlight.

Level Options lets you access the color and marker options described previously under "Single-Chart Options," p. 742. These options affect highlighted bars.

Label Options allow you to attach labels to your plots. In the **Label Options** menu, the first two options (**Show Label** and **Remove Label**) turn labels on and off. The last three options (**Label by Value**, **Label by Percent of Total Values**, **Label By Row**) specify what label should appear. Only one label can be shown at a time.

Thick Connecting Line toggles the connecting line in a line plot to be thick or thin.

Show Y Legend shows the Y legend of the plot. This option is on by default for overlaid charts.

Show Level Legend shows the level legend of the plot. This option is on by default for separated charts.

Separate Axes duplicate the axis notation for each chart when there are multiple charts. By default, the axis notation only shows for the last chart displayed if the charts are not overlaid.

Ungroup Charts creates a separate chart for each level of a specified grouping variable.

Script has a submenu of commands available to all platforms that let you redo the analysis, or save the JSL commands for the analysis to a window or a file.

Examples

The next sections give examples of different combinations of *X* variables and *Y* statistics. The examples use the Financial.jmp and the Companies.jmp data tables in the Sample Data folder.

Chart a Single Statistic

If you don't specify a variable for categories on the *x*-axis in the **Categories, X, Levels** box, most statistics produce a bar for each observation in the data table. In these cases, you usually specify **Data** as the statistic to plot. Each bar reflects the value of the *Y* variable, as illustrated here.

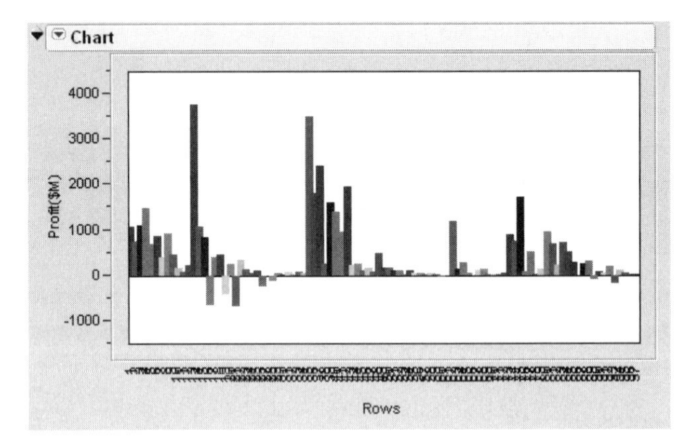

Plot a Line Chart of Several Statistics

To chart multiple statistics variables, highlight the variable in the variable selector list and select the statistic you want from the **Statistics** popup menu (See Figure 38.1).

Continue this process for as many statistics as you want to see on the same chart. If it is appropriate for your analysis, you can also have an *X* variable. If you don't have an *X* variable for grouping, the Chart platform gives a bar or point for each row in the data table.

Select **Y Options > Line Chart** from the platform drop-down menu to display the example shown on the right.

Figure 38.4 Overlaid Bar and Line Charts for an X Variable and Two Statistics

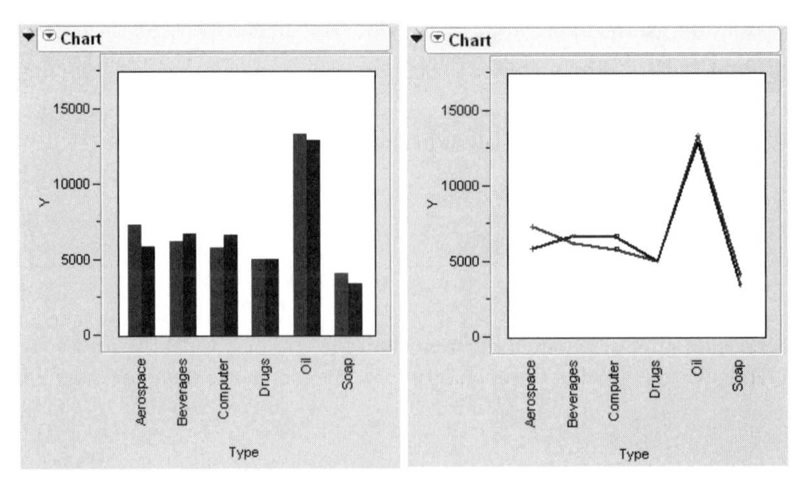

If **Overlay** is not checked and **Bar Chart** is selected, then bar charts are arranged vertically in the window, as shown in Figure 38.5.

Chart Multiple Statistics with Two X Variables

The next example, which shows two categorical variables, uses the Companies.jmp table in the Sample Data folder. This table has both company type (Type) and size of company (Size Co) as *x*-variables. When you choose two **Categories, X, Levels** variables in the Chart Launch dialog, the second variable forms subgroups. The Chart platform produces a chart for each level of the subgroup variable within each level of the grouping variable. The subgroup charts show side-by-side for each *Y*, as shown on the left in Figure 38.5. The chart on the right results when you use the overlay option. The *Y* variables show on a common axis, giving an overlaid bar chart for each subgroup.

Figure 38.5 Vertical and Overlaid Bar Charts for Multiple Responses

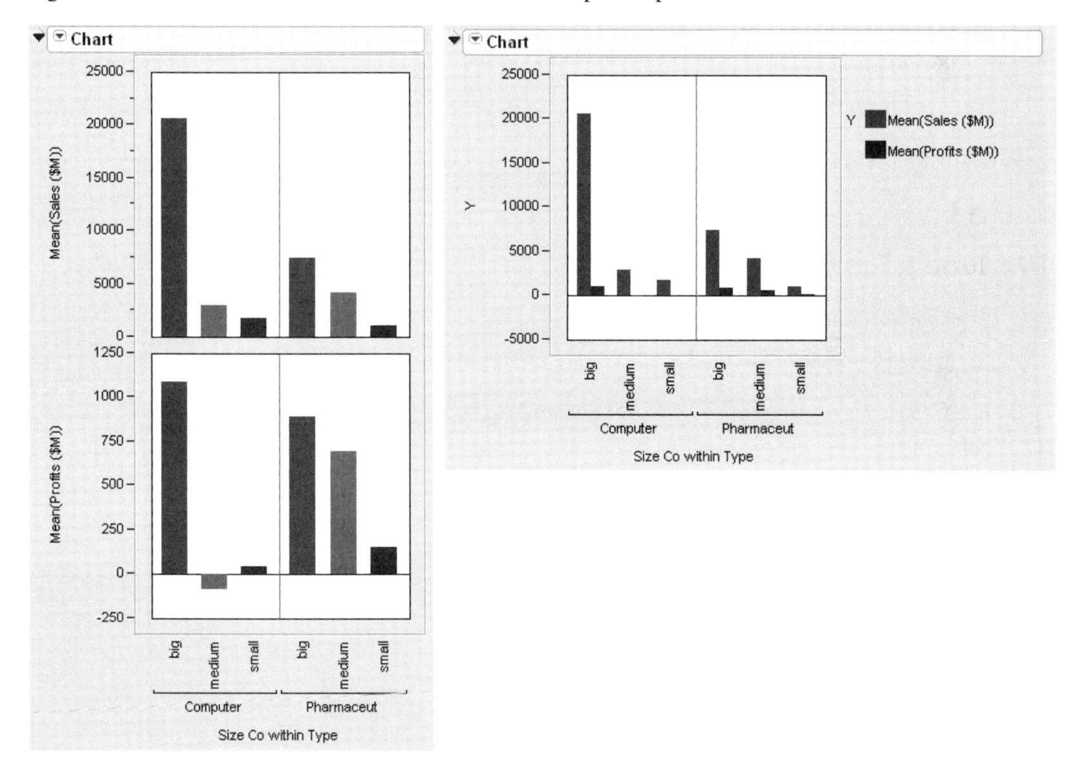

Plot a Stacked Bar Chart

When you have two *X* levels and a single *Y* variable, you can stack the bars by selecting the **Stack Bars** command from the platform menu. For example, create a chart with Mean (Sales($M)) as Y and Type and Size Co as X variables. When the chart appears, select **Stack Bars** from the platform menu. This produces the chart shown in Figure 38.6

Figure 38.6 Stacked Bar Chart

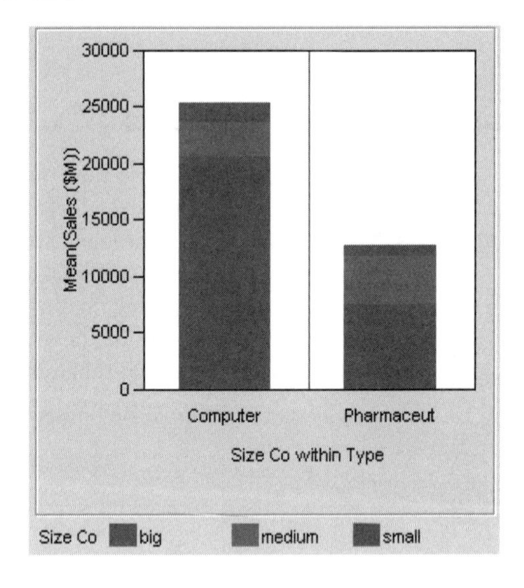

Produce a Pie Chart

Pie charts can be produced in two ways.

1 Select **Pie Chart** from the list of chart types in the launch dialog

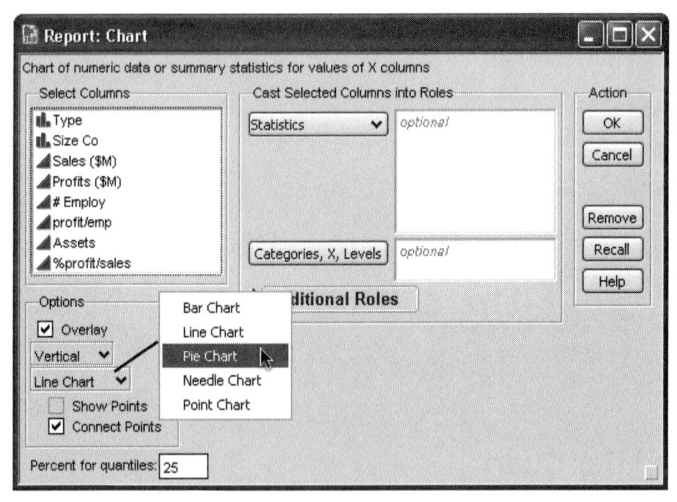

2 Select **Pie Chart** from the platform drop-down list

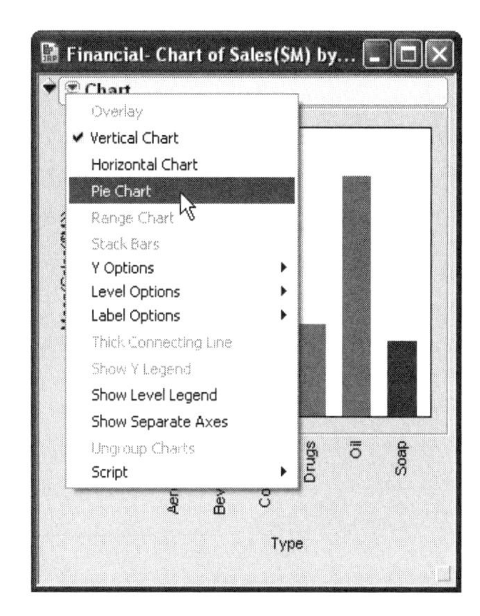

For example, using Companies.jmp, select **Graph > Chart** and assign Size Co as the **Categories, X, Levels** role. At this point, you can either

- change the chart type to **Pie** in the launch dialog and click **OK**.
- click **OK**, then select **Pie** from the platform drop-down menu.

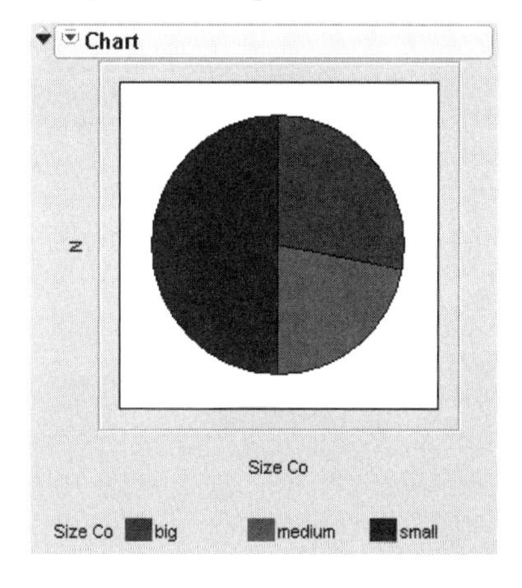

Create a Range Chart

Using Companies.JMP, use Mean(Profits($M)) and Mean(Sales($Mil)) as statistics to plot, and use **Size Co** as the **Categories, X, Levels**. Click **OK** to make a bar chart appear, then select **Range Chart** from the platform menu.

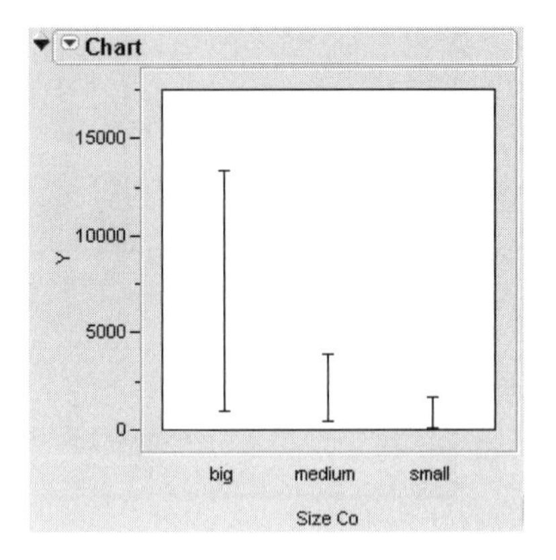

Chapter 39

Overlay Plots

The Overlay Plot Platform

The **Overlay Plot** command in the **Graph** menu produces plots of a single *X* column and one or more numeric *Y*'s. The curves can optionally be shown as separate plots for each *Y* with a common *x*-axis. Plots can be modified with range and needle options, color, log axes, and grid lines. Curves with two different scales can be overlaid on the same plot with the addition of a right axis.

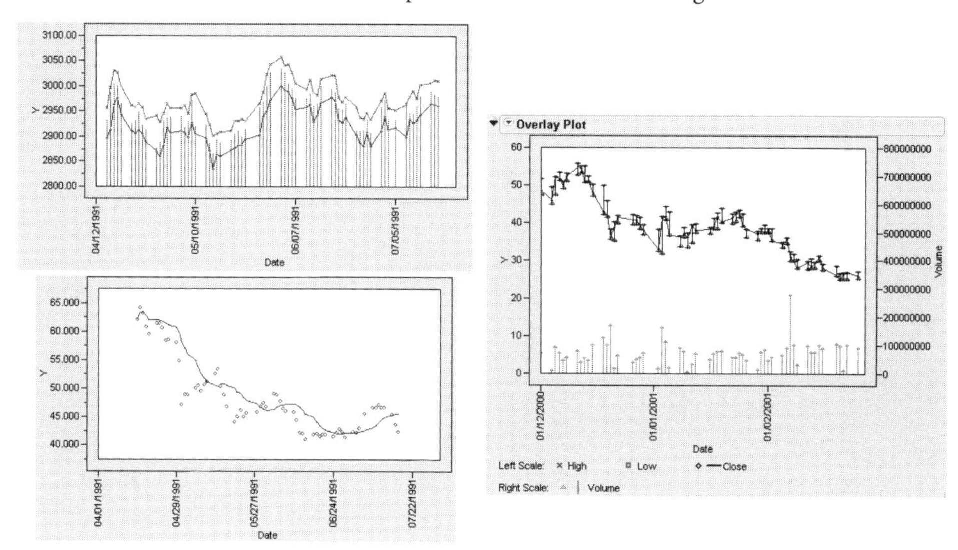

Contents

The Overlay Plot Platform

The **Graph > Overlay Plot** command overlays numeric *Y*'s with a single *X* variable. The Overlay Plot platform does not accept non-numeric variables for the *y*-axis.

Plotting Two Variables on a Single *y*-Axis

To illustrate the Overlay Plot platform, open the Spring.jmp data table in the sample data folder. The table has a row for each day in the month of April. The column called April is the numeric day of the month, and the remaining columns are various weather statistics.

Select the **Overlay Plot** command from the **Graph** menu and complete the Overlay Launch dialog as shown in Figure 39.1. The values in the column called April are the days of the month. April is assigned the **X** role. Daily humidity measures at 1:00 PM and 4:00 PM, Humid1:PM and Humid4:PM, are **Y** variables.

Note that the columns in the **Y** role have a left- or right-pointing arrow to the left of the column name. This arrow designates on which vertical axis (on the left or right of the plot) the variable appears. Change the designation by highlighting the column in the **Y** list and clicking the **Left Scale/Right Scale** button.

The **Sort X** option causes the points to be connected in order of ascending *X*-values. Otherwise, the points are connected in row order.

Figure 39.1 Launch Dialog for Overlay Plot

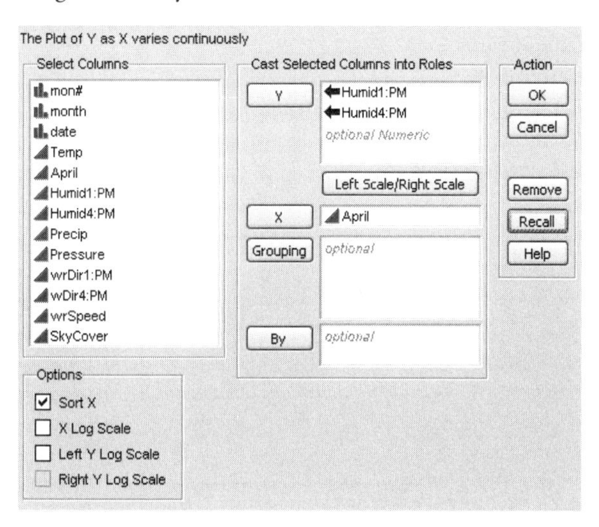

When you click **OK**, the plot shown in Figure 39.2 appears. Initially, this platform overlays all specified *Y* columns. The legend below the plot shows individual markers and colors that identify each *Y* column.

Figure 39.2 Plot with Overlay Options

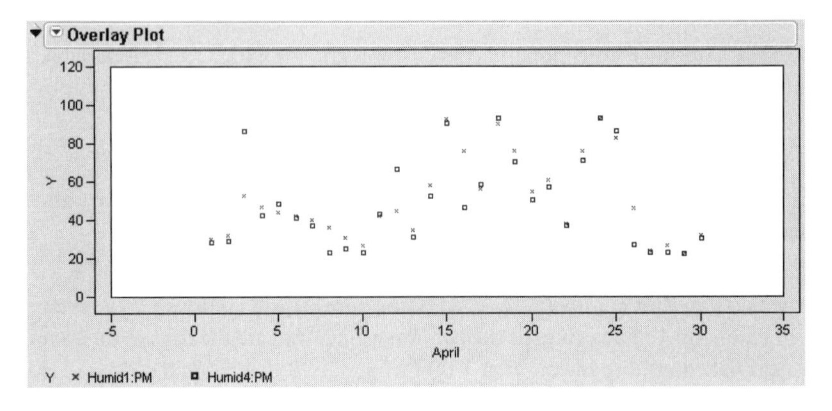

Plotting Two Variables with a Second *y*-axis

A second *Y*-axis is useful for plotting data with different scales on the same plot, such as a stock's closing price and its daily volume, or temperature and pressure. Consider, for example, plotting the selling price of an inexpensive stock against the Dow Jones Industrial Average. Data for this situation is found in the sample data file Stock Prices.JMP.

- Assign High, Low, Close, and Volume as **Y** variables.
- Select Volume in the **Y** list and click **Left Scale/Right Scale**.
- Assign Date as the **X** variable.
- Click **OK**.

The graph in Figure 39.3 appears.

Figure 39.3 Dual Axis Overlay Plot

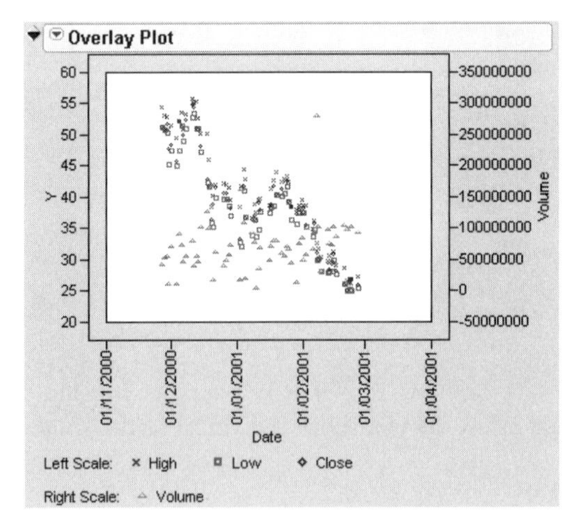

Grouping Variables

The Overlay Plot platform allows the production of several plots in one window through the use of grouping variables. With one grouping variable, a stacked vector of plots appears, with one plot for each level of the grouping variable. Two grouping variables result in a matrix of plots.

For example, use the **Big Class** data table to produce overlay plots of **height** (**Y**) vs. **weight** (**X**) with **age** and **sex** as grouping variables. A portion of this plot is shown in Figure 39.4.

Figure 39.4 Grouped Plots Without Separate Axes

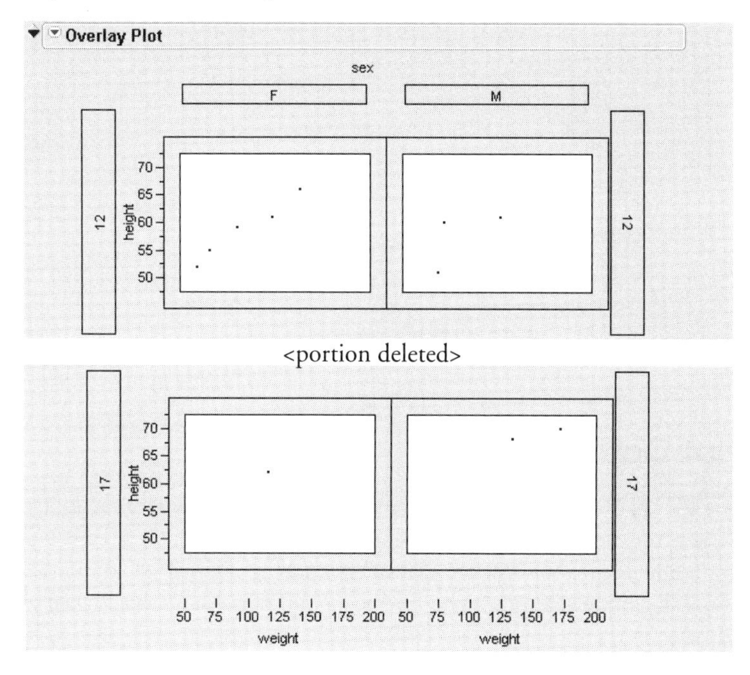

Check the **Separate Axes** option to produce plots that do not share axes, as shown in Figure 39.5.

Figure 39.5 Grouping Variables

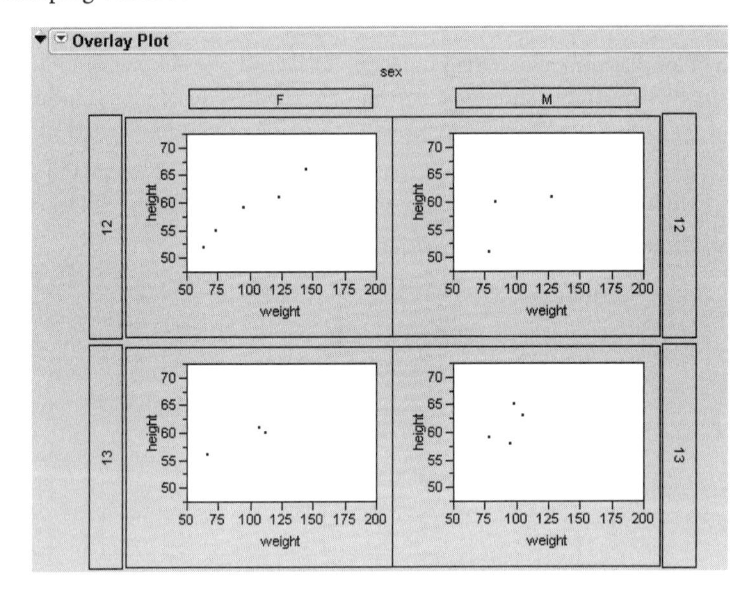

Overlay Plot Options

The Overlay Plot platform has plotting options accessed by the popup menu icon on the Overlay Plot title bar. When you select one of these options at the platform level, it affects all plots in the report if no legends are highlighted. If one or more plot legends are highlighted, the options affects only those plots. There is also a single-plot options menu for each *Y* variable, which show when you highlight a *Y* variable legend beneath the plot and Right-click.

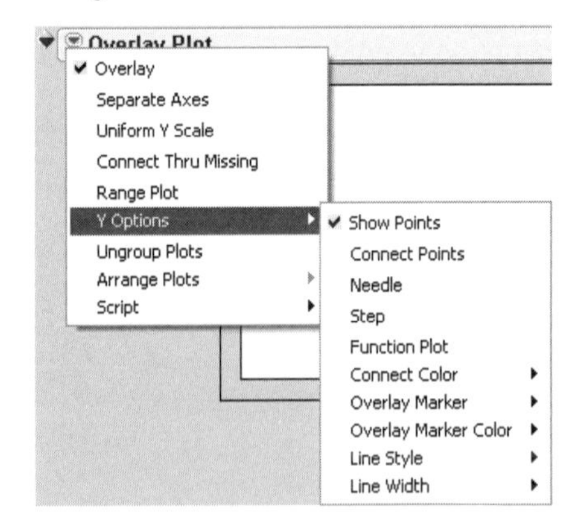

The individual plot options are the same as those in the **Y Options** submenu at the platform level.

Platform Options

Platform options affect every plot in the report window:

Overlay overlays plots for all columns assigned the *Y* role. Plots initially appear with the default **Overlay** option in effect. When you turn the **Overlay** option off, the plots appear separately. For example, the plots shown here result when you toggle off the **Overlay** option for the plot shown in Figure 39.2 (with points connected).

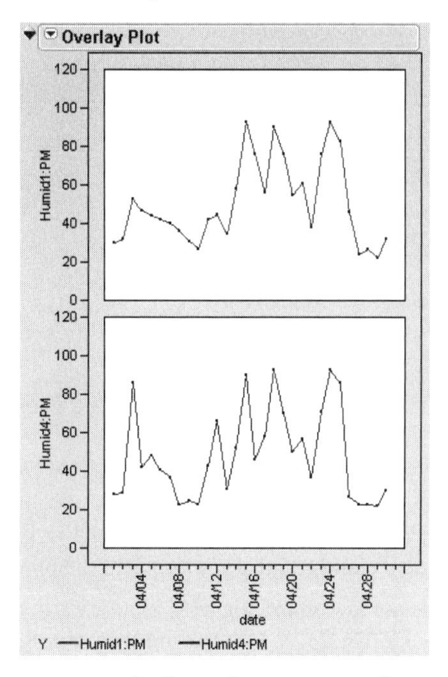

Separate Axes lets you associate each plot with its own set of *XY*-axes. If **Separate Axes** is off, the vertical axis is shared across the same row of plots and the horizontal axis is shared on the same column of plots, as shown here.

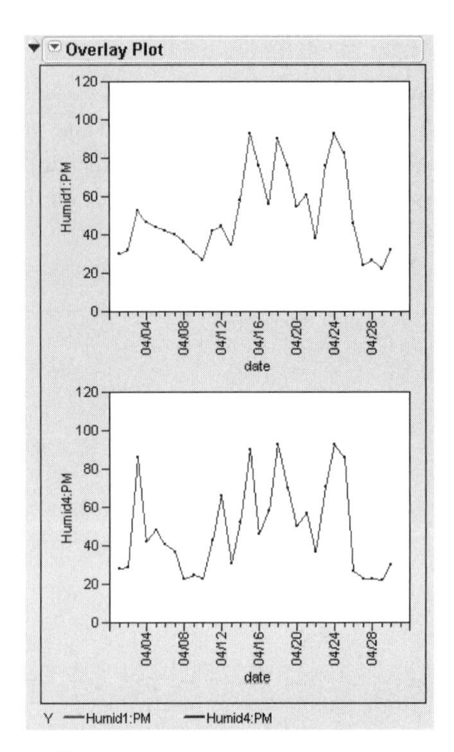

Uniform Y Scale makes the Y scales the same on grouped plots.

Connect Thru Missing connects adjacent points in the plot, regardless of missing values.

Range Plot connects the lowest and highest points at each *x* value with a line with bars at each end. The plot shown in Figure 39.7 "Overlaid Plot with Range Option and Assigned Markers," p. 760 is an example of the **Range Plot** option applied to the plot in Figure 39.2 "Plot with Overlay Options," p. 754.

Note: The **Needle** option, described in the next section, and **Range Plot** option are mutually exclusive.

Y Options apply to all plots in the report window when selected from the platform window or individual plots when selected from a plot legend. The **Y Options** menu is described next in the *Y Options* section.

Ungroup Plots creates a separate chart for each level of a grouping variable.

Arrange Plots allows you to specify the number of plots in each row.

Script has a submenu of commands available to all platforms that let you redo the analysis or save the JSL commands for the analysis to a window or a file.

Y Options

Each *Y* variable is labeled in a legend beneath the plot. The *Y* options are available from the **Y Options** submenu on the platform title bar as described above. Alternatively, you can right-click on any *Y* variable legend beneath the plot to see the **Y Options** menu. Use Shift-click to highlight multiple *Y* legends. Selecting an option affects all plots whose legends are highlighted.

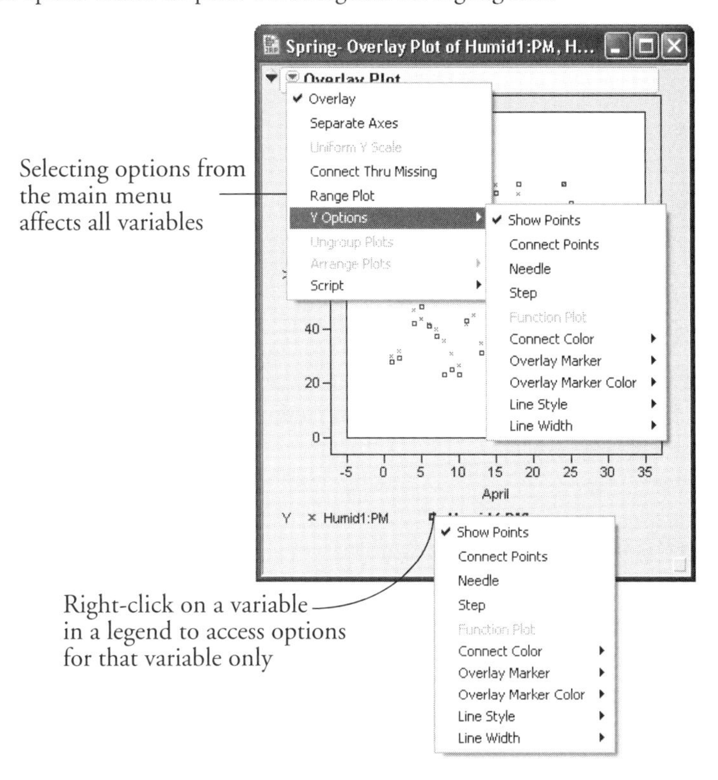

If you select **Y Options** from the main Overlay menu, each selection affects all Y variables in the plot.

Show Points alternately shows or hides points. The plot shown in Figure 39.7 uses the **Show Points** option and the **Range Plot** option described previously.

Connect Points is a toggle that alternately connects the points with lines. You can use **Connect** without showing points.

Needle draws a straight vertical line from each point to the *x*-axis.

Note: The **Connect**, **Step**, and **Needle** options are mutually exclusive. You can only have one selected at a time.

Step joins the position of the points with a discrete step by drawing a straight horizontal line from each point to the *x* value of the following point, and then a straight vertical line to that point. You can use **Step** without showing points, as in Figure 39.8.

Note: The **Connect**, **Step**, and **Needle** options are mutually exclusive. You can only have one selected at a time.

Function Plot plots a formula (stored in the Y column) as a smooth curve. To use this function, store a formula in a column that is a function of a single X column. Assign the formula to the Y role. For example, the data table in Figure 39.6 results in the overlay plot shown on the right. Use the **Line Style** and **Line Width** commands to alter the appearance of the plotted function.

Figure 39.6 Function Plot

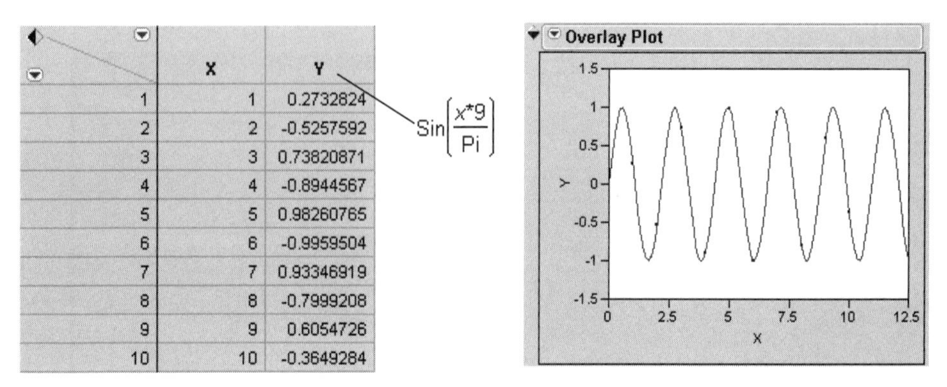

	X	Y
1	1	0.2732824
2	2	-0.5257592
3	3	0.73820871
4	4	-0.8944567
5	5	0.98260765
6	6	-0.9959504
7	7	0.93346919
8	8	-0.7999208
9	9	0.6054726
10	10	-0.3649284

Note: Overlay Plot normally assumes you want a function plot when the Y column contains a formula. However, formulas that contain random number functions are more frequently used with simulations, where function plotting is not often wanted. Therefore, the **Function Plot** option is off (by default) when a random number function is present, but on for all other functions.

Connect Color displays the standard JMP color palette for assigning colors to lines that connect points.

Overlay Marker assigns markers to plotted points using the standard JMP marker palette. The plot shown in Figure 39.7 uses assigned **Overlay Markers** and has all rows highlighted.

Overlay Marker Color assigns a color to all points of the selected variable. Selecting black causes the plot to use the color from the data table row states. If you want the points to be black, do not assign a color as a row state—black is the default.

Line Style allows the choice of dashed, dotted, or other line styles.

Line Width allows the choice of line widths.

Figure 39.7 Overlaid Plot with Range Option and Assigned Markers

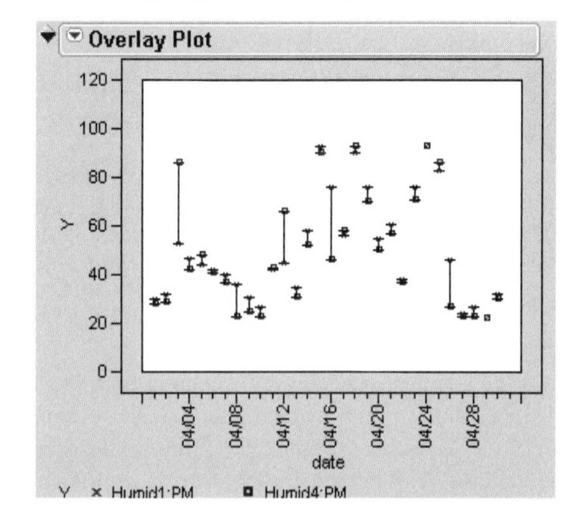

Figure 39.8 Step Plot with Hidden Points

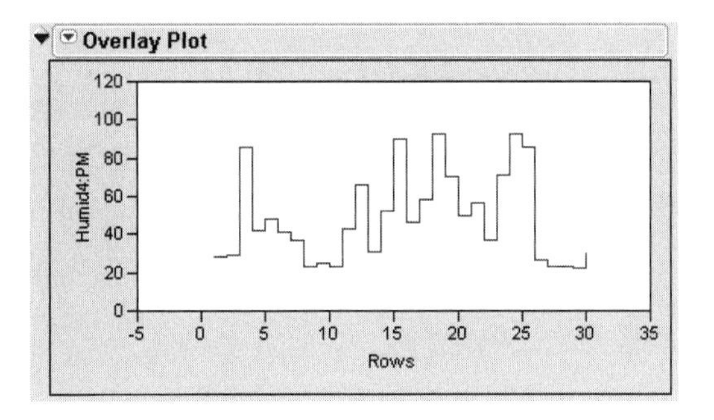

Chapter 40

Plotting Surfaces

The Surface Plot Platform

The Surface Plot platform functions both as a separate platform and as an option in model fitting platforms. Up to four dependent surfaces can be displayed in the same plot. The dependent variables section, below the plot, has four rows that correspond to the four surfaces. Depending on what you choose to view (sheets, points, isosurfaces, or density grids) and whether you supply a formula variable, different options appear in the dependent variables section.

Surface Plot is built using the 3-D scene commands from JMP's scripting language. Complete documentation of the OpenGL-style scene commands is found in the *JMP Scripting Guide*.

In this platform, you can

- use the mouse to drag the surface to a new position
- Right-click on the surface to change the background color or show the virtual ArcBall (which helps position the surface)
- enable hardware acceleration, which may increase performance if it is supported on your system
- drag lights to different positions, assign them colors, and turn them on and off.

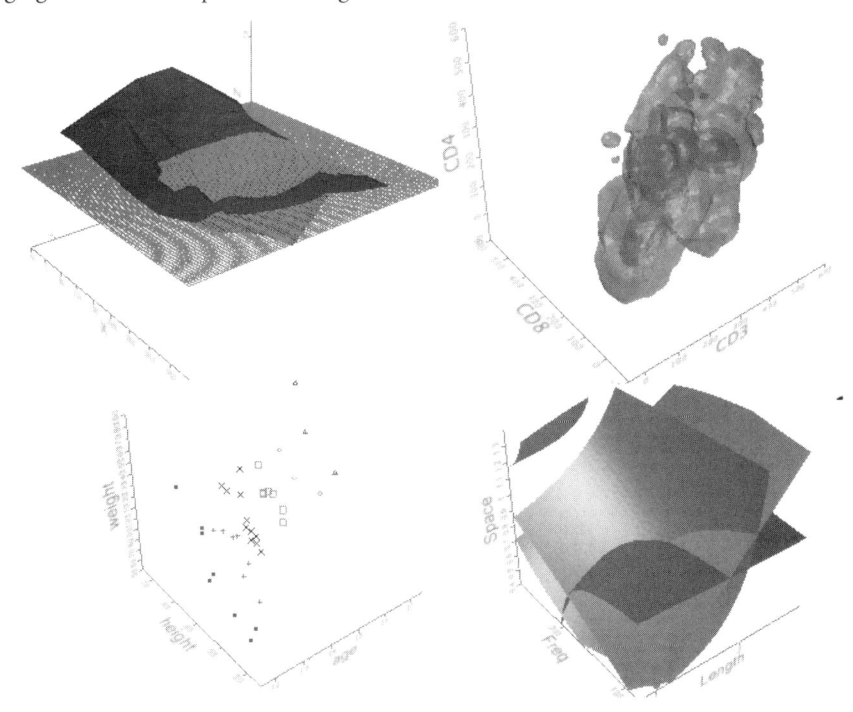

Contents

Surface Plots

The Surface Plot platform is used to plot points and surfaces in three dimensions.

Surface plots are available as a separate platform (**Graph > Surface Plot**) and as options in many reports (known as the **Surface Profiler**). Its functionality is similar wherever it appears.

The plots can be of points or surfaces. When the surface plot is used as a separate platform (that is, not as a profiler), the points are linked to the data table—they are clickable, respond to the brush tool, and reflect the colors and markers assigned in the data table. Surfaces can be defined by a mathematical equation, or through a set of points defining a polygonal surface. These surfaces can be displayed smoothly or as a mesh, with or without contour lines. Labels, axes, and lighting are fully customizable.

Launching the Platform

To launch the platform, select Surface Plot from the Graph menu. If there is a data table open, this displays the dialog in Figure 40.1. If there is no data table open, you are presented with the default surface plot shown in Figure 40.2.

Figure 40.1 Surface Plot Launch Dialog

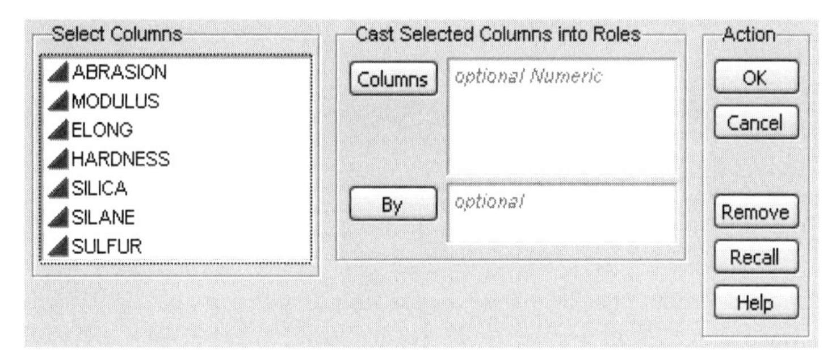

Specify the columns you want to plot by putting them in the **Columns** role. Variables in the **By** role produce a separate surface plot for each level of the **By** variable.

Plotting a Single Mathematical Function

To produce the graph of a mathematical function without any data points, do not fill in any of the roles on the launch dialog. Simply click **OK** to get a default plot, as shown in Figure 40.2.

Figure 40.2 Default Surface Plot

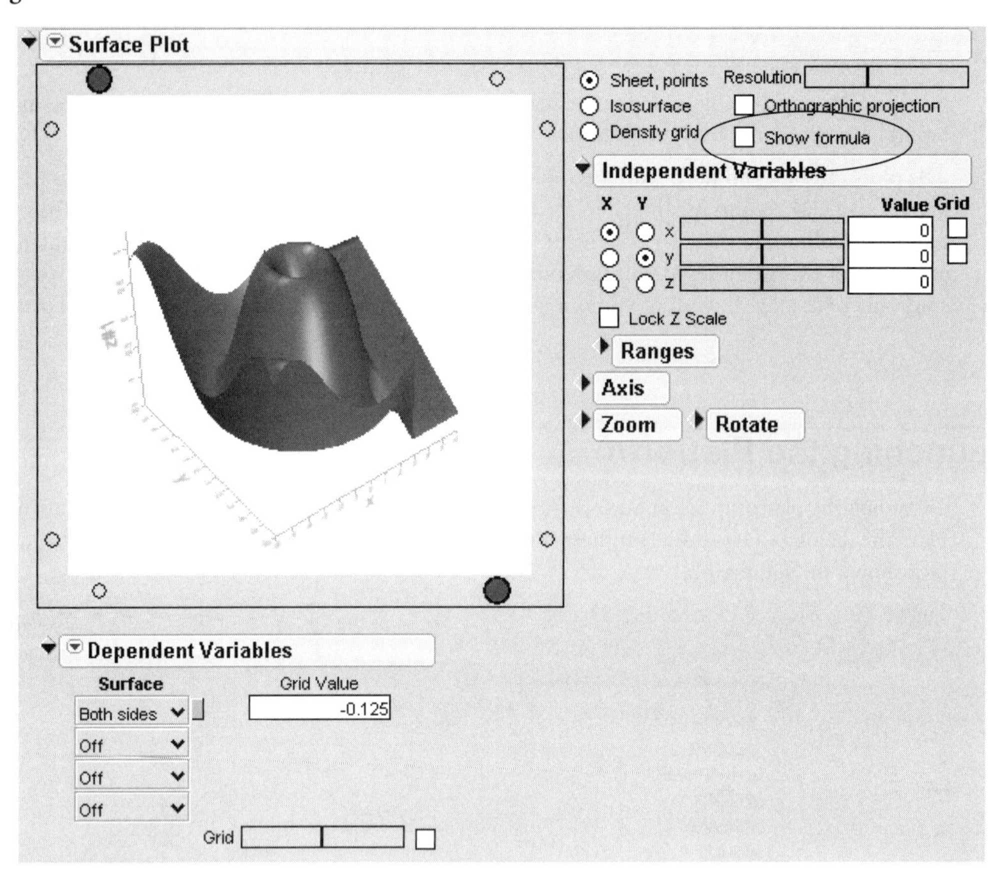

Select the **Show Formula** checkbox to show the four formulas spaces available for plotting. By default, only one surface is turned on, but you can activate more.

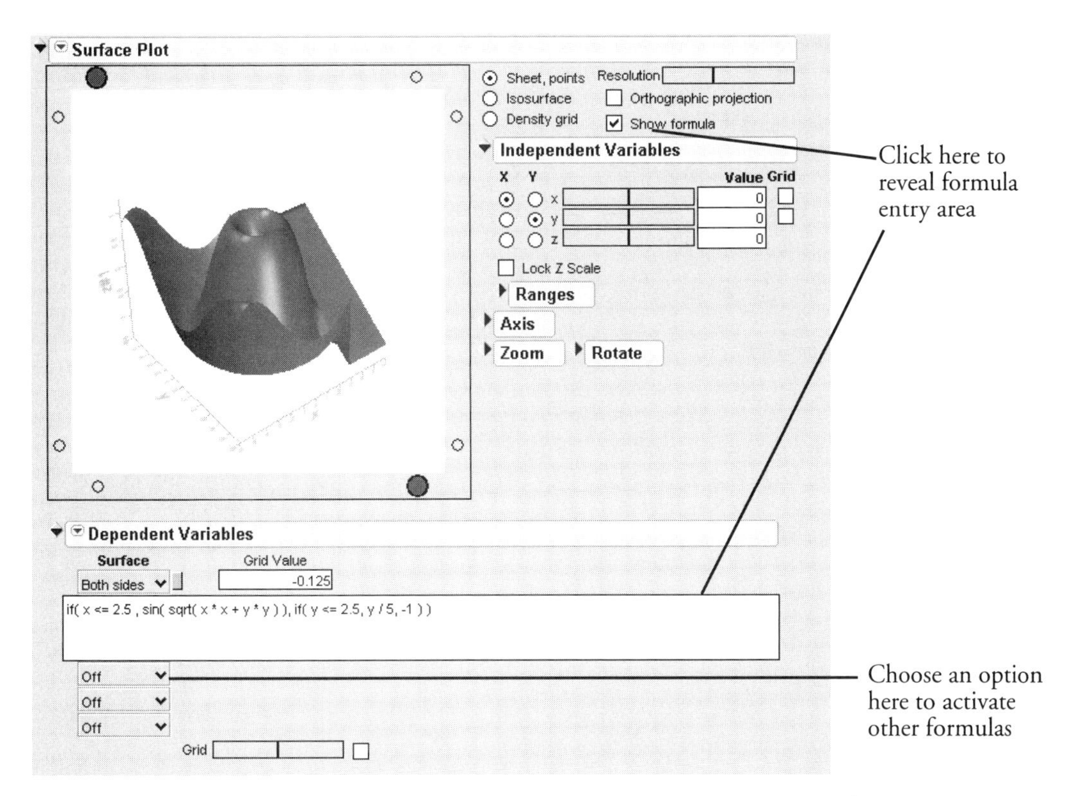

The default function shows in the box. To plot your own function, enter it in this box.

Plotting Points Only

To produce a 3-D scatterplot of points, select the *x*-, *y*-, and *z*-values from the columns list and click the **Columns** button. For example, using the Tiretread.jmp data, Figure 40.3 shows the launch dialog and resulting surface plot.

Figure 40.3 3-D Scatterplot Launch and Results

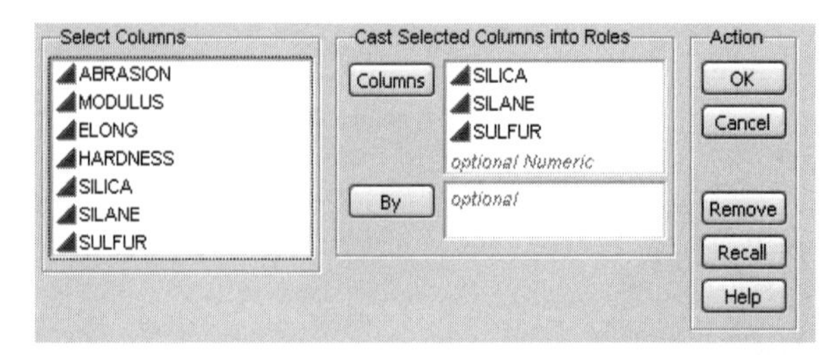

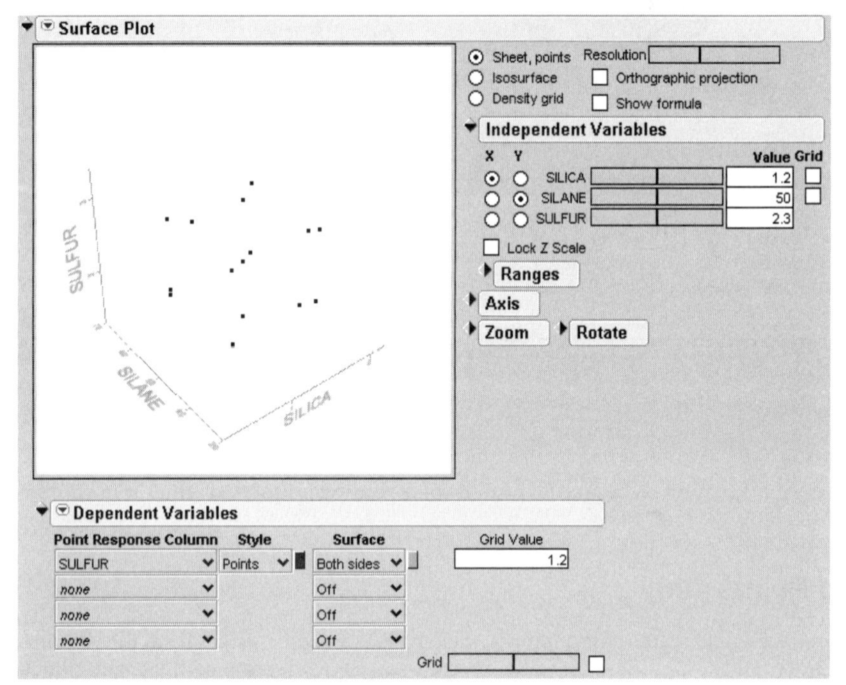

Plotting a Formula from a Data Column

To plot a formula (*i.e.* a formula from a column in the data table), select the column in the columns list, click the **Columns** button, and click **OK**. For example, suppose you have run the RSM script attached to the Tiretread.jmp data table and saved the prediction formulas to the data table. The launch dialog used to plot this surface (and its resulting output) is shown in Figure 40.4. You do not have to specify the factors for the plot, since the platform automatically extracts them from the formula.

Figure 40.4 Formula Launch and Output

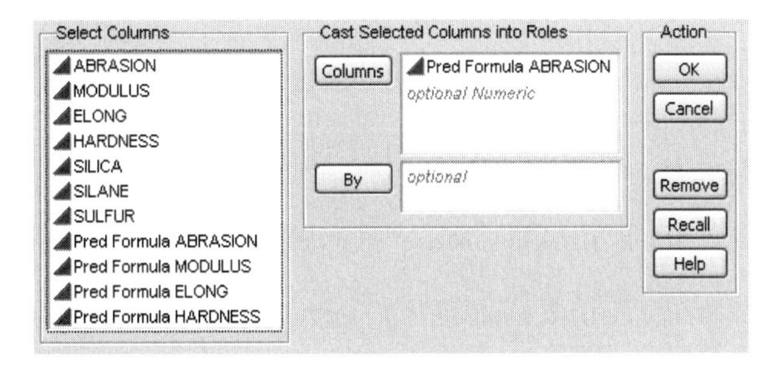

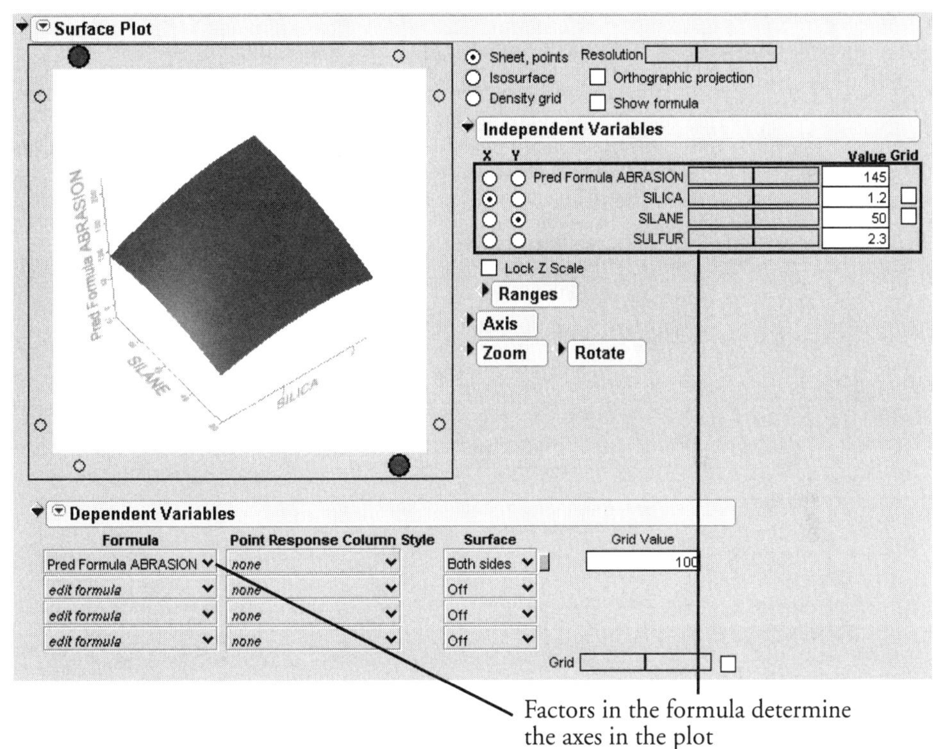

Note that this only plots the prediction surface. To plot the actual (observed) values in addition to the formula,

- Select **Graph > Surface Plot**.
- Include both the observed values (ABRASION) and the prediction formula (Pred Formula ABRASION) in the **Columns** role of the launch dialog.

Figure 40.5 shows the launch dialog and the completed results.

Figure 40.5 Formula and Data Points Launch and Output

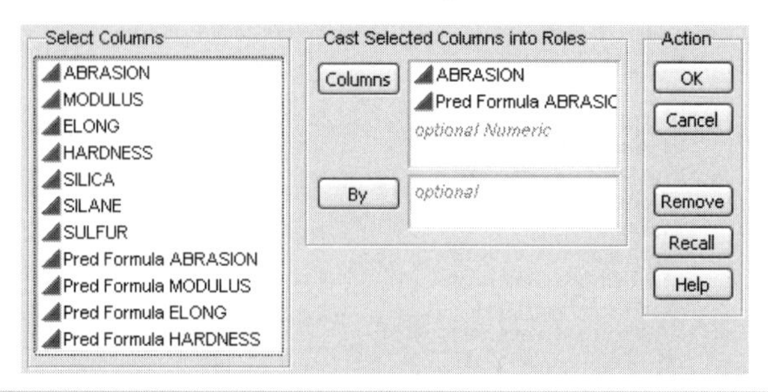

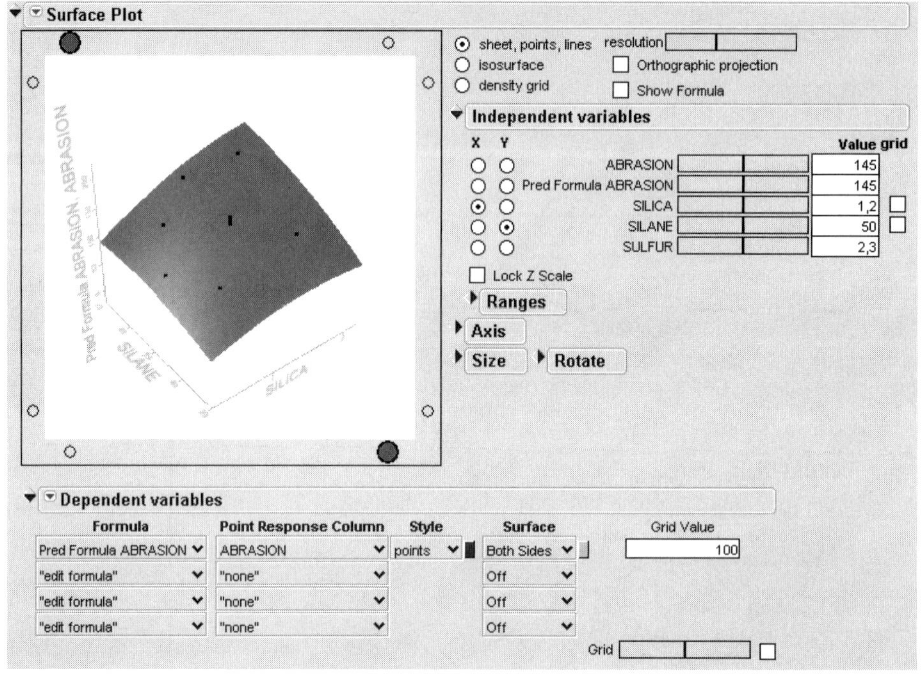

Density Grids

Density grids are shells defined by three variables. Denser clusters of points have higher densities, and the **Value** sliders in the Dependent Variables control panel vary the density that the shell encompasses. JMP can plot up to four nested density shells (represented by the four sets of controls under Dependent Variables, directly below the graph.

Figure 40.6 Density Grid with Three Shells

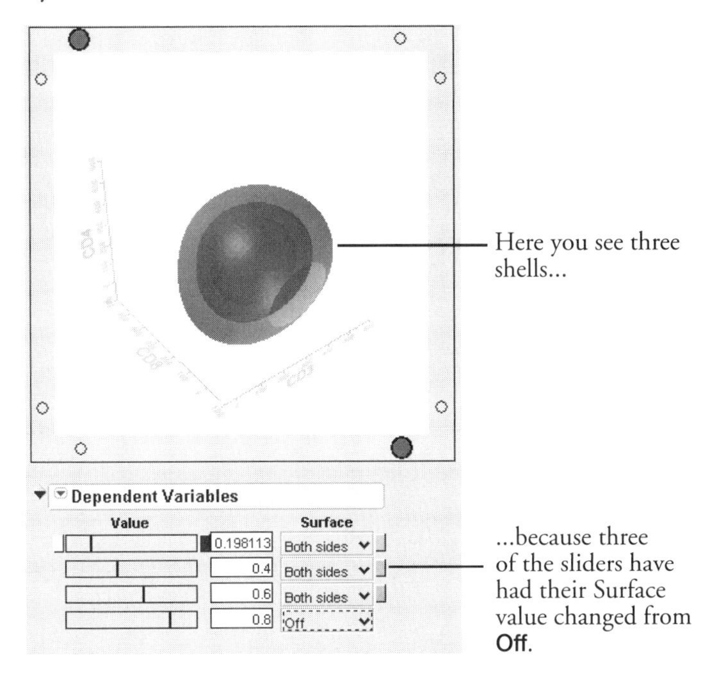

There are three important controls for getting a good picture. The first two are found at the top right of the control panel (see Figure 40.9 "Resolution Controls for Density Grids," p. 774.

- **Resolution** (a drop-down box) that determines the number of bins that the data is sorted into
- **Smooth** (a slider) that determines how much smoothing is applied to the bins after the data is loaded into them. Less smoothing gives a tighter, glove-like fit. Too much makes a loose, spherical fit.
- **Value** (a slider in the **Dependent Variables** section) that selects the desired density level. The highest density values are a tiny spot near the center of the data. The lowest density values are a large spherical region surrounding the data.

All three controls interact with each other, so for example, you may need to adjust the **Smooth** and **Value** sliders when changing resolutions.

To see an example of a density grid, open the Cytometry.jmp data set and select CD3, CD8, and CD4 as variables to plot.

When the report appears, select the Density Grid radio button. This draws a single density shell around the cloud of points. Move the Value slider under the graph to change the density encompassed by this shell.

To add more shells, change the Surface settings that are currently set to **Off** to **Both Sides**. Figure 40.6 shows three shells with resolution 64.

Isosurfaces

Isosurfaces are the 3-D analogy to a 2-D contour plot. An isosurface requires a formula with three independent variables. The Resolution slider determines the $n \times n \times n$ cube of points that the formula is evaluated over. The Value slider in the Dependent Variable section picks the isosurface (that is, the contour level) value.

As an example, open the Tiretread.jmp data table and run the RSM for 4 Responses script. This produces a response surface model with dependent variables ABRASION, MODULUS, ELONG, and HARDNESS. Since isosurfaces require formulas, select **Save Columns > Prediction Formula** for ABRASION, MODULUS, and ELONG.

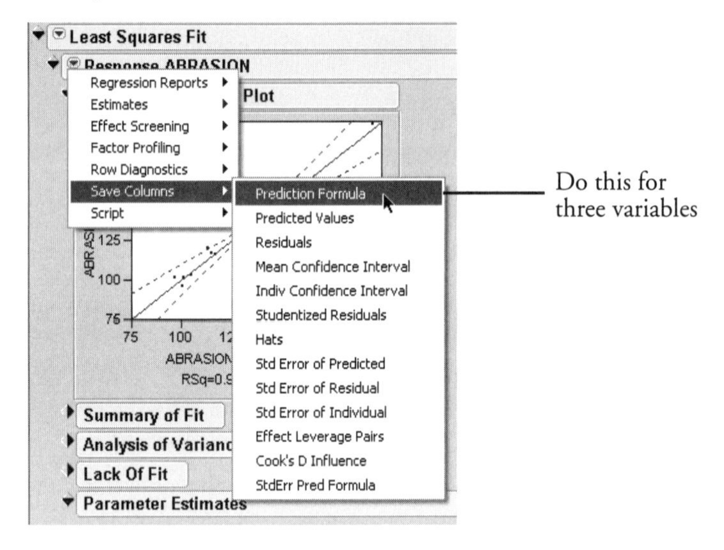

Now launch the Surface Plot platform and designate the three prediction columns as those to be plotted.

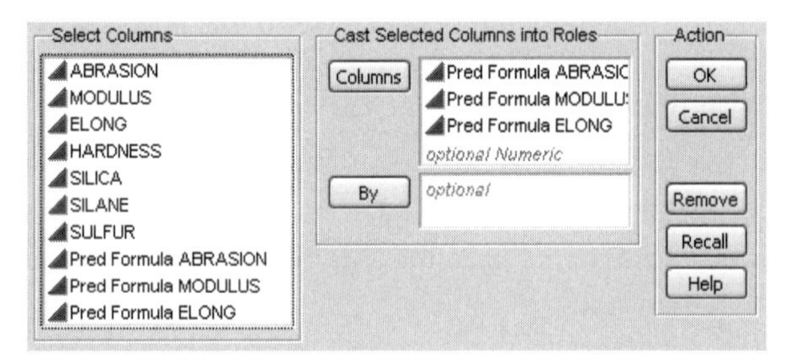

When the report appears, select the Isosurface radio button. Under the **Dependent Variables** outline node, select **Both Sides** for all three variables.

Figure 40.7 Isosurface of Three Variables

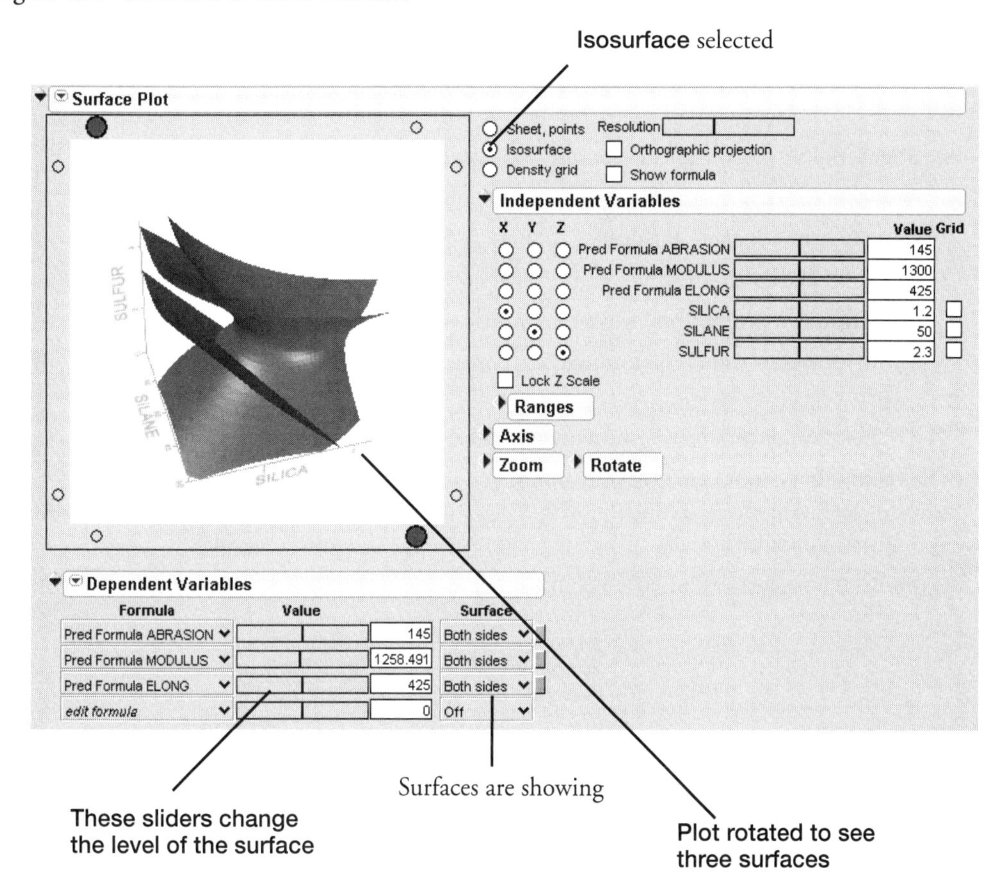

For the tire tread data, one might set the hardness at a fixed minimum setting and the elongation at a fixed maximum setting, then use the slider for modulus to see which values of modulus are inside the limits set by the other two surfaces.

The Surface Plot Control Panel

The drop-down menu in the main Surface Plot title bar has three entries. The first shows or hides the control panel. The second shows or hides the lights border. The third is the standard JMP **Script** menu.

The Control Panel consists of the following groups of options.

Appearance Controls

The first set of controls allows you to specify the overall appearance of the surface plot. They are shown in Figure 40.8.

Figure 40.8 Appearance Controls

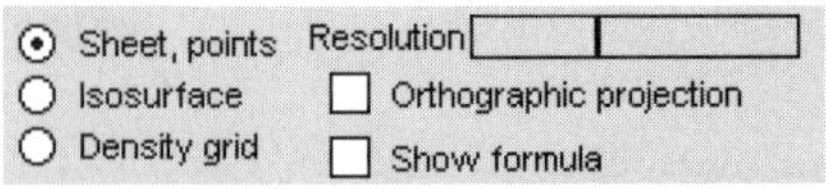

Sheet, points Resolution
Isosurface Orthographic projection
Density grid Show formula

Sheet, points is the setting for displaying sheets, points, and lines

Isosurface changes the display to show isosurfaces, described in "Isosurfaces," p. 772.

Density grid changes the display to show density grids, described in "Density Grids," p. 770.

The two checkboxes alter the way you see the plot.

Orthographic projection gives the orthographic projection of the surface.

Show formula shows the formula edit box under the plot, allowing you to enter a formula to be plotted.

The **Resolution** slider affects how many points are evaluated for a formula.

The **Resolution** control changes to a drop-down list of pre-defined resolutions for density grids. Too coarse a resolution means a function with a sharp change might not be represented very well, but setting the resolution high makes evaluating and displaying the surface slower.

Figure 40.9 Resolution Controls for Density Grids

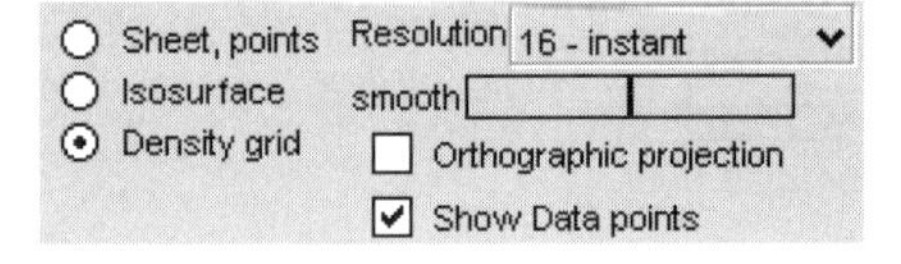

Independent Variables

The independent variables controls are displayed in Figure 40.10.

Figure 40.10 Variables Controls

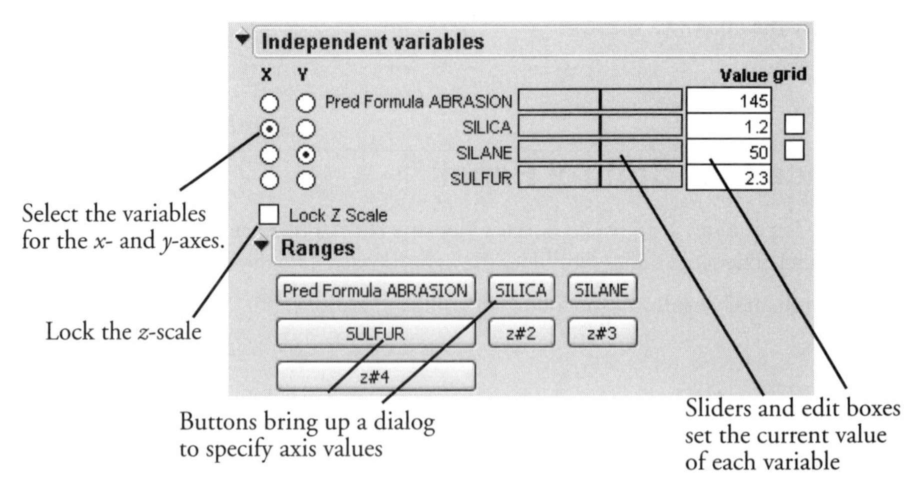

When there are more than two independent variables, you can select which two are displayed on the *x*- and *y*-axes using the radio buttons in this panel. The sliders and text boxes set the current values of each variable, which is most important for the variables that are not displayed on the axes. In essence, the plot shows the three-dimensional slice of the surface at the value shown in the text box. Move the slider to see different slices.

Lock Z Scale locks the *z*-axis to its current values. This is useful when moving the sliders that are not on an axis.

Grid check boxes activate a grid that is parallel to each axis. Sliders allow you to adjust the placement of each grid. To activate the grids, click the rectangle to the right of the sliders. The resolution of each grid can be controlled using the **Resolution** slider (found in the top right of the control panel). As an example, Figure 40.11 shows a surface with the X and Y grids activated.

Figure 40.11 Activated X and Y Grids

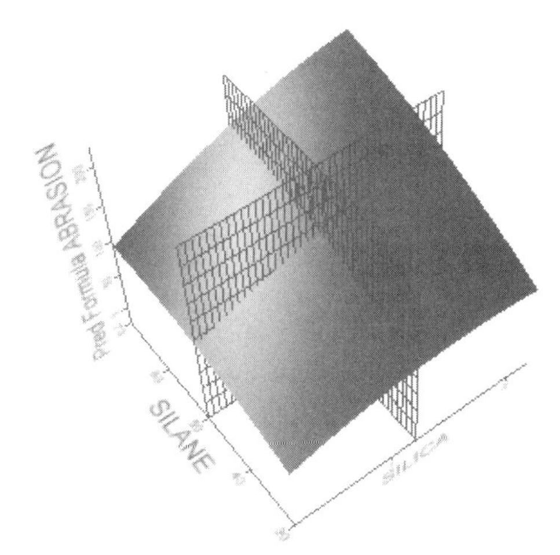

Dependent Variables

The Dependent Variables Controls change depending on whether you have selected **Sheet, Points**, **Isosurface**, or **Density Grid** in the Appearance Controls.

Controls for Sheet, Points

The Dependent Variables controls are shown in Figure 40.12 with its default menus.

Figure 40.12 Dependent Variable Controls

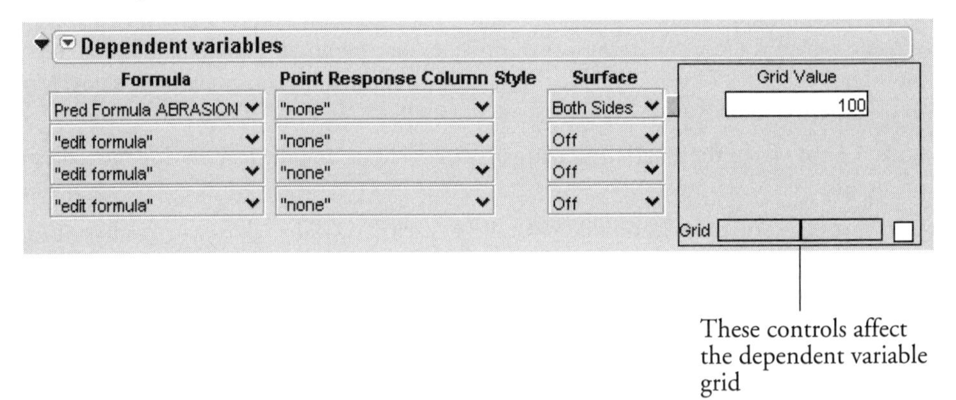

Formula lets you pick the formula(s) to be displayed in the plot as surfaces.

Point Response Column lets you pick the column that holds values to be plotted as points.

Style menus appear after you have selected a **Point Response Column**. The style menu lets you choose how those points are displayed, as **Points**, **Needles**, a **Mesh**, **Surface**, or **Off** (not at all). **Points** shows individual points, which change according to the color and marker settings of the row in the data table. **Needles** draws lines from the *x*-*y* plane to the points, or, if a surface is also plotted, connects the surface to the points. **Mesh** connects the points into a triangular mesh. **Surface** overlays a smooth, reflective surface on the points. An example using Little Pond.jmp is shown in Figure 40.13.

Figure 40.13 **Points**, **Needles**, **Mesh** and **Surface** Options Using Little Pond.jmp

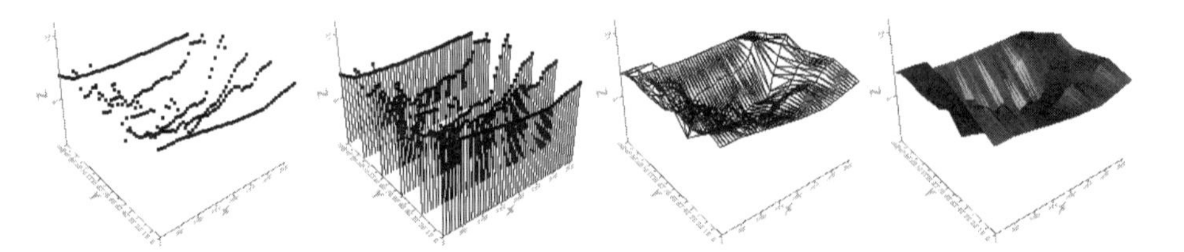

Surface lets you specify which side(s) of the surface are displayed. Choices are **Off, Both Sides, Above Only, Below Only.**

Grid slider and checkbox activate a grid for the dependent variable. Use the slider to adjust the value where the grid is drawn, or type the value into the Grid Value box above the slider.

Controls for Isosurface

Most of the controls for **Isosurface** are identical to those of **Sheet, Points.** Figure 40.14 shows the default controls, illustrating the slightly different presentation.

Figure 40.14 Dependent Variable Controls for Isosurfaces

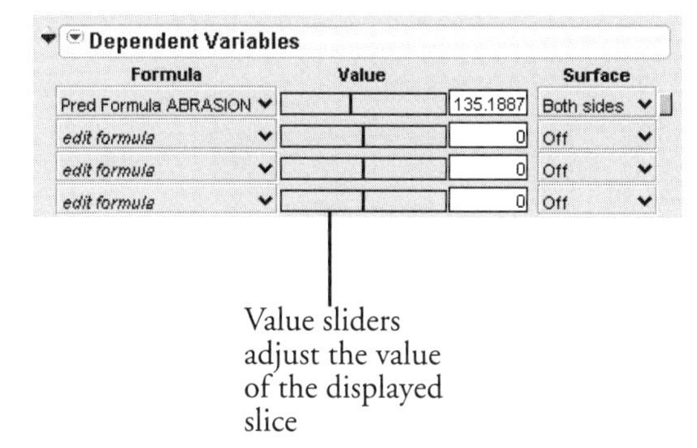

Controls for Density Grid

Most of the controls for **Density Grid** are identical to those of **Sheet, Points**. Figure 40.15 shows the default controls, illustrating the slightly different presentation.

Figure 40.15 Dependent Variable Controls for Density Grids

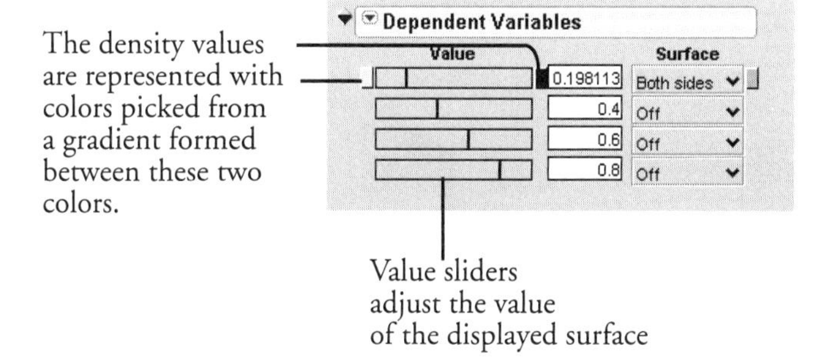

Dependent Variable Menu Options

There are several platform options for the Dependent Variable, accessed through the platform pop-up menu.

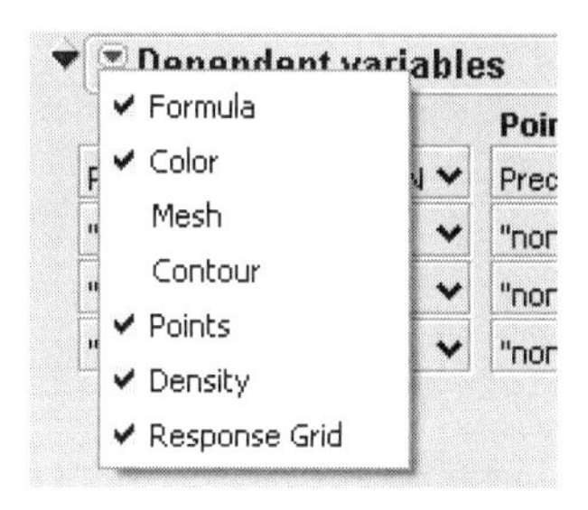

Formula toggles the edit box that allows you to enter a formula directly for each of the four surfaces.

Color toggles two controls. One lets you choose which sides of a surface (**Both Sides, Above Only, Below Only, Off**) are shown, the other allowing specification of the color of the shown surfaces.

Surface
Use this list to specify which sides are shown
Both sides
Off
Off
Off
Click here to choose color

Mesh reveals the menus that allow you to specify how the mesh should be drawn (**Off, X and Y, X,** or **Y**).

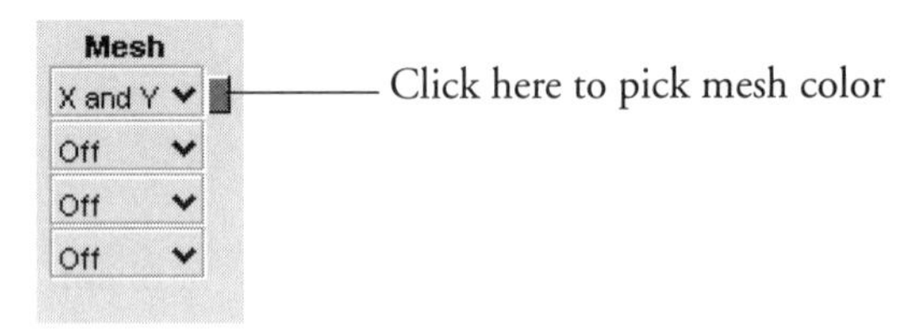

Figure 40.16 Surface with **Mesh** setting **X and Y** and color specified to black.

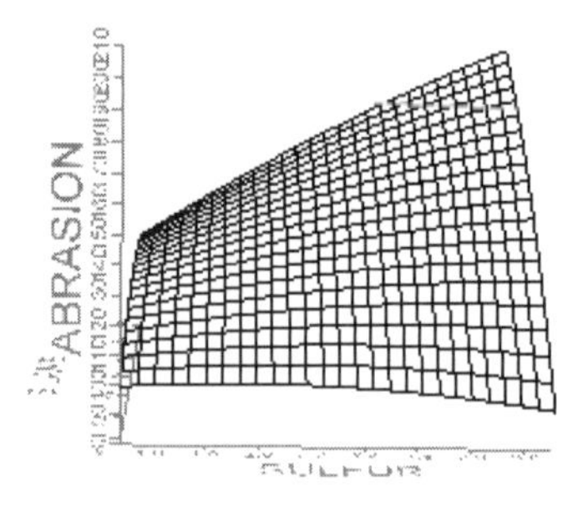

Contour reveals the menus that allow you to specify contour lines (sometimes known as *level curves*) of the surface. The **Contour** drop-down list allows you to select where the contours are drawn, either **Above, Below,** or **On Surface.** (See Figure 40.17). Click the colored rectangle to the right of the menu to change the display color of the contour. In Figure 40.17, the color of the contours on the left is black and on the right is white.

Figure 40.17 Contour lines below (left) and on (right) the surface

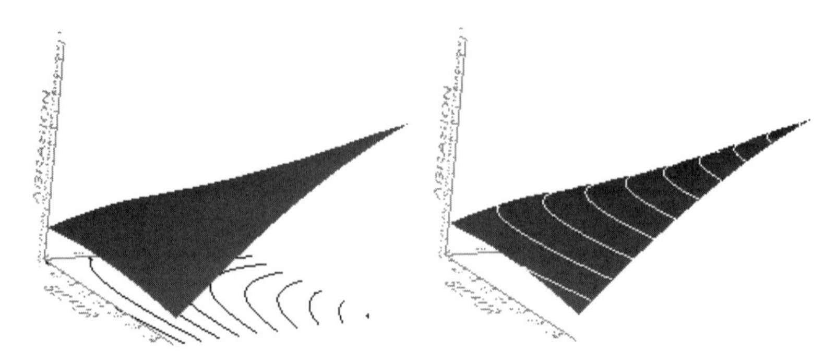

Points reveals or hides the **Point Response Column** and **Style** menus described above.

Density shows or hides the density value for a surface.

Response Grid shows or hides the grid controls detailed above and shown in Figure 40.12.

Axis

The Axis control panel is shown in Figure 40.18.It controls aspects of the appearance of axes.

Figure 40.18 Axis Control Panel

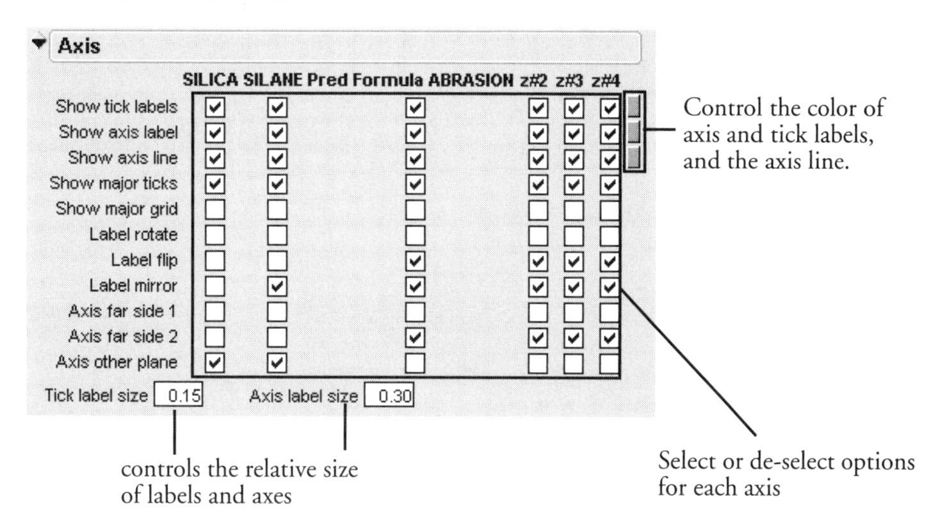

Show tick labels toggles the display of the corresponding tick labels.

Show axis labels toggles the display of the corresponding axis labels.

Show axis line toggles the display of the corresponding axis lines.

Show major ticks toggles the display of the major tick marks of the corresponding axis.

Show major grid toggles the display of the grid of the corresponding axis.

Label rotate toggles the rotation of the numerical labels (by 90 degrees) and the axis labels (180 degrees) of the corresponding axis.

Label flip flips the corresponding axis label.

Label mirror toggles the display of axis labels as their mirror image, which is useful when viewing plots from certain angles.

Axis far side 1 displays the corresponding axis labels and lines on the far side of the surface plot.

Axis far side 2 displays the corresponding axis labels and lines on the far side of the surface plot

Axis other plane displays the corresponding axis labels and lines in the plane of the axis or 90 degrees from the plane of the axis.

Tick label size sets the relative size of the corresponding tick label font.

Axis label size sets the relative size of the corresponding axis label font.

Zoom

The Zoom control panel is shown in Figure 40.19. Each slider adjusts the scaling of the corresponding axis.

Figure 40.19 Size Control Panel

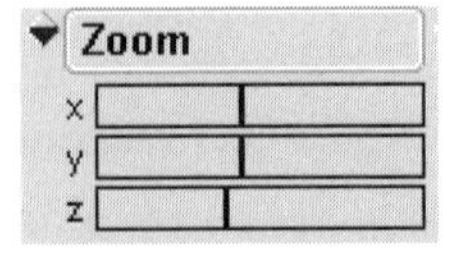

Rotate

The Rotate control panel is shown in Figure 40.20.Each slider allows precise control of rotation around the corresponding axis.

Figure 40.20 Rotate Control Panel

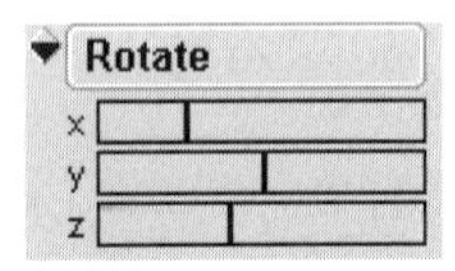

Keyboard Shortcuts

The following keyboard shortcuts can be used to manipulate the surface plot.

Table 40.1 Surface Plot Keyboard Shortcuts

Key	Function
left, right, up, and down arrows	spin
Page Up, Page Down, Home, End	diagonally spin
Enter	roll clockwise
Delete	roll counterclockwise
Control	boost spin speed 10X
Shift	allows spinning in background
Return	toggles ArcBall appearance

Chapter 41

Contour Plot

The Contour Plot Platform

The **Contour Plot** command in the **Graph** menu (or toolbar) constructs contours of a response in a rectangular coordinate system.

Some of the options available with the Contour platform are

- specify the number of contour levels
- choose a line or filled contours
- show or hide data points
- label contours with response values
- tailor a coloring scheme
- save contour construction information in a JMP table

Contents

A Simple Contour Plot Example

To create a contour plot, you need two variables for the *x*- and *y*-axes and at least one more variable for contours (although optionally, you can have several *y*-variables). For example, the data table in Figure 41.1 shows a partial listing of the Little Pond.jmp data table in the Sample Data folder.

The **Contour** command in the **Graph** menu displays the Contour Plot Launch dialog in Figure 41.1. For this example, the coordinate variables X and Y are assigned the **X** role for the plot, and Z (which is the pond depth) is the **Y**, or contour, variable.

Note: The function Z should be a function of *exactly* two variables. Those variables should be the *x*-variables entered in the contour plot launch dialog.

Figure 41.1 Data Table and Launch Dialog for Contour Plot

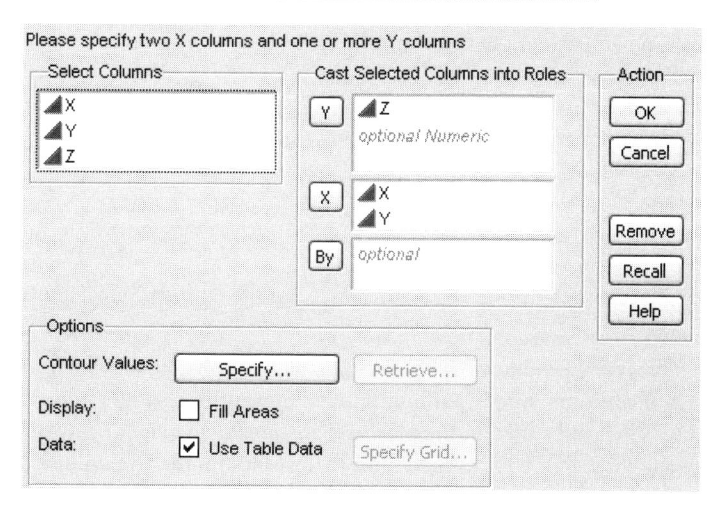

	X	Y	Z
1	0	0	5.9072265
2	0	30.609375	7.7841796
3	0	61.21875	9.6113281
4	0	91.828125	10
5	0	122.4375	6.8095703
6	0	150	3.7216796
7	7.24414063	0	5.8574218
8	7.24414063	30.609375	7.734375
9	7.24414063	61.21875	9.5800781
10	7.24414063	91.828125	6.8417968

When you click **OK** in the Contour Plot Launch dialog, the platform builds the plot as shown in Figure 41.2. By default, the contour levels are values computed from the data. You can specify your own number of levels and level increments with options in the Contour Plot Launch dialog before you create the plot, or options in the popup menu on the Contour Plot title bar. Also, you can use a column formula to compute the contour variable values.

There are more details about these options in the next section, "Contour Specification," p. 784.

Figure 41.2 Contour Plot with Legend

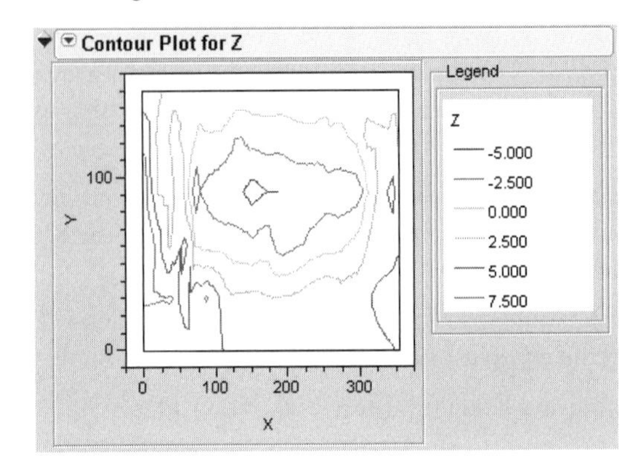

Contour Specification

If you don't choose options on the Contour Plot Launch dialog, the default plot spaces the contour levels equally within the range of the *Y* variable. The default colors are assigned from low to high level values by cycling from top to bottom through the middle column of the color palette. You can see the colors palette with the **Colors** command in the **Rows** menu or by right-clicking (Control-clicking on the Macintosh) on an item in the Contour plot legend.

The Contour Plot Launch dialog (see Figure 41.1) has the options shown above for specifying contour values:

Specify displays the Contour Specification Dialog (shown here), which enables you to change the number of contours, supply a minimum and maximum response value to define the range of response values to be used in the plot, and change the increment between contour values. You supply any three of the four values, and JMP computes the remaining value. Click on the check box to deselect one of the numbers and automatically select the remaining checkbox.

JMP: Contour Specification
Please provide any 3 of the 4 numbers requested below.
# of Contours 3
Minimum -5
Maximum 5
Fixed Increment 5
OK
Cancel
Help

Colors are automatically assigned, and are determined by the number of levels in the plot. After the plot appears, you can right-click (Control-click on the Macintosh) on any contour in the plot legend and choose from the JMP color palette to change that contour color.

Retrieve lets you retrieve the number of contours, an exact value for each level, and a color for each level from an open JMP data table.

For level value specification, the Contour Plot platform looks for a numeric column with the same name as the response column you specified in the Launch dialog. If it does not find a column with that name, it uses the first numeric column it finds. The number of rows is the number of levels that you want.

If there is a row state column with color information, those colors are used for the contour levels. Otherwise, the default platform colors are used.

Note: The **Retrieve** button is not active unless there is an open data table in addition to the table that has the contour plotting values. When you click **Retrieve**, the dialog shown here appears. You select the data table that contains the contour levels from the list of open data tables.

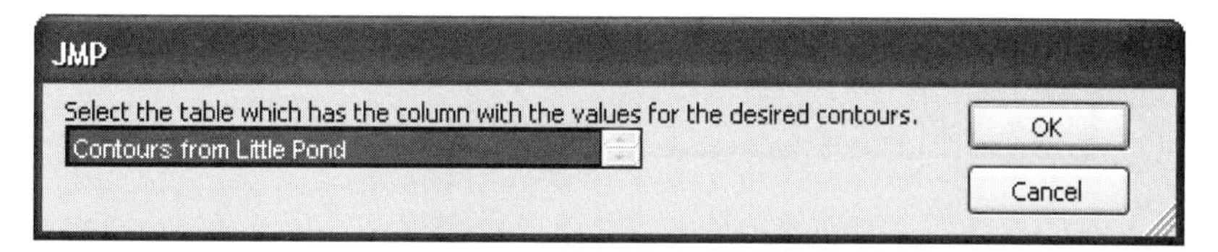

Fill Areas If you check the **Fill Areas** option box, the areas between contour lines are filled with the contour line colors. The **Fill Areas** option is also available in the platform popup menu described in the next section, “Platform Options,” p. 786.

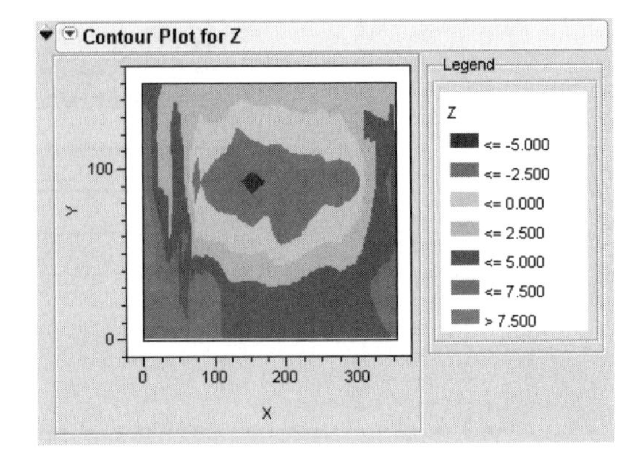

Use Table Data, Specify Grid Most often you construct a contour plot for a table of recorded response values such as the **Little Pond** table, shown previously. In that case, **Use Table Data** is checked and the **Specify Grid** button is dim. However, if a column has a formula and you specify that column as the response (**Z**), the **Specify Grid** button becomes active. When you click **Specify Grid**, you see the dialog shown in Figure 41.3. You can complete the Specify Grid dialog and define the contour grid in any way, regardless of the rows in the existing data table. This feature is also available with table templates that have one or more columns defined by formulas but no rows.

Figure 41.3 JMP Table of Exact Contour Specifications

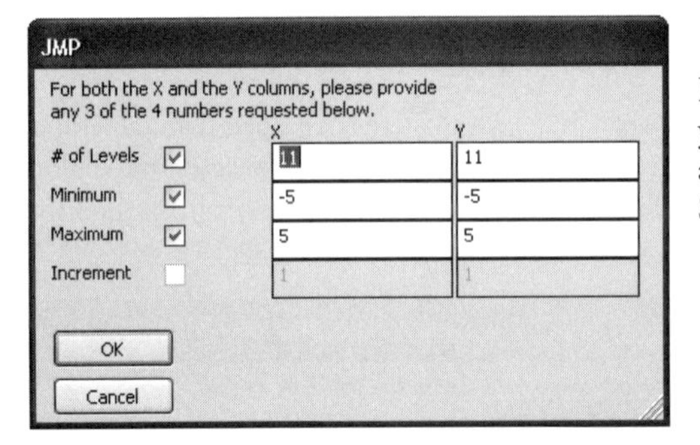

If the response has a formula, you can click **Specify Grid** and complete this Grid Specification dialog.

Platform Options

Options given by the platform popup menu on the Contour Plot title bar let you tailor the appearance of your contour plot and save information about its construction.

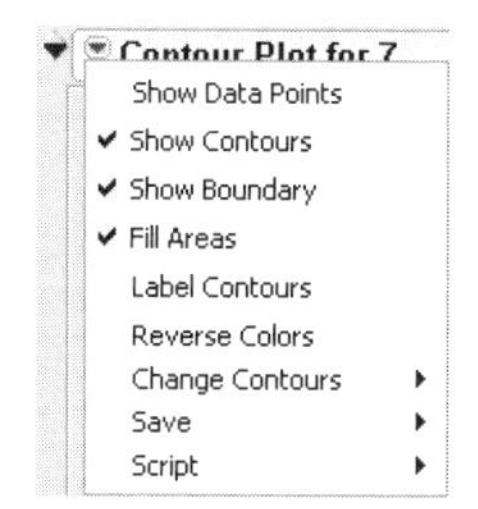

Show Data Points shows or hides (x, y) points. The points are hidden by default.

Show Contours shows or hides the contour lines or fills. The contours lines show by default.

Show Boundary shows or hides the boundary of the total contour area.

Fill Areas fills the areas between the contours with a solid color. It is the same option that is available on the Launch dialog, but it lets you fill after you see the line contour.

Label Contours shows or hides the label (z-value) of the contour line. By default, the labels are hidden.

Reverse Colors reverses the order of the colors assigned to the contour levels.

Change Contours gives you a submenu with the choices shown here, which has the same options as those on the Contour Plot Launch dialog. See "Contour Specification," p. 784, for a discussion of those options.

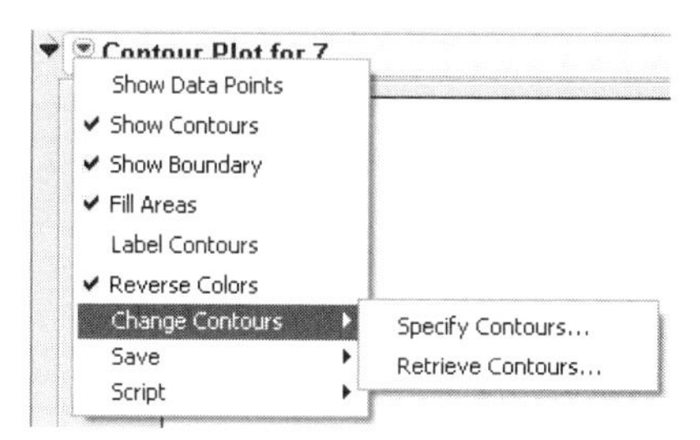

Save gives the submenu with the following selections:

Save Contours creates a new JMP table with columns for the x-and y-coordinate values generated by the Contour platform for each contour, the response computed for each coordinate set, and its contour level. The number of observations in this table depends on the number of contours you specified. You can use the coordinates and response values to look at the data with other JMP platforms as shown in Figure 41.4, which uses the Cowboy Hat.jmp sample data table and the Spinning Plot platform.

Generate Grid displays a dialog that prompts you for the grid size you want. When you click OK, the Contour platform creates a new JMP table with the number of grid coordinates you request, and contour values for the grid points computed from a linear interpolation.

Save Triangulation creates a new JMP table that lists coordinates of each triangle used to construct the contours.

Figure 41.4 Contour Plot and Corresponding Three-Dimensional Spinning Plot

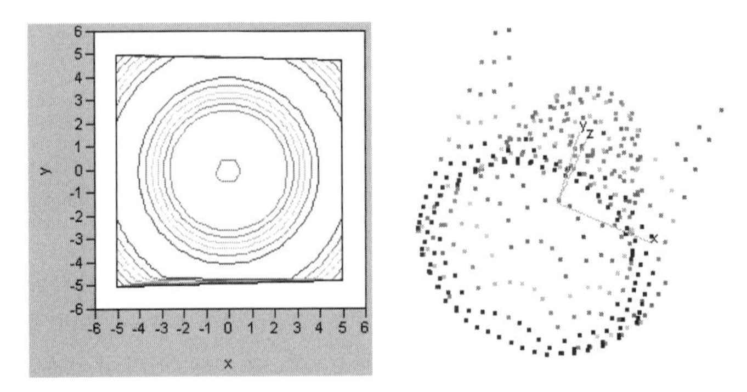

Script has a submenu of commands available to all platforms that let you redo the analysis or save the JSL commands for the analysis to a window or a file.

Chapter 42

Statistical Control Charts

The Control Chart Platform

This chapter contains descriptions of options that apply to all control charts.

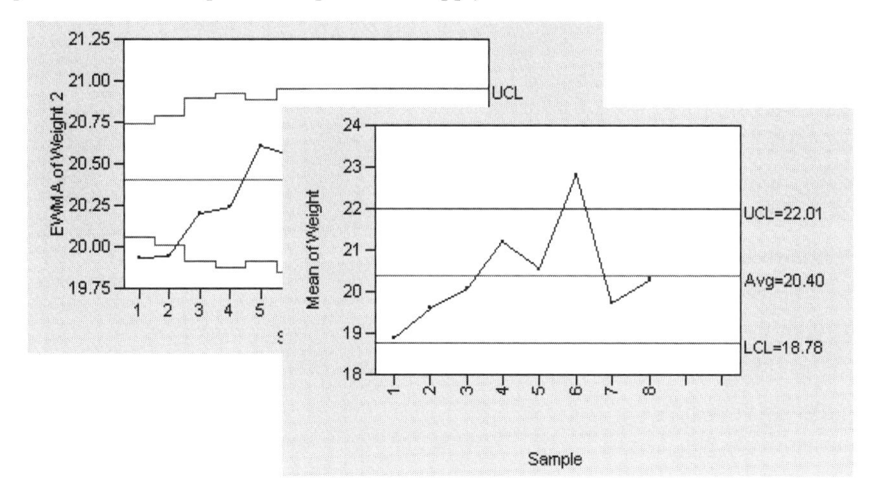

Contents

Statistical Quality Control with Control Charts

Control charts are a graphical and analytic tool for deciding whether a process is in a state of statistical control and for monitoring an in-control process.

Control charts have the following characteristics:

- Each point represents a summary statistic computed from a subgroup sample of measurements of a quality characteristic.
- The vertical axis of a control chart is scaled in the same units as the summary statistic.
- The horizontal axis of a control chart identifies the subgroup samples.
- The center line on a Shewhart control chart indicates the average (expected) value of the summary statistic when the process is in statistical control.
- The upper and lower control limits, labeled UCL and LCL, give the range of variation to be expected in the summary statistic when the process is in statistical control.
- A point outside the control limits (or the V-mask of a Cusum chart) signals the presence of a special cause of variation.
- **Graph > Control Chart** subcommands create control charts that can be updated dynamically as samples are received and recorded or added to the data table.

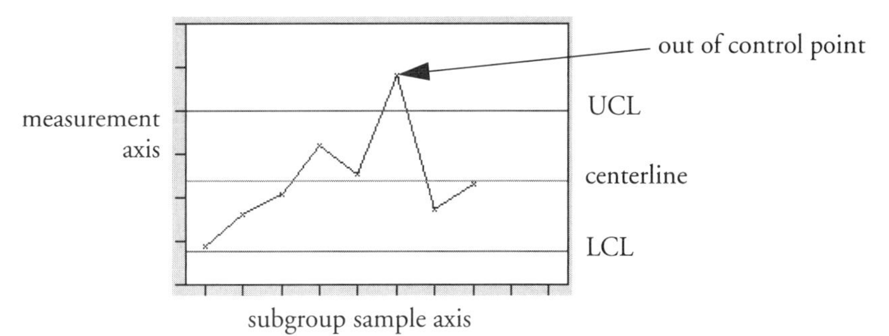

The Control Chart Launch Dialog

When you select a Control Chart from the **Graph > Control Chart** menu (Figure 42.1), you see a Control Chart Launch dialog similar to the one in Figure 42.2. (The exact controls vary depending on the type of chart you choose.) Initially, the dialog shows three kinds of information:

- process information, for measurement variable selection
- chart type information
- limits specification.

Figure 42.1 Control Chart Menu

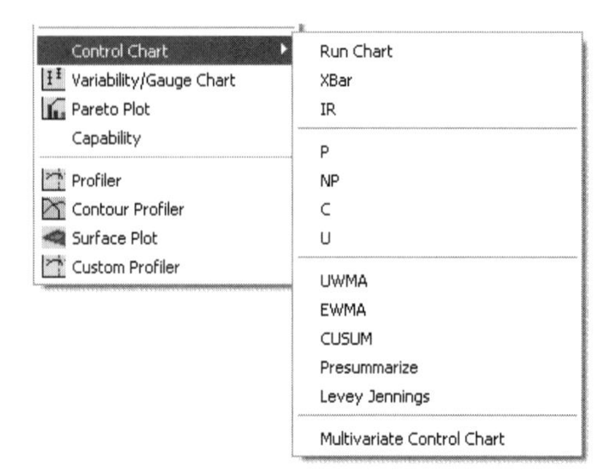

Specific information shown for each section varies according to the type of chart you request.

Figure 42.2 Control Chart Launch Dialog

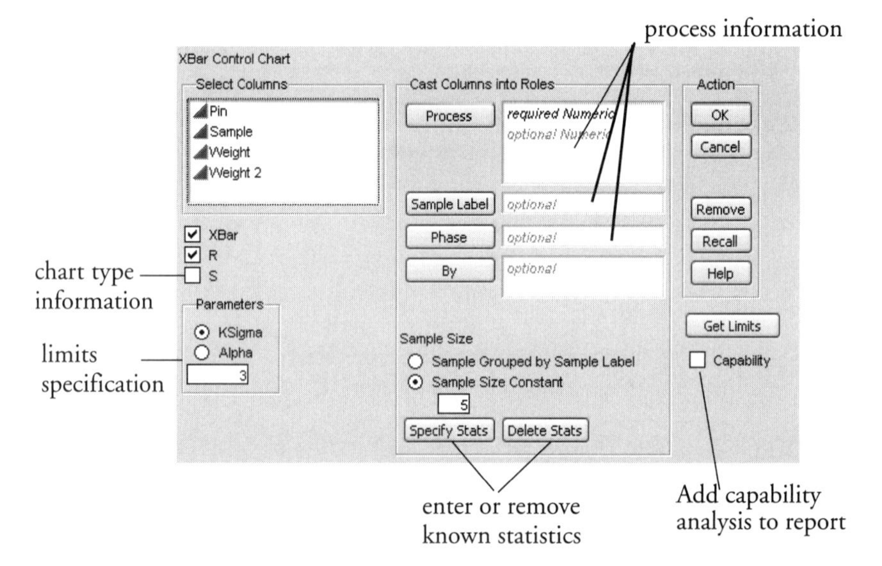

Through interaction with the Launch dialog, you specify exactly how you want your charts created. The following sections describe the panel elements.

Process Information

The Launch dialog displays a list of columns in the current data table. Here, you specify the variables to be analyzed and the subgroup sample size.

Process

selects variables for charting.

- For variables charts, specify measurements as the process.
- For attribute charts, specify the defect count or defective proportion as the process.

Sample Label

enables you to specify a variable whose values label the horizontal axis and can also identify unequal subgroup sizes. If no sample label variable is specified, the samples are identified by their subgroup sample number.

- If the sample subgroups are the same size, check the **Sample Size Constant** radio button and enter the size into the text box. If you entered a Sample Label variable, its values are used to label the horizontal axis.
- If the sample subgroups have an unequal number of rows or have missing values and you have a column identifying each sample, check the **Sample Grouped by Sample Label** radio button and enter the sample identifying column as the sample label.

For attribute charts (*p*-, *np*-, *c*-, and *u*-charts), this variable is the subgroup sample size. In Variables charts, it identifies the sample. When the chart type is **IR**, a **Range Span** text box appears. The *range span* specifies the number of consecutive measurements from which the moving ranges are computed.

The illustration in Figure 42.3 shows an $\bar{X}$-chart for a process with unequal subgroup sample sizes, using the Coating.jmp sample data from the Quality Control sample data folder.

Figure 42.3 Variables Charts with Unequal Subgroup Sample Sizes

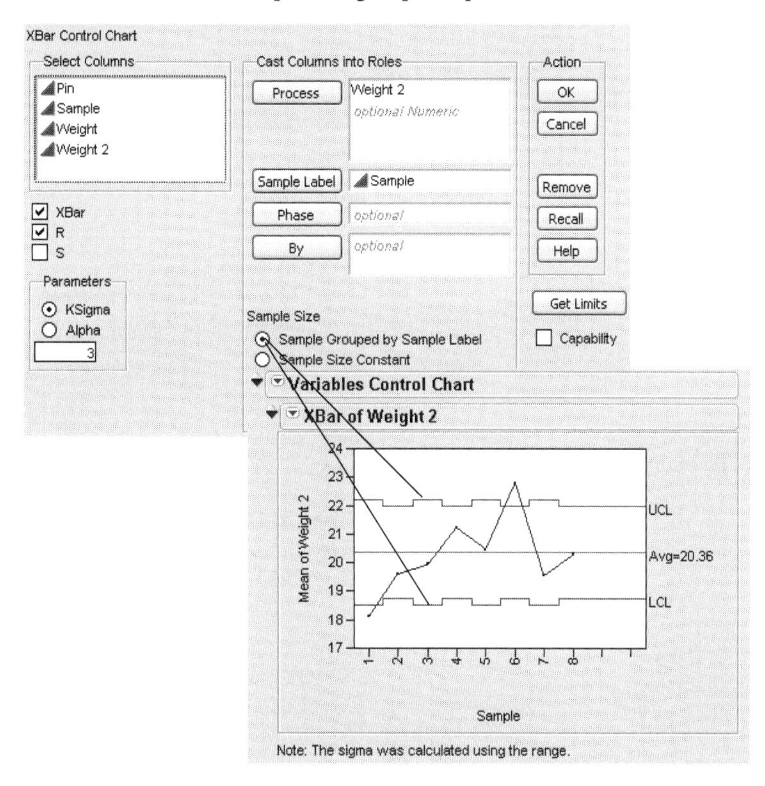

Phase

The **Phase** role enables you to specify a column identifying different phases, or sections. A *phase* is a group of consecutive observations in the data table. For example, phases might correspond to time periods during which a new process is brought into production and then put through successive changes. Phases generate, for each level of the specified Phase variable, a new sigma, set of limits, zones, and resulting tests. See "Phases," p. 836 IN THE "SHEWHART CONTROL CHARTS" CHAPTER for complete details of phases.

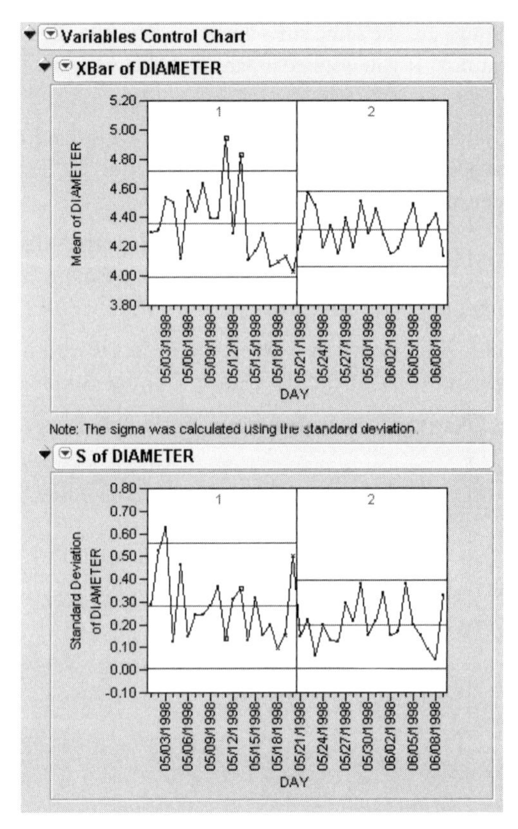

Chart Type Information

Shewhart control charts are broadly classified as *variables charts* and *attribute charts*. Moving average charts and cusum charts can be thought of as special kinds of variables charts.

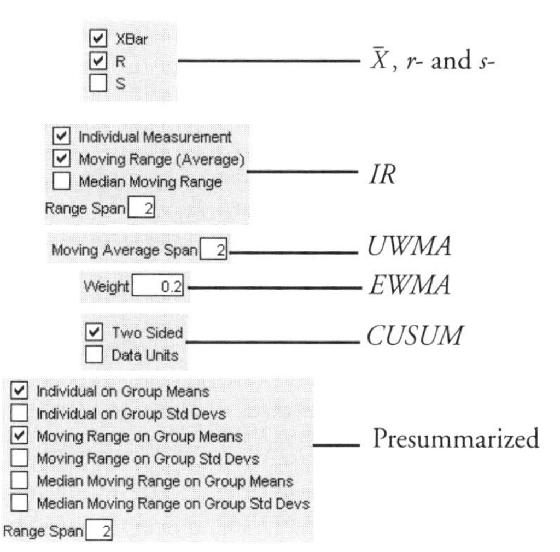

- **XBar** charts menu selection gives **XBar**, **R**, and **S** checkboxes.
- The **IR** menu selection has checkbox options for the Individual Measurement, Moving Range, and Median moving range charts.
- The uniformly weighted moving average (**UWMA**) and exponentially weighted moving average (**EWMA**) selections are special charts for means.
- The **Cusum** chart is a special chart for means or individual measurements.
- **Presummarize** allows you to specify information on pre-summarized statistics.
- **P**, **NP**, **C**, and **U** charts, **Run Charts**, and **Levey-Jennings** charts have no additional specifications.

The types of control charts are discussed in "Shewhart Control Charts," p. 815.

Parameters

You specify computations for control limits by entering a value for *k* (**K Sigma**), or by entering a probability for α(**Alpha**), or by retrieving a limits value from the process columns' properties or a previously created Limits Table. Limits Tables are discussed in the section "Saving and Retrieving Limits," p. 807, later in this chapter. There must be a specification of either **K Sigma** or **Alpha**. The dialog default for **K Sigma** is 3.

K Sigma

allows specification of control limits in terms of a multiple of the sample standard error. **K Sigma** specifies control limits at *k* sample standard errors above and below the expected value, which shows as the center line. To specify *k*, the number of sigmas, click **K Sigma** and enter a positive *k* value into the text-box. The usual choice for *k* is three, which is three standard deviations. The examples shown in Figure 42.4 compare the $\bar{X}$-chart for the Coating.jmp data with control lines drawn with **K Sigma** = 3 and **K Sigma** = 4.

Figure 42.4 K Sigma =3 (left) and K Sigma=4 (right) Control Limits

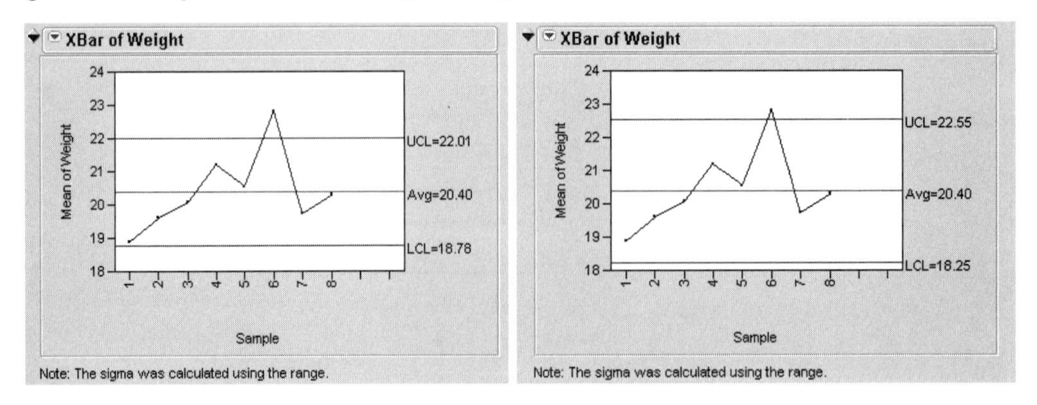

Alpha

specifies control limits (also called *probability limits*) in terms of the probability α that a single subgroup statistic exceeds its control limits, assuming that the process is in control. To specify alpha, click the **Alpha** radio button and enter the probability you want. Reasonable choices for α are 0.01 or 0.001

Using Specified Statistics

If you click the **Specify Stats** (when available) button on the Control Chart Launch dialog, a tab with editable fields is appended to the bottom of the launch dialog. This lets you enter historical statistics (statistics obtained from historical data) for the process variable. The Control Chart platform uses those entries to construct control charts. The example here shows 1 as the standard deviation of the process variable and 20 as the mean measurement.

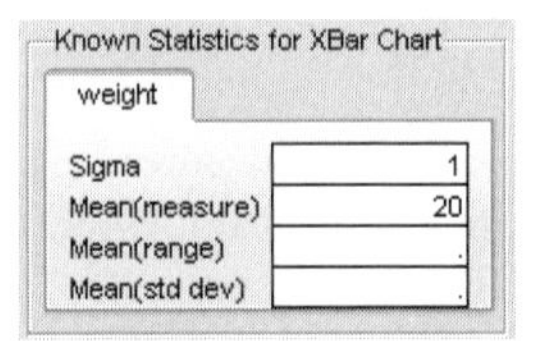

Note: When the mean is user-specified, it is labeled in the plot as $\mu 0$.

If you check the Capability option on the Control Chart launch dialog (see Figure 42.2), a dialog appears as the platform is launched asking for specification limits. The standard deviation for the control chart selected is sent to the dialog and appears as a Specified Sigma value, which is the default option. After entering the specification limits and clicking OK, capability output appears in the same window next to the control chart. For information on how the capability indices are computed, see Table 4.1 "Capability Index Names and Computations," p. 57.

Tailoring the Horizontal Axis

When you double-click the *x*-axis, the X Axis Specification dialog appears for you to specify the format, axis values, number of ticks, gridline and reference lines to display on the *x*-axis.

For example, the Pickles.JMP data lists eight measures a day for three days. In this example, by default, the *x*-axis is labeled at every other tick. Sometimes this gives redundant labels, as shown to the left in Figure 42.5. If you specify a label at an increment of eight, with seven ticks between them, the *x*-axis is labeled once for each day, as shown in the chart on the right.

Figure 42.5 Example of Labeled *x*-Axis Tick Marks

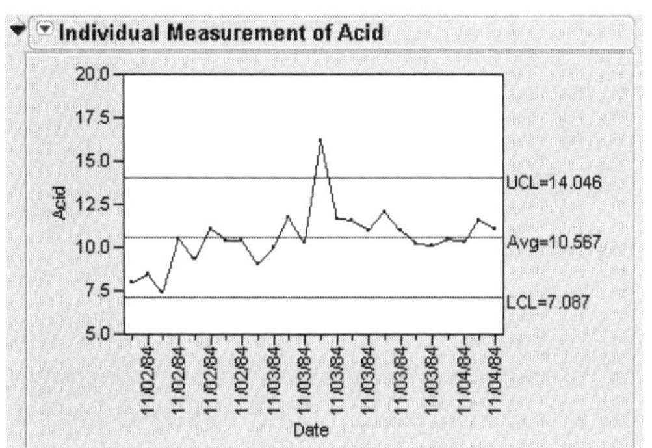

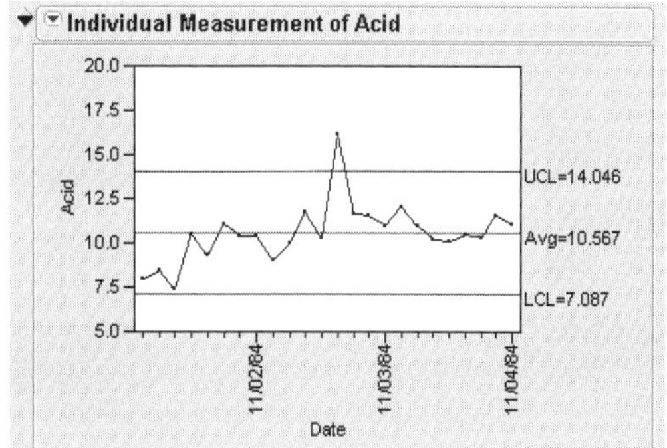

Display Options

Control Charts have popup menus that affect various parts of the platform:

- The menu on the top-most title bar affects the whole platform window. Its items vary with the type of chart you select.
- There is a menu of items on the chart type title bar with options that affect each chart individually.

Single Chart Options

The popup menu of chart options appears when you click the icon next to the chart name, or context-click the chart space (right-mouse click on Windows or Control-click on the Macintosh). The Cusum chart has different options that are discussed in "Cumulative Sum Control Charts," p. 839.

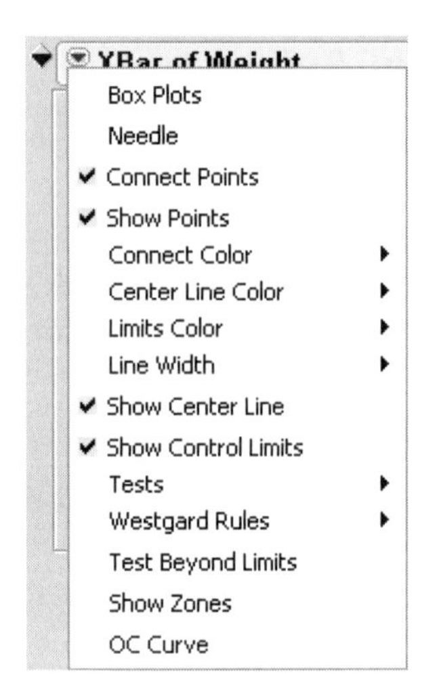

Box Plots

superimposes box plots on the subgroup means plotted in a Mean chart. The box plot shows the subgroup maximum, minimum, 75th percentile, 25th percentile, and median. Markers for subgroup means show unless you deselect the **Show Points** option. The control limits displayed apply only to the subgroup mean. The **Box Plots** option is available only for $\bar{X}$-charts. It is most appropriate for larger subgroup sample sizes (more than 10 samples in a subgroup).

Needle

connects plotted points to the center line with a vertical line segment.

Connect Points

toggles between connecting and not connecting the points.

Show Points

toggles between showing and not showing the points representing summary statistics. Initially, the points show. You can use this option to suppress the markers denoting subgroup means when the **Box Plots** option is in effect.

Figure 42.6 Box Plot Option and Needle Option for Airport.jmp Data

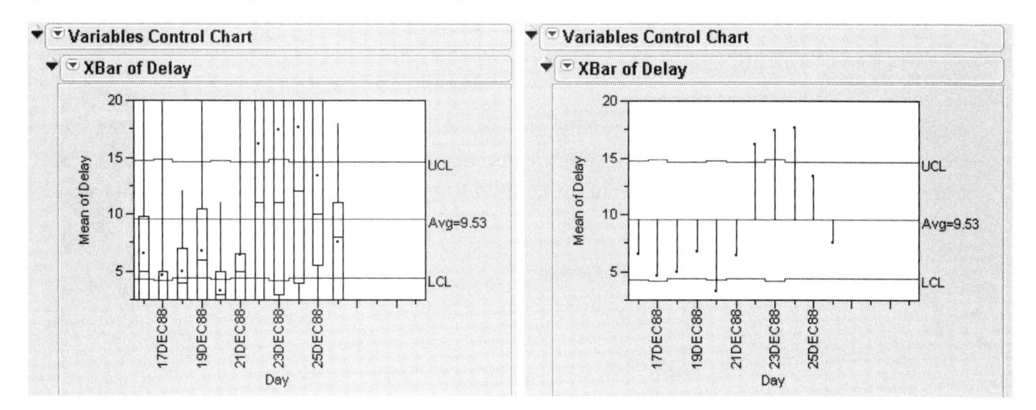

Connect Color

displays the JMP color palette for you to choose the color of the line segments used to connect points.

Center Line Color

displays the JMP color palette for you to choose the color of the line segments used to draw the center line.

Limits Color

displays the JMP color palette for you to choose the color of the line segments used in the upper and lower limits lines.

Line Width

allows you to pick the width of the control lines. Options are **Thin**, **Medium**, or **Thick**.

Show Center Line

initially displays the center line in green. Deselecting **Show Center Line** removes the center line and its legend from the chart.

Show Control Limits

toggles between showing and not showing the chart control limits and their legends.

Tests

shows a submenu that enables you to choose which tests to mark on the chart when the test is positive. Tests apply only for charts whose limits are 3σ limits. Tests 1 to 4 apply to Mean, Individual and attribute charts. Tests 5 to 8 apply to Mean charts and Individual Measurement charts only. If tests do not apply to a chart, the Tests option is dimmed. Tests apply, but will not appear for charts whose control limits vary due to unequal subgroup sample sizes, until the sample sizes become equal. These special tests are also referred to as the *Western Electric rules*. For more information on special causes tests, see *Tests for Special Causes* later in this chapter.

Show Zones

toggles between showing and not showing the *zone lines* with the tests for special causes. The zones are labeled A, B, and C as shown here in the Mean plot for **weight** in the **Coating.jmp** sample data. Control Chart tests use the zone lines as boundaries. The seven zone lines are set one sigma apart, centered on the center line.

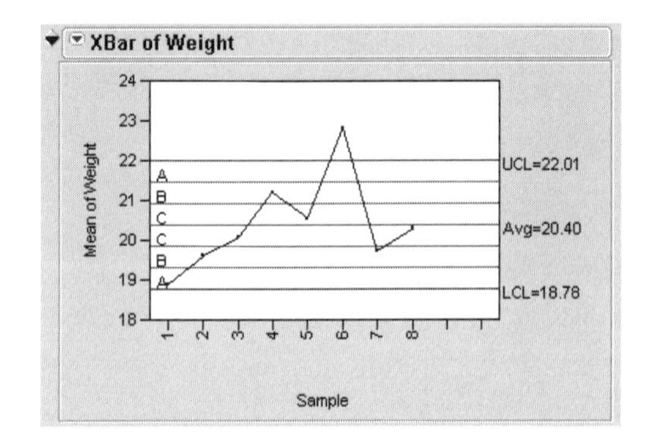

Westgard Rules

are detailed below. See the text and chart on p. 804.

Test Beyond Limits

flags as a "*" any point that is beyond the limits. This test works on all charts with limits, regardless of the sample size being constant, and regardless of the size of *k* or the width of the limits. For example, if you had unequal sample sizes, and wanted to flag any points beyond the limits of an *r*-chart, you could use this command.

OC Curve

gives Operating Characteristic (OC) curves for specific control charts. OC curves are defined in JMP only for $\bar{X}$-, *p*-, *np*-, *c*-, and *u*-charts. The curve shows how the probability of accepting a lot changes with the quality of the sample. When you choose the **OC Curve** option from the control chart option list, JMP opens a new window containing the curve, using all the calculated values directly from the active control chart. Alternatively, you can run an OC curve directly from the QC tab on the JMP Starter window. Select the chart on which you want the curve based, then a dialog prompts you for **Target**, **LCL**, **UCL**, **K**, **Sigma**, and **sample size**.

Window Options

The popup menu on the window title bar lists options that affect the report window. The example menu shown here appears if you request **XBar** and **R** at the same time. You can check each chart to show or hide it.

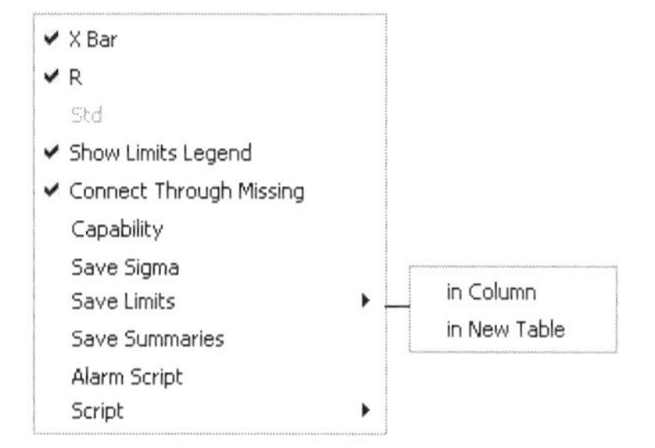

The specific options that are available depend on the type of control chart you request. Unavailable options show as grayed menu items.

The following options show for all control charts except Run charts:

Show Limits Legend shows or hides the Avg, UCL, and LCL values to the right of the chart.

Connect thru Missing connects points when some samples have missing values. The left-hand chart in Figure 42.7 is a control chart with no missing points. The middle chart has samples 8, 9, and 10 missing with the points not connected. The right-hand chart appears if you use the **Connect thru Missing** option, which is the default.

Figure 42.7 Example of **Connect thru Missing** Option

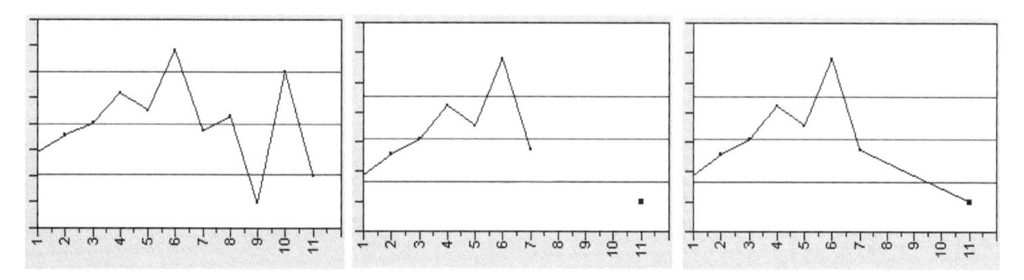

Save Sigma saves the computed value of sigma as a column property in the process variable column in the JMP data table.

Save Limits > in Column saves the computed values of sigma, center line, and the upper and lower limits as column properties in the process variable column in the JMP data table. These limits are later automatically retrieved by the Control Chart dialog and used in a later analysis.

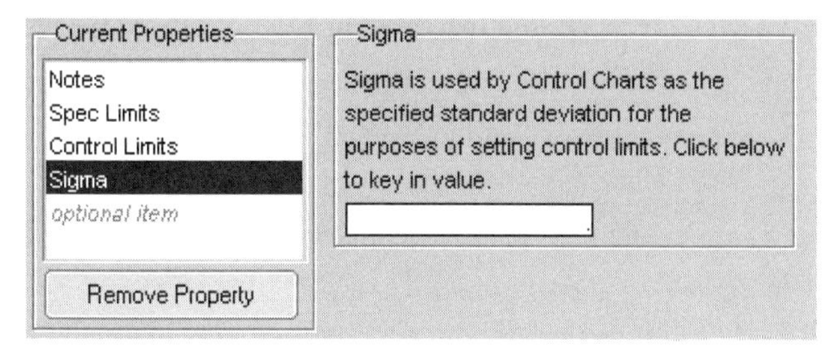

Save Limlts > in New Table saves all parameters for the particular chart type, including sigma and K Sigma, sample size, the center line, and the upper and lower control limits in a new JMP data table. These limits can be retrieved by the Control Chart dialog and used in a later analysis. See the section "Saving and Retrieving Limits," p. 807 for more information.

Save Summaries creates a new data table that contains the sample number, sample label (if there is one), and statistic being plotted for each plot within the window.

Alarm Script displays a dialog for choosing or entering a script or script name that executes whenever the tests for special causes is in effect and a point is out of range. See section "Tests for Special Causes," p. 802 for more information. See "Running Alarm Scripts," p. 806 for more information on writing custom Alarm Scripts.

Script has a submenu of commands available to all platforms that let you redo the analysis or save the JSL commands for the analysis to a window or a file.

Tests for Special Causes

The Tests option in the chart type popup menu displays a submenu for test selection. You can select one or more tests for special causes with the options popup menu. Nelson (1984) developed the numbering notation used to identify special tests on control charts.

If a selected test is positive, the last point in the test sequence is labeled with the test number, where the sequence is the moving set of points evaluated for that particular test. When you select several tests for display and more than one test signals at a particular point, the label of the numerically lowest test specified appears beside the point.

Western Electric Rules

Western Electric rules are implemented in the **Tests** submenu. Table 42.1 "Description and Interpretation of Special Causes Tests," p. 803 lists and interprets the eight tests, and Figure 42.8 illustrates the tests. The following rules apply to each test:

- The area between the upper and lower limits is divided into six zones, each with a width of one standard deviation.
- The zones are labeled A, B, C, C, B, A with zones C nearest the center line.
- A point lies in Zone B or beyond if it lies beyond the line separating zones C and B. That is, if it is more than one standard deviation from the centerline.
- Any point lying on a line separating two zones lines is considered belonging to the outermost zone.

Note: All Tests and zones require equal sample sizes in the subgroups of nonmissing data.

Tests 1 through 8 apply to Mean ($\bar{X}$) and individual measurement charts. Tests 1 through 4 can also apply to *p*-, *np*-, *c*-, and *u*-charts.

Tests 1, 2, 5, and 6 apply to the upper and lower halves of the chart separately. Tests 3, 4, 7, and 8 apply to the whole chart.

See Nelson (1984, 1985) for further recommendations on how to use these tests.

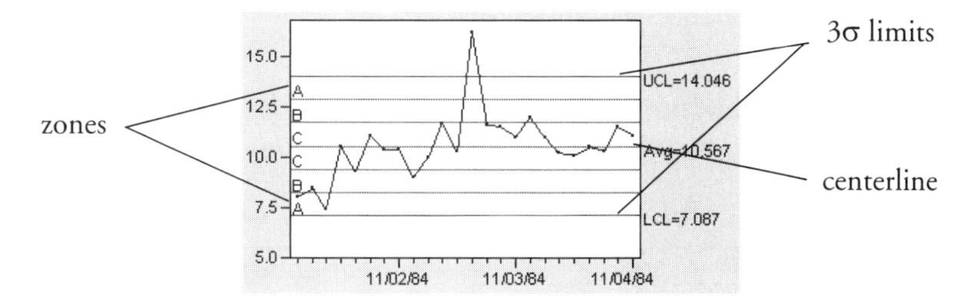

Nelson (1984, 1985)

Table 42.1 Description and Interpretation of Special Causes Tests

Test 1	One point beyond Zone A	detects a shift in the mean, an increase in the standard deviation, or a single aberration in the process. For interpreting Test 1, the *R*-chart can be used to rule out increases in variation.
Test 2	Nine points in a row in a single (upper or lower) side of Zone C or beyond	detects a shift in the process mean.
Test 3	Six points in a row steadily increasing or decreasing	detects a trend or drift in the process mean. Small trends will be signaled by this test before Test 1.
Test 4	Fourteen points in a row alternating up and down	detects systematic effects such as two alternately used machines, vendors, or operators.
Test 5	Two out of three points in a row in Zone A or beyond	detects a shift in the process average or increase in the standard deviation. Any two out of three points provide a positive test.
Test 6	Four out of five points in Zone B or beyond	detects a shift in the process mean. Any four out of five points provide a positive test.
Test 7	Fifteen points in a row in Zone C, above and below the center line	detects stratification of subgroups when the observations in a single subgroup come from various sources with different means.
Test 8	Eight points in a row on both sides of the center line with none in Zones C	detects stratification of subgroups when the observations in one subgroup come from a single source, but subgroups come from different sources with different means.

Figure 42.8 Illustration of Special Causes Tests

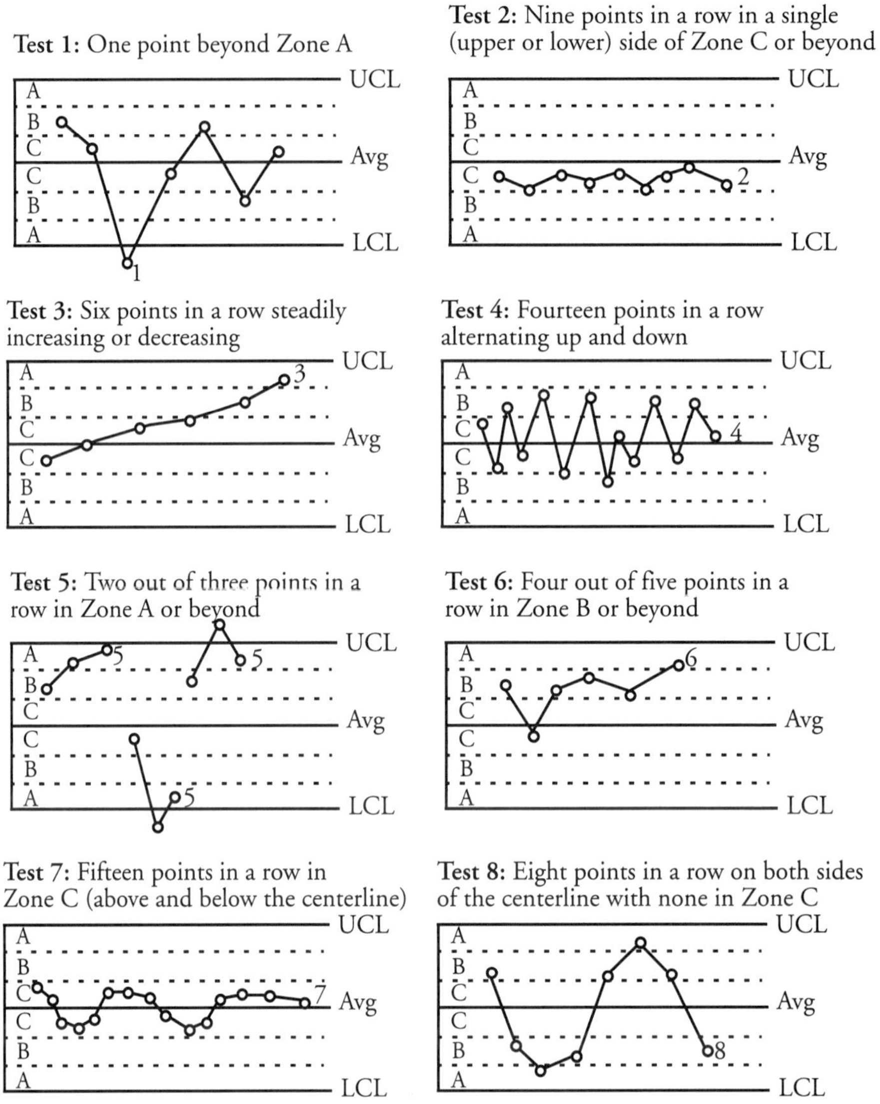
Test 1: One point beyond Zone A
UCL
Avg
LCL
Test 2: Nine points in a row in a single (upper or lower) side of Zone C or beyond
UCL
Avg
LCL
Test 3: Six points in a row steadily increasing or decreasing
UCL
Avg
LCL
Test 4: Fourteen points in a row alternating up and down
UCL
Avg
LCL
Test 5: Two out of three points in a row in Zone A or beyond
UCL
Avg
LCL
Test 6: Four out of five points in a row in Zone B or beyond
UCL
Avg
LCL
Test 7: Fifteen points in a row in Zone C (above and below the centerline)
UCL
Avg
LCL
Test 8: Eight points in a row on both sides of the centerline with none in Zone C
UCL
Avg
LCL

Nelson (1984, 1985)

Westgard Rules

Westgard rules are implemented under the **Westgard Rules** submenu of the Control Chart platform. The different tests are abbreviated with the decision rule for the particular test. For example, **1 2s** refers to a test that one point is two standard deviations away from the mean.

Because Westgard rules are based on sigma and not the zones, they can be computed without regard to constant sample size.

Table 42.2 Westgard Rules

Rule 1 2s is commonly used with Levey-Jennings plots, where control limits are set 2 standard deviations away from the mean. The rule is triggered when any one point goes beyond these limits.

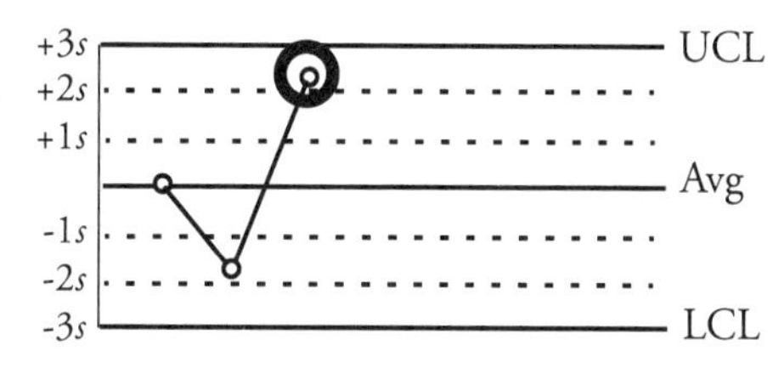

Rule 1 3s refers to a rule common to Levey-Jennings plots where the control limits are set 3 standard deviations away from the mean. The rule is triggered when any one point goes beyond these limits.

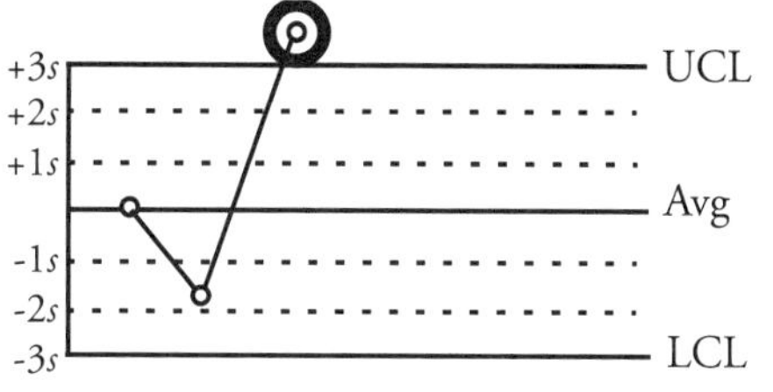

Rule 2 2s is triggered when two consecutive control measurements are farther than two standard deviations from the mean.

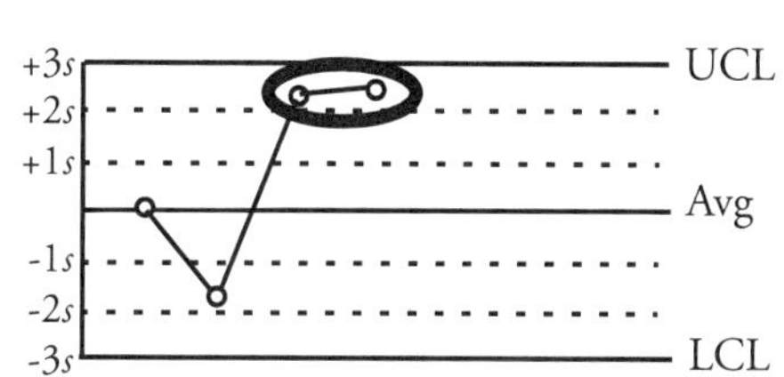

Rule 4s is triggered when one measurement in a group is two standard deviations above the mean and the next is two standard deviations below.

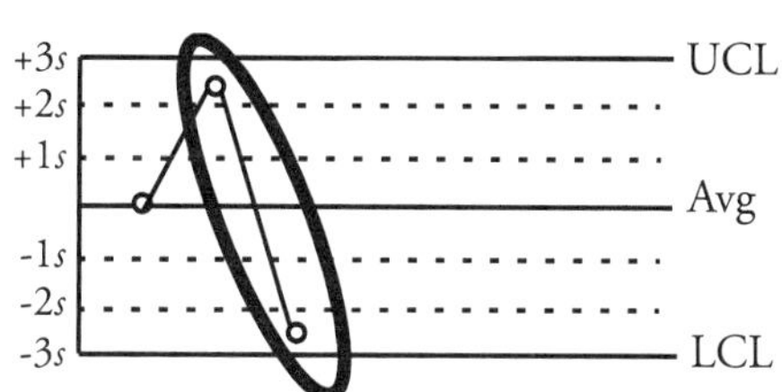

Rule 4 1s is triggered when four consecutive measurements are more than one standard deviation from the mean.

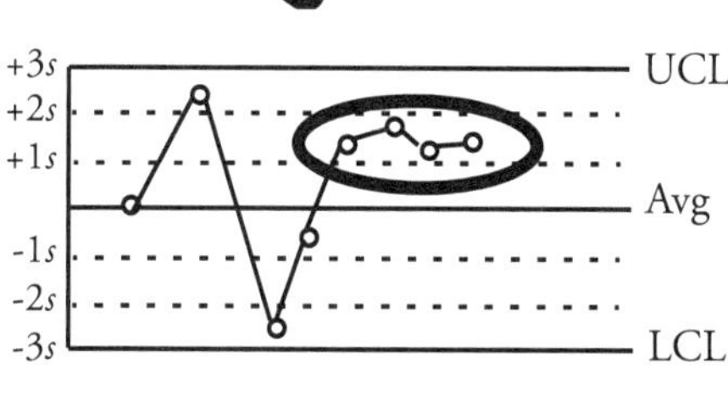

Rule 10 X is triggered when ten consecutive points are on one side of the mean.

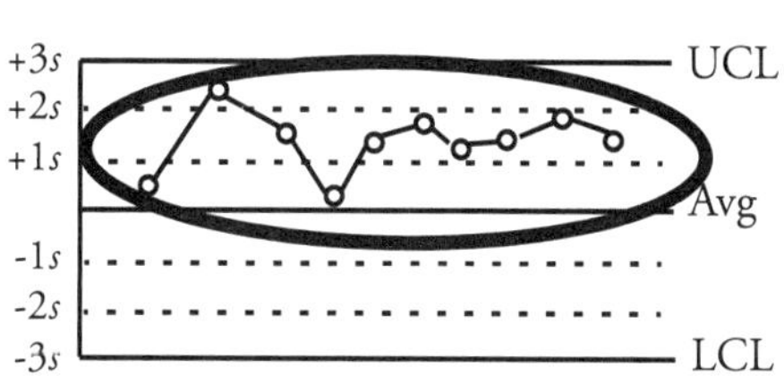

Running Alarm Scripts

If you want to run a script that alerts you when the data fail one or more tests, you can run an Alarm Script. As an Alarm Script is invoked, the following variables are available, both in the issued script and in subsequent JSL scripts:

`qc_col` is the name of the column

`qc_test` is the test that failed

`qc_sample` is the sample number

`qc_firstRow` is the first row in the sample

`qc_lastRow` is the last row in the sample

Example 1: Automatically writing to a log

One way to generate automatic alarms is to make a script and store it with the data table as a Data Table property named QC Alarm Script. To automatically write a message to the log whenever a test fails,

- Run the script below to save the script as a property to the data table,
- Run a control chart,
- Turn on the tests you're interested in. If there are any samples that failed, you'll see a message in the log.

```
CurrentData Table()<<Set Property("QC Alarm Script",
  Write(match(
    QC_Test,1,"One point beyond zone A",
          2,"Nine points in a row in zone C or beyond",
          3,"Six points in a row Steadily increasing or decreasing",
          4,"Fourteen points in a row alternating up and
            down",
          5,"Two out of three points in a row in Zone A or
            beyond",
          6,"Four out of five points in a row in Zone B or
            beyond",
          7,"Fifteen points in a row in Zone C",
          8,"Eight points in a row on both sides of the
            center line with none in Zone C" )));
```

Example 2: Running a chart with spoken tests

With the Coating.JMP data table open, submit the following script:

```
    Control Chart(Alarm Script(Speak(match(
      QC_Test,1, "One point beyond Zone A",
      QC_Test,2, "Nine points in a row in zone C or beyond",
      QC_Test,5, "Two out of three points in a row in Zone A
            or beyond"))),
  Sample Size( :Sample), Ksigma(3), Chart Col( :Weight,
  Xbar(Test 1(1), Test 2(1), Test 5(1)), R));
```

You can have either of these scripts use any of the JSL alert commands such as `Speak`, `Write` or `Mail`.

Note: Under Windows, in order to have sound alerts you must install the Microsoft Text-to-Speech engine, which is included as an option with the JMP product installation.

Saving and Retrieving Limits

JMP can use previously established control limits for control charts:

- upper and lower control limits, and a center line value
- parameters for computing limits such as a mean and standard deviation.

The control limits or limit parameter values must be either in a JMP data table, referred to as the *Limits Table* or stored as a column property in the process column. When you specify the **Control Chart** command, you can retrieve the Limits Table with the **Get Limits** button on the Control Chart launch dialog.

The easiest way to create a Limits Table is to save results computed by the Control Chart platform. The **Save Limits** command in the popup menu for each control chart automatically saves limits from the sample values. The type of data saved in the table varies according to the type of control chart in the analysis window. You can also use values from any source and create your own Limits Table. All Limits Tables must have

- a column of special key words that identify each row
- a column for each of the variables whose values are the known standard parameters or limits. This column name must be the same as the corresponding process variable name in the data table to be analyzed by the Control Chart platform.

You can save limits in a new data table or as properties of the response column. When you save control limits using the **in New Table** command, the limit key words written to the table depend on the current chart types displayed. A list of limit key words and their associated control chart is shown in Table 42.3 "Limits Table Keys with Appropriate Charts and Meanings," p. 808.

The data table shown next is the data table created when you **Save Limits** using the Clips1.jmp data analysis. The report window showed an Individual Measurement chart and a Moving Range chart, with a specified standard deviation of 0.2126. Note that there is a set containing a center line value and control limits for each chart. The rows with values _Mean, _LCL, and _UCL are for the individual measurement chart. The values _AvgR, _LCLR, and _UCLR are for the Moving Range chart. If you request these kinds of charts again using this Limits Table, the Control Chart platform identifies the appropriate limits from key words in the _LimitsKey column.

	_LimitsKey	Gap
1	_KSigma	3
2	_Alpha	0.0026998
3	_Range Span	2
4	_Std Dev	0.2126
5	_Mean	14.95
6	_LCL	14.3122
7	_UCL	15.5878
8	_AvgR	0.23989341
9	_LCLR	0
10	_UCLR	0.78361948

Note that values for _KSigma, _Alpha, and _Range Span can be specified in the Control Chart Launch dialog. JMP always looks at the values from the dialog first. Values specified in the dialog take precedence over those in an active Limits Table.

The **Control Chart** command ignores rows with unknown key words and rows marked with the excluded row state. Except for _Range Span, _KSigma, _Alpha, and _Sample Size, any needed values not specified are estimated from the data.

As an aid when referencing Table 42.3 "Limits Table Keys with Appropriate Charts and Meanings," p. 808, the following list summarizes the kinds of charts available in the Control Chart platform:

Run Charts

Variables charts are

- $\bar{X}$-chart (Mean)
- *R*-chart (range)
- *S*-chart (standard deviation)
- IM chart (individual measurement)
- MR chart (moving range)
- UWMA chart (uniformly weighted moving average)
- EWMA chart (exponentially weighted moving average)
- Cusum chart (cumulative sum).

Attribute charts are

- *p*-chart (proportion of nonconforming or defective items in a subgroup sample)
- *np*-chart (number of nonconforming or defective items in a subgroup sample)
- *c*-chart (number of nonconformities or defects in a subgroup sample)
- *u*-chart (number of nonconforming or defects per unit).

Table 42.3 Limits Table Keys with Appropriate Charts and Meanings

Key Words	For Charts...	Meaning
_Sample Size	$\bar{X}$, *R*-, *S*-, *p*-, *np*-, *c*-, *u*-, UWMA, EWMA, CUSUM	fixed sample size for control limits; set to missing if the sample size is not fixed. If specified in the Control Chart launch dialog, fixed sample size is displayed.

Table 42.3 Limits Table Keys with Appropriate Charts and Meanings (continued)

Key Words	For Charts...	Meaning
_Range Span	Individual Measurement, MR	specifies the number ($2 \le n \le 25$) of consecutive values for computation of moving range
_Span	UWMA	specifies the number ($2 \le n \le 25$) of consecutive subsample means for computation of moving average
_Weight	EWMA	constant weight for computation of EWMA
_KSigma	All	multiples of the standard deviation of the statistics to calculate the control limits; set to missing if the limits are in terms of the alpha level
_Alpha	All	Type I error probability used to calculate the control limits; used if multiple of the standard deviation is not specified in the launch dialog or in the Limits Table
_Std Dev	$\bar{X}$-, *R*-, *S*-, IM, MR, UWMA, EWMA, CUSUM	known process standard deviation
_Mean	$\bar{X}$-, IM, UWMA, EWMA, CUSUM	known process mean
_U	*c*-, *u*-	known average number of nonconformities per unit
_P	*np*-, *p*-	known value of average proportion nonconforming
_LCL, _UCL	$\bar{X}$-, IM, *p*-, *np*-, *c*-, *u*-	lower and upper control limit for Mean Chart, Individual Measurement chart, or any attribute chart
_AvgR	*R*-, MR	average range or average moving range
_LCLR, _UCLR	*R*-, MR	lower control limit for *R*- or MR chart upper control limit for *R*- or MR chart
_AvgS, _LCLS, _UCLS	*S*-Chart	average standard deviation, upper and lower control limits for *S*-chart
_HeadStart	Cusum	head start for one-sided scheme
_Two Sided, _Data Units	Cusum	type of chart
_H, _K	Cusum	alternative to alpha and beta; K is optional
_Delta _Beta	Cusum	Absolute value of the smallest shift to be detected as a multiple of the process standard deviation or standard error probability and are available only when _Alpha is specified.

Table 42.3 Limits Table Keys with Appropriate Charts and Meanings (continued)

Key Words	For Charts...	Meaning
_Avg R_PreMeans _Avg R_PreStdDev _LCL R_PreMeans _LCL R_PreStdDev _UCL R_PreMeans _UCL R_PreStdDev _Avg_PreMeans _Avg_PreStdDev _LCL_PreMeans _LCL_PreStdDev _UCL_PreMeans _UCL_PreStdDev	IM, MR	Mean, upper and lower control limits based on pre-summarized group means or standard deviations.

Real-Time Data Capture

In JMP, real-time data streams are handled with a `DataFeed` object set up through JMP Scripting Language (JSL) scripts. The `DataFeed` object sets up a concurrent thread with a queue for input lines that can arrive in real time, but are processed during background events. You set up scripts to process the lines and push data on to data tables, or do whatever else is called for. Full details for writing scripts are in the *JMP Scripting Language Guide*.

The Open Datafeed Command

To create a DataFeed object, use the `Open DataFeed` function specifying details about the connection, in the form

```
feedname = Open DataFeed( options... );
```

For example, submit this to get records from com1 and just list them in the log.

```
feed = OpenDataFeed(
     Connect( Port("com1:"),Baud(9600),DataBits(7)),
     SetScript(print(feed<<getLine)));
```

This command creates a scriptable object and starts up a thread to watch a communications port and collect lines. A reference to the object is returned, and you need to save this reference by assigning it to a global variable. The thread collects characters until it has a line. When it finishes a line, it appends it to the line queue and schedules an event to call the `On DataFeed` handler.

Commands for `DataFeed`

The scriptable `DataFeed` object responds to several messages. To send a message in JSL, use the << operator, aimed at the name of the variable holding a reference to the object.

To give it a script or the name of a global holding a script:

```
feedName << Set Script(script or script name);
```

To test DataFeed scripts, you can send it lines from a script:

```
feedName << Queue Line (character expression);
```

For the DataFeed script to get a line from the queue, use this message:

```
feedName << GetLine;
```

To get a list of all the lines to empty the queue, use this:

```
lineListName = feedName << GetLines;
```

To close the DataFeed, including the small window:

```
feedName << Close;
```

To connect to a live data source:

```
feedName << Connect(port specfication);
```

where the port specifications inside the Connect command are as follows. Each option takes only one argument, but they are shown below with the possible arguments separated by "|" with the default value shown first. The last three options take boolean values that specify which control characters are sent back and forth to the device indicating when it is ready to get data. Usually, at most, one of these three is used:

```
Port( "com1:" | "com2:" | "lpt1:" |...),
Baud( 9600 | 4800 | ...),
Data Bits( 8 | 7 ),
Parity( None | Odd | Even ),
Stop Bits( 1 | 0 | 2 ),
DTR_DSR( 0 | 1 ),   // DataTerminalReady
RTS_CTS( 0 | 1 ),   // RequestToSend/ClearToSend
XON_XOFF( 1 | 0 )
```

The Port specification is needed if you want to connect; otherwise, the object still works but is not connected to a data feed.

To disconnect from the live data source:

```
feedName << Disconnect;
```

To stop and later restart the processing of queued lines, either click the respective buttons, or submit the equivalent messages:

```
feedName << Stop;
feedName << Restart;
```

Operation

The script is specified with Set Script.

```
feedName << Set Script(myScript);
```

Here myScript is the global variable that you set up to contain the script to process the data feed. The script typically calls Get DataFeed to get a copy of the line, and does whatever it wants. Usually, it parses the line for data and adds it to some data table. In the example below, it expects to find a three-digit long number starting in column 11; if it does, it adds a row to the data table in the column called thickness.

```
mySCript= Expr(
```

```
        line = feed<<Get Line;
        if (Length(line)>=14,
          x = Num(SubString(line,11,3));
          if (x!=.,
            CurrentDataTable()<<Add Row({thickness=x})));
```

Setting up a script to start a new data table

Here is a sample script that sets up a new data table and starts a control chart based on the data feed:

```
// make a data table
  dt = NewTable("Gap Width");
// make a new column and setup control chart properties
  dc = dt<<NewColumn("gap",Numeric,
  SetProperty("Control Limits",
    {XBar(Avg(20),LCL(19.8),UCL(20.2))}),
  SetProperty("Sigma", 0.1));

// make the data feed
  feed = OpenDatafeed();
  feedScript = expr(
    line = feed<<get line;
    z = Num(line);
    Show(line,z); // if logging or debugging
    if (!IsMissing(z), dt<<AddRow({:gap = z}));
);
  feed<<SetScript(feedScript);
// start the control chart
  Control Chart(SampleSize(5),KSigma(3),ChartCol(gap,XBar,R));
// either start the feed:
// feed<<connect("com1:",Port("com1:"),Baud(9600));
// or test feed some data to see it work:
  For(i=1,i<20,i++,
    feed<<Queue Line(Char(20+RandomUniform()*.1)));
```

Setting Up a Script in a Data Table

In order to further automate the production setting, you can put a script like the one above into a data table property called On Open, which is executed when the data table is opened. If you further marked the data table as a template style document, a new data table is created each time the template table is opened.

Excluded, Hidden, and Deleted Samples

The following table summarizes the effects of various conditions on samples and subgroups:

Table 42.4 Excluded, Hidden, and Deleted Samples

Sample is excluded before creating the chart.	Sample is not included in the calculation of the limits, but it appears on the graph.
Sample is excluded after creating the chart.	Sample is included in the calculation of the limits, and it appears in the graph. Nothing will change on the output by excluding a sample with the graph open.
Sample is hidden before creating the chart.	Sample is included in the calculation of the limits, but does not appear on the graph.
Sample is hidden after creating the chart.	Sample is included in the calculation of the limits, but does not appear on the graph. The sample marker will disappear from the graph, the sample label will still appear on the axis, but limits remain the same.
Sample is both excluded and hidden before creating the chart.	Sample is not included in the calculation of the limits, and it does not appear on the graph.
Sample is both excluded and hidden after creating the chart.	Sample is included in the calculation of the limits, but does not appear on the graph. The sample marker will disappear from the graph, the sample label will still appear on the axis, but limits remain the same.
Data set is subsetted with Sample deleted before creating chart.	Sample is not included in the calculation of the limits, the axis will not include a value for the sample, and the sample marker does not appear on the graph.
Data set is subsetted with Sample deleted after creating chart.	Sample is not included in the calculation of the limits, and does not appear on the graph. The sample marker will disappear from the graph, the sample label will still be removed from the axis, the graph will shift, and the limits will change.

Some additional notes:

1 Exclude and Hide operate only on the rowstate of the first observation in the sample. For example, if the second observation in the sample is hidden, while the first observation is not hidden, the sample will still appear on the chart.

2 An exception to the exclude/hide rule: Tests for Special Causes can flag if a sample is excluded, but will not flag if a sample is hidden.

Chapter 43

Shewhart Control Charts

Variables and Attribute Control Charts

Control charts are a graphical and analytic tool for deciding whether a process is in a state of statistical control.

The concepts underlying the control chart are that the natural variability in any process can be quantified with a set of control limits and that variation exceeding these limits signals a special cause of variation. Out-of-control processes generally justify some intervention to fix a problem to bring the process back in control.

Shewhart control charts are broadly classified into control charts for variables and control charts for attributes. Moving average charts and cumulative sum (Cusum) charts are special kinds of control charts for variables.

The Control Chart platform in JMP implements a variety of control charts:

- $\bar{X}$-, *R*-, and *S*-charts,
- Individual and Moving Range charts,
- *p*-, *np*-, *c*-, and *u*-charts,
- UWMA and EWMA charts,
- Cusum charts
- Phase Control Charts for $\bar{X}$-, *r*-, *IR*-, *p*-, *np*-, *c*-, *u*-, Presummarized, and Levey-Jennings charts

This platform is launched by the **Control Chart** command in the **Graph** menu, by the toolbar or JMP Starter, or through scripting.

One feature special to Control Charts, different from other platforms in JMP, is that they update dynamically as data is added or changed in the table.

Contents

Shewhart Control Charts for Variables

Control charts for variables are classified according to the subgroup summary statistic plotted on the chart:

- $\bar{X}$-charts display subgroup means (averages)
- *R*-charts display subgroup ranges (maximum – minimum)
- *S*-charts display subgroup standard deviations.
- Run charts display data as a connected series of points.

The **IR** selection gives two additional chart types:

- **Individual Measurement** charts display individual measurements
- **Moving Range** charts display moving ranges of two or more successive measurements.

XBar-, *R*-, and *S*- Charts

For quality characteristics measured on a continuous scale, a typical analysis shows both the process mean and its variability with a mean chart aligned above its corresponding *R*- or *S*-chart. Or, if you are charting individual measurements, the individual measurement chart shows above its corresponding moving range chart.

Example. $\bar{X}$- and *R*-Charts

The following example uses the Coating.jmp data in the Quality Control sample data folder (taken from the *ASTM Manual on Presentation of Data and Control Chart Analysis*). The quality characteristic of interest is the Weight column. A subgroup sample of four is chosen. An $\bar{X}$-chart and an *R*-chart for the process are shown in Figure 43.1.

To replicate this example,

- Choose the **Graph > Control Chart > XBar** command.
- Note the selected chart types of **XBar** and **R**.
- Specify Weight as the **Process** variable.
- Change the **Sample Size Constant** from 5 to 4.
- Click **OK**.

Alternatively, you can also submit the following JSL for this example:

```
Control Chart(Sample Size(4), K Sigma(3), Chart Col( :Weight, XBar, R));
```

Sample six indicates that the process is not in statistical control. To check the sample values, click the sample six summary point on either control chart. The corresponding rows highlight in the data table.

Note: If an *S* chart is chosen with the $\bar{X}$-chart, then the limits for the $\bar{X}$-chart are based on the standard deviation. Otherwise, the limits for the $\bar{X}$-chart are based on the range.

Figure 43.1 Variables Charts for Coating Data

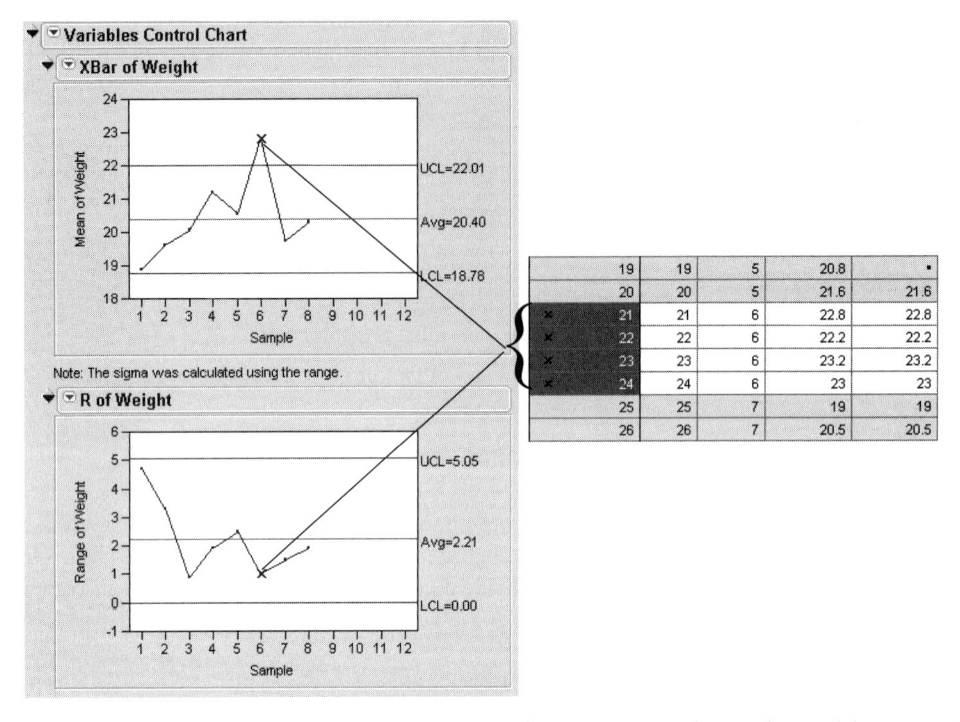

19	19	5	20.8	•
20	20	5	21.6	21.6
21	21	6	22.8	22.8
22	22	6	22.2	22.2
23	23	6	23.2	23.2
24	24	6	23	23
25	25	7	19	19
26	26	7	20.5	20.5

You can use **Fit Y by X** for an alternative visualization of the data. First, change the modeling type of Sample to Nominal. Specify the interval variable Weight as **Y** and the nominal variable Sample as **X**. The box plots in Figure 43.2 show that the sixth sample has a small range of high values.

Figure 43.2 Quantiles Option in Fit Y By X Platform

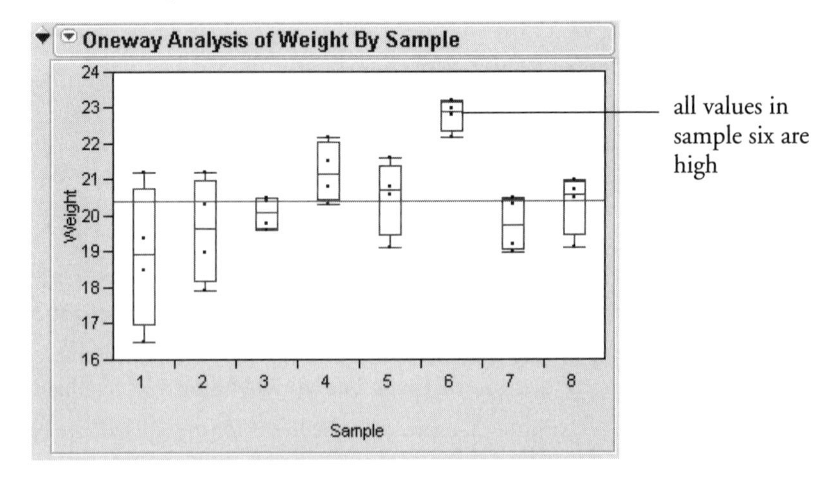

Control Limits for $\bar{X}$- and *R*-charts

JMP generates control limits for $\bar{X}$- and *R*-charts as follows.

LCL for $\bar{X}$ chart = $\bar{\bar{X}} - \frac{k\hat{\sigma}}{\sqrt{n_i}}$

UCL for $\bar{X}$ chart = $\bar{\bar{X}} + \frac{k\hat{\sigma}}{\sqrt{n_i}}$

LCL for *R*-chart = $\max\left(d_2(n_i)\hat{\sigma} - kd_3(n_i)\hat{\sigma}, 0\right)$

UCL for *R*-chart = $d_2(n_i)\hat{\sigma} + kd_3(n_i)\hat{\sigma}$

Center line for *R*-chart: By default, the center line for the ith subgroup indicates an estimate of the expected value of R_i, which is computed as $d_2(n_i)\hat{\sigma}$, where $\hat{\sigma}$ is an estimate of σ. If you specify a known value (σ_0) for σ, the central line indicates the value of $d_2(n_i)\hat{\sigma}_0$. Note that the central line varies with n_i.

The standard deviation of an $\bar{X}/R$ chart is estimated by

$$\hat{\sigma} = \frac{\frac{R_1}{d_2(n_1)} + \ldots + \frac{R_N}{d_2(n_N)}}{N}$$

where

$\bar{\bar{X}}$ = weighted average of subgroup means

σ = process standard deviation

n_i = sample size of ith subgroup

$d_2(n)$ is the expected value of the range of n independent normally distributed variables with unit standard deviation

$d_3(n)$ is the standard error of the range of n independent observations from a normal population with unit standard deviation

N is the number of subgroups for which $n_i \geq 2$

Example. $\bar{X}$- and *S*-charts with varying subgroup sizes

This example uses the same data as example 1, Coating.jmp, in the Quality Control sample data folder. This time the quality characteristic of interest is the Weight 2 column. An $\bar{X}$-chart and an *S* chart for the process are shown in Figure 43.3.

To replicate this example,

- Choose the **Graph > Control Chart > XBar** command.
- Select the chart types of **XBar** and **S**.
- Specify Weight 2 as the **Process** variable.

- Specify the column, **Sample** as the **Sample Label** variable.
- The **Sample Size** option should automatically change to **Sample Grouped by Sample Label.**
- Click **OK.**

Alternatively, you can also submit the following JSL for this example:

```
Control Chart(Sample Size( :Sample), KSigma(3), Chart Col( :Weight 2, XBar,
  S));
```

Figure 43.3 $\overline{X}$ and *S* charts for Varying Subgroup Sizes

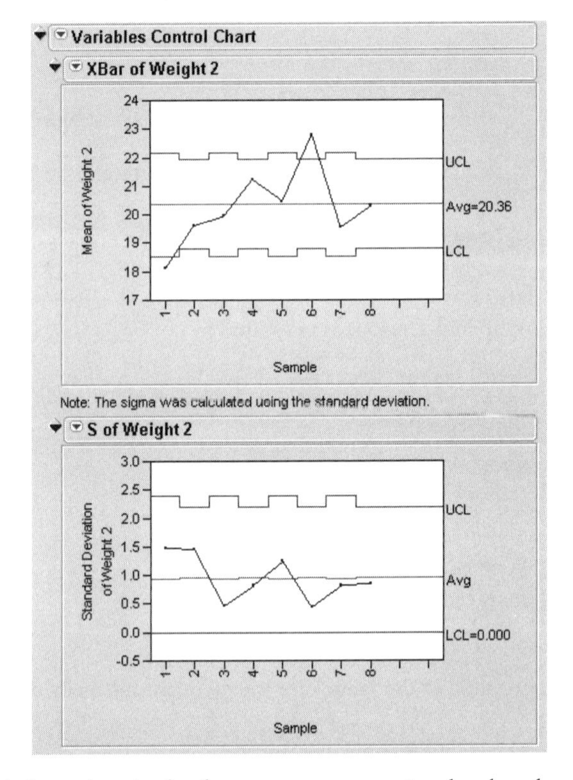

Weight 2 has several missing values in the data, so you may notice the chart has uneven limits. Although, each sample has the same number of observations, samples 1, 3, 5, and 7 each have a missing value.

Note: Although they will turn on and appear checked, no zones or tests will appear on the chart until all samples are equally sized, as neither are valid on charts with unequally sized samples. If the samples change while the chart is open and they become equally sized, and the zone and/or test option is selected, the zones and/or tests will be applied immediately and appear on the chart.

Control Limits for $\bar{X}$- and *S*-charts

JMP generates control limits for $\bar{X}$- and *S*-charts as follows.

LCL for $\bar{X}$ chart = $\bar{\bar{X}} - \frac{k\hat{\sigma}}{\sqrt{n_i}}$

UCL for $\bar{X}$ chart = $\bar{\bar{X}} + \frac{k\hat{\sigma}}{\sqrt{n_i}}$

LCL for S-chart = $\max\left(c_4(n_i)\sigma - kc_5(n_i)\sigma, 0\right)$

UCL for S-chart = $c_4(n_i)\sigma + kc_5(n_i)\sigma$

Center line for S-chart: By default, the center line for the ith subgroup indicates an estimate of the expected value of s_i, which is computed as $c_4(n_i)\hat{\sigma}$, where $\hat{\sigma}$ is an estimate of σ. If you specify a known value (σ_0) for σ, the central line indicates the value of $c_4(n_i)\hat{\sigma}_0$. Note that the central line varies with n_i.

The estimate for the standard deviation in an $\bar{X}/S$ chart is

$$\hat{\sigma} = \frac{\frac{s_1}{c_4(n_1)} + \ldots + \frac{s_n}{c_4(n_N)}}{N}$$

where

$\bar{\bar{X}}$ = weighted average of subgroup means

σ = process standard deviation

n_i = sample size of ith subgroup

$c_4(n)$ is the expected value of the standard deviation of n independent normally distributed variables with unit standard deviation

$c_5(n)$ is the standard error of the standard deviation of n independent observations from a normal population with unit standard deviation

N is the number of subgroups for which $n_i \geq 2$

s_i is the sample standard deviation of the ith subgroup

Run Charts

Run charts display a column of data as a connected series of points. The following example is a Run chart for the Weight variable from Coating.jmp.

Figure 43.4 Run Chart

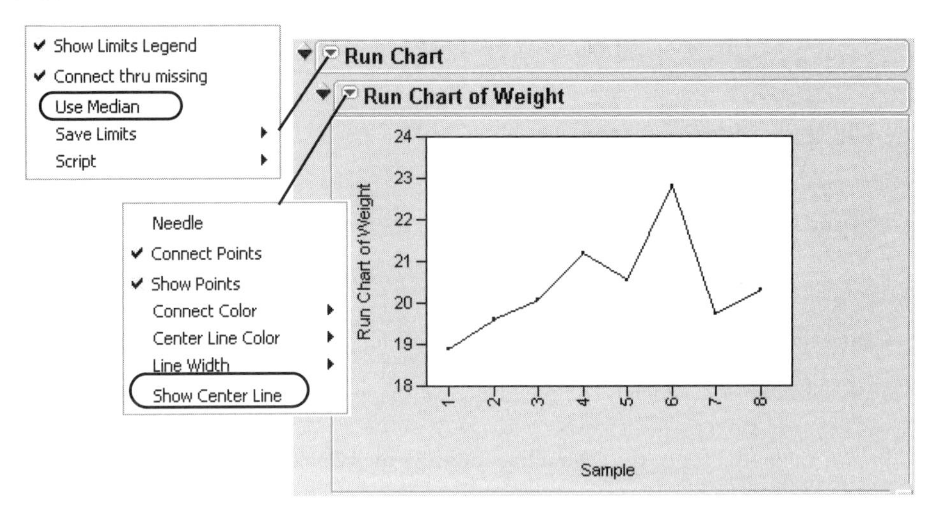

When you select the **Show Center Line** option in the Run Chart drop-down, a line is drawn through the center value of the column.The center line is determined by the **Use Median** setting of the platform drop-down. When **Use Median** is selected, the median is used as the center line. Otherwise, the mean is used. When saving limits to a file, both the overall mean and median are saved.

Run charts can also plot the group means when a sample label is given, either on the dialog or through a script.

Individual Measurement Charts

Individual Measurement Chart Type displays individual measurements. Individual Measurement charts are appropriate when only one measurement is available for each subgroup sample.

Moving Range Chart Type displays moving ranges of two or more successive measurements. Moving ranges are computed for the number of consecutive measurements you enter in the Range Span box. The default range span is 2. Because moving ranges are correlated, these charts should be interpreted with care.

Example. Individual Measurement and Moving Range Charts

The Pickles.jmp data in the Quality Control sample data folder contains the acid content for vats of pickles. Because the pickles are sensitive to acidity and produced in large vats, high acidity ruins an entire pickle vat. The acidity in four vats is measured each day at 1, 2, and 3 PM. The data table records day, time, and acidity measurements. The dialog in Figure 43.5 creates Individual Measurement and Moving Range charts with date labels on the horizontal axis.

Figure 43.5 Launch Dialog for Individual Measurement and Moving Range Chart

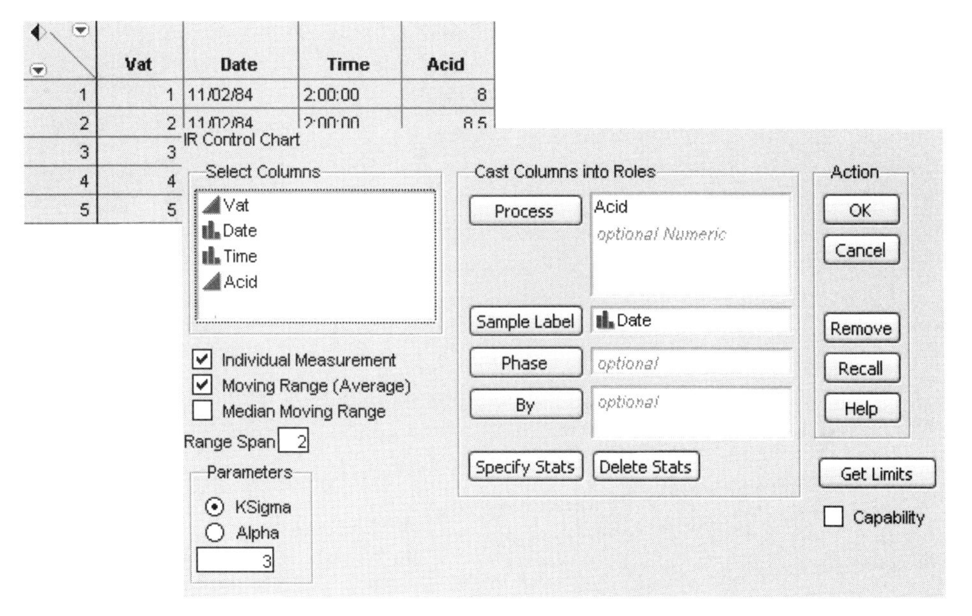

To complete this example,

- Choose the **Graph > Control Chart > IR** command.
- Select both **Individual Measurement** and **Moving Range** chart types.
- Specify Acid as the **Process** variable.
- Specify Date as the **Sample Label** variable.
- Click **OK**.

Alternatively, you can also submit the following JSL for this example:

```
Control Chart(Sample Label( :Date), KSigma(3), Chart Col( :Acid, Individual
  Measurement, Moving Range));
```

The individual measurement and moving range charts shown in Figure 43.6 monitor the acidity in each vat produced.

Figure 43.6 Individual Measurement and Moving Range Charts for Pickles Data

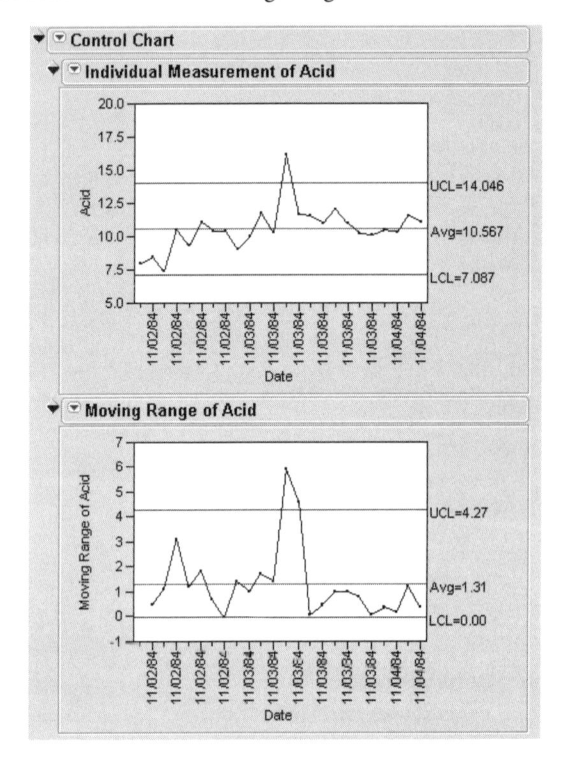

Control Limits for Individual Measurements and Moving Range Charts

LCL for Individual Measurement Chart = $\bar{X} - k\hat{\sigma}$

UCL for Individual Measurement Chart = $\bar{X} + k\hat{\sigma}$

LCL for Moving Range Chart = $\max\left(d_2(n)\sigma - kd_3(n)\sigma, 0\right)$

UCL for Moving Range Chart = $d_2(n)\sigma + kd_3(n)\sigma$

The standard deviation of an IR chart is estimated by

$$\hat{\sigma} = \frac{\bar{R}}{d_2(n)}$$

where

$\bar{X}$ = the mean of the individual measurements

$\bar{R}$ = the mean of the individual measurements

σ = the process standard deviation

$d_2(n)$ = expected value of the range of n independent normally distributed variables with unit standard deviation

$d_3(n)$ = standard error of the range of n independent observations from a normal population with unit standard deviation.

Note: If you choose a Median Moving range chart, the limits on the Individuals chart use the Median Moving Range as the sigma, rather than the Average Moving Range.

Pre-summarize Charts

If your data consists of repeated measurements of the same process unit, then you will want to combine these into one measurement for the unit. Pre-summarizing is not recommended unless the data has repeated measurements on each process or measurement unit.

Presummarize summarizes the process column into sample means and/or standard deviations, based either on the sample size or sample label chosen. Then it charts the summarized data based on the options chosen in the launch dialog. Optionally, you can append a capability analysis by checking the appropriate box in the launch dialog.

Example. Pre-Summarize Chart

For an example, using the Coating.jmp data table,

- Choose the **Graph > Control Chart > Presummarize** command.
- Choose Weight as the Process variable and Sample as the Sample Label.
- In the dialog check both **Individual on Group Means** and **Moving Range on Group Means**. The **Sample Grouped by Sample Label** button is automatically selected when you choose a Sample Label variable.
- Click **OK**.

Figure 43.7 Presummarize Dialog

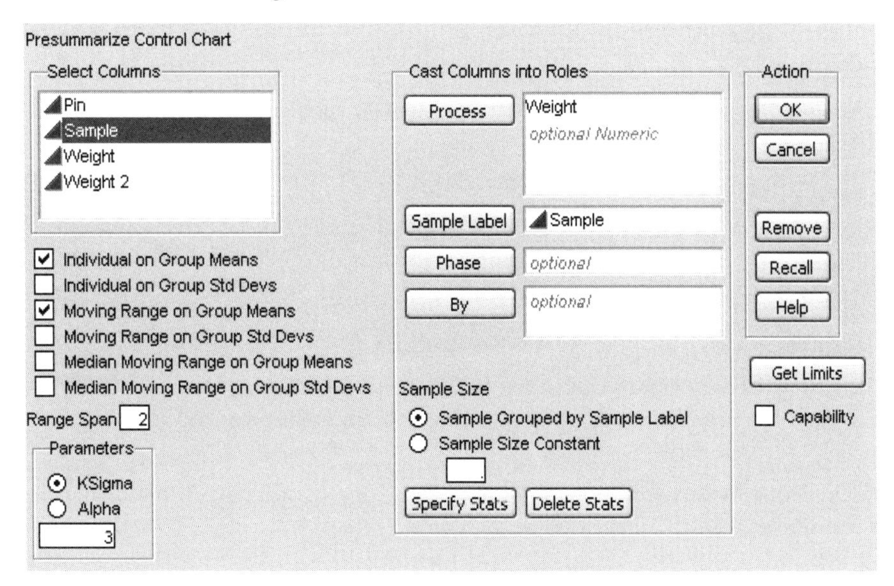

Figure 43.8 Example of Charting Pre-summarized Data

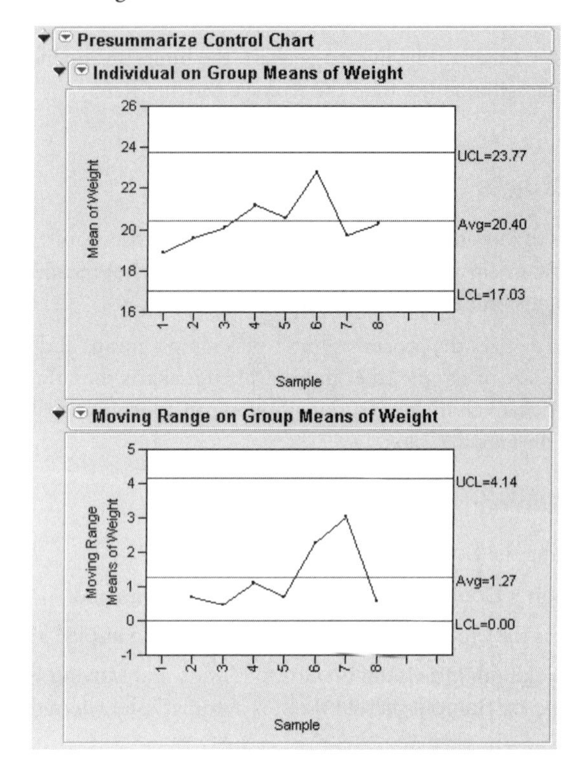

Although the points for $\bar{X}$- and S-charts are the same as the Individual on Group Means and Individual on Group Std Devs charts, the limits are different because they are computed as Individual charts.

Another way to generate the pre-summarized charts, with the Coating.jmp data table,

- Choose **Tables > Summary**.
- Assign Sample as the Group variable, then Mean(Weight) and Std Dev(Weight) as **Statistics**.
- Click **OK**.
- Select **Graph > Control Chart**.
- Set the chart type again to **IR** and choose both Mean(Weight) and Std Dev(Weight) as process variables.
- Click **OK**.

The resulting charts match the pre-summarize charts.

When using **Pre-summarize** charts, you can select either **On Group Means** or **On Group Std Devs** or both. Each option will create two charts (an Individual Measurement and a Moving Range), if both IR chart types are selected.

The **On Group Means** options compute each sample mean, then plot the means and create an Individual Measurement and a Moving Range chart on the means.

The **On Group Std Devs** options compute each sample standard deviation, then plot the standard deviations as individual points and create an Individual Measurement and a Moving Range chart on the standard deviations.

Moving Average Charts

The control charts previously discussed plot each point based on information from a single subgroup sample. The Moving Average chart is different from other types because each point combines information from the current sample and from past samples. As a result, the Moving Average chart is more sensitive to small shifts in the process average. On the other hand, it is more difficult to interpret patterns of points on a Moving Average chart because consecutive moving averages can be highly correlated (Nelson 1983).

In a Moving Average chart, the quantities that are averaged can be individual observations instead of subgroup means. However, a Moving Average chart for individual measurements is not the same as a control (Shewhart) chart for individual measurements or moving ranges with individual measurements plotted.

Uniformly Weighted Moving Average (UWMA) Charts

Each point on a Uniformly Weighted Moving Average (UWMA) chart, also called a Moving Average chart, is the average of the w most recent subgroup means, including the present subgroup mean. When you obtain a new subgroup sample, the next moving average is computed by dropping the oldest of the previous w subgroup means and including the newest subgroup mean. The constant, w, is called the *span* of the moving average, and indicates how many subgroups to include to form the moving average. The larger the span (w), the smoother the UWMA line, and the less it reflects the magnitude of shifts. This means that larger values of w guard against smaller shifts.

Example. UWMA Charts

As an example, consider Clips1.jmp. The measure of interest is the gap between the ends of manufactured metal clips. To monitor the process for a change in average gap, subgroup samples of five clips are selected daily and a UWMA chart with a moving average span of three is examined. To see this chart, complete the Control Chart launch dialog as shown in Figure 43.9, submit the JSL or follow the steps below.

```
Control Chart(Sample Size(5), KSigma(3), Moving Average Span(3), Chart Col(
   :Gap, UWMA));
```

- Choose the **Graph > Control Chart > UWMA** command.
- Change the **Moving Average Span** to 3.
- Choose Gap as the **Process** variable.
- Click **OK**.

Figure 43.9 Specification for UWMA Charts of Clips1.jmp Data

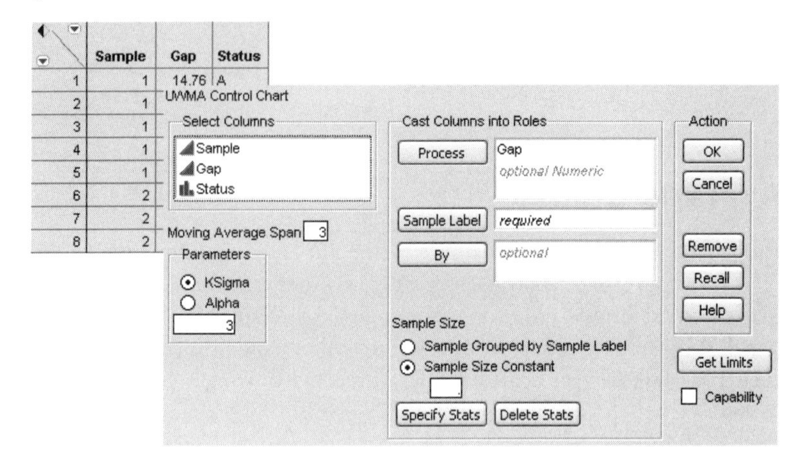

The result is the chart in Figure 43.10. The point for the first day is the mean of the five subgroup sample values for that day. The plotted point for the second day is the average of subgroup sample means for the first and second days. The points for the remaining days are the average of subsample means for each day and the two previous days.

The average clip gap appears to be decreasing, but no sample point falls outside the 3σ limits.

Like all control charts, the UWMA chart updates dynamically when you add rows to the current data table. The chart in Figure 43.11 shows the Clips1 chart with data added for five additional days.

Figure 43.10 UWMA Charts for the Clips1 data

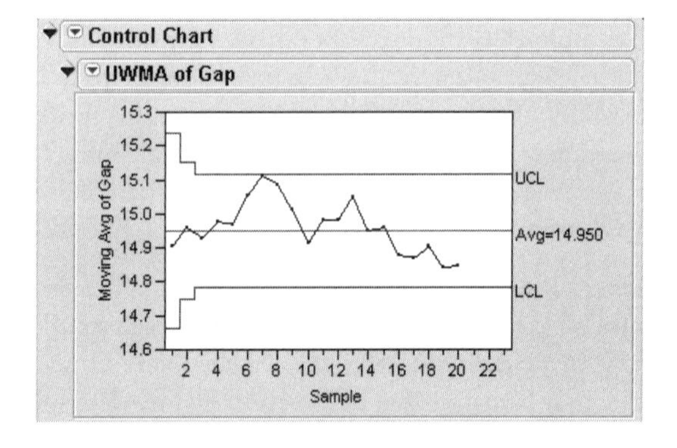

Figure 43.11 UWMA Chart with Five Points Added

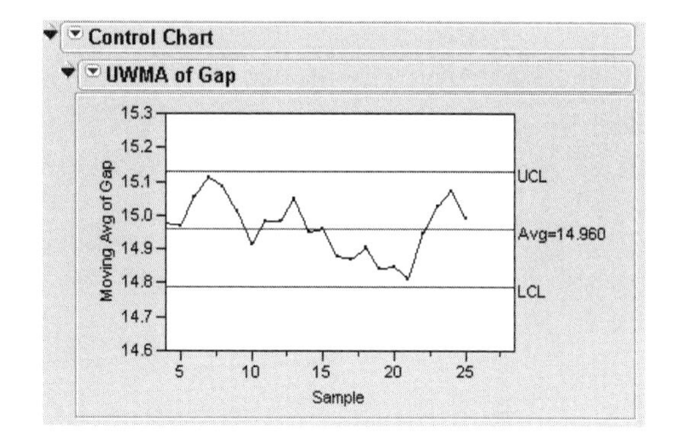

Control Limits for UWMA Charts

Control limits for UWMA charts are computed as follows. For each subgroup i,

$$\mathrm{LCL}_i = \left(\bar{p} - k\sqrt{\frac{\bar{p}(1-\bar{p})}{n_i}}\right)$$

$$\mathrm{UCL}_i = \left(\bar{p} + k\sqrt{\frac{\bar{p}(1-\bar{p})}{n_i}}\right)$$

where

w is the span parameter (number of terms in moving average)

n_i is the sample size of the ith subgroup

$\bar{\bar{X}}$ is the weighted average of subgroup means

Exponentially Weighted Moving Average (EWMA) Chart

Each point on an Exponentially Weighted Moving Average (EWMA) chart, also referred to as a Geometric Moving Average (GMA) chart, is the weighted average of all the previous subgroup means, including the mean of the present subgroup sample. The weights decrease exponentially going backward in time. The weight ($0 < \text{weight} \le 1$) assigned to the present subgroup sample mean is a parameter of the EWMA chart. Small values of weight are used to guard against small shifts.

Example. EWMA Charts

Using the Clips1.jmp data table, submit the JSL or follow the steps below.

```
Control Chart(Sample Size(5), KSigma(3), Weight(0.5), Chart Col( :Gap,
  EWMA));
```

- Choose the **Graph > Control Chart > EWMA** command.
- Change the **Weight** to 0.5.
- Choose **Gap** as the **Process** variable.

- Leave the **Sample Size Constant** as 5.
- Click **OK.**

The figure below shows the EWMA chart for the same data seen in Figure 43.10. This EWMA chart was generated for weight = 0.5.

Figure 43.12 EWMA Chart

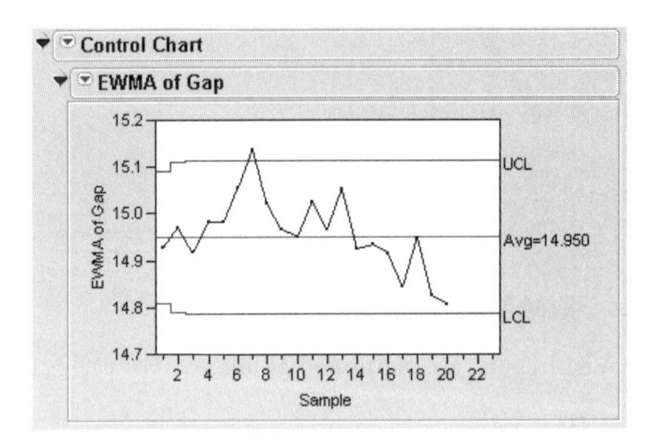

Control Limits for EWMA Charts

Control limits for EWMA charts are computed as follows.

$$\text{LCL} = \bar{\bar{X}} - k\hat{\sigma}r\sqrt{\sum_{j=0}^{i-1}(1-r)^{2j}/n_{i-j}}$$

$$\text{UCL} = \bar{\bar{X}} + k\hat{\sigma}r\sqrt{\sum_{j=0}^{i-1}(1-r)^{2j}/n_{i-j}}$$

where

r is the EWMA weight parameter ($0 < r \leq 1$)

x_{ij} is the jth measurement in the ith subgroup, with $j = 1, 2, 3,..., n_i$

n_i is the sample size of the ith subgroup

$\bar{\bar{X}}$ is the weighted average of subgroup means

Shewhart Control Charts for Attributes

In the previous types of charts, measurement data was the process variable. This data is often continuous, and the charts are based on continuous theory. Another type of data is count data, where the variable of interest is a discrete count of the number of defects or blemishes per subgroup. For discrete count data, attribute charts are applicable, as they are based on binomial and poisson models. Since the counts are measured per subgroup, it is important when comparing charts to determine whether you have similar number of items in the subgroups between the charts. Attribute charts, like variables charts, are classified according to the subgroup sample statistic plotted on the chart:

Table 43.1 Determining which Attribute Chart to use

Each item is judged as either conforming or non-conforming		For each item, the number of defects is counted	
The subgroups are a constant size	The subgroups vary in size	The subgroups are a constant size	The subgroups vary in size
np-chart	*p*-chart	*c*-Chart	*u*-chart

- *p*-charts display the proportion of nonconforming (defective) items in subgroup samples which can vary in size. Since each subgroup for a *p*-chart consists of N items, and an item is judged as either conforming or nonconforming, the maximum number of nonconforming items in a subgroup is N.
- *np*-charts display the number of nonconforming (defective) items in constant sized subgroup samples. Since each subgroup for a *np*-chart consists of N_i items, and an item is judged as either conforming or nonconforming, the maximum number of nonconforming items in subgroup i is N_i.
- *c*-charts display the number of nonconformities (defects) in a subgroup sample that usually consists of one *inspection* unit.
- *u*-charts display the number of nonconformities (defects) per unit in subgroup samples that can have a varying number of inspection units.

p- and *np*-Charts

Example. *np*-Charts

The Washers.jmp data in the Quality Control sample data folder contains defect counts of 15 lots of 400 galvanized washers. The washers were inspected for finish defects such as rough galvanization and exposed steel. If a washer contained a finish defect, it was deemed nonconforming or defective. Thus, the defect count represents how many washers were defective for each lot of size 400. To replicate this example, follow these steps or submit the JSL script below:

- Choose the **Graph > Control Chart > NP** command.
- Choose **# defects** as the **Process** variable.
- Change the **Constant Size** to 400.
- Click **OK**.

```
Control Chart(Sample Size(400), KSigma(3), Chart Col( :Name("# defects"),
  NP));
```

The example here illustrates an *np*-chart for the number of defects.

Figure 43.13 *np*-Chart

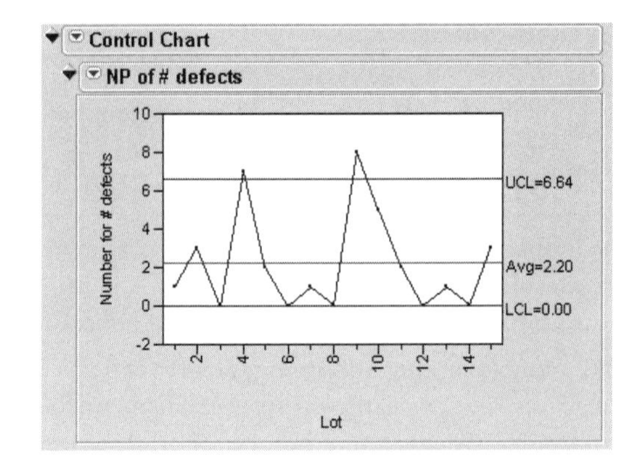

Example. *p*-Charts

Again, using the Washers.jmp data, we can specify a sample size variable, which would allow for varying sample sizes.

Note: This data contains all constant sample sizes. Follow these steps or submit the JSL script below:

- Choose the **Graph > Control Chart > P** command.
- Choose Lot as the **Sample Label** variable.
- Choose # defects as the **Process** variable.
- Choose Lot Size as the **Sample Size** variable.
- Click **OK**.

```
Control Chart(Sample Label( :Lot), Sample Size( :Lot Size), K Sigma(3),
  Chart Col(Name("# defects"), P))
```

The chart shown here illustrates a *p*-chart for the proportion of defects.

Figure 43.14 *p*-Chart

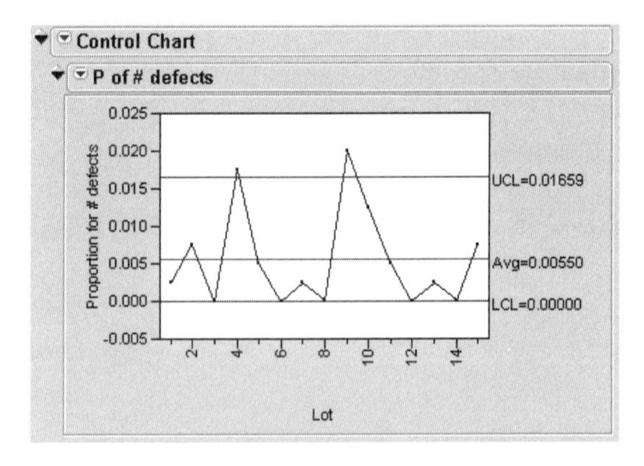

Note that although the points on the chart look the same as the *np*-chart, the *y*-axis, Avg and limits are all different since they are now based on proportions

Control Limits for *p*- and *np*- Charts

The lower and upper control limits, LCL and UCL, respectively, are computed as follows.

p-chart LCL = $\max(p - k\sqrt{p(1-\bar{p})/n_i}, 0)$

p-chart UCL = $\min(p + k\sqrt{p(1-\bar{p})/n_i}, 1)$

np-chart LCL = $\max(n_i p - k\sqrt{n_i p(1-\bar{p})}, 0)$

np-chart UCL = $\min(n_i p + k\sqrt{n_i p(1-\bar{p})}, n_i)$

where

$\bar{p}$ is the average proportion of nonconforming items taken across subgroups, i.e.

$$\bar{p} = \frac{n_1 p_1 + \ldots + n_N p_N}{n_1 + \ldots + n_n} = \frac{X_1 + \ldots + X_N}{n_1 + \ldots + n_N}$$

n_i is the number of items in the ith subgroup

N is the number of subgroups

u-Charts

The Braces.jmp data in the Quality Control sample data folder records the defect count in boxes of automobile support braces. A box of braces is one inspection unit. The number of boxes inspected (per day) is the subgroup sample size, which can vary. The *u*-chart, shown here, is monitoring the number of brace defects per subgroup sample size. The upper and lower bounds vary according to the number of units inspected.

Note: When you generate a *u*-chart, and select **Capability**, JMP launches the Poisson Fit in Distribution and gives a Poisson-specific capability analysis.

Figure 43.15 *u*-Chart

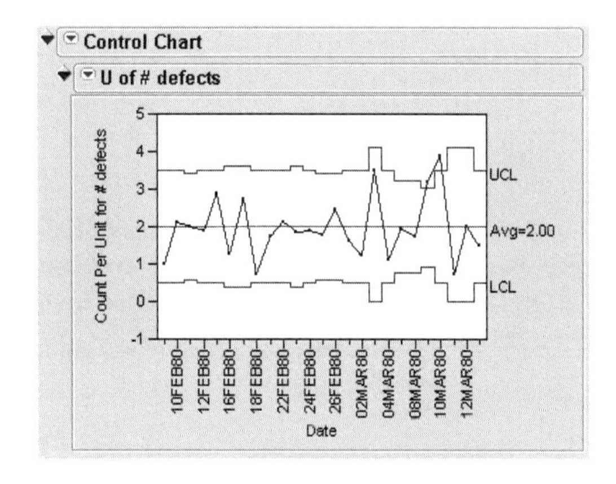

Example. *u*-Charts

To replicate this example, follow these steps or submit the JSL below.

- Open the Braces.jmp data in the Quality Control sample data folder.
- Choose the **Graph > Control Chart > U** command.
- Choose # defects as the **Process** variable.
- Choose Unit size as the **Unit Size** variable.
- Choose Date as the **Sample Label**.
- Click **OK**.

```
Control Chart(Sample Label( :Date), Unit Size( :Unit size), K Sigma(3),
  Chart Col( :Name("# defects"), U));
```

Control Limits on *u*-charts

The lower and upper control limits, LCL and UCL, are computed as follows

$$\text{LCL} = \max(\bar{u} - k\sqrt{\bar{u}/n_i}, 0)$$

$$\text{UCL} = \bar{u} + k\sqrt{\bar{u}/n_i}$$

The limits vary with n_i.

u is the expected number of nonconformities per unit produced by process

u_i is the number of nonconformities per unit in the ith subgroup. In general, $u_i = c_i/n_i$.

c_i is the total number of nonconformities in the ith subgroup

n_i is the number of inspection units in the ith subgroup

$\bar{u}$ is the average number of nonconformities per unit taken across subgroups. The quantity $\bar{u}$ is computed as a weighted average

$$\bar{u} = \frac{n_1 u_1 + \ldots + n_N u_N}{n_1 + \ldots + n_N} = \frac{c_1 + \ldots + c_N}{n_1 + \ldots + n_N}$$

N is the number of subgroups

c-Charts

c-charts are similar to *u*-charts in that they monitor the number of nonconformities in an entire subgroup, made up of one or more units. However, they require constant subgroup sizes. *c*-charts can also be used to monitor the average number of defects per inspection unit.

Note: When you generate a *c*-chart, and select **Capability**, JMP launches the Poisson Fit in Distribution and gives a Poisson-specific capability analysis.

Example 10. *c*-Charts for Noncomformities per Unit

In this example, a clothing manufacturer ships shirts in boxes of ten. Prior to shipment, each shirt is inspected for flaws. Since the manufacturer is interested in the average number of flaws per shirt, the number of flaws found in each box is divided by ten and then recorded. To replicate this example, follow these steps or submit the JSL below.

- Open the Shirts.jmp data in the Quality Control sample data folder.
- Choose the **Graph > Control Chart > C** command.
- Choose # Defects as the **Process** variable.
- Choose Box Size as the **Sample Size**.
- Choose Box as the **Sample Label**.
- Click **OK**.

```
Control Chart(Sample Label( :Box), Sample Size( :Box Size), K Sigma(3),
  Chart Col( :Name("# Defects"), C));
```

Figure 43.16 *c*-Chart

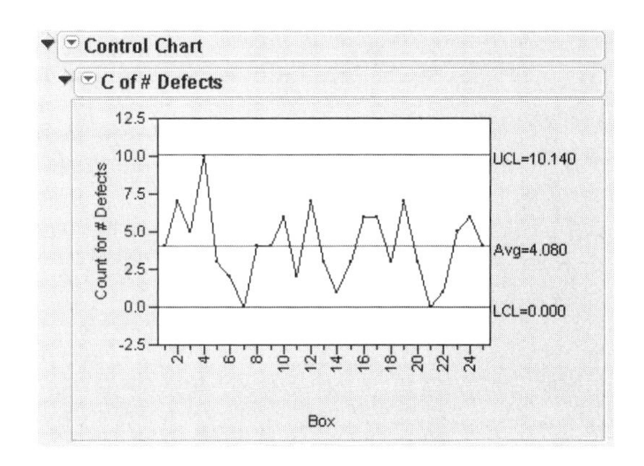

Control Limits on c-charts

The lower and upper control limits, LCL and UCL, are computed as follows.

$$\text{LCL} = \max(n_i\bar{u} - k\sqrt{n_i\bar{u}}, 0)$$

$$\text{UCL} = n_i\bar{u} + k\sqrt{n_i\bar{u}}$$

The limits vary with n_i.

u is the expected number of nonconformities per unit produced by process

u_i is the number of nonconformities per unit in the ith subgroup. In general, $u_i = c_i/n_i$.

c_i is the total number of nonconformities in the ith subgroup

n_i is the number of inspection units in the ith subgroup

$\bar{u}$ is the average number of nonconformities per unit taken across subgroups. The quantity $\bar{u}$ is computed as a weighted average

$$\bar{u} = \frac{n_1u_1 + \ldots + n_Nu_N}{n_1 + \ldots + n_N} = \frac{c_1 + \ldots + c_N}{n_1 + \ldots + n_N}$$

N is the number of subgroups

Levey-Jennings Plots

Levey-Jennings plots show a process mean with control limits based on a long-term sigma. The control limits are placed at $3s$ distance from the center line.

s for the Levey-Jennings plot is calculated the same way standard deviation is in the Distribution platform, *i.e.*

$$s = \sqrt{\sum_{i=1}^{N} \frac{w(\bar{y} - y_i)^2}{N-1}}$$

Figure 43.17 Levey Jennings Plot

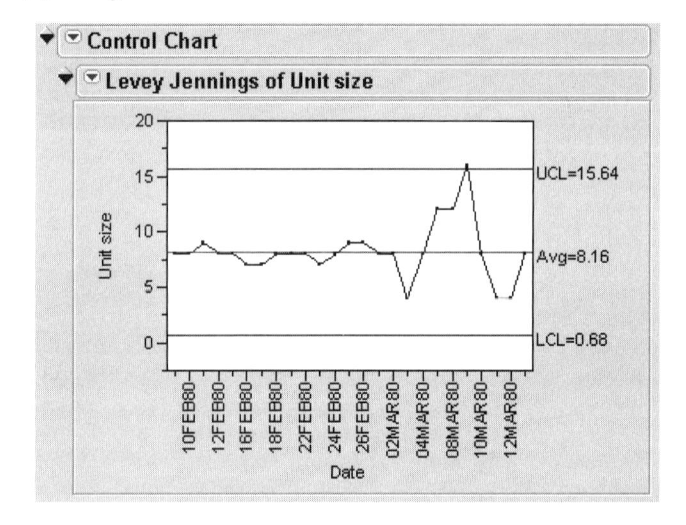

Phases

A *phase* is a group of consecutive observations in the data table. For example, phases might correspond to time periods during which a new process is brought into production and then put through successive changes. Phases generate, for each level of the specified Phase variable, a new sigma, set of limits, zones, and resulting tests.

On the dialog for $\bar{X}$-, *r*-, *s*-, *IR*-, *p*-, *np*-, *c*-, *u*-, Presummarized, and Levey-Jennings charts, a **Phase** variable button appears. If a phase variable is specified, the phase variable is examined, row by row, to identify to which phase each row belongs.

Saving to a limits file reveals the sigma and specific limits calculated for each phase.

Example

Open Diameter.JMP, found in the Quality Control sample data folder. This data set contains the diameters taken for each day, both with the first prototype (phase 1) and the second prototype (phase 2).

- Select **Graph > Control Chart > XBar.**
- Choose DIAMETER as the **Process**, DAY as the **Sample Label**, and Phase as the **Phase**.
- Click **OK**.

Figure 43.18 Launch Dialog for Phases

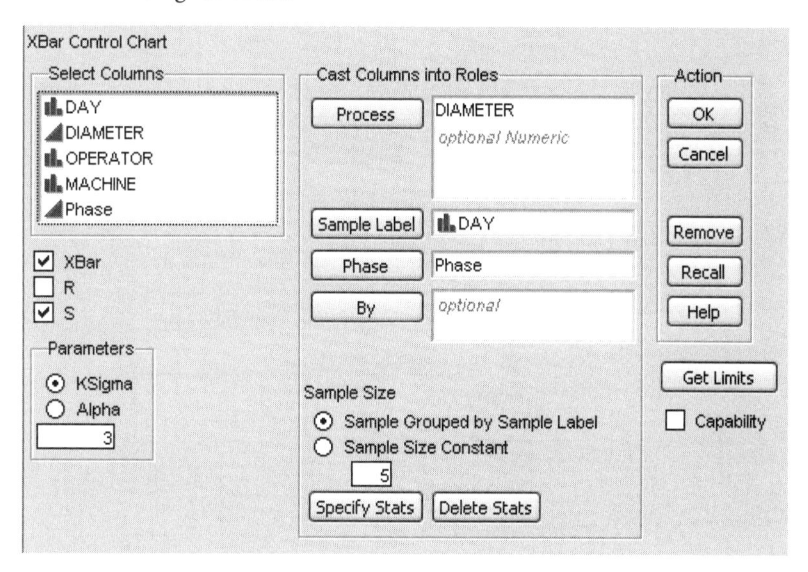

The resulting chart has different limits for each phase

Figure 43.19 Phase Control Chart

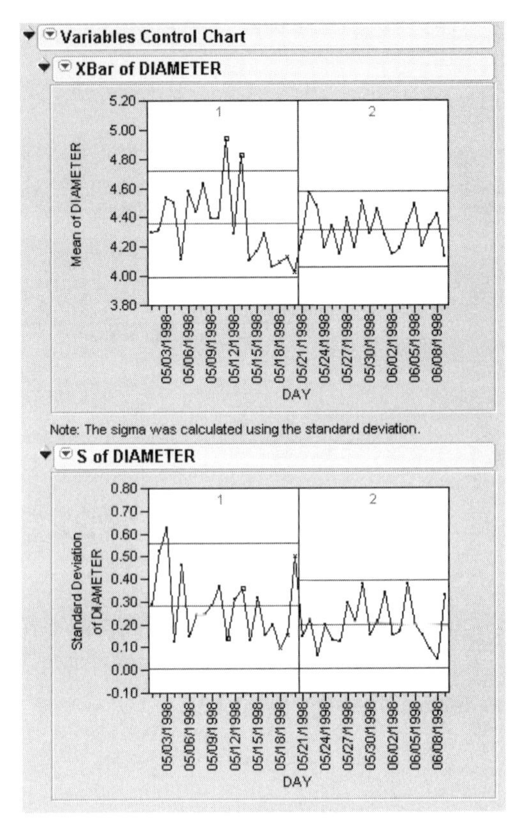

Chapter 44

Cumulative Sum Control Charts

CUSUM Charts

Cusum control charts are used when it is important to detect that a process has wandered away from a specified process mean. Although Shewhart $\bar{X}$-charts can detect if a process is moving beyond a two- or three-sigma shift, they are not effective at spotting a one-sigma shift in the mean. They will still appear in control because the cumulative sum of the deviations will wander farther and farther away from the specified target, and a small shift in the mean will appear very clearly and much sooner.

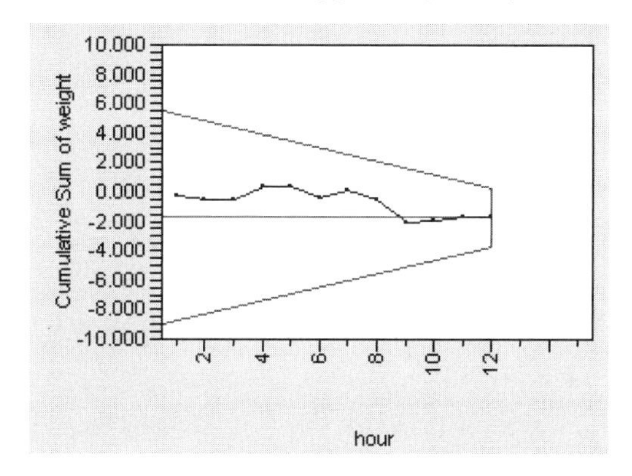

Contents

Cumulative Sum (Cusum) Charts

Cumulative Sum (Cusum) charts display cumulative sums of subgroup or individual measurements from a target value. Cusum charts are graphical and analytical tools for deciding whether a process is in a state of statistical control and for detecting a shift in the process mean.

JMP cusum charts can be one-sided, which detect a shift in one direction from a specified target mean, or two-sided to detect a shift in either direction. Both charts can be specified in terms of geometric parameters (h and k described in Figure 44.1); two-sided charts allow specification in terms of error probabilities α and β.

To interpret a two-sided Cusum chart, you compare the points with limits that compose a V-mask. A V-mask is formed by plotting V-shaped limits. The origin of a V-mask is the most recently plotted point, and the arms extended backward on the *x*-axis, as in Figure 44.1. As data are collected, the cumulative sum sequence is updated and the origin is relocated at the newest point.

Figure 44.1 Illustration of a V-Mask for a Two-Sided Cusum Chart

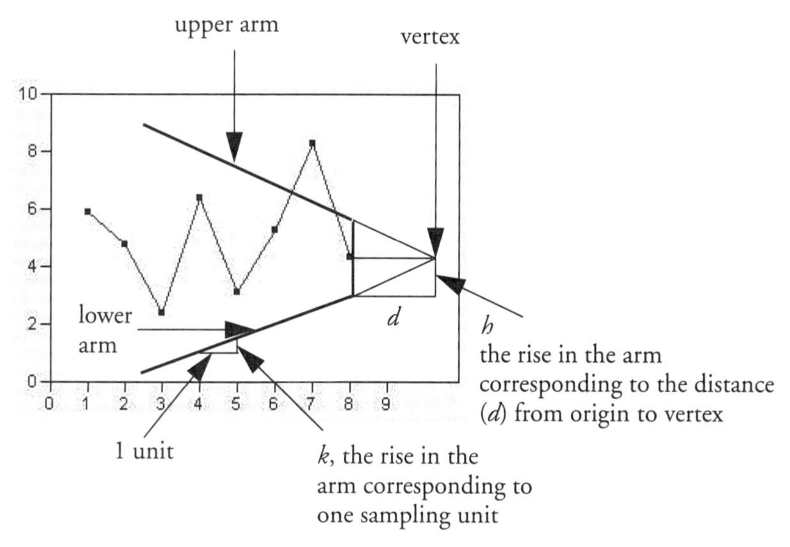

Shifts in the process mean are visually easy to detect on a cusum chart because they produce a change in the slope of the plotted points. The point where the slope changes is the point where the shift occurs. A condition is *out-of-control* if one or more of the points previously plotted crosses the upper or lower arm of the V-mask. Points crossing the lower arm signal an increasing process mean, and points crossing the upper arm signal a downward shift.

There are major differences between cusum charts and other control (Shewhart) charts:

- A Shewhart control chart plots points based on information from a single subgroup sample. In cusum charts, each point is based on information from all samples taken up to and including the current subgroup.
- On a Shewhart control chart, horizontal control limits define whether a point signals an out-of-control condition. On a cusum chart, the limits can be either in the form of a V-mask or a horizontal decision interval.

- The control limits on a Shewhart control chart are commonly specified as 3σ limits. On a cusum chart, the limits are determined from average run length, from error probabilities, or from an economic design.

A cusum chart is more efficient for detecting small shifts in the process mean. Lucas (1976) comments that a V-mask detects a 1σ shift about four times as fast as a Shewhart control chart.

Launch Options for Cusum Charts

When you choose **Graph > Control Charts > Cusum**, the Control Charts Launch dialog appears, including appropriate options and specifications as shown here.

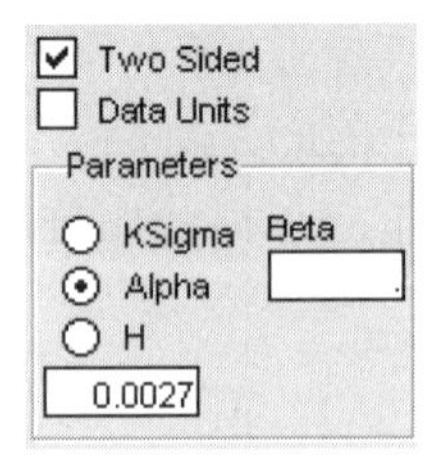

See "Parameters," p. 795 in the "Statistical Control Charts" chapter for a description of **K Sigma** and **Alpha**. The following items pertain only to cusum charts:

Two Sided

requests a two-sided cusum scheme when checked. If it is not checked, a one-sided scheme is used and no V-mask appears. If an H value is specified, a decision interval is displayed.

Data Units

specifies that the cumulative sums be computed without standardizing the subgroup means or individual values so that the vertical axis of the cusum chart is scaled in the same units as the data.

Note: Data Units requires that the subgroup sample size be designated as constant.

Beta

specifies the probability of failing to discover that the specified shift occurred. **Beta** is the probability of a Type II error and is available only when you specify **Alpha**.

H

is the vertical distance h between the origin for the V-mask and the upper or lower arm of the V-mask for a two-sided scheme. When you click **H**, the **Beta** entry box is labeled **K**. You also enter a value for the increase in the lower V-mask per unit change on the subgroup axis (See Figure 44.6). For a one-sided scheme, **H** is the decision interval. Choose **H** as a multiple of the standard error.

Specify Stats

appends the panel shown here to the Control Charts Launch dialog, which lets you enter the process variable specifications.

Known Statistics for CUSUM Chart
Weight
Target
Delta
Shift
Sigma
Head Start

Target is the target mean (goal) for the process or population. The target mean must be scaled in the same units as the data.

Delta specifies the absolute value of the smallest shift to be detected as a multiple of the process standard deviation or of the standard error, depending on whether the shift is viewed as a shift in the population mean or as a shift in the sampling distribution of the subgroup mean, respectively. Delta is an alternative to the Shift option (described next). The relationship between Shift and Delta is given by

$$\delta = \frac{\Delta}{(\sigma/(\sqrt{n}))}$$

where δ represents Delta, Δ represents the shift, σ represents the process standard deviation, and n is the (common) subgroup sample size.

Shift is the minimum value you want to detect on either side of the target mean. You enter the shift value in the same units as the data, and you interpret it as a shift in the mean of the sampling distribution of the subgroup mean. You can choose either Shift or Delta.

Sigma specifies a known standard deviation, σ_0, for the process standard deviation, σ. By default, the Control Chart platform estimates sigma from the data. You can use Sigma instead of the Alpha option on the Control Charts Launch dialog.

Head Start specifies an initial value for the cumulative sum, S_0, for a one-sided cusum scheme (S_0 is usually zero). Enter Head Start as a multiple of standard error.

Cusum Chart Options

Cusum charts have these options (in addition to standard chart options).

Show Points

shows or hides the sample data points.

Connect Points

connects the sample points with a straight line.

Mask Color

displays the JMP color palette for you to select a line color for the V-mask.

Connect Color

displays the JMP color palette for you to select a color for the connect line when the **Connect Points** option is in effect.

Center Line Color

displays the JMP color palette for you to select a color for the center line.

Show Shift

shows or hides the shift you entered, or center line.

Show V Mask

shows or hides the V-mask based on the parameters (statistics) specified on the Control Charts Launch dialog when Cusum is selected as the Chart Type.

Show Parameters

displays a Parameters table (see Figure 44.6) that summarizes the Cusum charting parameters.

Show ARL

displays the average run length (ARL) information.

Example 1. Two-Sided Cusum Chart with V-mask

To see an example of a two-sided cusum chart, open the Oil1 Cusum.jmp file from the Quality Control sample data folder. A machine fills 8-ounce cans of two-cycle engine oil additive. The filling process is believed to be in statistical control. The process is set so that the average weight of a filled can, μ_0, is 8.10 ounces. Previous analysis shows that the standard deviation of fill weights, σ_0, is 0.05 ounces.

Subgroup samples of four cans are selected and weighed every hour for twelve hours. Each observation in the Oil1 Cusum.jmp data table contains one value of weight along with its associated value of hour. The observations are sorted so that the values of hour are in increasing order. The Control Chart platform assumes that the data are sorted in increasing order.

A two-sided cusum chart is used to detect shifts of at least one standard deviation in either direction from the target mean of 8.10 ounces.

To create a Cusum chart for this example,

- Choose the **Graph > Control Chart > CUSUM** command.
- Click the **Two Sided** check box if it is not already checked.
- Specify weight as the Process variable.
- Specify hour as the Sample Label.
- Click the **H** radio button and enter 2 into the text box.
- Click **Specify Stats** to open the **Known Statistics for CUSUM chart** tab.
- Set **Target** to the average weight of 8.1.
- Enter a **Delta** value of 1.

- Set **Sigma** to the standard deviation of 0.05.

The finished dialog should look like the one in Figure 44.2.

Alternatively, you can bypass the dialog and submit the following JSL script:

```
Control Chart(Sample Size( :hour), H(2), Chart Col( :weight, CUSUM(Two
  sided(1), Target(8.1), Delta(1), Sigma(0.05))));
```

Figure 44.2 Dialog for Cusum Chart Example

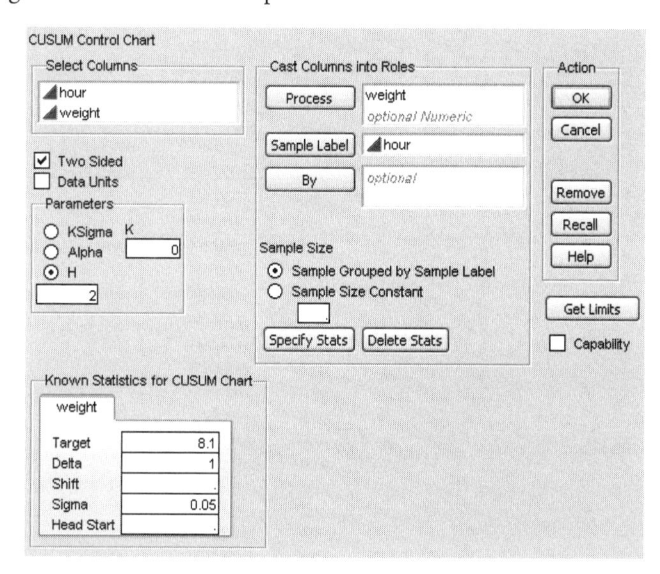

When you click **OK**, the chart in Figure 44.3 appears.

Figure 44.3 Cusum Chart for Oil1 Cusum.jmp Data

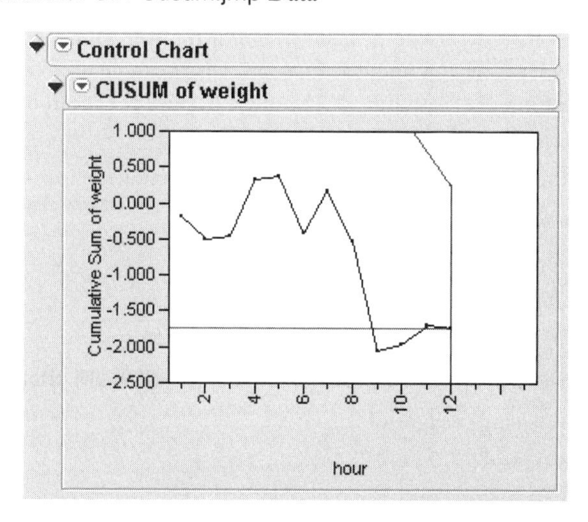

You can interpret the chart by comparing the points with the V-mask whose right edge is centered at the most recent point (hour=12). Because none of the points cross the arms of the V-mask, there is no evidence that a shift in the process has occurred.

A shift or out-of-control condition is signaled at a time *t* if one or more of the points plotted up to the time *t* cross an arm of the V-mask. An upward shift is signaled by points crossing the lower arm, and a downward shift is signaled by points crossing the upper arm. The time at which the shift occurred corresponds to the time at which a distinct change is observed in the slope of the plotted points.

The cusum chart automatically updates when you add new samples. The Cusum chart in Figure 44.4 is the previous chart with additional points. You can move the origin of the V-mask by using the hand to click a point. The center line and V-mask adjust to reflect the process condition at that point.

Figure 44.4 Updated Cusum Chart for the OIL Data

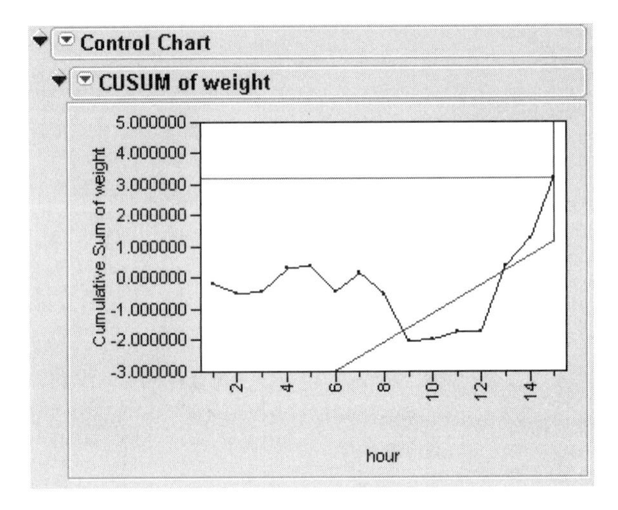

Example 2. One-Sided Cusum Chart with no V-mask

Consider the data used in Example 1, where the machine fills 8-ounce cans of engine oil. Consider also that the manufacturer is now concerned about significant over-filling in order to cut costs, and not so concerned about under-filling. A one-sided Cusum Chart can be used to identify data approaching or exceeding the side of interest. Anything 0.25 ounces beyond the mean of 8.1 is considered a problem. To do this example,

- Choose the **Graph > Control Chart > CUSUM** command.
- *Deselect* the **Two Sided** check box.
- Specify weight as the **Process** variable.
- Specify hour as the **Sample Label**.
- Click the **H** radio button and enter 0.25 into the text box.
- Click **Specify Stats** to open the **Known Statistics for CUSUM chart** tab.
- Set **Target** to the average weight of 8.1.
- Enter a **Delta** value of 1.
- Set **Sigma** to the standard deviation 0.05.

Alternatively, you can submit the following JSL script:

```
Control Chart(Sample Size( :hour), H(0.25), Show Limits Legend(0), Chart
   Col( :weight, CUSUM(Two Sided(0), Target(8.1), Delta(1), Sigma(0.05))));
```

The resulting output should look like the picture in Figure 44.5.

Figure 44.5 One-Sided Cusum Chart for the OIL Data

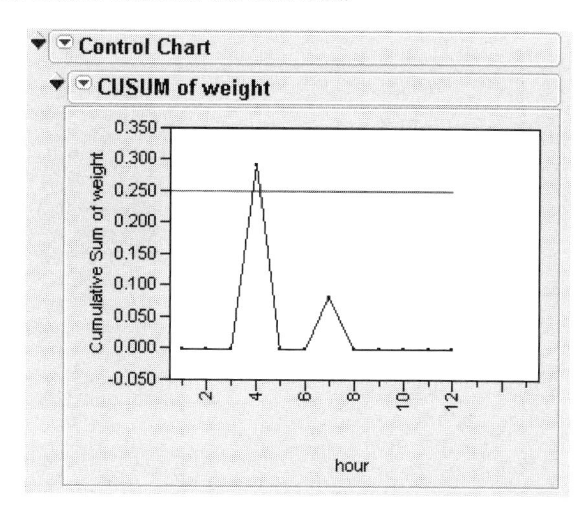

Notice that the *decision interval* or horizontal line is set at the H-value entered (0.25). Also note that no V-mask appears with One-Sided Cusum charts.

The **Show Parameters** option in the Cusum chart popup menu shows the Parameters report in Figure 44.6. The parameters report summarizes the charting parameters from the Known Statistics for CUSUM chart tab on the Control Chart Launch dialog. An additional chart option, **Show ARL**, adds the average run length (ARL) information to the report. The average run length is the expected number of samples taken before an out-of-control condition is signaled:

- ARL (Delta), sometimes denoted ARL1, is the average run length for detecting a shift the size of the specified Delta
- ARL(0), sometimes denoted ARL0, is the in-control average run length for the specified parameters (Montogomery (1985)).

Figure 44.6 Show Parameters and Show ARL Options

Parameters

Target	8.100000
Delta	1.000000
Shift	.
Sigma	0.050000
Head Start	0.000000

ARL

ARL (Delta)	1.658041
ARL (0)	4.351097

Formulas for CUSUM Charts

Notation

The following notation is used in these formulas:

μ denotes the mean of the population, also referred to as the process mean or the process level.

μ_0 denotes the target mean (goal) for the population. The symbol $\bar{X}_0$ is used for μ_0 in *Glossary and Tables for Statistical Quality Control.* You can provide μ_0 as the target on the dialog or through JSL.

σ denotes the population standard deviation.

σ_0 denotes a known standard deviation. You can provide σ_o as the target on the dialog or through JSL.

σ denotes an estimate of σ.

n denotes the nominal sample size for the cusum scheme.

δ denotes the shift in μ to be detected, expressed as a multiple of the standard deviation. You can provide δ as the target on the dialog or through JSL

Δ denotes the shift in μ to be detected, expressed in data units. If the sample size n is constant across subgroups, then

$$\Delta = \delta\sigma_{\bar{X}} = (\delta\sigma)/\sqrt{n}$$

Note that some authors use the symbol D instead of Δ. You can provide Δ as the target on the dialog or through JSL.

One-Sided CUSUM Charts

Positive Shifts

If the shift δ to be detected is positive, the CUSUM computed for the tth subgroup is

$$S_t = \max(0, S_{t-1} + (z_t - k))$$

for $t = 1, 2, \ldots, n$, where $S_0 = 0$, z_t is defined as for two-sided schemes, and the parameter k, termed the *reference value*, is positive. The cusum S_t is referred to as an *upper cumulative sum*. Since S_t can be written as

$$\max\left(0, S_{t-1} + \frac{\bar{X}_i - (\mu_0 + k\sigma_{\bar{X}_i})}{\sigma_{\bar{X}_i}}\right)$$

the sequence S_t cumulates deviations in the subgroup means greater than k standard errors from μ_0. If S_t exceeds a positive value h (referred to as the *decision interval*), a shift or out-of-control condition is signaled.

Negative Shifts

If the shift to be detected is negative, the cusum computed for the tth subgroup is

$$S_t = \max(0, S_{t-1} - (z_t + k))$$

for $t = 1, 2,..., n$, where $S_0 = 0$, z_t is defined as for two-sided schemes, and the parameter k, termed the *reference value*, is positive. The cusum S_t is referred to as a *lower cumulative sum*. Since S_t can be written as

$$\max\left(0, S_{t-1} - \frac{\bar{X}_i - (\mu_0 - k\sigma_{\bar{X}_i})}{\sigma_{\bar{X}_i}}\right)$$

the sequence S_t cumulates the absolute value of deviations in the subgroup means less than k standard errors from σ_0. If S_t exceeds a positive value h (referred to as the *decision interval*), a shift or out-of-control condition is signaled.

Note that S_t is always positive and h is always positive, regardless of whether δ is positive or negative. For schemes designed to detect a negative shift, some authors define a reflected version of S_t for which a shift is signaled when S_t is less than a negative limit.

Lucas and Crosier (1982) describe the properties of a fast initial response (FIR) feature for CUSUM schemes in which the initial CUSUM S_0 is set to a "headstart" value. Average run length calculations given by them show that the FIR feature has little effect when the process is in control and that it leads to a faster response to an initial out-of-control condition than a standard CUSUM scheme. You can provide headstart values on the dialog or through JSL.

Constant Sample Sizes

When the subgroup sample sizes are constant (= n), it may be preferable to compute cusums that are scaled in the same units as the data. Cusums are then computed as

$$S_t = \max(0, S_{t-1} + (\bar{X}_i - (\mu_0 + k\sigma/\sqrt{n})))$$

for $\delta > 0$ and the equation

$$S_t = \max(0, S_{t-1} - (\bar{X}_i - (\mu_0 - k\sigma/\sqrt{n})))$$

for $\delta < 0$. In either case, a shift is signaled if S_t exceeds $h' = h\sigma/\sqrt{n}$ .Some authors use the symbol H for h'.

Two-Sided Cusum Schemes

If the cusum scheme is two-sided, the cumulative sum S_t plotted for the tth subgroup is

$$S_t = S_{t-1} + z_t$$

for $t = 1, 2,..., n$. Here S_0=0, and the term z_t is calculated as

$$z_t = (\bar{X}_t - \mu_0)/(\sigma/\sqrt{n_t})$$

where $\bar{X}_t$ is the tth subgroup average, and n_t is the tth subgroup sample size. If the subgroup samples consist of individual measurements x_t, the term z_t simplifies to

$$z_t = (x_t - \mu_0)/\sigma$$

Since the first equation can be rewritten as

$$S_t = \sum_{i=1}^{t} z_i = \sum_{i=1}^{t} (\bar{X}_i - \mu_0)/\sigma_{\bar{X}_i}$$

the sequence S_t cumulates standardized deviations of the subgroup averages from the target mean μ_0.

In many applications, the subgroup sample sizes n_i are constant ($n_i = n$), and the equation for S_t can be simplified.

$$S_t = (1/\sigma_{\bar{X}}) \sum_{i=1}^{t} (\bar{X}_i - \mu_0) = (\sqrt{n}/\sigma) \sum_{i=1}^{t} (\bar{X}_i - \mu_0)$$

In some applications, it may be preferable to compute S_t as

$$S_t = \sum_{i=1}^{t} (\bar{X}_i - \mu_0)$$

which is scaled in the same units as the data. In this case, the procedure rescales the V-mask parameters h and k to $h' = h\sigma/\sqrt{n}$ and $k' = k\sigma/\sqrt{n}$, respectively. Some authors use the symbols F for k' and H for h'.

If the process is in control and the mean μ is at or near the target μ_0, the points will not exhibit a trend since positive and negative displacements from μ_0 tend to cancel each other. If μ shifts in the positive direction, the points exhibit an upward trend, and if μ shifts in the negative direction, the points exhibit a downward trend.

Chapter 45

Multivariate Control Charts

Quality Control for Multivariate Data

Multivariate control charts address process monitoring problems where several related variables are of interest.

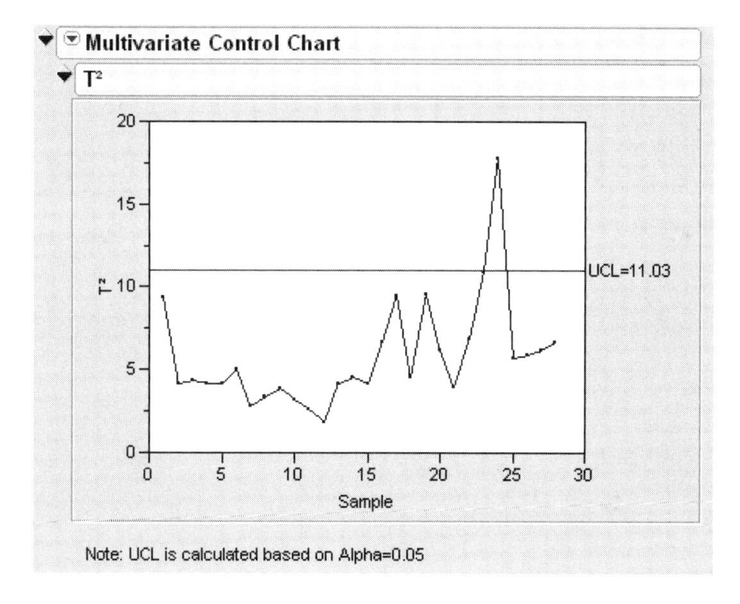

Contents

Launch the Platform

To generate a multivariate control chart, select **Graph > Control Charts > Multivariate Control Chart**.

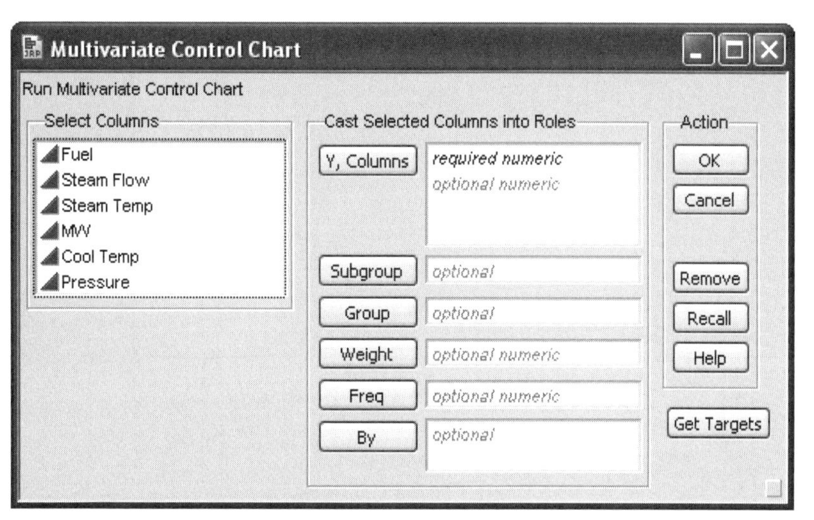

Y, Columns are the columns to be analyzed.

Subgroup is a column that specifies group membership. Hierarchically, this group is within **Group**.

Group is a column that specifies group membership at the highest hierarchical level.

In addition, there is a **Get Targets** button that allows you to pick a JMP table that contains historical targets for the process.

Control Chart Usage

There are two distinct phases in generating a multivariate control chart. Phase 1 is the use of the charts for establishing control. Historical limits are set in this stage, and initial tests show whether the process is in statistical control. The objective of phase 1 is to obtain an in-control set of observations so that control limits can be set for phase 2, which is the monitoring of future production.

Phase 1—Obtaining Targets

To illustrate the process of creating a multivariate control chart, we use data collected on steam turbines. Historical data, stored in Steam Turbine Historical.jmp, is used to construct the initial chart.

Launch the platform and assign all continuous variables to the **Y, Columns** role. When you click **OK**, Figure 45.1 appears.

Figure 45.1 Initial Multivariate Control Chart

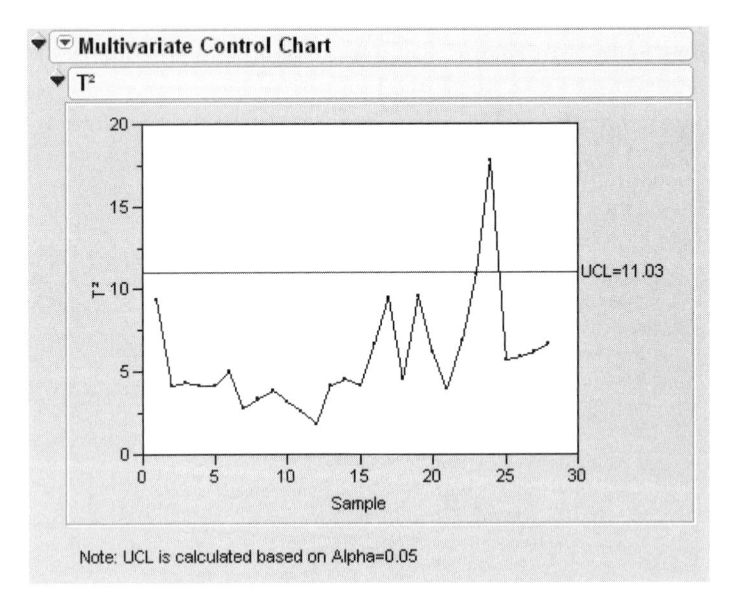

The process seems to be in reasonable statistical control, since there is only one out of control point. Therefore, we use it to establish targets. To do so, select **Save Target Statistics** from the platform menu. This creates a new data table containing target statistics for the process.

Figure 45.2 Target Statistics for Steam Turbine Data

	Ref_Stats	Fuel	Steam Flow	Steam Temp	MW	Cool Temp	Pressure
1	_SampleSize	28	28	28	28	28	28
2	_NumSample	1	1	1	1	1	1
3	_Mean	237595.786	179015.786	846.392857	20.6471429	53.8714286	29.1392857
4	_Std	7247.68598	4374.30638	2.9481857	0.53416503	0.20880106	0.04973475
5	_Corr_Fuel	1	0.87143829	-0.549875	0.85585708	-0.2700498	-0.4699285
6	_Corr_Steam Flow	0.87143829	1	-0.6290239	0.98525292	-0.223127	-0.5330562
7	_Corr_Steam Temp	-0.549875	-0.6290239	1	-0.5952146	0.24753872	0.21921473
8	_Corr_MW	0.85585708	0.98525292	-0.5952146	1	-0.2073058	-0.5044731
9	_Corr_Cool Temp	-0.2700498	-0.223127	0.24753872	-0.2073058	1	0.36174616
10	_Corr_Pressure	-0.4699285	-0.5330562	0.21921473	-0.5044731	0.36174616	1

Save these targets as Steam Turbine Targets.jmp so that they can be accessed in phase 2.

Phase 2—Monitoring the Process

With targets saved, we can create the multivariate control chart that monitors the process.

Open Steam Turbine Current.jmp, which holds recent observations from the process. Launch the Multivariate Control Chart platform, and again assign all variables to the **Y, Columns** role. This time, click the **Get Targets** button and select the Steam Turbine Targets.jmp table that was saved in phase 1.

Figure 45.3 Phase 2 Steam Turbine Control Chart

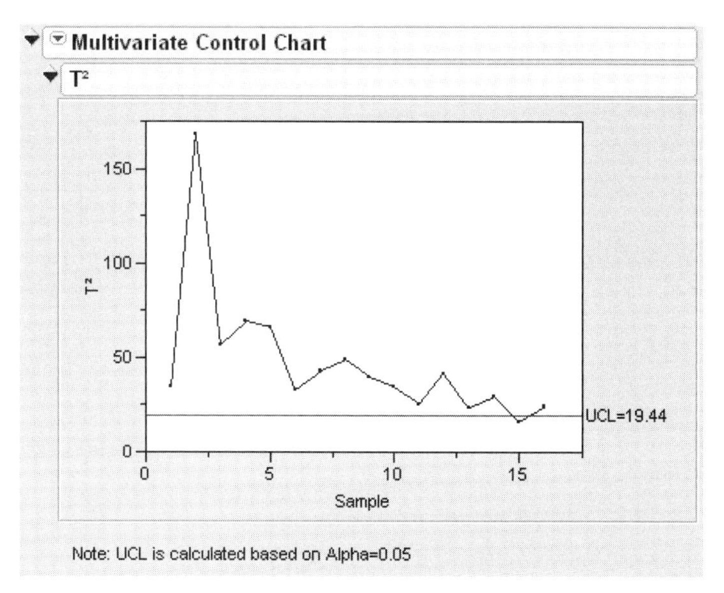

Monitoring a Grouped Process

The workflow for monitoring a multivariate process with grouped data is similar to the one for ungrouped data. An initial control chart is used to create target statistics, and these statistics are used in monitoring the process.

As an example, open Aluminum Pins Historical.jmp, which monitors a process of manufacturing aluminum pins. Enter all the Diameter and Length variables as **Y, Columns** and subgroup as the **Subgroup**. After clicking **OK**, you see the chart shown in Figure 45.4.

Figure 45.4 Grouped Multivariate Control Chart, Phase 1

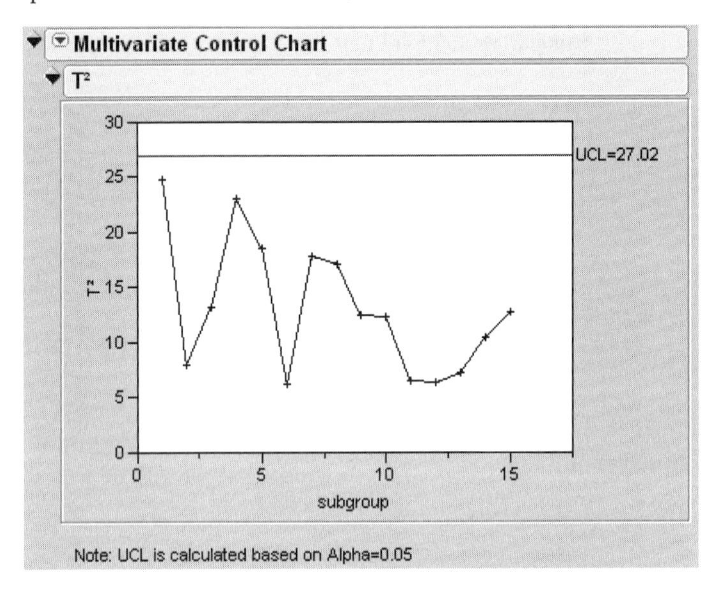

Again, the process seems to be in statistical control, making it appropriate to create targets. Select **Save Target Statistics** and save the resulting table as Aluminum Pins Targets.jmp.

Now, open Aluminum Pins Current.jmp to see current values for the process. To monitor the process, launch the Multivariate Control Chart platform, specifying the columns as in phase 1. Click **Get Targets** and select the saved targets file to produce the chart shown in Figure 45.5, which also has the **Show Means** option selected.

Figure 45.5 Grouped Multivariate Control Chart, Phase 2

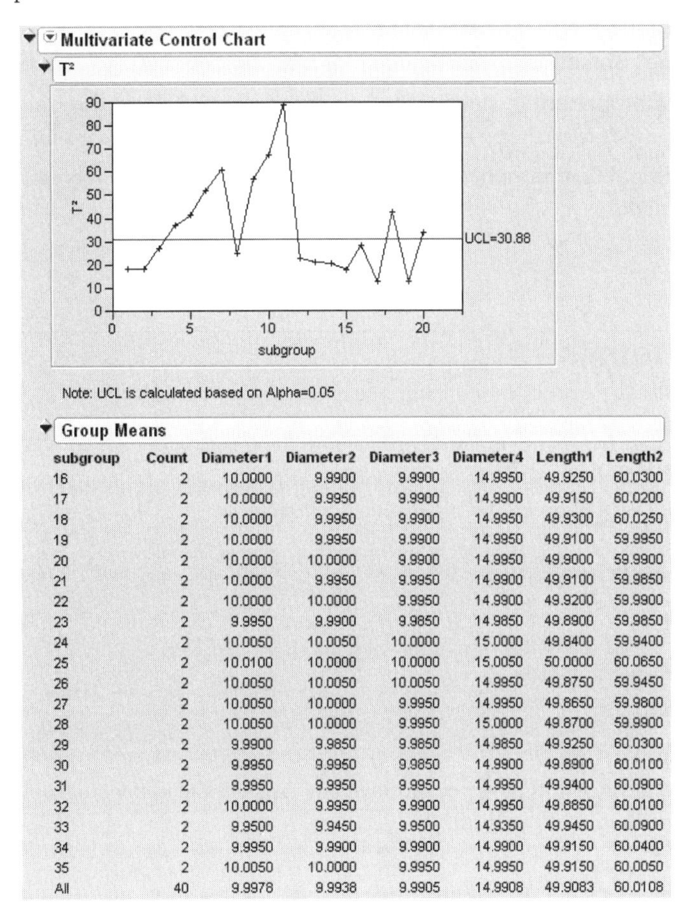

subgroup	Count	Diameter1	Diameter2	Diameter3	Diameter4	Length1	Length2
16	2	10.0000	9.9900	9.9900	14.9950	49.9250	60.0300
17	2	10.0000	9.9950	9.9900	14.9900	49.9150	60.0200
18	2	10.0000	9.9950	9.9900	14.9950	49.9300	60.0250
19	2	10.0000	9.9950	9.9900	14.9950	49.9100	59.9950
20	2	10.0000	10.0000	9.9900	14.9950	49.9000	59.9900
21	2	10.0000	9.9950	9.9950	14.9900	49.9100	59.9850
22	2	10.0000	10.0000	9.9950	14.9900	49.9200	59.9900
23	2	9.9950	9.9900	9.9850	14.9850	49.8900	59.9850
24	2	10.0050	10.0050	10.0000	15.0000	49.8400	59.9400
25	2	10.0100	10.0000	10.0000	15.0050	50.0000	60.0650
26	2	10.0050	10.0050	10.0050	14.9950	49.8750	59.9450
27	2	10.0050	10.0000	9.9950	14.9950	49.8650	59.9800
28	2	10.0050	10.0000	9.9950	15.0000	49.8700	59.9900
29	2	9.9900	9.9850	9.9850	14.9850	49.9250	60.0300
30	2	9.9950	9.9950	9.9850	14.9900	49.8900	60.0100
31	2	9.9950	9.9950	9.9950	14.9950	49.9400	60.0900
32	2	10.0000	9.9950	9.9900	14.9950	49.8850	60.0100
33	2	9.9500	9.9450	9.9500	14.9350	49.9450	60.0900
34	2	9.9950	9.9900	9.9900	14.9900	49.9150	60.0400
35	2	10.0050	10.0000	9.9950	14.9950	49.9150	60.0050
All	40	9.9978	9.9938	9.9905	14.9908	49.9083	60.0108

Platform Options

The following options are available from the platform drop-down menu

T^2 Chart shows or hides the T^2 chart.

Set Alpha Level sets the α-level used to calculate the control limit. The default is α=0.05.

Show Covariance shows or hides the covariance report.

Show Correlation shows or hides the correlation report.

Show Inverse Covariance shows or hides the inverse covariance report.

Show Inverse Correlation shows or hides the inverse correlation report.

Show Means shows or hides the group means.

Save T Square creates a new column in the data table containing T^2 values.

Save T Square Formula creates a new column in the data table, and stores a formula that calculates the T^2 values.

Save Target Statistics creates a new data table containing target statistics for the process.

Principal Components shows or hides a report showing a scaled version of the principal components on the covariances. The components are scaled so that their sum is the T^2 value.

Save Principal Components creates new columns in the data table that hold the scaled principal components.

Statistical Details

Ungrouped Data

The T^2 statistic is defined as

$$T^2 = (Y-\mu)'S^{-1}(Y-\mu)$$

where

S is the covariance matrix

μ is the true mean

Y represents the observations

During Phase 1 (when you have not specified any targets), the upper control limit (UCL) is a function of the beta distribution. Specifically,

$$UCL = \frac{(n-1)^2}{n}\beta\left(\alpha, \frac{p}{2}, \frac{n-p-1}{2}\right)$$

where

p is number of data points

n is the sample size for each subgroup

During phase 2, when targets are specified, the UCL is a function of the F-distribution, defined as

$$UCL = \frac{p(n+1)(n-1)}{n(n-p)}F(\alpha, p, n-p)$$

where

p is number of data points

n is the sample size for each subgroup

m is the number of subgroups

Grouped Data

The T^2 statistic is defined as

$$T^2 = n(\bar{Y} - \mu)'S^{-1}(\bar{Y} - \mu)$$

where

S is the covariance matrix

μ is the true mean

$\bar{Y}$ represents the observations

During Phase 1, the Upper Control Limit is

$$UCL = \frac{p(m-1)(n-1)}{mn-m-p+1}F(\alpha, p, mn-m-p+1)$$

where

p is number of data points

n is the sample size for each subgroup

m is the number of subgroups

During Phase 2, the Upper Control Limit is

$$UCL = \frac{p(m+1)(n-1)}{mn-m-p+1}F(\alpha, p, mn-m-p+1)$$

where

p is number of data points

n is the sample size for each subgroup

m is the number of subgroups

Additivity

When a sample of mn independent normal observations are grouped into m rational subgroups of size n, the distance between the mean $\bar{Y}_j$ of the jth subgroup and the expected value μ is T_M^2. Note that the components of the T^2 statistic are additive, much like sums of squares. That is,

$$T_{\text{all}}^2 = T_M^2 + T_D^2$$

Let T^2_M represent the distance from a target value,

$$T_M^2 = n(\bar{Y}_j - \mu)'S_P^{-1}(\bar{Y}_j - \mu)$$

The internal variability is

$$T_D^2 = \sum_{j=1}^{n}(Y_j - \bar{Y})'S_P^{-1}(Y_j - \bar{Y})$$

The overall variability is

$$T_{\text{all}}^2 = \sum_{j=1}^{n}(\bar{Y}_j - \mu)'S_P^{-1}(\bar{Y}_j - \mu)$$

Chapter 46

Variability Charts

Variability Chart and Gauge R&R Analysis

You launch the Variability Chart platform with the **Variability/Gauge Chart** command in the **Graph** menu (or toolbar), or with the **Variability Chart** or **Attribute Chart** buttons on the **Measure** tab of the JMP Starter.

A *variability chart* plots the mean for each level of a second factor, with all plots side by side. Along with the data, you can view the mean, range, and standard deviation of the data in each category, seeing how they change across the categories. The analysis options assume that the primary interest is how the mean and variance change across the categories.

- A traditional name for this chart is a *multivar* chart, but because that name is not well known, we use the more generic term variability chart.
- A variability chart shows data side-by-side like the Oneway platform, but it has been generalized to handle more than one grouping column.
- A common use of a variability chart is for measurement systems analysis such as gauge R&R, which analyzes how much of the variability is due to operator variation (reproducibility) and measurement variation (repeatability). Gauge R&R is available for many combinations of crossed and nested models, regardless of whether the model is balanced.
- Just as a control chart shows variation across time in a process, a variability chart shows the same kind of variation across categories such as parts, operators, repetitions, and instruments.
- The Variability Chart platform can compute variance components. Several models of crossed and nested factors of purely random models are available.
- Attribute (multi-level) data can also be analyzed with this platform.

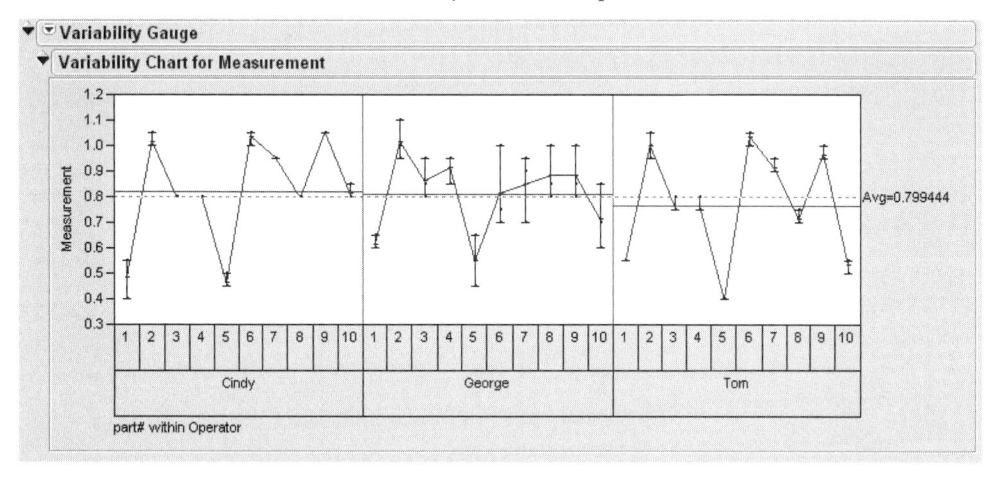

Contents

Variability Charts

A variability chart is built to study how a measurement varies across categories. Along with the data, you can view the mean, range, and standard deviation of the data in each category. The analysis options assume that the primary interest is how the mean and variance change across the categories.

A variability chart has the response on the *y*-axis and a multilevel categorized *x*-axis. The body of the chart can have the features illustrated in Figure 46.1.

Figure 46.1 Example of a Variability Chart

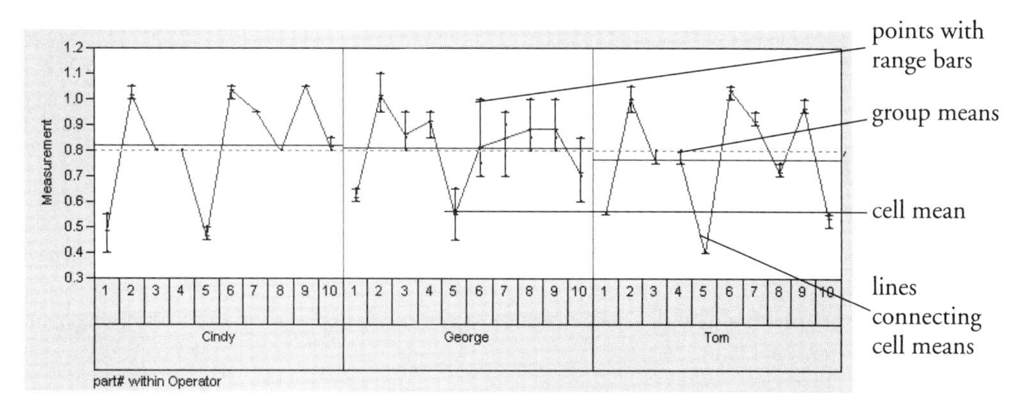

Note: The Variability/Gauge Chart platform invokes REML for computing the variance components if the data is unbalanced. To force it to do REML even if it is balanced, set the JSL global variable `Vari Chart Force REML`to 1 by typing the following into a script window and executing the script.

```
Vari Chart Force REML = 1;
```

Launch the Variability Platform

The **Variability /Gauge Chart** command on the **Graph** menu displays the Variability Chart Launch dialog shown in Figure 46.2. You specify the classification columns that group the measurements in the **X, Grouping** list. If the factors form a nested hierarchy, specify the higher terms first. If it is a gauge study, specify operator first and then the part. Specify the measurement column in the **Y, Response** list. If you specify more than one *Y* column, there will be a separate variability chart for each response.

Specifying a standard or reference column that contains the "true" or known values for the measured part enables the Bias and Linearity Study options. Both of these options perform analysis on the differences between the observed measurement and the reference or standard value.

You may decide on the model type (Main Effect, Crossed, Nested, etc.) here or by using the **Variance Components** command in the platform pop-up menu.

The following example uses the sample data table 2 Factors Crossed.jmp, found in the Variability Data folder.

Figure 46.2 The Variability/Gauge Chart Launch Dialog

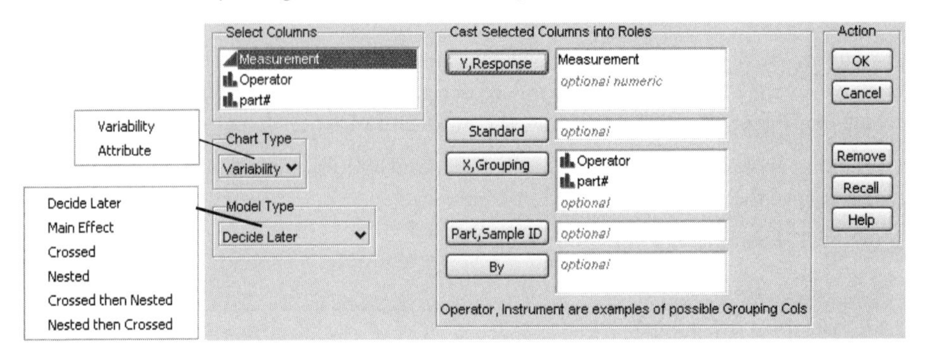

Variability Chart

When you complete the Launch dialog and click **OK**, the variability chart and the standard deviation chart shown in Figure 46.3 appear by default. This variability chart shows three measurements taken by each operator for parts numbered 1 to 10, with maximum and minimum bars to show the range of measurements. The standard deviation chart plots the standard deviation of measurements taken on each part by each operator.

Figure 46.3 Variability Charts for Two-Factors Crossed Data

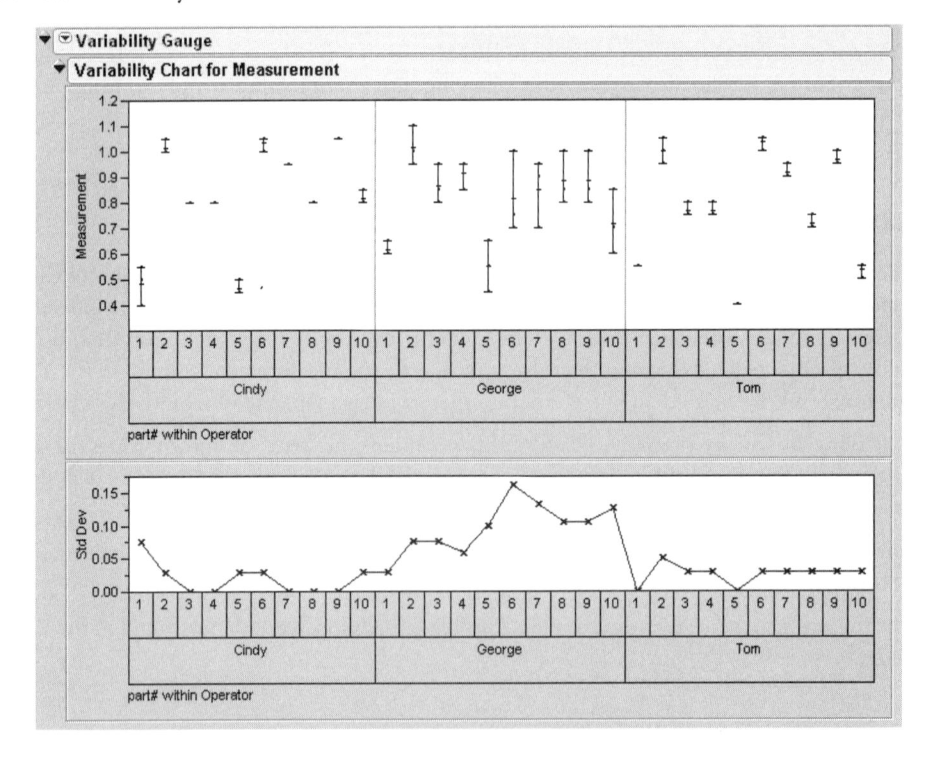

Variability Platform Options

The appearance of the variability chart and standard deviation chart in Figure 46.3 are a result of the checked options in the popup menu shown here. The following options let you tailor the appearance of a variability chart and generate a Gauge R&R analysis.

Variability Chart toggles the whole Gauge variability chart on or off.

Show Points shows the points for individual rows.

Show Range Bars shows the bar from the minimum to the maximum of each cell.

Show Cell Means shows the mean mark for each cell.

Connect Cell Means connects cell means within a group of cells.

Show Group Means shows the mean for groups of cells as a horizontal solid line.

Show Grand Mean shows the overall mean as a gray dotted line across the whole graph.

Show Grand Median shows the overall median as a blue dotted line across the whole graph.

Show Box Plots toggles box plots on and off.

Note that some users prefer not to show the bars on the tips of the plot's whiskers. They may be turned off using a global Platform preference.

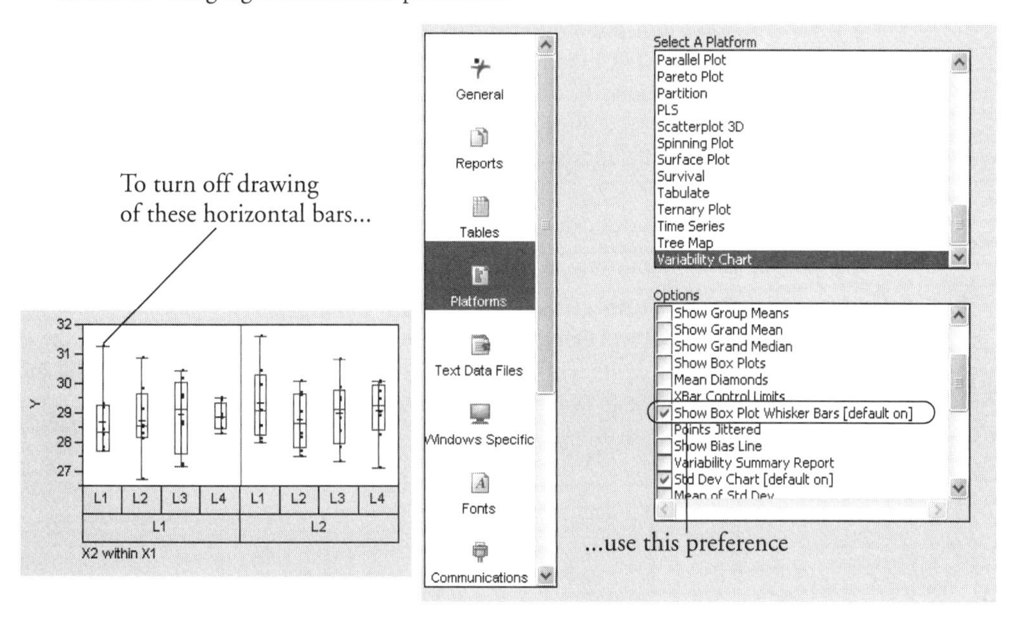

Means Diamonds turns the means diamonds on and off. The confidence intervals use the within-group standard deviation for each cell.

XBar Control Limits draws lines at the UCL and LCL on the Variability chart.

Points Jittered adds some random noise to the plotted points so that coincident points do not plot atop one another.

Show Bias Line toggles the bias line (in the main variability chart) on and off.

Variability Summary Report toggles a report that shows the mean, standard deviation, standard error of the mean, lower and upper 95% confidence intervals, and the minimum, maximum, and number of observations.

Std Dev Chart displays a separate graph that shows cell standard deviations across category cells.

Mean of Std Dev toggles a line at the mean standard deviation on the Std Dev chart.

S Chart Limits toggles lines showing the LCL and UCL in the Std Dev chart.

Variance Components estimates the variance components for a specific model. Variance components are computed for these models: nested, crossed, crossed then nested (three factors only), and nested then crossed (three factors only).

Gauge Studies interprets the first factors as grouping columns and the last as Part, and then it creates a gauge R&R report using the estimated variance components. (Note that there is also a **Part** field in the launch dialog). You are prompted to confirm a given *k* value to scale the results. You are also prompted for a tolerance interval or historical sigma, but these are optional and can be omitted.

Within this menu, you can request **Discrimination Ratio**, which characterizes the relative usefulness of a given measurement for a specific product. It compares the total variance of the measurement with the variance of the measurement error. **Bias Report** shows the average difference between the observed values and the standard. A graph of the average biases and a summary table are given for each X variable. **Linearity Study** performs a regression using the standard values as the X variable and the bias as the Y. This analysis examines the relationship between bias and the size of the part. Ideally, you want the slope to equal 0. A non-zero slope indicates your gauge performs differently with different sized parts. This option is only available when a standard variable is given.

AIAG Labels allows you to specify that quality statistics should be labeled in accordance with the AIAG standard, used extensively in automotive analyses.

A submenu for **Gauge Plots** lets you toggle **Gauge Mean** plots (the mean response by each main effect in the model) and **Gauge Std Dev** plots. If the model is purely nested, the graphs are displayed with a nesting structure. If the model is purely crossed, interaction graphs are shown. Otherwise, the graphs plot at each effect independently.

Figure 46.4 Gauge Mean plots for **2 Factors Crossed** example

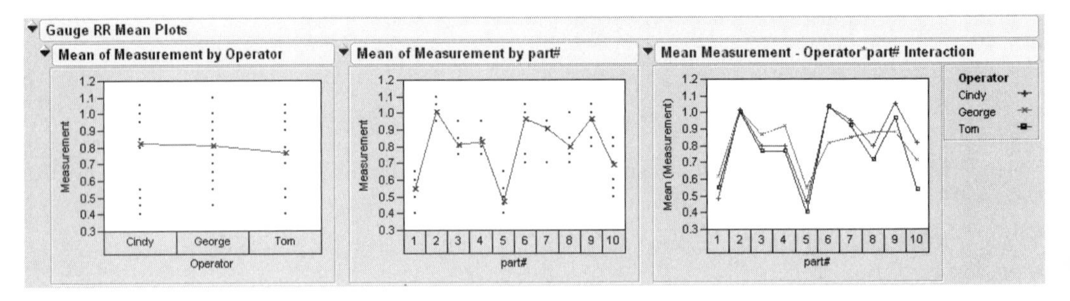

Figure 46.5 Gauge Std Dev plots for 2 Factors Crossed example

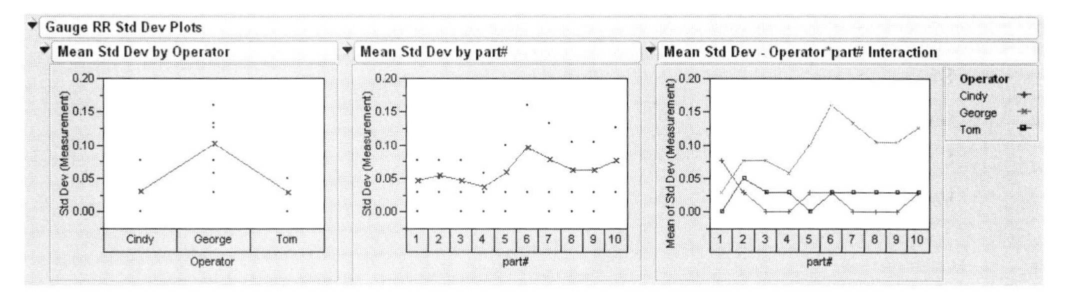

For the standard deviation plots, the red lines connect $\sqrt{\text{mean weighted variance}}$ for each effect.

Script has a submenu of commands available to all platforms that let you redo the analysis, or save the JSL commands for the analysis to a window or a file.

Variance Components

You can model the variation from measurement to measurement with a model, where the response is assumed to be a constant mean plus random effects associated with various levels of the classification. The exact model depends on how many new random values exist. For example, in a model where *B* is nested within *A*, multiple measurements are nested within both *B* and *A*, and there are *na•nb•nw* measurements. There are *na* random effects due to *A*, *na•nb* random effects due to each *nb B* levels within *A*, and *na•nb•nw* random effects due to each *nw* levels within *B* within *A*, which can be written:

$$y_{ijk} = u + Za_i + Zb_{ij} + Zw_{ijk}$$

The *Z*s are the random effects for each level of the classification. Each *Z* is assumed to have mean zero and to be independent from all other random terms. The variance of the response *y* is the sum of the variances due to each *z* component:

$$\text{Var}(y_{ijk}) = \text{Var}(Za_i) + \text{Var}(Zb_{ij}) + \text{Var}(Zw_{ijk})$$

The platform estimates all of these variances for models in the **Variance Components** command shown here.

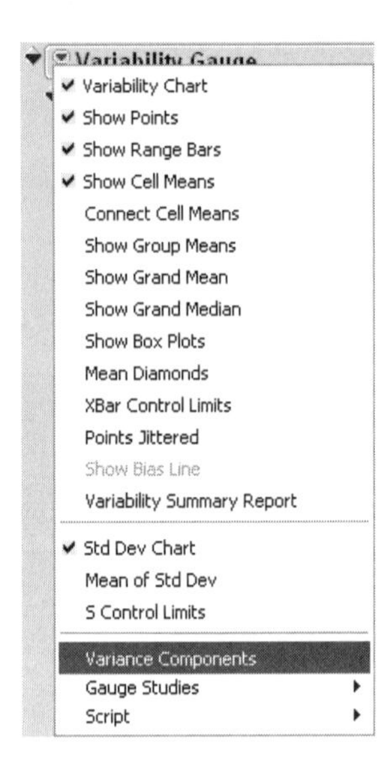

If you ask for **Variance Components** estimates and you have not specified the type of model in the launch dialog, then you get a dialog like the one shown in Figure 46.6.

Figure 46.6 Variance Component Dialog

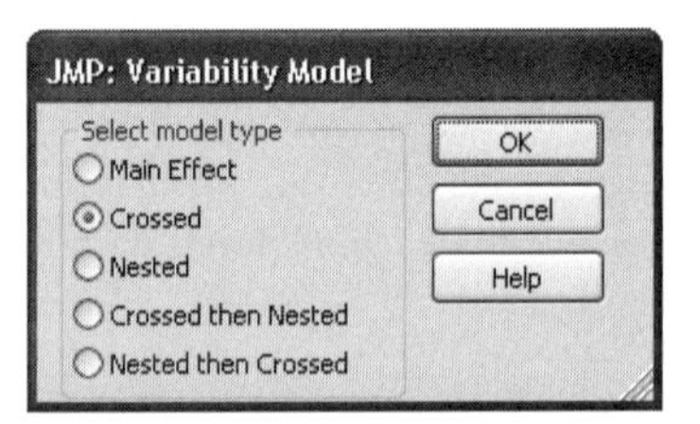

Table 46.1 "Models Supported by the Variability Charts Platform," p. 868, shows the models supported and what the effects in the model would be.'

Table 46.1 Models Supported by the Variability Charts Platform

Model	Factors	effects in the model
Crossed	1	A
	2	A, B, A*B
	3	A, B, A*B, C, A*C, B*C, A*B*C
	4	A, B, A*B, C, A*C, B*C, A*B*C, D, A*D, B*D, A*B*D, C*D, A*C*D, B*C*D, A*B*C*D,
	5	etc., for 5, 6 factors
Nested	1	A
	2	A, B(A)
	3	A, B(A), C(A,B)
	4	A, B(A), C(A,B), D(A,B,C)
Crossed then Nested	3	A, B, A*B, C(A,B)
Nested then Crossed	3	A, B(A), C, A*C, C*B(A)

Note: The Variability/Gauge Chart platform invokes REML for computing the variance components if the data is unbalanced. To force it to do REML even if it is balanced, set the JSL global variable `Vari Chart Force REML`to 1 by typint the following into a script window and executing the script.

Vari Chart Force REML = 1;

The Analysis of Variance shows the significance of each effect in the model. The Variance Components report shows the estimates themselves.Figure 46.7 shows these reports after selecting the **Crossed** selection in the dialog.

Figure 46.7 Analysis of Variance and Variance Components for Variability Analysis

Analysis of Variance

Source	DF	SS	Mean Square	F Ratio	Prob > F
Operator	2	0.054889	0.02744	1.3150	0.2931
part#	9	2.633583	0.29262	14.0209	<.0001*
Operator*part#	18	0.375667	0.02087	5.0425	<.0001*
Within	60	0.248333	0.00414		
Total	89	3.312472	0.03722		

Variance Components

Component	Var Component	% of Total	20 40 60 80	Sqrt(Var Con
Operator	0.00021914	0.5461		0.01
part#	0.03019444	75.2		0.17
Operator*part#	0.00557716	13.9		0.07
Within	0.00413889	10.3		0.06
Total	0.04012963	100.0		0.20

R&R Measurement Systems

Measurement systems analysis is an important step in any quality control application. Before you study the process itself, you need to make sure that the variation due to measurement errors is small relative to the variation in the process. The instruments that take the measurements are called gauges, and the analysis of their variation is a gauge study. If most of the variation you see comes from the measuring process itself, then you aren't learning reliably about the process. So, you do a measurement systems analysis, or gauge R&R study, to find out if the measurement system itself is performing well enough.

Gauge R&R results are available for all combinations of crossed and nested models, regardless of whether the model is balanced.

You collect a random sample of parts over the entire range of part sizes from your process. Select several operators randomly to measure each part several times. The variation is then attributed to the following sources:

- The *process variation*, from one part to another. This is the ultimate variation that you want to be studying if your measurements are reliable.
- The variability inherent in making multiple measurements—that is, *repeatability.* In Table 46.2 "Definition of Terms and Sums in Gauge R&R Analysis," p. 870, this is called the *within variation.*
- The variability due to having different operators measure parts—that is, *reproducibility.*

A Gauge R&R analysis then reports this attribution of the variation in terms of repeatability and reproducibility.

Table 46.2 Definition of Terms and Sums in Gauge R&R Analysis

Variances Sums	Term	Abbr.	Alternate Term
V(Within)	Repeatability	EV	Equipment Variation
V(Operator)+V(Oper*Part)	Reproducibility	AV	Appraiser Variation
V(Oper*Part)	Interaction	IV	Interaction Variation
V(Within)+V(Oper)+V(Oper*Part)	Gauge R&R	RR	Measurement Variation
V(Part)[a]		PV	Part Variation
V(Within)+V(Oper)+V(Oper*Part)+V(Part)	Total Variation	TV	Total Variation

a. A variance component attributed with Part that accounts for more than 90% of the variation indicates a good measurement system. This in turn suggests that working on fixing the part variability could be effective.

In the same way that a Shewhart control chart can identify processes that are going out of control over time, a variability chart can help identify operators, instruments, or part sources that are systematically different in mean or variance.

Some analysts study the variation using ranges instead of variances, because ranges are easier to hand calculate; but the calculations are easy when the computer is doing it. This is analogous to Shewhart $\bar{X}$-charts, which can be associated with Range charts or Std Dev charts. However, the JMP Variability Chart platform supports only the variance components approach.

Gauge R&R Variability Report

The Gauge R&R report shows measures of variation interpreted for a gauge study of operators and parts. When you select the **Gauge Studies > Gauge R R** option, the dialog shown here appears for you to change *K*, enter the tolerance for the process (which is the range of the specification limits, USL – LSL) and center the historical sigma.

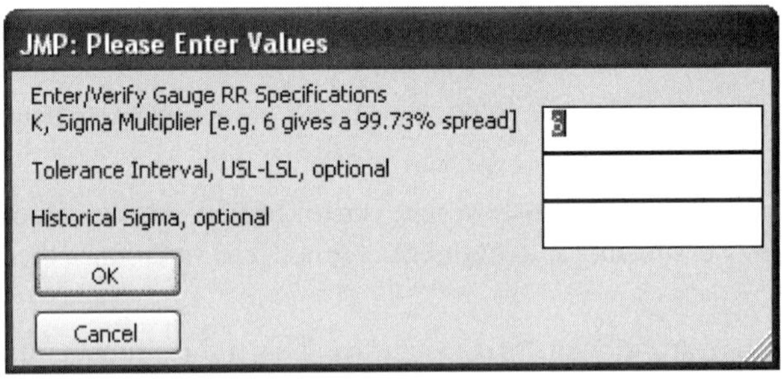

JMP: Please Enter Values
Enter/Verify Gauge RR Specifications
K, Sigma Multiplier [e.g. 6 gives a 99.73% spread]
Tolerance Interval, USL-LSL, optional
Historical Sigma, optional
OK
Cancel

Note that the tolerance interval and historical sigma are optional.

Also note that there is a platform preference (found in JMP's preferences) that allows you to set the default *K* that appears on this report.

In this example the report shows the statistics as a percentage of the tolerance interval (Upper Spec Limit minus Lower Spec Limit). The values are square roots of sums of variance components scaled by a value *k*, which is 6 in this example. Figure 46.8 shows the Gauge R&R report for the example shown previously.

Figure 46.8 Gauge R&R Report

Gauge R&R

Measurement Source		Variation (6*StdDev)		which is 6*sqrt of
Repeatability	(EV)	0.3860052	Equipment Variation	V(Within)
Reproducibility	(AV)	0.4568005	Appraiser Variation	V(Operator)+V(Operator*part#)
Operator		0.0888194		V(Operator)
Operator*part#		0.4480823		V(Operator*part#)
Gauge R&R	(RR)	0.5980524	Measurement Variation	V(Within)+V(Operator)+V(Operator*part#)
Part Variation	(PV)	1.0425929	Part Variation	V(part#)
Total Variation	(TV)	1.2019429	Total Variation	V(Within)+V(Operator)+V(Operator*part#)+V(part#)

6	k
49.7571	% Gauge R&R = 100*(RR/TV)
0.57362	Precision to Part Variation = RR/PV
2	Number of Distinct Categories = 1.41(PV/RR)

Using last column 'part#' for Part.

Variance Components for Gauge R&R

Component	Var Component	% of Total
Gauge R&R	0.00993519	24.76
Repeatability	0.00413889	10.31
Reproducibility	0.00579630	14.44
Part-to-Part	0.03019444	75.24

Barrentine (1991) suggests the following guidelines for an acceptable RR percent (percent measurement variation):

Figure 46.9 Acceptable Variation

< 10%	excellent
11% to 20%	adequate
21% to 30%	marginally acceptable
> 30%	unacceptable

- If a tolerance interval is given on the Gauge specifications dialog, a new column appears in the Gauge R&R report called '% of Tolerance". This column is computed as 100*(Variation/Tolerance). In addition, the Precision-to-Tolerance ratio is presented at the bottom of the report. It is useful in determining if precision is within the tolerance or capability interval. Ratios much larger than one are cause for concern. Those smaller than one indicate a system that measures well.
- If a historical sigma is given on the Gauge specifications dialog, a new column appears in the Gauge R&R report, named "% of Process". This column is defined as 100*(Variation/(K*Historical Sigma)).
- Number of distinct categories (NDC) is given in the summary table beneath the Gauge R&R report. NDC is defined as (1.41*(PV/RR)), rounded down to the nearest integer.

Bias

The **Gauge Studies > Bias** option shows a graph and summary table for each X variable, where the average bias, or differences between the observed values and the standard values, is given for each level of the X variable.

Note: The Bias option is only available when a Standard variable is given.

As an example, using the MSALinearity.JMP data table,

- Choose **Graph>Variability/Gauge Chart**.
- In the dialog, choose Response as the **Y, Response** variable.
- Choose Standard as the **Standard** variable.
- Choose Part as the **Grouping** variable.
- Click **OK**.
- From the platform menu, choose **Gauge Studies>Bias Report**.

The output in Figure 46.10 appears.

Figure 46.10 Bias Report

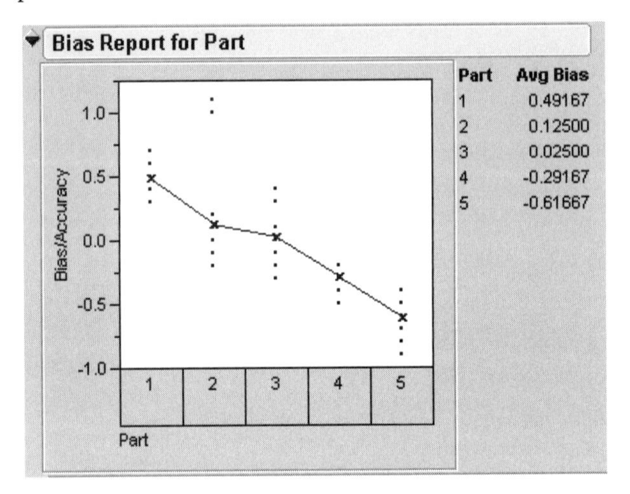

On the right is a table of average bias values for each part. Each of these bias averages is plotted on the graph along with the actual bias values for every part, so you can see the spread. In this example, Part number 2 has a wide bias range.

Linearity Study

The **Gauge Studies > Linearity Study** option performs a regression analysis using the standard variable as the X variable, and the bias as the Y. This analysis examines the relationship between bias and the size of the part. Ideally, you want a slope of 0. If the slope shows to be significantly different from zero, then you can conclude there is a significant relationship between the size of the part or variable measured as a standard, and the ability to measure.

The default ϖ-level for the linearity study is 0.05. To specify another α-level, hold down the Shift key when selecting **Linearity Study** from the **Gauge Studies** menu.

Note: The **Linearity Study** option is only available when a Standard variable is given.

Following the example above, after creating the Gauge output using the MSALinearity.JMP data table,

- From the platform menu, choose **Gauge Studies > Linearity Study**.
- In the dialog prompting **Specify Process Variation**, enter 16.5368.

The following output should appear:

Figure 46.11 Linearity Study

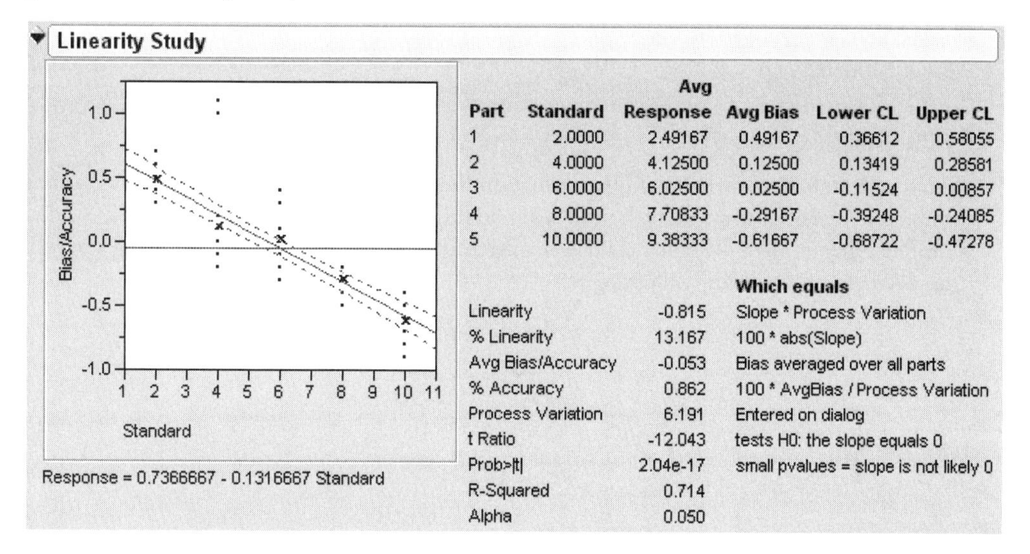

Part	Standard	Avg Response	Avg Bias	Lower CL	Upper CL
1	2.0000	2.49167	0.49167	0.36612	0.58055
2	4.0000	4.12500	0.12500	0.13419	0.28581
3	6.0000	6.02500	0.02500	-0.11524	0.00857
4	8.0000	7.70833	-0.29167	-0.39248	-0.24085
5	10.0000	9.38333	-0.61667	-0.68722	-0.47278

		Which equals
Linearity	-0.815	Slope * Process Variation
% Linearity	13.167	100 * abs(Slope)
Avg Bias/Accuracy	-0.053	Bias averaged over all parts
% Accuracy	0.862	100 * AvgBias / Process Variation
Process Variation	6.191	Entered on dialog
t Ratio	-12.043	tests H0: the slope equals 0
Prob>|t|	2.04e-17	small pvalues = slope is not likely 0
R-Squared	0.714	
Alpha	0.050	

Linearity represents how accurate measurements are over the range of (part) sizes.

% Accuracy shows how much bias from the gauge accounts for overall process variation.

t Ratio is the test statistic for testing that the slope = 0.

Prob>|t| is the *p*-value associated with testing that the slope = 0.

R-squared is a measure of the fit of the regression line, and represents the proportion of variation in the bias that is explained by the intercept and slope.

Alpha is the significance level.

Note: The equation of the line is shown directly beneath the graph.

Here, you see that the slope is -0.131667, and the *p*-value associated with the test on the slope is quite small (2.04e-17). From this, you can conclude that there is a significant relationship between the size of the parts and the ability to measure them. Looking at the output, the smaller parts appear to have a positive bias and are measured larger than they actually are, whereas the larger parts have a negative bias, with measurement readings being smaller than the actual parts.

Discrimination Ratio Report

The **Gauge Studies > Discrimination Ratio** option appends the **Discrim Ratio** table to the Variability report. The discrimination ratio characterizes the relative usefulness of a given measurement for a specific product. It compares the total variance of the measurement, M, with the variance of the measurement error, E. The discrimination ratio is computed for all main effects, including nested main effects. The Discrimination Ratio, D, is computed

$$D = \sqrt{2\left(\frac{M}{E}\right) - 1} = \sqrt{2\left(\frac{P}{E}\right) + 1} = \sqrt{2\left(\frac{P}{T-P}\right) + 1}$$

where

M = estimated measurement variance

P = estimated part variance

E = estimated variance of measurement error

T = estimated total variance

A rule of thumb is that when the Discrimination Ratio is less than 2, the measurement cannot detect product variation, so it would be best to work on improving the measurement process. A Discrimination Ratio greater than 4 adequately detects unacceptable product variation, implying a need for the improvement of the production process.

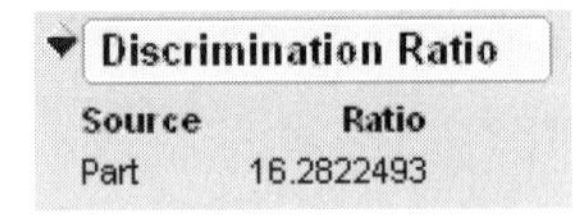
Discrimination Ratio

Source	Ratio
Part	16.2822493

Attribute Gauge Charts

Attribute gauge analysis gives measures of agreement across responses (raters, for example) in tables and graphs summarized by one or more X grouping variables. *Attribute data* is data where the variable of interest has a finite number of categories. Typically, data will have only two possible results (ex: pass/fail).

Data Organization

Data should be in the form where each rater is in a separate column, since agreement and effectiveness are both computed on these variables. In other words, if you want to compare agreement among raters, each rater needs to be in a separate (Y) column.

Any other variables of interest, (part, instrument, rep, and so on) should appear stacked in one column each. An optional standard column may be defined, which is then used in the Effectiveness Report. An example data table, contained in the sample data folder as AttributeGauge.jmp, is shown in Figure 46.12.

Figure 46.12 Attribute Data Example

AttributeGage
Attribute Chart
Columns (7/0)
Part
Standard
Code
A
B
C
RefValue
Rows
All rows 150

	Part	Standard	Code	A	B	C	RefValue
25	9	0	-	0	0	0	0.437817
26	9	0	-	0	0	0	0.437817
27	9	0	-	0	0	0	0.437817
28	10	1	+	1	1	1	0.515573
29	10	1	+	1	1	1	0.515573
30	10	1	+	1	1	1	0.515573
31	11	1	+	1	1	1	0.488905
32	11	1	+	1	1	1	0.488905
33	11	1	+	1	1	1	0.488905
34	12	0	x	0	0	0	0.559918
35	12	0	x	0	0	1	0.559918
36	12	0	x	0	0	0	0.559918
37	13	1	+	1	1	1	0.542704

Responses in the different Y columns may be character (Pass/Fail), numeric (0/1), or ordinal (low, medium, high).

Launching the Platform

To begin an attribute gauge analysis, select **Graph > Variability/ Gauge Chart**. For the Attribute-Gage.jmp example, fill in the dialog as shown in Figure 46.13.

Figure 46.13 Launching the Platform

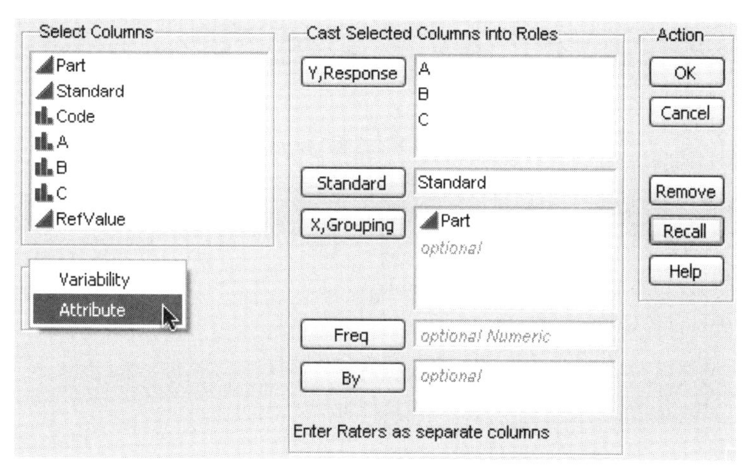

Attribute Gauge Plots

In the plots that appear, by default the % Agreement is plotted, where % Agreement is measured by comparing all pairs of rater by replicate combinations, for each part.

The first plot in Figure 46.14 uses all X (Grouping) variables on the x-axis, while the second plot contains all Y variables on the x-axis (typically the rater). For the top plot,

$$\%\text{Agreement for grouping variable } j = \frac{\sum_{i=1}^{\text{number of levels}} \binom{\text{number of responses for level } i}{2}}{\binom{(\text{number of raters}) \times (\text{number of reps})}{2}}$$

As an example, consider the following table of data for three raters, each having three replicates for one subject.

Table 46.3 Three replicates for raters A, B, and C

	A	B	C
1	1	1	1
2	1	1	0
3	0	0	0

Using this table,

$$\%\text{Agreement} = \frac{\binom{4}{2} + \binom{5}{2}}{\binom{9}{2}} = \frac{16}{36} = 0.444$$

For the bottom plot in Figure 46.14,

$$\%\text{Agreement for rater } k = \frac{\sum_{i=1}^{n}\left(\sum_{j=1}^{r_i} \begin{pmatrix}\text{number of uncounted matching levels} \\ \text{for this rater k within part i for rep j}\end{pmatrix}\right)}{\sum_{i=1}^{n}\left(\sum_{j=1}^{r_i} m_i \times r_i - j\right)}$$

where

n = number of subjects (grouping variables)

r_i = number of reps for subject i

m_i=number of raters for subject i

In the example using the data in Table 46.3,

$$\%\text{Agreement [A]} = \%\text{Agreement [B]} = \frac{4+3+3}{8+7+6} = \frac{10}{21} = 0.476$$

$$\%\text{Agreement[C]} = \frac{4+3+2}{8+7+6} = \frac{9}{21} = 0.4286$$

Figure 46.14 Agreement Percentage Plots

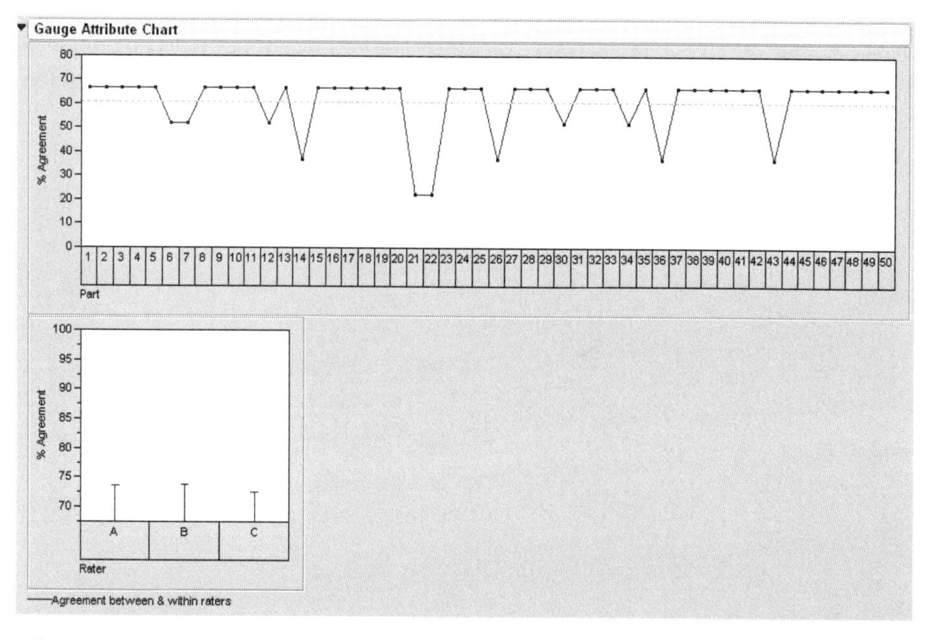

Note: If a standard column is defined, the option, **Show Effectiveness Means** is available, which shows the agreement to the standard (*i.e.* effectiveness).

Agreement

The Agreement Report gives agreement summarized for every rater as well as overall agreement.

The Agreement Comparisons give Kappa statistics for each *Y* variable compared with all other *Y* variables. In other words, each rater is compared with all other raters.

The Agreement within raters report shows the number of items that were inspected. The confidence intervals are Score confidence intervals, as suggested by Agresti, (1998). The Number Matched is defined as the sum of number of items inspected, where the rater agreed with him or herself on each inspection of an individual item. The Rater Score is Number Matched divided by Number Inspected.

The simple kappa coefficient is a measure of interrater agreement.

$$\hat{\kappa} = \frac{P_0 - P_e}{1 - P_e}$$

where

$$P_0 = \sum_i p_{ii}$$

and

$$P_e = \sum_i p_{i.} p_{.i}$$

Viewing the two response variables as two independent ratings of the n subjects, the kappa coefficient equals +1 when there is complete agreement of the raters. When the observed agreement exceeds chance agreement, the kappa coefficient is positive, with its magnitude reflecting the strength of agreement. Although unusual in practice, kappa is negative when the observed agreement is less than chance agreement. The minimum value of kappa is between -1 and 0, depending on the marginal proportions.

The asymptotic variance of the simple kappa coefficient is estimated by the following:

$$\text{var} = \frac{A+B-C}{(1-P_e)^2 n}$$

where

$$A = \sum_i p_{ii}\left[1-(p_{i.}+p_{.i})(1-\hat{\kappa})\right]$$

$$B = (1-\hat{\kappa})^2 \sum_{i \neq j}\sum p_{ij}(p_{.i}+p_{j.})^2$$

and

$$C = \left[\hat{\kappa} - P_e(1-\hat{\kappa})\right]^2$$

The Kappa's are plotted and the standard errors are also given.

Note: The Kappa statistic in the Attribute charts is given even when the levels of the variables are not the same.

Categorical Kappa statistics (Fleiss 1981) are found in the Agreement Across Categories report.

For

n = number of subjects reviewed

m = number of ratings per subject

k=number of categories in the response

x_{ij} = number of ratings on subject i (i = 1, 2,..., k) into category j (j=1, 2,...k)

The individual category Kappa is

$$\hat{\kappa}_j = \frac{\sum_{i=1}^{n} x_{ij}(m-x_{ij})}{nm(m-1)(\bar{p}_j\bar{q}_j)} \qquad \text{where} \qquad \bar{p}_j = \frac{\sum_{i=1}^{n} x_{ij}}{nm} \qquad \bar{q}_j = 1-\bar{p}_j$$

and the overall kappa is

$$\hat{\kappa} = \frac{\sum_{j=1}^{k} \bar{q}_j\bar{p}_j\hat{\kappa}_j}{\sum_{j=1}^{k} \bar{p}_j\bar{q}_j}$$

The variance of $\hat{\kappa}$ and $\hat{\bar{\kappa}}$ are

$$\mathrm{var}(\hat{\kappa}_j) = \frac{2}{nm(m-1)}$$

$$\mathrm{var}\hat{\bar{\kappa}} = \frac{2}{\left(\sum_{j=1}^{k} \bar{p}_j \bar{q}_j\right)^2 nm(m-1)} \times \left[\left(\sum_{j=1}^{k} \bar{p}_j \bar{q}_j\right)^2 - \sum_{j=1}^{k} \bar{p}_j \bar{q}_j (\bar{q}_j - \bar{p}_j)\right]$$

The standard errors of $\hat{\kappa}$ and $\hat{\bar{\kappa}}$ are only shown when there are an equal number to ratings per subject.

Figure 46.15 Agreement Reports

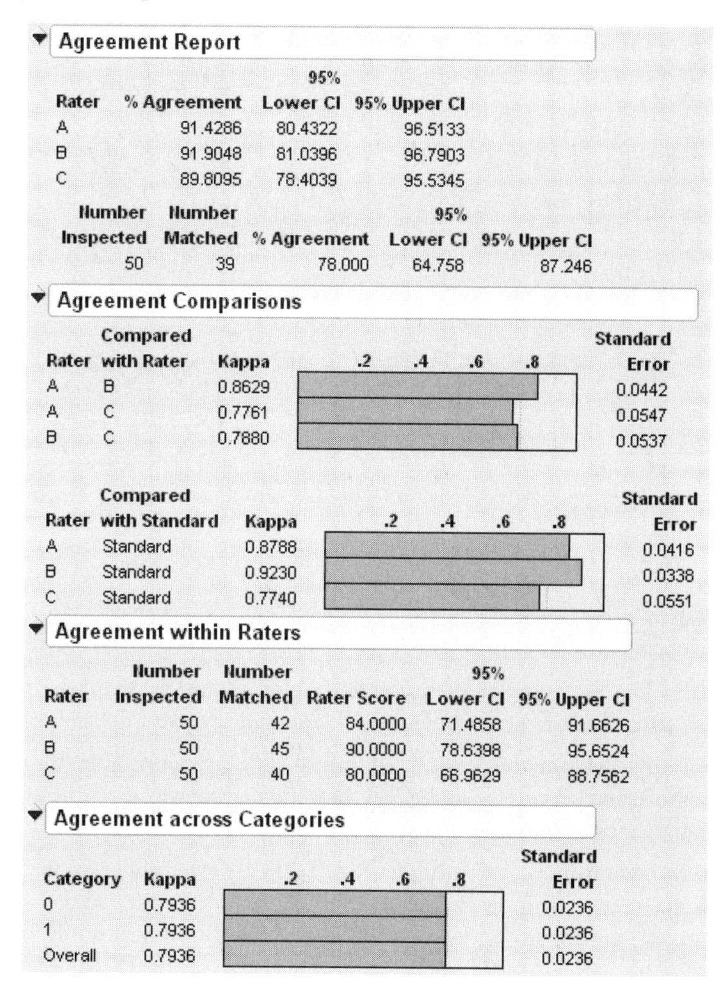

Agreement Report

Rater	% Agreement	95% Lower CI	95% Upper CI
A	91.4286	80.4322	96.5133
B	91.9048	81.0396	96.7903
C	89.8095	78.4039	95.5345

Number Inspected	Number Matched	% Agreement	95% Lower CI	95% Upper CI
50	39	78.000	64.758	87.246

Agreement Comparisons

Rater	Compared with Rater	Kappa	Standard Error
A	B	0.8629	0.0442
A	C	0.7761	0.0547
B	C	0.7880	0.0537

Rater	Compared with Standard	Kappa	Standard Error
A	Standard	0.8788	0.0416
B	Standard	0.9230	0.0338
C	Standard	0.7740	0.0551

Agreement within Raters

Rater	Number Inspected	Number Matched	Rater Score	95% Lower CI	95% Upper CI
A	50	42	84.0000	71.4858	91.6626
B	50	45	90.0000	78.6398	95.6524
C	50	40	80.0000	66.9629	88.7562

Agreement across Categories

Category	Kappa	Standard Error
0	0.7936	0.0236
1	0.7936	0.0236
Overall	0.7936	0.0236

If a standard variable is given, an additional Kappa report is given that compares every rater with the standard.

Effectiveness Report

The Effectiveness Report appears when a standard variable is given.

Figure 46.16 Effectiveness Report

Effectiveness Report

Agreement Counts

Rater	Correct(0)	Correct(1)	Total Correct	Incorrect(0)	Incorrect(1)	Grand Total
A	45	97	142	3	5	150
B	45	100	145	3	2	150
C	42	93	135	6	9	150

Effectiveness

Rater	Effectiveness	95% Lower CI	95% Upper CI	Error rate
A	94.6667	89.8296	97.2730	0.0533
B	96.6667	92.4348	98.5680	0.0333
C	90.0000	84.1565	93.8459	0.1000

Misclassifications

Standard Level	0	1
0	.	16
1	12	.
Other	0	0

Conformance Report

Rater	P(False Alarms)	P(Misses)
A	0.0490	0.0625
B	0.0196	0.0625
C	0.0882	0.1250

Assumptions	
NonConform =	0
Conform =	1

The Agreement Counts table gives cell counts on the number correct and incorrect for every level of the standard.

Effectiveness is defined as

$$\frac{\text{\# of correct decisions}}{\text{Total \# of opportunities for a decision}}$$

This means that if rater A sampled every part four times, and on the sixth part, one of the decisions did not agree, the other three decisions would still be counted as correct decisions.

Note: This is different from the MSA 3rd edition, as all four opportunities for rater A by part 6 would be counted as incorrect. We feel including all inspections separately gives the user more information about the overall inspection process.

In the **Effectiveness Report**, 95% confidence intervals are given about the effectiveness. JMP is using Score Confidence Intervals. In recent literature, it has been demonstrated that score confidence intervals provide more sensible answers, particularly where observations lie near the boundaries. (see Agresti, 2002)

The **Conformance Report** displays the probability of false alarms, and the probability of misses, where

- False Alarm = part is determined non-conforming, when it actually is conforming
- Miss = part is determined conforming, when it actually is not conforming.

The lowest sorted value is assumed to be the nonconforming value, while the second lowest sorted value is assumed to be the conforming value.

The conformance report is only displayed when the standard had only two levels.

The **Misclassification Table** shows the incorrect labeling with the *y*-axis representing the levels of the standard, or accepted reference value, and the *x*-axis containing the levels given by the raters.

Note: Missing values are treated as a separate category in this platform. If missing values are removed, different calculations are performed than if the missing values are excluded. We recommend excluding all rows containing missing values.

Chapter 47

Ishikawa Diagrams

The Diagram Platform

The **Diagram** platform is used to construct *Ishikawa charts*, also called *fishbone charts*, or *cause-and-effect diagrams*.

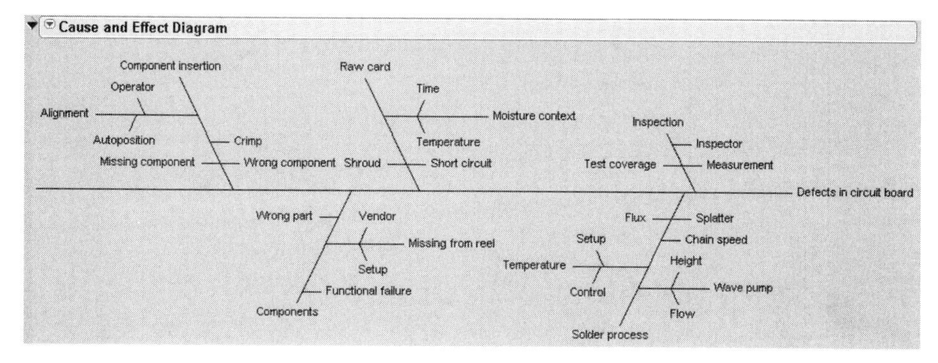

These charts are useful to organize the sources (causes) of a problem (effect), perhaps for brainstorming, or as a preliminary analysis to identify variables in preparation for further experimentation.

Contents

Preparing the Data for Diagramming

To produce the diagram, begin with data in two columns of a data table.

Figure 47.1 Ishikawa Diagram

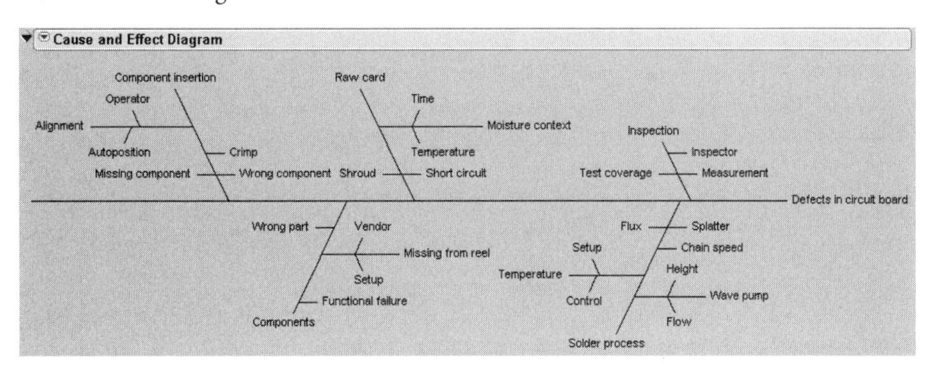

Some sample data (Montgomery, 1996) is shown in Figure 47.2, from the Ishikawa.jmp sample data table.

Figure 47.2 A Portion of the Ishikawa Sample Data

	Parent	Child
1	Defects in circuit board	Inspection
2	Defects in circuit board	Solder process
3	Defects in circuit board	Raw card
4	Defects in circuit board	Components
5	Defects in circuit board	Component insertion
6	Inspection	Measurement
7	Inspection	Test coverage
8	Inspection	Inspector
9	Solder process	Splatter
10	Solder process	Flux
11	Solder process	Chain speed
12	Solder process	Temperature
13	Solder process	Wave pump
14	Temperature	Setup

Notice that the Parent value “Defects in circuit board” (rows 1–5) has many causes, listed in the Child column. Among these causes is “Inspection”, which has its own children causes listed in rows 6–8.

Examine the plot in Figure 47.1 to see “Defects in circuit board” as the center line, with its children branching off above and below. “Inspection” is one of these branches, which has its own branches for its child causes.

Select **Diagram** to bring up the launch dialog (Figure 47.3). Provide the columns that represent the **X, Parent** (Parent in the sample data) and the **Y, Child** (Child in the sample data).

Figure 47.3 Diagram Launch Dialog

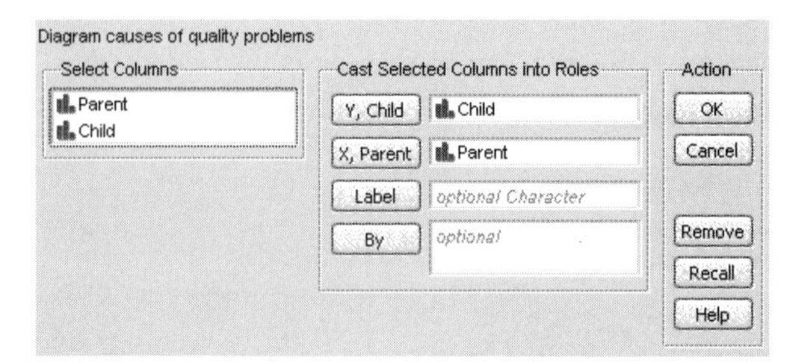

Including a variable in the **By** column produces separate diagrams for each value of the **By** variable. **Label** columns cause the text from them to be included in the nodes of the diagram.

Chart Types

The Diagram platform can draw charts of three types: Fishbone (Figure 47.4), Hierarchy (Figure 47.5), and Nested (Figure 47.6).

Figure 47.4 Fishbone Chart

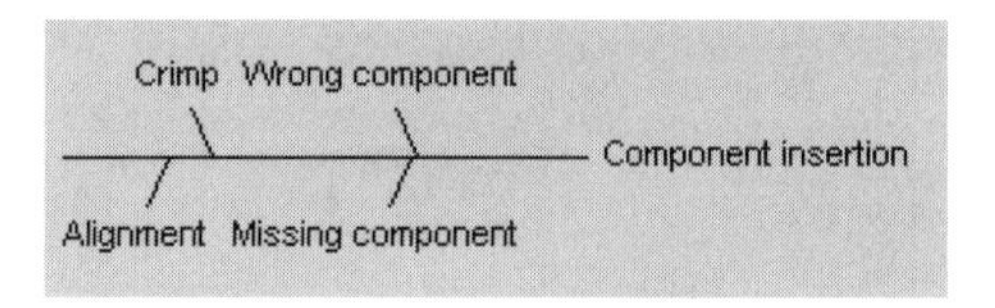

Figure 47.5 Hierarchy Chart

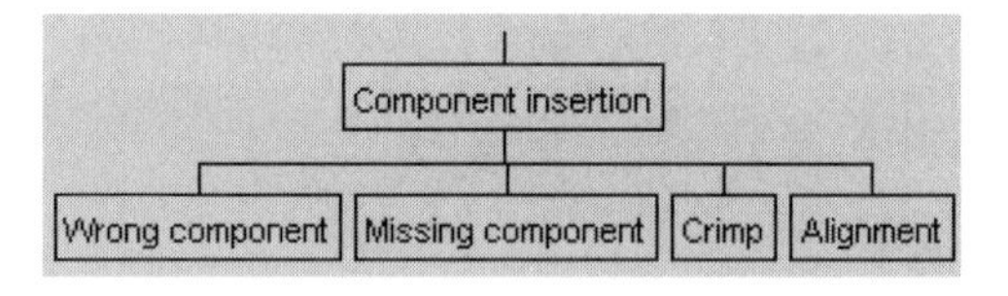

Figure 47.6 Nested Chart

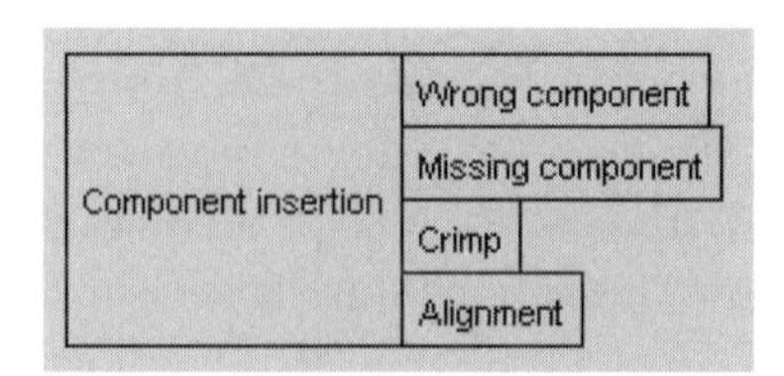

To change the chart type, right-click (control-click on the Macintosh) on any node line in the chart. The nodes highlight as the mouse passes over them. Then, select the desired chart from the **Chart Type** menu.

Building a Chart Interactively

Right click on any node in the chart to bring up a context menu (Figure 47.7) that allows a chart to be built piece-by-piece. You can edit new nodes into the diagram using context menus, accessible by right-clicking on the diagram itself. You can even create a diagram without a data table, which starts with the default diagram shown here.

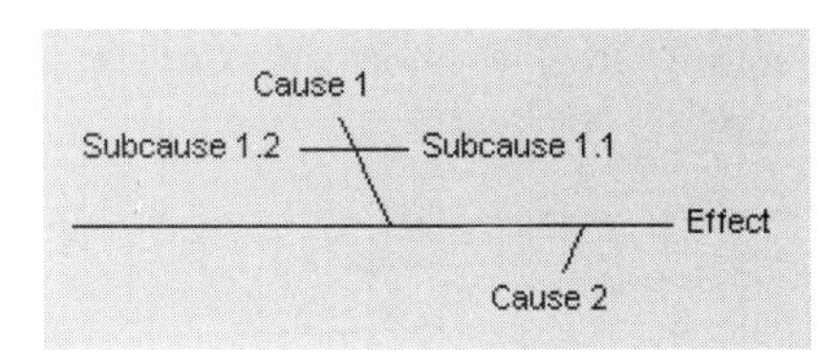

Figure 47.7 Diagram Platform Node Menu

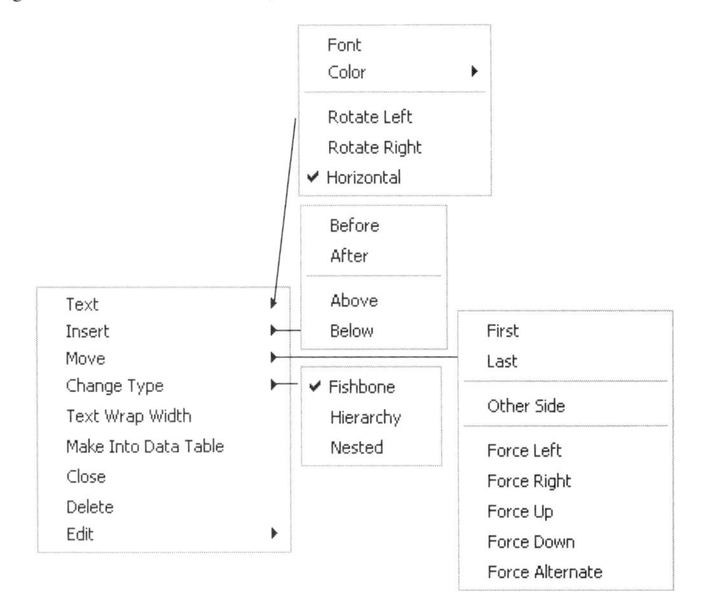

Text Menu

There are two ways to change the appearance of text in a diagram

- Right-click on highlighted node in the chart. This brings up the menu shown in Figure 47.7 that has the following options:

  Font brings up a dialog to select the font of the text.

  Color brings up a dialog to select the color of the text.

  Rotate Left, Rotate Right, Horizontal rotates the text to be horizontal, rotated 90 degrees left, or 90 degrees right (see Figure 47.8).

Figure 47.8 Rotated Text Example

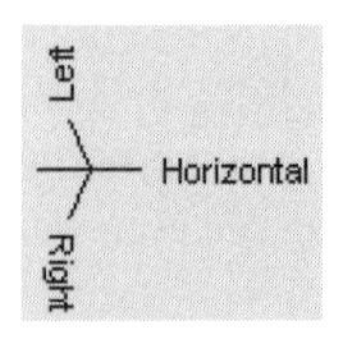

- Right-click on a word in the chart. This brings up a smaller menu that has the following options

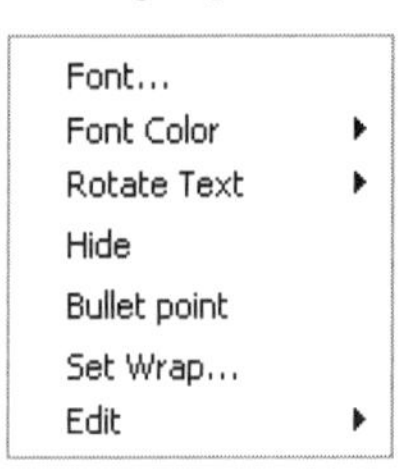

Font brings up a dialog to select the font of the text.

Font Color brings up a dialog to select the color of the text.

Rotate Text rotates the text to be **Horizontal**, rotated 90 degrees **Left**, or 90 degrees **Right** (see Figure 47.8).

Set Wrap brings up a dialog that allows you to set the text wrap width in pixels

Insert Menu

The **Insert** menu allows you to insert items onto existing nodes.

Insert > Before inserts a new node at the same level of the highlighted node. The new node appears before the highlighted node. For example, inserting "Child 1.5" before "Child 2" results in the following chart.

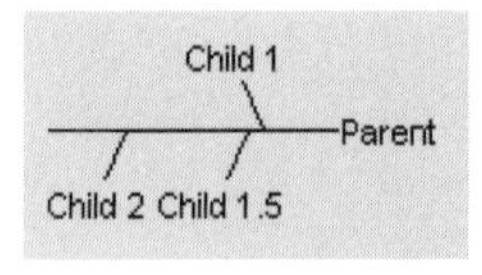

Insert > After inserts a new node at the same level of the highlighted node. The new node appears after the highlighted node. For example, inserting "Child 3" after "Child 2" results in the following chart.

Insert > Above inserts a new node at a level above the current node. For example, inserting "Grandparent" above "Parent" results in the following chart.

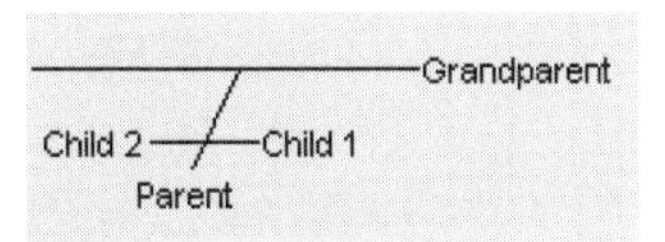

Insert > Below inserts a new node at a level below the current node. For example, inserting "Grandchild 1" below "Child 2" results in the following chart.

Child 1
Parent
Grandchild 1
Child 2

Move Menu

The Move menu allows you to customize the appearance of the diagram by giving you control over which side the branches appear on.

Move > First moves the highlighted node to the first position under its parent. For example, after switching sides, telling "Child 2" to **Move First** results in the following chart.

Child 1 Child 2
Parent

Move > Last moves the highlighted node to the last position under its parent.

Move > Other Side moves the highlighted node to the other side of its parent line. For example, telling "Child 2" to switch sides results in the following chart.

Child 2 Child 1
Parent

Force Left makes all horizontally-drawn elements appear to the left of their parent.

Force Right makes all horizontally-drawn elements appear to the right of their parent.

Force Up makes all vertically-drawn elements appear above their parent.

Force Down makes all vertically-drawn elements appear below their parent.

Force Alternate is the default setting, which draws siblings on alternate sides of the parent line.

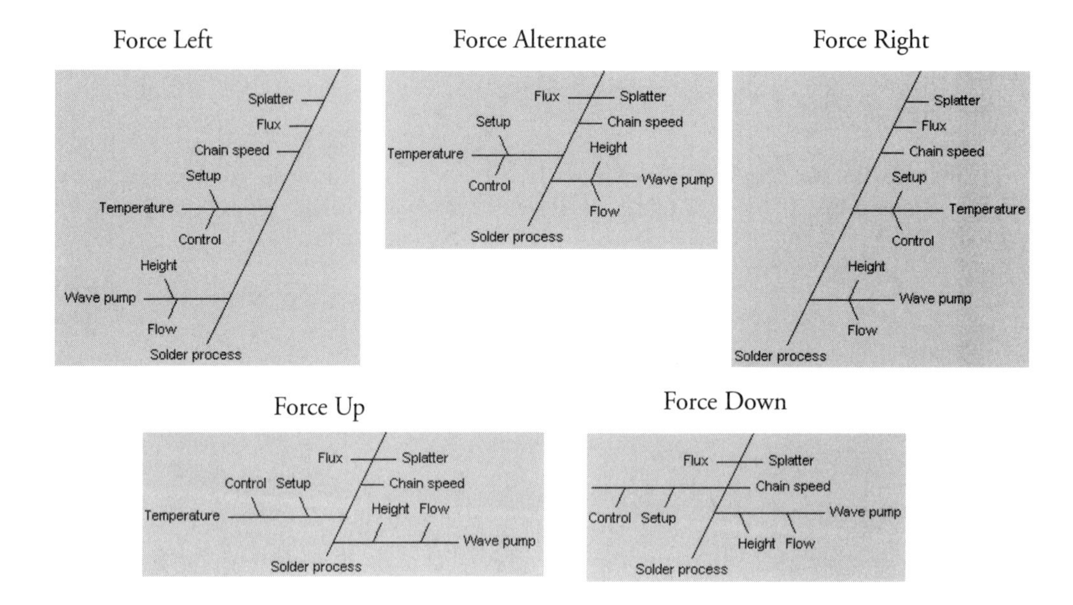

Other Menu Options

Change Type changes the entire chart type to **Fishbone**, **Hierarchy**, or **Nested**.

Text Wrap Width brings up a dialog that allows you to specify the width of labels where text wrapping occurs.

Make Into Data Table converts the currently highlighted node into a data table. Note that this can be applied to the whole chart by applying it to the uppermost level of the chart.

Close is a toggle to alternately show or hide the highlighted node.

Delete deletes the highlighted node and everything below it.

Drag and Drop

Nodes in a Diagram can be manipulated by drag and drop. Grab any outlined section of a Diagram and drag to a new location, using the highlighted bar as a guide to tell you where the node will appear after being dropped.

For example, the following picture shows the element Setup (initially a child of Temperature) being dragged to a position that makes it a child of Solder Process—in essence, equivalent to Flux, Splatter, and Chain Speed.

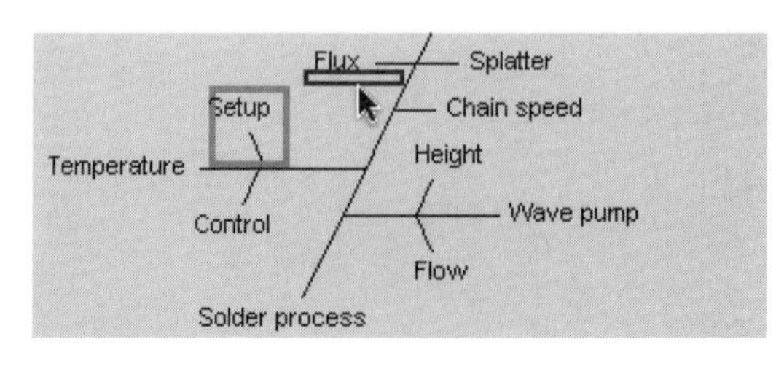

The next example shows two ways to make the Inspection tree a child of the Solder Process tree. Note that both drag operations result in the same final result.

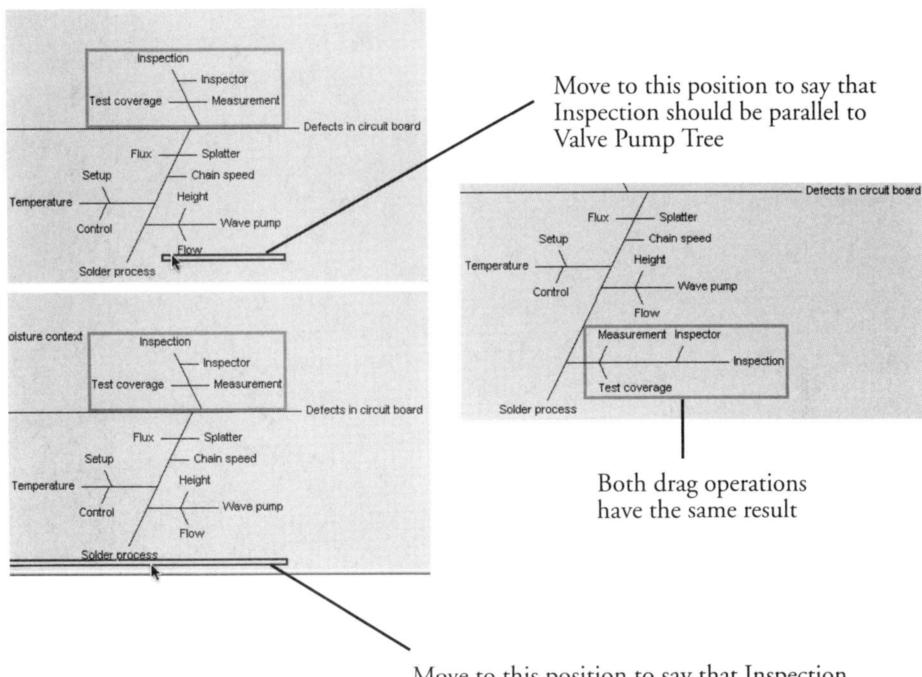

These principles extend to nested and Hierarchical charts. The following example shows the two ways to move Temperature from its initial spot (under Moisture Content) to a new position, under Missing from reel.

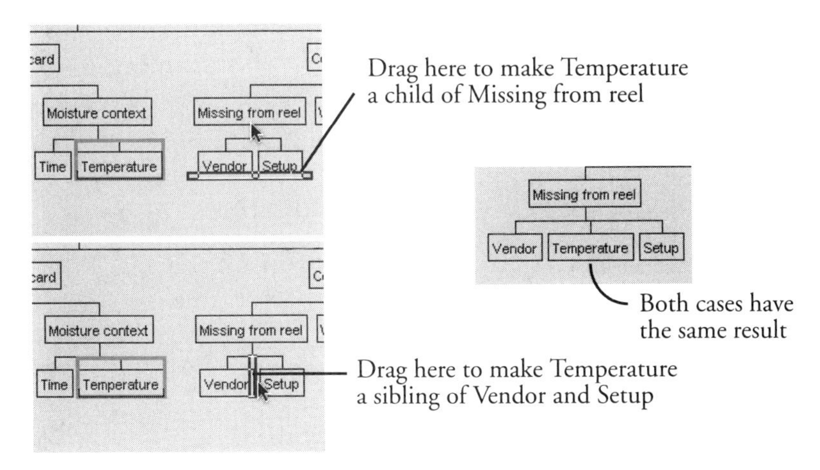

Chapter 48

Cell-Based Plots

Parallel Plots and Cell Plots

The plots in this chapter are all visualizations of each cell in a data table. Parallel plots draw connected line segments for each row. Cell plots, also known as heat maps, display the data table as a matrix of rectangular colors.

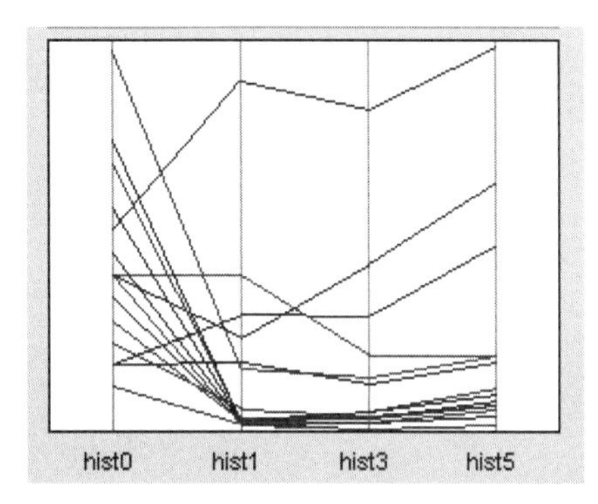

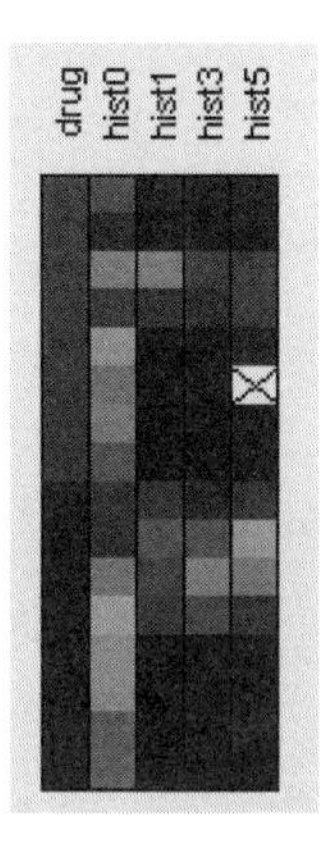

Contents

Introduction

Both styles of plots in this chapter display all the cells from a data table. Both are useful for seeing latent patterns in data.

Cell plots are direct representations of a data table, drawn as a rectangular array of cells with each cell corresponding to a data table entry. Representative colors are assigned to each column based on the range and type of values found in the column. The array is then drawn with a one-to-one correspondence of a colored cell representing each data table entry. Figure 48.1 shows a cell plot consisting of drug type and four histamine levels.

Figure 48.1 Cell plot of Dogs.jmp values

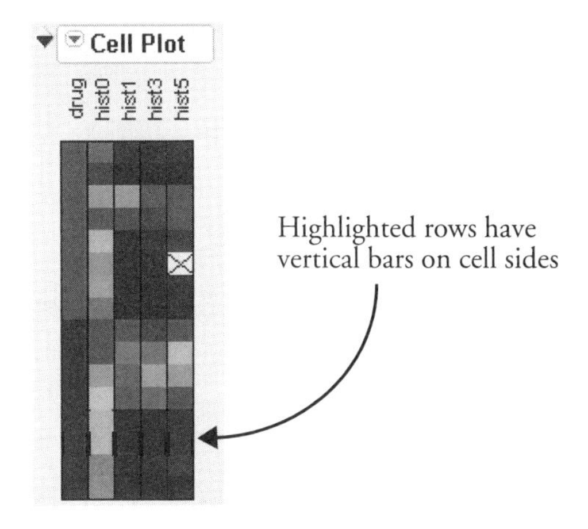

Parallel Coordinate Plots

Parallel Coordinate Plots show connected line segments representing each row of a data table. The parallel plot in Figure 48.2 shows a parallel plot of four histamine variables from the Dogs.jmp data table. Values from a single observation are connected.

Parallel coordinate plots are one of a few plots that show any number of columns in one plot. However, the drawback is that relationships between columns are only evident in neighboring columns, or if through color or selection it is possible to track groups of lines across several columns. They were initially developed by Inselberg (1985) and later popularized by Wegman (1990).

Figure 48.2 Parallel Plot of histamine variables from Dogs.jmp data table

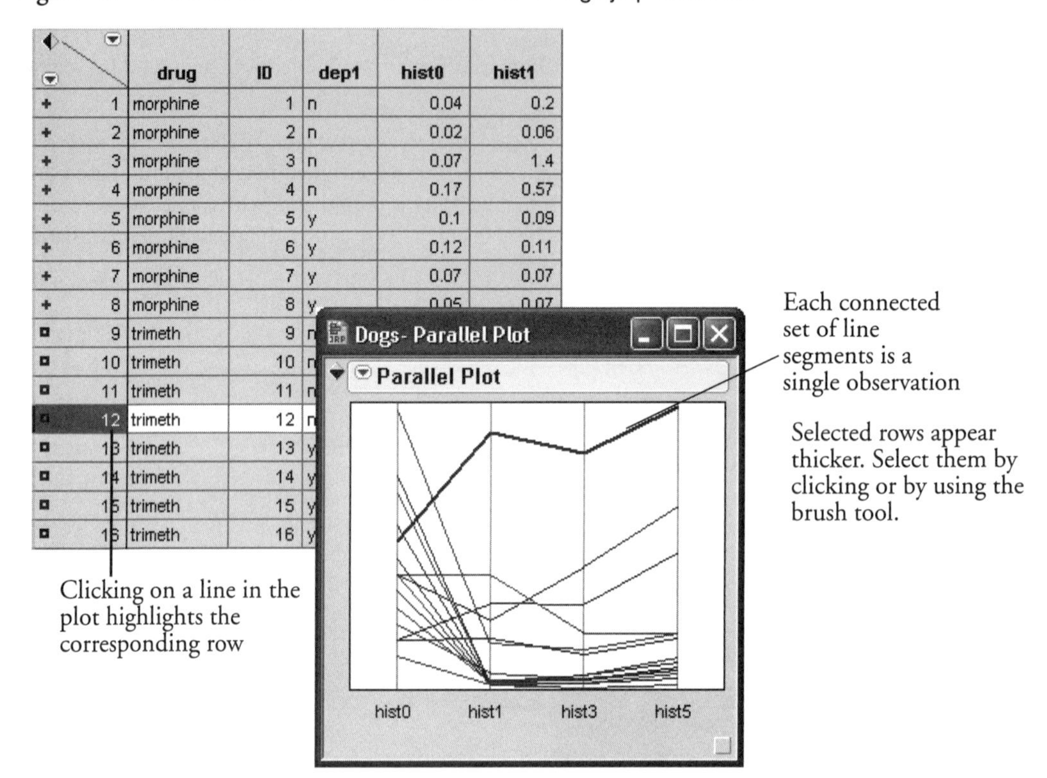

Launching the Platform

To draw a Parallel Coordinate Plot, select **Graph > Parallel Plot**. This produces the launch dialog shown in Figure 48.3. We have filled in scores from Decathlon.jmp.

Figure 48.3 Parallel Coordinate Plot Launch Dialog

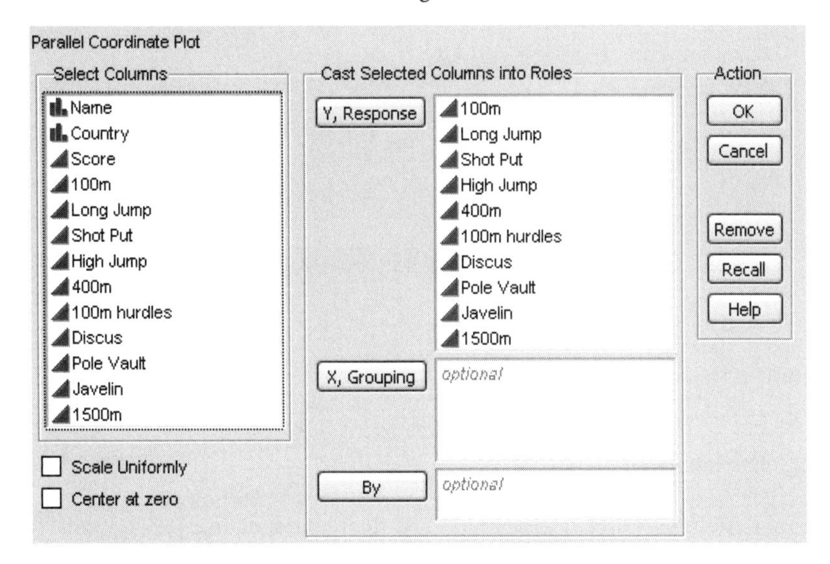

Y, Response variables appear on the horizontal axis of the plot. These values are the ones actually plotted and connected in the output. If you assign a value to the **X, Grouping** role, a separate plot is produced for each of its levels.

The following two options appear on the launch dialog.

Scale Uniformly causes all of the plots to be represented on the same scale. In other words, vertical distances represent the same quantities for each plotted column.This allows a *y*-axis to be displayed. Without this option, each variable is on a different scale.

Center at Zero requests that all the plots be centered at zero.

Platform Options

There are only two options available in the platform drop-down menu.

Show Reversing Checkboxes allows you to reverse the scale for one or more variables. In our Decathlon.jmp example, the jump and field events are larger-is-better value, whereas the running events are smaller-is-better. By reversing the running events, the better values are consistently on the high side.

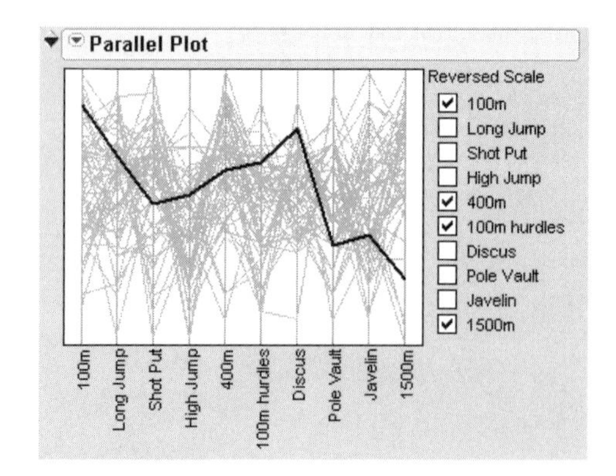

Script is the standard JMP script menu.

Examples

It is useful to see some familiar scatter plots and their corresponding parallel plots. In each of the following pictures, a parallel plot is shown on the left, with its scatterplot on the right.

The following relationship is a strong positive correlation. Notice the coherence of the lines in the parallel plot.

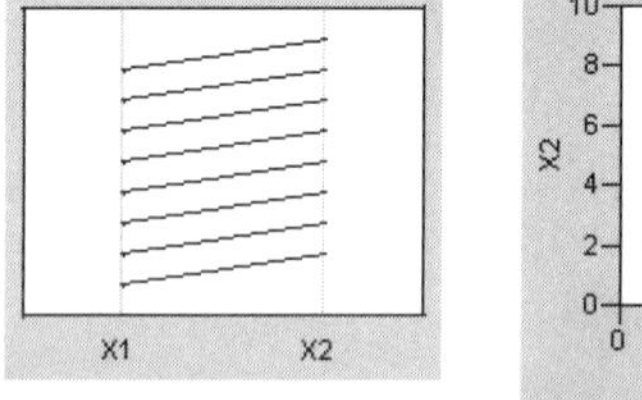

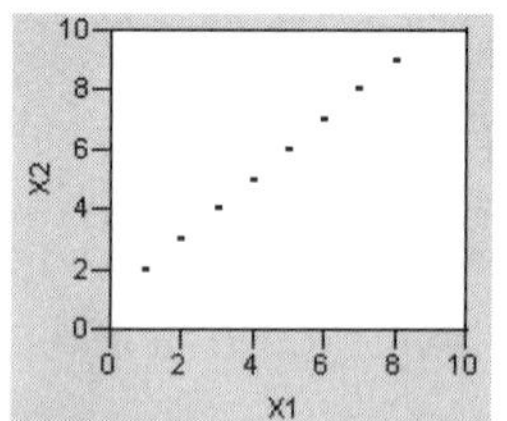

A strong negative correlation, by contrast, shows a narrow neck in the parallel plot.

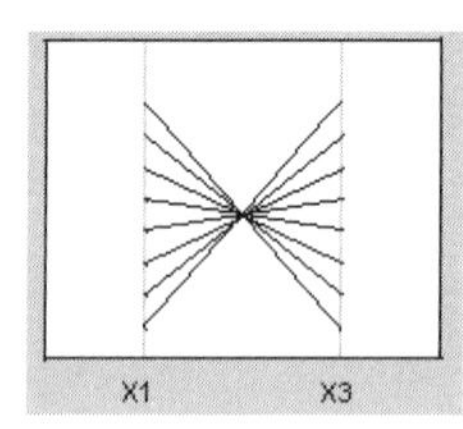

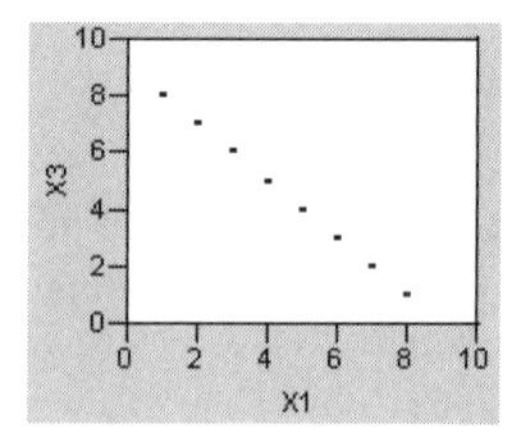

Now, consider a case that encompasses both situations: two groups, both strongly collinear. One has a positive slope, the other a negative. In the following picture, we have highlighted positively-sloped group.

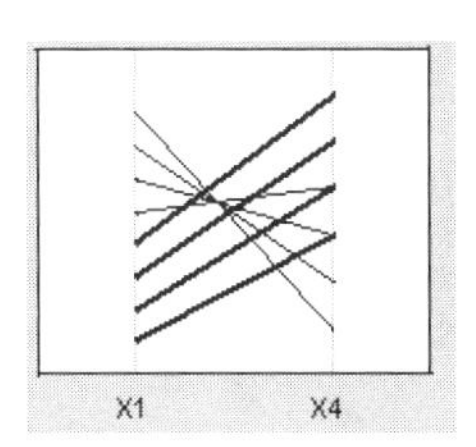

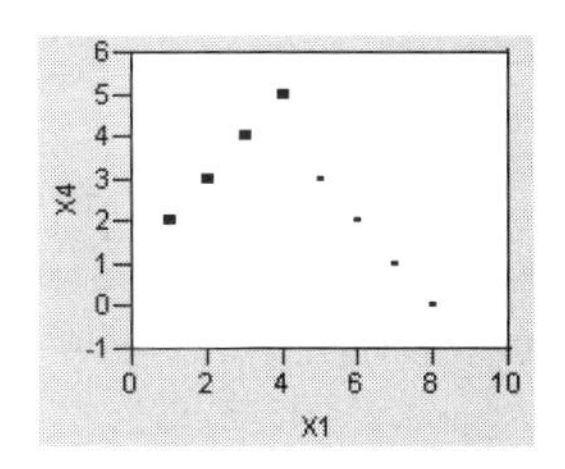

Finally, consider the case of a single outlier. The Parallel plot shows a general coherence among the lines, with a noticeable exception.

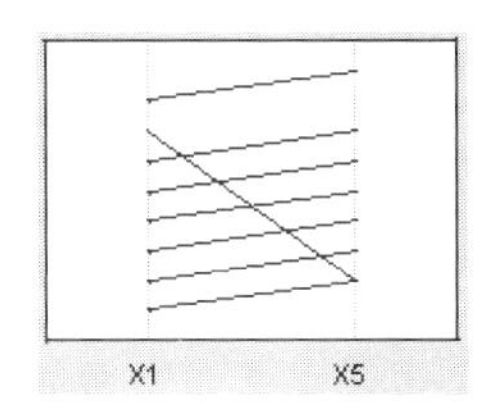

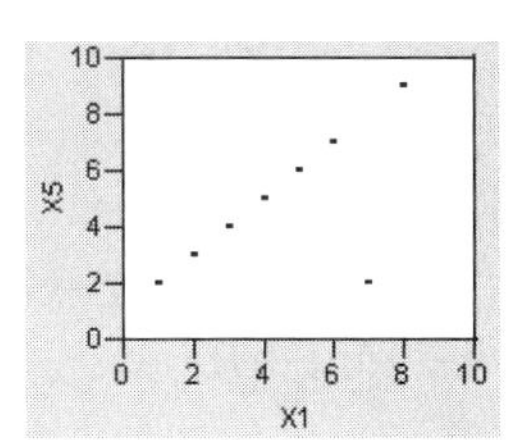

Spec Limits

If the columns in a Parallel Plot have **Spec Limits** properties, they are shown as red lines in the plot.

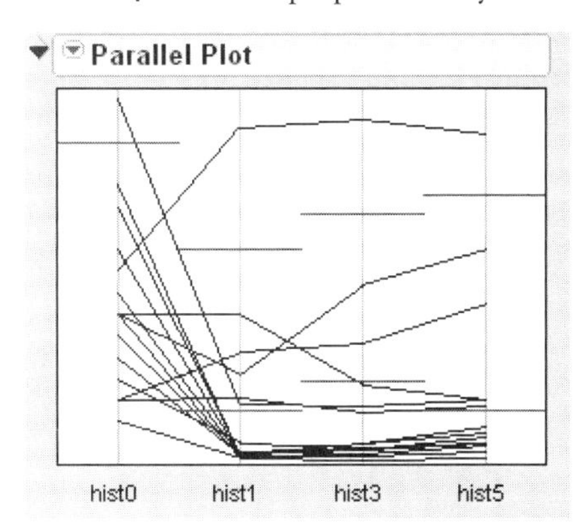

Cell Plots

Cell plots are direct representations of a data table, drawn as a rectangular array of cells with each cell corresponding to a data table entry. Colors are assigned to each cell based on the range and type of values found in the column.

- Nominal variables use a distinct color for each platform. As in other platforms, the colors cycle through twelve hues: red, green, blue, yellow, orange, blue-green, purple, yellow, cyan, magenta, yellow-green, blue-cyan, fuchsia. Nominal and ordinal colorings can be customized by using the **Value Colors** property of data columns, accessed through the **Column Info** command.

- Continuous variables are assigned a gradient of colors to show the smooth range of values of the variable.
- Ordinal variables are scaled like continuous, in order.

The array is then drawn with a one-to-one correspondence of a colored cell representing each data table entry. Figure 48.4 shows the cell plot for the Dogs.jmp data table using the drug, hist0, hist1, hist3, and hist5 columns.

Figure 48.4 Dogs.jmp cell plot

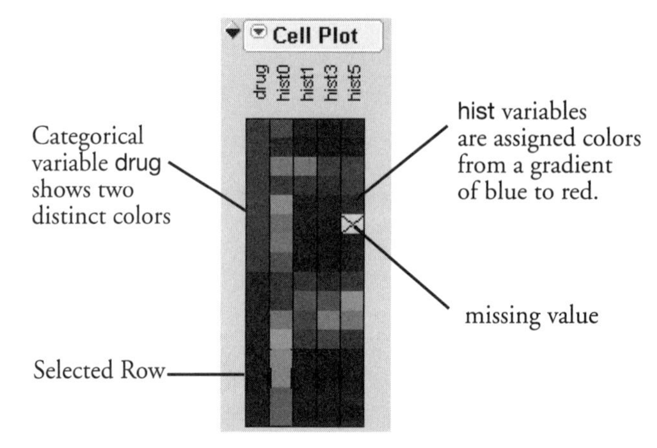

Cell plots were popularized by Genomics applications where large numbers of values for gene expression levels needed to be browsed.

The following picture shows SAT scores (split into verbal and mathematics portions) for all 50 US states for 2002, 2001, and 1997. Note that the cell plot, sorted by 2002 verbal score, shows Hawaii as an "outlier", since it has a strikingly different pattern for its math scores and its verbal scores.

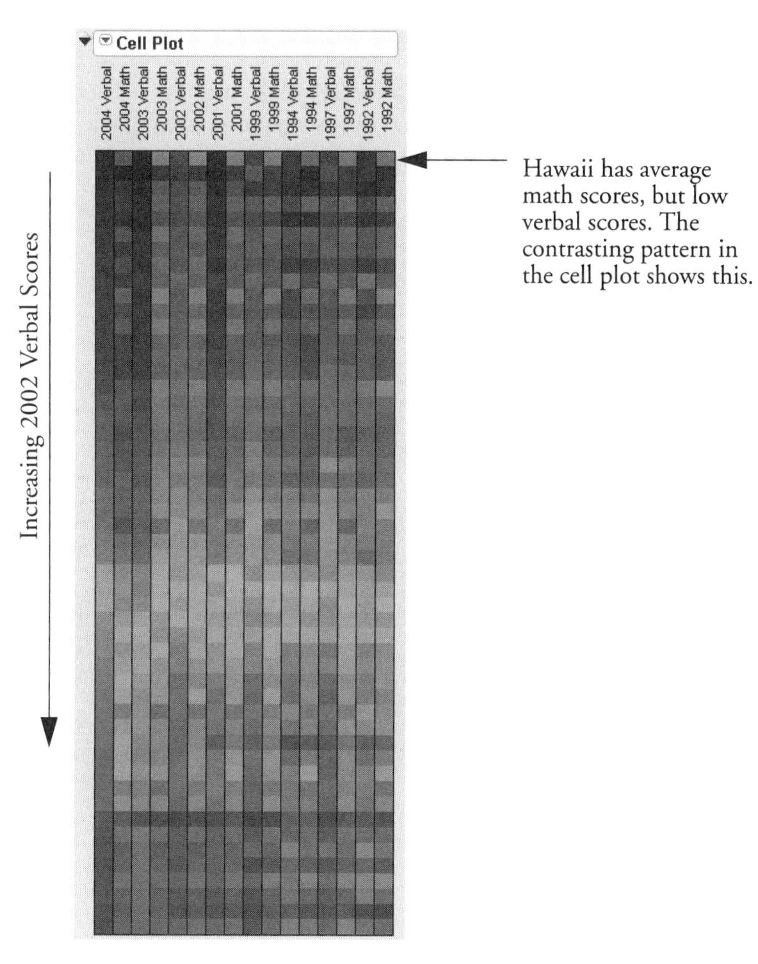

In addition, the Cell Plot platform supports label columns (one or more) for use in identifying the rows in the graph. In the following example, drug and dep1 are label columns for the Dogs.jmp data table example above. Hovering the mouse shows these labels.

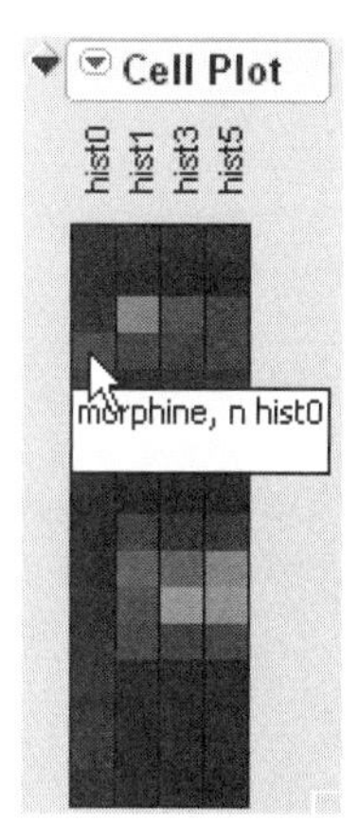

Launching the Platform

To launch the Cell Plot platform, select **Graph > Cell Plot**. A launch dialog like the one in Figure 48.5 appears. To duplicate the cell plot shown in Figure 48.4, specify the variables as shown in Figure 48.5.

The following two options appear on the launch dialog.

Scale Uniformly causes all of the plots to be represented on the same scale. In other words, vertical distances represent the same quantities for each plotted column.

Center at Zero requests that all the plots be scaled to have mean zero.

Figure 48.5 Cell Plot Launch Platform

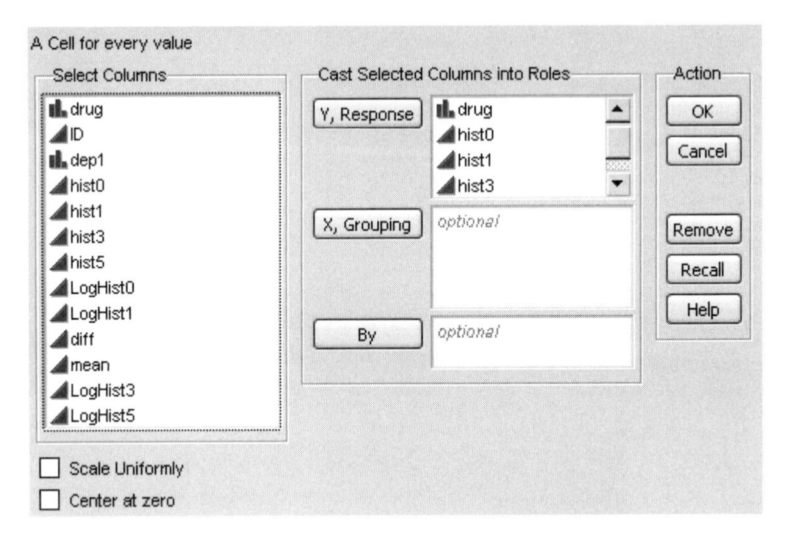

Platform Options

Platform options are accessed in two different ways.

Figure 48.6 Platform Option Locations

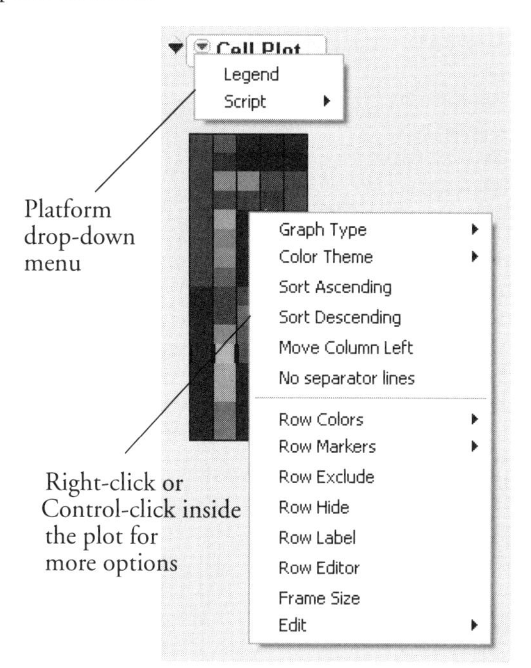

The platform drop-down menu contains two items.

Legend appends a legend to the right of the plot (Figure 48.7).

Script is the standard JMP script menu.

Figure 48.7 Cell Plot Legend

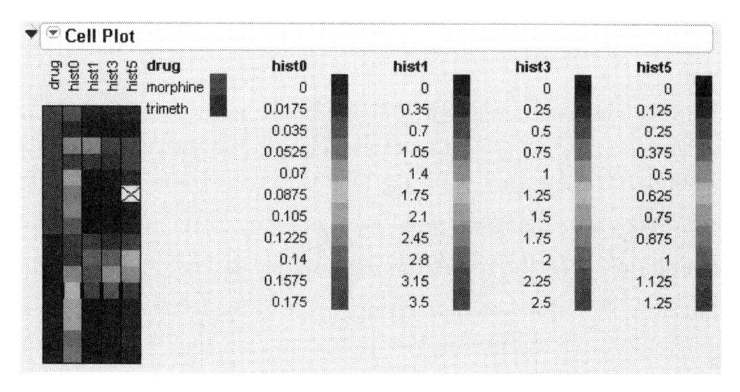

Right-click (Control-click on the Macintosh) to see the menu of options shown in Figure 48.8.

Figure 48.8 Cell Plot Options

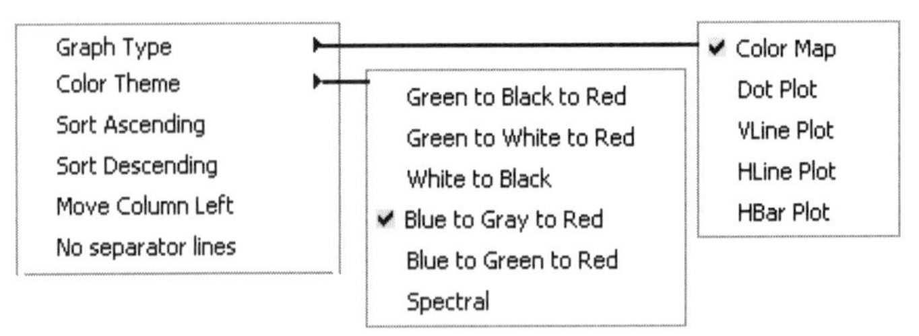

Graph Type determines the appearance of the graph. The options for graph type are illustrated in Figure 48.9.

Figure 48.9 Cell Plot Graph Types

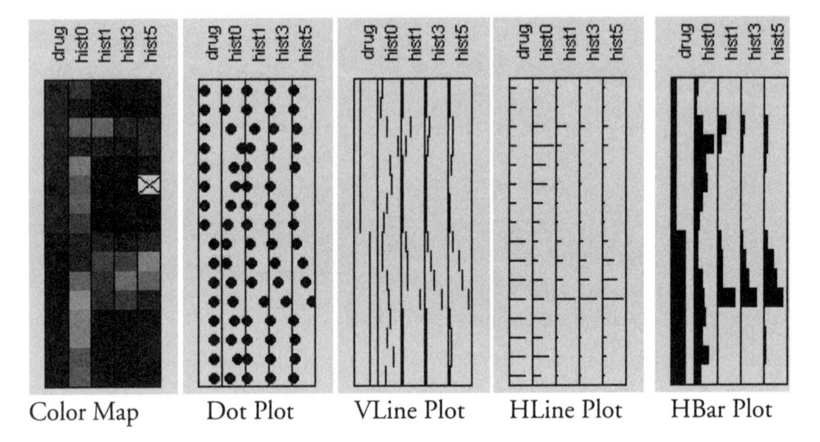

Color Theme reveals a list of color themes that affect continuous variables in Color Maps. Note that White to Black is useful when you need a gray-scale plot, frequently needed in print publications.

The following two commands sort the plot based on the values in a column. To display a sorted graph, right click (Control-click on the Macintosh) on the column you want to sort by and select a sort command.The entire plot is rearranged to correspond to the sorted column.

Sort Ascending sorts the rows from lowest to highest using the column that was clicked on.

Sort Descending sorts the rows from highest to lowest using the column that was clicked on.

Move Column Left moves a column once space to the left in a plot. To move a column, right-click (Control-click on the Macintosh) on the column and select **Move Column Left**.

No Separator Lines toggles the drawing of lines separating the columns. Separator lines are illustrated in Figure 48.10.

Figure 48.10 Separator Lines

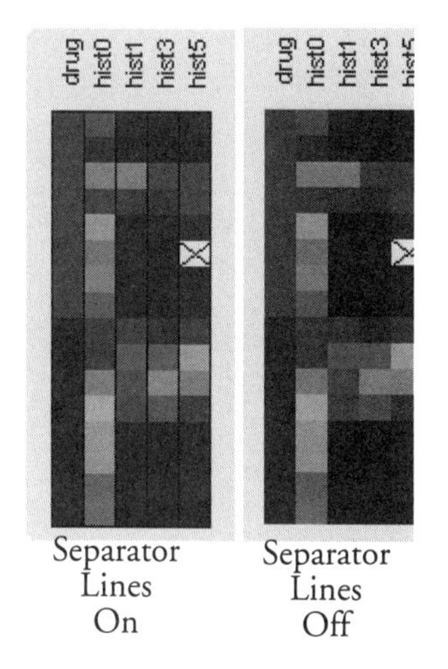

Chapter 49

Scatterplot Matrices

The Scatterplot Matrix Platform

The Scatterplot Matrix command allows quick production of scatterplot matrices. These matrices are orderly collections of bivariate graphs, assembled so that comparisons among many variables can be conducted visually. In addition, the plots can be customized and decorated with other analytical quantities (like density ellipses) to allow for further analysis.

These matrices can be square, showing the same variables on both sides of the matrix or triangular, showing only unique pairs of variables in either a lower or upper triangular fashion. In addition, you can specify that different variables be shown on the sides and bottom of the matrix, giving maximum flexibility for comparisons.

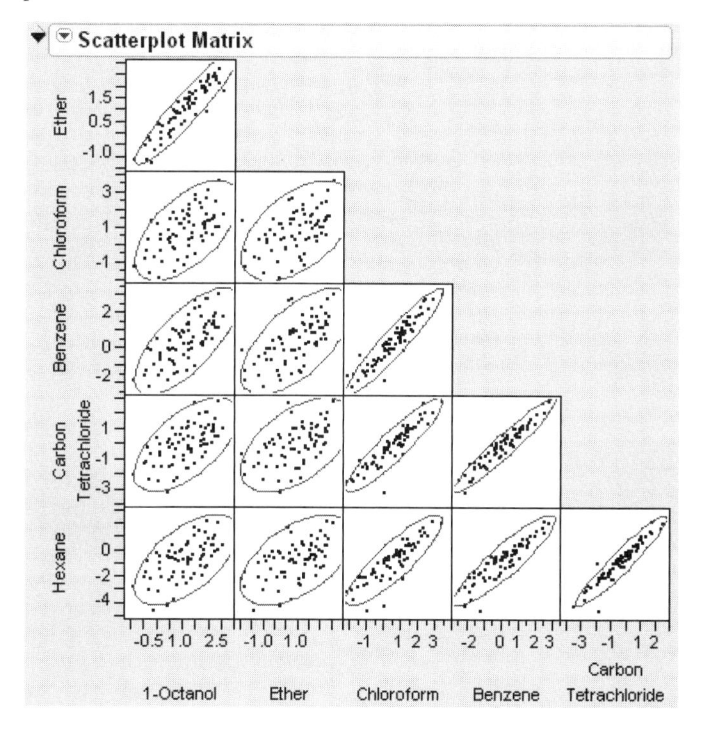

Contents

Launching the Platform

Select **Graphics > Scatterplot Matrix** to launch the platform. A dialog like the one shown in Figure 49.1 (which uses Solubility.jmp) appears.

Figure 49.1 Scatterplot Matrix Launch Dialog

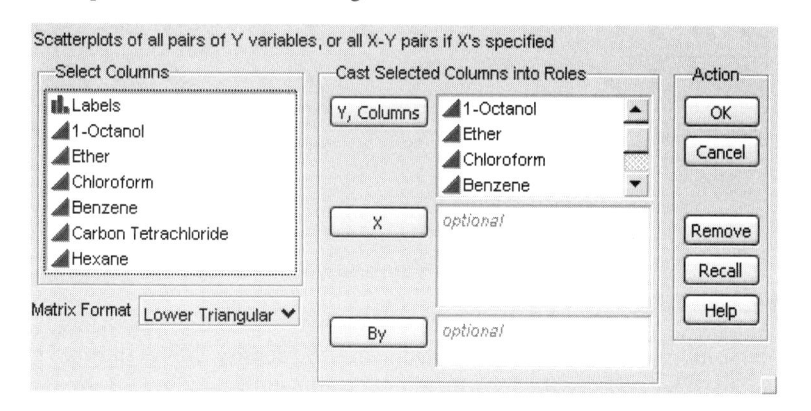

Here you choose the variables to appear in the matrix and the format of the matrix.

- If you assign variables to only the **Y, Columns** role, then these variables are used on both the horizontal and vertical axes of the resulting matrix.
- If you assign variables to both the **Y, Columns** and **X** role, then those assigned to **Y, Columns** are used on the vertical axis of the matrix, and those assigned as **X** are used on the horizontal axis of the matrix. In this way, you can produce rectangular matrices, or matrices that have different, yet overlapping, sets of variables forming the axes of the matrix.

For example, the following matrix was produced by assigning Ether and 1-Octanol to **Y** and Ether, Chloroform, and Benzene to **X**. Since Ether is in both **X** and **Y** roles, there is one matrix cell that shows the perfect Ether × Ether correlation.

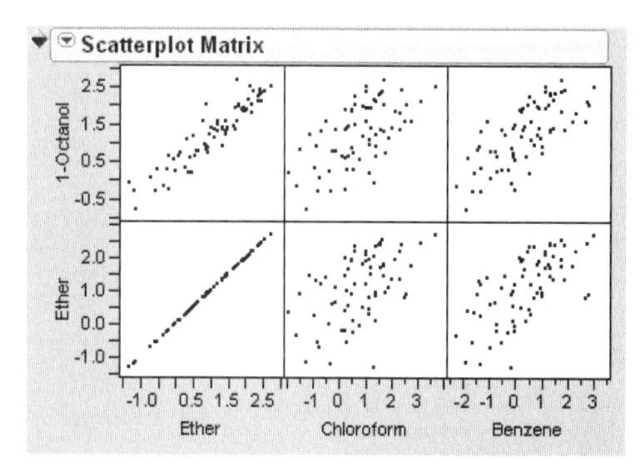

The matrix format can be one of three arrangements: **Upper Triangular**, **Lower Triangular**, or **Square**, as shown here.

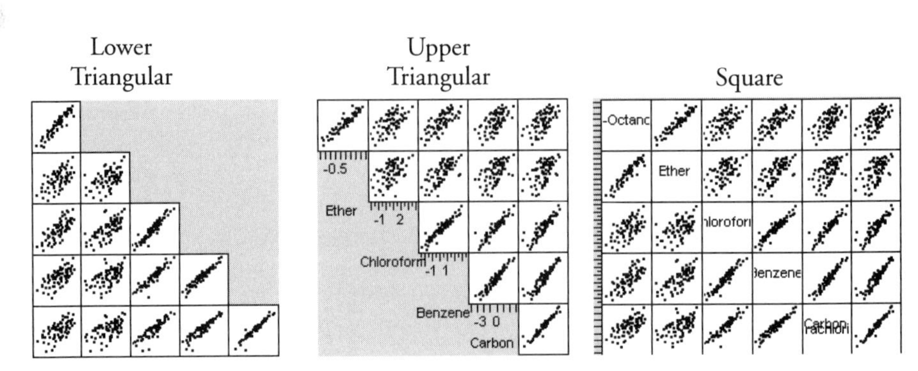

Platform Options

By default, Scatterplot Matrices are presented with 95% density ellipses shown. However, they can be toggled on or off by selecting the Density Ellipses command from the platform popup menu.

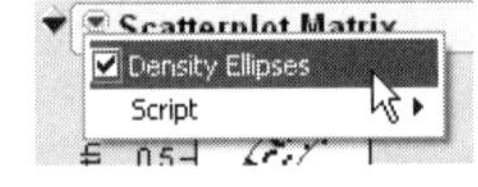

Chapter 50

Pareto Plots

The Pareto Plot Platform

A *Pareto plot* is a statistical quality improvement tool that shows frequency, relative frequency, and cumulative frequency of problems in a process or operation. It is a bar chart that displays severity (frequency) of problems in a quality-related process or operation. The bars are ordered by frequency in decreasing order from left to right, which makes a Pareto plot useful for deciding what problems should be solved first.

You launch the Pareto Plot platform with the **Pareto Plot** command in the **Graph** menu (or toolbar).

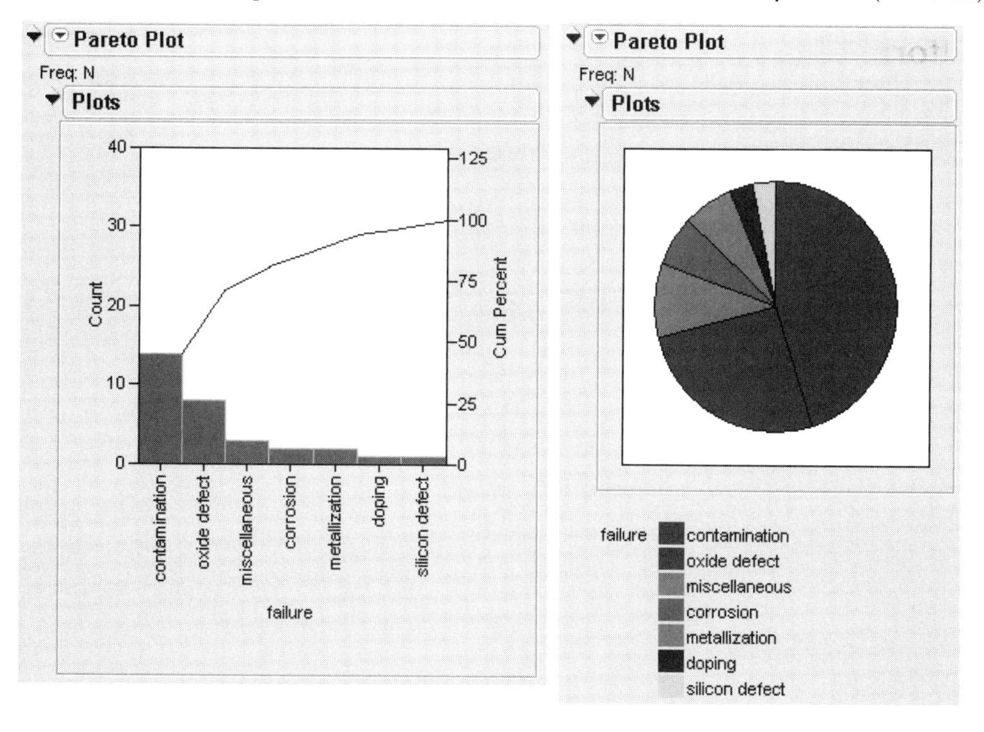

Contents

Pareto Plots

The **Pareto Plot** command produces charts to display the relative frequency or severity of problems in a quality-related process or operation. A Pareto plot is a bar chart that displays the classification of problems arranged in decreasing order. The column whose values are the cause of a problem is assigned as Y and is called the *process variable*. The column whose values hold the frequencies are assigned as **Freq**.

You can also request a *comparative Pareto Plot*, which is a graphical display that combines two or more **Pareto Plots** for the same process variable. **Pareto Plot** then produces a single graphical display with plots for each value in a column assigned the X role, or combination of levels from two *X* variables. Columns with the X role are called classification variables.

The **Pareto Plot** command can chart a single *Y* (process) variable with no *X* classification variables, with a single *X*, or with two *X* variables. The Pareto facility does not distinguish between numeric and character variables or between modeling types. All values are treated as discrete, and bars represent either counts or percentages. The following list describes the arrangement of the Pareto graphical display:

- A *Y* variable with no *X* classification variables produces a single chart with a bar for each value of the *Y* variable.
- A *Y* variable with one *X* classification variable produces a row of Pareto plots. There is a plot for each level of the *X* variable with bars for each *Y* level.
- A *Y* variable with two *X* variables produces rows and columns of Pareto plots. There is a row for each level of the first *X* variable listed and a column for each of the second *X* variable listed. The rows have a **Pareto Plot** for each value of the first *X* variable, as described previously.

The following sections illustrate each of these arrangements.

Assigning Variable Roles

The Failure.jmp table (Figure 50.1) from the Quality Control sample data folder lists causes of failure during the fabrication of integrated circuits. The N column in the table to the right lists the number of times each kind of defect occurred. It is a **Freq** variable in the Pareto Launch dialog. For the raw data table, shown on the left (Figure 50.1), causes of failure are not grouped. The **Pareto Plots** command produces the same results from either of these tables. The following example uses the failure data with a frequency column.

Figure 50.1 Partial Listing of the Failure.jmp and Failure Raw Data.jmp

	failure
1	corrosion
2	oxide defect
3	contamination
4	oxide defect
5	oxide defect
6	miscellaneous
7	oxide defect
8	contamination
9	metallization
10	oxide defect
11	contamination

	failure	N
1	contamination	14
2	corrosion	2
3	doping	1
4	metallization	2
5	miscellaneous	3
6	oxide defect	8
7	silicon defect	1

When you select the **Pareto Plot** command, you see the Pareto Plot Launch dialog shown in Figure 50.2. Select the failure column (causes of failure) as **Y, Cause**. It is the variable you want to inspect with Pareto plots. The N column in the data table is the **Freq** variable. When you click **OK**, you see the Pareto plot shown in Figure 50.3.

Figure 50.2 The Pareto Launch Dialog

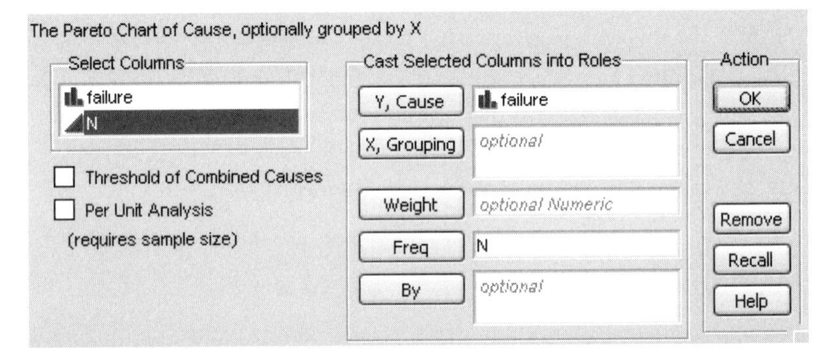

The height of each bar is the percent of failures in each category. For example, contamination accounts for 45% of the failures. The bars are in decreasing order with the most frequently occurring failure to the left. The curve indicates the cumulative percent of failures from left to right. If you place the crosshairs from the **Tools** menu on the point above the oxide defect bar, the cumulative percent axis shows that contamination and oxide defect together account for 71% of the failures.

The initial plot in Figure 50.3 shows the count scale on the left of the plot and the bars arranged in descending order with the cumulative curve. The type of scale and arrangement of bars are display options and are described in the next section. The options can be changed with the popup menu on the title bar of the window.

Figure 50.3 Simple Pareto Plot Showing Counts

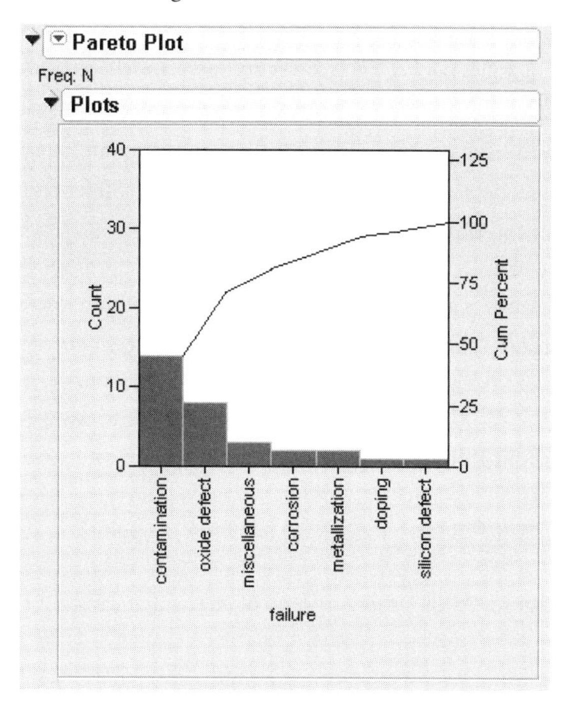

Pareto Plot Platform Commands

The popup menu on the Pareto plot title bar has commands that tailor the appearance of Pareto plots. It also has options in the **Causes** submenu that affect individual bars within a Pareto plot.

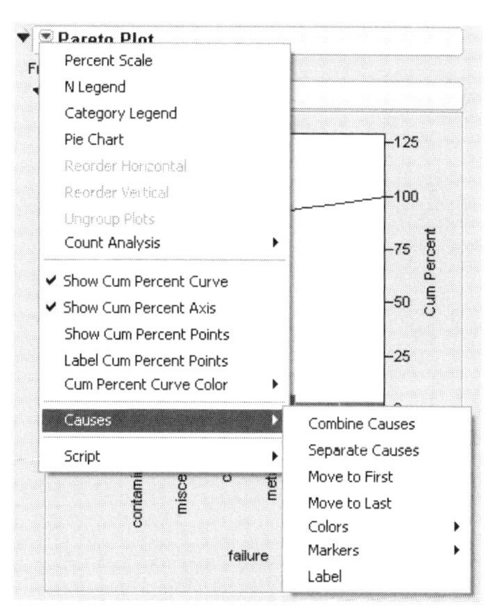

Initially, you see Pareto plots with the options checked as shown here.

The following commands affect the appearance of the Pareto plot as a whole:

Percent Scale toggles between the count and percent left vertical axis display.

N Legend toggles the legend in the plot frame on and off.

Category Legend toggles between labeled bars and separate category legend.

Pie Chart toggles between the bar chart and pie chart representation.

Reorder Horizontal, Reorder Vertical reorder grouped Pareto plots when there is one or more grouping variables.

Ungroup Plots allows a group of Pareto charts to be split up into separate plots.

Show Cum Percent Curve toggles the cumulative percent curve above the bars and the cumulative percent axis on the vertical right axis.

Show Cum Percent Axis toggles the cumulative percent axis on the vertical right axis.

Show Cum Percent Points toggles the points on the cumulative percent curve at each level of the Pareto plot.

Label Cum Percent Points toggles the labels on the points on the cumulative curve with cumulative percent values.

Cum Percent Curve Color shows palette of colors that can be assigned as the cumulative curve color.

Causes has options that affect one or more individual chart bars. See "Options for Bars," p. 917, for a description of these options.

Figure 50.4 shows the effect of various options. The left-hand plot has the **Percent Scale** option off, shows counts instead of percents, and uses the **Category Legend** on its horizontal axis. The vertical count axis is rescaled and has grid lines at the major tick marks. The plot on the right in Figure 50.4 has the **Percent Scale** option on and shows the **Cum Percent Curve**.

Figure 50.4 Pareto Plots with Display Options

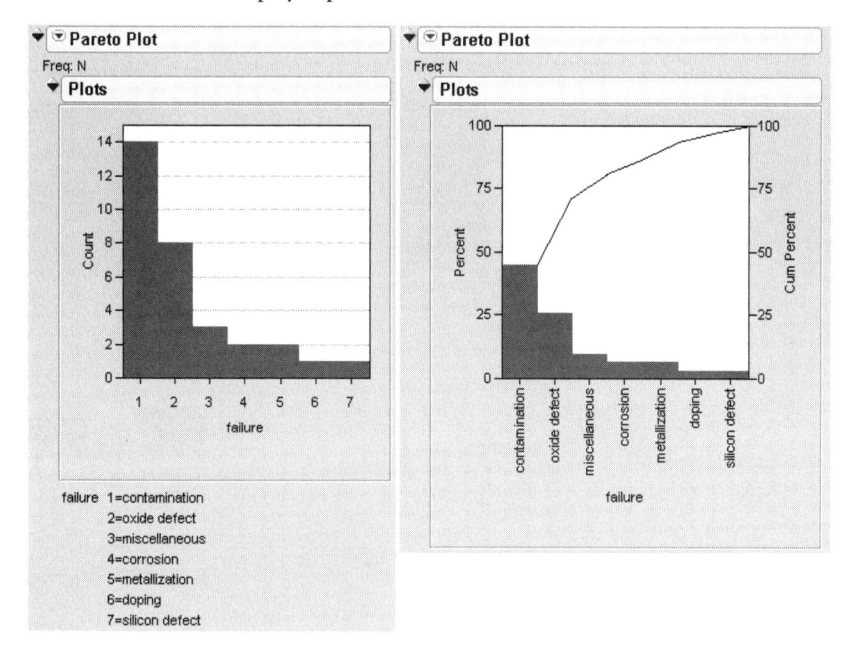

Options for Bars

You can highlight a bar by clicking on it, and you can extend the selection in standard fashion by Shift-clicking adjoining bars. Use Control-click (⌘-click on the Macintosh) to select multiple bars that are not contiguous. When you select bars, you can access the commands on the platform menu that affect Pareto plot bars. These options are also available with a right-click (Control-click on the Macintosh) anywhere in the plot area. The following options apply to highlighted bars within charts instead of to the chart as a whole:

Combine Causes combines selected (highlighted) bars.

Separate Causes separates selected bars into their original component bars.

Move to First moves one or more highlighted bars to the left (first) position.

Move to Last moves one or more highlighted bars to the right (last) position.

Colors shows the colors palette for coloring one or more highlighted bars.

Marker shows the markers palette for assigning a marker to the points on the cumulative percent curve corresponding to highlighted bars, when the **Show Cum Percent Points** command is in effect.

Label displays the bar value at the top of all highlighted bars.

The example Pareto plot on the left in Figure 50.5 is the default plot with a bar for each cause of failure. The smallest four bars are highlighted. The example on the right shows combined bars. The plot on the right results when bars are highlighted and you select the **Combine Causes** option in the popup menu. The combined bar is automatically labeled with the name of the largest of the selected bars, appended with a plus sign and the number of combined bars, or with the largest number of the selected bars if the **Category Legend** option is in effect. You can separate the highlighted bars into original categories with the **Separate Causes** option.

Figure 50.5 Example of Combining Bars

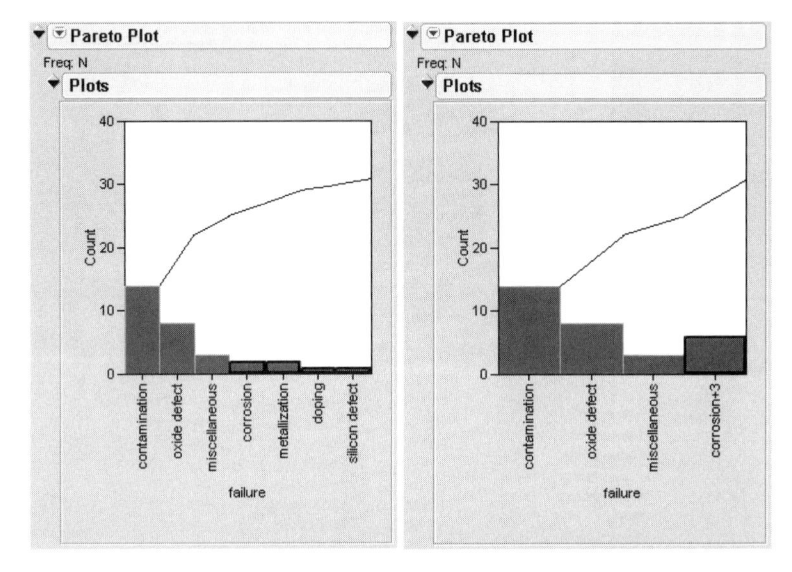

The plots in Figure 50.6 show the same data. The plot to the right results when you select the **Pie** display option.

Figure 50.6 Pareto with Bars and Corresponding Pie Representation

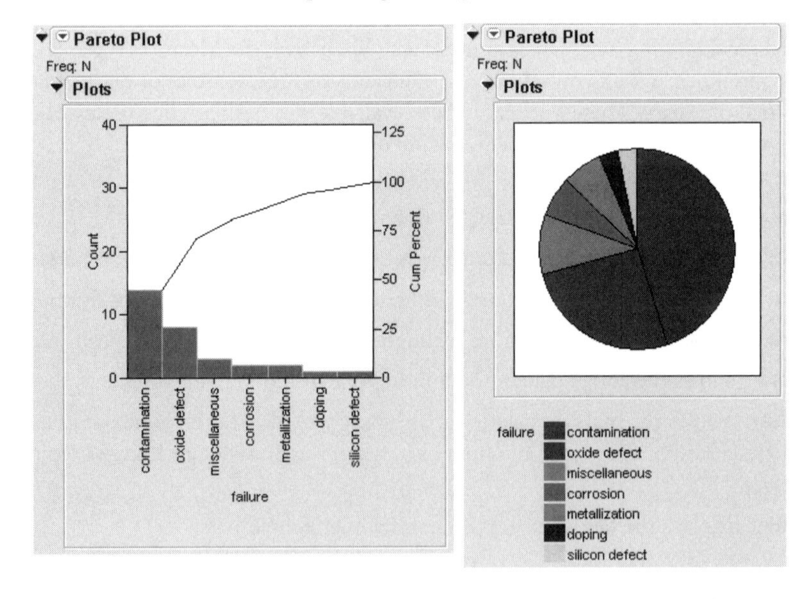

Launch Options

The following options are available on the Pareto Plot launch dialog.

Threshold of Combined Causes

This option allows you to specify a threshold for combining causes by specifying a minimum **Count** or a minimum **Rate**. To specify the threshold, select the **Threshold of Combined Causes** option on the launch dialog, as shown in Figure 50.7.

Figure 50.7 Threshold Menu

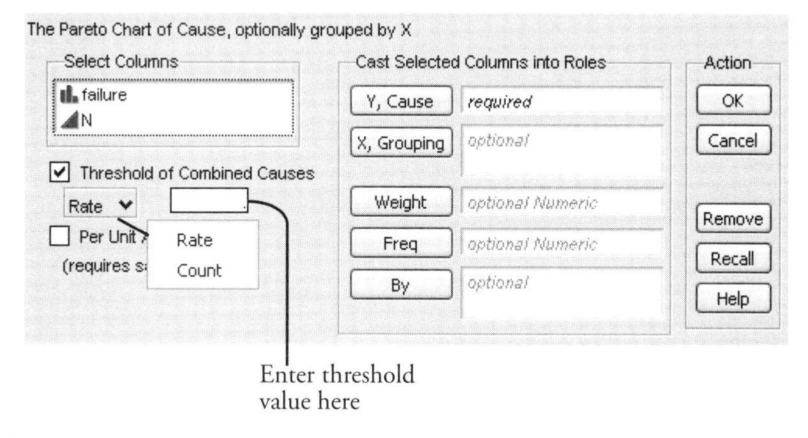

You then select **Rate** or **Count** in the menu that appears and enter the threshold value.

For example, using Failure.jmp, specifying a minimum count of 2 resulted in the following Pareto chart. All causes with counts 2 or fewer are combined into the final bar labeled corrosion+3. (Compare to Figure 50.5 to see which causes were combined).

Figure 50.8 Pareto Plot with Threshold Count = 2

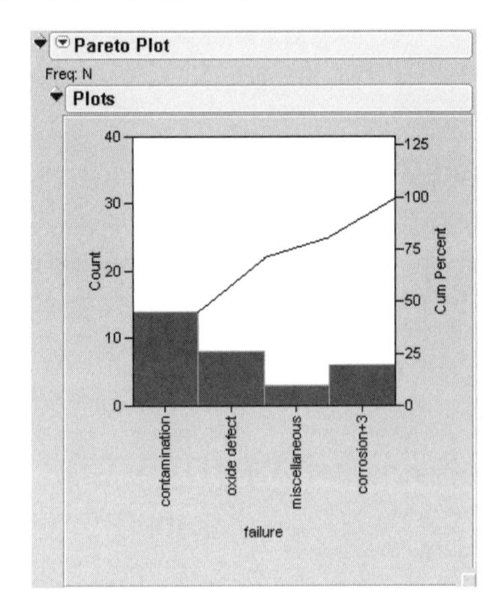

Before-and-After Pareto Plot

Figure 50.9 shows the contamination data called Failure2.jmp in the Quality Control sample data folder. This table records failures in a sample of capacitors manufactured before cleaning a tube in the diffusion furnace and in a sample manufactured after cleaning the furnace. For each type of failure, the variable clean identifies the samples with the values "before" or "after." It is a classification variable and has the X role in the Pareto Plot Launch dialog.

Figure 50.9 Failure2.jmp Data Table and Pareto Launch Dialog

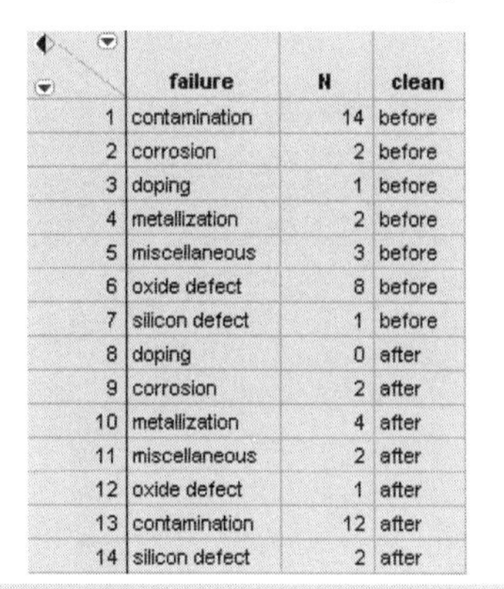

	failure	N	clean
1	contamination	14	before
2	corrosion	2	before
3	doping	1	before
4	metallization	2	before
5	miscellaneous	3	before
6	oxide defect	8	before
7	silicon defect	1	before
8	doping	0	after
9	corrosion	2	after
10	metallization	4	after
11	miscellaneous	2	after
12	oxide defect	1	after
13	contamination	12	after
14	silicon defect	2	after

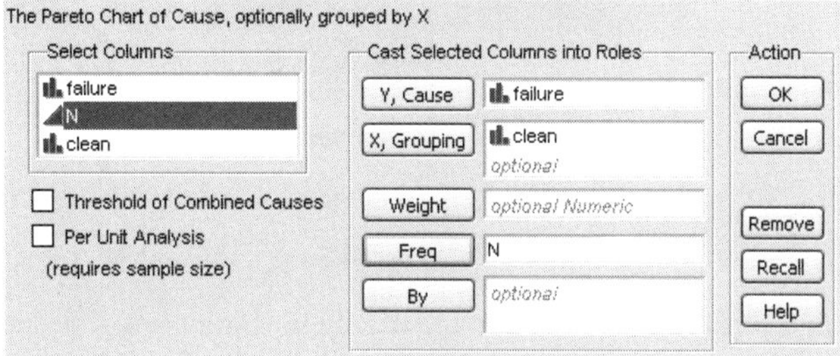

The grouping variable produces one Pareto plot window with side-by-side plots for each value of the **X, Grouping** variable, clean. You see the two Pareto plots in Figure 50.10.

These plots are referred to as the *cells* of a comparative Pareto plot. There is a cell for each level of the *X* (classification) variable. Because there is only one *X* variable, this is called a *one-way comparative Pareto plot.*

Figure 50.10 One-way Comparative Pareto Plot

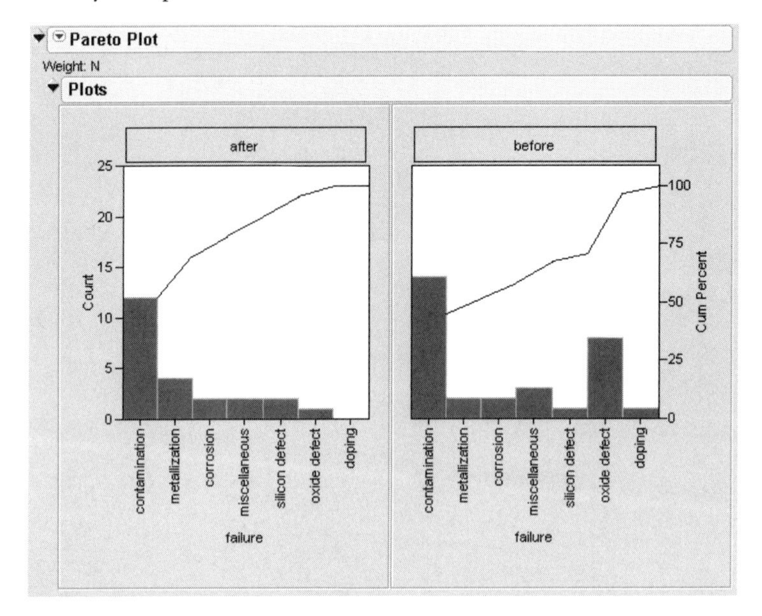

The horizontal and vertical axes are scaled identically for both plots. The bars in the first plot are in descending order of the *y*-axis values and determine the order for all cells.

The plots are arranged in alphabetical order of the classification variable levels. The levels ("after" and "before") of the classification variable, clean, show above each plot. You can rearrange the order of the plots by clicking on the title (level) of a classification and dragging it to another level of the same classification. For example, the comparative cells of the clean variable shown in Figure 50.10 are reversed. The "after" plot is first and is in descending order. The "before" plot is second and is reordered to conform with the "after" plot. A comparison of these two plots shows a reduction in oxide defects after cleaning, but the plots would be easier to interpret presented as the before-and-after plot shown in Figure 50.11.

Figure 50.11 One-way Comparative Pareto Plot with Reordered Cells

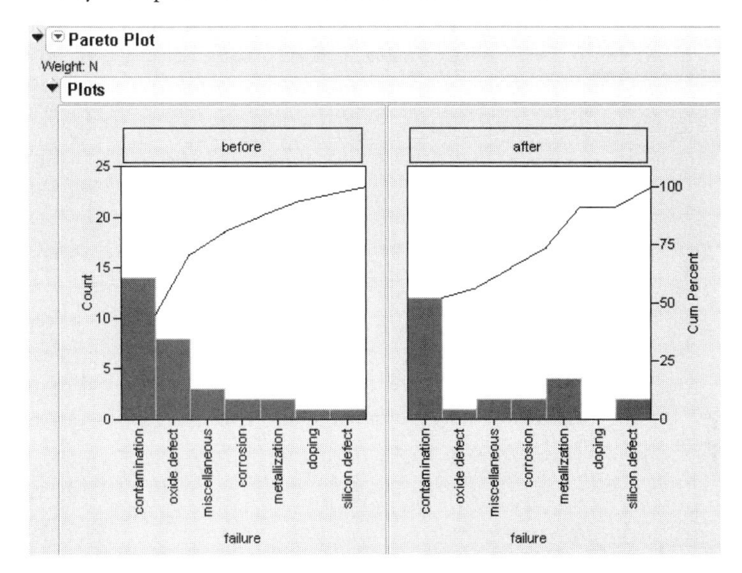

Note that the order of the categories have been changed to reflect the decreasing order based on the first cell.

Two-Way Comparative Pareto Plot

You can study the effects of two classification variables simultaneously with a *two-way comparative Pareto plot*. Suppose that the capacitor manufacturing process from the previous example monitors production samples before and after a furnace cleaning for three days. The Failure3.jmp table has the column called date with values OCT 1, OCT 2, and OCT 3. To see a two-way array of Pareto plots, select the **Pareto Plot** command and add both clean and date to the **X, Grouping** list as shown in Figure 50.12.

Figure 50.12 Pareto Plots with Two Grouping Variables

	failure	N	clean	date
1	contamination	14	before	OCT 1
2	corrosion	2	before	OCT 1
3	doping	1	before	OCT 1
4	metallization	2	before	OCT 1
5	miscellaneous	3	before	OCT 1
6	oxide defect	8	before	OCT 1
7	silicon defect	1	before	OCT 1
8	doping	0	after	OCT 1
9	corrosion	2	after	OCT 1
10	metallization	4	after	OCT 1

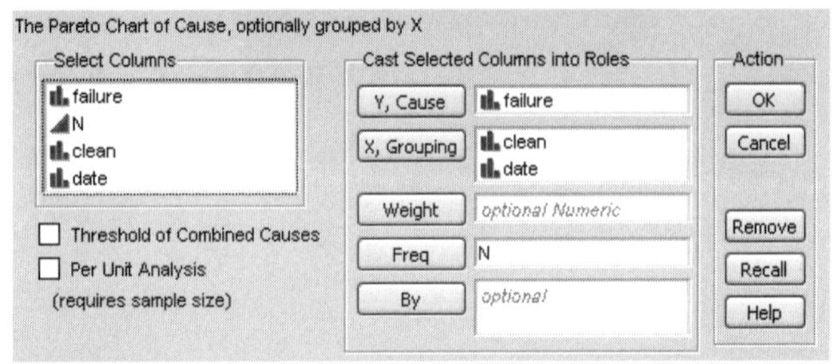

Two grouping variables produce one Pareto plot window with a two-way layout of plots that show each level of both *X* variables (Figure 50.13). The upper-left cell is called the *key cell.* Its bars are arranged in descending order. The bars in the other cells are in the same order as the key cell. You can reorder the rows and columns of cells with the **Reorder Vertical** or **Reorder Horizontal** option in the platform popup menu. The cell that moves to the upper-left corner becomes the new key cell and the bars in all other cells rearrange accordingly.

You can click bars in the key cell and the bars for the corresponding categories highlight in all other cells. Use Control-click (⌘-click on the Macintosh) to select nonadjacent bars.

The Pareto plot shown in Figure 50.13 illustrates highlighting the *vital few.* In each cell of the two-way comparative plot, the bars representing the two most frequently occurring problems are selected. Contamination and Metallization are the two vital categories in all cells, but they appear to be less of a problem after furnace cleaning.

Figure 50.13 Two-way Comparative Pareto Plot

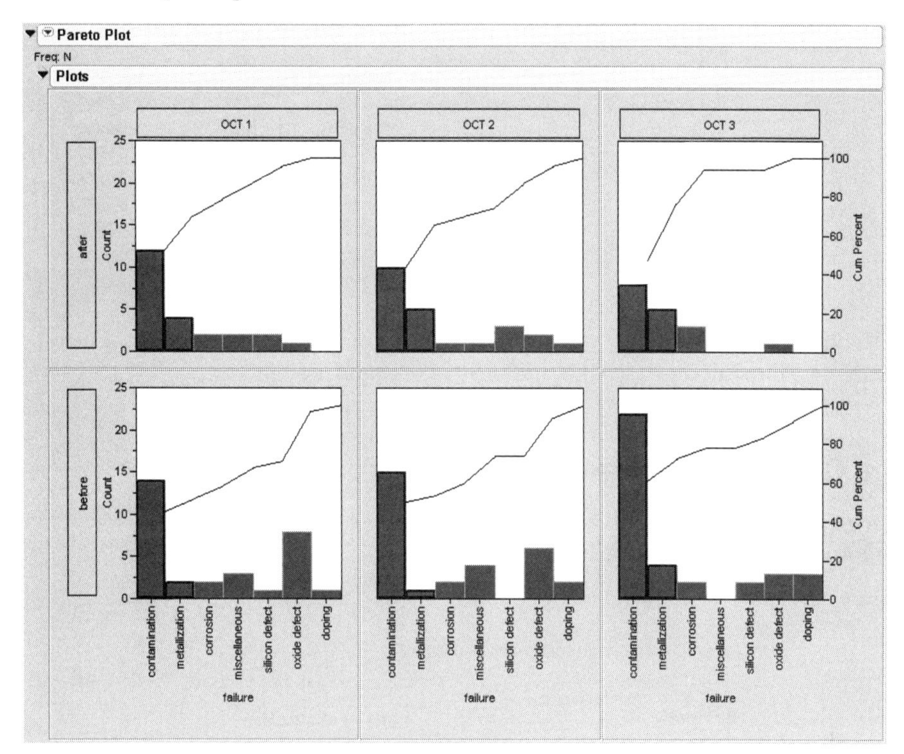

Defect Per Unit Analysis

The Defect Per Unit analysis allows you to compare defect rates across groups. JMP calculates the defect rate as well as 95% confidence intervals of the defect rate. You may optionally specify a constant sample size on the launch dialog.

Although causes are allowed to be combined in Pareto plots, the calculations for these analyses do not change correspondingly.

Using Number of Defects as Sample Size

As an example of calculating the rate per unit, use the Failures.jmp sample data table (Note that this is not Failure.jmp, but Failures.jmp). After assigning Causes to **Y, Cause** and Count to **Freq**, click **OK** to generate a Pareto plot.

When the chart appears, select **Count Analysis > Per Unit Rate** from the platform drop-down menu to get the Per Unit Rates table shown in Figure 50.14.

Figure 50.14 Per Unit Rates Table

Per Unit Rates

Cause	Count	Rate	Lower 95%	Upper 95%
Contamination	110	0.410448	0.3373	0.4947
Oxide Defect	86	0.320896	0.2567	0.3963
Miscellaneous	18	0.067164	0.0398	0.1061
Silicon Defect	17	0.063433	0.0370	0.1016
Corrosion	16	0.059701	0.0341	0.0970
Metallization	11	0.041045	0.0205	0.0734
Doping	10	0.037313	0.0179	0.0686
Total	268	1.000000	0.8838	1.1272

Since there was no sample size entered on the launch dialog, the total number of defect counts across causes is used to calculate each rate and their 95% confidence interval.

Using a Constant Sample Size Across Groups

Using **Failures.jmp**, fill in the dialog as shown in Figure 50.15 and click **OK**. Note that checking **Per Unit Analysis** causes options to appear.

Figure 50.15 Pareto Launch Dialog

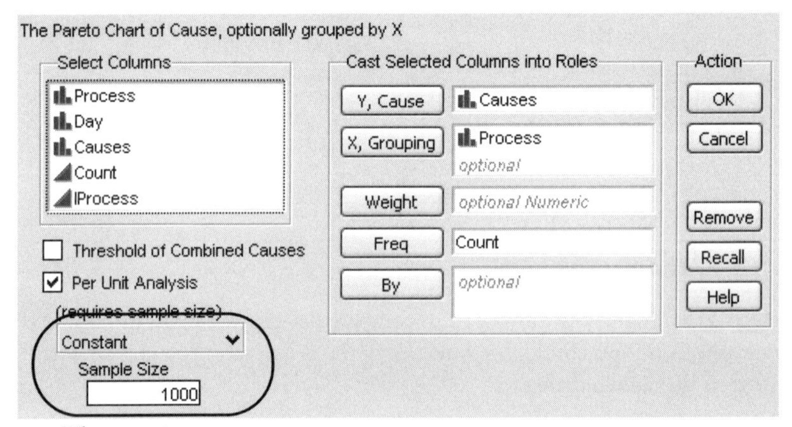

These options appear when **Per Unit Analysis** is checked

When the report appears, select **Count Analysis > Test Rates Across Groups**. This produces the tables shown in Figure 50.16.

Figure 50.16 Group Comparison Output

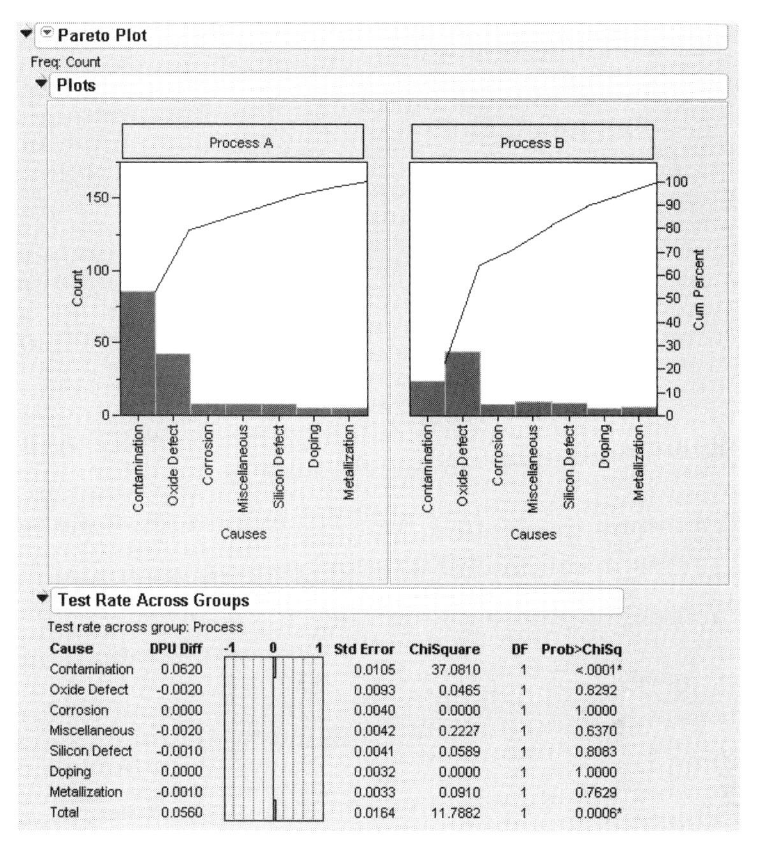

Cause	DPU Diff	-1 0 1	Std Error	ChiSquare	DF	Prob>ChiSq
Contamination	0.0620		0.0105	37.0810	1	<.0001*
Oxide Defect	-0.0020		0.0093	0.0465	1	0.8292
Corrosion	0.0000		0.0040	0.0000	1	1.0000
Miscellaneous	-0.0020		0.0042	0.2227	1	0.6370
Silicon Defect	-0.0010		0.0041	0.0589	1	0.8083
Doping	0.0000		0.0032	0.0000	1	1.0000
Metallization	-0.0010		0.0033	0.0910	1	0.7629
Total	0.0560		0.0164	11.7882	1	0.0006*

The statistical test (a likelihood-ratio chi square) tests whether the defects per unit (DPU) for each cause is the same for the two groups.

Using a Non-Constant Sample Size Across Groups

To specify a unique sample size for a group, add rows to the data table for each group, specifying a special cause code (e.g. “size”) to designate the rows as size rows. For example, look at the following excerpt from Failuressize.jmp. Among the other causes (Oxide Defect, Silicon Defect, etc.) you see a cause labeled size.

	Process	Day	Causes	Count
61	Process B	03/01/1991	Oxide Defect	10
62	Process B	03/02/1991	Oxide Defect	9
63	Process B	03/03/1991	Oxide Defect	10
64	Process B	03/04/1991	Oxide Defect	7
65	Process B	03/05/1991	Oxide Defect	8
66	Process B	03/01/1991	Silicon Defect	3
67	Process B	03/02/1991	Silicon Defect	2
68	Process B	03/03/1991	Silicon Defect	1
69	Process B	03/04/1991	Silicon Defect	1
70	Process B	03/05/1991	Silicon Defect	2
71	Process A	03/01/1991	size	100
72	Process A	03/02/1991	size	120
73	Process A	03/03/1991	size	130
74	Process A	03/04/1991	size	140
75	Process A	03/05/1991	size	150
76	Process B	03/01/1991	size	200
77	Process B	03/02/1991	size	220
78	Process B	03/03/1991	size	230
79	Process B	03/04/1991	size	240
80	Process B	03/05/1991	size	250

To conduct the analysis, fill in the Pareto launch dialog like the one shown in Figure 50.17.

Figure 50.17 Non-Constant Sample Size Launch

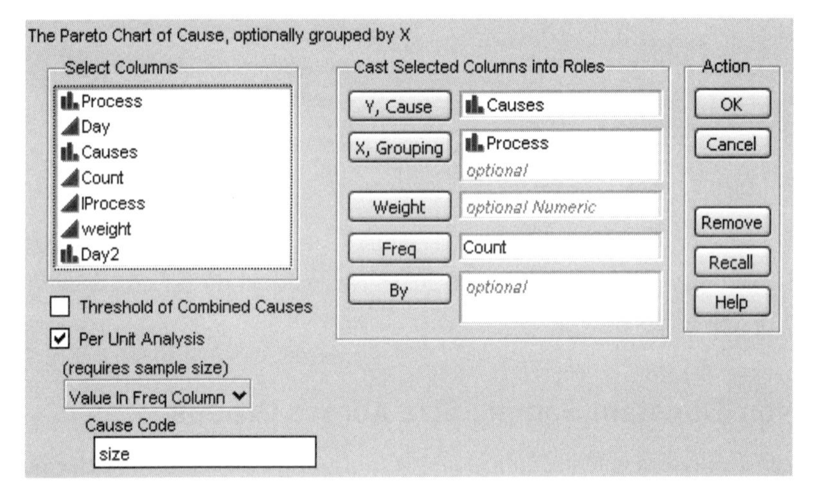

After clicking **OK**, select both **Per Unit Rates** and **Test Rates Across Groups**, found under **Count Analysis** in the platform drop-down. The resulting report is shown in Figure 50.18.

Figure 50.18 Pareto Analysis with Non-Constant Sample Sizes

Per Unit Rates

Process	Cause	Count	DPU	Lower 95%	Upper 95%
Process A	Contamination	86	0.8515	0.6811	1.0516
	Oxide Defect	42	0.4158	0.2997	0.5621
	Corrosion	8	0.0792	0.0342	0.1561
	Miscellaneous	8	0.0792	0.0342	0.1561
	Silicon Defect	8	0.0792	0.0342	0.1561
	Doping	5	0.0495	0.0161	0.1155
	Metallization	5	0.0495	0.0161	0.1155
	Pooled Total	162	0.2291	0.1952	0.2673
	size	101			
Process B	Contamination	24	0.1655	0.1061	0.2463
	Oxide Defect	44	0.3034	0.2205	0.4074
	Corrosion	8	0.0552	0.0238	0.1087
	Miscellaneous	10	0.0690	0.0331	0.1268
	Silicon Defect	9	0.0621	0.0284	0.1178
	Doping	5	0.0345	0.0112	0.0805
	Metallization	6	0.0414	0.0152	0.0901
	Pooled Total	106	0.1044	0.0855	0.1263
	size	145			

Test Rate Across Groups

Test rate across group: Process

Cause	DPU Diff	-1 0 1	Std Error	ChiSquare	DF	Prob>ChiSq
Contamination	0.6860		0.0978	63.0776	1	<.0001*
Oxide Defect	0.1124		0.0788	2.1195	1	0.1454
Corrosion	0.0240		0.0341	0.5202	1	0.4707
Miscellaneous	0.0102		0.0355	0.0847	1	0.7710
Silicon Defect	0.0171		0.0348	0.2500	1	0.6171
Doping	0.0150		0.0270	0.3251	1	0.5685
Metallization	0.0081		0.0278	0.0871	1	0.7679
Pooled Total	0.1247		0.0207	40.7524	1	<.0001*

Note that the sample size of 101 is used to calculate the DPU for the causes in group A, while the sample size of 145 is used to calculate the DPU for the causes in group B.

If there are two group variables (say, Day and Process), Per Unit Rates lists DPU or rates for every combination of Day and Process for each cause. However, Test Rate Across Groups only tests overall differences between groups.

Chapter 51

Three-Dimensional Scatterplots

The Scatterplot 3D Platform

The **Scatterplot 3D** command in the **Graph** menu (or toolbar) displays a three-dimensional view of data and an approximation of higher dimensions through principal components. The plot is a rotatable display of the values of numeric columns in the current data table. The Scatterplot 3D platform displays three variables at a time from the columns you select.

To help capture and visualize variation in higher dimensions, the Scatterplot 3Ddisplays a biplot representation of the points and variables when you request principal components.

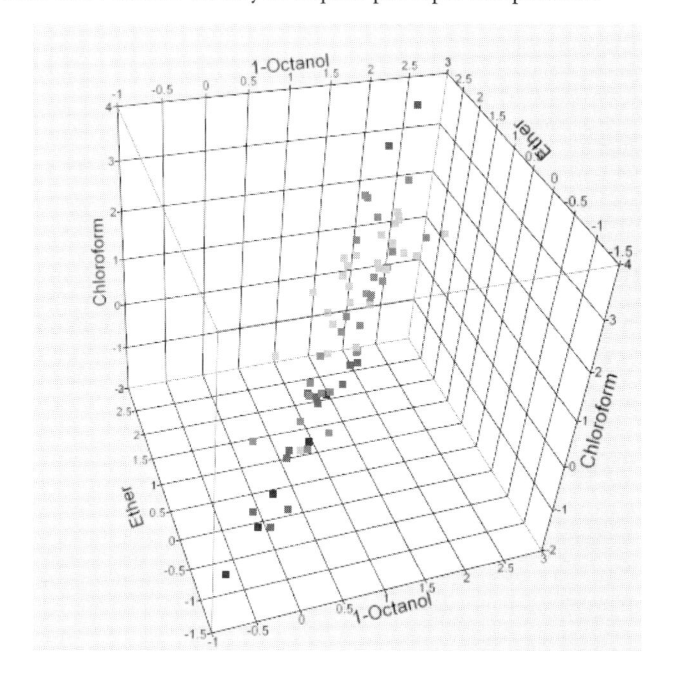

Contents

Launch the Platform

To generate a three-dimensional scatterplot, columns must be selected as components for plotting. When you select the **Scatterplot 3D** command, the launch dialog (using **Solubility.jmp**) appears.

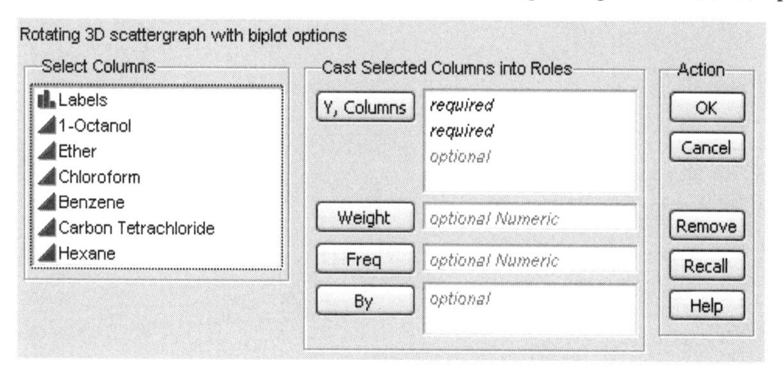

The Scatterplot 3D platform displays a three-dimensional spinnable plot, as shown in Figure 51.1 of the variables from the **Y, Columns** list. Three variables are plotted at a time, and controls are available below the plot to designate which three are shown on the plot. At launch time, therefore, you may include all the variables that you might want to show in the plot, since you can add more later.

Note: If you specify a Weight variable in the launch dialog, JMP draws the points as balls, scaled so that their volume represents the weight value. The report includes a slider control to scale the sizes.

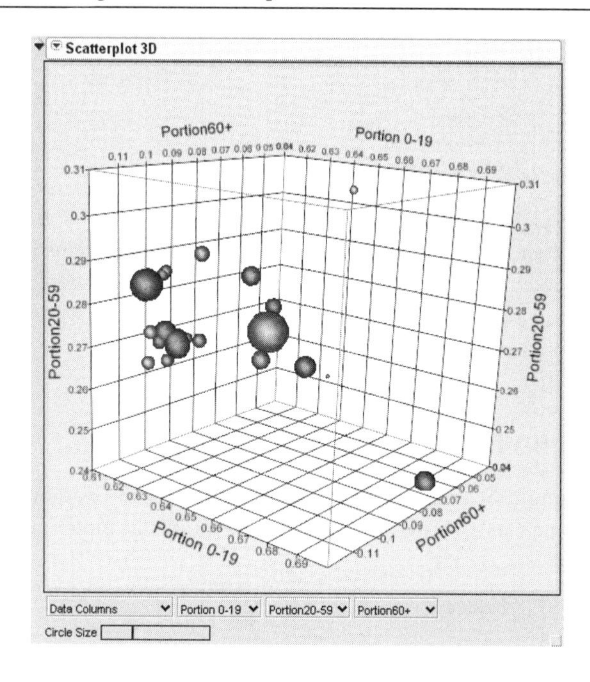

Figure 51.1 The Scatterplot 3D Platform

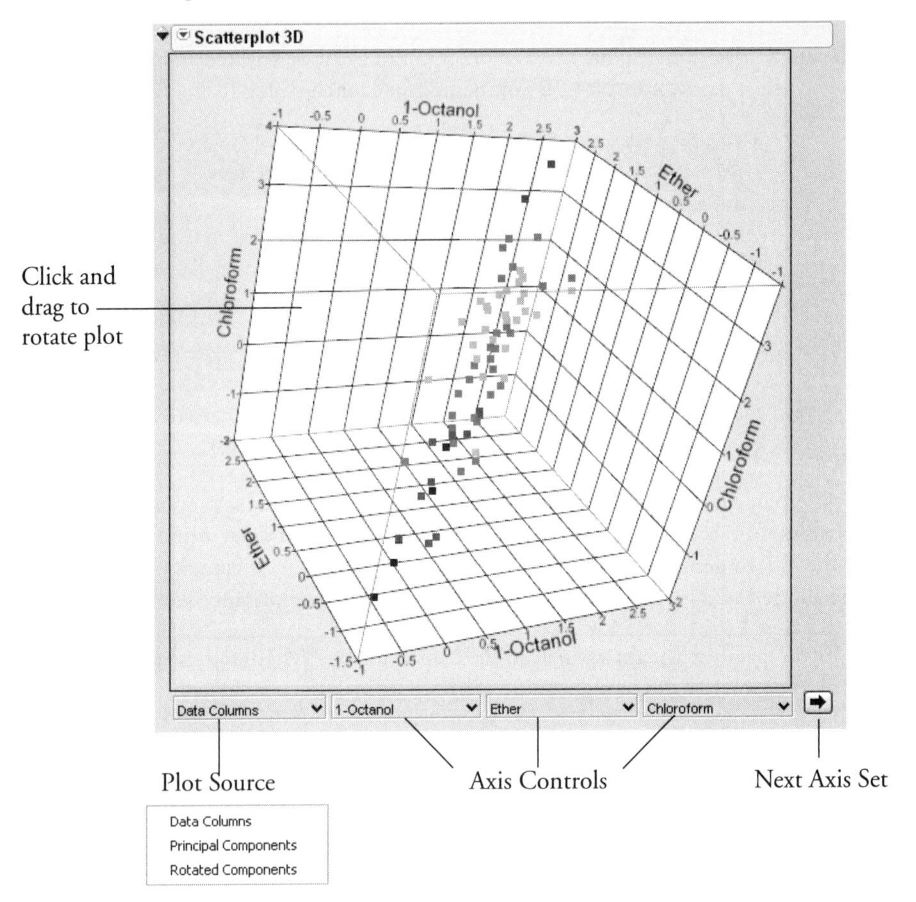

Initially, the Scatterplot 3D platform displays the first three variables in the **Y, Columns** list on the *x*-, *y*-, and *z*-axes of the plot (see Figure 51.1). The platform always assigns variables to three axes whenever there are three variables in the components list. To plot a different set of three variables, click one of the axis control drop-down lists and select another component. In this way, you can plot each combination of three variables in a list of components.

The Platform Popup Menu

The popup menu on the Scatterplot 3D title bar lists options to customize the display, and to compute, rotate, and save principal or rotated components. Additionally, biplot rays can be shown in the graph.

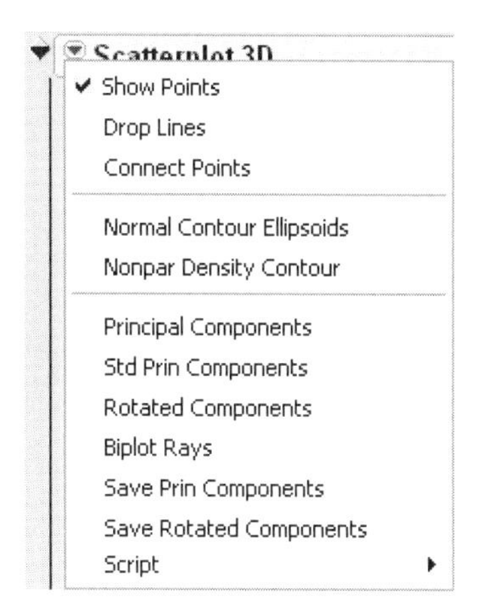

Show Points alternately shows and hides the points on the graph.

Drop Lines draws lines from the floor defined by the first and third variables to each point in the graph.

Connect Points connects the points with a line. Optionally, the points can be grouped.

Normal Contour Ellipsoids draws normal contour ellipsoids. After selecting this command, a dialog asks if you want a single shell drawn for the entire data set (the **ungrouped** option) or if there should be multiple contours based on the levels of a column (the **grouped** option).

Nonpar Density Contour draws a 95% kernel contour shell around the points. As with normal contour ellipsoids, the density contours can be for the entire data set (the **ungrouped** option) or for multiple contours based on the levels of a column (the **grouped** option).

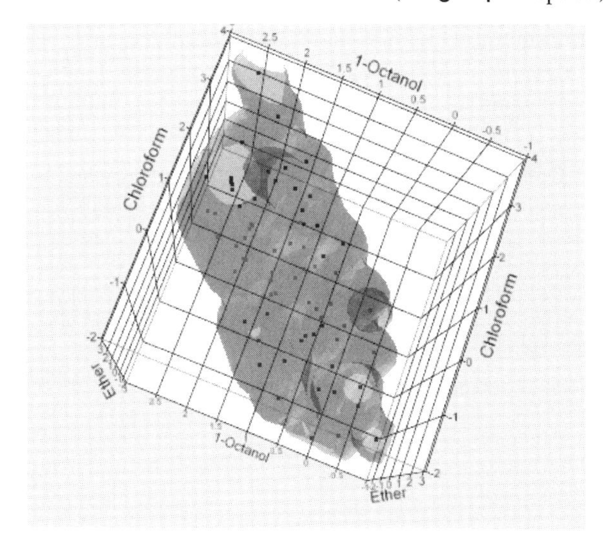

In addition, the **Nonpar Density Contour** command produces a control panel where you can tune the contour level, transparency, add contour levels, change the bandwidth, and change the resolution.

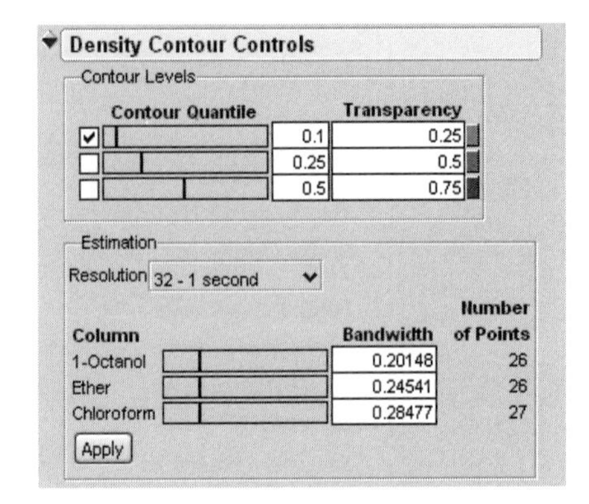

Principal Components calculates principal components on the set of variables in the components list and adds the principal components to the list. You can do this repeatedly with different sets of variables.

Std Prin Components calculates principal components the same as the **Principal Components** command but scales the principal component scores to have unit variance instead of canonical variance.

Note: Use this option if you want GH' rather than JK' biplots. The interpoint distance shown by GH' biplots is less meaningful, but the angles of the biplot rays measure correlations better.

Rotated Components prompts you to specify the number of factors you want to rotate and then computes that many rotated component scores.

Biplot Rays alternately shows or hides rays on the plot corresponding to the principal components on the variables or the variables on the principal components.

Save Prin Components saves the number of current principal component scores you request as columns in the current data table. These columns include their formulas and are locked. For *n* variables in the components list, *n* principal component columns are created and named Prin Comp 1, Prin Comp 2, …, Prin Comp n.

Save Rotated Components saves the rotated component scores as columns in the current data table. These columns include their formulas and are locked. If you requested *n* rotated components, then *n* rotated component columns are created and named Rot Comp 1, Rot Comp 2, …, Rot Comp n.

Script has a submenu of commands available to all platforms that lets you redo the analysis, or save the JSL commands for the analysis to a window or a file.

For details on principal components and rotated components, see the sections “Principal Components,” p. 515.

Spinning the Plot

After selecting three variables, you can spin the resulting point cloud in two ways:

- Click and drag the mouse

- Use the keypad arrows, or
- Use the arrow keys on the keyboard

to manually move or spin the plot about the origin in any direction

In each case, the spin persists as long as you hold down the mouse button or key. Use Shift-click to cause the spinning to continue after you release the mouse button.

Display Options

Right-click in the graph to produce the popup menu showing a variety of display options.

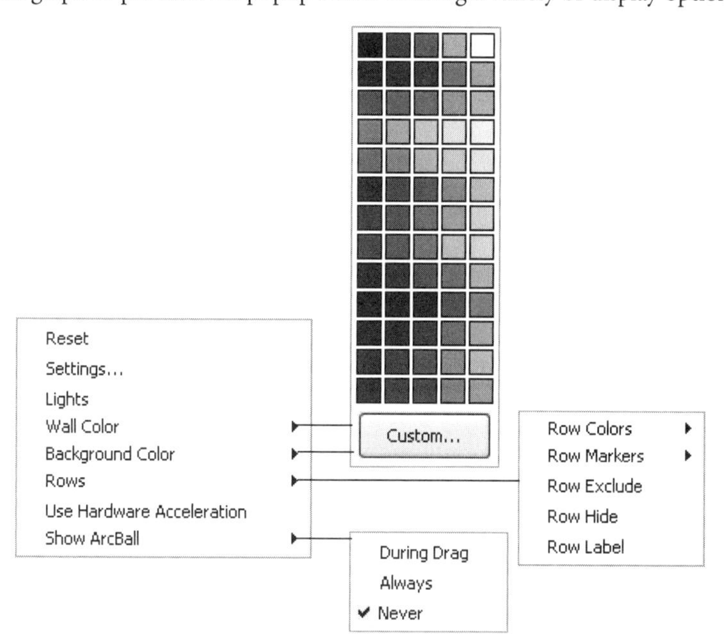

Reset returns the plot to its initial state.

Settings brings up a floating window with sliders to adjust the appearance of the plot.

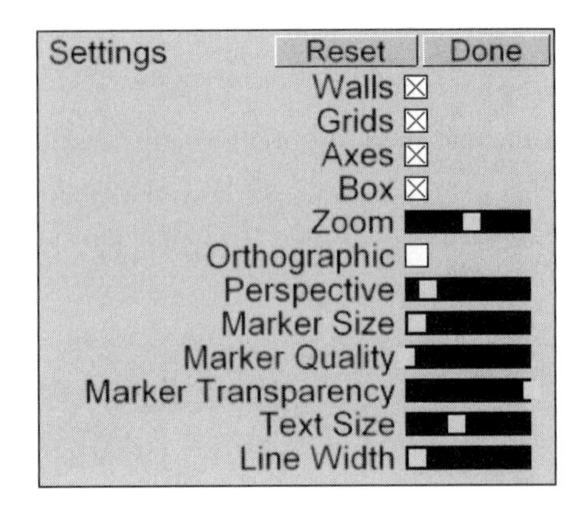

Reset resets the dialog to its initial settings.

Done dismisses the dialog.

Walls toggles the display of the graph frames.

Grids toggles the display of the graph grids.

Axes toggles the display of the axes.

Box toggles the display of the thin gray lines that show when a wall is hidden.

Zoom enlarges or shrinks the graph

Orthographic is a checkbox that toggles a view of the plot from an orthographic projection.

Perspective increases the perspective of the viewer. Large values of perspective create a field of view that is unnaturally large and visually disorientating.

Marker Size increases the size of the data markers, from small to large.

Marker Quality changes the marker quality. Hi-quality markers (roward the right side of the slider) are drawn more slowly.

Marker Transparency changes the marker transparency from completely transparent (left) to completely opaque (right).

Text Size changes the size of the text, from small to large.

Line Width changes the width of displayed lines (like axes and grids), from small to large.

Lights shows the lights border, allowing you to move the lights to new positions, or, using the right-click menu, turn them on or off, or alter their color.

Wall Color changes the color of the planes that frame the graph.

Background Color changes the color surrounding the graph.

Use Hardware Acceleration turns on hardware acceleration for machines that support it.

Show ArcBall shows the ArcBall that controls display rotation. It can be set to show Never, Always. or During Mouse Drag.

Highlighting Points

The arrow and brush tools found in the **Tools** menu highlight points on the plot and are described in the *JMP User Guide*. When you click on a point, it highlights on the plot, in the data table, and on all other plots that contain the point. Shift-click extends the highlighted selection. Alt-click (Option-click on the Macintosh) copies the label of the point to the clipboard.

Hovering over a point causes the row label to show. The example in Figure 51.2 shows the **Solubility.jmp** data with the **ETHANOL** point highlighted.

Figure 51.2 Highlighting a Labeled Point

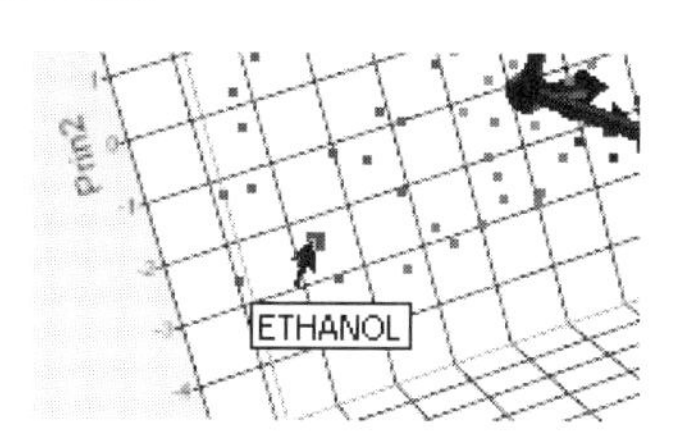

The brush tool in the **Tools** menu highlights all points within the brush rectangle and on the spreadsheet. Use Alt-click (or Option-click) to change the brush shape and Shift-click to extend a selection.

You can apply various *row state* attributes to highlighted points using commands in the **Rows** menu. Row states are color, marker, hide, exclude, label, and selection status. Row states are described in the"Data Tables" chapter of the *JMP Scripting Guide.*

Color and Marker Coding

To help visualize more dimensions on the spinning plot, you can code variable values with color and marker styles. There are several ways to do this:

- Use the **Color/Marker by Column** command in the Rows panel popup menu to the left of the spreadsheet. This command asks you to select a variable and then assigns a color and marker to each of its values.
- Use the **Distribution** command in the **Analyze** menu. Then highlight the points bar by bar, assigning colors or markers in sequence as you work through the levels or ranges of the variable.
- You can also use the formula editor to assign row state characteristics to rows according to the formula you construct. You can save these row states as a special row-state variable in the data table by creating a new variable with the **Row State** data type.

Rescaling the Axes

If you want to change the scaling of a variable in an axis, you can drag the axis interactively with the mouse, or double-click the axis to bring up a specification dialog. The dialog shown here gives you the maximum, minimum, and increment values for the axis. In addition, you are shown the axis type (linear or logarithmic) and the format used in displaying numerical axis labels. You can edit these values to best show the plot points of interest.

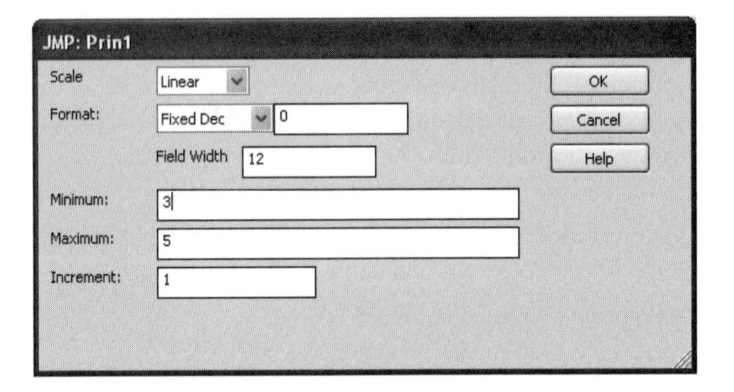
JMP: Prin1
Scale
Linear
Format:
Fixed Dec
0
Field Width
12
Minimum:
3
Maximum:
5
Increment:
1
OK
Cancel
Help

Chapter 52

Bubble Plots

The Bubble Plot Platform

A bubble plot is a scatter plot which represents its points as circles (bubbles). Optionally the bubbles can be sized according to a another column, colored by another column, aggregated across groups defined by one or more other columns, and dynamically indexed by a time column. With the opportunity to see up to five dimensions at once (*x* position, *y* position, size, color, and time), bubble plots can produce dramatic visualizations and make interesting discoveries easy.

Dynamic bubble plots were pioneered by Hans Rosling, Professor of International Health, Karolinska Institutet, and the people of the Gapminder.org project.

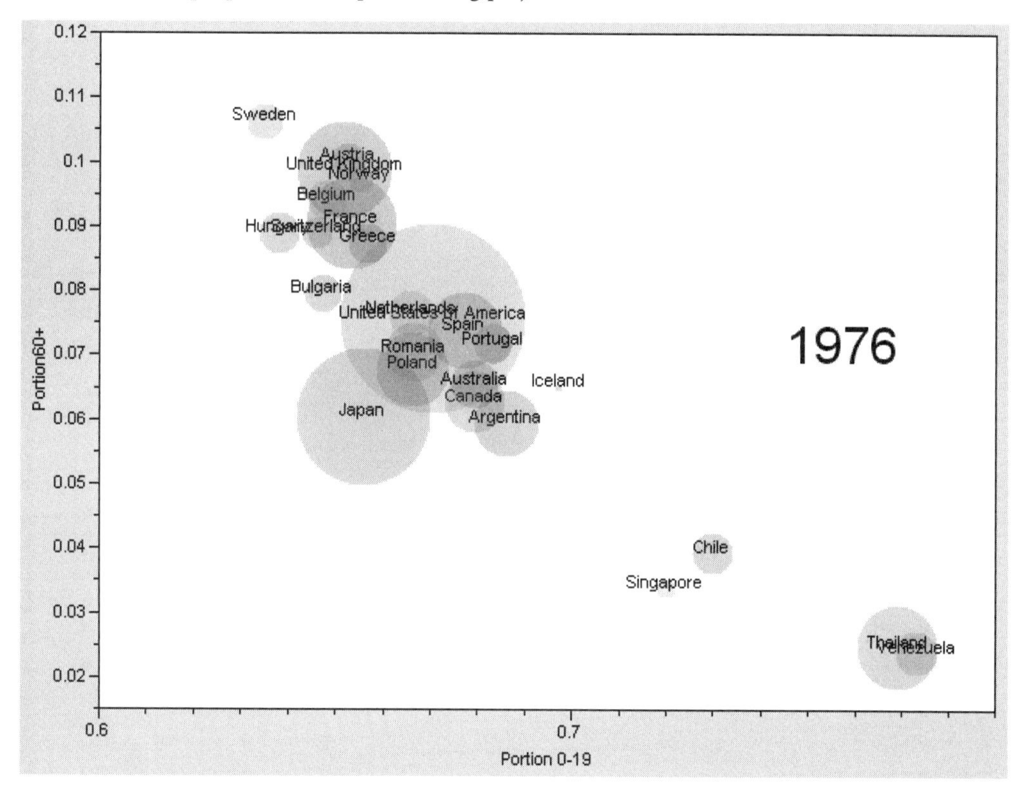

Contents

Launching the Platform

To launch the Bubble Plot Platform, Select **Graph > Bubble Plot** from the main menu. The launch dialog appears.

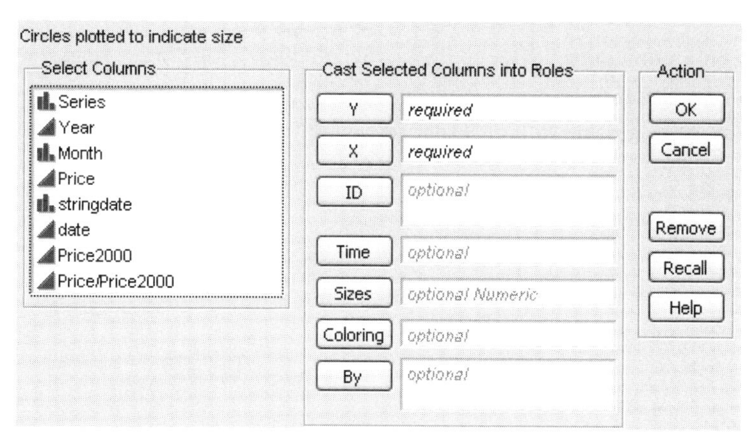

The following roles are used to generate the bubble plot.

Y, X columns become the (x, y) coordinates of the bubbles in the plot. These values can be continuous or categorical, where the bubbles are positioned by the category indices.

Sizes controls the size of the bubbles. The area of the bubble is proportional to the **Size** column's value. Bubble size has a lower limit to keep it visible, even when the size value is zero.

ID variables are optional and used to identify rows that should be aggregated and displayed as a single bubble. The coordinates of each bubble are the averaged x- and y-values, and the size of each bubble is the sum of the sizes of all aggregated members. A second **ID** variable provides a hierarchy of categories, but the bubbles are not split by the second category until they are selected and split interactively. If a second **ID** variable is specified, then Split and Merge buttons appear for this use.

For example, you may specify a country as the first **ID** variable, resulting in a separate aggregated bubble for each country. A second **ID** variable, perhaps designating regions within each country, would further split each country when the interactive **Split** button under the graph is pressed.

Time columns cause separate coordinates, sizes, and colors to be maintained for each unique time period. The bubble plot then shows these values for a single time period. For example, if the **Time** column contains years, the bubble plot shows a unique plot for each year's data. However, the displayed time can be adjusted through controls on the resulting graph.

If a **Time** column is specified, **Time** and **Speed** sliders, and **Step**, **Prev**, **Go**, and **Stop** buttons appear on the graph for interactive investigation.

When data is missing for values within the specified time column, a value is linearly interpolated. Values at the ends of the series are not extrapolated.

Coloring causes the bubbles to be colored. If the **Coloring** variable is categorical, each category is colored distinctly. If the **Coloring** variable is numeric, a continuous gradient of colors from blue to grey to red is used.

Using the Time and ID Variables

The Bubble Plot platform is used in one of two modes

- *static mode*, when you specify only **X**, **Y**, and **Size** variables.
- *dynamic mode*, where **Time** and **ID** variables are additionally specified.

Including only a time variable but no **ID** variable produces a plot with a single moving bubble that tracks the series as the **Time** value changes.

Figure 52.1 **Time** variable with no **ID** variable produces a single moving dot

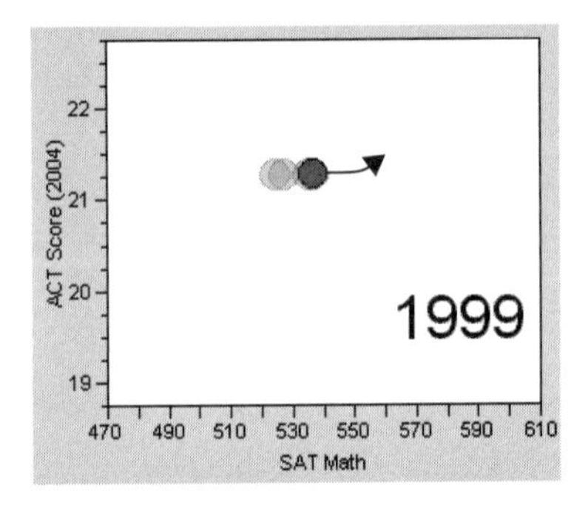

Including one or two **ID** but no **Time** variable produces a static plot with a bubble at each ID value. Although static, this summary graph supports splitting.

Figure 52.2 Splitting a Bubble using **SmogNBabies.jmp**

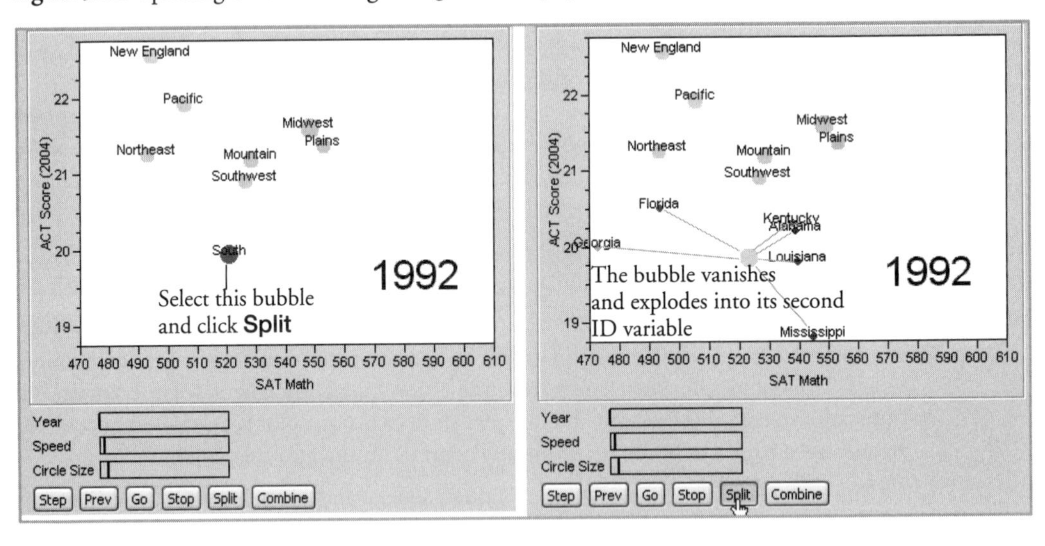

Driving the Platform

Bubble Plot is dynamic and provides smooth animation controls, allowing you to explore items across time interactively. Rapid motion of the bubbles, that is, those whose values are changing fastest, are easy to see.

Selection

Click the center of a bubble to select it. Visually, selected bubbles become more opaque, while non-selected bubbles become more transparent. In addition, if the bubble was initially not filled, selection fills it. With nothing selected, all bubbles are semi-transparent.

Selecting a bubble also selects all rows that the bubble represents. If the bubble is an aggregate based on an ID column, all rows for that ID are selected. Otherwise the one row represented by that bubble is selected.

Clicks in dynamic bubble plots select all times for that ID.

Rows selected in the data table by other means are reflected in the bubble plot.

Animation Controls

Depending on selections in the launch dialog, a variety of controls appear with the plot.

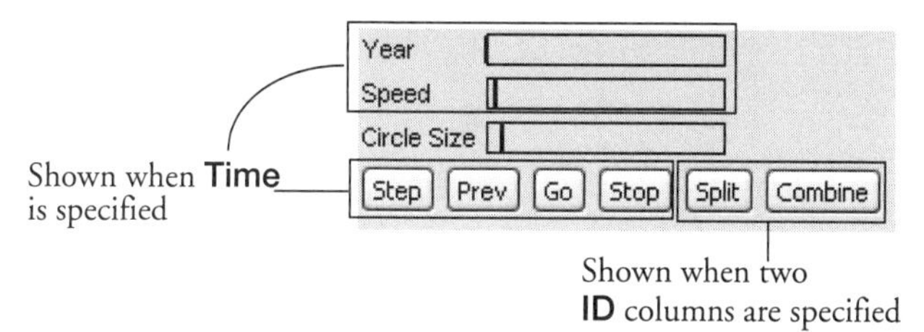

The Time slider (in this case **Year**) adjusts which year's data is displayed in the plot. You can manually drag the slider to see a progression of years.

The **Speed** slider is used in conjunction with the **Go** button, altering the speed that the animation progresses. Move the slider to the right to increase the speed, to the left to decrease the speed.

The **Circle Size** slider appears with every bubble plot, and is used to adjust the size of the bubbles themselves. The bubbles maintain their relative size, but their absolute size can be adjusted by moving this slider to the left for smaller circles and to the right for larger circles.

The **Step** and **Prev** buttons adjust the Time value by one unit. **Step** shows the next time, **Prev** shows the previous.

Go and **Stop** work like the Play and Stop buttons on a video player. **Go** moves through the time values, in order, looping back to the beginning when the last time period is reached. **Stop** stops the animation.

Split and **Combine** appear when two **ID** columns are specified. The second **ID** column should be nested within the first (*e.g.* country within continent). Choose **Split** to separate a bubble represented by the first, larger **ID** column into its smaller constituent parts, defined by the second **ID** column.

Figure 52.2 illustrates the splitting of a column. The **Combine** button undoes a split, *i.e.* recombines the smaller bubbles into their original bubble.

Platform Options

The following options are available from the report's drop-down menu.

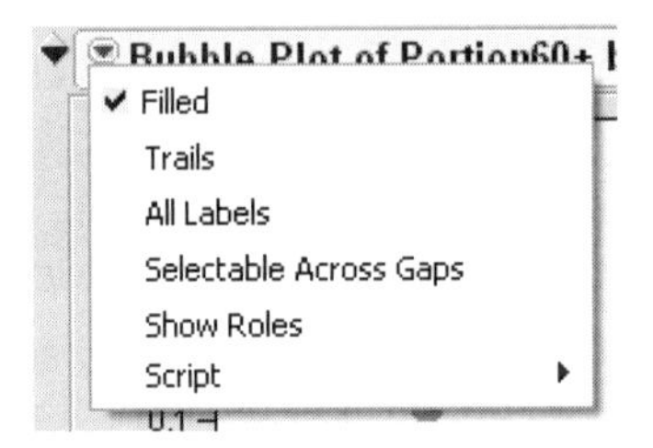

Filled toggles the fill of each bubble. Unfilled bubbles show as empty colored circles.

Trails are available when a Time column is specified. They show the past history of selected bubbles as a semi-transparent trail.

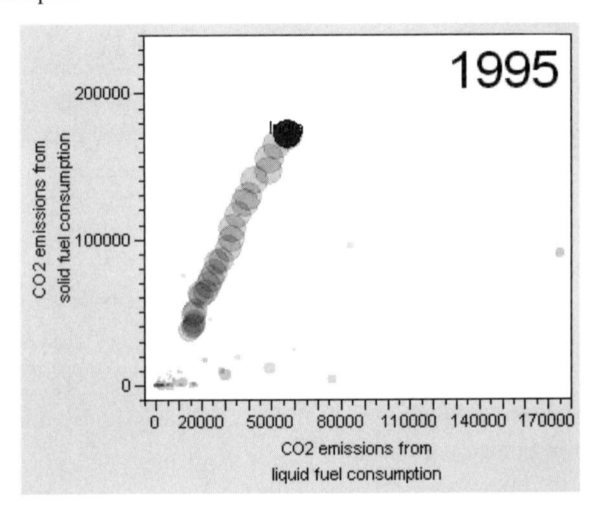

All Labels labels all bubbles in the plot, rather than just the selected bubbles.

Selectable Across Gaps allows a bubble to maintain its selection status, even if there are gaps in the actual data. Since data gaps are always interpolated by JMP, this option essentially toggles whether this interpolation is shown or not.

Show Roles shows the variables that are used in the plot. If you click the blue underlined role names, a dialog appears that allows you to choose another column. If you click OK without selecting a column, the existing column is removed. Note that X and Y roles can only be changed and cannot be removed. In addition, column names can be dragged from the data table to a role box.

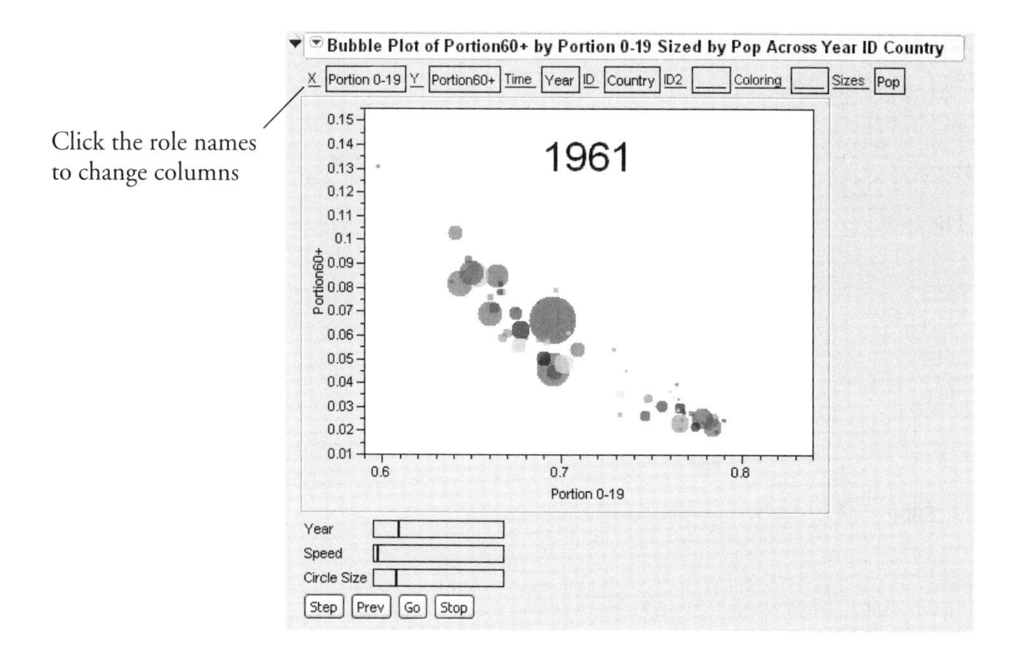

Brushing

A particularly useful property of dynamic bubble plots is that brushing is temporal, that is, it only selects and deselects for a given time period. This allows you to pounce on interesting bubbles at certain times, and JMP only highlights that bubble at that time.

To see an example of this, run the bubbles slowly with the brush tool selected, and quickly pounce on any behavior that's interesting. Then, replay the animation and note that the selection is highlighted in the animation and the data table, while points in other time periods are not. Additionally, you can manually adjust the time slider, watching for the selection and reviewing the discovery. For a granular examination of the details, use **Tables > Subset** or the Row Editor to examine interesting rows.

Examples

The following examples illustrate the features of the Bubble Plot platform.

Simple Static Example

A simple example (using SATByYear.jmp) shows a graph of SAT Verbal vs. SAT Math, with State as **ID** and % Taking as **Size**.

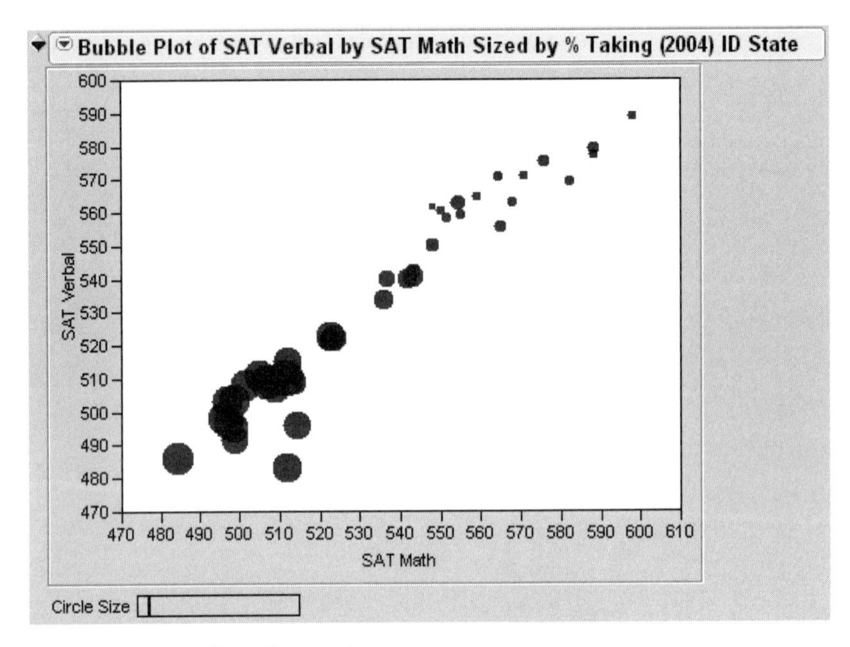

Two conclusions are obvious from this graph.

- there is a strong correlation between verbal and math scores
- States that have a large **% Taking** (large bubbles) are grouped together in the lower left part of the graph.

This data can also be shown according to region, only revealing the state-level information when needed. The following graph is identical to the previous one, except there are two ID variables: **Region**, then **State**. In addition, the Southwest region is highlighted.

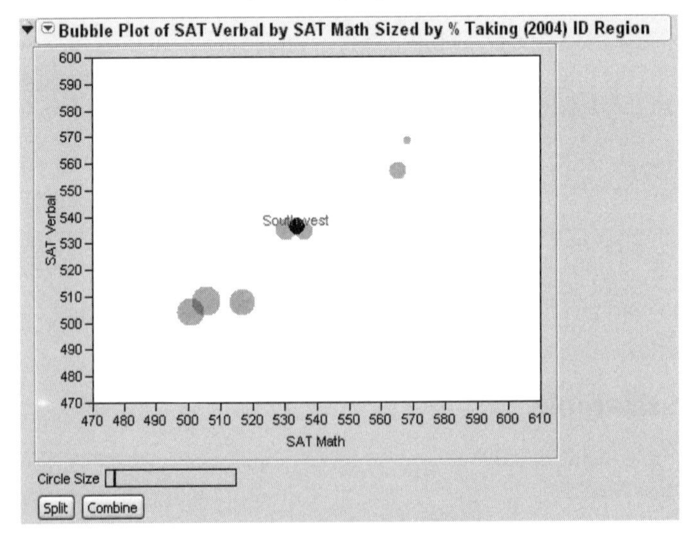

Click the **Split** button to see the states contributing to this bubble.

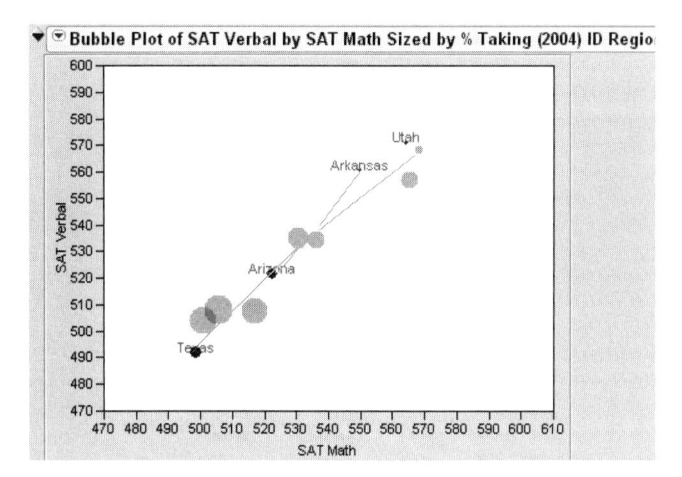

This shows the significant variation among the scores from the southwest states. After clicking **Combine**, you can select and Split the New England states to see an example of states that do not have as much variation.

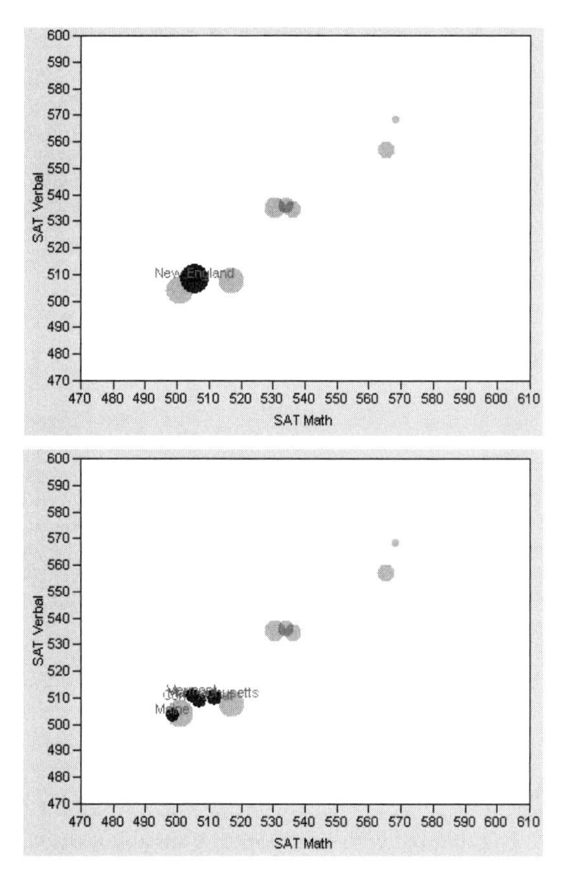

Dynamic Example

As an example of a dynamic bubble plot, open the PopAgeGroup.jmp data set, which contains population data for countries and regions around the world. Assign the variables as in the following dialog box.

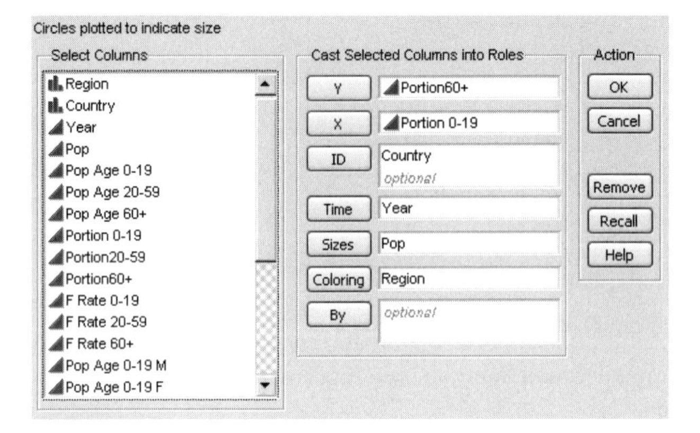

When you click OK, the following report appears. Since there is a Time variable, controls appear at the bottom of the report for progressing through the time levels. Click Step to move forward by one year. Click Go to see the animated, dynamic report.

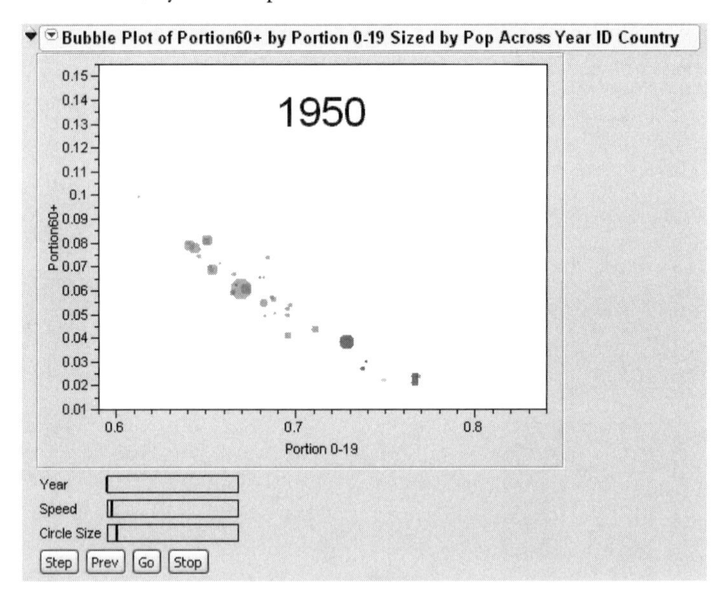

Categorical Y Example

Bubble plots can also be produced with a categorical Y variable. As an example, use blsPriceData.jmp, which shows the price of commodities for several years. Since the value of the US dollar changes over time, a column named Price/Price2000 is included that shows the ratio of each dollar amount compared to the price of the commodity in the year 2000.

Assign Price/Price2000 as **X**, Series as **Y** and date as **Time**. This produces a bubble plot that, when animated by clicking **Go**, shows the price bubbles moving side to side according to their price value.

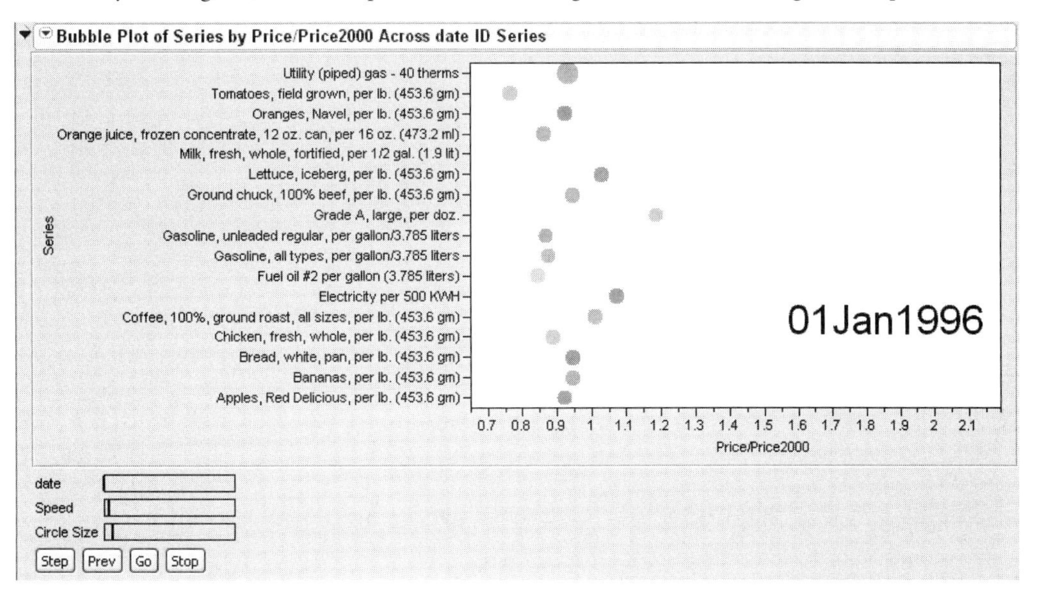

Tip: For easier readability, double-click a categorical axis and check a grid line option.

Chapter 53

Ternary Plots

The Ternary Plot Platform

You launch the Ternary platform with the **Ternary Plot** command in the **Graph** menu (or toolbar).

Ternary plots are a way of displaying the distribution and variability of three-part compositional data such as the proportion of sand, silt, and clay in soil or proportion of three chemical agents in a trial drug. You can use data expressed in proportions or use absolute measures.

The ternary display is a triangle with sides scaled from 0 to 1. Each side represents one of the three components. A point is plotted so that a line drawn perpendicular from the point to each leg of the triangle intersect at the component values of the point.

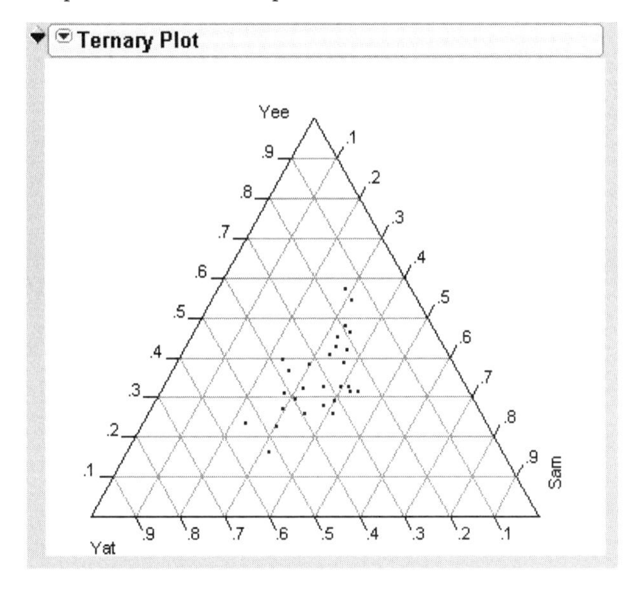

Contents

The Ternary Platform

To illustrate the features and options of the Ternary platform, consider the **Pogo Jumps.jmp** file in the **Sample Data** folder. Figure 53.1 shows a partial listing of the data. The data show measurements for pogo jumps of seven finalists in the 1985 Hong Kong Pogo-Jump Championship (Aitchison 1986). A pogo jump is the total jump distance in three consecutive *bounces*, referred to as *yat*, *yee*, and *sam*.

To look at each jump in a yat-yee-sam reference, select **Ternary Plot** from the **Graph** menu (or toolbar). Complete the dialog as shown in Figure 53.1.

Figure 53.1 Partial Listing of the Pogo Jumps Data Table and the Ternary Launch Dialog

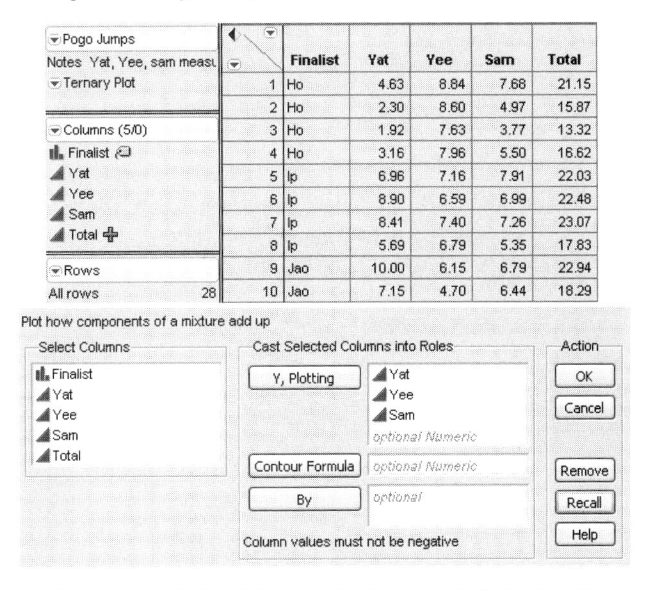

	Finalist	Yat	Yee	Sam	Total
1	Ho	4.63	8.84	7.68	21.15
2	Ho	2.30	8.60	4.97	15.87
3	Ho	1.92	7.63	3.77	13.32
4	Ho	3.16	7.96	5.50	16.62
5	Ip	6.96	7.16	7.91	22.03
6	Ip	8.90	6.59	6.99	22.48
7	Ip	8.41	7.40	7.26	23.07
8	Ip	5.69	6.79	5.35	17.83
9	Jao	10.00	6.15	6.79	22.94
10	Jao	7.15	4.70	6.44	18.29

When you click **OK** on the Launch dialog, Ternary displays the default plot shown to the left in Figure 53.2. The plot on the right results when you use the magnifier tool from the **Tools** menu to zoom so you can look closer at the cluster of points.

Figure 53.2 Simple Ternary Plot of Pogo Jumps

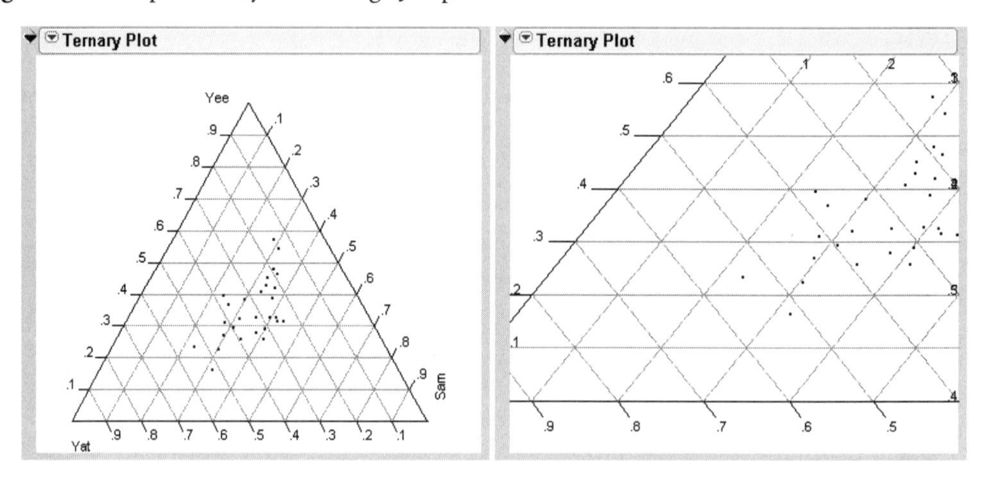

If you have variables in a Contour formula that are not listed as **Y, Plotting** variables, JMP appends sliders below the plot so that the values can be interactively adjusted. In the following picture, X1, X2, and X3 are plotted, with X4 and X5 shown as sliders.

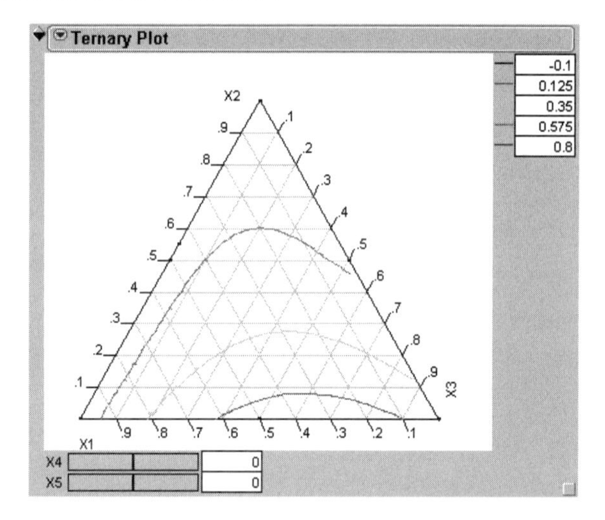

Use the crosshairs tool to determine exact coordinates of points within the plot.

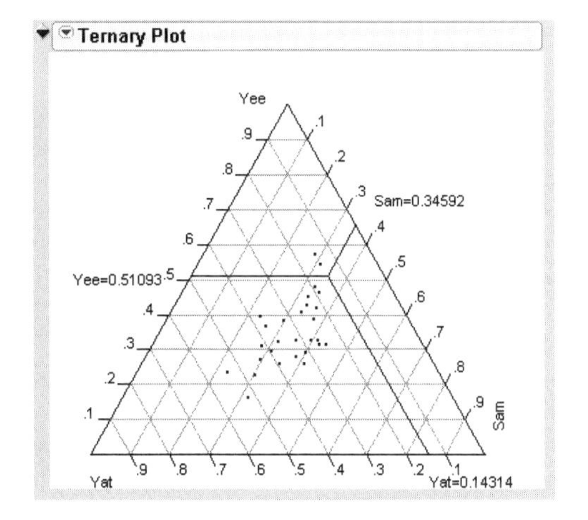

Display Options

The ternary plot has three display commands accessed by the popup menu on the Ternary Plot title bar:

Ref Labels shows or hides the labels on the three axes.

Ref Lines shows or hides the triangular grid lines. The default platform shows the grid.

Show Points shows or hides the plotted points. The default platform shows the points.

Contour Fill allows filling of contours if a contour formula is specified in the plot. It allows coloring as **Lines Only**, **Fill Above**, and **Fill Below**. The default platform shows lines only.

Script has a submenu of commands available to all platforms that let you redo the analysis, save the JSL commands for the analysis to a window or a file, or with the data table.

The Pogo Jump Example

To get a better feel for how the three bounces contribute to total scores, the Pogo Jumps.jmp table is summarized by Finalist using the **Summary** command in the **Tables** menu. To follow this example, complete the Summary dialog as shown in Figure 53.3. Create a new column (Grand Total) and use the formula editor to compute the sum of each finalist's scores. The data table in Figure 53.3 is sorted in ascending order, which shows the rank of the players.

Figure 53.3 Summary Table Showing Totals for Each Finalist and Grand Totals of the Pogo Champions

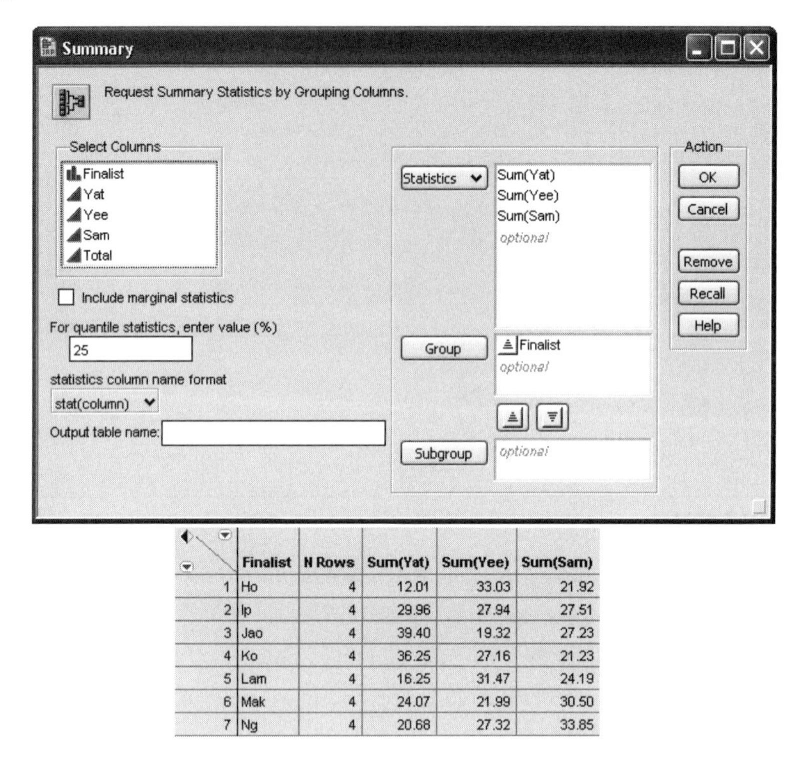

	Finalist	N Rows	Sum(Yat)	Sum(Yee)	Sum(Sam)
1	Ho	4	12.01	33.03	21.92
2	Ip	4	29.96	27.94	27.51
3	Jao	4	39.40	19.32	27.23
4	Ko	4	36.25	27.16	21.23
5	Lam	4	16.25	31.47	24.19
6	Mak	4	24.07	21.99	30.50
7	Ng	4	20.68	27.32	33.85

The sum of the three jumps can be displayed by the Ternary plot platform, as shown by the plot in Figure 53.4. Note that the winner and the loser appear with labels on the ternary plot. To show the value of one or more label variables on a plot, first identify the columns with **Cols > Label/Unlabel** whose values you want displayed as labels. Select the rows to be labeled in the data table as shown in the left in Figure 53.4 and select the **Label/Unlabel** command from the **Rows** menu.

Figure 53.4 Grand Totals of the Pogo Championship

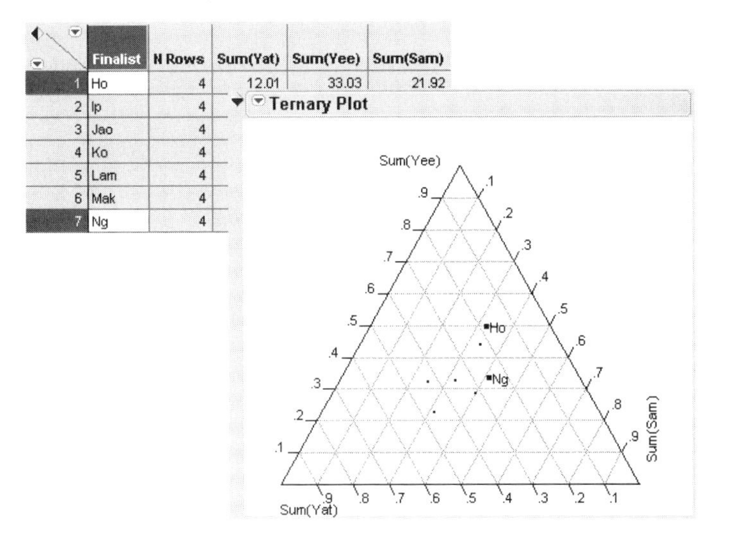

Chapter 54

Tree Maps

Visualizing Many Groups

Tree maps are like bar charts that have been folded over in two dimensions so that there is no unused space. It is useful when there are a lot of categories.

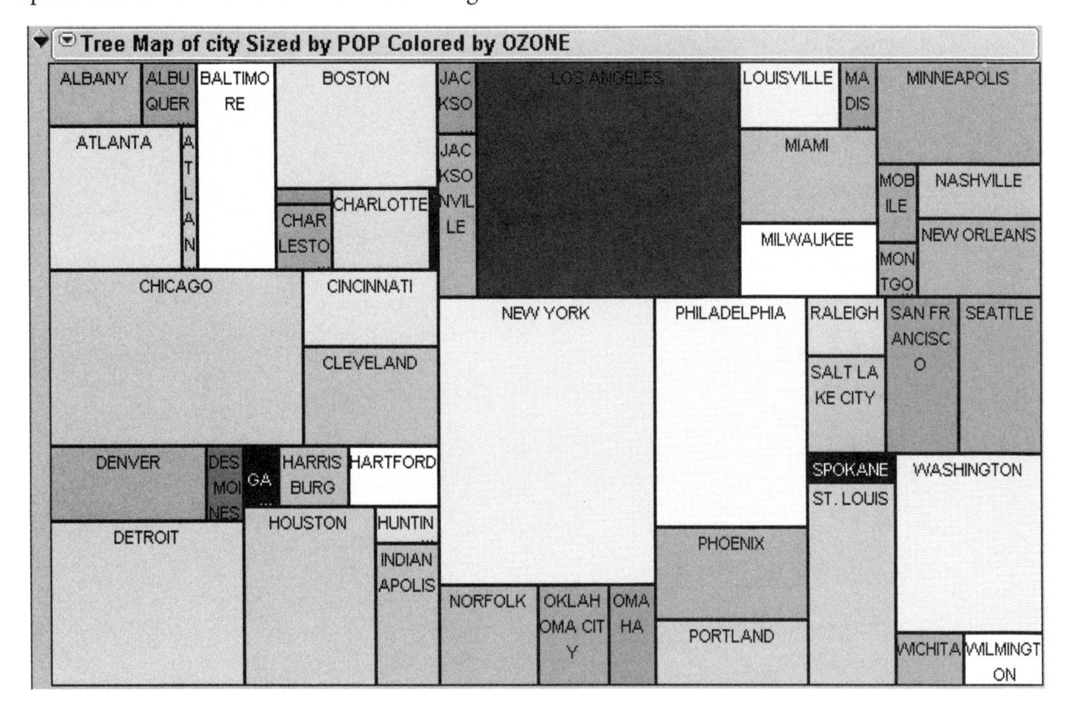

Contents

Tree Maps

Tree maps are a graphical technique of observing patterns among groups that have many levels. They are especially useful in cases where histograms are ineffective. For example, the Cities.jmp data set, containing various meteorological and demographic statistics for 52 cities. A bar chart of ozone levels shows the problem.

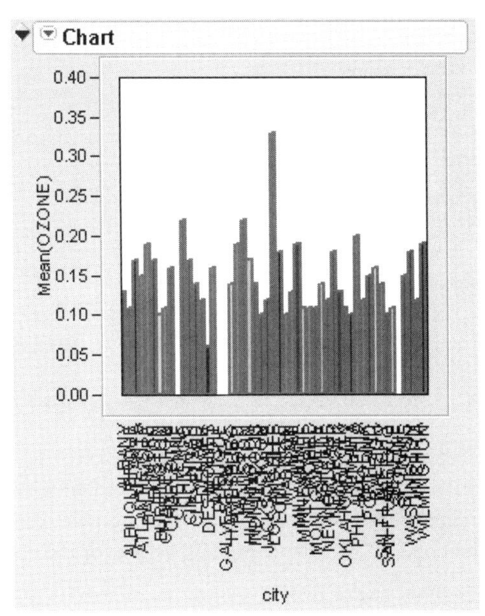

Although it is easy to see that there is a single large measurement, it is difficult to see what city it corresponds to.

A tree map approaches this problem by folding this data over two dimensions. Rather than drawing a single bar for each measurement, a tree map relays the magnitude of a measurement by varying the size or color of a rectangular area.

Tree maps were named and popularized by Ben Schneiderman, who has an extensive web site about the idea (http://www.cs.umd.edu/hcil/treemap-history/index.shtml).

An example shows this clearly. The corresponding tree map to the histogram above can be generated by selecting **Graph > Tree Map** and assigning City as the **Categories** variable and Ozone as the **Coloring** variable.

Figure 54.1 Tree Map of City Colored by Ozone

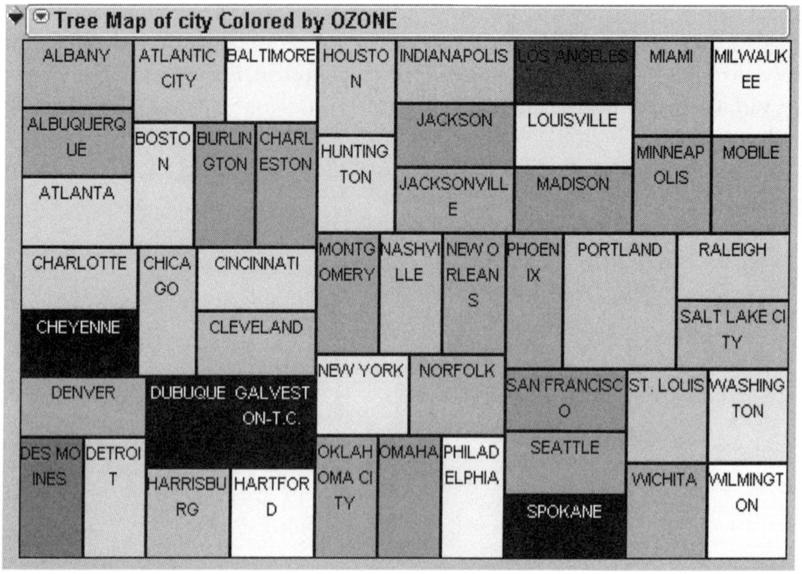

The magnitude of the ozone level for each city is shown by coloring a square from a continuous color spectrum, with deep green mapping to low ozone values and deep red mapping to high ozone values. More mediocre values decrease the intensity of the color, so pale green and pink are values closer to the mean ozone level. Cities colored black have missing values for the coloring variable.

Since we haven't specified any size or ordering options, the tree is split alphabetically.

Launching the Platform

After selecting **Graph > Tree Map**, you are presented with the following dialog box.

Figure 54.2 Tree Map Launch Dialog

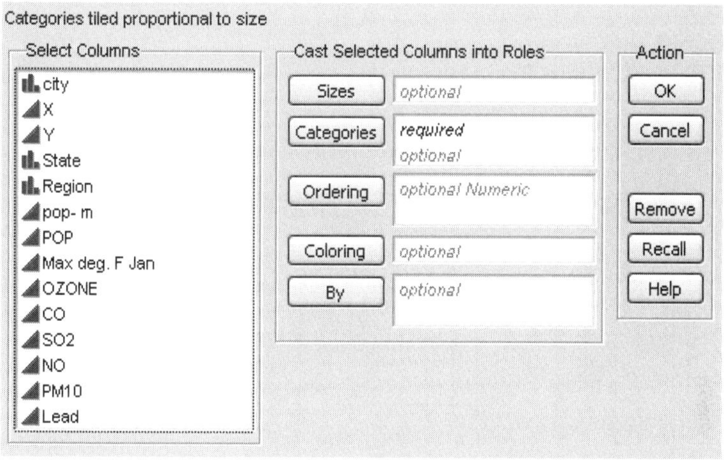

Categories tiled proportional to size

Select Columns: city, X, Y, State, Region, pop- m, POP, Max deg. F Jan, OZONE, CO, SO2, NO, PM10, Lead

Cast Selected Columns into Roles: Sizes (optional), Categories (required, optional), Ordering (optional Numeric), Coloring (optional), By (optional)

Action: OK, Cancel, Remove, Recall, Help

Categories

The only required variable role is that of **Categories**. When you specify only a **Categories** variable, JMP produces a tree map with the squares colored from a rotating color palette, arranged alphabetically, and sized by the number of occurrences of each group. Specifying **city** as a **Categories** variable produces a plot like the one in Figure 54.3. Most of the rectangles have equal area, because there is only one data point per city. Portland has two measurements, so has an area twice that of other cities.

Figure 54.3 Category = City Only

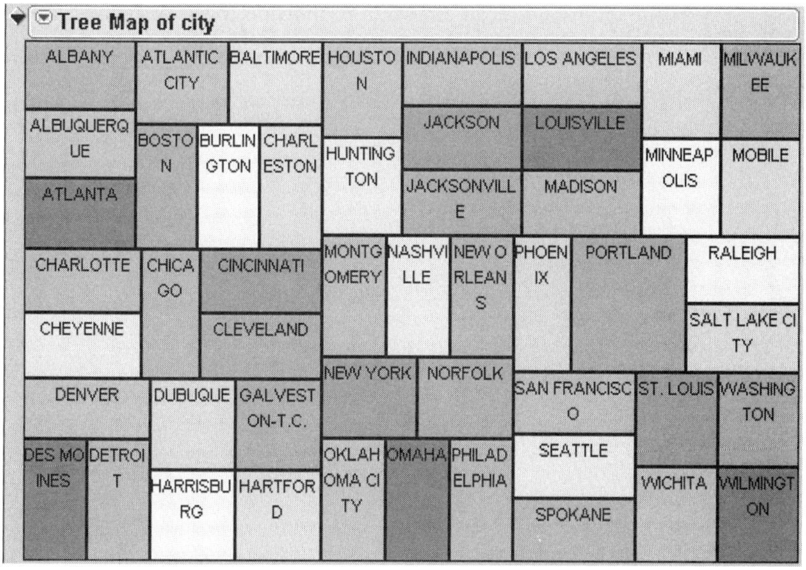

Tree Map of city

ALBANY, ATLANTIC CITY, BALTIMORE, HOUSTON, INDIANAPOLIS, LOS ANGELES, MIAMI, MILWAUKEE, ALBUQUERQUE, BOSTON, BURLINGTON, CHARLESTON, HUNTINGTON, JACKSON, LOUISVILLE, MINNEAPOLIS, MOBILE, ATLANTA, JACKSONVILLE, MADISON, CHARLOTTE, CHICAGO, CINCINNATI, MONTGOMERY, NASHVILLE, NEW ORLEANS, PHOENIX, PORTLAND, RALEIGH, CHEYENNE, CLEVELAND, SALT LAKE CITY, NEW YORK, NORFOLK, DENVER, DUBUQUE, GALVESTON-T.C., SAN FRANCISCO, ST. LOUIS, WASHINGTON, DES MOINES, DETROIT, OKLAHOMA CITY, OMAHA, PHILADELPHIA, SEATTLE, HARRISBURG, HARTFORD, WICHITA, WILMINGTON, SPOKANE

Assigning two variables to **Categories** tells JMP to group the tree map by the first variable, and to sort within groups by the second variable. For example, Figure 54.4 assigns Region and City (in that order) to the Categories role.

Figure 54.4 Two-Category Tree Map

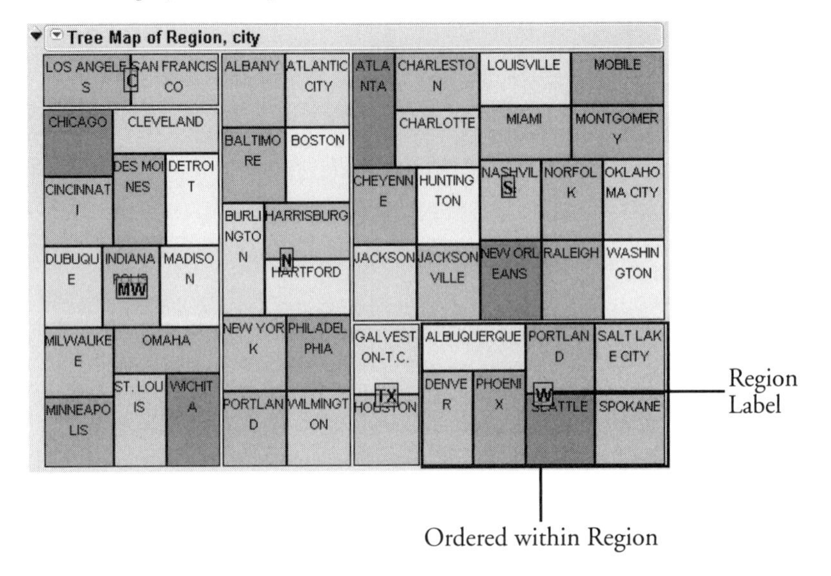

Sizes

If you want the size of the squares to correspond to the levels of a variable, assign that variable to the Sizes role. Areas of the resulting tiles are drawn in proportion to the designated variable.

For example, examine a tree map with city as **Categories** and POP (population) as the **Sizes** variable (Figure 54.5)

Figure 54.5 Tree Map with Sizes role

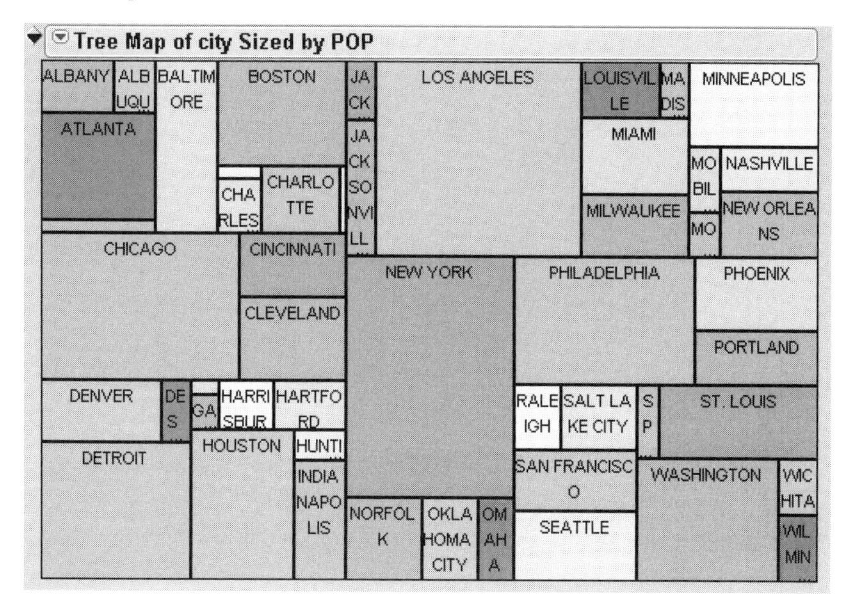

The large cities are represented by large rectangles, small cities by smaller rectangles.

Coloring

Specify a **Coloring** variable to have the colors of the rectangles correspond to the variable's levels. If the **Coloring** variable is continuous, the colors vary along a continuous scale from green (low) to red (high). The color is most intense at the extremes of the variable, with paler colors corresponding to levels close to the mean.

Figure 54.1 specifies city as the **Categories** and ozone as the **Coloring** variable. Note that the size of the squares is still based on the number of occurrences of the **Categories** variable, but the colors are mapped to ozone values. The high value for Los Angeles is clear.

Note: Missing values for the **Coloring** variable are drawn as black squares.

If the **Coloring** variable is categorical, colors are selected in order from JMP's color palette. In the following example, city is specified for **Categories** and region for **Coloring**. Default values are used for **Ordering** (alphabetical) and **Sizes** (number of occurrences).

Figure 54.6 Cities colored by Region

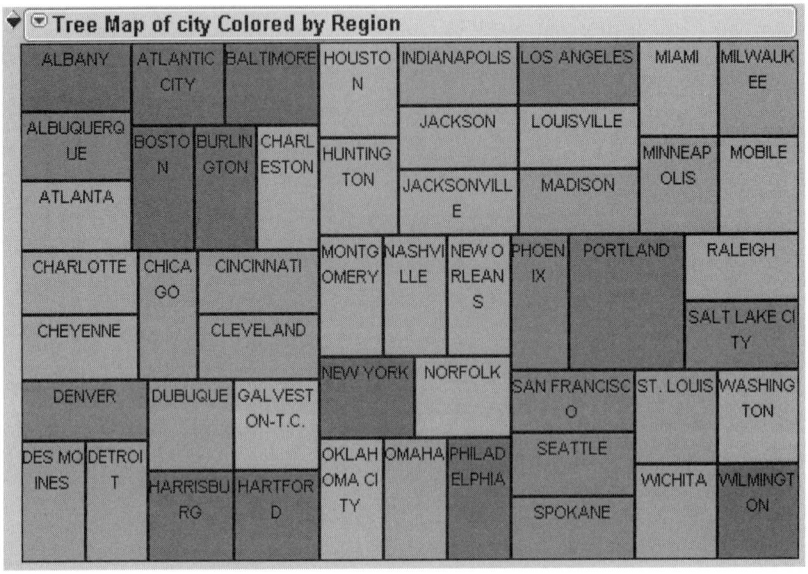

Ordering

By default, the tiles of a tree map are drawn in alphabetical order, with values progressing from the top left to the lower right. To change this ordering, use the **Ordering** role.

If you specify a single **Ordering** variable, tiles are drawn so that high levels of the variable cluster together, and low levels of the variable cluster together. For example, the following tree map uses city as **Categories** and POP (population) as the **Ordering** variable.

Figure 54.7 Tree Map win One Ordering Variable

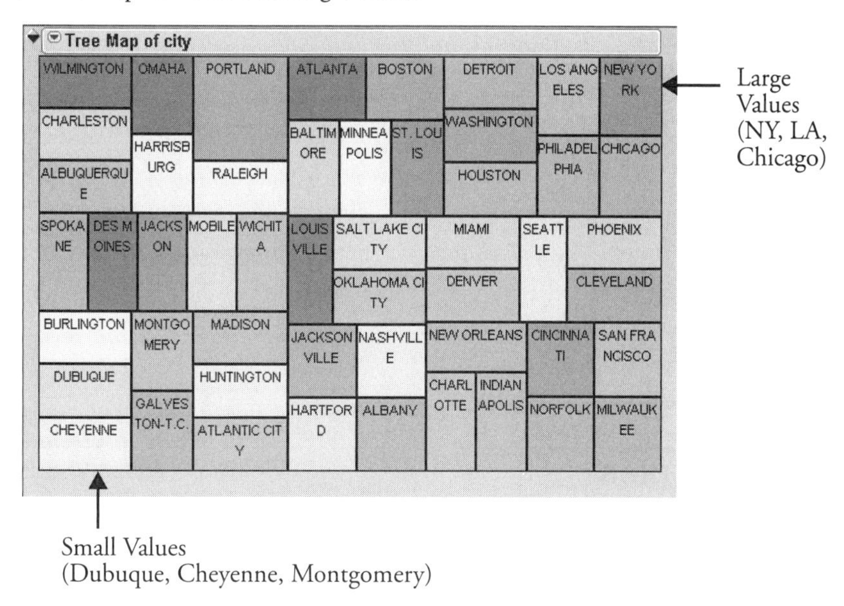

If you have two ordering variables, JMP produces a tree map that arranges the tiles horizontally by the first ordering variable and vertically by the second ordering variable. This is particularly useful for geographic data. For example, the variables X and Y in the Cities.jmp data set correspond to the geographic location of the cities. A tree map with X and Y as ordering variables is shown in Figure 54.8.

Figure 54.8 Tree Map with Two Ordering Variables

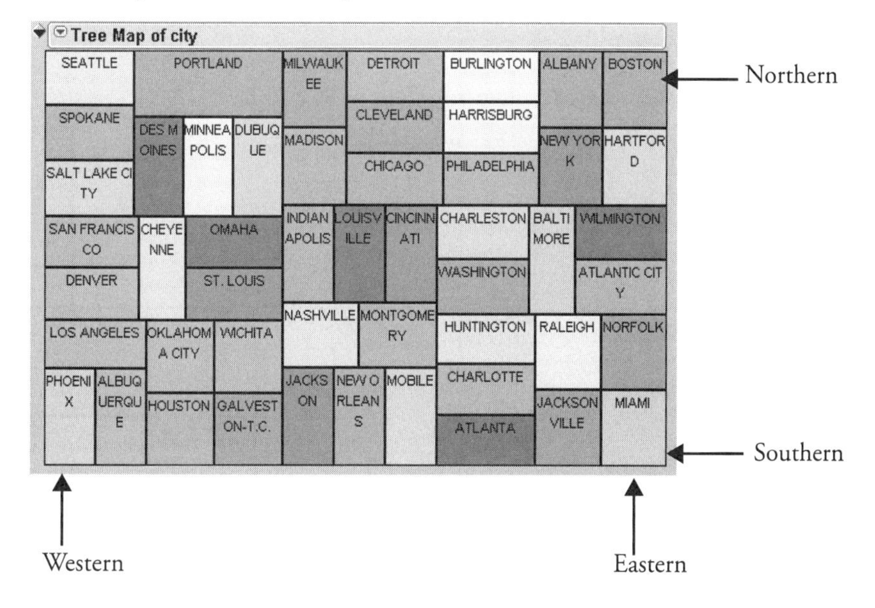

Platform Options

Aside from the standard JMP **Script** Menu, tree maps have the following options.

Change Color Column is available from the platform drop-down menu. It brings up a dialog that allows you to change the column used for coloring the squares. Choose a new column from the dialog and click **OK**.

The following two options are accessed by right-clicking on the tree map itself.

Suppress Box Frames suppresses the black lines outlining each box.

Ignore Group Hierarchy flattens the hierarchy and sorts by the ordering columns without using grouping except to define cells.

Tree Map Examples

The options described above can be combined together to form some interesting graphs.

Pollution

Using the same data set, we can examine the distribution of different pollution measurements across the United States, displaying geographical position and city population as well.

For example, to examine the ozone levels of the various cities, assign city to **Categories**, POP to **Sizes**, both X and Y to **Ordering**, and OZONE to **Coloring**. This results in the graph shown in Figure 54.9.

Figure 54.9 Ozone levels

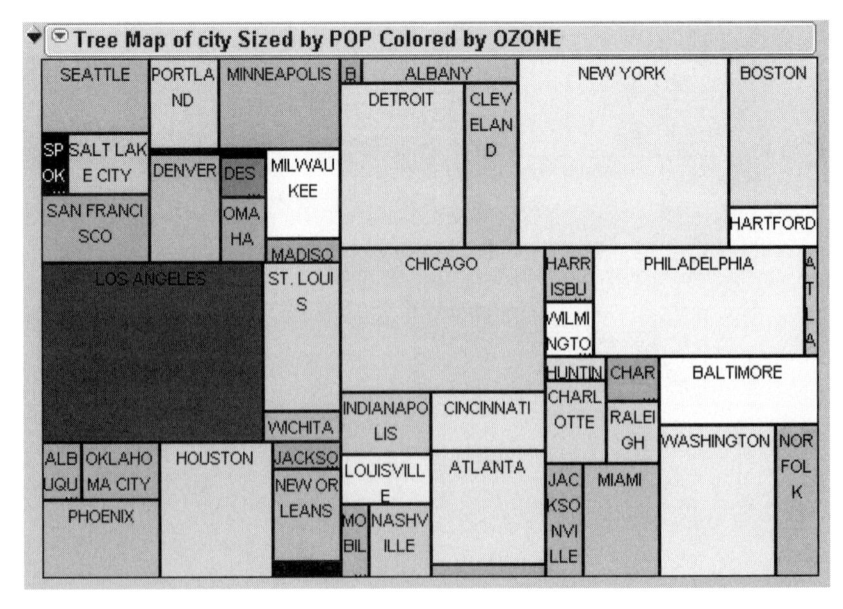

With this graph, you can see that Los Angeles, a western city, has a high level of ozone. Other large cities have slightly elevated ozone levels (Chicago and Houston), and some large cities have slightly lower-than-average ozone levels (New York and Washington).

Lead levels can be examined in a similar way. Another tree map with similar settings (city to **Categories**, POP to **Sizes**, both X and Y to **Ordering**) but with Lead as **Coloring** results in Figure 54.10.

Figure 54.10 Lead levels

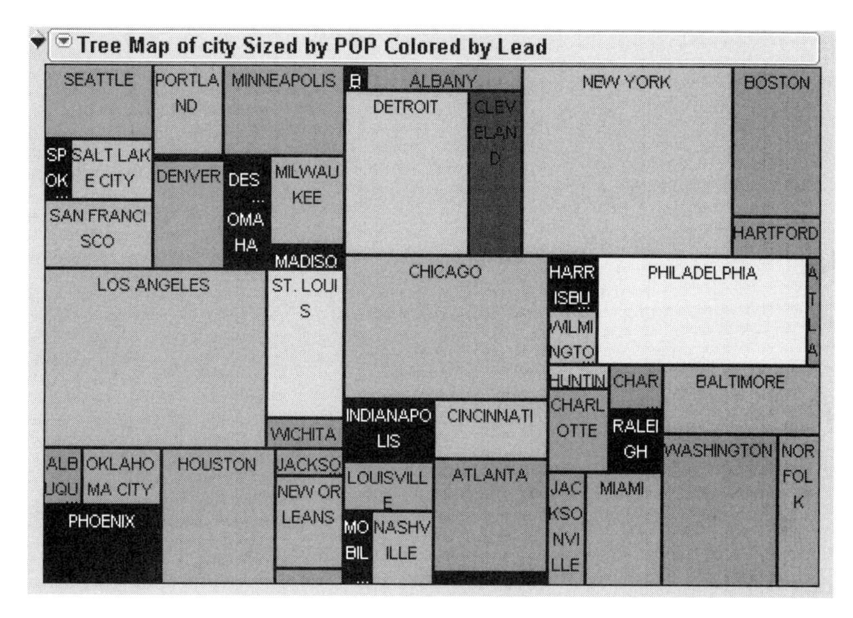

Seattle (a northwestern city) and Cleveland (a city toward the middle of the country) have high lead levels. Remarkably, most other cities have rather low lead levels. Note that several of the cities (Raleigh, Phoenix) have missing values for Lead measurements.

Failure Cause Counts

While the Pareto platform may be natural to use for a small number of categories, a tree map can be used to examine failure modes in quality control situations. For example, the following tree map uses the Failure3.jmp sample data file, found in the Quality Control sub-directory.

The tree map in Figure 54.11 uses failure and clean as **Categories**, N as **Sizes**, and clean as **Coloring**. By looking at the graph, it is obvious that contamination was the largest cause of failure, and contamination occurred at the "before" measurement more often than the "after" measurement.

Figure 54.11 Failure Tree Map

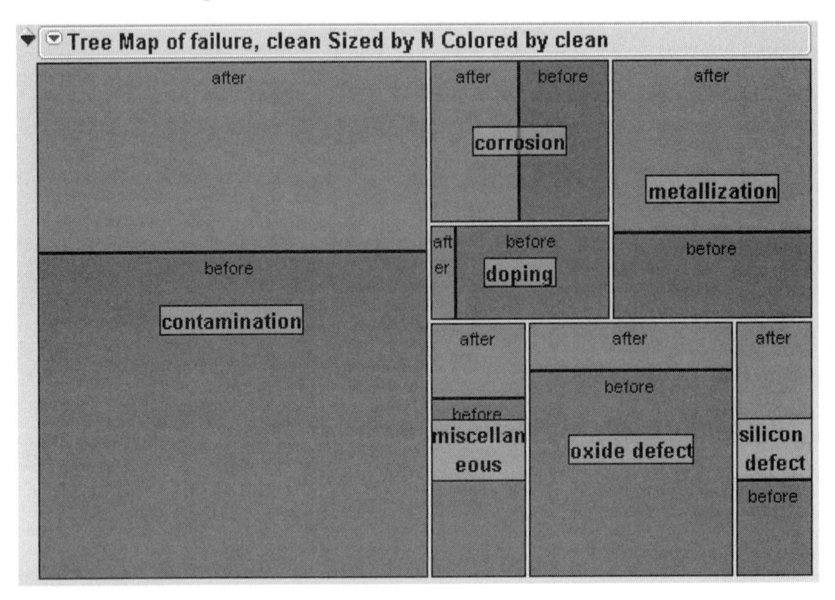

Patterns in Car Safety

The data file Cars.jmp contains measurements of various points of crash-test dummies in automobile safety tests. A tree map allows you to compare these measurements for different automobile makes and models.

In these examples, we used the Select Where and Subset commands to select only cars with year=90 and year=91.

Figure 54.12 Tree Map of Left Leg Deceleration

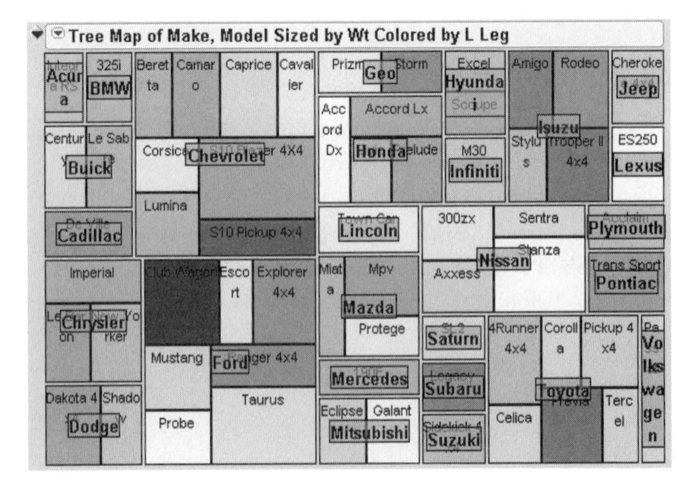

Figure 54.12 shows a map resulting from Make and Model as **Categories**, wt (weight) as **Sizes**, and L Leg (a measurement of the deceleration speed of the left leg, where more deceleration causes more injury) as **Coloring**.

The map shows that there are a few cars that seem to suffer from large values of left leg deceleration, most strongly the Club Wagon and the S10 Pickup.

You can quickly examine the other safety measurements by selecting **Change Color Column** from the platform drop-down menu and selecting another color column. Using this command lets you make comparisons among the measurements. For example, although the S10 had a bad rating for leg deceleration, it has a better measurement for head injuries. The Trooper B suffers from just the opposite problem, seen when comparing Figure 54.12 with Figure 54.13 (where the coloring column is Head IC).

Figure 54.13 Tree Map of Potential Head Injury

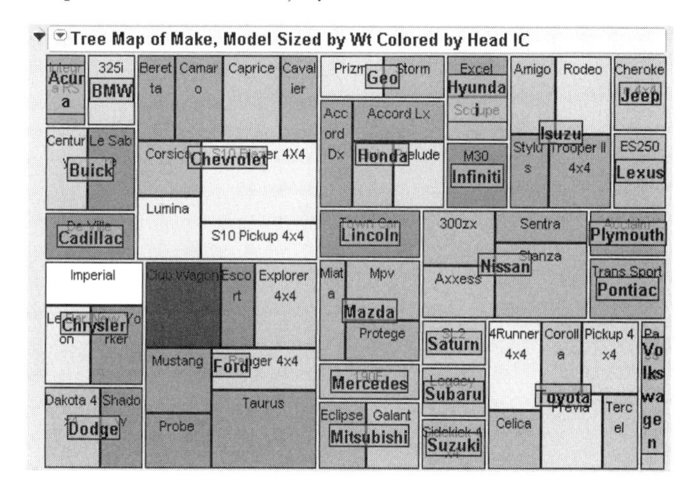

Statistical Details

Models in JMP

This appendix discusses the different types of response models, their factors, their design coding, and parameterization. It also includes many other details of methods described in the main text.

The JMP system fits linear models to three different types of response models that are labeled continuous, ordinal, and nominal. Many details on the factor side are the same between the different response models, but JMP only supports graphics and marginal profiles on continuous responses—not on ordinal and nominal.

Different computer programs use different design-matrix codings, and thus parameterizations, to fit effects and construct hypothesis tests. JMP uses a different coding than the GLM procedure in the SAS system, although in most cases JMP and SAS GLM procedure produce the same results. The following sections describe the details of JMP coding and highlight those cases when it differs from that of the SAS GLM procedure, which is frequently cited as the industry standard.

Contents

The Response Models

JMP fits linear models to three different kinds of responses: continuous, nominal, and ordinal.

Continuous Responses

JMP models continuous responses in the usual way through fitting a linear model by least squares. Continuous responses are much more richly supported than the other response types. Continuous responses are reported with leverage plots, least squares means, contrasts, and output formulas. None of these features has a corresponding feature for nominal or ordinal responses. Computationally, continuous responses are the easiest to fit.

Nominal Responses

Nominal responses are analyzed with a straightforward extension of the logit model. For a binary (two-level) response, a logit response model is

$$\log\left(\frac{P(y=1)}{P(y=2)}\right) = X\beta$$

which can be written

$$P(y=1) = F(X\beta)$$

where $F(x)$ is the cumulative distribution function of the standard logistic distribution

$$F(x) = \frac{1}{1+e^{-x}} = \frac{e^x}{1+e^x}$$

The extension for r responses is to relate each response probability to the rth probability and fit a separate set of design parameters to these $r-1$ models.

$$\log\left(\frac{P(y=j)}{P(y=r)}\right) = X\beta_{(j)} \text{ for } j = 1, \ldots, r-1$$

The fit is done with maximum likelihood using a modified Newton-Raphson iterative algorithm.

Ordinal Responses

Ordinal responses are modeled by fitting a series of parallel logistic curves to the cumulative probabilities. Each curve has the same design parameters but a different intercept and is written

$$P(y \le j) = F(\alpha_j + X\beta) \quad \text{for } j = 1, \ldots, r-1$$

where r response levels are present and $F(x)$ is the standard logistic cumulative distribution function

$$F(x) = \frac{1}{1+e^{-x}} = \frac{e^x}{1+e^x}$$

Another way to write this is in terms of an unobserved continuous variable, z, that causes the ordinal response to change as it crosses various thresholds

$$y = \begin{cases} r & \text{if } \alpha_{r-1} \le z \\ j & \text{if } \alpha_{j-1} \le z < \alpha_j \\ 1 & \text{if } z \le \alpha_1 \end{cases}$$

where z is an unobservable function of the linear model and error

$$z = X\beta + \varepsilon$$

and ε has the logistic distribution.

These models are attractive in that they recognize the ordinal character of the response, they need far fewer parameters than nominal models, and the computations are fast even though they involve iterative maximum likelihood calculation.

Continuous Factors

Continuous factors are placed directly into the design matrix as regressors. If a column is a linear function of other columns, then the parameter for this column is marked *zeroed* or *nonestimable*. Continuous factors are centered by their mean when they are crossed with other factors (interactions and polynomial terms). Centering is suppressed if the factor has a Column Property of **Mixture** or **Coding**, or if the centered polynomials option is turned off when specifying the model. If there is a coding column property, the factor is coded before fitting.

Nominal Factors

Nominal factors are transformed into indicator variables for the design matrix. SAS GLM constructs an indicator column for each nominal level. JMP constructs the same indicator columns for each nominal level except the last level. When the last nominal level occurs, a one is subtracted from all the other columns of the factor. For example, consider a nominal factor A with three levels coded for GLM and for JMP as shown below.

Figure A.1

	GLM			JMP	
A	A1	A2	A3	A13	A23
A1	1	0	0	1	0
A2	0	1	0	0	1
A3	0	0	1	–1	–1

In GLM, the linear model design matrix has linear dependencies among the columns, and the least squares solution employs a generalized inverse. The solution chosen happens to be such that the A3 parameter is set to zero.

In JMP, the linear model design matrix is coded so that it achieves full rank unless there are missing cells or other incidental collinearity. The parameter for the A effect for the last level is the negative sum of the other levels, which makes the parameters sum to zero over all the effect levels.

Interpretation of Parameters

Note: The parameter for a nominal level is interpreted as the differences in the predicted response for that level from the average predicted response over all levels.

The design column for a factor level is constructed as the zero-one indicator of that factor level minus the indicator of the last level. This is the coding that leads to the parameter interpretation above.

Figure A.2

JMP Parameter Report	How to Interpret	Design Column Coding
Intercept	mean over all levels	1′
A[1]	$\alpha_1 - 1/3(\alpha_1 + \alpha_2 + \alpha_3)$	(A==1) – (A==3)
A[2]	$\alpha_2 - 1/3(\alpha_1 + \alpha_2 + \alpha_3)$	(A==2) – (A==3)

Interactions and Crossed Effects

Interaction effects with both GLM and JMP are constructed by taking a direct product over the rows of the design columns of the factors being crossed. For example, the GLM code

```
PROC GLM;
   CLASS A B;
   MODEL A B A*B;
```

yields this design matrix:

Figure A.3

		A			B			AB								
A	**B**	1	2	3	1	2	3	11	12	13	21	22	23	31	32	33
A1	**B1**	1	0	0	1	0	0	1	0	0	0	0	0	0	0	0
A1	**B2**	1	0	0	0	1	0	0	1	0	0	0	0	0	0	0
A1	**B3**	1	0	0	0	0	1	0	0	1	0	0	0	0	0	0
A2	**B1**	0	1	0	1	0	0	0	0	0	1	0	0	0	0	0
A2	**B2**	0	1	0	0	1	0	0	0	0	0	1	0	0	0	0
A2	**B3**	0	1	0	0	0	1	0	0	0	0	0	1	0	0	0
A3	**B1**	0	0	1	1	0	0	0	0	0	0	0	0	1	0	0
A3	**B2**	0	0	1	0	1	0	0	0	0	0	0	0	0	1	0
A3	**B3**	0	0	1	0	0	1	0	0	0	0	0	0	0	0	1

Using the JMP **Fit Model** command and requesting a factorial model for columns A and B produces the following design matrix. Note that A13 in this matrix is A1–A3 in the previous matrix. However, A13B13 is A13*B13 in the current matrix.

Figure A.4

		A		B		A13	A13	A23	A23
A	B	13	23	13	23	B13	B23	B13	B23
A1	B1	1	0	1	0	1	0	0	0
A1	B2	1	0	0	1	0	1	0	0
A1	B3	1	0	–1	–1	–1	–1	0	0
A2	B1	0	1	1	0	0	0	1	0
A2	B2	0	1	0	1	0	0	0	1
A2	B3	0	1	–1	–1	0	0	–1	–1
A3	B1	–1	–1	1	0	–1	0	–1	0
A3	B2	–1	–1	0	1	0	–1	0	–1
A3	B3	–1	–1	–1	–1	1	1	1	1

The JMP coding saves memory and some computing time for problems with interactions of factors with few levels.

Nested Effects

Nested effects in GLM are coded the same as interaction effects because GLM determines the right test by what isn't in the model. Any effect not included in the model can have its effect soaked up by a containing interaction (or, equivalently, nested) effect.

Nested effects in JMP are coded differently. JMP uses the terms inside the parentheses as grouping terms for each group. For each combination of levels of the nesting terms, JMP constructs the effect on the outside of the parentheses. The levels of the outside term need not line up across the levels of the nesting terms. Each level of nest is considered separately with regard to the construction of design columns and parameters.

Figure A.5

				B(A)					
				A1	A1	A2	A2	A3	A3
A	B	A13	A23	B13	B23	B13	B23	B13	B23
A1	B1	1	0	1	0	0	0	0	0
A1	B2	1	0	0	1	0	0	0	0
A1	B3	1	0	–1	–1	0	0	0	0
A2	B1	0	1	0	0	1	0	0	0
A2	B2	0	1	0	0	0	1	0	0
A2	B3	0	1	0	0	–1	–1	0	0
A3	B1	–1	–1	0	0	0	0	1	0
A3	B2	–1	–1	0	0	0	0	0	1
A3	B3	–1	–1	0	0	0	0	–1	–1

Least Squares Means across Nominal Factors

Least squares means are the predicted values corresponding to some combination of levels, after setting all the other factors to some neutral value. The neutral value for direct continuous regressors is defined as the sample mean. The neutral value for an effect with uninvolved nominal factors is defined as the average effect taken over the levels (which happens to result in all zeroes in our coding). Ordinal factors use a different neutral value in "Ordinal Least Squares Means," p. 989. The least squares means might not be estimable, and if not, they are marked nonestimable. JMP's least squares means agree with GLM's (Goodnight and Harvey 1978) in all cases except when a weight is used, where JMP uses a weighted mean and GLM uses an unweighted mean for its neutral values.

The expected values of the cells in terms of the parameters for a three-by-three crossed model are:

Figure A.6

	B1	B2	B3
A1	$\mu+\alpha_1+\beta_1+\alpha\beta_{11}$	$\mu+\alpha_1+\beta_2+\alpha\beta_{12}$	$\mu+\alpha_1-\beta_1-\beta_2-\alpha\beta_{11}-\alpha\beta_{12}$
A2	$\mu+\alpha_2+\beta_1+\alpha\beta_{21}$	$\mu+\alpha_2+\beta_2+\alpha\beta_{22}$	$\mu+\alpha_2-\beta_1-\beta_2-\alpha\beta_{21}-\alpha\beta_{22}$
A3	$\mu-\alpha_1-\alpha_2+\beta_1$ $-\alpha\beta_{11}-\alpha\beta_{21}$	$\mu-\alpha_1-\alpha_2+\beta_2-\alpha\beta_{12}$ $-\alpha\beta_{22}$	$\mu+\alpha_1-\alpha_2-\beta_1-\beta_2-\alpha\beta_{11}-$ $\alpha\beta_{12}+\beta_{21}+\alpha\beta_{22}$

Effective Hypothesis Tests

Generally, the hypothesis tests produced by JMP agree with the hypothesis tests of most other trusted programs, such as SAS PROC GLM (Hypothesis types III and IV). The following two sections describe where there are differences.

In the SAS GLM procedure, the hypothesis tests for Types III and IV are constructed by looking at the general form of estimable functions and finding functions that involve only the effects of interest and effects contained by the effects of interest (Goodnight 1978).

In JMP, the same tests are constructed, but because there is a different parameterization, an effect can be tested (assuming full rank for now) by doing a joint test on all the parameters for that effect. The tests do not involve containing interaction parameters because the coding has made them uninvolved with the tests on their contained effects.

If there are missing cells or other singularities, the JMP tests are different than GLM tests. There are several ways to describe them:

- JMP tests are equivalent to testing that the least squares means are different, at least for main effects. If the least squares means are nonestimable, then the test cannot include some comparisons and, therefore, loses degrees of freedom. For interactions, JMP is testing that the least squares means differ by more than just the marginal pattern described by the containing effects in the model.
- JMP tests an effect by comparing the SSE for the model with that effect with the SSE for the model without that effect (at least if there are no nested terms, which complicate the logic slightly). JMP parameterizes so that this method makes sense.
- JMP implements the *effective hypothesis tests* described by Hocking (1985, 80–89, 163–166), although JMP uses structural rather than cell-means parameterization. Effective hypothesis tests

start with the hypothesis desired for the effect and include "as much as possible" of that test. Of course, if there are containing effects with missing cells, then this test will have to drop part of the hypothesis because the complete hypothesis would not be estimable. The effective hypothesis drops as little of the complete hypothesis as possible.

- The differences among hypothesis tests in JMP and GLM (and other programs) that relate to the presence of missing cells are not considered interesting tests anyway. If an interaction is significant, the test for the contained main effects are not interesting. If the interaction is not significant, then it can always be dropped from the model. Some tests are not even unique. If you relabel the levels in a missing cell design, then the GLM Type IV tests can change.

The following section continues this topic in finer detail.

Singularities and Missing Cells in Nominal Effects

Consider the case of linear dependencies among the design columns. With JMP coding, this does not occur unless there is insufficient data to fill out the combinations that need estimating, or unless there is some kind of confounding or collinearity of the effects.

With linear dependencies, a least squares solution for the parameters might not be unique and some tests of hypotheses cannot be tested. The strategy chosen for JMP is to set parameter estimates to zero in sequence as their design columns are found to be linearly dependent on previous effects in the model. A special column in the report shows what parameter estimates are zeroed and which parameter estimates are estimable. A separate *singularities* report shows what the linear dependencies are.

In cases of singularities the hypotheses tested by JMP can differ from those selected by GLM. Generally, JMP finds fewer degrees of freedom to test than GLM because it holds its tests to a higher standard of marginality. In other words, JMP tests always correspond to tests across least squares means for that effect, but GLM tests do not always have this property.

For example, consider a two-way model with interaction and one missing cell where A has three levels, B has two levels, and the A3B2 cell is missing

Figure A.7

A B	A1	A2	B1	A1B1	A2B1	
A1 B1	1	0	1	1	0	
A2 B1	0	1	1	0	1	
A3 B1	–1	–1	1	–1	–1	
A1 B2	1	0	–1	–1	0	
A2 B2	0	1	–1	0	–1	←
A3 B2	–1	–1	–1	1	1	suppose this missing

The expected values for each cell are:

Figure A.8

	B1	B2
A1	$\mu + \alpha_1 + \beta_1 + \alpha\beta_{11}$	$\mu + \alpha_1 - \beta_1 - \alpha\beta_{11}$
A2	$\mu + \alpha_2 + \beta_1 + \alpha\beta_{21}$	$\mu + \alpha_2 - \beta_1 - \alpha\beta_{21}$
A3	$\mu - \alpha_1 - \alpha_2 + \beta_1 - \alpha\beta_{11} - \alpha\beta_{21}$	$\mu - \alpha_1 - \alpha_2 - \beta_1 + \alpha\beta_{11} + \alpha\beta_{21}$

Obviously, any cell with data has an expectation that is estimable. The cell that is missing has an expectation that is nonestimable. In fact, its expectation is precisely that linear combination of the design columns that is in the singularity report

$$\mu - \alpha_1 - \alpha_2 - \beta_1 + \alpha\beta_{11} + \alpha\beta_{21}$$

Suppose that you want to construct a test that compares the least squares means of B1 and B2. In this example, the average of the rows in the above table give these least squares means.

$$\begin{aligned}\text{LSM(B1)} = (1/3)(&\mu + \alpha_1 + \beta_1 + \alpha\beta_{11} + \\ &\mu + \alpha_2 + \beta_1 + \alpha\beta_{21} + \\ &\mu - \alpha_1 - \alpha_2 + \beta_1 - \alpha\beta_{11} - \alpha\beta_{21}) \\ &= \mu + \beta_1\end{aligned}$$

$$\begin{aligned}\text{LSM(B2)} = (1/3)(&\mu + \alpha_1 + -\beta_1 - \alpha\beta_{11} + \\ &\mu + \alpha_2 + -\beta_1 - \alpha\beta_{21} + \\ &\mu - \alpha_1 - \alpha_2 - \beta_1 + \alpha\beta_{11} + \alpha\beta_{21}) \\ &= \mu - \beta_1\end{aligned}$$

$$\text{LSM(B1)} - \text{LSM(B2)} = 2\beta_1$$

Note that this shows that a test on the β_1 parameter is equivalent to testing that the least squares means are the same. But because β_1 is not estimable, the test is not testable, meaning there are no degrees of freedom for it.

Now, construct the test for the least squares means across the A levels.

$$\begin{aligned}\text{LSM(A1)} &= (1/2)(\mu + \alpha_1 + \beta_1 + \alpha\beta_{11} + \mu + \alpha_1 - \beta_1 - \alpha\beta_{11}) \\ &= \mu + \alpha_1\end{aligned}$$

$$\begin{aligned}\text{LSM(A2)} &= (1/2)(\mu + \alpha_2 + \beta_1 + \alpha\beta_{21} + \mu + \alpha_2 - \beta_1 - \alpha\beta_{21}) \\ &= \mu + \alpha_2\end{aligned}$$

$$\begin{aligned}\text{LSM(A3)} = (1/2)(&\mu - \alpha_1 - \alpha_2 + \beta_1 - \alpha\beta_{11} - \alpha\beta_{21} + \\ &\mu - \alpha_1 - \alpha_2 - \beta_1 + \alpha\beta_{11} + \alpha\beta_{21}) \\ &= \mu - \alpha_1 - \alpha_2\end{aligned}$$

$$\text{LSM(A1)} - \text{LSM(A3)} = 2\alpha_1 + \alpha_2$$

$$\text{LSM(A2)} - \text{LSM(A3)} = 2\alpha_2 + \alpha_1$$

Neither of these turn out to be estimable, but there is another comparison that is estimable; namely comparing the two A columns that have no missing cells.

$$\text{LSM(A1)} - \text{LSM(A2)} = \alpha_1 - \alpha_2$$

This combination is indeed tested by JMP using a test with 1 degree of freedom, although there are two parameters in the effect.

The estimability can be verified by taking its inner product with the singularity combination, and checking that it is zero:

Figure A.9

parameters	singularity combination	combination to be tested
m	1	0
a_1	–1	1
a_2	–1	–1
b_1	–1	0
ab_{11}	1	0
ab_{21}	1	0

It turns out that the design columns for missing cells for any interaction will always knock out degrees of freedom for the main effect (for nominal factors). Thus, there is a direct relation between the nonestimability of least squares means and the loss of degrees of freedom for testing the effect corresponding to these least squares means.

How does this compare with what GLM does? GLM and JMP do the same test when there are no missing cells. That is, they effectively test that the least squares means are equal. But when GLM encounters singularities, it focuses out these cells in different ways, depending on whether they are Type III or Type IV. For Type IV, it looks for estimable combinations that it can find. These might not be unique, and if you reorder the levels, you might get a different result. For Type III, it does some orthogonalization of the estimable functions to obtain a unique test. But the test might not be very interpretable in terms of the cell means.

The JMP approach has several points in its favor, although at first it might seem distressing that you might lose more degrees of freedom than with GLM:

1 The tests are philosophically linked to LSMs.

2 The tests are easy computationally, using reduction sum of squares for reparameterized models.

3 The tests agree with Hocking's "Effective Hypothesis Tests"

4 The tests are *whole marginal tests*, meaning they always go completely across other effects in interactions.

The last point needs some elaboration: Consider a graph of the expected values of the cell means in the previous example with a missing cell for A3B2.

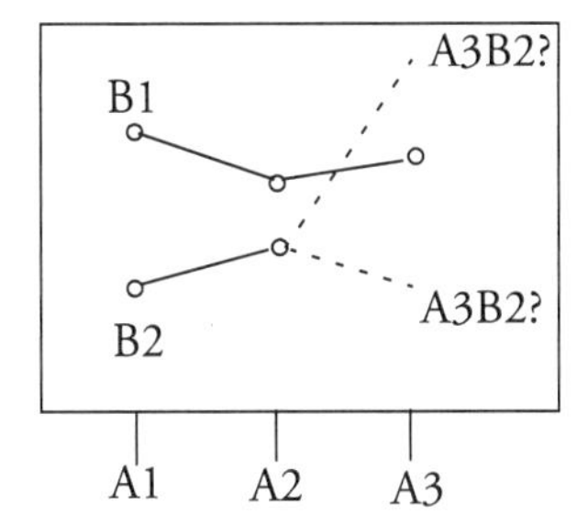

The graph shows expected cell means with a missing cell. The means of the A1 and A2 cells are profiled across the B levels. The JMP approach says you can't test the A main effect with a missing A2B3 cell, because the mean of the missing cell could be anything, as allowed by the interaction term. If the mean of the missing cell were the higher value shown, the A effect would test nonsignificant. If it were the lower, it would test significant. The point is that you don't know. That is what the least squares means are saying when they are declared nonestimable. That is what the hypotheses for the effects should be saying too—that you don't know.

If you want to test hypotheses involving margins for subsets of cells, then that is what GLM Type IV does. In JMP you would have to construct these tests yourself by partitioning the effects with a lot of calculations or by using contrasts.

Example of JMP and GLM differences

GLM works differently than JMP and produces different hypothesis tests in situations where there are missing cells. In particular, GLM does not recognize any difference between a nesting and a crossing in an effect, but JMP does. Suppose that you have a three-layer nesting of A, B(A), and C(A B) with different numbers of levels as you go down the nested design.

Table A.1 "Comparison of GLM and JMP Hypotheses," p. 985, shows the test of the main effect A in terms of the GLM parameters. The first set of columns is the test done by JMP. The second set of columns is the test done by GLM Type IV. The third set of columns is the test equivalent to that by JMP; it is the first two columns that have been multiplied by a matrix

$$\begin{bmatrix} 2 & 1 \\ 1 & 2 \end{bmatrix}$$

to be comparable to the GLM test. The last set of columns is the GLM Type III test. The difference is in how the test distributes across the containing effects. In JMP, it seems more top-down hierarchical. In GLM Type IV, the test seems more bottom-up. In practice, the test statistics are often similar.

Table A.1 Comparison of GLM and JMP Hypotheses

Parameter	JMP Test for A		GLM-IV Test for A		JMP Rotated Test		GLM-III Test for A	
u	0	0	0	0	0	0	0	0
a1	0.6667	-0.3333	1	0	1	0	1	0
a2	–0.3333	0.6667	0	1	0	1	0	1
a3	–0.3333	-0.3333	-1	-1	-1	-1	-1	-1

Table A.1 Comparison of GLM and JMP Hypotheses (continued)

a1b1	0.1667	-0.0833	0.2222	0	0.25	0	0.2424	0
a1b2	0.1667	-0.0833	0.3333	0	0.25	0	0.2727	0
a1b3	0.1667	-0.0833	0.2222	0	0.25	0	0.2424	0
a1b4	0.1667	-0.0833	0.2222	0	0.25	0	0.2424	0
a2b1	-0.1667	0.3333	0	0.5	0	0.5	0	.5
a2b2	-0.1667	0.3333	0	0.5	0	0.5	0	.5
a3b1	-0.1111	-0.1111	-0.3333	-0.3333	-0.3333	-0.3333	-0.3333	-0.3333
a3b2	-0.1111	-0.1111	-0.3333	-0.3333	-0.3333	-0.3333	-0.3333	-0.3333
a3b3	-0.1111	-0.1111	-0.3333	-0.3333	-0.3333	-0.3333	-0.3333	-0.3333
a1b1c1	0.0833	-0.0417	0.1111	0	0.125	0	0.1212	0
a1b1c2	0.0833	-0.0417	0.1111	0	0.125	0	0.1212	0
a1b2c1	0.0556	-0.0278	0.1111	0	0.0833	0	0.0909	0
a1b2c2	0.0556	-0.0278	0.1111	0	0.0833	0	0.0909	0
a1b2c3	0.0556	-0.0278	0.1111	0	0.0833	0	0.0909	0
a1b3c1	0.0833	-0.0417	0.1111	0	0.125	0	0.1212	0
a1b3c2	0.0833	-0.0417	0.1111	0	0.125	0	0.1212	0
a1b4c1	0.0833	-0.0417	0.1111	0	0.125	0	0.1212	0
a1b4c2	0.0833	-0.0417	0.1111	0	0.125	0	0.1212	0
a2b1c1	-0.0833	0.1667	0	0.25	0	0.25	0	0.25
a2b1c2	-0.0833	0.1667	0	0.25	0	0.25	0	0.25
a2b2c1	-0.0833	0.1667	0	0.25	0	0.25	0	0.25
a2b2c2	-0.0833	0.1667	0	0.25	0	0.25	0	0.25
a3b1c1	-0.0556	-0.0556	-0.1667	-0.1667	-0.1667	-0.1667	-0.1667	-0.1667
a3b1c2	-0.0556	-0.0556	-0.1667	-0.1667	-0.1667	-0.1667	-0.1667	-0.1667
a3b2c1	-0.0556	-0.0556	-0.1667	-0.1667	-0.1667	-0.1667	-0.1667	-0.1667
a3b2c2	-0.0556	-0.0556	-0.1667	-0.1667	-0.1667	-0.1667	-0.1667	-0.1667
a3b3c1	-0.0556	-0.0556	-0.1667	-0.1667	-0.1667	-0.1667	-0.1667	-0.1667
a3b3c2	-0.0556	-0.0556	-0.1667	-0.1667	-0.1667	-0.1667	-0.1667	-0.1667

Ordinal Factors

Factors marked with the ordinal modeling type are coded differently than nominal factors. The parameters estimates are interpreted differently, the tests are different, and the least squares means are different.

The theme for ordinal factors is that the first level of the factor is a control or baseline level, and the parameters measure the effect on the response as the ordinal factor is set to each succeeding level. The coding is appropriate for factors with levels representing various doses, where the first dose is zero:

Figure A.10

Term	Coded Column		
A	a2	a3	
A1	0	0	control level, zero dose
A2	1	0	low dose
A3	1	1	higher dose

From the perspective of the JMP parameterization, the tests for A are

Figure A.11

parameter	GLM–IV test		JMP test	
m	0	0	0	0
a13	2	1	1	0
a23	1	2	0	1
a1:b14	0	0	0	0
a1:b24	0.11111	0	0	0
a1:b34	0	0	0	0
a2:b12	0	0	0	0
a3:b13	0	0	0	0
a3:b23	0	0	0	0
a1b1:c12	0	0	0	0
a1b2:c13	0	0	0	0
a1b2:c23	0	0	0	0
a1b3:c12	0	0	0	0
a1b4:c12	0	0	0	0
a2b1:c13	0	0	0	0
a2b2:c12	0	0	0	0
a3b1:c12	0	0	0	0
a3b2:c12	0	0	0	0
a3b3:c12	0	0	0	0

So from JMP's perspective, the GLM test looks a little strange, putting a coefficient on the a1b24 parameter.

The pattern for the design is such that the lower triangle is ones with zeros elsewhere.

For a simple main-effects model, this can be written

$$y = \mu + \alpha_2 X_{(a \le 2)} + \alpha_3 X_{(a \le 3)} + \varepsilon$$

noting that μ is the expected response at $A = 1$, $\mu + \alpha_2$ is the expected response at $A = 2$, and $\mu + \alpha_2 + \alpha_3$ is the expected response at $A = 3$. Thus, α_2 estimates the effect moving from $A = 1$ to $A = 2$ and α_3 estimates the effect moving from $A = 2$ to $A = 3$.

If all the parameters for an ordinal main effect have the same sign, then the response effect is monotonic across the ordinal levels. The parameterization is done with a future JMP feature in mind: the ability to impose and test monotonicity on the estimates (Barlow et. al. 1972; Robertson et. al. 1988). We expect that monotonicity constraints could lead to tests of high power when compared to tests in unconstrained models.

Ordinal Interactions

The ordinal interactions, as with nominal effects, are produced with a horizontal direct product of the columns of the factors. Consider an example with two ordinal factors A and B, each with three levels JMP. JMP's ordinal coding produces the design matrix shown next. The pattern for the interaction is a block lower-triangular matrix of lower-triangular matrices of ones.

Figure A.12

						A*B			
						A2		A3	
A	B	A2	A3	B2	B3	B2	B3	B2	B3
A1	B1	0	0	0	0	0	0	0	0
A1	B2	0	0	1	0	0	0	0	0
A1	B3	0	0	1	1	0	0	0	0
A2	B1	1	0	0	0	0	0	0	0
A2	B2	1	0	1	0	1	0	0	0
A2	B3	1	0	1	1	1	1	0	0
A3	B1	1	1	0	0	0	0	0	0
A3	B2	1	1	1	0	1	0	1	0
A3	B3	1	1	1	1	1	1	1	1

Hypothesis Tests for Ordinal Crossed Models

To see what the parameters mean, examine this table of the expected cell means in terms of the parameters, where μ is the intercept, α_2 is the parameter for level A2, and so forth.

Figure A.13

	B1	B2	B3
A1	μ	$\mu + \alpha\beta_2 + \alpha\beta_{12}$	$\mu + \beta_2 + \beta_3$
A2	$\mu + \alpha_2$	$\mu + \alpha_2 + \beta_2 + \alpha\beta_{22}$	$\mu + \alpha_2 + \beta_2 + \beta_3 + \alpha\beta_{22} + \alpha\beta_{23}$

Figure A.13

A3	$\mu + \alpha_2 + \alpha_3$	$\mu + \alpha_2 + \alpha_3 + \beta_2 + \alpha\beta_{22} + \alpha\beta_{32} + \alpha\beta_{23} + \alpha\beta_{32} + \alpha\beta_{33}$	$\mu + \alpha_2 + \alpha_3 + \beta_2 + \beta_3 + \alpha\beta_{22} + \alpha\beta_{23} + \beta_{32} + \alpha\beta_{33}$

Note that the main effect test for A is really testing the A levels holding B at the first level. Similarly, the main effect test for B is testing across the top row for the various levels of B holding A at the first level. This is the appropriate test for an experiment where the two factors are both doses of different treatments. The main question is the efficacy of each treatment by itself, with fewer points devoted to looking for *drug interactions* when doses of both drugs are applied. In some cases it may even be dangerous to apply large doses of each drug.

Note that each cell's expectation can be obtained by adding all the parameters associated with each cell that is to the left and above it, inclusive of the current row and column. The expected value for the last cell is the sum of all the parameters.

Though the hypothesis tests for effects contained by other effects differs with ordinal and nominal codings, the test of effects not contained by other effects is the same. In the crossed design above, the test for the interaction would be the same no matter whether A and B were fit nominally or ordinally.

Ordinal Least Squares Means

As stated previously, least squares means are the predicted values corresponding to some combination of levels, after setting all the other factors to some neutral value. JMP defines the neutral value for an effect with uninvolved ordinal factors as the effect at the first level, meaning the *control* or *baseline* level.

This definition of least squares means for ordinal factors maintains the idea that the hypothesis tests for contained effects are equivalent to tests that the least squares means are equal.

Singularities and Missing Cells in Ordinal Effects

With the ordinal coding, you are saying that the first level of the ordinal effect is the baseline. It is thus possible to get good tests on the main effects even when there are missing cells in the interactions—even if you have no data for the interaction.

Example with Missing Cell

The example is the same as above, with two observations per cell except that the A3B2 cell has no data. You can now compare the results when the factors are coded nominally with results when they are coded ordinally. The model as a whole fits the same as seen in tables shown in Figure A.15.

Figure A.14

Y	A	B
12	1	1
14	1	1
15	1	2
16	1	2

Figure A.14

17	2	1
17	2	1
18	2	2
19	2	2
20	3	1
24	3	1

Figure A.15 Comparison of Summary Information for Nominal and Ordinal Fits

Summary of Fit (Nominal)

RSquare	0.891732
RSquare Adj	0.805118
Root Mean Square Error	1.48324
Mean of Response	17.2
Observations (or Sum Wgts)	10

Analysis of Variance (Nominal)

Source	DF	Sum of Squares	Mean Square	F Ratio
Model	4	90.60000	22.6500	10.2955
Error	5	11.00000	2.2000	**Prob > F**
C. Total	9	101.60000		0.0125

Summary of Fit (Ordinal)

RSquare	0.891732
RSquare Adj	0.805118
Root Mean Square Error	1.48324
Mean of Response	17.2
Observations (or Sum Wgts)	10

Analysis of Variance (Ordinal)

Source	DF	Sum of Squares	Mean Square	F Ratio
Model	4	90.60000	22.6500	10.2955
Error	5	11.00000	2.2000	**Prob > F**
C. Total	9	101.60000		0.0125

The parameter estimates are very different because of the different coding. Note that the missing cell affects estimability for all the nominal parameters but for none of the other ordinal parameters.

Figure A.16 Comparison of Parameter Estimates for Nominal and Ordinal Fits

Parameter Estimates (Nominal)

Term		Estimate	Std Error	t Ratio	Prob>|t|
Intercept	Biased	18.083333	0.74162	24.38	<.0001
A[1]	Biased	-3.833333	0.856349	-4.48	0.0065
A[2]	Biased	-0.333333	0.856349	-0.39	0.7131
B[1]	Biased	-0.75	0.74162	-1.01	0.3583
A[1]*B[1]	Biased	-0.5	1.048809	-0.48	0.6537
A[2]*B[1]	Zeroed	0	0	.	.

Parameter Estimates (Ordinal)

Term		Estimate	Std Error	t Ratio	Prob>|t|
Intercept		13	1.048809	12.40	<.0001
A[2-1]		4	1.48324	2.70	0.0429
A[3-2]		5	1.48324	3.37	0.0199
B[2-1]		2.5	1.48324	1.69	0.1527
A[2-1]*B[2-1]		-1	2.097618	-0.48	0.6537
A[3-2]*B[2-1]	Zeroed	0	0	.	.

The singularity details show the linear dependencies (and also identify the missing cell by examining the values).

Figure A.17 Comparison of Singularity Details for Nominal and Ordinal Fits

Singularity Details (Nominal)

Intercept = A[1] + A[2] + B[1] - A[1]*B[1] - A[2]*B[1]

Singularity Details (Ordinal)

A[3-2]*B[2-1] = 0

The effect tests lose degrees of freedom for nominal. In the case of B, there is no test. For ordinal, there is no loss because there is no missing cell for the *base* first level.

Figure A.18 Comparison of Effects Tests for Nominal and Ordinal Fits

Effect Tests (Nominal)

Source	Nparm	DF	Sum of Squares	F Ratio	Prob > F	
A	2	2	81.333333	18.4848	0.0049	
B	1	1	6.250000	2.8409	0.1527	
A*B	2	1	0.500000	0.2273	0.6537	LostDFs

Effect Tests (Ordinal)

Source	Nparm	DF	Sum of Squares	F Ratio	Prob > F	
A	2	1	24.500000	11.1364	0.0206	LostDFs
B	1	0	0.000000	.	.	LostDFs
A*B	2	1	0.500000	0.2273	0.6537	LostDFs

The least squares means are also different. The nominal LSMs are not all estimable, but the ordinal LSMs are. You can verify the values by looking at the cell means. Note that the A*B LSMs are the same for the two. Figure A.19 shows least squares means for an nominal and ordinal fits.

Figure A.19 Least Squares Means for Nominal and Ordinal Fits

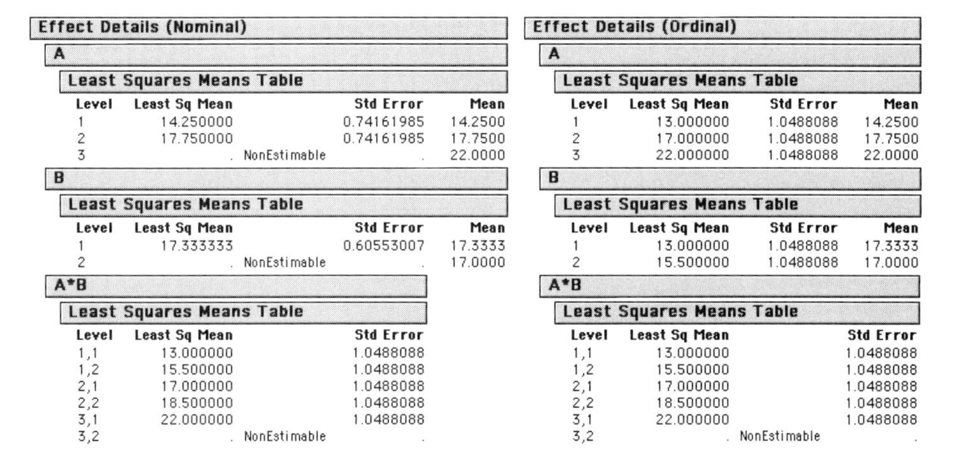

Effect Details (Nominal)

A — Least Squares Means Table

Level	Least Sq Mean	Std Error	Mean
1	14.250000	0.74161985	14.2500
2	17.750000	0.74161985	17.7500
3	. NonEstimable	.	22.0000

B — Least Squares Means Table

Level	Least Sq Mean	Std Error	Mean
1	17.333333	0.60553007	17.3333
2	. NonEstimable	.	17.0000

A*B — Least Squares Means Table

Level	Least Sq Mean	Std Error
1,1	13.000000	1.0488088
1,2	15.500000	1.0488088
2,1	17.000000	1.0488088
2,2	18.500000	1.0488088
3,1	22.000000	1.0488088
3,2	. NonEstimable	.

Effect Details (Ordinal)

A — Least Squares Means Table

Level	Least Sq Mean	Std Error	Mean
1	13.000000	1.0488088	14.2500
2	17.000000	1.0488088	17.7500
3	22.000000	1.0488088	22.0000

B — Least Squares Means Table

Level	Least Sq Mean	Std Error	Mean
1	13.000000	1.0488088	17.3333
2	15.500000	1.0488088	17.0000

A*B — Least Squares Means Table

Level	Least Sq Mean	Std Error
1,1	13.000000	1.0488088
1,2	15.500000	1.0488088
2,1	17.000000	1.0488088
2,2	18.500000	1.0488088
3,1	22.000000	1.0488088
3,2	. NonEstimable	.

Leverage Plot Details

Leverage plots for general linear hypotheses were introduced by Sall (1980). This section presents the details from that paper. Leverage plots generalize the partial regression residual leverage plots of Belsley, Kuh, and Welsch (1980) to apply to any linear hypothesis. Suppose that the estimable hypothesis of interest is

$$L\beta = 0$$

The leverage plot characterizes this test by plotting points so that the distance of each point to the sloped regression line displays the unconstrained residual, and the distance to the *x*-axis displays the residual when the fit is constrained by the hypothesis.

Of course the difference between the sums of squares of these two groups of residuals is the sum of squares due to the hypothesis, which becomes the main ingredient of the *F*-test.

The parameter estimates constrained by the hypothesis can be written

$$b_0 = b - (X'X)^{-1}L\lambda$$

where b is the least squares estimate

$$b = (X'X)^{-1}Xy'$$

and where lambda is the Lagrangian multiplier for the hypothesis constraint, which is calculated

$$\lambda = (L(X'X)^{-1}L')Lb$$

Compare the residuals for the unconstrained and hypothesis-constrained residuals, respectively.

$$r = y - Xb$$

$$r_0 = r + X(X'X)^{-1}L'\lambda$$

To get a leverage plot, the *x*-axis values $\mathbf{v_x}$ of the points are the differences in the residuals due to the hypothesis, so that the distance from the line of fit (with slope 1) to the *x*-axis is this difference. The *y*-axis values $\mathbf{v_y}$ are just the *x*-axis values plus the residuals under the full model as illustrated in Figure A.20. Thus, the leverage plot is composed of the points

$$v_x = X(X'X)^{-1}L'\lambda \text{ and } v_y = r + v_x$$

Figure A.20 Construction of Leverage Plot

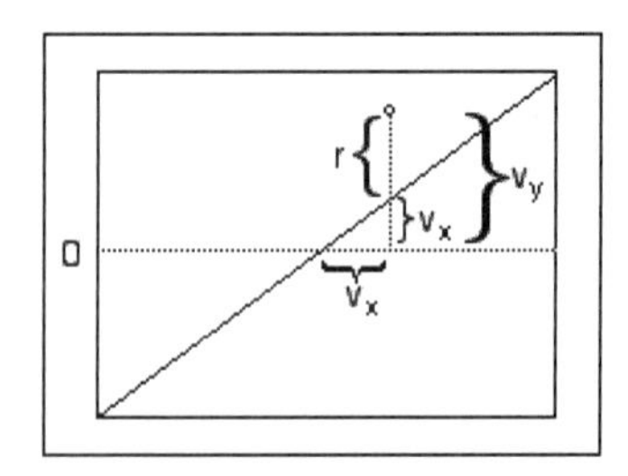

Superimposing a Test on the Leverage Plot

In simple linear regression, you can plot the confidence limits for the expected value as a smooth function of the regressor variable **x**

$$\text{Upper}(x) = xb + t_{\alpha/2}s\sqrt{x(X'X)^{-1}x'}$$

$$\text{Lower}(x) = xb - t_{\alpha/2}s\sqrt{x(X'X)^{-1}x'}$$

where $x = \begin{bmatrix} 1 & x \end{bmatrix}$ is the 2-vector of regressors.

This hyperbola is a useful significance-measuring instrument because it has the following nice properties:

- If the slope parameter is significantly different from zero, the confidence curve will cross the horizontal line of the response mean (left plot in Figure A.21).
- If the slope parameter is not significantly different from zero, the confidence curve will not cross the horizontal line of the response mean (plot at right in Figure A.21).

- If the t-test for the slope parameter is sitting right on the margin of significance, the confidence curve will have the horizontal line of the response mean as an asymptote (middle plot in Figure A.21).

This property can be verified algebraically or it can be understood by considering what the confidence region has to be when the regressor value goes off to plus or minus infinity.

Figure A.21 Cases of Significant, Borderline, and Nonsignificant Confidence Curves.

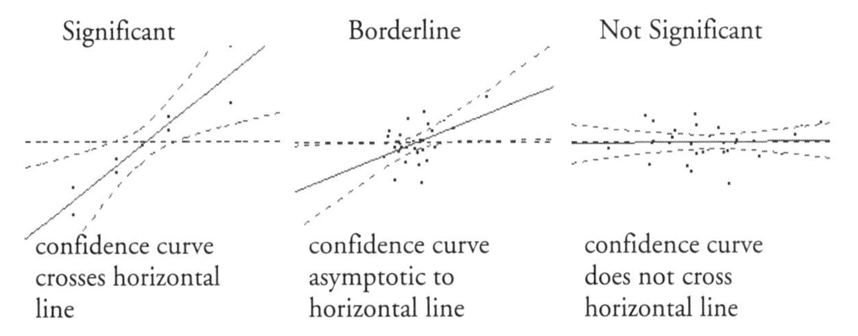

Leverage plots make use of the same device by calculating a confidence function with respect to one variable while holding all other variables constant at their sample mean value.

For general hypothesis tests, JMP can display a curve with the same properties: the confidence function shows at the mean value in the middle, with adjusted curvature so that it crosses the horizon if the F-test is significant at some α-level like 0.05.

Consider the functions

$$\text{Upper}(z) = z + \sqrt{s^2 t^2_{\alpha/2} \bar{h} + \frac{F_\alpha}{F} z^2}$$

and

$$\text{Lower}(z) = z - \sqrt{s^2 t^2_{\alpha/2} \bar{h} + \frac{F_\alpha}{F} z^2}$$

where F is the F-statistic for the hypothesis, F_α is the reference value for significance α, and

$\bar{h} = \bar{x}(X'X)^{-1}\bar{x}'$, where $\bar{x}$ is the regressors at a suitable middle value, such as the mean.

These functions serve the same properties listed above. If the F-statistic is greater than the reference value, the confidence functions cross the x-axis. If the F-statistic is less than the reference value, the confidence functions do not cross. If the F-statistic is equal to the reference value, the confidence functions have the x-axis as an asymptote. And the range between the confidence functions at the middle value is a valid confidence region of the predicted value at the point.

Multivariate Details

The following sections show computations used for multivariate tests and related, exact and approximate *F*-statistics, canonical details, and discriminant functions. In the following sections, **E** is the residual cross product matrix and $\frac{\mathbf{E}}{n-1}$ estimates the residual covariance matrix. Diagonal elements of **E** are the sum of squares for each variable. In discriminant analysis literature, this is often called **W**, where W stands for *within*.

Multivariate Tests

Test statistics in the multivariate results tables are functions of the eigenvalues λ of $E^{-1}H$. The following list describes the computation of each test statistic.

Note: After specification of a response design, the initial **E** and **H** matrices are premultiplied by M' and postmultiplied by **M**.

Figure A.22

Wilk's Lambda	$\Lambda = \frac{\det(E)}{\det(H+E)} = \prod\left(\frac{1}{1+\lambda_i}\right)$
Pillai's Trace	$V = \text{Trace}[H(H+E)^{-1}] = \sum\frac{\lambda_i}{1+\lambda_i}$
Hotelling-Lawley Trace	$U = \text{Trace}(E^{-1}H) = \sum\lambda_i$
Roy's Max Root	$\Theta = \lambda_1$, the maximum eigenvalue of $E^{-1}H$.

The whole model **L** is a column of zeros (for the intercept) concatenated with an identity matrix having the number of rows and columns equal to the number of parameters in the model. **L** matrices for effects are subsets of rows from the whole model **L** matrix.

Approximate *F*-Test

To compute *F*-values and degrees of freedom, let p be the rank of $H + E$. Let q be the rank of $L(\mathbf{X'X})^{-1}L'$, where the **L** matrix identifies elements of $\mathbf{X'X}$ associated with the effect being tested. Let v be the error degrees of freedom and s be the minimum of p and q. Also let $m = 0.5(|p-q|-1)$ and $n = 0.5(v-p-1)$.

Table A.2 "Approximate F-statistics," p. 994, gives the computation of each approximate F from the corresponding test statistic.

Table A.2 Approximate *F*-statistics

Test	Approximate F	Numerator DF	Denominator DF
Wilk's Lambda	$F = \left(\frac{1-\Lambda^{1/t}}{\Lambda^{1/t}}\right)\left(\frac{pq}{rt-2u}\right)$	pq	$rt-2u$

Table A.2 Approximate *F*-statistics

Pillai's Trace	$F = \left(\frac{V}{s-V}\right)\left(\frac{2n+s+1}{2m+s+1}\right)$	$s(2m+s+1)$	$s(2n+s+1)$
Hotelling-Lawley Trace	$F = \frac{2(sn+1)U}{s^2(2m+s+1)}$	$s(2m+s+1)$	$2(sn+1)$
Roy's Max Root	$F = \frac{\Theta(v-\max(p,q)+q)}{\max(p,q)}$	$\max(p,q)$	$v-\max(p,q)+q$

Canonical Details

The canonical correlations are computed as

$$\rho_i = \sqrt{\frac{\lambda_i}{1+\lambda_i}}$$

The canonical Y's are calculated as

$$\tilde{Y} = YMV$$

where **Y** is the matrix of response variables, **M** is the response design matrix, and **V** is the matrix of eigenvectors of $E^{-1}H$. Canonical Y's are saved for eigenvectors corresponding to eigenvalues larger than zero.

The total sample centroid is computed as

$$\text{Grand} = \bar{y}MV$$

where **V** is the matrix of eigenvectors of $E^{-1}H$.

The centroid values for effects are calculated as

$$m = (c'_1\bar{x}_j, c'_2\bar{x}_j, \ldots, c'_g\bar{x}_j) \quad \text{where } c_i = \left(v'_i\left(\frac{E}{N-r}\right)v_i\right)^{-1/2} v_i$$

and the vs are columns of **V**, the eigenvector matrix of $E^{-1}H$, $\bar{x}_j$ refers to the multivariate least squares mean for the *j*th effect, *g* is the number of eigenvalues of $E^{-1}H$ greater than 0, and *r* is the rank of the **X** matrix.

The centroid radii for effects are calculated as

$$d = \sqrt{\frac{\chi_g^2(0.95)}{L(X'X)^{-1}L}}$$

where *g* is the number of eigenvalues of $E^{-1}H$ greater than 0 and the denominator **L**'s are from the multivariate least squares means calculations.

Discriminant Analysis

The distance from an observation to the multivariate mean of the *i*th group is the Mahalanobis distance, D^2, and computed as

$$D^2 = (y - \bar{y}_i)'S^{-1}(y - \bar{y}_i) = y'S^{-1}y - 2y'S^{-1}\bar{y}_i + \bar{y}_i'S^{-1}\bar{y}_i$$

where

$$S = \frac{E}{N-1}$$

In saving discriminant columns, N is the number of observations and M is the identity matrix.

The **Save Discrim** command in the popup menu on the platform title bar saves discriminant scores with their formulas as columns in the current data table. Dist[0] is a quadratic form needed in all the distance calculations. It is the portion of the Mahalanobis distance formula that does not vary across groups. Dist[i] is the Mahalanobis distance of an observation from the *i*th centroid. Dist[0] and Dist[i] are calculated as

$$\text{Dist[0]} = y'S^{-1}y$$

and

$$\text{Dist[}i\text{]} = \text{Dist[0]} - 2y'S^{-1}\bar{y}_i + \bar{y}_i'S^{-1}\bar{y}_i$$

Assuming that each group has a multivariate normal distribution, the posterior probability that an observation belongs to the *i*th group is

$$\text{Prob[}i\text{]} = \frac{\exp(\text{Dist[i]})}{\text{Prob[0]}}$$

where

$$\text{Prob[0]} = \sum e^{-0.5\text{Dist[i]}}$$

Power Calculations

The next sections give formulas for computing LSV, LSN, power, and adjusted power.

Computations for the LSV

For one-degree-freedom tests, the LSV is easy to calculate. The formula for the *F*-test for the hypothesis $L\beta = 0$ is:

$$F = \frac{(Lb)'(L(X'X)^{-1}L')^{-1}(Lb)/r}{s^2}$$

Solving for **Lb**, a scalar, given an F for some α-level, like 0.05, and using a t as the square root of a one-degree-of-freedom F, making it properly two-sided, gives

$$(\mathrm{Lb})^{\mathrm{LSV}} = t_{\alpha/2} s \sqrt{\mathrm{L}(\mathrm{X'X})^{-1} c\mathrm{L}}$$

For **L** testing some β_i, this is

$$b_i^{\mathrm{LSV}} = t_{\alpha/2} s \sqrt{((\mathrm{X'X})^{-1})_{ii}}$$

which can be written with respect to the standard error as

$$b_i^{\mathrm{LSV}} = t_{\alpha/2} \mathrm{stderr}(b_i)$$

If you have a simple model of two means where the parameter of interest measures the difference between the means, this formula is the same as the LSD, least significant difference, from the literature

$$\mathrm{LSD} = t_{\alpha/2} s \sqrt{\frac{1}{n_1} + \frac{1}{n_2}}$$

In the JMP Fit Model platform, the parameter for a nominal effect measures the difference to the average of the levels, not to the other mean. So, the LSV for the parameter is half the LSD for the differences of the means.

Computations for the LSN

The LSN solves for n in the equation:

$$\alpha = 1 - \mathrm{ProbF}\left[\frac{\frac{\delta^2}{n}}{dfH\alpha^2}, dfH, n - dfH - 1\right]$$

where

ProbF is the central F-distribution function

dfH is the degrees of freedom for the hypothesis

δ^2 is the squared effect size, which can be estimated by $\frac{\mathrm{SS(H)}}{n}$

When planning a study, the LSN serves as the lower bound.

Computations for the Power

To calculate power, first get the critical value for the central F by solving for $\mathrm{F_{crit}}$ in the equation

$$\alpha = 1 - \mathrm{ProbF}[\mathrm{F_{Crit}}, dfH, n - dfH - 1]$$

Then obtain the probability of getting this critical value or higher

$$\mathrm{power} = 1 - \mathrm{ProbF}\left[\mathrm{F_{Crit}}, dfH, n - dfH - 1, \frac{n\delta^2}{\sigma^2}\right]$$

The last argument to ProbF is the noncentrality value

$$\lambda = \frac{n\delta^2}{\sigma^2}, \text{ which can be estimated as } \frac{SS(H)}{\hat{\sigma}^2} \text{ for retrospective power.}$$

Computations for Adjusted Power

The adjusted power is a function a noncentrality estimate that has been adjusted to remove positive bias in the original estimate (Wright and O'Brien 1988). Unfortunately, the adjustment to the noncentrality estimate can lead to negative values. Negative values are handled by letting the adjusted noncentrality estimate be

$$\hat{\lambda}_{adj} = \text{Max}\left[0, \frac{\hat{\lambda}(N - dfR - 1 - 2)}{N - dfH - 1} - dfH\right]$$

where N is the actual sample size, dfH is the degrees of freedom for the hypothesis, and dfR is the degrees of freedom for regression in the whole model.

The adjusted power is

$$\text{Power}_{adj} = 1 - \text{ProbF}[F_{Crit}, dfH, n - dfH - 1, \hat{\lambda}_{adj}]$$

where F_{crit} is calculated as above.

Confidence limits for the noncentrality parameter are constructed according to Dwass (1955) as

$$\text{Lower CL for } \lambda = dfH(\text{Max}[0, \sqrt{F_{sample}} - \sqrt{F_{crit}}])^2$$

$$\text{Upper CL for } \lambda = dfH(\sqrt{F_{sample}} - \sqrt{F_{crit}})^2$$

where F_{sample} is the F-value reported in the effect test report and F_{crit} is calculated as above. Power confidence limits are obtained by substituting confidence limits for λ into

$$\text{Power} = 1 - \text{ProbF}[F_{Crit}, dfH, n - dfH - 1, \lambda]$$

Inverse Prediction with Confidence Limits

Inverse prediction estimates values of independent variables. In bioassay problems, inverse prediction with confidence limits is especially useful. In JMP, you can request inverse prediction estimates of a single effect for continuous, nominal, and ordinal response models. If the response is a nominal variable and there is a single continuous effect, you can also request confidence limits for the inverse prediction estimate.

The confidence limits are computed using Fieller's theorem, which is based on the following logic. The goal is predicting the value of a single regressor and its confidence limits given the values of the other regressors and the response.

Let **b** estimate the parameters β so that we have **b** distributed as $N(\beta,V)$.

Let **x** be the regressor values of interest, with the *i*th value to be estimated.

Let **y** be the response value.

We desire a confidence region on the value of **x**[i] such that $\beta'x = y$ with all other values of **x** given.

The inverse prediction is just

$$x[i] = \frac{y - \beta'_{(i)}x_{(i)}}{\beta[i]}$$

where the parenthesized-subscript "$_{(i)}$" means that the *i*th component is omitted. A confidence interval can be formed from the relation:

$$(y - b'x)^2 < t^2 x'Vx$$

with specified confidence if $y = \beta'x$. A quadratic equation results of the form

$$z^2 g + zh + f = 0$$

where

$$g = b[i]^2 - t^2 V[i, i]$$

$$h = -2yb[i] + 2b[i]b'_{(i)}x_{(i)} - 2t^2 V[i, i]'x_{(i)}$$

$$f = y^2 - 2yb'_{(i)}x_{(i)} + (b_{(i)}x_{(i)})^2 - t^2 x_{(i)}'V_{(i)}x_{(i)}$$

It is possible for the quadratic equation to have only imaginary roots, in which case the confidence interval becomes the entire real line and missing values are returned.

Note: JMP uses *t* values when computing the confidence intervals for inverse prediction. SAS uses *z* values, which can give different results.

Details of Random Effects

The variance matrix of the fixed effects is always modified to include a Kackar-Harville correction. The variance matrix of the BLUPs and the covariances between the BLUPs and the fixed effects are not Kackar-Harville corrected because the corrections for these can be quite computationally intensive and use lots of memory when there are lots of levels of the random effects. In SAS, the Kackar-Harville correction is done for both fixed effects and BLUPs only when the `DDFM=KR` is set, so the standard errors from `PROC MIXED` with this option will differ from all the other options.

This implies:

- Standard errors for linear combinations involving only fixed effects parameters will match `PROC MIXED DDFM=KR`, assuming that one has taken care to transform between the different parameterizations used by `PROC MIXED` and JMP.
- Standard errors for linear combinations involving only BLUP parameters will match `PROC MIXED DDFM=SATTERTH`.

- Standard errors for linear combinations involving both fixed effects and BLUPS do not match PROC MIXED for any DDFM option if the data are unbalanced. However, these standard errors are between what you get with the DDFM=SATTERTH and DDFM=KR options. If the data are balanced, JMP matches SAS for balanced data, regardless of the DDFM option, since the Kackar-Harville correction is null.

The degrees of freedom for tests involving only linear combinations of fixed effects parameters are calculated using the Kenward and Roger correction, so JMP's results for these tests match PROC MIXED using the DDFM=KR option. If there are BLUPs in the linear combination, JMP uses a Satterthwaite approximation to get the degrees of freedom. So, the results follow a pattern similar to what is described for standard errors in the preceding paragraph.

For more details of the Kackar-Harville correction and the Kenward-Roger DF approach, see Kenward and Roger(1997). The Satterthwaite method is described in detail in the SAS PROC MIXED documentation.

Appendix B

References

Abernethy, Robert B. (1996) *The New Weibull Handbook.* Published by the author: 536 Oyster Road North Palm Beach, Florida 33408.

Agresti, A. (1984), *Analysis of Ordinal Categorical Data*, New York: John Wiley and Sons, Inc.

Agresti, A. (1990), *Categorical Data Analysis*, New York: John Wiley and Sons, Inc.

Agresti, A., and Coull, B. (1998), "Approximate is Better Than 'Exact' for Interval Estimation of Binomial Proportions," *The American Statistician*, 52, 119–126

Aitken, M. (1987) "Modelling Variance Heterogeneity in Normal Regression Using GLIM," *Applied Statistics* 36:3, 332–339.

Akaike, H. (1974), "Factor Analysis and AIC," *Pschychometrika*, 52, 317–332.

Akaike, H. (1987), "A new Look at the Statistical Identification Model," *IEEE Transactions on Automatic Control*, 19, 716–723.

Anderson, T.W. (1971), *The Statistical Analysis of Time Series*, New York: John Wiley and Sons.

Anderson, T. W. (1958) *An Introduction to Multivariate Statistical Analysis.* New York: John Wiley & Sons.

Andrews, D.F. and A. M. Herzberg (1985), *Data: A Collection of Problems from Many Fields for the Student and Research Worker.* New York: Springer-Verlag.

Ashton, W.D. (1972), "The Logit Transformation," *Griffin's Statistical Monographs*, New York: Hafner Publishing.

Atkinson, A.C. (1985), *Plots, Transformations and Regression*, Oxford: Clarendon Press.

Barlow, R.E., Bartholomew, D.J., Bremner, J.M., and Brunk, H.D. (1972), *Statistical Inference under Order Restrictions*, New York: John Wiley and Sons, Inc.

Barrentine (1991), *Concepts for R&R Studies*, Milwaukee, WI: ASQC Quality Press.

Bartlett, M.S. and D.G. Kendall (1946), "The Statistical Analysis of Variances–Heterogeneity and the Logarithmic Transformation," *JRSS* Suppl 8, 128–138.

Bartlett, M.S. (1966), *An Introduction to Stochastic Processes*, Second Edition, Cambridge: Cambridge University Press.

Bates, D.M. and Watts, D.G. (1988), *Nonlinear Regression Analysis & its Applications.* New York, John Wiley and Sons.

Beaton, A.E. and Tukey, J.W. (1974), "The Fitting of Power Series, Meaning Polynomials, Illustrated on Band–Spectroscopic Data," *Technometrics* 16, 147–180.

Becker, R.A. and Cleveland, W.S. (1987), "Brushing Scatterplots," *Technometrics*, 29, 2.

Berger, R.L., and Hsu, J.C. (1996), "Bioequivalence Trails, Intersection-Union Tests and Equivalence Confidence Sets," *Statistical Science*, 11, 283–319.

Belsley, D.A., Kuh, E., and Welsch, R.E. (1980), *Regression Diagnostics*, New York: John Wiley and Sons.

Benzecri, J.P. (1973), "L'Analyse des Donnees," *l'analyse des Correspondances*, Paris: Dunod.

Bowman, A. and Foster, P. (1992) "Density Based Exploration of Bivariate Data," Dept. of Statistics, Univ. of Glasgow, Tech Report No 92–1.

Bowman, A. and Schmee, J. (2004)"Estimating Sensitivity of Process Capability Modeled by a Transfer Function" Journal of Quality Technology, v36, no.2 (April)

Box, G. E. P. (1954). "Some Theorems on Quadratic Forms Applied in the Study of Analysis of Variance Problems, II: Effects of Inequality of Variance and Correlation Between Errors in the Two-Way Classification". *Annals of Mathematical Statistics*, 1, 69-82.

Box, G.E.P. (1988), "Signal–to–Noise Ratio, Performance Criteria, and Transformations," *Technometrics* 30, 1–40.

Box, G.E.P. and Cox, D.R. (1964), "An Analysis of Transformations," *JRSS* B26, 211–243.

Box, G.E.P. and Draper, N.R. (1969), *Evolutionary Operation: A Statistical Method for Process Improvement*, New York: John Wiley and Sons.

Box, G.E.P. and Draper, N.R. (1987), *Empirical Model–Building and Response Surfaces*, New York: John Wiley and Sons.

Box, G.E.P. and Meyer, R.D. (1986), "An analysis of Unreplicated Fractional Factorials," *Technometrics* 28, 11–18.

Box, G.E.P. and Meyer, R. D. (1993), "Finding the Active Factors in Fractionated Screening Experiments.", *Journal of Quality Technology* Vol.25 #2: 94–105.

Box, G.E.P., Hunter,W.G., and Hunter, J.S. (1978), *Statistics for Experimenters*, New York: John Wiley and Sons, Inc.

Brown, M.B. and Forsythe, A.B. (1974a), "The Small Sample Behavior of Some Statistics Which Test the Equality of Several Means," *Technometrics* 16:1, 129–132.

Brown, M.B. and Forsythe, A.B. (1974), "Robust tests for the equality of variances" *Journal of the American Statistical Association*, 69, 364–367.

Byrne, D.M. and Taguchi, G. (1986), *ASQC 40th Anniversary Quality Control Congress Transactions*, Milwaukee, WI: American Society of Quality Control, 168–177.

Carroll, R.J. and Ruppert, D. (1988), *Transformation and Weighting in Regression*, New York: Chapman and Hall.

Carroll, R.J., Ruppert, D. and Stefanski, L.A. (1995), *Measurement Error in Nonlinear Models*, New York: Chapman and Hall.

Cobb, G.W. (1998), *Introduction to Design and Analysis of Experiments*, Springer-Verlag: New York.

Cohen, J. (1960), "A coefficient of agreement for nominal scales," *Education Psychological Measurement*, 20: 37–46.

Cole, J.W.L. and Grizzle, J.E. (1966), "Applications of Multivariate Analysis of Variance to Repeated Measures Experiments," *Biometrics*, 22, 810–828.

Cochran, W.G. and Cox, G.M. (1957), *Experimental Designs*, Second Edition, New York: John Wiley and Sons.

Conover, W. J. (1972). "A Kolmogorov Goodness-of-fit Test for Discontinuous Distributions". *Journal of the American Statistical Association* 67: 591–596.

Conover, W.J. (1980), *Practical Nonparametric Statistics*, New York: John Wiley and Sons, Inc.

Cook, R.D. and Weisberg, S. (1982), *Residuals and Influence in Regression*, New York: Chapman and Hall.

Cook, R.D. and Weisberg, S. (1983), "Diagnostics for heteroscedasticity in regression" *Biometrika* 70, 1–10.

Cornell, J.A. (1990), *Experiments with Mixtures*, Second Edition New York: John Wiley and Sons.

Cox, D.R. (1970), *The Analysis of Binary Data*, London: Metheun.

Cox, D.R. (1972), "Regression Models And Life-tables", *Journal Of The Royal Statistical Society Series B-statistical Methodology.* 34 (2): 187–220, 1972.

Cronbach, L.J. (1951), "Coefficient Alpha and the Internal Structure of Tests," *Psychometrika*, 16, 297–334.

Daniel C. and Wood, F. (1980), *Fitting Equations to Data*, Revised Edition, New York: John Wiley and Sons, Inc.

Daniel, C. (1959), "Use of Half–normal Plots in Interpreting Factorial Two–level Experiments," *Technometrics*, 1, 311–314.

Davis, H.T. (1941), *The Analysis of Economic Time Series*, Bloomington, IN: Principia Press.

DeLong, E., Delong, D, and Clarke-Pearson, D.L. (1988), "Comparing the Areas Under Two or more Correlated Receiver Operating Characteristic Curves: A Nonparametric Approach," *Biometrics* 44, 837–845.

Derringer, D. and Suich, R. (1980), "Simultaneous Optimization of Several Response Variables," *Journal of Quality Technology*, 12:4, 214–219.

Devore, J. L. (1995), *Probability and Statistics for Engineering and the Sciences*, Duxbury Press, CA.

Do, K-A, and McLachlan G.J. (1984). Estimation of mixing proportions: a case study. *Journal of the Royal Statistical Society, Series C*, 33: 134-140.

Draper, N. and Smith, H. (1981), *Applied Regression Analysis*, Second Edition, New York: John Wiley and Sons, Inc.

Dunnett, C.W. (1955), "A multiple comparison procedure for comparing several treatments with a control" *Journal of the American Statistical Association*, 50, 1096–1121.

Dwass, M. (1955), "A Note on Simultaneous Confidence Intervals," *Annals of Mathematical Statistics* 26: 146–147.

Eppright, E.S., Fox, H.M., Fryer, B.A., Lamkin, G.H., Vivian, V.M., and Fuller, E.S. (1972), "Nutrition of Infants and Preschool Children in the North Central Region of the United States of America," World *Review of Nutrition and Dietetics*, 14.

Eubank, R.L. (1988), *Spline Smoothing and Nonparametric Regression*, New York: Marcel Dekker, Inc.

Farebrother, R.W. (1981), "Mechanical Representations of the L1 and L2 Estimation Problems," *Statistical Data Analysis*, 2nd Edition, Amsterdam, North Holland: edited by Y. Dodge.

Fleis, J.L., Cohen J., and Everitt, B.S. (1969), "Large-sample standard errors of kappa and weighted kappa," *Psychological Bulletin*, 72: 323–327.

Fisher, L. and Van Ness, J.W. (1971), "Admissible Clustering Procedures," *Biometrika*, 58, 91–104.

Fisherkeller, M.A., Friedman, J.H., and Tukey, J.W. (1974), "PRIM–9: An Interactive Multidimensional Data Display and Analysis System," SLAC–PUB–1408, Stanford, California: Stanford Linear Accelerator Center.

Fleiss, J. L. (1981). Statistical Methods for Rates and Proportions. New York: John Wiley and Sons.

Florek, K., Lukaszewicz, J., Perkal, J., and Zubrzycki, S. (1951a), "Sur La Liaison et la Division des Points d'un Ensemble Fini," *Colloquium Mathematicae*, 2, 282–285.

Foster, D.P., Stine, R.A., and Waterman, R.P. (1997), *Business Analysis Using Regression*, New York, Springer-Verlag.

Foster, D.P., Stine, R.A., and Waterman, R.P. (1997), *Basic Business Statistics*, New York, Springer-Verlag.

Friendly, M. (1991), "Mosaic Displays for Multiway Contingency Tables," *New York University Department of Psychology Reports*: 195.

Fuller, W.A. (1976), *Introduction to Statistical Time Series*, New York, John Wiley and Sons.

Fuller, W.A. (1987), *Measurement Error Models*, New York, John Wiley and Sons.

Gabriel, K.R. (1982), "Biplot," *Encyclopedia of Statistical Sciences*, Volume 1, eds. N.L.Johnson and S. Kotz, New York: John Wiley and Sons, Inc., 263–271.

Gallant, A.R. (1987), *Nonlinear Statistical Models*, New York, John Wiley and Sons.

Giesbrecht, F.G. and Gumpertz, M.L. (2004). *Planning, Construction, and Statistical Analysis of Comparative Experiments*. New York: John Wiley & Sons.

Goodnight, J.H. (1978), "Tests of Hypotheses in Fixed Effects Linear Models," *SAS Technical Report R–101*, Cary: SAS Institute Inc, also in Communications in Statistics (1980), A9 167–180.

Goodnight, J.H. and W.R. Harvey (1978), "Least Square Means in the Fixed Effect General Linear Model," *SAS Technical Report R–103*, Cary NC: SAS Institute Inc.

Greenacre, M.J. (1984), *Theory and Applications of Correspondence Analysis*, London: Academic Press.

Greenhouse, S. W. and Geiser, S. (1959). "On Methods in the Analysis of Profile Data." *Psychometrika*, 32, 95–112.

Gupta, S.S. (1965), On Some Multiple Decision (selection and ranking), Rules., *Technometrics* 7, 225–245.

Haaland, P.D. (1989), *Experimental Design in Biotechnology*, New York: Marcel Dekker, Inc.

Hahn, G.J. (1976), "Process Improvement Through Simplex EVOP," *Chemical Technology* 6, 243–345.

Hahn, G. J. and Meeker, W. Q. (1991) *Statistical Intervals: A Guide for Practitioners*. New York: Wiley.

Hajek, J. (1969), *A Course in Nonparametric Statistics*, San Francisco: Holden–Day.

Harrell, F. (1986), "The Logist Procedure," *SUGI Supplemental Library User's Guide*, Version 5 Edition, Cary, NC: SAS Institute Inc.

Harris, R.J. (1975), *A Primer of Multivariate Statistics*, New York: Academic Press.

Hartigan, J.A. (1975), *Clustering Algorithms*, New York, John Wiley and Sons.

Hartigan, J.A. (1981), "Consistence of Single Linkage for High–Density Clusters," *Journal of the American Statistical Association*, 76, 388–394.

Hartigan, J.A. and Kleiner, B. (1981), "Mosaics for Contingency Tables," *Proceedings of the 13th Symposium on the Interface between Computer Science and Statistics*, Ed. Eddy, W. F., New York: Springer–Verlag, 268–273.

Harvey, A.C. (1976), "Estimating Regression Models with Multiplicative Heteroscedasticity," *Econometrica* 44–3 461–465.

Hauck, W.W. and Donner, A. (1977), "Wald's Test as Applied to Hypotheses in Logit Analysis," *Journal of the American Statistical Association*, 72, 851–863.

Hawkins, D.M. (1974), "The Detection of Errors in Multivariate Data Using Principal Components," *Journal of the American Statistical Association*, 69.

Hayashi, C. (1950), "On the Quantification of Qualitative Data from the Mathematico–Statistical Point of View," *Annals of the Institute of Statistical Mathematics*, 2:1, 35–47.

Hayter, A.J. (1984), "A proof of the conjecture that the Tukey–Kramer multiple comparisons procedure is conservative," *Annals of Mathematical Statistics*, 12 61–75.

Henderson, C.R. (1984), *Applications of Linear Models in Animal Breeding*, Univ. of Guelph.

Hocking, R.R. (1985), *The Analysis of Linear Models*, Monterey: Brooks–Cole.

Hoeffding, W (1948), "A Non-Parametric Test of Independence", Annals of Mathematical Statistics, 19, 546–557.

Hoffman, Heike (2001). "Generalized Odds Ratios for Visual Modeling," Journal of Computational and Graphical Statistics, 10:4 pp. 628–640.

Holland, P.W. and Welsch, R.E. (1977), "Robust Regression Using Interactively Reweighted Least Squares," *Communications Statistics: Theory and Methods*, 6, 813–827.

Hooper, J. H. and Amster, S. J. (1990), "Analysis and Presentation of Reliability Data," *Handbook of Statistical Methods for Engineers and Scientists*, Harrison M. Wadsworth, editor. New York: McGraw Hill.

Hosmer, D.W. and Lemeshow, S. (1989), *Applied Logistic Regression*, New York: John Wiley and Sons.

"Hot Dogs," (1986), *Consumer Reports* (June), 364–367.

Hsu, J. (1981), "Simultaneous confidence intervals for all distances from the 'best'," *Annals of Statistics*, 9, 1026–1034.

Hsu, J. (1984), "Constrained two–sided simultaneous confidence intervals for multiple comparisons with the 'best'," *Annals of Statistics*, 12, 1136–1144.

Hsu, J. (1989), "Multiple Comparison Procedures" ASA Short Course notes, Columbus OH: Ohio State University.

Hsu, J. (1989), *Tutorial Notes on Multiple Comparisons*, American Statistical Association, Washington, DC.

Hunter, J.S. (1985), "Statistical Design Applied to Product Design," *Journal of Quality Technology*, 17, 210–221.

Huynh, H. and Feldt, L. S. (1970). "Conditions under which Mean Square Ratios in Repeated Measurements Designs have Exact *F*-Distributions." *Journal of the American Statistical Association*, 65, 1582–1589.

Huynh, H. and Feldt, L. S. (1976). "Estimation of the Box Correction for Degrees of Freedom from Sample Data in the Randomized Block Split Plot Designs." Journal of Educational Statistics, 1, 69–82.

Iman, R.L. and Conover, W.J. (1978), "Approximations of the Critical Regions for Spearman's Rho with and without Ties Present," *Communications in Statistics—Theory and Methods*, A5, 1335–1348 (5.2).

Inselberg, A. (1985) "The Plane with Parallel Coordinates." *Visual Computing* 1. pp 69–91.

Jardine, N. and Sibson, R. (1971), *Mathematical Taxonomy*, New York: John Wiley and Sons.

John, P.W.M. (1972), *Statistical Design and Analysis of Experiments*, New York: Macmillan Publishing Company, Inc.

Johnson, M.E. and Nachtsheim, C.J. (1983), "Some Guidelines for Constructing Exact D–Optimal Designs on Convex Design Spaces," *Technometrics* 25, 271–277.

Johnson, N.L. (1949). *Biometrika*, 36, 149-176.

Judge, G.G., Griffiths,W.E., Hill,R.C., and Lee, Tsoung–Chao (1980), *The Theory and Practice of Econometrics*, New York: John Wiley and Sons.

Kalbfleisch, J.D. and Prentice, R.L. (1980), *The Statistical Analysis of Failure Time Data*, New York: John Wiley and Sons.

Kackar, R.N. and Harville, D.A. (1984), Approximations for standard errors of estimators of fixed and random effects in mixed linear models, *Journal of the American Statistical Association*, 79, 853-862.

Kaiser, H.F. (1958), "The varimax criterion for analytic rotation in factor analysis" *Psychometrika*, 23, 187–200.

Kenward, M.G. and Roger, J.H. (1997). Small sample inference for fixed effects from restricted maximumþlikelihood. *Biometrics*, 53, 983-997.

Khuri, A.I. and Cornell J.A. (1987), *Response Surfaces: Design and Analysis*, New York: Marcel Dekker.

Koehler, M.G., Grigoras, S., and Dunn, J.D. (1988), "The Relationship Between Chemical Structure and the Logarithm of the Partition Coefficient," *Quantitative Structure Activity Relationships*, 7.

Kohonen, Teuvo. (1989) *Self-Organization and Associative Memory.* Berlin: Springer.

Kramer, C.Y. (1956), "Extension of multiple range tests to group means with unequal numbers of replications," *Biometrics*, 12, 309–310.

Lawless, J.E. (1982), *Statistical Models and Methods for Lifetime Data*, New York: John Wiley and Sons.

Lebart, L., Morineau, A., and Tabaard, N. (1977), *Techniques de la Description Statistique*, Paris: Dunod.

Lee, E.T. (1980), *Statistical Methods for Survival Data Analysis*, Belmont CA, Lifetime Learning Publications, a Division of Wadsworth, Inc.

Lenth, R.V. (1989), "Quick and Easy Analysis of Unreplicated Fractional Factorials," *Technometrics*, 31, 469–473.

Leven, J.R., Serlin, R.C., and Webne-Behrman, L. (1989), "Analysis of Variance Through Simple Correlation," *American Statistician*, 43, (1), 32–34.

Levene, H. (1960), "Robust tests for the equality of variances" In I. Olkin (ed), *Contributions to probability and statistics*, Stanford Univ. Press.

Linnerud: see Rawlings (1988)

Lucas, J.M. (1976), "The Design and Use of V–Mask Control Schemes," *Journal of Quality Technology*, 8, 1–12.

MacQueen, J.B. (1967) (1967) "Some Methods for Classification and Analysis of Multivariate Observations," *Proceedings of the fifth Berkeley Symposium on Mathematical Statistics and Probability*, 1, 281–297.

Mallows, C.L. (1967), "Choosing a Subset Regression," unpublished report, Bell Telephone Laboratories.

Mallows, C.L. (1973), "Some Comments on Cp," *Technometrics*, 15, 661–675.

Mardia, K.V., Kent, J.T., and Bibby J.M. (1979). *Multivariate Analysis*, New York: Academic Press.

Marsaglia, G. (1996) DIEHARD: A Battery of Tests of Randomness". http://stat.fsu.edu/~geo.

Matsumoto, M. and Nishimura, T. (1998)"Mersenne Twister: A 623-Dimensionally Equidistributed Uniform Pseudo-Random Number Generator", *ACM Transactions on Modeling and Computer Simulation*, Vol. 8, No. 1, January 1998, 3–*f*30.

McLachlan, G.J. and Krishnan, T. (1997), *The EM Algorithm and Extensions*, New York: John Wiley and Sons.

McCullagh, P. and Nelder, J.A. (1983), *Generalized Linear Models*, London: Chapman and Hall Ltd.

McQuitty, L.L. (1957), "Elementary Linkage Analysis for Isolating Orthogonal and Oblique Types and Typal Relevancies," *Educational and Psychological Measurement*, 17, 207–229.

Meeker, W.Q. and Escobar, L.A. (1998), *Statistical Methods for Reliability Data*, pp. 60–62, New York: John Wiley and Sons.

Myers, R.H. (1976), *Response Surface Methodology*, Boston: Allyn and Bacon.

Myers, R.H. (1988), *Response Surface Methodology*, Virginia Polytechnic and State University.

Myers, R.H. (1989), *Classical and Modern Regression with Applications*, Boston: PWS-KENT.

Meyer, R.D., Steinberg, D.M., and Box, G. (1996), "Follow-up Designs to Resolve Confounding in Multifactor Experiments," *Technometrics*, Vol. 38, #4, p307

Miller, A.J. (1990), *Subset Selection in Regression*, New York: Chapman and Hall.

Milligan, G.W. (1980), "An Examination of the Effect of Six Types of Error Perturbation on Fifteen Clustering Algorithms," *Psychometrika*, 45, 325–342.

Milliken, G.A. and Johnson, E.J. (1984), *Analysis of Messy Data Volume I: Design of Experiments*, New York: Van Nostrand Reinhold Company.

Montgomery, D.C. and Peck, E.A. (1982), *Introduction to Linear Regression Analysis*, New York: John Wiley and Sons.

Montgomery, D. C. (1991), "Using Fractional Factorial Designs for Robust Process Development," *Quality Engineering*, 3, 193–205.

Montgomery, D. C. (1996) *Introduction to Statistical Quality Control*, 3rd edition. New York: John Wiley.

Montgomery, D.C. (2001), *Introduction to Statistical Quality Control*, 4th Edition New York: John Wiley and Sons.

Moore, D.S. and McCabe, G. P. (1989), *Introduction to the Practice of Statistics*, New York and London: W. H. Freeman and Company.

Mosteller, F. and Tukey, J.W. (1977), *Data Analysis and Regression*, Reading, MA: Addison–Wesley.

Muller, K.E. and Barton, C.N. (1989), "Approximate Power for Repeated–measures ANOVA Lacking Sphericity," *Journal of the American Statistical Association*, 84, 549–555.

Myers, R. H. and Montgomery, D. C. (1995), *Response Surface Methodology*, New York: John Wiley and Sons.

Nelder, J.A. and Mead, R. (1965), "A Simplex Method for Function Minimization," *The Computer Journal*, 7, 308–313.

Nelder, J.A. and Wedderburn, R.W.M. (1983), "Generalized Linear Models," *Journal of the Royal Statistical Society*, Series A, 135, 370–384.

Nelson, F. (1976), "On a General Computer Algorithm for the Analysis of Model with Limited Dependent Variables," *Annals of Economic and Social Measurement*, 5/4.

Nelson, L. (1984), "The Shewhart Control Chart—Tests for Special Causes," *Journal of Quality Technology*, 15, 237–239.

Nelson, L. (1985), "Interpreting Shewhart X Control Charts," *Journal of Quality Technology*, 17, 114–116.

Nelson, W. (1982), *Applied Life Data Analysis*, New York: John Wiley and Sons.

Nelson, W. (1990), *Accelerated Testing: Statistical Models, Test Plans, and Data analysis*, New York: John Wiley and Sons.

Neter, J., Wasserman, W. and Kutner, M.H. (1990), *Applied Linear Statistical Models*, Third Edition, Boston: Irwin, Inc.

Nunnaly, J. C. (1978) Psychometric Theory, 2nd Ed., New York: McGraw-Hill.

O'Brien, R.G. (1979), "A general ANOVA method for robust tests of additive models for variances," *Journal of the American Statistical Association*, 74, 877–880.

O'Brien, R., and Lohr, V. (1984), "Power Analysis For Linear Models: The Time Has Come," *Proceedings of the Ninth Annual SAS User's Group International Conference*, 840–846.

Odeh, R. E. and Owen, D. B. (1980) *Tables for Normal Tolerance Limits, Sampling Plans, and Screening*. New York: Marcel Dekker, Inc.

Olejnik, S.F. and Algina, J. (1987), "Type I Error Rates and Power Estimates of Selected Parametric and Nonparametric Tests of Scale," *Journal of Educational Statistics* 12, 45–61.

Olson, C.L. (1976), "On Choosing a Test Statistic in MANOVA," *Psychological Bulletin* 83, 579–586.

Patterson, H. D. and Thompson, R. (1974). Maximum likelihood estimation of components of variance. *Proc. Eighth International Biochem. Conf.*, 197–209.

Piepel, G.F. (1988), "Programs for Generating Extreme Vertices and Centroids of Linearly Constrained Experimental Regions," *Journal of Quality Technology* 20:2, 125–139.

Plackett, R.L. and Burman, J.P. (1947), "The Design of Optimum Multifactorial Experiments," *Biometrika*, 33, 305–325.

Ratkowsky, D.A. (1990), *Handbook of Nonlinear Regression Models*, New York, Marcel Dekker, Inc.

Rawlings, J.O. (1988), *Applied Regression Analysis: A Research Tool*, Pacific Grove CA: Wadsworth and Books/Cole.

Reinsch, C.H. (1967), *Smoothing by Spline Functions*, Numerische Mathematik, 10, 177–183.

Robertson, T., Wright, F.T., and R.L. Dykstra, R.L (1988), *Order Restricted Statistical Inference*, New York: John Wiley and Sons, Inc.

Rodriguez, R.N. (1990), "Selected SAS/QC Software Examples, Release 6.06," SAS Institute Inc., Cary, NC.

Rodriguez, R.N. (1991), "Applications of Computer Graphics to Two Basic Statistical Quality Improvement Methods," National Computer Graphics Association Proceedings, 17–26.

Rosenbaum, P.R. (1989), "Exploratory Plots for Paired Data," *American Statistician*, 108–109.

Rousseuw, P.J. and Leroy, A.M. (1987), *Robust Regression and Outlier Detection*, New York: John Wiley and Sons.

Royston, J.P. (1982), "An Extension of Shapiro and Wilk's W Test for Normality to Large Samples," *Applied Statistics* 31, 115–124.

Sall, J.P. (1990), "Leverage Plots for General Linear Hypotheses," *American Statistician*, 44, (4), 303–315.

SAS Institute Inc. (1995), *SAS/QC Software: Usage and References, Version* 6, 1st Ed., Vol. 1, SAS Institute Inc., Cary, NC.

SAS Institute Inc. (1986), *SAS/QC User's Guide*, Version 5 Edition, SAS Institute Inc., Cary, NC.

SAS Institute Inc. (1987), *SAS/STAT Guide for Personal Computers*, Version 6 Edition, Cary NC: SAS Institute Inc.

SAS Institute Inc. (1999), *SAS/ETS User's Guide, Version 8*, Cary NC: SAS Institute Inc.

SAS Institute Inc. (1989), "SAS/ Technical Report P–188: SAS/QC Software Examples, Version 6 Edition," SAS Institute Inc., Cary, NC.

SAS Institute Inc. (1999), *SAS/STAT User's Guide*, Version 8, Cary, NC: SAS Institute Inc.

SAS Institute Inc. (1996), "SAS/STAT Software, Changes and Enhancements through Version 6.11, The Mixed Procedure, Cary, NC: SAS Institute Inc.

SAS Institute Inc. (2004), *SAS/STAT User's Guide*, Version 9.1, Cary, NC: SAS Institute Inc.

Satterthwaite, F.E., (1946), "An approximate distribution of Estimates of Variance Components," *Biometrics Bulletin*, 2, 110–114.

Scheffé, H. (1958) "Experiments with Mixtures", *Journal of the Royal Statistical Society B* v20, 344–360.

Searle, S. R, Casella, G. and McCulloch, C. E. (1992) *Variance Components*. New York: John Wiley and Sons.

Seber, G.A.F. (1984), *Multivariate Observations*, New York: John Wiley and Sons, 413–416.

Seder, L.A. (1950) "Diagnosis with Diagrams, Part I and Part II, Industrial Quality Control," 6, 11–19, 7–11 reprinted in Quality Engineering 2, 505–530 (1990).

Shapiro, S.S. and Wilk, M.B. (1965), "An Analysis of Variance Test for Normality (complete samples)," *Biometrika* 52, 591–611.

Slifker, J. F. and Shapiro, S. S. (1980). *Technometrics*, 22, 239-246.

Sneath, P.H.A. (1957) "The Application of Computers to Taxonomy," *Journal of General Microbiology*,17, 201–226.

Snedecor, G.W. and Cochran, W.G. (1967), *Statistical Methods*, Ames, Iowa: Iowa State University Press.

Snee, R.D. (1979), "Experimental Designs for Mixture Systems with Multicomponent Constraints," *Commun. Statistics*, A8(4), 303–326.

Snee, R.D. and Marquardt, D.W. (1974), "Extreme Vertices Designs for Linear Mixture Models," *Technometrics* 16, 391–408.

Snee, R.D. and Marquardt D.W. (1975), "Extreme vertices designs for linear mixture models," *Technometrics* 16 399–408.

Sokal, R.R. and Michener, C.D. (1958), "A Statistical Method for Evaluating Systematic Relationships," *University of Kansas Science Bulletin*, 38, 1409–1438.

Spendley, W., Hext, G.R., and Minsworth, F.R. (1962), "Sequential Application of Simplex Designs in Optimization and Evolutionary Operation," *Technometrics* 4, 441–461.

Stevens, J.P. (1979), "Comment on Olson: Choosing a Test Statistic in Multivariate Analysis of Variance," *Psychological Bulletin*, 86, 355–360.

Stevens, J.P. (1986), *Applied Multivariate Statistics for the Social Sciences*, New Jersey: Laurence Erlbaum Associates.

Tan, Charles Y., and Iglewicz, Boris (1999), "Measurement-methods Comparisons and Linear Statistical Relationship," *Technometrics*, 41:3, 192–201.

Taguchi, G. (1976), "An Introduction to Quality Control," Nagoya, Japan: Central Japan Quality Control Association.

Tukey, J. (1953), "A problem of multiple comparisons," Dittoed manuscript of 396 pages, Princeton University.

Tobin, J. (1958), "Estimation of Relationships for Limited Dependent Variables," *Econometrica*, 26 24–36.

Tamhane, A. C. and Dunlop, D. D. (2000) Statistics and Data Analysis. Prentice Hall.

Tukey, J. (1991), "The Philosophy of Multiple Comparisons," *Statistical Science*, 6, 100–116.

Wadsworth, H. M., Stephens, K., Godfrey, A.B. (1986) *Modern Methods for Quality Control and Improvement*. John Wiley & Sons.

Walker, S.H. and Duncan, D.B. (1967), "Estimation of the Probability of an Event as a Function of Several Independent Variables," *Biometrika* 54.

Wegman, E. J. (1990) "Hyperdimensional Data Analysis using Parallel Coordinates." *Journal of the American Statistical Association* 85, pp. 664–675.

Welch, B.L. (1951), "On the comparison of several mean values: an alternative approach," *Biometrika* 38, 330–336.

Western Electric Company (1956), *Statistical Quality Control Handbook*, currently referred to as the *AT&T Statistical Quality Control Handbook*.

Westgard, J. O (2002), *Basic QC Practices, 2nd Edition*. Madison, Wisconsin: Westgard QC Inc.

Winer, B.J. (1971), *Statistical Principals in Experimental Design*, Second Edition, New York: McGraw–Hill, Inc.

Wolfinger, R., Tobias, R., and Sall, J. (1994). Computing Gaussian likelihoods and their derivatives for general linear mixed models. *SIAM J. Sci. Comput.* 15, 6 (Nov. 1994), 1294-1310.

Wright, S.P. and R.G. O'Brien (1988), "Power Analysis in an Enhanced GLM Procedure: What it Might Look Like," *SUGI 1988, Proceedings of the Thirteenth Annual Conference*, 1097–1102, Cary NC: SAS Institute Inc.

Ye, K. Q., and Hamada, M.(2001)Critical Values of the Lenth Method for Unreplicated Factorial Designs. *Journal of Quality Technology*: 33(2).

Index

JMP Statistics and Graphics Guide

Symbols

Numerics

A

B

C

D

E

F

G

H

I

J

K

L

M

N

O

P

Q

R

S

T

U

V

W-Z